Mathematical Techniques
for Engineers and Scientists

Mathematical Techniques
for Engineers and Scientists

Larry C. Andrews
Ronald L. Phillips

SPIE PRESS
A Publication of SPIE—The International Society for Optical Engineering
Bellingham, Washington USA

Library of Congress Cataloging-in-Publication Data

Andrews, Larry C.
    Mathematical techniques for engineers and scientists / Larry C. Andrews, Ronald L. Phillips.
        p. cm. – (SPIE Press monograph ; PM118)
    Includes bibliographical references and index.
    ISBN 0-8194-4506-1
    1. Mathematical analysis. I. Phillips, Ronald L. II. Title. III. Series.

QA300 .A5585 2002
515–dc21   2002070783

Published by

SPIE—The International Society for Optical Engineering
P.O. Box 10
Bellingham, Washington 98227-0010 USA
Phone: 360.676.3290
Fax: 360.647.1445
Email: spie@spie.org
www.spie.org

Printed in the United States of America.

Contents

Preface / xi
Symbols and Notation / xv

1 Ordinary Differential Equations / 1

1.1 Introduction / 2
1.2 Classifications / 3
1.3 First-Order Equations / 6
1.4 Second-Order Linear Equations / 17
1.5 Power Series Method / 34
1.6 Solutions Near an Ordinary Point / 35
1.7 Legendre's Equation / 40
1.8 Solutions Near a Singular Point / 43
1.9 Bessel's Equation / 50
Suggested Reading / 57
Exercises / 58

2 Special Functions / 61

2.1 Introduction / 62
2.2 Engineering Functions / 63
2.3 Functions Defined by Integrals / 67
2.4 Orthogonal Polynomials / 76
2.5 Family of Bessel Functions / 83
2.6 Family of Hypergeometric-like Functions / 94
2.7 Summary of Notations for Special Functions / 103
Suggested Reading / 104
Exercises / 105

3 Matrix Methods and Linear Vector Spaces / 109

3.1 Introduction / 110
3.2 Basic Matrix Concepts and Operations / 110
3.3 Linear Systems of Equations / 114
3.4 Linear Systems of Differential Equations / 121
3.5 Linear Vector Spaces / 133
Suggested Reading / 140
Exercises / 140

4   Vector Analysis / 143

4.1   Introduction / 145
4.2   Cartesian Coordinates / 146
4.3   Tensor Notation / 156
4.4   Vector Functions of One Variable / 161
4.5   Scalar and Vector Fields / 170
4.6   Line and Surface Integrals / 179
4.7   Integral Relations Between Line, Surface, Volume Integrals / 194
4.8   Electromagnetic Theory / 206
        Suggested Reading / 210
        Exercises / 211

5   Tensor Analysis / 215

5.1   Introduction / 216
5.2   Tensor Notation / 216
5.3   Rectilinear Coordinates / 218
5.4   Base Vectors / 226
5.5   Vector Algebra / 231
5.6   Relations Between Tensor Components / 238
5.7   Reduction of Tensors to Principal Axes / 241
5.8   Tensor Calculus: Rectilinear Coordinates / 243
5.9   Curvilinear Coordinates / 245
5.10   Tensor Calculus: Curvilinear Coordinates / 250
5.11   Riemann-Christoffel Curvature Tensor / 259
5.12   Applications / 260
        Suggested Reading / 266
        Exercises / 266

6   Complex Variables / 271

6.1   Introduction / 273
6.2   Basic Concepts: Complex Numbers / 273
6.3   Complex Functions / 281
6.4   The Complex Derivative / 287
6.5   Elementary Functions—Part I / 295
6.6   Elementary Functions—Part II / 300
6.7   Mappings by Elementary Functions / 306
        Exercises / 316

7   Complex Integration, Laurent Series, and Residues / 319

7.1   Introduction / 320
7.2   Line Integrals in the Complex Plane / 320

7.3   Cauchy's Theory of Integration / 325
7.4   Infinite Series / 339
7.5   Residue Theory / 357
7.6   Evaluation of Real Integrals—Part I / 363
7.7   Evaluation of Real Integrals—Part II / 371
7.8   Harmonic Functions Revisited / 376
7.9   Heat Conduction / 383
7.10   Two-Dimensional Fluid Flow / 386
7.11   Flow Around Obstacles / 393
       Suggested Reading / 399
       Exercises / 400

8   **Fourier Series, Eigenvalue Problems, and Green's Function / 403**

8.1   Introduction / 405
8.2   Fourier Trigonometric Series / 405
8.3   Power Signals: Exponential Fourier Series / 416
8.4   Eigenvalue Problems and Orthogonal Functions / 420
8.5   Green's Function / 438
       Suggested Reading / 449
       Exercises / 450

9   **Fourier and Related Transforms / 453**

9.1   Introduction / 454
9.2   Fourier Integral Representation / 454
9.3   Fourier Transforms in Mathematics / 458
9.4   Fourier Transforms in Engineering / 461
9.5   Properties of the Fourier Transform / 466
9.6   Linear Shift-Invariant Systems / 471
9.7   Hilbert Transforms / 473
9.8   Two-Dimensional Fourier Transforms / 477
9.9   Fractional Fourier Transform / 483
9.10   Wavelets / 487
       Suggested Reading / 492
       Exercises / 493

10   **Laplace, Hankel, and Mellin Transforms / 495**

10.1   Introduction / 496
10.2   Laplace Transform / 496
10.3   Initial Value Problems / 508
10.4   Hankel Transform / 513
10.5   Mellin Transform / 519
10.6   Applications Involving the Mellin Transform / 526
10.7   Discrete Fourier Transform / 529

10.8   *Z*-Transform / 533
10.9   Walsh Transform / 538
       Suggested Reading / 542
       Exercises / 542

11   Calculus of Variations / 545

11.1   Introduction / 546
11.2   Functionals and Extremals / 547
11.3   Some Classical Variational Problems / 552
11.4   Variational Notation / 555
11.5   Other Types of Functionals / 559
11.6   Isoperimetric Problems / 564
11.7   Rayleigh-Ritz Approximation Method / 567
11.8   Hamilton's Principle / 572
11.9   Static Equilibrium of Deformable Bodies / 579
11.10 Two-Dimensional Variational Problems / 581
       Suggested Reading / 584
       Exercises / 584

12   Partial Differential Equations / 589

12.1   Introduction / 591
12.2   Classification of Second-Order PDEs / 591
12.3   The Heat Equation / 592
12.4   The Wave Equation / 600
12.5   The Equation of Laplace / 604
12.6   Generalized Fourier Series / 611
12.7   Applications Involving Bessel Functions / 617
12.8   Transform Methods / 621
       Suggested Reading / 631
       Exercises / 632

13   Probability and Random Variables / 637

13.1   Introduction / 638
13.2   Random Variables and Probability Distributions / 640
13.3   Examples of Density Functions / 646
13.4   Expected Values / 649
13.5   Conditional Probability / 655
13.6   Functions of One Random Variable / 658
13.7   Two Random Variables / 665
13.8   Functions of Two or More Random Variables / 677
13.9   Limit Distributions / 690
       Suggested Reading / 692
       Exercises / 693

14 Random Processes / 697

14.1 Introduction / 698
14.2 Probabilistic Description of Random Process / 698
14.3 Autocorrelation and Autocovariance Functions / 700
14.4 Cross-Correlation and Cross-Covariance / 708
14.5 Power Spectral Density Functions / 711
14.6 Transformations of Random Processes / 716
14.7 Stationary Gaussian Processes / 722
    Suggested Reading / 729
    Exercises / 729

15 Applications / 733

15.1 Introduction / 734
15.2 Mechanical Vibrations and Electric Circuits / 734
15.3 Buckling of a Long Column / 742
15.4 Communication Systems / 745
15.5 Applications in Geometrical Optics / 756
15.6 Wave Propagation in Free Space / 762
15.7 *ABCD* Matrices for Paraxial Systems / 767
15.8 Zernike Polynomials / 773
    Exercises / 780

References / 783

Index / 785

Preface

Modern engineers and scientists are frequently faced with difficult mathematical problems to solve. As technology continues to move ahead, some of these problems will require a greater understanding of advanced mathematical concepts than ever before. Unfortunately, the mathematical training in many engineering and science undergraduate university programs ends with an introductory course in differential equations. Even in those engineering and science curriculums that require some mathematics beyond differential equations, the required advanced mathematics courses often do not make a clear connection between abstract mathematical concepts and practical engineering applications.

This mathematics book is designed as a self-study text for practicing engineers and scientists, and as a useful reference source to complement more comprehensive publications. In particular, the text might serve as a supplemental text for certain undergraduate or graduate mathematics courses designed primarily for engineers and/or scientists. It takes the reader from ordinary differential equations to more sophisticated mathematics—Fourier analysis, vector and tensor analysis, complex variables, partial differential equations, and random processes. The assumed formal training of the reader is at the undergraduate or beginning graduate level with possible extended experience on the job. We present the exposition in a way that is intended to bridge the gap between the formal education of the practitioner and his/her experience. The emphasis in this text is on the use of mathematical tools and techniques. In that regard it should be useful to those who have little or no experience in the subjects, but should also provide a useful review for readers with some background in the various topics.

Some special features of the text that may be of interest to readers include the following:

- Historical comments appear in a box at the beginning of many chapters to identify some of the major contributors to the subject.
- The most important equations in each section are enclosed in a box to help the reader identify key results.
- Boxes are also used to enclose important lists of identities and sometimes to summarize special results.
- Numbered examples are given in every chapter, each of which appears between horizontal lines.
- Exercise sets are included at the end of each chapter except for Chapter 15. Most of the problems in these exercise sets have answers provided.
- Remark boxes are occasionally introduced to provide some additional comments about a given point.
- At the end of each chapter is a "Suggested Reading" section which contains a brief list of textbooks that generally provide a deeper treatment of the mathematical concepts.
- A more comprehensive numbered set of references is also provided at the end of the text to which the reader is directed throughout the text, e.g., (see [10]).
- We have included a Symbols and Notation page for easy reference to some of the acronyms and special symbols as well as a list of Special Function notation (at the end of Chapter 2).

The text is composed of 15 chapters, each of which is presented independently of other chapters as much as possible. Thus, the particular ordering of the chapters is not necessarily crucial to the user with few exceptions. We begin Chapter 1 with a review of ordinary differential equations, concentrating on second-order linear equations. Equations of this type arise in simple mechanical oscillating systems and in the analysis of electric circuits. Special functions such as the gamma function, orthogonal polynomials, Bessel functions, and hypergeometric functions are introduced in Chapter 2. Our presentation also includes useful engineering functions like the step function, rectangle function, and delta (impulse) function. An introduction to matrix methods and linear vector spaces is presented in Chapter 3, the ideas of which are used repeatedly throughout the text. Chapters 4 and 5 are devoted to vector and tensor analysis, respectively. Vectors are used in the study of electromagnetic theory and to describe the motion of an object moving through space. Tensors are useful in studies of continuum mechanics like elasticity, and in describing various properties of anisotropic materials like crystals. In Chapters 6 and 7 we present a fairly detailed discussion of analytic functions of a complex variable. The Cauchy-Riemann equations are developed in Chapter 6 along with the mapping properties associated with analytic functions. The Laurent series representation of complex functions and the residue calculus presented in Chapter 7 are powerful tools that can be used in a variety of applications, such as the evaluation of nonelementary integrals associated with various integral transforms.

Fourier series and eigenvalue problems are discussed in Chapter 8, followed by an introduction to the Fourier transform in Chapter 9. Generally speaking, the Fourier series representation is useful in describing spectral properties of power signals, whereas the Fourier transform is used in the same fashion for energy signals. However, through the development of formal properties associated with the impulse function, the Fourier transform can also be used for power signals. Other integral transforms are discussed in Chapter 10— the Laplace transform associated with initial value problems, the Hankel transform for circularly symmetric functions, and the Mellin transform for more specialized applications. A brief discussion of discrete transforms ends this chapter. We present some of the classical problems associated with the calculus of variations in Chapter 11, including the famous brachistochrone problem which is similar to Fermat's principle for light. In Chapter 12 we give an introductory treatment of partial differential equations, concentrating primarily on the separation of variables method and transform methods applied to the heat equation, wave equation, and Laplace's equation. Basic probability theory is introduced in Chapter 13, followed by a similar treatment of random processes in Chapter 14. The theory of random processes is essential to the treatment of random noise as found, for example, in the study of statistical communication systems. Chapter 15 is a collection of applications that involve a number of the mathematical techniques introduced in the first 14 chapters. Some additional applications are also presented throughout the text in the various chapters.

In addition to the classical mathematical topics mentioned above, we also include a cursory introduction to some more specialized areas of mathematics that are of growing interest to engineers and scientists. These other topics include the fractional Fourier transform (Chapter 9), wavelets (Chapter 9), and the Walsh transform (Chapter 10).

Except for Chapter 15, each chapter is a condensed version of a subject ordinarily expanded to cover an entire textbook. Consequently, the material found here is necessarily less comprehensive, and also generally less formal (i.e., it is presented in somewhat of a tutorial style). We discuss the main ideas that we feel are essential to each chapter topic and try to relate the mathematical techniques to a variety of applications, many of which are commonly associated with electrical and optical engineering—e.g., communications, imaging, radar, antennas, and optics, among others. Nonetheless, we believe the general exposition and choice of topics should appeal to a wide audience of applied practitioners. Last, we wish to thank our reviewers Christopher Groves-Kirkby and Andrew Tescher for their careful review of the manuscript and helpful suggestions.

Larry C. Andrews
Ronald L. Phillips
Orlando, Florida (USA)
2003

Symbols and Notation

$\mathbf{a}, \mathbf{x}, \mathbf{A}, \ldots$	Vector, matrix, or random variable
AM	Amplitude modulation
arg, Arg	Argument
BC	Boundary condition
BVP	Boundary value problem
CDF	Cumulative distribution function
CNR	Carrier-to-noise ratio
Cov	Covariance
CTF	Coherent transfer function
CW	Continuous wave
DE	Differential equation
DFT	Discrete Fourier transform
DWT	Discrete Walsh transform
E[.]	Expectation operator
EO	Electro-optics
$\mathscr{F}\{.\}$	Fourier transform operator
FFT	Fast Fourier transform
FM	Frequency modulation
FRFT	Fractional Fourier transform
FWT	Fast Walsh transform
GRIN	Graded index
$\mathcal{H}\{.\}$	Hankel transform operator
IC	Initial condition
IF	Intermediate frequency
Im	Imaginary part of
ISP	Irregular singular point
IVP	Initial value problem
J	Jacobian
$\mathscr{L}\{.\}$	Laplace transform operator
LO	Local oscillator
LRC	Inductor-resistor-capacitor network
LSI	Linear shift-invariant
M{.}	Mellin transform operator
Mod	Modulus
ODE	Ordinary differential equation
PDE	Partial differential equation
PDF	Probability density function

Pr	Probability
PV	Principal value
RC	Resistor-capacitor network
Re	Real part of
Res	Residue
RSP	Regular singular point
SNR	Signal-to-noise ratio
$u_x, u_{xt}, \ldots$	Partial derivative
U.H.P.	Upper half-plane
Var	Variance
z^*	Complex conjugate of z
$\delta_{jk}, \delta_j^k, \delta^{jk}$	Kronecker delta
$\delta y, \delta F$	First variation
∇	Del (or gradient) operator
e_{ijk}, e^{ijk}	Permutation symbols
$\in$	Belonging to
$\notin$	Not belonging to
$\int_C$	Path or contour integral
$\oint_C$	Closed path (contour) integral
$\iint_S$	Surface integral
$\oiint_S$	Closed surface integtral
$\iiint_V$	Volume integral
$\cap$	Intersection
$<>$	Ensemble average
$<\mathbf{x}, \mathbf{y}>$	Inner product
$<a,b,c>$	Vector components

NOTE: Notation for special functions of mathematics is provided in Table 2.5 at the end of Chapter 2.

Mathematical Techniques
for Engineers and Scientists

Chapter 1

Ordinary Differential Equations

1.1 Introduction / 2
1.2 Classifications / 3
    1.2.1   Solutions of differential equations / 4
1.3 First-Order Equations / 6
    1.3.1   Separation of variables / 7
    1.3.2   Linear equations / 8
    1.3.3   Initial condition / 11
    1.3.4   Applications / 13
1.4 Second-Order Linear Equations / 17
    1.4.1   Homogeneous equations: fundamental solution sets / 18
    1.4.2   Constant coefficient equations / 20
    1.4.3   Nonhomogeneous equations—Part I / 24
    1.4.4   Nonhomogeneous equations—Part II / 28
    1.4.5   Cauchy-Euler equations / 31
1.5 Power Series Method / 34
    1.5.1   Review of power series / 34
1.6 Solutions Near an Ordinary Point / 35
    1.6.1   Ordinary and singular points / 36
    1.6.2   General method for ordinary points / 37
1.7 Legendre's Equation / 40
    1.7.1   Legendre polynomials: $P_n(x)$ / 41
    1.7.2   Legendre functions of the second kind: $Q_n(x)$ / 42
1.8 Solutions Near a Singular Point / 43
    1.8.1   Method of Frobenius / 44
1.9 Bessel's Equation / 50
    1.9.1   The gamma function: $\Gamma(x)$ / 50
    1.9.2   Bessel functions of the first kind: $J_v(x)$ / 52
    1.9.3   Bessel functions of the second kind: $Y_v(x)$ / 55
    1.9.4   Differential equations related to Bessel's equation / 56
Suggested Reading / 57
Exercises / 58

Historical Comments: Historically, the study of differential equations originated with the introduction of the calculus by Sir Isaac Newton (1642-1727) and Gottfried Wilhelm von Leibniz (1646-1716). Although mathematics began as a recreation for Newton, he became known as a great mathematician by the age of 24 after his invention of the calculus, discovery of the law of universal gravitation, and experimental proof that white light is composed of all colors. Leibniz completed his doctorate in philosophy by the age of 20 at the University of Altdorf. Afterward he studied mathematics under the supervision of Christian Huygens (1629-1695) and, independently of Newton, helped develop the calculus. Leibniz corresponded regularly with other mathematicians concerning differential equations, and he developed several methods for solving first-order equations.

Other prominent mathematicians who contributed to the development of differential equations and their applications were members of the famous Bernoulli family of Switzerland, the most famous of which are James (1654-1705) and John (1667-1748). Over the years there have been a host of mathematicians who contributed to the general development of differential equations.

The objective of this chapter is to review the basic ideas found in a first course in ordinary differential equations (ODEs). In doing so, we will concentrate primarily on those we deem most important in engineering applications. Because DEs are considered the most fundamental models that are used in a wide variety of physical phenomena, they play a central role in many of the following chapters of this text.

1.1 Introduction

Differential equations (DEs) play a fundamental role in engineering and science because they can be used in the formulation of many physical laws and relations. The development of the theory of DEs is closely interlaced with the development of mathematics in general, and it is indeed difficult to separate the two. In fact, most of the famous mathematicians from the time of Newton and Leibniz had a part in the cultivation of this fascinating subject. The first problems studied that involved the notion of DE came from the field of mechanics. Consequently, some of the terminology that persists today (like "forcing function") had its beginning in these early mechanics problems.

At its most basic level, *Newton's second law of motion* is commonly expressed by the simple algebraic formulation

$$F = ma.$$

For a "particle" or body in motion, F denotes the force acting on the body, m is the mass of the body (generally assumed to be constant), and a is its *acceleration*. In practice, however, it is the *velocity* and *position* of the body as a function of time that may be more useful. Recalling that the velocity (speed) v is related to acceleration by $a = dv/dt$, Newton's second law may also be expressed by

$$F = m\frac{dv}{dt}.$$

Last, the position y of the body is related to velocity by $v = dy/dt$, and by substituting this last expression into the above equation, we get another variation of Newton's law given by

$$F = m\frac{d^2y}{dt^2}.$$

These last two expressions of Newton's second law are considered DEs because they involve derivatives of unknown functions. And although the study of DEs grew out of certain kinds of problems in mechanics, their use today is far more widespread. For example, they occur in various branches of engineering and physical science to study problems in the following areas (among others):

- the study of particle motion
- the analysis of electric circuits and servomechanisms
- continuum and quantum mechanics
- the theory of diffusion processes and heat flow
- electromagnetic theory

Other disciplines such as economics and the biological sciences are also using DEs to investigate problems like the following:

- interest rates
- population growth
- the ecological balance of systems
- the spread of epidemics

1.2 Classifications

Definition 1.1 By *differential equation* we mean an equation that is composed of
- a single unknown function y, and
- a finite number of its derivatives.

A simple example of a DE found in the calculus is to find all functions y for which

$$y' = f(x), \tag{1}$$

where $y' = dy/dx$ and $f(x)$ is a given function. The formal solution to this equation is

$$y = \int f(x)dx + C, \tag{2}$$

where C is an arbitrary constant.

Most of the DEs that concern us are not of the simple type illustrated by (1). Typical examples of the types of equations found here include

$$y' = x^2 y^3 \tag{3}$$

$$y'' + cy' + k^2 y = \sin x \tag{4}$$

$$y'' + b \sin y = 0 \tag{5}$$

$$(y')^2 + 3xy = 1 \tag{6}$$

In order to provide a framework in which to discuss various solution techniques for DEs, it is helpful to first introduce *classification schemes* for the equations. For example, some important classifications are the following:

▸ **Order:** The *order* of a DE is the order of its highest derivative.

▸ **Linear:** The DE is said to be *linear* if it is linear in the unknown function y and all derivatives. If a DE is not linear, it is called *nonlinear*.

▸ **Ordinary DE:** A DE is called *ordinary* when the unknown function depends on only one independent variable. Otherwise, it is a *partial* DE.

Based on these definitions, Eqs. (3)-(6) are *ordinary*, Eqs. (3) and (6) are *first-order* equations, and (4) and (5) are *second-order*. Also, only Eq. (4) is *linear*—the others are *nonlinear*.

1.2.1 Solutions of differential equations

Definition 1.2 A *solution* $y = y(x)$ of a DE on an interval I is a continuous function possessing all derivatives occurring in the equation that, when substituted into the DE, reduces it to an identity for all x in the interval I.

EXAMPLE 1 Verify that $y = e^{-x}$ is a *solution* of the first-order DE

$$y' + y = 0.$$

Solution: To verify that a given (differentiable) function is a solution of a DE, we simply substitute it directly into the DE. Note that the function $y = e^{-x}$ is continuous and has continuous derivatives for all x. Furthermore,

$$y' + y = -e^{-x} + e^{-x} = 0$$

for all values of x. It also follows in the same manner that $y = C_1 e^{-x}$ is a solution for all values of x and all values of the constant C_1.

EXAMPLE 2 Verify that $y_1 = C_1 \cos 2x$ and $y_2 = C_2 \sin 2x$ are both solutions of the second-order DE

$$y'' + 4y = 0$$

for any values of the constants C_1 and C_2.

Solution: Both functions are continuous and have continuous derivatives for all x. For y_1, we have

$$y_1'' + 4y_1 = -4C_1 \cos 2x + 4C_1 \cos 2x = 0,$$

for all values of x. Similarly, for y_2 it follows that

$$y_2'' + 4y_2 = -4C_1 \sin 2x + 4C_1 \sin 2x = 0.$$

In Example 2, we illustrated that both y_1 and y_2 are solutions of the given DE. Moreover, it is easy to verify that the functions

$$y = C_1 \cos 2x + C_2 \sin 2x,$$

$$y = C_3 \sin x \cos x,$$

are also solutions of the same DE.

Solutions are classified in the following manner:

▸ **Particular solution:** If a solution contains no arbitrary constants, it is called a *particular solution* of the DE.

▸ **General solution:** A function that contains all particular solutions of the DE is called a *general solution*.

Because of arbitrary constants, the above DEs (Examples 1 and 2) have infinitely many solutions. However, the *number of arbitrary constants* that appears in a general solution is always *equal to the order of the DE*.

In solving DEs it can be important to know in advance: "Does a solution *exist*?" If so, we may then also want to know: "Is it *unique*?" In general, questions concerning the existence and uniqueness of solutions can be very difficult to answer.

1.3 First-Order Equations

First-order DEs arise naturally in problems involving the determination of the velocity of free-falling bodies subject to a resistive force, finding the current or charge in an electric circuit, finding curves of population growth, and in radioactive decay, among other applications. Each type of first-order DE that arises in practice may demand a different method of solution. And although others exist, we will introduce only two methods of solution—*separation of variables* and *linear equations*—both of which are applicable to a wide variety of practical problems involving first-order DEs.

First-order DEs are typically written in either the *derivative form*

$$y' = F(x,y) \tag{7}$$

or, through formal manipulations, in the *differential form*

$$M(x,y)dx + N(x,y)dy = 0. \tag{8}$$

In applications, the solution of (7) or (8) is usually required to also satisfy an *auxiliary condition* of the form

$$y(x_0) = y_0, \tag{9}$$

which geometrically specifies that the graph of the solution pass through the point (x_0,y_0) of the *xy*-plane. Because x_0 is often the beginning point in the interval of interest, the condition (9) is also called an *initial condition* (IC). Hence, solving (7) or (8) subject to the auxiliary condition (9) is called an *initial value problem* (IVP).

Many first-order DEs that routinely arise in applications are *nonlinear*. In some cases these may be very difficult or impossible to solve by known methods. To ensure that the DE together with its initial condition (9) has a solution, we have the following *existence-uniqueness theorem* which we state without proof (see [17]).

Theorem 1.1 If $F(x,y)$ is a continuous function in a rectangular domain $a<x<b$, $c<y<d$ containing the point (x_0,y_0), then the IVP

$$y' = F(x,y), \quad y(x_0) = y_0$$

has at least one solution in some interval $|x - x_0| < h$, $(h > 0)$ embedded in $a<x<b$. If, in addition, the partial derivative $\partial F/\partial y$ is continuous in that rectangle, then the IVP has a *unique* solution.

Remark: The conditions stated in Theorem 1.1 are only *sufficient conditions*—not *necessary conditions*. That is, if these conditions are not satisfied, the problem may have *no solution*, but in some cases may have *more than one solution* or even a *unique solution*!

1.3.1 Separation of variables

Perhaps the easiest method to apply when it is appropriate is that called *separation of variables*. The first-order DE written in differential form

$$M(x,y)dx + N(x,y)dy = 0 \tag{10}$$

is said to be "separable" if it can be rearranged in the form

$$\boxed{f(y)dy = g(x)dx.} \tag{11}$$

Observe that the left-hand side of (11) is a function of y alone and the right-hand side is a function of x alone. A family of solutions can be obtained by simple integration of each side to yield the form

$$\int f(y)dy = \int g(x)dx + C, \tag{12}$$

where C is a constant of integration (only one such constant is required because the arbitrary constants from each side of the equation can be combined).

EXAMPLE 3 Solve the DE
$$(1 - x)dy + ydx = 0.$$

Solution: We see that division by y and $(1 - x)$ leads to the "separated form"

$$\frac{dy}{y} = -\frac{dx}{1 - x}, \quad x \neq 1, y \neq 0.$$

Thus, by integrating each side independently, we arrive at

$$\ln|y| = \ln|1 - x| + C,$$

which, by writing $C = \ln|C_1|$ and using properties of logarithms, yields

$$y = C_1(1 - x).$$

Remark: Recall from calculus that $\int \frac{du}{u} = \ln|u| + C$, where the absolute value is retained except in those cases in which we know $u > 0$.

EXAMPLE 4 Solve the IVP

$$(x^2 + 1)y' + y^2 + 1 = 0, \quad y(0) = 1.$$

Solution: By rearranging terms, we have

$$\frac{dy}{y^2 + 1} = -\frac{dx}{x^2 + 1},$$

which, upon integration, leads to

$$\tan^{-1} y = -\tan^{-1} x + C.$$

If we apply the prescribed initial condition ($x = 0$, $y = 1$), we see that $C = \pi/4$, and consequently,

$$\tan^{-1} y + \tan^{-1} x = \frac{\pi}{4}.$$

A more convenient form of the solution can be obtained by using the trigonometric identity

$$\tan(A + B) = \frac{\tan A + \tan B}{1 - \tan A \tan B}.$$

Thus, we find

$$\tan(\tan^{-1} y + \tan^{-1} x) = \tan(\pi/4),$$

or, equivalently,

$$\frac{y + x}{1 - xy} = 1.$$

Now, solving explicitly for y, we get

$$y = \frac{1 - x}{1 + x}, \quad x \neq -1.$$

1.3.2 Linear equations

A *linear first-order DE* is any equation that can be put in the form

$$A_1(x)y' + +A_0(x)y = F(x). \tag{13}$$

The functions $A_0(x)$ and $A_1(x)$ are the *coefficients* of the DE (which do not depend on y) and $F(x)$ is the *input function*, also called the *forcing function*. In practice, the solution of (13) is usually required to satisfy the initial condition (IC)

$$y(x_0) = y_0. \tag{14}$$

For developing the solution of a linear DE it is customary to first put (13) in the *normal form*

$$y' + a_0(x)y = f(x), \qquad (15)$$

obtained by dividing each term of (13) by $A_1(x)$. Hence, $a_0(x) = A_0(x)/A_1(x)$ and $f(x) = F(x)/A_1(x)$.

To develop a general solution of (15), we first consider the associated *homogeneous equation*

$$y' + a_0(x)y = 0, \qquad (16)$$

in which the input function $f(x)$ is identically zero. One feature of a homogeneous equation is that $y = 0$ is always a solution, called the *trivial solution*. However, our interest concerns *nontrivial* (nonzero) *solutions*. We do note that (16) can be formally solved by the method of separation of variables, which leads to

$$\frac{dy}{y} = -a_0(x)dx, \quad y \neq 0.$$

The direct integration of this expression then yields

$$\ln|y| = -\int a_0(x)dx + C, \qquad (17)$$

where C is a constant of integration. By solving (17) directly for y, we obtain the family of solutions

$$y_H = C_1 y_1(x) = C_1 \exp\left[-\int a_0(x)dx\right], \qquad (18)$$

where we write y_H to denote that (18) is a *solution* of the *homogeneous* equation. We also introduced the notation $C_1 = e^C$, and the function $y_1(x)$ is defined by

$$y_1(x) = \exp\left[-\int a_0(x)dx\right]. \qquad (19)$$

The solution function (18) is a *general solution* of the *homogeneous equation* (16). A *general solution* of the *nonhomogeneous equation* (15) is defined by the sum $y = y_H + y_P$, where y_P is any *particular solution* and y_H is defined above. To construct the particular solution, we employ a method called *variation of parameters* that was developed by J. L. Lagrange (1736-1813). Specifically, we look for a solution of the form

$$y_P = u(x)y_1(x), \qquad (20)$$

where $u(x)$ is a function to be determined. The technique derives its name from the fact that the arbitrary constant C_1 in the homogeneous solution (18) is replaced by the unknown function $u(x)$.

The direct substitution of (20) into the left-hand side of the DE in (15) gives us

$$y_P' + a_0(x)y_P = \frac{d}{dx}[u(x)y_1(x)] + a_0(x)u(x)y_1(x)$$
$$= u'(x)y_1(x) + u(x)[y'(x) + a_0(x)y_1(x)] \qquad (21)$$
$$= u'(x)y_1(x) + 0,$$

where we are using the fact that y_1 satisfies the associated homogeneous DE (16). Now, if (20) is indeed a particular solution, then the result of (21) requires that $u(x)$ be a solution of $u'(x)y_1(x) = f(x)$. Upon integration, we determine

$$u(x) = \int \frac{f(x)}{y_1(x)} dx,$$

where we can ignore the arbitrary constant of integration (i.e., any particular solution is good enough!). Hence, the particular solution is

$$\boxed{y_P = y_1(x)\int \frac{f(x)}{y_1(x)} dx,} \qquad (22)$$

and the general solution of (15) becomes

$$\boxed{y = y_H + y_P = C_1 y_1(x) + y_1(x)\int \frac{f(x)}{y_1(x)} dx.} \qquad (23)$$

EXAMPLE 5 Find the solution of

$$xy' + (1 - x)y = xe^x, \quad y(1) = 3e.$$

Solution: We first rewrite the DE in normal form, which yields

$$y' + \left(\frac{1}{x} - 1\right)y = e^x, \quad x \neq 0.$$

Thus, using (19) we see that

$$y_1(x) = \exp\left[-\int\left(\frac{1}{x} - 1\right)dx\right] = \frac{1}{x}e^x,$$

and, consequently, the homogeneous solution (18) is $y_H = C_1 e^x/x$.

From the normal form above we note that $f(x) = e^x$, and thus, the particular solution is $y_P = u(x)e^x/x$, where

$$u(x) = \int \frac{e^x}{x^{-1}e^x}dx = \int x\,dx = \frac{x^2}{2}.$$

Hence, $y_P = \frac{1}{2}xe^x$, and the general solution we seek can be expressed in the form

$$y = y_H + y_P = \frac{C_1 e^x}{x} + \frac{1}{2}xe^x, \quad x \neq 0.$$

Last, by imposing the IC, we are led to

$$y(1) = C_1 e + \frac{1}{2}e = 3e,$$

or $C_1 = 5/2$. Thus,

$$y = \frac{1}{2}\left(\frac{5}{2x} + x\right)e^x, \quad x \neq 0.$$

1.3.3 Initial condition

When solving (15) subject to the IC (14), it can be useful for physical interpretation to split the problem into two simpler problems, defined by

PROBLEM (A): $\quad y' + a_0(x)y = 0, \quad y(x_0) = y_0$ \hfill (24)

PROBLEM (B): $\quad y' + a_0(x)y = f(x), \quad y(x_0) = 0.$ \hfill (25)

If we subject the general solution (18) of the homogeneous equation in PROBLEM (A) to the prescribed IC, we are led to

$$y_H(x_0) = C_1 y_1(x_0) = y_0,$$

and hence, the IVP (24) has the unique solution

$$y_H = \frac{y_0 y_1(x)}{y_1(x_0)}. \tag{26}$$

Equation (26) physically represents the "response" of the system (15) to the IC (14) in the absence of an input function $f(x)$. Similarly, the solution of (25) is considered the response of the system when y_0 in the IC is zero, i.e., the system is "at rest." To obtain this solution, we now define the specific function

$$u(x) = \int_{x_0}^x \frac{f(s)}{y_1(s)}\,ds, \tag{27}$$

where we have introduced the dummy variable of integration s; thus,

$$y_P = y_1(x) \int_{x_0}^{x} \frac{f(s)}{y_1(s)} ds. \tag{28}$$

The function $y_1(x)/y_1(s)$ represents what is commonly called a *Green's function* (Section 8.5) or *impulse response function* (Section 9.6). Note that, by selecting the limits of integration as shown, the solution (28) necessarily satisfies the IC in PROBLEM (B), viz.,

$$y_P(x_0) = y_1(x_0) \int_{x_0}^{x_0} \frac{f(s)}{y_1(s)} ds = 0.$$

Summarizing the above results for y_H and y_P, the solution of the original IVP (14) and (15) is given by the sum of solutions

$$y = y_H + y_P = \frac{y_0 y_1(x)}{y_1(x_0)} + y_1(x) \int_{x_0}^{x} \frac{f(s)}{y_1(s)} ds, \tag{29}$$

where $y_1(x)$ is defined by (19).

EXAMPLE 6 Solve the IVP
$$y' - 2xy = 1, \quad y(0) = 3.$$

Solution: Based on the discussion presented above, we start by splitting the problem into the following two problems:

PROBLEM (A): $y' - 2xy = 0, \quad y(0) = 3,$

PROBLEM (B): $y' - 2xy = 1, \quad y(0) = 0.$

The general solution of the homogeneous DE in PROBLEM (A) is

$$y_H = C_1 e^{x^2},$$

and by imposing the IC, we see that $C_1 = 3$; thus,

$$y_H = 3e^{x^2}.$$

On the other hand, PROBLEM (B) above has the particular solution

$$y_P = e^{x^2} \int_0^x e^{-t^2} dt.$$

However, the integral in this case is nonelementary, but is related to a special function called the *error function* (see Section 2.3.4) and defined by

$$\text{erf}(x) = \frac{2}{\sqrt{\pi}} \int_0^x e^{-t^2} dt.$$

Therefore, we can express the solution of our IVP in the form

$$y = 3e^{x^2} + \frac{\sqrt{\pi}}{2} e^{x^2} \text{erf}(x).$$

1.3.4 Applications

DEs have always been applied to problems in mechanics, most of which involve quantities that change in a *continuous* manner such as distance, velocity, acceleration, and force. However, in problems in the life sciences, for example, the quantity of interest may be the population size of a particular community that changes by *discrete amounts* rather than continuously. For this reason we may not be able to model this latter class of problems by a DE, because derivatives require continuous functions. However, if the population size is "sufficiently large," it can often be modeled as a continuous system with a DE whose solution may describe general characteristics of the population quite well. In this section, we look at some typical problems in different application areas that involve first-order equations.

In the application of DEs we are generally concerned with more than simply solving a particular mathematical problem. Indeed, the starting point is usually some real-world problem that must be mathematically formulated before it can be solved. The complete *solution process*, therefore, consists primarily of the following three steps (see also Fig. 1.1).

---

1. *Construction of a mathematical model*:

   The variables involved must be carefully defined and the governing physical (or biological) laws identified. The mathematical model is usually some differential equation(s) representing an idealization of the laws, taking into account some simplifying assumptions in order to make the model tractable.

2. *Solution of the mathematical model*:

   When permitted, exact solutions are desired, but in many cases one must rely on approximate solutions. It can sometimes be useful to establish the existence and/or uniqueness of the solution. If a solution doesn't exist, it is probably a poor model.

3. *Interpretation of the results*:

   The solutions obtained from the model should be consistent with physical evidence. If a good model has been constructed, the solution should describe many of the essential characteristics of the system under study.

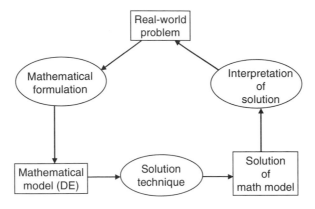

Figure 1.1

EXAMPLE 7 Newton's second law of motion for a "particle" or body of mass m can be written as

$$\frac{d}{dt}(mv) = F,$$

where v is the velocity and the product mv can be interpreted as momentum. The function F denotes the sum of external forces acting on the body. If this DE is applied to a skydiver weighing 150 lb who opens his/her parachute when his/her downward velocity (speed) is 100 ft/s and the force of resistance is $5v$,

(a) what is his/her velocity as a function of time t?
(b) What is his/her limiting velocity (after a long time)?
(c) If the skydiver opens his/her parachute at 2000 ft above ground, how close to the ground will he/she be after 1 min?
(d) If the air resistance is $5v^2$, what is the limiting velocity of the skydiver?

Solution: The forces acting on the skydiver are his/her weight $W = mg$, where $g = 32$ ft/s², and the force of resistance is $F_R = 5v$. Therefore, the governing IVP is

$$m\frac{dv}{dt} = mg - 5v, \quad v(0) = 100.$$

(a) The DE is linear and the mass of the skydiver is $m = 150/32 = 75/16$ slugs (a "slug" is a unit of mass). In normal form, the IVP becomes

$$\frac{dv}{dt} + \frac{16}{15}v = 32, \quad v(0) = 100.$$

The homogeneous solution (26) is

$$v_H(t) = 100e^{-16t/15},$$

and the corresponding particular solution (28) is

$$v_P(t) = 32e^{-16t/15} \int_0^t e^{16s/15} ds = 30 - 30e^{-16t/15},$$

from which we deduce

$$v(t) = v_H(t) + v_P(t) = 30 + 70e^{-16t/15}.$$

(b) The limiting or terminal velocity of the skydiver is

$$v_\infty = \lim_{t \to \infty} v(t) = 30\text{ft/s}.$$

(c) The distance y of the skydiver above the ground is related to his/her velocity by

$$\frac{dy}{dt} = -v(t) = -\left(30 + 70e^{-16t/15}\right),$$

where the negative sign indicates that y is decreasing. Direct integration of this result leads to

$$y(t) = -30t + \frac{525}{8}e^{-16t/15} + C,$$

and by imposing the condition that $y(0) = 2000$, we see that $C \cong 1934$. Because 1 minute corresponds to 60 seconds, the position of the skydiver above the ground at this time is

$$y(60) \cong 134\,\text{ft}.$$

(d) In this case the IVP is described by

$$m\frac{dv}{dt} = mg - 5v^2, \quad v(0) = 100.$$

Here, the governing equation is nonlinear but can be solved by separating the variables. Doing so yields

$$\frac{dv}{30 - v^2} = \frac{16}{15}dt,$$

which leads to

$$\ln\left|\frac{v + \sqrt{30}}{v - \sqrt{30}}\right| = \frac{32\sqrt{30}}{15}t + \ln C_1,$$

or

$$\frac{v + \sqrt{30}}{v - \sqrt{30}} = C\exp\left(\frac{32\sqrt{30}}{15}t\right),$$

where C is a new arbitrary constant. Solving for v we obtain

$$v(t) = \sqrt{30}\left[\frac{C + \exp\left(-32\sqrt{30}\,t/15\right)}{C - \exp\left(-32\sqrt{30}\,t/15\right)}\right].$$

By imposing the initial condition $v(0) = 100$, we find $C \cong 1.12$. The limiting velocity this time is $(t \to \infty)$

$$v_\infty = \sqrt{30} \cong 5.48 \text{ ft/s}.$$

Example 7 is typical of problems involving the motion of a body. Many other related problems can be similarly modeled and solved by comparable methods. For example, the charge q in a *simple electric circuit* satisfies the linear equation

$$R\frac{dq}{dt} + \frac{1}{C}q = E(t), \tag{30}$$

where R is resistance in the circuit, C is capacitance, and $E(t)$ is a voltage source. *Radioactive decay* or *population growth* may both be modeled by the simple DE

$$\frac{dy}{dt} = ky, \tag{31}$$

under suitable conditions.

Below we present another type of problem involving an epidemic, which is similar to some of the models used in describing the growth of certain populations.

EXAMPLE 8 A student carrying a flu virus returns after Christmas vacation to an isolated college campus of 2000 healthy students and begins to infect the other students. Suppose the rate at which the virus spreads among the students is proportional to the number of infected students P and is proportional as well to the number of students not infected, i.e., $2000 - P$. The governing DE is therefore given by

$$\frac{dP}{dt} = kP(2000 - P),$$

where k is an unknown proportionality constant. If 20 students are infected with the virus after 3 days, how many students were infected after 1 day and how many will be infected by the end of a week?

Solution: Because only one student is infected with the virus at the beginning, we formulate the IVP we wish to solve by

$$\frac{dP}{dt} = kP(2000 - P), \quad P(0) = 1.$$

This equation is nonlinear but can be solved by separating the variables similar to part (*d*) of Example 7. The solution we obtain can be put in the form

$$P(t) = \frac{2000}{1 + 1999e^{-2000kt}}.$$

Using the condition $P(3) = 20$, we obtain $20 = 2000/[1 + 1999(e^{-2000k})^3]$, or

$$e^{-2000k} = \left(\frac{99}{1999}\right)^{1/3}.$$

Rather than solve this last equation for k, we substitute this expression directly into the above solution to get

$$P(t) = \frac{2000}{1 + 1999\left(\frac{99}{1999}\right)^{t/3}}.$$

Now, by setting $t = 1$ in this last equation, we find $P(1) \cong 3$ students infected at the end of the first day, and by the end of the week ($t = 7$ days), we find $P(7) \cong 715$ students infected. In both cases, we have rounded up to the nearest whole number.

1.4 Second-Order Linear Equations

Our treatment in this section will be limited to linear DEs of the second order, which are the most common type of DE in practice.

▸ **Linear DE:** A *linear DE* is any second-order equation that can be put in the form

$$\boxed{A_2(x)y'' + A_1(x)y' + A_0(x)y = F(x).} \tag{32}$$

We call $A_0(x)$, $A_1(x)$, and $A_2(x)$ the *coefficients* of the equation, which are functions of x alone. Unless otherwise stated, we generally assume the coefficients and the *input function* $F(x)$ are continuous functions on some interval I, and further, that $A_2(x) \neq 0$ on I. When the input function $F(x)$ is identically zero, Eq. (32) reduces to the *associated homogeneous equation*

$$A_2(x)y'' + A_1(x)y' + A_0(x)y = 0. \tag{33}$$

▸ **Normal form:**   The *normal form* of Eq. (32) is given by

$$\boxed{y'' + a_1(x)y' + a_0(x)y = f(x),} \tag{34}$$

obtained after division of each term in (32) by the lead coefficient $A_2(x)$.

Because the general solution of a second-order DE will contain two arbitrary constants, we find in applications the need for *two* auxiliary conditions. These are often in the form of ICs where $y(x_0)$ and $y'(x_0)$ are both specified at the same point x_0. The following existence-uniqueness theorem is an extension of Theorem 1.1 to second-order linear DEs.

Theorem 1.2 (*Existence-uniqueness*) If $a_0(x)$, $a_1(x)$, and $f(x)$ are continuous functions on an open interval I containing the point x_0, then the IVP

$$y'' + a_1(x)y' + a_0(x)y = f(x), \quad y(x_0) = \alpha, \quad y'(x_0) = \beta,$$

has a unique solution on I for every α and β.

1.4.1 Homogeneous equations: fundamental solution sets

Our development of the theory of second-order equations starts with the associated *homogeneous DE*

$$y'' + a_1(x)y' + a_0(x)y = 0. \tag{35}$$

One of the most useful properties of this DE is the fact that the sum of two (or more) solutions is also a solution, as is any multiple of a solution. This remarkable property is known as the *linearity property*, or *superposition principle*.

Theorem 1.3 (*Superposition principle*) If y_1 and y_2 are both solutions (on some interval I) of the homogeneous DE

$$y'' + a_1(x)y' + a_0(x)y = 0,$$

then

$$y = C_1 y_1(x) + C_2 y_2(x)$$

is also a solution on I for any constants C_1 and C_2.

Proof: Because the proof is short and instructive we will give it. The substitution of $y = C_1 y_1(x) + C_2 y_2(x)$ into the DE yields

$$\begin{aligned}
y'' + a_1(x)y' + a_0(x)y &= (C_1 y'' + C_2 y_2'') + a_1(x)(C_1 y' + C_2 y_2') \\
&\quad + a_0(x)(C_1 y + C_2 y_2) \\
&= C_1 [y_1'' + a_1(x)y_1' + a_0(x)y_1] \\
&\quad + C_2 [y_2'' + a_1(x)y_2' + a_0(x)y_2] \\
&= C_1 \cdot 0 + C_2 \cdot 0 = 0.
\end{aligned}$$

In the last step we are recognizing that both y_1 and y_2 are solutions of the homogeneous DE. ❑

Theorem 1.3 tells us that $y = C_1 y_1(x) + C_2 y_2(x)$ is a solution of the homogeneous DE (35), provided both y_1 and y_2 are solutions. To ensure that this represents the *general solution* of (35), we need to establish that y_1 and y_2 are linearly independent solutions.

> **Definition 1.3** We say two functions y_1 and y_2 are *linearly independent* functions on some interval I if neither one is a constant multiple of the other on I.

- ▸ **Fundamental solution set:** Linearly independent solutions y_1 and y_2 of a second-order homogeneous equation on some interval I are said to form a *fundamental set of solutions* on I, and in this case the general solution is

$$y = C_1 y_1(x) + C_2 y_2(x). \tag{36}$$

For theoretical purposes, and also to extend the notion of linear independence of solutions to higher-order equations, we now offer an equivalent test of linear independence that involves a determinant called the Wronskian. First, observe that if constants C_1 and C_2 (not both zero) can be found such that $C_1 y_1(x) + C_2 y_2(x) = 0$ for all x in some interval I, then y_1 and y_2 are linearly dependent on I (i.e., one must be a multiple of the other). Let us now consider the simultaneous equations

$$\begin{aligned} C_1 y_1(x) + C_2 y_2(x) &= 0, \\ C_1 y_1{}'(x) + C_2 y_2{}'(x) &= 0, \end{aligned} \tag{37}$$

the second of which is the derivative of the first. If we consider the constants C_1 and C_2 as unknowns in (37), then *nonzero* values for C_1 and C_2 are possible only when the coefficient determinant of (37) is zero for every x on some interval I (see Section 3.3). That is, y_1 and y_2 are linearly dependent on I only when

$$\begin{vmatrix} y_1(x) & y_2(x) \\ y_1{}'(x) & y_2{}'(x) \end{vmatrix} \equiv 0 \quad \text{on } I. \tag{38}$$

This coefficient determinant is called the *Wronskian*, named after the Polish mathematician J. M. H. Wronski (1778-1853), and is denoted by

$$W(y_1, y_2)(x) = \begin{vmatrix} y_1(x) & y_2(x) \\ y_1{}'(x) & y_2{}'(x) \end{vmatrix} = y_1(x)y_2{}'(x) - y_1{}'(x)y_2(x). \tag{39}$$

Thus far we have only shown that, if the functions y_1 and y_2 are *linearly dependent* on the interval I, the Wronskian (39) is zero everywhere on I. Although we won't present the proof, it has been shown that if y_1 and y_2 are any continuous solutions of the homogeneous equation (35), we can draw the following conclusions:

1. If $W(y_1,y_2)(x_0) \neq 0$ at any point x_0 on I, then y_1 and y_2 are *linearly independent* on I and thus form a *fundamental solution set*.

2. If $W(y_1,y_2)(x_0) = 0$ at any point x_0 on I, then $W(y_1,y_2)(x) \equiv 0$ on I, and y_1 and y_2 are *linearly dependent* on I.

1.4.2 Constant coefficient equations

Differential equations with *constant coefficients* are the easiest class of linear DEs to solve because the solution technique is primarily algebraic. The *homogeneous* second-order equation of this class is given by

$$ay'' + by' + cy = 0, \quad -\infty < x < \infty, \tag{40}$$

where a, b, and c are constants and $a \neq 0$. The form of the equation suggests that the solution must be a function that reproduces itself upon differentiation. Thus, we seek solutions of the general form

$$y = e^{mx} \tag{41}$$

in which we must determine values of the parameter m. Direct substitution of (41) into the left-hand side of (40) leads to

$$ay'' + by' + cy = am^2 e^{mx} + bme^{mx} + ce^{mx}$$
$$= (am^2 + bm + c)e^{mx} = 0,$$

which will vanish for all x if and only if

$$am^2 + bm + c = 0. \tag{42}$$

▸ **Auxiliary equation:** Equation (42) is called the *auxiliary equation* of (40). It clearly has the two solutions defined by

$$m_1 = \frac{-b + \sqrt{b^2 - 4ac}}{2a}, \quad m_2 = \frac{-b - \sqrt{b^2 - 4ac}}{2a}. \tag{43}$$

In developing linearly independent solutions of (40), we must consider three distinct cases corresponding to the discriminant of (43), viz., $b^2 - 4ac > 0$, $b^2 - 4ac = 0$, and $b^2 - 4ac < 0$.

CASE I $(b^2 - 4ac > 0)$: m_1 and m_2 real and distinct roots

When m_1 and m_2 are real and $m_1 \neq m_2$, we obtain two solutions of the form (41), viz.,

$$y_1 = e^{m_1 x}, \quad y_2 = e^{m_2 x}.$$

In this case neither solution is a multiple of the other, i.e., $y_1/y_2 = e^{(m_1 - m_2)x} \neq$ const., so we conclude that these are linearly independent solutions. The general solution is

$$y = C_1 e^{m_1 x} + C_2 e^{m_2 x}, \quad -\infty < x < \infty, \tag{44}$$

where C_1 and C_2 are arbitrary constants.

EXAMPLE 9 Find a general solution of the DE

$$2y'' + 5y' - 3y = 0.$$

Solution: By inspection, we see that the auxiliary equation is

$$2m^2 + 5m - 3 = (2m - 1)(m + 3) = 0,$$

which we have factored. Thus, the roots are $m_1 = 1/2$ and $m_2 = -3$, leading to the general solution

$$y = C_1 e^{x/2} + C_2 e^{-3x}.$$

EXAMPLE 10 Solve the IVP

$$y'' + 3y' = 0, \quad y(0) = 2, \ y'(0) = 1.$$

Solution: The auxiliary equation is

$$m^2 + 3m = m(m + 3) = 0,$$

with roots $m_1 = 0$ and $m_2 = -3$. Therefore, the general solution is

$$y = C_1 + C_2 e^{-3x}.$$

The derivative yields $y' = -3C_2 e^{-3x}$, and hence, imposing the ICs we get

$$y(0) = C_1 + C_2 = 2,$$
$$y'(0) = -3C_2 = 1.$$

It follows that $C_2 = -1/3$ and $C_1 = 7/3$, giving us the solution

$$y = \frac{1}{3}(7 - e^{-3x}).$$

CASE II $(b^2 - 4ac = 0)$: $m_1 = m_2 = m$

Here we initially obtain only one solution

$$y_1 = e^{mx}.$$

It has been shown, and easily verified, that

$$y_2 = xe^{mx}$$

is also a solution of the DE that is linearly independent of y_1. Hence, the general solution when the roots are equal is

$$y = C_1 e^{mx} + C_2 xe^{mx}$$
$$= (C_1 + C_2 x)e^{mx}, \quad -\infty < x < \infty. \tag{45}$$

EXAMPLE 11 Find a general solution of

$$y'' + 4y' + 4y = 0.$$

Solution: The auxiliary equation $m^2 + 4m + 4 = (m + 2)^2 = 0$ has a double root $m = -2, -2$. Therefore, the general solution is

$$y = (C_1 + C_2 x)e^{-2x}.$$

EXAMPLE 12 Solve the IVP

$$y'' - 2y' + y = 0, \quad y(0) = 3, \; y'(0) = 1.$$

Solution: The auxiliary equation $m^2 - 2m + 1 = (m - 1)^2 = 0$ has the double root $m = 1, 1$, from which we deduce

$$y = (C_1 + C_2 x)e^x.$$

The derivative of y is $y' = C_1 e^x + C_2(1 + x)e^x$. Consequently, the ICs demand that

$$y(0) = C_1 = 3,$$
$$y'(0) = C_1 + C_2 = 1.$$

Hence, $C_1 = 3$ and $C_2 = -2$, and the solution of the IVP we seek is

$$y = (3 - 2x)e^x.$$

CASE III ($b^2 - 4ac < 0$): $m = p \pm iq$, where both p and q are real

---

When the roots are complex, they are complex conjugates. Formal solutions are given by

$$\begin{aligned}
y_1 &= e^{(p+iq)x} = e^{px}(\cos qx + i\sin qx), \\
y_2 &= e^{(p-iq)x} = e^{px}(\cos qx - i\sin qx),
\end{aligned} \tag{46}$$

where we have invoked Euler's formula $e^{\pm i\theta} = \cos\theta \pm i\sin\theta$. However, to avoid using complex solutions, we use the superposition principle to obtain real solutions y_3 and y_4 defined by

$$y_3 = \frac{1}{2}(y_1 + y_2) = e^{px}\cos qx,$$

$$y_4 = \frac{1}{2i}(y_1 - y_2) = e^{px}\sin qx.$$

We leave it to the reader to verify that these are linearly independent solutions of (40) under the assumed conditions. In this case, therefore, the general solution can be expressed in the form (by definition we choose q to be positive)

$$y = e^{px}(C_1\cos qx + C_2\sin qx), \quad -\infty < x < \infty. \tag{47}$$

EXAMPLE 13 Find the general solution of

$$y'' - 4y' + 13y = 0.$$

Solution: The auxiliary equation is $m^2 - 4m + 13 = 0$, which, by way of the quadratic formula, leads to the complex roots $m = 2 \pm 3i$. Hence, we identify $p = 2$ and $q = 3$, from which we deduce the general solution

$$y = e^{2x}(C_1\cos 3x + C_2\sin 3x).$$

In summary, we have derived the following solution forms for the three cases of solutions of the homogeneous DE (40).

CASE I. $m_1 \neq m_2$ (both real): $y + C_1 e^{m_1 x} + C_2 e^{m_2 x}$

CASE II. $m_1 = m_2 = m$: $y = (C_1 + C_2 x)e^{mx}$

CASE III. $m_1, m_2 = p \pm iq$: $y = e^{px}(C_1\cos qx + C_2\sin qx)$

Remark: The above method is easily extended to higher-order DEs. For example, if the DE is of order n, we must find n roots of a polynomial

$$a_n m^n + a_{n-1} m^{n-1} + \cdots + a_1 m + a_0 = 0,$$

where $a_0, a_1, \ldots a_n$ are constants. By grouping roots according to the above three cases, we seek n corresponding fundamental solutions $y_1(x), \ldots, y_n(x)$ to form the general solution

$$y = C_1 y_1(x) + \ldots + C_n y_n(x).$$

1.4.3 Nonhomogeneous equations—Part I

In this section we consider a technique for finding the general solution of the constant-coefficient *nonhomogeneous* DE

$$ay'' + by' + cy = F(x), \tag{48}$$

where a, b, c are constants. Similar to first-order linear equations discussed in Section 1.3.2, it has been shown that every solution of Eq. (48) is contained in the sum

$$y = y_H + y_P, \tag{49}$$

where y_H denotes the general solution of the associated homogeneous equation [obtained by setting $F(x) \equiv 0$] and y_P is any particular solution of the nonhomogeneous equation (48). Particular solutions are not unique—any particular solution that we can find will work for producing the general solution! There are essentially two distinct methods for constructing the particular solution y_P, one of which we discuss here (method of undetermined coefficients) and another method (variation of parameters) discussed in Section 1.4.4.

When the input function $F(x)$ is composed of functions whose nth derivative is zero or else reproduces a multiple of $F(x)$, we can use a simple technique called the method of *undetermined coefficients* to construct y_P. That is, to apply this method we require $F(x)$ to be composed of functions of the following type:

- A polynomial in x
- e^{px}
- $\cos qx$ or $\sin qx$
- A finite sum and/or product of these functions

Functions not suitable for this solution technique are $\ln x$, $\tan x$, e^{x^2}, and so on.

To illustrate the method, suppose we wish to find a particular solution of

$$y'' + y = 9e^{2x}. \tag{50}$$

Because differentiation of an exponential function reproduces itself, we "guess" a particular solution of the form

$$y_P = Ae^{2x}, \tag{51}$$

where A is an "undetermined coefficient." We substitute (51) into (50), getting

$$4Ae^{2x} + Ae^{2x} = 9e^{2x},$$

from which we deduce $A = 9/5$. Thus, a suitable particular solution of (50) is

$$y_P = \frac{9}{5}e^{2x}. \tag{52}$$

Recognizing that $y_H = C_1 \cos x + C_2 \sin x$ is the general solution of the associated homogeneous equation, we obtain the general solution of (50) in the form

$$y = y_H + y_P = C_1 \cos x + C_2 \sin x + \frac{9}{5}e^{2x}. \tag{53}$$

As a second example, let us consider

$$y'' + y = 3x^2. \tag{54}$$

This time we see that derivatives of x^2 lead to new functions involving x and x^0. Thus, we start with the more general form

$$y_P = Ax^2 + Bx + C. \tag{55}$$

If we substitute this function into (54) and simplify, we are led to

$$Ax^2 + Bx + (2A + C) = 3x^2.$$

By equating like coefficients in this last equation, we find $A = 3$, $B = 0$, $C = -6$. Hence, the particular solution of (54) is

$$y_P = 3x^2 - 6, \tag{56}$$

and the general solution is

$$y = y_H + y_P = C_1 \cos x + C_2 \sin x + 3x^2 - 6. \tag{57}$$

The general rule illustrated by these examples is the following:

Rule 1 Assume y_P has the same functional form as the nonhomogeneous term $F(x)$, plus all linearly independent derivatives of $F(x)$.

EXAMPLE 14 Find the general solution of

$$y'' + 2y' = 3\cos 2x.$$

Solution: Because the derivative of $\cos 2x$ leads to $\sin 2x$, the use of Rule 1 suggests the choice

$$y_P = A\cos 2x + B\sin 2x.$$

By substituting this assumed form for y_P into the DE and simplifying the results, we have

$$4(B - A)\cos 2x - 4(A + B)\sin 2x = 3\cos 2x,$$

and by comparing like coefficients, we see that $A = -B$ and $B = 3/8$. The homogeneous equation has general solution $y_H = C_1 + C_2 e^{-3x}$, and thus,

$$y = y_H + y_P = C_1 + C_2 e^{-3x} + \frac{3}{8}\sin 2x - \frac{3}{8}\cos 2x.$$

Suppose we now wish to solve

$$y'' + y = 3\cos x, \tag{58}$$

which resembles the nonhomogeneous term in Example 14. However, following Rule 1 and assuming the particular solution

$$y_P = A\cos x + B\sin x,$$

we find, after substituting this into (58) and simplifying, $0 = 3\cos x$. Clearly, we cannot satisfy this last equation for all x. To remedy this situation, we must select y_P to be a function that is linearly independent of any function in the homogeneous solution $y_H = C_1\cos x + C_2\sin x$. One way to do this is to assume

$$y_P = x(A\cos x + B\sin x). \tag{59}$$

Now, when we substitute this new form for y_P into (58), we are led to

$$-2A\sin x + 2B\cos x = 3\cos x.$$

Hence, $A = 0$ and $B = 3/2$, and $y_P = \frac{3}{2}x\sin x$. The general solution of (58) is

$$y = y_H + y_P = C_1\cos x + C_2\sin x + \frac{3}{2}x\sin x. \tag{60}$$

Based on this last example, we are led to our second and last rule. If the input function is $F(x) = F_1(x) + F_2(x)$, then Rule 2 applies to $F_1(x)$ and $F_2(x)$ separately.

> **Rule 2** If $F(x)$ or any of its derivatives satisfy the associated homogeneous equation, then the choice of y_P following Rule 1 must be multiplied by x. For a double root of the homogeneous equation, we multiply by x^2.

EXAMPLE 15 Solve the IVP

$$y'' + y = x\cos x - \cos x + 2e^x, \quad y(0) = 2, \ y'(0) = \frac{1}{4}.$$

Solution: The homogeneous equation $y'' + y = 0$ has the general solution

$$y_H = C_1 \cos x + C_2 \sin x.$$

By using both Rule 1 and Rule 2, we assume a particular solution of the form

$$
\begin{aligned}
y_P &= x[(Ax + B)\cos x + (Cx + D)\sin x] + Ee^x \\
&= (Ax^2 + Bx)\cos x + (Cx^2 + Dx)\sin x + Ee^x.
\end{aligned}
$$

In arriving at y_P, we must assume a particular solution for each distinct type of function in $F(x)$. The direct substitution of this function into the DE leads to (after some algebraic simplification)

$$
\begin{aligned}
4Cx\cos x + 2(A + D)\cos x - 4Ax\sin x + 2(C - B)\sin x + 2Ee^x \\
= x\cos x - \cos x + 2e^x.
\end{aligned}
$$

By equating like coefficients and solving simultaneously for the constants, we arrive at $A = 0$, $B = 1/4$, $C = 1/4$, $D = -1/2$, and $E = 1$. Thus,

$$y = C_1 \cos x + C_2 \sin x + \frac{1}{4}x^2\sin x + \frac{1}{4}x\cos x - \frac{1}{2}x\sin x + e^x.$$

Last, by imposing the prescribed ICs, we get

$$
\begin{aligned}
y(0) &= C_1 + 1 = 2, \\
y'(0) &= C_2 + \frac{1}{4} = \frac{1}{4},
\end{aligned}
$$

from which it follows $C_1 = 1$ and $C_2 = 0$. The solution we seek is therefore

$$
\begin{aligned}
y &= \cos x + \frac{1}{4}x^2\sin x + \frac{1}{4}x\cos x - \frac{1}{2}x\sin x + e^x \\
&= (1 + \frac{1}{4}x)\cos x + \frac{1}{2}x(\frac{1}{2}x - 1)\sin x + e^x.
\end{aligned}
$$

1.4.4 Nonhomogeneous equations—Part II

When the input function $F(x)$ is not of the form assumed in Section 1.4.3, or when the DE has variable coefficients, a more general method of constructing a particular solution is needed. This more general method, called *variation of parameters*, is a generalization of the method used in Section 1.3.2 with the same name.

To begin, we need to put the DE once again in the *normal form*

$$y'' + a_1(x)y' + a_0(x)y = f(x). \tag{61}$$

If the associated homogeneous DE has the general solution

$$y_H = C_1 y_1(x) + C_2 y_2(x), \tag{62}$$

then we seek a *particular solution* of the form

$$y_P = u(x)y_1(x) + v(x)y_2(x), \tag{63}$$

where $u(x)$ and $v(x)$ are functions to be determined in such a way that (63) satisfies (61).

From (63), we first calculate

$$y_P' = u(x)y_1'(x) + v(x)y_2'(x) + u'(x)y_1(x) + v'(x)y_2(x),$$

which can be simplified if we impose the condition

$$u'(x)y_1(x) + v'(x)y_2(x) = 0. \tag{64}$$

By using (64) and differentiating again, we get

$$y_P'' = u(x)y_1''(x) + v(x)y_2''(x) + u'(x)y_1'(x) + v'(x)y_2'(x).$$

If we now substitute the above expressions for y_P, y_P', and y_P'' into (61), we obtain

$$\begin{aligned} y_P'' + a_1(x)y_P' + a_0(x)y_P &= u(x)[y_1'' + a_1(x)y_1' + a_0(x)y_1] \\ &\quad + v(x)[y_2'' + a_1(x)y_2' + a_0(x)y_2] \\ &\quad + u'(x)y_1'(x) + v'(x)y_2'(x). \end{aligned}$$

However, because y_1 and y_2 are solutions of the homogeneous equation, it follows that the terms in brackets are zero and this expression reduces to

$$y_P'' + a_1(x)y_P' + a_0(x)y_P = u'(x)y_1'(x) + v'(x)y_2'(x) = f(x). \tag{65}$$

That is, if y_P is indeed a solution, then the left-hand side of (65) must equal $f(x)$. Combined with (64), we are therefore led to the set of equations

$$u'(x)y_1(x) + v'(x)y_2(x) = 0.$$
$$u'(x)y_1'(x) + v'(x)y_2'(x) = f(x). \tag{66}$$

Equations (66) are simultaneous equations that must be satisfied by the derivatives $u'(x)$ and $v'(x)$. It is known that nontrivial solutions for $u'(x)$ and $v'(x)$ exist, provided the coefficient determinant (e.g., see Section 3.3)

$$\begin{vmatrix} y_1(x) & y_2(x) \\ y_1'(x) & y_2'(x) \end{vmatrix} \neq 0.$$

Because this determinant is the Wronskian (39), it can never be zero and, hence, the simultaneous solution of (66) leads to

$$u'(x) = -\frac{y_2(x)f(x)}{W(y_1,y_2)(x)}, \qquad v'(x) = \frac{y_1(x)f(x)}{W(y_1,y_2)(x)}. \tag{67}$$

Any indefinite integral of the expressions in (67) provides suitable choices for the functions $u(x)$ and $v(x)$. We conclude, therefore, that the particular solution (63) can be expressed in the form

$$y_P = -y_1(x)\int \frac{y_2(x)f(x)}{W(y_1,y_2)(x)}dx + y_2(x)\int \frac{y_1(x)f(x)}{W(y_1,y_2)(x)}dx. \tag{68}$$

In summary, by using (62) and (68), we can express the general solution of Eq. (61) as

$$y = y_H + y_P. \tag{69}$$

EXAMPLE 16 Find the general solution of

$$2y'' + 18y = \csc 3x.$$

Solution: To proceed, we first write the DE in normal form

$$y'' + 9y = \frac{1}{2}\csc 3x,$$

which identifies the input function $f(x) = 1/(2\csc 3x)$. The solution of the associated homogeneous equation

$$y'' + 9y = 0$$

is given by

$$y_H = C_1 \cos 3x + C_2 \sin 3x.$$

The Wronskian of these solutions is $W(\cos 3x, \sin 3x) = 3$.

We now construct a particular solution of the form

$$y_P = u(x)\cos 3x + v(x)\sin 3x.$$

Thus, from (67) we see that

$$u'(x) = -\frac{1}{6}\sin 3x \csc 3x = -\frac{1}{6},$$

from which we deduce

$$u(x) = -\frac{1}{6}\int dx = -\frac{x}{6}.$$

Similarly,

$$v'(x) = \frac{1}{6}\cos 3x \csc 3x = -\frac{1}{6}\cot 3x,$$

and

$$v(x) = -\frac{1}{6}\int \cot 3x\, dx = -\frac{1}{18}\ln|\sin 3x|,$$

leading to the particular solution

$$y_P = u(x)\cos 3x + v(x)\sin 3x$$
$$= -\frac{x}{6}\cos 3x + \frac{1}{18}\ln|\sin 3x|\sin 3x.$$

The general solution is $y = y_H + y_P$, which in this case is

$$y = \left(C_1 - \frac{x}{6}\right)\cos 3x + \left(C_2 + \frac{1}{18}\ln|\sin 3x|\right)\sin 3x.$$

EXAMPLE 17 Find the general solution of

$$y'' - 3y' + 2y + \frac{1}{1 + e^{-x}}.$$

Solution: The associated homogeneous equation has the general solution

$$y_H = C_1 e^x + C_2 e^{2x},$$

which we leave for the reader to show. Here, we find the Wronskian is given by $W(e^x, e^{2x}) = e^{3x}$. Hence,

$$u'(x) = -\frac{e^{2x}}{e^{3x}(1 + e^{-x})} = -\frac{e^{-x}}{1 + e^{-x}}$$

and consequently,

$$u(x) = -\int \frac{e^{-x}}{1 + e^{-x}}dx = \ln(1 + e^{-x}).$$

Also,

$$v'(x) = \frac{e^x}{e^{3x}(1 + e^{-x})} = \frac{e^{-2x}}{1 + e^{-x}} = e^{-x} - \frac{e^{-x}}{1 + e^{-x}},$$

which yields

$$v(x) = \int \left(e^{-x} - \frac{e^{-x}}{1 + e^{-x}} \right) dx = -e^{-x} + \ln(1 + e^{-x}).$$

The particular solution is therefore

$$y_P = u(x)e^x + v(x)e^{2x}$$
$$= e^x(1 + e^x)\ln(1 + e^{-x}) - e^x,$$

from which we get, by summing y_H and y_P,

$$y = C_1 e^x + C_2 e^{2x} + e^x(1 + e^x)\ln(1 + e^{-x}) - e^x$$
$$= C_3 e^x + C_2 e^{2x} + e^x(1 + e^x)\ln(1 + e^{-x}),$$

where $C_3 = C_1 - 1$.

1.4.5 Cauchy-Euler equations

Thus far, our solution techniques apply only to constant-coefficient DEs, although the general theory of fundamental solution sets is applicable to general variable-coefficient equations. Unfortunately, variable-coefficient DEs cannot be solved as easily as constant-coefficient DEs, and we must usually resort to some sort of power series solution as discussed in Section 1.5. One exception to this is the class of *Cauchy-Euler equations*, which have the general form

$$\boxed{ax^2 y'' + bxy' + cy = F(x),} \tag{70}$$

where a, b, and c are constants. The significant feature here is that the power of x in each coefficient corresponds to the order of the derivative of y. For this reason, the equation (70) is sometimes called an *equidimensional equation*.

To solve (70), we can transform it to a constant-coefficient DE by the simple change of variable $x = e^t$, where we are assuming $x > 0$. From the chain rule, we have

$$y' = \frac{dy}{dt}\frac{dt}{dx} = \frac{1}{x}\frac{dy}{dt}$$

and

$$y'' = \frac{d}{dx}\left(\frac{dy}{dx} \right) = \frac{d}{dx}\left(\frac{1}{x}\frac{dy}{dt} \right) = \frac{1}{x}\frac{d^2 y}{dt^2}\frac{dt}{dx} - \frac{1}{x^2}\frac{dy}{dt} = \frac{1}{x^2}\left(\frac{d^2 y}{dt^2} - \frac{dy}{dt} \right).$$

By substituting these expressions into (70) and simplifying the results, we are led to the constant-coefficient equation

$$a \frac{d^2 y}{dt^2} + (b - a) \frac{dy}{dt} + cy = F(e^t). \tag{71}$$

EXAMPLE 18 Find the general solution of

$$4x^2 y'' + y = 0, \quad x > 0.$$

Solution: Under the transformation $x = e^t$, we obtain the constant-coefficient DE

$$4 \frac{d^2 y}{dt^2} - 4 \frac{dy}{dt} + y = 0.$$

The auxiliary equation $4m^2 - 4m + 1 = (2m - 1)^2 = 0$ has a double root $m = 1/2, 1/2$. Hence, the general solution of the transformed DE is

$$y(t) = (C_1 + C_2 t)e^{t/2}.$$

Transforming back to the original variable by setting $t = \ln x$, we get

$$y = (C_1 + C_2 \ln x)\sqrt{x}.$$

EXAMPLE 19 Find the general solution of

$$x^2 y'' - xy' + 5y = 0, \quad x > 0.$$

Solution: The transformed constant-coefficient DE is

$$\frac{d^2 y}{dt^2} - 2 \frac{dy}{dt} + 5y = 0$$

with $m = 1 \pm 2i$ as the roots of the auxiliary equation. Thus,

$$y(t) = e^t(C_1 \cos 2t + C_2 \sin 2t),$$

and, in terms of the original variable x,

$$y = x[C_1 \cos(2\ln x) + C_2 \sin(2\ln x)].$$

EXAMPLE 20 Find a general solution of

$$x^2 y'' - 3xy' + 3y = 2x^4 e^x, \quad x > 0.$$

Solution: Because the DE is nonhomogeneous, we must find both y_H and y_P. Under the transformation $x = e^t$, the associated homogeneous DE becomes

$$\frac{d^2 y}{dt^2} - 4\frac{dy}{dt} + 3y = 0.$$

Here, we find the general solution

$$y_H(t) = C_1 e^t + C_2 e^{3t},$$

or, in terms of the original variable x,

$$y_H = C_1 x + C_2 x^3.$$

The Wronskian of the linearly independent solutions $y_1 = x$ and $y_2 = x^3$ is given by

$$W(x, x^3) = \begin{vmatrix} x & x^3 \\ 1 & 3x^2 \end{vmatrix} = 2x^3.$$

We now seek a particular solution of the form

$$y_P = u(x)x + v(x)x^3.$$

By putting the Cauchy-Euler equation in normal form,

$$y'' - \frac{3}{x}y' + \frac{3}{x^2}y = 2x^2 e^x,$$

we identify $f(x) = 2x^2 e^x$. Thus, it follows that

$$u'(x) = -\frac{x^3(2x^2 e^x)}{2x^3} = -x^2 e^x,$$

$$v'(x) = \frac{x(2x^2 e^x)}{2x^3} = e^x,$$

integrations of which yield

$$u(x) = -x^2 e^x + 2xe^x - 2e^x$$
$$v(x) = e^x.$$

The particular solution is

$$\begin{aligned} y_P &= u(x)x + v(x)x^3 \\ &= 2x(x-1)e^x, \end{aligned}$$

from which we deduce the general solution

$$y = y_H + y_P = C_1 x + C_2 x^3 + 2x(x - 1)e^x.$$

1.5 Power Series Method

The solution of homogeneous linear differential equations (DEs) with constant coefficients can be reduced to the algebraic problem of finding the roots of a polynomial. The solutions in such cases can be expressed in terms of elementary functions like exponentials, polynomials, and trigonometric functions. Even the class of Cauchy-Euler equations, which have variable coefficients, can be solved by purely algebraic methods. However, this is not the case for most other variable-coefficient DEs, because their solutions typically involve nonelementary functions. The methods discussed in the following sections produce solutions in the form of *power series*, so the general procedure is usually called the power series method. Although it can be extended to higher-order DEs, we restrict our attention primarily to second-order equations.

1.5.1 Review of power series

In this short section we review a few basic properties concerning power series as studied in calculus. A *power series* is an infinite series of the form

$$c_0 + c_1(x - x_0) + c_2(x - x_0)^2 + \cdots = \sum_{n=0}^{\infty} c_n(x - x_0)^n, \tag{72}$$

where the c's are the *coefficients* of the series, and x_0 is a fixed constant called the *center*. Note that each term of the series (72) is raised to the "power" n—hence, the name "power series." When $x = x_0$, the series has the sum c_0, but generally we are interested in whether the series has a finite sum for other values of x.

Theorem 1.4 To every power series (72) is a number R, $0 \le R < \infty$, called the radius of convergence, with the property that the series converges absolutely for $|x - x_0| < R$ and diverges for $|x - x_0| > R$. If the series converges for all values of x, then $R \to \infty$. However, if the series converges only at x_0, then $R = 0$.

The radius of convergence R in Theorem 1.4 is usually found by the *ratio test*. Note, however, that Theorem 1.4 does not include values for which $|x - x_0| = R$. In general, these endpoints must be separately analyzed for possible convergence.

Because a power series converges for all x in the interval $|x - x_0| < R$, it defines a function $f(x)$ on this interval. One of the most important properties of a convergent power series (with $R > 0$) in our work is that such series can be differentiated termwise to produce another convergent series. That is, if $f(x)$ is a function that has a convergent power series described by

$$f(x) = \sum_{n=0}^{\infty} c_n (x - x_0)^n, \tag{73}$$

then its derivative is

$$f'(x) = \sum_{n=0}^{\infty} n c_n (x - x_0)^{n-1}. \tag{74}$$

The new series (74) is again a power series that can now be differentiated, and so on. For our purposes we can stop with the second derivative.

1.6 Solutions Near an Ordinary Point

Given some DE, we look for solutions of it in a power series of the form

$$y = \sum_{n=0}^{\infty} c_n x^n, \quad |x| < R. \tag{75}$$

If this is the case, we can substitute the series (75) and its derivatives directly into the DE and try to determine the unknown constants $c_0, c_1, c_2, \dots$.

To illustrate the method, let us consider the simple first-order DE

$$y' - y = 0, \tag{76}$$

for which we know the solution $y = Ce^x$, where C is an arbitrary constant. The substitution of (75) into the DE yields

$$y' - y = \sum_{n=0}^{\infty} n c_n x^{n-1} - \sum_{n=0}^{\infty} c_n x^n = 0. \tag{77}$$

To combine these two series, we make the change of index $n \Rightarrow n - 1$ in the second summation and write

$$\sum_{n=0}^{\infty} n c_n x^{n-1} - \sum_{n=1}^{\infty} c_{n-1} x^{n-1} = 0, \tag{78}$$

or

$$0 \cdot c_0 x^{-1} + \sum_{n=1}^{\infty} (n c_n - c_{n-1}) x^{n-1} = 0, \tag{79}$$

where we have now placed the first term of the first sum outside so that the summation limits are the same for each sum. Because x can take on various values, the coefficient of x^{n-1} must vanish in order to satisfy (79), i.e., $n c_n - c_{n-1} = 0$, $n = 1,2,3,\dots$; whereas c_0 can remain *arbitrary* because the term in which it appears is already zero. This relationship is called a *recurrence formula* for the unknown coefficients, which we write as

$$c_n = \frac{c_{n-1}}{n}, \quad n = 1,2,3,\dots . \tag{80}$$

Successively setting $n = 1, 2, 3, \ldots$, into the recurrence formula (80), leads to

$$
\begin{aligned}
n = 1: & \quad c_1 = c_0, \\
n = 2: & \quad c_2 = \frac{c_1}{2} = \frac{c_0}{2!}, \\
n = 3: & \quad c_3 = \frac{c_2}{3} = \frac{c_0}{3!}, \ldots
\end{aligned}
\tag{81}
$$

from which we obtain the series solution

$$
\begin{aligned}
y &= c_0 + c_1 x + c_2 x^2 + c_3 x^3 + \cdots \\
  &= c_0 \left(1 + x + \frac{x^2}{2!} + \frac{x^3}{3!} + \cdots \right),
\end{aligned}
\tag{82}
$$

or

$$
y = c_0 \sum_{n=0}^{\infty} \frac{x^n}{n!} = c_0 e^x.
\tag{83}
$$

Although this example is quite elementary, it still illustrates the basic manipulations required to solve any DE by the power series method. Even when the coefficients are polynomials rather than constants, the required manipulations differ little from those above. The power series method works for many variable-coefficient DEs but not all. To understand when it can be used we need to introduce the notion of ordinary and singular points.

1.6.1 Ordinary and singular points

Consider the homogeneous linear DE

$$
A_2(x)y'' + A_1(x)y' + A_0(x)y = 0,
\tag{84}
$$

in which the coefficients are *polynomials* with no common factors.

> **Definition 1.4** Those points for which $A_2(x) = 0$ are classified as *singular points* of the DE. All other points are called *ordinary points*.

In the classification of singular points, we take into account complex values as well as real values of the variable x.

EXAMPLE 21 Identify the singular points of the DEs:

(a) $(1 - x^2)y'' - 2xy' + 2y = 0.$

(b) $(x^2 + 4)y'' + 3xy' - 7y = 0.$

Solution:
   (a) Here $A_2(x) = 1 - x^2 = 0$ at $x = \pm 1$, which are the singular points. All other points are considered ordinary points.

   (b) In this case $A_2(x) = x^2 + 4 = 0$ at $x = \pm 2i$, which constitute the singular points. Again, all other points are ordinary points

---

1.6.2 General method for ordinary points

Although in theory we can develop a power series about any ordinary point of the DE, we are generally interested only in the point $x = 0$ because most DEs are formulated in such a way that the origin is a special point of interest. The following theorem is based on this special point.

Theorem 1.5 If $x = 0$ is an ordinary point of the DE

$$A_2(x)y'' + A_1(x)y' + A_0(x)y = 0,$$

then the general solution is given by

$$y = \sum_{n=0}^{\infty} c_n x^n = c_0 y_1(x) + c_1 y_2(x),$$

where c_0 and c_1 are arbitrary constants. Moreover, the series for linearly independent solutions y_1 and y_2 will converge at least in the interval $|x| < R$, where R is the distance from the origin to the nearest singular point (real or complex).

EXAMPLE 22 Solve the initial value problem

$$y'' + xy' + y = 0, \quad y(0) = 3, \quad y'(0) = -7.$$

Solution: We first note that this DE has no singular points. Thus, the series solution about $x = 0$ will converge for all x. By substituting the series for y and its derivatives into the given DE, we find

$$\sum_{n=0}^{\infty} n(n-1)c_n x^{n-2} + \sum_{n=0}^{\infty} nc_n x^n + \sum_{n=0}^{\infty} c_n x^n = 0.$$

After making the change of index $n \Rightarrow n-2$ in the last two summations and combining the series, we obtain

$$0 \cdot c_0 x^{-2} + 0 \cdot c_1 x^{-1} + \sum_{n=2}^{\infty} \left[n(n-1)c_n + (n-2)c_{n-2} + c_{n-2} \right] x^{n-2} = 0.$$

Clearly, both c_0 and c_1 are arbitrary and the remaining constants are found from the recurrence formula

$$c_n = -\frac{c_{n-2}}{n}, \quad n = 2, 3, 4, \ldots,$$

from which we deduce (by successively setting $n = 2, 3, 4, \ldots$ in the recurrence formula)

$$c_2 = -\frac{1}{2}c_0, \; c_3 = -\frac{1}{3}c_1, \; c_4 = -\frac{1}{4}c_2 = \frac{1}{2 \cdot 4}c_0, \; c_5 = -\frac{1}{5}c_3 = \frac{1}{3 \cdot 5}c_1, \; \ldots.$$

The general solution is therefore

$$
\begin{aligned}
y &= c_0 + c_1 x + c_2 x^2 + c_3 x^3 + \cdots \\
&= c_0 \left(1 - \frac{x^2}{2} + \frac{x^4}{2 \cdot 4} - \cdots \right) + c_1 \left(x - \frac{x^3}{3} + \frac{x^5}{3 \cdot 5} - \cdots \right).
\end{aligned}
$$

By imposing the first initial condition, we find

$$y(0) = c_0 = 3,$$

and the second initial condition yields

$$y'(0) = c_1 = -7.$$

The desired solution now takes the form

$$y = 3 \left(1 - \frac{x^2}{2} + \frac{x^4}{2 \cdot 4} - \cdots \right) - 7 \left(x - \frac{x^3}{3} + \frac{x^5}{3 \cdot 5} - \cdots \right).$$

Notice that we left the constants in our solution in Example 22 in product form rather than multiplying them. In practice this approach lends itself more easily to the determination of the general term of the series so that the entire series can be represented. Here, however, we are content to merely list the first few nonzero terms of the series rather than attempt to find the general term. Finally, it is interesting to also note in solving for the arbitrary constants in Example 22, we found that $c_0 = y(0)$ and $c_1 = y'(0)$. This is not a peculiarity of just this particular problem, but is the case with all initial value problems solved by the power series method.

EXAMPLE 23 Use the power series method to solve

$$y'' + (1+x)y = 0.$$

Solution: Once again we see that the given DE has no singularities, so the solutions obtained will converge for all values of x. Proceeding as above, we are eventually led to

$$0 \cdot c_0 x^{-2} + 0 \cdot c_1 x^{-1} + (2c_2 + c_0)x^0 + \sum_{n=3}^{\infty} \left[n(n-1)c_n + c_{n-2} + c_{n-3} \right] x^{n-2} = 0.$$

Thus, both c_0 and c_1 are arbitrary, whereas $c_2 = -c_0/2$. The recurrence formula for the remaining constants is

$$c_n = -\frac{c_{n-2} + c_{n-3}}{n(n-1)}, \quad n = 3,4,5,\dots .$$

To simplify matters here and still obtain two linearly independent solutions, let us first set $c_0 = 1$ and $c_1 = 0$ (i.e., because they are arbitrary constants, we can set them to any value). This will produce one solution $y_1(x)$. Then, we set $c_0 = 0$ and $c_1 = 1$, which will lead to a second solution $y_2(x)$.

With $c_0 = 1$ and $c_1 = 0$, we find that

$$c_2 = -\frac{c_0}{2} = -\frac{1}{2}, \quad c_3 = -\frac{c_1 + c_0}{3 \cdot 2} = -\frac{1}{3!}, \quad c_4 = -\frac{c_2 + c_1}{4 \cdot 3} = \frac{1}{4!}, \dots,$$

from which we obtain

$$y_1(x) = 1 - \frac{x^2}{2!} - \frac{x^3}{3!} + \frac{x^4}{4!} + \cdots .$$

In the same fashion, with $c_0 = 0$ and $c_1 = 1$, we get

$$c_2 = -\frac{c_0}{2} = 0, \quad c_3 = -\frac{c_1 + c_0}{3 \cdot 2} = -\frac{1}{3!}, \quad c_4 = -\frac{c_2 + c_1}{4 \cdot 3} = -\frac{2}{4!}, \dots,$$

which yields

$$y_2(x) = x - \frac{x^3}{3!} - \frac{2x^4}{4!} + \cdots .$$

The general solution is

$$y = Ay_1(x) + By_2(x),$$

where A and B are arbitrary constants.

1.7 Legendre's Equation

Some variable-coefficient DEs occur so frequently in practice that they have been named and their solutions extensively studied. One such equation is *Legendre's equation*

$$(1 - x^2)y'' - 2xy' + \alpha(\alpha + 1)y = 0, \tag{85}$$

where α is a constant.[1] This DE arises in various applications involving spherical symmetry, as in finding the steady-state temperature within a solid spherical ball whose surface temperature is prescribed.

Legendre's DE (85) has singularities at $x = \pm 1$. All other points are ordinary points, including $x = 0$. Thus, we expect to find a power series solution

$$y = \sum_{k=0}^{\infty} c_k x^k, \quad -1 < x < 1, \tag{86}$$

where we use the index k here (for reasons to appear later) and the known interval of convergence is shown. By substituting the series and its derivatives into (85) and simplifying, we are led to

$$0 \cdot c_0 x^{-2} + 0 \cdot c_1 x^{-1}$$
$$+ \sum_{k=2}^{\infty} \left[k(k - 1)c_k - (k - 2)(k - 1)c_{k-2} + \alpha(\alpha + 1)c_{k-2} \right] x^{k-2} = 0. \tag{87}$$

Consequently, both c_0 and c_1 are arbitrary, and the remaining constants are determined from the recurrence formula

$$c_k = \frac{(k - 2)(k - 1) - \alpha(\alpha + 1)}{k(k - 1)} c_{k-2}$$
$$= \frac{(k - \alpha - 2)(k + \alpha - 1)}{k(k - 1)} c_{k-2}, \quad k = 2, 3, 4, \ldots, \tag{88}$$

which we have written in factored form in the last step. From the recurrence formula, we see that

$$k = 2: \quad c_2 = -\frac{\alpha(\alpha + 1)}{2!} c_0,$$

$$k = 3: \quad c_3 = -\frac{(\alpha - 1)(\alpha + 2)}{3!} c_1,$$

$$k = 4: \quad c_4 = -\frac{(\alpha - 2)(\alpha + 3)}{4 \cdot 3} c_2 = \frac{(\alpha - 2)\alpha(\alpha + 1)(\alpha + 3)}{4!} c_0, \tag{89}$$

$$k = 5: \quad c_5 = -\frac{(\alpha - 3)(\alpha + 4)}{5 \cdot 4} c_3 = \frac{(\alpha - 3)(\alpha - 1)(\alpha + 2)(\alpha + 4)}{4!} c_1,$$

and so forth. Thus, the general solution is

---

[1]We write the constant in Legendre's equation as $\alpha(\alpha + 1)$ for mathematical convenience in finding the solution.

$$y = c_0 y_1(x) + c_1 y_2(x), \tag{90}$$

where

$$y_1(x) = 1 - \frac{\alpha(\alpha+1)}{2!}x^2 + \frac{(\alpha-2)\alpha(\alpha+1)(\alpha+3)}{4!}x^4 - \cdots \tag{91}$$

and

$$y_2(x) = x - \frac{(\alpha-1)(\alpha+2)}{3!}x^3 + \frac{(\alpha-3)(\alpha-1)(\alpha+2)(\alpha+4)}{5!}x^5 - \cdots . \tag{92}$$

1.7.1 Legendre polynomials: $P_n(x)$

The convergence of the series (91) and (92) is guaranteed by the power series method only for the open interval $(-1,1)$. In fact, standard convergence tests reveal that, for general α, the above two series diverge at the endpoints $x = \pm 1$ (e.g., see [5]). However, in most applications we require that the solution be bounded throughout the interval $[-1,1]$, including the endpoints. To remedy this situation, we note that by carefully selecting values of α, one series or the other will truncate. In particular, when $\alpha = n$, $n = 0,1,2,\ldots$, then either y_1 or y_2 will reduce to a *polynomial* of degree n. If n is even, then y_1 reduces to a polynomial, and, if n is odd, y_2 reduces to a polynomial. That multiple of the polynomial of degree n that has the value unity when $x = 1$ is called the nth *Legendre polynomial* and is denoted by the symbol $P_n(x)$. We conclude, therefore, that the only bounded solutions of Legendre's equation are multiples of

$$y = P_n(x), \quad n = 0,1,2,\ldots, \tag{93}$$

where $P_n(x)$ is a polynomial defined by

$$P_n(x) = \begin{cases} y_1(x)/y_1(1), & n \text{ even}, \\ y_2(x)/y_2(1), & n \text{ odd}. \end{cases} \tag{94}$$

The first few Legendre polynomials obtained from (94) are

$$\begin{aligned}
P_0(x) &= 1, \\
P_1(x) &= x, \\
P_2(x) &= \frac{1}{2}(3x^2 - 1), \\
P_3(x) &= \frac{1}{2}(5x^3 - 3x),
\end{aligned} \tag{95}$$

whereas in general it has been shown that

$$P_n(x) = \sum_{n=0}^{[n/2]} \frac{(-1)^k(2n-2k)!x^{n-2k}}{2^n k!(n-k)!(n-2k)!}, \quad n = 0,1,2,\ldots, \tag{96}$$

where we have introduced the bracket notation

$$[n/2] = \begin{cases} n/2, & n \text{ even} \\ (n-1)/2, & n \text{ odd.} \end{cases} \tag{97}$$

A graph of the first few Legendre polynomials is given in Fig. 2.14. These polynomials are remarkable in that they satisfy a large number of useful properties. Foremost among such properties is the integral relation

$$\int_{-1}^{1} P_n(x)P_k(x)dx = 0, \quad n \neq k, \tag{98}$$

called the *orthogonality property*. The orthogonality property also extends to other similar polynomial sets such as the *Hermite, Laguerre*, and *Chebyshev polynomials* (see Section 2.4). This particular property has far-reaching consequences that will be explored in detail in Chapter 8 with regards to sets of functions (in addition to polynomial sets) that commonly arise in practice, such as sines and cosines. Another useful relation follows directly from the series definition (96), viz.,

$$P_n(-x) = (-1)^n P_n(x). \tag{99}$$

This last property shows that $P_n(x)$ is an *even function* for $n = 0, 2, 4, \ldots$, and is an *odd function* for $n = 1, 3, 5, \ldots$. Additional properties of the Legendre polynomials are presented in Section 2.4.1.

1.7.2 Legendre functions of the second kind: $Q_n(x)$

For $\alpha = n$, $n = 0, 1, 2, \ldots$, Legendre's DE (85) becomes

$$(1 - x^2)y'' - 2xy' + n(n+1)y = 0, \quad -1 < x < 1. \tag{100}$$

The Legendre polynomials $P_n(x)$ denote the solutions of (100) obtained from the truncated series. On the open interval $(-1,1)$, the remaining power series in the general solution (90)-(92) is still a valid linearly independent solution and can be used to introduce another Legendre function that has significance in certain specialized applications. We represent this *Legendre function of the second kind* by the symbol $Q_n(x)$, where

$$Q_n(x) = \begin{cases} y_1(1)y_2(x), & n \text{ even,} \\ -y_2(1)y_1(x), & n \text{ odd.} \end{cases} \tag{101}$$

Thus, the general solution of (100) is

$$y = A_n P_n(x) + B_n Q_n(x), \quad n = 0, 1, 2, \ldots, \tag{102}$$

where A_n and B_n are arbitrary constants.

Specifically, for $n = 0$, we find that $y_1(x) = 1$ and

$$y_2(x) = x + \frac{x^3}{3} + \frac{x^4}{4} + \frac{x^5}{5} + \cdots = \frac{1}{2} \ln \frac{1-x}{1+x}, \quad |x| < 1. \qquad (103)$$

Hence, from (101) it follows that

$$Q_0(x) = \frac{1}{2} \ln \frac{1-x}{1+x}. \qquad (104)$$

Note that the logarithm in (104) has singularities at both endpoints $x = \pm 1$. By some involved manipulation, it can also be shown that

$$Q_1(x) = xQ_0(x) - 1,$$

$$Q_2(x) = P_2(x)Q_0(x) - \frac{3}{2}x, \qquad (105)$$

and so on. In general, the function $Q_0(x)$ will appear in all higher order Legendre functions so the singularities at $x = \pm 1$ are contained in each of these. Although the appearance of the Legendre functions $Q_n(x)$ in applications is more rare than that for the Legendre polynomials $P_n(x)$, the former are useful in certain applications involving curvilinear coordinate systems like the prolate spheroidal and prolate ellipsoidal coordinate systems, among others.

1.8 Solutions Near a Singular Point

In the neighborhood of a singular point the behavior of the solution of a DE, and hence that of the physical system described by the DE, is usually quite different from that near an ordinary point. For this reason, solutions in the neighborhood of singular points are often of special interest. Unfortunately, when $x = 0$ is a singular point, for example, it may not be possible to find a power series solution like those in Section 1.6. We find instead that solutions may exist of the more general form

$$y = x^s \sum_{n=0}^{\infty} c_n x^n, \qquad (106)$$

where s is a parameter to be determined along with the c's. To develop the theory asssociated with such solutions, we need to further classify the singularities of the DE. Once again, we restrict ourselves to DEs of the form

$$A_2(x)y'' + A_1(x)y' + A_0(x)y = 0, \qquad (107)$$

in which the coefficients are polynomials with no common factors. For classification of the singular points, we rewrite (107) in *normal form*

$$y'' + a(x)y' + b(x)y = 0, \qquad (108)$$

where we have written $a(x) = A_1(x)/A_2(x)$ and $b(x) = A_0(x)/A_2(x)$.

Definition 1.5 If x_0 is a singular point of (108), it is classified as a *regular singular point* (RSP) if both of the following limits are finite:

$$\lim_{x \to x_0} (x - x_0)a(x) \quad \text{and} \quad \lim_{x \to x_0} (x - x_0)^2 b(x)$$

If either limit, or both, fail to exist or become infinite, we say that x_0 is an *irregular singular point* (ISP).

EXAMPLE 24 Classify the singular points of

$$x(x - 1)^2 y'' + 5x^2 y' + (2x^2 + 1)y = 0.$$

Solution: The singular points are $x = 0, 1$. Putting the DE in normal form we get

$$a(x) = \frac{5x}{(x - 1)^2}, \quad b(x) = \frac{2x^2 + 1}{x(x - 1)^2}.$$

For $x = 0$, we see that

$$\lim_{x \to 0} x a(x) = 0 \quad \text{and} \quad \lim_{x \to 0} x^2 b(x) = 0,$$

so we conclude that $x = 0$ is an RSP. At $x = 1$, this leads to

$$\lim_{x \to 1} (x - 1)a(x) = \infty \quad \text{and} \quad \lim_{x \to 1} (x - 1)^2 b(x) = 3,$$

so clearly $x = 1$ is an ISP.

1.8.1 Method of Frobenius

In 1873, a method was published by the German mathematician Georg Frobenius (1849-1917) for finding a solution of a DE about an RSP. The method is based on the following theorem which we state specifically for the case when $x = 0$ is an RSP.

Theorem 1.6 If $x = 0$ is an RSP of the DE

$$A_2(x)y'' + A_1(x)y' + A_0(x)y = 0,$$

then there exists at least one solution of the form

$$y = x^s \sum_{n=0}^{\infty} c_n x^n, \quad c_0 \neq 0, \quad x > 0.$$

Moreover, the series will converge at least in the interval $0 < x < R$, where R is the distance from the origin to the nearest other singular point.

In finding a solution of the form suggested in Theorem 1.6, it is necessary to determine the values of the parameter s in addition to the c's. The restriction $x > 0$ is required to prevent complex solutions that might arise otherwise. By writing the series as

$$y = \sum_{n=0}^{\infty} c_n x^{n+s},$$

(109)

and formally differentiating it termwise, we substitute these results into the DE (108) to find

$$\sum_{n=0}^{\infty} \left[(n+s)(n+s-1) + xa(x)(n+s) + x^2 b(x)\right] c_n x^{n+s-2} = 0.$$

(110)

If we assume that $x = 0$ is an RSP of the DE, then it follows that $xa(x)$ and $x^2 b(x)$ have Maclaurin series expansions of the form

$$\begin{aligned}
xa(x) &= \alpha_0 + \alpha_1 x + \alpha_2 x^2 + \cdots, \\
x^2 b(x) &= \beta_0 + \beta_1 x + \beta_2 x^2 + \cdots.
\end{aligned}$$

(111)

If we now substitute these last expressions into (110) and group the resulting terms in like powers of x, we get

$$\begin{aligned}
&[s(s-1) + \alpha_0 s + \beta_0] c_0 x^{s-2} \\
&\quad + [(s+1)s c_1 + (s+1)\alpha_0 c_1 + s\alpha_1 c_0 + \beta_0 c_1 + \beta_1 c_0] x^{s-1} + \cdots = 0.
\end{aligned}$$

(112)

This last expression can be satisfied only if the coefficients of all powers of x vanish independently. In particular, the first term involving the constant c_0 can vanish for all x if and only if

$$\boxed{s(s-1) + \alpha_0 s + \beta_0 = 0.}$$

(113)

This important quadratic equation in s is called the *indicial equation* of the DE. By forcing s to always be a solution of this equation, we are essentially demanding that c_0 be our arbitrary constant.

The two roots of the indicial equation s_1 and s_2 are sometimes referred to as the *exponents of the singularity*. It can be shown that one solution of the form given in Theorem 1.6 always exists corresponding to the *larger* root between s_1 and s_2. Finding a second linearly independent solution, however, will depend upon the nature of these two roots. Specifically, there are three cases to consider:

CASE I:	Roots s_1 and s_2 differ by a noninteger
CASE II:	Roots s_1 and s_2 are equal
CASE III:	Roots s_1 and s_2 differ by a nonzero integer

For CASE I, we have the following theorem.

Theorem 1.7 If $x = 0$ is an RSP of the DE and the roots of the indicial equation s_1 and s_2 differ by a noninteger, then the general solution of the DE is given by

$$y = Ay_1(x) + By_2(x), \quad x > 0,$$

where A and B are arbitrary constants and where

$$y_1(x) = x^{s_1} \sum_{n=0}^{\infty} c_n(s_1)x^n, \quad c_0(s_1) = 1,$$

$$y_2(x) = x^{s_2} \sum_{n=0}^{\infty} c_n(s_2)x^n, \quad c_0(s_2) = 1.$$

Note that it is not necessary to set $c_0 = 1$ in solutions y_1 and y_2 in Theorem 1.7 but merely a convenience. That is why we simply introduce new arbitrary constants A and B in forming the general solution.

EXAMPLE 25 Use the method of Frobenius to find the general solution of

$$2xy'' + (1 - 2x)y' - y = 0, \quad x > 0.$$

Solution: We first check and find that $x = 0$ is an RSP. Thus, by substituting the series (109) into the DE and simplifying the results, we obtain

$$[2s(s - 1) + s]c_0 x^{s-1}$$
$$+ \sum_{n=1}^{\infty} [2(n + s)(n + s - 1)c_n + (n + s)c_n$$
$$- 2(n + s - 1)c_{n-1} - c_{n-1}]x^{n+s-1} = 0.$$

The coefficient of c_0 equated to zero gives us the indicial equation

$$2s^2 - s = s(2s - 1) = 0,$$

with solutions $s_1 = 0$ and $s_2 = 1/2$. Thus, Theorem 1.7 is applicable. For general values of s, the recurrence formula is

$$c_0 = 1, \quad c_n(s) = \frac{c_{n-1}(s)}{n + s}, \quad n = 1, 2, 3, \dots .$$

Now putting $s = s_1 = 0$ in the above recurrence formula, we are led to

$$s_1 = 0: \quad c_n = \frac{c_{n-1}}{n}, \quad n = 1, 2, 3, \dots,$$

$$y_1(x) = x^0(c_0 + c_1 x + c_2 x^2 + \cdots)$$

$$= 1 + x + \frac{x^2}{2!} + \frac{x^3}{3!} + \cdots.$$

Similarly, for $s = s_2 = 1/2$ in the above recurrence formula, we have

$$s_2 = \frac{1}{2}: \quad c_n = \frac{2c_{n-1}}{2n+1}, \quad n = 1,2,3,\ldots,$$

$$y_2(x) = x^{1/2}(c_0 + c_1 x + c_2 x^2 + \cdots)$$

$$= x^{1/2}\left(1 + \frac{2x}{3} + \frac{4x^2}{15} + \frac{8x^3}{105} + \cdots\right).$$

Finally, the general solution is

$$y = A\left(1 + x + \frac{x^2}{2!} + \frac{x^3}{3!} + \cdots\right) + Bx^{1/2}\left(1 + \frac{2x}{3} + \frac{4x^2}{15} + \frac{8x^3}{105} + \cdots\right)$$

$$= Ae^x + Bx^{1/2}\left(1 + \frac{2x}{3} + \frac{4x^2}{15} + \frac{8x^3}{105} + \cdots\right),$$

where we recognize the first series as that for e^x. The constants A and B are arbitrary constants, and, because $x = 0$ is the only singular point, the series converges for all positive x

---

When the roots of the indicial equation are equal, i.e., $s_1 = s_2$, the procedure used for CASE I will produce only one solution. To obtain two linearly independent solutions in CASE II, we introduce the function

$$y(x,s) = x^s\left[1 + \sum_{n=1}^{\infty} c_n(s)x^n\right], \quad c_0(s) = 1. \tag{114}$$

The two linearly independent solutions are then defined by

$$y_1(x) = y(x,s_1)$$

$$= x^{s_1}\left[1 + \sum_{n=1}^{\infty} c_n(s_1)x^n\right], \quad c_0(s_1) = 1, \tag{115}$$

$$y_2(x) = \left.\frac{\partial y(x,s)}{\partial s}\right|_{s=s_1}$$

$$= y_1(x)\ln x + x^{s_1}\sum_{n=1}^{\infty} c_n'(s_1)x^n. \tag{116}$$

EXAMPLE 26 Find the general solution of

$$xy'' + y' + xy = 0, \quad x > 0.$$

Solution: Once again, $x = 0$ is an RSP of the DE and all other points are ordinary points. The Frobenius method leads to

$$[s(s-1) + s]c_0 x^{s-1} + [(s+1)s + s + 1]c_1 x^s$$
$$+ \sum_{n=2}^{\infty} \left[(n+s)(n+s-1)c_n + (n+s)c_n + c_{n-2}\right]x^{n+s-1} = 0.$$

Setting the coefficient of c_0 to zero yields the indicial equation

$$s(s-1) + s = s^2 = 0,$$

from which it follows that $s_1 = s_2 = 0$. Hence, c_0 is arbitrary and the remaining constants must satisfy

$$[(s+1)s + s + 1]c_1 = 0,$$
$$[(n+s)(n+s-1) + (n+s)]c_n + c_{n-2} = 0, \quad n = 2,3,4,\dots.$$

For $s = 0$, the coefficient of c_1 does not vanish so we are forced to set $c_1 = 0$. The remaining relations lead to the recurrence formula

$$c_0 = 1, \quad c_n(s) = -\frac{c_{n-2}(s)}{(s+n)^2}, \quad n = 2,3,4,\dots.$$

The first few coefficients calculated from the recurrence formula gives us

$$c_2(s) = -\frac{1}{(s+2)^2}, \quad c_3(s) = 0, \quad c_4(s) = \frac{1}{(s+2)^2(s+4)^2}, \quad \dots,$$

and hence,

$$y(x,s) = 1 - \frac{x^2}{(s+2)^2} + \frac{x^4}{(s+2)^2(s+4)^2} - \dots.$$

By setting $s = 0$ in this last expression, we obtain the first solution

$$y_1(x) = 1 - \frac{x^2}{4} + \frac{x^4}{64} - \dots.$$

For the second solution, we first differentiate the expression $y(x,s)$ with respect to s and then set $s = 0$. This action leads to

$$y_2(x) = y_1(x)\ln x + \frac{x^2}{4} - \frac{x^4}{64}\left(1 + \frac{1}{2}\right) + \frac{x^6}{288}\left(1 + \frac{1}{2} + \frac{1}{3}\right) - \dots,$$

from which we deduce the general solution

$$y = Ay_1(x) + By_2(x),$$

where A and B are arbitrary constants.

When s_1 and s_2 differ by a nonzero integer, we have CASE III which in general can be handled similar to CASE II, leading to a logarithm solution. However, in some special cases we find that two solutions, neither of which involves a logarithmic term, can be found by selecting the smaller root between s_1 and s_2. Typically, this will happen when the larger root is a positive constant and then both c_0 and c_r turn out to be arbitrary constants, where r denotes the larger root. To test this case, we usually substitute the smaller root directly into the recurrence formula in the hope that two arbitrary constants show up. Let us illustrate this particular case with an example.

EXAMPLE 27 (*Nonlog case*) Find the general solution of

$$xy'' - (4 + x)y' + 2y = 0, \quad x > 0.$$

Solution: Based on the Frobenius method, we are led to

$$[s(s-1) - 4s]c_0 x^{s-1}$$
$$+ \sum_{n=1}^{\infty} \big[(n+s)(n+s-1)c_n - 4(n+s)c_n$$
$$- (n-s-3)c_{n-1}\big]x^{n+s-1} = 0.$$

Therefore, the indicial equation is

$$s(s-1) - 4s = s(s-5) = 0,$$

with roots $s = 0, 5$ that differ by a nonzero integer. For the smaller root $s = 0$, the recurrence formula simplifies to

$$n(n-5)c_n = (n-3)c_{n-1}, \quad n = 1,2,3,...,$$

where we write the recurrence formula in this fashion because division by the factor $(n - 5)$ is not permitted when $n = 5$. From the recurrence formula, we find

$$c_1 = \frac{1}{2}c_0, \quad c_2 = \frac{1}{12}c_0, \quad c_3 = c_4 = 0,$$

and, for $n = 5$, we have $0 \cdot c_5 = 0$, so c_5 is arbitrary. Additional constants are

$$c_6 = \frac{1}{2}c_5, \qquad c_7 = \frac{1}{7}, \ ...,$$

and thus, we obtain the solution

$$y = c_0\left(1 + \frac{1}{2}x + \frac{1}{12}x^2\right) + c_5\left(x^5 + \frac{1}{2}x^6 + \frac{1}{7}x^7 + \cdots\right), \quad x > 0.$$

1.9 Bessel's Equation

Solutions of *Bessel's equation*

$$\boxed{x^2y'' + xy' + (x^2 - v^2)y = 0, \quad x > 0,} \tag{117}$$

are called *Bessel functions of order* v, where we assume $v \geq 0$. The equation is named in honor of the German astronomer F. W. Bessel (1784-1846) who investigated its solutions in connection with planetary motion. Bessel's equation occurs in a wide variety of applications, including the free vibrations of a circular membrane, finding the temperature distribution in a solid cylinder, and in electromagnetic theory. In fact, Bessel functions occur so frequently in engineering and physics applications that they are undoubtedly the most important functions beyond the study of elementary functions in the calculus (e.g., see [34]) .

1.9.1 The gamma function: $\Gamma(x)$

Solutions of Bessel's DE (117) for general values of the parameter v are usually expressed in terms of the *gamma function*, defined by the improper integral[2]

$$\boxed{\Gamma(x) = \int_0^\infty e^{-t}t^{x-1}\,dt, \quad x > 0.} \tag{118}$$

The most important property of this function is its *recurrence formula*

$$\Gamma(x+1) = x\Gamma(x), \tag{119}$$

obtained by replacing x with $x + 1$ in (118) and integrating by parts.

For $x = 1$, it follows from definition that

$$\Gamma(1) = \int_0^\infty e^{-t}\,dt = 1. \tag{120}$$

By using this result and the recurrence formula (119), it can also be readily seen that for

---

[2]See also Section 2.3.1 for a graph of the gamma function and further discussion of its properties.

integer values of x we obtain the interesting formula

$$\Gamma(n + 1) = n!, \quad n = 1,2,3,\dots . \tag{121}$$

That is, the gamma function is a generalization of the factorial function from discrete values to a continuum of values.

The gamma function is not defined by the integral (118) for $x < 0$. However, by use of the recurrence formula (119), which is valid for all values of x for which the gamma function is defined, we see that

$$\Gamma(x) = \frac{\Gamma(x + 1)}{x}, \quad x \neq 0. \tag{122}$$

Hence, we can use (122) to formally extend the definition of $\Gamma(x)$ to negative values of x. In particular, (122) can be used to infer the behavior of the gamma function in the vicinity of the origin. For example, it follows that

$$
\begin{aligned}
\lim_{x \to 0^+} \Gamma(x) &= \lim_{x \to 0^+} \frac{\Gamma(x + 1)}{x} = +\infty, \\
\lim_{x \to 0^-} \Gamma(x) &= \lim_{x \to 0^-} \frac{\Gamma(x + 1)}{x} = -\infty.
\end{aligned}
\tag{123}
$$

By repeated use of (122), it can be established that the gamma function is defined for all $x < 0$ except for the negative integers. That is, for $x = -n$, we find

$$\frac{1}{\Gamma(-n)} = 0, \quad n = 0,1,2,\dots . \tag{124}$$

Evaluation of the gamma function for values of x other than an integer is a difficult task without some numerical assistance. An exception is the case $x = 1/2$ and odd multiples of $1/2$ [see Example 28 and Eq. (125) below].

EXAMPLE 28 Show that $\Gamma\left(\dfrac{1}{2}\right) = \sqrt{\pi}$.

Solution: By setting $x = 1/2$ in (118) and making the change of variable $t = u^2$, we have

$$\Gamma\left(\frac{1}{2}\right) = 2\int_0^\infty e^{-u^2}\, du.$$

To evaluate this integral, we first consider its square, i.e.,

$$\left[\Gamma\left(\frac{1}{2}\right)\right]^2 = 2\int_0^\infty e^{-u^2}\, du \cdot 2\int_0^\infty e^{-v^2}\, dv = 4\int_0^\infty \int_0^\infty e^{-(u^2 + v^2)}\, du\, dv,$$

where we have expressed the product of two integrals as an iterated integral. The term $u^2 + v^2$ suggests a change of variables to polar coordinates. Thus, if we let

$$u = r\cos\theta, \quad v = r\sin\theta,$$

the resulting integral is

$$\left[\Gamma\left(\frac{1}{2}\right)\right]^2 = 4\int_0^{\pi/2}\int_0^\infty re^{-r^2}\,dr\,d\theta = \pi.$$

Finally, by taking the positive square root, we obtain the desired result

$$\Gamma\left(\frac{1}{2}\right) = \sqrt{\pi}.$$

The result of Example 28 combined with the recurrence formula leads to the relation

$$\Gamma\left(n + \frac{1}{2}\right) = \frac{(2n)!}{2^{2n}n!}\sqrt{\pi}, \quad n = 0,1,2,\ldots . \tag{125}$$

There are a number of additional relations involving the gamma function that are also useful in practice but we will not list them here (see Chapter 2).

1.9.2 Bessel functions of the first kind: $J_\nu(x)$

Bessel's equation (117) has a regular singular point at $x = 0$; thus, we seek a Frobenius solution of the form

$$y = x^s\sum_{n=0}^\infty c_n x^n = \sum_{n=0}^\infty c_n x^{n+s}. \tag{126}$$

The substitution of (126) and its derivatives into (117), followed by some algebraic manipulation, leads to

$$(s^2 - v^2)c_0 x^s + [(s+1)^2 - v^2]c_1 x^{s+1}$$
$$+ \sum_{n=2}^\infty \left\{[(n+s)^2 - v^2]c_n + c_{n-2}\right\} x^{n+s} = 0. \tag{127}$$

From the coefficient of c_0, we obtain the indicial equation $s^2 - v^2 = 0$, with solutions $s = \pm v$. For this choice of roots, we must set $c_1 = 0$ and the recurrence formula can be written as

$$c_n = -\frac{c_{n-2}}{n^2 \pm 2nv}, \quad n = 2,3,4,\ldots . \tag{128}$$

Because $c_1 = 0$, it is clear from (128) that all c_n with odd index also vanish.

Unless we know the value of v, we do not know which of the three cases of the method of Frobenius we are dealing with. Therefore, to be precise and produce at least one solution of Bessel's equation, we select $s = v$, and then (128) for even values of n $(n = 2m)$ takes the form

$$c_{2m} = -\frac{c_{2m-1}}{2^2 m(m+v)}, \quad m = 1,2,3,\dots. \tag{129}$$

By repeated use of this recurrence formula, it follows that

$$\begin{aligned} c_{2m} &= \frac{(-1)^m c_0}{2^{2m} m!(v+1)(v+2)\cdots(v+m)} \\ &= \frac{(-1)^m \Gamma(v+1) c_0}{2^{2m} m!\Gamma(m+v+1)}, \quad m = 1,2,3,\dots, \end{aligned} \tag{130}$$

where we have used properties of the gamma function. Finally, by choosing

$$c_0 = \frac{1}{2^v \Gamma(v+1)},$$

we obtain the solution

$$\boxed{y_1(x) = J_v(x) = \sum_{m=0}^{\infty} \frac{(-1)^m (x/2)^{2m+v}}{m!\Gamma(m+v+1)}.} \tag{131}$$

The series solution (131) is called the *Bessel function of the first kind and order* v (see also Section 2.5.1).

For $s_1 = v$, we have shown that $y_1(x) = J_v(x)$ is a solution of Bessel's equation. If we define $s_2 = -v$, then

$$y_2(x) = J_{-v}(x) = \sum_{m=0}^{\infty} \frac{(-1)^m (x/2)^{2m-v}}{m!\Gamma(m-v+1)} \tag{132}$$

is another solution, obtained by the formal replacement of v with $-v$ in (131). In fact, if v is not an integer, it can be shown that $J_{-v}(x)$ is always linearly independent of $J_v(x)$. Therefore, we conclude that

$$\boxed{y = C_1 J_v(x) + C_2 J_{-v}(x), \quad v \ne 0,1,2,\dots,} \tag{133}$$

is a general solution of Bessel's equation for noninteger values of v.

For small arguments, the two solutions in (133) have the asymptotic behavior

$$\begin{aligned} J_v(x) &\sim \frac{(x/2)^v}{\Gamma(1+v)}, \quad x \to 0^+, \\ J_{-v}(x) &\sim \frac{(x/2)^{-v}}{\Gamma(1-v)}, \quad x \to 0^+. \end{aligned} \tag{134}$$

Clearly, for v not an integer, the function $J_{-v}(x)$ is unbounded at the origin.

EXAMPLE 29 Find the general solution of

$$x^2 y'' + xy' + \left(x^2 - \frac{1}{4} \right) y = 0, \quad x > 0.$$

Solution: By comparison with (117), we see that $v = 1/2$ (and $-v = -1/2$). Thus, even though $v - (-v) = 1$ is an integer, it follows from above that the general solution is

$$y = C_1 J_{1/2}(x) + C_2 J_{-1/2}(x), \quad x > 0.$$

The most commonly occurring Bessel functions in practice are those of integer order. By setting $v = n$, $n = 0,1,2,...$, in (131), we obtain

$$J_n(x) = \sum_{m=0}^{\infty} \frac{(-1)^m (x/2)^{2m+n}}{m!\,\Gamma(m+n+1)}, \quad n = 0,1,2,.... \tag{135}$$

All Bessel functions of the first kind have an oscillatory behavior somewhat like that of a simple sinusoid, but the zeros of the Bessel functions are not evenly spaced and the amplitude steadily decreases with increasing values of the argument. The graphs of some of these functions is provided in Fig. 2.19.

By replacing n with $-n$ in (135), we obtain

$$\begin{aligned} J_{-n}(x) &= \sum_{m=0}^{\infty} \frac{(-1)^m (x/2)^{2m-n}}{m!\,\Gamma(m-n+1)} \\ &= \sum_{m=n}^{\infty} \frac{(-1)^m (x/2)^{2m-n}}{m!\,\Gamma(m-n+1)}, \end{aligned} \tag{136}$$

where we are recognizing that $1/\Gamma(m-n+1) = 0$, $m = 0,1,...,n-1$. If we now set $k = m - n$, then (136) becomes (upon simplification)

$$J_{-n}(x) = (-1)^n \sum_{k=0}^{\infty} \frac{(-1)^m (x/2)^{2k+n}}{k!\,\Gamma(k+n+1)}, \quad n = 0,1,2,..., \tag{137}$$

or

$$J_{-n}(x) = (-1)^n J_n(x), \quad n = 0,1,2,.... \tag{138}$$

Hence, although $J_{-n}(x)$ is also a solution of Bessel's equation, the integer order Bessel functions $J_n(x)$ and $J_{-n}(x)$ are not linearly independent functions.

1.9.3 Bessel functions of the second kind: $Y_\nu(x)$

To produce a solution of Bessel's equation (117) that is linearly independent of $J_\nu(x)$ for all values of ν, including integer values, it is customary to introduce the function

$$Y_\nu(x) = \frac{J_\nu(x)\cos \nu\pi - J_{-\nu}(x)}{\sin \nu\pi}, \tag{139}$$

called the *Bessel function of the second kind* (see also Section 2.5.2). For ν not an integer, this function is linearly independent of $J_\nu(x)$ because it is a linear combination of two linearly independent functions. For integer values of ν, (139) reduces to the indeterminate form $0/0$. In these latter cases, we define

$$Y_n(x) = \lim_{\nu \to n} Y_\nu(x), \quad n = 0,1,2,..., \tag{140}$$

the evaluation of which involves L'Hôpital's rule from calculus.[3] For instance, when $n = 0$ this leads to

$$Y_0(x) = \frac{2}{\pi}J_0(x)\left(\ln\frac{x}{2}+\gamma\right) - \frac{2}{\pi}\sum_{k=1}^{\infty}\frac{(-1)^k(x/2)^{2k}}{(k!)^2}\left(1+\frac{1}{2}+\cdots+\frac{1}{k}\right), \tag{141}$$

where γ is Euler's constant defined by

$$\gamma = \lim_{n \to \infty}\left(\sum_{k=1}^{n}\frac{1}{k} - \ln n\right) = 0.577215.... \tag{142}$$

Thus, for arbitrary values of ν, we write the *general solution* of Bessel's equation as

$$\boxed{y = C_1 J_\nu(x) + C_2 Y_\nu(x), \quad \nu \geq 0,} \tag{143}$$

where C_1 and C_2 are arbitrary constants.

For small arguments, the asymptotic form of the Bessel function (141) is

$$Y_0(x) \sim \frac{2}{\pi}\ln x, \quad x \to 0^+, \tag{144}$$

which becomes unbounded at the origin. For Bessel functions of the second kind of order $\nu > 0$, it can be shown that the asymptotic behavior near the origin is described by

$$Y_\nu(x) \sim -\frac{\Gamma(\nu)}{\pi}\left(\frac{2}{x}\right)^\nu, \quad \nu > 0, \quad x \to 0^+. \tag{145}$$

---

[3]L'Hôpital's rule is used to evaluate indeterminate forms like $0/0$ or ∞/∞. For example, if $\lim_{x \to a} f(x)/g(x) = 0/0$, then $\lim_{x \to a} f(x)/g(x) = \lim_{x \to a} f'(x)/g'(x)$.

Hence, like $J_{-v}(x)$, the Bessel functions $Y_v(x)$ are unbounded at the origin. For this reason, these functions must often be eliminated as potential solutions of problems in practice where bounded solutions at the origin are required.

1.9.4 Differential equations related to Bessel's equation

Elementary problems involving DEs are regarded as solved when their solutions can be expressed in terms of elementary functions such as trigonometric and exponential functions. The same can be said of many problems of a more complicated nature when their solutions can be expressed in terms of Bessel functions.

A fairly large number of DEs occurring in physics and engineering problems are specializations of

$$x^2y'' + (1 - 2a)xy' + \left[b^2c^2x^{2c} + (a^2 - c^2v^2)\right]y = 0, \quad v \geq 0, \quad b > 0, \tag{146}$$

where a, b, c, and v are constants that can assume various values. The general solution of (146) can be expressed as

$$y = x^a\left[C_1 J_v(bx^c) + C_2 Y_v(bx^c)\right], \tag{147}$$

where C_1 and C_2 are arbitrary constants. To derive this solution requires the following transformations:

$$\begin{aligned} y &= x^a z, \\ t &= x^c, \\ s &= bt. \end{aligned} \tag{148}$$

The combination of these transformations reduces (146) to

$$s^2\frac{d^2z}{ds^2} + s\frac{dz}{ds} + (s^2 - v^2)z = 0, \tag{149}$$

which is Bessel's equation with general solution $z = C_1 J_v(s) + C_2 Y_v(s)$. By using the transformations (148), this general solution can be written in the form given by (147).

EXAMPLE 30 Find the general solution of *Airy's equation*

$$y'' + xy = 0.$$

Solution: To compare Airy's equation with the general form (146), we first multiply through by x^2 to get

$$x^2y'' + x^3y = 0.$$

Thus, we see that $a = 1/2$, $b = 2/3$, $c = 3/2$, and $v = 1/3$. We conclude, therefore, that the general solution is

$$y = \sqrt{x}\left[C_1 J_{1/3}\left(\frac{2x^{3/2}}{3}\right) + C_2 Y_{1/3}\left(\frac{2x^{3/2}}{3}\right)\right],$$

or, because v is not an integer, we can also express this solution as

$$y = \sqrt{x}\left[C_1 J_{1/3}\left(\frac{2x^{3/2}}{3}\right) + C_2 J_{-1/3}\left(\frac{2x^{3/2}}{3}\right)\right].$$

Suggested Reading

L. C. Andrews, *Introduction to Differential Equations with Boundary Value Problems* (HarperCollins, New York, 1991).

W. Boyce and R. DiPrima, *Elementary Differential Equations*, 5th ed. (Wiley, New York, 1992).

C. H. Edwards, Jr., and D.E. Penny, *Differential Equations and Boundary Value Problems*, 2nd ed. (Prentice-Hall, Englewood Cliffs, 2000).

E. D. Rainville, P. E. Bedient, and R. E. Bedient, *Elementary Differential Equations*, 8th ed. (Prentice-Hall, Upper Saddle River, 1997).

D. Zill, *A First Course in Differential Equations with Applications*, 5th ed. (Brooks/Cole, Pacific Grove, 2001).

D. Zwillinger, *Handbook of Differential Equations*, 2nd ed. (Academic Press, New York, 1992).

For additional reading about the history of mathematics, the following textbooks are recommended.

1. E. T. Bell, *Men of Mathematics* (Simon and Schuster, New York, 1965).

2. C. B. Boyer and U. C. Merzbackh, *A History of Mathematics*, 2nd ed. (Wiley & Sons, New York, 1991).

3. F. Cajori, *A History of Mathematics*, 3rd ed. (Chelsea, New York, 1980).

4. H. Eves, *An Introduction to the History of Mathematics* (Saunder, Philadelphia, 1983).

Chapter 1

Exercises

In Problems 1-4, solve the given equation by separating the variables.

1. $y' = y^2 x^3$

Ans. $x^4/4 + 1/y = C$

2. $\dfrac{dy}{dx} = \dfrac{2x}{1+2y}$, $y(2) = 0$

Ans. $y^2 + y = x^2 - 4$

3. $4x\,dy - y\,dx = x^2\,dy$

Ans. $y = \pm\left(\dfrac{C_1 x}{x - 4}\right)^{1/4}$

4. $y + xy' = \dfrac{2x^2 y}{x^2 - x^2 y}$

Ans. $x(y^2 + 1) = C_1 y$

In Problems 5-8, solve the given linear equation.

5. $\dfrac{dy}{dx} - y\sin x = 0$

Ans. $y = C_1 e^{-\cos x}$

6. $y' + y = 4\cos 2x$

Ans. $y = C_1 e^{-x} + \frac{8}{5}\sin 2x + \frac{4}{5}\cos 2x$

7. $y' = x^2 + y$, $y(0) = -3$

Ans. $y = -(e^x + x^2 + 2x + 2)$

8. $y' - 3y = 6$, $y(1) = 0$

Ans. $y = 2e^{3(x-1)} - 2$

9. A given radioactive material is known to decay at a rate proportional to the amount present. If 500 mg of the material are initially present, and after 3 years 20% of the original mass has decayed, how much is present after

(*a*) 200 weeks?
(*b*) 25 years?

Ans. (*a*) 375.6 mg   (*b*) 77.9 mg

10. At time $t = 0$, half of a population of 100,000 persons have a cold that is spreading among the population. If the number of people who have a cold is increased by 1000 persons at the end of the first day, how long will it take until 80% of the population have the cold?

Ans. Approximately 35 days

In Problems 11-20, solve the given DE.

11. $9y'' + 6y' + y = 0$
Ans. $y = (C_1 + C_2 x)e^{-x/3}$

12. $y'' - y' - 2y = 0$
Ans. $y = C_1 e^{-x} + C_2 e^{2x}$

13. $y'' + 8y' + 25y = 0$
Ans. $y = e^{-4x}(C_1 \cos 3x + C_2 \sin 3x)$

14. $y'' - 2y' - 3y = 0$,
$y(0) = 0$, $y'(0) = -4$
Ans. $y = e^{-x} - e^{3x}$

15. $y'' + 2y' + 5y = 0$
$y(0) = 1$, $y'(0) = 5$

Ans. $y = e^{-x}(\cos 2x + 3\sin 2x)$

16. $y'' + 6y' + 9y = 0$
$y(0) = -4$, $y'(0) = 14$
Ans. $y = (2x - 4)e^{-3x}$

17. $y'' + y' = -\cos x$
Ans. $y = C_1 + C_2 e^{-x} + \frac{1}{2}(\cos x - \sin x)$

18. $y'' - 3y' - 4y = e^{-x}$,
    $y(0) = 3$, $y'(0) = 0$

Ans. $y = \dfrac{1}{25}(59 - 5x)e^{-x} + \dfrac{16}{25}e^{4x}$

19. $y'' + y = \sec x$

  Ans. $y = (C_1 + \ln|\cos x|)\cos x$
     $+ (C_2 + x)\sin x$

20. $y'' + 2y' + y = e^{-x}\ln x$
  Ans. $y = (C_1 + C_2 x$
     $+ \dfrac{1}{2}x^2\ln|x| - \dfrac{3}{4}x^2)e^{-x}$

In Problems 21 and 22, solve the given Cauchy-Euler DE for $x > 0$.

21. $x^2 y'' - 5xy' + 25y = 0$

  Ans. $y = x^3[C_1\cos(4\ln x)$
     $+ C_2\sin(4\ln x)]$

22. $x^2 y'' - 4xy' + 6y = 0$
    $y(2) = 0$, $y'(2) = 4$

  Ans. $y = x^3 - 2x^2$

In Problems 23 and 24, find a power series solution about $x = 0$.

23. $(1 - x^2)y'' - 2xy' + 2y = 0$

  Ans. $y = c_0(1 - x^2 - \dfrac{1}{3}x^4 - \cdots) + c_1 x$,
     $-1 < x < 1$

24. $y'' - xy' - y = 0$, $y(0) = 1$, $y'(0) = 0$

  Ans. $y = 1 + \dfrac{1}{2}x^2 + \dfrac{1}{8}x^4 + \cdots$, $|x| < \infty$

In Problems 25 and 26, use the method of Frobenius to solve the DE for $x > 0$.

25. $2xy'' + (1 - 2x)y' - y = 0$

Ans. $y = A\left(1 + x + \dfrac{1}{2!}x^2 + \cdots\right)$
    $+ Bx^{1/2}\left(1 + \dfrac{2}{3}x + \dfrac{4}{15}x^2 + \cdots\right)$

26. $xy'' + y' - 4y = 0$
  Ans. $y = (A + B\ln x)(1 + 4x + 4x^2 + \cdots)$
     $- B\left(8x + 12x^2 + \dfrac{176}{27}x^3 + \cdots\right)$

In Problems 27 and 28, find the general solution in terms of Bessel functions.

27. $x^2 y'' + 5xy' + (9x^2 - 12)y = 0$

  Ans. $y = \dfrac{1}{x^2}[C_1 J_4(3x) + C_2 Y_4(3x)]$

28. $4x^2 y'' + (1 + 4x)y = 0$

Ans. $y = \sqrt{x}[C_1 J_1(2\sqrt{x}) + C_2 Y_1(2\sqrt{x})]$

29. By comparing series, show that

  (a) $J_{1/2}(x) = \sqrt{\dfrac{2}{\pi x}}\sin x$.

  (b) $J_{-1/2}(x) = \sqrt{\dfrac{2}{\pi x}}\cos x$.

  (c) Use the results of parts (a) and (b) to deduce that

$$[J_{1/2}(x)]^2 + [J_{-1/2}(x)]^2 = \dfrac{2}{\pi x}.$$

30. Based on the result of Problem 29, show that the general solution of Example 29 can also be written as

$$y = \dfrac{1}{\sqrt{x}}(C_1\cos x + C_2\sin x).$$

Chapter 2

Special Functions

2.1 Introduction / 62
2.2 Engineering Functions / 63
    2.2.1   Step and signum (sign) functions: $\text{step}(x)$, $\text{sgn}(x)$ / 64
    2.2.2   Rectangle and triangle functions: $\text{rect}(x)$, $\text{tri}(x)$ / 65
    2.2.3   Sinc and Gaussian functions: $\text{sinc}(x)$, $\text{Gaus}(x)$ / 65
    2.2.4   Delta and comb functions: $\delta(x)$, $\text{comb}(x)$ / 65
2.3 Functions Defined by Integrals / 67
    2.3.1   Gamma functions: $\Gamma(x)$, $\gamma(a,x)$, $\Gamma(a,x)$ / 67
    2.3.2   Beta function: $B(x,y)$ / 70
    2.3.3   Digamma and polygamma functions: $\psi(x)$, $\psi^{(m)}(x)$ / 71
    2.3.4   Error functions and Fresnel integrals: $\text{erf}(x), \text{erfc}(x), C(x), S(x)$ / 72
2.4 Orthogonal Polynomials / 76
    2.4.1   Legendre polynomials: $P_n(x)$ / 76
    2.4.2   Hermite polynomials: $H_n(x)$ / 78
    2.4.3   Laguerre polynomials: $L_n(x)$, $L_n^{(m)}(x)$ / 80
    2.4.4   Chebyshev polynomials: $T_n(x)$, $U_n(x)$ / 81
2.5 Family of Bessel Functions / 83
    2.5.1   Standard Bessel functions: $J_p(x)$, $Y_p(x)$ / 84
    2.5.2   Modified Bessel functions: $I_p(x)$, $K_p(x)$ / 88
    2.5.3   Other Bessel functions / 92
2.6 Family of Hypergeometric-like Functions / 94
    2.6.1   Pochhammer symbol: $(a)_n$ / 94
    2.6.2   Hypergeometric function of Gauss: $_2F_1(a,b;c;x)$ / 95
    2.6.3   Confluent hypergeometric functions: $_1F_1(a;c;x)$, $U(a;c;x)$ / 97
    2.6.4   Generalized hypergeometric functions: $_pF_q$ / 98
    2.6.5   Applications involving hypergeometric functions / 100
2.7 Summary of Notations for Special Functions / 103
Suggested Reading / 104
Exercises / 105

Historical Comments: The study of special functions involves too many names to mention them all, so we concentrate on a few of the more well known people. The famous mathematician Leonhard Euler (1707-1783) was concerned with a generalization of the factorial function now called the gamma function. Other mathematicians involved with the gamma function were Adrien M. Legendre (1752-1833) and Karl Weierstrass (1815-1897). Legendre is also credited with developing the polynomial set now bearing his name in connection with his development of the gravitational potential into a power series. Additional polynomial sets with similar properties carry the name of Charles Hermite (1822-1901) and Edmund Laguerre (1834-1886), among others.

The German astronomer Friedrich W. Bessel (1784-1846) first achieved fame by computing the orbit of Halley's comet. In addition to many other accomplishments in connection with his studies of planetary motion, he is credited with deriving the differential equation now bearing his name and carrying out the first systematic study of the general properties of Bessel functions in his famous memoir of 1824. Nonetheless, Bessel functions were first (unknowingly) discovered in 1732 by Daniel Bernoulli (1700-1782) who provided a series representation. There now exists a wide variety of types of Bessel functions that also bear such famous names as Hankel, Kelvin, Lommel, Struve, Airy, Anger, and Weber.

The major development of the properties of the hypergeometric function was carried out in 1812 by Carl F. Gauss (1777-1855), and a similar analysis involving the confluent hypergeometric functions was carried out in 1836 by Ernst E. Kummer (1810-1893). Generalized hypergeometric functions were developed in the late 1800s and early 1900s by people such as Clausen, Appel, Lauricella, Whittaker,

---

Our objective in this chapter is to familiarize the reader with many of the special functions that arise in advanced engineering applications. In addition to defining these functions, we provide lists of their most important properties for easy reference. We start, however, with a review of several of the standard engineering-type functions that are widely used in practice, like the step function and Dirac delta function.

2.1 Introduction

Functions introduced in introductory calculus and engineering courses are called *elementary functions*—i.e., algebraic functions, trigonometric functions, exponential functions, logarithm functions, and so on. In more advanced courses the need often arises to consider other types of functions known as *special functions*. Such functions occur naturally in a variety of engineering problems in electromagnetic theory, communication theory, electro-optics, wave propagation, electric circuit theory, and heat conduction, among many others. Although somewhat difficult to generate numerical values of some of the special functions a few years ago, today it is quite easy owing to the development of a variety of standard software programs which incorporate many of these functions. The term "special function" originally referred to only *higher transcendental functions*, such as Bessel functions and hypergeometric functions, but the term is more inclusive now and also includes many *engineering-type functions* such as the step function, delta function, sinc function, and so on. Our treatment of special functions here is limited to real variables, but many of the special functions such as Bessel functions and hypergeometric functions can also be defined with complex arguments.

2.2 Engineering Functions

In the solution of engineering problems it is often helpful to employ the use of special notation to identify certain functions, particularly those that must be prescribed in a piecewise manner. Some authors use a single letter to denote these functions and others use some sort of abbreviation. However, because the same letters or abbreviations are not consistently used by all authors, some care must be exercised when using various reference sources. Although some of these functions are also defined in two or three dimensions, we limit our discussion below to only functions in one dimension.

2.2.1 Step and signum (sign) functions: $\text{step}(x)$, $\text{sgn}(x)$

The *unit step function* was first introduced by the English engineer Oliver Heaviside (1850-1925) in circuit analysis problems to handle finite jump discontinuities, acting like a "switch." This function is defined by (see Fig. 2.1)

$$\text{step}(x) \;=\; U(x) \;=\; \begin{cases} 0, & x < 0 \\ 1/2, & x = 0 \\ 1, & x > 0 \end{cases} \tag{1}$$

where the value ½ at $x = 0$ is optional. That is, in many texts the value of this function at the jump discontinuity is either omitted or taken to be either zero or one.

As indicated above, the unit step function is often designated by the symbol $U(x)$, among others. When the jump takes place at an arbitrary point $x = a$, we write $\text{step}(x - a)$, $U(x - a)$, and so on. A more general version of the step function is sometimes given by $\text{step}[(x - a)/b]$, in which the jump occurs at $x = a$ and the constant $b = \pm 1$ permits reflection about the line $x = a$ by selecting $b = -1$.

The *sign* or *signum function* defined by (see Fig. 2.2)

$$\text{sgn}(x) \;=\; \begin{cases} -1, & x < 0 \\ 0, & x = 0 \\ 1, & x > 0 \end{cases} \tag{2}$$

is used to reverse the polarity of another function at some point. It is directly related to the step function by $\text{sgn}(x) = 2\text{step}(x) - 1$.

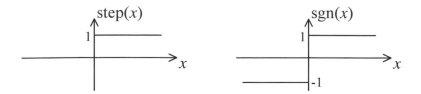

Figure 2.1 Unit step function.   **Figure 2.2** Signum function.

2.2.2 Rectangle and triangle functions: $\text{rect}(x)$, $\text{tri}(x)$

One of the most common and useful engineering functions is the *rectangle function* (see Fig. 2.3)

$$\text{rect}(x) = \Pi(x) = \begin{cases} 1, & |x| < 1/2 \\ 0, & |x| > 1/2. \end{cases} \tag{3}$$

In some instances it is also called the "window" function and is directly related to the step function by $\text{rect}(x) = \text{step}(x + 1/2) - \text{step}(x - 1/2)$. In the time domain, this function can be used to represent a gating operation in an electrical circuit, for example. Like the step function, the rectangle function is sometimes defined by the value $1/2$ at $|x| = 1/2$. The rectangle function (3) has a height of unity and is centered at $x = 0$. More generally, the rectangle function is written as $\text{rect}[(x - a)/b]$, in which case it is centered at $x = a$ and $|b|$ is equal to both its width and area.

The *triangle function* (see Fig. 2.4)

$$\text{tri}(x) = \Lambda(x) = \begin{cases} 1 - |x|, & |x| < 1 \\ 0, & |x| > 1 \end{cases} \tag{4}$$

is sometimes used as a model for the correlation function of a stochastic variable. Written as $\text{tri}[(x - a)/b]$, the width of the base is $2|b|$, its area is $|b|$, and it is centered at $x = a$.

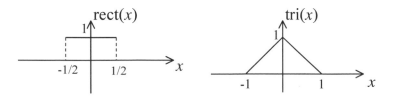

Figure 2.3 Rectangle function.   **Figure 2.4** Triangle function.

2.2.3 Sinc and Gaussian functions: $\mathrm{sinc}(x)$, $\mathrm{Gaus}(x)$

The *sinc function* is defined by (see Fig. 2.5)

$$\mathrm{sinc}(x) = \frac{\sin \pi x}{\pi x}. \tag{5}$$

This function consistently shows up in studies of linear systems. The π factor in the sine function is introduced so that the zeros of the sinc function are at integer values of x. The sinc function is directly related to the Fourier transform of the rectangle function and its square is directly related to the Fourier transform of the triangle function (see Table 9.2 in Chapter 9). Again, this function can be defined more generally by $\mathrm{sinc}[(x - a)/b]$, in which the zeros occur at $a \pm nb$, $n = 1, 2, 3, \ldots$ and it is centered at $x = a$.

The *Gaussian function* (see Fig. 2.6)

$$\mathrm{Gaus}(x) = e^{-\pi x^2} \tag{6}$$

is encountered in the field of statistics and is a common model used to characterize the output of a laser operating in the lowest-order mode. To center this function at $x = a$, we write $\mathrm{Gaus}[(x - a)/b]$, which still has a height of unity and an area equal to $|b|$.

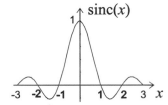

Figure 2.5 Sinc function.   **Figure 2.6** Gaussian function.

2.2.4 Delta and comb functions: $\delta(x)$, $\mathrm{comb}(x)$

The *Dirac delta function*, or simply the *delta function*, is named after Paul A. M. Dirac (1902-1984) who was awarded a Nobel prize (with E. Schrödinger) in 1933 for his work in quantum mechanics. The delta function is now widely accepted as a standard tool in engineering analysis, and in applications involving linear systems it is customary to call it the *impulse function*. In the field of optics, the delta function is often used as a model for a point source.

The defining properties of the delta function are given by (see Fig. 2.7)

$$\delta(x - a) = 0, \quad x \neq a,$$

$$\int_{x_1}^{x_2} f(x)\delta(x - a)\,dx = \begin{cases} f(a), & x_1 < a < x_2 \\ 0, & \text{otherwise} \end{cases} \tag{7}$$

The second expression in (7) is commonly called the "sifting" property of this function. Clearly, this is not a function in the usual sense of the word! Nonetheless, it is common practice to assign certain function properties to this symbol, the primary ones being the following:

$$
\begin{aligned}
&(a) \quad \delta(-x) = \delta(x) \\
&(b) \quad \delta(ax) = \frac{1}{|a|}\delta(x) \\
&(c) \quad f(x)\delta(x - a) = f(a)\delta(x - a) \\
&(d) \quad \frac{d}{dx}\text{step}(x - a) = \delta(x - a)
\end{aligned}
\tag{8}
$$

The derivative of the step function occurring in property (d) of (8) is not a derivative in the usual sense because the step function is not continuous at $x = a$. We refer to this as a "generalized derivative," which describes the jump discontinuity in the step function.

In the strict sense, the delta function has significance only as part of an integrand. Thus, it is best in dealing with this function to avoid the idea of assigning "functional values" and instead refer to its integral property given in (7). Following more rigorous lines, the delta function can be defined as a limit of an infinite sequence of well-behaved functions (called a delta sequence); for example,

$$
\delta(x) = \lim_{N \to \infty} N\sqrt{\pi}\,\text{Gaus}(Nx).
\tag{9}
$$

In this more rigorous context, it is termed a *generalized function*.

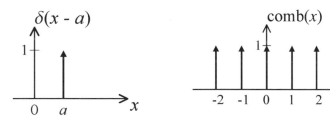

Figure 2.7 Delta function.   **Figure 2.8** Comb function.

The *comb function* is an infinite series of delta functions that is useful in representing sampled data (see Fig. 2.8). It is typically defined by

$$
\text{comb}(x) = \sum_{n=-\infty}^{\infty} \delta(x - n).
\tag{10}
$$

Among other uses, the delta function (7) and the comb function (10) are both very expedient in the spectral representations of periodic functions that commonly appear in the Fourier analysis of various power and energy signals. To place the delta functions in (10) at $a \pm nb$, $n = 1, 2, 3, \ldots$ rather than at the integers, we introduce the notation $\text{comb}[(x - a)/b]$.

2.3 Functions Defined by Integrals

Special functions defined by an integral occur in a variety of applications. In many cases this is because of the widespread use of Fourier transforms for which closed-form solutions often lead to functions not commonly encountered in the undergraduate curriculum, such as Bessel functions, Fresnel integrals, and hypergeometric functions.

2.3.1 Gamma functions: $\Gamma(x)$, $\gamma(a,x)$, $\Gamma(a,x)$

The gamma function, briefly introduced in Section 1.9.1, is widely used in calculations in probability theory. It is also used in the kinetic theory of gases and statistical mechanics because of its connection to the Maxwell-Boltzmann distribution; it is used in problems in condensed-matter physics involving Fermi-Dirac and Einstein-Bose statistics; it arises in a variety of problems related to calculating areas, volumes, moments of inertia, an so on; and it is used in the definition of other special functions like Bessel functions (recall Section 1.9.2) and hypergeometric functions. It was originally discovered by Leonhard Euler (1707–1783) in 1729 as a means of interpolating between the numbers

$$n! = \int_0^\infty e^{-t} t^n dt, \quad n = 0, 1, 2, \ldots$$

with nonintegral values of n. His studies eventually led him to the *gamma function* defined by

$$\Gamma(x) = \int_0^\infty e^{-t} t^{x-1} dt, \quad x > 0. \tag{11}$$

The gamma function is defined for all real numbers x except for zero and the negative integers [see property (G7) below]. Although (11) is valid for only positive numbers, the gamma function can be extended to negative (real) arguments by the use of the recurrence formula (G1) (except for zero and all negative integers). The graph of this function is illustrated in Fig. 2.9.

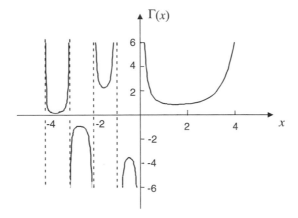

Figure 2.9 Gamma function.

Some of the important properties associated with the gamma function are listed below:

(G1): $\Gamma(x + 1) = x\Gamma(x)$

(G2): $\Gamma(n + 1) = n!, \quad n = 0, 1, 2, \ldots$

(G3): $\Gamma(\tfrac{1}{2}) = \sqrt{\pi}$

(G4): $\Gamma(n + \dfrac{1}{2}) = \dfrac{(2n)!}{2^{2n} n!}\sqrt{\pi}, \quad n = 0, 1, 2, \ldots$

(G5): $\Gamma(x)\Gamma(1 - x) = \dfrac{\pi}{\sin \pi x}, \quad (x \text{ nonintegral}).$

(G6): $\Gamma\!\left(\dfrac{1}{2} + x\right)\Gamma\!\left(\dfrac{1}{2} - x\right) = \dfrac{\pi}{\cos \pi x}, \quad (x \text{ nonintegral})$

(G7): $\dfrac{1}{\Gamma(-n)} = 0, \quad n = 0, 1, 2, \ldots$

(G8): $\sqrt{\pi}\,\Gamma(2x) = 2^{2x-1}\Gamma(x)\Gamma\!\left(x + \dfrac{1}{2}\right)$

(G9): $\Gamma(x + 1) \sim \sqrt{2\pi x}\, x^{x} e^{-x}\!\left(1 + \dfrac{1}{12x} + \cdots\right), \quad x \to \infty$

(G10): $n! \sim \sqrt{2\pi n}\,(n/e)^{n}, \quad n = 1, 2, 3, \ldots, \quad n \gg 1$

EXAMPLE 1 Evaluate the following integral:

$$I = \int_{0}^{\infty} e^{-pt}\sqrt{t}\,dt, \quad p > 0.$$

Solution: We can interpret this integral as the Laplace transform of $f(t) = \sqrt{t}$, although this is not necessary. We simply make the substitution $x = pt$ to find

$$I = \frac{1}{p\sqrt{p}}\int_{0}^{\infty} e^{-x}\sqrt{x}\,dx = \frac{\Gamma(3/2)}{p\sqrt{p}} = \frac{1}{2p}\sqrt{\frac{\pi}{p}}.$$

In the last step we compared our integral with that in (11) and used property (G4).

A function directly related to the gamma function is the *incomplete gamma function* defined by

$$\gamma(a,x) = \int_{0}^{x} e^{-t} t^{a-1}\,dt, \quad a > 0. \tag{12}$$

This function occurs in calculating the cumulative distribution associated with a gamma or chi-square probability density function, among others. The companion function

$$\Gamma(a,x) = \int_x^\infty e^{-t} t^{a-1} dt, \quad a > 0. \tag{13}$$

is called the *complementary incomplete gamma function.*

Some identities associated with these functions are listed below:

(IG1): $\gamma(a,x) + \Gamma(a,x) = \Gamma(a)$.

(IG2): $\gamma(a + 1,x) = a\gamma(a,x) - x^a e^{-x}$

(IG3): $\Gamma(a + 1,x) = a\Gamma(a,x) + x^a e^{-x}$

(IG4): $\dfrac{d}{dx}[x^{-a}\gamma(a,x)] = -x^{-a-1}\gamma(a + 1,x)$

(IG5): $\dfrac{d}{dx}[x^{-a}\Gamma(a,x)] = -x^{-a-1}\Gamma(a + 1,x)$

(IG6): $\gamma(a,x) = x^a \displaystyle\sum_{n=1}^{\infty} \dfrac{(-1)^n x^n}{n!(n + a)}$

(IG7): $\gamma(\frac{1}{2},x^2) = \sqrt{\pi}\,\mathrm{erf}(x)$ (See Section 2.3.4.)

(IG8): $\Gamma(\frac{1}{2},x^2) = \sqrt{\pi}\,\mathrm{erfc}(x)$ (See Section 2.3.4.)

(IG9): $\gamma(a,x) = \dfrac{x^a}{a}\,{}_1F_1(a;a + 1; -x)$ (See Section 2.6.3.)

(IG10): $\Gamma(a,x) = e^{-x}U(1 - a;1 - a;x)$ (See Section 2.6.3.)

(IG11): $\Gamma(a,x) \sim \Gamma(a)x^{a-1}e^{-x}\displaystyle\sum_{n=0}^{\infty} \dfrac{1}{\Gamma(a - n)x^n}, \quad a > 0, \quad x \to \infty$

EXAMPLE 2 Given the *gamma probability density function* (see Chapter 13)

$$f_x(x) = \frac{\alpha(\alpha x)^{\alpha-1}}{\Gamma(\alpha)} e^{-\alpha x} U(x),$$

where α is a positive parameter and $U(x)$ is the unit step function (Section 2.2.1),
(*a*) calculate the *n*th statistical moment of the distribution.
(*b*) calculate the cumulative distribution function $F_x(x) = \mathrm{Pr}(\mathbf{x} \le x)$.

Solution:
(*a*) The *n*th moment is defined by the ensemble average

$$\langle \mathbf{x}^n \rangle = \int_0^\infty x^n f_x(x)dx = \frac{\alpha^\alpha}{\Gamma(\alpha)}\int_0^\infty x^{n+\alpha-1} e^{-\alpha x} dx.$$

By making the change of variable $t = \alpha x$, we obtain

$$<\mathbf{x}^n> = \frac{1}{\alpha^n \Gamma(\alpha)} \int_0^\infty t^{n+\alpha-1} e^{-t} dt = \frac{\Gamma(\alpha + n)}{\alpha^n \Gamma(\alpha)}, \quad n = 1, 2, 3, \dots.$$

Note that, for $n = 1$, the *mean* is unity and for $n = 2$, we obtain $E[\mathbf{x}^2] = 1 + \dfrac{1}{\alpha}$. Consequently, the variance is $\sigma^2 = E[\mathbf{x}^2] - E^2[\mathbf{x}] = \dfrac{1}{\alpha}$.

(*b*) The cumulative distribution is defined by

$$F_\mathbf{x}(x) = \int_0^x f_\mathbf{x}(u) du = \frac{\alpha^\alpha}{\Gamma(\alpha)} \int_0^x u^{\alpha-1} e^{-\alpha u} du.$$

Here, we make the change of variable $t = \alpha u$ and use the incomplete gamma function (12) to deduce that

$$F_\mathbf{x}(x) = \frac{1}{\Gamma(\alpha)} \int_0^{\alpha x} t^{\alpha-1} e^{-t} dt = \frac{\gamma(\alpha, \alpha x)}{\Gamma(\alpha)}.$$

For the special case in which $\alpha = 1$, the reader can verify that $F_\mathbf{x}(x) = 1 - e^{-x}$.

2.3.2 Beta function: $B(x,y)$

A particular combination of gamma functions defines the *beta function*

$$B(x,y) = \frac{\Gamma(x)\Gamma(y)}{\Gamma(x + y)} = \int_0^1 t^{x-1}(1 - t)^{y-1} dt, \quad x, y > 0. \tag{14}$$

From (14), it follows that the beta function is symmetric in x and y; i.e., $B(x,y) = B(y,x)$. By introducing the change of variables $t = u/(1 + u)$ and $t = \cos^2\theta$, respectively, the beta function assumes the other useful integral forms

$$B(x,y) = \int_0^\infty \frac{u^{x-1}}{(1 + u)^{x+y}} du, \quad x, y > 0, \tag{15}$$

$$B(x,y) = 2 \int_0^{\pi/2} \cos^{2x-1}\theta \, \sin^{2y-1}\theta \, d\theta, \quad x, y > 0. \tag{16}$$

2.3.3 Digamma and polygamma functions: $\psi(x)$, $\psi^{(m)}(x)$

Closely associated with the derivative of the gamma function is the *digamma function* (also called the *psi function*) defined by

$$\psi(x) = \frac{d}{dx} \ln \Gamma(x) = \frac{\Gamma'(x)}{\Gamma(x)}, \quad x \neq 0, -1, -2, \dots . \tag{17}$$

A series representation of this function is given by

$$\psi(x) = -\gamma + \sum_{n=0}^{\infty} \left(\frac{1}{n+1} - \frac{1}{n+x} \right), \tag{18}$$

where *Euler's constant* $\gamma = -\psi(1) = -\Gamma'(1)$ has the integral representation

$$\gamma = -\int_0^{\infty} e^{-t} \ln t \, dt. \tag{19}$$

The Euler constant has numerical value $\gamma = 0.577215\dots$.

Continued derivatives of the gamma function lead to the *polygamma functions* given by

$$\psi^{(m)}(x) = \frac{d^{m+1}}{dx^{m+1}} \ln \Gamma(x), \quad m = 1, 2, 3, \dots . \tag{20}$$

Both the digamma and polygamma functions are used for summing series involving rational functions with the power of the denominator at least 2 greater than that in the numerator. When $x = 1$, we obtain the special value

$$\psi^{(m)}(1) = (-1)^{n+1} m! \, \zeta(m+1), \tag{21}$$

where $\zeta(p)$ is the *Riemann zeta function* defined by

$$\zeta(p) = \sum_{n=0}^{\infty} \frac{1}{n^p}, \quad p > 1. \tag{22}$$

The basic properties of the digamma and polygamma functions are listed below:

(DP1): $\psi(x+1) = \psi(x) + \dfrac{1}{x}$

(DP2): $\psi(1-x) - \psi(x) = \pi \cot \pi x$

(DP3): $\psi(n+1) = -\gamma + \displaystyle\sum_{k=1}^{n} \frac{1}{k}, \quad n = 1, 2, 3, \dots .$

(DP4): $\psi(x+1) = -\gamma + \displaystyle\sum_{n=1}^{\infty} (-1)^{n+1} \zeta(n+1) x^n, \quad -1 < x < 1$

(DP5): $\psi^{(m)}(x+1) = (-1)^{m+1} \sum_{n=0}^{\infty} (-1)^{n+1} \frac{(m+n)!}{n!} \zeta(m+n+1)x^n,$

$$-1 < x < 1$$

(DP6): $\psi(x+1) \sim \ln x + \dfrac{1}{2x} - \dfrac{1}{12x^3} + \cdots, \quad x \to \infty$

2.3.4 Error functions and Fresnel integrals: $\mathrm{erf}(x)$, $\mathrm{erfc}(x)$, $C(x)$, $S(x)$

The *error function*, which derives its name from its importance in the theory of errors, also occurs in probability theory and in certain heat conduction problems, among other areas. It is defined by the integral

$$\mathrm{erf}(x) = \frac{2}{\sqrt{\pi}} \int_0^x e^{-t^2} dt, \quad -\infty < x < \infty. \tag{23}$$

In some applications it is useful to also introduce the *complementary error function*

$$\mathrm{erfc}(x) = \frac{2}{\sqrt{\pi}} \int_x^{\infty} e^{-t^2} dt, \quad -\infty < x < \infty. \tag{24}$$

The graph of the error function (23) is shown in Fig. 2.10, and the primary properties associated with both error functions (23) and (24) are the following:

(E1): $\mathrm{erf}(-x) = -\mathrm{erf}(x)$

(E2): $\mathrm{erf}(0) = 0$

(E3): $\mathrm{erf}(\infty) = 1$

(E4): $\mathrm{erf}(x) = \dfrac{2}{\sqrt{\pi}} \sum_{n=0}^{\infty} \dfrac{(-1)^n x^{2n+1}}{n!(2n+1)}, \quad |x| < \infty$

(E5): $\mathrm{erfc}(x) = 1 - \mathrm{erf}(x)$

(E6): $\mathrm{erf}(x) = \dfrac{2x}{\sqrt{\pi}}\; {}_1F_1\left(\dfrac{1}{2}; \dfrac{3}{2}; -x^2\right)$ (See Section 2.6.3.)

(E7): $\mathrm{erfc}(x) = \dfrac{1}{\sqrt{\pi}} e^{-x^2} U\left(\dfrac{1}{2}; \dfrac{1}{2}; x^2\right)$ (See Section 2.6.3.)

(E8): $\mathrm{erfc}(x) \sim \dfrac{e^{-x^2}}{\sqrt{\pi}\,x}, \quad x \to \infty$

(E9): $\dfrac{d}{dx}\mathrm{erf}(x) = \dfrac{2}{\sqrt{\pi}} e^{-x^2}$

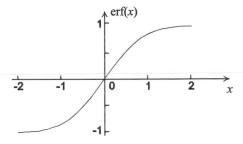

Figure 2.10 Error function.

EXAMPLE 3 Given the *normal distribution function* (see Chapter 13)

$$f_{\mathbf{x}}(x) = \frac{1}{\sigma\sqrt{2\pi}} e^{-(x-m)^2/2\sigma^2}, \quad -\infty < x < \infty,$$

where m is the *mean value* and σ^2 is the *variance*, calculate the cumulative distribution function.

Solution: From definition, the cumulative distribution function is

$$F_{\mathbf{x}}(x) = \frac{1}{\sigma\sqrt{2\pi}} \int_{-\infty}^{x} e^{-(u-m)^2/2\sigma^2} \, du$$

$$= \frac{1}{\sigma\sqrt{2\pi}} \int_{-\infty}^{\infty} e^{-(u-m)^2/2\sigma^2} \, du - \frac{1}{\sigma\sqrt{2\pi}} \int_{x}^{\infty} e^{-(u-m)^2/2\sigma^2} \, du.$$

By making the change of variable $t = (u-m)/\sigma\sqrt{2}$, and using the complementary error function (24), we obtain

$$F_{\mathbf{x}}(x) = 1 - \frac{1}{2}\operatorname{erfc}\left(\frac{x-m}{\sigma\sqrt{2}} \right),$$

which can also be written as

$$F_{\mathbf{x}}(x) = \Pr(\mathbf{x} \le x) = \frac{1}{2}\left[1 + \operatorname{erf}\left(\frac{x-m}{\sigma\sqrt{2}} \right) \right].$$

Closely associated with the error function are the *Fresnel integrals*

$$\boxed{C(x) = \int_{0}^{x} \cos\left(\frac{1}{2}\pi t^2 \right) dt,} \qquad (25)$$

$$S(x) = \int_0^x \sin\left(\frac{1}{2}\pi t^2\right) dt, \qquad\qquad (26)$$

the graphs of which are illustrated in Fig. 2.11. From definition, it follows that $C(0) = S(0) = 0$. Also, the derivatives are

$$C'(x) = \cos\left(\frac{1}{2}\pi x^2\right), \quad S'(x) = \sin\left(\frac{1}{2}\pi x^2\right), \qquad (27)$$

from which we deduce that both $C(x)$ and $S(x)$ are *oscillatory*. In particular, the maxima and minima of these functions are determined by the zeros of the simple trigonometric functions given in (27). In the limit $x \to \infty$, it has been shown that

$$C(\infty) = S(\infty) = \frac{1}{2}. \qquad\qquad (28)$$

Among other applications, the Fresnel integrals occur naturally in optics in calculating the diffraction pattern arising from a square aperture (see Example 4 below).

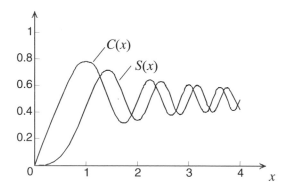

Figure 2.11 Fresnel integrals.

Some basic properties of the Fresnel integrals are listed below:

(F1): $C(-x) = -C(x)$

(F2): $S(-x) = -S(x)$

(F3): $C(x) = \displaystyle\sum_{n=0}^{\infty} \frac{(-1)^n (\pi/2)^{2n}}{(2n)!(4n+1)} x^{4n+1}$

(F4): $S(x) = \displaystyle\sum_{n=0}^{\infty} \frac{(-1)^n (\pi/2)^{2n+1}}{(2n+1)!(4n+3)} x^{4n+3}$

(F5): $C(x) - iS(x) = \dfrac{1}{\sqrt{2i}} \mathrm{erf}\left(x\sqrt{\dfrac{i\pi}{2}}\right)$

EXAMPLE 4 The Fresnel diffraction pattern of monochromatic light by a finite aperture is determined by the Huygens-Fresnel principle (see [16]). If $U(x,y,0)$ denotes the distribution of the field immediately behind the finite aperture, then, under a paraxial approximation, the field distribution at distance z behind the aperture is given by

$$U(x,y,z) = \frac{\exp(ikz)}{i\lambda z} \iint U(\xi,\eta,0)\exp\left\{\frac{ik}{2z}\left[(x-\xi)^2 + (y-\eta)^2\right]\right\}d\xi\,d\eta,$$

where λ is optical wavelength and $k = 2\pi/\lambda$. Given that the finite aperture is square with width $2a$, use the above integral to determine the diffraction pattern of an incident plane wave.

Solution: The distribution of the optical field immediately behind the square aperture may be written

$$U(x,y,0) = \text{rect}\left(\frac{x}{2a}\right)\text{rect}\left(\frac{y}{2a}\right).$$

The Huygens-Fresnel integral then takes the form

$$U(x,y,z) = \frac{\exp(ikz)}{i\lambda z}\int_{-a}^{a}\int_{-a}^{a}\exp\left\{\frac{ik}{2z}\left[(x-\xi)^2 + (y-\eta)^2\right]\right\}d\xi\,d\eta$$

$$= \frac{\exp(ikz)}{i\lambda z}u(x)u(y),$$

where

$$u(x) = \int_{-a}^{a}\exp\left[\frac{ik}{2z}(\xi-x)^2\right]d\xi, \quad u(y) = \int_{-a}^{a}\exp\left[\frac{ik}{2z}(\eta-y)^2\right]d\eta.$$

Now, by making the change of variable $s = \sqrt{k/\pi z}(\xi - x)$ and $t = \sqrt{k/\pi z}(\eta - y)$, it follows from definition of the Fresnel integrals that

$$u(x) = \sqrt{\frac{\pi z}{k}}\left\{\left[C(s_2) - C(s_1)\right] + i\left[S(s_2) - S(s_1)\right]\right\},$$

$$u(y) = \sqrt{\frac{\pi z}{k}}\left\{\left[C(t_2) - C(t_1)\right] + i\left[S(t_2) - S(t_1)\right]\right\},$$

where

$$s_1 = -\sqrt{\frac{k}{\pi z}}(a + x), \quad s_2 = \sqrt{\frac{k}{\pi z}}(a - x),$$

$$t_1 = -\sqrt{\frac{k}{\pi z}}(a + y), \quad t_2 = \sqrt{\frac{k}{\pi z}}(a - y).$$

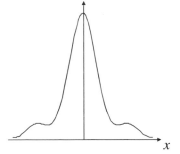

Figure 2.12 Cross section of Fresnel diffraction pattern of a rectangular aperture.

Last, the intensity pattern is given by (see Fig. 2.12 for a cross-section view)

$$I(x,y,z) = |U(x,y,z)|^2 = \frac{|u(x)|^2\,|u(y)|^2}{(\lambda z)^2}.$$

2.4 Orthogonal Polynomials

There are several special polynomial sets with properties that distinguish them from ordinary polynomials. In particular, each polynomial set satisfies a three-term recurrence relation between consecutive polynomials and each set is a mutually orthogonal set of functions (see also Chapter 8). We say a set of functions $\{\varphi_n(x)\}$, $n = 0,1,2,...$, is *orthogonal* on an interval $a < x < b$ with respect to a weighting function $r(x) > 0$ if it is true for all members of the set that

$$\int_a^b r(x)\varphi_n(x)\varphi_k(x)\,dx = 0, \quad k \neq n. \tag{29}$$

In the treatment below we assume $\{\varphi_n(x)\}$ denotes one of the following sets—*Legendre polynomials, Hermite polynomials, Laguerre polynomials*, or *Chebyshev polynomials*.

2.4.1 Legendre polynomials: $P_n(x)$

The *Legendre polynomial* $y = P_n(x)$ was briefly introduced in Section 1.7.1 as a bounded solution of

$$(1 - x^2)y'' - 2xy' + n(n+1)y = 0, \quad -1 < x < 1. \tag{30}$$

Equation (30), called *Legendre's equation*, arises in problems in mechanics, quantum mechanics, electromagnetic theory, and heat conduction, all with spherical symmetry.

Historically, the Legendre polynomials first arose in the problem of expressing the Newtonian potential of a conservative force field for a point mass at P in a series involving

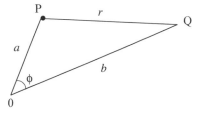

Figure 2.13 Geometry used for the Newtonian potential $V(r) = K/r$.

the distance variables a and b of two points P and Q and their included central angle ϕ (see Fig. 2.13). The force field potential function given by $V(r) = K/r$, where K is a physical constant, was developed in 1785 by A. M. Legendre (1752-1833) into the series

$$
\begin{aligned}
V(r) = \frac{K}{r} &= \frac{K}{b}(1 - 2xt + t^2)^{-1/2} \\
&= \frac{K}{b}\sum_{k=0}^{\infty} P_n(x)t^n, \quad |x| \le 1, \quad |t| < 1,
\end{aligned}
\tag{31}
$$

where $r^2 = a^2 + b^2 - 2ab\cos\phi$, $x = \cos\phi$, and $t = a/b$ $(a < b)$. In this setting, the coefficient functions $P_n(x)$ are the Legendre polynomials defined by (see Fig. 2.14)

$$
P_n(x) = \sum_{n=0}^{[n/2]} \frac{(-1)^k(2n-2k)!\,x^{n-2k}}{2^n k!(n-k)!(n-2k)!}, \quad n = 0,1,2,\dots,
\tag{32}
$$

where we have introduced the special notation

$$
[n/2] = \begin{cases} n/2, & n \text{ even}, \\ (n-1)/2, & n \text{ odd}. \end{cases}
\tag{33}
$$

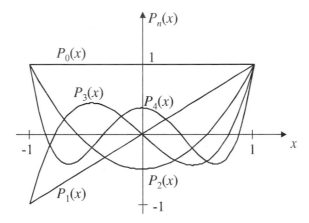

Figure 2.14 Legendre polynomials $P_n(x)$, $n = 0,1,2,3,4$.

The function $w(x,t) = (1 - 2xt + t^2)^{-1/2}$ given in (31) is called a *generating function* because of its connection to the Legendre polynomials. The first few Legendre polynomials are listed in Table 2.1.

Table 2.1 Legendre Polynomials

$n = 0$	$P_0(x) = 1$
$n = 1$	$P_1(x) = x$
$n = 2$	$P_2(x) = \dfrac{1}{2}(3x^2 - 1)$
$n = 3$	$P_3(x) = \dfrac{1}{2}(5x^3 - 3x)$
$n = 4$	$P_4(x) = \dfrac{1}{8}(35x^4 - 30x^2 + 3)$
$n = 5$	$P_5(x) = \dfrac{1}{8}(63x^5 - 70x^3 + 15x)$

Some properties associated with the Legendre polynomial set are given below:

(P1): $P_n(-x) = (-1)^n P_n(x)$

(P2): $P_n(x) = \dfrac{1}{2^n n!} \dfrac{d^n}{dx^n}[(x^2 - 1)^n]$

(P3): $P_n(1) = 1; \quad P_n(-1) = (-1)^n$

(P4): $P_{2n}(0) = \dfrac{(-1)^n (2n)!}{2^{2n} (n!)^2}; \quad P_{2n+1}(0) = 0$

(P5): $(n + 1)P_{n+1}(x) - (2n + 1)xP_n(x) + nP_{n-1}(x) = 0$

(P6): $(1 - x^2)P_n{}'(x) = nP_{n-1}(x) - nxP_n(x)$

(P7): $P_n(x) = \dfrac{1}{\pi}\int_0^\pi [x + (x^2 - 1)^{1/2}\cos\theta]^n d\theta$

(P8): $\int_{-1}^1 P_n(x)P_k(x)dx = 0, \quad k \neq n$

(P9): $\int_{-1}^1 [P_n(x)]^2 dx = \dfrac{2}{2n + 1}$

(P10): $P_n(x) = {}_2F_1\left(-n, n + 1; 1; \dfrac{1-x}{2}\right)$ (See Section 2.6.2.)

2.4.2 Hermite polynomials: $H_n(x)$

The *Hermite polynomials* $H_n(x)$, $n = 0, 1, 2, ...$, play an important role in quantum mechanics, probability theory, and they occur in the rectangular coordinates model used to characterize the output of a laser operating in one of its higher-order modes. For a given value of n, the function $y = H_n(x)$ is a solution of the second-order linear DE

$$y'' - 2xy' + 2ny = 0, \quad -\infty < x < \infty. \tag{34}$$

The series representation of the Hermite polynomial of degree n is

$$H_n(x) = \sum_{k=0}^{[n/2]} \frac{(-1)^k n!}{k!(n-2k)!} (2x)^{n-2k}, \quad n = 0,1,2,.... \tag{35}$$

The series (35) reveals that, like the Legendre polynomial $P_n(x)$, the Hermite polynomial $H_n(x)$ is a polynomial of degree n and, further, is an even function when n is even and an odd function when n is odd (see Fig. 2.15). The first few Hermite polynomials are listed in Table 2.2.

Table 2.2 Hermite Polynomials

$n = 0$	$H_0(x) = 1$
$n = 1$	$H_1(x) = 2x$
$n = 2$	$H_2(x) = 4x^2 - 2$
$n = 3$	$H_3(x) = 8x^3 - 12x$
$n = 4$	$H_4(x) = 16x^4 - 48x^2 + 12$
$n = 5$	$H_5(x) = 32x^5 - 160x^3 + 120x$

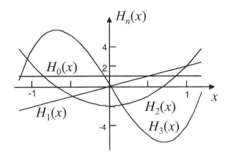

Figure 2.15 Hermite polynomials.

Some properties associated with the set of Hermite polynomials are given below:

(H1): $H_n(-x) = (-1)^n H_n(x)$

(H2): $H_n(x) = (-1)^n e^{x^2} \dfrac{d^n}{dx^n} (e^{-x^2})$

(H3): $H_{2n}(0) = (-1)^n \dfrac{(2n)!}{n!}, \quad H_{2n+1}(0) = 0$

(H4): $H_{n+1}(x) - 2xH_n(x) + 2nH_{n-1}(x) = 0$

(H5): $H_n'(x) = 2nH_{n-1}(x)$

(H6): $\displaystyle\int_{-\infty}^{\infty} e^{-x^2} H_n(x) H_k(x)\, dx = 0, \quad k \neq n$

(H7): $\displaystyle\int_{-\infty}^{\infty} e^{-x^2} [H_n(x)]^2\, dx = 2^n n! \sqrt{\pi}$

(H8): $H_{2n}(x) = (-1)^n \dfrac{(2n)!}{n!}\, {}_1F_1\left(-n; \dfrac{1}{2}; x^2\right)$ (See Section 2.6.3.)

$$H_{2n+1}(x) = (-1)^n \frac{(2n+1)!}{n!} 2x \ _1F_1\left(-n; \frac{3}{2}; x^2\right)$$

2.4.3 Laguerre polynomials: $L_n(x)$, $L_n^{(m)}(x)$

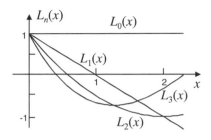

Figure 2.16 Laguerre polynomials.

The standard *Laguerre polynomials* $L_n(x)$, $n = 0, 1, 2, ...$, (see Fig. 2.16) occur in quantum mechanics, probability theory, and they occur in the circular symmetry model (polar coordinates) used to characterize the output of a laser operating in one of its higher-order modes. For a given value of n, the function $y = L_n(x)$ is a solution of the linear differential equation

$$xy'' + (1 - x)y' + ny = 0, \ 0 < x < \infty. \quad (36)$$

The series representation of the Laguerre polynomial of degree n is

$$L_n(x) = \sum_{k=0}^{n} \frac{(-1)^k n!}{(n-k)!(k!)^2} x^k, \quad n = 0, 1, 2, ..., \quad (37)$$

some of which are listed in Table 2.3.

Table 2.3 Laguerre Polynomials

$n = 0$	$L_0(x) = 1$
$n = 1$	$L_1(x) = -x + 1$
$n = 2$	$L_2(x) = \frac{1}{2!}(x^2 - 4x + 2)$
$n = 3$	$L_3(x) = \frac{1}{3!}(-x^3 + 9x^2 - 18x + 6)$
$n = 4$	$L_4(x) = \frac{1}{4!}(x^4 - 16x^3 + 72x^2 - 96x + 24)$

A generalization of the Laguerre polynomials, called the *associated Laguerre polynomials*, occurs in some applications. These more general polynomials can be derived from the standard Laguerre polynomials through the derivative relation

$$L_n^{(m)}(x) = (-1)^m \frac{d^m}{dx^m}[L_{n+m}(x)], \quad m = 0, 1, 2, ... \quad (38)$$

The series representation for the associated Laguerre polynomials is

$$L_n^{(m)}(x) = \sum_{k=0}^{n} \frac{(-1)^k (n + m)!}{(n - k)!(k + m)!k!} x^k. \tag{39}$$

For m not zero nor a positive integer, the associated Laguerre polynomials can be generalized even further by

$$L_n^{(a)}(x) = \sum_{k=0}^{n} \frac{(-1)^k \Gamma(n + a + 1)}{(n - k)! \Gamma(k + a + 1)k!} x^k, \quad a > -1. \tag{40}$$

Some properties associated with these sets of polynomials are given below. Those properties associated with the standard Laguerre polynomials are readily deduced from the property of the associated Laguerre polynomials by setting $m = 0$.

(L1): $L_n^{(m)}(x) = \dfrac{e^x x^{-m}}{n!} \dfrac{d^n}{dx^n}(x^{n+m} e^{-x})$

(L2): $L_n^{(m)}(0) = \dfrac{(n + m)!}{n!m!}$

(L3): $(n + 1)L_{n+1}^{(m)}(x) + (x - 1 - 2n - m)L_n^{(m)}(x) + (n + m)L_{n-1}^{(m)}(x) = 0$

(L4): $xL_{n-1}^{(m)}{}''(x) + (m + 1 - x)L_n^{(m)}{}'(x) + nL_n^{(m)}(x) = 0$

(L5): $\displaystyle\int_0^\infty e^{-x} x^m L_n^{(m)}(x) L_k^{(m)}(x)\,dx = 0, \quad k \neq n$

(L6): $\displaystyle\int_0^\infty e^{-x} x^m [L_n^{(m)}(x)]^2\,dx = \dfrac{(n + m)!}{n!}$

(L7): $L_n^{(m)}(x) = \dfrac{(n + m)!}{n!m!}\,{}_1F_1(-n; m + 1; x)$ (See Section 2.6.3.)

2.4.4 Chebyshev polynomials: $T_n(x)$, $U_n(x)$

Chebyshev[1] polynomials arise in problems concerning polynomial approximation methods. Specifically, it has been shown that a series of Chebyshev polynomials converges more rapidly than any other series of polynomials, including a power series. The *Chebyshev polynomials of the first kind* are defined by

$$T_n(x) = \frac{n}{2} \sum_{k=0}^{[n/2]} \frac{(-1)^k (n - k - 1)!}{k!(n - 2k)!}(2x)^{n-2k}, \quad n = 0,1,2,\dots. \tag{41}$$

---

[1] There are numerous spellings of Chebyshev that occur throughout the literature. Some of these are Tchebysheff, Tchebycheff, Tchebichef, and Chebysheff, among others.

A related set, called *Chebyshev polynomials of the second kind*, are given by

$$U_n(x) = \sum_{k=0}^{[n/2]} \binom{n-k}{k} (-1)^k (2x)^{n-2k}, \quad n = 0, 1, 2, \dots, \tag{42}$$

where $\binom{n}{k} = n!/[k!(n-k)!]$ is the binomial coefficient. The first few polynomials of each kind are listed in Tables 2.4 and 2.5 and some are shown in Figs. 2.17 and 2.18.

Table 2.4 Chebyshev Polynomials of the First Kind

$n = 0$	$T_0(x) = 1$
$n = 1$	$T_1(x) = x$
$n = 2$	$T_2(x) = 2x^2 - 1$
$n = 3$	$T_3(x) = 4x^3 - 3x$
$n = 4$	$T_4(x) = 8x^4 - 8x^2 + 1$
$n = 5$	$T_5(x) = 16x^5 - 20x^3 + 5x$

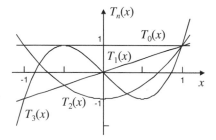

Figure 2.17 Chebyshev polynomials of the first kind.

Table 2.5 Chebyshev Polynomials of the Second Kind

$n = 0$	$U_0(x) = 1$
$n = 1$	$U_1(x) = 2x$
$n = 2$	$U_2(x) = 4x^2 - 1$
$n = 3$	$U_3(x) = 8x^3 - 4x$
$n = 4$	$U_4(x) = 16x^4 - 12x^2 + 1$
$n = 5$	$U_5(x) = 32x^5 - 32x^3 + 6x$

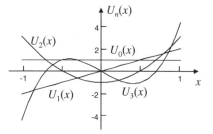

Figure 2.18 Chebyshev polynomials of the second kind.

The polynomial $y = T_n(x)$ is a solution of the second-order differential equation

$$(1 - x^2)y'' - xy' + n^2 y = 0, \quad -1 < x < 1, \tag{43}$$

and $y = U_n(x)$ is a solution of

$$(1 - x^2)y'' - 3xy' + n(n + 2)y = 0, \quad -1 < x < 1. \tag{44}$$

Some additional properties associated with both kinds of Chebyshev polynomials are listed below:

Type I:

(T1): $T_n(-x) = (-1)^n T_n(x)$

(T2): $T_n(1) = 1$

(T3): $T_{2n}(0) = (-1)^n$; $T_{2n+1}(0) = 0$

(T4): $T_{n+1}(x) - 2xT_n(x) + T_{n-1}(x) = 0$

(T5): $T_n(x) = U_n(x) - xU_{n-1}(x)$

(T6): $\int_{-1}^{1} (1-x^2)^{-1/2} T_n(x)T_k(x)dx = 0, \quad k \neq n$

(T7): $\int_{-1}^{1} (1-x^2)^{-1/2} [T_n(x)]^2 dx = \begin{cases} \pi, & n = 0 \\ \pi/2, & n \geq 1 \end{cases}$

Type II:

(U1): $U_n(-x) = (-1)^n U_n(x)$

(U2): $U_n(1) = n + 1$

(U3): $U_{2n}(0) = (-1)^n$; $U_{2n+1}(0) = 0$

(U4): $U_{n+1}(x) - 2xU_n(x) + U_{n-1}(x) = 0$

(U5): $(1-x^2)U_{n-1}(x) = xT_n(x) - T_{n+1}(x)$

(U6): $\int_{-1}^{1} (1-x^2)^{1/2} U_n(x)U_k(x)dx = 0, \quad k \neq n$

(U7): $\int_{-1}^{1} (1-x^2)^{1/2} [U_n(x)]^2 dx = \dfrac{\pi}{2}$

2.5 Family of Bessel Functions

When discovering the *Bessel functions*, the German astronomer Friedrich W. Bessel (1784-1846) was working on a problem associated with elliptic planetary motion originally proposed by Johannes Kepler (1571-1630), using the powerful method of Fourier series that had been recently introduced by Joseph Fourier (1768-1830).

Bessel functions also occur in problems in electricity, heat, hydrodynamics, elasticity, wave motion, Fresnel integrals, etc. In particular, they are closely associated with problems possessing circular or cylindrical symmetry and, thus, are sometimes called *cylinder functions*. There are many functions belonging to the general family of Bessel functions, including the *standard Bessel functions*, *modified Bessel functions*, *Hankel functions*, *spherical Bessel functions*, *Kelvin's functions*, *Lommel functions*, *Struve functions*, *Airy functions*, and *Anger* and *Weber functions*. These are often divided further into functions of the first and second kinds, and so on. Bessel functions of one type or another occur so frequently in physics and engineering applications that they are undoubtedly the most important functions beyond the elementary functions (see [34]).

2.5.1 Standard Bessel functions: $J_p(x)$, $Y_p(x)$

The standard Bessel functions, denoted by the symbols $J_p(x)$ and $Y_p(x)$, are linearly independent solutions of *Bessel's equation* (recall Section 1.9)

$$x^2 y'' + xy' + (x^2 - p^2)y = 0, \quad p \geq 0, \ x > 0. \tag{45}$$

The *Bessel function of the first kind* has the power series representations

$$J_p(x) = \sum_{k=0}^{\infty} \frac{(-1)^k (x/2)^{2k+p}}{k! \Gamma(k + p + 1)}, \quad p \geq 0, \tag{46}$$

$$J_{-p}(x) = \sum_{k=0}^{\infty} \frac{(-1)^k (x/2)^{2k-p}}{k! \Gamma(k - p + 1)}, \quad p > 0, \tag{47}$$

where $\Gamma(x)$ is the gamma function. Equation (47) is formally obtained from (46) by replacing p with $-p$. The special cases when $p = n$, $n = 0, 1, 2, \ldots$, lead to

$$J_n(x) = \sum_{k=0}^{\infty} \frac{(-1)^k (x/2)^{2k+n}}{k! (k + n)!}, \quad J_{-n}(x) = (-1)^n J_n(x). \tag{48}$$

The second relation in (48) reveals that a Bessel function of negative integer order $-n$ is basically the same as the Bessel function of positive integer order n. These functions arise in the *generating function* relation

$$\exp\left[\frac{x}{2}\left(t - \frac{1}{t}\right)\right] = \sum_{n=-\infty}^{\infty} J_n(x) t^n, \quad t \neq 0. \tag{49}$$

The graphs of $J_0(x)$, $J_1(x)$, and $J_2(x)$ are shown in Fig. 2.19. Observe that these functions exhibit an oscillatory behavior somewhat like that of the sine and cosine functions, except that the amplitude of each Bessel function diminishes as the argument increases, and the (infinitely many) zeros are not evenly spaced.

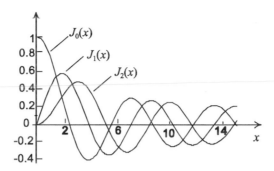

Figure 2.19 Bessel functions of the first kind.

The Bessel functions (46) and (47) satisfy a large number of identities and integral relations, some of which are provided below:

(BJ1): $J_0(0) = 1; \quad J_p(0) = 0, \; p > 0$

(BJ2): $e^{\pm iz\cos\theta} = J_0(z) + 2\sum_{n=0}^{\infty} (\pm i)^n J_n(z)\cos n\theta$

(BJ3): $J_n(x + y) = \sum_{k=-\infty}^{\infty} J_k(x)J_{n-k}(y)$

(BJ4): $J_0(x) = \dfrac{1}{2\pi} \displaystyle\int_0^{2\pi} e^{\pm ix\cos\theta} d\theta$

(BJ5): $J_n(x) = \dfrac{1}{\pi} \displaystyle\int_0^{\pi} \cos(n\theta - x\sin\theta)d\theta, \; n = 0,1,2,....$

(BJ6): $J_{1/2}(x) = \sqrt{\dfrac{2}{\pi x}}\sin x; \quad J_{-1/2}(x) = \sqrt{\dfrac{2}{\pi x}}\cos x$

(BJ7): $\dfrac{d}{dx}[x^p J_p(x)] = x^p J_{p-1}(x)$

(BJ8): $\dfrac{d}{dx}[x^{-p} J_p(x)] = -x^{-p} J_{p+1}(x)$

(BJ9): $J_p{}'(x) + \dfrac{p}{x}J_p(x) = J_{p-1}(x)$

(BJ10): $J_p{}'(x) - \dfrac{p}{x}J_p(x) = -J_{p+1}(x)$

(BJ11): $J_{p-1}(x) - J_{p+1}(x) = 2J_p{}'(x)$

(BJ12): $J_{p-1}(x) + J_{p+1}(x) = \dfrac{2p}{x}J_p(x)$

(BJ13): $\displaystyle\int x^p J_{p-1}(x)dx = x^p J_p(x) + C$

(BJ14): $\displaystyle\int x^{-p} J_{p+1}(x)dx = -x^{-p} J_p(x) + C$

(BJ15): $\displaystyle\int_0^b x J_p(k_m x) J_p(k_n x)\, dx = 0,\ m \neq n;$

$$J_p(k_n b) = 0,\ n = 1,2,3,\ldots$$

(BJ16): $\displaystyle\int_0^b x [J_p(k_n x)]^2\, dx = \frac{1}{2} b^2 [J_{p+1}(k_n b)]^2;$

$$J_p(k_n b) = 0,\ n = 1,2,3,\ldots$$

(BJ17): $\displaystyle J_p(x) \sim \frac{(x/2)^p}{\Gamma(1+p)},\quad p \neq -1,-2,-3,\ldots,\quad x \to 0^+$

(BJ18): $\displaystyle J_p(x) \sim \sqrt{\frac{2}{\pi x}} \cos\left[x - (p + 1/2)\frac{\pi}{2} \right],\quad x \to \infty,$

EXAMPLE 5 Show that

$$I = \int_0^\infty x e^{-a^2 x^2} J_0(bx)\, dx = \frac{1}{2a^2} \exp\left(-\frac{b^2}{4a^2} \right).$$

Solution: If we replace the Bessel function with its series representation (46) and interchange the order of summation and integration, then

$$I = \sum_{n=0}^\infty \frac{(-1)^n b^{2n}}{(n!)^2 2^{2n}} \int_0^\infty x^{2n+1} e^{-a^2 x^2}\, dx$$

$$= \frac{1}{2a^2} \sum_{n=0}^\infty \frac{(-1)^n b^{2n}}{n!} \left(\frac{1}{4a^2} \right)^n$$

$$= \frac{1}{2a^2} \exp\left(-\frac{b^2}{4a^2} \right),$$

where we have used properties of the gamma function to evaluate the termwise integrals and summed the remaining series.

A linearly independent solution of Bessel's equation (45) is $Y_p(x)$, called the *Bessel function of the second kind* of nonintegral order p and defined by

$$Y_p(x) = \frac{(\cos p\pi) J_p(x) - J_{-p}(x)}{\sin p\pi}. \tag{50}$$

When $p = n$, $n = 0,1,2,\ldots$, Eq. (50) requires further investigation because the right-hand side approaches the indeterminate form 0/0. By application of L'Hôpital's rule, it has been shown that for $p = 0$,

$$Y_0(x) = \frac{2}{\pi}J_0(x)\left(\ln\frac{x}{2} + \gamma\right) - \frac{2}{\pi}\sum_{k=1}^{\infty}\frac{(-1)^k(x/2)^{2k}}{(k!)^2}\left(1 + \frac{1}{2} + \cdots + \frac{1}{k}\right), \quad x > 0, \tag{51}$$

where γ is Euler's constant (Section 2.3.3), and for $n = 1, 2, 3, \ldots,$

$$Y_n(x) = \frac{2}{\pi}J_n(x)\ln\frac{x}{2} - \frac{1}{\pi}\sum_{k=0}^{n-1}\frac{(n-k-1)!}{k!}\left(\frac{x}{2}\right)^{2k-n}$$
$$- \frac{1}{\pi}\sum_{k=0}^{\infty}\frac{(-1)^k(x/2)^{2k+n}}{k!(k+n)!}[\psi(k+n+1) + \psi(k+1)], \quad x > 0. \tag{52}$$

Graphs of $Y_0(x)$, $Y_1(x)$, and $Y_2(x)$ are shown in Fig. 2.20. Note that away from the origin these functions have oscillatory characteristics similar to those of $J_n(x)$, but their asymptotic behavior near the origin is quite different. That is, their behavior near the origin closely resembles that of a logarithm function. Also, Bessel functions of the second kind satisfy the same recurrence formulas as those given by (BJ7)-(BJ12) above.

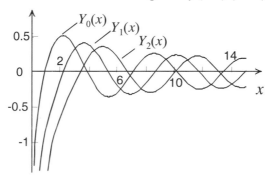

Figure 2.20 Bessel functions of the second kind.

EXAMPLE 6 A classical problem involving Bessel functions concerns the small oscillations of a flexible hanging chain governed by

$$\frac{\partial}{\partial x}\left(T\frac{\partial y}{\partial x}\right) = \rho\frac{\partial^2 y}{\partial t^2}, \quad |y(0)| < \infty, \quad y(L) = 0,$$

where T is the tension in the chain and ρ is mass per unit length. In this formulation we assume the fixed end is at $x = L$ (the ceiling) and the free end of the chain is at $x = 0$ (see Fig. 2.21). Given that the tension in the chain at x is due only to the weight of the chain below x, then $T = \rho g x$, where g is the gravitational constant. In this case, find the lowest frequency of oscillation for a 10-ft long chain.

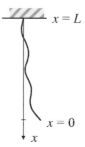

Figure 2.21

Solution: If we assume the oscillations are essentially sinusoidal, then we set

$$y(x,t) = R(x)\cos(\omega t - \varphi),$$

where R is the amplitude, ω is angular frequency, and φ is an unimportant phase. The substitution of this assumed solution form into the governing equation above, together with the prescribed boundary conditions (BCs), reduces it to an ordinary differential equation with BCs

$$xR'' + R' + k^2 R = 0, \quad |R(0)| < \infty, \ R(L) = 0, \quad (k^2 = \omega^2/g).$$

The general solution of this last equation can be expressed in terms of Bessel functions as [recall Eqs. (146) and (147) in Chapter 1]

$$R(x) = C_1 J_0(2k\sqrt{x}) + C_2 Y_0(2k\sqrt{x}),$$

where C_1 and C_2 are arbitrary constants.

To satisfy the first BC above, we must set $C_2 = 0$ because the Bessel function Y_0 becomes unbounded at the origin (see Fig. 2.20). The remaining BC at $x = L$ leads to

$$R(L) = C_1 J_0(2k\sqrt{L}) = 0.$$

The first zero of $J_0(x)$ occurs at $x = 2.405$, from which we find $2k\sqrt{10} = 2.405$. Thus, $k = 0.380$, and using the numerical value $g = 32$ ft/s^2, we obtain for the lowest frequency

$$f = 2\pi\omega = 2\pi k\sqrt{g} = 13.5 \text{ Hz}.$$

2.5.2 Modified Bessel functions: $I_p(x)$, $K_p(x)$

The modified Bessel functions $I_p(x)$ and $K_p(x)$ are solutions of Bessel's modified equation

$$x^2 y'' + xy' - (x^2 + p^2)y = 0, \quad p \geq 0, \ x > 0. \tag{53}$$

The series representation for the *modified Bessel function of the first kind* is

$$I_p(x) = i^{-p} J_p(ix) = \sum_{k=0}^{\infty} \frac{(x/2)^{2k+p}}{k!\Gamma(k+p+1)}, \quad p \geq 0, \tag{54}$$

$$I_{-p}(x) = i^p J_{-p}(ix) = \sum_{k=0}^{\infty} \frac{(x/2)^{2k-p}}{k!\Gamma(k-p+1)}, \quad p > 0. \tag{55}$$

When $p = n$, $n = 0, 1, 2, \ldots$, we have that

$$I_{-n}(x) = I_n(x). \tag{56}$$

That is, the modified Bessel functions with negative integer index are the same as those with positive integer index. The graphs of $I_0(x)$, $I_1(x)$, and $I_2(x)$ are shown in Fig. 2.22. Observe that these functions do not exhibit the oscillatory behavior like that of the standard Bessel functions. The modified Bessel functions are commonly associated with probability density functions in probability theory like the Rician distribution arising in the analysis of the envelope of a signal plus narrowband Gaussian noise in statistical communication theory (e.g., see Example 7 and Sections 14.6 and 14.7).

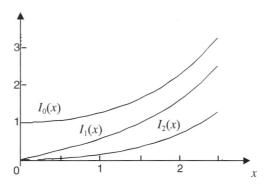

Figure 2.22 Modified Bessel functions of the first kind.

Many of the identities associated with the modified Bessel function of the first kind are similar to those of the standard Bessel function, some of which are displayed below:

(BI1): $I_0(0) = 1$; $I_p(0) = 0$, $p > 0$

(BI2): $e^{\pm x \cos \theta} = \displaystyle\sum_{n=-\infty}^{\infty} I_n(x) \cos n\theta$

(BI3): $I_n(x + y) = \displaystyle\sum_{k=-\infty}^{\infty} I_k(x) I_{n-k}(y)$

(BI4): $I_0(x) = \dfrac{1}{2\pi} \displaystyle\int_0^{2\pi} e^{\pm x \cos \theta} d\theta$

(BI5): $I_n(x) = \dfrac{1}{\pi} \displaystyle\int_0^{\pi} e^{x \cos \theta} \cos n\theta \, d\theta$, $n = 0, 1, 2, \ldots$

(BI6): $I_{1/2}(x) = \sqrt{\dfrac{2}{\pi x}} \sinh x$; $I_{-1/2}(x) = \sqrt{\dfrac{2}{\pi x}} \cosh x$

(BI7): $\dfrac{d}{dx}[x^p I_p(x)] = x^p I_{p-1}(x)$

(BI8): $\dfrac{d}{dx}[x^{-p} I_p(x)] = x^{-p} I_{p+1}(x)$

(BI9): $I_p'(x) + \dfrac{p}{x} I_p(x) = I_{p-1}(x)$

(BI10): $I_p'(x) - \dfrac{p}{x} I_p(x) = I_{p+1}(x)$

(BI11): $I_{p-1}(x) + I_{p+1}(x) = 2I_p'(x)$

(BI12): $I_{p-1}(x) - I_{p+1}(x) = \dfrac{2p}{x} I_p(x)$

(BI13): $I_p(x) \sim \dfrac{(x/2)^p}{\Gamma(1+p)}, \quad p \neq -1, -2, -3, \ldots, \quad x \to 0^+$

(BI14): $I_p(x) \sim \dfrac{e^x}{\sqrt{2\pi x}}, \quad x \to \infty$

A linearly independent solution of Bessel's modified equation (53) is the *modified Bessel function of the second kind* of nonintegral order p defined by

$$K_p(x) = \frac{\pi}{2} \frac{I_{-p}(x) - I_p(x)}{\sin p\pi}. \tag{57}$$

When $n = 0$, Eq. (57) leads to

$$K_0(x) = -I_0(x)\left(\ln\frac{x}{2} + \gamma \right) + \sum_{k=1}^{\infty} \frac{(x/2)^{2k}}{(k!)^2}\left(1 + \frac{1}{2} + \cdots + \frac{1}{k} \right), \quad x > 0, \tag{58}$$

whereas in general when $p = n, \ n = 1, 2, 3, \ldots,$

$$K_n(x) = (-1)^n I_n(x) \ln\frac{x}{2} + \frac{1}{2}\sum_{k=0}^{n-1} \frac{(-1)^k (n-k-1)!}{k!}\left(\frac{x}{2}\right)^{2k-n}$$

$$- \frac{(-1)^n}{2}\sum_{k=0}^{\infty} \frac{(x/2)^{2k+n}}{k!(k+n)!}[\psi(k+n+1) + \psi(k+1)], \quad x > 0. \tag{59}$$

Graphs of $K_0(x)$, $K_1(x)$, and $K_2(x)$ are shown in Fig. 2.23. Some of the basic properties analogous to those of the other Bessel functions are listed below:

(BK1): $K_{-p}(x) = K_p(x)$

(BK2): $K_{1/2}(x) = \sqrt{\dfrac{\pi}{2x}} e^{-x}$

(BK3): $\dfrac{d}{dx}[x^p K_p(x)] = -x^p K_{p-1}(x)$

(BK4): $\dfrac{d}{dx}[x^{-p} K_p(x)] = -x^{-p} K_{p+1}(x)$

(BK5): $K_p{}'(x) + \dfrac{p}{x} K_p(x) = -K_{p-1}(x)$

(BK6): $K_p{}'(x) - \dfrac{p}{x} K_p(x) = -K_{p+1}(x)$

(BK7): $K_{p-1}(x) + K_{p+1}(x) = -2K_p{}'(x)$

(BK8): $K_{p-1}(x) - K_{p+1}(x) = -\dfrac{2p}{x} K_p(x)$

(BK9): $K_0(x) \sim -\ln x; \quad K_p(x) \sim \dfrac{\Gamma(p)}{2}\left(\dfrac{2}{x}\right)^p, \quad p > 0, \quad x \to 0^+$

(BK10): $K_p(x) \sim \sqrt{\dfrac{\pi}{2x}}\, e^{-x}, \quad x \to \infty$

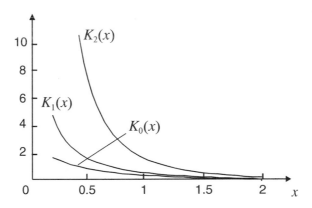

Figure 2.23 Modified Bessel functions of the second kind.

EXAMPLE 7 The joint probability distribution of the envelope and phase of a sinusoidal signal embedded in narrowband zero-mean Gaussian random noise is given by (see Section 13.9.3 and [22])

$$f_{r\theta}(r,\theta) = \frac{r}{2\pi N} \exp\left[-\frac{(r^2 + A^2 - 2A\cos\theta)}{2N}\right] U(r), \quad -\pi < \theta \leq \pi,$$

where A is the signal amplitude and N is the noise variance.
(a) Calculate the marginal distribution $f_r(r)$.
(b) Calculate the marginal distribution $f_\theta(\theta)$.

Solution:

(a) The marginal distribution of the envelope r is found by integrating the joint distribution over θ (modulo 2π). That is,

$$\begin{aligned}
f_r(r) &= \int_{-\pi}^{\pi} f_{r\theta}(r,\theta)\, d\theta \\
&= \frac{rU(r)}{2\pi N} e^{-(r^2 + A^2)/2N} \int_{-\pi}^{\pi} \exp\left(\frac{2Ar\cos\theta}{2N}\right) d\theta,
\end{aligned}$$

from which, using (BI4), we deduce

$$f_{\mathbf{r}}(r) = \frac{r}{N} e^{-(r^2 + A^2)/2N} I_0\left(\frac{Ar}{N}\right) U(r).$$

This probability density function is widely known as the *Rician distribution*. In the absence of signal (i.e., $A = 0$), it reduces to the more familiar *Rayleigh distribution*

$$f_{\mathbf{r}}(r) = \frac{r}{N} e^{-r^2/2N} U(r).$$

(*b*) The marginal phase distribution is likewise determined by evaluating the integral

$$f_{\theta}(\theta) = \int_0^{\infty} f_{\mathbf{r}\theta}(r,\theta)\,dr$$
$$= \frac{1}{2\pi N} e^{-A^2/2N} \int_0^{\infty} r e^{-(r^2 - 2Ar\cos\theta)/2N}\,dr.$$

To perform the integration, we complete the square in the argument of the exponential function to find

$$f_{\theta}(\theta) = \frac{1}{2\pi N} e^{-s} e^{s\cos^2\theta} \int_0^{\infty} r e^{-(r/\sqrt{2N} - \sqrt{s}\cos\theta)^2}\,dr,$$

where $s = A^2/2N$ is the *signal-to-noise ratio* (SNR). Next, we make the change of variable $t = r/\sqrt{2N} - \sqrt{s}\cos\theta$ and integrate the resulting expression to find

$$f_{\theta}(\theta) = \frac{1}{2\pi} e^{-s}\left\{1 + \sqrt{\pi s}\,e^{s\cos^2\theta}\cos\theta\,[1 + \mathrm{erf}(\sqrt{s}\cos\theta)]\right\}, \quad -\pi < \theta \le \pi,$$

where erf(x) is the error function (Section 2.3.4). Note that in the absence of signal (i.e., $s = 0$), this distribution reduces to the *uniform distribution*

$$f_{\theta}(\theta) = \frac{1}{2\pi}, \quad -\pi < \theta \le \pi.$$

2.5.3 Other Bessel functions

There are a variety of other Bessel functions that arise in certain specialized applications. In this section we briefly introduce some of these functions.

Spherical Bessel functions are commonly associated with problems featuring spherical symmetry. These functions are directly related to half-integral order Bessel functions according to

$$j_n(x) = \sqrt{\frac{\pi}{2x}}\, J_{n+1/2}(x), \quad n = 0, 1, 2, \ldots, \tag{60}$$

$$y_n(x) = \sqrt{\frac{\pi}{2x}}\, Y_{n+1/2}(x), \quad n = 0, 1, 2, \ldots, \tag{61}$$

and called, respectively, *spherical Bessel functions of the first* and *second kinds* of order n. The series representation for $j_n(x)$ can be deduced directly from that of $J_p(x)$, which leads to

$$j_n(x) = (2x)^n \sum_{k=0}^{\infty} \frac{(-1)^k (k+n)! \, x^{2k}}{k! \, (2k+2n+1)!}. \tag{62}$$

The spherical Bessel functions (60) and (61) are linearly independent solutions of the linear differential equation

$$x^2 y'' + 2xy' + [x^2 - n(n+1)]y = 0, \quad n = 0, 1, 2, \ldots. \tag{63}$$

These Bessel functions are significant in that they are directly related to the trigonometric functions. For example, it can readily be shown that

$$j_0(x) = \frac{\sin x}{x}, \quad j_1(x) = \frac{\sin x}{x^2} - \frac{\cos x}{x}, \tag{64}$$

$$y_0(x) = -\frac{\cos x}{x}, \quad y_1(x) = -\frac{\cos x}{x^2} - \frac{\sin x}{x}. \tag{65}$$

The Hankel functions are related to the standard Bessel functions by

$$H_p^{(1)}(x) = J_p(x) + i Y_p(x), \tag{66}$$

$$H_p^{(2)}(x) = J_p(x) - i Y_p(x). \tag{67}$$

Equations (66) and (67) define *Hankel functions of the first* and *second kinds*, and are defined analogous to the Euler formulas

$$e^{\pm ix} = \cos x \pm i \sin x. \tag{68}$$

Among other applications, Hankel functions are often used in the theoretical treatment of optical fibers and in certain wave scattering problems (e.g., see Section 12.7.2).

Solutions of *Airy's equation*

$$y'' - xy = 0 \tag{69}$$

are important in several applications, including the diffraction of radio waves around the Earth's surface and the construction of asymptotic results near a two-dimensional caustic in the study of optical fields. The general solution of (69) can be expressed as

$$y = C_1 \operatorname{Ai}(x) + C_2 \operatorname{Bi}(x), \tag{70}$$

where

$$\text{Ai}(x) = \frac{1}{\pi}\sqrt{\frac{x}{3}}K_{1/3}\left(\frac{2}{3}x^{3/2}\right),$$

$$\text{Bi}(x) = \sqrt{\frac{x}{3}}\left[I_{-1/3}\left(\frac{2}{3}x^{3/2}\right) + I_{1/3}\left(\frac{2}{3}x^{3/2}\right)\right]. \tag{71}$$

The functions $\text{Ai}(x)$ and $\text{Bi}(x)$ are called *Airy functions of the first* and *second kinds*, respectively. When the argument is negative, the Airy functions are defined by

$$\text{Ai}(-x) = \frac{1}{3}\sqrt{x}\left[J_{-1/3}\left(\frac{2}{3}x^{3/2}\right) + J_{1/3}\left(\frac{2}{3}x^{3/2}\right)\right],$$

$$\text{Bi}(-x) = \sqrt{\frac{x}{3}}\left[J_{-1/3}\left(\frac{2}{3}x^{3/2}\right) - J_{1/3}\left(\frac{2}{3}x^{3/2}\right)\right]. \tag{72}$$

This latter set of functions (72) represent linearly independent solutions of

$$y'' + xy = 0. \tag{73}$$

2.6 Family of Hypergeometric-type Functions

Many of the special functions introduced above are specializations of more general functions referred to as *functions of the hypergeometric type*. There are several varieties of these functions, the most common of which are the *hypergeometric function* and the *confluent hypergeometric functions*. Hypergeometric-type functions arise in engineering applications such as the analysis of random noise (e.g., see Section 15.5.4 and [22]) and laser propagation through random media (see [2] and [28]), among many other areas.

2.6.1 Pochhammer symbol: $(a)_n$

In the discussion of hypergeometric-type functions, it is useful to employ special notation known as the *Pochhammer symbol*, the defining properties of which are given by

$$\boxed{\begin{aligned}(a)_0 &= 1,\\ (a)_n &= a(a+1)\cdots(a+n-1) = \frac{\Gamma(a+n)}{\Gamma(a)},\quad n = 1,2,3,\dots.\end{aligned}} \tag{74}$$

Because of its close relation to the gamma function, it is clear the Pochhammer symbol satisfies a number of identities directly related to the gamma function. Some of those relations are listed below:

(Poch1): $(0)_0 = 1$

(Poch2): $(1)_n = n!$

(Poch3): $(a)_{n+k} = (a)_k(a+k)_n$

(Poch4): $(a)_{-n} = \dfrac{(-1)^n}{(1-a)_n}$

(Poch5): $(-k)_n = \begin{cases} \dfrac{(-1)^n k!}{(k-n)!}, & 0 \le n \le k \\ 0, & n > k \end{cases}$

(Poch6): $\dbinom{a}{n} = \dfrac{(-1)^n}{n!}(-a)_n$

(Poch7): $(2n)! = 2^{2n}(1/2)_n n!$

(Poch8): $(2n+1)! = 2^{2n}(3/2)_n n!$

(Poch9): $(a)_{2n} = 2^{2n}\left(\dfrac{1}{2}a\right)_n\left(\dfrac{1}{2}+\dfrac{1}{2}a\right)_n.$

EXAMPLE 8 Verify the identity

$$\frac{1}{2n+a} = \frac{(a/2)_n}{a(1+a/2)_n}.$$

Solution: Expressing the quotient as a ratio of gamma functions, we have

$$\frac{1}{2n+a} = \frac{1}{2(n+a/2)} = \frac{\Gamma(n+a/2)}{2\Gamma(n+a/2+1)}.$$

Now using the definition (74), it follows that

$$\frac{1}{2n+a} = \frac{\Gamma(a/2)(a/2)_n}{2\Gamma(1+a/2)(1+a/2)_n} = \frac{(a/2)_n}{a(1+a/2)_n}.$$

2.6.2 Hypergeometric function of Gauss: $_2F_1(a,b;c;x)$

The major development of the theory of the hypergeometric function was carried out by Carl F. Gauss (1777–1855) and published in his famous memoir of 1812. Specializations of this function include various elementary functions like the arcsine and natural log, and several orthogonal polynomial sets like the Chebyshev polynomials and Legendre polynomials. The series definition of the *hypergeometric function* is given by

$$
{}_2F_1(a,b;c;x) = \sum_{n=0}^{\infty} \frac{(a)_n (b)_n}{(c)_n} \frac{x^n}{n!}, \quad |x| < 1, \tag{75}
$$

where $c \neq 0, -1, -2, \ldots$. The subscripts 2 and 1 refer to the number of numerator and denominator parameters, respectively, in its series representation. The hypergeometric function is also commonly denoted by the symbol $F(a,b;c;x)$ where the subscripts are omitted. If either a or b is zero or a negative integer, the series (75) truncates. Otherwise, the series (75) is restricted to values of the argument for which $|x| < 1$. There is also a hypergeometric function of the second kind, which together with (75) represent linearly independent solutions of the same linear differential equation [see Eq. (77)]. However, this latter function is not very useful in applications.

If we replace x in the argument of the hypergeometric function with the complex variable $z = x + iy$, the resulting function is analytic in various domains of the complex plane (see Chapter 6). In particular, the hypergeometric function is defined for values outside the unit circle in the complex plane. For example, if $|z| > 1$, it has been shown that the hypergeometric function with argument z can be expressed as combinations of hypergeometric functions with argument $1/z$, viz.,

$$
\begin{aligned}
{}_2F_1(a,b;c;z) &= \frac{\Gamma(c)\Gamma(b-a)}{\Gamma(b)\Gamma(c-a)}(-z)^{-a} \, {}_2F_1\left(a,1-c+a;1-b+a;\frac{1}{z}\right) \\
&+ \frac{\Gamma(c)\Gamma(a-b)}{\Gamma(a)\Gamma(c-b)}(-z)^{-b} \, {}_2F_1\left(b,1-c+b;1-a+b;\frac{1}{z}\right), \\
&\qquad\qquad\qquad\qquad\qquad |\arg(-z)| < \pi.
\end{aligned} \tag{76}
$$

The hypergeometric function (75) is one of 24 solutions of this type that satisfy the second-order equation

$$
x(1-x)y'' + [c - (a+b+1)x]y' - aby = 0. \tag{77}
$$

The six functions defined by ${}_2F_1(a \pm 1, b; c; x)$, ${}_2F_1(a, b \pm 1; c; x)$, and ${}_2F_1(a, b; c \pm 1; x)$ are called *contiguous functions*. Gauss showed that there exist 15 linear recurrence relations with coefficients at most linear in x, each of which is between (75) and two contiguous functions. Some additional relations involving the hypergeometric function are listed below:

(HY1): $\quad {}_2F_1(a,b;c;x) = {}_2F_1(b,a;c;x)$

(HY2): $\quad \dfrac{d^k}{dx^k} \, {}_2F_1(a,b;c;x)$

$$
= \frac{(a)_k (b)_k}{(c)_k} \, {}_2F_1(a+k, b+k; c+k; x), \quad k = 1, 2, 3, \ldots
$$

(HY3): $_2F_1(a,b;c;x)$

$$= \frac{\Gamma(c)}{\Gamma(b)\Gamma(c-b)} \int_0^1 t^{b-1}(1-t)^{c-b-1}(1-xt)^{-a}\,dt, \quad c > b > 0$$

(HY4): $_2F_1(a,b;c;1) = \dfrac{\Gamma(c)\Gamma(c-a-b)}{\Gamma(c-a)\Gamma(c-b)}$

(HY5): $_2F_1(a,b;c;-x) = (1+x)^{-a}\, _2F_1\!\left(a,c-b;c;\dfrac{x}{1+x}\right)$

2.6.3 Confluent hypergeometric functions: $_1F_1(a;c;x)$, $U(a;c;x)$

Ernst E. Kummer (1810–1893) is the name most closely associated with the confluent hypergeometric functions. For this reason, these functions are also known as Kummer's functions. The series representation for the *confluent hypergeometric function of the first kind* is given by

$$_1F_1(a;c;x) = \sum_{n=0}^{\infty} \frac{(a)_n}{(c)_n}\frac{x^n}{n!}, \quad |x| < \infty, \tag{78}$$

where $c \neq 0, -1, -2, \ldots$. Specializations of this function include the exponential function, Hermite and Laguerre polynomial sets, and several Bessel functions, among others. The *confluent hypergeometric function of the second kind* is a linear combination of functions of the first kind that can be expressed as

$$U(a;c;x) = \frac{\Gamma(1-c)}{\Gamma(1+a-c)}\, _1F_1(a;c;x)$$
$$+ \frac{\Gamma(c-1)}{\Gamma(a)} x^{1-c}\, _1F_1(1+a-c;2-c;x). \tag{79}$$

The confluent hypergeometric functions (78) and (79) are linearly independent solutions of the second order equation

$$xy'' + (c-x)y' - ay = 0. \tag{80}$$

Some useful properties associated with the confluent hypergeometric functions of the first and second kinds are listed below:

(CH1): $\dfrac{d^k}{dx^k}\, _1F_1(a;c;x) = \dfrac{(a)_k}{(c)_k}\, _1F_1(a+k;c+k;x), \quad k = 1,2,3,\ldots$

(CH2): $_1F_1(a;c;x)$

$$= \frac{\Gamma(c)}{\Gamma(a)\Gamma(c-a)} \int_0^1 e^{xt} t^{a-1}(1-t)^{c-a-1}\,dt, \quad c > a > 0$$

(CH3): $_1F_1(a;c;x) = e^x \ _1F_1(c-a;c;-x)$

(CH4): $_1F_1(a;c;-x) \sim \begin{cases} 1 - \dfrac{ax}{c}, & |x| \ll 1 \\[3mm] \dfrac{\Gamma(c)}{\Gamma(c-a)} x^{-a}, & x \gg 1 \end{cases}$

(CH5): $\dfrac{d}{dx} U(a;c;x) = -a\,U(a+1;c+1;x)$

(CH6): $U(a;c;x)$
$\qquad = \dfrac{1}{\Gamma(a)} \displaystyle\int_0^\infty e^{-xt} t^{a-1}(1+t)^{c-a-1}\,dt, \quad a > 0, \ x > 0$

(CH7): $U(a;c;x) = x^{1-c} U(1+a-c;2-c;x)$

(CH8): $U(a;c;x) \sim \begin{cases} \dfrac{\Gamma(1-c)}{\Gamma(1+a-c)} + \dfrac{\Gamma(c-1)}{\Gamma(a)} x^{1-c}, & |x| \ll 1 \\[3mm] x^{-a}, & x \gg 1 \end{cases}$

2.6.4 Generalized hypergeometric functions: $_pF_q$

During the last 70 years there has been considerable interest in working with generalized hypergeometric functions, of which the hypergeometric and confluent hypergeometric functions are special cases. In general, we say a power series $\sum A_n x^n$ is a *series of hypergeometric type* if the ratio A_{n+1}/A_n is a rational function of n. A particular series of this type is

$$_pF_q(a_1,...,a_p;c_1,...,c_q;x) = \sum_{n=0}^\infty \frac{(a_1)_n \cdots (a_p)_n}{(c_1)_n \cdots (c_q)_n} \frac{x^n}{n!}, \tag{81}$$

where p and q are nonnegative integers and no c_k $(k = 1, 2, ..., q)$ is zero or a negative integer. This function, which we denote by simply $_pF_q$, is called a *generalized hypergeometric function*. Provided the series (81) does not terminate, it can be established by the ratio test of calculus that

  1. If $p < q + 1$, the series *converges* for all $|x| < \infty$.
  2. If $p = q + 1$, the series *converges* for $|x| < 1$ and diverges for $|x| > 1$.
  3. If $p > q + 1$, the series *diverges* for all x except $x = 0$.

Many elementary functions, as well as special functions, are specializations of a certain generalized hypergeometric function.[2] Some of these relations are listed below, which can generally be verified by comparing series on each side:

(GH1): $\quad {}_0F_0(-;-;x) = e^x$

(GH2): $\quad {}_1F_0(a;-;x) = (1 - x)^{-a}$

(GH3): $\quad {}_0F_1\left(-;\dfrac{1}{2};-\dfrac{x^2}{4}\right) = \cos x$

(GH4): $\quad {}_0F_1\left(-;\dfrac{3}{2};-\dfrac{x^2}{4}\right) = \dfrac{\sin x}{x}$

(GH5): $\quad {}_0F_1\left(-;1;-\dfrac{x^2}{4}\right) = J_0(x)$

(GH6): $\quad {}_0F_1\left(-;a+\dfrac{1}{2};-\dfrac{x^2}{4}\right) = \left(\dfrac{x}{2}\right)^{1/2-a} \Gamma\left(a + \dfrac{1}{2}\right) J_{a-1/2}(x), \ x > 0$

(GH7): $\quad {}_1F_1(a;a;x) = {}_0F_0(-;-;x) = e^x$

(GH8): $\quad {}_1F_1\left(-n;\dfrac{1}{2};x^2\right) = (-1)^n \dfrac{n!}{(2n)!} H_{2n}(x)$

(GH9): $\quad {}_1F_1\left(-n;\dfrac{3}{2};x^2\right) = (-1)^n \dfrac{n!}{(2n + 1)!} \dfrac{1}{2x} H_{2n+1}(x)$

(GH9): $\quad {}_1F_1(-n;a + 1;x) = \dfrac{n!}{(a + 1)_n} L_n^{(a)}(x)$

(GH10): $\quad {}_1F_1\left(\dfrac{1}{2};\dfrac{3}{2};-x^2\right) = \dfrac{\sqrt{\pi}}{2x} \operatorname{erf}(x)$

(GH11): $\quad {}_1F_1\left(\dfrac{1}{2};1;-x\right) = e^{-x/2} I_0(x/2)$

(GH12): $\quad {}_1F_1\left(\dfrac{1}{2};2;-x\right) = e^{-x/2}\left[I_0(x/2) + I_1(x/2)\right]$

(GH13): $\quad {}_1F_1(1;2;-x) = \dfrac{1}{x}(1 - e^{-x})$

(GH14): $\quad {}_1F_1(2;3;-x) = \dfrac{1}{x^2}(1 - e^{-x} - xe^{-x})$

---

[2]The absence of a parameter in ${}_pF_q$ is emphasized by a dash.

(GH15): $\quad {}_2F_1\left(\dfrac{1}{2},1;\dfrac{3}{2};x^2\right) = \dfrac{1}{2x}\ln\dfrac{1+x}{1-x}$

(GH16): $\quad {}_2F_1\left(\dfrac{1}{2},1;\dfrac{3}{2};-x^2\right) = \dfrac{1}{x}\tan^{-1}x$

(GH17): $\quad {}_2F_1\left(\dfrac{1}{2},\dfrac{1}{2};\dfrac{3}{2};x^2\right) = \dfrac{1}{x}\sin^{-1}x$

(GH18): $\quad {}_2F_1\left(-n,n+1;1;\dfrac{1-x}{2}\right) = P_n(x)$

2.6.5 Applications involving hypergeometric functions

Hypergeometric functions frequently arise in applications, particularly in problems dealing with integral transforms, combinatorics, and probability theory, among others. Regardless of the application area, the problem often requires the evaluation of some series or integral that can not be found in integral tables or solved by one of the readily available software packages. By way of some examples, we wish to illustrate a technique that can help in many of these situations without the need for more advanced mathematical techniques such as residue theory from complex variables.

EXAMPLE 9 Sum the finite series $\displaystyle\sum_{k=0}^{n}\binom{n}{k}^2$.

Solution: A standard approach to summing series of the above form is to convert every term in the summand to Pochhammer notation in order to recognize a particular hypergeometric function. In the present case, this action leads to [using (Poch6) in Section 2.6.1]

$$\binom{n}{k} = (-1)^k\frac{(-n)_k}{k!}.$$

We can now sum the resulting series as

$$\sum_{k=0}^{n}\binom{n}{k}^2 = \sum_{k=0}^{n}\frac{(-n)_k(-n)_k}{(1)_k k!} = {}_2F_1(-n,-n;1;1),$$

where the last step follows from Eq. (75). Last, based on property (HY4), we can evaluate the hypergeometric function with unit argument to obtain

$$\sum_{k=0}^{n}\binom{n}{k}^2 = \frac{(2n)!}{(n!)^2} = \binom{2n}{n}.$$

EXAMPLE 10 Express the following series as a generalized hypergeometric function:

$$S = \sum_{n=1}^{\infty} \frac{(-1)^{n-1}(n+2)}{(n+1)!} x^n.$$

Solution: We first make the change of index $n \to n+1$ to find

$$S = \sum_{\substack{n=1 \\ n \to n+1}}^{\infty} \frac{(-1)^{n-1}(n+2)}{(n+1)!} x^n = x \sum_{n=0}^{\infty} \frac{(-1)^n(n+3)}{(n+2)!} x^n.$$

Then, by noting that (see Example 8)

$$\frac{n+3}{(n+2)!} = \frac{(n+3)}{(n+2)(n+1)n!} = \frac{3}{2} \frac{(1)_n(2)_n(4)_n}{(2)_n(3)_n(3)_n n!} = \frac{3}{2} \frac{(1)_n(4)_n}{(3)_n(3)_n n!},$$

we have

$$S = \frac{3x}{2} \sum_{n=0}^{\infty} \frac{(1)_n(4)_n}{(3)_n(3)_n} \frac{(-x)^n}{n!} = \frac{3x}{2} \,_2F_2(1,4;3,3;-x).$$

A similar approach can be used in the evaluation of many integrals like the following two examples.

EXAMPLE 11 Evaluate the integral

$$I = \int_0^x \frac{t^{\mu-1}}{(1+\beta t)^\nu} dt, \quad 0 < x < 1, \ 0 < \beta < 1.$$

Solution: We begin with the binomial series expansion

$$\frac{1}{(1+\beta t)^\nu} = \sum_{n=0}^{\infty} \binom{-\nu}{n} (\beta t)^n = \sum_{n=0}^{\infty} (\nu)_n \frac{(-\beta t)^n}{n!}, \quad |\beta t| < 1,$$

where we have used property (Poch6) in Section 2.6.1. Interchanging the order of integration and summation, we get

$$I = \sum_{n=0}^{\infty} (\nu)_n \frac{(-\beta)^n}{n!} \int_0^x t^{n+\mu-1} dt = \sum_{n=0}^{\infty} (\nu)_n \frac{(-\beta)^n}{n!} \frac{x^{n+\mu}}{n+\mu}.$$

Next, by writing (see Example 8)

$$\frac{1}{n+\mu} = \frac{\Gamma(\mu+n)}{\Gamma(1+\mu+n)} = \frac{(\mu)_n}{\mu(1+\mu)_n},$$

the above summation becomes

$$I = \frac{x^{\mu}}{\mu} \sum_{n=0}^{\infty} \frac{(v)_n (\mu)_n}{(1+\mu)_n} \frac{(-\beta x)^n}{n!},$$

from which we deduce

$$\int_0^x \frac{t^{\mu-1}}{(1+\beta t)^{v}} dt = \frac{x^{\mu}}{\mu} {}_2F_1(v, \mu; 1+\mu; -\beta x).$$

EXAMPLE 12 Evaluate the integral

$$I = \int_0^{\infty} e^{-t} t^{n} J_0(2\sqrt{xt}) dt.$$

Solution: Integrals of this type are common in the use of the Laplace transform (Chapter 10). To begin, we start with the series representation (48) for the Bessel function

$$J_0(2\sqrt{xt}) = \sum_{k=0}^{\infty} \frac{(-1)^k x^k t^k}{(k!)^2}.$$

Next, we substitute this series directly into the integral, followed by an interchange in the order of integration and summation to find

$$I = \sum_{k=0}^{\infty} \frac{(-1)^k x^k}{(k!)^2} \int_0^{\infty} e^{-t} t^{n+k} dt = \sum_{k=0}^{\infty} \frac{(-1)^k (n+k)!}{(k!)^2} x^k.$$

By using the Pochhammer identities $(n+k)! = n!(n+1)_k$ and $k! = (1)_k$, we can express our series in terms of Pochhammer symbols and deduce that

$$I = n! \sum_{k=0}^{\infty} \frac{(n+1)_k}{(1)_k} \frac{(-x)^k}{k!} = n! {}_1F_1(n+1; 1; -x).$$

Finally, with the help of Kummer's transformation (CH3), we arrive at [using GH9) in Section 2.6.4]

$$\int_0^{\infty} e^{-t} t^{n} J_0(2\sqrt{xt}) dt = n! e^{-x} {}_1F_1(-n; 1; x) = n! e^{-x} L_n(x),$$

where $L_n(x)$ is the nth Laguerre polynomial (Section 2.4.3).

There are numerous examples similar to those above that routinely arise in practice—some of these examples will appear in various chapters throughout the text. What we are trying to illustrate here is the power of the notation of hypergeometric-type functions in attacking a class of problems from an analytical approach. When analytic expressions can be obtained, they generally provide greater insight into the problem as compared with purely numerical approaches. In particular, analytic expressions can help identify the essential parameters of the problem under study and how the solution will change as certain parameters of the system are changed. It is difficult to achieve the same degree of understanding by running a number of numerical experiments.

2.7 Summary of Notations for Special Functions

Listed below in Table 2.5 is a summary of notation used for special functions.

Table 2.5 Notation for Special Functions

Notation	Name of Function
$(a)_n$	Pochhammer symbol
$Ai(x)$	Airy function of the first kind
$Bi(x)$	Airy function of the second kind
$B(x,y)$	Beta function
$C(x)$	Fresnel cosine integral
$comb(x)$	Comb function
$\delta(x)$	Delta function
$erf(x)$, $erfc(x)$	Error functions
$_1F_1(a;c;z)$	Confluent hypergeometric function of the first kind
$_2F_1(a,b;c;z)$	Hypergeometric function
$_pF_q(a_1, .., a_p; c_1,...,c_q;z)$	Generalized hypergeometric function
$Gaus(x)$	Gaussian function
γ	Euler's constant
$\gamma(a,x)$, $\Gamma(a,x)$	Incomplete gamma functions
$\Gamma(x)$	Gamma function
$H_n(x)$	Hermite polynomial

Table 2.5 Cont'd.

$I_p(x)$	Modified Bessel function of the first kind
$j_n(x)$	Spherical Bessel function of the first kind
$J_p(x)$	Bessel function of the first kind
$K_p(x)$	Modified Bessel function of the second kind
$L_n(x)$	Laguerre polynomial
$L_n^{(m)}(x)$	Generalized Laguerre polynomial
$P_n(x)$	Legendre polynomial
$\psi(x)$	Digamma (psi) function
$\psi^{(m)}(x)$	Polygamma function
$\text{rect}(x), \Pi(x)$	Rectangle function
$S(x)$	Fresnel sine integral
$\text{sgn}(x)$	Signum function
$\text{sinc}(x)$	Sinc function
$\text{step}(x), U(x)$	Step function
$T_n(x)$	Chebyshev polynomial of the first kind
$\text{tri}(x), \Lambda(x)$	Triangle function
$U(a;c;z)$	Confluent hypergeometric function of the second kind
$U_n(x)$	Chebyshev polynomial of the second kind
$y_n(x)$	Spherical Bessel function of the second kind
$Y_p(x)$	Bessel function of the second kind
$\zeta(p)$	Riemann zeta function

Suggested Reading

M. Abramowitz and I. A. Stegun (eds.), *Handbook of Mathematical Functions* (Dover, New York, 1965).

L. C. Andrews, *Special Functions of Mathematics for Engineers*, 2nd ed. (SPIE Engineering Press, Bellingham; Oxford University Press, Oxford, 1998).

G. Arfken, *Mathematical Methods for Physicists*, 3rd ed. (Academic, New York, 1985).

W. W. Bell, *Special Functions for Scientists and Engineers* (Van Nostrand, London, 1968).

Y. L. Luke, *The Special Functions and Their Applications*, 2 vols. (Academic, New York, 1969).

G. N. Watson, *A Treatise on the Theory of Bessel Functions*, 2nd ed. (Cambridge University Press, London, 1952).

Chapter 2

Exercises

In Problems 1-4, use gamma functions and beta functions to evaluate the given integral.

1. $\int_0^\infty e^{-st} t^{x-1} dt$, $s > 0$ *Ans.* $\Gamma(x)/s^x$

2. $\int_0^\infty x^4 e^{-x^3} dx$

Ans. $\dfrac{1}{3}\Gamma\left(\dfrac{5}{3}\right)$

3. $\int_0^\infty \dfrac{dx}{\sqrt{x}(1+x)^2}$ *Ans.* $\dfrac{\pi}{2}$

4. $\int_0^{\pi/2} \cos^6\theta\, d\theta$ *Ans.* $\dfrac{5\pi}{32}$

5. Take the logarithmic derivative of $\Gamma(x)\Gamma(1-x) = \pi \csc \pi x$ to deduce

$$\psi(1-x) - \psi(x) = \pi \cot \pi x.$$

In Problems 6-8, use the error functions to evaluate the given integral.

6. $\int_{-a}^a e^{-t^2} dt$ *Ans.* $\sqrt{\pi}\,\mathrm{erf}(a)$

7. $\int_0^\infty e^{-x^2 t^2} dt$, $x > 0$ *Ans.* $\dfrac{\sqrt{\pi}}{2x}$

8. $\int_0^\infty e^{-st - t^2/4} dt$, $s > 0$

Ans. $\sqrt{\pi}\, e^{s^2} \mathrm{erfc}(s)$

9. Show that the Fresnel integrals have the series representations.

(*a*) $C(x) = \displaystyle\sum_{n=0}^\infty \frac{(-1)^n(\pi/2)^{2n} x^{4n+1}}{(2n)!(4n+1)}$

(*b*) $S(x) = \displaystyle\sum_{n=0}^\infty \frac{(-1)^n(\pi/2)^{2n+1} x^{4n+3}}{(2n+1)!(4n+3)}$

10. Show that

$$C(\infty) = S(\infty) = \frac{1}{2}.$$

11. Given the generating function relation for the Legendre polynomials

$$w(x,t) = (1 - 2xt + t^2)^{-1/2} = \sum_{n=0}^\infty P_n(x) t^n,$$

(*a*) show that

$$(1 - 2xt + t^2)\frac{\partial w}{\partial t} + (t - x)w = 0.$$

(*b*) Use part (*a*) and the generating function relation given above to deduce the recurrence formula (P5).

12. An electric dipole consists of electric charges q and $-q$ located along the x-axis as shown in the figure below. The potential induced at point P due to the charges is known to be ($r > a$)

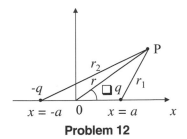

Problem 12

$$V = kq\left(\frac{1}{r_1} - \frac{1}{r_2}\right), \quad k > 0.$$

Use the first term of a generating function series to show that

$$V \cong \frac{2akq}{r^2}\cos\theta, \quad r \gg a.$$

13. Given that the Hermite polynomial $y = H_n(x)$ is a solution of the DE (34), show that $\varphi(x) = H_n(x)e^{-x^2/2}$, is a solution of the Schrödinger equation

$$\varphi'' + (2n + 1 - x^2)\varphi = 0, \quad -\infty < x < \infty.$$

14. Show that the Laplace transform of the Laguerre polynomial $L_n(t)$ leads to

$$\int_0^\infty e^{-pt} L_n(t)dt = \frac{1}{p}\left(1 - \frac{1}{p}\right)^n, \quad p > 1.$$

15. Define

$$w(x,t) = \exp\left[\frac{x}{2}\left(t - \frac{1}{t}\right)\right],$$

and use the generating function relation [Eq. (49)] to show that the product $w(x,t)w(-x,t)$ leads to

(a) $1 = [J_0(x)]^2 + 2\sum_{n=1}^\infty [J_n(x)]^2$.

(b) From (a), show that for all x

$$|J_0(x)| \le 1 \text{ and } |J_n(x)| \le \frac{1}{\sqrt{2}},$$

for $n = 1, 2, 3, \dots$.

16. Show that the function $w(x,t)$ in Problem 15 satisfies the identity $w(x + y,t) = w(x,t)w(y,t)$ and deduce the addition formula (BJ3).

17. Use the generating function relation (49) to derive (BJ2).

18. By comparing Maclaurin series, verify the identities in (BJ6).

19. Use the integral representation

$$\frac{1}{\sqrt{x^2 + a^2}} = \frac{1}{\sqrt{\pi}}\int_0^\infty \frac{e^{-(x^2 + a^2)t}}{\sqrt{t}}dt$$

and interchange the order of integration to deduce that ($a \ge 0$, $b > 0$)

$$\int_0^\infty \frac{x J_0(x)}{\sqrt{x^2 + a^2}}dx = \frac{1}{b}e^{-ab}.$$

20. Show, by expanding the Bessel function in an infinite series followed by termwise integration, that

(a) $\int_0^{\pi/2} J_0(x\cos\theta)\cos\theta\, d\theta = \frac{\sin x}{x}$.

(b) $\int_0^{\pi/2} J_1(x\cos\theta)\, d\theta = \frac{1 - \cos x}{x}$.

21. Show that

(a) $I_{1/2}(x) = \sqrt{\frac{2}{\pi x}}\sinh x$.

(b) $I_{-1/2}(x) = \sqrt{\frac{2}{\pi x}}\cosh x$.

(c) $K_{1/2}(x) = \sqrt{\frac{\pi}{2x}}e^{-x}$.

22. Use Eqs. (146) and (147) in Chapter 1 to deduce that the general solution of

$$y'' + xy = 0$$

is given by the Airy functions

$$y = C_1 \, \text{Ai}(-x) + C_2 \, \text{Bi}(-x).$$

23. When the intensity $\mathbf{E}$ of a radar signal is governed by the modified Rician distribution

$$p_{\mathbf{E}}(E) = \frac{1}{N} e^{-(A^2 + E)/N} I_0\left(\frac{2A\sqrt{E}}{N}\right) U(E),$$

show that the statistical moments are

$$\langle \mathbf{E}^n \rangle = N^n n! L_n\left(-\frac{A^2}{N}\right), \quad n = 1,2,3,\dots.$$

24. Verify the following identities for the Pochhammer symbol.

(a) $(a)_{n+1} = a(a+1)_n$

(b) $(n+k)! = n!(n+1)_k$

(c) $(2n)! = 2^{2n}(1/2)_n n!$

(d) $(2n+1)! = 2^{2n}(3/2)_n n!$

25. By assuming a power series solution

$$y = \sum_{n=1}^{\infty} c_n x^n,$$

verify that $y = {}_2F_1(a,b;c;x)$ is a solution of the hypergeometric equation (77).

26. By comparing Maclaurin series, show that

(a) $\arcsin x = x \, {}_2F_1(\frac{1}{2},\frac{1}{2};\frac{3}{2};x^2).$

(b) $\ln(1+x) = x \, {}_2F_1(1,1;2;-x).$

27. By comparing Maclaurin series, show that

(a) ${}_1F_1(1;2;-x) = \frac{1}{x}(1 - e^{-x}).$

(b) ${}_1F_1(\frac{1}{2};1;-x) = e^{-x/2} I_0(x/2).$

(c) ${}_1F_1(-n;1;x) = L_n(x).$

28. By assuming a power series solution

$$y = \sum_{n=1}^{\infty} c_n x^n,$$

verify that $y = {}_1F_1(a;c;x)$ is a solution of the confluent hypergeometric equation (80).

29. Show that

(a) ${}_1F_0(a;\text{—};-x) = \dfrac{1}{(1+x)^a}.$

(b) ${}_0F_1(\text{—};\frac{1}{2};-\dfrac{x^2}{4}) = \cos x.$

(c) ${}_0F_1(\text{—};1;-\dfrac{x^2}{4}) = J_0(x).$

(d) ${}_0F_1(\text{—};\frac{3}{2};-\dfrac{x^2}{4}) = \dfrac{\sin x}{x}.$

30. Show that

$$\sum_{k=0}^{n} \binom{2n}{2k} = 2^{2n-1}, \quad n > 0.$$

31. Show that

$$\sum_{k=0}^{n} (-1)^k \binom{2n}{2k}\binom{-1/2}{k} = \frac{2^{2n}\Gamma(2n+\frac{1}{2})}{\sqrt{\pi}\,(2n)!},$$
$$n = 0,1,2,\dots.$$

32. By expanding $J_0(at)$ in a series,
(a) show that $(p > a)$

$$\int_0^{\infty} e^{-pt} J_0(at)\,dt = \frac{1}{p} \, {}_1F_0\left(\frac{1}{2};\text{—};-\frac{a^2}{p^2}\right).$$

(b) From (a), deduce that

$$\int_0^{\infty} e^{-pt} J_0(at)\,dt = \frac{1}{\sqrt{p^2 + a^2}}.$$

33. Show that (for $s > 1$)

$$\int_0^{\infty} e^{-st} \, {}_1F_1(a;c;t)\,dt = \frac{1}{s} \, {}_2F_1\left(a,1;c;\frac{1}{s}\right).$$

34. Show that

$$\int_0^{x} {}_0F_1[\text{—};1;-\tfrac{1}{4}t(x-t)]\,dt = 2\sin\frac{x}{2}.$$

Chapter 3

Matrix Methods and Linear Vector Spaces

3.1 Introduction / 110
3.2 Basic Matrix Concepts and Operations / 110
    3.2.1   Algebraic properties / 111
    3.2.2   Determinants / 112
    3.3.3   Special matrices / 113
3.3 Linear Systems of Equations / 114
    3.3.1   Matrix eigenvalue problems / 116
    3.3.2   Real symmetric and skew-symmetric matrices / 119
3.4 Linear Systems of Differential Equations / 121
    3.4.1   Homogeneous systems / 122
    3.4.2   Homogeneous systems with constant coefficients / 123
    3.4.3   Stability of linear systems / 129
    3.4.4   Nonhomogeneous systems / 131
3.5 Linear Vector Spaces / 133
    3.5.1   Linear independence and basis vectors / 134
    3.5.2   Inner product spaces / 135
    3.5.3   Orthonormal basis and the expansion theorem / 137
    3.5.4   Hilbert spaces / 138
Suggested Reading / 140
Exercises / 140

Because matrix methods play such an important role in solving systems of linear equations, we devote this chapter to a review of the basic concepts associated with matrix operations. In addition, we also provide a brief treatment of the notion of linear vector space, which arises again in later chapters. The linear vector space concept provides a unifying approach to several common areas of applied mathematics.

3.1 Introduction

The development of engineering mathematics during the past few decades has been greatly affected by the increasing role of *linear analysis* and extensive use of computers to solve engineering problems. The subject matter of general linear analysis is vast, including differential equations, matrices, linear algebra, vector and tensor analysis, integral equations, and linear vector spaces. In fact, virtually all of the topics covered in this text can be included under the general title of linear analysis; however, in this chapter we concentrate only on matrix methods and linear vector spaces, the latter including some discussion of Hilbert spaces. Like most chapters throughout this text, our treatment is not intended to be exhaustive.

Linear algebra is basically the theory and application of vectors and matrices in connection with the solution of linear transformations, linear systems of differential equations, eigenvalue problems, and so on. Such systems of equations may arise, for example, from the analysis of electrical networks and various frameworks in mechanics such as describing the motion of a system of particles. Matrices are useful in that they permit us to consider a rectangular array of numbers as a single entity; hence, we can perform various operations in a systematic and compact fashion.

3.2 Basic Matrix Concepts and Operations

The solution of systems of linear equations can be greatly systematized by the use of matrix methods, particularly when the number of equations is greater than two. We will denote a matrix here by a bold letter such as $\mathbf{A}, \mathbf{B}, \mathbf{X}, \mathbf{Y}, \mathbf{x}, \mathbf{y}, \dots$.

▸   **Matrix:** A *matrix* is a rectangular array of numbers, or elements, arranged in m rows and n columns; that is,

$$
\mathbf{A} = \begin{pmatrix} a_{11} & a_{12} & \cdots & a_{1n} \\ a_{21} & a_{22} & \cdots & a_{2n} \\ \vdots & \vdots & & \vdots \\ a_{m1} & a_{m2} & \cdots & a_{mn} \end{pmatrix}. \tag{1}
$$

We call $\mathbf{A}$ an $m \times n$ matrix with elements a_{ij} $(i = 1, 2, \dots, m;\ j = 1, 2, \dots, n)$. The first subscript denotes the row, while the second subscript identifies the column. It is also customary to write simply $\mathbf{A} = [a_{ij}]$. If $m = n$, we say that $\mathbf{A}$ is a *square matrix*.

3.2.1 Algebraic properties

Various algebraic properties can be identified with certain matrices related to each other primarily through their size. These properties are listed below.

▸ **Equality:** Two $m \times n$ matrices $\mathbf{A} = [a_{ij}]$ and $\mathbf{B} = [b_{ij}]$ are said to be *equal* if and only if $a_{ij} = b_{ij}$ for each i and j. In this case we write $\mathbf{A} = \mathbf{B}$.

▸ **Addition:** The sum of two $m \times n$ matrices $\mathbf{A} = [a_{ij}]$ and $\mathbf{B} = [b_{ij}]$ is the $m \times n$ matrix $\mathbf{C} = [c_{ij}]$, where $c_{ij} = a_{ij} + b_{ij}$.

▸ **Multiplication by a number:** The product of an $m \times n$ matrix $\mathbf{A} = [a_{ij}]$ and a number k is defined by $k\mathbf{A} = [ka_{ij}]$. In particular, if $k = -1$, we write $(-1)\mathbf{A} = -\mathbf{A}$.

▸ **Multiplication of matrices:** Let $\mathbf{A} = [a_{ij}]$ be an $m \times n$ matrix and $\mathbf{B} = [b_{ij}]$ be an $n \times r$ matrix. The product $\mathbf{AB} = \mathbf{C}$ is an $m \times r$ matrix $\mathbf{C} = [c_{ij}]$, where

$$c_{ij} = \sum_{k=1}^{n} a_{ik} b_{kj}.$$

Remark: In order to define a matrix product $\mathbf{AB}$, it is necessary that the number of columns in $\mathbf{A}$ be equal to the number of rows in $\mathbf{B}$. Thus, the product $\mathbf{AB}$ may be defined, whereas $\mathbf{BA}$ may not be defined.

EXAMPLE 1 Calculate the matrix product $\mathbf{AB}$ given that

$$\mathbf{A} = \begin{pmatrix} 1 & 6 \\ -4 & 2 \\ 5 & 0 \end{pmatrix}, \quad \mathbf{B} = \begin{pmatrix} -5 & 4 & 1 \\ 0 & -1 & 3 \end{pmatrix}.$$

Solution: We first note that the number of columns in $\mathbf{A}$ is two, equal to the number of rows in $\mathbf{B}$ as required. Hence, by following the above rule of matrix multiplication, we have

$$\mathbf{AB} = \begin{pmatrix} 1 & 6 \\ -4 & 2 \\ 5 & 0 \end{pmatrix} \begin{pmatrix} -5 & 4 & 1 \\ 0 & -1 & 3 \end{pmatrix}$$

$$= \begin{pmatrix} (1)(-5) + (6)(0) & (1)(4) + (6)(-1) & (1)(1) + (6)(3) \\ (-4)(-5) + (2)(0) & (-4)(4) + (2)(-1) & (-4)(1) + (2)(3) \\ (5)(-5) + (0)(0) & (5)(4) + (0)(-1) & (5)(1) + (0)(3) \end{pmatrix},$$

which, upon simplifying, yields

$$\mathbf{AB} = \begin{pmatrix} -5 & -2 & 19 \\ 20 & -18 & 2 \\ -25 & 20 & 5 \end{pmatrix}.$$

When defined, matrix multiplication obeys the following properties:

$$
\begin{aligned}
(\mathbf{AB})\mathbf{C} &= \mathbf{A}(\mathbf{BC}) \\
\mathbf{A}(\mathbf{B} + \mathbf{C}) &= \mathbf{AB} + \mathbf{AC} \\
\mathbf{AB} &\neq \mathbf{BA} \quad \text{(in general)}
\end{aligned}
$$

3.2.2 Determinants

Associated with every square matrix $\mathbf{A}$ is a number called the *determinant,* denoted by $\det(\mathbf{A})$. For example, if $\mathbf{A}$ is a 2×2 matrix

$$\mathbf{A} = \begin{pmatrix} a_{11} & a_{12} \\ a_{21} & a_{22} \end{pmatrix}, \tag{2}$$

its determinant is defined by

$$\det(\mathbf{A}) = \begin{vmatrix} a_{11} & a_{12} \\ a_{21} & a_{22} \end{vmatrix} = a_{11}a_{22} - a_{12}a_{21}. \tag{3}$$

Among other uses, determinants can be used to solve systems of equations like

$$
\begin{aligned}
a_{11}x + a_{12}y &= b_1, \\
a_{21}x + a_{22}y &= b_2,
\end{aligned}
\tag{4}
$$

whose solution can be written as $x = \Delta_x/\Delta$, $y = \Delta_y/\Delta$, where

$$\Delta = \begin{vmatrix} a_{11} & a_{12} \\ a_{21} & a_{22} \end{vmatrix}, \quad \Delta_x = \begin{vmatrix} b_1 & a_{12} \\ b_2 & a_{22} \end{vmatrix}, \quad \Delta_y = \begin{vmatrix} a_{11} & b_1 \\ a_{21} & b_2 \end{vmatrix}. \tag{5}$$

Finding the solution of the linear system (4) by the method of determinants is called *Cramer's rule,* named after the Swiss mathematician Gabriel Cramer (1704-1752).

A *determinant of third order* can be defined by

$$\det(\mathbf{A}) = \begin{vmatrix} a_{11} & a_{12} & a_{13} \\ a_{21} & a_{22} & a_{23} \\ a_{31} & a_{32} & a_{33} \end{vmatrix} = a_{11} \begin{vmatrix} a_{22} & a_{23} \\ a_{32} & a_{33} \end{vmatrix} - a_{12} \begin{vmatrix} a_{21} & a_{23} \\ a_{31} & a_{33} \end{vmatrix} + a_{13} \begin{vmatrix} a_{21} & a_{22} \\ a_{31} & a_{32} \end{vmatrix}. \quad (6)$$

Note that the signs of the terms in front of the 2×2 determinants alternate. Each of the second-order determinants in (6) is called a *minor*, obtained by deleting the column and row of the element in front. We may expand a determinant by any row or column—doing so will yield the same number.

Listed below are some of the basic properties associated with determinants:

- The value of a determinant is not altered if the rows and columns of the square matrix are interchanged.
- If all elements of a row or column are zero, the determinant is zero.
- If two rows (or two columns) are proportional, the determinant is zero.
- If two rows (or two columns) of a square matrix are interchanged, the sign of the determinant is changed.
- If all elements of one row (or one column) of a square matrix are multiplied

3.2.3 Special matrices

Presented below are several special matrices and a few of their properties that are of some importance in practice.

▸ **Transpose:** Associated with each matrix $\mathbf{A}$ is the *transpose* matrix $\mathbf{A}^T$, obtained from $\mathbf{A}$ by interchanging the rows and columns of $\mathbf{A}$.

The transpose of a product (when defined) equals the product of transposes, taken in reverse order. That is,

$$(\mathbf{AB})^T = \mathbf{B}^T \mathbf{A}^T.$$

▸ **Symmetric matrix:** If $\mathbf{A}$ is a square matrix and $\mathbf{A} = \mathbf{A}^T$, we say that $\mathbf{A}$ is a *symmetric matrix*.

▸ **Identity:** The *identity matrix* is a square matrix for which all off-diagonal elements are zero and all diagonal elements a_{jj} are unity; for example,

$$\mathbf{I} = \begin{pmatrix} 1 & 0 \\ 0 & 1 \end{pmatrix}, \quad \mathbf{I} = \begin{pmatrix} 1 & 0 & 0 \\ 0 & 1 & 0 \\ 0 & 0 & 1 \end{pmatrix}, \cdots$$

▸ **Inverse:** Given a square matrix $\mathbf{A}$, we say that $\mathbf{A}^{-1}$ is its inverse provided that

$$\mathbf{A}\mathbf{A}^{-1} = \mathbf{A}^{-1}\mathbf{A} = \mathbf{I}.$$

If the inverse $\mathbf{A}^{-1}$ exists, we say that $\mathbf{A}$ is *nonsingular*; otherwise, it is *singular*. It can be shown that a square matrix $\mathbf{A}$ is nonsingular if and only if $\det(\mathbf{A}) \neq 0$. For example, if $\mathbf{A}$ is a 2×2 matrix with nonzero determinant

$$\mathbf{A} = \begin{pmatrix} a_{11} & a_{12} \\ a_{21} & a_{22} \end{pmatrix},$$

its inverse is given by

$$\mathbf{A}^{-1} = \frac{1}{\det(\mathbf{A})} \begin{pmatrix} a_{22} & -a_{12} \\ -a_{21} & a_{11} \end{pmatrix}, \quad \det(\mathbf{A}) = a_{11}a_{22} - a_{12}a_{21} \neq 0. \tag{7}$$

In the case of an $n \times n$ nonsingular matrix $\mathbf{A}$, its inverse is defined by

$$\boxed{\mathbf{A}^{-1} = \frac{1}{\det(\mathbf{A})}[A_{ij}]^T,} \tag{8}$$

where A_{ij} is the *cofactor* of a_{ij} in $\det(\mathbf{A})$. That is, if M_{ij} is the $(n-1) \times (n-1)$ determinant obtained by deleting the ith row and the jth column from $\det(\mathbf{A})$, then $A_{ij} = (-1)^{i+j} M_{ij}$. To illustrate, a 3×3 matrix $\mathbf{A}$ and its inverse are given by

$$\mathbf{A} = \begin{pmatrix} a_{11} & a_{12} & a_{13} \\ a_{21} & a_{22} & a_{23} \\ a_{31} & a_{32} & a_{33} \end{pmatrix}, \quad \mathbf{A}^{-1} = \frac{1}{\det(\mathbf{A})} \begin{pmatrix} A_{11} & A_{12} & A_{13} \\ A_{21} & A_{22} & A_{23} \\ A_{31} & A_{32} & A_{33} \end{pmatrix}, \tag{9}$$

where $\det(\mathbf{A}) = a_{11}A_{11} + a_{12}A_{12} + a_{13}A_{13}$ and

$$A_{11} = \begin{vmatrix} a_{22} & a_{23} \\ a_{32} & a_{33} \end{vmatrix}, \quad A_{12} = -\begin{vmatrix} a_{21} & a_{23} \\ a_{31} & a_{33} \end{vmatrix}, \quad A_{13} = \begin{vmatrix} a_{21} & a_{22} \\ a_{31} & a_{32} \end{vmatrix}, \quad \dots.$$

Similar to the transpose, the inverse of a product of square matrices is equal to the product of inverses of each matrix factor in reverse order, namely,

$$(\mathbf{AB})^{-1} = \mathbf{B}^{-1}\mathbf{A}^{-1}.$$

3.3 Linear Systems of Equations

Probably the most important practical use of matrices is in the solution of linear systems of equations. A *linear system* of m equations and n unknowns $x_1, \dots, x_n$ is a set of equations of the form:

$$\begin{aligned}
a_{11}x_1 + \cdots + a_{1n}x_n &= b_1 \\
a_{21}x_1 + \cdots + a_{2n}x_n &= b_2 \\
&\cdots\cdots\cdots\cdots\cdots\cdots\cdots \\
a_{m1}x_1 + \cdots + a_{mn}x_n &= b_m
\end{aligned} \tag{10}$$

Using the above definition of matrix multiplication, we can write the linear system (10) as a single matrix equation

$$\mathbf{Ax} = \mathbf{b}. \tag{11}$$

In this case, the *coefficient matrix* $\mathbf{A}$, and *column vectors* $\mathbf{x}$ and $\mathbf{b}$ are defined by

$$\mathbf{A} = \begin{pmatrix} a_{11} & a_{12} & \cdots & a_{1n} \\ a_{21} & a_{22} & \cdots & a_{2n} \\ \vdots & \vdots & \cdots & \vdots \\ a_{m1} & a_{m2} & \cdots & a_{mn} \end{pmatrix}, \quad \mathbf{x} = \begin{pmatrix} x_1 \\ x_2 \\ \vdots \\ x_n \end{pmatrix}, \quad \mathbf{b} = \begin{pmatrix} b_1 \\ b_2 \\ \vdots \\ b_m \end{pmatrix}.$$

We note that $\mathbf{x}$ has n elements whereas $\mathbf{b}$ has m elements. In many cases of interest, we find that $m = n$ and then a formal solution of (11) is

$$\mathbf{x} = \mathbf{A}^{-1}\mathbf{b}. \tag{12}$$

Suppose, for example, we consider the simple case where we wish to find values of x and y such that

$$\begin{aligned}
a_{11}x + a_{12}y &= b_1 \\
a_{21}x + a_{22}y &= b_2.
\end{aligned} \tag{13}$$

By the use of Eq. (12), we see that the solution is given by

$$\begin{pmatrix} x \\ y \end{pmatrix} = \frac{1}{\det(\mathbf{A})} \begin{pmatrix} a_{22} & -a_{12} \\ -a_{21} & a_{11} \end{pmatrix} \begin{pmatrix} b_1 \\ b_2 \end{pmatrix} = \frac{1}{\det(\mathbf{A})} \begin{pmatrix} b_1 a_{22} - b_2 a_{12} \\ -b_1 a_{21} + b_2 a_{11} \end{pmatrix}. \tag{14}$$

Geometrically, the system (13) represents two lines in the xy-plane. The solution is the point of intersection of the two lines. We say the system is *consistent* if it possesses at least one solution. If the two lines have different slopes, exactly one solution exists [see Fig. 3.1(*a*) and (*d*)]. If the two lines are parallel, there will be no solution [Fig. 3.1(*b*)], and we say the system is *inconsistent*. Last, if the two lines are the same, the solution set becomes infinite [Fig. 3.1(*c*)].

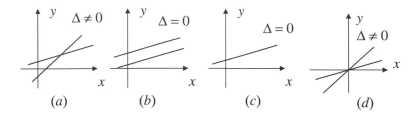

Figure 3.1 Solution of (6) illustrating (*a*) a unique solution, (*b*) no solution, (*c*) infinitely many solutions, and (*d*) a unique (trivial) solution for a homogeneous system. The symbol Δ is the determinant of the coefficient matrix, i.e., $\Delta = \det(\mathbf{A})$.

When $b_1 = b_2 = 0$, the system (11) is called *homogeneous*, and *nonhomogeneous* otherwise. For a homogeneous system it is clear that $x = y = 0$ is always a solution [see Fig. 3.1(*d*)], called the *trivial solution*. Thus, a homogeneous system is always consistent. From a geometrical point of view, the homogeneous system represents two lines passing through the origin. Only if one line lies directly on top of the other will a nontrivial solution exist, and in this case there will be infinitely many solutions (every point on either line). In summary, we have the following conclusions about linear systems in which $m = n$:[1]

Homogeneous system: $(b_1 = b_2 = \cdots = 0)$
    (*a*) If $\det(\mathbf{A}) \neq 0$, only the trivial solution exists.
    (*b*) If $\det(\mathbf{A}) = 0$, infinitely many solutions exist.

Nonhomogeneous system:
    (*a*) If $\det(\mathbf{A}) \neq 0$, a unique solution exists (i.e., the related homogeneous system has only the trivial solution.
    (*b*) If $\det(\mathbf{A}) = 0$, either no solution exists or infinitely many solutions exist (i.e., the related homogeneous system has a nontrivial solution.

3.3.1 Matrix eigenvalue problems

Eigenvalue problems are among some of the most important engineering problems that arise in applications. Such problems involve the solution of certain types of matrix equation in which we seek values of the scalar λ that permit nontrivial solutions of

$$\mathbf{A}\mathbf{x} = \lambda\mathbf{x}, \tag{15}$$

where $\mathbf{A}$ is an $n \times n$ matrix and $\mathbf{x}$ is an $n \times 1$ column vector. From inspection we see that $\mathbf{x} = \mathbf{0}$ is a solution for any value of λ, called the *trivial solution*. The values of λ that permit nontrivial solutions are called *eigenvalues* (or *characteristic values*) of the square matrix

---

[1]Although Fig. 3.1 applies to the case $m = n = 2$, the conclusions about linear systems are also valid for cases when $m = n > 2$.

A and the corresponding nontrivial solutions **x** are called *eigenvectors* (or *characteristic vectors*). To solve this problem it is customary to rewrite (15) in the form

$$(\mathbf{A} - \lambda \mathbf{I})\mathbf{x} = \mathbf{0}. \tag{16}$$

From our discussion above, we recognize this homogeneous system of equations has a nontrivial solution if and only if

$$\det(\mathbf{A} - \lambda \mathbf{I}) = 0. \tag{17}$$

For an $n \times n$ matrix **A**, Eq. (17) is a polynomial in λ of degree n and therefore must have n zeros (not necessarily real or distinct), denoted by $\lambda_1, \lambda_2, \ldots, \lambda_n$. The set of n eigenvalues is called the *spectrum* of **A**. Each distinct eigenvalue λ_j gives rise to a nontrivial eigenvector $\mathbf{e}_j$. The set of n eigenvectors, together with **0**, constitutes what is called an *eigenspace*. That there is at least one eigenvalue of **A** and at most n follows from the theory of polynomials.

EXAMPLE 2 Find the eigenvalues and eigenvectors of

$$\mathbf{A} = \begin{pmatrix} 1 & 2 \\ 3 & 2 \end{pmatrix}.$$

Solution: Based on Eq. (17), we have

$$\det(\mathbf{A} - \lambda \mathbf{I}) = \begin{vmatrix} 1 - \lambda & 2 \\ 3 & 2 - \lambda \end{vmatrix} = 0,$$

which leads to

$$\lambda^2 - 3\lambda - 4 = (\lambda - 4)(\lambda + 1) = 0.$$

Hence, the eigenvalues are $\lambda_1 = -1$, $\lambda_2 = 4$.

To find the corresponding eigenvectors, we look for solutions of

$$\begin{pmatrix} 1 - \lambda & 2 \\ 3 & 2 - \lambda \end{pmatrix} \begin{pmatrix} x_1 \\ x_2 \end{pmatrix} = 0,$$

for each of the eigenvalues. For $\lambda_1 = -1$, this becomes the system of equations

$$2x_1 + 2x_2 = 0,$$
$$3x_1 + 3x_2 = 0,$$

from which we deduce (from either equation) $x_2 = -x_1$; thus,

$$\mathbf{e}_1 = \begin{pmatrix} x_1 \\ x_2 \end{pmatrix} = c_1 \begin{pmatrix} 1 \\ -1 \end{pmatrix},$$

where c_1 is an arbitrary constant. In the same manner we find for $\lambda_2 = 4$ that

$$\mathbf{e}_2 = c_2 \begin{pmatrix} 2 \\ 3 \end{pmatrix}.$$

Remark: Example 2 illustrates that eigenvectors are unique only to within a (nonzero) multiplicative constant.

EXAMPLE 3 Find the eigenvalues and eigenvectors of the matrix

$$\mathbf{A} = \begin{pmatrix} -4 & -1 \\ 1 & -2 \end{pmatrix}.$$

Solution: Here the characteristic determinant yields

$$\det(\mathbf{A} - \lambda \mathbf{I}) = \begin{vmatrix} -(4 + \lambda) & -1 \\ 1 & -(2 + \lambda) \end{vmatrix} = \lambda^2 + 6\lambda + 9 = 0,$$

which has the double root $\lambda_1 = \lambda_2 = -3$. For this case, there is only one eigenvector found by solving

$$x_1 + x_2 = 0.$$

Thus, $x_2 = -x_1$, and consequently,

$$\mathbf{e}_1 = \mathbf{e}_2 = \begin{pmatrix} x_1 \\ x_2 \end{pmatrix} = c_1 \begin{pmatrix} 1 \\ -1 \end{pmatrix},$$

EXAMPLE 4 Find the eigenvalues of the matrix

$$\mathbf{A} = \begin{pmatrix} 6 & -1 \\ 5 & 4 \end{pmatrix}.$$

Solution: The characteristic determinant yields

$$\det(\mathbf{A} - \lambda \mathbf{I}) = \begin{vmatrix} 6 - \lambda & -1 \\ 1 & 4 - \lambda \end{vmatrix} = \lambda^2 - 10\lambda + 29 = 0.$$

In this case we find complex eigenvalues $\lambda_1 = 5 + 2i$ and $\lambda_2 = 5 - 2i$. We leave it to the reader to find the complex eigenvectors in this case (e.g. see Example 8).

In general, if $\mathbf{A}$ is an $n \times n$ matrix, the characteristic determinant can be expressed in the factored form

$$\det(\mathbf{A} - \lambda\mathbf{I}) = (\lambda_1 - \lambda)(\lambda_2 - \lambda)\cdots(\lambda_n - \lambda) \tag{18}$$

for all λ. In particular, when $\lambda = 0$ we see that

$$\det(\mathbf{A}) = \lambda_1\lambda_2\cdots\lambda_n. \tag{19}$$

That is, the determinant of $\mathbf{A}$ is simply the product of its eigenvalues. Hence, if one of the eigenvalues is zero, we have $\det(\mathbf{A}) = 0$, which implies that $\mathbf{A}$ is singular.

3.3.2 Real symmetric and skew-symmetric matrices

A *real symmetric matrix* $\mathbf{A}$ is a square matrix with real elements satisfying $a_{ij} = a_{ji}$. Alternatively, we say $\mathbf{A}$ is symmetric if $\mathbf{A}^T = \mathbf{A}$. We say the matrix is *skew-symmetric* if the real elements satisfy $a_{ij} = -a_{ji}$, or $\mathbf{A}^T = -\mathbf{A}$. These particular classes of matrices have the following important property concerning their eigenvalues.

Theorem 3.1 The eigenvalues of
- a symmetric matrix are real.
- a skew-symmetric matrix are pure imaginary or zero.

Additional properties concerning symmetric matrices will be discussed later under the general topic of linear vector spaces. That is, the properties associated with symmetric matrices are applicable to a much broader class of problems. We end this section with an application concerning the elastic deformation of a thin membrane.

EXAMPLE 5 A circular elastic membrane of unit radius in the xy-plane is stretched in such a way that the point $P(x,y)$ is moved to the point $Q(X,Y)$ under the linear transformation

$$\begin{aligned} X &= 5x + 3y, \\ Y &= 3x + 5y. \end{aligned}$$

If the boundary curve of the unstretched membrane is $x^2 + y^2 = 1$ before deformation, find the principal directions and shape of the membrane after stretching (see also Section 5.7).

Solution: If we define column vectors

$$\mathbf{x} = \begin{pmatrix} x \\ y \end{pmatrix}, \quad \mathbf{X} = \begin{pmatrix} X \\ Y \end{pmatrix},$$

we can write the equation of transformation in matrix form $\mathbf{X} = \mathbf{A}\mathbf{x}$, where

$$\mathbf{A} = \begin{pmatrix} 5 & 3 \\ 3 & 5 \end{pmatrix}.$$

The principal directions associated with this matrix are defined by the eigenvectors. Hence, we first solve

$$\begin{vmatrix} 5 - \lambda & 3 \\ 3 & 5 - \lambda \end{vmatrix} = \lambda^2 - 10\lambda + 16 = (\lambda - 2)(\lambda - 8) = 0.$$

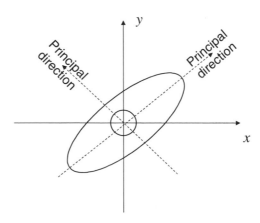

Figure 3.2 Elastic membrane before and after deformation.

The eigenvalues are therefore given by $\lambda_1 = 2$ and $\lambda_2 = 8$. We can interpret these values as representing the "stretching factors" of the membrane in the principal directions (see Fig. 3.2).

For $\lambda_1 = 2$, the corresponding eigenvector is obtained by solving

$$3x_1 + 3x_2 = 0,$$

from which we deduce

$$\mathbf{e}_1 = \begin{pmatrix} x_1 \\ x_2 \end{pmatrix} = c_1 \begin{pmatrix} 1 \\ -1 \end{pmatrix}.$$

In a similar manner, for the eigenvalue $\lambda_2 = 8$, we are led to

$$-3x_1 + 3x_2 = 0,$$

and

$$\mathbf{e}_2 = c_2 \begin{pmatrix} 1 \\ 1 \end{pmatrix}.$$

Note that the vectors $\mathbf{e}_1$ and $\mathbf{e}_2$ are rotated through $\pi/4$ radians with respect to the original axes. If we let (u,v) denote the axes in the principal directions, it now follows that the shape of the stretched membrane is that of an ellipse described by

$$\frac{u^2}{8^2} + \frac{v^2}{2^2} = 1.$$

3.4 Linear Systems of Differential Equations

In solving matrix differential equations the need arises to perform calculus operations on a matrix function $\mathbf{A}(t)$, where t is the independent variable. The derivative and integral of $\mathbf{A}(t) = [a_{ij}(t)]$ are defined, respectively, by

$$\mathbf{A}'(t) = [a_{ij}'(t)],$$
$$\int_{t_0}^{t} \mathbf{A}(s)ds = \left[\int_{t_0}^{t} a_{ij}(s)ds\right]. \tag{20}$$

The theory and solution techniques associated with solving linear systems of differential equations are well suited for matrix methods. In part, this follows from the fact that a system of n equations such as

$$\begin{aligned}
y_1' &= a_{11}(t)y_1 + a_{12}(t)y_2 + \cdots + a_{1n}(t)y_n + f_1(t) \\
y_2' &= a_{21}(t)y_1 + a_{22}(t)y_2 + \cdots + a_{2n}(t)y_n + f_2(t) \\
&\cdots\cdots\cdots\cdots\cdots\cdots\cdots\cdots\cdots\cdots\cdots\cdots\cdots \\
y_n' &= a_{n1}(t)y_1 + a_{n2}(t)y_2 + \cdots + a_{nn}(t)y_n + f_n(t)
\end{aligned} \tag{21}$$

can be represented by the single matrix equation

$$\boxed{\mathbf{Y}' = \mathbf{A}(t)\mathbf{Y} + \mathbf{F}(t),} \tag{22}$$

where $\mathbf{A}(t) = [a_{ij}(t)]$ is the $n \times n$ coefficient matrix

$$\mathbf{A}(t) = \begin{pmatrix}
a_{11}(t) & a_{12}(t) & \cdots & a_{1n}(t) \\
a_{21}(t) & a_{22}(t) & \cdots & a_{2n}(t) \\
\vdots & \vdots & & \vdots \\
a_{n1}(t) & a_{n2}(t) & \cdots & a_{nn}(t)
\end{pmatrix}, \tag{23}$$

and $\mathbf{Y}(t)$ and $\mathbf{F}(t)$ are column vectors defined, respectively, by

$$\mathbf{Y}(t) = \begin{pmatrix} y_1(t) \\ y_2(t) \\ \vdots \\ y_n(t) \end{pmatrix}, \quad \mathbf{F}(t) = \begin{pmatrix} f_1(t) \\ f_2(t) \\ \vdots \\ f_n(t) \end{pmatrix}. \tag{24}$$

Note that (22) assumes the same form, regardless of the number of equations n in the system.

Equation (22) is called a *homogeneous* equation when the column vector $\mathbf{F}(t) = \mathbf{0}$ for all t in some interval I (the interval I should not to be confused with the identity matrix $\mathbf{I}$); otherwise, we say (22) is *nonhomogeneous*.

3.4.1 Homogeneous systems

We first investigate the nature of solutions of the associated *homogeneous system*

$$\mathbf{Y}' = \mathbf{A}(t)\mathbf{Y}. \tag{25}$$

The theory attached to this class of equations closely follows that presented in Chapter 1 for linear equations. In the interest of notational simplicity, however, we will develop the theory only for the case when $n = 2$. Generalizations to systems of order $n > 2$ follows in a direct manner for many results. To start, if $\mathbf{Y}^{(1)}(t)$ and $\mathbf{Y}^{(2)}(t)$ are solutions of (25), then by the superposition principle it follows that

$$\mathbf{Y}(t) = C_1 \mathbf{Y}^{(1)}(t) + C_2 \mathbf{Y}^{(2)}(t) = C_1 \begin{pmatrix} y_{11}(t) \\ y_{21}(t) \end{pmatrix} + C_2 \begin{pmatrix} y_{12}(t) \\ y_{22}(t) \end{pmatrix} \tag{26}$$

is also a solution for any choice of constants C_1 and C_2. Moreover, if $\mathbf{Y}^{(1)}(t)$ and $\mathbf{Y}^{(2)}(t)$ are not multiples of one another, they are considered *linearly independent* solutions and Eq. (26) represents the *general solution* of (25). The linearly independent solutions $\mathbf{Y}^{(1)}(t)$ and $\mathbf{Y}^{(2)}(t)$ are also called a *fundamental set of solution vectors*.

From the general solution (26), we can generate the matrix

$$\mathbf{\Phi}(t) = \begin{pmatrix} y_{11}(t) & y_{12}(t) \\ y_{21}(t) & y_{22}(t) \end{pmatrix}, \tag{27}$$

called the *fundamental matrix* of the system (25). Note that the columns of (27) are simply the solution vectors $\mathbf{Y}^{(1)}(t)$ and $\mathbf{Y}^{(2)}(t)$. The determinant of the fundamental matrix is called the *Wronskian*, defined here by

$$\det[\mathbf{\Phi}(t)] = W(t) = \begin{vmatrix} y_{11}(t) & y_{12}(t) \\ y_{21}(t) & y_{22}(t) \end{vmatrix}. \tag{28}$$

Because the solution vectors $\mathbf{Y}^{(1)}(t)$ and $\mathbf{Y}^{(2)}(t)$ are assumed to be linearly independent, it can be shown that the Wronskian (28) cannot vanish on the interval I. Hence, it follows that $\mathbf{\Phi}(t)$ is nonsingular and therefore has an inverse defined by

$$\mathbf{\Phi}(t)^{-1} = \frac{1}{\det[\mathbf{\Phi}(t)]} \begin{pmatrix} y_{22}(t) & -y_{12}(t) \\ -y_{21}(t) & y_{11}(t) \end{pmatrix}. \tag{29}$$

In terms of the fundamental matrix, the general solution of (25) can also be written as

$$
\mathbf{Y}(t) = \mathbf{\Phi}(t)\mathbf{C} = \begin{pmatrix} y_{11}(t) & y_{12}(t) \\ y_{21}(t) & y_{22}(t) \end{pmatrix} \begin{pmatrix} C_1 \\ C_2 \end{pmatrix}. \tag{30}
$$

If the solution (30) is subject to the *initial condition* $\mathbf{Y}(0) = \mathbf{Y}_0$, then $\mathbf{\Phi}(0)\mathbf{C} = \mathbf{Y}_0$ and solving for the constant vector $\mathbf{C}$ yields

$$
\mathbf{C} = \mathbf{\Phi}^{-1}(0)\mathbf{Y}_0. \tag{31}
$$

3.4.2 Homogeneous systems with constant coefficients

We now restrict our attention to the 2×2 linear system of equations

$$
\begin{aligned}
x' &= ax + by, \\
y' &= cx + dy,
\end{aligned} \qquad (ad - bc \neq 0) \tag{32}
$$

in which the coefficients a, b, c, d are *constants*. To understand the theory presented below in this section, we first solve the second equation for x, finding

$$
x = \frac{1}{c}y' - \frac{d}{c}y.
$$

Next, we substitute this expression into the first equation (and simplify the algebra) to obtain

$$
y'' - (a + d)y' + (ad - bc)y = 0. \tag{33}
$$

Hence, by this "elimination process" we have shown the equivalence of the linear system (32) with a second-order linear DE of the type discussed in Section 1.4.2. Consequently, the theory that we develop below is equivalent to our previous treatment of second-order DEs. In particular, seeking solutions of the form e^{mt} leads to the *auxiliary equation*

$$
m^2 - (a + d)m + (ad - bc) = 0. \tag{34}
$$

Formulated in matrix notation, we rewrite the system (32) as

$$
\mathbf{Y}' = \mathbf{A}\mathbf{Y}, \tag{35}
$$

in which the matrices $\mathbf{Y}$ and $\mathbf{A}$ are defined by

$$
\mathbf{Y} = \begin{pmatrix} x \\ y \end{pmatrix}, \qquad \mathbf{A} = \begin{pmatrix} a & b \\ c & d \end{pmatrix}. \tag{36}
$$

The advantage of the matrix formulation is that it is readily expanded to systems of n

equations and we can use the power of matrix operations for solution purposes.

By analogy with our solution treatment in Section 1.4.2 for constant-coefficient DEs, we seek solutions of (35) of the form

$$\mathbf{Y} = \mathbf{K}e^{\lambda t} = \begin{pmatrix} k_1 \\ k_2 \end{pmatrix} e^{\lambda t}, \tag{37}$$

where λ, k_1, k_2 are scalar constants to be determined. The substitution of (37) into (35) yields

$$\lambda \mathbf{K}e^{\lambda t} = \mathbf{A}\mathbf{K}e^{\lambda t},$$

which, after canceling the common factor $e^{\lambda t}$ and turning the equation around, becomes $\mathbf{A}\mathbf{K} = \lambda \mathbf{K}$, or

$$(\mathbf{A} - \lambda \mathbf{I})\mathbf{K} = \mathbf{0}. \tag{38}$$

We recognize (38) as the standard form of an eigenvalue problem for which the eigenvalues λ_1 and λ_2 of the matrix $\mathbf{A}$ are solutions of (see Section 3.3.1)

$$\det(\mathbf{A} - \lambda \mathbf{I}) = \Delta(\lambda) = 0, \tag{39}$$

where

$$\Delta(\lambda) = \lambda^2 - (a + d)\lambda + (ad - bc). \tag{40}$$

Thus, finding the eigenvalues of the matrix $\mathbf{A}$ is equivalent to solving the auxiliary equation (34) of the second-order DE (33). The corresponding eigenvectors of matrix $\mathbf{A}$ are therefore $\mathbf{K}^{(1)}$ and $\mathbf{K}^{(2)}$. To proceed, we must separately consider cases in which the eigenvalues are either *real and distinct*, *real and equal*, or *complex conjugates*.

| CASE I: | λ_1 and λ_2 real and distinct eigenvalues |

When the eigenvalues are real and distinct, there are two distinct eigenvectors $\mathbf{K}^{(1)}$ and $\mathbf{K}^{(2)}$ belonging, respectively, to λ_1 and λ_2. The general solution is therefore

$$\mathbf{Y} = C_1 \mathbf{K}^{(1)} e^{\lambda_1 t} + C_2 \mathbf{K}^{(2)} e^{\lambda_2 t}. \tag{41}$$

EXAMPLE 6 Use matrix methods to solve the system of equations

$$\begin{aligned} x' &= x + 2y, & x(0) &= 5 \\ y' &= 3x + 2y, & y(0) &= 0. \end{aligned}$$

Solution: In matrix notation, this system becomes

$$\mathbf{Y}' = \begin{pmatrix} 1 & 2 \\ 3 & 2 \end{pmatrix} \mathbf{Y}, \quad \mathbf{Y}(0) = \begin{pmatrix} 5 \\ 0 \end{pmatrix}.$$

The characteristic polynomial of the coefficient matrix is

$$\Delta(\lambda) = \begin{vmatrix} 1 - \lambda & 2 \\ 3 & 2 - \lambda \end{vmatrix} = \lambda^2 - 3\lambda - 4 = 0.$$

Hence, factoring the polynomial yields $\lambda^2 - 3\lambda - 4 = (\lambda - 4)(\lambda + 1)$, from which we deduce the eigenvalues

$$\lambda_1 = -1, \quad \lambda_2 = 4.$$

The eigenvector $\mathbf{K}^{(1)}$ corresponding to eigenvalue $\lambda_1 = -1$ is determined by solving the algebraic equation $2k_1 + 2k_2 = 0$ (recall Example 2) which leads to (among other choices)

$$\mathbf{K}^{(1)} = \begin{pmatrix} 1 \\ -1 \end{pmatrix}.$$

In the same way for eigenvalue $\lambda_2 = 4$ we solve $-3k_1 + 2k_2 = 0$, giving us

$$\mathbf{K}^{(2)} = \begin{pmatrix} 2 \\ 3 \end{pmatrix}.$$

Hence, the fundamental matrix is

$$\Phi(t) = \begin{pmatrix} e^{-t} & 2e^{4t} \\ -e^{-t} & 3e^{4t} \end{pmatrix}.$$

By setting $t = 0$ in the fundamental matrix, we obtain

$$\Phi(0) = \begin{pmatrix} 1 & 2 \\ -1 & 3 \end{pmatrix}, \quad \det[\Phi(0)] = W(0) = 5,$$

from which we calculate

$$\Phi^{-1}(0) = \frac{1}{5} \begin{pmatrix} 3 & -2 \\ 1 & 1 \end{pmatrix}.$$

The solution we seek is therefore

$$\mathbf{Y} = \begin{pmatrix} e^{-t} & 2e^{4t} \\ -e^{-t} & 3e^{4t} \end{pmatrix} \frac{1}{5} \begin{pmatrix} 3 & -2 \\ 1 & 1 \end{pmatrix} \begin{pmatrix} 5 \\ 0 \end{pmatrix}$$

$$= \begin{pmatrix} 3e^{-t} + 2e^{4t} \\ -3e^{-t} + 3e^{4t} \end{pmatrix} = \begin{pmatrix} 3 \\ -3 \end{pmatrix} e^{-t} + \begin{pmatrix} 2 \\ 3 \end{pmatrix} e^{4t}.$$

Last, in scalar form we can also express the solution as

$$x(t) = 3e^{-t} + 2e^{4t},$$
$$y(t) = -3e^{-t} + 3e^{4t}.$$

CASE II:	λ_1 and λ_2 are real and equal eigenvalues

If the two roots λ_1 and λ_2 of the characteristic equation are real and equal, then setting $\lambda_1 = \lambda_2 = \lambda$ produces only one solution vector, viz.,

$$\mathbf{Y}^{(1)}(t) = \mathbf{K}^{(1)}e^{\lambda t}.$$

By analogy with second-order DEs in Section 1.4.2, we seek a second solution vector in the form

$$\mathbf{Y}^{(2)}(t) = [\mathbf{K}^{(2)}t + \mathbf{L}^{(2)}]e^{\lambda t},$$

where $\mathbf{K}^{(2)}$ and $\mathbf{L}^{(2)}$ are vectors that must both be determined. The substitution of $\mathbf{Y}^{(2)}(t)$ into the matrix equation $\mathbf{Y}' = \mathbf{AY}$ yields the condition

$$[\lambda \mathbf{K}^{(2)}t + \lambda \mathbf{L}^{(2)} + \mathbf{K}^{(2)}]e^{\lambda t} = [\mathbf{AK}^{(2)}t + \mathbf{AK}^{(2)}]e^{\lambda t}.$$

By equating like terms in $e^{\lambda t}$ and $te^{\lambda t}$, we are led to the matrix equations

$$(\mathbf{A} - \lambda \mathbf{I})\mathbf{K}^{(2)} = \mathbf{0},$$
$$(\mathbf{A} - \lambda \mathbf{I})\mathbf{L}^{(2)} = \mathbf{K}^{(2)}. \tag{42}$$

The first of these last two equations shows that $\mathbf{K}^{(2)}$ is an eigenvector of $\mathbf{A}$. Hence, in many cases we simply set $\mathbf{K}^{(2)} = \mathbf{K}^{(1)}$. We can then solve the nonhomogeneous equation in (42) for the remaining vector $\mathbf{L}^{(2)}$. The general solution in this case is

$$\boxed{\mathbf{Y} = C_1 \mathbf{K}^{(1)}e^{\lambda t} + C_2[\mathbf{K}^{(2)}t + \mathbf{L}^{(2)}]e^{\lambda t}.} \tag{43}$$

EXAMPLE 7 Use matrix methods to solve the system of equations

$$x' = -4x - y,$$
$$y' = x - 2y.$$

Solution: The characteristic polynomial of the coefficient matrix is

$$\Delta(\lambda) = \begin{vmatrix} -(4 + \lambda) & -1 \\ 1 & -(2 + \lambda) \end{vmatrix} = \lambda^2 + 6\lambda + 9 = 0,$$

which yields the double root $\lambda_1 = \lambda_2 = -3$. The eigenvector $\mathbf{K}^{(1)}$ corresponding to eigenvalue $\lambda = -3$ is determined by solving the algebraic equation $-k_1 - k_2 = 0$, which leads to

$$\mathbf{K}^{(1)} = \begin{pmatrix} 1 \\ -1 \end{pmatrix}.$$

Hence, the first solution vector is

$$\mathbf{Y}^{(1)}(t) = \begin{pmatrix} 1 \\ -1 \end{pmatrix} e^{-3t}.$$

To obtain a second solution vector, we choose

$$\mathbf{K}^{(2)} = \mathbf{K}^{(1)} = \begin{pmatrix} 1 \\ -1 \end{pmatrix},$$

and determine $\mathbf{L}^{(2)}$ by solving [see (42)]

$$\begin{pmatrix} -(4+\lambda) & -1 \\ 1 & -(2+\lambda) \end{pmatrix} \begin{pmatrix} l_1 \\ l_2 \end{pmatrix} = \begin{pmatrix} 1 \\ -1 \end{pmatrix}.$$

Upon setting $\lambda = -3$, we see that $l_1 + l_2 = -1$; thus, choosing $l_1 = 1$ and $l_2 = -2$, we obtain the second solution

$$\mathbf{Y}^{(2)}(t) = \begin{pmatrix} 1 \\ -1 \end{pmatrix} te^{-3t} + \begin{pmatrix} 1 \\ -2 \end{pmatrix} e^{-3t} = \begin{pmatrix} t+1 \\ -(t+2) \end{pmatrix} e^{-3t}.$$

For arbitrary constants C_1 and C_2, we deduce that

$$\mathbf{Y} = C_1 \begin{pmatrix} 1 \\ -1 \end{pmatrix} e^{-3t} + C_2 \begin{pmatrix} t+1 \\ -(t+2) \end{pmatrix} e^{-3t},$$

which, in scalar form, becomes

$$x(t) = C_1 e^{-3t} + C_2(t+1)e^{-3t}$$
$$y(t) = -C_1 e^{-3t} - C_2(t+2)e^{-3t}.$$

CASE III:	λ_1 and λ_2 complex conjugate eigenvalues

When the characteristic roots of (40) are complex, they will appear as complex conjugates, i.e., $\lambda_1 = p + iq$ and $\lambda_2 = p - iq$, where p and q are real numbers. Following the general approach in CASE I with real distinct roots, it follows that the resulting solution vectors

$$\mathbf{Y}^{(1)}(t) = \mathbf{K}^{(1)} e^{(p+iq)t},$$

$$\mathbf{Y}^{(2)}(t) = \mathbf{K}^{(2)} e^{(p-iq)t}, \tag{44}$$

are linearly independent and thus can be used in the construction of a general solution. However, in most applications we desire *real* solutions, not complex solutions.

In general we expect $\mathbf{K}^{(1)}$ and $\mathbf{K}^{(2)}$ to be complex eigenvectors, viz.,

$$\mathbf{K}^{(1)} = \begin{pmatrix} a_1 \\ a_2 \end{pmatrix} + i \begin{pmatrix} b_1 \\ b_2 \end{pmatrix} = \mathbf{a} + i\mathbf{b},$$

and $\mathbf{K}^{(2)} = \mathbf{a} - i\mathbf{b}$. Therefore, we can express (44) in the form

$$
\begin{aligned}
\mathbf{Y}^{(1)}(t) &= (\mathbf{a} + i\mathbf{b})e^{(p+iq)t} = \mathbf{U}(t) + i\mathbf{V}(t), \\
\mathbf{Y}^{(2)}(t) &= (\mathbf{a} - i\mathbf{b})e^{(p-iq)t} = \mathbf{U}(t) - i\mathbf{V}(t),
\end{aligned}
\tag{45}
$$

where

$$
\begin{aligned}
\mathbf{U}(t) &= \frac{1}{2}\left[\mathbf{Y}^{(1)}(t) + \mathbf{Y}^{(2)}(t)\right] = e^{pt}(\mathbf{a}\cos qt - \mathbf{b}\sin qt), \\
\mathbf{V}(t) &= \frac{1}{2i}\left[\mathbf{Y}^{(1)}(t) - \mathbf{Y}^{(2)}(t)\right] = e^{pt}(\mathbf{a}\sin qt + \mathbf{b}\cos qt).
\end{aligned}
\tag{46}
$$

We conclude, therefore, that the general solution in this case is

$$\boxed{\mathbf{Y} = C_1\mathbf{U}(t) + C_2\mathbf{V}(t).} \tag{47}$$

EXAMPLE 8 Use matrix methods to solve the system of equations

$$
\begin{aligned}
x' &= 6x - y, \\
y' &= 5x + 4y.
\end{aligned}
$$

Solution: The characteristic polynomial of the coefficient matrix is

$$\Delta(\lambda) = \begin{vmatrix} 6 - \lambda & -1 \\ 5 & 4 - \lambda \end{vmatrix} = \lambda^2 - 10\lambda + 29 = 0,$$

the roots of which are $\lambda_1 = 5 + 2i$ and $\lambda_2 = 5 - 2i$. The eigenvector $\mathbf{K}^{(1)}$ corresponding to eigenvalue λ_1 is determined by solving

$$\begin{pmatrix} 6 - \lambda & -1 \\ 5 & 4 - \lambda \end{pmatrix}\begin{pmatrix} a_1 + ib_1 \\ a_2 + ib_2 \end{pmatrix} = \begin{pmatrix} 1 - 2i & -1 \\ 5 & -1 - 2i \end{pmatrix}\begin{pmatrix} a_1 + ib_1 \\ a_2 + ib_2 \end{pmatrix} = \begin{pmatrix} 0 \\ 0 \end{pmatrix},$$

from which we obtain (from the first equation)

$$(1 - 2i)(a_1 + ib_1) - (a_2 + ib_2) = 0.$$

Among other choices, we can set $a_1 + ib_1 = 1$ and $a_2 + ib_2 = 1 - 2i$, which requires that $a_1 = 1$, $b_1 = 0$, $a_2 = 1$, and $b_2 = -2$. Because the eigenvector $\mathbf{K}^{(2)}$ is the complex conjugate of $\mathbf{K}^{(1)}$, we deduce that

$$\mathbf{Y} = C_1 e^{5t}\left[\begin{pmatrix} 1 \\ 1 \end{pmatrix} \cos 2t - \begin{pmatrix} 0 \\ -2 \end{pmatrix} \sin 2t\right] + C_2 e^{5t}\left[\begin{pmatrix} 1 \\ 1 \end{pmatrix} \sin 2t + \begin{pmatrix} 0 \\ -2 \end{pmatrix} \cos 2t\right],$$

which in scalar form is

$$x(t) = e^{5t}(C_1 \cos 2t + C_2 \sin 2t),$$

$$y(t) = e^{5t}\left[(C_1 - 2C_2)\cos 2t + (2C_1 + C_2)\sin 2t\right].$$

3.4.3 Stability of linear systems

In this section we briefly discuss the notions of *critical point*, *stability*, and *phase plane* for the linear systems studied in the previous section. Curves in the phase plane known as trajectories provide much information about the general behavior of the system. Of course, this type of analysis is most crucial in dealing with *nonlinear systems*, but the theory for linear systems provides much insight into how to treat nonlinear systems.

The general linear system we discuss here is

$$\begin{aligned} x' &= ax + by, \\ y' &= cx + dy, \end{aligned} \qquad (ad - bc \neq 0) \tag{48}$$

in which the coefficients are constants. It is customary in practice to also specify a set of initial conditions (ICs) of the form

$$x(t_0) = \alpha, \quad y(t_0) = \beta, \tag{49}$$

where α and β are constants. In many cases, the value $t_0 = 0$. In the discussion here it is useful to think of the solution functions $x = x(t)$ and $y = y(t)$ of (48) and (49) as *parametric equations* of an arc in the xy-plane that passes through the point (α, β). From this point of view it is customary to refer to the xy-plane as the *phase plane*. Any arc described parametrically by a solution of (48) is called a *trajectory* (or *path* or *orbit*), with the positive direction along the trajectory defined by the direction of increasing t.

The trajectories can be found by eliminating the parameter t. For example, by taking the ratio of the two equations in (48), we obtain

$$\frac{y'}{x'} = \frac{dy}{dx} = \frac{cx + dy}{ax + by}. \tag{50}$$

The trajectories are then simply the solutions of (50).

▸ **Critical point:** *Critical points* (also called *equilibrium points*) are those points in the phase plane for which

$$x'(t) = y'(t) = 0.$$

For the linear system (48) the only critical point is the origin (0,0). In the case of nonlinear systems, however, there may be several such points. A critical point in mechanics in which $x(t)$ and $y(t)$ denote the position and velocity, respectively, of a particle in motion is a point where both velocity and acceleration vanish, and hence, the motion stops. Alternatively, the two functions $x(t)$ and $y(t)$ could represent the populations of two species of animals living in the same environment and competing for the same food supply (or one animal may prey on the other). In this latter case a critical point specifies the constant populations of these two competing species that can coexist in the same environment.

If the solution set $\{x(t), y(t)\}$ has a finite limit as t tends to infinity, that limit point must be a critical point. Because of the above interpretations of critical points, it is natural to ask what happens to a particle (or population) that is slightly displaced from a critical point. Specifically, we want to know if the particle (or population)

     1. Returns to the critical point
     2. Moves away from the critical point
     3. Moves about the critical point but does not approach it

Each of the above types of behavior leads to a different classification of stability.

▸ **Asymptotically stable:** We say the critical point is *asymptotically stable* if a particle slightly displaced from its critical point returns to the critical point.

▸ **Unstable:** We say a critical point is *unstable* if a displaced particle moves away from the critical point.

▸ **Stable:** We say a critical point is *stable* if a displaced particle neither moves toward or away from the critical point.

The above definitions of stability are for general systems—either linear or nonlinear. In the case of the linear system (48), however, we can describe the above notions of stability more precisely in terms of the roots of the characteristic equation $\Delta(\lambda) = 0$. Once again, we consider cases.

CASE I:	λ_1, λ_2 real, distinct, and both positive

In this case the two solution vectors are $\mathbf{Y}^{(1)} = \mathbf{K}^{(1)} e^{\lambda_1 t}$ and $\mathbf{Y}^{(2)} = \mathbf{K}^{(2)} e^{\lambda_2 t}$. Note that if both λ_1 and λ_2 are *positive*, then both solution vectors become infinite as $t \to \infty$. Hence, we say the critical point (0,0) is *unstable* (also called an unstable *node*).

CASE II:	λ_1, λ_2 real, distinct, and both negative

If both λ_1 or λ_2 are *negative*, then the solution vectors both approach the critical point (0,0). Hence, we then say the origin is an *asymptotically stable node*.

CASE III:	λ_1, λ_2 real, distinct, and opposite signs

In this case one solution approaches the origin whereas the other moves away from it. We call such a critical point an *unstable saddle point*.

> CASE IV: λ_1, λ_2 real and equal (i.e., $\lambda_1 = \lambda_2 = \lambda$)

Here, when $\lambda_1 = \lambda_2 = \lambda < 0$, the origin is *asymptotically stable*. On the other hand, when $\lambda_1 = \lambda_2 = \lambda > 0$, the origin is *unstable*. There are some subcases here, but we will not discuss them.

> CASE V: λ_1, λ_2 complex roots $p \pm iq$

For $\lambda_1 = p + iq$ and $\lambda_2 = p - iq$, we need only look at p to decide stability. That is, if $p > 0$, the origin is *unstable*, but if $p < 0$ the origin is *asymptotically stable*. If $p = 0$, i.e., the eigenvalues are pure imaginary, the origin is said to be *stable* but not asymptotically stable. The motion of a particle in this case is a periodic rotation about the origin. Hence, the origin is also called a *center*.

In Table 3.1 below we summarize the analysis of the stability of linear systems of the form given by Eqs. (48) near a critical point at the origin.

3.4.4 Nonhomogeneous systems

To complete this discussion we now briefly consider the *nonhomogeneous system* of equations

$$\mathbf{Y}' = \mathbf{A}(t)\mathbf{Y} + \mathbf{F}(t), \tag{51}$$

where $\mathbf{F}(t)$ is considered an external input to the system. The general solution of (51) can be expressed in the form

Table 3.1 Stability of Linear Systems

Eigenvalues	Stability
$0 < \lambda_1 < \lambda_2 \quad (0 < \lambda_2 < \lambda_1)$	unstable
$\lambda_1 < \lambda_2 < 0 \quad (\lambda_2 < \lambda_1 < 0)$	asymptotically stable
$\lambda_1 < 0 < \lambda_2 \quad (\lambda_2 < 0 < \lambda_1)$	unstable
$\lambda_1 = \lambda_2 > 0$	unstable
$\lambda_1 = \lambda_2 < 0$	asymptotically stable
$\lambda_{1,2} = p \pm iq, \, p > 0$	unstable
$\lambda_{1,2} = p \pm iq, \, p < 0$	asymptotically stable
$\lambda_{1,2} = \pm iq, \, p = 0$	stable

$$\mathbf{Y} = \mathbf{Y}_H + \mathbf{Y}_P$$
$$= C_1 \mathbf{Y}^{(1)}(t) + C_2 \mathbf{Y}^{(2)}(t) + \mathbf{Y}_P(t), \tag{52}$$

where $\mathbf{Y}_P$ is any *particular solution* of (51) and $\mathbf{Y}_H$ is the general solution of the associated homogeneous system.

Once we have determined $\mathbf{Y}_H$, we can construct $\mathbf{Y}_P$ by the method of *variation of parameters*. To start, we assume that

$$\mathbf{Y}_P = \mathbf{\Phi}(t)\mathbf{Z}(t), \tag{53}$$

where $\mathbf{\Phi}(t)$ is the fundamental matrix and $\mathbf{Z}(t)$ is an unknown vector function. The direct substitution of (53) into (51) leads to

$$\mathbf{\Phi}'(t)\mathbf{Z}(t) + \mathbf{\Phi}(t)\mathbf{Z}'(t) = \mathbf{A}(t)\mathbf{\Phi}(t)\mathbf{Z}(t) + \mathbf{F}(t).$$

However, $\mathbf{\Phi}'(t) = \mathbf{A}(t)\mathbf{\Phi}(t)$, and consequently,

$$\mathbf{\Phi}(t)\mathbf{Z}'(t) = \mathbf{F}(t)$$

or

$$\mathbf{Z}'(t) = \mathbf{\Phi}^{-1}(t)\mathbf{F}(t). \tag{54}$$

The integral of (54) gives us our result

$$\mathbf{Z}(t) = \int \mathbf{\Phi}^{-1}(t)\mathbf{F}(t)dt, \tag{55}$$

where the constant vector of integration has been conveniently set to zero. The general solution (52) can now be written as

$$\boxed{\mathbf{Y} = \mathbf{\Phi}(t)\mathbf{C} + \mathbf{\Phi}(t)\int \mathbf{\Phi}^{-1}(t)\mathbf{F}(t)dt,} \tag{56}$$

where $\mathbf{C}$ is an arbitrary constant vector. Lastly, for an initial condition of the form

$$\mathbf{Y}(0) = \mathbf{Y}_0, \tag{57}$$

the solution (56) becomes [e.g., see Eq. (29) in Section 1.3.3]

$$\mathbf{Y} = \mathbf{\Phi}(t)\mathbf{\Phi}^{-1}(0)\mathbf{Y}_0 + \mathbf{\Phi}(t)\int_0^t \mathbf{\Phi}^{-1}(\tau)\mathbf{F}(\tau)d\tau. \tag{58}$$

EXAMPLE 9 Find the general solution of the matrix equation

$$\mathbf{Y}' = \begin{pmatrix} -4 & 2 \\ 2 & -1 \end{pmatrix} \mathbf{Y} + \begin{pmatrix} 1/t \\ 4 + 2/t \end{pmatrix}.$$

Solution: Following the approach used in Example 7, we find the eigenvalues are given by $\lambda_1 = 0$ and $\lambda_2 = -5$. Corresponding to each eigenvalue are the eigenvectors

$$\mathbf{K}^{(1)} = \begin{pmatrix} 1 \\ 2 \end{pmatrix}, \qquad \mathbf{K}^{(2)} = \begin{pmatrix} 2 \\ -1 \end{pmatrix},$$

from which we deduce

$$\mathbf{Y}_H = C_1 \begin{pmatrix} 1 \\ 2 \end{pmatrix} + C_2 \begin{pmatrix} 2 \\ -1 \end{pmatrix} e^{-5t}.$$

The fundamental matrix and its inverse are given by

$$\mathbf{\Phi}(t) = \begin{pmatrix} 1 & 2e^{-5t} \\ 2 & -e^{-5t} \end{pmatrix}, \qquad \mathbf{\Phi}^{-1}(t) = \frac{1}{5} \begin{pmatrix} 1 & 2 \\ 2e^{5t} & -e^{5t} \end{pmatrix},$$

and thus, by using (53) and (55), we obtain

$$\mathbf{Y}_P = \begin{pmatrix} 1 & 2e^{-5t} \\ 2 & -e^{-5t} \end{pmatrix} \int \frac{1}{5} \begin{pmatrix} 1 & 2 \\ 2e^{5t} & -e^{5t} \end{pmatrix} \begin{pmatrix} 1/t \\ 4 + 2/t \end{pmatrix} dt$$

$$= \begin{pmatrix} \dfrac{8}{5}t + \ln t - \dfrac{8}{25} \\[2mm] \dfrac{16}{5}t + 2\ln t + \dfrac{4}{25} \end{pmatrix}.$$

Combining results, we have the general solution

$$\mathbf{Y} = C_1 \begin{pmatrix} 1 \\ 2 \end{pmatrix} + C_2 \begin{pmatrix} 2 \\ -1 \end{pmatrix} e^{-5t} + \begin{pmatrix} \dfrac{8}{5}t + \ln t - \dfrac{8}{25} \\[2mm] \dfrac{16}{5}t + 2\ln t + \dfrac{4}{25} \end{pmatrix}.$$

3.5 Linear Vector Spaces

Certain quantities, like force and velocity, may be mathematically represented by *vectors* (see Chapter 4). Other physical quantities like stress tensor and moments of inertia can be represented by arrays of numbers, e.g., *matrices*. Both vectors and matrices have many properties in common with each other and, in particular, in common with *functions*. For this reason we find that linear systems like matrix equations, differential equations, integral equations, and so on, have many properties in common. The analogy between vectors, matrices, and functions can be best pursued in terms of *inner products*, *orthogonality*, *series representations*, and so on. To do this it is instructive to focus upon their common properties from a unified point of view called *linear vector spaces*. Our treatment below will be mostly of an introductory nature, and will be limited to only real vector spaces, i.e., where the vectors and scalars are all real.

> **Definition 3.1** A set of vectors is called a *linear vector space V* if the following axioms are satisfied by all vectors **x**, **y**, **z**, and **0** in *V* and all real scalars *a* and *b*.
>
> *Addition Axioms:* If **x** and **y** are any pair of vectors in *V*, so is **x** + **y**. Furthermore,
> - $\mathbf{x} + \mathbf{y} = \mathbf{y} + \mathbf{x}$ (commutative law)
> - $(\mathbf{x} + \mathbf{y}) + \mathbf{z} = \mathbf{x} + (\mathbf{y} + \mathbf{z})$ (associative law)
> - $\mathbf{x} + \mathbf{0} = \mathbf{x}$
> - $\mathbf{x} + (-\mathbf{x}) = \mathbf{0}$
>
> *Scalar Multiplication Axioms:* If **x** is in *V* and *a* is any scalar, then *a***x** is in *V*. Also,
> - $a(\mathbf{x} + \mathbf{y}) = a\mathbf{x} + a\mathbf{y}$
> - $(a + b)\mathbf{x} = a\mathbf{x} + b\mathbf{x}$
> - $a(b\mathbf{x}) = (ab)\mathbf{x}$
> - $1\mathbf{x} = \mathbf{x},\ 0\mathbf{x} = \mathbf{0}$

We recognize that the axioms listed in Definition 3.1 are simply the elementary properties associated with three-dimensional vectors used in the study of analytic geometry. Hence, the space of three-dimensional vectors, denoted by R^3, is an example of a linear vector space. Also, the space where the vectors are *n-tuples*, i.e.,

$$\mathbf{x} = (x_1, x_2, \ldots, x_n), \tag{59}$$

is a vector space provided we define addition of **x** and $\mathbf{y} = (y_1, y_2, \ldots, y_n)$, and scalar multiplication, respectively, by

$$\mathbf{x} + \mathbf{y} = (x_1 + y_1, x_2 + y_2, \ldots, x_n + y_n),$$
$$a\mathbf{x} = (ax_1, ax_2, \ldots, ax_n).$$

It is customary to use the notation R^n to denote the linear vector space consisting of the totality of ordered *n*-tuples like (59).

3.5.1 Linear independence and basis vectors

Fundamental to linear vector spaces is the concept of a set of *basis vectors*. This is well known for vectors in R^3 (see Chapter 4) in which every vector **x** has the representation

$$\mathbf{x} = x_1\hat{\mathbf{i}} + x_2\hat{\mathbf{j}} + x_3\hat{\mathbf{k}}, \tag{60}$$

where $\hat{\mathbf{i}}$, $\hat{\mathbf{j}}$, $\hat{\mathbf{k}}$ are *unit base vectors* in rectangular coordinates. Representations of the form (60) are useful in that they enable us to establish properties of vectors **x** quite simply in terms of properties of the corresponding base vectors. Thus, we can study a particular vector space by focusing on its basis vectors.

Definition 3.2 Let $e_1, ..., e_n$ be a set of n vectors of a linear vector space V. If every vector $\mathbf{x}$ of V can be expressed by a linear combination

$$\mathbf{x} = c_1 \mathbf{e}_1 + c_2 \mathbf{e}_2 + \cdots + c_n \mathbf{e}_n$$

for some set of scalars $c_1, c_2, ..., c_n$, we say the vectors $\mathbf{e}_1, ..., \mathbf{e}_n$ form a *basis* for V.

The key concept in selecting a set of basis vectors $\mathbf{e}_1, ..., \mathbf{e}_n$ is that the set of vectors be linearly independent.

Definition 3.3 A set of n vectors $\mathbf{e}_1, ..., \mathbf{e}_n$ is said to be *linearly dependent* if there exists a set of scalars $c_1, c_2, ..., c_n$, not all zero, such that

$$c_1 \mathbf{e}_1 + c_2 \mathbf{e}_2 + \cdots + c_n \mathbf{e}_n = \mathbf{0};$$

otherwise, the set of vectors $\mathbf{e}_1, ..., \mathbf{e}_n$ is *linearly independent*.

Remark: The set of vectors $\mathbf{e}_1, ..., \mathbf{e}_n$ is linearly dependent if and only if at least one of them can be expressed as a linear combination of the remaining $n - 1$ vectors.

If a vector space V contains a linearly independent set of n vectors for every n (no matter how large), then V is an *infinite-dimensional space*. A vector space of dimension n is called a *finite dimensional space*. It follows that any set of $n + 1$ vectors in an n-dimensional space is linearly dependent—otherwise, the set could form a basis. It also follows that a set of basis vectors must be linearly independent.

3.5.2 Inner product spaces

Thus far we have discussed only the algebraic structure associated with linear vector spaces. In order to discuss geometric properties, such as distance and direction, we must introduce the notion of a "metric."

Definition 3.4 A set of vectors forms a *metric space* if, for all vectors $\mathbf{x}$, $\mathbf{y}$, $\mathbf{z}$, there is associated the real number d satisfying the properties

- ▸ $d(\mathbf{x},\mathbf{y}) \geq 0$
- ▸ $d(\mathbf{x},\mathbf{y}) = 0$ if and only if $\mathbf{x} = \mathbf{y}$
- ▸ $d(\mathbf{x},\mathbf{y}) = d(\mathbf{y},\mathbf{x})$
- ▸ $d(\mathbf{x},\mathbf{z}) \leq d(\mathbf{x},\mathbf{y}) + d(\mathbf{y},\mathbf{z})$

The "function" d in Definition 3.4 is called the distance function, or *metric*, of the linear vector space. This definition is based on the idea that the "distance between two points" must be a nonnegative real number that is zero only if the two points coincide, it must be independent of the order in which the points are considered, and finally, the triangle inequality (last property in Def. 3.4) must be satisfied.

Fundamental to much of the following development is the concept of an *inner product*, also called a *scalar product*, which is a generalization of the familiar "dot product" between three-dimensional vectors (Chapter 4).

Definition 3.5 An *inner product* is said to be defined on a real vector space V if, for each pair of vectors **x** and **y** in V, there is a real number $<\mathbf{x},\mathbf{y}>$ such that

▸ $<\mathbf{x},\mathbf{y}> = <\mathbf{y},\mathbf{x}>$
▸ $<a\mathbf{x},\mathbf{y}> = a<\mathbf{x},\mathbf{y}>$
▸ $<\mathbf{x},\mathbf{x}> \geq 0$
▸ $<\mathbf{x},\mathbf{x}> = 0$ if and only if $\mathbf{x} = \mathbf{0}$

A vector space in which an inner product is defined is called a *Euclidean space* (also called an *inner product space*). A finite dimensional complex vector space in which an inner product is defined is called a *unitary space*. For two (n-tuple) vectors $\mathbf{x} = (x_1, x_2, ..., x_n)$, $\mathbf{y} = (y_1, y_2, ..., y_n)$ in R^n, the Euclidean inner product is[2]

$$<\mathbf{x},\mathbf{y}> = \sum_{j=1}^{n} x_j y_j. \tag{61}$$

An important use of the inner product is to define the "norm" of a vector, which we can use as a metric in inner product spaces.

Definition 3.6 The *norm* of a vector **x** in an inner product space is defined by

$$\|\mathbf{x}\| = \sqrt{<\mathbf{x},\mathbf{x}>}.$$

A linear vector space featuring both an inner product and a norm is called a *normed inner product space*. Such a space is automatically a metric space with metric defined by $d(\mathbf{x},\mathbf{y}) = \|\mathbf{x} - \mathbf{y}\|$. When the norm of a vector is unity, i.e., $\|\mathbf{x}\| = 1$, we say the vector is a

---

[2]When considering *complex* vector spaces, we define the inner product by

$$<\mathbf{x},\mathbf{y}^*> = \sum_{j=1}^{n} x_j y_j^*,$$

where * denotes the complex conjugate.

unit vector, or is *normalized*.

The following two inequalities are of fundamental importance in theoretical developments involving inner product spaces.

► **Schwarz inequality:** If **x** and **y** are two vectors in an inner product space, then (equality holds if and only if **x** and **y** are linearly dependent)

$$|<\mathbf{x},\mathbf{y}>| \le \|\mathbf{x}\| \|\mathbf{y}\|. \tag{62}$$

► **Triangle inequality:** If **x** and **y** are two vectors in an inner product space, then

$$\|\mathbf{x} + \mathbf{y}\| \le \|\mathbf{x}\| + \|\mathbf{y}\|. \tag{63}$$

The Schwarz inequality (62) is recognized as a natural generalization of the dot product relation between two vectors **x** and **y** in R^3, viz., given $\mathbf{x} \cdot \mathbf{y} = \|\mathbf{x}\| \|\mathbf{y}\| \cos \theta$, where θ is the angle between the vectors, then $|\mathbf{x} \cdot \mathbf{y}| \le \|\mathbf{x}\| \|\mathbf{y}\|$. The triangle inequality (63) is a generalization of the statement that the length of one side of a triangle is less than or equal to the sum of lengths of the remaining sides.

3.5.3 Orthonormal basis and the expansion theorem

In three-dimensional vector space we say two vectors are orthogonal if neither vector is the zero vector and their dot product vanishes. Orthogonal vectors play an important role in simplifying vector calculations and they are also useful in theoretical developments.

We say a set of nonzero vectors $\mathbf{e}_1, ..., \mathbf{e}_n$ is an *orthogonal set of vectors* if all vectors in the set satisfy

$$<\mathbf{e}_j, \mathbf{e}_k> = 0, \quad j \ne k. \tag{64}$$

If it is also true that $<\mathbf{e}_j, \mathbf{e}_j> = 1$ for each j, the set is then called *orthonormal*.

It is easy to show that every orthogonal set of n vectors in an n-dimensional Euclidean vector space is linearly independent. Moreover, every orthogonal set of n vectors can be a basis for the vector space. As a consequence, we conclude this section with the following far-reaching *expansion theorem*.

Theorem 3.2 If $\mathbf{e}_1, ..., \mathbf{e}_n$ forms an orthonormal basis for the Euclidean vector space V, and if **x** is any vector in V that has the representation

$$\mathbf{x} = \sum_{j=1}^{n} x_j \mathbf{e}_j,$$

then the numbers (components) x_j are uniquely determined by

$$x_j = <\mathbf{x}, \mathbf{e}_j>, \quad j = 1, 2, ..., n.$$

Proof: Let us form the inner product of each side of the given expansion with the unit vector $\mathbf{e}_k$ to obtain

$$<\mathbf{x},\mathbf{e}_k> = \sum_{j=1}^{n} x_j <\mathbf{e}_j,\mathbf{e}_k> = x_k, \quad k = 1,2,...,n,$$

the last step of which follows from the orthonormal properties of the base vectors. ❏

3.5.4 Hilbert spaces

In this section we discuss a class of vector spaces that is particularly useful in the theoretical analysis of signals $f(t)$ (e.g., a current, a voltage, an optical wave, and so on). Such a class of spaces, known as *Hilbert spaces*, is named in honor of David Hilbert, who initiated their theory in his famous work on integral equations in 1912.

A vector space, which is not necessarily of finite dimension but on which an inner product is defined, is called a *pre-Hilbert space*. If a certain "completeness" condition is also satisfied, we obtain a *Hilbert space*. A Hilbert space is ordinarily defined as "a complete metric space with an inner product." An example of a complete metric space is the real line R^1. Omitting one point from the line makes it incomplete. Another example of Hilbert space is the set of all real (or possibly complex) vectors $\mathbf{x} = (x_1,x_2,...,x_n,...)$, where

$$\sum_{j=1}^{\infty} |x_j|^2 < \infty, \tag{65}$$

the sum $\mathbf{x} + \mathbf{y}$ is defined as $(x_1+y_1,x_2+y_2,...,x_n+y_n,...)$, the product $a\mathbf{x}$ is defined as $(ax_1,ax_2,...,ax_n,...)$, and the inner product is defined as

$$<\mathbf{x},\mathbf{y}> = \sum_{j=1}^{\infty} x_j y_j. \tag{66}$$

An energy signal is a function for which the total energy is finite (see Chapters 8 and 9). To ensure finite energy of a given signal, we must place a certain "growth" restriction on the signal function. Doing so leads to the following definition.

Definition 3.7 For an interval $a \le t \le b$, the space $L^2([a,b])$ is the set of all *square-integrable functions* $f(t)$ defined on $a \le t \le b$, i.e.,

$$\int_a^b |f(t)|^2 dt < \infty.$$

For finite a and b it is clear that continuous functions are square-integrable. However, in dealing with energy signals we are often interested in the extended interval $(-\infty,\infty)$. In this case we find that not all continuous functions satisfy the condition of being square-integrable (i.e., have finite energy). On the other hand, certain functions with a finite number of discontinuities can be square-integrable on $(-\infty,\infty)$. Also, the space of L^2 functions implied by Definition 3.7 allows for functions that cannot be integrated by the conventional Riemann integral— in such cases a more general integral (called a Lebesgue integral) is used (see [12]). Nonetheless, we will restrict the permissible class of functions in our analysis to only those for which the integral in Definition 3.7 is the standard Riemann integral from calculus.

▸ **Inner product on L^2:** The *inner product* on the space of $L^2([a,b])$ functions is defined by

$$<f(t),g(t)> = \int_a^b f(t)g(t)dt. \tag{67}$$

If the inner product (67) of two (nonzero) functions $f(t)$ and $g(t)$ is zero, we say the functions are *orthogonal*.

▸ **Norm on L^2:** The *norm* on the space of $L^2([a,b])$ functions is defined by the nonnegative quantity

$$\|f(t)\| = \sqrt{<f(t),f(t)>} = \sqrt{\int_a^b |f(t)|^2 \, dt}. \tag{68}$$

EXAMPLE 10 Show that the functions $f(t) = \sin t$ and $g(t) = \cos t$ are orthogonal on the interval $[-\pi,\pi]$ and find the norm of each.

Solution: From Eq. (67), we have

$$<f(t),g(t)> = \int_{-\pi}^{\pi} \sin t \cos t \, dt = \frac{1}{2}\sin^2 t \Big|_{-\pi}^{\pi} = 0,$$

which establishes the orthogonality of the functions. Also, their respective norms are

$$\|f(t)\| = \sqrt{\int_{\pi}^{\pi} \sin^2 t \, dt} = \sqrt{\pi}, \quad \|g(t)\| = \sqrt{\int_{\pi}^{\pi} \cos^2 t \, dt} = \sqrt{\pi}.$$

Beyond what we have presented here, the general theory of Hilbert spaces takes us too far afield for our purposes (see [12] for a more extensive treatment). However, we do want to point out that much of the theory developed for Euclidean inner product spaces can be readily extended to Hilbert spaces. In fact, in many of the following chapters the reader may recognize various specializations of the notions of linear vector space, inner product, and orthogonal bases as applied to Hilbert spaces. In particular, the Fourier series analysis that we consider in detail in Chapter 8 is an extension of Theorem 3.2 to the infinite dimensional vector space consisting of all piecewise continuous functions with piecewise continuous derivatives. The eigenfunctions associated with various Sturm-Liouville systems (Section 8.4) will form the orthogonal basis used for such expansions. Thus, the basis vectors may be sets of simple sines and/or cosines, or may involve higher level functions such as orthogonal polynomial sets (Legendre, Hermite, Laguerre, etc.) or even sets of Bessel functions.

Suggested Reading

R. Bellman, *Introduction to Matrix Analysis*, 2nd ed. (McGraw-Hill, New York, 1970).

R. Courant and D. Hilbert, *Methods of Mathematical Physics*, Vol I (Wiley-Interscience, New York, 1953).

D. L. Kreider, R. G. Kuller, D. R. Ostberg, and F. W. Perkins, *An Introduction to Linear Analysis* (Addison-Wesley, Reading, 1966).

J. Schwartz, *Introduction to Matrices and Vectors* (McGraw-Hill, New York, 1961).

Chapter 3

Exercises

In Problems 1 and 2, use the given matrices to calculate (*a*) $\mathbf{A} + \mathbf{B}$ and (*b*) $3\mathbf{B} - 2\mathbf{A}$.

1. $\mathbf{A} = \begin{pmatrix} 3 & 4 \\ 5 & 6 \end{pmatrix}$, $\mathbf{B} = \begin{pmatrix} -1 & 2 \\ 0 & 3 \end{pmatrix}$

Ans. (*a*) $\begin{pmatrix} 2 & 6 \\ 5 & 9 \end{pmatrix}$, (*b*) $\begin{pmatrix} -9 & -2 \\ -10 & -3 \end{pmatrix}$

2. $\mathbf{A} = \begin{pmatrix} -2 & 0 \\ 4 & 1 \\ 7 & 3 \end{pmatrix}$, $\mathbf{B} = \begin{pmatrix} 3 & -1 \\ 0 & 2 \\ -4 & -2 \end{pmatrix}$

Ans. (*a*) $\begin{pmatrix} 1 & -1 \\ 4 & 3 \\ 3 & 1 \end{pmatrix}$, (*b*) $\begin{pmatrix} 13 & -3 \\ -8 & 4 \\ -26 & -12 \end{pmatrix}$

3. Calculate both $\mathbf{AB}$ and $\mathbf{BA}$, where

$$\mathbf{A} = \begin{pmatrix} 2 & 1 \\ 3 & 4 \end{pmatrix}, \quad \mathbf{B} = \begin{pmatrix} 1 & -2 \\ 5 & 3 \end{pmatrix}.$$

Ans. $\mathbf{AB} = \begin{pmatrix} 7 & -1 \\ 23 & 6 \end{pmatrix}$, $\mathbf{BA} = \begin{pmatrix} -4 & -7 \\ 19 & 17 \end{pmatrix}$

In Problems 4 and 5, calculate the inverse matrix $\mathbf{A}^{-1}$.

4. $\mathbf{A} = \begin{pmatrix} -4 & -8 \\ 3 & 5 \end{pmatrix}$ *Ans.* $\begin{pmatrix} 5/4 & 2 \\ -3/4 & -1 \end{pmatrix}$

5. $\mathbf{A} = \begin{pmatrix} 2 & 1 & 0 \\ -2 & 1 & 1 \\ 3 & 0 & 1 \end{pmatrix}$

Ans. $\dfrac{1}{7} \begin{pmatrix} 1 & -1 & 1 \\ 5 & 2 & 2 \\ -3 & 3 & 4 \end{pmatrix}$

6. Show that $\mathbf{A}(t)$ is nonsingular for all t and find its inverse $\mathbf{A}^{-1}(t)$, where

$$\mathbf{A}(t) = \begin{pmatrix} e^{-t} & 2e^{4t} \\ -e^{-t} & 3e^{4t} \end{pmatrix}.$$

Ans. $\dfrac{1}{5} \begin{pmatrix} 3e^{t} & -2e^{t} \\ -e^{-4t} & e^{-4t} \end{pmatrix}$

In Problems 7-10, find the eigenvalues and eigenvectors of the given matrix.

7. A = $\begin{pmatrix} -5 & 2 \\ 2 & -2 \end{pmatrix}$

Ans. $\lambda_1 = -1$, $\mathbf{e}_1 = \begin{pmatrix} 1 \\ 2 \end{pmatrix}$;

$\lambda_2 = -6$, $\mathbf{e}_2 = \begin{pmatrix} 2 \\ -1 \end{pmatrix}$

8. A = $\begin{pmatrix} 0 & 1 \\ -1 & 0 \end{pmatrix}$

Ans. $\lambda_1 = i$, $\mathbf{e}_1 = \begin{pmatrix} 1 \\ i \end{pmatrix}$;

$\lambda_2 = -i$, $\mathbf{e}_2 = \begin{pmatrix} 1 \\ -i \end{pmatrix}$

9. A = $\begin{pmatrix} 1 & 0 & 1 \\ 1 & 1 & 0 \\ 0 & 0 & 1 \end{pmatrix}$

Ans. $\lambda_1 = \lambda_2 = \lambda_3 = 1$, $\mathbf{e} = \begin{pmatrix} 0 \\ 1 \\ 0 \end{pmatrix}$

10. A = $\begin{pmatrix} 6 & 10 & 6 \\ 0 & 8 & 12 \\ 0 & 0 & 2 \end{pmatrix}$

Ans. $\lambda_1 = 2$, $\mathbf{e}_1 = \begin{pmatrix} 7 \\ -4 \\ 2 \end{pmatrix}$;

$\lambda_2 = 6$, $\mathbf{e}_2 = \begin{pmatrix} 1 \\ 0 \\ 0 \end{pmatrix}$; $\lambda_3 = 8$, $\mathbf{e}_3 = \begin{pmatrix} 5 \\ 1 \\ 0 \end{pmatrix}$

In Problems 11-14, solve the given homogeneous system of equations.

11. $x' = 2x - y$
$y' = 3x - 2y$

Ans. $\mathbf{Y} = C_1 \begin{pmatrix} 1 \\ 1 \end{pmatrix} e^t + C_2 \begin{pmatrix} 1 \\ 3 \end{pmatrix} e^{-t}$

12. $x' = 4x - 3y$
$y' = 8x - 6y$

Ans. $\mathbf{Y} = C_1 \begin{pmatrix} 1 \\ 4/3 \end{pmatrix} + C_2 \begin{pmatrix} 1 \\ 2 \end{pmatrix} e^{-2t}$

13. $x' = x - 4y$, $x(0) = 1$
$y' = x + y$, $y(0) = 0$

Ans. $\mathbf{Y} = \begin{pmatrix} 1 \\ 0 \end{pmatrix} e^t \cos 2t + \begin{pmatrix} 0 \\ 1/2 \end{pmatrix} e^t \sin 2t$

14. Y′ = $\begin{pmatrix} 2 & -5 \\ 4 & -2 \end{pmatrix}$ **Y**, **Y**(0) = $\begin{pmatrix} 2 \\ 3 \end{pmatrix}$

Ans. $\mathbf{Y} = \dfrac{2}{5} \begin{pmatrix} \cos 4t \\ 2\cos 4t + 4\sin 4t \end{pmatrix}$

$\qquad - \dfrac{11}{20} \begin{pmatrix} 5\sin 4t \\ 2\sin 4t - 4\cos 4t \end{pmatrix}$

In Problems 15-18, use matrix methods to solve the given nonhomogeneous system of equations.

15. $x' = 2x + y + e^t$
$y' = 4x - y - e^t$

Ans.
$\mathbf{Y} = C_1 \begin{pmatrix} 1 \\ -4 \end{pmatrix} e^{-2t} + C_2 \begin{pmatrix} 1 \\ 1 \end{pmatrix} e^{3t} - \dfrac{1}{6} \begin{pmatrix} 1 \\ 5 \end{pmatrix} e^t$

16. $x' = 4x - 2y + e^t$
    $y' = 5x + 2y - t$

Ans. $\mathbf{Y} = C_1 \begin{pmatrix} 2\cos 3t \\ \cos 3t + 3\sin 3t \end{pmatrix} e^{3t}$

$$+ C_2 \begin{pmatrix} 2\sin 3t \\ \sin 3t - 3\cos 3t \end{pmatrix} e^{3t}$$

$$+ \frac{1}{9} \begin{pmatrix} t + 1/3 \\ 2t + 1/5 \end{pmatrix} + \begin{pmatrix} -1/3 \\ 5/13 \end{pmatrix} e^t$$

17. $\mathbf{Y}' = \begin{pmatrix} -10 & 6 \\ -12 & 7 \end{pmatrix} \mathbf{Y} + \begin{pmatrix} 10e^{-3t} \\ 18e^{-3t} \end{pmatrix}$,

$$\mathbf{Y}(0) = \begin{pmatrix} -1 \\ 2 \end{pmatrix}$$

Ans.

$$\mathbf{Y} = 17 \begin{pmatrix} 2 \\ 3 \end{pmatrix} e^{-t} - 13 \begin{pmatrix} 3 \\ 4 \end{pmatrix} e^{-2t} + \begin{pmatrix} 4 \\ 3 \end{pmatrix} e^{-3t}$$

18. $\mathbf{Y}' = \begin{pmatrix} 4 & 2 \\ 3 & -1 \end{pmatrix} \mathbf{Y} - \begin{pmatrix} 15 \\ 4 \end{pmatrix} te^{-2t}$,

$$\mathbf{Y}(0) = \begin{pmatrix} 1 \\ -1 \end{pmatrix}$$

Ans. $\mathbf{Y} = \frac{1}{7} \begin{pmatrix} 2 \\ 1 \end{pmatrix} e^{5t} - \frac{2}{7} \begin{pmatrix} -1 \\ 3 \end{pmatrix} e^{-2t}$

$$+ \frac{1}{14} \begin{pmatrix} 4 + 28t - 7t^2 \\ 2 + 14t + 21t^2 \end{pmatrix} e^{-2t}$$

In Problems 19-22, discuss the stability of the critical point (0,0) of the given system.

19. $x' = -x$
    $y' = -2y$

   Ans. Asymptotically stable

20. $x' = 4x - y$
    $y' = 6x - 3y$

   Ans. Saddle point

21. $x' = -4x - y$
    $y' = x - 2y$

   Ans. Asymptotically stable

22. $x' = x - 3y$
    $y' = x - y$

   Ans. Center

Chapter 4

Vector Analysis

4.1 Introduction / 145
4.2 Cartesian Coordinates / 146
    4.2.1   Base vectors / 148
    4.2.2   Products of vectors / 148
    4.2.3   Vector identities / 152
    4.2.4   Applications / 153
4.3 Tensor Notation / 156
    4.3.1   Einstein summation convention / 156
    4.3.2   Kronecker delta and permutation symbols: δ_{jk}, e_{ijk} / 157
    4.3.3   Products of vectors and identities / 159
4.4 Vector Functions of One Variable / 161
    4.4.1   Space curves / 162
    4.4.2   Frenet-Serret formulas / 164
    4.4.3   Velocity and acceleration / 166
    4.4.4   Planar motion in polar coordinates / 168
4.5 Scalar and Vector Fields / 170
    4.5.1   Gradient: ∇f / 171
    4.5.2   Divergence: $\nabla \cdot \mathbf{F}$ / 173
    4.5.3   Physical interpretation of divergence / 174
    4.5.4   Curl: $\nabla \times \mathbf{F}$ / 176
    4.5.5   Vector differential operators: tensor notation / 177
4.6 Line and Surface Integrals / 179
    4.6.1   Line integrals / 179
    4.6.2   Conservative fields / 183
    4.6.3   Surface integrals / 185
4.7 Integral Relations Between Line, Surface and Volume Integrals / 194
    4.7.1   Green's theorem in the plane / 194
    4.7.2   Harmonic functions / 197
    4.7.3   Divergence theorem and Stokes' theorem / 200
4.8 Electromagnetic Theory / 206
    4.8.1   Maxwell's equations / 206
    4.8.2   Poisson's equation / 209
    4.8.3   Electromagnetic wave equation / 209
    Suggested Reading / 210
    Exercises / 211

Historical Comments: The word "vector" comes from a Latin word meaning "to carry," but actually entered mathematics through astronomy (where it had a different meaning). The word vector and the idea that a force is a vector quantity was known to Aristotle. Also, Galileo Galilei (1564-1642) explicitly stated the parallelogram law for the vector addition of forces. Nonetheless, vector algebra and vector analysis are both considered to be a product of the nineteenth century.

It was known in 1830 that complex numbers could be used to represent vectors in a plane. The search for a "three-dimensional complex number" eventually led to the invention of the "quaternion" in 1843 by William Rowan Hamilton (1805-1865). A quaternion is something of a cross between a complex number and a vector. For example, if the real part of a quaternion is zero, the part that is left can be identified with an ordinary vector. Hamilton felt that quaternions were well suited for solving problems in physics—in fact, strong efforts were made by several mathematicians over half a century to introduce quaternions into physics. Eventually, the quaternion concept was considered unacceptable for applications.

The next significant development in vector analysis took place when the idea of "curl" appeared in 1873 in the revolutionary work of James C. Maxwell (1831-79) on electromagnetic waves. Shortly afterward, W. L. Clifford (1845-79) coined the term "divergence." The notation that is commonly used today in vector analysis is due primarily to the work of J. Willard Gibbs (1839-1903) in 1893, but also to some extent the work of Oliver Heaviside (1850-1925) on electromagnetic theory. The two basic integral theorems—that of Stokes and the divergence theorem—were actually developed before Gibbs. In 1831, for example, the Russian M. Ostogradsky (1801-61) converted a volume integral into a surface integral (equivalent to the divergence theorem). The divergence theorem can also be found in the work of Carl F. Gauss (1777-1855) and in that of the English mathematician George Green (1793-1841). Oddly enough, the theorem attributed to George G. Stokes (1819-1903) was used as a question in a prize examination at Cambridge in 1854.

The words "scalar," "vector," and "tensor" were all used by Hamilton. However, tensor calculus was perfected in 1889 and 1926-27, respectively, by the Italian mathematicians G. Ricci (1853-1925) and T. Levi-Civita (1873-1941). Albert Einstein (1879-1955) used the term "tensor" only in connection with the transformation laws, but he was first in applying the generalized calculus of tensors to problems in gravitation.

> The objective of this chapter is to provide a review of the fundamental vector operations that are essential to application areas like electromagnetic fields. Our discussion involves the treatment of vector calculus in addition to the standard algebraic approach. Thus, we will introduce important theorems like the divergence theorem and Stokes' theorem that link line, surface, and volume integrals.

4.1 Introduction

In geometry, physics, and engineering applications it is useful to distinguish between two kinds of quantities called scalars and vectors.

- **Scalar:** Quantities completely identified by only a *magnitude* (number) are *scalars*; examples of scalar are distance, mass, voltage, and temperature.

- **Vector:** Quantities that require both a *magnitude* and a *direction* for their identification are *vectors*; force, velocity, and electric field are typical examples of vectors.

It is common to define a "vector" as a collection of equivalent directed line segments (all with the same magnitude and direction)—therefore, vectors have a "tail," called the *initial point*, and a "tip," called the *terminal point*. We will denote vectors by a bold letter like **a** and scalars by simply *a*.

- **Norm:** The *length* or *magnitude* $\|\mathbf{a}\|$ of a vector **a** is also called the *norm*.

> **Remark:** Because parallel vectors with the same norm and sense (direction) are considered to be the same vector, we generally take the standard position of a vector to be with its initial point at the origin (see Fig. 4.1).

Vectors and vector functions are important in almost every facet of applied mathematics. In particular, they are routinely used in mechanics, quantum mechanics, hydrodynamics, optics, etc. Motivated by applications involving forces, early users of vectors wanted to develop an algebra and calculus of vectors and vector functions that permitted them to calculate in much the same fashion as they could with scalars and scalar functions. To begin our treatment, we first introduce two basic kinds of vectors.

- **Zero vector:** The vector of zero norm and arbitrary direction is the *zero vector* **0**.

- **Unit vector:** Any vector of unit norm is called a *unit vector*. A unit vector parallel to **a** in our work will be identified by the symbol **â**.

4.2 Cartesian Coordinates

Operations with vectors are greatly facilitated by the introduction of *components* relative to a coordinate system. Of particular importance is the *rectangular* or *Cartesian coordinate system* with mutually perpendicular axes designated by x, y, and z. If P(x,y,z) is any point in space, then the directed line segment from the origin 0 to P can be used to define a vector **a** (see Fig. 4.1). In particular, we write

$$\mathbf{a} = \overline{OP} = <x,y,z>. \tag{1}$$

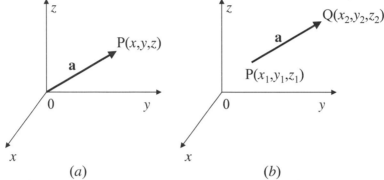

Figure 4.1 Rectangular coordinate system with (*a*) vector **a** in standard position, (*b*) vector **a** defined by line segment from P to Q.

In this case the coordinates (x,y,z) uniquely define the vector **a** in Fig. 4.1(*a*) and are referred to as the *components* of **a**, also written $\mathbf{a} = <a_1, a_2, a_3>$. That is, we identify the components as $a_1 = x$, $a_2 = y$, $a_3 = z$. The angle brackets $<\ >$ are commonly used, although not universally, to distinguish vector components from coordinates. If we define the vector **a** by the directed line segment between P(x_1,y_1,z_1) and Q(x_2,y_2,z_2), as shown in Fig. 4.1(*b*), the components of $\mathbf{a} = \overline{PQ} = <a_1, a_2, a_3>$ are then defined by

$$a_1 = x_2 - x_1, \quad a_2 = y_2 - y_1, \quad a_3 = z_2 - z_1. \tag{2}$$

The *norm* of vector **a** is simply the distance between P and Q, which leads to

$$\|\mathbf{a}\| = \sqrt{(x_2 - x_1)^2 + (y_2 - y_1)^2 + (z_2 - z_1)^2} = \sqrt{a_1^2 + a_2^2 + a_3^2}. \tag{3}$$

▸ **Equality:** We say two vectors $\mathbf{a} = <a_1,a_2,a_3>$ and $\mathbf{b} = <b_1,b_2,b_3>$ are *equal* if and only if their respective components are equal, i.e.,

$$\mathbf{a} = \mathbf{b} \quad \text{if and only if} \quad a_1 = b_1, \ a_2 = b_2, \ a_3 = b_3.$$

▸ **Addition:** The *addition* of two vectors **a** + **b** satisfies the *parallelogram law* (see Fig. 4.2).

This definition of addition is motivated by the sum of two displacements (or forces) **a** and **b** as illustrated in Fig. 4.2. In terms of components, vector addition is defined by

$$\mathbf{a} + \mathbf{b} = <a_1 + b_1, a_2 + b_2, a_3 + b_3>, \qquad (4)$$

and in general, vector addition has the following properties:

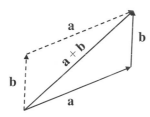

Figure 4.2 Parallelogram law for vector addition.

1. $\mathbf{a} + \mathbf{b} = \mathbf{b} + \mathbf{a}$

2. $\mathbf{a} + (\mathbf{b} + \mathbf{c}) = (\mathbf{a} + \mathbf{b}) + \mathbf{c}$

3. $K(\mathbf{a} + \mathbf{b}) = K\mathbf{a} + K\mathbf{b}$

4. $\mathbf{a} + \mathbf{0} = \mathbf{a}$

5. $\mathbf{a} + (-\mathbf{a}) = \mathbf{0}$

6. $\mathbf{a} + \mathbf{a} = 2\mathbf{a}$

Note that Properties 3 and 6 listed above for vector addition involve the notion of multiplication by a scalar. The multiplication of a vector $\mathbf{a} = <a_1, a_2, a_3>$ by a scalar K is defined by

$$K\mathbf{a} = <Ka_1, Ka_2, Ka_3>, \qquad (5)$$

obtained by multiplying each component of **a** by the scalar K. Geometrically, if **a** is a nonzero vector and $K > 0$, then $K\mathbf{a}$ has the same direction as **a**. If $K < 0$, then $K\mathbf{a}$ has the opposite direction of **a**. In particular, the vector $-\mathbf{a}$ has the same norm as **a** but opposite sense or direction.

▸ **Direction angles:** The direction of a vector **a** can be specified by the three angles α, β, γ called *direction angles* (see Fig. 4.3). They are defined by the components of **a** indirectly through the relations

$$\cos \alpha = \frac{a_1}{\|\mathbf{a}\|}, \quad \cos \beta = \frac{a_2}{\|\mathbf{a}\|}, \quad \cos \gamma = \frac{a_3}{\|\mathbf{a}\|}. \qquad (6)$$

The cosines appearing in (6) are a special set of direction numbers called the *direction cosines*. From definition, it is clear that the direction cosines satisfy the identity

$$\cos^2 \alpha + \cos^2 \beta + \cos^2 \gamma = 1. \qquad (7)$$

4.2.1 Base vectors

Another, but equivalent, representation of vector is by the use of *unit base vectors* $\hat{\mathbf{i}}, \hat{\mathbf{j}}, \hat{\mathbf{k}}$ defined along the positive x, y, and z axes, respectively (see Fig. 4.4). In terms of these unit vectors, we can use the definition of vector addition to deduce that

$$\mathbf{a} = a_1\hat{\mathbf{i}} + a_2\hat{\mathbf{j}} + a_3\hat{\mathbf{k}}, \tag{8}$$

where a_1, a_2, a_3 are the same components as defined above. That is, the representation (8) and the representation $\mathbf{a} = <a_1, a_2, a_3>$ can be used interchangeably, depending upon which is more convenient at the time.

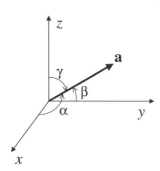

Figure 4.3 Direction angles.

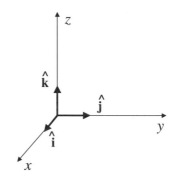

Figure 4.4 Rectangular coordinate system with base vectors.

4.2.2 Products of vectors

Multiplication between vectors can take on two basic forms—one called the scalar product and the other called the vector product. Each definition was originally motivated by its utility in solving certain problems in mechanics.

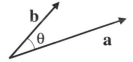

Figure 4.5

▸ **Scalar product:** The *scalar product* or *dot product* between two vectors $\mathbf{a}$ and $\mathbf{b}$ is a scalar quantity defined by (see Fig. 4.5)

$$\mathbf{a} \cdot \mathbf{b} = \|\mathbf{a}\| \|\mathbf{b}\| \cos\theta, \tag{9}$$

where θ is the angle between the vectors.

▸ **Orthogonal:** We say the vectors $\mathbf{a}$ and $\mathbf{b}$ are *orthogonal* in the special case when $\theta = \pm\pi/2$, and then $\mathbf{a} \cdot \mathbf{b} = 0$.

The scalar product has the following properties:

1. $\mathbf{a} \cdot \mathbf{b} = \mathbf{b} \cdot \mathbf{a}$

2. $\mathbf{a} \cdot K\mathbf{b} = K\mathbf{a} \cdot \mathbf{b} = K(\mathbf{a} \cdot \mathbf{b})$

3. $\mathbf{a} \cdot (\mathbf{b} + \mathbf{c}) = \mathbf{a} \cdot \mathbf{b} + \mathbf{a} \cdot \mathbf{c}$

4. $\mathbf{a} \cdot \mathbf{a} = \|\mathbf{a}\|^2$

5. $\mathbf{a} \cdot \mathbf{b} = 0$ if and only if $\mathbf{a}$ is orthogonal to $\mathbf{b}$ $(\mathbf{a} \neq \mathbf{0}, \mathbf{b} \neq \mathbf{0})$

6. $\mathbf{a} \cdot \mathbf{0} = 0$

From definition, it also follows that

$$\hat{\mathbf{i}} \cdot \hat{\mathbf{i}} = \hat{\mathbf{j}} \cdot \hat{\mathbf{j}} = \hat{\mathbf{k}} \cdot \hat{\mathbf{k}} = 1,$$

$$\hat{\mathbf{i}} \cdot \hat{\mathbf{j}} = \hat{\mathbf{j}} \cdot \hat{\mathbf{k}} = \hat{\mathbf{i}} \cdot \hat{\mathbf{k}} = 0. \tag{10}$$

By using the relations in (10), we can readily show that the component form of the dot product is

$$\mathbf{a} \cdot \mathbf{b} = (a_1 \hat{\mathbf{i}} + a_2 \hat{\mathbf{j}} + a_3 \hat{\mathbf{k}}) \cdot (b_1 \hat{\mathbf{i}} + b_2 \hat{\mathbf{j}} + b_3 \hat{\mathbf{k}})$$

$$= a_1 b_1 + a_2 b_2 + a_3 b_3. \tag{11}$$

EXAMPLE 1 Given $\mathbf{a} = \langle 2, 2, -1 \rangle$ and $\mathbf{b} = \langle 3, 0, 4 \rangle$, find $\mathbf{a} \cdot \mathbf{b}$ and the angle θ between the vectors $\mathbf{a}$ and $\mathbf{b}$.

Solution: Using (11), we see that

$$\mathbf{a} \cdot \mathbf{b} = (2)(3) + (2)(0) + (-1)(4) = 2.$$

Also, $\|\mathbf{a}\| = \sqrt{4 + 4 + 1} = 3$ and $\|\mathbf{b}\| = \sqrt{9 + 16} = 5$, from which we deduce

$$\cos \theta = \frac{\mathbf{a} \cdot \mathbf{b}}{\|\mathbf{a}\| \|\mathbf{b}\|} = \frac{2}{15}, \quad \theta \cong 1.437 \text{ radians.}$$

Typical applications that use the scalar or dot product involve the concept of *scalar component*, or *scalar projection*, of a given vector $\mathbf{a}$ in the direction of another vector $\mathbf{b}$. We define this scalar projection by (see Fig. 4.6)

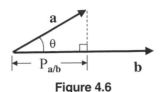

Figure 4.6

$$P_{a/b} = \|\mathbf{a}\| \cos\theta = \frac{\mathbf{a}\cdot\mathbf{b}}{\|\mathbf{b}\|} = \mathbf{a}\cdot\hat{\mathbf{b}}. \qquad (12)$$

Some applications involving the notion of scalar projection will be briefly discussed in Section 4.2.4.

Remark: The components of a vector are projections onto the coordinate axes.

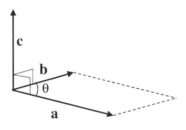

Figure 4.7

▶ **Vector product:** The *vector product* or *cross product* between vectors **a** and **b** is defined by (see Fig. 4.7)

$$\mathbf{a}\times\mathbf{b} = \mathbf{c}, \qquad (13)$$

where **c** is a vector with the properties:
1. $\|\mathbf{c}\| = \|\mathbf{a}\|\,\|\mathbf{b}\|\sin\theta$
2. **c** is perpendicular (orthogonal) to both **a** and **b**
3. the sense of **c** is given by the right-hand rule[1]

The vector product has the following properties:

1. $\mathbf{a}\times\mathbf{b} = -(\mathbf{b}\times\mathbf{a})$

2. $\mathbf{a}\times K\mathbf{b} = K\mathbf{a}\times\mathbf{b} = K(\mathbf{a}\times\mathbf{b})$

3. $\mathbf{a}\times(\mathbf{b}+\mathbf{c}) = \mathbf{a}\times\mathbf{b} + \mathbf{a}\times\mathbf{c}$

4. $\mathbf{a}\times\mathbf{a} = \mathbf{0}$

5. $\mathbf{a}\times\mathbf{b} = \mathbf{0}$ if and only if **a** is parallel to **b** $(\mathbf{a}\neq\mathbf{0}, \mathbf{b}\neq\mathbf{0})$

6. $\mathbf{a}\times\mathbf{0} = \mathbf{0}$

From the definition of cross product, it can readily be established that the base vectors satisfy the properties

$$\begin{aligned}
\hat{\mathbf{i}}\times\hat{\mathbf{i}} &= \mathbf{0}, \quad \hat{\mathbf{i}}\times\hat{\mathbf{j}} = \hat{\mathbf{k}}, \quad \hat{\mathbf{i}}\times\hat{\mathbf{k}} = -\hat{\mathbf{j}}, \\
\hat{\mathbf{j}}\times\hat{\mathbf{i}} &= -\hat{\mathbf{k}}, \quad \hat{\mathbf{j}}\times\hat{\mathbf{j}} = \mathbf{0}, \quad \hat{\mathbf{j}}\times\hat{\mathbf{k}} = \hat{\mathbf{i}}, \\
\hat{\mathbf{k}}\times\hat{\mathbf{i}} &= \hat{\mathbf{j}}, \quad \hat{\mathbf{k}}\times\hat{\mathbf{j}} = -\hat{\mathbf{i}}, \quad \hat{\mathbf{k}}\times\hat{\mathbf{k}} = \mathbf{0}.
\end{aligned} \qquad (14)$$

---

[1] A coordinate system is called *right-handed* if the axes x, y, z are arranged as shown in Fig. 4.4, or a cyclic permutation of such. The *right-hand rule* for vectors assumes the vectors **a**, **b**, **c** are similarly related to each other (see Fig. 4.7).

By use of (14), we are led to the component form for the cross product given by

$$
\begin{aligned}
\mathbf{a} \times \mathbf{b} &= (a_1\hat{\mathbf{i}} + a_2\hat{\mathbf{j}} + a_3\hat{\mathbf{k}}) \times (b_1\hat{\mathbf{i}} + b_2\hat{\mathbf{j}} + b_3\hat{\mathbf{k}}) \\
&= (a_2 b_3 - a_3 b_2)\hat{\mathbf{i}} - (a_1 b_3 - a_3 b_1)\hat{\mathbf{j}} + (a_1 b_2 - a_2 b_1)\hat{\mathbf{k}},
\end{aligned}
\tag{15}
$$

the result of which can be also expressed by the "determinant"

$$
\mathbf{a} \times \mathbf{b} = \begin{vmatrix} \hat{\mathbf{i}} & \hat{\mathbf{j}} & \hat{\mathbf{k}} \\ a_1 & a_2 & a_3 \\ b_1 & b_2 & b_3 \end{vmatrix}.
\tag{16}
$$

Equation (16) is a symbolic determinant in that we use the rules of a determinant only as a convenience for obtaining the result in Eq. (15). Also, it serves as a reminder that if the order of **a** and **b** is reversed, the cross product **b** × **a** produces the negative of **a** × **b**. (That is, interchanging two rows of a determinant changes only the sign of the result.)

EXAMPLE 2 Calculate **a** × **b**, where **a** = <3, 4, 0> and **b** = <1, −2, 5>.

Solution: Using (16), we have

$$
\mathbf{a} \times \mathbf{b} = \begin{vmatrix} \hat{\mathbf{i}} & \hat{\mathbf{j}} & \hat{\mathbf{k}} \\ 3 & 4 & 0 \\ 1 & -2 & 5 \end{vmatrix} = 20\hat{\mathbf{i}} - 15\hat{\mathbf{j}} - 10\hat{\mathbf{k}} = <20, -15, -10>.
$$

The cross product is useful in applications involving the moment of a force, finding the linear velocity of a rotating body, and in certain problems in analytic geometry. We will briefly examine some of these below in Section 4.2.4.

▸ **Scalar triple product:** The *scalar triple product* (also called the *mixed triple product*) of vectors **a**, **b**, **c** is defined by

$$
(\mathbf{a}\,\mathbf{b}\,\mathbf{c}) = \mathbf{a} \cdot (\mathbf{b} \times \mathbf{c}).
\tag{17}
$$

The use of parentheses on the right-hand side in (17) is usually omitted because there is only one possible interpretation of the symbols [i.e., $(\mathbf{a} \cdot \mathbf{b}) \times \mathbf{c}$ is not defined]. By use of (16) for the cross product, it readily follows that the scalar triple product can be defined by the determinant

$$(\mathbf{a\,b\,c}) = \begin{vmatrix} a_1 & a_2 & a_3 \\ b_1 & b_2 & b_3 \\ c_1 & c_2 & c_3 \end{vmatrix}. \tag{18}$$

Thus, based on properties of determinants, we have the following properties:

1. $(\mathbf{a\,b\,c}) = (\mathbf{c\,a\,b}) = (\mathbf{b\,c\,a})$
2. $(\mathbf{a\,b}\,K\mathbf{c}) = K(\mathbf{a\,b\,c})$
3. $(\mathbf{a\,b\,c}) = -(\mathbf{b\,a\,c})$
4. $(\mathbf{a\,a\,b}) = 0$

In addition, it can also be shown that the locations of the dot and cross in the definition (17) can be interchanged (property of determinants), i.e.,

$$(\mathbf{a\,b\,c}) = \mathbf{a} \cdot (\mathbf{b} \times \mathbf{c}) = (\mathbf{a} \times \mathbf{b}) \cdot \mathbf{c}, \tag{19}$$

the proof of which we leave to the reader. Last, based on Property 4 above, we note that if any two vectors in the scalar triple product are parallel, the product is zero.

EXAMPLE 3 Calculate the product $(\mathbf{a\ b\ c})$, where $\mathbf{a} = <2, 0, 1>$, $\mathbf{b} = <3, 1, 1>$, and $\mathbf{c} = <1, 1, 4>$.

Solution: By use of (18), we deduce that (expanding by the first row)

$$(\mathbf{a\,b\,c}) = \begin{vmatrix} 2 & 0 & 1 \\ 3 & 1 & 1 \\ 1 & 1 & 4 \end{vmatrix} = 6 - 0 + 2 = 8.$$

4.2.3 Vector identities

In the algebraic manipulation of various vector expressions it is helpful to have a source of identities that relate the various vector products to each other. The following identities are particularly useful in this regard:

$$\mathbf{a} \times (\mathbf{b} \times \mathbf{c}) = (\mathbf{a} \cdot \mathbf{c})\mathbf{b} - (\mathbf{a} \cdot \mathbf{b})\mathbf{c} \tag{20}$$

$$(\mathbf{a} \times \mathbf{b}) \times \mathbf{c} = (\mathbf{a} \cdot \mathbf{c})\mathbf{b} - (\mathbf{b} \cdot \mathbf{c})\mathbf{a} \qquad (21)$$

$$(\mathbf{a} \times \mathbf{b}) \times (\mathbf{c} \times \mathbf{d}) = (\mathbf{a}\,\mathbf{b}\,\mathbf{d})\mathbf{c} - (\mathbf{a}\,\mathbf{b}\,\mathbf{c})\mathbf{d} \qquad (22)$$

$$(\mathbf{a} \times \mathbf{b}) \cdot (\mathbf{c} \times \mathbf{d}) = (\mathbf{a} \cdot \mathbf{c})(\mathbf{b} \cdot \mathbf{d}) - (\mathbf{a} \cdot \mathbf{d})(\mathbf{b} \cdot \mathbf{c}) \qquad (23)$$

$$(\mathbf{a} \times \mathbf{b}) \cdot (\mathbf{b} \times \mathbf{c}) \times (\mathbf{c} \times \mathbf{a}) = (\mathbf{a}\,\mathbf{b}\,\mathbf{c})^2 \qquad (24)$$

4.2.4 Applications

In this section we will briefly look at some applications of vector algebra to problems in geometry and basic physics.

Equation of a Line:

Let $P_0(x_0, y_0, z_0)$ be a point on a line that is parallel to a given vector $\mathbf{v} = <a, b, c>$. If $\mathbf{R} = <x, y, z>$ is the *position vector* of any other point on the line (see Fig. 4.8) and $\mathbf{R}_0 = <x_0, y_0, z_0>$ is the position vector of the original point P_0, then the equation of the line can be represented by the vector function

$$\mathbf{R}(t) = \mathbf{R}_0 + t\mathbf{v}, \qquad (25)$$

where t is a parameter. Equation (25) is a simple consequence of vector addition. In component form, Eq. (25) becomes

$$\begin{aligned} x &= x_0 + at, \\ y &= y_0 + bt, \\ z &= z_0 + ct. \end{aligned} \qquad (26)$$

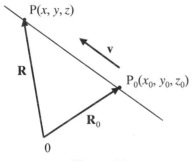

Figure 4.8

Equations written in this latter form are called *parametric equations*. By allowing the parameter t to vary over all real numbers, all points on the line are described by either (25) or (26).

Last, if we eliminate the parameter t in (26), the line can also be represented in the *symmetric form*

$$\frac{x - x_0}{a} = \frac{y - y_0}{b} = \frac{z - z_0}{c}. \qquad (27)$$

The numbers a, b, c, which represent the components of the vector $\mathbf{v}$, are also called *direction numbers* in this setting.

Equation of a Plane:

Let $P_0(x_0, y_0, z_0)$ be a point in some plane which is perpendicular to the vector $\mathbf{n} = \langle a, b, c \rangle$ as illustrated in Fig. 4.9. We wish to find the equation of the plane. To do so, let $P(x, y, z)$ be any other point in the plane which defines the vector

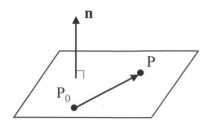

Figure 4.9

$$\overline{P_0 P} = \langle x - x_0, y - y_0, z - z_0 \rangle. \qquad (28)$$

It follows that the vector $\overline{P_0 P}$ is perpendicular to the *normal vector* $\mathbf{n}$, i.e., $\overline{P_0 P} \cdot \mathbf{n} = 0$, or

$$a(x - x_0) + b(y - y_0) + c(z - z_0) = 0. \qquad (29)$$

Equation (29) represents the equation of the plane.

EXAMPLE 4 Find the equation of the plane passing through $P_0(1, 3, -6)$ that is perpendicular to the line

$$\frac{x - 1}{4} = \frac{y}{5} = \frac{z + 5}{6}.$$

Solution: The direction numbers of the line define the normal vector $\mathbf{n} = \langle 4, 5, 6 \rangle$ which is perpendicular to the plane. Hence, based on (29), the equation of the plane is

$$4(x - 1) + 5(y - 3) + 6(z + 6) = 0,$$

or, upon simplification,

$$4x + 5y + 6z + 17 = 0.$$

EXAMPLE 5 Find the distance from the point $P_0(x_0, y_0, z_0)$ to the plane

$$ax + by + cz + d = 0.$$

Solution: Choose any point $P(x, y, z)$ in the plane as shown in Fig. 4.10 and form the vector

$$\overline{PP_0} = \langle x_0 - x, y_0 - y, z_0 - z \rangle.$$

The normal to the plane is $\mathbf{n} = <a,b,c>$ and hence, by forming the scalar projection of $\overline{PP_0}$ onto $\mathbf{n}$, we obtain the distance formula

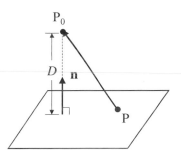

$$D = \frac{|\overline{PP_0} \cdot \mathbf{n}|}{\|\mathbf{n}\|}$$

$$= \frac{|a(x_0 - x) + b(y_0 - y) + c(z_0 - z)|}{\sqrt{a^2 + b^2 + c^2}},$$

or, upon simplification,

Figure 4.10

$$D = \frac{|ax_0 + by_0 + cz_0 + d|}{\sqrt{a^2 + b^2 + c^2}}.$$

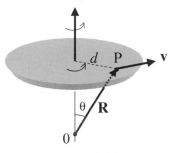

Rotating Disk:

Consider the circular disk illustrated in Fig. 4.11 that is rotating with constant *angular velocity* $\boldsymbol{\omega}$. Let us assume that $\boldsymbol{\omega}$ has direction along the axis of rotation as determined by the right-hand rule. The *linear speed* $\|\mathbf{v}\|$ of a point P with position vector $\mathbf{R}$ from the origin is given by

$$\|\mathbf{v}\| = \|\boldsymbol{\omega}\| d = \|\boldsymbol{\omega}\| \|\mathbf{R}\| \sin\theta, \qquad (30)$$

Figure 4.11

where d is the perpendicular distance from P to the axis of rotation. Hence, from this result and the direction of $\boldsymbol{\omega}$, we deduce that the *linear velocity* is related to the angular velocity by the vector product

$$\mathbf{v} = \boldsymbol{\omega} \times \mathbf{R}. \qquad (31)$$

In particular, if $\boldsymbol{\omega} = <\omega_1, \omega_2, \omega_3>$ and $\mathbf{R} = <x,y,z>$, then clearly

$$\begin{aligned} \mathbf{v} &= \boldsymbol{\omega} \times \mathbf{R} \\ &= (\omega_2 z - \omega_3 y)\hat{\mathbf{i}} + (\omega_3 x - \omega_1 z)\hat{\mathbf{j}} + (\omega_1 y - \omega_2 x)\hat{\mathbf{k}}. \end{aligned} \qquad (32)$$

If the axis of rotation coincides with the z-axis, then (32) reduces to

$$\mathbf{v} = \omega_3(-y\hat{\mathbf{i}} + x\hat{\mathbf{j}}). \qquad (33)$$

These relations make it easy for determining velocity $\mathbf{v}$ at any point on the rotating body.

4.3 Tensor Notation

Vectors and scalars belong to a larger class of quantities known as *tensors*. In particular, vectors are considered *tensors of order one* whereas scalars are *tensors of order zero*. We define the concept of tensor more completely in Chapter 5, but find it convenient in this chapter to introduce the basic notation of tensors without a formal definition. Doing so can simplify certain results and derivations. That is, the tensor notation can be interpreted as a expedient notation for vectors and certain vector operations.

4.3.1 Einstein summation convention

In *tensor notation* we represent a vector $\mathbf{a} = <a_1, a_2, a_3>$ by the symbol

$$a_i, \; i = 1,2,3 \quad \text{or} \quad a_n, \; n = 1,2,3.$$

The subscript i (or n) is called an *index*. As suggested above, the choice of dummy index is somewhat arbitrary, but the letters i, j, k or m, n, p, and so on, are commonly used. This brings us to Rule 1:

> **Rule 1** (*Range*) When a particular index i, j, k, ... appears in a term but is not repeated, it is called a *free index* and is considered to have the range of values 1, 2, 3 unless otherwise stated.

Using Rule 1, the statement $\mathbf{a} = \mathbf{b}$ becomes $a_i = b_i$. Similarly, the statement $\mathbf{a} + \mathbf{b} = \mathbf{c}$ takes the form $a_i + b_i = c_i$. Continuing in this fashion, let us redefine the base vectors

$$\hat{\mathbf{I}}_1 = \hat{\mathbf{i}}, \quad \hat{\mathbf{I}}_2 = \hat{\mathbf{j}}, \quad \hat{\mathbf{I}}_3 = \hat{\mathbf{k}}. \tag{34}$$

The vector $\mathbf{a} = <a_1, a_2, a_3>$ can then be expressed as

$$\mathbf{a} = \sum_{i=1}^{3} a_i \hat{\mathbf{I}}_i = \sum_{k=1}^{3} a_k \hat{\mathbf{I}}_k. \tag{35}$$

Sums analogous to the form of (35) appear so frequently in tensor analysis that it is convenient to adopt a notation that eliminates the need to write the summation symbol. This we address in Rule 2:

> **Rule 2** (*Summation*) If the same index i, j, k, ... occurs exactly twice in a single term, we take an automatic sum on it over values 1, 2, 3. If the same index occurs more than twice, no sum is taken.

If we now invoke Rule 2, Eq. (35) takes the simpler form

$$\mathbf{a} = a_k \hat{\mathbf{I}}_k. \tag{36}$$

Once again, we point out that the index k in (36) is a dummy index. Hence, we make no distinctions between the various forms

$$a_i \hat{\mathbf{I}}_i = a_k \hat{\mathbf{I}}_k = a_n \hat{\mathbf{I}}_n. \tag{37}$$

Rule 2 is widely known as the *Einstein summation convention.*

Thus far we have only introduced symbols (vectors) with a single index. Later on we will also be concerned with symbols like a_{jk} or a_{ijk}, and so on.

EXAMPLE 6 Write out the equation(s) $a_{jk} b_k = c_j$ in complete detail.

Solution: First applying Rule 2, we obtain

$$a_{j1} b_1 + a_{j2} b_2 + a_{j3} b_3 = c_j,$$

where $j = 1, 2, 3$. Now invoking Rule 1, we obtain the three statements

$$a_{11} b_1 + a_{12} b_2 + a_{13} b_3 = c_1,$$

$$a_{21} b_1 + a_{22} b_2 + a_{23} b_3 = c_2,$$

$$a_{31} b_1 + a_{32} b_2 + a_{33} b_3 = c_3.$$

Remark: The reader may recognize that the final result of Example 6 is equivalent to the matrix equation $\mathbf{AB} = \mathbf{C}$, where $\mathbf{A}$ is a 3×3 matrix and $\mathbf{B}$ and $\mathbf{C}$ are 3×1 column vectors.

4.3.2 Kronecker delta and permutation symbols

The use of tensor notation is greater facilitated with the introduction of two special symbols—the Kronecker delta and the permutation symbol.

▸ **Kronecker delta:** The *Kronecker delta* is a double-indexed symbol defined by

$$\delta_{jk} = \begin{cases} 1, & j = k \\ 0, & j \neq k. \end{cases} \tag{38}$$

From definition, we recognize $\delta_{jk} = \delta_{kj}$. Also, it follows that

$$
\begin{aligned}
\delta_{kk} &= 3 \\
a_j \delta_{jk} &= a_k, \\
A_{ij} \delta_{jk} &= A_{ik}.
\end{aligned}
$$

(39)

‣ **Permutation symbol:** The *permutation symbol* is a triple-indexed quantity defined by

$$
e_{ijk} = \begin{cases} +1, & \text{if } (ijk) \text{ either } (123),\ (231),\ \text{or } (312) \\ -1, & \text{if } (ijk) \text{ either } (132),\ (321),\ \text{or } (213) \\ 0, & \text{otherwise} \end{cases}
$$

(40)

Note that the circular order in (40) remains the same for $+1$, but a reverse circular order is used for -1. Making an observation, we see that the permutation symbol satisfies

$$e_{ijk} = 0 \text{ if any two indices are alike}$$

$$e_{ijk} = e_{jki} = e_{kij}$$

$$e_{ijk} = -e_{ikj}.$$

In addition, the manipulation of certain expressions involving the permutation symbol are often aided by use of the following identities:

$$
\begin{aligned}
e_{ijk} e_{imn} &= \delta_{jm}\delta_{kn} - \delta_{jn}\delta_{km} \\
e_{ijk} e_{ijm} &= 2\delta_{km} \\
e_{ijk} e_{ijk} &= 6
\end{aligned}
$$

(41)

The permutation symbol is particularly useful in the expansion of determinant expansions. For example, we can write the determinant

$$
\begin{vmatrix} a_1 & a_2 & a_3 \\ b_1 & b_2 & b_3 \\ c_1 & c_2 & c_3 \end{vmatrix} = e_{ijk} a_i b_j c_k.
$$

(42)

To verify the result of (42), we simply expand the right-hand side to get

$$e_{ijk}a_ib_jc_k = e_{1jk}a_1b_jc_k + e_{2jk}a_2b_jc_k + e_{3jk}a_3b_jc_k$$

$$= (e_{12k}a_1b_2c_k + e_{13k}a_1b_3c_k) + (e_{21k}a_2b_1c_k$$
$$+ e_{23k}a_2b_3c_k) + (e_{31k}a_3b_1c_k + e_{32k}a_3b_2c_k)$$

$$= e_{123}a_1b_2c_3 + e_{132}a_1b_3c_2 + e_{213}a_2b_1c_3 \tag{43}$$
$$+ e_{231k}a_2b_3c_1 + e_{312}a_3b_1c_2 + e_{321}a_3b_2c_1$$

$$= a_1(b_2c_3 - c_2b_3) - a_2(b_1c_3 - b_3c_1) + a_3(b_1c_2 - b_2c_1).$$

In arriving at this result, we eliminated right away all permutation symbols for which any index is repeated, e.g., $e_{11k} = e_{1k1} = e_{k11} = \cdots = 0$, and also have recognized the particular permutation values

$$e_{123} = e_{312} = e_{231} = 1,$$
$$e_{132} = e_{213} = e_{321} = -1.$$

By definition, the last line in (43) is the value of the determinant in (42), obtained by expanding the determinant by the elements in its first row.

4.3.3 Products of vectors and identities

The various products between vectors introduced in Section 4.2.2 take on a different appearance when using tensor notation. For example, the scalar or dot product becomes

$$\boxed{\mathbf{a} \cdot \mathbf{b} = (a_j\hat{\mathbf{I}}_j) \cdot (b_k\hat{\mathbf{I}}_k) = \delta_{jk}a_jb_k = a_kb_k,} \tag{44}$$

where we are recognizing the orthogonality property of the base vectors, i.e.,

$$\hat{\mathbf{I}}_j \cdot \hat{\mathbf{I}}_k = \delta_{jk}. \tag{45}$$

Also, the last step in (44) shows simply that $\mathbf{a} \cdot \mathbf{b} = a_kb_k$. The norm of a vector $\mathbf{a}$ is

$$\|\mathbf{a}\| = (\delta_{jk}a_ja_k)^{1/2} = (a_ka_k)^{1/2}. \tag{46}$$

Remark: Note that $a_ka_k \neq a_k^2$, because the left-hand side is summed and the right-hand side is not. In general, we do not use powers with tensor notation.

Because the vector product and scalar triple product are both obtained by the rules of determinant expansion, we can write these expressions in tensor notation as

$$\mathbf{a} \times \mathbf{b} = e_{ijk} a_i b_j \hat{\mathbf{I}}_k \qquad\qquad (47)$$

$$(\mathbf{a}\,\mathbf{b}\,\mathbf{c}) = e_{ijk} a_i b_j c_k. \qquad\qquad (48)$$

To verify the scalar triple product in tensor notation, we merely observe that

$$
\begin{aligned}
(\mathbf{a}\,\mathbf{b}\,\mathbf{c}) = \mathbf{a} \times \mathbf{b} \cdot \mathbf{c} &= (e_{ijk} a_i b_j \hat{\mathbf{I}}_k) \cdot (c_m \hat{\mathbf{I}}_m) \\
&= (e_{ijk} a_i b_j c_m) \delta_{km} \\
&= e_{ijk} a_i b_j c_k.
\end{aligned}
$$

Remark: As introduced, the tensor notation does not aid in making vector calculations. It can be very useful, however, in the analysis of complicated expressions and in establishng identities like that below in Example 7.

EXAMPLE 7 Use tensor notation to prove the vector identity

$$\mathbf{a} \times (\mathbf{b} \times \mathbf{c}) = (\mathbf{a} \cdot \mathbf{c})\mathbf{b} - (\mathbf{a} \cdot \mathbf{b})\mathbf{c}.$$

Solution: Let us first introduce the vector $\mathbf{B} = \mathbf{b} \times \mathbf{c}$ and recall from above that

$$\mathbf{B} = B_k \hat{\mathbf{I}}_k = e_{ijk} b_i c_j \hat{\mathbf{I}}_k,$$

where we identify $B_k = e_{ijk} b_i c_j$. It then follows that

$$
\begin{aligned}
\mathbf{a} \times (\mathbf{b} \times \mathbf{c}) = \mathbf{a} \times \mathbf{B} &= e_{mkn} a_m B_k \hat{\mathbf{I}}_n \\
&= e_{mkn} e_{ijk} a_m b_i c_j \hat{\mathbf{I}}_n.
\end{aligned}
$$

By use of the identity $e_{mkn} e_{ijk} = e_{knm} e_{kij} = \delta_{ni}\delta_{mj} - \delta_{nj}\delta_{mi}$, we find

$$
\begin{aligned}
\mathbf{a} \times (\mathbf{b} \times \mathbf{c}) &= (\delta_{ni}\delta_{mj} - \delta_{nj}\delta_{mi}) a_m b_i c_j \hat{\mathbf{I}}_n \\
&= \delta_{ni}\delta_{mj} a_m b_i c_j \hat{\mathbf{I}}_n - \delta_{nj}\delta_{mi} a_m b_i c_j \hat{\mathbf{I}}_n \\
&= a_j b_i c_j \hat{\mathbf{I}}_i - a_i b_i c_j \hat{\mathbf{I}}_j \\
&= (a_j c_j) b_i \hat{\mathbf{I}}_i - (a_i b_i) c_j \hat{\mathbf{I}}_j,
\end{aligned}
$$

from which we deduce

$$\mathbf{a} \times (\mathbf{b} \times \mathbf{c}) = (\mathbf{a} \cdot \mathbf{c})\mathbf{b} - (\mathbf{a} \cdot \mathbf{b})\mathbf{c}.$$

4.4 Vector Functions of One Variable

A *vector function* of one variable is defined in terms of three scalar functions $F_1(t)$, $F_2(t)$, and $F_3(t)$, viz.,

$$\mathbf{F}(t) = F_1(t)\hat{\mathbf{i}} + F_2(t)\hat{\mathbf{j}} + F_3(t)\hat{\mathbf{k}} = F_k(t)\hat{\mathbf{I}}_k. \tag{49}$$

Therefore, general concepts of *limit* and *continuity* for vector functions follow directly from that for the scalar component functions. The same is true for the derivative, which we formally state below.

Definition 4.1 (*Derivative*) The *derivative* of a vector function $\mathbf{F}(t) = F_k(t)\hat{\mathbf{I}}_k$ at a fixed point t is defined by the following limit, provided the limit exists:

$$\mathbf{F}'(t) = \lim_{\Delta t \to 0} \frac{\mathbf{F}(t + \Delta t) - \mathbf{F}(t)}{\Delta t} = \mathbf{I}_k \lim_{\Delta t \to 0} \frac{F_k(t + \Delta t) - F_k(t)}{\Delta t}.$$

From above, we have $\mathbf{F}'(t) = F_k'(t)\hat{\mathbf{I}}_k$, and the following rules readily follow:

1. $\dfrac{d}{dt}[\mathbf{F}(t) + \mathbf{G}(t)] = \mathbf{F}'(t) + \mathbf{G}'(t)$

2. $\dfrac{d}{dt}[a(t)\mathbf{F}(t)] = a(t)\mathbf{F}'(t) + a'(t)\mathbf{F}(t)$

3. $\dfrac{d}{dt}[\mathbf{F}(t)\cdot\mathbf{G}(t)] = \mathbf{F}'(t)\cdot\mathbf{G}(t) + \mathbf{F}(t)\cdot\mathbf{G}'(t)$

4. $\dfrac{d}{dt}[\mathbf{F}(t)\times\mathbf{G}(t)] = \mathbf{F}'(t)\times\mathbf{G}(t) + \mathbf{F}(t)\times\mathbf{G}'(t)$

EXAMPLE 8 Show that if $\mathbf{F}(t)$ has a constant (nonzero) norm, then

$$\mathbf{F}(t)\cdot\mathbf{F}'(t) = 0.$$

Solution: Because

$$\|\mathbf{F}(t)\|^2 = \mathbf{F}(t)\cdot\mathbf{F}(t) = \text{constant},$$

we simply differentiate this result to obtain

$$\frac{d}{dt}[\mathbf{F}(t)\cdot\mathbf{F}(t)] = \mathbf{F}'(t)\cdot\mathbf{F}(t) + \mathbf{F}(t)\cdot\mathbf{F}'(t)$$
$$= 2\mathbf{F}(t)\cdot\mathbf{F}'(t) = 0.$$

Example 8 reveals that the derivative of a vector function with constant norm is always orthogonal to the original vector function. We will find this property useful in some of our work, particularly concerning other sets of base vectors.

4.4.1 Space curves

A vector function of one variable $\mathbf{R}(t) = \langle x(t), y(t), z(t)\rangle$ may be geometrically associated with a *curve C in space*, where t is an arbitrary parameter that may or may not represent time. The coordinates at a particular point are given by the parametric equations

$$C: \quad x = x(t), \quad y = y(t), \quad z = z(t). \tag{50}$$

For example, in the previous section we found that $\mathbf{R}(t) = \mathbf{R}_0 + t\mathbf{v}$ represents the equation of a line in space that passes through the point (x_0, y_0, z_0) and is parallel to $\mathbf{v}$. Curves in space are important in a variety of applications in engineering and physics—a space curve in physics, for instance, may represent the path of an object moving through space or a ray of light, and the parameter t in these cases will usually represent time.

From the definition of derivative, it can be shown that $\mathbf{R}'(t)$ has the direction of a tangent vector $\mathbf{T}$ to the space curve C (see Fig. 4.12). To understand this, consider the vector equation

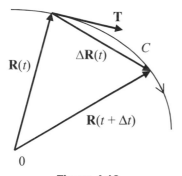

Figure 4.12

$$\mathbf{R}(t + \Delta t) - \mathbf{R}(t) = \Delta\mathbf{R}(t),$$

obtained through simple vector addition. By dividing both sides by Δt and passing to the limit $\Delta t \to 0$, it is clear that the secant vector $\Delta\mathbf{R}(t)$ approaches the direction of the tangent vector $\mathbf{T}$. The *unit tangent vector* to the space curve is therefore defined by

$$\hat{\mathbf{T}} = \frac{\mathbf{R}'(t)}{\|\mathbf{R}'(t)\|}. \tag{51}$$

A space curve C that has a continuous, nonzero tangent vector at each point is called a *smooth curve*. If the space curve is made up of several sections of smooth curves, we say it is *piecewise smooth*. The border of a rectangle is an example of a piecewise smooth curve.

The *length L* of a curve C with $a \le t \le b$ is a positive constant. We introduce this notion in calculus by dividing ("partitioning") the total curve into segments determined by the points $t_0 = a$, t_1, t_2, ... , $t_n = b$ (e.g., see Section 4.6). If we let Δs_k denote the length of each "small" segment, then the total length is

$$L = \sum_{k=1}^{n} \Delta s_k. \tag{52}$$

Now, by approximating the lengths of each arc segment by the secant lines (chords)

$$\Delta s_k \cong \| \mathbf{R}_k(t) - \mathbf{R}_{k-1}(t) \| = \| \Delta \mathbf{R}_k(t) \|,$$

the total length L can be expressed exactly by the limit

$$L = \lim_{\substack{n \to \infty \\ \max |\Delta t_k| \to 0}} \sum_{k=1}^{n} \left\| \frac{\Delta \mathbf{R}_k(t)}{\Delta t_k} \right\| \Delta t_k,$$

which reduces to

$$L = \int_a^b \| \mathbf{R}'(t) \| \, dt = \int_a^b \sqrt{\left(\frac{dx}{dt} \right)^2 + \left(\frac{dy}{dt} \right)^2 + \left(\frac{dz}{dt} \right)^2} \, dt. \tag{53}$$

If the endpoint b of the curve is a variable point t, then we designate the resulting arc length $s(t)$ by

$$s(t) = \int_a^t \| \mathbf{R}'(\tau) \| \, d\tau. \tag{54}$$

Here we introduce the dummy variable τ to avoid confusion. By differentiating both sides of (54) with respect to t, we deduce the important relation

$$\frac{ds}{dt} = \| \mathbf{R}'(t) \|, \tag{55}$$

which shows that the rate of change of the arc length s with respect to the parameter t is simply the norm of the tangent vector. Moreover, by use of the chain rule, we find that

$$\frac{d\mathbf{R}}{ds} = \frac{d\mathbf{R}}{dt} \frac{dt}{ds} = \frac{\mathbf{R}'(t)}{\| \mathbf{R}'(t) \|}. \tag{56}$$

This last relation identifies the derivative $d\mathbf{R}/ds$ as a unit tangent vector. In this special case we write

$$\hat{\mathbf{T}}(s) = \mathbf{R}'(s). \tag{57}$$

Remark: Only when the parameter t can be identified with the arc length s will the derivative of the vector function $\mathbf{R}(t)$ yield the unit tangent vector.

4.4.2 Frenet-Serret formulas

In applications such as that describing the motion of a particle (object) along a space curve it is useful to "attach" an orthogonal coordinate system to the particle (see Fig. 4.13). For *planar motion* the unit tangent vector $\hat{\mathbf{T}}$ and unit normal vector $\hat{\mathbf{N}}$ to the curve are often used as a basis for this attached coordinate system. In three-dimensional motion there is a third unit vector $\hat{\mathbf{B}}$ called the unit binormal that identifies the remaining axis for the basis $(\hat{\mathbf{T}}, \hat{\mathbf{N}}, \hat{\mathbf{B}})$. In this section we wish to develop some geometric relations between these vectors that are useful in applications as well as in theoretical work.

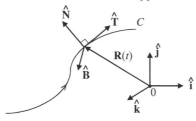

To begin, we start with the relation for the *unit tangent vector* [see Eq. (57)]

$$\hat{\mathbf{T}}(s) = \mathbf{R}'(s),$$

where s is arc length and $\mathbf{R}(s)$ denotes a space curve. If we take the derivative of each side of this equation, we are led to

$$\hat{\mathbf{T}}'(s) = \mathbf{R}''(s) = \mathbf{N}(s), \tag{58}$$

Figure 4.13 Orthogonal basis attached to particle on curve *C*.

where $\mathbf{N}$ is a vector orthogonal to $\hat{\mathbf{T}}$ by virtue of the result of Example 8. A *normal vector* is defined as one that is orthogonal to the tangent vector. Because this definition does not uniquely determine a normal vector, we will call $\mathbf{N}$ the *principal* normal vector. The norm or magnitude of this vector is

$$\|\mathbf{N}(s)\| = \|\hat{\mathbf{T}}'(s)\| = \kappa(s), \tag{59}$$

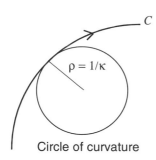

Circle of curvature

Figure 4.14

where κ is called the *curvature* of the space curve. In general, the curvature will vary from point to point along the curve, but is constant for a circle. The reciprocal quantity $\rho = 1/\kappa$ defines the *radius of curvature* (see Fig. 4.14). Thus, we write

$$\hat{\mathbf{T}}'(s) = \kappa(s)\hat{\mathbf{N}}(s), \tag{60}$$

where $\hat{\mathbf{N}}$ is the *principal unit normal vector*.

To define a third vector $\hat{\mathbf{B}}$ orthogonal to both $\hat{\mathbf{T}}$ and $\hat{\mathbf{N}}$, we simply rely on the vector product relation

$$\hat{\mathbf{B}} = \hat{\mathbf{T}} \times \hat{\mathbf{N}}, \tag{61}$$

where the vector $\hat{\mathbf{B}}$ is the *unit binormal vector*. In planar motion, the vectors $\hat{\mathbf{T}}$ and $\hat{\mathbf{N}}$ lie in the plane and the binormal vector $\hat{\mathbf{B}}$ is always orthogonal to the plane of motion.

Derivatives of the unit base vectors $\hat{\mathbf{T}}$, $\hat{\mathbf{N}}$, and $\hat{\mathbf{B}}$ with respect to arc length s lead to the *Frenet-Serret formulas* (see also Section 15.5.2 and [32])

$$\boxed{\begin{aligned}
\hat{\mathbf{T}}'(s) &= \kappa\hat{\mathbf{N}}(s), \\
\hat{\mathbf{N}}'(s) &= -\kappa\hat{\mathbf{T}}(s) + \tau\hat{\mathbf{B}}(s), \\
\hat{\mathbf{B}}'(s) &= -\tau\hat{\mathbf{N}}(s).
\end{aligned}} \tag{62}$$

The quantity τ, called the *torsion*, can be defined by

$$\tau(s) = \pm\|\hat{\mathbf{B}}'(s)\|. \tag{63}$$

Geometrically, the torsion is a measure of the rate of "twisting" of the space curve from the plane defined by the tangent and normal vectors. The derivations of Eqs. (62) follow from differentiating (60), (61) and $\hat{\mathbf{N}} = \hat{\mathbf{B}} \times \hat{\mathbf{T}}$.

Last, we point out that the curvature and torsion can both be calculated directly from the position vector $\mathbf{R}(t)$. Namely,

$$\kappa(t) = \frac{\|\mathbf{R}'(t) \times \mathbf{R}''(t)\|}{\|\mathbf{R}'(t)\|^3} \tag{64}$$

and

$$\tau(t) = \frac{|\mathbf{R}'(t) \times \mathbf{R}''(t) \cdot \mathbf{R}'''(t)|}{\|\mathbf{R}'(t) \times \mathbf{R}''(t)\|^2}, \tag{65}$$

the proofs of which we omit. In (64) and (65), the parameter t is not restricted to arc length.

EXAMPLE 9 Given the space curve $x = t$, $y = t^2$, $z = \frac{2}{3}t^3$, find the following quantities at $t = 1$:

    (*a*) $\hat{\mathbf{T}}$ (*b*) $\hat{\mathbf{N}}$ (*c*) $\hat{\mathbf{B}}$ (*d*) κ and τ

Solution: (*a*) Writing the space curve in vector notation yields

$$\mathbf{R}(t) = \langle t, t^2, \frac{2}{3}t^3 \rangle,$$

from which we deduce $\mathbf{R}'(t) = \langle 1, 2t, 2t^2 \rangle$ and $\|\mathbf{R}'(t)\| = 1 + 2t^2$. Hence, the unit tangent vector is

$$\hat{\mathbf{T}}(t) = \frac{\mathbf{R}'(t)}{\|\mathbf{R}'(t)\|} = \frac{\hat{\mathbf{i}} + 2t\hat{\mathbf{j}} + 2t^2\hat{\mathbf{k}}}{1 + 2t^2},$$

from which we obtain

$$\hat{\mathbf{T}}(1) = \langle \frac{1}{3}, \frac{2}{3}, \frac{2}{3} \rangle.$$

(*b*) Here we first calculate

$$\mathbf{N}(t) = \hat{\mathbf{T}}'(t)\frac{dt}{ds} = \frac{\hat{\mathbf{T}}'(t)}{\|\mathbf{R}'(t)\|} = \frac{2[-2t\hat{\mathbf{i}} + (1-2t)\hat{\mathbf{j}} + 2t\hat{\mathbf{k}}]}{(1+2t^2)^3},$$

which leads to

$$\hat{\mathbf{N}}(1) = \frac{\mathbf{N}(t)}{\|\mathbf{N}(t)\|}\bigg|_{t=1} = \frac{-2t\hat{\mathbf{i}} + (1-2t)\hat{\mathbf{j}} + 2t\hat{\mathbf{k}}}{1+2t^2}\bigg|_{t=1} = <-\frac{2}{3}, -\frac{1}{3}, \frac{2}{3}>.$$

(*c*) Based on parts (*a*) and (*b*), we obtain

$$\hat{\mathbf{B}}(1) = \hat{\mathbf{T}}(1)\times\hat{\mathbf{N}}(1) = <\frac{2}{3}, -\frac{2}{3}, \frac{1}{3}>.$$

(*d*) Here we can use the result above for $\mathbf{N}(t)$ to deduce $\kappa(t)$, or simply calculate $\mathbf{R}'(1) = <1,2,2>$, $\mathbf{R}''(1) = <0,2,4>$, and $\mathbf{R}'''(1) = <0,0,4>$, and then use Eqs. (64) and (65) to obtain

$$\kappa(1) = \tau(1) = \frac{2}{9}.$$

4.4.3 Velocity and acceleration

The motion of an object (particle) in space traces out a space curve. Given the position vector $\mathbf{R}(t)$ of the object as a function of time t enables us to completely describe the motion, including its velocity and acceleration at all times.

Let the position of the object be described by the space curve

$$\mathbf{R}(t) = x(t)\hat{\mathbf{i}} + y(t)\hat{\mathbf{j}} + z(t)\hat{\mathbf{k}}. \tag{66}$$

The *velocity* of the object is defined by the time rate change of position, or

$$\mathbf{v}(t) = \mathbf{R}'(t) = x'(t)\hat{\mathbf{i}} + y'(t)\hat{\mathbf{j}} + z'(t)\hat{\mathbf{k}}. \tag{67}$$

Similarly, *acceleration* of the object is the time rate change of velocity, i.e., $\mathbf{a}(t) = \mathbf{v}'(t)$, which leads to

$$\mathbf{a}(t) = \mathbf{R}''(t) = x''(t)\hat{\mathbf{i}} + y''(t)\hat{\mathbf{j}} + z''(t)\hat{\mathbf{k}}. \tag{68}$$

From (67), it is clear that the velocity vector always points in the direction of the *instantaneous tangent* to the space curve. As such, we can also represent velocity by

$$\mathbf{v}(t) = \frac{ds}{dt}\hat{\mathbf{T}}(t) = v(t)\hat{\mathbf{T}}(t), \tag{69}$$

where $v(t) = ds/dt$ denotes the *speed* of the object. By differentiating (69), we find that

$$\begin{aligned}
\mathbf{a}(t) &= v'(t)\hat{\mathbf{T}}(t) + v(t)\hat{\mathbf{T}}'(t) \\
&= v'(t)\hat{\mathbf{T}}(t) + \kappa(t)v(t)\frac{ds}{dt}\hat{\mathbf{N}}(t) \\
&= v'(t)\hat{\mathbf{T}}(t) + \kappa(t)v^2(t)\hat{\mathbf{N}}(t).
\end{aligned} \tag{70}$$

It follows, therefore, that the acceleration vector can always be partitioned into two components along the tangential and normal directions, i.e.,

$$\boxed{\mathbf{a}(t) = a_T(t)\hat{\mathbf{T}}(t) + a_N(t)\hat{\mathbf{N}}(t),} \tag{71}$$

where $a_T(t)$ and $a_N(t)$ are the *tangential* and *normal components* of acceleration defined, respectively, by (ρ is the radius of curvature)

$$a_T(t) = v'(t) = \frac{d^2 s}{dt^2}, \qquad a_N(t) = \kappa(t)v^2(t) = \frac{v^2(t)}{\rho}. \tag{72}$$

The tangential component measures the *rate of change of speed* of the object. Thus, $a_T(t) > 0$ when the object is speeding up, and $a_T(t) < 0$ when it is slowing down. The normal component is also called the *centripetal acceleration*. It is always directed toward the center of curvature of the path and measures the *change in direction of the velocity* (or motion) of the object. The magnitude of acceleration can be found from either representation (68) or (71); thus,

$$\|\mathbf{a}(t)\| = \sqrt{\left(\frac{d^2 x}{dt^2}\right)^2 + \left(\frac{d^2 y}{dt^2}\right)^2 + \left(\frac{d^2 z}{dt^2}\right)^2} = \sqrt{a_T^2(t) + a_N^2(t)}. \tag{73}$$

Although the acceleration vector can be defined in terms of the original basis $(\hat{\mathbf{i}},\hat{\mathbf{j}},\hat{\mathbf{k}})$, it is physically more meaningful in most applications to use the basis $(\hat{\mathbf{T}},\hat{\mathbf{N}},\hat{\mathbf{B}})$ in which the component along the binormal direction is zero.

> **Remark:** In order to avoid abrupt changes in acceleration, airplane wings, railroad tracks, high-speed cams, etc., must be carefully designed to provide a continuous change in acceleration.

EXAMPLE 10 (*Coriolis acceleration*) Consider an object moving along a line with speed v_0 from the center of a rotating disk with constant angular speed ω (see Fig. 4.15). Determine the acceleration vector acting on the object in terms of the unit vector $\hat{\mathbf{b}}$ pointing in the direction of motion.

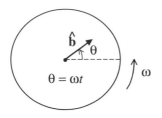

Figure 4.15

Solution: The position of the object as measured from the center of the disk is

$$\mathbf{R}(t) = v_0 t \hat{\mathbf{b}},$$

where

$$\hat{\mathbf{b}} = \cos \omega t \, \hat{\mathbf{i}} + \sin \omega t \, \hat{\mathbf{j}}.$$

Calculating the first derivative, we obtain

$$\mathbf{v}(t) = \mathbf{R}'(t) = v_0 [\hat{\mathbf{b}}(t) + t \hat{\mathbf{b}}'(t)],$$

whereas for the acceleration vector,

$$\mathbf{a}(t) = \mathbf{R}''(t) = v_0 [2\hat{\mathbf{b}}'(t) + t \hat{\mathbf{b}}''(t)].$$

However,

$$\hat{\mathbf{b}}''(t) = -\omega^2 \cos \omega t \, \hat{\mathbf{i}} - \omega^2 \sin \omega t \, \hat{\mathbf{j}} = -\omega^2 \hat{\mathbf{b}},$$

from which we deduce

$$\mathbf{a}(t) = v_0 [2\hat{\mathbf{b}}'(t) - t\omega^2 \hat{\mathbf{b}}(t)].$$

The motion in this case is a combination of two motions—the *rectilinear motion* toward the edge of the disk and the *circular motion* due to the rotating disk. The second term $-t\omega^2 v_0 \hat{\mathbf{b}}(t)$ is toward the center of the disk and can therefore be identified with the centripetal acceleration. The first term $2v_0 \hat{\mathbf{b}}'(t)$ is caused by a combination of the two motions and is known as the *Coriolis acceleration.*

4.4.4 Planar motion in polar coordinates

Particle motion that is restricted to a plane (or is two-dimensional) is called *planar motion*. For certain planar motion problems it may be more convenient to use polar coordinates rather than standard rectangular coordinates. This is particularly true for problems in planetary motion in which the motion of the planets is the consequence of an inverse square law of gravitational attraction. One of the great advances in the history of astronomy took place in the early 1600s when the German astronomer Johannes Kepler (1571-1630) introduced the three basic laws of planetary motion now called *Kepler's laws* (see below). It was later shown by Newton that these three laws actually reduce to a simple single law concerning central forces, and this ultimately led to his remarkable generalization known as "the law of universal gravitation."

> **Kepler's Laws**
> 1. Each planet moves in an elliptical orbit with the sun at one of its foci.
> 2. The radius vector drawn from the sun to a planet sweeps out equal areas in equal times.
> 3. The squares of the periodic times of the planets are proportional to the cubes of the semimajor axes of their orbits.

Let us consider planar motion shown in Fig. 4.16 and introduce unit vectors $\hat{\mathbf{e}}_r$ and $\hat{\mathbf{e}}_\theta$ defined by the rectangular coordinate components

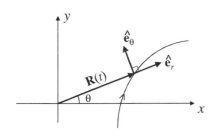

$$\hat{\mathbf{e}}_r = \langle \cos\theta, \sin\theta \rangle, \tag{74}$$

$$\hat{\mathbf{e}}_\theta = \frac{d\hat{\mathbf{e}}_r}{d\theta} = \langle -\sin\theta, \cos\theta \rangle. \tag{75}$$

Note that the unit vector $\hat{\mathbf{e}}_r$ is defined parallel to the position vector $\mathbf{R}(t)$ and both vectors change direction along the curve as the object moves. Moreover, the unit vectors $\hat{\mathbf{e}}_r$ and $\hat{\mathbf{e}}_\theta$ are mutually orthogonal, i.e.,

Figure 4.16

$$\hat{\mathbf{e}}_r \cdot \hat{\mathbf{e}}_\theta = 0. \tag{76}$$

The position vector $\mathbf{R}(t)$ of a object moving along the curve in Fig. 4.16 can be expressed as

$$\mathbf{R}(t) = R(t)\hat{\mathbf{e}}_r, \tag{77}$$

where $R(t) = \|\mathbf{R}(t)\|$ is the distance to the object. Hence, the velocity vector is

$$\begin{aligned}
\mathbf{v}(t) = \mathbf{R}'(t) &= R'(t)\hat{\mathbf{e}}_r + R(t)\frac{d\hat{\mathbf{e}}_r}{d\theta}\theta'(t) \\
&= R'(t)\hat{\mathbf{e}}_r + R(t)\theta'(t)\hat{\mathbf{e}}_\theta,
\end{aligned} \tag{78}$$

where we have used the chain rule and Eq. (75). Note that if $R'(t) \equiv 0$, then the distance $R(t)$ to the object is constant. This means that the object is moving along a *circular path* with tangential component $R(t)\theta'(t)$.

The acceleration vector is likewise given by

$$\begin{aligned}
\mathbf{a}(t) = \mathbf{R}''(t) &= R''(t)\hat{\mathbf{e}}_r + R'(t)\frac{d\hat{\mathbf{e}}_r}{d\theta}\theta'(t) + R'(t)\theta'(t)\hat{\mathbf{e}}_\theta \\
&\quad + R(t)\theta''(t)\hat{\mathbf{e}}_\theta + R(t)\frac{d\hat{\mathbf{e}}_\theta}{d\theta}[\theta'(t)]^2.
\end{aligned} \tag{79}$$

However, from definition it follows that

$$\frac{d\hat{\mathbf{e}}_\theta}{d\theta} = -\hat{\mathbf{e}}_r, \tag{80}$$

and consequently, (79) reduces to

$$\mathbf{a}(t) = \left\{ R''(t) - R(t)[\theta'(t)]^2 \right\} \hat{\mathbf{e}}_r + \left[2R'(t)\theta'(t) + R(t)\theta''(t) \right] \hat{\mathbf{e}}_\theta. \tag{81}$$

The acceleration term $2R'(t)\theta'(t)$ can sometimes be identified with the Coriolis acceleration. Once again, if we take the special case $R'(t) \equiv 0$ (i.e., circular path motion), the resulting acceleration vector (81) reduces to

$$\mathbf{a}(t) = -R(t)[\theta'(t)]^2 \hat{\mathbf{e}}_r + R(t)\theta''(t)\hat{\mathbf{e}}_\theta, \tag{82}$$

where the component in the radial direction is recognized as the centripetal acceleration. In this special case of circular motion, the base vectors $\hat{\mathbf{e}}_r$ and $\hat{\mathbf{e}}_\theta$ are, respectively, the vectors $\hat{\mathbf{T}}$ and $-\hat{\mathbf{N}}$. If we now consider the motion of an object in the plane under the influence of the force field

$$\mathbf{F} = F_r \hat{\mathbf{e}}_r + F_\theta \hat{\mathbf{e}}_\theta, \tag{83}$$

then, based on *Newton's second law of motion* $\mathbf{F} = m\mathbf{a}$, where m is the mass of the object, it follows from Eq. (81) that the components of the given force field are

$$F_r(t) = m\left\{ R''(t) - R(t)[\theta'(t)]^2 \right\},$$
$$F_\theta(t) = m\left[2R'(t)\theta'(t) + R(t)\theta''(t) \right]. \tag{84}$$

By multiplying the second equation in (84) by $R(t)$, we obtain

$$RF_\theta = m\left[2R(t)R'(t)\theta'(t) + R^2(t)\theta''(t) \right] = m\frac{d}{dt}\left(R^2 \frac{d\theta}{dt} \right), \tag{85}$$

which we can interpret as a *torque*. That is, the torque applied to the object equals the time rate change of angular momentum. For motion such that $F_\theta = 0$, the governing force is always in the radial direction (called a *central force*). This is the case for planetary motion. Under a central force field, Eq. (85) may be integrated to yield $mR^2\, d\theta/dt = C$, where C is constant. Thus, this leads to Kepler's second law because the time derivative of area A swept out by the radius vector $\mathbf{R}(t)$ is given by

$$\frac{dA}{dt} = \frac{1}{2}R^2 \frac{d\theta}{dt}. \tag{86}$$

4.5 Scalar and Vector Fields

Functions that commonly arise in applications are generally classified as either scalar fields or vector fields. The rate of change of these types of functions leads to the concepts of gradient, divergence, and curl.

A *scalar field* is simply a scalar valued function $f(x,y,z)$ of three variables such that every point in a region is assigned a scalar value. If $f(x,y,z) = K$, where K is a constant, we obtain a *surface*. By selecting several values for the constant K, we obtain a layered family of surfaces called *isotimic surfaces*, or *level surfaces*. Other common names for these surfaces are *equipotential surface* if f denotes an electric or gravitational field potential, *isothermal surface* if f denotes temperature, and *isobaric surface* if f denotes pressure.

4.5.1 Gradient

Given a continuous scalar field f we consider the rate of change of f in a given direction. For example, $\partial f/\partial x$ is the rate of change of f along the x axis. Let point P lie on a ray-line C which has direction parallel to the unit vector $\hat{\mathbf{u}}$ (see Fig. 4.17). If Q is another point on the ray-line, we consider the limit of $f(Q) - f(P)$ divided by the distance Δs as $\Delta s \to 0$. If this limit exists, we call it the *directional derivative* of f (in the direction $\hat{\mathbf{u}}$) and denote it by

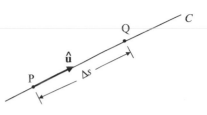

Figure 4.17

$$\frac{df}{ds} = \lim_{\Delta s \to 0} \frac{f(Q) - f(P)}{\Delta s}. \tag{87}$$

The notation $df/d\hat{\mathbf{u}}$ is also commonly used for df/ds, but both are ambiguous and confusing. We do note that if $\hat{\mathbf{u}} = \hat{\mathbf{i}}$, the directional derivative (87) is exactly $\partial f/\partial x$.

For computing directional derivatives, it is useful to introduce the notion of a gradient. To do so, let us define the ray-line C in Fig. 4.17 by the vector function

$$C: \quad \mathbf{R}(s) = x(s)\hat{\mathbf{i}} + y(s)\hat{\mathbf{j}} + z(s)\hat{\mathbf{k}}, \tag{88}$$

where s is arc length. It then follows that

$$\hat{\mathbf{u}} = \frac{d\mathbf{R}}{ds} = \frac{dx}{ds}\hat{\mathbf{i}} + \frac{dy}{ds}\hat{\mathbf{j}} + \frac{dz}{ds}\hat{\mathbf{k}}, \tag{89}$$

and from the chain rule applied to the continuous function $f(x,y,z)$, we also have

$$\frac{df}{ds} = \frac{\partial f}{\partial x}\frac{dx}{ds} + \frac{\partial f}{\partial y}\frac{dy}{ds} + \frac{\partial f}{\partial z}\frac{dz}{ds}. \tag{90}$$

Now, by defining the vector quantity

$$\boxed{\operatorname{grad} f \equiv \nabla f = \frac{\partial f}{\partial x}\hat{\mathbf{i}} + \frac{\partial f}{\partial y}\hat{\mathbf{j}} + \frac{\partial f}{\partial z}\hat{\mathbf{k}},} \tag{91}$$

called the *gradient*, we recognize that the directional derivative (90) can be expressed as a simple dot product, viz.,

$$\boxed{\frac{df}{ds} = \nabla f \cdot \hat{\mathbf{u}}.} \tag{92}$$

From (92), it follows that the directional derivative is the *scalar projection* of the gradient vector ∇f onto the unit vector $\hat{\mathbf{u}}$.

Theorem 4.1 If the gradient ∇f is not the zero vector, it has the direction of maximum increase of the scalar function f.

Proof: From definition of the dot product, we have

$$\frac{df}{ds} = \nabla f \cdot \hat{\mathbf{u}} = \|\nabla f\| \cos\theta,$$

where θ is the angle between vectors (see Fig. 4.18). Clearly, this quantity is a maximum (minimum) only when the vector $\hat{\mathbf{u}}$ has the same (opposite) direction as the gradient ∇f, i.e., only when $\theta = 0$ (or $\theta = \pi$). ❏

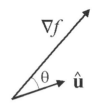

Figure 4.18

Based on Theorem 4.1, we conclude that $\|\nabla f\|$ equals the *maximum rate of increase* (*decrease*) of the function f per unit distance. Last, we have the following important property associated with the gradient vector.

Theorem 4.2 If f is a continuous scalar field such that through any point P there exists exactly one equipotential surface S, then for $\nabla f \neq \mathbf{0}$, the gradient ∇f has the direction of the normal vector to S, i.e.,

$$\hat{\mathbf{N}} = \pm \frac{\nabla f}{\|\nabla f\|}.$$

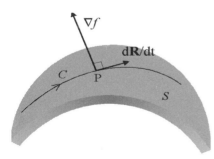

Figure 4.19

Proof: The surface S is defined by $f = K$. Let C be a curve on the surface S (see Fig. 4.19), i.e.,

$$C: \quad \mathbf{R}(t) = x(t)\hat{\mathbf{i}} + y(t)\hat{\mathbf{j}} + z(t)\hat{\mathbf{k}}.$$

Because f is a constant on S, it follows that

$$\frac{df}{dt} = \nabla f \cdot \frac{d\mathbf{R}}{dt} = 0,$$

which shows that the gradient ∇f and $d\mathbf{R}/dt$ are orthogonal. But, because $d\mathbf{R}/dt$ is tangent to the curve C, the direction of ∇f is normal to C. Since this must hold true for every curve C on the surface S, we conclude that the gradient ∇f is everywhere normal to S. ❏

EXAMPLE 11 Find the directional derivative of $f(x,y,z) = xy - x^3z^3$ at the point P(1,3,−1) in the direction defined by the vector $\mathbf{v} = \langle -1, 2, 2 \rangle$.

Solution: First calculating the norm $\|\mathbf{v}\| = \sqrt{(-1)^2 + 2^2 + 2^2} = 3$, the unit vector defining the direction is

$$\hat{\mathbf{u}} = \frac{\mathbf{v}}{\|\mathbf{v}\|} = <-\frac{1}{3}, \frac{2}{3}, \frac{2}{3}>.$$

The gradient of the scalar function f at the point P is

$$\nabla f = <y - 3x^2z^3, x, -3x^3z^2>\Big|_{(1,3,-1)} = <6, 1, -3>,$$

and consequently,

$$\frac{df}{ds} = \nabla f \cdot \hat{\mathbf{u}} = -2 + \frac{2}{3} - 2 = -\frac{10}{3}.$$

EXAMPLE 12 At the point $(2,3,13)$, find the equation of the tangent plane to the surface

$$z = x^2 + y^2$$

Solution: The equation of the tangent plane at the given point has the form

$$A(x - 2) + B(y - 3) + C(z - 13) = 0,$$

where $\mathbf{N} = <A, B, C>$ is a normal vector. By defining the function

$$f(x,y,z) = x^2 + y^2 - z,$$

we find that the normal vector $\mathbf{N}$ becomes

$$\mathbf{N} = \nabla f = <2x, 2y, -1>\Big|_{(2,3,13)} = <4, 6, -1>,$$

which leads to (upon simplification)

$$4x + 6y - z = 13.$$

4.5.2 Divergence

A *vector field* $\mathbf{F}(x,y,z)$ is a vector-valued function of three variables—in essence, a collection of three scalar fields that form the components. Thus, every point in a region is assigned a vector. Common examples of vector field include *gravity force field*, *velocity field*, *electric field*, *magnetic field*, and *gradient*. Whereas the rate of change of a scalar field is measured by the gradient, the rate of change of a vector field is measured by either the divergence or curl. In this section we consider the divergence, which is a scalar field, and in the following section we introduce the curl, which is a vector field.

Given the vector field

$$\mathbf{F}(x,y,z) = F_1(x,y,z)\hat{\mathbf{i}} + F_2(x,y,z)\hat{\mathbf{j}} + F_3(x,y,z)\hat{\mathbf{k}}, \tag{93}$$

we define the *divergence* of $\mathbf{F}$ by the scalar field

$$\boxed{\operatorname{div}\mathbf{F} \equiv \nabla\cdot\mathbf{F} = \frac{\partial F_1}{\partial x} + \frac{\partial F_2}{\partial y} + \frac{\partial F_3}{\partial z}.} \tag{94}$$

The symbol $\nabla\cdot\mathbf{F}$ for the divergence is only a convenient mathematical notation—the divergence is not a true dot product (see Section 4.5.5).

EXAMPLE 13 Find the divergence of $\mathbf{F} = \langle x, y^2z, xz^3\rangle$.

Solution: From the partial derivatives

$$\frac{\partial F_1}{\partial x} = 1, \quad \frac{\partial F_2}{\partial y} = 2yz, \quad \frac{\partial F_3}{\partial z} = 3xz^2,$$

it follows that

$$\nabla\cdot\mathbf{F} = 1 + 2yz + 3xz^2.$$

4.5.3 Physical interpretation of divergence

To obtain a physical interpretation of divergence, we consider the motion of a compressible fluid through a region of space modeled as a small rectangular box of dimensions $\Delta x \times \Delta y \times \Delta z$ (see Fig. 4.20). Other physical interpretations are also possible. We assume the box contains no sources nor sinks (i.e., fluid is not being created nor destroyed within the box). Physically, we will show that divergence measures the *rate of change of matter at a point per unit volume per unit time*.

If $\rho(x,y,z,t)$ is the mass density (mass/unit volume) of the fluid and $\mathbf{v}(x,y,z,t)$ is the velocity of the flow, then the *mass flow rate density* (or *flux*) of the fluid is given by

$$\mathbf{F}(x,y,z,t) = \rho(x,y,z,t)\,\mathbf{v}(x,y,z,t)$$

$$= \rho(x,y,z,t)\langle v_1(x,y,z,t), v_2(x,y,z,t), v_3(x,y,z,t)\rangle.$$

The physical principle that we adopt is the following statement based on experimental data:

"... the net efflux rate of mass through the control surface is equal to the rate of decrease of mass inside the control surface."

The "flux" across a boundary is defined as the total loss of mass leaving the control surface per unit time. This is the flow of mass $\mathbf{F}$ normal to the surface times the area ΔA of the surface. For the left face marked #1 in Fig. 4.20 the mass flowing cross the rectangular face is $\mathbf{F} \cdot \hat{\mathbf{j}} = (\rho \mathbf{v}) \cdot \mathbf{j}$ and the surface area is $\Delta x \Delta z$. Hence, the flux across this face is

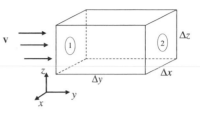

Figure 4.20 Divergence.

$$-(\rho \mathbf{v} \cdot \mathbf{j}) \Delta x \Delta z = -\rho v_2 \Delta x \Delta z,$$

where the minus sign is conventional for inward flow. The flux out the opposite face at #2 is

$$[\rho v_2 + \Delta(\rho v_2)] \Delta x \Delta z,$$

and thus, the net efflux is $\Delta(\rho v_2) \Delta x \Delta z$. The same is true for the other faces of the box so that the net efflux per unit of time across the control surface is the sum

$$\Delta(\rho v_1) \, \Delta y \Delta z + \Delta(\rho v_2) \, \Delta x \Delta z + \Delta(\rho v_3) \, \Delta x \Delta y.$$

The loss of mass per unit of time is simply

$$-\frac{\partial \rho}{\partial t} \Delta x \Delta y \Delta z.$$

By equating these last results and dividing by the volume of the box, we are led to

$$\frac{\Delta(\rho v_1)}{\Delta x} + \frac{\Delta(\rho v_2)}{\Delta y} + \frac{\Delta(\rho v_3)}{\Delta z} = -\frac{\partial \rho}{\partial t}. \tag{95}$$

Last, by passing to the limit as $\Delta x, \Delta y, \Delta z \to 0$ we obtain the *continuity equation*

$$\frac{\partial \rho}{\partial t} + \nabla \cdot (\rho \mathbf{v}) = 0. \tag{96}$$

The continuity equation, fundamental in the field of fluid mechanics, is simply the *law of conservation of mass.*

For the special case in which the mass density ρ is independent of time t, the flow is called *steady* and the continuity equation (96) reduces to

$$\nabla \cdot (\rho \mathbf{v}) = 0. \tag{97}$$

Moreover, if the mass density ρ is constant, we obtain the equation for *incompressible flow*

$$\nabla \cdot \mathbf{v} = 0. \tag{98}$$

Of course, the assumption of no sources nor sinks is essential to our argument.

Remark: Basically, the above example illustrates that *divergence* is, roughly speaking, a measure of outflow minus inflow.

4.5.4 Curl

Another measure of the rate of change of a vector field is the curl, which characterizes the "rotational effects" of the field. Given the vector field

$$\mathbf{F}(x,y,z) = F_1(x,y,z)\hat{\mathbf{i}} + F_2(x,y,z)\hat{\mathbf{j}} + F_3(x,y,z)\hat{\mathbf{k}}, \tag{99}$$

we define the *curl* by

$$\mathrm{curl}\,\mathbf{F} \equiv \nabla \times \mathbf{F}$$
$$= \left(\frac{\partial F_3}{\partial y} - \frac{\partial F_2}{\partial z}\right)\hat{\mathbf{i}} - \left(\frac{\partial F_3}{\partial x} - \frac{\partial F_1}{\partial z}\right)\hat{\mathbf{j}} + \left(\frac{\partial F_2}{\partial x} - \frac{\partial F_1}{\partial y}\right)\hat{\mathbf{k}}. \tag{100}$$

Analogous to the divergence, the curl is not a true cross product—the cross symbol is only a convenient mathematical notation. In particular, to aid in remembering the components of this expression, we can rely on the symbolic determinant used for the cross product, i.e.,

$$\nabla \times \mathbf{F} = \begin{vmatrix} \hat{\mathbf{i}} & \hat{\mathbf{j}} & \hat{\mathbf{k}} \\ \dfrac{\partial}{\partial x} & \dfrac{\partial}{\partial y} & \dfrac{\partial}{\partial z} \\ F_1 & F_2 & F_3 \end{vmatrix}. \tag{101}$$

In this case, of course, we must properly interpret products involving differential operators as differentiation operations, so "order" is important.

EXAMPLE 14 Find the curl of $\mathbf{F} = \langle x, y^2z, xz^3 \rangle$.

Solution: By using (101), we have

$$\nabla \times \mathbf{F} = \begin{vmatrix} \hat{\mathbf{i}} & \hat{\mathbf{j}} & \hat{\mathbf{k}} \\ \dfrac{\partial}{\partial x} & \dfrac{\partial}{\partial y} & \dfrac{\partial}{\partial z} \\ x & y^2z & xz^3 \end{vmatrix}$$

$$= \hat{\mathbf{i}}\left[\frac{\partial}{\partial y}(xz^3) - \frac{\partial}{\partial z}(y^2z)\right] - \hat{\mathbf{j}}\left[\frac{\partial}{\partial x}(xz^3) - \frac{\partial}{\partial z}(x)\right] + \hat{\mathbf{k}}\left[\frac{\partial}{\partial x}(y^2z) - \frac{\partial}{\partial y}(x)\right],$$

or

$$\nabla \times \mathbf{F} = -y^2\hat{\mathbf{i}} - z^3\hat{\mathbf{j}} = \langle -y^2, -z^3, 0 \rangle.$$

To obtain a physical interpretation of the curl, let us recall the rotating disk problem in Section 4.2.4 (see Fig. 4.11). There we showed that the linear velocity $\mathbf{v}$ of a point P on the disk and the angular velocity $\boldsymbol{\omega} = \langle \omega_1, \omega_2, \omega_3 \rangle$ are related by

$$\mathbf{v} = \boldsymbol{\omega} \times \mathbf{R}, \tag{102}$$

where $\mathbf{R}$ is the position vector of the point P. By calculating the curl of $\mathbf{v}$, we obtain

$$\nabla \times \mathbf{v} = \begin{vmatrix} \hat{\mathbf{i}} & \hat{\mathbf{j}} & \hat{\mathbf{k}} \\ \dfrac{\partial}{\partial x} & \dfrac{\partial}{\partial y} & \dfrac{\partial}{\partial z} \\ \omega_2 z - \omega_3 y & \omega_3 x - \omega_1 z & \omega_1 y - \omega_2 x \end{vmatrix} = 2\boldsymbol{\omega}. \tag{103}$$

Hence, the curl of the velocity vector $\mathbf{v}$ is just twice the angular velocity $\boldsymbol{\omega}$, suggesting that in some sense the curl is related to the rotation of the field.

Both the divergence and curl are difficult concepts to interpret physically from the infinitesimal point of view that we have used thus far. These concepts become more meaningful when we consider the *divergence theorem* and *Stokes' theorem* in Section 4.7.3.

4.5.5 Vector differential operators: tensor notation

The notation ∇ used in defining the gradient, divergence, and curl is an example of what we call a *differential operator*. The common name for ∇ is the *del operator*. Differential operators such as $D = d/dx$ are commonly used in calculus and differential equations. The del operator, a generalized version of the simple operator D, is defined by

$$\nabla = \hat{\mathbf{i}} \frac{\partial}{\partial x} + \hat{\mathbf{j}} \frac{\partial}{\partial y} + \hat{\mathbf{k}} \frac{\partial}{\partial z}. \tag{104}$$

In general, we can think of a differential operator as a rule that transforms functions in one domain into other functions. Although we wrote the divergence and curl in terms of the del operator, using a "dot product" and "cross product," we recognize that these are symbolic products; moreover, it is clear that

$$\nabla \cdot \mathbf{F} \neq \mathbf{F} \cdot \nabla, \qquad \nabla \times \mathbf{F} \neq \mathbf{F} \times \nabla.$$

Another related operator, called the *Laplacian*, is defined by

$$\nabla \cdot \nabla = \nabla^2 = \frac{\partial^2}{\partial x^2} + \frac{\partial^2}{\partial y^2} + \frac{\partial^2}{\partial z^2}. \tag{105}$$

The Laplacian is a second-order differential operator that leads to expressions like

$$\nabla^2 u = \frac{\partial^2 u}{\partial x^2} + \frac{\partial^2 u}{\partial y^2} + \frac{\partial^2 u}{\partial z^2}. \tag{106}$$

One of the most important partial differential equations arising in practice is $\nabla^2 u = 0$, called

Laplace's equation. This equation is fundamental in studying steady-state heat conduction, gravitational or electrostatic potentials, steady flow of currents in solid conductors, and fluid mechanics, among other areas (see Section 12.5).

There are various interconnecting relations, or identities, between the gradient, divergence, curl, and so on, that are useful in practice. For reference purposes, we list some of those below.

$$
\begin{aligned}
&1.\ \nabla(fg) \;=\; g\,\nabla f + f\nabla g \\[4pt]
&2.\ \nabla\cdot(\varphi\mathbf{F}) \;=\; \varphi(\nabla\cdot\mathbf{F}) + \nabla\varphi\cdot\mathbf{F} \\[4pt]
&3.\ \nabla\times(\varphi\mathbf{F}) \;=\; \varphi(\nabla\times\mathbf{F}) + \nabla\varphi\times\mathbf{F} \\[4pt]
&4.\ \nabla\times\nabla\varphi \;=\; \mathbf{0} \\[4pt]
&5.\ \nabla\cdot\nabla\times\mathbf{F} \;=\; 0 \\[4pt]
&6.\ \nabla\cdot(\mathbf{F}\times\mathbf{G}) \;=\; \mathbf{G}\cdot(\nabla\times\mathbf{F}) - \mathbf{F}\cdot(\nabla\times\mathbf{G}) \\[4pt]
&7.\ \nabla\times(\mathbf{F}\times\mathbf{G}) \;=\; (\mathbf{G}\cdot\nabla)\mathbf{F} - (\mathbf{F}\cdot\nabla)\mathbf{G} \\
&\qquad\qquad\qquad + (\nabla\cdot\mathbf{G})\mathbf{F} - (\nabla\cdot\mathbf{F})\mathbf{G} \\[4pt]
&8.\ \nabla\times(\nabla\times\mathbf{F}) \;=\; \nabla(\nabla\cdot\mathbf{F}) - \nabla^2\mathbf{F} \\[4pt]
&9.\ \nabla(\mathbf{F}\cdot\mathbf{G}) \;=\; (\mathbf{F}\cdot\nabla)\mathbf{G} + (\mathbf{G}\cdot\nabla)\mathbf{F} \\
&\qquad\qquad\qquad + \mathbf{F}\times(\nabla\times\mathbf{G}) + \mathbf{G}\times(\nabla\times\mathbf{F}) \\[4pt]
&10.\ \nabla f(u) \;=\; \frac{df}{du}\nabla u
\end{aligned}
\tag{107}
$$

To write the gradient, divergence, and curl in tensor form, we introduce the "comma notation"

$$
\frac{\partial\varphi}{\partial x} \;=\; \varphi_{,1}; \quad \frac{\partial\varphi}{\partial y} \;=\; \varphi_{,2}; \quad \frac{\partial\varphi}{\partial z} \;=\; \varphi_{,3};
\tag{108}
$$

or, in general, $\varphi_{,k}$. Hence, using this notation for derivative, we can define

$$
\nabla\varphi \;=\; \varphi_{,k}\hat{\mathbf{I}}_k
\tag{109}
$$

$$
\nabla\cdot\mathbf{F} \;=\; \delta_{jk}F_{k,j} \;=\; F_{k,k}
\tag{110}
$$

$$
\nabla\times\mathbf{F} \;=\; e_{ijk}F_{j,i}\hat{\mathbf{I}}_k.
\tag{111}
$$

EXAMPLE 15 Use tensor notation to prove the identity

$$
\nabla\times(\mathbf{F}\times\mathbf{G}) \;=\; (\mathbf{G}\cdot\nabla)\mathbf{F} - (\mathbf{F}\cdot\nabla)\mathbf{G} + (\nabla\cdot\mathbf{G})\mathbf{F} - (\nabla\cdot\mathbf{F})\mathbf{G}.
$$

Solution: First, by defining the vector $\mathbf{E} = \mathbf{F}\times\mathbf{G}$, or $E_j = e_{mnj}F_m G_n$ we can write

$$\begin{aligned}
\nabla \times (\mathbf{F} \times \mathbf{G}) &= e_{ijk} E_{j,i} \hat{\mathbf{I}}_k \\
&= e_{ijk} e_{mnj} (F_m G_n)_{,i} \hat{\mathbf{I}}_k \\
&= (\delta_{km}\delta_{in} - \delta_{kn}\delta_{im})(F_m G_n)_{,i} \hat{\mathbf{I}}_k \\
&= (F_k G_i)_{,i} \hat{\mathbf{I}}_k - (F_i G_k)_{,i} \hat{\mathbf{I}}_k \\
&= F_{k,i} G_i \hat{\mathbf{I}}_k + F_k G_{i,i} \hat{\mathbf{I}}_k - F_{i,i} G_k \hat{\mathbf{I}}_k - F_i G_{k,i} \hat{\mathbf{I}}_k.
\end{aligned}$$

We now make the observation that

$$\mathbf{G}\cdot\nabla = G_1 \frac{\partial}{\partial x} + G_2 \frac{\partial}{\partial y} + G_3 \frac{\partial}{\partial z}$$

and hence,

$$(\mathbf{G}\cdot\nabla)\mathbf{F} = G_i F_{j,i} \hat{\mathbf{I}}_j.$$

Similarly, we find

$$(\nabla\cdot\mathbf{G})\mathbf{F} = G_{i,i} \left(F_j \hat{\mathbf{I}}_j \right).$$

Comparing the above expression for $\nabla \times (\mathbf{F} \times \mathbf{G})$ with these last relations, we see that our intended result follows, viz.,

$$\nabla \times (\mathbf{F} \times \mathbf{G}) = (\mathbf{G}\cdot\nabla)\mathbf{F} - (\mathbf{F}\cdot\nabla)\mathbf{G} + (\nabla\cdot\mathbf{G})\mathbf{F} - (\nabla\cdot\mathbf{F})\mathbf{G}.$$

4.6 Line and Surface Integrals

In this section we are concerned with extending the calculus of vector functions to integration. This will involve generalizations of the ordinary Riemann integral

$$\int_a^b f(x)\,dx$$

along the *x*-axis to more general types of integrals called *line integrals* and *surface integrals*. Relations between line, surface, and volume integrals will be addressed in Section 4.7.

4.6.1 Line integrals

The concept of line integral is a natural generalization of the definite integral, but integration takes place along a space curve C rather than along a coordinate axis. Let us consider the curve shown in Fig. 4.21 with initial point A and terminal point B. The vector representation of this curve is

$$C: \quad \mathbf{R}(s) = x(s)\hat{\mathbf{i}} + y(s)\hat{\mathbf{j}} + z(s)\hat{\mathbf{k}}, \qquad (112)$$

where s denotes arc length. We assume that $\mathbf{R}(s)$ is continuous with continuous first derivative (i.e., a smooth curve), and, further, that $f(x,y,z)$ is a continuous scalar field defined at each point along the curve C. Let us subdivide the curve C into intervals by the points $P_0, P_1, \ldots, P_{k-1}, P_k, \ldots, P_n$, where $P_0 = A$ and $P_n = B$. In each interval of length Δs_k we let $\xi_k = (x_k, y_k, z_k)$, $k = 1, 2, \ldots, n$ designate an arbitrary point at which we define $f(\xi_k)$. We then define the *line integral* of f along the curve C by the limit (provided it exists)

$$\int_C f(x,y,z)\,ds \;=\; \lim_{\substack{n \to \infty \\ \max|\Delta s_k| \to 0}} \sum_{k=1}^{n} f(\xi_k)\,\Delta s_k. \tag{113}$$

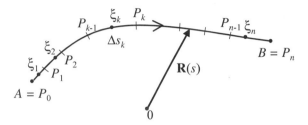

Figure 4.21 A space curve C.

Observe that if the curve C (also called a path) for the line integral (113) is some portion of the x-axis, the line integral reduces to the ordinary Riemann integral. Thus, the Riemann integral is a subset of the more general line integral.

Although it was convenient in arriving at (113) to use arc length s as the parameter for the position vector $\mathbf{R}$, in most problems this is not practical. Suppose, instead, that we represent the curve C by

$$C: \quad \mathbf{R}(t) \;=\; x(t)\hat{\mathbf{i}} + y(t)\hat{\mathbf{j}} + z(t)\hat{\mathbf{k}}, \quad a \le t \le b, \tag{114}$$

where t is any convenient parameter. By the chain rule,

$$ds \;=\; \frac{ds}{dt}\,dt \;=\; \|\mathbf{R}'(t)\|\,dt, \tag{115}$$

and hence, the integral in (113) takes the form

$$\boxed{\int_C f(x,y,z)\,ds \;=\; \int_a^b f[x(t),y(t),z(t)]\,\|\mathbf{R}'(t)\|\,dt.} \tag{116}$$

Because the right-hand side of Eq. (116) is a Riemann integral, this actually provides us with a scheme for evaluating line integrals (see Example 16 below). Also, from this relation we recognize that line integrals share all the familiar properties of the Riemann integral.

EXAMPLE 16 Evaluate the line integral $\int_C (x + \sqrt{y})ds$ from $(0,0)$ to $(1,1)$ along the following curves in the xy-plane:

(a) C_1: $y = x$, $0 \le x \le 1$

(b) C_2: $y = x^2$, $0 \le x \le 1$

(c) C_3: along $y = 0$, $0 \le x \le 1$ and then along $x = 1$, $0 < y \le 1$

Solution: Each of the curves is shown in Fig. 4.22.

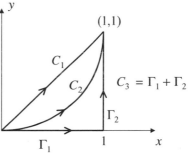

Figure 4.22

(a) For C_1, we let $x = t$, $y = t$, and thus,

$$C_1: \quad \mathbf{R}(t) = <t, t>, \quad 0 \le t \le 1.$$

Also, $\mathbf{R}'(t) = <1, 1>$ and $\|\mathbf{R}'(t)\| = \sqrt{2}$, from which we deduce

$$\int_{C_1} (x + \sqrt{y})ds = \int_0^1 (t + \sqrt{t})\sqrt{2}\,dt = \frac{7\sqrt{2}}{6}.$$

(b) Along C_2, we let $x = t$, $y = t^2$, leading to

$$C_2: \quad \mathbf{R}(t) = <t, t^2>, \quad 0 \le t \le 1.$$

Also, $\mathbf{R}'(t) = <1, 2t>$ and $\|\mathbf{R}'(t)\| = \sqrt{1 + 4t^2}$, from which we deduce

$$\int_{C_2} (x + \sqrt{y})ds = \int_0^1 2t\sqrt{1 + 4t^2}\,dt = \frac{1}{6}(5^{3/2} - 1).$$

(c) Here the curve is composed of two parts, i.e., $C_3 = \Gamma_1 + \Gamma_2$, where

$$\Gamma_1: \quad x = t, \ y = 0, \ \text{or} \ \mathbf{R}(t) = <t, 0>, \ 0 \le t \le 1$$
$$\Gamma_2: \quad x = 1, \ y = t, \ \text{or} \ \mathbf{R}(t) = <1, t>, \ 0 < t \le 1.$$

Hence,

$$\int_{C_3} (x + \sqrt{y})ds = \int_{\Gamma_1} (x + \sqrt{y})ds + \int_{\Gamma_2} (x + \sqrt{y})ds$$
$$= \int_0^1 t\,dt + \int_0^1 (1 + \sqrt{t})dt = \frac{13}{6}.$$

Example 16 illustrates an important concept—i.e., the value of a line integral between two points depends (in general) upon the path C connecting the points. Later on, we will identify a special class of integrals for which the value of the line integral is independent of path. Integrals of this latter category will generally lead to simpler evaluations.

There are various interpretations of line integral depending upon the definition of the scalar field $f(x,y,z)$. For example, if $f(x,y,z) = 1$, the resulting line integral (116) simplifies to

$$I = \int_C ds = \int_a^b \|\mathbf{R}'(t)\| \, dt, \tag{117}$$

which we recognize as that defining the *arc length* of C [recall Eq. (53)]. If $f(x,y,z) = \mathbf{F} \cdot \hat{\mathbf{T}}$, where $\hat{\mathbf{T}}$ is the unit tangent vector to the curve, then

$$I = \int_C (\mathbf{F} \cdot \hat{\mathbf{T}}) ds = \int_C \mathbf{F} \cdot d\mathbf{R}, \tag{118}$$

where we have used the relation $\hat{\mathbf{T}} ds = (d\mathbf{R}/ds) ds = d\mathbf{R}$. When the vector field $\mathbf{F}$ is a force field, then the integral (118) represents the *work* done in moving an object along the path C. When the integral (118) is around a closed path, it represents the *circulation*, such as that associated with fluid flow where $\mathbf{F}$ is the velocity field (e.g., see Section 7.11.1). Other interpretations are also possible.

EXAMPLE 17 If an object is acted on by the force

$$\mathbf{F}(x,y,z) = \langle 3x^2, 2xz - y, z \rangle,$$

find the work done in moving the object along the curve described by $x = 2t^2$, $y = t$, $z = 4t^2 - t$, $0 \le t \le 1$.

Solution: In this case we employ a variation of the approach used in Example 16. First, we write

$$\int_C \mathbf{F} \cdot d\mathbf{R} = \int_C 3x^2 \, dx + (2xz - y) dy + z \, dz$$
$$= \int_C 3x^2 \, dx + \int_C (2xz - y) dy + \int_C z \, dz.$$

In the first and last integrals in the second step on the right-hand side, we can let x and z be the parameter of integration (since no other variables occur), whereas in the middle integral we use the given parametric representation. This leads to the integrals

$$\int_C \mathbf{F} \cdot d\mathbf{R} = \int_0^2 3x^2 \, dx + \int_0^1 [2(8t^4 - 2t^3) - t] dt + \int_0^3 z \, dz = \frac{71}{5}.$$

4.6.2 Conservative fields

To begin, we introduce some terminology. We say a region R is *open* if every point of R is an interior point (not including the boundary). The open region R is further said to be *connected* if, given any two points P and Q in R, a smooth curve (arc) C exists in R connecting P and Q. It is *simply-connected* if there are no "holes" in the region (i.e., "donut shapes"). Last, an open and connected region is called a *domain D*.

▸ **Conservative field:** We say a vector field $\mathbf{F}$ is *conservative* in a domain D if and only if there exists a scalar field φ such that $\mathbf{F} = \nabla\varphi$.

▸ **Potential function:** If $\mathbf{F} = \nabla\varphi$, we call the scalar field φ a *potential function*.

Theorem 4.3 (*Independence of path*) If the vector field $\mathbf{F}$ is continuous in a domain D, then $\mathbf{F}$ is conservative if and only if the line integral of the tangential component of $\mathbf{F}$ along every simple curve (i.e., curves that do not cross themselves) in D depends only on the endpoints (i.e., is independent of path).

Proof: Because of the "if and only if" statement, the proof consists of two parts.

Part I: Here we will show that if $\mathbf{F} = \nabla\varphi$, then

$$\int_P^Q \mathbf{F} \cdot d\mathbf{R} = \text{independent of path } C.$$

Because $\mathbf{F} = \nabla\varphi$, it follows that $\mathbf{F} \cdot d\mathbf{R}$ is an exact differential, i.e.,

$$\mathbf{F} \cdot d\mathbf{R} = \nabla\varphi \cdot d\mathbf{R} = \frac{\partial\varphi}{\partial x}dx + \frac{\partial\varphi}{\partial y}dy + \frac{\partial\varphi}{\partial z}dz = d\varphi,$$

and hence,

$$\int_P^Q \mathbf{F} \cdot d\mathbf{R} = \int_P^Q d\varphi = \varphi(Q) - \varphi(P).$$

This shows that the integral only depends on the endpoints of the path C.

Part II: Now we want to show that if the integral $\int_P^Q \mathbf{F} \cdot d\mathbf{R}$ is independent of path, then necessarily $\mathbf{F} = \nabla\varphi$.

Let P be a fixed point in D and $Q = (x,y,z)$ a variable point. Because the integral is assumed independent of path, there exists a function φ such that

$$\varphi(x,y,z) - \varphi(P) = \int_P^{(x,y,z)} \left(\mathbf{F} \cdot \frac{d\mathbf{R}}{ds} \right) ds;$$

and by differentiating both sides of this expression, we obtain

$$\frac{d\varphi}{ds} = \mathbf{F} \cdot \frac{d\mathbf{R}}{ds}.$$

Also, from definition it is known that

$$\frac{d\varphi}{ds} = \nabla\varphi \cdot \frac{d\mathbf{R}}{ds},$$

from which it follows by subtracting these last two results,

$$(\nabla\varphi - \mathbf{F}) \cdot \frac{d\mathbf{R}}{ds} = 0.$$

We now argue that $\nabla\varphi - \mathbf{F}$ cannot always be orthogonal to $d\mathbf{R}/ds$ (because $\mathbf{F}$ can change), so we must conclude $\nabla\varphi - \mathbf{F} = \mathbf{0}$, or $\nabla\varphi = \mathbf{F}$. ❏

Theorem 4.4 A continuous vector field $\mathbf{F}$ is conservative in a domain D if and only if, for all simple closed curves C in D, it can be shown that

$$\oint_C \mathbf{F} \cdot d\mathbf{R} = 0$$

The proof of Theorem 4.4 follows directly from Theorem 4.3 so we do not present it. Also, we use the special notation $\oint_C$ to define a line integral around a closed path C. Last, we have the following theorem.

Theorem 4.5 If $\mathbf{F}$ is a continuous vector field in a simply-connected domain D and has continuous first partial derivatives in D, then $\mathbf{F}$ is conservative in D if and only if

$$\nabla \times \mathbf{F} = \mathbf{0}.$$

If we assume $\mathbf{F} = \nabla\varphi$, then the first part of the proof of Theorem 4.5 follows immediately from the vector identity $\nabla \times \nabla\varphi = \mathbf{0}$. To show that $\mathbf{F}$ is conservative, given $\nabla \times \mathbf{F} = \mathbf{0}$, is a little more complicated and will be omitted.

EXAMPLE 18 Given the vector field

$$\mathbf{F}(x,y,z) = \langle 2xy + z^3, x^2, 3xz^2 \rangle,$$

find the amount of work done in moving the object from the point $(1, -2, 1)$ to the point $(3, 1, 4)$ along any curve.

Solution: We first note that

$$\nabla \times \mathbf{F} = \begin{vmatrix} \hat{\mathbf{i}} & \hat{\mathbf{j}} & \hat{\mathbf{k}} \\ \dfrac{\partial}{\partial x} & \dfrac{\partial}{\partial y} & \dfrac{\partial}{\partial z} \\ 2xy + z^3 & x^2 & 3xz^2 \end{vmatrix} = \mathbf{0},$$

thereby showing that $\mathbf{F}$ is a conservative field. Hence, based on Theorem 4.3 we know that the line integral for work is independent of the path. To find the associated scalar potential function, we must solve the simultaneous equations

$$\frac{\partial \varphi}{\partial x} = 2xy + z^3,$$

$$\frac{\partial \varphi}{\partial y} = x^2,$$

$$\frac{\partial \varphi}{\partial z} = 3xz^2.$$

By integrating the first of these, we get $\varphi(x,y,z) = x^2 y + xz^3 + f(y,z)$, where we must now determine $f(y,z)$. Based on the two remaining equations above, we get

$$\frac{\partial \varphi}{\partial y} = x^2 + \frac{\partial f}{\partial y} = x^2,$$

$$\frac{\partial \varphi}{\partial z} = 3xz^2 + \frac{\partial f}{\partial z} = 3xz^2,$$

from which we deduce $f(y,z) = $ constant. By selecting $f(y,z) = 0$, we have

$$\varphi(x,y,z) = x^2 y + xz^3.$$

Because the work done is independent of path, we now deduce that

$$W = \int_{(1,-2,1)}^{(3,1,4)} \mathbf{F} \cdot d\mathbf{R} = \int_{(1,-2,1)}^{(3,1,4)} d\varphi$$

$$= \varphi(3,1,4) - \varphi(1,-2,1) = 202.$$

4.6.3 Surface integrals

We have previously shown that space curves are characterized by tangent vectors. Surfaces, on the other hand, are characterized by the *normal vector*. In particular, a surface S is said to be *smooth* if it has a unit normal vector $\hat{\mathbf{N}}$ that changes direction in a continuous manner over the surface.

We can define a surface in a three-dimensional space by specifying

$$z = f(x,y) \text{ or } g(x,y,z) = K \text{ (constant)}. \tag{119}$$

For example, $z = x^2 + y^2$ represents a paraboloid of revolution, whereas $x^2 + y^2 + z^2 = a^2$ denotes a sphere of radius a. We can define a unit normal in each case by using the gradient. That is, by writing $g(x,y,z) = f(x,y) - z = K$ for a surface defined by the first equation in (119), we get in either case the *unit normal* to the surface given by

$$\hat{\mathbf{N}} = \pm \frac{\nabla g}{\|\nabla g\|}. \tag{120}$$

In dealing with space curves we found it convenient in many cases to use parametric equations. For surfaces, the parametric representation requires two parameters (two degrees of freedom), which we denote by u and v. Thus, we write

$$S: \quad \mathbf{R}(u,v) = \langle x(u,v), y(u,v), z(u,v) \rangle, \tag{121}$$

where $\mathbf{R}(u,v)$ is a position vector to any point on the surface. In representing a surface by the vector equation (121), the unit normal vector to the surface is defined by a particular cross product. To derive this result, we note that families of curves are traced out on the surface S by fixing one parameter and varying the other. For example, these families are defined by $u = C_1$ and $v = C_2$, where C_1 and C_2 are any constants. If we consider any curve $v = C_2$, we can obtain the tangent vector

$$\frac{\partial \mathbf{R}}{\partial u} = \langle \frac{\partial x}{\partial u}, \frac{\partial y}{\partial u}, \frac{\partial z}{\partial u} \rangle,$$

and similarly, along the curve $u = C_1$, we obtain the tangent vector

$$\frac{\partial \mathbf{R}}{\partial v} = \langle \frac{\partial x}{\partial v}, \frac{\partial y}{\partial v}, \frac{\partial z}{\partial v} \rangle.$$

Because both these vectors are tangent to the surface, it follows from the definition of cross product that the unit normal to the surface is given by the expression

$$\hat{\mathbf{N}} = \pm \frac{\dfrac{\partial \mathbf{R}}{\partial u} \times \dfrac{\partial \mathbf{R}}{\partial v}}{\left\| \dfrac{\partial \mathbf{R}}{\partial u} \times \dfrac{\partial \mathbf{R}}{\partial v} \right\|}. \tag{122}$$

EXAMPLE 19 Use both (120) and (122) to calculate the unit normal for the sphere

$$x^2 + y^2 + z^2 = a^2$$

Solution: The easiest way to calculate the unit normal in this case is to define the function $g(x,y,z) = x^2 + y^2 + z^2 - a^2$, and use Eq. (120). Doing so, we see that $\nabla g = \langle 2x, 2y, 2z \rangle$ and $\|\nabla g\| = 2a$. Hence, the *outward* unit normal is

$$\hat{\mathbf{N}} \ = \ <\frac{x}{a}, \frac{y}{a}, \frac{z}{a}>.$$

To calculate the unit normal directly from Eq. (122) requires more effort. First, we note that the standard (but not only) parameterization for a sphere is defined by

$$S: \quad x = a \cos u \cos v, \quad y = a \sin u \cos v, \quad z = a \sin v,$$

$$0 \le u < 2\pi, \quad -\frac{\pi}{2} < v < \frac{\pi}{2}.$$

We can interpret this parameterization as a mapping from the rectangular domain defined in the *uv*-plane to the sphere. Note that, although similar, this representation is *not* the same as spherical coordinates. From this representation, we find that

$$\frac{\partial \mathbf{R}}{\partial u} \ = \ <-a \sin u \cos v, \, a \cos u \cos v, \, 0>,$$

$$\frac{\partial \mathbf{R}}{\partial v} \ = \ <-a \cos u \sin v, \, -a \sin u \sin v, \, a \cos v>.$$

Next, using the relations

$$\frac{\partial \mathbf{R}}{\partial u} \times \frac{\partial \mathbf{R}}{\partial v} \ = \ <a^2 \cos u \cos^2 v, \, a^2 \sin u \cos^2 v, \, a^2 \sin v \cos v>,$$

$$\left\| \frac{\partial \mathbf{R}}{\partial u} \times \frac{\partial \mathbf{R}}{\partial v} \right\| \ = \ a^2 \left| \cos v \right|,$$

we arrive at

$$\hat{\mathbf{N}} \ = \ <\cos u \cos v, \, \sin u \cos v, \, \sin v>.$$

Surface integrals are a generalization of the double integral introduced in calculus for finding area between curves. In particular, in a surface integral the integration takes place over a surface *S*. We have essentially two methods for developing and solving a surface integral, both of which we present here. The first method involves the notion of "projection" of the surface area onto one of the coordinate planes, which in effect reduces the surface integral to a double integral in the plane. The second method is based on a parametric representation of the surface.

METHOD I:

Let $f(x,y,z) = K$ define a surface *S* that lies completely above the *xy*-plane as illustrated in Fig. 4.23. The projection of the surface *S* onto the *xy*-plane identifies the area denoted by A_{xy} in the figure. Also, let $G(x,y,z)$ be a continuous scalar field defined on *S*. If we partition the area A_{xy} in the *xy*-plane into *m* small elements $\Delta A_p, p = 1, 2, ..., m$ corresponding to the projection of the surface area element ΔS_p onto the xy-plane and (x_p,y_p,z_p) an arbitrary point on ΔS_p, then we define the *surface integral* of G over the surface S by the limit (provided it exists)

$$\iint_S G(x,y,z)\,dS \;=\; \lim_{\substack{m \to \infty \\ \max|\Delta S_p| \to 0}} \sum_{p=1}^{m} G(x_p,y_p,z_p)\Delta S_p. \tag{123}$$

Because we have assumed the surface S has the projection A_{xy} in the xy-plane, we can transform the surface integral in (123) into a double integral in the xy-plane for evaluation. To do so, we write the projection of ΔS_p onto the element ΔA_p by

$$\Delta A_p \;\cong\; \Delta S_p \cos\gamma,$$

where γ is the angle between the unit normal $\hat{\mathbf{N}}_p$ to ΔS_p and the unit normal $\hat{\mathbf{k}}$ to ΔA_p. Hence,

$$\Delta A_p \;\cong\; \Delta S_p \|\hat{\mathbf{N}}_p \cdot \hat{\mathbf{k}}\|.$$

If the area ΔA_p is rectangular so that $\Delta A_p = \Delta x_p \Delta y_p$, then in the limit (123), we obtain

$$dS \;=\; \frac{dx\,dy}{\|\hat{\mathbf{N}} \cdot \hat{\mathbf{k}}\|},$$

and therefore, the surface integral in (123) becomes

$$\boxed{\iint_S G(x,y,z)\,dS \;=\; \iint_{A_{xy}} G(x,y,z)\frac{dx\,dy}{\|\hat{\mathbf{N}} \cdot \hat{\mathbf{k}}\|}.} \tag{124}$$

The integral on the right-hand side of (124) is now simply a double integral in the xy-plane that can be evaluated through conventional methods from calculus.

If the projection is in either the yz-plane or the xz-plane, rather than in the xy-plane, then it follows through similar arguments that, respectively,

$$\boxed{\iint_S G(x,y,z)\,dS \;=\; \iint_{A_{yz}} G(x,y,z)\frac{dy\,dz}{\|\hat{\mathbf{N}} \cdot \hat{\mathbf{i}}\|},} \tag{125}$$

$$\boxed{\iint_S G(x,y,z)\,dS \;=\; \iint_{A_{xz}} G(x,y,z)\frac{dx\,dz}{\|\hat{\mathbf{N}} \cdot \hat{\mathbf{j}}\|}.} \tag{126}$$

Like line integral, there are several interpretations of surface integral depending on the nature of the scalar field $G(x,y,z)$. For example, if $G(x,y,z) = 1$, the resulting surface integral yields the surface area of S, i.e.,

$$S = \iint_S dS.$$

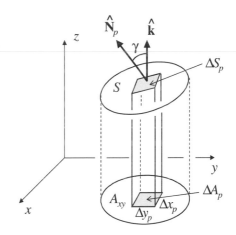

Also, if $G(x,y,z) = \mathbf{F} \cdot \hat{\mathbf{N}}$, where $\hat{\mathbf{N}}$ is the unit normal to the surface, the surface integral represents the flux of the vector quantity $\mathbf{F}$ through the surface (see Example 20 below). For example, this flux integral might represent the mass or volume of fluid (or electromagnetic energy) flowing across the surface in the normal direction.

Figure 4.23

EXAMPLE 20 Given the vector field $\mathbf{F}(x,y,z) = <z, x, -3y^2z>$, evaluate the surface integral

$$I = \iint_S (\mathbf{F} \cdot \hat{\mathbf{N}}) dS$$

over the portion of the cylinder in the first octant defined by

$$S: \ x^2 + y^2 = 16, \quad 0 \le x \le 4, \ 0 \le y \le 4, \ 0 \le z \le 5.$$

Solution: We interpret the vector $\hat{\mathbf{N}}$ to be the outward unit normal to the surface which we calculate first. If we write the surface as

$$f(x,y,z) = x^2 + y^2 - 16 = 0,$$

we find the unit normal

$$\hat{\mathbf{N}} = \frac{\nabla f}{\|\nabla f\|} = \frac{<2x, 2y, 0>}{\sqrt{4x^2 + 4y^2}} = <\frac{x}{4}, \frac{y}{4}, 0>,$$

which is parallel to the xy-plane. Using this expression, it follows that

$$\mathbf{F} \cdot \hat{\mathbf{N}} = \frac{1}{4}(xz + xy).$$

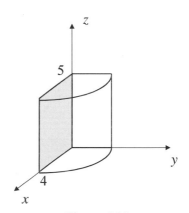

Figure 4.24

Because the normal vector $\hat{\mathbf{N}}$ is parallel to the xy-plane, the surface doesn't have a projection in the xy-plane, but it does have a projection in either of the other coordinate planes. If we choose the xz-plane as shown in Fig. 4.24, then

$$dS = \frac{dx\,dz}{\|\hat{\mathbf{N}}\cdot\hat{\mathbf{j}}\|} = \frac{4}{y}dx\,dz;$$

and, by writing $y = \sqrt{16 - x^2}$, we see that

$$I = \int_0^5\int_0^4\left(\frac{xz + xy}{y}\right)dx\,dz$$
$$= \int_0^5\int_0^4\frac{xz}{\sqrt{16 - x^2}}dx\,dz + \int_0^5\int_0^4 x\,dx\,dz,$$

or

$$I = \iint_S (\mathbf{F}\cdot\hat{\mathbf{N}})dS = 90.$$

Had we projected the cylindrical surface onto the yz-plane in Example 20, the calculations would be virtually identical. However, in some cases it may take more than one projection to obtain the double integrals in the plane of projection. That is, there may exist a projection of the spatial surface onto the "front" surface of the plane and another onto the "back" surface of the same plane. For instance, consider the following example.

EXAMPLE 21 Evaluate the surface integral

$$I = \iint_S (2x^2 - y^3)dS$$

over the entire cylinder described by

$$S: \quad x^2 + y^2 = 4, \ 0 \le z \le 3.$$

Solution: As in Example 20, we consider the projection onto the xz-plane. However, because we want to integrate over the entire cylinder this time, we must consider *two* projections. The projection from the positive y-axis side is considered a positive projection in which we write $+dx\,dz$, whereas that from the negative y-axis side is negative, i.e., $-dx\,dz$. In addition, we must replace y by $-y$ on the negative side of the projection.

Based on Example 20, we know

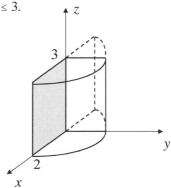

Figure 4.25 Projection from the positive y-axis side.

$$\hat{\mathbf{N}} = <\frac{x}{2}, \frac{y}{2}, 0>$$

and thus, $\|\hat{\mathbf{N}} \cdot \hat{\mathbf{j}}\| = y/2$. The surface integral projected onto both sides of the xz-plane leads to the expression

$$I = 4\int_0^3 \int_0^2 \left(\frac{2x^2 - y^3}{y}\right) dx\,dz - 4\int_0^3 \int_0^2 \left(\frac{2x^2 + y^3}{-y}\right) dx\,dz$$

$$= 16\int_0^3 \int_0^2 \frac{x^2}{\sqrt{4-x^2}} dx\,dz,$$

which, upon evaluation, yields

$$I = \iint_S (2x^2 - y^3)\,dS = 48\pi.$$

METHOD II:

The method of projections (METHOD I) requires a reasonable sketch of the surface to identify the appropriate planes for the projection(s). In addition, it is not always well-suited for problems in which multiple projections are required. Here we present another method that uses the vector representation of the surface and does not require any projection.

Let us suppose the surface integral we wish to evaluate has a single projection in the xy-plane so that we may write

$$\iint_S G(x,y,z)\,dS = \iint_{A_{xy}} G(x,y,z)\frac{dx\,dy}{\|\hat{\mathbf{N}} \cdot \hat{\mathbf{k}}\|}. \tag{127}$$

If the surface has the vector representation

$$S: \quad \mathbf{R}(u,v) = <x(u,v), y(u,v), z(u,v)>, \tag{128}$$

then we simply make a transformation of variables from the xy-plane to the uv-plane, which leads to

$$\iint_{A_{xy}} G(x,y,z)\frac{dx\,dy}{\|\hat{\mathbf{N}} \cdot \hat{\mathbf{k}}\|} = \iint_{A_{uv}} G[x(u,v),y(u,v),z(u,v)]\left|\frac{\partial(x,y)}{\partial(u,v)}\right|\frac{du\,dv}{\|\hat{\mathbf{N}} \cdot \hat{\mathbf{k}}\|}, \tag{129}$$

where $J = \left|\dfrac{\partial(x,y)}{\partial(u,v)}\right|$ is the *Jacobian of transformation* defined by

$$J = \left|\frac{\partial(x,y)}{\partial(u,v)}\right| = \begin{vmatrix} \dfrac{\partial x}{\partial u} & \dfrac{\partial x}{\partial v} \\[2mm] \dfrac{\partial y}{\partial u} & \dfrac{\partial y}{\partial v} \end{vmatrix}. \tag{130}$$

For example, given the polar coordinate transformation $x = r\cos\theta$, $y = r\sin\theta$, the Jacobian of transformation is

$$J = \begin{vmatrix} \dfrac{\partial x}{\partial r} & \dfrac{\partial x}{\partial \theta} \\[2mm] \dfrac{\partial y}{\partial r} & \dfrac{\partial y}{\partial \theta} \end{vmatrix} = \begin{vmatrix} \cos\theta & -r\sin\theta \\ \sin\theta & r\cos\theta \end{vmatrix} = r. \tag{131}$$

At this point we need the following theorem to simplify Eq. (129).

Theorem 4.6 The Jacobian of transformation (130) satisfies the identity

$$\left| \frac{\partial(x,y)}{\partial(u,v)} \right| \frac{1}{\|\hat{\mathbf{N}}\cdot\hat{\mathbf{k}}\|} = \left\| \frac{\partial \mathbf{R}}{\partial u} \times \frac{\partial \mathbf{R}}{\partial v} \right\|.$$

Proof: From Eq. (122), we recall

$$\hat{\mathbf{N}} = \pm \frac{\dfrac{\partial \mathbf{R}}{\partial u} \times \dfrac{\partial \mathbf{R}}{\partial v}}{\left\| \dfrac{\partial \mathbf{R}}{\partial u} \times \dfrac{\partial \mathbf{R}}{\partial v} \right\|},$$

and therefore,

$$\|\hat{\mathbf{N}}\cdot\hat{\mathbf{k}}\| = \frac{\left\| \dfrac{\partial \mathbf{R}}{\partial u} \times \dfrac{\partial \mathbf{R}}{\partial v}\cdot\hat{\mathbf{k}} \right\|}{\left\| \dfrac{\partial \mathbf{R}}{\partial u} \times \dfrac{\partial \mathbf{R}}{\partial v} \right\|}.$$

Now, by forming the triple scalar product

$$\frac{\partial \mathbf{R}}{\partial u} \times \frac{\partial \mathbf{R}}{\partial v}\cdot\hat{\mathbf{k}} = \begin{vmatrix} \dfrac{\partial x}{\partial u} & \dfrac{\partial y}{\partial u} & \dfrac{\partial z}{\partial u} \\[2mm] \dfrac{\partial x}{\partial v} & \dfrac{\partial y}{\partial v} & \dfrac{\partial z}{\partial v} \\[2mm] 0 & 0 & 1 \end{vmatrix} = \begin{vmatrix} \dfrac{\partial x}{\partial u} & \dfrac{\partial y}{\partial u} \\[2mm] \dfrac{\partial x}{\partial v} & \dfrac{\partial y}{\partial v} \end{vmatrix},$$

we recognize that the last determinant is simply the Jacobian with rows and columns interchanged. Hence, we see that

$$\|\hat{\mathbf{N}}\cdot\hat{\mathbf{k}}\| = \frac{\left\| \dfrac{\partial \mathbf{R}}{\partial u} \times \dfrac{\partial \mathbf{R}}{\partial v}\cdot\hat{\mathbf{k}} \right\|}{\left\| \dfrac{\partial \mathbf{R}}{\partial u} \times \dfrac{\partial \mathbf{R}}{\partial v} \right\|} = \frac{1}{\left\| \dfrac{\partial \mathbf{R}}{\partial u} \times \dfrac{\partial \mathbf{R}}{\partial v} \right\|} \left| \frac{\partial(x,y)}{\partial(u,v)} \right|,$$

from which our intended result follows. ❏

Last, by using the identity in Theorem 4.6, we deduce that

$$\iint_S G(x,y,z)\,dS = \iint_{A_{uv}} G[x(u,v),y(u,v),z(u,v)]\left\|\frac{\partial \mathbf{R}}{\partial u}\times\frac{\partial \mathbf{R}}{\partial v}\right\|du\,dv. \tag{132}$$

Although our derivation of (132) was based on an initial projection into the *xy*-plane, this result is independent of which plane contains the initial projection. One of the main advantages in using (132) over the projection method is that the integral "setups" in the *uv*-plane are often more straightforward (although sometimes tedious).

EXAMPLE 22 Solve Example 21 by METHOD II.

Solution: The cylindrical surface $x^2 + y^2 = 4$, $0 \le z \le 3$, has the parametric equation representation S: $x = 2\cos u$, $y = 2\sin u$, $z = v$, or

$$S: \quad \mathbf{R}(u,v) = <2\cos u, 2\sin u, v>,$$
$$0 \le u \le 2\pi, \qquad 0 \le v \le 3.$$

From this result we calculate

$$\frac{\partial \mathbf{R}}{\partial u}\times\frac{\partial \mathbf{R}}{\partial v} = \begin{vmatrix} \hat{\mathbf{i}} & \hat{\mathbf{j}} & \hat{\mathbf{k}} \\ -2\sin u & 2\cos u & 0 \\ 0 & 0 & 1 \end{vmatrix} = <2\cos u, 2\sin u, 0>,$$

and find $\left\|\dfrac{\partial \mathbf{R}}{\partial u}\times\dfrac{\partial \mathbf{R}}{\partial v}\right\| = 2$. The given integral in this case reduces to

$$I = 16\int_0^3\int_0^{2\pi}(\cos^2 u - \sin^3 u)\,du\,dv$$
$$= 16\pi\int_0^3 dv = 48\pi.$$

Observe that in Example 22 we set up only one integral over the entire surface whereas in Example 21 we had to consider separate projections over the positive side and negative side of the projection plane.

EXAMPLE 23 Calculate the surface area of a sphere of radius a.

Solution: The vector representation for a sphere is given by (see Example 19)

$$S: \quad \mathbf{R}(u,v) \; = \; <a \cos u \cos v, a \sin u \cos v, a \sin v>.$$

Moreover, in Example 19 we showed that

$$\left\| \frac{\partial \mathbf{R}}{\partial u} \times \frac{\partial \mathbf{R}}{\partial v} \right\| \; = \; a^2 \, |\cos v| \, ,$$

and therefore,

$$S \; = \; \iint_S dS \; = \; a^2 \int_{-\pi/2}^{\pi/2} \int_0^{2\pi} |\cos v| \, du \, dv.$$

Upon further evaluation, we get the familiar result

$$S \; = \; 4\pi a^2 \int_0^{\pi/2} \cos v \, dv \; = \; 4\pi a^2.$$

4.7 Integral Relations Between Line, Surface, and Volume Integrals

Developing relations between various line integrals, surface integrals, and volume integrals can be important for our theoretical understanding of these concepts, but also practical in that such relations may permit us to exchange a difficult surface integral, for example, with a simpler line integral, or, in some cases, the other way around. In this section we introduce the classic theorem of Green (Theorem 4.7), the divergence theorem of Gauss (Theorem 4.10), and Stokes' theorem (Theorem 4.11), all of which have far-reaching consequences in various branches of mathematics, physics, and engineering.

4.7.1 Green's Theorem in the Plane

We begin with the following theorem (*Green's theorem in the plane*) relating certain line integrals in the plane around closed paths with equivalent double integrals over the enclosed area.

Theorem 4.7 (*Green's theorem in the plane*) Let D be a bounded simply-connected domain in the xy-plane with a boundary curve C consisting of finitely-many smooth curves. If $P(x,y)$ and $Q(x,y)$, and their first-partial derivatives, are all continuous in D, then

$$\oint_C P(x,y)dx + Q(x,y)dy \; = \; \iint_D \left(\frac{\partial Q}{\partial x} - \frac{\partial P}{\partial y} \right) dx\,dy,$$

where integration is in the counterclockwise direction around C.

Partial Proof: We will present the proof only for a special domain D which can be represented by both of the following [see Figs. 4.26(*a*) and (*b*)] :

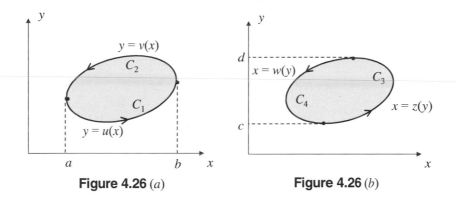

Figure 4.26 (*a*)   **Figure 4.26** (*b*)

$$a \le x \le b, \quad u(x) \le y \le v(x)$$

$$c \le y \le d, \quad w(y) \le x \le z(y)$$

We first wish to show that

$$\iint_D \frac{\partial P}{\partial y} dx \, dy = -\oint_C P(x,y) dx.$$

From Fig. 4.26(*a*), it follows that

$$\iint_D \frac{\partial P}{\partial y} dx \, dy = \int_a^b \int_{u(x)}^{v(x)} \frac{\partial P}{\partial y} dy \, dx$$

$$= \int_a^b P[x,v(x)] \, dx - \int_a^b P[x,u(x)] \, dx$$

or

$$\iint_D \frac{\partial P}{\partial y} dx \, dy = -\int_{C_2} P(x,y) dx - \int_{C_1} P(x,y) dx$$

$$= -\oint_C P(x,y) dx,$$

where $C = C_1 + C_2$. In a similar manner, referring to Fig. 4.26(*b*), we can deduce that

$$\iint_D \frac{\partial Q}{\partial x} dx \, dy = \oint_C Q(x,y) dy,$$

and, thus, by combining results, we have proved the theorem. ❏

Remark: Green's theorem is also valid in multiply-connected domains, but we will not separately discuss this case.

EXAMPLE 24 Use Green's theorem to evaluate the line integral

$$I = \oint_C (3x + 4y)dx + (2x - 3y)dy,$$

where C is the circle $x^2 + y^2 = 4$.

Solution: With $P(x,y) = 3x + 4y$ and $Q(x,y) = 2x - 3y$, the application of Green's theorem leads to

$$I = -2\iint_{x^2+y^2<4} dx\,dy = -2\times(\text{area of circle}),$$

or

$$I = -2(4\pi) = -8\pi.$$

For the special case in which $P(x,y) = -y/2$ and $Q(x,y) = x/2$, the double integral in Green's theorem reduces to calculating the area of the domain D, i.e., $\iint(\partial Q/\partial x - \partial P/\partial y)dx\,dy = \iint dx\,dy = A$. Hence, we see that the area of a plane region can be written as the line integral

$$A = \frac{1}{2}\oint_C (x\,dy - y\,dx). \qquad (133)$$

EXAMPLE 25 Find the area of the ellipse

$$\frac{x^2}{a^2} + \frac{y^2}{b^2} = 1.$$

Solution: Let us represent the boundary curve of the ellipse by

$$C: \quad x = a\cos t, \quad y = b\sin t, \quad 0 \le t \le 2\pi.$$

From (133), it follows that

$$A = \frac{1}{2}\oint_C (x\,dy - y\,dx)$$
$$= \frac{1}{2}\int_0^{2\pi} [a\cos t(b\cos t\,dt) - b\sin t(-a\sin t\,dt)]$$

which reduces to

$$A = \frac{1}{2}\int_0^{2\pi} (ab)dt = \pi ab.$$

For later work it is useful to formulate Green's theorem in vector notation. To begin, we define the vector field

$$\mathbf{F}(x,y,z) = <P(x,y), Q(x,y)>, \tag{134}$$

from which we make the observation

$$\nabla \times \mathbf{F} \cdot \hat{\mathbf{k}} = \frac{\partial Q}{\partial x} - \frac{\partial P}{\partial y}. \tag{135}$$

If we define the position vector $\mathbf{R} = <x, y>$, then Green's theorem can be reformulated as

$$\oint_C \mathbf{F} \cdot d\mathbf{R} = \iint_D (\nabla \times \mathbf{F} \cdot \hat{\mathbf{k}}) \, dx \, dy. \tag{136}$$

Note that if $\mathbf{F}$ is conservative, the right-hand side is zero.

Another useful vector formulation arises by first defining the vector

$$\boldsymbol{\psi} = \mathbf{F} \times \hat{\mathbf{k}}, \tag{137}$$

where $\mathbf{F}$ is the vector field defined by (134). If $\hat{\mathbf{T}}$ and $\hat{\mathbf{N}}$ denote the unit tangent vector and outward unit normal vector, respectively, to the curve C, then

$$\boldsymbol{\psi} \cdot \hat{\mathbf{N}} = \mathbf{F} \times \hat{\mathbf{k}} \cdot \hat{\mathbf{N}} = \mathbf{F} \cdot \hat{\mathbf{k}} \times \hat{\mathbf{N}},$$

or

$$\boldsymbol{\psi} \cdot \hat{\mathbf{N}} = \mathbf{F} \cdot \hat{\mathbf{T}}. \tag{138}$$

Also, noting that

$$\nabla \cdot \boldsymbol{\psi} = \nabla \cdot \mathbf{F} \times \hat{\mathbf{k}} = \nabla \times \mathbf{F} \cdot \hat{\mathbf{k}},$$

and using (118), we find that Green's theorem (136) takes the form

$$\oint_C (\boldsymbol{\psi} \cdot \hat{\mathbf{N}}) \, ds = \iint_D (\nabla \cdot \boldsymbol{\psi}) \, dx \, dy. \tag{139}$$

4.7.2 Harmonic functions

There are a number of consequences of Green's theorem in the plane that are important in the general theory of harmonic functions (see also Sections 6.4.3 and 7.8). Here we discuss some simple notions about harmonic functions and their connection to conservative vector fields, but limit our analysis to two-dimensional domains D.

To begin, let us define two vector fields

$$\begin{aligned}
\mathbf{F}(x,y) &= <v(x,y), u(x,y)>, \\
\mathbf{G}(x,y) &= <u(x,y), -v(x,y)>.
\end{aligned} \tag{140}$$

From inspection, it is clear that the vector fields in (14) are orthogonal, i.e., $\mathbf{F} \cdot \mathbf{G} = 0$. Using the vector formulation of Green's theorem given by (136), we obtain

$$\oint_C \mathbf{F} \cdot d\mathbf{R} = \iint_D (\nabla \times \mathbf{F} \cdot \hat{\mathbf{k}}) \, dx \, dy = \iint_D \left(\frac{\partial u}{\partial x} - \frac{\partial v}{\partial y} \right) dx \, dy,$$

$$\oint_C \mathbf{G} \cdot d\mathbf{R} = \iint_D (\nabla \times \mathbf{G} \cdot \hat{\mathbf{k}}) \, dx \, dy = \iint_D \left(-\frac{\partial v}{\partial x} - \frac{\partial u}{\partial y} \right) dx \, dy.$$

If we now specify that both $\mathbf{F}$ and $\mathbf{G}$ are *conservative*, then the above integrals must vanish, which leads to the set of equations

$$\begin{aligned} \frac{\partial u}{\partial x} &= \frac{\partial v}{\partial y} \\ \frac{\partial u}{\partial y} &= -\frac{\partial v}{\partial x}. \end{aligned} \tag{141}$$

Equations (141) are the well-known *Cauchy-Riemann equations* for analytic functions from complex variables (see Section 6.4.1). Through differentiation of these results it is easy to show that

$$\nabla^2 u = 0, \qquad \nabla^2 v = 0. \tag{142}$$

Each partial differential equation appearing in (142) is widely known as *Laplace's equation*, or the *potential equation* (see Section 12.5). Continuous solutions of this equation with continuous first- and second-order derivatives in a domain D are called *harmonic functions*. The above analysis reveals that the scalar components of a conservative vector field are harmonic functions.

The two most important problems associated with Laplace's equation are known as the *Dirichlet problem* and *Neumann problem*. These are defined, respectively, by

$$\text{Dirichlet Problem:} \qquad \begin{aligned} \nabla^2 u &= 0 \quad \text{in } D, \\ u &= f \quad \text{on } C, \end{aligned} \tag{143}$$

$$\text{Neumann Problem:} \qquad \begin{aligned} \nabla^2 u &= 0 \quad \text{in } D, \\ \frac{\partial u}{\partial \hat{\mathbf{N}}} &= f \quad \text{on } C. \end{aligned} \tag{144}$$

In both cases, the function f prescribed on the boundary C is assumed to be a continuous function. Although we won't solve these problems here (see Section 12.5), we can prove some properties about their solution.

To begin, we reformulate Green's theorem given by Eq. (139). Let us introduce the vector function $\boldsymbol{\psi} = v \nabla u$, which leads to

$$\nabla \cdot \boldsymbol{\psi} = \nabla \cdot (v \nabla u) = v \nabla^2 u + \nabla u \cdot \nabla v,$$

$$\boldsymbol{\psi} \cdot \hat{\mathbf{N}} = v (\nabla u \cdot \hat{\mathbf{N}}) = v \frac{\partial u}{\partial \hat{\mathbf{N}}},$$

where the first result follows from the second identity in Eqs. (107). The substitution of these equations into (139) gives us *Green's first identity*

$$\oint_C v \frac{\partial u}{\partial \hat{\mathbf{N}}} ds = \iint_D (v \nabla^2 u + \nabla u \cdot \nabla v) dx\, dy. \tag{145}$$

Using (145), we now prove the following lemma.

Lemma 4.1 If u is harmonic in a domain D and $u = 0$ on the boundary C, then $u = 0$ in D.

Proof: Let $v = u$ in (145), which yields

$$\oint_C u \frac{\partial u}{\partial \hat{\mathbf{N}}} ds = \iint_D (u \nabla^2 u + \|\nabla u\|^2) dx\, dy.$$

Because $u = 0$ on C and $\nabla^2 u = 0$ in D, this equation reduces to

$$\iint_D (\|\nabla u\|^2) dx\, dy = 0.$$

For this to be true, ∇u must be the zero vector which implies that u is constant. Last, because $u = 0$ on C, it must be zero everywhere. ❑

Using Lemma 4.1, we can now prove that if the Dirichlet problem (143) has a solution, that solution is *unique*.

Theorem 4.8 If u_1 and u_2 are both harmonic functions in D, and if $u_1 = u_2$ on the boundary C, then $u_1 = u_2$ in D.

Proof: Here we simply set $u = u_1 - u_2$ and use Lemma 4.1 to prove our result. ❑

The uniqueness of the Dirichlet problem implied by Theorem 4.8 is based on the *existence* of a solution. However, proving the existence of a solution for the Dirichlet problem is more difficult and we will not pursue it. Also, assuming a solution exists, it can be established by arguments similar to those above that the Neumann problem is unique only to within an additive constant. For a solution of the Neumann problem to *exist*, however, a certain *compatibility condition* must be satisfied by the boundary data.

Theorem 4.9 The Neumann problem (144) has a solution only if

$$\oint_C f ds = 0.$$

Proof: Let us set $v = 1$ in Green's first identity (145) to get

$$\oint_C \frac{\partial u}{\partial \hat{\mathbf{N}}} ds = \iint_D \nabla^2 u \, dx \, dy.$$

But, $\nabla^2 u = 0$ and $\partial u / \partial \hat{\mathbf{N}} = f$, which reduces the above expression to our intended compatibility condition. ❑

4.7.3 Divergence theorem and Stokes' theorem

In this section we want to present generalizations of Green's theorem in the plane to three-dimensional domains. In doing so, we find there are essentially two generalizations that arise, known as the divergence theorem and Stokes' theorem. Before introducing these theorems, however, it may be instructive to review the notion of volume integral.

Let $f(x,y,z)$ be a continuous scalar field throughout a closed and bounded domain V of space. Integrals of the form

$$I = \iiint_V f(x,y,z) \, dV$$

are commonly called *volume integrals* because of the domain of integration. Various interpretations of volume integrals are possible. For example, if $f(x,y,z) \equiv 1$, the integral simply reduces to the enclosed volume V. Also, the differential dV takes on different appearances depending upon coordinate system. In rectangular coordinates, for instance, the volume element becomes $dV = dx \, dy \, dz$.

EXAMPLE 26 Find the volume of the solid in the first octant under the surface $z = 4 - x^2 - y$ (see Fig. 4.27).

Solution: The traces (curves in the coordinate planes) are sketched in Fig. 4.27. Setting up the integral in rectangular coordinates yields

$$V = \int_0^2 \int_0^{4-x^2} \int_0^{4-x^2-y} dz \, dy \, dx.$$

Now performing successive integrations, we obtain

$$V = \int_0^2 \int_0^{4-x^2} (4 - x^2 - y) \, dy \, dx$$

$$= \frac{1}{2} \int_0^2 (4 - x^2)^2 \, dx,$$

which simplifies to $V = \dfrac{128}{15}$.

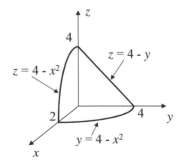

Figure 4.27

For coordinate systems other than rectangular, it is necessary to obtain the Jacobian of transformation in order to transform the volume integral from one coordinate system to another. The *Jacobian* is defined in three-dimensional space by the determinant

$$J = \left| \frac{\partial(x,y,z)}{\partial(u,v,w)} \right| = \begin{vmatrix} \dfrac{\partial x}{\partial u} & \dfrac{\partial x}{\partial v} & \dfrac{\partial x}{\partial w} \\[2mm] \dfrac{\partial y}{\partial u} & \dfrac{\partial y}{\partial v} & \dfrac{\partial y}{\partial w} \\[2mm] \dfrac{\partial z}{\partial u} & \dfrac{\partial z}{\partial v} & \dfrac{\partial z}{\partial w} \end{vmatrix}. \tag{146}$$

For example, if we introduce the coordinate transformation

$$x = x(u,v,w), \quad y = y(u,v,w), \quad z = z(u,v,w), \tag{147}$$

then the volume integral becomes

$$\iiint_{V_{xyz}} f(x,y,z)\,dx\,dy\,dz = \iiint_{V_{uvw}} f(u,v,w) \left| \frac{\partial(x,y,z)}{\partial(u,v,w)} \right| du\,dv\,dw. \tag{148}$$

EXAMPLE 27 Find the volume of a sphere of radius a.

Solution: In rectangular coordinates, the equation of the sphere is

$$x^2 + y^2 + z^2 = a^2.$$

Converting to spherical coordinates by the transformation

$$x = \rho \sin\varphi \cos\theta, \quad y = \rho \sin\varphi \sin\theta, \quad z = \rho \cos\varphi,$$

$$0 \le \rho \le a, \quad 0 \le \theta \le 2\pi, \quad 0 \le \varphi \le \pi.$$

the equation of the sphere reduces to simply $\rho = a$ and the Jacobian of transformation is $J = \rho^2 \sin\varphi$. Based on (148), the resulting volume integral becomes

$$V = \int_0^\pi \int_0^{2\pi} \int_0^a \rho^2 \sin\varphi \, d\rho\, d\theta\, d\varphi = \frac{4}{3}\pi a^3.$$

Theorem 4.10 (*Divergence theorem*) If V is a closed, bounded, and simply-connected domain of space whose boundary surface S consists of a finite number of piecewise smooth surfaces, and $\mathbf{F}(x,y,z)$ is a continuous vector field with continuous first partial derivatives in V, then

$$\oiint_S (\mathbf{F} \cdot \hat{\mathbf{N}})\, dS = \iiint_V (\nabla \cdot \mathbf{F})\, dV,$$

where $\hat{\mathbf{N}}$ is the outward unit normal to S.

The proof of Theorem 4.10 is similar to that of Green's theorem in the plane (Theorem 4.7), and is therefore omitted. Note, however, that if we expand the indicated vector operations under the integrals in rectangular coordinates, the divergence theorem can also be expressed as

$$\oiint_S (F_1\, dy\, dz + F_2\, dx\, dz + F_3\, dx\, dy) = \iiint_V \left(\frac{\partial F_1}{\partial x} + \frac{\partial F_2}{\partial y} + \frac{\partial F_3}{\partial z} \right) dx\, dy\, dz. \quad (149)$$

EXAMPLE 28 Given the vector function $\mathbf{F}(x,y,z) = \langle 4x, -2y^2, z^2 \rangle$, verify the divergence theorem over the surface (see Fig. 4.28)

$$S: \quad x^2 + y^2 = 4, \quad z = 0, \quad z = 3.$$

Solution: To verify the divergence theorem, we need to separately evaluate the surface integral and the volume integral, showing they lead to the same value.

Note that the surface consists of three piecewise smooth surfaces (top, lateral surface, and bottom) denoted by

$$S_1: \quad z = 3$$
$$S_2: \quad x^2 + y^2 = 4, \ 0 < z < 3$$
$$S_3: \quad z = 0$$

Hence, the surface integral we need to evaluate can be expressed as a sum of three surface integrals, viz.,

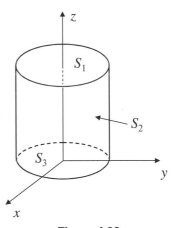

Figure 4.28

$$I = \oiint_S (\mathbf{F} \cdot \hat{\mathbf{N}})\, dS$$
$$= \iint_{S_1} (\mathbf{F} \cdot \hat{\mathbf{N}})\, dS + \iint_{S_2} (\mathbf{F} \cdot \hat{\mathbf{N}})\, dS + \iint_{S_3} (\mathbf{F} \cdot \hat{\mathbf{N}})\, dS.$$

S_1: $\hat{\mathbf{N}} = \hat{\mathbf{k}}$; $\mathbf{F} \cdot \hat{\mathbf{N}} = 9$

Here we find that

$$\iint_{S_1} (\mathbf{F} \cdot \hat{\mathbf{N}}) dS = 9 \iint_{S_1} dy \, dx = 9 \times (\text{area of top}) = 36\pi.$$

S_2: $\hat{\mathbf{N}} = <x/2, y/2, 0>$; $\mathbf{F} \cdot \hat{\mathbf{N}} = 2x^2 - y^3$

In this case it is best to convert to the vector representation of the cylindrical surface, i.e.,

$$S_2: \quad \mathbf{R}(u,v) = <2\cos u, 2\sin u, v>, \quad 0 \le u \le 2\pi, \quad 0 \le v \le 3$$

$$\left\| \frac{\partial \mathbf{R}}{\partial u} \times \frac{\partial \mathbf{R}}{\partial v} \right\| = 2, \quad \mathbf{F} \cdot \hat{\mathbf{N}} = 8\cos^2 u - 8\sin^3 u$$

Using the vector representation, we find

$$\iint_{S_2} (\mathbf{F} \cdot \hat{\mathbf{N}}) dS = 16 \int_0^{2\pi} \int_0^3 (\cos^2 u - \sin^3 u) \, dv \, du = 48\pi.$$

S_3: $\hat{\mathbf{N}} = -\hat{\mathbf{k}}$; $\mathbf{F} \cdot \hat{\mathbf{N}} = 0$

Because the dot product is zero, it follows that

$$\iint_{S_3} (\mathbf{F} \cdot \hat{\mathbf{N}}) dS = 0.$$

By summing results we deduce that the surface integral yields

$$I = \oiint_S (\mathbf{F} \cdot \hat{\mathbf{N}}) dS = 36\pi + 48\pi + 0 = 84\pi.$$

For the volume integral, we write

$$I = \iiint_V (\nabla \cdot \mathbf{F}) dV,$$

where $\nabla \cdot \mathbf{F} = 4 - 4y + 2z$. The shape of the volume suggests a change to cylindrical coordinates

$$x = r \cos\theta, \quad y = r\sin\theta, \quad z = z,$$

$$0 \le r \le 2, \; 0 \le \theta \le 2\pi, \; 0 \le z \le 3.$$

The Jacobian of transformation is $J = r$, leading to

$$I = 2\int_0^{2\pi}\int_0^2\int_0^3 (2 - 2r\sin\theta + z)r\,dz\,dr\,d\theta$$
$$= \int_0^{2\pi}\int_0^2 (21 - 12r\sin\theta)r\,dr\,d\theta$$
$$= \int_0^{2\pi} (42 - 32\sin\theta)\,d\theta = 84\pi.$$

Whereas the divergence theorem of Gauss provides a transformation link between surface integrals and volume integrals, the following theorem provides a similar transformation link between line integrals and surface integrals.

Theorem 4.11 (*Stokes' theorem*) If S is a piecewise smooth surface whose boundary C is a smooth simple curve, and if $\mathbf{F}(x,y,z)$ is a continuous vector field with continuous first partial derivatives on S and C, then

$$\oint_C \mathbf{F}\cdot d\mathbf{R} = \iint_S (\nabla\times\mathbf{F}\cdot\hat{\mathbf{N}})\,dS,$$

where $\hat{\mathbf{N}}$ is the unit normal consistent with the positive orientation of the curve.

In Theorem 4.11 it is clear that the value of the surface integral is dependent upon the choice of unit normal vector $\hat{\mathbf{N}}$. That is, instead of $\hat{\mathbf{N}}$ we could choose $-\hat{\mathbf{N}}$. Thus, by making a selection between these normal vectors we then say the integral is over an *oriented surface*. This orientation determines what we mean by the *positive direction* along C. We can state this by the following rule:

Right-hand rule: An observer on the positive side of the surface (i.e., the side on which $\hat{\mathbf{N}}$ emerges) is moving in the positive direction along C if the surface is at his/her left.

EXAMPLE 29 Given the vector function $\mathbf{F}(x,y,z) = <2y, z, 3y>$ and the line integral

$$I = \oint_C \mathbf{F}\cdot d\mathbf{R},$$

(a)   evaluate I over the intersection of the cylinder $x^2 + y^2 = 4$ and the plane $z = 2$.
(b)   Use Stokes' theorem to evaluate I as a surface integral over the disk formed by C.
(c)   Use Stokes' theorem to evaluate I as a surface integral over the hemisphere above C with C as the boundary curve (see Fig. 4.29).

Solution: (*a*) The curve *C* is a circle of radius 2 around the *z*-axis. Thus, we write

$$C: \quad \mathbf{R}(t) = <2\cos t, 2\sin t, 2>, \quad 0 \le t \le 2\pi,$$

where *t* in this case represents a central angle on the shaded circular disk *S*. Also, we see that $\mathbf{F} \cdot d\mathbf{R} = 4(-2\sin^2 t + \cos t)dt$, and we are led to

$$I = 4\int_0^{2\pi}(-2\sin^2 t + \cos t)dt = -8\pi.$$

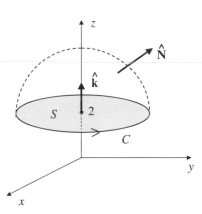

Figure 4.29

(*b*) Because the disk *S* in Fig. 4.29 is the circular area bounded by the circle *C*, we can write

$$\nabla \times \mathbf{F} \cdot \hat{\mathbf{N}} = \nabla \times \mathbf{F} \cdot \hat{\mathbf{k}} = -2,$$

and, based on Stokes' theorem, it follows that

$$I = \iint_S (\nabla \times \mathbf{F} \cdot \hat{\mathbf{N}})dS = -2\iint_S dS = -8\pi.$$

(*c*) For the hemisphere shown in Fig. 4.29, we have

$$\hat{\mathbf{N}} = <x/2, y/2, (z-2)/2>,$$

$$\nabla \times \mathbf{F} \cdot \hat{\mathbf{N}} = x + z - 2.$$

Because there is a projection onto the *xy*-plane, we write $dS = dx\,dy/\|\hat{\mathbf{N}} \cdot \hat{\mathbf{k}}\|$, which leads to

$$I = 2\iint_{A_{xy}}\left(\frac{x}{\sqrt{4 - x^2 - y^2}} - 1\right)dx\,dy,$$

where A_{xy} is the projected circle $x^2 + y^2 = 4$ in the *xy*-plane. Last, changing to polar coordinates then yields

$$I = 2\int_0^2\int_0^{2\pi}\left(\frac{r^2\cos\theta}{\sqrt{4 - r^2}} - r\right)d\theta\,dr = -8\pi.$$

4.8 Electromagnetic Theory

Although vectors are commonly associated with quantities from mechanics like force and acceleration, vector analysis was not used in the original development of mechanics. Instead, the need for vector analysis really emerged with the introduction of Maxwell's electromagnetic equations for the electric field $\mathbf{E}$ and magnetic field $\mathbf{H}$ vectors. The theory of electromagnetic fields and waves makes use of the divergence and curl concepts in a concise manner that aids in understanding the underlying physics. We illustrate below how the classical Maxwell equations can be deduced from the divergence theorem and Stokes' theorem under suitable conditions.

4.8.1 Maxwell's equations

Consider a point electric charge within a material occupying some region of space. Any other charged object within this region will experience a force of attraction (opposite charge) or repulsion (same charge) caused by the original charged body. In general, such a force may vary with both position of the object in the field and with the time of observation. The amount of force acting upon a unit charged object in the region is defined as the *electric field* at that point, which we denote by the vector $\mathbf{E}$.

The region around an electric charge is usually represented by *lines of flux*, directed away from the positively charged body to the negatively charged one. The number of lines per unit area defines what is called the *flux density* or *electric displacement* $\mathbf{D}$. For a given material with constant *dielectric permittivity* ε, the electric displacement and electric field are related by

$$\mathbf{D} = \varepsilon\mathbf{E}. \tag{150}$$

Treating the permittivity ε as a constant means that we are assuming the medium is *linear* and *isotropic*. As a consequence of Eq. (150), the electric displacement $\mathbf{D}$ is also commonly referred to as the electric field. That is, the only distinction between electric fields $\mathbf{D}$ and $\mathbf{E}$ is that they are measured in different units.

A magnetic field is defined in a manner analogous to that of the electric field, except that the lines of force are magnetic instead of electrical. The *magnetic field* $\mathbf{H}$ and *magnetic flux density* $\mathbf{B}$ are related to each other in a manner similar to $\mathbf{E}$ and $\mathbf{D}$, viz.,

$$\mathbf{B} = \mu\mathbf{H}, \tag{151}$$

where the constant μ is the *magnetic permeability* of the material.

Electric current is the flow of electric charge. The total electric current flowing through any surface S extending through the electric field at a given time t is composed of two parts—the *conduction current* I_c and the *displacement current* I_d. The conduction current I_c can be obtained from the surface flux integral

$$I_c = \iint_S \mathbf{J}_c \cdot \hat{\mathbf{N}} dS, \tag{152}$$

where $\mathbf{J}_c = \sigma\mathbf{E}$ is the *conduction current density* per unit area, $\hat{\mathbf{N}}$ is the outward unit normal to S, and the constant σ is the *specific conductivity*.[2] In a similar manner, the displacement current I_d is defined by

$$I_d = \iint_S \mathbf{J}_d \cdot \hat{\mathbf{N}} dS. = \iint_S \frac{\partial \mathbf{D}}{\partial t} \cdot \hat{\mathbf{N}} dS, \tag{153}$$

where $\mathbf{J}_d = \partial \mathbf{D}/\partial t$ is the *displacement* (electric flux) *current* per unit area. The total *current density* is $\mathbf{J} = \mathbf{J}_c + \mathbf{J}_d = \rho\mathbf{u}$, where $\mathbf{u}$ is the mean velocity of the electric charge and ρ denotes the *charge density* within a given volume V.

The magnetic flux passing through the surface S at a given time t is described by the integral of the normal component of the magnetic flux density over the surface, or

$$\Phi = \iint_S \mathbf{B} \cdot \hat{\mathbf{N}} dS. \tag{154}$$

Thus, the time rate change of magnetic flux density is described by

$$\frac{\partial \Phi}{\partial t} = \frac{\partial}{\partial t} \iint_S \mathbf{B} \cdot \hat{\mathbf{N}} dS = \iint_S \frac{\partial \mathbf{B}}{\partial t} \cdot \hat{\mathbf{N}} dS. \tag{155}$$

Faraday experimentally discovered a connection between a time-varying magnetic field and an electric field. To obtain Faraday's law, let C denote a boundary curve for the surface S. A key relationship introduced by Faraday is given by

$$\frac{\partial \Phi}{\partial t} = -\oint_C \mathbf{E} \cdot d\mathbf{R} = -\iint_S \nabla \times \mathbf{E} \cdot \hat{\mathbf{N}} dS, \tag{156}$$

where the integral on the far right is a consequence of invoking Stokes' theorem (see Theorem 4.11). Hence, by equating (155) and (156) we are led to

$$\iint_S \left(\nabla \times \mathbf{E} + \frac{\partial \mathbf{B}}{\partial t} \right) \cdot \hat{\mathbf{N}} dS = 0 \quad \text{for all } S, \tag{157}$$

from which we deduce *Faraday's law*

$$\boxed{\nabla \times \mathbf{E} = -\frac{\partial \mathbf{B}}{\partial t} \quad \text{(Faraday's law).}} \tag{158}$$

In words, Faraday's law says that a time-varying magnetic field induces an electric field $\mathbf{E}$ whose curl is equal to the negative of the time derivative of $\mathbf{B}$.

The converse of Faraday's law, known as Ampère's law, states that a time-varying electric field gives rise to a magnetic field. Again relying on Stokes' theorem, a key relationship by Ampère involving the total current $I = I_c + I_d$ can be expressed by

---

[2] The quantities ε, μ, and σ are all assumed to be constant scalars in the present discussion, although in general they may be varying tensor quantities (e.g., see Section 5.12).

$$I_c + I_d = \oint_C \mathbf{H} \cdot d\mathbf{R} = \iint_S \nabla \times \mathbf{H} \cdot \hat{\mathbf{N}} dS, \tag{159}$$

which, coupled with Eqs. (152) and (153), yields

$$\iint_S \mathbf{J}_c \cdot \hat{\mathbf{N}} dS + \iint_S \frac{\partial \mathbf{D}}{\partial t} \cdot \hat{\mathbf{N}} dS = \iint_S \nabla \times \mathbf{H} \cdot \hat{\mathbf{N}} dS. \tag{160}$$

In differential form, this last result leads to *Ampère's law*

$$\boxed{\nabla \times \mathbf{H} = \mathbf{J}_c + \frac{\partial \mathbf{D}}{\partial t} \qquad \text{(Ampère's law).}} \tag{161}$$

Next we wish to develop Gauss's law. Given that ρ is the charge density within a given volume V, the total charge Q contained within the volume is then given by the volume integral

$$Q = \iiint_V \rho \, dV. \tag{162}$$

The total charge is also given by the total electric flux through the surface S enclosing the volume V, which is defined by

$$Q = \oiint_S \mathbf{D} \cdot \hat{\mathbf{N}} dS = \iiint_V \nabla \cdot \mathbf{D} dV, \tag{163}$$

the last step following an application of the divergence theorem (see Theorem 4.10). By equating (162) and (163), we obtain *Gauss's law*

$$\boxed{\nabla \cdot \mathbf{D} = \rho \qquad \text{(Gauss's law).}} \tag{164}$$

The magnetic analogue to a point charge is a magnetic pole. However, whereas electric charges can exist in isolation, magnetic poles always occur in pairs. Thus, there is no magnetic equivalent to total charge Q. Analogous to (163), therefore,

$$0 = \oiint_S \mathbf{B} \cdot \hat{\mathbf{N}} dS = \iiint_V \nabla \cdot \mathbf{B} dV, \tag{165}$$

which states that the total magnetic flux passing through any closed surface S is zero. The last integral on the right follows from an application of the divergence theorem. Based on (165), we can deduce Gauss's law for magnetic fields given by

$$\boxed{\nabla \cdot \mathbf{B} = 0 \qquad \text{(Gauss's law for magnetic fields).}} \tag{166}$$

The four relations (158), (161), (164), and (166) constitute what are known as *Maxwell's equations*.

4.8.2 Poisson's equation

By writing $\mathbf{D} = \varepsilon\mathbf{E}$, we can express Gauss's law (164) as

$$\nabla \cdot \mathbf{E} = \frac{\rho}{\varepsilon}. \tag{167}$$

In applications where the electric field $\mathbf{E}$ is a conservative field, there exists a scalar potential function φ (called the *electric potential*) related to $\mathbf{E}$ by

$$\mathbf{E} = -\nabla\varphi \tag{168}$$

where the minus sign is conventional. By inserting Eq. (168) into (167) and making use of the identity $\nabla \cdot (\nabla\varphi) = \nabla^2\varphi$, we find that (167) becomes

$$\nabla^2\varphi = -\frac{\rho}{\varepsilon}, \tag{169}$$

which is known as *Poisson's equation*. If the medium under consideration contains no free charges, then $\rho = 0$ and (169) reduces to *Laplace's equation* (see Chapter 12)

$$\nabla^2\varphi = 0. \tag{170}$$

4.8.3 Electromagnetic wave equation

In this section we wish to derive the electromagnetic wave equation for a charge-free medium. To do so we start with Eqs. (158) and (161) written as

$$\nabla \times \mathbf{E} = -\frac{\partial \mathbf{B}}{\partial t},$$
$$\nabla \times \mathbf{B} = \sigma\mu\mathbf{E} + \varepsilon\mu\frac{\partial \mathbf{E}}{\partial t}, \tag{171}$$

where we have used Eq. (151). By taking the curl of both sides of the first equation above, we obtain

$$\nabla \times (\nabla \times \mathbf{E}) = -\nabla \times \frac{\partial \mathbf{B}}{\partial t}, \tag{172}$$

and the time derivative of the second equation yields

$$\nabla \times \frac{\partial \mathbf{B}}{\partial t} = \sigma\mu\frac{\partial \mathbf{E}}{\partial t} + \varepsilon\mu\frac{\partial^2 \mathbf{E}}{\partial t^2}. \tag{173}$$

Now, by use of the vector identity [see #8 in (107)]

$$\nabla \times (\nabla \times \mathbf{E}) = \nabla(\nabla \cdot \mathbf{E}) - \nabla^2\mathbf{E}, \tag{174}$$

and the results of (164) (with $\rho = 0$) and (173), we arrive at

$$\nabla^2 \mathbf{E} = \sigma\mu\frac{\partial \mathbf{E}}{\partial t} + \varepsilon\mu\frac{\partial^2 \mathbf{E}}{\partial t^2}, \tag{175}$$

which is *Maxwell's equation for the electric field intensity* **E**. In the same fashion, if we take the curl of the second equation in (171), it can be shown that

$$\nabla^2 \mathbf{H} = \sigma\mu\frac{\partial \mathbf{H}}{\partial t} + \varepsilon\mu\frac{\partial^2 \mathbf{H}}{\partial t^2}, \tag{176}$$

called *Maxwell's equation for the magnetic field intensity* **H**.

   Note that when $\sigma = 0$ (i.e., in a nonconducting medium like an insulator), Maxwell's equations (175) and (176) reduce, respectively, to

$$\nabla^2 \mathbf{E} = \varepsilon\mu\frac{\partial^2 \mathbf{E}}{\partial t^2}, \tag{177}$$

$$\nabla^2 \mathbf{H} = \varepsilon\mu\frac{\partial^2 \mathbf{H}}{\partial t^2}. \tag{178}$$

These equations are called the *electromagnetic vector wave equation* for the electric field **E** and magnetic field **H**. By expressing each field vector in Cartesian coordinates, these equations separate into three standard scalar wave equations, each of the form (177) or (178) for the respective rectangular components. On the other hand, in a good conductor, the second term on the right in (175) and (176) vanishes and Maxwell's equations reduce to the three-dimensional heat equation

$$\nabla^2 \mathbf{E} = \sigma\mu\frac{\partial \mathbf{E}}{\partial t}, \tag{179}$$

$$\nabla^2 \mathbf{H} = \sigma\mu\frac{\partial \mathbf{H}}{\partial t}. \tag{180}$$

Last, for static fields (time-independent) the curls of the electric and magnetic fields lead to the uncoupled equations

$$\nabla \times \mathbf{E} = \mathbf{0},$$
$$\nabla \times \mathbf{H} = \mathbf{J}_c. \tag{181}$$

Suggested Reading

G. Arfken, *Mathematical Methods for Physicists*, 3rd ed. (Academic, New York, 1985).

A. I. Borisenko and I. E. Tarapov, *Vector and Tensor Analysis with Applications*, trans. from the Russian by R. A. Silverman (Dover, New York, 1968).

F. Chorlton, *Vector and Tensor Methods* (Ellis Horwood Limited, Chichester; Wiley & Sons, New York, 1976).

H. F. Davis and A. D. Snider, *Introduction to Vector Analysis*, 7th ed. (Hawkes Publishing, Tallahassee, 1999).

M. Rahman and I. Mulolani, *Applied Vector Analysis* (CRC Press, Boca Raton, 2001).

J. G. Simmonds, *A Brief on Tensor Analysis*, 2nd ed. (Springer, New York, 1994).

Chapter 4

Exercises

In Problems 1 and 2, find the components of the vector **a** between the initial point P and terminal point Q. Also find $\|\mathbf{a}\|$.

1. P:$(4,0,-1)$, Q:$(1,0,2)$

    Ans. $<-3,0,3>$, $\|\mathbf{a}\| = 3\sqrt{2}$

2. P:$(3,1,4)$, Q:$(1,-2,4)$

    Ans. $<-2,-3,0>$, $\|\mathbf{a}\| = \sqrt{13}$

In Problems 3-5, find the indicated operations for vectors $\mathbf{a} = <2,1,3>$, $\mathbf{b} = <1,0,-4>$, and $\mathbf{c} = <3,-1,2>$.

3. $\mathbf{a}\cdot\mathbf{b}$ *Ans.* -10

4. $\mathbf{a}\cdot(\mathbf{b}-\mathbf{c})$ *Ans.* -21

5. $\|\mathbf{a}+\mathbf{b}+\mathbf{c}\|$ *Ans.* $\sqrt{37}$

6. For what value of C are the planes
$x + 2y + 3z = 6$ and $x + Cy + z = 2$ orthogonal?

In Problems 7-9, find the indicated operations for vectors $\mathbf{a} = <1,1,0>$, $\mathbf{b} = <-1,2,0>$, and $\mathbf{c} = <2,3,1>$.

7. $\mathbf{a}\times\mathbf{b}$ *Ans.* $<0,0,3>$

8. $(\mathbf{a}+\mathbf{b})\times\mathbf{c}$ *Ans.* $<3,0,-6>$

9. $(\mathbf{abc})$ *Ans.* 3

10. Determine the scalars x and y so that $\mathbf{D} = \mathbf{C} - x\mathbf{A} + y\mathbf{B}$ is perpendicular to both **A** and **B**, given the vectors $\mathbf{A} = <1,1,2>$, $\mathbf{B} = <2,-1,1>$, and $\mathbf{C} = <2,-1,4>$.

11. Find the distance of the point P_0:$(6,-4,4)$ to the line through points A:$(2,1,2)$ and B:$(3,-1,4)$.

    Ans. 3

12. Find the equation of the line through the point $(4,2,1)$ that is parallel to the intersection of the planes $3x + y + z = 5$, $x - 2y + 3z = 1$.

    Ans. $\dfrac{x-4}{5} = \dfrac{y-2}{-8} = \dfrac{z-1}{-7}$

In Problems 13 and 14, use *tensor notation* to prove the identity.

13. $e_{ijk} e_{ijm} = 2\delta_{km}$

14. $(\mathbf{a} \times \mathbf{b}) \cdot (\mathbf{c} \times \mathbf{d})$
$\qquad = (\mathbf{a} \cdot \mathbf{c})(\mathbf{b} \cdot \mathbf{d}) - (\mathbf{a} \cdot \mathbf{d})(\mathbf{b} \cdot \mathbf{c})$

15. Given the space curve

$$\mathbf{R}(t) = 8\cos^3 t\, \hat{\mathbf{i}} + 8\sin^3 t\, \hat{\mathbf{j}}$$
$$+ 3\cos 2t\, \hat{\mathbf{k}}, \ 0 \le t \le 2\pi,$$

(*a*) Find the triad of unit vectors:
$\hat{\mathbf{T}}(t), \ \hat{\mathbf{N}}(t), \ $ and $ \ \hat{\mathbf{B}}(t)$.

(*b*) Find the *curvature* $\kappa(t)$ and the *torsion* $\tau(t)$.

Ans. (*a*) $\hat{\mathbf{T}} = \dfrac{1}{\sqrt{5}} <-2\cos t, 2\sin t, -1>$

$\qquad \hat{\mathbf{N}} = <\sin t, \cos t, 0>$

$\qquad \hat{\mathbf{B}} = \dfrac{1}{\sqrt{5}} <\cos t, -\sin t, -2>$

(*b*) $\kappa = 2\tau = \dfrac{1}{15\sin 2t}$

16. A particle moves with constant radial speed of 2 cm/s away from the center of a platform rotating counterclockwise with uniform angular speed of 30 rev/min.

(*a*) What is its radial acceleration?

(*b*) What is its Coriolis acceleration?

Ans. (*a*) $\pi^2 r$ cm/s toward center
$\qquad$ (*b*) $4\pi\, \hat{\mathbf{e}}_\theta$

17. The path of a particle is given by

$$\mathbf{R}(t) = \cos \pi t\, \hat{\mathbf{i}} + \sin \pi t\, \hat{\mathbf{j}} + t\, \hat{\mathbf{k}}.$$

(*a*) Determine the velocity and acceleration vectors at time $t = 1$.

(*b*) Determine the tangential and normal components of acceleration.

Ans. (*a*) $\mathbf{v} = -\pi \hat{\mathbf{i}} + \hat{\mathbf{k}}, \ \mathbf{a} = \pi^2 \hat{\mathbf{i}}$
$\qquad$ (*b*) $a_T = 0, \ a_N = \pi^2$

18. Calculate the gradient ∇f.

(*a*) $f = e^x \sin y$

(*b*) $f = \dfrac{z}{x^2 + y^2}$

Ans.

(*a*) $<e^x \sin y, e^x \cos y>$

(*b*) $<-\dfrac{2xz}{(x^2 + y^2)^2}, -\dfrac{2yz}{(x^2 + y^2)^2}, \dfrac{1}{x^2 + y^2}>$

19. Find $\nabla \varphi$, where $\mathbf{r} = <x, y, z>$ and

(*a*) $\varphi = \ln \|\mathbf{r}\|$.

(*b*) $\varphi = \dfrac{1}{r}$.

Ans. (*a*) $\nabla \varphi = \dfrac{\mathbf{r}}{r^2}$ $\quad$ (*b*) $\nabla \varphi = -\dfrac{\mathbf{r}}{r^3}$

20. Find an equation for the tangent plane to the surface $2xz^2 - 3xy - 4x = 7$ at the point $(1, -1, 2)$.

Ans. $7x - 3y + 8z = 26$

21. Find the directional derivative of

$$f(x, y, z) = 2xy^2 + z^2 xy + x^2$$

at the point $(1, 1, 2)$ in the direction $\mathbf{u} = <3, -2, 5>$.

Ans. $\dfrac{28}{\sqrt{38}}$

22. Given the point $(2, 1, -1)$ and the function $f(x, y, z) = x^2 y z^3$,

(*a*) find the direction from the point in which the directional derivative is a maximum.

(*b*) What is the magnitude of this maximum?

Ans. (*a*) $<-4, -4, 12>$ $\quad$ (*b*) $4\sqrt{11}$

23. Calculate the divergence $\nabla \cdot \mathbf{F}$.

(*a*) $\mathbf{F} = <y^2 e^z, 0, x^2 z^2>$

(*b*) $\mathbf{F} = <x^2 z, -2y^3 z^2, xy^2 z>$

Ans. (a) $2x^2z$ (b) $2xz - 6y^2z^2 + xy^2$

24. Find grad(div **v**), where
$$\mathbf{v} = x^2\hat{\mathbf{i}} + y^2\hat{\mathbf{j}} - 4z^2\hat{\mathbf{k}}.$$

Ans. $\langle 2, 2, -8 \rangle$

25. Find the charge density at $(1,1,2)$ if the electric field displacement vector is
$$\mathbf{D} = \langle x^2y, z^2, z^3 \rangle.$$

Ans. 14

26. Calculate the curl $\nabla \times \mathbf{F}$.

(a) $\mathbf{F} = \langle yz, 3zx, z \rangle$

(b) $\mathbf{F} = \langle \ln(x^2 + y^2), \tan^{-1}(y/x), 0 \rangle$

Ans. (a) $\langle -3x, y, 2z \rangle$
(b) $\langle 0, 0, -3y/(x^2 + y^2) \rangle$

27. Given the vector field
$$\mathbf{F} = \langle 2xy, x^2 + 2yz, y^2 + 1 \rangle,$$

(a) show that **F** is *conservative*.
(b) find the scalar potential function $\varphi(x,y,z)$ such that $\mathbf{F} = \nabla\varphi$.

Ans. $\varphi(x,y,z) = x^2y + y^2z + z + C$

28. Use *tensor notation* to prove that
$$\nabla \cdot (\mathbf{F} \times \mathbf{G}) = \mathbf{G} \cdot (\nabla \times \mathbf{F}) - \mathbf{F} \cdot (\nabla \times \mathbf{G}).$$

29. Evaluate the line integral of the tangential component of the vector field
$$\mathbf{F} = \langle x^2, x + y, 0 \rangle$$
along the curve $y = x^2$ from $(0,0)$ to $(2,4)$.
Ans. 16

In Problems 30-32, evaluate the given line integral.

30. $\int_C (x^2 - y^2)ds$, $C: y = 2x$, $z = 0$,
from $(0,0,0)$ to $(1,2,0)$

Ans. $-\sqrt{5}$

31. $\int_C \mathbf{F} \cdot d\mathbf{R}$, $\mathbf{F} = \langle x^2y, y^2z, z^2x \rangle$,

(a) C: line from $(0,0,0)$ to $(1,3,3)$.
(b) C: $\mathbf{R}(t) = \langle t, t^2, t^3 \rangle$
from $t = 0$ to $t = 1$.

Ans. (a) $\dfrac{111}{4}$ (b) $\dfrac{13}{18}$

32. $\int_C 2xyz^2 d\mathbf{R}$,
$C: x = t^2$, $y = 2t$, $z = t^3$,
from $t = 0$ to $t = 1$

Ans. $\dfrac{8}{11}\hat{\mathbf{i}} + \dfrac{4}{5}\hat{\mathbf{j}} + \hat{\mathbf{k}}$

33. For the vector field given in Problem 27, find the *work* done in moving a particle from $(0,0,0)$ to $(1,1,1)$ along any path.

Ans. 3

34. Verify *Green's Theorem in the Plane* for
$$I = \oint_C (3x^2 - 8y^2)dx + (4y - 6xy)dy,$$

where C is the boundary of the region enclosed by $y = \sqrt{x}$, $y = x^2$. (That is, evaluate the line integral directly and also by converting it to an appropriate double integral.)

35. Evaluate the surface integral
$$I = \iint_S x^2 z\, dS,$$
where $S: z = \sqrt{1 - x^2 - y^2}$

(a) by the *projection method*.
(b) by the *method of parametric representation*.

Ans. $\dfrac{\pi}{4}$

36. Use the *Divergence Theorem* to show that

$$\oiint_S R^2 (\mathbf{R \cdot \hat{N}}) dS = 5 \iiint_V R^2 \, dV,$$

where S is a *closed* smooth surface, $\mathbf{R}$ is the position vector, $R = \|\mathbf{R}\|$, and $\mathbf{\hat{N}}$ is the outward unit normal to the surface.

37. Use the *Divergence Theorem* to evaluate the surface integral below as a volume integral:

$$I = \oiint_S (2xz\,dydz + y\,dzdx + z^2\,dxdy)$$

where the volume is that enclosed by the surfaces

S: $z = \sqrt{1 - x^2 - y^2}$, $z = 0$.

Ans. $\dfrac{5\pi}{3}$

38. Verify *Stoke's Theorem* for the vector field $\mathbf{F} = <xy, -y, x^2 y>$, where S is the first-octant portion of the plane $2x + y + 2z = 8$. (That is, evaluate both the line integral and the surface integral.)

Ans. $\dfrac{32}{3}$

Chapter 5

Tensor Analysis

5.1 Introduction / 216
5.2 Tensor Notation / 216
    5.2.1 Special symbols / 217
5.3 Rectilinear Coordinates / 218
    5.3.1 Definition of tensor / 221
    5.3.2 Tensor operations / 224
    5.3.3 Symmetric and skew-symmetric tensors / 225
5.4 Base Vectors / 226
    5.4.1 Covariant base vectors / 226
    5.4.2 Contravariant base vectors: reciprocal basis / 227
    5.4.3 Metric tensor / 228
5.5 Vector Algebra / 231
    5.5.1 Permutation symbols in rectilinear coordinates / 232
    5.5.2 Dot product / 235
    5.5.3 Cross product and mixed triple product / 235
5.6 Relations Between Tensor Components / 238
    5.6.1 Raising and lowering indices / 238
    5.6.2 Physical components / 239
5.7 Reduction of Tensors to Principal Axes / 241
    5.7.1 Two-dimensional case / 241
    5.7.2 Three-dimensional case / 242
5.8 Tensor Calculus: Rectilinear Coordinates / 243
    5.8.1 Gradient, divergence, and curl / 244
5.9 Curvilinear Coordinates / 245
    5.9.1 Differentials as tensors / 247
    5.9.2 Tensor fields and base vectors / 248
    5.9.3 Metric tensors / 248
5.10 Tensor Calculus: Curvilinear Coordinates / 250
    5.10.1 Christoffel symbols / 251
    5.10.2 Orthogonal systems / 253
    5.10.3 Covariant derivative / 255
    5.10.4 Absolute derivative / 258
5.11 Riemann-Christoffel Curvature Tensor / 259
5.12 Applications / 260
    5.12.1 Gradient, divergence, and curl / 261
    5.12.2 Dynamics of a particle: Newton's second law / 262
    5.12.3 Dielectric tensor of an anisotropic medium / 265
Suggested Reading / 266
Exercises / 266

In this chapter we introduce the basic transformation laws of tensors that are used in various engineering application areas like elasticity and general relativity, among others. Because many of the tensors of interest are of second order, we find the matrix operations introduced in Chapter 3 to be particularly useful in our treatment here.

5.1 Introduction

A *scalar* is a quantity that can be specified (in any coordinate system) by just one number, whereas the specification of a *vector* requires three numbers (see Chapter 4). Both scalars and vectors are special cases of a more general concept called a *tensor*. To specify a tensor of *order n* in a coordinate system requires 3^n numbers, called the *components* of the tensor.[1] Scalars are tensors of order 0 and vectors are tensors of order 1.

A tensor of order n is more than just a set of 3^n numbers. Only when these numbers satisfy a particular *transformation law* do they represent the components of a tensor. The transformation law describes how the tensor components in one coordinate system are related to those in another coordinate system. Because they have useful properties that are independent of coordinate system, tensors are used to represent various fundamental laws of physics, engineering, science, and mathematics. In particular, tensors are an important tool in general relativity, elasticity, hydrodynamics, and electromagnetic theory. In these areas of application the elastic, optical, electrical, and magnetic properties must often be described by tensor quantities. For example, this is the case if the medium is anisotropic, like in many crystals, or is a plasma in the presence of a magnetic field.

As a final comment, we alert the reader to the change in the use of certain punctuation marks in the present chapter to avoid confusion with the "comma" notation commonly used for partial derivative in this material. For example, when deemed necessary, a semicolon (;) is often used in place of a comma.

5.2 Tensor Notation

The use of vector symbols like **a** and **a** + **b** to denote a vector or vector operation provides a convenient notation for expressing relationships in geometry and physics (see Chapter 4). For actually performing the various operations between vectors, such as addition and multiplication, it is the components of the vector that are most useful. Shorthand notation based on vector components that is widely used in advanced works is that involving tensor. We briefly introduced tensor notation for Cartesian coordinates in Chapter 4, but now we wish to build upon that notation for more general coordinate systems that require both *upper* and *lower indices*.

In *tensor notation* we can represent a triad of numbers (x_1, x_2, x_3) by the symbol x_j, where $j = 1$, 2, or 3. We refer to the subscript j in this context as a *dummy index*. In particular, if a vector **a** can be represented in component form **a** $= <a_1, a_2, a_3>$, then by use of the index notation we may also represent these components by

$$a_j \quad (j = 1,2,3) \quad \text{or} \quad a_n \quad (n = 1,2,3).$$

---

[1] The *order* of a tensor is also called the *rank* of a tensor in some texts.

The choice of dummy index is somewhat arbitrary as illustrated in the above example. The letters i, j, k or m, n, p, and so on, are commonly used.

In a *Cartesian (rectangular) coordinate system* it is customary to use only subscripts to denote vector or tensor components, but in more general *rectilinear* or *curvilinear coordinate systems* the need arises to also consider superscripts as well. This need arises because, in general rectilinear or curvilinear coordinates, there exist two sets of base vectors. For coordinate systems other than Cartesian, we always use superscripts like (y^1, y^2, y^3) to denote the axes. In Cartesian coordinates, however, we could continue to let (x_1, x_2, x_3) denote the coordinates (x, y, z), but it is conventional to also use superscripts (x^1, x^2, x^3) for these coordinates, which we denote by x^j. We remind the reader that superscripts that appear in tensor notation should not be confused with powers since the latter is almost never used in tensor notation.

The *Einstein summation convention* was introduced in Section 4.3.1. Namely, an automatic sum is taken over 1, 2, 3 when any index is repeated *exactly twice* in a single term. If the index appears more than two times, no sum is taken. Now when using the Einstein summation convention, we modify our rule somewhat by always requiring *one upper and one lower index*; for example, we now write expressions like

$$a_k x^k = a_1 x^1 + a_2 x^2 + a_3 x^3$$
$$a^k x_k = a^1 x_1 + a^2 x_2 + a^3 x_3.$$

However, because we make no distinction between vector components a_j and a^j in Cartesian coordinates, we will occasionally move the indices up or down whenever necessary to conform with this new summation convention. That is, we can simply replace components a_j with a^j in order to conform with this convention.

5.2.1 Special symbols

Tensor notation is greatly facilitated by use of the following two special symbols called, respectively, the Kronecker delta and the permutation symbol.

▸ **Kronecker delta:** We denote the *Kronecker delta* by any one of the symbols:

$$\delta_{jk} = \delta^{jk} = \delta^j_k = \begin{cases} 1, & j = k \\ 0, & j \ne k \end{cases} \tag{1}$$

Thus, the Kronecker delta is either 1 or 0, depending upon whether the indices are the same or distinct. It follows from (1) that $\delta_{jk} = \delta_{kj}$, $\delta^k_j = \delta^j_k$, $\delta^{jk} = \delta^{kj}$. Finally, from our summation convention it also follows that

$$\delta_{kk} = \delta^k_k = \delta^{kk} = 3$$
$$a^j \delta_{jk} = a_k; \quad a^j \delta^k_j = a^k; \quad a_j \delta^{jk} = a^k$$
$$A^{ij} \delta_{jk} = A^i_k; \quad A^{ij} \delta^k_j = A^{ik} \tag{2}$$

▶ **Permutation symbol:** The *permutation symbol* for Cartesian coordinates is a triple-indexed quantity defined by

$$e_{ijk} = e^{ijk} = \begin{cases} +1, & \text{if } (ijk) \text{ either (123), (231), or (312)} \\ -1, & \text{if } (ijk) \text{ either (132), (321), or (213)} \\ 0, & \text{otherwise} \end{cases} \qquad (3)$$

Making an observation, we see from definition of the permutation symbol that

$$e_{ijk} = e^{ijk} = 0 \text{ if any two indices are alike}$$

$$e_{ijk} = e_{jki} = e_{kij}; \quad e^{ijk} = e^{jki} = e^{kij}$$

$$e_{ijk} = - e_{ikj}; \quad e^{ijk} = - e^{ikj}.$$

In addition, the permutation symbol satisfies the following identities:

$$e^{ijk} e_{imn} = \delta_m^j \delta_n^k - \delta_n^j \delta_m^k$$

$$e^{ijk} e_{ijm} = 2\delta_m^k \qquad (4)$$

$$e^{ijk} e_{ijk} = 6$$

$$e_{ijk} = \delta_i^m \delta_j^n \delta_k^p e_{mnp}; \qquad e^{ijk} = \delta_m^i \delta_n^j \delta_p^k e^{mnp}$$

$$e_{ijk} = \delta_{im} \delta_{jn} \delta_{kp} e^{mnp}; \qquad e^{ijk} = \delta^{im} \delta^{jn} \delta^{kp} e_{mnp} \qquad (5)$$

5.3 Rectilinear Coordinates

Let us represent the Cartesian coordinate system (x,y,z) by the symbols (x^1,x^2,x^3), or simply x^j, as shown in Fig. 5.1. If we define a new set of variables (y^1,y^2,y^3) by the *linear transformation*

$$x^1 = \alpha_1^1 y^1 + \alpha_2^1 y^2 + \alpha_3^1 y^3$$
$$x^2 = \alpha_1^2 y^1 + \alpha_2^2 y^2 + \alpha_3^2 y^3 \qquad (6)$$
$$x^3 = \alpha_1^3 y^1 + \alpha_2^3 y^2 + \alpha_3^3 y^3$$

then we can interpret y^j as the axes of a *rectilinear coordinate system*, also shown in Fig. 5.1. In tensor notation, we write the transformation (6) more compactly as

$$x^j = \alpha_k^j y^k \qquad (7)$$

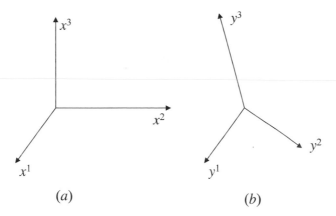

Figure 5.1 (*a*) Cartesian coordinate system and (*b*)
rectilinear coordinate system. Both systems, and any
cyclic permutation of these systems, are right-handed.

where α_k^j denotes the coefficients in (6) defined by the partial derivatives $\alpha_k^j = \partial x^j / \partial y^k$.

We recognize that we can represent the linear transformation (6) by the use of matrix representations. In particular, x^j and y^j can be represented by 3×1 column vectors $[x^j]$ and $[y^j]$, and α_k^j can be represented by a 3×3 matrix $[\alpha_k^j]$. In particular,

$$[x^j] = \begin{pmatrix} x^1 \\ x^2 \\ x^3 \end{pmatrix}; \quad [y^j] = \begin{pmatrix} y^1 \\ y^2 \\ y^3 \end{pmatrix}; \quad [\alpha_k^j] = \begin{pmatrix} \alpha_1^1 & \alpha_2^1 & \alpha_3^1 \\ \alpha_1^2 & \alpha_2^2 & \alpha_3^2 \\ \alpha_1^3 & \alpha_2^3 & \alpha_3^3 \end{pmatrix}; \tag{8}$$

and thus, Eq. (6) can be written in matrix notation as (see Chapter 3)

$$\begin{pmatrix} x^1 \\ x^2 \\ x^3 \end{pmatrix} = \begin{pmatrix} \alpha_1^1 & \alpha_2^1 & \alpha_3^1 \\ \alpha_1^2 & \alpha_2^2 & \alpha_3^2 \\ \alpha_1^3 & \alpha_2^3 & \alpha_3^3 \end{pmatrix} \begin{pmatrix} y^1 \\ y^2 \\ y^3 \end{pmatrix}. \tag{9}$$

We denote the determinant of the matrix $[\alpha_k^j]$ by the symbol J, called the *Jacobian of transformation*, which we express as [recall Eq. (42) in Chapter 4]

$$J = \det[\alpha_k^j] = e_{ijk} \alpha_1^i \alpha_2^j \alpha_3^k. \tag{10}$$

If $J > 0$, a right-handed system is transformed into another right-handed system (see Fig. 5.1), and a left-handed system is transformed into another left-handed system. In both cases we call this a *proper transformation*. When $J < 0$, the transformation is said to be *improper* because it transforms right-handed systems into left-handed systems, and vice versa. If $J = 0$, the transformation is *singular*.

Assuming $J \neq 0$, there exists a unique *inverse linear transformation* to (7) given by

$$\boxed{y^m = \beta_k^m x^k} \tag{11}$$

where $\beta_k^m = \partial y^m / \partial x^k$. We recognize $[\beta_k^m]$ as the inverse matrix of $[\alpha_k^j]$, for which

$$\frac{1}{J} = \det[\beta_k^j] = e^{ijk} \beta_i^1 \beta_j^2 \beta_k^3. \tag{12}$$

To verify that $[\beta_k^m]$ is the inverse matrix, whose determinant is the reciprocal of the Jacobian J, we simply differentiate both sides of (11) with respect to y^n to obtain

$$\frac{\partial y^m}{\partial y^n} = \beta_k^m \frac{\partial x^k}{\partial y^n} = \beta_k^m \alpha_n^k.$$

Thus, because $\partial y^m / \partial y^n = \delta_n^m$, we deduce that

$$\beta_k^m \alpha_n^k = \delta_n^m \tag{13}$$

where the matrix representation of the Kronecker delta is the *identity matrix*. We conclude, therefore, that $[\alpha_k^j]$ and $[\beta_k^m]$ are indeed inverses of each other and the determinant of one is the reciprocal determinant of the other. Of course, if $\alpha_k^j = \delta_k^j$, we obtain the *identity transformation* $x^j = \delta_k^j y^k = y^j$.

EXAMPLE 1 Given the linear transformation

$$\begin{aligned} x^1 &= y^1 + y^2 + y^3 \\ x^2 &= y^1 - y^3 \\ x^3 &= 2y^2 - 3y^3 \end{aligned}$$

(a)   find the Jacobian of transformation.

(b)   find the inverse transformation and the matrix representation of $[\beta_k^m]$.

Solution:

(a) The Jacobian of transformation is

$$J = \det[\alpha_k^j] = \begin{vmatrix} 1 & 1 & 1 \\ 1 & 0 & -1 \\ 0 & 2 & -3 \end{vmatrix} = 7$$

which illustrates that this is a proper transformation.

(b) By use of the method of determinants (i.e., Cramer's rule from algebra) we obtain the inverse transformation

$$y^1 = \frac{1}{7} \begin{vmatrix} x^1 & 1 & 1 \\ x^2 & 0 & -1 \\ x^3 & 2 & -3 \end{vmatrix} = \frac{1}{7}(2x^1 + 5x^2 - x^3)$$

$$y^2 = \frac{1}{7} \begin{vmatrix} 1 & x^1 & 1 \\ 1 & x^2 & -1 \\ 0 & x^3 & -3 \end{vmatrix} = \frac{1}{7}(3x^1 - 3x^2 + 2x^3)$$

$$y^3 = \frac{1}{7} \begin{vmatrix} 1 & 1 & x^1 \\ 1 & 0 & x^2 \\ 0 & 2 & x^3 \end{vmatrix} = \frac{1}{7}(2x^1 - 2x^2 - x^3).$$

Thus, the matrix representation of $[\beta_k^m]$ is

$$[\beta_k^m] = \frac{1}{7} \begin{pmatrix} 2 & 5 & -1 \\ 3 & -3 & 2 \\ 2 & -2 & -1 \end{pmatrix}.$$

5.3.1 Definition of tensor

Basically, a tensor is a system of numbers or functions of various orders whose components obey a certain transformation law when the variables undergo a linear transformation. To describe these transformation laws we will follow the convention that a *lowercase letter* denotes the tensor components in a *Cartesian coordinate system* and the corresponding *uppercase letter* denotes the tensor components in a *rectilinear coordinate system*. Hence, if a_j denotes the components of a vector **a** in Cartesian coordinates, then A_j denotes the corresponding components of the vector **a** in rectilinear coordinates.

Let us assume the basic transformation equations between a Cartesian coordinate system and a rectilinear coordinate system are given by

$$x^j = \alpha_k^j y^k; \quad \det[\alpha_k^j] = J \neq 0$$

$$y^j = \beta_k^j x^k; \quad \det[\beta_k^j] = 1/J.$$

(14)

In terms of these equations of transformation, we define tensor of various orders by the following:

Systems of Order Zero: A system of order zero has one component a in the variable x^j and one component A in the variable y^j. If

$$A = a \quad \text{for all} \quad x^j, y^j,$$

then a and A are components of a *tensor of order zero*, also called a *scalar* or *invariant*.

Systems of Order One: Let a_j and A_j be components in the variables x^j and y^j, respectively. If

$$A_j = \alpha_j^k a_k,$$

then a_j and A_j are components of a *covariant tensor of order one*, also called a *covariant vector*. Similarly, if a^j and A^j are components in the variables x^j and y^j, respectively, and if

$$A^j = \beta_k^j a^k,$$

then a^j and A^j are components of a *contravariant tensor of order one*, also called a *contravariant vector*.

Systems of Order Two: Let a_{jk} and A_{jk} be components in the variables x^j and y^j, respectively. If

$$A_{jk} = \alpha_j^m \alpha_k^n a_{mn},$$

then a_{jk} and A_{jk} are components of a *covariant tensor of order two*. Similarly, if a^{jk} and A^{jk} are components in the variables x^j and y^j, respectively, and if

$$A^{jk} = \beta_m^j \beta_n^k a^{mn},$$

then a^{jk} and A^{jk} are components of a *contravariant tensor of order two*. Last, if a_k^j and A_k^j are components in the variables x^j and y^j, respectively, and if

$$A_k^j = \beta_m^j \alpha_k^n a_n^m,$$

then a_k^j and A_k^j are components of a *mixed tensor of order two*.

From the above definitions, we see that "covariant components" of a tensor are identified by a "subscript" and "contravariant components" by a "superscript." In general, these are not the same components (e.g., see Example 2 below). Higher order tensors transform in a similar fashion. For instance, tensors of order three satisfy the following transformation laws:

$$A_{ijk} = \alpha_i^m \alpha_j^n \alpha_k^p a_{mnp},$$
$$A_{ij}^{k} = \alpha_i^m \alpha_j^n \beta_p^k a_{mn}^{p},$$
$$A^{ijk} = \beta_m^i \beta_n^j \beta_p^k a^{mnp}$$

If physical laws are to be valid, they must be independent of any particular coordinate system used in the mathematical formulation of them. Thus, a law stated by a mathematical equation in one coordinate system must hold true even under a transformation of coordinate systems. The consequence of this concept forms what is called *tensor analysis*.

From the above discussion we see that tensors are certain systems of numbers (or functions) that obey specific linear transformation laws. Because the transformations are linear (homogeneous) equations, it follows that if all components are zero in one coordinate system, they must be zero in all coordinate systems.

It is useful to use matrix notation and matrix operations in calculating various transformations between first-order and second-order tensor components. For this reason, we must take note of which index represents a row and which represents a column. For example, a *row* is indicated by the *first index* and a *column* by the *second index* in second-order tensor components A_{jk} and A^{jk}. Also, we associate the *upper index* of the quantity α_k^j with a *row* and the *lower index* with a *column* [e.g., see Eq. (8)].

EXAMPLE 2 Given the vector $\mathbf{a} = <-1,3,0>$ in Cartesian coordinates and the linear transformation described in Example 1, find

(a) the covariant components A_j in rectilinear coordinates.

(b) the contravariant components A^j in rectilinear coordinates.

Solution:

(a) The covariant components are defined by $A_j = \alpha_j^k a_k$, which in matrix notation takes the form $[A_j] = [\alpha_j^k]^T [a_k]$, where T denotes the transpose. Therefore, we see that

$$\begin{pmatrix} A_1 \\ A_2 \\ A_3 \end{pmatrix} = \begin{pmatrix} 1 & 1 & 1 \\ 1 & 0 & -1 \\ 0 & 2 & -3 \end{pmatrix}^T \begin{pmatrix} -1 \\ 3 \\ 0 \end{pmatrix} = \begin{pmatrix} 1 & 1 & 0 \\ 1 & 0 & 2 \\ 1 & -1 & -3 \end{pmatrix} \begin{pmatrix} -1 \\ 3 \\ 0 \end{pmatrix} = \begin{pmatrix} 2 \\ -1 \\ -4 \end{pmatrix}.$$

Alternatively, we could write $[A_j] = [a_k][\alpha_j^k]$, which yields

$$\begin{pmatrix} A_1 \\ A_2 \\ A_3 \end{pmatrix}^T = (-1, 3, 0) \begin{pmatrix} 1 & 1 & 1 \\ 1 & 0 & -1 \\ 0 & 2 & -3 \end{pmatrix} = \begin{pmatrix} 2 \\ -1 \\ -4 \end{pmatrix}^T.$$

(*b*) To evaluate the contravariant components, we first note that $A^j = \beta_k^j a^k$, which has the matrix formulation $[A^j] = [\beta_k^j][a^k]$. Consequently, because $[\beta_k^j]$ is given in Example 1, we deduce that

$$\begin{pmatrix} A^1 \\ A^2 \\ A^3 \end{pmatrix} = \frac{1}{7} \begin{pmatrix} 2 & 5 & -1 \\ 3 & -3 & 2 \\ 2 & -2 & -1 \end{pmatrix} \begin{pmatrix} -1 \\ 3 \\ 0 \end{pmatrix} = \frac{1}{7} \begin{pmatrix} 13 \\ -12 \\ -8 \end{pmatrix} = \begin{pmatrix} 13/7 \\ -12/7 \\ -8/7 \end{pmatrix}.$$

5.3.2 Tensor operations

The following are some important tensor operations.

- **Addition:** The sum of two or more tensors of the same order and type is a tensor of the same order and type; for example,

$$A_{ij}^k + B_{ij}^k = C_{ij}^k.$$

- **Outer Product:** The outer product of two tensors is a tensor whose order is the sum of orders of the given tensors forming the product; for example,

$$A_k^{ij} B_n^m = C_{kn}^{ijm}.$$

- **Contraction:** If one contravariant and one covariant index of a tensor component are set equal, the resulting sum is a tensor of order two less than the original system; for example,

$$A_{jk}^j = B_k.$$

- **Quotient Law:** Let B^{jk} and C^i be arbitrary tensor components of order two and one. Then A_{jk}^i is a tensor of order three if it is true that

$$A_{jk}^i B^{jk} = C^i.$$

The quotient law given above can be generalized to tensors of any order. To prove the assertion given above, it is necessary to show that A_{jk}^i satisfies the appropriate transformation law $A_{jk}^i = \beta_m^i \alpha_j^n \alpha_k^p a_{np}^m$ (see Example 3 below).

EXAMPLE 3 Given the relationship $A_{jk}^i B^{jk} = C^i$, where it is known that B^{jk} and C^i are tensors of order two and one, respectively, show that A_{jk}^i is a tensor of order three.

Solution: In Cartesian coordinates we assume that

$$a_{jk}^{i} b^{jk} = c^{i}$$

where it is known that b^{jk} and c^{i} satisfy the transformation laws

$$b^{jk} = \alpha_m^j \alpha_n^k B^{mn}; \qquad c^i = \alpha_r^i C^r.$$

By substituting these relations into the first equation, we arrive at

$$a_{jk}^i \alpha_m^j \alpha_n^k B^{mn} = \alpha_r^i C^r = \alpha_r^i A_{mn}^r B^{mn},$$

which, since B^{mn} is common to both sides, we can rewrite as

$$\left(a_{jk}^i \alpha_m^j \alpha_n^k - \alpha_r^i A_{mn}^r \right) B^{mn} = 0.$$

If we now multiply the term in parentheses by β_i^p, we find

$$a_{jk}^i \alpha_m^j \alpha_n^k \beta_i^p - \beta_i^p \alpha_r^i A_{mn}^r = a_{jk}^i \alpha_m^j \alpha_n^k \beta_i^p - \delta_r^p A_{mn}^r$$
$$= a_{jk}^i \alpha_m^j \alpha_n^k \beta_i^p - A_{mn}^p,$$

which leads to

$$\left(a_{jk}^i \alpha_m^j \alpha_n^k \beta_i^p - A_{mn}^p \right) B^{mn} = 0.$$

Because B^{mn} is an arbitrary tensor, it follows that $A_{mn}^p = a_{jk}^i \alpha_m^j \alpha_n^k \beta_i^p$, which is a proper tensor transformation. Therefore, A_{jk}^i is a tensor of order three, which is what we intended to prove.

5.3.3 Symmetric and skew-symmetric tensors

A tensor of order n is said to be *symmetric* in two of its indices if the components are unchanged when the indices are interchanged. For example, the following tensors are symmetric in the indices j and k provided:

$$A_{jk} = A_{kj}; \qquad T_i^{jk} = T_i^{kj}.$$

The Kronecker delta δ_{jk}, δ_j^k, δ^{jk} is another example of a symmetric tensor.

We say a tensor of order n is *skew-symmetric* in two of its indices if the components change sign when the indices are interchanged. Examples of skew-symmetric tensors are

$$A_{jk} = -A_{kj}; \qquad T_i^{jk} = -T_i^{kj}.$$

If a tensor A_{jk} is skew-symmetric in two indices, an immediate consequence is that the corresponding tensor diagonal components are zero; i.e., $A_{11} = A_{22} = A_{33} = 0$.

5.4 Base Vectors

In Cartesian coordinates we denote the *unit base vectors* by $\hat{\mathbf{I}}_j$, which lie along the coordinate axes as shown in Figure 5.2. As such, they are mutually *orthogonal*; i.e.,

$$\hat{\mathbf{I}}_j \cdot \hat{\mathbf{I}}_k = \delta_{jk}. \tag{15}$$

Also, any vector **a** with contravariant components a^j and covariant components a_j can be expressed as

$$\mathbf{a} = a^j \hat{\mathbf{I}}_j = a_j \hat{\mathbf{I}}^j. \tag{16}$$

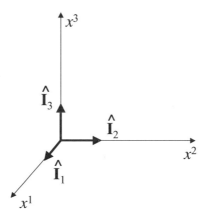

Figure 5.2 Base vectors in Cartesian coordinates.

Of course, in Eq. (16) we make no distinction between contravariant and covariant components in Cartesian coordinates; that is, $\hat{\mathbf{I}}_j = \hat{\mathbf{I}}^j$.

5.4.1 Covariant base vectors

To introduce the notion of base vectors into a rectilinear coordinate system, we start with the equations of transformation

$$x^j = \alpha_k^j y^k; \quad \det[\alpha_k^j] = J \neq 0$$
$$y^j = \beta_k^j x^k; \quad \det[\beta_k^j] = 1/J. \tag{17}$$

If A^j represents the contravariant components of a vector **a**, these components are related to the Cartesian components a^j by $a^j = A^k \alpha_k^j$, and consequently,

$$\mathbf{a} = a^j \hat{\mathbf{I}}_j = A^k \left(\alpha_k^j \hat{\mathbf{I}}_j \right). \tag{18}$$

Equation (18) suggests that we define a set of *covariant base vectors* $\mathbf{E}_k$ in rectilinear coordinates by

$$\mathbf{E}_k = \alpha_k^j \hat{\mathbf{I}}_j. \tag{19}$$

By doing so, we can represent a given vector $\mathbf{a}$ by its contravariant components in either Cartesian coordinates or rectilinear coordinates, viz.,

$$\mathbf{a} = a^j \hat{\mathbf{I}}_j = A^j \mathbf{E}_j. \tag{20}$$

Similar to the Cartesian base vectors $\hat{\mathbf{I}}_j$, the rectilinear base vectors $\mathbf{E}_j$ defined by (19) lie along the coordinate axes y^j as shown in Fig. 5.3. To verify that this is so, let us denote a position vector of some point x^j in Cartesian coordinates by

$$\mathbf{R} = x^j \hat{\mathbf{I}}_j = \alpha_k^j y^k \hat{\mathbf{I}}_j, \tag{21}$$

where y^k denotes the same point in rectilinear coordinates. By taking a partial derivative of (21) to obtain a tangent vector, we observe that

$$\frac{\partial \mathbf{R}}{\partial y^k} = \alpha_k^j \hat{\mathbf{I}}_j = \mathbf{E}_k, \tag{22}$$

which shows that $\mathbf{E}_k$ is indeed tangent to the y^k axis.

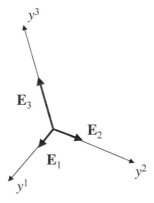

Figure 5.3 Covariant base vectors in rectilinear coordinates.

5.4.2 Contravariant base vectors: reciprocal basis

The Cartesian base vectors $\hat{\mathbf{I}}_j$ are a mutually *orthogonal set of unit vectors* (also called an *orthonormal set*), which we denote by

$$\hat{\mathbf{I}}_j \cdot \hat{\mathbf{I}}_k = \delta_{jk}. \tag{23}$$

However, this is generally not the case for the corresponding covariant base vectors $\mathbf{E}_j$ in rectilinear coordinates. That is, the base vectors $\mathbf{E}_j$ are not required to be mutually orthogonal nor to have unit norms. One way to remedy this situation is to introduce a *reciprocal* set of base vectors defined by

$$\mathbf{E}^j = \beta_k^j \hat{\mathbf{I}}^k. \qquad (24)$$

The reciprocal set of base vectors $\mathbf{E}^j$ are also called *contravariant base vectors*. Here we now see that

$$\mathbf{E}_j \cdot \mathbf{E}^k = \alpha_j^m \beta_n^k \hat{\mathbf{I}}_m \cdot \hat{\mathbf{I}}^n = \alpha_j^m \beta_n^k \delta_m^n = \alpha_j^m \beta_m^k, \qquad (25)$$

which reduces to

$$\mathbf{E}_j \cdot \mathbf{E}^k = \delta_j^k. \qquad (26)$$

Hence, we conclude that the reciprocal sets of base vectors $\mathbf{E}_j$ and $\mathbf{E}^k$ are mutually orthogonal, even though neither covariant nor contravariant sets of base vectors may have this property. Geometrically, it now also follows that the covariant base vectors $\mathbf{E}^k$ are orthogonal to the coordinate planes $y^k = 0$.

Both the covariant and contravariant base vectors satisfy the transformation law for tensors of order one. To generate the inverse transformation, we simply multiply both sides of (19) by β_m^k to find

$$\beta_m^k \mathbf{E}_k = \alpha_k^j \beta_m^k \hat{\mathbf{I}}_j = \delta_m^j \hat{\mathbf{I}}_j \qquad (27)$$

from which we deduce

$$\hat{\mathbf{I}}_m = \beta_m^k \mathbf{E}_k. \qquad (28)$$

By a similar analysis imposed on (24), it can be shown that

$$\hat{\mathbf{I}}^m = \alpha_k^m \mathbf{E}^k. \qquad (29)$$

Last, we note that a given vector $\mathbf{a}$ can be represented in rectilinear coordinates by using either covariant or contravariant base vectors; i.e.,

$$\mathbf{a} = A^j \mathbf{E}_j = A_j \mathbf{E}^j. \qquad (30)$$

This result illustrates that the contravariant components A^j belong to the covariant base vectors and the covariant components A_j belong to the contravariant base vectors.

5.4.3 Metric tensors

As shown above, nonorthogonal rectilinear coordinate systems are distinguished from Cartesian coordinate systems by the existence of two sets of base vectors. Although neither set of base vectors may be mutually orthogonal, the dot products between the covariant base vectors or between the contravariant base vectors form the components of important second-order tensors.

To begin, let us consider the dot product given by

$$\mathbf{E}_j \cdot \mathbf{E}_k = \alpha_j^m \alpha_k^n \left(\hat{\mathbf{I}}_m \cdot \hat{\mathbf{I}}_n \right) = \alpha_j^m \alpha_k^n \delta_{mn}. \tag{31}$$

The last term in (31) defines the second-order tensor

$$g_{jk} = \alpha_j^m \alpha_k^n \delta_{mn} = \alpha_j^m \alpha_k^m \tag{32}$$

called the *covariant metric tensor*. Because it satisfies the tensor transformation law (32), the covariant metric tensor g_{jk} is the rectilinear coordinate analogue of the Kronecker delta δ_{jk}, i.e.,

$$\hat{\mathbf{I}}_j \cdot \hat{\mathbf{I}}_k = \delta_{jk}; \qquad \mathbf{E}_j \cdot \mathbf{E}_k = g_{jk}. \tag{33}$$

Unlike the Kronecker delta, however, the components of the metric tensor g_{jk} are not restricted to zeros and ones.

Similarly, from the contravariant base vectors we are led to

$$\mathbf{E}^j \cdot \mathbf{E}^k = g^{jk}, \tag{34}$$

where g^{jk} are components of the *contravariant metric tensor* defined by

$$g^{jk} = \beta_m^j \beta_n^k \delta^{mn} = \beta_m^j \beta_m^k. \tag{35}$$

If follows from definition that both g_{jk} and g^{jk} are *symmetric tensors*; i.e., $g_{jk} = g_{kj}$ and $g^{jk} = g^{kj}$. Also, from the dot product between reciprocal base vectors (26), we deduce that the mixed metric tensor $g_j^{\ k} = \delta_j^k$.

The norms of reciprocal base vectors are typically not unity. By use of the metric tensors, we find that the norms of these base vectors are given by

$$\| \mathbf{E}_k \| = \sqrt{g_{\underline{kk}}}; \qquad \| \mathbf{E}^k \| = \sqrt{g^{\underline{kk}}}, \tag{36}$$

where the underline indicates *no sum is taken*.

EXAMPLE 4 Under the transformation of coordinates (see Example 1)

$$\begin{aligned}
x^1 &= y^1 + y^2 + y^3 \\
x^2 &= y^1 - y^3 \\
x^3 &= 2y^2 - 3y^3
\end{aligned}$$

(a) find the covariant base vectors $\mathbf{E}_j$ and their norms.

(b) find the contravariant base vectors $\mathbf{E}^j$ and their norms.

(c) find the metric tensors g_{jk} and g^{jk}.

Solution:

(a) From (19), the covariant base vectors are defined by one of the matrix products

$$[\mathbf{E}_j] = [\hat{\mathbf{I}}_k]^T[\alpha_j^k] = [\alpha_j^k]^T[\hat{\mathbf{I}}_k],$$

where T denotes matrix transpose. By recalling the matrix representation of α_j^k from Example 1 and using the first product above, we have

$$\begin{pmatrix} \mathbf{E}_1 \\ \mathbf{E}_2 \\ \mathbf{E}_3 \end{pmatrix}^T = (\mathbf{I}_1, \mathbf{I}_2, \mathbf{I}_3) \begin{pmatrix} \alpha_1^1 & \alpha_2^1 & \alpha_3^1 \\ \alpha_1^2 & \alpha_2^2 & \alpha_3^2 \\ \alpha_1^3 & \alpha_2^3 & \alpha_3^3 \end{pmatrix} = (\mathbf{I}_1, \mathbf{I}_2, \mathbf{I}_3) \begin{pmatrix} 1 & 1 & 1 \\ 1 & 0 & -1 \\ 0 & 2 & -3 \end{pmatrix}$$

from which we deduce

$$\begin{aligned} \mathbf{E}_1 &= \mathbf{I}_1 + \mathbf{I}_2 = \;<1,1,0> \\ \mathbf{E}_2 &= \mathbf{I}_1 + 2\mathbf{I}_3 = \;<1,0,2> \\ \mathbf{E}_3 &= \mathbf{I}_1 - \mathbf{I}_2 - 3\mathbf{I}_3 = \;<1,-1,-3>. \end{aligned}$$

Thus, we see that the columns of the matrix $[\alpha_j^k]$ correspond to the components of the base vectors $\mathbf{E}_1$, $\mathbf{E}_2$, and $\mathbf{E}_3$. From these results, the respective norms are readily found to be

$$\|\mathbf{E}_1\| = \sqrt{2}; \quad \|\mathbf{E}_2\| = \sqrt{5}; \quad \|\mathbf{E}_3\| = \sqrt{11}.$$

(b) In a similar manner, we recall from (24) that

$$[\mathbf{E}^j] = [\beta_k^j][\hat{\mathbf{I}}^k]$$

and by using the result from Example 1 for β_j^k, we obtain

$$\begin{pmatrix} \mathbf{E}^1 \\ \mathbf{E}^2 \\ \mathbf{E}^3 \end{pmatrix} = \begin{pmatrix} \beta_1^1 & \beta_2^1 & \beta_3^1 \\ \beta_1^2 & \beta_2^2 & \beta_3^2 \\ \beta_1^3 & \beta_2^3 & \beta_3^3 \end{pmatrix} \begin{pmatrix} \mathbf{I}^1 \\ \mathbf{I}^2 \\ \mathbf{I}^3 \end{pmatrix} = \frac{1}{7} \begin{pmatrix} 2 & 5 & -1 \\ 3 & -3 & 2 \\ 2 & -2 & -1 \end{pmatrix} \begin{pmatrix} \mathbf{I}^1 \\ \mathbf{I}^2 \\ \mathbf{I}^3 \end{pmatrix}$$

or

$$\mathbf{E}^1 \;=\; \frac{1}{7}(2\mathbf{I}^1 \;+\; 5\mathbf{I}^2 \;-\; \mathbf{I}^3) \;=\; \left\langle \frac{2}{7}, \frac{5}{7}, -\frac{1}{7} \right\rangle$$

$$\mathbf{E}^2 \;=\; \frac{1}{7}(3\mathbf{I}^1 \;-\; 3\mathbf{I}^3 \;+\; 2\mathbf{I}^3) \;=\; \left\langle \frac{3}{7}, -\frac{3}{7}, \frac{2}{7} \right\rangle$$

$$\mathbf{E}^3 \;=\; \frac{1}{7}(2\mathbf{I}_1 \;-\; 2\mathbf{I}_2 \;-\; \mathbf{I}_3) \;=\; \left\langle \frac{2}{7}, -\frac{2}{7}, -\frac{1}{7} \right\rangle.$$

Thus, the components of the contravariant base vectors $\mathbf{E}^1$, $\mathbf{E}^2$, and $\mathbf{E}^3$ are given by the rows of the matrix $[\beta_j^k]$. Here, we find the respective norms

$$\|\mathbf{E}^1\| \;=\; \frac{\sqrt{30}}{7}; \quad \|\mathbf{E}^2\| \;=\; \frac{\sqrt{22}}{7}; \quad \|\mathbf{E}^3\| \;=\; \frac{3}{7}.$$

(*c*) The components of the metric tensors g_{jk} and g^{jk} can be found directly from the base vectors using (33) and (34), or we can use Eqs. (32) and (35), respectively. In matrix form, Eq. (32) becomes $[g_{jk}] = [\alpha_j^m]^T[\alpha_k^m]$, which leads to

$$[g_{jk}] \;=\; \begin{pmatrix} 1 & 1 & 0 \\ 1 & 0 & 2 \\ 1 & -1 & -3 \end{pmatrix} \begin{pmatrix} 1 & 1 & 1 \\ 1 & 0 & -1 \\ 0 & 2 & -3 \end{pmatrix} \;=\; \begin{pmatrix} 2 & 1 & 0 \\ 1 & 5 & -5 \\ 0 & -5 & 11 \end{pmatrix}.$$

Similarly, from Eq. (35) we see that $[g^{jk}] = [\beta_m^j][\beta_m^k]^T$, and consequently,

$$[g^{jk}] \;=\; \frac{1}{49} \begin{pmatrix} 2 & 5 & -1 \\ 3 & -3 & 2 \\ 2 & -2 & -1 \end{pmatrix} \begin{pmatrix} 2 & 3 & 2 \\ 5 & -3 & -2 \\ -1 & 2 & -1 \end{pmatrix} \;=\; \frac{1}{49} \begin{pmatrix} 30 & -11 & -5 \\ -11 & 22 & 10 \\ -5 & 10 & 9 \end{pmatrix}.$$

Remark: The reader can easily verify that the reciprocal base vectors found in Example 4 satisfy the orthogonality relation $\mathbf{E}_j \cdot \mathbf{E}^k = \delta_j^k$.

5.5 Vector Algebra

In this section we examine various algebraic operations like the dot product, cross product, and mixed triple product of vectors in rectilinear coordinates. To do so, we first recall the definitions of the vector operations in Cartesian coordinates.

Let **a**, **b**, and **c** denote three arbitrary vectors. In terms of their Cartesian components written in tensor notation, we define the *dot product*, *cross product*, and *mixed triple* *product*,[2] respectively, by (see Chapter 4)

$$\mathbf{a} \cdot \mathbf{b} = \delta_{jk} a^j b^k = \delta^{jk} a_j b_k \tag{37}$$

$$\mathbf{a} \times \mathbf{b} = e_{ijk} a^i b^j \hat{\mathbf{I}}^k = e^{ijk} a_i b_j \hat{\mathbf{I}}_k \tag{38}$$

$$\mathbf{a} \times \mathbf{b} \cdot \mathbf{c} = (\mathbf{abc}) = e_{ijk} a^i b^j c^k = e^{ijk} a_i b_j c_k. \tag{39}$$

5.5.1 Permutation symbols in rectilinear coordinates

The Kronecker delta is associated with dot products between the Cartesian coordinate base vectors, i.e., $\delta_{jk} = \hat{\mathbf{I}}_j \cdot \hat{\mathbf{I}}_k$ and $\delta^{jk} = \hat{\mathbf{I}}^j \cdot \hat{\mathbf{I}}^k$. Similarly, the permutation symbol defined by Eq. (3) can be expressed in terms of the mixed triple product according to (assuming a right-handed coordinate system)

$$e_{ijk} = \hat{\mathbf{I}}_i \times \hat{\mathbf{I}}_j \cdot \hat{\mathbf{I}}_k = 1; \qquad e^{ijk} = \hat{\mathbf{I}}^i \times \hat{\mathbf{I}}^j \cdot \hat{\mathbf{I}}^k = 1. \tag{40}$$

Thus, if we interchange two indices, there is a sign change in the result; and if two indices are equal, then $e_{ijk} = e^{ijk} = 0$ (i.e., if two columns or two rows of a determinant are the same, the determinant is zero).

We have already introduced the metric tensors g_{jk} and g^{jk} as the rectilinear coordinate analogues of the corresponding Kronecker deltas. For calculating cross products and mixed triple products in rectilinear components, it is useful to introduce the rectilinear coordinate *permutation symbols*

$$\varepsilon_{ijk} = \mathbf{E}_i \times \mathbf{E}_j \cdot \mathbf{E}_k = e_{ijk} \alpha_1^i \alpha_2^j \alpha_3^k = J,$$

$$\varepsilon^{ijk} = \mathbf{E}^i \times \mathbf{E}^j \cdot \mathbf{E}^k = e^{ijk} \beta_i^1 \beta_j^2 \beta_k^3 = \frac{1}{J}, \tag{41}$$

where J is the Jacobian of transformation. The verification that these mixed triple products are related to the Jacobian in this fashion follows immediately from the fact that the covariant and contravariant base vectors represent columns and rows, respectively, of the matrices $[\alpha_m^n]$ and $[\beta_m^n]$.

By using the transformation law $\mathbf{E}_i = \alpha_i^m \hat{\mathbf{I}}_m$, the permutation symbol ε_{ijk} can be expressed in terms of the Cartesian coordinate permutation symbol e_{ijk}; i.e.,

$$\varepsilon_{ijk} = \alpha_i^m \alpha_j^n \alpha_k^p (\hat{\mathbf{I}}_m \times \hat{\mathbf{I}}_n \cdot \hat{\mathbf{I}}_p),$$

from which we deduce

$$\varepsilon_{ijk} = \alpha_i^m \alpha_j^n \alpha_k^p e_{mnp}. \tag{42}$$

---

[2]Recall that the placement of dot and cross in a mixed triple product is arbitrary, i.e., $\mathbf{a} \times \mathbf{b} \cdot \mathbf{c} = \mathbf{a} \cdot \mathbf{b} \times \mathbf{c}$.

In a similar manner, it can be shown that

$$\varepsilon^{ijk} = \beta^i_m \beta^j_n \beta^k_p e^{mnp}. \tag{43}$$

It follows from properties of determinants that the permutation symbols ε_{ijk} and ε^{ijk} are also defined similar to (3) in Cartesian coordinates. Namely,

$$\varepsilon_{ijk} = \begin{cases} J & (i,j,k \;\; \text{cyclic}) \\ -J, & (i,j,k \;\; \text{noncyclic}) \\ 0, & (\text{two or more indices alike}) \end{cases} \tag{44}$$

$$\varepsilon^{ijk} = \begin{cases} J^{-1} & (i,j,k \;\; \text{cyclic}) \\ -J^{-1} & (i,j,k \;\; \text{noncyclic}) \\ 0, & (\text{two or more indices alike}) \end{cases} \tag{45}$$

In addition, we note the following properties:

$$\boxed{\begin{aligned} \varepsilon^{ijk}\varepsilon_{imn} &= \delta^j_m \delta^k_n - \delta^j_n \delta^k_m \\ \varepsilon^{ijk} &= g^{im} g^{jn} g^{kp} \varepsilon_{mnp} \\ \varepsilon_{ijk} &= g_{im} g_{jn} g_{kp} \varepsilon^{mnp} \end{aligned}} \tag{46}$$

$$\boxed{\begin{aligned} \varepsilon^{ijk}\varepsilon^{mnp} g_{im} g_{jn} &= 2g^{kp} \\ \varepsilon_{ijk}\varepsilon_{mnp} g^{im} g^{jn} &= 2g_{kp} \end{aligned}} \tag{47}$$

$$\boxed{\begin{aligned} \varepsilon^{ijk}\varepsilon^{mnp} g_{im} g_{jn} g_{kp} &= 6 \\ \varepsilon_{ijk}\varepsilon_{mnp} g^{im} g^{jn} g^{kp} &= 6 \end{aligned}} \tag{48}$$

In Example 5 below, we prove one of the identities in (46).

EXAMPLE 5 Verify the identity $\varepsilon_{ijk} = g_{im} g_{jn} g_{kp} \varepsilon^{mnp}$.

Solution: Starting with the right-hand side of the identity, we have

$$\begin{aligned} g_{im} g_{jn} g_{kp} \varepsilon^{mnp} &= \alpha^r_i \alpha^r_m \alpha^s_j \alpha^s_n \alpha^t_k \alpha^t_p \varepsilon^{mnp} \\ &= \alpha^r_i \alpha^r_m \alpha^s_j \alpha^s_n \alpha^t_k \alpha^t_p (\beta^m_u \beta^n_v \beta^p_w) e^{uvw} \\ &= \alpha^r_i \alpha^s_j \alpha^t_k (\alpha^r_m \beta^m_u)(\alpha^s_n \beta^n_v)(\alpha^t_p \beta^p_w) e^{uvw} \end{aligned}$$

which simplifies to

$$g_{im} g_{jn} g_{kp} \varepsilon^{mnp} = \alpha_i^r \alpha_j^s \alpha_k^t \delta_u^r \delta_v^s \delta_w^t e^{uvw}$$
$$= \alpha_i^r \alpha_j^s \alpha_k^t e_{rst}$$
$$= \varepsilon_{ijk}.$$

In arriving at our result, we used the identity $e^{rst} = e_{rst}$.

From definition, the metric tensors and the Jacobian are related by

$$\det[g_{jk}] = \{\det[\alpha_j^k]\}^2 = J^2$$
$$\det[g^{jk}] = \{\det[\beta_k^j]\}^2 = J^{-2}$$
$$\tag{49}$$

from which we deduce

$$\varepsilon_{ijk} = J e_{ijk}; \quad \varepsilon^{ijk} = \frac{1}{J} e^{ijk}. \tag{50}$$

Using these last results, we are now able to show that reciprocal base vectors $\mathbf{E}_k$ and $\mathbf{E}^k$ are related by

$$\mathbf{E}_i \times \mathbf{E}_j = J e_{ijk} \mathbf{E}^k$$
$$\mathbf{E}^i \times \mathbf{E}^j = \frac{1}{J} e^{ijk} \mathbf{E}_k. \tag{51}$$

To prove the first relation in (51), we start with the cross product

$$\mathbf{E}_i \times \mathbf{E}_j = e_{mnp} \alpha_i^m \alpha_j^n \hat{\mathbf{I}}^p ;$$

but since $\hat{\mathbf{I}}^p = \alpha_k^p \mathbf{E}^k$, this becomes

$$\mathbf{E}_i \times \mathbf{E}_j = e_{mnp} \alpha_i^m \alpha_j^n \alpha_k^p \mathbf{E}^k.$$

To complete the proof, we combine (42) and (50) to obtain

$$e_{mnp} \alpha_i^m \alpha_j^n \alpha_k^p = J e_{ijk}. \tag{52}$$

The proof of the second relation in (51) is similar.

We previously showed that the contravariant base vectors $\mathbf{E}^k$ are orthogonal to the coordinate planes $y^k = 0$. Based on Eqs. (51), we now see that

$$\mathbf{E}_2 \times \mathbf{E}_3 = J \mathbf{E}^1$$
$$\mathbf{E}_3 \times \mathbf{E}_1 = J \mathbf{E}^2$$
$$\mathbf{E}_1 \times \mathbf{E}_2 = J \mathbf{E}^3. \tag{53}$$

5.5.2 Dot product

Given the vector **a**, its covariant and contravariant components in Cartesian coordinates are related to those in rectilinear coordinates by

$$A_j = \alpha_j^k a_k; \qquad a_j = \beta_j^k A_k$$
$$A^j = \beta_k^j a^k; \qquad a^j = \alpha_k^j A^k. \tag{54}$$

Using these relations, the dot product (37) between vectors **a** and **b** becomes

$$\mathbf{a} \cdot \mathbf{b} = \delta_{jk} a^j b^k = \delta_{jk} \alpha_m^j \alpha_n^k A^m B^n$$
$$\mathbf{a} \cdot \mathbf{b} = \delta^{jk} a_j b_k = \delta^{jk} \beta_j^m \beta_k^n A_m B_n \tag{55}$$

from which we deduce

$$\boxed{\mathbf{a} \cdot \mathbf{b} = g_{jk} A^j B^k = g^{jk} A_j B_k.} \tag{56}$$

Note that (56) has the same form as (37) for Cartesian coordinates with the metric tensors replacing the Kronecker deltas and the rectilinear vector components replacing Cartesian vector components. Finally, by using one covariant and one contravariant component in the dot product, we find that

$$\mathbf{a} \cdot \mathbf{b} = g_k^j A_j B^k = A_k B^k; \qquad \mathbf{a} \cdot \mathbf{b} = g_j^k A^j B_k = A^k B_k \tag{57}$$

where $g_k^j = \delta_k^j$. The advantage in using (57) over (56) is that (57) has the same ease of computation as a dot product in Cartesian coordinates.

5.5.3 Cross product and mixed triple product

In the case of cross products, we replace the Cartesian components by rectilinear components to get

$$\boxed{\begin{aligned} \mathbf{a} \times \mathbf{b} &= e_{ijk} a^i b^j \hat{\mathbf{I}}^k = \varepsilon_{ijk} A^i B^j \mathbf{E}^k \\ \mathbf{a} \times \mathbf{b} &= e^{ijk} a_i b_j \hat{\mathbf{I}}_k = \varepsilon^{ijk} A_i B_j \mathbf{E}_k \end{aligned}} \tag{58}$$

where $\hat{\mathbf{I}}^k = \alpha_m^k \mathbf{E}^m$, $\hat{\mathbf{I}}_k = \beta_k^m \mathbf{E}_m$, and where we have used (42) and (43). The mixed triple product between three vectors **a**, **b**, **c** given by (39) in Cartesian coordinates is similarly defined in rectilinear coordinates by

$$\boxed{(\mathbf{abc}) = \varepsilon_{ijk} A^i B^j C^k = \varepsilon^{ijk} A_i B_j C_k.} \tag{59}$$

We omit the verification of (59), but point out that (59) can be formally obtained from the Cartesian relation (39) by simple symbol replacement.

EXAMPLE 6 Given vectors **a** and **b** with Cartesian components

$$\mathbf{a} = <1, -3, 2>; \quad \mathbf{b} = <0, 5, 1>$$

and the equations of transformation

$$y^1 = x^1 + x^2 + x^3$$
$$y^2 = x^1 - x^3$$
$$y^3 = 2x^2 - 3x^3$$

(a) find the contravariant components A^k and B^k under the transformation given above.

(b) Calculate the dot product $\mathbf{a \cdot b}$ in both Cartesian and rectilinear coordinates.

(c) Calculate the cross product $\mathbf{a} \times \mathbf{b}$ and $\|\mathbf{a} \times \mathbf{b}\|$ in both Cartesian and rectilinear coordinates (relative to $\mathbf{E}^k$ basis only).

Solution: The above transformation, similar to that in Example 1 with the roles of x^j and y^j interchanged, is characterized by the matrix representation

$$[\beta_k^j] = \begin{pmatrix} 1 & 1 & 1 \\ 1 & 0 & -1 \\ 0 & 2 & -3 \end{pmatrix}.$$

(a) Under the transformation laws $A^j = \beta_k^j a^k$, $B^j = \beta_k^j b^k$, we obtain

$$\begin{pmatrix} A^1 \\ A^2 \\ A^3 \end{pmatrix} = \begin{pmatrix} 1 & 1 & 1 \\ 1 & 0 & -1 \\ 0 & 2 & -3 \end{pmatrix} \begin{pmatrix} 1 \\ -3 \\ 2 \end{pmatrix} = \begin{pmatrix} 0 \\ -1 \\ -12 \end{pmatrix}$$

$$\begin{pmatrix} B^1 \\ B^2 \\ B^3 \end{pmatrix} = \begin{pmatrix} 1 & 1 & 1 \\ 1 & 0 & -1 \\ 0 & 2 & -3 \end{pmatrix} \begin{pmatrix} 0 \\ 5 \\ 1 \end{pmatrix} = \begin{pmatrix} 6 \\ -1 \\ 7 \end{pmatrix}.$$

(b) In Cartesian coordinates, the dot product leads to

$$\mathbf{a \cdot b} = a_k b^k = 0 - 15 + 2 = -13.$$

When using rectilinear coordinates, we have a choice between (56) or (57). We choose to take advantage of the simpler result of (57), which necessitates first finding the covariant components A_j. From the given transformation, we can readily find

$$[\alpha_k^j] = \frac{1}{7}\begin{pmatrix} 2 & 5 & -1 \\ 3 & -3 & 2 \\ 2 & -2 & -1 \end{pmatrix}.$$

Hence, recalling $[A_j] = [\alpha_j^k]^T[a_k]$, we then calculate

$$\begin{pmatrix} A_1 \\ A_2 \\ A_3 \end{pmatrix} = \frac{1}{7}\begin{pmatrix} 2 & 3 & 2 \\ 5 & -3 & -2 \\ -1 & 2 & -1 \end{pmatrix}\begin{pmatrix} 1 \\ -3 \\ 2 \end{pmatrix} = \frac{1}{7}\begin{pmatrix} -3 \\ 10 \\ -9 \end{pmatrix},$$

from which we obtain our result

$$\mathbf{a \cdot b} = A_k B^k$$
$$= \frac{1}{7}<-3, 10, -9> \cdot <6, -1, 7> = \frac{-91}{7} = -13.$$

[We leave it to the reader to show that using one of the forms in (56) leads to the same result.]

(c) In Cartesian coordinates, the cross product $\mathbf{a} \times \mathbf{b} = \mathbf{c}$ produces a vector $\mathbf{c}$, which leads directly to

$$\mathbf{a} \times \mathbf{b} = c_k \hat{\mathbf{I}}^k = \begin{vmatrix} 1 & -3 & 2 \\ 0 & 5 & 1 \\ \hat{\mathbf{I}}^1 & \hat{\mathbf{I}}^2 & \hat{\mathbf{I}}^3 \end{vmatrix} = -13\hat{\mathbf{I}}^1 - \hat{\mathbf{I}}^2 + 5\hat{\mathbf{I}}^3.$$

In rectilinear coordinates, we use the transformation equations (58) with (50) to obtain

$$\mathbf{a} \times \mathbf{b} = \varepsilon_{ijk}A^i B^j \mathbf{E}^k = Je_{ijk}A^i B^j \mathbf{E}^k,$$

where $J = \det[\alpha_k^j] = 1/7$. Therefore,

$$\mathbf{a} \times \mathbf{b} = C_k \mathbf{E}^k = \frac{1}{7}\begin{vmatrix} 0 & -1 & -12 \\ 6 & -1 & 7 \\ \mathbf{E}^1 & \mathbf{E}^2 & \mathbf{E}^3 \end{vmatrix}$$

or

$$\mathbf{a} \times \mathbf{b} = \frac{1}{7}\left(-19\mathbf{E}^1 - 72\mathbf{E}^2 + 6\mathbf{E}^3\right).$$

To calculate the magnitude in Cartesian coordinates, we note that

$$\| \mathbf{a} \times \mathbf{b} \| = \sqrt{(-13)^2 + (-1)^2 + 5^2} = \sqrt{195}.$$

In rectilinear coordinates, we let $\mathbf{a} \times \mathbf{b} = \mathbf{c}$, and then use $\| \mathbf{a} \times \mathbf{b} \| = \sqrt{\mathbf{c} \cdot \mathbf{c}}$, which leads to (this time we use the metric tensor for illustrative purposes)

$$
\begin{aligned}
\mathbf{c} \cdot \mathbf{c} &= C_j g^{jk} C_k \\
&= \frac{1}{49} (-19, \; -72, \; 6)
\begin{pmatrix} 3 & 0 & -1 \\ 0 & 2 & 3 \\ -1 & 3 & 13 \end{pmatrix}
\begin{pmatrix} -19 \\ -72 \\ 6 \end{pmatrix} \\
&= 195.
\end{aligned}
$$

Thus, $\| \mathbf{a} \times \mathbf{b} \| = \sqrt{195}$ in agreement with that for Cartesian coordinates. Here we have calculated the metric tensor g^{jk} using the technique illustrated in Example 4.

5.6 Relations Between Tensor Components

We have provided relations between tensor components in Cartesian coordinates and those in rectilinear coordinates, but we have not yet given any direct relations between covariant and contravariant components in the same coordinate system. In Cartesian coordinates the Kronecker delta is instrumental in finding such relations, whereas in rectilinear coordinates it is the metric tensor we need.

5.6.1 Raising and lowering indices

Although we recognize that the vector covariant and contravariant components in Cartesian coordinates satisfy $a_j = a^j$, it can be suggestive to use the Kronecker delta to rewrite this relation as

$$a_j = \delta_{jk} a^k; \qquad a^j = \delta^{jk} a_k. \tag{60}$$

For second-order tensor components, the comparable relations are

$$
\begin{aligned}
a_{jk} &= \delta_{jm} \delta_{kn} a^{mn}; & a^{jk} &= \delta^{jm} \delta^{kn} a_{mn} \\
a_{\bullet k}^{j} &= \delta^{jm} \delta_{kn} a_m^{\bullet n}; & a_j^{\bullet k} &= \delta_{jm} \delta^{kn} a_{\bullet n}^{m}.
\end{aligned}
\tag{61}
$$

The "dots" in (62) are used here to distinguish between the first and second index of mixed tensors, which may not be symmetric. We will continue the use of this "dot" notation in the following discussion of mixed tensor expressions in rectilinear coordinates.

Similar expressions for tensor components in rectilinear coordinates can be derived by using the above results combined with the proper transformation laws. That is, the representation of a vector **a** in rectilinear coordinates leads to

$$A_j \mathbf{E}^j = A^j \mathbf{E}_j. \tag{62}$$

By forming the dot product of each side of this expression, first with respect to $\mathbf{E}_k$ and then with respect to $\mathbf{E}^k$, we find that

$$A^j \mathbf{E}_j \cdot \mathbf{E}^k = A_j \mathbf{E}^j \cdot \mathbf{E}^k$$
$$A^j \mathbf{E}_j \cdot \mathbf{E}_k = A_j \mathbf{E}^j \cdot \mathbf{E}_k$$

which reduce to the analog of Eqs. (61) given by

$$A_j = g_{jk} A^k; \qquad A^j = g^{jk} A_k. \tag{63}$$

Thus, the metric tensors permit the transformation between covariant and contravariant components of a vector. This process is known as *raising and lowering indices*. For higher-order tensor components, the relations comparable to (63) are

$$A_{jk} = g_{jm} g_{kn} A^{mn}; \qquad A^{jk} = g^{jm} g^{kn} A_{mn}$$
$$A_{\cdot k}^{j} = g_{km} A^{jm} = g^{jm} A_{mk}; \qquad A_j^{\cdot k} = g_{jm} A^{mk} = g^{mk} A_{jm}. \tag{64}$$

Clearly, for higher-order tensor components we again rely on the metric tensors as illustrated above to raise or lower various tensor component indices.

5.6.2 Physical components

Because the base vectors $\mathbf{E}_k$ and $\mathbf{E}^k$ are not unit vectors in general, not all vector components A_k or A^k will have the same physical dimensions. In problems in physics and engineering it is the physical components of the vector or tensor (with appropriate physical dimensions) that must ultimately be defined. These physical components are obtained by taking the parallel projections of vectors on unit vectors lying along the coordinate curves. Thus, we define the *physical components* $A^{(k)}$ of a vector **a** by the relation

$$\mathbf{a} = A^k \mathbf{E}_k = A^{(k)} \hat{\mathbf{E}}_k \tag{65}$$

where the *unit* base vector in (65) is defined by

$$\hat{\mathbf{E}}_k = \frac{1}{\sqrt{g_{kk}}} \mathbf{E}_k. \tag{66}$$

The physical components are directly related to the contravariant tensor components by

$$A^{(k)} = A^k \sqrt{g_{kk}}. \tag{67}$$

The physical components can also be defined in terms of the parallel projections on unit vectors lying along the reciprocal base vectors $\mathbf{E}^k$, but we will not do so here. However, we can relate the physical components to the covariant tensor components by

$$A^{(k)} = g^{jk}A_j\sqrt{g_{kk}}. \tag{68}$$

Several different types of physical components can arise in tensors of order greater than one. However, the identification of these physical components gets more complicated, so we will not discuss them here.

As a final comment here, we point out that the physical components do not transform as tensors; that is, they do not obey linear laws of transformation. The fact that they do not transform as tensors is the primary reason that transforming equations from one coordinate system to another in terms of the physical components becomes such a difficult task in general. If we choose to write our equations in terms of tensor components, then the transformation of coordinate systems becomes a relatively straightforward operation, regardless of coordinate system. Formally, the conversion of tensor components from one coordinate system to another is simple "symbol replacement." That is, Cartesian tensor components are replaced by rectilinear tensor components, Kronecker deltas are replaced by metric tensors, and so on. It is this invariance of form of the equations that is the real advantage of tensor notation.

EXAMPLE 7 Given the vector $\mathbf{a} = \langle 2, -1, 3\rangle$ in Cartesian coordinates and the linear transformation of coordinates

$$\begin{aligned}
y^1 &= x^1 + x^2 + x^3 \\
y^2 &= x^1 - x^3 \\
y^3 &= 2x^2 - 3x^3
\end{aligned}$$

find the physical components of the vector $\mathbf{a}$ in the defined rectilinear coordinate system.

Solution: The coordinate transformation is the same as that introduced in Example 6; thus, the coefficient matrix $[\beta_j^k]$ of the transformation is the same as that given there. Using this matrix representation, the contravariant components of vector $\mathbf{a}$ are defined by

$$\begin{pmatrix} A^1 \\ A^2 \\ A^3 \end{pmatrix} = \begin{pmatrix} 1 & 1 & 1 \\ 1 & 0 & -1 \\ 0 & 2 & -3 \end{pmatrix}\begin{pmatrix} 2 \\ -1 \\ 3 \end{pmatrix} = \begin{pmatrix} 4 \\ -1 \\ -11 \end{pmatrix}.$$

To calculate the physical components from (67) requires knowledge of the covariant metric tensors, defined by the matrix $[g_{jk}] = [\alpha_j^m]^T[\alpha_k^m]$. The inverse matrix $[\alpha_j^m] = [\beta_j^m]^{-1}$ is determined by solving the given transformation equations for the Cartesian coordinates x^j, which leads to

$$[\alpha_k^j] = \frac{1}{7}\begin{pmatrix} 2 & 5 & -1 \\ 3 & -3 & 2 \\ 2 & -2 & -1 \end{pmatrix}.$$

Hence, the diagonal elements of the covariant metric tensor are

$$g_{11} = \frac{17}{49}; \qquad g_{22} = \frac{38}{49}; \qquad g_{33} = \frac{6}{49};$$

from which it follows that

$$A^{(1)} = A^1\sqrt{g_{11}} = \frac{4}{7}\sqrt{17}$$

$$A^{(2)} = A^2\sqrt{g_{22}} = -\frac{1}{7}\sqrt{38}$$

$$A^{(3)} = A^3\sqrt{g_{33}} = -\frac{11}{7}\sqrt{6}.$$

5.7 Reduction of Tensors to Principal Axes

A problem of great physical importance concerns finding the principal axes associated with various second-order tensors like a stress tensor, dielectric tensor, etc. To begin, let us consider the second-order tensor T_{jk} and any vector **A** such that

$$T_{jk}A^k = B_j,$$

where **B** is the resulting vector. In general, the vector **B** differs from **A** in both direction and magnitude. Suppose now we wish to find all vectors **A** for which their direction is not altered by such a transformation, i.e., vectors that satisfy the relation

$$T_{jk}A^k = \lambda A_j, \tag{69}$$

where λ is a scalar. Such vectors, if they exist, are called *characteristic vectors*, or *eigenvectors* of the tensor T_{jk}. The axes determined by the directions of these vectors are called *principal axes*. Clearly, there are at most three such axes (recall Section 3.3.1). The values of the components of T_{jk} in the coordinate system determined by the principal axes are the *eigenvalues* of the tensor T_{jk}.

5.7.1 Two-dimensional case

Let us first consider the simpler problem when the number of dimensions is only two. Also, we take the special case in which the tensor T_{jk} is symmetric so that $T_{12} = T_{21}$. Thus, if the vector **A** lies along a principal axis of the tensor T_{jk}, it follows that

$$(T_{11} - \lambda)A^1 + T_{12}A^2 = 0,$$

$$T_{21}A^1 + (T_{22} - \lambda)A^2 = 0, \tag{70}$$

which has nonzero solutions only if the coefficient determinant is zero; i.e.,

$$\begin{vmatrix} T_{11} - \lambda & T_{12} \\ T_{21} & T_{22} - \lambda \end{vmatrix} = 0, \tag{71}$$

or

$$\lambda^2 - (T_{11} + T_{22})\lambda + (T_{11}T_{22} - T_{12}T_{12}) = 0. \tag{72}$$

The two solutions of (72) are readily found to be

$$\lambda_1 = \frac{T_{11} + T_{22}}{2} + \sqrt{\left(\frac{T_{11} - T_{22}}{2}\right)^2 + T_{12}T_{12}},$$

$$\lambda_2 = \frac{T_{11} + T_{22}}{2} - \sqrt{\left(\frac{T_{11} - T_{22}}{2}\right)^2 + T_{12}T_{12}}. \tag{73}$$

Clearly, if $T_{12} = T_{21} = 0$, the original Cartesian axes x^j are principal axes. Otherwise, there are two distinct principal axes denoted by X^j with directions given by

$$\tan \varphi_1 = \frac{X^2}{X^1} = \frac{\lambda_1 - T_{11}}{T_{12}}, \tag{74}$$

and $\varphi_2 = \varphi_1 + \pi/2$. In the principal axes system X^j, the tensor T_{jk} has components

$$[T_{jk}] = \begin{pmatrix} \lambda_1 & 0 \\ 0 & \lambda_2 \end{pmatrix}. \tag{75}$$

5.7.2 Three-dimensional case

Based on the result of Eq. (75), we see that identifying the principal axes reduces the number of components of a tensor T_{jk} to only those along the diagonal of the matrix. Hence, in the case of a three-dimensional system, this means that we can reduce the nine components of a general second-order tensor to only *three* components associated with the principal axes.

The standard procedure in the three-dimensional case is essentially the same as that in the two-dimensional case. That is, we look for vectors **A** such that

$$(T_{jk} - \lambda\delta_{jk})A^k = 0, \tag{76}$$

from which we determine the eigenvalues by solving $\det(T_{jk} - \lambda\delta_{jk}) = 0$, or

$$\begin{vmatrix} T_{11} - \lambda & T_{12} & T_{13} \\ T_{21} & T_{22} - \lambda & T_{23} \\ T_{31} & T_{32} & T_{33} - \lambda \end{vmatrix} = 0. \tag{77}$$

In this case the eigenvalues $\lambda_1, \lambda_2, \lambda_3$ are solutions of a polynomial of degree three. If the tensor T_{jk} is symmetric, the eigenvalues are all real. We then follow the procedure discussed in Section 3.3 for determining the principal axes. Last, in terms of the coordinate system based on the principal axes, the tensor components are

$$[T_{jk}] = \begin{pmatrix} \lambda_1 & 0 & 0 \\ 0 & \lambda_2 & 0 \\ 0 & 0 & \lambda_3 \end{pmatrix}. \tag{78}$$

5.8 Tensor Calculus: Rectilinear Coordinates

In order to introduce the notions of derivative and integral into our analysis it is necessary to consider tensor functions, or *tensor fields* (see also Chapter 4). In particular, let $A_j = A_j(y^1, y^2, y^3) \equiv A_j(\mathbf{y})$ denote a covariant vector field defined in rectilinear coordinates by the transformation law

$$A_j = \alpha_j^k a_k, \tag{79}$$

where $a_k = a_k(x^1, x^2, x^3) \equiv a_k(\mathbf{x})$ is the corresponding vector field in Cartesian coordinates. To emphasize the coordinate dependency of the vector components, the transformation law (79) can also be expressed as

$$A_j(\mathbf{y}) = \alpha_j^k a_k(\mathbf{x}), \tag{80}$$

where we are illustrating that the coefficient elements α_j^k are constant.

If we differentiate (80) with respect to y^n, this leads to

$$\frac{\partial A_j(\mathbf{y})}{\partial y^n} = \alpha_j^k \frac{\partial a_k(\mathbf{x})}{\partial y^n} = \alpha_j^k \frac{\partial a_k(\mathbf{x})}{\partial x^m} \frac{\partial x^m}{\partial y^n}, \tag{81}$$

where we have applied the chain rule in the last step. By first recognizing that $\partial x^m / \partial y^n = \alpha_n^m$ and then introducing the "comma notation" for partial derivative

$$\frac{\partial A_j}{\partial y^n} \equiv A_{j,n}; \qquad \frac{\partial a_k}{\partial x^m} \equiv a_{k,m} \tag{82}$$

we can write Eq. (81) in tensor shorthand as

$$A_{j,n} = \alpha_j^k \alpha_n^m a_{k,m}$$
(83)

By a similar analysis, it can be shown that

$$A^j_{,n} = \beta_k^j \alpha_n^m a^k_{,m}$$
(84)

Equations (83) and (84) are valid transformation laws that reveal that the quantities $A_{j,n}$ and $A^j_{,n}$ are *second-order tensors*. That is, by differentiating a first-order tensor we end up with a second-order tensor. In this case, (83) is a second-order covariant tensor and (84) is a second-order mixed tensor. Higher-order derivatives can be denoted in similar fashion. For example, the second derivative of the vector component A_j leads to the expression

$$A_{j,np} = \alpha_j^k \alpha_n^m \alpha_p^r a_{k,mr}$$
(85)

and so on.

5.8.1 Gradient, divergence, and curl

In Section 4.5.5 we presented the gradient, divergence, and curl in tensor notation for Cartesian coordinates. The *gradient* provides information on the maximum rate of increase in a scalar field, whereas the *divergence* and *curl* are fundamental measures of the rate of change of a vector field. We begin here by recalling those expressions in Cartesian coordinates and then extending them to the case of rectilinear coordinates.

Let φ denote a scalar field and $\mathbf{F}$ a vector field, the latter represented in Cartesian coordinates by $\mathbf{F} = f_k \hat{\mathbf{I}}^k = f^k \hat{\mathbf{I}}_k$. The gradient, divergence, and curl are defined, respectively, in Cartesian coordinates by

$$\nabla\varphi = \varphi_{,k} \hat{\mathbf{I}}^k$$
$$\nabla \cdot \mathbf{F} = f^k_{,k}$$
$$\nabla \times \mathbf{F} = e^{ijk} f_{j,i} \hat{\mathbf{I}}_k$$
(86)

In addition, the Laplacian is defined by

$$\nabla^2 \varphi = \varphi_{,kk}.$$
(87)

To obtain the corresponding representations of these quantities in rectilinear coordinates, we can formally replace each symbol in (86) and (87) by the corresponding symbol in rectilinear coordinates. In particular, we first write the vector field as $\mathbf{F} = F_k \mathbf{E}^k = F^k \mathbf{E}_k$. The rectilinear coordinate representations of the gradient, divergence, and curl are therefore given by

$$\nabla\varphi = \varphi_{,k}\mathbf{E}^k$$
$$\nabla\cdot\mathbf{F} = F^k_{,k}$$
$$\nabla\times\mathbf{F} = \varepsilon^{ijk}F_{j,i}\mathbf{E}_k$$

(88)

and that for the Laplacian is still given by (87). Direct verification of these results can be implemented by using the equations of transformation between the various tensor components.

5.9 Curvilinear Coordinates

Up to now we have restricted our attention to *rectilinear coordinates*—those for which the equations of transformation from Cartesian to non-Cartesian coordinates y^j are *linear*. When the equations of transformation are *nonlinear*, they can be described in general by

$$x^j = x^j(z^1,z^2,z^3) = x^j(\mathbf{z})$$

(89)

where the functions $x^j(\mathbf{z})$ are assumed to be differentiable. We define the *Jacobian of transformation* by

$$J = \det\left[\frac{\partial x^j(\mathbf{z})}{\partial z^k}\right] = \det[\alpha^j_k(\mathbf{z})].$$

(90)

Note that the Jacobian is now a function of $\mathbf{z}$ rather than constant. This means that there may be points in the domain of interest for which $J = 0$, known as *singular points* of the transformation. By considering only points for which the Jacobian is nonzero, there exists a unique *inverse transformation* given by

$$z^j = z^j(x^1,x^2,x^3) = z^j(\mathbf{x}).$$

(91)

In this case, we find that

$$\frac{1}{J} = \det\left[\frac{\partial z^j(\mathbf{x})}{\partial x^k}\right] = \det[\beta^j_k(\mathbf{x})].$$

(92)

From the transformation equations (89) and (91) we see that to each point (x^1,x^2,x^3) in Cartesian coordinates there is a unique representation of that point given by (z^1,z^2,z^3) relative to the variables z^j. We then say the variables z^j define a set of *curvilinear coordinates*. Geometrically, the expressions

$$z^j(\mathbf{x}) = C \quad \text{(constant)}$$

represent the *coordinate surfaces* of the curvilinear coordinate system and the intersection of these surfaces define the *coordinate curves*. A typical curvilinear coordinate system is illustrated below in Fig. 5.4. Familiar examples from calculus of such coordinate systems are *cylindrical* and *spherical coordinate systems*.

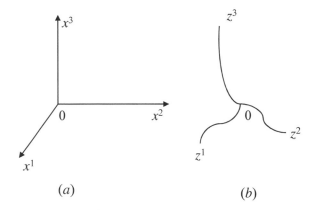

$$(a) \qquad\qquad\qquad\qquad (b)$$

Figure 5.4 (*a*) Cartesian coordinate system and (*b*) curvilinear coordinate system.

EXAMPLE 8 Given the equations of transformation in cylindrical coordinates

$$x = r\cos\theta; \quad y = r\sin\theta; \quad z = z;$$

(*a*) find the Jacobian of transformation.
(*b*) find the inverse transformation.
(*c*) determine the coordinate surfaces.

Solution: For notational ease, let (x,y,z) represent the Cartesian coordinates (x^1,x^2,x^3) and (r,θ,z) denote the curvilinear coordinates (z^1,z^2,z^3).

(*a*) The Jacobian of transformation is defined by

$$J = \det\left[\frac{\partial x^j}{\partial z^k}\right] = \begin{vmatrix} \dfrac{\partial x}{\partial r} & \dfrac{\partial x}{\partial \theta} & \dfrac{\partial x}{\partial z} \\[2mm] \dfrac{\partial y}{\partial r} & \dfrac{\partial y}{\partial \theta} & \dfrac{\partial y}{\partial z} \\[2mm] \dfrac{\partial z}{\partial r} & \dfrac{\partial z}{\partial \theta} & \dfrac{\partial z}{\partial z} \end{vmatrix}$$

which leads to

$$J = \begin{vmatrix} \cos\theta & -r\sin\theta & 0 \\ \sin\theta & r\cos\theta & 0 \\ 0 & 0 & 1 \end{vmatrix} = r.$$

(*b*) Solving simultaneously the transformation equations for r, θ, and z, we obtain the well-known relations

$$r = \sqrt{x^2 + y^2}; \quad \theta = \text{Arctan} \frac{y}{x}; \quad z = z.$$

(c) From the results given in (b), we see that $r = C$ (constant) yields

$$x^2 + y^2 = C^2,$$

which represent a family of cylinders with common axes of symmetry given by the the z axis. For $\theta = C$, the corresponding coordinate surfaces are

$$y = kx,$$

where $k = \tan C$. These surfaces are a family of planes containing the z axis. Finally, the surfaces

$$z = C$$

are simply a family of planes parallel to the xy plane.

5.9.1 Differentials as tensors

The transformation equations (14) for a rectilinear coordinate system indicate that the coordinates x^j and y^j transform as tensors. The transformation equations (89) and (91) for curvilinear coordinates show that the corresponding coordinates x^j and z^j do *not* transform as tensors. The reason for this is that the coefficients $\alpha_j^k(\mathbf{z})$ in curvilinear coordinates are not constants but functions of $\mathbf{z}$. The consequence of this is that the notion of "tensor" in curvilinear coordinates exists only in the *local sense* (i.e., only at a point).

To better understand these comments, let us take the differential of Eq. (89) to obtain

$$dx^j = \frac{\partial x^j}{\partial z^k} dz^k = \alpha_k^j(\mathbf{z}) dz^k. \tag{93}$$

By taking the differential of Eq. (91), we are similarly led to

$$dz^j = \frac{\partial z^j}{\partial x^k} dx^k = \beta_k^j(\mathbf{x}) dx^k. \tag{94}$$

Here we see that, although the coordinates x^j and z^j do not transform as tensor components, the differentials dx^j and dz^k do conform to tensor transformations—hence the reason we consider only local tensor transformations in curvilinear coordinates. For completeness, we also define covariant differential tensors dx_j and dz_k according to

$$dx_j = \beta_j^k(\mathbf{x}) dz_k; \qquad dz_j = \alpha_j^k(\mathbf{z}) dx_k. \tag{95}$$

5.9.2 Tensor fields and base vectors

If a vector $\mathbf{a}$ has covariant components a_j in Cartesian coordinates and covariant components A_j in curvilinear coordinates, they are related at each point of space for which $J \neq 0$ by

$$A_j = \alpha_j^k(\mathbf{z})a_k; \qquad a_j = \beta_j^k(\mathbf{x})A_k. \tag{96}$$

If these equations of transformation are satisfied at every point of a particular domain, such an aggregate of tensors is called a *tensor field*.

Let

$$\mathbf{R}(\mathbf{x}) = x^j\hat{\mathbf{I}}_j = x_j\hat{\mathbf{I}}^j$$

represent a position vector in Cartesian coordinates. By forming the differential of each side, we see that

$$d\mathbf{R} = dx^j\hat{\mathbf{I}}_j = dx_j\hat{\mathbf{I}}^j$$

$$= \alpha_k^j(\mathbf{z})dz^k\hat{\mathbf{I}}_j = \beta_j^k(\mathbf{x})dz_k\hat{\mathbf{I}}^j. \tag{97}$$

Thus, we use these tensor relations to define *covariant* and *contravariant base vectors*

$$\boxed{\mathbf{G}_k = \alpha_k^j(\mathbf{z})\hat{\mathbf{I}}_j; \qquad \mathbf{G}^k = \beta_j^k(\mathbf{x})\hat{\mathbf{I}}^j.} \tag{98}$$

From the above definitions, we see once again that the covariant base vectors $\mathbf{G}_j$ can be identified with the columns of the matrix $[\alpha_j^k(\mathbf{z})]$, whereas the contravariant base vectors $\mathbf{G}^j$ can be identified with the rows of the matrix $[\beta_j^k(\mathbf{x})]$. Also, the geometric interpretation of these sets of base vectors is the same as that for the rectilinear base vectors $\mathbf{E}_j$ and $\mathbf{E}^j$. Specifically, the set of covariant base vectors $\mathbf{G}_j$ are tangent to the coordinate curves, whereas the contravariant base vectors $\mathbf{G}^j$ are orthogonal to the coordinate surfaces. These sets of base vectors form reciprocal bases for which

$$\mathbf{G}_j \cdot \mathbf{G}^k = \delta_j^k. \tag{99}$$

This result follows directly from Eqs. (98).

5.9.3 Metric tensors

Continuing our analogous treatment to that for rectilinear coordinates, we are now led to the notion of covariant and contravariant *metric tensors* defined, respectively, by

$$g_{jk} = \mathbf{G}_j \cdot \mathbf{G}_k = \alpha_j^m(\mathbf{z})\alpha_k^n(\mathbf{z})\delta_{mn} = \alpha_j^m(\mathbf{z})\alpha_k^m(\mathbf{z})$$

$$g^{jk} = \mathbf{G}^j \cdot \mathbf{G}^k = \beta_m^j(\mathbf{x})\beta_n^k(\mathbf{x})\delta^{mn} = \beta_m^j(\mathbf{x})\beta_m^k(\mathbf{x}). \tag{100}$$

Once again, we have the familiar mixed metric tensor $g_j{}^k = \delta_j^k$. Among other uses, the metric tensors in curvilinear coordinates serve to raise and lower indices in much the same fashion as shown for rectilinear coordinates. In particular, it can readily be deduced that

$$dz^j = g^{jk}dz_k; \qquad dz_j = g_{jk}dz^k. \tag{101}$$

EXAMPLE 9 Given the equations of transformation in cylindrical coordinates

$$x = r\cos\theta; \quad y = r\sin\theta; \quad z = z;$$

(*a*) determine the covariant and contravariant base vectors.
(*b*) determine the covariant and contravariant metric tensors.

Solution: We first notice this is the same transformation considered in Example 8. Thus, we recall from Example 8 the matrix representation

$$[\alpha_j^k(\mathbf{z})] = \begin{pmatrix} \dfrac{\partial x}{\partial r} & \dfrac{\partial x}{\partial \theta} & \dfrac{\partial x}{\partial z} \\[2mm] \dfrac{\partial y}{\partial r} & \dfrac{\partial y}{\partial \theta} & \dfrac{\partial y}{\partial z} \\[2mm] \dfrac{\partial z}{\partial r} & \dfrac{\partial z}{\partial \theta} & \dfrac{\partial z}{\partial z} \end{pmatrix} = \begin{pmatrix} \cos\theta & -r\sin\theta & 0 \\ \sin\theta & r\cos\theta & 0 \\ 0 & 0 & 1 \end{pmatrix}.$$

Similarly, the inverse matrix is defined by

$$[\beta_j^k(\mathbf{x})] = \begin{pmatrix} \dfrac{\partial r}{\partial x} & \dfrac{\partial r}{\partial y} & \dfrac{\partial r}{\partial z} \\[2mm] \dfrac{\partial \theta}{\partial x} & \dfrac{\partial \theta}{\partial y} & \dfrac{\partial \theta}{\partial z} \\[2mm] \dfrac{\partial z}{\partial x} & \dfrac{\partial z}{\partial y} & \dfrac{\partial z}{\partial z} \end{pmatrix} = \begin{pmatrix} \cos\theta & \sin\theta & 0 \\ -\dfrac{\sin\theta}{r} & \dfrac{\cos\theta}{r} & 0 \\ 0 & 0 & 1 \end{pmatrix}.$$

(*a*) From the first expression in (98) and the columns of matrix $[\alpha_j^k(\mathbf{z})]$, we recognize that

$$\mathbf{G}_1 = \langle\cos\theta, \sin\theta, 0\rangle$$

$$\mathbf{G}_2 = \langle -r\sin\theta, r\cos\theta, 0\rangle$$

$$\mathbf{G}_3 = \langle 0, 0, 1\rangle.$$

Similarly, using the second expression in (98) and the rows of $[\beta_j^k(\mathbf{x})]$, we readily see that

$$\mathbf{G}^1 = <\cos\theta, \sin\theta, 0>$$

$$\mathbf{G}^2 = <-\frac{\sin\theta}{r}, \frac{\cos\theta}{r}, 0>$$

$$\mathbf{G}^3 = <0, 0, 1>.$$

(b) The metric tensors can be deduced directly from the base vectors or by using the matrix representations of $[\alpha_j^k(\mathbf{z})]$ and $[\beta_j^k(\mathbf{x})]$. Either way, it follows that

$$[g_{jk}] = \begin{pmatrix} 1 & 0 & 0 \\ 0 & r^2 & 0 \\ 0 & 0 & 1 \end{pmatrix}; \quad [g^{jk}] = \begin{pmatrix} 1 & 0 & 0 \\ 0 & \dfrac{1}{r^2} & 0 \\ 0 & 0 & 1 \end{pmatrix}.$$

The fact that only the diagonal elements of the metric tensors in Example 9 are nonzero indicates that both sets of base vectors are mutually orthogonal. That is, cylindrical coordinates represent an *orthogonal* system of coordinates. Although we don't present the details, it can be similarly shown that spherical coordinates are also an *orthogonal* system.

5.10 Tensor Calculus: Curvilinear Coordinates

In rectilinear coordinates a tensor of order two can be generated from a tensor of order one by partial differentiation [recall Eqs. (83) and (84)]. This is not so, however, in general curvilinear coordinates, because the base vectors are not constant. For example, let us consider the tensor transformation

$$A^i = \beta_j^i(\mathbf{x})a^j. \tag{102}$$

By differentiating each side of this equation with respect to z^k we obtain

$$\frac{\partial A^i}{\partial z^k} = \beta_j^i(\mathbf{x})\frac{\partial a^j}{\partial z^k} + \frac{\partial \beta_j^i(\mathbf{x})}{\partial z^k}a^j.$$

However, using the chain rule relations

$$\frac{\partial a^i}{\partial z^k} = \frac{\partial a^i}{\partial x^m}\frac{\partial x^m}{\partial z^k} = \alpha_k^m(\mathbf{z})a^i_{,m}$$

$$\frac{\partial \beta_j^i(\mathbf{x})}{\partial z^k} = \frac{\partial \beta_j^i(\mathbf{x})}{\partial x^m}\frac{\partial x^m}{\partial z^k} = \alpha_k^m(\mathbf{z})\beta_{j,m}^i(\mathbf{x})$$

where

$$\beta^i_{j,m}(\mathbf{x}) = \frac{\partial^2 z^i}{\partial x^m \partial x^j},$$ (103)

we get

$$A^i_{,k} = \alpha^m_k(\mathbf{z})\beta^i_j(\mathbf{x})a^j_{,m} + \beta^i_{j,m}(\mathbf{x})\alpha^m_k(\mathbf{z})a^j.$$ (104)

In rectilinear coordinates the last term in (104) is zero and we are left with a proper tensor transformation law. In curvilinear coordinates this term is not zero, and therefore $A^i_{,k}$ is not a tensor!

To rectify the above situation involving the partial derivatives of tensor components it will be necessary to generalize the notion of partial differentiation so that tensor fields evolve from the operation. Such a generalization is called covariant differentiation, but it can help motivate our definition if we first investigate derivatives of the base vectors.

5.10.1 Christoffel symbols

The derivatives of the base vectors in rectilinear coordinates are zero because they are constant. In curvilinear coordinates the base vectors $\mathbf{G}_i(\mathbf{z})$ and $\mathbf{G}^i(\mathbf{z})$ are functions of the coordinates z^j and, hence, do not generally have zero derivatives. From the transformation laws

$$\mathbf{G}_i(\mathbf{z}) = \alpha^m_i(\mathbf{z})\hat{\mathbf{I}}^m = \alpha^m_i(\mathbf{z})\hat{\mathbf{I}}_m$$

$$\hat{\mathbf{I}}^m = \alpha^m_k(\mathbf{z})\mathbf{G}^k(\mathbf{z})$$ (105)

$$\hat{\mathbf{I}}_m = \beta^k_m(\mathbf{x})\mathbf{G}_k(\mathbf{z})$$

we find through partial differentiation that

$$\mathbf{G}_{i,j}(\mathbf{z}) = \alpha^m_{i,j}(\mathbf{z})\hat{\mathbf{I}}^m = \alpha^m_{i,j}(\mathbf{z})\alpha^m_k(\mathbf{z})\mathbf{G}^k(\mathbf{z})$$

$$\mathbf{G}_{i,j}(\mathbf{z}) = \alpha^m_{i,j}(\mathbf{z})\hat{\mathbf{I}}_m = \alpha^m_{i,j}(\mathbf{z})\beta^k_m(\mathbf{x})\mathbf{G}_k(\mathbf{z})$$ (106)

where

$$\alpha^m_{i,j}(\mathbf{z}) = \frac{\partial^2 x^m}{\partial z^j \partial z^i}.$$ (107)

▸ **Christoffel symbols:** The *Christoffel symbols of the first and second kinds,* respectively, are special symbols defined by

$$\boxed{\Gamma_{ijk} = \alpha^m_{i,j}(\mathbf{z})\alpha^m_k(\mathbf{z}); \qquad \Gamma^k_{ij} = \alpha^m_{i,j}(\mathbf{z})\beta^k_m(\mathbf{x}).}$$ (108)

By use of the Christoffel symbols we find that the expressions in (106) yield

$$\mathbf{G}_{i,j}(\mathbf{z}) = \Gamma_{ijk}\mathbf{G}^k(\mathbf{z}) = \Gamma_{ij}^k\mathbf{G}_k(\mathbf{z}). \tag{109}$$

Also, it can be shown that

$$\mathbf{G}^k_{,i}(\mathbf{z}) = -\Gamma_{ij}^k\mathbf{G}^j(\mathbf{z}), \tag{110}$$

the proof of which is given below in Example 11.

The Christoffel symbols (108) are rich in identities, some of which we list below. In particular, for Christoffel symbols of the first kind, we have

$$\begin{array}{ll}
(a) & \Gamma_{ijk} = \Gamma_{jik} \\
(b) & \Gamma_{ijk} + \Gamma_{kji} = g_{ik,j} \\
(c) & \Gamma_{ijk} = \dfrac{1}{2}(g_{ik,j} + g_{jk,i} - g_{ij,k})
\end{array} \tag{111}$$

whereas comparable identities for the Christoffel symbol of the second kind are

$$\begin{array}{ll}
(a) & \Gamma_{ij}^k = g^{km}\Gamma_{ijm} \\
(b) & \Gamma_{ij}^k = \Gamma_{ji}^k \\
(c) & \Gamma_{ij}^k = \mathbf{G}^k\cdot\mathbf{G}_{i,j} = -\mathbf{G}_i\cdot\mathbf{G}^k_{,j} \\
(d) & \Gamma_{kj}^k = \dfrac{1}{2}g^{mn}g_{mn,j}
\end{array} \tag{112}$$

Some of these identities follow directly from definition of the Christoffel symbols, but others are more involved. For illustrative purposes, we will solve only part (*c*) in (112) above (see Example 10), which we use in Example 11.

EXAMPLE 10 Prove that $\Gamma_{ij}^k = \mathbf{G}^k\cdot\mathbf{G}_{i,j} = -\mathbf{G}_i\cdot\mathbf{G}^k_{,j}$.

Solution: If we form the dot product of each side of (109) with $\mathbf{G}^k$, we obtain

$$\mathbf{G}^k\cdot\mathbf{G}_{i,j} = \Gamma_{ij}^m\mathbf{G}^k\cdot\mathbf{G}_m = \Gamma_{ij}^m\delta_m^k = \Gamma_{ij}^k,$$

which proves the first half of the identity. To prove the second half of the identity we start with the known identity $\mathbf{G}_j\cdot\mathbf{G}^k = \delta_j^k$. By taking the partial derivative of each side we find

$$(\mathbf{G}_j\cdot\mathbf{G}^k)_{,i} = \mathbf{G}_j\cdot\mathbf{G}^k_{,i} + \mathbf{G}_{j,i}\cdot\mathbf{G}^k = 0.$$

By now using the first half of the identity, this last expression yields

$$\mathbf{G}_j \cdot \mathbf{G}^k_{,i} + \Gamma^m_{ji} \mathbf{G}_m \cdot \mathbf{G}^k = \mathbf{G}_j \cdot \mathbf{G}^k_{,i} + \Gamma^m_{ji} \delta^k_m = \mathbf{G}_j \cdot \mathbf{G}^k_{,i} + \Gamma^k_{ji} = 0.$$

Hence, our intended result now follows.

EXAMPLE 11 Verify the identity $\mathbf{G}^k_{,i}(\mathbf{z}) = -\Gamma^k_{ij} \mathbf{G}^j(\mathbf{z})$.

Solution: We start with the relation

$$\mathbf{G}^k(\mathbf{z}) = \beta^k_m(\mathbf{x}) \hat{\mathbf{I}}^m$$

and differentiate both sides with respect to z^i to get

$$\mathbf{G}^k_{,i}(\mathbf{z}) = \beta^k_{m,n}(\mathbf{x}) \alpha^n_i(\mathbf{z}) \hat{\mathbf{I}}^m$$

where we used the chain rule. We can now replace the Cartesian base vector by use of the second equation in (105), which gives us

$$\mathbf{G}^k_{,i}(\mathbf{z}) = \beta^k_{m,n}(\mathbf{x}) \alpha^n_i(\mathbf{z}) \alpha^m_j(\mathbf{z}) \mathbf{G}^j.$$

To complete the proof we need to establish the identity

$$\beta^k_{m,n}(\mathbf{x}) \alpha^n_i(\mathbf{z}) \alpha^m_j(\mathbf{z}) = -\Gamma^k_{ij}.$$

By using the result of Example 10, we see that

$$\begin{aligned}
-\Gamma^k_{ij} &= \mathbf{G}_i(\mathbf{z}) \cdot \mathbf{G}^k_{,j}(\mathbf{z}) \\
&= [\alpha^p_i(\mathbf{z}) \hat{\mathbf{I}}_p] \cdot [\beta^k_{m,n}(\mathbf{x}) \alpha^n_j(\mathbf{z}) \hat{\mathbf{I}}^m] \\
&= \beta^k_{m,n}(\mathbf{x}) \alpha^n_j(\mathbf{z}) \alpha^p_j(\mathbf{z}) \delta^m_p \\
&= \beta^k_{m,n}(\mathbf{x}) \alpha^n_i(\mathbf{z}) \alpha^m_j(\mathbf{z}).
\end{aligned}$$

Then, by use of this last identity and (*b*) in (112), we have our intended result.

5.10.2 Orthogonal systems

A curvilinear coordinate system is said to be *orthogonal* if and only if the base vectors satisfy $\mathbf{G}_j \cdot \mathbf{G}_k = 0$ and $\mathbf{G}^j \cdot \mathbf{G}^k = 0$ for $j \neq k$. In other words, the metric tensors all have diagonal matrix representations. If the coordinate system is orthogonal (like cylindrical or spherical coordinates), there are some additional identities satisfied by the Christoffel symbols. For Christoffel symbols of the first kind, we have

$$
\begin{array}{lll}
(a) & \Gamma_{ijk} = -\Gamma_{jki} = 0 & (i \ne j \ne k) \\
(b) & \Gamma_{iik} = -\dfrac{1}{2} g_{ii,k} & (i \ne k) \\
(c) & \Gamma_{iki} = \Gamma_{kii} = \dfrac{1}{2} g_{ii,k} & (i \ne k)
\end{array}
\tag{113}
$$

and for Christoffel symbols of the second kind, similar properties are

$$
\begin{array}{lll}
(a) & \Gamma_{ij}^{k} = 0 & (i \ne j \ne k) \\
(b) & \Gamma_{ii}^{k} = -\dfrac{1}{2 g_{kk}} g_{ii,k} & (i \ne k) \\
(c) & \Gamma_{ik}^{i} = \Gamma_{ki}^{i} = \dfrac{1}{2 g_{ii}} g_{ii,k} & (i \ne k)
\end{array}
\tag{114}
$$

The Christoffel symbols Γ_{ijk} and Γ_{ij}^{k} have 27 components each, but these are not tensor components because they do not satisfy tensor transformation laws. Nonetheless, they are useful because they provide us with the *rate of change* of the local basis. For this reason they will appear in the generalization of a partial derivative to be discussed in Section 5.10.3.

EXAMPLE 12 Given the equations of transformation in cylindrical coordinates

$$ x = r\cos\theta; \quad y = r\sin\theta; \quad z = z; $$

(a) find the Christoffel symbols of the first kind.
(b) find the Christoffel symbols of the second kind.

Solution: Because this system is orthogonal (recall Example 9) and we already know the metric tensors, we can use properties in (113) and (114) to calculate these symbols.

(a) Starting with identity (a) in (113), we see that

$$ \Gamma_{123} = \Gamma_{132} = \Gamma_{213} = \Gamma_{231} = \Gamma_{312} = \Gamma_{321} = 0. $$

Also, from the covariant metric tensor representation (see Example 9)

$$
[g_{jk}] = \begin{pmatrix} 1 & 0 & 0 \\ 0 & r^2 & 0 \\ 0 & 0 & 1 \end{pmatrix}
$$

and identity (b) in (113), we see that only g_{22} has nonzero derivative. Furthermore, $g_{22,1} = 2r$, whereas $g_{22,2} = g_{22,3} = 0$. Hence, we deduce that

$$\Gamma_{112} = \Gamma_{113} = \Gamma_{223} = \Gamma_{331} = \Gamma_{332} = 0$$
$$\Gamma_{221} = -r.$$

Similarly, from identity (c) in (113) it follows that

$$\Gamma_{121} = \Gamma_{211} = \Gamma_{131} = \Gamma_{311} = \Gamma_{232} = 0$$
$$\Gamma_{322} = \Gamma_{313} = \Gamma_{133} = \Gamma_{323} = \Gamma_{233} = 0$$
$$\Gamma_{212} = \Gamma_{122} = r.$$

Last, from identity (b) in (111), we see that

$$\Gamma_{iii} = \frac{1}{2} g_{ii,i}$$

and conclude that

$$\Gamma_{111} = \Gamma_{222} = \Gamma_{333} = 0.$$

(b) By a similar analysis, here we have

$$\Gamma_{22}^1 = -r; \quad \Gamma_{12}^2 = \Gamma_{21}^2 = \frac{1}{r},$$

whereas all others are zero.

5.10.3 Covariant derivative

Equation (114) for partial derivative shows that this is not a tensor transformation law. Analogously to rectilinear coordinates, it is useful to have derivatives of tensor fields in curvilinear coordinates leading to higher-order tensor fields.

Given a vector **a** in curvilinear coordinates it can be represented in terms of covariant and contravariant base vectors by

$$\mathbf{a} = A^i \mathbf{G}_i = A_i \mathbf{G}^i. \tag{115}$$

Although both sets of base vectors are vector fields, not constant vectors, what we desire is a generalization of partial derivative that treats the base vectors "as if they were constants." Let us introduce a special notation with the property that

$$\frac{\partial \mathbf{a}}{\partial z^j} = A^i\big|_j \mathbf{G}_i(\mathbf{z}) = A_i\big|_j \mathbf{G}^i(\mathbf{z}) \tag{116}$$

and where $A^i\big|_j$ and $A_i\big|_j$ are tensors of order two. Such a derivative is called a covariant derivative.

We can derive a definition of the covariant derivative by applying the standard product rule to the vector (115) with covariant base vector to find

$$\frac{\partial \mathbf{a}}{\partial z^j} = A^i_{,j} \mathbf{G}_i(\mathbf{z}) + A^i \mathbf{G}_{i,j}(\mathbf{z})$$

$$= A^i_{,j} \mathbf{G}_i(\mathbf{z}) + A^i \Gamma^m_{ij} \mathbf{G}_m(\mathbf{z}) \qquad (117)$$

$$= (A^i_{,j} + A^m \Gamma^i_{mj}) \mathbf{G}_i(\mathbf{z}).$$

In the second term of the last step we have interchanged the roles of i and m.

▸ **Covariant derivative:** We define the *covariant derivative* of the contravariant component and covariant component, respectively, of a first-order tensor by

$$\boxed{A^i\big|_j = A^i_{,j} + A^k \Gamma^i_{jk}} \qquad (118)$$

$$\boxed{A_i\big|_j = A_{i,j} - A_k \Gamma^k_{ij}.} \qquad (119)$$

The notion of covariant derivative can be extended to tensors of higher order. For example, the *covariant derivatives* of the components of a second-order tensor are given by

$$\boxed{\begin{aligned} A^{ij}\big|_k &= A^{ij}_{,k} + A^{im} \Gamma^j_{mk} + A^{mj} \Gamma^i_{mk} \\ A_j^i\big|_k &= A_{j,k}^i - A_m^i \Gamma^m_{jk} + A_j^m \Gamma^i_{mk} \\ A_{ij}\big|_k &= A_{ij,k} - A_{im} \Gamma^m_{jk} - A_{mj} \Gamma^m_{ik}. \end{aligned}} \qquad (120)$$

We will prove only the first of these results, the others having similar proofs. We begin by introducing the first-order tensor T^i by

$$A^{ij} V_j = T^i$$

and then forming the covariant derivative of each side to get

$$(A^{ij} V_j)\big|_k = T^i\big|_k.$$

Upon expanding each side,

$$A^{ij}\big|_k V_j + A^{ij} V_j\big|_k = T^i_{,k} + T^m \Gamma^i_{mk}$$

and further expanding the term $V_j\big|_k = V_{j,k} - V_m \Gamma^m_{jk}$, we find

$$A^{ij}\big|_k V_j + A^{ij}(V_{j,k} - V_m \Gamma^m_{jk}) = T^i_{,k} + T^m \Gamma^i_{mk}.$$

By making the replacements

$$T^m = A^{mj} V_j ;$$
$$T^i_{,k} = (A^{ij} V_j)_{,k} = A^{ij}_{,k} V_j + A^{ij} V_{j,k}$$

the above expression leads to (after eliminating common terms)

$$A^{ij}\big|_k V_j - A^{ij} V_m \Gamma^m_{jk} = A^{ij}_{,k} V_j + A^{mj} V_j \Gamma^i_{mk} .$$

Finally, by rearranging terms, we obtain

$$A^{ij}\big|_k V_j = A^{ij}_{,k} V_j + A^{ij} V_m \Gamma^m_{jk} + A^{mj} V_j \Gamma^i_{mk}$$
$$= (A^{ij}_{,k} + A^{im} \Gamma^j_{mk} + A^{mj} \Gamma^i_{mk}) V_j ,$$

from which we can now deduce the first result in (120).

> **Remark:** Note in the above proof that we tacitly made the assumption that the covariant derivative satisfied the product rule of ordinary differentiation. In general, it can be shown that the usual rules of differentiation carry over to the covariant derivative, but we will not pursue this.

Schemes similar to those illustrated above in (120) can be used to define the covariant derivative of higher-order tensors. Also, higher-order covariant derivatives follow in a natural way as illustrated above. For example, it is easy to show that

$$
\begin{aligned}
A_i\big|_{jk} &= (A_i\big|_j)\big|_k \\
&= (A_i\big|_j)_{,k} - A_m\big|_j \Gamma^m_{ik} - A_i\big|_m \Gamma^m_{jk} \\
&= (A_{i,j} - A_m \Gamma^m_{ij})_{,k} - (A_{m,j} - A_n \Gamma^n_{mj}) \Gamma^m_{ik} \\
&\quad - (A_{i,m} - A_n \Gamma^n_{im}) \Gamma^m_{jk} .
\end{aligned}
\tag{121}
$$

If we consider the fundamental metric tensors g_{ij}, g^{ij}, and g^i_j in rectilinear coordinates, we know all components are constant and therefore the partial derivatives must all be zero. Consequently, in every coordinate system this must be so and, in particular, we are led to the following theorem due to Ricci.

> **Theorem 5.1** (*Ricci's theorem*) The covariant derivatives of the metric tensors all vanish, i.e.,
>
> $$g_{ij}\big|_k = g^i_j\big|_k = g^{ij}\big|_k = 0.$$

Proof (partial): From the definition of covariant derivative, we have

$$g_{ij}\big|_k = g_{ij,k} - g_{mj} \Gamma^m_{ik} - g_{im} \Gamma^m_{jk}$$

and by lowering indices on the Christoffel symbols, this yields

$$g_{ij}|_k = g_{ij,k} - \Gamma_{ikj} - \Gamma_{jki}$$
$$= g_{ij,k} - g_{ij,k} = 0.$$

In the last step we used identity (b) in (111). Proving the other covariant derivatives are zero follows in the same fashion. ❑

As a consequence of Ricci's theorem, we see that

$$A^i|_j = (g^{ik}A_k)|_j = g^{ik}A_k|_j$$
$$A_i|_j = (g_{ik}A^k)|_j = g_{ik}A^k|_j. \tag{122}$$

But this is precisely what we would expect if the metric tensors are to be used for raising and lowering indices of all tensors, including those that arise from covariant differentiation.

5.10.4 Absolute derivative

In the study of particle motion we are concerned with various curves in space. Such a curve, which we represent by the parametric equations $z^i = z^i(t)$, may denote the path of a moving particle. In this case the parameter t is taken as time. Along this curve the velocity and acceleration vectors of the particle are also functions of time and are related to derivatives of $z^i(t)$.

Rather than restrict our attention to particle motion, let us consider the generic case of an arbitrary vector function $\mathbf{F}(t)$ of parameter t in some curvilinear coordinate system defined by its components according to

$$\mathbf{F}(t) = F^i(t)\mathbf{G}_i(\mathbf{z}) = F_i(t)\mathbf{G}^i(\mathbf{z}). \tag{123}$$

By taking the derivative of the first term on the right in (123) with respect to t, we obtain

$$\frac{d\mathbf{F}(t)}{dt} = \frac{dF^i(t)}{dt}\mathbf{G}_i(\mathbf{z}) + F^i(t)\mathbf{G}_{i,j}(\mathbf{z})\frac{dz^j}{dt}$$
$$= \frac{dF^i(t)}{dt}\mathbf{G}_i(\mathbf{z}) + F^j(t)\mathbf{G}_{j,k}(\mathbf{z})\frac{dz^k}{dt} \tag{124}$$
$$= \left(\frac{dF^i(t)}{dt} + \Gamma^i_{jk}F^j(t)\frac{dz^k}{dt}\right)\mathbf{G}_i(\mathbf{z}),$$

where we have used the identity $\mathbf{G}_{j,k}(\mathbf{z}) = \Gamma^i_{jk}\mathbf{G}_i(\mathbf{z})$. The last expression in parentheses motivates us to introduce the notation

$$\frac{\delta F^i}{\delta t} = \frac{dF^i(t)}{dt} + \Gamma^i_{jk}F^j(t)\frac{dz^k}{dt} \tag{125}$$

called the *absolute derivative* (also called the *intrinsic derivative*). In terms of the absolute derivative, we can rewrite (124) as

$$\frac{d\mathbf{F}(t)}{dt} = \frac{\delta F^i}{\delta t}\mathbf{G}_i(\mathbf{z}).$$

(126)

By the use of this notation, the base vector $\mathbf{G}_i(\mathbf{z})$ appears to be constant under the derivative. Thus, the absolute derivative in curvilinear coordinates is the counterpart of the ordinary derivative in rectilinear coordinates. If we choose to use the second representation of $\mathbf{F}(t)$ in (123), the absolute derivative is then defined by

$$\frac{\delta F_i}{\delta t} = \frac{dF_i(t)}{dt} - \Gamma_{ij}^k F_k(t)\frac{dz^j}{dt}$$

(127)

and, consequently,

$$\frac{d\mathbf{F}(t)}{dt} = \frac{\delta F_i}{\delta t}\mathbf{G}^i(\mathbf{z}).$$

(128)

The absolute derivative is related to the covariant derivative defined in Section 5.10.3 when the contravariant vector components F^i are defined throughout a region of space containing the curve C. In this case, F^i is a function of the curvilinear coordinates $z^k(\mathbf{z})$ and thus, it follows that

$$\frac{dF^i}{dt} = F^i_{,k}\frac{dz^k}{dt}.$$

(129)

The absolute derivative (125) can then be expressed in the form

$$\frac{\delta F^i}{\delta t} = \frac{dF^i(t)}{dt} + \Gamma_{jk}^i F^j(t)\frac{dz^k}{dt}$$
$$= \left(F^i_{,k} + \Gamma_{jk}^i F^j\right)\frac{dz^k}{dt}$$

(130)

or, in terms of the covariant derivative,

$$\frac{\delta F^i}{\delta t} = F^i|_k\frac{dz^k}{dt}.$$

(131)

Similarly, it can be shown that

$$\frac{\delta F_i}{\delta t} = F_i|_k\frac{dz^k}{dt}.$$

(132)

5.11 Riemann-Christoffel Curvature Tensor

From calculus we recall that if a scalar function $\varphi(x^1, x^2)$ is continuous with continuous first partial derivatives, that is a sufficient condition for the equality of mixed partial derivatives, viz.,

$$\frac{\partial^2 \varphi}{\partial x^1 \partial x^2} = \frac{\partial^2 \varphi}{\partial x^2 \partial x^1}.$$

We might wonder now what conditions are sufficient for equality of the mixed covariant derivatives, i.e., under what conditions can we write

$$A_i\big|_{jk} = A_i\big|_{kj}. \tag{133}$$

Based on Eq. (121), we have

$$
\begin{aligned}
A_i\big|_{jk} &= (A_{i,j} - A_m \Gamma^m_{ij})_{,k} - (A_{m,j} - A_n \Gamma^n_{mj})\Gamma^m_{ik} \\
&\quad - (A_{i,m} - A_n \Gamma^n_{im})\Gamma^m_{jk} \\
&= A_{i,jk} - A_m \Gamma^m_{ij,k} - A_{m,k}\Gamma^m_{ij} - (A_{m,j} - A_n \Gamma^n_{mj})\Gamma^m_{ik} \\
&\quad - (A_{i,m} - A_n \Gamma^n_{im})\Gamma^m_{jk}
\end{aligned}
\tag{134}
$$

and by interchanging the roles of j and k in the last step, we also obtain

$$
\begin{aligned}
A_i\big|_{kj} &= A_{i,kj} - A_m \Gamma^m_{ik,j} - A_{m,j}\Gamma^m_{ik} - (A_{m,k} - A_n \Gamma^n_{mk})\Gamma^m_{ij} \\
&\quad - (A_{i,m} - A_n \Gamma^n_{im})\Gamma^m_{kj}.
\end{aligned}
\tag{135}
$$

Now subtracting (135) from (134), it follows that

$$A_i\big|_{jk} - A_i\big|_{kj} = R^m_{ijk} A_m, \tag{136}$$

where R^m_{ijk} is the *Riemann-Christoffel tensor* defined by

$$\boxed{R^m_{ijk} = \Gamma^m_{ik,j} - \Gamma^m_{ij,k} + \Gamma^n_{ik}\Gamma^m_{nj} - \Gamma^n_{ij}\Gamma^m_{nk}.} \tag{137}$$

We see from (136) that the vanishing of the Riemann-Christoffel tensor is a necessary and sufficient condition for equating mixed covariant derivatives (133). Through the development of a number of important identities, it has been shown that this tensor plays a basic role in the investigations of differential geometry, dynamics of rigid and deformable bodies, electrodynamics, and relativity. However, such analysis is beyond the scope of this text. We do make a final note here that the Riemann-Christoffel tensor is a measure of the "curvature" of a particular space. That is, in spaces for which this tensor is identically zero (Euclidean spaces), we say that space is "flat."

5.12 Applications

Tensor analysis can be a useful tool in a variety of applications. We will briefly examine some of these applications below, but first develop expressions for the gradient, divergence, and curl in curvilinear coordinates.

5.12.1 Gradient, divergence, and curl

In Section 5.8.1 we examined expressions for the gradient, divergence, and curl expressed in rectilinear coordinates. Here we formally extend those expressions to the case of curvilinear coordinates. Once again we let φ denote a scalar field and $\mathbf{F}$ a vector field with representation $\mathbf{F} = F_k \mathbf{G}^k = F^k \mathbf{G}_k$. By simple symbol replacement, the curvilinear coordinate representations of the gradient, divergence, and curl are therefore given by

$$
\begin{aligned}
\nabla\varphi &= \varphi|_k \mathbf{G}^k = \varphi_{,k} \mathbf{G}^k \\
\nabla\cdot\mathbf{F} &= F^k|_k \\
\nabla\times\mathbf{F} &= \varepsilon^{ijk} F_j|_i \mathbf{G}_k
\end{aligned}
\tag{138}
$$

Similarly, the Laplacian takes the form

$$
\nabla^2\varphi = \varphi|_{kk}.
\tag{139}
$$

The direct verification of these results from the equations of transformation is left to the reader. Also, we are emphasizing the fact that the covariant derivative of a scalar field is the same as a partial derivative.

The gradient, divergence, and curl all have important physical interpretations. For example, the gradient $\nabla\varphi$ at a point $\mathbf{z}$ is the vector having the direction and magnitude of the maximum increase of the scalar field φ with respect to distance away from $\mathbf{z}$ (see Section 4.5.1). In the theory of heat transfer, Fourier's law of heat conduction for a materially isotropic body states that the heat flux $\mathbf{q}$ is proportional to the negative of the temperature gradient, i.e., $\mathbf{q} = -k\nabla T$, where k is a positive constant called the *heat conductivity* and T is temperature. In ideal fluid flow, the velocity vector $\mathbf{v}$ at any point in the fluid is equal to the gradient of a potential function, $\mathbf{v} = \nabla\varphi$.

The divergence $\nabla\cdot\mathbf{q}$ of the heat flux $\mathbf{q}$ at a point P in the theory of heat transfer measures the rate of heat outflow from a neighborhood of P. Similarly, the quantity $\nabla\cdot(\rho\mathbf{v})$ at a point P in fluid flow measures the rate of decrease of mass in a neighborhood of P, where ρ is the mass density and $\mathbf{v}$ is the fluid velocity (see Section 4.5.3). Also, the curl $\mathbf{\Omega} = \nabla\times\mathbf{v}$ of the fluid velocity $\mathbf{v}$ in fluid dynamics is the *vorticity* of the flow in the field. This same quantity $\nabla\times\mathbf{v}$ for a solid rotating body with linear velocity $\mathbf{v}$ at a point P represents twice the *angular velocity* of the rotating body (see Section 4.5.4).

EXAMPLE 13 Given the equations of transformation in cylindrical coordinates

$$
x = r\cos\theta; \quad y = r\sin\theta; \quad z = z;
$$

(a) find the divergence of a vector field $\mathbf{F}$.
(b) find the Laplacian of a scalar field φ.

Solution:

(*a*) Writing out the divergence in terms of Christoffel symbols, we get

$$\nabla \cdot \mathbf{F} = F^k\big|_k = F^k_{,k} + \Gamma^k_{jk}F^j$$
$$= \frac{\partial F^1}{\partial r} + \frac{\partial F^2}{\partial \theta} + \frac{\partial F^3}{\partial z} + \Gamma^k_{1k}F^1 + \Gamma^k_{2k}F^2 + \Gamma^k_{3k}F^3.$$

Based on results derived in Example 12, we know

$$\Gamma^k_{2k} = \Gamma^k_{3k} = 0; \qquad \Gamma^k_{1k} = \Gamma^2_{12} = \frac{1}{r}.$$

Hence,

$$\nabla \cdot \mathbf{F} = \frac{\partial F^1}{\partial r} + \frac{\partial F^2}{\partial \theta} + \frac{\partial F^3}{\partial z} + \frac{F^1}{r}.$$

Because we have not used the physical components of the vector $\mathbf{F}$, the above expression differs from that usually found for the divergence. Using the relation for the physical components

$$F^{(k)} = F^k \sqrt{g_{kk}}$$

we see that $F^{(1)} = F^1$; $F^{(2)} = rF^2$; and $F^{(3)} = F^3$. The above expression for divergence now takes the more familiar form

$$\nabla \cdot \mathbf{F} = \frac{1}{r}\frac{\partial}{\partial r}\left(rF^{(1)}\right) + \frac{1}{r}\frac{\partial F^{(2)}}{\partial \theta} + \frac{\partial F^{(3)}}{\partial z}.$$

(*b*) By writing the Laplacian as $\nabla^2\varphi = \nabla \cdot \nabla\varphi$; we can first define $\mathbf{F} = \nabla\varphi$ and then use the above expression for the divergence. Thus, by identifying

$$F^1 = \frac{\partial \varphi}{\partial r}; \quad F^2 = \frac{\partial \varphi}{\partial \theta}; \quad F^3 = \frac{\partial \varphi}{\partial z};$$

and relating these to the physical components, we obtain

$$\nabla^2\varphi = \nabla \cdot \nabla\varphi = \frac{1}{r}\frac{\partial}{\partial r}\left(r\frac{\partial \varphi}{\partial r}\right) + \frac{1}{r^2}\frac{\partial^2 \varphi}{\partial \theta^2} + \frac{\partial^2 \varphi}{\partial z^2}.$$

This last result is the familiar expression usually found through application of the chain rule from Cartesian coordinates to cylindrical coordinates.

5.12.2 Dynamics of a particle: Newton's second law

In Cartesian coordinates, Newton's second law of motion $\mathbf{F} = m\mathbf{a}$ becomes $f^k = ma^k$. The velocity and acceleration components of a particle in motion are related to the position x^k

by

$$v^k = \frac{dx^k}{dt}; \qquad a^k = \frac{dv^k}{dt} = \frac{d^2x^k}{dt^2}. \tag{140}$$

In terms of generalized curvilinear coordinates, Newton's law of motion reads

$$F^k = mA^k, \tag{141}$$

where the tensor components F^k are *generalized force components* (see Example 13) which may represent torques, etc., rather than actual physical components. To identify the corresponding velocity components V^k and acceleration components A^k that are counterparts of (140), we start with the generalized position coordinates z^k. The velocity components are obtained from the standard time derivative

$$V^k = \frac{dz^k}{dt}, \tag{142}$$

whereas the acceleration components involve the *absolute derivative* [see (127)]

$$A^k = \frac{\delta V^k}{\delta t} = \frac{d^2z^k}{dt^2} + \Gamma_{ij}^k \frac{dz^i}{dt} \frac{dz^j}{dt}. \tag{143}$$

Thus, Newton's second law of motion for the contravariant components in generalized curvilinear coordinates takes the form

$$F^k = m\left(\frac{d^2z^k}{dt^2} + \Gamma_{ij}^k \frac{dz^i}{dt} \frac{dz^j}{dt} \right) = m\left(\frac{dV^k}{dt} + \Gamma_{ij}^k V^i V^j \right). \tag{144}$$

Similarly, by use of the identity $A_i = g_{ik} A^k$ it can be shown that the covariant components are

$$F_k = m\left(g_{ik} \frac{dV^i}{dt} + \Gamma_{ijk} V^i V^j \right). \tag{145}$$

Because these last equations involve the Christoffel symbols (which have 27 components) it is sometimes more convenient to express Newton's second law in a form independent of these symbols. To obtain such a form we start with the identity

$$\frac{d}{dt}\left(g_{ik} V^i \right) = g_{ik} \frac{dV^i}{dt} + g_{ik,j} V^i V^j. \tag{146}$$

Also, recalling property (c) in (111), we have

$$\begin{aligned}
\Gamma_{ijk} V^i V^j &= \frac{1}{2}\left(g_{ik,j} V^i V^j + g_{jk,i} V^i V^j - g_{ij,k} V^i V^j \right) \\
&= g_{ik,j} V^i V^j - \frac{1}{2} g_{ij,k} V^i V^j
\end{aligned} \tag{147}$$

where the last step arises from making the observation

$$g_{ik,j} \, V^i V^j = g_{jk,i} \, V^i V^j.$$

By combining (146) and (147), it follows that

$$A_k = g_{ik}\frac{dV^i}{dt} + \Gamma_{ijk} V^i V^j = \frac{d}{dt}\left(g_{ik}V^i\right) - \frac{1}{2}g_{ij,k} V^i V^j, \tag{148}$$

and, consequently, Newton's law for the covariant force components becomes

$$F_k = m\left[\frac{d}{dt}\left(g_{ik}V^i\right) - \frac{1}{2}g_{ij,k} V^i V^j\right]. \tag{149}$$

EXAMPLE 14 Determine the physical components of acceleration in cylindrical coordinates (r,θ,z).

Solution: From Example 12 we found all covariant metric tensors are zero except for

$$g_{11} = g_{33} = 1; \quad g_{22} = r^2.$$

Based on (148), it follows that for $k = 1$, the covariant component of acceleration is

$$A_1 = \frac{d}{dt}\left(g_{i1}\frac{dr}{dt}\right) - \frac{1}{2}g_{ij,1} V^i V^j = \frac{d^2 r}{dt^2} - r\left(\frac{d\theta}{dt}\right)^2.$$

Similarly,

$$A_2 = \frac{d}{dt}\left(g_{i2}\frac{d\theta}{dt}\right) - \frac{1}{2}g_{ij,2} V^i V^j = \frac{d}{dt}\left[r^2\left(\frac{d\theta}{dt}\right)\right]$$

$$A_3 = \frac{d}{dt}\left(g_{i3}\frac{dz}{dt}\right) - \frac{1}{2}g_{ij,3} V^i V^j = \frac{d^2 z}{dt^2}.$$

The physical components are related by

$$A^{(i)} = g^{ij}A_j\sqrt{g_{ii}},$$

where $g^{ii} = 1/g_{ii}$. Hence, we now deduce that

$$A^{(1)} = g^{1j}A_j\sqrt{g_{11}} = A_1 = \frac{d^2 r}{dt^2} - r\left(\frac{d\theta}{dt}\right)^2$$

$$A^{(2)} = g^{2j}A_j\sqrt{g_{22}} = \frac{1}{r}A_2 = \frac{1}{r}\frac{d}{dt}\left[r^2\left(\frac{d\theta}{dt}\right)\right]$$

$$A^{(3)} = g^{3j}A_j\sqrt{g_{33}} = A_3 = \frac{d^2 z}{dt^2}.$$

5.12.3 Dielectric tensor of an anisotropic medium

In Section 4.8 we discussed the basic equations of Maxwell for the *electric* and *magnetic field vectors*, **E** and **H**, respectively. These equations connect **E** and **H** to three other quantities **B**, **D**, and **J**$_c$, for a total of five basic quantities. The quantity **B** is the *magnetic flux density* (also called the *magnetic induction*), **D** is the *electric flux density* (also called the *electric displacement*), and **J**$_c$ is the *electric* or *conduction current density*.

Maxwell's equations must be supplemented by additional relations known as *material equations* (or *constitutive relations*) that describe the behavior of substances under the influence of the field. In general these material equations can be rather complicated, but, under the assumptions that the field is time-harmonic (i.e., represented by a simple sinusoid in time), the bodies are at rest, and the material is *isotropic* (i.e., the physical properties are independent of direction), they take on the particularly simple form

$$\begin{aligned}
\mathbf{J}_c &= \sigma\mathbf{E} \\
\mathbf{D} &= \varepsilon\mathbf{E} \\
\mathbf{B} &= \mu\mathbf{H}.
\end{aligned} \tag{150}$$

The scalar σ is the *specific conductivity*, ε is the *dielectric permittivity*, and μ is the *magnetic permeability*. Substances in which $\sigma \neq 0$ are known as *conductors,* and those for which $\sigma \cong 0$ are called *insulators* or *dielectrics*. In this latter case, the magnetic and electric properties of the substance are completely described by ε and μ.

For a number of substances it is sufficient to assume the material is isotropic and rely on the relations given in Eqs. (150). However, it becomes necessary to account for the *anisotropy* of the substance in dealing with crystals, for example. In this section we will briefly consider the case of a crystal in which the medium is nonconducting ($\sigma = 0$) and magnetically isotropic, but is electrically anisotropic. Thus, the vector **D** will no longer have the same direction as **E**. In place of the middle equation in (150) we will assume the electric flux density in Cartesian coordinates is related to the electric field according to

$$\begin{aligned}
D_1 &= \varepsilon_{11}E_1 + \varepsilon_{12}E_2 + \varepsilon_{13}E_3 \\
D_2 &= \varepsilon_{21}E_1 + \varepsilon_{22}E_2 + \varepsilon_{23}E_3 \\
D_3 &= \varepsilon_{31}E_1 + \varepsilon_{32}E_2 + \varepsilon_{33}E_3
\end{aligned} \tag{151}$$

which in tensor notation we can write as

$$D_j = \varepsilon_{jk}E_k. \tag{152}$$

Because we are using Cartesian coordinates, it is not necessary to use both subscripts and superscripts in our notation here. (Note also that ε_{jk} is *not* the permutation symbol.)

The nine quantities ε_{jk} are constants of the medium which now constitute the components of the *dielectric tensor*. It can be shown that the dielectric tensor is symmetric, i.e., $\varepsilon_{jk} = \varepsilon_{kj}$, and consequently, there are only six independent tensor components. In this case, it is possible to simplify the relations (151) by a change of coordinates to the *principal dielectric axes*. This we do by finding the eigenvalues and eigenvectors of the dielectric tensor (recall Section 5.7). That is, the eigenvalues λ are all real and are solutions of the equation

$$\det(\varepsilon_{jk} - \lambda\delta_{jk}) = 0, \tag{153}$$

which we denote by $\lambda_1 = \varepsilon_x$, $\lambda_2 = \varepsilon_y$, $\lambda_3 = \varepsilon_z$. The quantities ε_x, ε_y, ε_z are called the *principal dielectric constants* (or *principal permittivities*). The principal dielectric axes are subsequently found by substituting these values of λ (one at a time) into the matrix equation

$$(\varepsilon - \lambda\mathbf{I})\mathbf{x} = \mathbf{0}, \tag{154}$$

where $\mathbf{I}$ is the identity matrix, and solving for $\mathbf{x}$. This action essentially reduces the set of material equations (151) to the simpler set given by

$$\begin{aligned} D_x &= \varepsilon_x E_x \\ D_y &= \varepsilon_y E_y \\ D_z &= \varepsilon_z E_z. \end{aligned} \tag{155}$$

Suggested Reading

A. J. McConnell, *Applications of Tensor Analysis* (Dover, New York, 1957).

A. I. Borisenko and I. E. Tarapov, *Vector and Tensor Analysis with Applications*, trans. from the Russian by R. A. Silverman (Dover, New York, 1968).

D. A. Danielson, *Vectors and Tensors in Engineering and Physics*, 2nd ed. (Addison-Wesley, Reading, 1997).

J. G. Simmonds, *A Brief on Tensor Analysis*, 2nd ed. (Springer, New York, 1994).

I. S. Sokolnikoff, *Tensor Analysis Theory and Applications to Geometry and Mechanics of Continua* (Wiley, New York, 1963).

Chapter 5

Exercises

1. Use *tensor notation* to prove the following identities:

(a) $\nabla\cdot(\mathbf{F}\times\mathbf{G})$
$= \mathbf{G}\cdot(\nabla\times\mathbf{F}) - \mathbf{F}\cdot(\nabla\times\mathbf{G})$

(b) $\nabla\times(\nabla\times\mathbf{F}) = \nabla(\nabla\cdot\mathbf{F}) - \nabla^2\mathbf{F}$

2. Given the linear transformation $y^j = \beta_k^j x^k$ described by

$$\begin{aligned} y^1 &= 3x^1 + x^2 + 2x^3 \\ y^2 &= x^1 + x^2 + 3x^3 \\ y^3 &= 2x^1 - 3x^2 + x^3 \end{aligned}$$

find the following:

(a) The *Jacobian* of transformation, the matrix $[\alpha_k^j]$ for the inverse transformation, and explicitly write out the inverse transformation $x^j = \alpha^j_k y^k$.

(b) The *covariant* base vectors $\mathbf{E}_j$.

Ans. (a) $J = 25$

$$[\alpha^j_k] = \frac{1}{25}\begin{pmatrix} 10 & -7 & 1 \\ 5 & -1 & -7 \\ -5 & 11 & 2 \end{pmatrix}$$

(b) $\mathbf{E}_1 = \frac{1}{25}<10,5,-5>$

$\mathbf{E}_2 = \frac{1}{25}<-7,-1,11>$

$\mathbf{E}_3 = \frac{1}{25}<1,-7,2>$

3. For the transformation given in Problem 2, calculate the

(a) *contravariant* base vectors $\mathbf{E}^j$.
(b) metric tensors g_{jk} and g^{jk}.
(c) Find $\mathbf{E}_2 \times \mathbf{E}^3$.

Ans. (a) $\mathbf{E}^1 = <3,1,2>$
$\mathbf{E}^2 = <1,1,3>$
$\mathbf{E}^3 = <2,-3,1>$

(b)

$$[g_{jk}] = \frac{1}{625}\begin{pmatrix} 150 & -130 & -35 \\ -130 & 171 & 22 \\ -35 & 22 & 54 \end{pmatrix}$$

$$[g^{jk}] = \begin{pmatrix} 14 & 10 & 5 \\ -10 & 11 & 2 \\ -5 & 2 & 14 \end{pmatrix}$$

4. Given the vector $\mathbf{a} = <3,3,6>$ in Cartesian coordinates and the retilinear base vectors

$\mathbf{E}_1 = <1,-1,2>$, $\mathbf{E}_2 = <0,1,1>$,
$\mathbf{E}_3 = <-1,-2,1>$,

(a) find contravariant components A^k.

(b) find the reciprocal base vectors.
(c) find covariant components A_k.

Ans. (a) $<A^1,A^2,A^3> = <2,3,-1>$
(b) $\mathbf{E}^1 = \frac{1}{6}<3,-1,1>$,

$\mathbf{E}^2 = \frac{1}{2}<-1,1,1>$,

$\mathbf{E}^3 = \frac{1}{6}<-3,-1,1>$
(c) $<A_1,A_2,A_3> = <12,9,-3>$

5. Given the linear transformation $y^j = \beta^j_k x^k$ described by

$$y^1 = x^1 + x^2$$
$$y^2 = -x^1 + 3x^2$$
$$y^3 = x^3$$

(a) find $g = \det(g_{jk})$.

(b) If $\mathbf{a} = <1, 1, 0>$, $\mathbf{b} = <2, 0, -3>$ are vectors in a Cartesian coordinate system, find *contravariant* components A^k and B^k in the rectilinear system defined above.

(c) Find *covariant* components A_k and B_k for the same rectilinear system.

Ans. (a) $g = \frac{1}{16}$
(b) $<A^1,A^2,A^3> = <2,2,0>$
$<B^1,B^2,B^3> = <2,-2,-3>$
(c) $<A_1,A_2,A_3> = <1,0,0>$
$<B_1,B_2,B_3> = <\frac{3}{2},-\frac{1}{2},-3>$

6. Given the vectors

$$\mathbf{A} = 2\mathbf{E}_1 - \mathbf{E}_2 + 4\mathbf{E}_3$$
$$\mathbf{B} = -3\mathbf{E}^1 + 2\mathbf{E}^2 - 2\mathbf{E}^3$$

find $\mathbf{A} \cdot \mathbf{B}$.

Ans. -16

7. Given the vectors

$$\mathbf{A} = 2\mathbf{E}_1 - \mathbf{E}_2 + 4\mathbf{E}_3$$
$$\mathbf{B} = 2\mathbf{E}_1 + 3\mathbf{E}_2 - \mathbf{E}_3$$

find $\mathbf{A} \times \mathbf{B}$.

Ans. $<-66,60,48>$

8. Prove the following identities:

(a) $\varepsilon^{ijk}\varepsilon^{mnp}g_{im}g_{jn}g_{kp} = 6$

(b) $\varepsilon^{ijk}\varepsilon^{mnp}g_{im}g_{jn} = 2g^{kp}$

9. A second-order tensor in Cartesian coordinates is defined by

$$[a_{jk}] = [a^{jk}] = [a_j^{\cdot k}] = [a_{\cdot k}^j] = \begin{pmatrix} 2 & 1 & 3 \\ 2 & 3 & 4 \\ 1 & 2 & 1 \end{pmatrix}$$

Given that a rectilinear coordinate system is defined by base vectors

$$\mathbf{E}_1 = \hat{\mathbf{I}}_1,\ \mathbf{E}_2 = \hat{\mathbf{I}}_1 + \hat{\mathbf{I}}_2,\ \mathbf{E}_3 = \hat{\mathbf{I}}_1 + \hat{\mathbf{I}}_2 + \hat{\mathbf{I}}_3,$$

(a) find the rectilinear components $[A_{jk}]$ and $[A_j^{\cdot k}]$.

(b) find the rectilinear components $[A^{jk}]$ and $[A_{\cdot k}^j]$.

Ans. (a) $[A_{jk}] = \begin{pmatrix} 2 & 3 & 6 \\ 4 & 8 & 15 \\ 5 & 11 & 19 \end{pmatrix}$

$[A_j^{\cdot k}] = \begin{pmatrix} 1 & -2 & 3 \\ 0 & -3 & 7 \\ -1 & -2 & 8 \end{pmatrix}$

(b) $[A^{jk}] = \begin{pmatrix} 2 & -1 & -1 \\ 0 & -2 & 3 \\ -1 & 1 & 1 \end{pmatrix}$

$[A_{\cdot k}^j] = \begin{pmatrix} 0 & -2 & -3 \\ 1 & 2 & 5 \\ 1 & 3 & 4 \end{pmatrix}$

10. Find the components of A_{jk} and A^{jk} from Problem 9 along the principal axes.

11. Show that

(a) $\varepsilon^{ijk} = \beta_m^i \beta_n^j \beta_p^k e^{mnp}$.

(b) $\varepsilon^{ijk} = g^{im}g^{jn}g^{kp}\varepsilon_{mnp}$.

(c) $\varepsilon^{ijk}\varepsilon^{mnp}g_{im}g_{jn} = 2g^{kp}$.

(d) $\varepsilon^{ijk}\varepsilon^{mnp}g_{im}g_{jn}g_{kp} = 6$.

(e) $\varepsilon_{ijk}\varepsilon_{mnp}g^{im}g^{jn} = 2g_{kp}$.

(f) $\varepsilon_{ijk}\varepsilon_{mnp}g^{im}g^{jn}g^{kp} = 6$.

12. Show that

(a) $\det[g_{jk}] = J^2$.

(b) $\det[g^{jk}] = \dfrac{1}{J^2}$.

13. In standard notation, spherical coordinates are defined by:

$$x = r\cos\theta\sin\varphi$$
$$y = r\sin\theta\sin\varphi$$
$$z = r\cos\varphi.$$

(a) Calculate metric tensors: g_{jk} and g^{jk}.

(b) Calculate Christoffel symbols: Γ_{323}, Γ_{232}, and Γ_{22}^1.

Ans. (a) $[g_{jk}] = \begin{pmatrix} 1 & 0 & 0 \\ 0 & r^2\sin^2\varphi & 0 \\ 0 & 0 & r^2 \end{pmatrix}$

(b) $\Gamma_{323} = 0,\ \Gamma_{22}^1 = -r\sin^2\varphi$

$\Gamma_{232} = r^2\sin\varphi\cos\varphi$

14. *Bipolar coordinates* in two dimensions are related to Cartesian coordinates (x^1, x^2) by

$$x^1 = \frac{\sinh\theta^1}{\cosh\theta^1 + \cos\theta^2},$$

$$x^2 = \frac{\sin\theta^2}{\cosh\theta^1 + \cos\theta^2}.$$

(a) find the covariant base vectors $\mathbf{G}_1$ and $\mathbf{G}_2$.

(b) Are the axes of the curvilinear coordinate system (θ^1, θ^2) orthogonal?

(c) Calculate the covariant metric tensor components g_{jk}.

Ans. (*a*) $\mathbf{G}_1 = \dfrac{(1+\cosh\theta^1 \cos\theta^2)}{(\cosh\theta^1 + \cos\theta^2)^2}\mathbf{I}_1$

$\qquad\qquad - \dfrac{\sinh\theta^1 \sin\theta^2}{(\cosh\theta^1 + \cos\theta^2)^2}\mathbf{I}_2$

$\quad \mathbf{G}_2 = \dfrac{\sinh\theta^1 \sin\theta^2}{(\cosh\theta^1 + \cos\theta^2)^2}\mathbf{I}_1$

$\qquad\qquad + \dfrac{(1+\cosh\theta^1 \cos\theta^2)}{(\cosh\theta^1 + \cos\theta^2)^2}\mathbf{I}_2$

(*b*) Yes

(*c*) $\begin{pmatrix} \dfrac{1}{(\cosh\theta^1 + \cos\theta^2)^2} & 0 \\[2mm] 0 & \dfrac{1}{(\cosh\theta^1 + \cos\theta^2)^2} \end{pmatrix}$

15. Calculate the six Christoffel symbols of the two-dimensional *uv*-system defined by

$$x = 2e^{u-v}, \quad y = -e^{3u+2v}.$$

Chapter 6

Complex Variables

6.1 Introduction / 273
6.2 Basic Concepts: Complex Numbers / 273
    6.2.1   Geometric interpretation: the complex plane / 275
    6.2.2   Polar coordinate representation / 276
    6.2.3   Euler formulas / 278
    6.2.4   Powers and roots of complex numbers / 279
6.3 Complex Functions / 281
    6.3.1   Loci and terminology / 282
    6.3.2   Functions as mappings / 283
    6.3.3   Limits and continuity / 284
6.4 The Complex Derivative / 287
    6.4.1   Cauchy-Riemann equations / 289
    6.4.2   Analytic functions / 291
    6.4.3   Harmonic functions / 293
6.5 Elementary Functions—Part I / 295
    6.5.1   Complex exponential function / 295
    6.5.2   Trigonometric functions / 297
    6.5.3   Hyperbolic functions / 299
6.6 Elementary Functions—Part II / 300
    6.6.1   Complex logarithm / 301
    6.6.2   Complex powers / 303
    6.6.3   Inverse trigonometric and hyperbolic functions / 304
6.7 Mappings by Elementary Functions / 306
    6.7.1   Orthogonal families / 306
    6.7.2   Simple polynomials / 307
    6.7.3   Reciprocal mapping / 308
    6.7.4   Bilinear transformations / 311
    6.7.5   Conformal mapping / 313
    Exercises / 316

Historical Comments: The study of analytic functions of a complex variable is synonymous with the name of Augustin-Louis Cauchy (1789-1857), one of the great mathematicians. Cauchy entered the Ecole Polytechnic to study engineering, but, because of poor health, was advised by Lagrange and Laplace to study mathematics instead. One of the most prolific mathematicians of all time, he published more than 700 papers. Among the subjects on which he worked were determinants, ordinary differential equations, partial differential equations, and complex variable theory. Cauchy provided the first systematic study of the theory of limits and the first rigorous proof of the existence of solutions to first-order differential equations. He also developed the concept of convergence of an infinite series and the theory of functions of a complex variable. In fact, most of the important theorems of the complex integral are associated with his name.

Other prominent names associated with complex variables are Georg Friedrich Bernhard Riemann (1826-1866) and Karl Weierstrass (1815-1897). Riemann entered Göttingen to study theology, but instead received his Ph.D. in mathematics under Gauss. He became professor of mathematics there in 1859 but died in this post when he was only 39 years old. Riemann introduced the concept of the integral, as it is used in basic calculus courses, in connection with his work on Fourier series. He also developed the mathematical basis for Einstein's theory of relativity known now as Riemannian geometry. After studying law for four years at Bonn, Weierstrass likewise turned to mathematics. Weierstrass developed complex variable theory based upon the power series representation of functions, but his "reverse" approach to the theory failed to attract many followers. Today we approach complex variables by first laying the groundwork of the complex differential and integral calculus, after which power series and Laurent series are developed.

Our objective in this chapter is to provide a fairly detailed treatment of the differential calculus of complex variables up through the Cauchy-Riemann equations. In Chapter 7 we extend the analysis to include the complex integral and Laurent series. Because complex variables are used so extensively in engineering and physics applications, our discussion here and in Chapter 7 is generally more detailed than it is in other subjects. The analysis of elementary complex functions provides a natural means of mapping certain two-dimensional regions into other two-dimensional regions—a powerful concept that is further explored in Chapter 7 in connection with steady-state heat conduction and fluid flow.

6.1 Introduction

Complex variables and their applications have become a standard component of the undergraduate curriculum in engineering and physical science at many universities. In large part, the inclusion of complex variables in the curriculum stems from the need for complex numbers and complex functions in the study of electric circuits, mechanical vibrations, fluid flow, optical wave propagation, and electrostatics, among other areas.

Functions of a complex variable have many properties in common with functions of a real variable studied in calculus. However, there is a special class of complex functions, called *analytic functions*, which have remarkable properties that are very different from those of real variables. Most importantly, these remarkable properties are the very properties that prove useful in a variety of engineering and science applications. For example, the real and imaginary parts of an analytic function are *harmonic functions*, the theory of which is used in fluid flow and electrostatics. Complex variable methods are particularly useful in problems that require the evaluation of certain real integrals that cannot be handled by standard calculus techniques. For example, the Fourier transform may necessitate the evaluation of such an integral when the desired transform cannot be found in tables (see Chapter 9). Also, the inverse Laplace transform can be formulated as a contour integral in the complex plane (see Chapter 10), although this is rarely discussed at the undergraduate level.

Historically, complex numbers were introduced several hundred years ago to provide solutions to equations of the form

$$x^2 = -2,$$

$$x^2 - 10x + 40 = 0,$$

which have no solution in terms of real numbers. In general, complex numbers are important in finding solutions of any polynomial equation. In this chapter we will begin by reviewing some of the basic concepts and algebraic operations associated with complex numbers before addressing the notion of complex function.

6.2 Basic Concepts: Complex Numbers

There are two conventional ways in which to introduce a complex number—both of which illustrate a close connection with vectors in the plane. For example, many texts use the *ordered pair* concept in which a *complex number z* can be defined by an ordered pair of real numbers x and y, i.e.,

$$z = (x,y). \tag{1}$$

Using the notation of Eq. (1), all *real numbers x* are associated with the ordered pair $(x,0)$.

The ordered pair $(0,1)$ is given the special symbol i.[1] Thus, we can write

$$(x,0) = x \text{ and } (0,1) = i.$$

---

[1] In electrical engineering it is customary to use the symbol j in place of i.

The real numbers x and y are called the *real* and *imaginary parts* of the complex number z, also designated by

$$x = \text{Re}(z), \quad y = \text{Im}(z). \tag{2}$$

The ordered pair notation $z = (x,y)$ is similar to the ordered pair notation used to define vectors (see Chapter 4). Its connection with a complex number was introduced in the mid-nineteenth century by the Irish mathematician William Rowan Hamilton (1805-1865). Prior to that time, complex numbers were expressed in the form

$$\boxed{z = x + iy,} \tag{3}$$

where the significance of the symbol i, which was invented in 1779 by the Swiss mathematician Leonhard Euler (1707-1783), will soon become clear. The ordered pair notation (x,y) and representation $x + iy$ are often used interchangeably, depending upon which is convenient at the time. Motivated by the representation (3), however, we define the following operations between the complex numbers $z_1 = a + ib$ and $z_2 = c + id$:

▸ **Equality:** $z_1 = z_2$ if and only if $a = c$ and $b = d$

▸ **Addition:** $z_1 + z_2 = a + c + i(b + d)$

▸ **Multiplication:** $z_1 z_2 = ac - bd + i(ad + bc)$

▸ **Quotient:** $\dfrac{z_1}{z_2} = \dfrac{ac + bd}{c^2 + d^2} + i\dfrac{(bc - ad)}{c^2 + d^2}$

Based on the multiplication property, if we choose $a = c = 0$ and $b = d = 1$, we find that

$$i^2 = -1. \tag{4}$$

Of course, we recognize that (4) was actually the motivation for the above properties. Thus, the symbol i denotes a complex number whose square equals -1. Since no real number has a square that is negative, the term *pure imaginary number* was used by early mathematicians to define the quantity iy. Also, from (4) we see continued multiplication by positive powers of i leads to the following pattern:

$$\boxed{\begin{aligned} i &= i, \\ i^2 &= -1, \\ i^3 &= -i, \\ i^4 &= 1, \\ i^5 &= i, \\ &\vdots \end{aligned}} \tag{5}$$

Complex addition and multiplication obey many of the familiar rules for real numbers, such as the *commutative, associative,* and *distributive laws.*

(*Commutative law*):	$z_1 + z_2 = z_2 + z_1, \quad z_1 z_2 = z_2 z_1$
(*Associative law*):	$z_1 + (z_2 + z_3) = (z_1 + z_2) + z_3,$
	$z_1(z_2 z_3) = (z_2 z_1)z_3$
(*Distributive law*):	$z_1(z_2 + z_3) = z_1 z_2 + z_1 z_3$

6.2.1 Geometric interpretation: the complex plane

For many purposes it is useful to represent the complex number $z = (x,y)$ as a point in the *xy*-plane. Each complex number corresponds to just one point in the plane, and each point in the plane corresponds to just one complex number. In this case it is customary to refer to this plane as the *complex plane* or, simply, the *z-plane* (see Fig. 6.1). The *x*-axis is called the *real axis*, whereas the *y*-axis is the *imaginary axis*. In this setting we see the obvious similarity between complex numbers and vectors in the plane—in fact, many properties associated with vectors in the plane have the same counterpart among the complex numbers, which is effective in some applications (e.g., see Section 7.10).

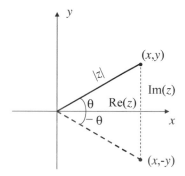

Figure 6.1 Complex plane.

▸   **Complex conjugate:** The *complex conjugate* of $z = (x,y) = x + iy$ is the number $(x,-y)$, also denoted by[2]

$$z^* = x - iy \quad \text{or} \quad \bar{z} = x - iy. \tag{6}$$

Geometrically, the complex conjugate z^* is the reflection of z across the real axis (see Fig. 6.1). It follows directly from definition that

$$x = \text{Re}(z) = \frac{1}{2}(z + z^*),$$
$$y = \text{Im}(z) = \frac{1}{2i}(z - z^*). \tag{7}$$

---

[2]In many textbooks it is customary to represent the complex conjugate by $\bar{z}$ rather than by z^*. Here, however, we will use only z^*.

▸   **Modulus**: The *modulus* of the complex number $z = (x,y) = x + iy$ is the real number

$$|z| = \text{mod}(z) = \sqrt{x^2 + y^2}. \tag{8}$$

The modulus (also called the *absolute value*) geometrically represents the distance of the complex number $z = (x,y)$ from the origin (see Fig. 6.1). It follows, therefore, that the quantity $|z_1 - z_2|$ represents the distance between the complex numbers z_1 and z_2. Clearly, the modulus of a (nonzero) complex number is a positive number. Although we cannot say that one complex number is greater than (or less than) another, we can say that the modulus of one number exceeds that of another.

There are numerous identities involving the complex conjugate and modulus of a complex number. Some of these identities are quite useful in practice, which we list below for easy reference:

$$
\begin{aligned}
&(a) \quad (z_1 \pm z_2)^* = z_1^* \pm z_2^* \\[2mm]
&(b) \quad (z_1 z_2)^* = z_1^* z_2^* \\[2mm]
&(c) \quad \left(\frac{z_1}{z_2}\right)^* = \frac{z_1^*}{z_2^*}
\end{aligned}
\tag{9}
$$

$$
\begin{aligned}
&(a) \quad zz^* = x^2 + y^2 = |z|^2 \\
&(b) \quad |z^*| = |z| \\
&(c) \quad |z|^2 = |z^2| = |z^*|^2 \\
&(d) \quad |z_1 z_2| = |z_1||z_2| \\
&(e) \quad \left|\frac{z_1}{z_2}\right| = \frac{|z_1|}{|z_2|}
\end{aligned}
\tag{10}
$$

In addition to these, we have the important inequalities

$$|z_1 + z_2| \leq |z_1| + |z_2|,$$

$$|z_1 - z_2| \geq ||z_1| - |z_2||, \tag{11}$$

the first of which is called the *triangle inequality*.

6.2.2 Polar coordinate representation

In certain calculations it may be more convenient to use *polar coordinates* in place of rectangular coordinates to describe complex numbers. By introducing polar coordinates (r,θ) according to the standard equations of transformation

$$x = r\cos\theta,$$

$$y = r\sin\theta, \tag{12}$$

we can write the complex number $z = x + iy$ in the form

$$z = r\cos\theta + ir\sin\theta$$

$$= r(\cos\theta + i\cos\theta). \tag{13}$$

In polar coordinates, therefore, we make the associations

$$\boxed{\begin{aligned} r &= \mathrm{mod}(z) = |z|, \\ \theta &= \arg(z) = \tan^{-1}\frac{y}{x}. \end{aligned}} \tag{14}$$

That is, r is the *modulus* of z and the angle θ is called the *argument* of z. Clearly, $\arg(z)$ is not uniquely determined, but is a multiple-valued quantity which we may express by

$$\arg(z) = \theta + 2k\pi, \quad k = 0,\pm 1,\pm 2,\pm 3,\pm \cdots . \tag{15}$$

To avoid the multiple values associated with $\arg(z)$, we frequently use the notion of a principal angle for θ.

▸ **Principal value (angle):** The symbol $\theta_p = \mathrm{Arg}(z)$ is called the *principal value* or *principal angle*, which is restricted to a 2π interval such as

$$\boxed{\begin{aligned} -\pi &< \theta_p \le \pi, \\ 0 &\le \theta_p < 2\pi, \end{aligned}} \tag{16}$$

In most of our work to follow we will use the first interval in (16) for the principal angle. Note that $\arg(0)$ is not defined.

Finally, using polar coordinates, we see that

$$z^* = r(\cos\theta - i\sin\theta), \tag{17}$$

from which we deduce

$$\arg(z^*) = -\arg(z). \tag{18}$$

EXAMPLE 1 Write $z = 3 - 3i$ in polar form.

Solution: We first note that

$$r = |3 - 3i| = 3\sqrt{2}, \qquad \theta_p = -\frac{\pi}{4}.$$

Hence, using (13), we can write

$$z = 3 - 3i = 3\sqrt{2}\left(\cos\frac{\pi}{4} - i\sin\frac{\pi}{4}\right).$$

In polar form, the product and quotient of two complex numbers can be expressed, respectively, by (see Section 6.3.2)

$$z_1 z_2 = r_1 r_2 \left[\cos(\theta_1 + \theta_2) + i \sin(\theta_1 + \theta_2) \right] \tag{19}$$

$$\frac{z_1}{z_2} = \frac{r_1}{r_2} \left[\cos(\theta_1 - \theta_2) + i \sin(\theta_1 - \theta_2) \right]. \tag{20}$$

EXAMPLE 2 Express the following complex number in polar form:

$$z = \frac{1 + i}{\sqrt{3} - i}$$

Solution: We first observe that

$$1 + i = \sqrt{2} \left(\cos \frac{\pi}{4} + i \sin \frac{\pi}{4} \right),$$

$$\sqrt{3} - i = 2 \left[\cos \left(-\frac{\pi}{6} \right) + i \sin \left(-\frac{\pi}{6} \right) \right].$$

Hence, based on Eq. (20), we deduce that

$$z = \frac{1 + i}{\sqrt{3} - i} = \frac{\sqrt{2}}{2} \left[\cos \left(\frac{\pi}{4} + \frac{\pi}{6} \right) + i \sin \left(\frac{\pi}{4} + \frac{\pi}{6} \right) \right]$$

$$= \frac{\sqrt{2}}{2} \left[\cos \left(\frac{5\pi}{12} \right) + i \sin \left(\frac{5\pi}{12} \right) \right].$$

> **Remark:** The product or quotient of complex numbers may lead to an argument that is not a principal value.

6.2.3 Euler formulas

Rather than use sines and cosines in the polar representation, it is usually more convenient to define the symbol[3]

$$e^{i\theta} = \cos \theta + i \sin \theta, \tag{21}$$

known as *Euler's formula*. From (21) we note that $|e^{i\theta}| = 1$. Also, by use of this symbol, we can now express the polar representation of a complex number in the form

---

[3] Some authors use cis(θ) as an abbreviation for $\cos\theta + i\sin\theta$.

$$z = re^{i\theta}.$$ (22)

The important *additive property*

$$e^{i\theta_1} e^{i\theta_2} = e^{i(\theta_1 + \theta_2)}$$

is merely a restatement of Eq. (19) for the case $r_1 = r_2 = 1$. Also, if $\theta_1 = \theta$ and $\theta_2 = -\theta$, then we obtain $e^{i\theta} e^{-i\theta} = 1$, from which we deduce

$$e^{-i\theta} = \frac{1}{e^{i\theta}}.$$

Hence, it follows that [see also Eqs. (19) and (20)]

$$\frac{1}{z} = \frac{1}{r} e^{-i\theta},$$ (23)

$$z_1 z_2 = r_1 r_2 e^{i(\theta_1 + \theta_2)},$$ (24)

$$\frac{z_1}{z_2} = \frac{r_1}{r_2} e^{i(\theta_1 - \theta_2)}.$$ (25)

We will further investigate properties of the function $e^{i\theta}$ in Section 6.3 and find that it is a natural generalization of the real exponential function for complex arguments.

6.2.4 Powers and roots of complex numbers

The polar representation (22) is particularly useful in finding powers and roots of various complex numbers. For example, we see from (24) that

$$z^2 = r^2 e^{2i\theta},$$

whereas, repeatedly multiplying by z and $re^{i\theta}$, we obtain

$$z^n = r^n (e^{i\theta})^n = r^n e^{in\theta}, \quad n = 1,2,3,....$$ (26)

From this last result, we easily infer the identity

$$\left(e^{i\theta}\right)^n = e^{in\theta}, \quad n = 1,2,3,...,$$

which in terms of trigonometric functions becomes the famous *DeMoivre's formula*

$$\boxed{(\cos\theta + i\sin\theta)^n = \cos n\theta + i\sin n\theta, \quad n = 1,2,3,....}$$ (27)

One of the principal uses of Demoivre's formula is finding fractional powers of complex numbers. For example, suppose we wish to find the solutions of the equation

$$z^n = z_0. \tag{28}$$

Formally, we represent the solution as $z = z_0^{1/n}$, but we don't know how to find the nth root of a complex number. To do so, we first write $z_0 = r_0 e^{i\theta_0}$ and then express Eq. (28) in the polar representation

$$r^n e^{in\theta} = r_0 e^{i\theta_0}.$$

From this relation it now follows that $r^n = r_0$ and $n\theta = \theta_0$, from which we deduce

$$r = \sqrt[n]{r_0}, \quad \theta = \frac{\theta_0 + 2k\pi}{n}, \quad k = 0, \pm 1, \pm 2, \dots .$$

Thus, the solution to (28) that we seek can be written as

$$z = \sqrt[n]{r_0} \exp\left[i\left(\frac{\theta_0 + 2k\pi}{n}\right)\right], \quad k = 0, \pm 1, \pm 2, \dots , \tag{29}$$

which leads to the following result.

▸   **nth root:** The nth root of any complex number z can expressed as

$$z^{1/n} = z_{k+1} = \sqrt[n]{|z|} \exp\left[i\left(\frac{\theta_p + 2k\pi}{n}\right)\right], \quad k = 0, 1, 2, \dots, n-1, \tag{30}$$

where $|z| = r$ and $\theta_p = \text{Arg}(z)$.

Although the number k in (30) can assume any integer value, there are only n unique values for the nth root of z corresponding to z_{k+1}, $k = 0, 1, 2, \dots , n - 1$.

The geometric interpretation of (30) is important since it permits the n roots $z_1, z_2, \dots, z_n$ to be quickly plotted in the complex plane. That is, since the moduli of all roots given in (30) are the same, the roots are representable by uniformly spaced points on a circle having radius $\sqrt[n]{|z|}$. For example, the roots of $z^{1/2}$ are on a diameter of the circle of radius $\sqrt{|z|}$ and the nth roots of unity lie equally spaced around the unit circle.

EXAMPLE 3 Find the cube roots of unity.

Solution: For $z = 1$, it follows that $\theta_p = 0$. Therefore, we write

$$1^{1/3} = z_{k+1} = \exp\left(\frac{2k\pi i}{3}\right), \quad k = 0, 1, 2.$$

Successively, by writing $e^{i\theta} = \cos\theta + i \sin\theta$, we find

$k = 0$: $z_1 = \cos 0 + i \sin 0 = 1$

$k = 1$: $z_2 = \cos \dfrac{2\pi}{3} + i \sin \dfrac{2\pi}{3} = -\dfrac{1}{2} + i \dfrac{\sqrt{3}}{2}$

$k = 2$: $z_3 = \cos \dfrac{4\pi}{3} + i \sin \dfrac{4\pi}{3} = -\dfrac{1}{2} - i \dfrac{\sqrt{3}}{2}$

Last, we can generalize the result of Eq. (30) to include the case where z is raised to any rational power. Doing so leads to

$$z^{m/n} = \left(\sqrt[n]{|z|}\right)^m \exp\left[\frac{im}{n}\left(\theta_p + 2k\pi\right)\right]; \quad m = 1,2,3,..., \quad k = 0,1,...,n-1. \qquad (31)$$

EXAMPLE 4 Find all values of $(-1 + i\sqrt{3})^{3/2}$.

Solution: With $z = -1 + i\sqrt{3}$, we see that

$$|z| = \sqrt{1 + 3} = 2,$$

$$\theta_p = \tan^{-1}\left(\frac{\sqrt{3}}{-1}\right) = \frac{2\pi}{3}.$$

Hence, it follows that the roots are given by

$$z_{k+1} = 2\sqrt{2}\exp\left[i\frac{3}{2}\left(\frac{2\pi}{3} + 2k\pi\right)\right]$$

$$= 2\sqrt{2}\exp[i(\pi + 3k\pi)], \quad k = 0,1,$$

or

$k = 0$: $z_1 = 2\sqrt{2}(\cos \pi + i \sin \pi) = -2\sqrt{2},$

$k = 1$: $z_2 = 2\sqrt{2}(\cos 4\pi + i \sin 4\pi) = 2\sqrt{2}.$

6.3 Complex Functions

The notion of a complex function closely parallels that of a real function. However, in the case of real functions we have graphical representations that provide nice geometrical interpretations of *continuity, derivative,* and *integral,* but no similar graphical representation exists in the case of complex functions. Nonetheless, our definitions of continuity, derivative and integral are virtually the same as those in the calculus of real functions. The primary distinction occurs in certain types of complex functions, called *analytic functions*, which have extraordinary properties that prove useful in physical applications. To begin, however, it is useful to briefly review some notions about sets and related terminology.

6.3.1 Loci and terminology

The modulus $|z|$ represents the *distance* of a point z from the origin. Similarly, $|z_1 - z_2|$ denotes the distance between two complex points z_1 and z_2. If ρ denotes a fixed real number and z_0 a fixed complex number, then the locus of points described by $|z - z_0| = \rho$ is a circle of radius ρ in the complex plane centered at z_0. When $z_0 = 0$, the circle is centered at the origin and any point on this circle has the polar form

$$z = \rho e^{i\theta}, \tag{32}$$

where the angle θ is restricted to a 2π interval. All points on the circle $|z - z_0| = \rho$ are likewise described in polar form by

$$z = z_0 + \rho e^{i\theta}, \quad 0 \le \theta < 2\pi. \tag{33}$$

In order to describe various points and collections of points (sets) in the complex plane, it is helpful to develop a standard vocabulary of certain terms. We will let S represent a set of complex numbers and define the following terms.

> **1. Neighborhood:** By *neighborhood* of z_0, we mean all points within a small circle of radius δ centered at z_0, i.e., $|z - z_0| < \delta$. A *deleted neighborhood* of z_0 consists of all points in the neighborhood of z_0 except the point itself, i.e., $0 < |z - z_0| < \delta$.
> **2. Limit point:** A point z_0 is called a *limit point* of a set S if every deleted neighborhood of z_0 contains points in S. (Note that a limit point does not have to belong to the set S.)
> **3. Bounded set:** A set S is called *bounded* if there exists a constant M such that $|z| < M$ for every point z in S.
> **4. Interior, exterior and boundary points:** A point z_0 is called an *interior point* of a set S if there exists a neighborhood of z_0 contained entirely in S. If every δ neighborhood of z_0 contains points in S and points not in S, then z_0 is called a *boundary point*. All other points are called *exterior points*.
> **5. Open set:** An *open set* of points S is one that contains only interior points.
> **6. Connected set:** If every combination of two points of the set S can be joined by a curve lying entirely in S, we say that S is a *connected set*.
> **7. Domain:** An open-connected set is a *domain*.
> **8. Region:** A *region* is a domain plus some or all of its boundary points. By *closed region*, we mean a domain plus all of its boundary points.

6.3.2 Functions as mappings

Let S denote the set of complex numbers $z = x + iy$. If to each z in S (denoted by $z \in S$) there exists a number w defined by

$$w = f(z), \quad z \in S, \tag{34}$$

we say this is a *complex function* of the variable z. For example, if $f(z) = z^2$, then $f(3i) = -9$. The set S of complex numbers is called the *domain of definition* of the function $f(z)$ and the set of values w is the *range of values* of the function $w = f(z)$. We also refer to Eq. (34) as a *mapping*, and w is sometimes called the *image* of z. We will discuss the mapping idea in more detail below in Section 6.7.

Because $z = x + iy$ can be decomposed into real and imaginary parts, we find that $w = f(z)$ can also be similarly decomposed. That is, by substituting $z = x + iy$ into (34), we can express it in the alternate form

$$w = u(x,y) + iv(x,y), \tag{35}$$

where $u(x,y)$ and $v(x,y)$ are real functions of x and y. In this fashion, all complex functions can be expressed as the sum of two real functions of two real variables.

Unfortunately, a graphical representation of a complex function (34) analogous to that of a real function does not exist. Moreover, to provide a graphical representation displaying two real functions $u(x,y)$ and $v(x,y)$ would require four dimensions. Graphical information about complex functions is obtained by considering two complex planes—the z-plane and the w-plane—as illustrated in Fig. 6.2. Corresponding to each point (x,y) in the z-plane (domain) is an *image point* (u,v) in the w-plane (range).

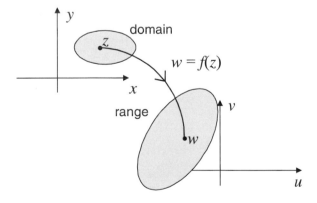

Figure 6.2 A complex function illustrated as a mapping from the z-plane to the w-plane.

EXAMPLE 5 Write the function $f(z) = z^2$ in the form of Eq. (35).

Solution: By writing $z = x + iy$, we have

$$w = (x + iy)^2 = x^2 - y^2 + i\,2xy,$$

from which we identify the real functions

$$u(x,y) = x^2 - y^2, \quad v(x,y) = 2xy.$$

EXAMPLE 6 Determine the range of the complex function

$$f(z) = x^2 + 2i, \quad |z| \leq 1.$$

Solution: By inspection, we identify $u(x,y) = x^2$, $v(x,y) = 2$. Thus, for $|z| \leq 1$, it is clear that $-1 \leq x \leq 1$. Consequently,

$$0 \leq u \leq 1, \quad v = 2,$$

so the range is along the line segment from $w = 2i$ to $w = 1 + 2i$ (see Fig. 6.3).

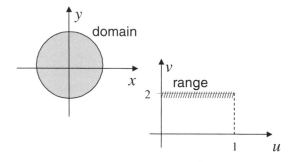

Figure 6.3 Domain and range of $f(z)$.

6.3.3 Limits and continuity

The general concept of limit as used in dealing with complex functions is the same as it is for real functions. However, when we compute the limit

$$\lim_{x \to x_0} f(x) = y_0$$

for real functions we are concerned only with values of x lying either to the left or right of the point x_0. That is, there are only *two ways* in which we can approach the point x_0 in real variables. In the complex plane the concept of limit is more complicated because z can approach z_0 from any direction in the complex plane [e.g., see Fig. 6.4 (a)].

To begin, let $w = f(z)$ be defined at all points in some neighborhood of z_0, except possibly at the point z_0. Then, the limit of this function as z approaches z_0 is written as

$$\lim_{z \to z_0} f(z) = w_0. \tag{36}$$

In precise words, the limit (36) means that for every positive ε there is a positive δ such that

$$|f(z) - w_0| < \varepsilon$$

whenever

$$0 < |z - z_0| < \delta.$$

A graphical representation of limit is shown in Fig. 6.4.

To relate the concept of limit of a complex function to limit notions introduced in calculus, we have the following theorem which we state without proof.

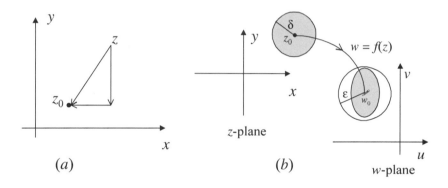

Figure 6.4 Illustration of a limit (*a*) along two possible paths in which $z \to z_0$ and (*b*) as given by (36) relating the z-plane to the w-plane.

Theorem 6.1 If $w = f(z) = u(x,y) + iv(x,y)$ and $z_0 = (x_0,y_0)$, then

$$\lim_{z \to z_0} f(z) = u_0 + iv_0$$

if and only if

$$\lim_{x \to x_0, y \to y_0} u(x,y) = u_0 \quad \text{and} \quad \lim_{x \to x_0, y \to y_0} v(x,y) = v_0.$$

In view of the above theorem, we see that the limit of a complex function is formally the same as the sum of limits of two real functions of two real variables. Because the concept of limit involving a complex function is simply the limit of its real and imaginary parts (which are both real functions), it follows that many of the familiar properties of limits involving real functions carry over to the complex plane.

Theorem 6.2 If $\lim\limits_{z \to z_0} f(z) = A$ and $\lim\limits_{z \to z_0} g(z) = B$, then

- $\lim\limits_{z \to z_0} [f(z) \pm g(z)] = A \pm B$.

- $\lim\limits_{z \to z_0} [f(z)g(z)] = AB$.

- $\lim\limits_{z \to z_0} \dfrac{f(z)}{g(z)} = \dfrac{A}{B}$.

EXAMPLE 7 Evaluate the limit $\lim\limits_{z \to i} \dfrac{z - i}{z^2 + 1}$.

Solution: By factoring the term in the denominator, we see that

$$\lim_{z \to i} \frac{z - i}{z^2 + 1} = \lim_{z \to i} \frac{z - i}{(z + i)(z - i)}$$

$$= \lim_{z \to i} \frac{1}{z + i} = \frac{1}{2i}.$$

From the limit concept introduced in Theorem 6.1, we see that the continuity of the function $w = f(z) = u(x,y) + i\,v(x,y)$ is determined by the continuity of the real functions $u(x,y)$ and $v(x,y)$. More precisely, we have the following definition.

Definition 6.1 (*Continuity*) A function $w = f(z)$ is said to be *continuous* at $z = z_0$, provided:

  ▸ $f(z_0)$ is defined

  ▸ $\lim\limits_{z \to z_0} f(z)$ exists

  ▸ $\lim\limits_{z \to z_0} f(z) = f(z_0)$.

We say that $f(z)$ is *continuous in a domain D* if it is continuous at every point of D. The following theorem is another analogue of a theorem concerning real functions.

Theorem 6.3 If $f(z)$ and $g(z)$ are continuous at z_0, then so are $f(z) \pm g(z)$ and $f(z)g(z)$. The quotient $f(z)/g(z)$ is continuous at z_0 provided $g(z_0) \neq 0$.

It is easy to verify that constant functions and powers of z are continuous functions everywhere in the complex plane. Thus, with the aid of Theorem 6.3, it follows that *polynomial functions* such as

$$f(z) = a_0 + a_1 z + a_2 z^2 + \cdots + a_n z^n$$

are also continuous everywhere in the z-plane. Similarly, *rational functions*, which are quotients of polynomials, are continuous at all points except where the denominator is zero. Other complex functions will be discussed separately in Sections 6.5 and 6.6.

6.4 Complex Derivative

The derivative of a real function is defined as a certain limit of a continuous function. Even if the real function is continuous, it still may not have a derivative. For complex functions, the restriction on which functions have derivatives at a given point is even more severe than it is for real functions.

Definition 6.2 (*Derivative*) The *derivative* of the complex function $w = f(z)$ at a fixed point z is defined by the following limit, provided the limit exists:

$$f'(z) = \lim_{\Delta z \to 0} \frac{f(z + \Delta z) - f(z)}{\Delta z}.$$

As in the case of real functions, differentiability implies continuity of the function, but the converse is not necessarily true. For those functions in the complex plane that have a derivative, all the familiar rules of real differential calculus carry over to complex functions (e.g., see Theorem 6.4 below).

Theorem 6.4 If $f(z)$ and $g(z)$ are differentiable functions at a given point z, then

- $\dfrac{d}{dz}[f(z) \pm g(z)] = f'(z) \pm g'(z).$

- $\dfrac{d}{dz}[cf(z)] = cf'(z),$ where c is a constant.

- $\dfrac{d}{dz}[f(z)g(z)] = f(z)g'(z) + f'(z)g(z).$

- $\dfrac{d}{dz}\left[\dfrac{f(z)}{g(z)}\right] = \dfrac{g(z)f'(z) - f(z)g'(z)}{[g(z)]^2},$ if $g(z) \neq 0.$

EXAMPLE 8 Given $f(z) = z^2$, show that $f'(z) = 2z$.

Solution: From definition,

$$f'(z) = \lim_{\Delta z \to 0} \frac{(z + \Delta z)^2 - z^2}{\Delta z} = \lim_{\Delta z \to 0} (2z + \Delta z)$$

$$= 2z.$$

By a similar analysis as given in Example 8, we can show that the derivative of $f(z) = z^n$ is given by the well-known relationship $f'(z) = nz^{n-1}$, $n = \pm 1, \pm 2, \pm 3, \dots$. However, for negative exponents we must restrict $z \neq 0$. Generally speaking, whenever a complex function $f(z)$ has a derivative, we find it in much the same manner as we do in the case of $f(x)$, where x is real. Of course, not all complex functions have a derivative as illustrated by the next simple example.

EXAMPLE 9 Show that $f(z) = z^*$ is not differentiable at any point in the complex plane.

Solution: We start with the limit

$$f'(z) = \lim_{\Delta z \to 0} \frac{(z + \Delta z)^* - z^*}{\Delta z} = \lim_{\Delta z \to 0} \frac{(\Delta z)^*}{\Delta z}.$$

To evaluate this last limit, we convert to x and y coordinates to obtain

$$\lim_{\Delta z \to 0} \frac{(\Delta z)^*}{\Delta z} = \lim_{\Delta x \to 0, \Delta y \to 0} \frac{\Delta x - i \Delta y}{\Delta x + i \Delta y}.$$

If we allow $\Delta y \to 0$ first, we have

$$\lim_{\Delta z \to 0} \frac{(\Delta z)^*}{\Delta z} = \lim_{\Delta x \to 0} \frac{\Delta x}{\Delta x} = 1;$$

whereas when $\Delta x \to 0$ first, the above limit yields

$$\lim_{\Delta z \to 0} \frac{(\Delta z)^*}{\Delta z} = \lim_{\Delta y \to 0} \frac{-i \Delta y}{i \Delta y} = -1.$$

Because the limit value is not unique (i.e., does not exist), we conclude that $f(z) = z^*$ does not have a derivative.

6.4.1 Cauchy-Riemann equations

Unlike real functions, it is quite common to find complex functions differentiable only at isolated points and/or only along some curve in the complex plane. If a complex function has a continuous derivative throughout some domain of the complex plane, it has a number of additional properties that are very important in applications. Before discussing such properties, we must first establish necessary conditions for $f'(z)$ to exist.

Let us consider the function $w = f(z)$ defined in the neighborhood of some point z. If it has a derivative, it must be defined by the limit

$$f'(z) = \lim_{\Delta z \to 0} \frac{\Delta f}{\Delta z} = \lim_{\Delta x \to 0, \Delta y \to 0} \frac{\Delta u + i \Delta v}{\Delta x + i \Delta y}, \tag{37}$$

where we are using the notation $\Delta f = f(z + \Delta z) - f(z)$ and $f(z) = u(x,y) + iv(x,y)$. In order for the limit (37) to exist, it must be independent of the order in which Δx and Δy approach zero. Thus, if we first let $\Delta y \to 0$, we find that

$$f'(z) = \lim_{\Delta x \to 0} \left(\frac{\Delta u}{\Delta x} + i \frac{\Delta v}{\Delta x} \right)$$

or, upon taking the remaining limits,

$$f'(z) = \frac{\partial u}{\partial x} + i \frac{\partial v}{\partial x}. \tag{38}$$

By reversing the order of the limits in Eq. (37), we are led to

$$f'(z) = \lim_{\Delta y \to 0} \left(\frac{1}{i} \frac{\Delta u}{\Delta y} + \frac{\Delta v}{\Delta y} \right)$$

or

$$f'(z) = \frac{\partial v}{\partial y} - i \frac{\partial u}{\partial y}. \tag{39}$$

▸ **Cauchy-Riemann equations:** Under the assumption that $f'(z)$ exists, we can equate the real and imaginary parts of Eqs. (38) and (39) to deduce the *Cauchy-Riemann equations*

$$\boxed{\begin{aligned} \frac{\partial u}{\partial x} &= \frac{\partial v}{\partial y}, \\ \frac{\partial u}{\partial y} &= -\frac{\partial v}{\partial x}, \end{aligned}} \tag{40}$$

Thus, we have shown that the Cauchy-Riemann equations are a *necessary condition* for the existence of a derivative (see also Theorem 6.5 below). If these equations fail to be satisfied for some value of z, say z_0, then $f'(z)$ does not exist at this point.

> **Theorem 6.5** (*Cauchy-Riemann*) If $u(x,y)$ and $v(x,y)$ and their first partial derivatives are continuous throughout some neighborhood of a point z, then the Cauchy-Riemann equations (40) are both a necessary and sufficient condition that $f(z) = u(x,y) + iv(x,y)$ have a derivative $f'(z)$ at the point z.

When a function is known to have a derivative, we can usually find it in the same way as for a real function. However, Eqs. (38) and (39) provide an alternate method for computing a derivative in some cases.

EXAMPLE 10 Use Eq. (38) or (39) to calculate the derivative of $f(z) = z^2$.

Solution: We first check the Cauchy-Riemann equations to see that the function has a derivative. With

$$u(x,y) = x^2 - y^2, \quad v(x,y) = 2xy,$$

we see that the conditions of Theorem 6.5 are satisfied and further, that

$$\frac{\partial u}{\partial x} = 2x = \frac{\partial v}{\partial y},$$
$$\frac{\partial u}{\partial y} = -2y = -\frac{\partial v}{\partial x}.$$

Thus, the Cauchy-Riemann equations are satisfied everywhere in the finite complex plane, i.e., $f'(z)$ exists evewhere. Using Eq. (38), we now obtain

$$f'(z) = \frac{\partial u}{\partial x} + i\frac{\partial v}{\partial x} = 2x + i2y,$$

or $f'(z) = 2z$. We leave it to the reader to verify that the same result is obtained by the use of Eq. (39).

On certain occasions it is useful to express the Cauchy-Riemann equations in polar coordinates. We leave it to the reader to verify that

$$f'(z) = (\cos\theta - i\sin\theta)\left(\frac{\partial u}{\partial r} + i\frac{\partial v}{\partial r}\right)$$

$$= e^{-i\theta}\left(\frac{\partial u}{\partial r} + i\frac{\partial v}{\partial r}\right) \tag{41}$$

and that the Cauchy-Riemann equations assume the form

$$\frac{\partial u}{\partial r} = \frac{1}{r}\frac{\partial v}{\partial \theta}$$

$$\frac{1}{r}\frac{\partial u}{\partial \theta} = -\frac{\partial v}{\partial r}.$$

(42)

> **Remark:** We can derive the Cauchy-Riemann equations (42) directly from (40) by use of the chain rule, i.e., $\partial u/\partial x = (\partial u/\partial r)(\partial r/\partial x) + (\partial u/\partial \theta)(\partial \theta/\partial x)$, or
>
> $$\frac{\partial u}{\partial x} = \frac{\cos\theta}{r}\frac{\partial u}{\partial r} - \frac{\sin\theta}{r}\frac{\partial u}{\partial \theta},$$
>
> and so on.

6.4.2 Analytic functions

While appearing to be a simple notion, the concept of analytic function is at the very core of complex variable theory and its applications.

> **Definition 6.3** (*Analytic function*) A function $w = f(z)$ is said to be *analytic at a point* z_0 if it is defined at the point and $f'(z)$ exists at every point in some neighborhood of z_0.

From Definition 6.3 we see that a function cannot be analytic at an isolated point or just along some curve. Moreover, it can be shown that $f(z)$ is *analytic in a domain D* if and only if the Cauchy-Riemann equations are satisfied at each point in D. Thus, the concept of analytic function is stronger than that of differentiability and, in fact, *analyticity implies differentiability.*

EXAMPLE 11 For what values of z is the function $f(z) = 3x^2 + iy^2$ analytic?

Solution: The functions $u(x,y) = 3x^2$ and $v(x,y) = y^2$ and their first partial derivatives are continuous everywhere. However, we see that

$$\frac{\partial u}{\partial x} = 6x, \quad \frac{\partial v}{\partial y} = 2y,$$

$$\frac{\partial u}{\partial y} = 0 = -\frac{\partial v}{\partial x},$$

and hence, the Cauchy-Riemann equations are satisfied only along the line $y = 3x$ in the z-plane. We deduce, therefore, that the function $f(z) = 3x^2 + iy^2$ is *differentiable* along the line $y = 3x$, but is *analytic nowhere.*

Theorem 6.6 Let $f(z)$ and $g(z)$ be analytic functions in some domain D. Then,

- $f(z) \pm g(z)$ and $f(z)g(z)$ are analytic functions in D.

- $f(z)/g(z)$ is analytic in D except where $g(z) = 0$.

- $f[g(z)]$ and $g[f(z)]$ are analytic functions in D.

The third property of Theorem 6.6 says that an *analytic function of an analytic function is analytic*. A function that is analytic at every point of the finite z-plane is said to be *entire*. An example of an entire function is any polynomial.

EXAMPLE 12 For what values of z is the following function analytic:

$$f(z) = e^x \cos y + i e^x \sin y$$

Solution: From inspection we see that $u(x,y) = e^x \cos y$ and $v(x,y) = e^x \sin y$ together with their first partial derivatives are all continuous. Moreover,

$$\frac{\partial u}{\partial x} = e^x \cos y = \frac{\partial v}{\partial y},$$

$$\frac{\partial u}{\partial y} = -e^x \sin y = -\frac{\partial v}{\partial x},$$

and thus the Cauchy-Riemann equations are satisfied everywhere in the finite z-plane. The given function is therefore an *entire function*. If we use Eq. (38) to calculate $f'(z)$, we find the "surprising" result

$$f'(z) = \frac{\partial u}{\partial x} + i\frac{\partial v}{\partial x} = e^x \cos y + i e^x \sin y,$$

or

$$f'(z) = f(z).$$

If $f(z)$ is analytic in some domain D except for a finite number of points, those points where $f(z)$ fails to be analytic are called *singular points*. An example of a function with singular points is given by the rational function

$$f(z) = \frac{2z}{z^2 + 1}.$$

Clearly, this function is not analytic at $z = \pm i$ because the function is not defined at these points. Hence, the points $z = \pm i$ are *singular points*.

6.4.3 Harmonic functions

Perhaps the most important partial differential equation that occurs in mathematical physics is the two-dimensional *equation of Laplace* or *potential equation* (see also Section 12.5). For the function $\varphi(x,y)$, the potential equation is defined by

$$\frac{\partial^2 \varphi}{\partial x^2} + \frac{\partial^2 \varphi}{\partial y^2} = 0. \tag{43}$$

This equation arises in steady-state heat conduction problems involving homogeneous solids. It is also the governing equation for the gravitational potential in free space, the electrostatic potential in a uniform dielectric, the magnetic potential in free space, and the velocity potential of inviscid, irrotational fluids. Not all solutions of (43) are continuous, but the *continuous solutions* are of particular importance in our work on complex variables.

Definition 6.4 A real-valued function $\varphi(x,y)$ that is continuous in a domain D, has continuous first- and second-order derivatives in D, and satisfies the potential equation (43) is called a *harmonic function* in D.

The connection between harmonic functions and complex variables is explained in the following theorem.

Theorem 6.7 If $f(z) = u(x,y) + iv(x,y)$ is analytic in some domain D, then both $u(x,y)$ and $v(x,y)$ are harmonic functions in D.

Proof: The proof is straightforward but will depend on the notion that an analytic function has at least continuous second-order derivatives. (Later we will show that an analytic function has *continuous derivatives of all orders*.)

If $f(z)$ is analytic in D, then $u(x,y)$ and $v(x,y)$ satisfy the Cauchy-Riemann equations

$$\frac{\partial u}{\partial x} = \frac{\partial v}{\partial y}, \quad \frac{\partial u}{\partial y} = -\frac{\partial v}{\partial x}.$$

If we differentiate the first of these equations with respect to x and the second with respect to y, we obtain

$$\frac{\partial^2 u}{\partial x^2} = \frac{\partial^2 v}{\partial x \partial y}, \quad \frac{\partial^2 u}{\partial y^2} = -\frac{\partial^2 v}{\partial y \partial x}.$$

Assuming these second-order derivatives are continuous in D, it follows that

$$\frac{\partial^2 u}{\partial x^2} + \frac{\partial^2 u}{\partial y^2} = \frac{\partial^2 v}{\partial x \partial y} - \frac{\partial^2 v}{\partial y \partial x} = 0,$$

which shows that $u(x,y)$ is harmonic in D. A similar calculation shows that $v(x,y)$ is harmonic in D. ❏

When $f(z) = u(x,y) + iv(x,y)$ is analytic in some domain D, we refer to $u(x,y)$ and $v(x,y)$ as *conjugate harmonic functions*. It can be shown that if we are given a function $u(x,y)$ that is harmonic in some domain D, then we can always find a conjugate harmonic function $v(x,y)$ such that $f(z) = u(x,y) + iv(x,y)$ or $f(z) = v(x,y) + iu(x,y)$ is analytic in D. Consider the next example.

EXAMPLE 13 Construct an analytic function whose real part is

$$u(x,y) \ = \ y^3 \ - \ 3x^2y.$$

Solution: Let us first demonstrate that $u(x,y)$ is harmonic. Calculating derivatives, we find

$$\frac{\partial^2 u}{\partial x^2} \ + \ \frac{\partial^2 u}{\partial y^2} \ = \ -6y \ + \ 6y \ = \ 0.$$

Next, using the Cauchy-Riemann equations, we obtain

$$\frac{\partial u}{\partial x} \ = \ -6xy \ = \ \frac{\partial v}{\partial y},$$
$$\frac{\partial u}{\partial y} \ = \ 3y^2 \ - \ 3x^2 \ = \ -\frac{\partial v}{\partial x}.$$

Hence, by integrating the first relation with respect to y, we get

$$v(x,y) \ = \ -3xy^2 \ + \ g(x),$$

where $g(x)$ is a "constant" of integration. Using the second Cauchy-Riemann equation above leads to

$$\frac{\partial v}{\partial x} \ = \ -3y^2 \ + \ g'(x) \ = \ -3y^2 \ + \ 3x^2.$$

From this result we deduce that $g'(x) = 3x^2$, or $g(x) = x^3 + C$, where C is any constant. Regardless of the value of C, the conjugate function

$$v(x,y) \ = \ -3xy^2 \ + \ x^3 \ + \ C$$

is a harmonic function everywhere in the z-plane. It is conventional to set the constant $C = 0$ so that the analytic function we seek takes the form

$$\begin{aligned}
f(z) \ &= \ u(x,y) \ + \ iv(x,y) \\
&= \ (y^3 \ - \ 3x^2y) \ + \ i(x^3 \ - \ 3xy^2).
\end{aligned}$$

Although not obvious, this analytic function can also be expressed in the form $f(z) = iz^3$. To see this, we can let

$$x \ = \ \frac{1}{2}(z + z^*), \quad y \ = \ \frac{1}{2i}(z - z^*),$$

and simplify the result. A function which is analytic must reduce to a function of

z alone (recall that z^* is not analytic).

6.5 Elementary Functions—Part I

In introducing functions of a complex variable we wish to define them in such a way that the definition is consistent with the real function that arises when z is replaced by the real variable x. Thus, all basic properties of the real function should be retained. In this section we restrict our attention to single-valued functions. That is, if to each complex number z there is but one value $w = f(z)$, we say that $f(z)$ is *single-valued*. Multiple-valued functions will be separately discussed in Section 6.6.

6.5.1 Complex exponential function

We define the *complex exponential function* by

$$\boxed{\begin{aligned} e^z &= e^{x+iy} \\ &= e^x(\cos y + i\sin y), \end{aligned}}$$

(44)

where y must be used as the radian measure of the angle defining the sine and cosine functions. One of our reasons for choosing this definition is that when $y = 0$, it reduces to the real exponential function e^x. Also, when $x = 0$, Eq. (44) reduces to the Euler formula

$$e^{iy} = \cos y + i\sin y.$$

(45)

> **Remark:** Like in the case of the real exponential function, we will have occasion to use the symbol $\exp(z)$ to denote e^z when the argument is complicated.

The complex exponential function is an *entire function* (recall Example 12). Also, its derivative

$$\frac{d}{dz}e^z = e^z$$

(46)

is consistent with that for the real exponential function. Additional properties of e^z are listed below, the verification of which is left to the reader:

$$\boxed{\begin{aligned} &(a) \quad e^{z_1}e^{z_2} = e^{z_1+z_2} \\ &(b) \quad |e^z| = e^x \\ &(c) \quad |e^{iy}| = 1, \quad y \text{ real} \\ &(d) \quad e^z \neq 0 \quad \text{for all } z \\ &(e) \quad [\exp(z)]^* = \exp(z^*) \end{aligned}}$$

(47)

EXAMPLE 14 Show that

$$\left|e^{-3iz+5i}\right| = e^{3y}.$$

Solution: Based on (b) and (c) in (47), we can write

$$\left|e^{-3iz+5i}\right| = \left|e^{3y+i(5-3x)}\right|$$

$$= e^{3y}\left|e^{i(5-3x)}\right| = e^{3y}.$$

By the use of Euler's formula (45), we see that

$$e^{2\pi i} = \cos 2\pi + i \sin 2\pi = 1.$$

Hence, it follows that

$$e^{z+2\pi i} = e^z e^{2\pi i} = e^z,$$

which reveals that the complex exponential function is a *periodic function* with fundamental *period* $2\pi i$. That is,

$$e^{z+2k\pi i} = e^z, \quad k = 0,\pm 1,\pm 2,\dots. \tag{48}$$

In general we say a function $f(z)$ is *periodic* if $f(z+P) = f(z)$ for all z, and if P is the smallest number for which this is true, it is called the *period* (see also Section 8.2). In the case of e^z the period is pure imaginary.

A single-valued function is said to be *one-to-one* on a given set S if the relation $f(a) = f(b)$ implies that $a = b$. Certainly the real exponential function satisfies this relation, but not the complex exponential function because of its periodic nature. However, if we divide the z-plane up into horizontal strips, each width 2π, then the complex exponential function will be one-to-one in each horizontal strip.

EXAMPLE 15 Find all complex numbers z that satisfy $e^z = -1$.

Solution: We simply write

$$e^z = e^x \cos y + i e^x \sin y = -1,$$

from which we deduce (by equating real and imaginary parts)

$$e^x \cos y = -1, \quad e^x \sin y = 0.$$

Clearly, the last relation requires $y = n\pi$, $n = 0, \pm 1, \pm 2, \dots$. But, since $e^x > 0$ and $\cos n\pi = (-1)^n$, we see that we must set $x = 0$ in the first relation above and restrict the choice of y to only odd multiples of π; i.e.,

$$y = (2n + 1)\pi, \quad n = 0, \pm 1, \pm 2, \ldots.$$

Combining results, the solutions we seek are the pure imaginary numbers

$$z = x + iy = (2n + 1)\pi i, \quad n = 0, \pm 1, \pm 2, \ldots.$$

6.5.2 Trigonometric functions

If we add and subtract the Euler formulas

$$e^{iy} = \cos y + i \sin y,$$
$$e^{-iy} = \cos y - i \sin y,$$

we are led to the (real) trigonometric functions

$$\cos y = \frac{1}{2}\left(e^{iy} + e^{-iy}\right),$$

$$(49)$$

$$\sin y = \frac{1}{2i}\left(e^{iy} - e^{-iy}\right).$$

By using (49) to define the real sine and cosine functions, it seems natural then to define the *complex trigonometric functions*

$$\boxed{\begin{aligned} \cos z &= \frac{1}{2}\left(e^{iz} + e^{-iz}\right), \\[2mm] \sin z &= \frac{1}{2i}\left(e^{iz} - e^{-iz}\right), \end{aligned}}$$

$$(50)$$

which reduce to Eqs. (49) when z is real.

Because $\cos z$ and $\sin z$ are linear combinations of entire functions, it follows that they themselves are *entire functions*. Also, based on the differentiation formula for the complex exponential, it can easily be shown that the derivative of the complex trigonometric functions agrees with that for real trigonometric functions; i.e.,

$$\frac{d}{dz}\cos z = -\sin z,$$

$$(51)$$

$$\frac{d}{dz}\sin z = \cos z.$$

Without providing the details, it can also be shown that the basic trigonometric identities for real trigonometric functions carries over to the complex counterparts. Among these are

$$\cos(z + 2\pi) = \cos z, \quad \sin(z + 2\pi) = \sin z,$$

$$(52)$$

$$\cos(-z) = \cos z, \quad \sin(-z) = -\sin z,$$

$$\cos^2 z + \sin^2 z = 1.$$

$$(53)$$

Equations (52) show simply that these functions are periodic with real period 2π, and also that $\cos z$ is an even function, whereas $\sin z$ is an odd function.

In addition to the above familiar identities, we list the following properties:

$$
\begin{array}{ll}
(a) & \cos z = \cos x \cosh y - i \sin x \sinh y \\
(b) & \sin z = \sin x \cosh y + i \cos x \sinh y \\
(c) & \cos iz = \cosh z \\
(d) & \sin iz = i \sinh z
\end{array}
\tag{54}
$$

$$
\begin{array}{ll}
(a) & |\cos z|^2 = \cos^2 x + \sinh^2 y \\
(b) & |\sin z|^2 = \sin^2 x + \sinh^2 y
\end{array}
\tag{55}
$$

$$
\begin{array}{lll}
(a) & \cos z = 0 \quad \Rightarrow & z = \pm \dfrac{(2n-1)\pi}{2}, \quad n = 1,2,3,\dots \\[2mm]
(b) & \sin z = 0 \quad \Rightarrow & z = \pm n\pi, \quad n = 0,1,2,3,\dots
\end{array}
\tag{56}
$$

EXAMPLE 16 Prove property (a) in (54) above.

Solution: By the use of Euler's formulas, we obtain

$$
\begin{aligned}
\cos z &= \frac{1}{2}\left[e^{i(x+iy)} + e^{-i(x+iy)}\right] \\
&= \frac{1}{2}\left[e^{-y}(\cos x + i\sin x) + e^{y}(\cos x - i\sin x)\right] \\
&= \frac{1}{2}\left(e^{y} + e^{-y}\right)\cos x - i\frac{1}{2}\left(e^{y} - e^{-y}\right)\sin x,
\end{aligned}
$$

from which we deduce

$$
\cos z = \cos x \cosh y - i \sin x \sinh y.
$$

For the sake of completeness, we sometimes find it convenient to use the additional four trigonometric functions

$$
\tan z = \frac{\sin z}{\cos z}, \quad \cot z = \frac{\cos z}{\sin z},
$$

$$
\sec z = \frac{1}{\cos z}, \quad \csc z = \frac{1}{\sin z}.
\tag{57}
$$

Here we see that tan z and sec z are analytic in any domain for which cos $z \neq 0$, and cot z and csc z are analytic in any domain for which sin $z \neq 0$. The derivatives and other properties of these functions are analogous to those of the corresponding real functions.

6.6.3 Hyperbolic functions

We define *complex hyperbolic functions* in the same fashion as the real hyperbolic functions, viz.,

$$
\cosh z = \frac{1}{2}\left(e^{z} + e^{-z}\right),
$$

$$
\sinh z = \frac{1}{2}\left(e^{z} - e^{-z}\right).
$$

(58)

From definition it follows that both cosh z and sinh z are *entire functions*, and their derivatives are

$$
\frac{d}{dz}\cosh z = \sinh z,
$$

$$
\frac{d}{dz}\sinh z = \cosh z.
$$

(59)

As in the case of trigonometric functions, we find that the standard properties of real hyperbolic functions carry over to the complex case. However, because the complex exponential function is periodic, we see that both cosh z and sinh z are periodic functions with period $2\pi i$. Analogously to Euler's formula, we have the useful identity

$$
e^{z} = \cosh z + \sinh z.
$$

(60)

In addition, we have the following properties:

$$
\begin{array}{ll}
(a) & \cosh z = \cosh x \cos y + i \sinh x \sin y \\
(b) & \sinh z = \sinh x \cos y + i \cosh x \sin y \\
(c) & \cosh iz = \cos z \\
(d) & \sinh iz = i \sin z
\end{array}
$$

(61)

$$
\begin{array}{ll}
(a) & |\cosh z|^2 = \sinh^2 x + \cos^2 y \\
(b) & |\sinh z|^2 = \sinh^2 x + \sin^2 y
\end{array}
$$

(62)

EXAMPLE 17 Find all solutions of $\cosh z = -2$.

Solution: Through use of (*a*) in (61), we have

$$\cosh z = \cosh x \cos y + i \sinh x \sin y = -2,$$

which, on equating real and imaginary parts, yields

$$\cosh x \cos y = -2,$$

$$\sinh x \sin y = 0.$$

The possibility $\sinh x = 0$ requires both $x = 0$ and $\cos y = -2$, the latter of which is impossible. Therefore, we must set $\sin y = 0$ in the second equation, which has solutions $y = k\pi, k = 0, \pm 1, \pm 2, \ldots$. Hence, the first equation now requires

$$\cosh x \cos k\pi = (-1)^k \cosh x = -2,$$

but, because $\cosh x > 0$, we must restrict k to odd integers. The solution we seek can then be expressed as

$$z = x + iy = \cosh^{-1}(2) + ik\pi, \quad k = \pm 1, \pm 3, \pm 5, \ldots \ .$$

6.6 Elementary Functions—Part II

The functions discussed in Section 6.5 are examples of *single-valued functions*. That is, for each complex value z, there is only one value $f(z)$. Functions which are not single-valued are called *multiple-valued functions*. For example, $w = \arg(z)$ has infinitely-many values, whereas $w = z^{1/2}$ has two values.

Like their analogue in real functions, we often have the need to introduce the concept of the "inverse of a function." By definition, we say that $g(z)$ is the *inverse* of $f(z)$ provided

$$f[g(z)] = g[f(z)] = z. \tag{63}$$

However, in dealing with complex functions we often find there may be more than one value that satisfies such a relation. For that reason, we ordinarily don't make a restriction that the inverse must be a single-valued function, but in some instances we may want to restrict the function to its principal branch. In the following we will formally address the problem of finding a value of z for which

$$w = f(z), \tag{64}$$

and simply introduce a function that can be expressed as

$$z = f^{-1}(w). \tag{65}$$

6.6.1 Complex logarithm

We define the complex logarithm function $\log z$ to be the inverse of the exponential function, the same as for real functions. That is, we define

$$w = \log z \quad \text{if} \quad z = e^w. \tag{66}$$

Because $e^w \neq 0$, we must assume that $z \neq 0$. In order to write the logarithm in terms of its real and imaginary parts, let us write $w = u + iv$ and express z in polar coordinates. In this case, the equation $z = e^w$ becomes

$$re^{i\theta} = e^{u+iv} = e^u e^{iv}.$$

Thus, we have $r = e^u$ or $u = \ln r$, which is the real natural logarithm. Similarly, $\theta = v$, or, because of the multivalued nature of $\theta = \arg(z)$, we write $v = \theta_p + 2k\pi$, $k = 0, \pm 1, \pm 2, \ldots$, $-\pi < \theta_p \leq \pi$. Based on these results, we define the *complex logarithm* by

$$\boxed{\log z = \ln r + i(\theta_p + 2k\pi), \quad k = 0, \pm 1, \pm 2, \ldots.} \tag{67}$$

Clearly, Eq. (67) defines a multiple-valued function.

If we restrict the argument of z to the principal value θ_p, we obtain what is called the *principal branch* of the function $w = \log z$. In the following work, we designate the principal branch of the logarithm by the symbol Log z, using a capital letter; i.e.,

$$\boxed{\text{Log } z = \ln r + i\theta_p, \quad -\pi < \theta_p \leq \pi.} \tag{68}$$

The principal branch represents a single-valued function that is the inverse of $w = e^z$ in the horizontal strip $-\pi < y \leq \pi$. Other branches of $\log z$ can be similarly defined by restricting the argument θ to a 2π interval.

EXAMPLE 18 Show that $\text{Log}(-1) = i\pi$.

Solution: From Eq. (68) we see that $r = |-1| = 1$ and $\theta_p = \pi$. Because $\ln 1 = 0$, the intended result follows.

EXAMPLE 19 Find all values of $\log(-1 - i)$.

Solution: For $z = -1 - i$, we see that

$$r = |z| = \sqrt{2}, \qquad \theta_p = -\frac{3\pi}{4}.$$

Hence, we obtain

$$\log(-1 - i) = \ln\sqrt{2} + i\left(2k\pi - \frac{3\pi}{4}\right), \quad k = 0, \pm 1, \pm 2, \ldots.$$

Notice that the principal value

$$\text{Log}(-1 - i) = \frac{1}{2}\ln 2 - \frac{3\pi i}{4}$$

is obtained by setting $k = 0$.

The concept of analytic function was defined only for single-valued functions. Thus, we can discuss this property only with respect to one of the branches of log z. For discussion purposes, let us consider only the principal branch defined by (68). Recall that the Cauchy-Riemann equations in polar coordinates take the form

$$\frac{\partial u}{\partial r} = \frac{1}{r}\frac{\partial v}{\partial \theta}$$

$$\frac{1}{r}\frac{\partial u}{\partial \theta} = -\frac{\partial v}{\partial r}. \tag{69}$$

The derivatives of $\text{Log } z = \ln r + i\theta_p$ are

$$\frac{\partial u}{\partial r} = \frac{1}{r}, \qquad \frac{\partial v}{\partial \theta} = 1, \qquad \frac{\partial u}{\partial \theta} = 0, \qquad \frac{\partial v}{\partial r} = 0,$$

which, for $z \neq 0$, are all continuous functions. Thus, we deduce that the function $w = \text{Log } z$ is analytic everywhere $-\pi < \arg(z) < \pi$, except for $z = 0$. The strict inequality on $\arg(z)$ imples that no point on the negative x-axis can be a point where $\text{Log } z$ is analytic. That is, on the negative x-axis the function $\text{Log } z$ is not even continuous because every point on this axis has a neighborhood that contains points close to $\theta = \pi$ as well as points close to $\theta = -\pi$. Finally, using Eq. (41) we find that

$$\frac{d}{dz}\text{Log } z = e^{-i\theta}\left(\frac{\partial u}{\partial r} + i\frac{\partial v}{\partial r}\right) = \frac{1}{re^{i\theta}},$$

from which we conclude

$$\frac{d}{dz}\text{Log } z = \frac{1}{z}, \quad z \neq 0, \quad -\pi < \arg(z) < \pi. \tag{70}$$

By selecting any other *branch* of the multiple-valued function $w = \log z$ we can obtain an analytic function. For the principal branch, the function (70) is analytic in the z-plane with the negative real axis ($\theta = \pi$) removed. The ray-line $\theta = \pi$, including the origin, is called a *branch cut* (or *branch line*). If we form another branch cut $\theta = \theta_0$, another branch of $w = \log z$ can be formed that is also single-valued and analytic in the resulting domain. The singular point $z = 0$, which is common to all branch cuts, is called a *branch point*. In all branches the derivative of the complex logarithm is $1/z$, the same as for the principal branch.

The basic properties of the complex logarithmic function are listed below, the verification of which we leave to the reader:

$$
\begin{aligned}
(a) &\quad e^{\log z} = z \\
(b) &\quad \log(e^z) = z + 2n\pi i, \quad n = 0, \pm 1, \pm 2, \ldots \\
(c) &\quad \log(z_1 z_2) = \log z_1 + \log z_2 \\
(d) &\quad \log(z_1/z_2) = \log z_1 - \log z_2
\end{aligned}
\tag{72}
$$

Remark: Properties (c) and (d) in (71) are to be interpreted that each value of one side is also contained among the values of the other side.

6.6.2 Complex powers

Thus far we have discussed powers of z like z^α only in the case where α is a rational number (i.e., the ratio of two integers). We now wish to consider the case where α is *any* complex number. To do so, we will use the complex logarithmic function.

For $z \neq 0$, we define *complex powers* of z according to

$$
w = z^\alpha = \exp(\alpha \log z).
\tag{73}
$$

This means that a particular value of $\log z$ will lead to a particular value of $w = z^\alpha$, and because $\log z$ is multiple-valued, so too is $w = z^\alpha$. Thus, we write

$$
z^\alpha = \exp\{\alpha[\ln r + i(\theta_p + 2k\pi)]\}, \quad k = 0, \pm 1, \pm 2, \ldots.
\tag{74}
$$

EXAMPLE 20 Find all values of $(-i)^i$.

Solution: By the use of (73), we find

$$
(-i)^i = \exp\left\{ i\left[\ln |-i| + i\left(2k\pi - \frac{\pi}{2}\right)\right]\right\}
$$

$$
= \exp\left(\frac{\pi}{2} - 2k\pi\right), \quad k = 0, \pm 1, \pm 2, \ldots.
$$

Notice that there are infinitely-many values and that they are *all real*. In particular, the principal value is $(-i)^i = e^{\pi/2}$.

If $\log z$ occurring in Eq. (72) is restricted to one of its branches, the resulting function $w = z^\alpha$ will be single-valued and analytic in the same domain as the logarithmic function.

Furthermore, by using properties of log z, it can be shown that

$$\frac{d}{dz}z^{\alpha} = \alpha z^{\alpha-1}. \tag{74}$$

Finally, based on Eq. (72) it follows that we can also define for any complex number a,

$$\boxed{a^z = \exp(z\log a).} \tag{75}$$

Note that for $a = e$, this last function reduces to the standard complex exponential function.

6.6.3 Inverse trigonometric and hyperbolic functions

Once again we can use logarithmic functions to introduce inverses of the trigonometric and hyperbolic functions. This is so because both trigonometric and hyperbolic functions are defined in terms of complex exponential functions.

From the relation

$$z = \sin w = \frac{e^{iw} - e^{-iw}}{2i},$$

we define the *inverse sine function* by

$$w = \sin^{-1}z. \tag{76}$$

To find an explicit expression for the inverse sine function, we rewrite the first equation for $\sin w$ in the form

$$e^{2iw} - 2ize^{iw} - 1 = 0. \tag{77}$$

We recognize (77) as quadratic in form and hence, has the solution

$$e^{iw} = iz + (1 - z^2)^{1/2},$$

where $(1 - z^2)^{1/2}$ is a double-valued function. By taking the complex logarithm of both sides of this last expression, we deduce that

$$\boxed{w = \sin^{-1}z = -i\log[iz + (1 - z^2)^{1/2}],} \tag{78}$$

which is a multiple-valued function with infinitely-many values. Once again, if we restrict the square root and logarithm to specific branches, the inverse sine function becomes single-valued and analytic because it is a composite of analytic functions.

In a similar manner we can define

$$\boxed{w = \cos^{-1}z = -i\log[z + (z^2 - 1)^{1/2}],} \tag{79}$$

$$w = \tan^{-1} z = \frac{i}{2} \log \frac{i + z}{i - z},$$ (80)

the details of which we leave to the reader. However, we notice that, because

$$\frac{1}{z - \sqrt{z^2 - 1}} = z + \sqrt{z^2 - 1},$$

changing the sign associated with the square root in the inverse cosine function is equivalent to changing the sign of the logarithm. A similar statement applies to the inverse sine function.

By introducing the notion of an inverse function, we can avoid solving problems by the method illustrated in Examples 15 and 17. Consider the following example.

EXAMPLE 21 Find all values of z for which $\sin z = 2$.

Solution: For the real sine function this would be an absurd problem. However, in the context of complex trigonometric functions, there are infinitely-many solutions. Using the definition of the inverse sine function (78) leads to

$$z = \sin^{-1} 2 = -i \log\left[(2 \pm \sqrt{3})i\right].$$

By taking the positive square root first, we have

$$z = -i\left[\ln(2 + \sqrt{3}) + i\left(\frac{\pi}{2} + 2k\pi\right)\right]$$

$$= \frac{\pi}{2} + 2k\pi - i\ln(2 + \sqrt{3}), \quad k = 0, \pm 1, \pm 2, \ldots .$$

For the negative square root, we find in a similar manner

$$z = -i\left[\ln(2 - \sqrt{3}) + i\left(\frac{\pi}{2} + 2k\pi\right)\right]$$

$$= \frac{\pi}{2} + 2k\pi - i\ln\left(\frac{1}{2 + \sqrt{3}}\right)$$

$$= \frac{\pi}{2} + 2k\pi + i\ln(2 + \sqrt{3}), \quad k = 0, \pm 1, \pm 2, \ldots .$$

Combining results yields

$$z = \frac{\pi}{2} + 2k\pi \pm i\ln(2 + \sqrt{3}), \quad k = 0, \pm 1, \pm 2, \ldots .$$

Lastly, the inverses of the hyperbolic functions can be defined in a corresponding manner. Doing so leads to

$$w = \sinh^{-1} z = \log\left[z + (z^2 + 1)^{1/2}\right], \tag{81}$$

$$w = \cosh^{-1} z = \log\left[z + (z^2 - 1)^{1/2}\right], \tag{82}$$

$$w = \tanh^{-1} z = \frac{1}{2}\log\frac{1 + z}{1 - z}. \tag{83}$$

As in the case of the inverse trigonometric functions, these functions are multiple-valued unless restricted to particular branches. For each branch, these functions are single-valued and analytic.

6.7 Mappings by Elementary Functions

In Section 6.3.2 we briefly introduced the notion that a function $w = f(z)$ is a *mapping* from the z-plane to the w-plane (see Fig. 6.2). The ability to map one region into another is a very useful concept in various applications. To begin, we will discuss the notion of *orthogonal families* of curves associated with an analytic function. We follow this by examining several specific mappings, and then confine our attention to a particular type of transformation called a *conformal mapping*.

6.7.1 Orthogonal families

Let $f(z) = u(x,y) + i\,v(x,y)$ represent an analytic function in some domain D, and consider the families of *level curves* defined by

$$\begin{aligned} u(x,y) &= C_1, \\ v(x,y) &= C_2, \end{aligned} \tag{84}$$

where C_1 and C_2 are real constants that can take on any range of values (see Fig. 6.5). The gradients ∇u and ∇v are each normal to their respective family of level curves (recall Section 4.5.1). Moreover, we see that

$$\nabla u \cdot \nabla v = \frac{\partial u}{\partial x}\frac{\partial v}{\partial x} + \frac{\partial u}{\partial y}\frac{\partial v}{\partial y} = -\frac{\partial u}{\partial x}\frac{\partial u}{\partial y} + \frac{\partial u}{\partial y}\frac{\partial u}{\partial x} = 0, \tag{85}$$

the last step a consequence of the Cauchy-Riemann equations. The relation (85) illustrates that the normals to the two families of level curves (84) are everywhere *mutually orthogonal* within the domain D. Consequently, the level curves (84) are *mutually orthogonal* families (see Fig. 6.5). This is an important property associated with the intersection angles of harmonic conjugates of analytic functions that has special significance in applications involving mappings.

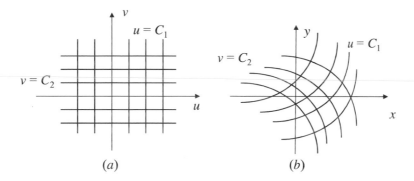

Figure 6.5 Orthogonal families of curves (*a*) plotted in the *uv*-plane and (*b*) plotted in the *xy*-plane. In each case we assume C_1 and C_2 take on several values.

6.7.2 Simple polynomials

To begin, we consider several specializations of the general *linear mapping* $w = Bz + C$, where B and C are complex constants.

▸ **Identity mapping:** The function $w = z$ is the *identity mapping*.

▸ **Translation:** The mapping $w = z + C$ is simply a *translation* of every point z in the z-plane to the new point $z + C$ in the w-plane.

▸ **Rotation and expansion (contraction):** The function $w = Bz$ represents both a *rotation* and *expansion* (or *contraction*) of every region in the z-plane.

To understand this last mapping, it is best to use the polar coordinate representation. Thus, we write $z = re^{i\theta}$ and $w = \rho e^{i\varphi}$. If we also represent the complex constant B in polar form

$$B = be^{i\beta},$$

then

$$w = \rho e^{i\varphi} = bre^{i(\theta + \beta)}. \tag{86}$$

By equating the modulii and phases in (86), we deduce that $\rho = br$ and $\varphi = \theta + \beta$. Hence, we see that the point (r,θ) in the z-plane maps into the point $(br,\theta + \beta)$ in the w-plane. The radial distance r is expanded (contracted) by the constant amount b and the radial vector rotated by the angle β. By combining the above mappings, we obtain the general *linear mapping*

$$\boxed{w = Bz + C.} \tag{87}$$

Under this mapping every region in the z-plane is mapped into a geometrically similar region in the w-plane, but is rotated, expanded (contracted), and translated (see Fig. 6.6).

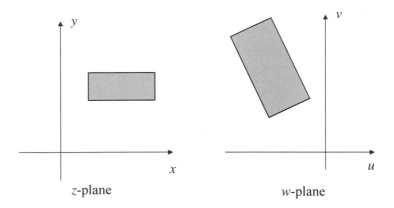

Figure 6.6 Linear mapping $w = Bz + C$.

Next, let us consider the quadratic function

$$w = z^2. \tag{88}$$

In polar coordinates, we write $z = re^{i\theta}$ and $w = \rho e^{i\varphi}$, which leads to

$$\rho e^{i\varphi} = r^2 e^{2i\theta}.$$

Hence, we see that the image of the point (r,θ) in the z-plane is the point $(r^2, 2\theta)$ in the w-plane. For example, under the quadratic mapping function (88), the first quadrant in the z-plane maps onto the entire upper half of the w-plane.

In rectangular coordinates Eq. (88) becomes

$$u + iv = x^2 - y^2 + i2xy,$$

from which we deduce

$$u = x^2 - y^2, \qquad v = 2xy. \tag{89}$$

It is instructive to plot the lines $u = C_1 = \text{const.}$ and $v = C_2 = \text{const.}$ in the w-plane as well as their images in the z-plane (e.g., see Fig. 6.5). Clearly, the line $u = 0$ (i.e., the v axis) in the w-plane maps into the lines $y = \pm x$ in the z-plane. Other vertical lines $u = C_1$ in the w-plane map into the hyperbolas $x^2 - y^2 = C_1$ in the z-plane. Similarly, the line $v = 0$ (i.e., the u axis) maps into the x and y axes in the z-plane and lines described by $v = C_2$ map into the hyperbolas $2xy = C_2$ in the z-plane (see also Section 7.9.2).

6.7.3 Reciprocal mapping

The *reciprocal function*

$$\boxed{w = \frac{1}{z}, \quad z \neq 0,} \tag{90}$$

sets up a one-to-one correspondence between points in the z-plane and points in the w-plane, except for the points $z = 0$ and $w = 0$ which have no images in the finite plane. In polar coordinates, the function (90) becomes

$$\rho e^{i\varphi} = \frac{1}{r} e^{-i\theta},$$

which we may interpret as successive transformations (mappings) given by an inversion and a conjugation, viz.,

$$\zeta = \frac{1}{z^*} = \frac{1}{r} e^{i\theta}, \qquad w = \zeta^*. \qquad (91)$$

The first transformation in (91) is an inversion with respect to the unit circle $r = 1$ (see Fig. 6.7), whereas the second transformation in (91) is a reflection across the real axis.

By expressing (90) in terms of its real and imaginary parts, we obtain

$$u + iv = \frac{x}{x^2 + y^2} - i\frac{y}{x^2 + y^2}. \qquad (92)$$

We now wish to show that all circles and lines in one plane are mapped into circles and lines in the other plane. For example, the quadratic equation

$$A(u^2 + v^2) + Bu + Cv + D = 0 \qquad (93)$$

represents all circles ($A \neq 0$) and lines ($A = 0$) in the w-plane. Under the mapping (92), this equation reduces to (after some algebraic manipulation)

$$A + Bx - Cy + D(x^2 + y^2) = 0, \qquad (94)$$

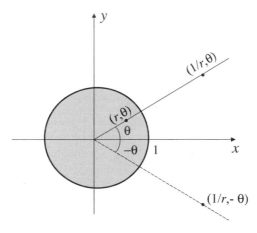

Figure 6.7 Reciprocal mapping.

which represents circles and lines in the z-plane. If $D = 0$, $A \neq 0$ in (93), then the image of all such circles in the w-plane are lines in the z-plane. Similarly, if $A = 0$, $D \neq 0$ in (94), then the image of all such circles in the z-plane are lines in the w-plane. More specifically, we can draw the following conclusions:

- All circles *not* passing through the origin in one plane transform into circles in the other plane *not* passing through the origin.

- Any circle *through* the origin in one plane transforms into a line in the other plane *not* through the origin.

It is often convenient to use the concept of the "point" at infinity, denoted by $z = \infty$. In the complex plane we do not distinguish between $+\infty$ and $-\infty$ as we do on the real axis. Instead, we define $z = \infty$ as the image of the point $w = 0$ under the reciprocal mapping function (90). Likewise, the point $z = 0$ is the image of the point $w = \infty$.

The conventional complex plane is somewhat inconvenient when discussing the concept of infinity for complex numbers because the point at infinity can be reached in all directions away from the origin. Perhaps the best means of visualizing the "point" ∞ is to use stereographic projection as suggested by Riemann, whereby all points in the complex plane correspond to a unique point on a sphere. Let us consider a sphere of unit diameter "sitting" on the complex plane at the origin (see Fig. 6.8;, i.e., the "South Pole" coincides with the origin. In three dimensions, the center of the sphere is $C(0,0,1/2)$. A line joining any point z in the complex plane with the "North Pole" will pierce the sphere at a unique point, which becomes the *stereographic projection* of the point z. The unit circle in the z-plane maps onto the "equator" of the sphere. Finally, the North Pole represents the "point" infinity.

Following the procedure described above, every point in the z-plane gets mapped into a unique point on the surface of the sphere; i.e., the mapping is one-to-one. Without proof, we list the following observations concerning stereographic projection:

- The unit circle $|z| = 1$ corresponds to the equator of the sphere. Points interior to the unit circle are mapped onto the southern hemisphere, and points exterior to the unit circle are mapped onto the northern hemisphere.

- A line of the form $y = kx$ corresponds to a circle on the sphere that passes through both the north and the south poles.

- All points such that $|z| > R \gg 1$ map onto a spherical cap about the north pole.

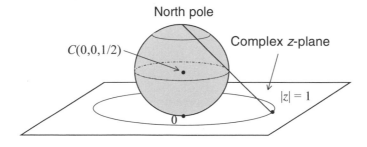

Figure 6.8 Stereographic projection.

6.7.4 Bilinear transformation

The mapping described by

$$w = \frac{Az + B}{Cz + D}, \quad AD - BC \neq 0,$$

(95)

where A, B, C, and D are complex constants, is called the *linear fractional* or *bilinear transformation*. The special case $C = 0$ reduces it to the linear transformation (87). The bilinear mapping (95) assigns to each point in the z-plane a unique point in the w-plane. In particular, the point $z = -D/C$ $(C \neq 0)$ maps into $w = \infty$. For $C \neq 0$, Eq. (95) can be rewritten in the form

$$w = \frac{A(z + D/C) + (B - AD/C)}{C(z + D/C)}$$
$$= \frac{A}{C} + \frac{B - AD/C}{Cz + D},$$

(96)

which shows that it is simply a succession of elementary transformations already discussed. That is, the three successive transformations are the *linear transformation*

$$\zeta_1 = Cz + D,$$

(97)

followed by a *reciprocal transformation*

$$\zeta_2 = \frac{1}{\zeta_1} = \frac{1}{Cz + D},$$

(98)

and, finally, followed by another *linear transformation*

$$w = \left(\frac{BC - AD}{C} \right) \zeta_2 + \frac{A}{C}, \quad BC - AD \neq 0.$$

(99)

Based on these results, we can draw the following conclusion:

- The bilinear transformation (95) maps circles and lines in the z-plane into circles and lines in the w-plane, and vice-versa.

Remark: Equation (99) reveals why we make the restriction $AD \neq BC$. That is, if we allow $AD = BC$, then (99) reduces to the constant mapping $w = A/C$ which maps all points in the z-plane into the single point A/C in the w-plane.

Observe that if (95) is cleared of fractions, we can write it as

$$Cwz + Dw - Az - B = 0,$$

(100)

which is *linear* in both z and w—hence, the name "bilinear." If we now solve (100) for z, we obtain another bilinear transformation

$$z = \frac{-Dw + B}{Cw - A}, \quad AD - BC \neq 0. \tag{101}$$

A *fixed point* of a mapping is one in which the image is the same complex number. We can determine such points by letting $w = z$ in (95), viz.,

$$z = \frac{Az + B}{Cz + D}, \tag{102}$$

or, upon simplifying,

$$Cz^2 + (D - A)z - B = 0. \tag{103}$$

Because (103) is quadratic, it leads to at most two fixed points. If $C = 0$ and $D \neq A$, there is only one fixed point, and if $C = 0$ and $D = A$, there are no fixed points. (It can be shown that if there are three or more fixed points of a mapping, that mapping must be the identity mapping.)

Although we used four constants in both (95) and (101), these transformations are not altered by dividing the numerator and denominator by any one of the constants which is not zero, thus leaving only three essential constants. This would seem to indicate that if we know the images of three points, say $z_1, z_2,$ and z_3, in the z-plane are given by $w_1, w_2,$ and w_3, respectively, in the w-plane, then (95) [or (101)] is uniquely determined. In this regard, consider the following theorems.

> **Remark:** Recall that a circle is uniquely determined by three points, not all on the same line.

Theorem 6.8 Given three distinct points $\{z_1, z_2, z_3\}$ in the z-plane, there exists a bilinear transformation that maps these points into $\{0, 1, \infty\}$, respectively, in the w-plane.

Proof: The proof requires that we consider four cases.

CASE I: If the points z_1, z_2, z_3 are all finite, the transformation is given by

$$w = \frac{(z - z_1)(z_2 - z_3)}{(z - z_3)(z_2 - z_1)}.$$

By setting $z = z_1, z_2, z_3$ into the above equation, we get $w = 0, 1, \infty$, respectively.

CASE II: If $z_1 = \infty$ whereas z_2, z_3 remain finite, then

$$w = \frac{z_2 - z_3}{z - z_3}.$$

CASE III: If $z_2 = \infty$ whereas z_1, z_3 remain finite, then

$$w = \frac{z - z_1}{z - z_3}.$$

CASE IV: If $z_3 = \infty$ whereas z_1, z_2 remain finite, then

$$w = \frac{z - z_1}{z_2 - z_1}.$$

❑

Theorem 6.9 Given three distinct points $z_1, z_2,$ and z_3, in the z-plane (not all necessarily finite) and three distinct points $w_1, w_2,$ and w_3, in the w-plane, there exists precisely one bilinear transformation which maps $z_1, z_2,$ and z_3, into $w_1, w_2,$ and w_3, respectively, given by

$$\frac{(w - w_1)(w_2 - w_3)}{(w - w_3)(w_2 - w_1)} = \frac{(z - z_1)(z_2 - z_3)}{(z - z_3)(z_2 - z_1)}.$$

Remark: If any one of the three points cited in Theorem 6.9 in either plane is ∞, the quotient of the two differences containing this point must be replaced by unity.

EXAMPLE 22 Find a bilinear transformation that maps the points $\{0, -i, -1\}$ in the z-plane into the points $\{i, 1, 0\}$ in the w-plane.

Solution: By substituting the given values into the equation given in Theorem 6.8, we are led to

$$\frac{(w - i)(1 - 0)}{(w - 0)(1 - i)} = \frac{(z - 0)(-i + 1)}{(z + 1)(-i - 0)}.$$

Solving this equation algebraically for w yields the result

$$w = -i\left(\frac{z + 1}{z - 1}\right).$$

6.7.5 Conformal mapping

The geometric character of a mapping under an analytic function $f(z)$ is of special interest in that the angle of intersection between two curves in the z-plane is preserved under the mapping $w = f(z)$. Before discussing this further, however, let us again introduce the notion of a function that is "one-to-one."

Let us consider the analytic function $w = f(z)$ which defines the transformation

$$\begin{aligned} u &= u(x,y), \\ v &= v(x,y). \end{aligned} \tag{104}$$

In order to obtain a unique inverse transformation of the form

$$x = x(u,v),$$
$$y = y(u,v), \tag{105}$$

the *Jacobian of transformation* must satisfy (recall the discussions in Section 5.4 and 5.9)

$$J = \frac{\partial(u,v)}{\partial(x,y)} = \begin{vmatrix} \dfrac{\partial u}{\partial x} & \dfrac{\partial u}{\partial y} \\[2mm] \dfrac{\partial v}{\partial x} & \dfrac{\partial v}{\partial y} \end{vmatrix} \neq 0. \tag{106}$$

Because we have assumed that $w = f(z)$ is analytic, it follows that

$$f'(z) = \frac{\partial u}{\partial x} + i\frac{\partial v}{\partial x} = \frac{\partial v}{\partial y} - i\frac{\partial u}{\partial y},$$

and, thus, we see that the Jacobian J is related to $f'(z)$ by

$$J = \frac{\partial u}{\partial x}\frac{\partial v}{\partial y} - \frac{\partial u}{\partial y}\frac{\partial v}{\partial x} = |f'(z)|^2. \tag{107}$$

Based on this last result, it follows that the Jacobian will be nonzero at all points of the mapping $w = f(z)$ for which $f'(z) \neq 0$.

Theorem 6.10 If $f(z)$ is analytic at z_0 and $f'(z_0) \neq 0$, then $w = f(z)$ provides a one-to-one mapping of a neighborhood of z_0.

Remark: Points where $f'(z) = 0$ are called *critical points* of the mapping $w = f(z)$. Among others, these are the points where the families of level curves associated with the real and imaginary parts of $f(z)$ may fail to be orthogonal (recall Section 6.7.1).

▸ **Conformal mapping:** A *conformal mapping* is a transformation that "preserves angles" between oriented curves, both in magnitude and in sense (e.g., orthogonal families of curves map into orthogonal families).

Theorem 6.11 An analytic function $w = f(z)$ is a conformal mapping at every point z_0 for which $f'(z_0) \neq 0$.

EXAMPLE 23 Determine the image of the semi-infinite strip in the upper half-plane defined by the lines $x = \pm\pi/2$ and $y = 0$ (see Fig. 6.9) under the mapping $f(z) = \sin z$.

Solution: We first note that the mapping $f(z) = \sin z$ has critical points at $z = \pm\pi/2$ because $f'(z) = \cos z$ vanishes at these points. These are points where the angles may not be preserved. Also, the mapping is defined by the equations of transformation

$$u(x,y) = \sin x \cosh y,$$

$$v(x,y) = \cos x \sinh y.$$

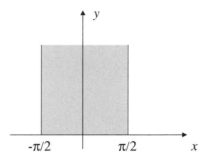

Figure 6.9 Semi-infinite strip.

Thus, if $x = \pi/2$, it follows that $u(x,y) = \cosh y$ and $v(x,y) = 0$;   similarly, if $x = -\pi/2$, it follows that $u(x,y) = -\cosh y$ and $v(x,y) = 0$. We deduce, therefore, that two vertical ray-lines shown in Fig. 6.9 for $y > 0$ are mapped into portions of the u-axis described by $u > 1$ and $u < -1$. In addition, the part of the x-axis described by $-\pi/2 < x < \pi/2$ maps into that part of the u-axis described by $-1 < u < 1$. The image of the semi-infinite strip is clearly the upper-half plane defined by $v > 0$ (see Fig. 6.10).

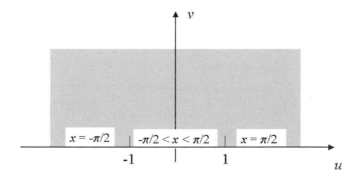

Figure 6.10 Image of semi-infinite strip in Fig. 6.9.

NOTE: Suggesting reading for this chapter is the same as that listed in Chapter 7.

Chapter 6

Exercises

1. Perform the necessary operations to put the number in the form $a + ib$.

 (a) $(2 + 3i)(-2 - 3i)$

 (b) $\dfrac{1}{3 + 2i}$

 (c) $\dfrac{1 + i}{1 - i}$

 Ans. (a) $5 - 12i$ (b) $\dfrac{3}{13} - \dfrac{2}{13}i$

 (c) i

2. Find the *modulus* and *argument* of the following numbers.

 (a) $z = -\sqrt{12} - 2i$

 (b) $z = -5 + i\sqrt{75}$

 Ans. (a) $|z| = 4$, $\arg(z) = -\dfrac{5\pi}{6} + 2k\pi$

 (b) $|z| = 10$, $\arg(z) = \dfrac{2\pi}{3} + 2k\pi$

3. Write the number in polar form.

 (a) $z = -2 + 2i$

 (b) $z = -\sqrt{27} - 3i$

 Ans. (a) $z = \sqrt{8}\,e^{i3\pi/4}$
       (b) $z = 6e^{i7\pi/6}$

4. In each case, find the locus of points in the plane.

 (a) $|z - 5| = 6$

 (b) $\mathrm{Re}(z + 2) = -1$

 (c) $|z + i| = |z - i|$

 Ans. (a) Circle of radius 6 with center at $z = 5$.

   (b) Vertical line: $x = -3$
   (c) Real axis: $y = 0$

5. Find the three cube roots of i.

 Ans. $z_1 = e^{i\pi/6} = \dfrac{\sqrt{3}}{2} + \dfrac{i}{2}$

   $z_2 = e^{i5\pi/6} = -\dfrac{\sqrt{3}}{2} + \dfrac{i}{2}$

   $z_3 = e^{i3\pi/2} = -i$

6. Solve the equation $z^2 + i = 0$ and use your answer to solve $z^4 + 2iz^2 - 1 = 0$.

7. Determine the points at which the derivative exists.

 (a) $f(z) = x + iy^2$
 (b) $f(z) = \sin x \cosh y + i \cos x \sinh y$
 (c) $f(z) = 2x^2 + 3iy^3$

 Ans. (a) $y = 1/2$ (b) all z
   (c) $9y^2 = 4x$

8. Given $f(z) = x^2 + iy^2$, where is $f(z)$
 (a) differentiable?
 (b) analytic?

 Ans. (a) $y = x$ (b) nowhere

9. For what values of a, b, c, d is the following function analytic?

 $$f(z) = x^2 + axy + by^2$$
 $$+ i(cx^2 + dxy + y^2)$$

 Ans. $a = d = 2$, $b = c = -1$

10. Given the harmonic function

 $$u(x,y) = 2x(1 - y),$$

 (a) find $v(x,y)$ so that $f(z) = u + iv$ is analytic.

 (b) Express $f(z)$ as a function of z alone.

 Ans. (a) $v(x,y) = x^2 - y^2 + 2y$
   (b) $f(z) = 2z + iz^2$

11. Express each of the following in the form $a + ib$.

 (a) e^{2+i} (b) $e^{(1+i)^2}$ (c) $e^{1/(1+i)}$
 Ans. (a) $3.99 + i6.22$
   (b) $-0.42 + i0.91$
   (c) $1.45 - i0.79$

12. Express each of the following in the form $u + iv$.

(a) e^{z^2} (b) e^{e^z}

Ans. (a) $e^{x^2-y^2}(\cos 2xy + i\sin 2xy)$
(b) $\exp(e^x\cos y)$
$\times [\cos(e^x\sin y) + i\sin(e^x\sin y)]$

13. Find the *principal value* of
(a) $|\sin(\pi/3 - i)|^2$
(b) $(1 - i)^{1+i}$
(c) $\log(\log i)$

Ans. (a) $\dfrac{3}{4} + \sinh^2 1$
(b) $e^{\pi/2}[\cos(\ln 2) + i\sin(\ln 2)]$
(c) $\ln(\pi/2) + i\pi/2$

14. Find all values of
(a) $\cosh[\log(-2i)]$
(b) $e^z = 3^{2i}$
(c) $(e^z - 1)^2 = 1$

Ans. (a) $-\dfrac{3}{4}i$
(b) $-4k\pi + 2i\ln 3, k = 0, \pm 1, \pm 2, ...$
(c) $\ln 2 + 1 + i2k\pi, k = 0, \pm 1, \pm 2, ...$

15. Show that the image of the infinite line $\text{Im} z = 1$ under the mapping $f(z) = 1/z$ is a circle. Find its center and radius.

16. Describe the image of the following domains under the mapping $f(z) = \cos z$.
(a) The infinite strip $1 \le \text{Re} z \le 2$.
(b) The semi-infinite strip $\text{Im} z \ge 0$, $0 \le \text{Re} z \le \pi$.

Ans. (a) Region lying on and between two parabolas.
(b) Half-plane: $\text{Im} w \le 0$

17. Where are the following mappings conformal?
(a) $w = z + 1/z$
(b) $w = z^2 + 1/z^2$

Ans. (a) $z \ne 0, \pm 1$
(b) $z \ne 0, \pm 1, \pm i$

18. Find a bilinear transformation that

maps the points $\{1, i, 0\}$ in the z-plane into $\{0, \infty, -i\}$ in the w-plane.

Ans. $w = \dfrac{1 - z}{i - z}$

19. Find a bilinear transformation that maps the points $\{1, 2, 3\}$ in the z-plane into $\{0, \infty, 1\}$ in the w-plane.

Ans. $w = \dfrac{1}{2}\left(\dfrac{z - 1}{z - 2}\right)$

20. Find a bilinear transformation that has fixed points
(a) -1 and $+1$.
(b) $-i$ and $+i$.

Ans. (a) $w = \dfrac{Az + B}{Bz + A}$
(b) $w = \dfrac{Az - B}{Bz + A}$

21. Find a bilinear transformation that has fixed points -1 and $+1$, and which maps the upper-half z-plane onto the interior of the unit circle.

Ans. $w = \dfrac{iz + 1}{z + i}$

22. Given the bilinear mapping

$$w = \frac{z - 1}{z - 2},$$

find the image of
(a) the circle $|z + 1| = 1$.
(b) the circle $|z + 1| = 3$.
(c) the line $|z - 1| = |z - 2|$.

Ans. (a) $(u - 5/8)^2 + v^2 = \dfrac{1}{64}$
(b) $u = 5/6$
(c) $u^2 + v^2 = 1$

Chapter 7

Complex Integration, Laurent Series, and Residues

7.1   Introduction / 320
7.2   Line Integrals in the Complex Plane / 320
    7.2.1   Bounded integrals / 323
7.3   Cauchy's Theory of Integration / 325
    7.3.1   Deformation of contours / 326
    7.3.2   Integrals independent of path / 328
    7.3.3   Cauchy's integral formula / 332
    7.3.4   Cauchy's generalized formula / 334
    7.3.5   Bounds on analytic functions / 337
7.4   Infinite Series / 339
    7.4.1   Sequences and series of constants / 339
    7.4.2   Power series / 340
    7.4.3   Laurent series / 346
    7.4.4   Zeros and singularities / 352
7.5   Residue Theory / 357
    7.5.1   Residues / 357
7.6   Evaluation of Real Integrals—Part I / 363
    7.6.1   Rational functions of cos Θ and/or sin Θ / 363
    7.6.2   Improper integrals of rational functions / 365
    7.6.3   Fourier transform integrals / 369
7.7   Evaluation of Real Integrals—part II / 371
7.8   Harmonic Functions Revisited / 376
    7.8.1   Harmonic functions in the half-plane / 377
    7.8.2   Harmonic functions in circular domains / 380
    7.8.3   Invariance of Laplace's equation / 382
7.9   Heat Conduction / 383
    7.9.1   Steady-state temperatures in the plane / 383
    7.9.2   Conformal mapping / 384
7.10 Two-Dimensional Fluid Flow / 386
    7.10.1 Complex potential / 387
    7.10.2 Source, sink, and doublet flows / 391
7.11 Flow Around Obstacles / 393
    7.11.1 Circulation and lift / 394
    7.11.2 Flow around a cylinder / 396
    Suggested Reading / 399
    Exercises / 400

In this chapter we extend the discussion of complex variables to complex integrals and Laurent series. Complex integrals in general behave similar to the line integrals found in vector analysis. However, the method of residue calculus used for evaluating integrals that involve an analytic function is more powerful than any counterpart in vector analysis—i.e., the residue calculus can be used to evaluate inverse Laplace transforms, and is also useful in calculating Fourier and Mellin transform integrals. Laurent series are a natural generalization of Taylor series that provide the basis for developing the residue calculus.

7.1 Introduction

The two-dimensional nature of a complex variable required us in Chapter 6 to generalize our notion of derivative in the complex plane. This was a consequence of the fact that a complex variable can approach its limit value from infinitely-many directions, rather than just two directions as in the case of a real variable. This two-dimensional aspect of a complex variable will also influence the theory of integration in the complex plane, requiring us to consider integrals along general curves in the plane rather than simply along segments of the x-axis. Because of this, we find that complex integrals behave more like *line integrals* from vector analysis instead of like standard Riemann integrals.

The most practical application of the complex integral is in the evaluation of certain real integrals, including those that commonly appear in the use of integral transforms. A secondary reason why the complex integral is important is that the method of complex integration (also called contour integration) yields simple proofs of some basic properties of analytic functions that would otherwise be very difficult to prove. In particular, we can use complex integration to prove that an analytic function has higher-order derivatives—in fact, *derivatives of all orders*. Thus far, however, we have only shown that an analytic function has a first derivative.

Most of the theory of integration in the complex plane involving analytic functions is due to Augustin L. Cauchy (1789-1857), the famous French mathematician. His main result, now known as *Cauchy's integral theorem*, or simply *Cauchy's theorem*, states that the integral of an analytic function around a simple closed curve is zero. Although sounding almost trivial, most of the theory of complex integration is a consequence of this very important result.

7.2 Line Integrals in the Complex Plane

Once again, we find it necessary to introduce some special terminology that arises in the discussion of integration in the complex plane. However, there are variations of these terms that appear in some textbooks.

Let us imagine that the real and imaginary parts of the complex variable $z = x + iy$ depend upon a real parameter t, i.e., $x = x(t)$ and $y = y(t)$. In this case we can write

$$C: \quad z(t) = x(t) + iy(t), \quad a \le t \le b.$$

(1)

If t is a continuous parameter, then Eq. (1) for increasing t represents a continuous oriented curve (or arc) C in the complex plane, where $t = a$ and $t = b$ correspond to the endpoints of the curve (see Fig. 7.1). This parameterization of the curve C is completely analogous to the notion of such curves in a standard two-dimensional vector space.

- *simple curve* or *arc*—a curve that does not intersect itself. If the endpoints coincide it is a *simple closed curve*.

- *smooth arc*—$z'(t)$ exists (nonzero) and is continuous on the interval $a \leq t \leq b$.

- *contour*—a curve consisting of a finite number of smooth arcs joined end to end.

The concept of integral in the complex plane is closely related to the concept of line integral in the xy-plane. Let us consider the oriented contour C defined by Eq. (1), and let $f(z)$ be a continuous function at each point of C. Suppose we partition the contour C into parts by the points $z_0, z_1, z_2, ..., z_n$ as shown in Fig. 7.1. We then choose arbitrary points ζ_k ($k = 1, 2, ... , n$) on C between the points z_{k-1} and z_k, and form the sum

$$S_n = \sum_{k=1}^{n} f(\zeta_k)\Delta z_k, \qquad \Delta z_k = z_k - z_{k-1}. \tag{2}$$

If the limit of this expression exists as $n \to \infty$ in such a way that the largest of the chord lengths $|\Delta z_k|$ approaches zero, this limit is called the *complex integral* of $f(z)$ along C and is designated by

$$\int_C f(z)dz = \lim_{\substack{n \to \infty \\ \max|\Delta z| \to 0}} \sum_{k=1}^{n} f(\zeta_k)\Delta z_k. \tag{3}$$

▸ **Parameterization method:** By using the parameterization given by Eq. (1), we reduce the integral in (3) to a standard integral of a real variable t, viz.,

$$\boxed{\int_C f(z)dz = \int_a^b f[z(t)]z'(t)dt.} \tag{4}$$

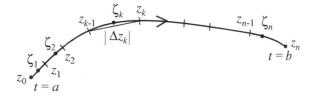

Figure 7.1 Contour for complex line integral.

▸ **Relation to real line integrals:** Another useful form of the integral (3) is possible by expressing $f(z) = u(x,y) + iv(x,y)$ and $dz = dx + idy$, viz.,

$$\int_C f(z)dz = \int_C [u(x,y) + iv(x,y)](dx + idy)$$

$$= \int_C u(x,y)dx - v(x,y)dy + i\int_C v(x,y)dx + u(x,y)dy.$$

(5)

Equation (5) reveals that the integral of a complex function in the complex plane is directly related to the sum of two real integrals in the xy-plane. Because of this relationship, basic properties of the complex integral follow immediately from the well-known properties of real line integrals.

Unless otherwise stated, we will generally assume all integration paths are contours and hence, the complex integral is also called a *contour integral*. Because much of the subsequent discussion concerns integrals over closed contours, it is convenient in these cases to adopt the special notation

$$\oint_C f(z)dz.$$

▸ **Positive direction:** The *positive direction* along a closed contour is always taken as *counterclockwise*; we generally assume that a closed contour integral is in the positive direction unless otherwise stated.

EXAMPLE 1 Evaluate the integral I along the specified contours:

$$I = \int_C (xy + ix^2)dz$$

(a) C_1: $z(t) = t + it$, $0 \le t \le 1$.

(b) C_2: $z(t) = t + it^2$, $0 \le t \le 1$.

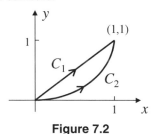

Figure 7.2

Solution: The curves C_1 and C_2 are both shown in Fig. 7.2.

(a) For the curve C_1, we have $x = y = t$ and $z'(t) = 1 + i$. Hence, using (4), we obtain

$$I = \int_0^1 (t^2 + it^2)(1 + i)dt = (1 + i)^2 \int_0^1 t^2 dt$$

$$= \frac{2i}{3}.$$

(b) Along C_2 we have $x = t$, $y = t^2$, and $z'(t) = 1 + 2it$. Using (4) again, we get

$$I = \int_0^1 (t^3 + it^2)(1 + 2it)dt = -\int_0^1 t^3 dt + i\int_0^1 (t^2 + 2t^4)dt$$

$$= -\frac{1}{4} + i\frac{11}{15}.$$

Example 1 illustrates that, similar to the case of real line integrals, complex integrals are dependent upon the choice of contour connecting the two points. The next example has far-reaching consequences in later developments.

EXAMPLE 2 Evaluate the integral I around a circle of radius R that is centered at $z = z_0$:

$$I = \oint_C (z - z_0)^n \, dz, \quad n = 0, \pm 1, \pm 2, \dots .$$

Solution: Our standard parameterization for the circle $|z - z_0| = R$ is

$$C: \quad z(t) = z_0 + Re^{it}, \quad 0 \le t \le 2\pi .$$

Thus, $z - z_0 = Re^{it}$ and $z'(t) = iRe^{it}$, which leads to

$$I = \int_0^{2\pi} (Re^{it})^n iRe^{it} \, dt = iR^{n+1} \int_0^{2\pi} e^{i(n+1)t} \, dt .$$

From this last result, we consider cases $(n \ne -1, \ n = -1)$. For $n \ne -1$, we see that

$$\int_0^{2\pi} e^{i(n+1)t} \, dt = \left. \frac{e^{i(n+1)t}}{i(n+1)} \right|_0^{2\pi} = 0,$$

whereas for $n = -1$, we have

$$\int_0^{2\pi} dt = 2\pi ;$$

consequently, it follows that

$$\boxed{I = \oint_C (z - z_0)^n \, dz = \begin{cases} 0, & n \ne -1 \\ 2\pi i, & n = -1. \end{cases}}$$

7.2.1 Bounded integrals

In developing some of the subsequent theoretical results it is often enough to obtain an *upper bound* for a complex integral rather than know its exact value. By using the analogous property of real integrals, we start with the inequality

$$\left| \int_C f(z) \, dz \right| \le \int_C |f(z)| \, |dz| . \tag{6}$$

Now, if M is a real number such that $|f(z)| \le M$, $z \in C$, then Eq. (6) becomes

$$\left| \int_C f(z)\,dz \right| \le M \int_C |dz|.$$

Also, for $z \in C$, we interpret the symbol $|dz|$ as

$$|dz| = |z'(t)|\,dt = \sqrt{[x'(t)]^2 + [y'(t)]^2}\,dt,$$

which we recognize as the standard symbol for the differential of arc length, i.e., $|dz| = ds$. Hence, we deduce that for a rectifiable[1] curve C, the quantity

$$L = \int_C |dz|$$

represents the *length* of the contour C. In view of the above results, the inequality (6) reduces to

$$\boxed{\left| \int_C f(z)\,dz \right| \le ML.} \tag{7}$$

EXAMPLE 3 Use the result of Eq. (7) to find an upper bound for the contour integral

$$I = \int_C \frac{dz}{z^4},$$

where C is the line segment from $z = i$ to $z = 1$ (see Fig. 7.3).

Solution: By use of the Pythagorean theorem, we note from inspection that the length of the contour C is $L = \sqrt{2}$. To find an upper bound M, it is useful to parameterize the curve C by the linear function

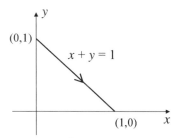

Figure 7.3

$$C: \quad z(t) = t + i(1 - t),\ 0 \le t \le 1.$$

The denominator in the integrand then becomes

$$|z^4| = |z|^4 = (x^2 + y^2)^2,$$

which in parametric form is

---

[1]A rectifiable curve is one that has finite length.

$$|z^4| = \left[t^2 + (1 - t)^2\right]^2$$

$$= \left[2\left(t - \frac{1}{2}\right)^2 + \frac{1}{2}\right]^2 \geq \frac{1}{4}.$$

The last step is a consequence of recognizing a minimum value of $|z^4|$ occurs for $t = 1/2$. [In some cases, however, it may be useful to define $g(t) = t^2 + (1 - t)^2$ and set $g'(t) = 0$ to find the maximum/minimum.] Hence, the integrand is bounded by

$$\left|\frac{1}{z^4}\right| \leq 4 = M,$$

and we deduce that

$$\left|\int_C z^{-4}\,dz\right| \leq ML = 4\sqrt{2}.$$

7.3 Cauchy's Theorem of Integration

The following theorem from vector analysis, known as *Green's theorem in the plane* (recall Theorem 4.7), is basic to real functions integrated around a closed contour.

> **Theorem 7.1** (*Green's theorem in the plane*) If $P(x,y)$ and $Q(x,y)$, together with their first partial derivatives, are continuous functions throughout a region R consisting of a simple closed contour C and its interior, then
>
> $$\oint_C P(x,y)dx + Q(x,y)dy = \iint_R \left(\frac{\partial Q}{\partial x} - \frac{\partial P}{\partial y}\right) dx\,dy.$$

One of the most significant theorems in all of mathematics is the following theorem of Cauchy, the proof of which (using Green's theorem in the plane) was first presented by Cauchy in the early eighteenth century. Although Cauchy's proof relies on the continuity of the derivative $f'(z)$, it was later shown by Goursat that this restriction is not necessary. Goursat's proof, however, is much more difficult and lengthy.

> **Theorem 7.2** (*Cauchy's integral theorem*) If $f(z)$ is analytic on and inside a simple closed contour C, then
>
> $$\oint_C f(z)dz = 0.$$

Proof: Our proof (like that of Cauchy) will rely on the assumed continuity of the derivative $f'(z)$. Recalling Eq. (5), we have

$$\oint_C f(z)\,dz = \oint_C u(x,y)\,dx - v(x,y)\,dy + i\oint_C v(x,y)\,dx + u(x,y)\,dy.$$

Now, because $f(z)$ is analytic, it is also *continuous*, and since we assume $f'(z)$ is continuous, it follows that the real and imaginary parts of $f(z) = u(x,y) + iv(x,y)$ are likewise continuous as are the first partial derivatives of the real functions $u(x,y)$ and $v(x,y)$. Thus, the conditions of Green's theorem (Theorem 7.1) are satisfied and we can write

$$\oint_C u(x,y)\,dx - v(x,y)\,dy = \iint_R \left(-\frac{\partial v}{\partial x} - \frac{\partial u}{\partial y} \right) dx\,dy,$$

$$\oint_C v(x,y)\,dx + u(x,y)\,dy = \iint_R \left(\frac{\partial u}{\partial x} - \frac{\partial v}{\partial y} \right) dx\,dy,$$

where R is the region containing the contour C and the interior of the contour. Because $f(z)$ is analytic, the Cauchy-Riemann equations are satisfied and, hence, the double integrals on the right are both zero, which leads to

$$\oint_C f(z)\,dz = 0.$$

❏

Cauchy's theorem (Theorem 7.2) has a number of consequences, which we study throughout the remainder of this chapter. To start, we discuss the deformation of contour principle in the following section.

7.3.1 Deformation of contours

Lemma 7.1 below is sometimes called the "deformation of contour principle" since it enables us in certain situations to replace a complicated shaped contour C with a simpler shaped contour C_1 (such as a circle) wholly inside C.

Lemma 7.1 If $f(z)$ is analytic on simple closed contours C and C_1, with C_1 inside C, and analytic also in the annular region between them (see Fig. 7.4), then

$$\oint_C f(z)\,dz = \oint_{C_1} f(z)\,dz,$$

Proof:

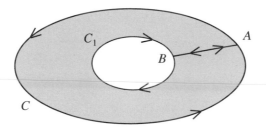

Figure 7.4 Annular domain.

Let A denote a point on contour C and B a point on C_1 as shown in Fig. 7.4, and connect them with a line. We then consider the contour integral described by

$$I = \oint_C f(z)\,dz + \int_A^B f(z)\,dz - \oint_{C_1} f(z)\,dz + \int_B^A f(z)\,dz,$$

where the direction of integration is shown in the figure. Because $f(z)$ is analytic on and inside the closed path described by the integral I above, it follows from Cauchy's integral theorem that $I = 0$. Moreover,

$$\int_A^B f(z)\,dz + \int_B^A f(z)\,dz = 0,$$

and we deduce that $I = \oint_C f(z)\,dz - \oint_{C_1} f(z)\,dz = 0$, or

$$\oint_C f(z)\,dz = \oint_{C_1} f(z)\,dz.$$

❏

The next theorem is a generalization of Lemma 7.1 to the case where there are several closed contours C_1, C_2, ... , C_n, all lying inside a larger closed contour C. The proof follows from continued application of Lemma 7.1 to the contours shown in Fig. 7.5.

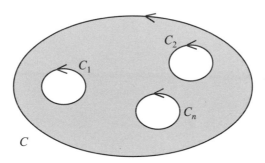

Figure 7.5 Multiannular domain.

Theorem 7.3 If $f(z)$ is analytic on the simple closed contour C and on the simple closed contours C_1, C_2, ... , C_n, all of which lie within C, and if $f(z)$ is analytic in the annular domain between the curves, then

$$\oint_C f(z)dz = \oint_{C_1} f(z)dz + \cdots + \oint_{C_n} f(z)dz.$$

7.3.2 Integrals independent of path

One of the important results in the theory of complex integration is the extension of the *fundamental theorem of calculus* to contour integrals (Theorem 7.6 below). In general, when we integrate a complex function over a path connecting two points, the numerical value of the integral will depend on the choice of path. However, under proper conditions, we can show the complex integral is *independent of path*.

Theorem 7.4 If $f(z)$ is analytic in a simply-connected domain D, and C is any contour contained entirely in D, then the integral

$$I = \int_C f(z)dz$$

is independent of the contour C.

Proof:

Let C_1 and C_2 be two contours in D connecting points z_0 and z as shown in Fig. 7.6 above. If we define $C' = C_1 - C_2$, then, as a consequence of Cauchy's theorem (Theorem 7.2), we have

$$\oint_{C'} f(z)dz = \int_{C_1} f(z)dz - \int_{C_2} f(z)dz = 0,$$

from which we deduce (for any two curves in D connecting the same two points)

$$\int_{C_1} f(z)dz = \int_{C_2} f(z)dz.$$

❏

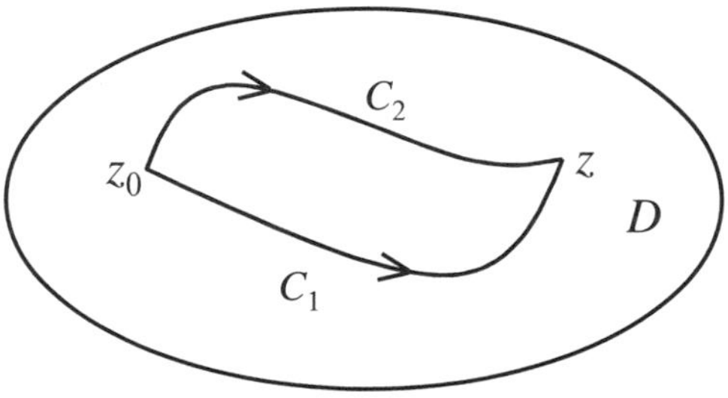

Figure 7.6

Theorem 7.5 If z_0 is a point in a simply-connected domain D, and if

- $f(z)$ is continuous in D,

- $\int_{z_0}^{z} f(z)dz$ is independent of path,

- $\int_{z_0}^{z} f(z)dz = F(z)$,

then $F(z)$ is analytic in D and $F'(z) = f(z)$.

Proof: Let $f(z) = u(x,y) + iv(x,y)$ and $F(z) = U(x,y) + iV(x,y)$ so that the third hypothesis in the theorem becomes

$$\int_{(x_0,y_0)}^{(x,y)} [u(x,y) + iv(x,y)](dx + idy) = U(x,y) + iV(x,y).$$

If we separate the left-hand side into real and imaginary parts, we see that

$$U(x,y) = \int_{(x_0,y_0)}^{(x,y)} u(x,y)dx - v(x,y)dy,$$

$$V(x,y) = \int_{(x_0,y_0)}^{(x,y)} v(x,y)dx + u(x,y)dy.$$

Because of the first two hypotheses, it follows that $u(x,y)$ and $v(x,y)$ are continuous functions in D and that the real line integrals above for $U(x,y)$ and $V(x,y)$ are independent of path. Hence, there exist two real functions $g(x,y)$ and $h(x,y)$ such that

$$u(x,y)dx - v(x,y)dy = dg = \frac{\partial g}{\partial x}dx + \frac{\partial g}{\partial y}dy,$$

$$v(x,y)dx + u(x,y)dy = dh = \frac{\partial h}{\partial x}dx + \frac{\partial h}{\partial y}dy.$$

By using these results and integrating the above line integrals, we deduce that

$$U(x,y) = \int_{(x_0,y_0)}^{(x,y)} dg = g(x,y) - g(x_0,y_0),$$

$$V(x,y) = \int_{(x_0,y_0)}^{(x,y)} dh = h(x,y) - h(x_0,y_0).$$

To prove now that $F(z)$ is analytic, we merely observe

$$\frac{\partial U}{\partial x} = \frac{\partial g}{\partial x} = u(x,y), \quad \frac{\partial V}{\partial y} = \frac{\partial h}{\partial y} = u(x,y),$$

$$\frac{\partial U}{\partial y} = \frac{\partial g}{\partial y} = -v(x,y), \quad \frac{\partial V}{\partial x} = \frac{\partial h}{\partial x} = v(x,y).$$

Hence, the Cauchy-Riemann equations are satisfied, i.e., $\partial U/\partial x = \partial V/\partial y$, $\partial U/\partial y = -\partial V/\partial x$. Finally, from the fact that $F(z)$ is analytic, it follows that

$$F'(z) = \frac{\partial U}{\partial x} + i\frac{\partial V}{\partial x} = u(x,y) + iv(x,y) = f(z),$$

and the theorem is proved. ❏

Last, the combined results of Theorems 7.4 and 7.5 lead to the following fundamental theorem of calculus for complex variables.

Theorem 7.6 (*Fundamental theorem of calculus*) If $f(z)$ is analytic in a simply-connected domain D, and if α and β are any two points in D, then

$$\int_\alpha^\beta f(z)dz = F(\beta) - F(\alpha),$$

where $F'(z) = f(z)$.

To illustrate the use of the above theorems in evaluating integrals, let us consider the following examples.

EXAMPLE 4 Evaluate $I = \int_C \sin z\, dz$, where C is composed of the circular arc

C_1 and straight line segment C_2 shown in Fig. 7.7 that connects the points $z = 0$ and $z = i\pi$.

Solution: The integrand $f(z) = \sin z$ is an entire function. Hence, its integral is independent of path so we may write

$$\int_C \sin z\, dz = \int_0^{i\pi} \sin z\, dz$$

$$= -\cos z \Big|_0^{i\pi}$$

$$= 1 - \cos i\pi,$$

or, finally,

$$\int_C \sin z\, dz = 1 - \cosh \pi.$$

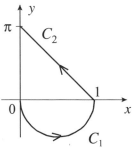

Figure 7.7

EXAMPLE 5 Evaluate $I = \int_{-1}^{1} z^{1/2} dz$.

Solution: Because the function $f(z) = z^{1/2} = \sqrt{r} e^{i\theta/2}$ is multiple-valued (i.e., two values), it is not analytic in the entire z-plane. However, if we restrict it to one of its branches illustrated below in Fig. 7.8, it becomes an analytic function for which we can then apply Theorem 7.6.

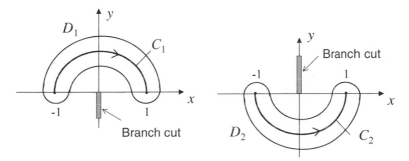

Figure 7.8 Contours of integration for different branches.

For example, by restricting $-\pi/2 < \theta < 3\pi/2$, the integrand is analytic in the domain D_1 shown above in the figure. Because of this restriction, our evaluations will eventually require the use of polar coordinates. Thus, along any contour C_1 lying inside D_1, the integral leads to

$$I = \int_{-1}^{1} z^{1/2} dz = \int_{C_1} z^{1/2} dz$$

$$= \frac{2}{3} z^{3/2} \Big|_{-1}^{1} = \frac{2}{3} r\sqrt{r} e^{3i\theta/2} \Big|_{(1,\pi)}^{(1,0)},$$

where only the last step is expressed in polar coordinates. Hence,

$$I = \frac{2}{3}\left(1 - e^{3\pi i/2}\right) = \frac{2}{3}(1 + i).$$

On the other hand, the integrand is also analytic in the domain D_2 shown above in the figure where $-3\pi/2 < \theta < \pi/2$. In this case, we find that

$$I = \int_{-1}^{1} z^{1/2} dz = \int_{C_2} z^{1/2} dz$$

$$= \frac{2}{3} r\sqrt{r} e^{3i\theta/2} \Big|_{(1,-\pi)}^{(1,0)},$$

or,

$$I = \frac{2}{3}\left(1 - e^{-3\pi i/2}\right) = \frac{2}{3}(1 - i).$$

In summary, we deduce that

$$\int_{-1}^{1} z^{1/2}\,dz = \begin{cases} \dfrac{2}{3}(1\,+\,i), & \text{in } D_1 \quad (-\pi/2 < \theta < 3\pi/2) \\[2mm] \dfrac{2}{3}(1\,-\,i), & \text{in } D_2 \quad (-3\pi/2 < \theta < \pi/2). \end{cases}$$

7.3.3 Cauchy's integral formula

If $f(z)$ is analytic on and inside a simple closed contour C, we know from Cauchy's integral theorem that $\oint_C f(z)\,dz = 0$. However, if z_0 is a point inside C, then in general

$$\oint_C \frac{f(z)}{z - z_0}\,dz \neq 0.$$

To deal with integrals of this latter form, we use *Cauchy's integral formula* given below as Theorem 7.7.

> **Theorem 7.7** (*Cauchy's integral formula*) If $f(z)$ is analytic on and inside a simple closed contour C, and if z_0 is any point inside C, then
>
> $$\oint_C \frac{f(z)}{z - z_0}\,dz = 2\pi i f(z_0).$$

Proof: Let us begin by writing $f(z) = f(z_0) + [f(z) - f(z_0)]$, giving us

$$\oint_C \frac{f(z)}{z - z_0}\,dz = \oint_C \frac{f(z_0)}{z - z_0}\,dz + \oint_C \frac{f(z) - f(z_0)}{z - z_0}\,dz$$

$$= 2\pi i f(z_0) + I(z),$$

where the first integral on the right has been evaluated using Lemma 7.1 and the result of Example 2. Using the continuity of $f(z)$ and Lemma 7.1, we wish to show that the second integral $I(z)$ is bounded by an arbitrarily small positive real number.

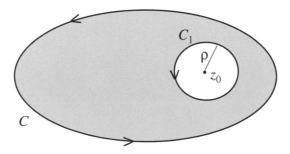

Figure 7.9

Let C_1 be a circle of arbitrarily small radius ρ, centered at z_0 (i.e., $|z - z_0| = \rho$), which lies inside the closed contour C as illustrated in Fig. 7.9. By applying the deformation of contour principle, we obtain

$$I(z) = \oint_{C_1} \frac{f(z) - f(z_0)}{z - z_0} dz.$$

From the assumed continuity of $f(z)$, we can choose the radius ρ of the circle C_1 so small that

$$|f(z) - f(z_0)| < \varepsilon, \quad z \in C_1,$$

where ε is any positive real number. Also, because $|z - z_0| = \rho$, we see that using Eq. (7) leads to

$$\begin{aligned}
|I(z)| &= \left| \oint_{C_1} \frac{f(z) - f(z_0)}{(z - z_0)} dz \right| \\
&\leq \oint_{C_1} \frac{|f(z) - f(z_0)|}{|z - z_0|} |dz| \\
&\leq \frac{\varepsilon}{\rho} 2\pi\rho = 2\pi\varepsilon.
\end{aligned}$$

This last result implies that $|I(z)| \to 0$ as $\varepsilon \to 0$, and the theorem is proved. ❏

EXAMPLE 6 Evaluate $I = \oint_C \dfrac{dz}{z(3z + 1)(z - 3)}$, where C: $|z| = 1$.

Solution: The singularities of the integrand are clearly $z = 0$, $-1/3$, and 3, but $z = 3$ lies outside the contour C (see Fig. 7.10). To begin, we use the deformation of contour principle (Theorem 7.3) and write the given integral as the sum

$$I = \oint_{C_1} \frac{1/[(3z + 1)(z - 3)]}{z} dz + \frac{1}{3} \oint_{C_2} \frac{1/[z(z - 3)]}{z + 1/3} dz,$$

where we have expressed each integral in a form satisfying the conditions of Cauchy's integral formula (Theorem 7.7). Hence, by use of this formula, we deduce

$$\begin{aligned}
I &= 2\pi i \left[\frac{1}{(3z + 1)(z - 3)} \right]\Bigg|_{z=0} + \frac{2\pi i}{3} \left[\frac{1}{z(z - 3)} \right]\Bigg|_{z=-1/3} \\
&= -\frac{2\pi i}{3} + \frac{3\pi i}{5},
\end{aligned}$$

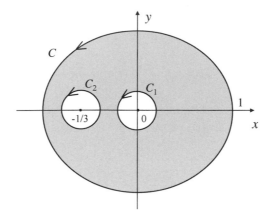

Figure 7.10 Integration contours around singularities.

or, upon simplification, we obtain

$$\oint_C \frac{dz}{z(3z + 1)(z - 3)} = -\frac{\pi i}{15}.$$

7.3.4 Cauchy's generalized formula

Based on Cauchy's integral formula, we can now prove that if $f(z)$ is analytic at a point, it has *derivatives of all orders* at that point. To start, we rewrite Cauchy's integral formula as

$$f(z) = \frac{1}{2\pi i} \oint_C \frac{f(\zeta)}{\zeta - z} d\zeta, \tag{8}$$

where z represents any point inside C and we use ζ as the dummy variable of integration. By formally differentiating both sides of (8), we arrive at

$$f'(z) = \frac{1}{2\pi i} \oint_C \frac{f(\zeta)}{(\zeta - z)^2} d\zeta. \tag{9}$$

Although Eq. (9) turns out to be a correct result, our formal derivation of it lacks rigor. To rigorously prove the validity of Eq. (9), we first observe that

$$\frac{f(z + \Delta z) - f(z)}{\Delta z} = \frac{1}{2\pi i} \oint_C \left(\frac{1}{\zeta - z - \Delta z} - \frac{1}{\zeta - z} \right) \frac{f(\zeta)}{\Delta z} d\zeta$$

$$= \frac{1}{2\pi i} \oint_C \frac{f(\zeta)}{(\zeta - z - \Delta z)(\zeta - z)} d\zeta,$$

which can be rearranged as

$$\frac{f(z + \Delta z) - f(z)}{\Delta z} = \frac{1}{2\pi i} \oint_C \frac{f(\zeta)}{(\zeta - z)^2} d\zeta$$

$$+ \frac{1}{2\pi i} \left[\oint_C \frac{f(\zeta)}{(\zeta - z - \Delta z)(\zeta - z)} d\zeta - \oint_C \frac{f(\zeta)}{(\zeta - z)^2} d\zeta \right]$$

$$= \frac{1}{2\pi i} \oint_C \frac{f(\zeta)}{(\zeta - z)^2} d\zeta + I(z),$$

where

$$I(z) = \frac{1}{2\pi i} \left[\oint_C \frac{f(\zeta)}{(\zeta - z - \Delta z)(\zeta - z)} d\zeta - \oint_C \frac{f(\zeta)}{(\zeta - z)^2} d\zeta \right]$$

$$= \frac{\Delta z}{2\pi i} \oint_C \frac{f(\zeta)}{(\zeta - z - \Delta z)(\zeta - z)^2} d\zeta. \tag{10}$$

To finish proving the validity of Eq. (9), we only need show that $|I(z)| \to 0$ as $\Delta z \to 0$. This we do by establishing an upper bound for $I(z)$ by the use of Eq. (7).

Using the deformation of contour principle, we first replace C by a small circle C_1 of radius ρ and centered at z (see Fig. 7.11). If we take Δz small enough so that $|\Delta z| < \rho$, the second inequality in (11) in Chapter 6 yields

$$|\zeta - z - \Delta z| \geq |\, |\zeta - z| - |\Delta z| \,|$$

$$\geq \rho - |\Delta z|, \quad |\Delta z| < \rho.$$

If we assume $|f(\zeta)| \leq M$ on C_1, then the integral in (10) satisfies the inequality

$$|I(z)| = \left| \frac{\Delta z}{2\pi i} \oint_{C_1} \frac{f(\zeta)}{(\zeta - z - \Delta z)(\zeta - z)^2} d\zeta \right|$$

$$\leq \frac{|\Delta z| M \rho}{(\rho - |\Delta z|)\rho^2}$$

$$\leq \frac{|\Delta z| M}{(\rho - |\Delta z|)\rho}.$$

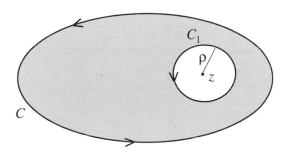

Figure 7.11

Thus, we see that $|I(z)| \to 0$ as $\Delta z \to 0$, so that Eq. (9) is now rigorously established.

By continued formal differentiation of (9), we obtain the additional results

$$f''(z) = \frac{2}{2\pi i} \oint_C \frac{f(\zeta)}{(\zeta - z)^3} d\zeta,$$

$$f'''(z) = \frac{3 \cdot 2}{2\pi i} \oint_C \frac{f(\zeta)}{(\zeta - z)^4} d\zeta,$$

whereas, in general $[f^{(n)}(z) = d^n f/dz^n]$,

$$f^{(n)}(z) = \frac{n!}{2\pi i} \oint_C \frac{f(\zeta)}{(\zeta - z)^{n+1}} d\zeta, \quad n = 0, 1, 2, \dots . \tag{11}$$

Of course, we have only formally derived Eq. (11). To rigorously prove it, we need to use an argument similar to that used in arriving at the validity of (9) combined with mathematical induction. Nonetheless, Eq. (11) is a correct and important result that is summed up in the following theorems.

Theorem 7.8 If $f(z)$ is analytic at a point z_0, then $f(z)$ has derivatives of all orders at z_0, each of which is again an analytic function at z_0.

Theorem 7.9 (*Cauchy's generalized integral formula*) If $f(z)$ is analytic on and inside a simple closed contour C, and z_0 is a point inside C, then

$$\oint_C \frac{f(z)}{(z - z_0)^{n+1}} dz = \frac{2\pi i}{n!} f^{(n)}(z_0), \quad n = 0, 1, 2, \dots .$$

EXAMPLE 7 Evaluate $I = \oint_C \frac{e^{2z}}{(z + 1)^4} dz$, where $C: |z| = 3$.

Solution: The singularity of the integrand occurs at $z = -1$, which is enclosed by the contour C. Hence, Theorem 7.9 is satisfied with $n = 3$, so that

$$\oint_C \frac{e^{2z}}{(z + 1)^4} dz = \frac{2\pi i}{3!} \frac{d^3}{dz^3} (e^{2z}) \Big|_{z = -1}$$

$$= \frac{\pi i}{3} (8 e^{2z}) \Big|_{z = -1},$$

or

$$\oint_C \frac{e^{2z}}{(z+1)^4} dz = \frac{8\pi i}{3e^2}.$$

From the above results concerning derivatives of analytic functions, we can now prove the converse of Cauchy's theorem, known as *Morera's theorem.*

Theorem 7.10 (*Morera's theorem*) If $f(z)$ is continuous in a simply-connected domain D; and if

$$\oint_C f(z)dz = 0$$

for all simple closed contours C in D, then $f(z)$ is analytic in D.

Proof: Consider any point z_0 inside D. Because $\oint_C f(z)dz = 0$ for all closed paths in D, it follows that the integral $\int_{z_0}^z f(\zeta)d\zeta$ is independent of the path in D. Hence, we can write

$$\int_{z_0}^z f(\zeta)d\zeta = F(z),$$

where $F(z)$ is analytic in D and $F'(z) = f(z)$. Consequently, because the derivative of an analytic function is also analytic, we see that $f(z)$ is analytic in D. ❑

7.3.5 Bounds for analytic functions

It is sometimes useful to obtain upper bounds on the modulii of analytic functions and their derivatives. Through use of the generalized formula of Cauchy, we can obtain the following theorem, also known as *Cauchy's inequality.*

Theorem 7.11 (*Cauchy's inequality*) Let $f(z)$ be analytic on and inside a circle C_R of radius R and centered at z_0. If $|f(z)| \leq M$ for all $z \in C_R$, then the derivatives of $f(z)$ evaluated at $z = z_0$ satisfy

$$\left| f^{(n)}(z_0) \right| \leq \frac{Mn!}{R^n}, \quad n = 1,2,3,\dots .$$

Proof: Based on Cauchy's generalized integral formula, it follows that

$$f^{(n)}(z_0) = \frac{n!}{2\pi i} \oint_C \frac{f(z)}{(z - z_0)^{n+1}} dz, \quad n = 1,2,3,\dots .$$

On the contour C_R we see that $|z - z_0| = R$ and $|f(z)| \leq M$ by hypothesis. Thus,

$$\left| f^{(n)}(z_0) \right| = \frac{n!}{2\pi} \left| \oint_C \frac{f(z)}{(z - z_0)^{n+1}} dz \right|$$

$$\leq \frac{n!}{2\pi} \oint_C \frac{|f(z)|}{|z - z_0|^{n+1}} |dz|$$

$$\leq \frac{n!}{2\pi} \cdot \frac{M}{R^{n+1}} \cdot 2\pi R, \quad n = 1,2,3,\dots ,$$

where we have recalled Eq. (7). Upon simplification, we obtain

$$\left| f^{(n)}(z_0) \right| \leq \frac{Mn!}{R^n}, \quad n = 1,2,3,\dots .$$

❏

There are a number of interesting results concerning bounded functions that are consequences of Theorem 7.11. In particular, we offer the following two theorems.

> **Theorem 7.12** (*Liouville's theorem*) If $f(z)$ is entire and bounded everywhere, then $f(z)$ is constant.

Proof: By assumption we have $|f(z)| \leq M$ for all z. From Cauchy's inequality (Theorem 7.11), we see that $|f'(z_0)| \leq M/R$, which is true for every R in this case (entire function). Hence, we can take R as large as we want and conclude that $f'(z_0) = 0$. Because z_0 is arbitrary, it follows that $f'(z_0) = 0$ for every z_0, and consequently, $f(z)$ is constant.❏

> **Theorem 7.13** (*Maximum-minimum modulus theorem*) If $f(z)$ is analytic and non-constant on and inside a simple closed contour C, then the maximum value of $|f(z)|$ occurs on C. If also, $f(z) \neq 0$ inside C, then the minimum of $|f(z)|$ occurs on C.

See Section 12.5.3 for further discussion of Theorem 7.13 and its consequences on the solution of Laplace's equation.

7.4 Infinite Series

Infinite series that are studied in calculus are important in almost all areas of pure and applied mathematics. In addition to other uses, they are used to define functions and to calculate accurate numerical estimates of the values of transcendental functions. Our treatment below will mostly review basic concepts from calculus associated with infinite sequences and series of *real* numbers and functions, but we now permit the numbers and functions to take on *complex* values.

7.4.1 Sequences and series

The ordered arrangement of complex numbers

$$S_1, S_2, \dots, S_n, \dots$$

is called an *infinite sequence* with *general term* S_n. If it should happen that

$$\lim_{n \to \infty} S_n = S, \tag{12}$$

where $|S|$ is finite, the sequence is said to *converge* to S, and is otherwise said to *diverge*.

An *infinite series* of (complex) constants has the form

$$z_1 + z_2 + \cdots + z_n + \cdots = \sum_{n=1}^{\infty} z_n. \tag{13}$$

In this case, z_n is called the *general term* of the series. Closely associated with the series (13) is a particular sequence defined by

$$\begin{aligned} S_1 &= z_1, \\ S_2 &= z_1 + z_2, \\ &\vdots \\ S_n &= z_1 + z_2 + \cdots + z_n = \sum_{k=1}^{n} z_k, \end{aligned} \tag{14}$$

called the *sequence of partial sums*. If the sequence of partial sums (14) converges to a finite limit S, we say the infinite series (13) *converges*, or sums, to S. The series *diverges* when the limit of partial sums fails to exist.

Because we can write a complex series in the form

$$\sum_{n=1}^{\infty} z_n = \sum_{n=1}^{\infty} x_n + i \sum_{n=1}^{\infty} y_n, \tag{15}$$

it becomes clear that the complex series (13) converges if and only if the real series in (15) each converge. Hence, all of the usual tests of convergence for real series carry over to complex series. In particular, if $\sum z_n$ converges, then $z_n \to 0$ as $n \to \infty$. Finally, if the series $\sum_{n=1}^{\infty} |z_n|$ converges, we say that the series $\sum_{n=1}^{\infty} z_n$ *converges absolutely*.

▸ **Geometric series:** For complex number z, the particular series

$$1 + z + z^2 + \cdots + z^n + \cdots = \sum_{n=0}^{\infty} z^n \qquad (16)$$

is called a *geometric series*.

The geometric series is one of the simplest and yet most important of all series. By considering the limit of partial sums, it can be shown that

$$\sum_{n=0}^{\infty} z^n = \frac{1}{1-z}, \quad |z| < 1. \qquad (17)$$

EXAMPLE 8 Sum the geometric series: $\displaystyle\sum_{n=0}^{\infty} \left(\frac{i}{3}\right)^n$.

Solution: Because $|i/3| = 1/3 < 1$, it follows that the series converges. Moreover, from Eq. (17) we see that

$$\sum_{n=0}^{\infty} \left(\frac{i}{3}\right)^n = \frac{1}{1 - i/3} = \frac{9}{10}\left(1 + \frac{i}{3}\right).$$

7.4.2 Power series

Of special interest to us are those series that arise when the general term $g_n(z)$ is a function of z. In this case the resulting series has the form $\sum_{n=0}^{\infty} g_n(z)$, and the general question is: For what values of z does this series have a sum, if any? Those values constitute the domain of a function defined by the series. To prove convergence, we first fix z and then rely on the convergence definition of series of constants. In particular, we define the nth partial sum by

$$S_n(z) = \sum_{k=0}^{n} g_k(z). \qquad (18)$$

▸ **Pointwise convergence:** If, for a fixed value of z, it happens that

$$\lim_{n \to \infty} S_n(z) = f(z), \qquad (19)$$

we say the series $\sum_{n=0}^{\infty} g_n(z)$ *converges pointwise* to $f(z)$.

▸ **Uniform convergence:** If, for $\varepsilon > 0$, there exists a real number N, independent of z (but depending on ε), such that

$$\left| S_n(z) - f(z) \right| < \varepsilon$$

for all z in some domain D and for all $n > N$, we say that $\sum_{n=0}^{\infty} g_n(z)$ *converges uniformly* to $f(z)$ in D.

Uniform convergence is a stronger type of convergence than pointwise convergence, and is what we need to perform operations like termwise differentiation and integration.

The most commonly used test for establishing uniform convergence of an infinite series is the *Weierstrass M-test*, which we state below without proof.

Theorem 7.14 (*Weierstrasss M-test*) If $\sum_{n=0}^{\infty} M_n$ is a convergent series of real positive constants such that $\left| g_n(z) \right| \le M_n$ $(n = 1, 2, 3, ...)$ for all z in some domain D, then the series $\sum_{n=0}^{\infty} g_n(z)$ is uniformly (and absolutely) convergent in D.

Some immediate consequences of the uniform convergence of an infinite series are contained in the following theorem.

Theorem 7.15 If each function $g_n(z)$, $n = 1, 2, 3, ...$, is continuous in a domain D and the series

$$f(z) = \sum_{n=1}^{\infty} g_n(z)$$

is uniformly convergent in D, then
- $f(z)$ is a continuous function.
- the integral of $f(z)$ along any contour C of finite length in D is

$$\int_C f(z) dz = \sum_{n=1}^{\infty} \int_C g_n(z) dz.$$

- If, in addition, each $g_n'(z)$, $n = 1, 2, 3, ...$, is a continuous function in D, and $\sum_{n=1}^{\infty} g_n(z)$ and $\sum_{n=1}^{\infty} g_n'(z)$ are uniformly convergent series in D, then

$$f'(z) = \sum_{n=1}^{\infty} g_n'(z).$$

▸ **Power series:**   A *power series* is one of the general form

$$A_0 + A_1(z - z_0) + \cdots + A_n(z - z_0)^n + \cdots = \sum_{n=0}^{\infty} A_n(z - z_0)^n. \tag{20}$$

Power series have a number of important properties that are stated below as theorems.

Theorem 7.16 Every power series has a radius of convergence ρ such that the series converges absolutely whenever $|z - z_0| < \rho$ and diverges whenever $|z - z_0| > \rho$. If the series converges for all z, then $\rho = \infty$, but if the series converges only at $z = z_0$, then $\rho = 0$.

Notice that Theorem 7.16 provides no information about convergence when $|z - z_0| = \rho \neq 0$. Such values of z must be analyzed independently to determine convergence. Geometrically, the curve $|z - z_0| = \rho$ in the complex plane is a circle of radius ρ that is centered at z_0. Hence, the power series converges for all z within this *circle of convergence* and diverges for all z outside the circle.

Theorem 7.17 A power series converges uniformly and absolutely in any domain which lies entirely within the circle of convergence.

This last theorem shows that power series have the desired property of uniform convergence everywhere within their circle of convergence. In addition, it can be shown that power series always converge to analytic functions (Theorem 7.18) and, moreover, analytic functions always have power series representations (Theorem 7.19).

Theorem 7.18 If a power series $f(z) = \sum_{n=0}^{\infty} A_n(z - z_0)^n$ converges within some circle of convergence it converges to an analytic function in the circle.

Proof: We first note that if the power series $f(z) = \sum_{n=0}^{\infty} A_n(z - z_0)^n$ converges within any domain D containing the contour C, then, as a consequence of Theorem 7.15, $f(z)$ is continuous in D and

$$\int_C f(z)\,dz = \sum_{n=0}^{\infty} A_n \int_C (z - z_0)^n\,dz.$$

In particular, if C is any closed contour within the circle of convergence of the series, then it follows that $\oint_C (z - z_0)^n\,dz = 0$, $n = 0, 1, 2, \ldots$ (recall Example 2) and, therefore,

$$\oint_C f(z)\,dz = 0.$$

Hence, based on Morera's theorem (Theorem 7.10) we deduce that $f(z)$ is analytic in the circle of convergence, and the theorem is proved. ❏

To prove the converse of Theorem 7.18, let C be a circle of radius ρ centered at z_0 which lies entirely within some domain D for which $f(z)$ is an analytic function (see Fig. 7.12). Then, a valid representation for $f(z)$ inside C is given by the Cauchy integral formula

$$f(z) = \frac{1}{2\pi i} \oint_C \frac{f(\zeta)}{\zeta - z} d\zeta, \tag{21}$$

where C: $\ |\zeta - z_0| = \rho$ and $|z - z_0| < \rho$.

Now consider the identity

$$\frac{1}{\zeta - z} = \frac{1}{(\zeta - z_0) - (z - z_0)} = \frac{1}{(\zeta - z_0)\left[1 - \left(\dfrac{z - z_0}{\zeta - z_0}\right)\right]}. \tag{23}$$

Because $|z - z_0| < |\zeta - z_0|$, we can use the geometric series representation

$$\frac{1}{1 - u} = \sum_{n=0}^{\infty} u^n, \quad |u| < 1 \tag{24}$$

to rewrite the identity (22) as

$$\frac{1}{\zeta - z} = \frac{1}{\zeta - z_0} \sum_{n=0}^{\infty} \left(\frac{z - z_0}{\zeta - z_0}\right)^n, \quad \left|\frac{z - z_0}{\zeta - z_0}\right| < 1, \tag{25}$$

and therefore Eq. (21) can be rewritten in the form

$$f(z) = \frac{1}{2\pi i} \oint_C \frac{f(\zeta)}{\zeta - z_0} \sum_{n=0}^{\infty} \left(\frac{z - z_0}{\zeta - z_0}\right)^n d\zeta. \tag{26}$$

The geometric series in (25) converges uniformly and, thus, we can interchange the order of summation and integration to get

$$f(z) = \frac{1}{2\pi i} \sum_{n=0}^{\infty} (z - z_0)^n \oint_C \frac{f(\zeta)}{(\zeta - z_0)^{n+1}} d\zeta = \sum_{n=0}^{\infty} A_n (z - z_0)^n, \tag{27}$$

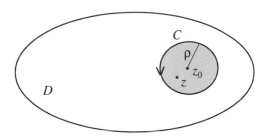

Figure 7.12 Domain D where $f(z)$ is analytic.

where

$$A_n = \frac{1}{2\pi i} \oint_C \frac{f(\zeta)}{(\zeta - z_0)^{n+1}} d\zeta = \frac{f^{(n)}(z_0)}{n!}, \quad n = 0, 1, 2, \dots \ . \tag{28}$$

The last step in (27) is a consequence of the generalized Cauchy integral formula.

In summary, we have the following result known as *Taylor's theorem*.

Theorem 7.19 (*Taylor's theorem*) If $f(z)$ is analytic in a domain D and z_0 is a point inside D, then there exists precisely one power series with center at z_0 that represents $f(z)$, viz.,

$$f(z) = \sum_{n=0}^{\infty} A_n (z - z_0)^n,$$

where $A_n = f^{(n)}(z_0)/n!$, $n = 0, 1, 2, \dots \ .$

Taylor's theorem says that, given any analytic function $f(z)$, a power series representation is possible about any point where $f(z)$ is *analytic*. The special case of Taylor's theorem with $z_0 = 0$ leads to a *Maclaurin series*. Most of the series we encounter in practice are of the Maclaurin variety.

The radius of convergence of a Taylor (or Maclaurin) series can always be found by applying the ratio test (or root test). However, this is usually not necessary since the radius of convergence is always the shortest distance between the center z_0 of the series and the nearest *singularity* of $f(z)$. [Recall that a singularity is a point where $f(z)$ is *not* analytic. See also Section 7.4.4.]

For example, the function

$$f(z) = \frac{1}{1 - z}$$

has a singularity at $z = 1$. If we expand $f(z)$ in a Taylor series about $z_0 = -i$, the radius of convergence will be given by $\rho = |1 - (-i)| = \sqrt{2}$. To actually find its Taylor series, we can use the technique utilized in the proof of Taylor's theorem. Namely, we start with the identity

$$\frac{1}{1 - z} = \frac{1}{(1 + i) - (z + i)} = \frac{1}{(1 + i)\left[1 - \left(\dfrac{z + i}{1 + i}\right)\right]},$$

and, through properties of a geometric series, deduce that

$$\frac{1}{1 - z} = \frac{1}{1 + i} \sum_{n=0}^{\infty} \left(\frac{z + i}{1 + i}\right)^n, \quad |z + i| < \sqrt{2}.$$

Of course, if a function is entire, its power series has an infinite radius of convergence about any point.

By application of Taylor's theorem (Theorem 7.19), we can readily derive the following well-known power series representations:

$$
\begin{aligned}
&(a) \quad e^z = \sum_{n=0}^{\infty} \frac{z^n}{n!}, \quad |z| < \infty \\[2mm]
&(b) \quad \cos z = \sum_{n=0}^{\infty} \frac{(-1)^n z^{2n}}{(2n)!}, \quad |z| < \infty \\[2mm]
&(c) \quad \sin z = \sum_{n=0}^{\infty} \frac{(-1)^n z^{2n+1}}{(2n+1)!}, \quad |z| < \infty
\end{aligned}
\tag{28}
$$

$$
\begin{aligned}
&(a) \quad \frac{1}{1-z} = \sum_{n=0}^{\infty} z^n, \quad |z| < 1 \\[2mm]
&(b) \quad \frac{1}{z} = \sum_{n=0}^{\infty} (-1)^n (z-1)^n, \quad |z-1| < 1 \\[2mm]
&(c) \quad \operatorname{Log} z = \sum_{n=0}^{\infty} \frac{(-1)^n}{n+1} (z-1)^{n+1}, \quad |z-1| < 1
\end{aligned}
\tag{29}
$$

Once we have derived several such series as those above, other related series can often be found through formal operations on these known series. To illustrate the generation of series from known series, we can replace z with z^2 in (28a) to find

$$
e^{z^2} = \sum_{n=0}^{\infty} \frac{z^{2n}}{n!}, \quad |z| < \infty.
\tag{30}
$$

Similarly, by replacing z with $-z$ in (29a) yields

$$
\frac{1}{1+z} = \sum_{n=0}^{\infty} (-1)^n z^n, \quad |z| < 1.
\tag{31}
$$

Last, by writing

$$
\frac{1}{2+4z} = \frac{1/2}{1+2z} = \frac{1}{2} \sum_{n=0}^{\infty} (-1)^n (2z)^n, \quad |2z| < 1,
$$

we deduce that

$$\frac{1}{2+4z} = \sum_{n=0}^{\infty} (-1)^n 2^{n-1} z^n$$

$$= \frac{1}{2} - z + 2z^2 - 4z^3 + \dots, \quad |z| < \frac{1}{2}. \tag{32}$$

Another method of producing new series from known series involves either differentiating or integrating the known series. For instance, observe that

$$\frac{d}{dz}\left(\frac{1}{1-z}\right) = \sum_{n=0}^{\infty} \frac{d}{dz}(z^n),$$

or

$$\frac{1}{(1-z)^2} = \sum_{n=1}^{\infty} n z^{n-1} = \sum_{n=0}^{\infty} (n+1)z^n, \quad |z| < 1. \tag{33}$$

Also, by integrating the series (31) termwise from $z = 0$ to any point z within the circle of convergence, we see that

$$\text{Log}(1+z) = \sum_{n=0}^{\infty} \frac{(-1)^n}{n+1} z^{n+1}, \quad |z| < 1. \tag{34}$$

EXAMPLE 9 Find the Maclaurin series for $f(z) = \tan^{-1} z$.

Solution: We first note that

$$f'(z) = \frac{1}{1+z^2} = \sum_{n=0}^{\infty} (-1)^n z^{2n}, \quad |z| < 1,$$

where the series is deduced from (31) by replacing z with z^2. If we now integrate this series termwise, we have (using the dummy variable ζ)

$$\int_0^z \frac{1}{1+\zeta^2} d\zeta = \sum_{n=0}^{\infty} (-1)^n \int_0^z \zeta^{2n} d\zeta,$$

from which we deduce

$$\tan^{-1} z = \sum_{n=0}^{\infty} \frac{(-1)^n}{2n+1} z^{2n+1}, \quad |z| < 1.$$

7.4.3 Laurent series

Taylor series provide valid representations of functions which are analytic at some point $z = z_0$ and throughout a circle centered at z_0. In many applications it is necessary to expand functions around points at which (or near which) these functions are not analytic. Such series, called *Laurent series*,[2] are distinguished by the fact they include both positive and negative powers of $z - z_0$. Laurent series, therefore, furnish us with a series representation for a function $f(z)$ in the annular ring between two concentric circles (see Fig. 7.13), provided $f(z)$ is analytic everywhere on and between the two circles.

Let us consider the annular domain D between the concentric circles C_1 and C_2 shown in Fig. 7.13. We assume z_0 is the center of the circles and that the respective radii are ρ_1 and ρ_2, with $\rho_1 < \rho_2$. Suppose $f(z)$ is analytic everywhere on the circles and in the annular domain D between them, but not necessarily inside C_1. Consider a contour consisting of the two circles and the straight line segment AB connecting them as shown in Fig. 7.13. If z is any point in the domain D, it follows from Cauchy's integral formula (Theorem 7.7) that

$$
\begin{aligned}
f(z) &= \frac{1}{2\pi i}\oint_{C_2}\frac{f(\zeta)}{\zeta - z}\,d\zeta + \frac{1}{2\pi i}\int_A^B\frac{f(\zeta)}{\zeta - z}\,d\zeta \\
&\quad - \frac{1}{2\pi i}\oint_{C_1}\frac{f(\zeta)}{\zeta - z}\,d\zeta - \frac{1}{2\pi i}\int_A^B\frac{f(\zeta)}{\zeta - z}\,d\zeta \\
&= \frac{1}{2\pi i}\oint_{C_2}\frac{f(\zeta)}{\zeta - z}\,d\zeta - \frac{1}{2\pi i}\oint_{C_1}\frac{f(\zeta)}{\zeta - z}\,d\zeta.
\end{aligned}
\tag{35}
$$

Because $|z - z_0| < |\zeta - z_0|$, where $\zeta \in C_2$, it follows from using (24) that the first integral in (35) reduces to

$$
A_n = \frac{1}{2\pi i}\oint_{C_2}\frac{f(\zeta)}{(\zeta - z_0)^{n+1}}\,d\zeta = \frac{1}{2\pi i}\oint_C\frac{f(\zeta)}{(\zeta - z_0)^{n+1}}\,d\zeta, \quad n = 0, 1, 2, \ldots ,
\tag{36}
$$

In the last step, we interchanged the order of integration and summation, and defined where C is any simple closed contour in the domain D enclosing the point z_0. To evaluate the second integral in (35), we first observe that the integrand can be written as

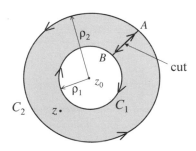

Figure 7.13 Annular domain for Laurent series about z_0.

[2]Named for the French mathematician Hermann Laurent (1841-1908).

$$-\frac{f(\zeta)}{\zeta - z} = \frac{f(\zeta)}{(z - z_0) - (\zeta - z_0)} = \frac{f(\zeta)}{(z - z_0)\left[1 - \left(\dfrac{\zeta - z_0}{z - z_0}\right)\right]}.$$

But, along the contour C_1 we note that $|\zeta - z_0| < |z - z_0|$, so we can once again use the geometric series to write

$$-\frac{f(\zeta)}{\zeta - z} = \frac{f(\zeta)}{z - z_0}\sum_{n=0}^{\infty}\left(\frac{\zeta - z_0}{z - z_0}\right)^n = f(\zeta)\sum_{n=1}^{\infty}\frac{(\zeta - z_0)^{n-1}}{(z - z_0)^n}, \quad |z - z_0| > \rho_1. \tag{37}$$

Hence, the second integral in (35) leads to

$$-\frac{1}{2\pi i}\oint_{C_1}\frac{f(\zeta)}{\zeta - z}d\zeta = \frac{1}{2\pi i}\oint_{C_1}f(\zeta)\sum_{n=1}^{\infty}\frac{(\zeta - z_0)^{n-1}}{(z - z_0)^n}d\zeta$$

$$= \sum_{n=1}^{\infty}B_n(z - z_0)^{-n}, \quad \rho_1 < |z - z_0| < \rho_2, \tag{38}$$

where we have interchanged the order of integration and summation, and defined

$$B_n = \frac{1}{2\pi i}\oint_{C_1}\frac{f(\zeta)}{(\zeta - z_0)^{-n+1}}d\zeta = \frac{1}{2\pi i}\oint_{C}\frac{f(\zeta)}{(\zeta - z_0)^{-n+1}}d\zeta, \quad n = 1, 2, 3, \dots. \tag{39}$$

In the last integral, C can be the same contour as that in (37), but this is not necessary. By further comparing (40) with (37), we recognize the relationship $B_n = A_{-n}$ for $n = 1, 2, 3, \dots$ Thus, by combining results, we have

$$f(z) = \sum_{n=0}^{\infty}A_n(z - z_0)^n + \sum_{n=1}^{\infty}A_{-n}(z - z_0)^{-n},$$

which is usually written more briefly as

$$\boxed{f(z) = \sum_{n=-\infty}^{\infty}A_n(z - z_0)^n, \quad \rho_1 < |z - z_0| < \rho_2,} \tag{40}$$

obtained from the change of index $n \to -n$ in the second summation above.

Series like (41) featuring both positive and negative powers of $z - z_0$ are generalizations of Taylor series called *Laurent series*. Such series will converge for all z lying in the annular domain D shown in Fig. 7.13. Also, because the values of A_n and A_{-n} are independent of the paths of integration, the path C can be any path around z_0 between the contours C_1 and C_2. In summary, we have the following theorem.

Theorem 7.20 (*Laurent's theorem*) If $f(z)$ is analytic in the annular domain defined by $\rho_1 < |z - z_0| < \rho_2$, then $f(z)$ can be represented by the convergent series

$$f(z) = \sum_{n=-\infty}^{\infty} A_n (z - z_0)^n, \quad \rho_1 < |z - z_0| < \rho_2.$$

The coefficients of the series are defined by

$$A_n = \frac{1}{2\pi i} \oint_C \frac{f(z)}{(z - z_0)^{n+1}} dz, \quad n = 0, \pm 1, \pm 2, \ldots,$$

where C is any simple closed contour lying in D and containing z_0 in its interior.

Based on the above theorem, we deduce that $f(z)$ can be expressed as the sum of two functions, i.e.,

$$f(z) = f_1(z) + f_2(z),$$

with $f_1(z)$ analytic on and inside the circle C_2 in Fig. 7.13 and $f_2(z)$ analytic outside the circle C_1. The sum is then analytic in the annular domain between C_1 and C_2. Although we won't prove it, it can be shown that the domain of convergence of the Laurent series of a given function $f(z)$ is the broadest annular domain about z_0 that contains no singular points of $f(z)$.

We should point out that, unlike Taylor series, the integral in Theorem 7.20 for the coefficients A_n *does not* reduce to the derivative relation $A_n = f^{(n)}(z_0)/n!$. Although these coefficients can, in principle, be calculated by use of the defining integral in Theorem 7.20, this formula is rarely used in practice. Mostly, we use various manipulations involving known series and techniques such as partial fractions to obtain the desired series from known series. Some of these techniques are illustrated in the following examples.

EXAMPLE 10 Find the Laurent series for $f(z) = z^2 e^{1/z}$ about the singular point $z = 0$.

Solution: Using the known series [see Eq. (28)]

$$e^z = \sum_{n=0}^{\infty} \frac{z^n}{n!}, \quad |z| < \infty$$

we can simply replace z by $1/z$ to obtain

$$z^2 e^{1/z} = z^2 \sum_{n=0}^{\infty} \frac{(1/z)^n}{n!}$$

$$= z^2 + z + \frac{1}{2!} + \frac{1}{3!z} + \frac{1}{4!z^2} + \cdots, \quad |z| > 0.$$

The domain of convergence is therefore all $|z| > 0$.

EXAMPLE 11 Find all Laurent series about the singular points of

$$f(z) = \frac{1}{z^2(z-1)}.$$

Solution: The singularities of $f(z)$ are clearly given by $z = 0$ and $z = 1$. Thus, centered at the singular point $z_0 = 0$ there are two possible Laurent series—one valid for $0 < |z| < 1$ and one for $|z| > 1$ (see Fig. 7.14).

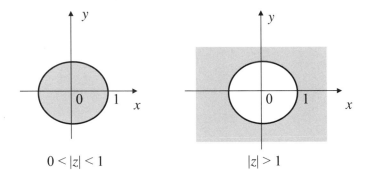

$$0 < |z| < 1 \qquad\qquad |z| > 1$$

Figure 7.14 Domains for Laurent series about $z_0 = 0$.

In the domain $0 < |z| < 1$ we can use the geometric series for $1/(1-u)$ to write

$$\frac{1}{z^2(z-1)} = -\frac{1}{z^2}\sum_{n=0}^{\infty} z^n = -\sum_{n=0}^{\infty} z^{n-2}.$$

By making the change of index $n \to n+2$, we then obtain

$$f(z) = \frac{1}{z^2(z-1)} = -\sum_{n=-2}^{\infty} z^n$$

$$= -\frac{1}{z^2} - \frac{1}{z} - 1 - z - z^2 - \cdots, \quad 0 < |z| < 1.$$

For the second domain in Fig. 7.14, we use a similar approach to write

$$\frac{1}{z^2(z-1)} = \frac{1}{z^3(1-1/z)} = \frac{1}{z^3}\sum_{n=0}^{\infty}\left(\frac{1}{z}\right)^n,$$

which, following the index change $n \to -(n+3)$, we can write in the form

$$f(z) = \frac{1}{z^2(z-1)} = \sum_{n=-\infty}^{-3} z^n, \quad |z| > 1.$$

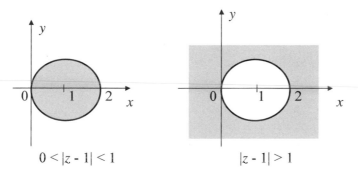

Figure 7.15 Domains for Laurent series about $z_0 = 1$.

At the other singular point $z = 1$, there are again two possible Laurent series corresponding to the domains $0 < |z - 1| < 1$ and $|z - 1| > 1$ as illustrated above in Fig. 7.15. In this case we begin by writing the given function as a function of $u = z - 1$, which leads to

$$\frac{1}{z^2(z - 1)} = \frac{1}{(u + 1)^2 u}.$$

Then, we proceed as before to find $(0 < |u| < 1)$

$$\frac{1}{z^2(z - 1)} = \frac{1}{u(1 + u)^2} = \frac{1}{u} \sum_{n=0}^{\infty} (-1)^n (n + 1)u^n \Bigg|_{u = z - 1}$$

$$= \frac{1}{z - 1} \sum_{n=0}^{\infty} (-1)^n (n + 1)(z - 1)^n,$$

or

$$\frac{1}{z^2(z - 1)} = \sum_{n=0}^{\infty} (-1)^n (n + 1)(z - 1)^{n-1}.$$

In arriving at the series above, we have used the result of Eq. (33). Making the index change $n \to n + 1$, we now obtain the more standard form

$$f(z) = \frac{1}{z^2(z - 1)} = \sum_{n=-1}^{\infty} (-1)^{n+1} (n + 2)(z - 1)^n, \quad 0 < |z - 1| < 1.$$

Last, for $|z - 1| > 1$, we write $(|u| > 1)$

$$\frac{1}{z^2(z - 1)} = \frac{1}{u^3(1 + 1/u)^2} = \frac{1}{u^3} \sum_{n=0}^{\infty} (-1)^n (n + 1)u^{-n} \Bigg|_{u = z - 1}$$

$$= \frac{1}{(z - 1)^3} \sum_{n=0}^{\infty} (-1)^n (n + 1)(z - 1)^{-n}$$

$$= \sum_{n=0}^{\infty} (-1)^n (n + 1)(z - 1)^{-n-3},$$

or, after some algebraic manipulation,

$$f(z) = \frac{1}{z^2(z-1)} = \sum_{n=-\infty}^{-3} (-1)^n (n+2)(z-1)^n, \quad |z-1| > 1.$$

As mentioned above, we don't ordinarily resort to the integral definition of A_n given in Theorem 7.20, but sometimes it is useful to do so. Consider the next example.

EXAMPLE 12 Find a Laurent series about $z_0 = 1$ for the function

$$f(z) = \frac{e^{2z}}{(z-1)^3}.$$

Solution: Since the only singularity occurs at $z = 1$, we seek a series of the form

$$f(z) = \frac{e^{2z}}{(z-1)^3} = \sum_{n=-\infty}^{\infty} A_n (z-1)^n, \quad 0 < |z-1| < \infty,$$

where

$$A_n = \frac{1}{2\pi i} \oint_C \frac{f(z)}{(z-1)^{n+1}} dz = \frac{1}{2\pi i} \oint_C \frac{e^{2z}}{(z-1)^{n+4}} dz, \quad n = 0, \pm 1, \pm 2, \ldots,$$

and where C is any simple closed curve around the point $z = 1$.

To evaluate the above integral, we first note that for $n \leq -4$, the integral vanishes as a consequence of Cauchy's theorem, i.e.,

$$A_n = 0, \quad n = -4, -5, -6, \ldots.$$

For $n \geq -3$, we can use the generalized Cauchy integral formula to obtain

$$A_n = \frac{1}{(n+3)!} \frac{d^{n+3}}{dz^{n+3}} (e^{2z}) \Big|_{z=1} = \frac{e^2 2^{n+3}}{(n+3)!}, \quad n \geq -3.$$

Hence, the desired Laurent series is

$$f(z) = \frac{e^{2z}}{(z-1)^3} = 8e^2 \sum_{n=-3}^{\infty} \frac{2^n}{(n+3)!} (z-1)^n, \quad 0 < |z-1| < \infty.$$

7.4.4 Zeros and singularities

One of the firsthand uses of Laurent series is in the classification of various types of *singularities* of analytic functions. We start, however, with the notion of *zero*.

‣ **Zero:** If $f(z)$ is analytic at a point z_0 and $f(z_0) = 0$, we say z_0 is a *zero* of $f(z)$. We say z_0 is a *zero of order m* if $f(z)$ is analytic at z_0 and

$$f(z_0) = f'(z_0) = f''(z_0) = \cdots = f^{(m-1)}(z_0) = 0, \quad f^{(m)}(z_0) \neq 0. \tag{42}$$

For example, the function $f(z) = z^2$ has a zero of *order two* at $z = 0$ because $f(0) = f'(0) = 0$, whereas $f''(0) = 2 \neq 0$.

‣ **Isolated singular point:** If $f(z)$ is analytic throughout any deleted neighborhood $0 < |z - z_0| < \rho$ of a given singular point z_0, then z_0 is called an *isolated singularity* of $f(z)$.

We previously defined a "singular point" z_0 as a point where $f(z)$ is not analytic. More precisely, we find in the isolated singularity case, the function $f(z)$ has a Laurent series about z_0 that converges in the deleted neighborhood $0 < |z - z_0| < \rho$. However, there are *three* possible types of behavior of $f(z)$ in the neighborhood of an isolated singularity z_0, which are further classified by characteristics of the Laurent series for $f(z)$.

Definition 7.1 If $f(z)$ has the Laurent series representation

$$f(z) = \sum_{n=-\infty}^{\infty} A_n(z - z_0)^n, \quad 0 < |z - z_0| < \rho,$$

where z_0 is an isolated singularity of $f(z)$, then we say that z_0 is:

‣   a *removable singularity* if $A_n = 0$ for all $n < 0$.
‣   a *pole of order m* if $A_{-m} \neq 0$, but $A_n = 0$ for all $n < -m$.
‣   an *essential singularity* if $A_n \neq 0$ for an infinite number of negative values of n.

Remark: If a function $f(z)$ has only a finite number of singular points in a domain, they are necessarily isolated singularities.

Removable singularities of a function $f(z)$ are distinguished by the absence of negative powers in its Laurent series about the singular point. The most familiar example of a function with this type of singularity is given by

$$f(z) = \frac{\sin z}{z}.$$

Clearly, $z = 0$ is a singular point of this function because $f(0) = (\sin 0)/0$ is not defined. However, by using the Taylor series for $\sin z$, we immediately see that

$$\frac{\sin z}{z} = \frac{1}{z}\left(z - \frac{z^3}{3!} + \frac{z^5}{5!} - \frac{z^7}{7!} + \cdots\right)$$

$$= 1 - \frac{z^2}{3!} + \frac{z^4}{5!} - \frac{z^6}{7!} + \cdots,$$

and so, based on Definition 7.1, we see that $z = 0$ is indeed a removable singularity. That is, the singularity can be "removed" by defining the function so that it is analytic at the singular point. In the above example, we do this by simply defining $f(0) = 1$ to obtain the related function

$$g(z) = \begin{cases} \dfrac{\sin z}{z}, & z \neq 0 \\ 1, & z = 0. \end{cases}$$

Clearly, the function $g(z)$ is analytic for all z and coincides with the function $f(z)$ above for $z \neq 0$. Because we can always do this with removable singularities, they are of little concern in our work. Mostly, we have to distinguish them from poles, which is what they often appear to be without analysis.

Based on Definition 7.1, the Laurent series for a function with a *pole* of order m at z_0 appears like

$$f(z) = \sum_{n=-m}^{\infty} A_n (z - z_0)^n$$

$$= \frac{A_{-m}}{(z - z_0)^m} + \frac{A_{-m+1}}{(z - z_0)^{m-1}} + \cdots + \frac{A_{-1}}{z - z_0} + A_0 + A_1(z - z_0) + \cdots, \tag{43}$$

where $A_{-m} \neq 0$. A pole of order one is also called a *simple pole*. The types of functions that commonly have poles are *rational functions* and some *trigonometric functions*. In the case of rational functions, the poles are simply the zeros of its denominator polynomial.

To determine the order of a pole in general when its Laurent series is not available, we rely on the following theorem.

Theorem 7.21 If $f(z)$ is not finite at some point $z = z_0$, but for a positive integer m the product function $(z - z_0)^m f(z)$ has a removable singularity at $z = z_0$, then $f(z)$ has a *pole of order m* at z_0, provided m is the smallest integer for which this is true.

Proof: Because $(z - z_0)^m f(z)$ has a removable singularity at $z = z_0$, it follows that

$$(z - z_0)^m f(z) = a_0 + a_1(z - z_0) + a_2(z - z_0)^2 + \cdots.$$

Dividing this result by $(z - z_0)^m$ leads to

$$f(z) = \frac{a_0}{(z - z_0)^m} + \frac{a_1}{(z - z_0)^{m-1}} + \frac{a_2}{(z - z_0)^{m-2}} + \cdots,$$

which, by comparison with (43), identifies $z = z_0$ as a pole of order m. ◻

EXAMPLE 13 Classify the poles of

$$f(z) = \frac{z - 2}{(z^2 + 1)(z - 1)^3}.$$

Solution: By inspection of the denominator, we see that the poles are located at $z = \pm i$ and $z = 1$. Both $z = \pm i$ are simple poles (because they are factors one time each), whereas $z = 1$ is a pole of order 3.

EXAMPLE 14 Classify the singularities of

$$f(z) = \frac{1}{(2 \sin z - 1)^2}.$$

Solution: Clearly, the singularities of $f(z)$ are located at

$$z = \begin{cases} \dfrac{\pi}{6} + 2k\pi, & k = 0, \pm 1, \pm 2, \ldots \\ \dfrac{5\pi}{6} + 2k\pi, & k = 0, \pm 1, \pm 2, \ldots \end{cases}.$$

To show that these are all poles of order 2, we use L'Hôpital's rule. For $z = \pi/6$, for example, we find

$$\lim_{z \to \pi/6} \frac{(z - \pi/6)^2}{(2 \sin z - 1)^2} = \lim_{z \to \pi/6} \frac{2(z - \pi/6)}{4 \cos z (2 \sin z - 1)}$$

$$= \lim_{z \to \pi/6} \left(\frac{z - \pi/6}{2 \sin z - 1} \right) \frac{1}{2 \cos z}$$

$$= \frac{1}{2 \cos \pi/6} \lim_{z \to \pi/6} \frac{1}{2 \cos z}$$

$$= \left(\frac{1}{2 \cos \pi/6} \right)^2,$$

which is finite, but nonzero. A similar treatment shows that $(z - \pi/6)f(z)$ still has a singularity, so we deduce that this is indeed a pole of order 2. The analysis is virtually the same for all other singular points.

Zeros and poles are related in that if $f(z)$ has a zero of order m at $z = z_0$, then the function $1/f(z)$ has a pole of order m at $z = z_0$. Conversely, if $f(z)$ has a pole of order m at $z = z_0$, then the function $1/f(z)$ has a zero of order m at $z = z_0$.

According to Definition 7.1, an *essential singularity* of a function $f(z)$ is characterized by an infinite number of negative powers in its Laurent series about the singularity. For instance, by replacing z with $1/z$ in the power series for e^z, we see that

$$e^{1/z} = \sum_{n=0}^{\infty} \frac{1}{z^n n!} = 1 + \frac{1}{z} + \frac{1}{2z^2} + \frac{1}{6z^3} + \cdots, \quad |z| > 0.$$

Clearly, $z = 0$ is an essential singularity of $e^{1/z}$, because the above series representation has an infinite number of negative powers of the form $1/z^n$, $n \geq 1$. Basically, essential singular points are the isolated singular points that are neither poles nor removable singularities. The behavior of an analytic function in the neighborhood of an essential singularity is quite different from that of a pole. An important result concerning the behavior of a function near an essential singularity is the following theorem due to Picard, which we state without proof.

Theorem 7.22 (*Picard's theorem*) In every neighborhood of an essential singular point the function $f(z)$ takes on every conceivable finite value, with one possible exception, an infinite number of times.

Note, for example, that $\left| e^{1/z} \right| \neq 0$ for *any* value of z, so zero is the exceptional value for this function alluded to in Picard's theorem.

Based on the above discussion concerning essential singular points, we see that a convenient test for a pole at z_0 is verification that

$$\lim_{z \to z_0} |f(z)| = \infty.$$

That is, as a consequence of Picard's theorem, such a limit cannot be true for an essential singularity. Using the concept of limit, therefore, we can identify the three types of isolated singularities according to the limit behavior cited in the following theorem.

Theorem 7.23 An isolated singularity z_0 of the function $f(z)$ is said to be

- a *removable singularity* if $\lim\limits_{z \to z_0} f(z)$ exists (finitely).

- a *pole* if $\lim\limits_{z \to z_0} |f(z)| = \infty$.

- an *essential singularity* if $\lim\limits_{z \to z_0} f(z)$ does not exist (finitely or infinitely).

7.5 Residue Theory

Thus far, we have shown that Laurent series are useful in classifying singularities. In the following sections we will illustrate that Laurent series can also be very useful in the evaluation of certain kinds of complex integrals as well as certain real integrals that could not be evaluated by conventional techniques from calculus.

7.5.1 Residues

When Cauchy's integral theorem is extended to functions having isolated singularities, the value of the integral in general is not zero. In fact, each singularity of the integrand contributes a term, called a "residue," to the total value of the integral.

Another way to introduce the notion of residue is to examine the Laurent series of a function $f(z)$. That is, let $f(z)$ be analytic on and inside a circle C except at an interior point $z = z_0$. Then, as we have seen in Section 7.4, the function $f(z)$ has a Laurent series in a deleted neighborhood about $z = z_0$ given by

$$f(z) = \sum_{n=-\infty}^{\infty} A_n (z - z_0)^n, \quad 0 < |z - z_0| < \rho,$$

$$A_n = \frac{1}{2\pi i} \oint_C \frac{f(z)}{(z - z_0)^{n+1}} dz, \quad n = 0, \pm 1, \pm 2, \dots ,$$

(44)

where C lies within the deleted neighborhood of convergence. Suppose we formally integrate the function $f(z)$ around the circle C by termwise integration of the series (44). This action leads to

$$\oint_C f(z) dz = \sum_{n=-\infty}^{\infty} A_n \oint_C (z - z_0)^n dz,$$

(45)

which, by use of the result of Example 2, i.e.,

$$\oint_C (z - z_0)^n dz = \begin{cases} 0, & n \neq -1 \\ 2\pi i, & n = -1, \end{cases}$$

reduces (45) to

$$\oint_C f(z) dz = 2\pi i A_{-1}.$$

(46)

▸ **Residue:** The coefficient A_{-1} is called the *residue* of $f(z)$ at the point $z = z_0$. Standard notation that we also use for residue is $A_{-1} \equiv \mathrm{Res}\{f(z); z_0\}$, or, when there is no confusion $A_{-1} \equiv \mathrm{Res}(z_0)$.

To obtain the residue of a function $f(z)$ at $z = z_0$, we can (theoretically) always examine its Laurent series (44) to find the value of A_{-1}. For example, if $f(z) = z^2 e^{1/z}$, then the Laurent series about the singularity $z = 0$ is (see Example 3)

$$z^2 e^{1/z} = z^2 \sum_{n=0}^{\infty} \frac{(1/z)^n}{n!}$$

$$= z^2 + z + \frac{1}{2!} + \frac{1}{3!z} + \frac{1}{4!z^2} + \cdots , \quad |z| > 0.$$

The residue of this function at $z = 0$ is the coefficient of $1/z$, viz., $\text{Res}(0) = 1/3! = 1/6$.

Finding the Laurent series as in the above example is not always an easy task, so other means of calculating residues are desired. In the case where the isolated singularity $z = z_0$ is a pole of order m, we can derive a simple formula for calculating the residue. Suppose first that $f(z)$ has a *simple pole* at $z = z_0$. Its Laurent series is therefore of the form

$$f(z) = \frac{A_{-1}}{z - z_0} + A_0 + A_1(z - z_0) + A_2(z - z_0)^2 + \cdots .$$

Let us multiply both sides of this expression by $(z - z_0)$ to obtain

$$(z - z_0)f(z) = A_{-1} + A_0(z - z_0) + A_1(z - z_0)^2 + A_2(z - z_0)^3 + \cdots .$$

Hence, in the limit $z \to z_0$, we see that

$$\lim_{z \to z_0} (z - z_0)f(z) = A_{-1},$$

and deduce that, at a simple pole,

$$\boxed{\text{Res}\{f(z); z_0\} = \lim_{z \to z_0} (z - z_0)f(z).} \qquad (47)$$

EXAMPLE 15 Calculate the residues at the poles of

$$f(z) = \frac{e^{-2z}}{z(z + 3)}.$$

Solution: By inspection, we see that $f(z)$ has simple poles at $z = 0$ and $z = -3$. Therefore, from (47) we have

$$\text{Res}\{f(z); 0\} = \lim_{z \to 0} zf(z) = \lim_{z \to 0} \frac{e^{-2z}}{z + 3} = \frac{1}{3},$$

and, similarly,

$$\text{Res}\{f(z); -3\} = \lim_{z \to -3} (z + 3)f(z) = \lim_{z \to -3} \frac{e^{-2z}}{z} = -\frac{e^6}{3}.$$

As a consequence of Eq. (47), we can derive another formula for calculating residues at a simple pole $z = z_0$ when the function $f(z)$ has the quotient form $f(z) = P(z)/Q(z)$, where $P(z)$ and $Q(z)$ are both analytic at $z = z_0$, $Q(z)$ has a *simple zero* at $z = z_0$, but $P(z_0) \neq 0$. Then, from Eq. (47) we see that the residue

$$\text{Res}\{f(z); z_0\} = \lim_{z \to z_0} (z - z_0)\frac{P(z)}{Q(z)} = P(z_0) \lim_{z \to z_0} \frac{(z - z_0)}{Q(z)}$$

leads to the indeterminate form 0/0. By an application of L'Hôpital's rule, we obtain

$$\boxed{(z = z_0 \text{ is a simple pole}): \quad \text{Res}\left\{\frac{P(z)}{Q(z)}; z_0\right\} = \frac{P(z_0)}{Q'(z_0)}.} \tag{48}$$

EXAMPLE 16 Calculate the residue at each singular point of

$$f(z) = \frac{\sin z}{z^2 + 4}.$$

Solution: Because $z^2 + 4 = (z - 2i)(z + 2i)$, we see that $f(z)$ has simple poles at $z = \pm 2i$. Therefore, using (48), we obtain

$$\text{Res}\{f(z); 2i\} = \frac{\sin z}{2z}\bigg|_{z=2i} = \frac{\sin 2i}{4i},$$

which simplifies to

$$\text{Res}\{f(z); 2i\} = \frac{1}{4} \sinh 2.$$

Similarly,

$$\text{Res}\{f(z); -2i\} = \frac{\sin z}{2z}\bigg|_{z=-2i} = \frac{\sin(-2i)}{(-4i)},$$

or,

$$\text{Res}\{f(z); -2i\} = \frac{1}{4} \sinh 2.$$

For calculating the residue of a function $f(z)$ at a *pole of order m*, we use the following theorem.

Theorem 7.24 If $f(z)$ has a pole of order m at $z = z_0$, then

$$\text{Res}\{f(z); z_0\} = \lim_{z \to z_0} \frac{1}{(m-1)!} \frac{d^{m-1}}{dz^{m-1}} \left[(z - z_0)^m f(z)\right].$$

Proof: Given that $f(z)$ has a pole of order m at $z = z_0$, its Laurent series has the form

$$f(z) = \frac{A_{-m}}{(z - z_0)^m} + \frac{A_{-m+1}}{(z - z_0)^{m-1}} + \cdots$$
$$+ \frac{A_{-2}}{(z - z_0)^2} + \frac{A_{-1}}{z - z_0} + A_0 + A_1(z - z_0) + \cdots .$$

Thus,

$$(z - z_0)^m f(z) = A_{-m} + A_{-m+1}(z - z_0) + \cdots + A_{-2}(z - z_0)^{m-2}$$
$$+ A_{-1}(z - z_0)^{m-1} + A_0(z - z_0)^m + A_1(z - z_0)^{m+1} + \cdots ,$$

and by differentiating both sides $(m - 1)$ times, we obtain

$$\frac{d^{m-1}}{dz^{m-1}}\left[(z - z_0)^m f(z)\right] = 0 + 0 + \cdots + (m-1)! A_{-1}$$
$$+ m! A_0(z - z_0) + \frac{(m+1)!}{2!} A_1(z - z_0)^2 + \cdots ,$$

where all terms preceding A_{-1} are zero. In the limit $z \to z_0$, we then obtain

$$\frac{d^{m-1}}{dz^{m-1}}\left[(z - z_0)^m f(z)\right]\Bigg|_{z = z_0} = (m-1)! A_{-1},$$

which is what we needed to prove. ❑

EXAMPLE 17 Calculate the residue at the singular points of

$$f(z) = \frac{\sin z}{z^4}.$$

Solution: If we write

$$f(z) = \left(\frac{\sin z}{z}\right) \frac{1}{z^3},$$

the first factor has a removable singularity at $z = 0$ and we conclude the function $f(z)$

has a pole of order 3 at $z = 0$. Hence, using Theorem 7.23, we see that

$$\text{Res}\,\{f(z); 0\} = \frac{1}{2!} \lim_{z \to 0} \frac{d^2}{dz^2} \left(\frac{\sin z}{z} \right)$$

$$= \frac{1}{2} \lim_{z \to 0} \left(\frac{2 \sin z - z^2 \sin z - 2z \cos z}{z^3} \right),$$

from which we deduce, through application of L'Hôpital's rule,

$$\text{Res}\,\{f(z); 0\} = -\frac{1}{3!} = -\frac{1}{6}.$$

Of course, in this case we can more easily find the residue by inspection of the $1/z$ coefficient of the Laurent series

$$\frac{\sin z}{z^4} = \frac{1}{z^3} - \frac{1}{3!z} + \frac{1}{5!}z - \frac{1}{7!}z^2 + \cdots .$$

Thus far we have shown that if $f(z)$ has a singularity at $z = z_0$ inside the simple closed contour C, then

$$\oint_C f(z)dz = 2\pi i \,\text{Res}\,\{f(z); z_0\},$$

when only one singularity of $f(z)$ is contained inside the simple closed contour C. For the more general case where C encloses a finite number of isolated singularities at $z = a_k$, $k = 1, 2, 3, ..., n$, we can use the deformation of contour principle to write

$$\oint_C f(z)dz = \sum_{k=1}^{n} \oint_{C_k} f(z)dz,$$

where each simple closed contour C_k encloses only one of the singularities. From this observation, we deduce the important *residue theorem*.

Theorem 7.25 (*Residue theorem*) If $f(z)$ is analytic on and inside a simple closed contour C, except for finitely many isolated singular points $a_1, a_2, ..., a_n$ inside C, then

$$\oint_C f(z)dz = 2\pi i \sum_{k=1}^{n} \text{Res}\,\{f(z); a_k\}.$$

EXAMPLE 18 Evaluate the integral

$$I = \oint_C \frac{dz}{z(z-1)^2(z+3)}, \quad C: |z| = 2.$$

Solution: The function $f(z) = 1/z(z-1)^2(z+3)$ has a simple pole at $z = 0$ and a pole of order 2 at $z = 1$ inside the contour C. The pole at $z = -3$ lies outside C. Hence, from the residue theorem, we see that

$$I = 2\pi i[\text{Res}(0) + \text{Res}(1)].$$

Calculating residues, we obtain

$$\text{Res}(0) = \frac{1}{(z-1)^2(z+3)}\bigg|_{z=0} = \frac{1}{3},$$

$$\text{Res}(1) = \lim_{z \to 1} \frac{d}{dz}\left[(z-1)^2 f(z)\right]$$

$$= -\frac{2z+3}{z^2(z+3)^2}\bigg|_{z=1} = -\frac{5}{16},$$

and thus, our result is

$$\oint_C \frac{dz}{z(z-1)^2(z+3)} = 2\pi i\left(\frac{1}{3} - \frac{5}{16}\right) = \frac{\pi i}{24}.$$

To summarize our treatment of complex integrals, let us define the integral

$$I = \int_C f(z)\,dz.$$

- If $f(z)$ is *analytic* in a bounded domain containing a simple contour C (not closed), then I can be evaluated by indefinite integration.

- If C is a *simple closed contour* and $f(z)$ has *no singular points* on C or its interior, then $I = 0$.

- If C is a *simple closed contour* and $f(z)$ has a *finite number of isolated singularities* in the interior of C, some of which may be essential singularities, then the *residue theorem* must be used to evaluate the integral I.

7.6 Evaluation of Real Integrals—Part I

The residue theorem (Theorem 7.25) is a powerful tool in the evaluation of various types of real integrals. In this section we will consider several classes of real integrals that are either difficult or impossible to evaluate by conventional methods from calculus.

7.6.1 Rational functions of cos θ and/or sin θ

To begin, let us consider integrals of the general form

$$I = \int_0^{2\pi} F(\cos\theta, \sin\theta)\, d\theta, \tag{49}$$

where $F(\cos\theta, \sin\theta)$ is a *rational function* of $\cos\theta$ and/or $\sin\theta$. Such integrals can be evaluated by methods of elementary calculus, but the calculus of residues is often much easier to employ. Our approach here is to convert (49) into a contour integral about the unit circle $|z| = 1$.

If we start with the substitution

$$z = e^{i\theta}, \quad 0 \le \theta < 2\pi, \tag{50}$$

we see that

$$\cos\theta = \frac{1}{2}\left(z + \frac{1}{z}\right), \quad \sin\theta = \frac{1}{2i}\left(z - \frac{1}{z}\right), \quad d\theta = \frac{dz}{iz}, \tag{51}$$

where the differential relation follows from $dz = ie^{i\theta}\,d\theta$. As θ ranges in values from 0 to 2π, the variable $z = e^{i\theta}$ traverses the unit circle $|z| = 1$ in the counterclockwise direction (see Fig. 7.16). The integrand of I will then end up a rational function $f(z) = P(z)/Q(z)$, which has either removable singularities (these can be ignored in the evaluation of integrals) or poles. Consequently, by the residue theorem, the integral I is $2\pi i$ times the sum of the residues at the poles of $f(z) = P(z)/Q(z)$ *inside* $|z| = 1$. If one of the poles occurs on the unit circle, however, the method will fail. Thus, the integral (49) becomes

$$I = \oint_{|z|=1} f(z)\,dz = 2\pi i \sum \text{Res}\{f(z);\ \text{all poles inside } |z| = 1\}. \tag{52}$$

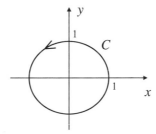

Figure 7.16 Contour of integration for the integral (52).

EXAMPLE 19 Evaluate the integral

$$I = \int_0^{2\pi} \frac{d\theta}{5 + 4\sin\theta}.$$

Solution: Based on Eqs. (50) and (51), we have $d\theta = dz/iz$ and

$$5 + 4\sin\theta = 5 + \frac{2}{i}\left(z - \frac{1}{z}\right) = \frac{2z^2 + 5iz - 2}{iz}.$$

Hence, in the complex plane our integral becomes

$$I = \oint_{|z|=1}\left(\frac{iz}{2z^2 + 5iz - 2}\right)\frac{dz}{iz} = \oint_{|z|=1}\left(\frac{dz}{2z^2 + 5iz - 2}\right).$$

By setting $2z^2 + 5iz - 2 = 0$, we find simple poles at $z = -2i$ and $z = -i/2$, which identifies $z = -i/2$ as the only singular point inside $|z| = 1$. Based on the residue theorem, it follows that

$$I = 2\pi i\operatorname{Res}(-i/2) = \frac{1}{4z + 5i}\bigg|_{z=-i/2} = 2\pi i\left(\frac{1}{3i}\right),$$

and thus,

$$\int_0^{2\pi}\frac{d\theta}{5 + 4\sin\theta} = \frac{2\pi}{3}.$$

EXAMPLE 20 Evaluate the integral

$$I = \int_0^{\pi}\frac{d\theta}{2 - \cos\theta}.$$

Solution: By recognizing that the cosine is an *even function*, it follows that

$$I = \frac{1}{2}\int_{-\pi}^{\pi}\frac{d\theta}{2 - \cos\theta}.$$

Because we now have a 2π interval, the substitution $z = e^{i\theta}$ will map the segment $-\pi < \theta \leq \pi$ into the complete unit circle. If we utilize the relations (50) and (51), the integral I transforms to

$$I = \frac{1}{2}\oint_{|z|=1}\left[\frac{1}{2 - \frac{1}{2}(z + 1/z)}\right]\frac{dz}{iz} = i\oint_{|z|=1}\frac{dz}{z^2 - 4z + 1}.$$

From the quadratic formula, we find the zeros of the denominator in the last integral to be $z = 2 \pm \sqrt{3}$, with $z = 2 - \sqrt{3}$ the only one inside $|z| = 1$. The singularity $z = 2 - \sqrt{3}$ is a simple pole of the integrand, which leads to

$$\text{Res}(2 - \sqrt{3}) = \left. \frac{1}{2z - 4} \right|_{z = 2 - \sqrt{3}} = -\frac{1}{2\sqrt{3}},$$

and consequently,

$$\int_0^\pi \frac{d\theta}{2 - \cos\theta} = i(2\pi i)\left(-\frac{1}{2\sqrt{3}} \right) = \frac{\pi}{\sqrt{3}}.$$

In some cases, we need to make the transformation on trigonometric functions of a more general argument, such as double angles, etc. For these situations, we use the more general formulas

$$\cos n\theta = \frac{1}{2}\left(e^{in\theta} + e^{-in\theta} \right) = \frac{1}{2}\left(z^n + \frac{1}{z^n} \right), \quad n = 1, 2, 3, \ldots$$

$$\sin n\theta = \frac{1}{2i}\left(e^{in\theta} - e^{-in\theta} \right) = \frac{1}{2i}\left(z^n + \frac{1}{z^n} \right), \quad n = 1, 2, 3, \ldots .$$

(53)

7.6.2 Improper integrals of rational functions

We now turn our attention to improper integrals of the types

$$\int_0^\infty f(x)\,dx, \quad \int_{-\infty}^\infty f(x)\,dx,$$

where $f(x)$ is a real function that is continuous on the interval of integration. These are classified as *improper integrals* because of the infinite limits of integration. If such integrals exist, they are defined by the limits

$$\int_0^\infty f(x)\,dx = \lim_{b \to \infty} \int_0^b f(x)\,dx,$$

$$\int_{-\infty}^\infty f(x)\,dx = \lim_{a \to -\infty} \int_a^0 f(x)\,dx + \lim_{b \to \infty} \int_0^b f(x)\,dx.$$

(54)

In place of the second expression in (54), we sometimes define the value of the improper integral by the limit

$$\int_{-\infty}^\infty f(x)\,dx = \lim_{R \to \infty} \int_{-R}^R f(x)\,dx,$$

called the *Cauchy principal value* of the integral. We usually designate this limiting procedure by the special notation

$$PV\int_{-\infty}^{\infty} f(x)dx = \lim_{R\to\infty} \int_{-R}^{R} f(x)dx. \qquad (55)$$

If the limit of the second expression in (54) exists, it always equals its principal value given by (55). However, it may happen that the principal value (55) exists but one or both of the limits in the second expression in (54) fail to exist. For example,

$$PV\int_{-\infty}^{\infty} x\,dx = 0,$$

whereas the integral does not converge in the usual sense.

In this section we will be interested in how the theory of residues can be used to calculate the principal values of certain types of integrals. Specifically, we will initially restrict the function $f(x)$ to those with the following characteristics:

- $f(x) = P(x)/Q(x)$, where $P(x)$ and $Q(x)$ are polynomials
- $Q(x)$ is at least *two* degrees higher than $P(x)$
- $Q(x) \neq 0$ on the interval $-\infty < x < \infty$

To help, we will need the following definition and theorem.

Definition 7.2 We say that $g(z)$ *tends to zero uniformly* on a circular arc C_R of radius R as $R\to\infty$ (or as $R\to 0$) if and only if $|g(z)| \le M_R$ along C_R and $M_R\to 0$ as $R\to\infty$ (or $R\to 0$).

Theorem 7.26 If, on a circular arc C_R of radius R and center at $z = 0$, we have $zf(z)\to 0$ uniformly as $R\to\infty$, then

$$\lim_{R\to\infty} \int_{C_R} f(z)dz = 0.$$

Proof: Let C_R be a circular arc with subtended angle α (see Fig. 7.17). Based on Definition 7.2 with $g(z) = zf(z)$, there exists a bound M_R such that $|zf(z)| \le M_R$, or

$$|f(z)| \le \frac{M_R}{R}.$$

The length of the circular arc C_R is $L = \alpha R$ so that

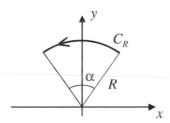

Figure 7.17

$$\left| \int_{C_R} f(z)dz \right| \leq \frac{M_R}{R} \alpha R = \alpha M_R.$$

Because $zf(z) \to 0$ uniformly as $R \to \infty$, it follows that $M_R \to 0$ as $R \to \infty$, and we get our intended result. ❏

Let $f(z)$ be a complex rational function satisfying the three conditions imposed above upon the real function $f(x)$. We then consider the integral $\oint_C f(z)dz$ around the simple closed path illustrated in Fig. 7.18. Because $f(z)$ is a rational function, it has finitely many poles in the upper half-plane (U.H.P.) denoted by $a_1, a_2, ..., a_n$. Thus, by choosing the radius R sufficiently large, the closed path C, consisting of the semicircular arc C_R and the straight line segment from $z = -R$ to $z = R$, will enclose all such poles. Based on the residue theorem, we then have

$$\oint_C f(z)dz = 2\pi i \sum_{k=1}^{n} \text{Res} \{f(z); a_k \text{ in U.H.P.}\}. \tag{56}$$

We now need to relate (56) to the real integral $\text{PV} \int_{-\infty}^{\infty} f(x)dx$. By breaking C up into the straight line segment along the x-axis and the semicircular arc C_R, we can write

$$\oint_C f(z)dz = \int_{-R}^{R} f(x)dx + \int_{C_R} f(z)dz. \tag{57}$$

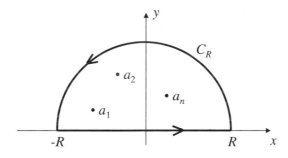

Figure 7.18

Next, we investigate what happens to these integrals as the radius R becomes unbounded. In particular, we wish to show that the contour integral around C_R vanishes in this limit.

Because $f(z) = P(z)/Q(z)$, where the denominator is at least two degrees higher than the numerator, it follows that

$$|zf(z)| \leq \left| Kz \frac{z^n + \cdots}{z^{n+2} + \cdots} \right| \leq \frac{|K|}{R}, \quad R \gg 1, \tag{58}$$

where K is some constant. Thus, $zf(z) \to 0$ as $R \to \infty$, and by Theorem 7.26 we have

$$\lim_{R \to \infty} \int_{C_R} f(z)\,dz = 0. \tag{59}$$

Hence, by combining (57) and (59), we deduce in the limit

$$\text{PV} \int_{-\infty}^{\infty} f(x)\,dx = 2\pi i \sum_{k=1}^{n} \text{Res}\,\{f(z); a_k \text{ in U.H.P.}\}. \tag{60}$$

Remark: If $f(z)$ is an *odd function*, then

$$\text{PV} \int_{-\infty}^{\infty} f(x)\,dx = 0,$$

whereas if $f(z)$ is an *even function*,

$$\text{PV} \int_{-\infty}^{\infty} f(x)\,dx = 2 \int_{0}^{\infty} f(x)\,dx.$$

EXAMPLE 21 Evaluate the integral

$$I = \int_{-\infty}^{\infty} \frac{dx}{x^4 + 1}.$$

Solution: We first observe that $f(x) = 1/(x^4 + 1)$ is a rational function with the denominator at least two degrees greater than the numerator. Also the related complex function $f(z) = 1/(z^4 + 1)$ has simple poles at

$$a_{k+1} = (-1)^{1/4} = \exp\left[\frac{(2k+1)\pi i}{4} \right], \quad k = 0, 1, 2, 3,$$

or, explicitly,

$$a_1 = e^{\pi i/4} = (1 + i)/\sqrt{2}$$
$$a_2 = e^{3\pi i/4} = (-1 + i)/\sqrt{2}$$
$$a_3 = e^{5\pi i/4} = (-1 - i)/\sqrt{2}$$
$$a_4 = e^{7\pi i/4} = (1 - i)/\sqrt{2}.$$

Clearly, the only poles in the U.H.P. are $z = a_1$ and $z = a_2$; thus,

$$\text{PV} \int_{-\infty}^{\infty} \frac{dx}{x^4 + 1} = 2\pi i \left[\text{Res}(a_1) + \text{Res}(a_2) \right].$$

By the use of (48), the residues at these poles are

$$\text{Res}(a_1) = \frac{1}{4z^3} \bigg|_{z = e^{\pi i/4}} = \frac{1}{4} e^{-3\pi i/4} = -\frac{\sqrt{2}}{8}(1 + i),$$

$$\text{Res}(a_2) = \frac{1}{4z^3} \bigg|_{z = e^{3\pi i/4}} = \frac{1}{4} e^{-\pi i/4} = \frac{\sqrt{2}}{8}(1 - i),$$

which leads to

$$\text{PV} \int_{-\infty}^{\infty} \frac{dx}{x^4 + 1} = 2\pi i \left(-i \frac{\sqrt{2}}{4} \right) = \frac{\pi\sqrt{2}}{2}.$$

7.6.3 Fourier transform integrals

The technique illustrated in Section 7.6.2 can readily be extended to more general integrands such as those that arise in Fourier transform integrals (see Chapter 9). Such integrals usually appear as one of the following types:

$$\int_{-\infty}^{\infty} f(x) \cos mx \, dx, \qquad \int_{-\infty}^{\infty} f(x) \sin mx \, dx, \qquad \int_{-\infty}^{\infty} f(x) e^{imx} \, dx, \quad (m > 0). \tag{61}$$

Each of the integrals depicted in (61) is an improper integral where the parameter m is taken as a positive real number. Once again we assume $f(x)$ is a rational function, i.e., $f(x) = P(x)/Q(x)$, but now the polynomial $Q(x)$ need only be *one* degree higher than the polynomial $P(x)$. In our analysis, we only treat the third integral in (61) because of the connection

$$\text{PV} \int_{-\infty}^{\infty} f(x) e^{imx} \, dx = \text{PV} \int_{-\infty}^{\infty} f(x) \cos mx \, dx + i \, \text{PV} \int_{-\infty}^{\infty} f(x) \sin mx \, dx. \tag{62}$$

Our approach to evaluating integrals of this type will closely parallel the method leading to Eq. (60). Here, however, we require the following theorem, which we state without proof.

Theorem 7.27 If, on a circular arc C_R of radius R and center at $z = 0$, we have $f(z) \to 0$ uniformly as $R \to \infty$, then

$$\lim_{R \to \infty} \int_{C_R} f(z) e^{imz} dz = 0, \quad m > 0,$$

provided that C_R is confined to the first and/or second quadrants.

Based on the same contour illustrated in Fig. 7.18, we have that

$$\oint_C f(z) e^{imz} dz = \int_{-R}^{R} f(x) e^{imx} dx + \int_{C_R} f(z) e^{imz} dz$$

$$= 2\pi i \sum_{k=1}^{n} \operatorname{Res} \{ f(z) e^{imz}; a_k \text{ in U.H.P.} \}, \tag{63}$$

where $a_1, a_2, \ldots, a_n$ denote the poles of $f(z)$ in the U.H.P. Using Theorem 7.27, it can be shown that the integral along the circular arc C_R in (63) vanishes in the limit $R \to \infty$, and hence,

$$\operatorname{PV} \int_{-\infty}^{\infty} f(x) e^{imx} dx = 2\pi i \sum_{k=1}^{n} \operatorname{Res} \{ f(z) e^{imz}; a_k \text{ in U.H.P.} \}. \tag{64}$$

By simply matching the real and imaginary parts of (64), we can obtain results for $\int_{-\infty}^{\infty} f(x) \cos mx \, dx$ and $\int_{-\infty}^{\infty} f(x) \sin mx \, dx$, respectively.

EXAMPLE 22 Evaluate the integral

$$I = \int_{0}^{\infty} \frac{\cos x}{x^2 + 1} dx.$$

Solution: Because the integrand is an even function, we can use the relation

$$I = \frac{1}{2} \int_{-\infty}^{\infty} \frac{\cos x}{x^2 + 1} dx = \frac{1}{2} \operatorname{Re} \int_{-\infty}^{\infty} \frac{e^{ix}}{x^2 + 1} dx.$$

Now, the only pole of $f(z) = 1/(z^2 + 1)$ in the U.H.P. is $z = i$, so by the use of Eq. (64), we have

$$PV\int_{-\infty}^{\infty}\frac{e^{ix}}{x^2+1}dx = 2\pi i\,\mathrm{Res}\left\{\frac{e^{iz}}{z^2+1};i\right\}.$$

Clearly, $z = i$ is a simple pole leading to the residue

$$\mathrm{Res}(i) = \left.\frac{e^{iz}}{2z}\right|_{z=i} = \frac{1}{2ei},$$

from which we obtain

$$\frac{1}{2}PV\int_{-\infty}^{\infty}\frac{e^{ix}}{x^2+1}dx = \frac{2\pi i}{2}\left(\frac{1}{2ei}\right) = \frac{\pi}{2e}.$$

Consequently,

$$\int_0^{\infty}\frac{\cos x}{x^2+1}dx = \frac{\pi}{2e}.$$

7.7 Evaluation of Real Integrals—Part II

In Sections 7.6.2 and 7.6.3 we limited our treatment of evaluating real integrals to those cases in which the function $f(x)$ was continuous over the whole interval of integration, i.e., the corresponding complex function $f(z)$ had no singularities on the real axis. In some cases, the presence of a pole or branch point along the contour of integration (generally the real axis) may necessitate a small indentation of the contour to avoid integration through the singularity. This technique, called the method of *indented contours*, is a small modification of the treatment presented in Section 7.6.2. Namely, we introduce an arc of a small circle of radius ρ around each such singularity and then consider the limit of the integral over this arc as $\rho \to 0$.

The indented contour technique illustrated below requires the use of the following two theorems which we state without proof.

Theorem 7.28 Let C_ρ denote a circular arc with radius ρ, center at $z = a$, and subtending an angle α (see Fig. 7.19). If $(z - a)f(z) \to 0$ uniformly on C_ρ as $\rho \to 0$, then

$$\lim_{\rho \to 0}\int_{C_\rho}f(z)dz = 0.$$

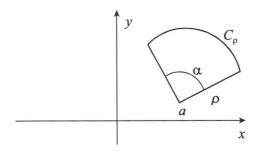

Figure 7.19 Circular arc for Theorems 7.28 and 7.29.

Theorem 7.29 Let C_ρ denote a circular arc with radius ρ, center at $z = a$, and subtending an angle α (see Fig. 7.19). If the function $f(z)$ has a simple pole at $z = a$, then

$$\lim_{\rho \to 0} \int_{C_\rho} f(z)dz = \pm\alpha \operatorname{Res}\{f(z); a\},$$

where the plus sign is used when integration is counterclockwise over C_ρ and the negative sign is used when integration is clockwise.

If the real function $f(x)$ is continuous on the entire real axis except at $x = b$, the Cauchy principal value of its integral over $(-\infty,\infty)$ is defined by

$$\mathrm{PV}\int_{-\infty}^{\infty} f(x)dx = \lim_{\substack{R \to \infty \\ \rho \to 0}} \left[\int_{-R}^{b-\rho} f(x)dx + \int_{b+\rho}^{R} f(x)dx \right], \tag{65}$$

provided these limits exist independently. We can extend this idea to the case of several discontinuities $b_1, b_2, ..., b_m$ occurring on the real axis.

Suppose we once again consider integrals of the form

$$\int_{-\infty}^{\infty} f(x)dx, \qquad \int_{-\infty}^{\infty} f(x)e^{imx}dx, \quad (m > 0),$$

in which $f(x) = P(x)/Q(x)$ is a rational function satisfying the conditions stated in Sections 7.6.2 and 7.6.3. Here, however, we include the case in which $Q(x)$ has *simple zeros* at the m points $b_1, b_2, ..., b_m$, all located on the real axis. These zeros correspond to *simple poles* of the complex function $f(z)$. As before, the function $f(z)$ may also contain n poles in the U.H.P. located at $z = a_1, a_2, ..., a_n$.

Let us consider the semicircular domain shown below in Fig. 7.20 where the radius R of the large semicircle is large enough so that all poles of $f(z)$ in the U.H.P. are contained in the domain and the interval $-R \le x \le R$ includes all poles of $f(z)$ on the real axis. In addition, we indent the contour around each of the poles on the real axis with a small-radius semicircle denoted by $C_1, C_2, ..., C_m$. By use of the residue theorem, we deduce that

$$\oint_C f(z)dz = 2\pi i \sum_{k=1}^{n} \text{Res}\{f(z); a_k \text{ in U.H.P.}\}, \tag{66}$$

where $C = C_R + L_1 + C_1 + L_2 + C_2 + ... + C_m + L_{m+1}$. If we now split the integral around the closed path C into its various subcontours, we obtain

$$\oint_C f(z)dz = \int_{C_R} f(z)dz + \sum_{k=1}^{m+1} \int_{L_k} f(x)dx + \sum_{k=1}^{m} \int_{C_k} f(z)dz.$$

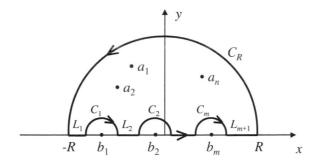

Figure 7.20

As $R \to \infty$, the integral along C_R tends to zero (Theorem 7.26), whereas the remaining integrals lead to

$$\lim_{\substack{\rho \to 0 \\ R \to \infty}} \left[\sum_{k=1}^{m+1} \int_{L_k} f(x)dx + \sum_{k=1}^{m} \int_{C_k} f(z)dz \right] = \int_{-\infty}^{\infty} f(x)dx - \pi i \sum_{k=1}^{m} \text{Res}\{f(z); b_k\}, \tag{67}$$

where we have used Theorem 7.29. Thus, by combining (66) and (67), we deduce that

$$\boxed{\text{PV} \int_{-\infty}^{\infty} f(x)dx = \pi i \sum_{k=1}^{m} \text{Res}\{f(z); b_k\} + 2\pi i \sum_{k=1}^{n} \text{Res}\{f(z); a_k\}.} \tag{68}$$

Following similar arguments, if the integrand is $f(x)e^{imx}$, we obtain

$$\boxed{\begin{aligned} \text{PV} \int_{-\infty}^{\infty} f(x)e^{imx}dx = &\pi i \sum_{k=1}^{m} \text{Res}\{f(z)e^{imz}; b_k\} \\ &+ 2\pi i \sum_{k=1}^{n} \text{Res}\{f(z)e^{imz}; a_k\}, \quad (m > 0). \end{aligned}} \tag{69}$$

EXAMPLE 23 Evaluate the integral

$$I = \int_0^\infty \frac{\sin x}{x}\,dx.$$

Solution:

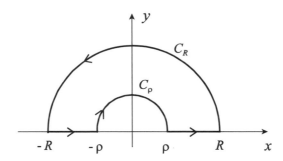

Figure 7.21

Because the integrand is an even function, we first observe that

$$I = \int_0^\infty \frac{\sin x}{x}\,dx = \frac{1}{2}\mathrm{PV}\int_{-\infty}^\infty \frac{\sin x}{x}\,dx.$$

Next, we integrate the complex function e^{iz}/z around the closed contour C shown in Fig. 7.21 where we have indented around the simple pole at $z = 0$. Inside this closed contour the function e^{iz}/z is analytic, so from Cauchy's theorem we see that

$$\oint_C \frac{e^{iz}}{z}\,dz = \int_{C_R} \frac{e^{iz}}{z}\,dz + \int_{-R}^{-\rho} \frac{e^{ix}}{x}\,dx + \int_{C_\rho} \frac{e^{iz}}{z}\,dz + \int_\rho^R \frac{e^{ix}}{x}\,dx = 0.$$

As $R \to \infty$ and $\rho \to 0$, we now deduce that

$$\mathrm{PV}\int_{-\infty}^\infty \frac{e^{ix}}{x}\,dx - \pi i\,\mathrm{Res}\left\{\frac{e^{iz}}{z};0\right\} = 0,$$

or

$$\mathrm{PV}\int_{-\infty}^\infty \frac{e^{ix}}{x}\,dx = \pi i.$$

Last, by comparing real and imaginary parts, we are led to

$$I = \frac{1}{2}\mathrm{PV}\left(\mathrm{Im}\int_{-\infty}^\infty \frac{e^{ix}}{x}\,dx\right) = \int_0^\infty \frac{\sin x}{x}\,dx = \frac{\pi}{2}.$$

As a final example here, let us consider the integral of a multiple-valued function. To do so, we need to modify our procedure by taking into account *branch points* and *branch cuts* along with isolated singularities.

EXAMPLE 24 Evaluate the integral

$$I = \int_0^\infty \frac{x^{m-1}}{1+x} dx, \quad 0 < m < 1.$$

Solution: Because of the restriction on the real parameter m, it follows that the complex function

$$f(z) = \frac{z^{m-1}}{1+z}$$

has a branch point at $z = 0$ as well as a simple pole at $z = -1$.

To evaluate the integral I, we integrate $f(z) = z^{m-1}/(1+z)$ around the closed contour C illustrated in Fig. 7.22, taking into account the branch cut along the positive real axis. Based on the residue theorem, we have

$$\oint_C \frac{z^{m-1}}{1+z} dz = 2\pi i \operatorname{Res}(-1).$$

That is, the simple pole at $z = -1$ is the only isolated singularity inside C. If we write $z = x$ along the upper boundary of the branch cut, then $z = xe^{2\pi i}$ along the lower boundary of the cut. Thus, the integral around C becomes

$$\oint_C f(z)dz = -\int_{C_\rho} f(z)dz + \int_\rho^R f(x)dx$$
$$+ \int_{C_R} f(z)dz + \int_R^\rho f(xe^{2\pi i})dx.$$

Based on Theorems 7.26 and 7.28, it follows that the integrals over C_R and C_ρ both tend to zero as $R \to \infty$, $\rho \to 0$, and therefore, the above result reduces to

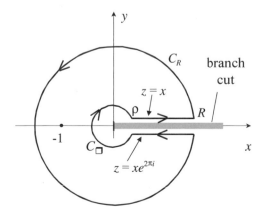

Figure 7.22

$$\oint_C \frac{z^{m-1}}{1+z}\,dz = \int_0^\infty \frac{x^{m-1}}{1+x}\,dx + \int_\infty^0 \frac{(xe^{2\pi i})^{m-1}}{1+x}\,dx = 2\pi i\,\mathrm{Res}(-1),$$

or, upon combining integrals,

$$(1 - e^{2m\pi i})\int_0^\infty \frac{x^{m-1}}{1+x}\,dx = 2\pi i\,\mathrm{Res}(-1).$$

For the simple pole, we find

$$\mathrm{Res}(-1) = \lim_{z \to -1} z^{m-1} = (-1)^{m-1}.$$

However,

$$(-1)^{m-1} = e^{i\pi(m-1)} = -e^{im\pi},$$

and, consequently,

$$\int_0^\infty \frac{x^{m-1}}{1+x}\,dx = \frac{-2\pi i\,e^{im\pi}}{1 - e^{2m\pi i}} = \frac{-2\pi i}{e^{-im\pi} - e^{im\pi}},$$

from which we deduce

$$\int_0^\infty \frac{x^{m-1}}{1+x}\,dx = \frac{\pi}{\sin m\pi}, \quad 0 < m < 1.$$

What makes problems like Example 24 more difficult than previous examples is that each such problem may dictate a different shaped contour to avoid the branch point and branch cut. A general discussion of such contours, however, is beyond the scope of our text.

7.8 Harmonic Functions Revisited

All analytic functions have real and imaginary components that are *harmonic functions*. Furthermore, there is a close connection between harmonic functions and two-dimensional physical problems involving heat conduction, fluid flow, and electrostatics. Such problems are frequently formulated as *Dirichlet problems* or *Neumann problems*, which constitute what are also called *boundary value problems of the first and second kinds* (see Section 12.5). In this section we will solve some of these classic problems in the *half-plane* and in a *circular domain*. Then, using properties of *conformal mappings* (Section 6.7.5), we can extend such results to other shaped domains.

We have previously shown (see Section 6.4.3) that the real and imaginary parts of an analytic function $f(z) = u(x,y) + i\,v(x,y)$ are harmonic functions; i.e.,

$$\frac{\partial^2 u}{\partial x^2} + \frac{\partial^2 u}{\partial y^2} = 0, \quad \frac{\partial^2 v}{\partial x^2} + \frac{\partial^2 v}{\partial y^2} = 0. \tag{70}$$

Also, Cauchy's integral formula

$$f(z) = \frac{1}{2\pi i} \oint_C \frac{f(\zeta)}{\zeta - z} d\zeta \tag{71}$$

gives values of the analytic function $f(z)$ at points interior to a closed contour C in terms of values of the function on the contour itself. The combination of these properties provides us with powerful tools for solving certain important boundary value problems formulated in the plane.

The *Dirichlet problem* is to find a function $\varphi(x,y)$ that is harmonic in a domain D and which takes on prescribed values on the boundary curve C of D. It is mathematically characterized by

$$\frac{\partial^2 \varphi}{\partial x^2} + \frac{\partial^2 \varphi}{\partial y^2} = 0 \quad \text{in } D,$$

$$\text{BC:} \quad \varphi = g \quad \text{on } C, \tag{72}$$

where g is a prescribed function. The other boundary value problem, known as the *Neumann problem*, is the same except that the normal derivative of φ is prescribed along the boundary curve C. Its mathematical formulation is given by

$$\frac{\partial^2 \varphi}{\partial x^2} + \frac{\partial^2 \varphi}{\partial y^2} = 0 \quad \text{in } D,$$

$$\text{BC:} \quad \frac{\partial \varphi}{\partial n} = g \quad \text{on } C. \tag{73}$$

Through application of the Cauchy integral formula, we can derive integral formulas for the solution of (73) and (74) when the domain D is either the half-plane $y > 0$ or a circle centered at the origin.

7.8.1 Harmonic functions in the half-plane

To begin, let us consider the semicircular domain in the ζ-plane, where $\zeta = \xi + i\eta$, as shown in Fig. 7.23. The boundary contour is $C = C_R + C_1$, where C_R is a semicircle of radius R centered at the origin and C_1 is the straight line segment from $\zeta = -R$ to $\zeta = R$. Parameterizations of these curves are given by

$$C_1: \quad \zeta(t) = t, \quad -R \le t \le R$$

$$C_R: \quad \zeta(t) = Re^{it}, \quad 0 < t < \pi. \tag{74}$$

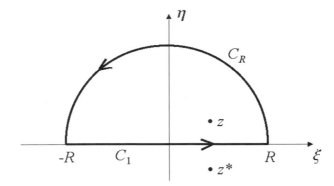

Figure 7.23 Semicircular domain in the ζ-plane.

If z is any point inside the contour C, then the point z^* is necessarily outside C (see Fig. 7.23). By Cauchy's theorems, we therefore deduce that

$$f(z) = \frac{1}{2\pi i} \oint_C \frac{f(\zeta)}{\zeta - z} d\zeta,$$

$$0 = \frac{1}{2\pi i} \oint_C \frac{f(\zeta)}{\zeta - z^*} d\zeta, \tag{75}$$

where $f(z)$ is assumed analytic on and inside C. The difference of these results leads to

$$f(z) = \frac{1}{2\pi i} \oint_C f(\zeta) \left(\frac{1}{\zeta - z} - \frac{1}{\zeta - z^*} \right) d\zeta = \frac{y}{\pi} \oint_C \frac{f(\zeta)}{(\zeta - z)(\zeta - z^*)} d\zeta, \tag{76}$$

where we have used $z - z^* = 2iy$. If we now separate the contour C into its segments C_1 and C_R, we obtain

$$f(z) = \frac{y}{\pi} \int_{-R}^{R} \frac{f(t)}{(t - x)^2 + y^2} dt + \frac{y}{\pi} \int_{C_R} \frac{f(\zeta)}{(\zeta - z)(\zeta - z^*)} d\zeta, \tag{77}$$

We next wish to show that the second integral in (77) vanishes in the limit $R \to \infty$. To do so, we will establish an appropriate bound on the integral. From the second inequality in Eqs. (11) in Chapter 6, we have that

$$|\zeta - z| \geq R - |z|,$$

$$|\zeta - z^*| \geq R - |z|,$$

along the contour C_R, and we also know that $|f(z)| \leq M$ on C_R. Thus, it follows that

$$\left| \frac{f(\zeta)}{(\zeta - z)(\zeta - z^*)} \right| \leq \frac{M}{(R - |z|)^2}, \tag{78}$$

and because the length of C_R is πR, we see that

$$\left| \int_{C_R} \frac{f(\zeta)}{(\zeta - z)(\zeta - z^*)} d\zeta \right| \le \frac{M\pi R}{(R - |z|)^2}. \tag{79}$$

Clearly, as $R \to \infty$ this result tends to zero. It follows, therefore, that (77) reduces to

$$f(z) = \frac{y}{\pi} \int_{-\infty}^{\infty} \frac{f(t)}{(t - x)^2 + y^2} dt. \tag{80}$$

Last, because $f(z) = u(x,y) + i v(x,y)$ and $f(t) = u(t,0) + i v(t,0)$, it follows by equating real and imaginary parts that

$$
\begin{aligned}
u(x,y) &= \frac{y}{\pi} \int_{-\infty}^{\infty} \frac{u(t,0)}{(t - x)^2 + y^2} dt, \\[2mm]
v(x,y) &= \frac{y}{\pi} \int_{-\infty}^{\infty} \frac{v(t,0)}{(t - x)^2 + y^2} dt,
\end{aligned}
\tag{81}
$$

each of which is called *Poisson's formula for the half-plane*. Either integral represents a formal solution of the Dirichlet problem for the half-plane formulated, for example, by

$$\frac{\partial^2 u}{\partial x^2} + \frac{\partial^2 u}{\partial y^2} = 0, \quad -\infty < x < \infty, \ y > 0 \tag{82}$$

$$\text{BC:} \quad u(x,0) = g(x), \quad -\infty < x < \infty.$$

In Section 7.9 we will illustrate the use of Poisson's integral formula in solving problems like (82) in the half-plane.

Although Poisson's integral formula (81) is a solution to only the Dirichlet problem (82) in the half-plane, we can still use it to derive a similar integral formula for the Neumann problem in the half-plane. To see this, let us replace the Dirichlet problem in (82) by the Neumann problem

$$\frac{\partial^2 u}{\partial x^2} + \frac{\partial^2 u}{\partial y^2} = 0, \quad -\infty < x < \infty, \ y > 0 \tag{83}$$

$$\text{BC:} \quad \frac{\partial u}{\partial y}(x,0) = g(x), \quad -\infty < x < \infty,$$

where the normal derivative in this case is in the y direction. Physically, specifying the normal derivative tells us, for example, the heat flux or flow at this boundary (an insulated boundary has zero heat flow).

Let us introduce a new function $\varphi(x,y) = \partial u/\partial y$, where $u(x,y)$ is a solution of the Neumann problem. Hence, the new function satisfies

$$\frac{\partial^2 \varphi}{\partial x^2} + \frac{\partial^2 \varphi}{\partial y^2} = \frac{\partial}{\partial y}\left(\frac{\partial^2 u}{\partial x^2} + \frac{\partial^2 u}{\partial y^2}\right) = 0, \quad -\infty < x < \infty,\ y > 0$$

$$\text{(84)}$$

$$\text{BC:} \quad \varphi(x,0) = \frac{\partial u}{\partial y}(x,0) = g(x), \quad -\infty < x < \infty,$$

which reveals that $\varphi(x,y)$ is a solution of the Dirichlet problem described by (82). It follows that the solution $u(x,y)$ we are seeking can be obtained from $\varphi(x,y)$ by a simple indefinite integration, i.e.,

$$u(x,y) = \int \varphi(x,y)\,dy$$

$$= \frac{1}{\pi}\int_{-\infty}^{\infty} g(t)\int \frac{y}{(t-x)^2 + y^2}\,dy\,dt,$$

where we have interchanged the order of integration in the last step. Last, by completing the above indefinite integration, we get

$$u(x,y) = \frac{1}{2\pi}\int_{-\infty}^{\infty} g(t)\ln\left[(t-x)^2 + y^2\right]dt, \quad y > 0 \tag{85}$$

as the formal solution of the Neumann problem (84) in the half-plane. Of course, since we actually performed an indefinite integration, an arbitrary constant can be added to the solution (85). In fact, it has been shown that the solution to any Neumann problem is unique only to within an additive constant.

7.8.2 Harmonic functions in circular domains

Let us now consider a circle C of unit radius in the ζ plane as shown in Fig. 7.24. Let z be any point inside the circle and define $z_1 = 1/z^*$, which lies outside the circle C. That is, because $|z| = |z^*| < 1$, it follows that $|z_1| = 1/|z^*| > 1$. By Cauchy's integral theorem we can deduce that

$$f(z) = \frac{1}{2\pi i}\oint_C \frac{f(\zeta)}{\zeta - z}\,d\zeta,$$

$$0 = \frac{1}{2\pi i}\oint_C \frac{f(\zeta)}{\zeta - 1/z^*}\,d\zeta, \tag{86}$$

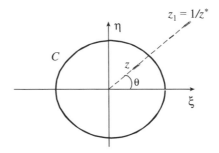

Figure 7.24 Circular domain of unit radius.

the difference of which yields

$$f(z) = \frac{1}{2\pi i} \oint_C f(\zeta) \left(\frac{1}{\zeta - z} - \frac{1}{\zeta - 1/z^*} \right) d\zeta = \frac{1}{2\pi i} \oint_C f(\zeta) \left[\frac{z - 1/z^*}{(\zeta - z)(\zeta - 1/z^*)} \right] d\zeta. \quad (87)$$

In polar coordinates we write

$$\zeta = e^{i\varphi}, \ z = re^{i\theta}, \text{ and } \frac{1}{z^*} = \frac{1}{r} e^{i\theta},$$

and thus, find that (87) can be expressed in the form

$$f(re^{i\theta}) = \frac{1}{2\pi i} \int_0^{2\pi} \frac{f(e^{i\varphi})(r^2 - 1)(ie^{i\theta})e^{i\varphi}}{re^{i\varphi} - re^{i\theta}\left(e^{i\varphi} - \frac{1}{r} e^{i\theta} \right)} d\varphi$$

$$= \frac{1 - r^2}{2\pi} \int_0^{2\pi} \frac{f(e^{i\varphi})}{1 - 2r\cos(\theta - \varphi) + r^2} d\varphi. \quad (88)$$

Once again, setting $f(re^{i\theta}) = u(r,\theta) + iv(r,\theta)$ and equating real and imaginary parts of (88), we obtain

$$u(r,\theta) = \frac{1 - r^2}{2\pi} \int_0^{2\pi} \frac{u(1,\varphi)}{1 - 2r\cos(\theta - \varphi) + r^2} d\varphi,$$

$$v(r,\theta) = \frac{1 - r^2}{2\pi} \int_0^{2\pi} \frac{v(1,\varphi)}{1 - 2r\cos(\theta - \varphi) + r^2} d\varphi. \quad (89)$$

Each of these relations is called *Poisson's integral formula for a unit circle*. By replacing r with r/a, we can formally develop a similar formula for a circle of radius a given by

$$u(r,\theta) = \frac{a^2 - r^2}{2\pi} \int_0^{2\pi} \frac{u(a,\varphi)}{a^2 - 2ar\cos(\theta - \varphi) + r^2} d\varphi. \quad (90)$$

Poisson's integral formula (90) gives the values of a harmonic function inside a circle in terms of its values on the boundary curve of the circle. As such, it represents a formal solution of the Dirichlet problem

$$\frac{\partial^2 u}{\partial r^2} + \frac{1}{r} \frac{\partial u}{\partial r} + \frac{1}{r^2} \frac{\partial^2 u}{\partial \theta^2} = 0, \quad 0 < r < a, \ -\pi < \theta < \pi \quad (91)$$

$$\text{BC: } u(a,\theta) = g(\theta), \quad -\pi < \theta < \pi,$$

where we are using the relation

$$\frac{\partial^2 u}{\partial x^2} + \frac{\partial^2 u}{\partial y^2} = \frac{\partial^2 u}{\partial r^2} + \frac{1}{r}\frac{\partial u}{\partial r} + \frac{1}{r^2}\frac{\partial^2 u}{\partial \theta^2}, \tag{92}$$

which follows a transformation to polar coordinates $x = r\cos\theta$, $y = r\sin\theta$.

7.8.3 Invariance of Laplace's equation

Harmonic functions represent the solution to many two-dimensional physical problems. One of the complexities that arises in such problems is that the solution function must satisfy certain prescribed conditions along the boundary of the domain of interest, called *boundary conditions*. Poisson's integral formulas derived above can be useful in finding harmonic functions in half-plane or circular domains in which the boundary conditions are prescribed along the finite boundaries of the domain.

When the domain of interest D is not a half-plane or circle, we may be able to find a *conformal mapping* that maps the domain D into one of these two specialized domains (e.g., see Section 7.9.2). One of the most significant consequences of conformal mappings is the invariance of Laplace's equation under such a transformation. Specifically, we have the following theorem.

Theorem 7.30 Every harmonic function of x and y transforms into a harmonic function of u and v under a conformal mapping $w = f(z)$.

In order to directly prove Theorem 7.30, we apply the chain rule which leads to

$$\frac{\partial^2 \varphi}{\partial x^2} + \frac{\partial^2 \varphi}{\partial y^2} = |f'(z)|^2 \left(\frac{\partial^2 \varphi}{\partial u^2} + \frac{\partial^2 \varphi}{\partial v^2} \right). \tag{93}$$

Hence, if $\varphi(x,y)$ is harmonic and $f'(z) \neq 0$ (a requirement of conformal mappings), it follows from (93) that $\varphi(u,v)$ is also harmonic.

In addition to Laplace's equation, there are certain types of boundary conditions that remain unaltered under a conformal mapping. In particular, boundary conditions of the form

$$\varphi = \text{constant}, \quad \frac{\partial \varphi}{\partial n} = 0, \tag{94}$$

which are characteristic of those belonging to Dirichlet and Neumann problems, respectively, remain unchanged under a conformal mapping. Boundary conditions of other forms, however, may transform into conditions substantially different under various mappings.

In the following section we will illustrate an example of a Dirichlet problem in which we can map a specific domain into the half-plane by a conformal transformation. If the prescribed boundary conditions are of the variety given in (94), we can often solve the transformed problem via Poisson's integral formula. The solution of the original Dirichlet problem is then obtained through the inverse mapping.

7.9 Heat Conduction

In this section we apply some of the ideas of analytic functions to problems of heat conduction. The same ideas can also be easily applied to problems in electrostatics and fluid flow (e.g., see Section 7.10).

The conduction of heat in two-dimensional solids (like flat plates) is basically the study of the *two-dimensional heat equation*

$$\frac{\partial^2 T}{\partial x^2} + \frac{\partial^2 T}{\partial y^2} = \frac{1}{a^2}\frac{\partial T}{\partial t}, \tag{95}$$

where $T(x,y,t)$ is the temperature at a point (x,y) and time t, and a^2 is a physical constant known as the *diffusivity*. In many cases of interest the temperature in the solid eventually becomes independent of time, in which case we say it approaches a *steady-state condition*. In such instances, the steady-state temperature $T(x,y)$ is a solution of Laplace's equation

$$\frac{\partial^2 T}{\partial x^2} + \frac{\partial^2 T}{\partial y^2} = 0. \tag{96}$$

Hence, the temperature $T(x,y)$ is a *harmonic function* in the domain occupied by the solid body under consideration (see also Section 6.4.3).

7.9.1 Steady-state temperatures in the plane

The curves $T(x,y) = C_1$, where C_1 is any constant are called *isotherms* (curves of constant temperature) or *level curves*. The gradient of $T(x,y)$, which is orthogonal to the isotherms, is in the direction of the heat flux vector describing heat flow. Thus, if $S(x,y)$ is the *harmonic conjugate* of $T(x,y)$, the curves $S(x,y) = C_2$ represent the lines of flow or flux lines orthogonal to the isotherms (recall Section 6.7.1). In such cases it can be useful to introduce the concept of the *complex temperature* defined by the analytic function

$$F(z) = T(x,y) + i S(x,y). \tag{97}$$

Let us assume the steady-state temperature $T(x,y)$ in a very large rectangular plate (or slab), whose flat surfaces are insulated and the temperature along one edge is prescribed by the function g but otherwise tends to zero along the other edges, is governed by the Dirichlet problem

$$\frac{\partial^2 T}{\partial x^2} + \frac{\partial^2 T}{\partial y^2} = 0, \quad -\infty < x < \infty, \ y > 0$$

$$\text{BC:} \quad T(x,0) = g(x) = \begin{cases} T_0, & |x| < b \\ 0, & |x| > b. \end{cases} \tag{98}$$

The solution to this problem is formally given by the Poisson integral formula in Eq. (81). In particular, if we substitute $u(x,y) = T(x,y)$ and $u(t,0) = g(t)$ directly into the top equation in (81), we are led to

$$T(x,y) = \frac{T_0 y}{\pi} \int_{-b}^{b} \frac{dt}{(t - x)^2 + y^2}$$

$$= \frac{T_0}{\pi} \left[\tan^{-1}\left(\frac{x + b}{y} \right) - \tan^{-1}\left(\frac{x - b}{y} \right) \right]. \tag{99}$$

However, for interpretation purposes it is useful to simplify our answer. By use of the trigonometric identity

$$\tan(A - B) = \frac{\tan A - \tan B}{1 + \tan A \tan B},$$

we can express the solution (99) in the more convenient form

$$T(x,y) = \frac{T_0}{\pi} \tan^{-1}\left(\frac{2by}{x^2 + y^2 - b^2} \right). \tag{100}$$

For the solution given by Eq. (100), we find that the *isotherms* (curves of constant temperature) are defined by the family of curves

$$x^2 + y^2 - Cy = b^2, \tag{101}$$

where C is a constant that can take on various values. We recognize this family of curves as a family of circular arcs that have centers on the y axis and endpoints on the x axis at $x = \pm b$ as shown in Fig. 7.25.

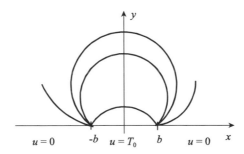

Figure 7.25 Family of isotherms.

7.9.2 Conformal mapping

Let us illustrate the conformal mapping technique (Section 6.7.5) applied to heat conduction by finding the steady-state temperature $T(x,y)$ in a semi-infinite strip (or slab) bounded by the planes $x = \pm\pi/2$ and $y = 0$ (see Fig. 7.26). Suppose the boundaries at $x = \pm\pi/2$ are kept at temperature zero while that along $y = 0$ is kept at T_0. (Discontinuities at the corners are of little concern in heat conduction because of the "smoothing" process of heat flow.) The steady-state temperature is a solution of the Dirichlet problem

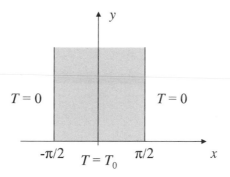

Figure 7.26 Semi-infinite strip.

$$\frac{\partial^2 T}{\partial x^2} + \frac{\partial^2 T}{\partial y^2} = 0, \quad -\pi/2 < x < \pi/2, \ y > 0$$

B.C.:
$$\begin{cases} T(-\pi/2,y) = 0, \ T(\pi/2,y) = 0, \\ T(x,0) = T_0. \end{cases}$$
(102)

Our objective is to find a mapping function $w = f(z)$ that maps the above semi-infinite strip onto the half-plane in the w plane, which can then be solved by Poisson's integral formula. A mapping function suitable for our needs is the function $w = \sin z$ (recall Example 23 in Section 6.7.5). Writing $w = u + iv$, the corresponding Dirichlet problem in the w plane is therefore characterized by

$$\frac{\partial^2 T}{\partial u^2} + \frac{\partial^2 T}{\partial v^2} = 0, \quad -\infty < u < \infty, \ v > 0$$

BC: $T(u,0) = \begin{cases} T_0, & |u| < 1 \\ 0, & |u| > 1. \end{cases}$
(103)

Based on Poisson's integral formula for the half-plane and the result given by Eq. (100), the solution of (103) in the w-plane is

$$T(u,v) = \frac{T_0}{\pi} \tan^{-1}\left(\frac{2v}{u^2 + v^2 - 1} \right).$$
(104)

To transform the solution (104) back into the z plane, we use the mapping function $w = \sin z$, which leads to

$$u(x,y) = \sin x \cosh y,$$
$$v(x,y) = \cos x \sinh y.$$
(105)

Hence, Eq. (104) becomes

$$T(x,y) = \frac{T_0}{\pi} \tan^{-1}\left(\frac{2\cos x \sinh y}{\sin^2 x + \sinh^2 y - 1}\right) = \frac{T_0}{\pi} \tan^{-1}\left(\frac{2\cos x \sinh y}{\sinh^2 y - \cos^2 x}\right). \qquad (106)$$

With some algebraic manipulation we can express this solution in a more convenient form. First, by defining

$$\alpha = \tan^{-1}\left(\frac{2\cos x \sinh y}{\sinh^2 y - \cos^2 x}\right), \qquad (107)$$

it follows that

$$\tan \alpha = \frac{2\cos x \sinh y}{\sinh^2 y - \cos^2 x} = \frac{2\left(\dfrac{\cos x}{\sinh y}\right)}{1 - \left(\dfrac{\cos x}{\sinh y}\right)^2} = \tan 2\beta,$$

where $\tan 2\beta = (2 \tan \beta)/(1 - \tan^2 \beta)$ and

$$\tan \beta = \frac{\cos x}{\sinh y}.$$

Now, because $\alpha = 2\beta$, we see that our solution (106) becomes

$$T(x,y) = \frac{T_0}{\pi} \tan^{-1}(\tan 2\beta) = \frac{2T_0}{\pi} \tan^{-1}\left(\frac{\cos x}{\sinh y}\right). \qquad (108)$$

7.10 Two-Dimensional Fluid Flow

A fluid flow in three-dimensional space is called *two-dimensional* if the velocity vector **V** is always parallel to a fixed plane (*xy*-plane), and if the velocity components (*u*,*v*) parallel to this plane along with the pressure *p* and fluid density ρ are all constant along any normal to the plane. This permits us to confine our attention to just a single plane which we interpret as a cross section of the three-dimensional region under consideration. Under the assumption that the flow is nonviscous (see below), the four functions *u*, *v*, *p*, and ρ are connected through a system of differential equations and appropriate boundary conditions:

$$\text{(Continuity equation):} \qquad \frac{\partial \rho}{\partial t} + \nabla \cdot (\rho \mathbf{V}) = 0 \qquad (109)$$

$$\text{(Euler equation):} \qquad \frac{\partial \mathbf{V}}{\partial t} + (\mathbf{V} \cdot \nabla) \mathbf{V} = -\frac{1}{\rho} \nabla p. \qquad (110)$$

The continuity equation is merely a statement of the conservation of mass, whereas the Euler equation is a consequence of Newton's second law of motion.

There are a number of terms that occur in the study of fluid flow, some of which we present below:

1. **Nonviscous fluid:** one in which the stress on an element of area is wholly normal and independent of the orientation of the area (fluid at rest, etc.).
2. **Incompressible fluid:** one in which the density ρ is constant.
3. **Ideal fluid:** a nonviscous, incompressible fluid (also called a *perfect fluid* or *inviscid fluid*).
4. **Steady flow:** one in which the velocity, pressure, and density are time-independent.
5. **Irrotational flow:** one in which curl $\mathbf{V} = \mathbf{0}$ ($\nabla \times \mathbf{V} = \mathbf{0}$).
6. **Solenoidal fluid:** one in which div $\mathbf{V} = 0$ ($\nabla \cdot \mathbf{V} = 0$); i.e., there are no *sources* or *sinks*.
7. **Streamline:** a curve for which the velocity vector $\mathbf{V}$ is everywhere tangent.

7.10.1 Complex potential

To formulate fluid flow in the complex plane we define the velocity vector $\mathbf{V}$ by the complex function

$$V = u + iv, \tag{111}$$

where u and v are the x and y components of velocity (note that we use u and v in a new context here). The "divergence" and "curl" then satisfy the relations

$$\text{div}(V) = \frac{\partial u}{\partial x} + \frac{\partial v}{\partial y}, \tag{112}$$

$$|\text{curl}(V)| = \omega = \left| \frac{\partial v}{\partial x} - \frac{\partial u}{\partial y} \right|, \tag{113}$$

where ω is called the *rotation* of the fluid.

If the flow is *irrotational*, then $\omega = |\text{curl}(V)| = 0$, or

$$\frac{\partial v}{\partial x} - \frac{\partial u}{\partial y} = 0. \tag{114}$$

Under the assumption of a *steady* irrotational flow with uniform density ρ, it can be shown that the fluid pressure p also satisfies *Bernoulli's equation*

$$\frac{p}{\rho} + \frac{1}{2}|V|^2 = \text{const.} \tag{115}$$

Thus, pressure is greatest where the speed $|V|$ is least. In the case of an *incompressible fluid* (i.e., $\rho = \text{const.}$), the continuity equation (109) reduces to $\text{div}(V) = 0$, or

$$\frac{\partial u}{\partial x} + \frac{\partial v}{\partial y} = 0; \tag{116}$$

Equations (114) and (116) are easily recognized as the Cauchy-Riemann equations for the *analytic function* $V^* = u - iv$. That is, the complex conjugate of velocity is an analytic function, but *not* the velocity itself. From vector analysis (see Chapter 4) we know that if a vector field is irrotational (i.e., conservative), there exists a scalar potential function $\Phi(x,y)$ such that $V = \nabla\Phi$, or

$$u = \frac{\partial\Phi}{\partial x}, \qquad v = \frac{\partial\Phi}{\partial y}. \tag{117}$$

By substituting these last results into (116), we see that

$$\frac{\partial^2\Phi}{\partial x^2} + \frac{\partial^2\Phi}{\partial y^2} = 0, \tag{118}$$

which shows that $\Phi(x,y)$ is *harmonic*. Hence, there exists a conjugate harmonic function $\Psi(x,y)$, called the *stream function*, such that the complex function

$$F(z) = \Phi(x,y) + i\Psi(x,y) \tag{119}$$

is *analytic* in the region of interest. The functions $\Phi(x,y)$ and $\Psi(x,y)$ are related through the Cauchy-Riemann equations, i.e.,

$$\frac{\partial\Phi}{\partial x} = \frac{\partial\Psi}{\partial y}, \qquad \frac{\partial\Phi}{\partial y} = -\frac{\partial\Psi}{\partial x}. \tag{120}$$

The analytic function $F(z)$, called the *complex potential function*, has the derivative

$$F'(z) = \frac{\partial\Phi}{\partial x} + i\frac{\partial\Psi}{\partial x} = u - iv,$$

which shows it is related to the velocity of the flow by

$$F'(z) = V^*(z). \tag{121}$$

Thus, knowledge of the complex potential permits us to calculate the velocity of the field at all points in the flow through use of Eq. (121). In particular, points in the flow where the velocity is zero are called *stagnation points*. Based on Eq. (121), the stagnation points can be found by solving $F'(z) = 0$. For example, if $F(z) = V_0(z + a^2/z)$, we find that $F'(z) = V_0(1 - a^2/z^2)$, and thus deduce that the stagnation points are $z = \pm a$. This particular complex potential is associated with a cylinder in a uniform flow (see Section 7.10.2)

From Section 6.7.1, we know the curves $\Psi(x,y) = C_1$ are orthogonal to the equipotential curves $\Phi(x,y) = C_2$, where C_1 and C_2 are any constants. Hence, because $V = \nabla\Phi$, we see that $\Psi(x,y) = C_1$ represents a family of curves parallel to the velocity vector and can therefore be identified with the *streamlines* of the flow. Thus, the complex potential $F(z)$ can also be used to determine the streamlines of the flow.

EXAMPLE 25 Find the complex potential $F(z)$ for a fluid moving with constant speed V_0, making an angle α with the positive x-axis (see Fig. 7.27).

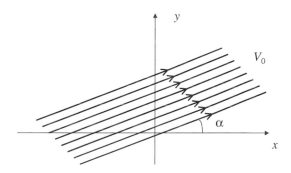

Figure 7.27 Fluid moving with constant speed V_0.

Solution: Because the angle α of the flow lines with the horizontal is constant, we see that the velocity components are

$$u = V_0 \cos \alpha,$$

$$v = V_0 \sin \alpha,$$

from which we deduce

$$V = V_0 (\cos \alpha + i \sin \alpha) = V_0 e^{i\alpha}.$$

Based on (120), it follows that the derivative of the complex potential function $F(z)$ is

$$F'(z) = V^*(z) = V_0 e^{-i\alpha},$$

which leads to

$$F(z) = V_0 e^{-i\alpha} z.$$

(A constant of integration is not needed for potential functions.) By separating this potential function into its real and imaginary parts [see (118)] we find that

$$\Phi(x,y) = V_0 (x \cos \alpha + y \sin \alpha),$$

$$\Psi(x,y) = V_0 (y \cos \alpha - x \sin \alpha).$$

EXAMPLE 26 Under the mapping $z = \zeta^2$, determine the images of the flow lines in the upper half of the z-plane in Example 25 for the special case $\alpha = 0$. Also, determine the speed of the flow in the ζ-plane.

Solution: The complex potential from Example 25 with $\alpha = 0$ is

$$F(z) = V_0 z = V_0 x + i V_0 y,$$

which, under the mapping $z = \zeta^2$, becomes $F(\zeta) = V_0 \zeta^2$. The mapping function $z = \zeta^2$ maps the upper half of the z-plane into the first quadrant of the ζ-plane (recall the discussion in Section 6.7.2). If we write $\zeta = \xi + i\eta$, then

$$x = \xi^2 - \eta^2,$$

$$y = 2\xi\eta.$$

Hence, the streamlines $\Psi(x,y) = V_0 y = C_1$ in the z-plane are mapped into branches of rectangular hyperbolas in the ζ-plane, viz.,

$$\Psi(\xi,\eta) = 2V_0 \xi\eta = C_1.$$

In particular, the boundary line $y = 0$ in the z-plane maps into the boundary lines $\xi = 0$ and $\eta = 0$ in the ζ-plane. We see, therefore, that the flow in the ζ-plane is that of a fluid flowing into a corner (see Fig. 7.28).

Last, the complex potential in the ζ-plane is

$$F(\zeta) = V_0 \zeta^2,$$

from which we deduce the speed of the flow [see Eq. (120)]

$$|V| = |V^*| = |F'(\zeta)| = 2V_0 \sqrt{\xi^2 + \eta^2}.$$

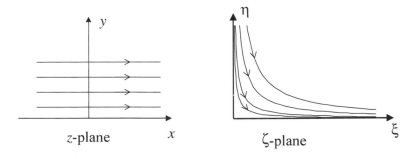

Figure 7.28 Flow of fluid into a corner under the mapping $z = \zeta^2$.

7.10.2 Source, sink, and doublet flows

Our development to this point has been based on the assumption there are no sources or sinks (i.e., places at which the fluid appears or disappears) within the region of interest. Such points, when they exist, introduce singularities into the complex potential $F(z)$ that we have not accounted for thus far. In this section we wish to briefly examine what happens when a source (sink) of heat or fluid is placed in a domain whose boundaries are maintained in some fashion. For simplicity, we employ what is called a line source (sink)—one that is unchanging and of infinite extent perpendicular to the complex plane.

The complex potential associated with a line source always displays a singularity at the point of intersection with the complex plane. To begin, let us consider the complex potential described by

$$F(z) = k \operatorname{Log} z, \tag{122}$$

where k is a constant. In polar coordinates, we write

$$F(z) = k \ln r + ik\theta_p. \tag{123}$$

Thus, the streamlines $\Psi(r,\theta) = k\theta_p = C_1$ correspond to lines radiating from the origin $(k > 0)$ or into the origin $(k < 0)$. The equipotential curves $\Phi(r,\theta) = k \ln r = C_2$ correspond to concentric circles around the origin. If we generalize the potential function (122) to

$$F(z) = k \operatorname{Log}(z - a), \tag{124}$$

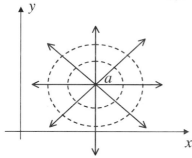

Figure 7.29 Source flow.

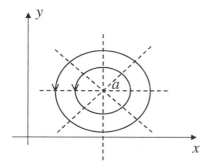

Figure 7.30 Vortex flow.

the streamlines radiate away from (or into) the point $z = a$ and the equipotential curves are concentric circles with centers at $z = a$ (see Fig. 7.29). Thus, if $k > 0$, the flow represents a *source* of strength k and if $k < 0$, it represents a *sink* of strength k.

Similarly, the flow described by the complex potential

$$F(z) = -ik \operatorname{Log}(z - a), \tag{125}$$

simply interchanges the roles of the potential function $\Phi(r,\theta)$ and the streamlines $\Psi(r,\theta)$. That is, the flow lines are now the concentric circles and the equipotential curves are the lines through the point $z = a$. Such a flow pattern is called a *vortex flow*, or *flow with circulation* (see Fig. 7.30).

Last, we consider the flow due to a sink of strength k located at the origin and a source of strength k located at a nearby point $z = a$. The complex potential in this case is given by

$$F(z) = -k \operatorname{Log} z + k \operatorname{Log}(z - a) = k \operatorname{Log}\left(\frac{z-a}{z}\right), \quad k > 0. \tag{126}$$

If we consider the limit of (126) as $a \to 0$, the complex potential will vanish unless at the same time the strength k is continuously increased in inverse proportion to the separation $|a|$. For this reason we write

$$k = \frac{m}{|a|}, \quad a = |a|e^{i\alpha}, \tag{127}$$

and then the complex potential assumes the form

$$F(z) = \frac{m}{|a|} \operatorname{Log}\left(\frac{z-a}{z}\right) = \frac{me^{i\alpha}}{a} \operatorname{Log}\left(\frac{z-a}{z}\right). \tag{128}$$

Consequently, in the limit $a \to 0$, we obtain

$$F(z) = -me^{i\alpha}\left(\frac{1}{z}\right). \tag{129}$$

The limiting concept of a source and sink as described by (129) is called a *doublet* of strength m. The angle α denotes the direction along which the coincidence was effected (see Fig. 7.31), and is called the orientation of the doublet. The same phenomenon occurs when two charges of equal magnitude but opposite type are placed close together. In this setting it is called a *dipole*.

By writing (129) in polar coordinates, we have

$$F(z) = -\frac{m}{r}\big[\cos(\theta - \alpha) - i\sin(\theta - \alpha)\big],$$

so that the streamlines are circles described by

$$\psi(r,\theta) = \frac{m}{r}\sin(\theta - \alpha) = C_1. \tag{130}$$

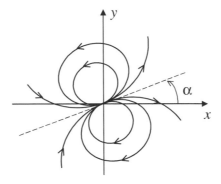

Figure 7.31 Doublet.

7.11 Flow around Obstacles

An important problem in fluid flow is that of determining the flow pattern of a fluid initially moving with uniform speed V_0 in which an obstacle has been placed. To begin, we consider a uniform flow in the ζ-plane described by the complex potential

$$F(\zeta) = V_0 \zeta. \tag{131}$$

The streamlines $\Psi(\xi,\eta) = V_0 \eta = C$ (const.) are the lines parallel to the ξ-axis as shown in Fig. 7.32(*a*). The basic problem is to find a mapping function

$$\zeta = f(z) \tag{132}$$

such that one of the streamlines in the ζ-plane [e.g., $\Psi(\xi,\eta) = 0$] maps onto the boundary of the obstacle in the z-plane. The resulting complex potential in the z-plane is therefore

$$F(z) = V_0 f(z), \tag{133}$$

and the corresponding streamlines in the z-plane are now defined by

$$\Psi(x,y) = V_0 \, \mathrm{Im}[f(z)] = C. \tag{134}$$

Using this approach, it can be shown that no streamline passes through the obstacle; rather, the streamlines (134) represent flow around the obstacle as illustrated in Fig. 7.32(*b*).

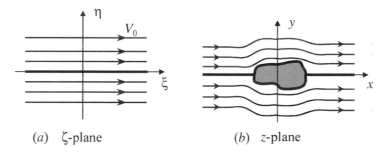

(*a*) ζ-plane   (*b*) z-plane

Figure 7.32 Flow around an obstacle.

A general principle involved in the selection of the mapping function (132) is to construct the complex potential (133) in the form

$$F(z) = V_0 z + G(z), \tag{135}$$

where $F'(z) = V_0 + G'(z)$ and the function $G(z)$ is chosen so that

$$\lim_{|z| \to \infty} G'(z) = 0.$$

Physically, this restriction on the function $G(z)$ tells us that at distances far from the obstacle the velocity of the flow has constant magnitude V_0. We conclude, therefore, that $F'(z) = V_0 + G'(z)$ is *analytic* in the region of the z-plane exterior to the obstacle itself, and thus, has the Laurent series representation

$$F'(z) = V^*(z) = V_0 + \sum_{n=1}^{\infty} A_n (z - z_0)^{-n}, \qquad (136)$$

where z_0 is a point inside the obstacle. Although (136) does not completely identify the function $G(z)$, it does give us information on what type of behavior is acceptable.

7.11.1 Circulation and lift

Before investigating the flow around a specific obstacle, we need to introduce an important concept in fluid flow called *circulation*, which is comparable to the notion of "work" around a closed path in mechanics (e.g., see Section 4.6.1).

▸ **Circulation:** The *circulation* Γ of a fluid is the line integral of the tangential component of velocity taken around a closed contour in the positive sense, i.e.,

$$\Gamma = \oint_C \mathbf{V} \cdot \hat{\mathbf{T}} ds = \oint_C u\,dx + v\,dy. \qquad (137)$$

Through the use of Green's theorem in the plane (Theorem 7.1) we can rewrite the line integral for circulation in the form of a double integral over the domain D enclosed by C, i.e., the circulation can also be expressed in the form

$$\Gamma = \iint_D \left(\frac{\partial v}{\partial x} - \frac{\partial u}{\partial y} \right) dx\,dy.$$

Yet, another method for calculating circulation directly from the complex potential function $F(z)$ follows from integrating

$$F'(z) = V^*(z) = u - iv$$

around a closed contour C within the flow. Doing so yields

$$\begin{aligned}
\oint_C F'(z)dz &= \oint_C (u - iv)(dx + i\,dy) \\
&= \oint_C (u\,dx + v\,dy) - i\oint_C (v\,dx - u\,dy),
\end{aligned} \qquad (138)$$

which we write as

$$\Gamma + i\Lambda = \oint_C F'(z)dz. \qquad (139)$$

That is, the first integral in (138) is recognized as the circulation (137) and the second integral Λ defines the *volume flow of fluid* (or *flux*) through C, which is the normal component of the velocity V_n integrated around C. By the use of Green's theorem in the plane once again, the volume flow of fluid Λ can be written as

$$\Lambda = -\oint_C (v\,dx - u\,dy) = \iint_D \left(\frac{\partial u}{\partial x} + \frac{\partial v}{\partial y} \right) dx\,dy. \tag{140}$$

Based on this last result, we see that an incompressible fluid is one for which $\Lambda = 0$. Consequently, through the use of Eq. (139), we can define the circulation for incompressible flows by the complex integral

$$\Gamma = \oint_C F'(z)\,dz, \tag{141}$$

which can be solved by the method of residues, if necessary.

EXAMPLE 27 Compute the circulation for the flow described by

$$F(z) = -ik\,\text{Log}(z - a)$$

around any simple closed contour enclosing the point $z = a$.

Solution: The complex potential is that associated with a vortex flow for which

$$F'(z) = -\frac{ik}{z - a}.$$

In this case, it can be shown

$$\Gamma + i\Lambda = \oint_C F'(z)\,dz = -ik \oint_C \frac{dz}{z - a} = 2\pi k.$$

It follows from this result that $\Lambda = 0$ and, consequently, the circulation is defined by

$$\Gamma = 2\pi k.$$

Hence, the constant k is directly proportional to the circulation.

In the flow of fluid around an obstacle there exists a net *aerodynamic force* due to fluid pressure on the surface of the obstacle. The existence of this aerodynamic force, for example, is essential for the motion of an aircraft through the atmosphere.

The net aerodynamic force acting on an obstacle in the flow can be reduced to two orthogonal force components, D and L, and a moment M. Specifically, the orthogonal force components D and L can be derived by the *theorem of Blasius* for steady and irrotational flows by calculating

$$D - iL = \frac{1}{2} i\rho \oint_C [F'(z)]^2 \, dz, \tag{142}$$

where ρ is the fluid density and C is a simple closed curve enclosing the obstacle. The component D is parallel to the direction of flow and is called the *drag*, whereas L represents the force component perpendicular to the flow direction and is called the *lift*. The direction of the lift is determined by rotating the velocity vector through a right angle in the sense opposite to that of the circulation. If M is the moment about the origin of the pressure forces on the obstacle, then the theorem of Blasius states that

$$M = \text{Re}\left\{-\frac{1}{2} i\rho \oint_C z[F'(z)]^2 \, dz\right\}. \tag{143}$$

7.11.2 Flow around a cylinder

As a specific example of a flow around an obstacle, let us consider the complex potential defined by

$$F(z) = V_0\left(z + \frac{a^2}{z}\right), \quad a > 0. \tag{144}$$

By introducing the mapping function

$$\zeta = f(z) = z + \frac{a^2}{z}, \tag{145}$$

we can interpret $F(\zeta) = V_0\zeta$ as a uniform flow in the ζ-plane with constant velocity V_0 parallel to the horizontal axis [see Fig. 7.32(a)].

The transformation (145) in rectangular coordinates leads to

$$\xi = x + \frac{a^2 x}{x^2 + y^2}, \quad \eta = y - \frac{a^2 y}{x^2 + y^2}. \tag{146}$$

By expressing the second relation in (146) in the form

$$\eta = \frac{y(x^2 + y^2 - a^2)}{x^2 + y^2},$$

we see that the ξ-axis ($\eta = 0$) maps onto the x-axis ($y = 0$) plus the circle $x^2 + y^2 = a^2$. We conclude from this that the circle (or cylinder in three dimensions) represents the obstacle in the path of the fluid flow in the z-plane.

To determine the flow lines around the cylinder for the complex potential (145), we observe that the streamlines $\Psi(\xi, \eta) = V_0\eta = C$ in the ζ-plane are mapped into the streamlines in the z-plane described in polar coordinates by

$$\Psi(r,\theta) = V_0\left(r - \frac{a^2}{r}\right)\sin\theta = C. \tag{147}$$

We show some representative streamlines in Fig. 7.33(a).

By definition, stagnation points are points where the fluid velocity is zero. Note that the derivative of (144) leads to

$$F'(z) = V_0\left(1 - \frac{a^2}{z^2}\right) = 0;$$

consequently, the stagnation points of the flow occur on the cylinder at the front and back points defined by $z = \pm a$ [see Fig. 7.33(a)]. Also, from Eq. (140) we can show that Eq. (144) represents flow around a cylinder without circulation, i.e.,

$$\Gamma = V_0 \oint_C \left(1 - \frac{a^2}{z^2}\right) dz = 0. \tag{148}$$

Let us now consider the flow described by the complex potential

$$F(z) = V_0\left(z + \frac{a^2}{z}\right) + \frac{\Gamma}{2\pi i}\text{Log}\,z, \quad \Gamma \geq 0, \tag{149}$$

which has the effect of superimposing a circulation Γ (see Section 7.9.3) on the flow around the cylinder described by (144). In writing the second term in (149), we have taken into account the result of Example 27.

To understand the effect of introducing the circulation term in the complex potential (149), we can examine the stagnation points on the cylinder (away from the cylinder there can be no such points). Thus, by setting

$$V^*(z) = F'(z) = V_0\left(1 - \frac{a^2}{z^2}\right) + \frac{\Gamma}{2\pi i z} = 0, \tag{150}$$

we find that the stagnation points are solutions of the quadratic equation

$$V_0 z^2 + \frac{\Gamma}{2\pi i}z - a^2 V_0 = 0,$$

which leads to

$$z = \frac{i\Gamma}{4\pi V_0} \pm \frac{\sqrt{16\pi^2 a^2 V_0^2 - \Gamma^2}}{4\pi V_0}. \tag{151}$$

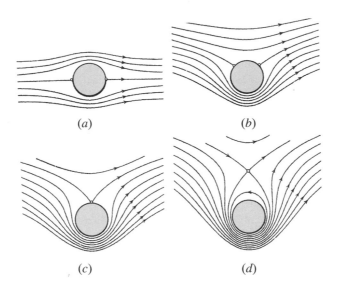

Figure 7.33 Flow around a cylinder (a) without circulation $(\Gamma = 0)$, (b) with circulation $0 \le \Gamma < 4\pi a V_0$, (c) with circulation $\Gamma = 4\pi a V_0$, and (d) with circulation $\Gamma > 4\pi a V_0$.

Based on (151), it can be shown there are *two* stagnation points on the cylinder when $0 \le \Gamma < 4\pi a V_0$, there is *one* stagnation point on the cylinder when $\Gamma = 4\pi a V_0$, and there are *no* stagnation points on the cylinder when $\Gamma > 4\pi a V_0$. That is, for the case when $\Gamma = 4\pi a V_0$, we note that the single stagnation point is at $z = ia$, located at the top of the cylinder. When $0 \le \Gamma < 4\pi a V_0$, the two stagnation points are both also on the cylinder because in this case

$$|z| = \sqrt{\frac{\Gamma^2}{16\pi^2 V_0^2} \pm a^2 - \frac{\Gamma^2}{16\pi^2 V_0^2}} = a.$$

Last, when $\Gamma > 4\pi a V_0$, the stagnation points are pure imaginary—one lies above the cylinder and the other is inside the cylinder. Recall that in the absence of circulation $(\Gamma = 0)$, there are two stagnation points located at $z = \pm a$. These four cases are illustrated in Fig. 7.33.

The general effect of the circulation on the cylinder is to increase the speed of the fluid on the bottom of the cylinder and decrease it on top. Based on Bernoulli's equation (136), the pressure below the cylinder is therefore diminished and the pressure above is increased. Consequently, there will be a downward force (lift) acting on the cylinder due to the pressure differences.

To determine the flow lines associated with the complex potential (149), we note that the stream function is described by

$$\Psi(x,y) = \frac{V_0 y(x^2 + y^2 - a^2)}{x^2 + y^2} - \frac{\Gamma}{2\pi} \text{Log}(x^2 + y^2), \tag{152}$$

or, in polar coordinates, by

$$\Psi(r,\theta) = V_0\left(r - \frac{a^2}{r}\right)\sin\theta - \frac{\Gamma}{2\pi}\operatorname{Log} r. \tag{153}$$

Some representative flow lines deduced from (153) are also featured in Fig. 7.33.

The lift component of the aerodynamic force on the cylinder perpendicular to the specified velocity can be deduced from Eq. (142), viz.,

$$\begin{aligned} D - iL &= \frac{1}{2}i\rho\oint_C [F'(z)]^2\,dz \\ &= \frac{1}{2}i\rho\oint_C\left[V_0^2 - \frac{iV_0\Gamma}{\pi z} - \left(2a^2V_0^2 + \frac{\Gamma^2}{4\pi^2}\right)\frac{1}{z^2} + \frac{ia^2V_0\Gamma}{\pi z^3} + \frac{a^4V_0^2}{z^4}\right]dz, \end{aligned} \tag{154}$$

which, upon application of the residue theory, yields $D = 0$ and the classic result

$$L = -\rho V_0\Gamma. \tag{155}$$

Equation (155) is the famous *Kutta-Joutkowski lift formula*, which relates the lift on the obstacle directly to the density, circulation, and flow velocity (at infinity). Following the stated convention for our example, the lift on the cylinder is "downward." Also, because the stream function (152) is symmetrical about the y-axis, there can be no resultant drag force in the direction of the flow.

Last, the pitching moment (143) about the origin leads to the result

$$\begin{aligned} M &= \operatorname{Re}\left\{\frac{1}{2}i\rho\oint_C z[F'(z)]^2\,dz\right\} \\ &= \operatorname{Re}\left\{\frac{1}{2}i\rho\oint_C\left[V_0^2 z - \frac{iV_0\Gamma}{\pi} - \left(2a^2V_0^2 + \frac{\Gamma^2}{4\pi^2}\right)\frac{1}{z} + \frac{ia^2V_0\Gamma}{\pi z^2} + \frac{a^4V_0^2}{z^3}\right]dz\right\}, \end{aligned} \tag{156}$$

from which we deduce

$$M = \frac{\rho}{4\pi}(8\pi^2 a^2 V_0^2 + \Gamma^2). \tag{157}$$

Suggested Reading

R. V. Churchill and J. W. Brown, *Complex Variables and Applications*, 5th ed. (McGraw-Hill, New York, 1990).

W. R. LePage, *Complex Variables and the Laplace Transform for Engineers* (Dover, New York, 1961).

J. D. Paliouras and D. S. Meadows, *Complex Variables for Scientists and Engineers*, 2nd ed. (Macmillan, New York, 1990).

R. Roth, F. Ollendorf, and K. Pohlhausen, *Theory of Functions as Applied to Engineering Problems* (Dover, New York, 1961).

E. B. Saff and A. D. Snider, *Fundamentals of Complex Analysis for Mathematics, Science, and Engineering*, 2nd ed. (Prentice-Hall, Upper Saddle River, 1993).

A. D. Wunsch, *Complex Variables with Applications*, 2nd ed. (Addison-Wesley, Reading, 1994).

Chapter 7

Exercises

In Problems 1 and 2, evaluate the given integral along the specified curve.

1. $\oint_C |z^*|^2 \, dz$, C: $|z + 2i| = 4$

Ans. 64π

2. $\int_C (z - 3i)\, dz$, C: $x = t^2$, $y = t + 2$,
$$0 \le t \le 1$$

Ans. 1

In Problems 3 and 4, find an upper bound ML for the integral.

3. $\oint_C \frac{dz}{(z^2 + 1)^2}$, C: $|z| = 5$

Ans. $\frac{5\pi}{288}$

4. $\int_C \operatorname{Log} z \, dz$, C: line from 1 to $1 + i$

Ans. $\ln \sqrt{2} + \frac{\pi}{4}$

In Problems 5-7, use *Cauchy's Integral Formulas* to evaluate the given integral.

5. $\oint_C \frac{(z + 2)\,dz}{z^3 - 3z^2}$, C: $|z - 4| = 2$

Ans. $\frac{10\pi i}{9}$

6. $\oint_C \frac{dz}{(z^2 + z + 1)^2}$, C: $4x^2 + 9y^2 = 9$

Ans. 0

7. $\oint_C \frac{z\,dz}{(z^2 + 9)(z + 5)^2}$, C: $|z| = \pi$

Ans. $\frac{8\pi i}{289}$

8. Given the function
$$f(z) = \frac{z + 1}{z^2 + 2z + 3}$$
find a Taylor series around $z = -1$ and give its radius of convergence.

Ans. $\frac{1}{2} \sum_{n=0}^{\infty} \frac{(-1)^n}{2^n} (z + 1)^{2n+1}$
$$\rho = \sqrt{2}$$

9. Given the function
$$f(z) = \frac{z}{z^2 - z - 2}$$
find a Taylor series around $z = 1$ and give its radius of convergence.

Ans. $\displaystyle\sum_{n=0}^{\infty}\left[\frac{1}{6}\left(-\frac{1}{2}\right)^{n}-\frac{2}{3}\right](z-1)^{n}$

$\rho = 1$

10. Find a *Laurent series* for

$$f(z) = \frac{(z+1)^2}{z(z+2)}$$

that converges for

(a) $0 < |z| < 2$.

(b) $|z+1| > 1$.

Ans. (a) $\displaystyle 1 - \frac{1}{4}\sum_{n=-1}^{\infty}\frac{(-1)^n}{2^n}z^n$

(b) $\displaystyle\sum_{n=-\infty}^{0}(z+1)^{2n}$

11. Classify all the isolated singularities of the following functions:

(a) $f(z)$

$$= \frac{z\cos z}{(z^2+1)^2(z-1)(z^2+3z+2)}$$

(b) $f(z) = \dfrac{e^z - 1}{z^2}$

Ans. (a) simple poles: $z = 1, -1, -2$
poles of order 2: $z = \pm i$
(b) simple pole: $z = 0$

In Problems 12-15, compute the residue at each finite singular point.

12. $f(z) = \dfrac{z}{(z-1)(z+1)^2}$

Ans. $\text{Res}(1) = \dfrac{1}{4}$, $\text{Res}(-1) = \dfrac{1}{4}$

13. $f(z) = \dfrac{e^{1/z}}{1-z}$

Ans. $\text{Res}(0) = e - 1$, $\text{Res}(1) = -e$

14. $f(z) = \dfrac{e^{-z}}{z^5}$

Ans. $\text{Res}(0) = \dfrac{1}{24}$

In Problems 15-22, use residue theory to evaluate the given real integral.

15. $\displaystyle\int_0^{2\pi}\frac{d\theta}{a+b\sin\theta}$, $a > |b|$

Ans. $\dfrac{2\pi}{\sqrt{a^2 - b^2}}$

16. $\displaystyle\int_0^{2\pi}\frac{\cos 2\theta\, d\theta}{5 - 4\sin\theta}$

Ans. $-\dfrac{\pi}{6}$

17. $\displaystyle\int_0^{2\pi}\frac{d\theta}{(5 - 3\cos\theta)^2}$

Ans. $\dfrac{5\pi}{32}$

18. $\displaystyle\int_0^{\infty}\frac{dx}{x^6 + 1}$

Ans. $\dfrac{\pi}{3}$

19. $\displaystyle\int_{-\infty}^{\infty}\frac{dx}{(x+1)(x^2+2)}$

Ans. $\dfrac{\sqrt{2}\pi}{6}$

20. $\displaystyle\int_{-\infty}^{\infty}\frac{dx}{x^4 + x^2 + 1}$

Ans. $\dfrac{\sqrt{3}\pi}{6}$

21. $\displaystyle\int_{-\infty}^{\infty}\frac{x\cos\pi x\, dx}{x^2 + 2x + 5}$

Ans. $\dfrac{\pi}{2}e^{-2\pi}$

22. $\displaystyle\int_0^{\infty}\frac{\sin^2 x\, dx}{x^2}$

Ans. $\dfrac{\pi}{2}$

23. Use the contour shown in Fig. 7.22 to evaluate the integral:

$$I = \int_0^{\infty}\frac{x^{1/5}}{1+x^3}\, dx$$

Hint: Let $t = x^3$.

Ans. $\dfrac{\pi}{3\sin(2\pi/5)}$

24. Using *Poisson's Integral Formula* and the transformation $w = \sin^2 z$, show that the solution of the Dirichlet problem

$$T_{xx} + T_{yy} = 0, \ \ 0 < x < \pi/2, 0 < y < \infty$$

BC: $T(0,y) = 1, \ \ T(\pi/2,y) = 0,$
   $T(x,0) = 0$

can be expressed in the form

$$T(x,y) = \frac{2}{\pi}\tan^{-1}\left(\frac{\tanh y}{\tan x}\right).$$

25. Given the potential problem for the stream function described by

$$\psi_{xx} + \psi_{yy} = 0, \ \ -\infty < x < \infty, y > 0$$

BC: $\psi_x(x,0) = \begin{cases} -V_0, \ |x| < b \\ \ \ 0, \ |x| > b, \end{cases}$

rewrite the problem in terms of the potential function $\varphi(x,y)$ and determine the velocity components $u(x,y), \ v(x,y)$.

Ans. $u = \dfrac{V_0}{\pi}\ln\left[\dfrac{(x+b)^2 + y^2}{(x-b)^2 + y^2}\right]$

$v = \dfrac{V_0}{\pi}\tan^{-1}\left(\dfrac{2by}{x^2 + y^2 - b^2}\right)$

Chapter 8

Fourier Series, Eigenvalue Problems, and Green's Function

8.1 Introduction / 405
8.2 Fourier Trigonometric Series / 405
    8.2.1 Periodic functions as power signals / 406
    8.2.2 Convergence of the series / 409
    8.2.3 Even and odd functions: cosine and sine series / 411
    8.2.4 Nonperiodic functions: extensions to other intervals / 413
8.3 Power Signals: Exponential Fourier Series / 416
    8.3.1 Parseval's theorem and the power spectrum / 418
8.4 Eigenvalue Problems and Orthogonal Functions / 420
    8.4.1 Regular Sturm-Liouville systems / 425
    8.4.2 Generalized Fourier series / 428
    8.4.3 Periodic Sturm-Liouville systems / 431
    8.4.4 Singular Sturm-Liouville systems / 432
8.5 Green's Function / 438
    8.5.1 One-sided Green's function / 439
    8.5.2 Boundary value problems / 442
    8.5.3 Bilinear formula / 446
    Suggested Reading / 449
    Exercises / 450

Historical Comments: Joseph Fourier (1768-1830) was one of several famous French mathematicians who flourished during the time of Napoleon. In 1794, Napoleon offered Fourier the chair of mathematics at the Ecole Normale in Paris. He left this position in 1798 to accompany Napoleon and a group of other scientists to Egypt, where he remained for four years, establishing the scientific institute of Cairo. Fourier returned to France in 1802 to become prefect of the department of Isere at Grenoble in the French Alps.

The theory widely known as Fourier series is credited to Fourier, who came across such representations in his classic studies of heat conduction. His basis papers, presented to the Academy of Sciences in Paris in 1807 and 1811, were criticized by the referees (most strongly by Lagrange) for a lack of rigor and consequently were not then published. Fourier was called to Paris by the Academy of Sciences in 1816, whereupon he succeeded Laplace as president of the board of the Ecole Polytechnique. When publishing the classic *Théorie analytique de la Chaleur* in 1822, he also incorporated his earlier work that was previously rejected. Fourier died in Paris on May 16, 1830.

Leonhard Euler (1707-1783) solved the first eigenvalue problem when he developed a model for describing the "buckling" modes of a vertical column. However, the general theory of eigenvalue problems for second-order DEs, commonly called the Sturm-Liouville theory, originated in the work of Jacques C. F. Sturm (1803-1855), a professor of mechanics at the Sorbonne, and Joseph Liouville (1809-1882), a professor of mathematics at the College de France.

The method of Green's function is named for George Green (1793-1841), who gained recognition for his work on reflection and refraction of sound and light waves. He also extended the work of Poisson in the theory of electricity and magnetism.

In this chapter we introduce the method of Fourier series for the analysis of periodic waveforms (e.g., power signals). This approach reduces the signal being studied to a spectral representation in which the distribution of power is found to be concentrated at specific frequencies that are harmonically related to a fundamental frequency. In addition, we discuss the related notion of eigenvalue problem for homogeneous boundary value problems, the eigenfunctions of which are used to develop generalized Fourier series. Last, the method of Green's function is introduced for solving nonhomogeneous problems (including eigenvalue problems). By representing the Green's function in a "bilinear" representation, we amalgamate the theory of Fourier series and eigenvalue problems with that of the Green's function method.

8.1 Introduction

There are numerous engineering problems that involve the notions of Fourier series and Fourier transforms. For example, Fourier series and Fourier transforms can both arise in the analysis of signals such as those that occur in communications, radar, and imaging applications. In particular, *Fourier series* are closely associated with *power signals* (e.g., periodic signals), whereas the related *Fourier transform* (see Chapter 9) is closely associated with *energy signals* (e.g., pulses).

Standard Fourier series are those defined by simple sinusoids leading to trigonometric series. In addition to the analysis of power signals, Fourier trigonometric series can be used in the general analysis of periodic functions and they also arise in the solution of partial differential equations (see Chapter 12).

More general forms of the Fourier series can arise in connection with various applications for which the solution of a differential equation and/or its boundary conditions depend on a parameter λ. Problems of this kind are widely known as *eigenvalue problems*— the values of the parameter λ that permit nontrivial (i.e., nonzero) solutions are called *eigenvalues* and the corresponding solutions are the *eigenfunctions*. In applications the eigenvalues and eigenfunctions have many different physical interpretations. For example, in vibration problems the eigenvalues are proportional to the squares of the natural frequencies of vibration, whereas the eigenfunctions provide the natural configuration modes. The eigenvalues denote the possible energy states in quantum mechanics, and the eigenfunctions are the associated wave functions. Of particular importance here is that eigenfunctions are mutually *orthogonal* and, as such, are useful in developing *generalized Fourier series*, also called *eigenfunction expansions*.

When the boundary value problem (or eigenvalue problem) arises from a nonhomogeneous equation, the method of *Green's function* can be a powerful tool for solution and interpretation. If the nonhomogeneous equation arises from an initial value problem, we are led to the *one-sided Green's function* for solution and interpretation. This latter function is directly related to the *impulse response function* of a linear shift-invariant system (e.g., see Section 9.6).

8.2 Fourier Trigonometric Series

In the field of communications the *classification* of a signal (function) $f(t)$ is important for deciding which representation is appropriate for analysis. There are basically two classifications—"energy signals" and "power signals."

The *energy* of a signal $f(t)$ over all time is defined by the integral

$$E = \int_{-\infty}^{\infty} |f(t)|^2 \, dt, \tag{1}$$

whereas the *average power* of the signal $f(t)$ is defined by

$$P_{\text{ave}} = \lim_{T \to \infty} \frac{1}{T} \int_{-T/2}^{T/2} |f(t)|^2 \, dt. \tag{2}$$

▸ **Energy signal:** If the energy E is finite, i.e., if

$$E = \int_{-\infty}^{\infty} |f(t)|^2 \, dt < \infty,$$

we say that $f(t)$ is an *energy signal* (also called a *square-integrable* function). Physically realizable waveforms are energy signals, usually with a finite duration. However, we often model them by infinite-duration waveforms, e.g., $e^{-|t|}$ and e^{-t^2}, among others.

▸ **Power signal:** If the average power P_{ave} is finite but nonzero, we say that $f(t)$ is a *power signal*. Examples of power signals are periodic functions.

Clearly, if $f(t)$ is an energy signal, its average power is zero. However, if the energy is infinite it is possible to have a nonzero average power.

8.2.1 Periodic functions as power signals

▸ **Periodic function:** A function $f(t)$ is called *periodic* if there exists a constant $T > 0$ for which $f(t + T) = f(t)$ for all t. The smallest value of T for which the property is true is called the *period*.

A periodic function $f(t)$ is an example of a *power signal* if its energy over one period is finite. In this case the average power (2) need only be calculated over one period of the signal, i.e.,

$$P_{ave} = \frac{1}{T} \int_0^T |f(t)|^2 \, dt = \frac{1}{T} \int_{-T/2}^{T/2} |f(t)|^2 \, dt. \tag{3}$$

Periodic functions appear in a variety of physical problems, such as those concerning vibrating springs and membranes, planetary motion, a swinging pendulum, and musical sounds, to name a few. Because trigonometric functions like sines and cosines are the simplest examples of periodic functions, it may be reasonable to look for series representations in terms of these functions.

The simple sinusoid

$$f(t) = \cos \omega_0 t$$

is a well-known example of a periodic function. Its period is clearly $T = 2\pi/\omega_0$, i.e., $f(t + 2\pi/\omega_0) = f(t)$. The reciprocal quantity $f_0 = 1/T = \omega_0/2\pi$ is called the *fundamental frequency* of the signal, whereas the quantity $\omega_0 = 2\pi f_0$ is the *angular frequency*.

Observe that all members of the set $\{1, \cos(n\omega_0 t), \sin(n\omega_0 t)\}$ for $n = 1, 2, 3, ...,$ have the same period $T = 2\pi/\omega_0$.[1] Therefore, if $f(t)$ is any periodic function with period $T = 2\pi/\omega_0$, we look for series representations of the form

$$f(t) = A_0 + \sum_{n=1}^{\infty} \left(a_n \cos n\omega_0 t + b_n \sin n\omega_0 t \right), \tag{4}$$

---

[1]A constant is a periodic function with arbitrary period.

where A_0, a_n, and b_n are all constants to be determined. Series of this nature are called *Fourier trigonometric series,* or, simply, *Fourier series*.

In order to determine the constants in (4), let us begin by integrating both sides of Eq. (4) over one period. This action yields

$$\int_{-T/2}^{T/2} f(t)\,dt = A_0 \int_{-T/2}^{T/2} dt + \sum_{n=1}^{\infty} \left(a_n \int_{-T/2}^{T/2} \cos n\omega_0 t\,dt + b_n \int_{-T/2}^{T/2} \sin n\omega_0 t\,dt \right)$$
$$= TA_0,$$

where we have tacitly interchanged the order of summation and integration and recognized that all integrals under the summation are zero. Consequently, we have

$$A_0 = \frac{1}{T}\int_{-T/2}^{T/2} f(t)\,dt. \tag{5}$$

The constant A_0 defined by (5) represents the *average (power) value* of the function $f(t)$ over one period. Next, we multiply the original series (4) by $\cos(k\omega_0 t)$, $k = 1,2,3,...$, and again integrate over one period to find

$$\int_{-T/2}^{T/2} f(t)\cos k\omega_0 t\,dt = A_0 \int_{-T/2}^{T/2} \cos k\omega_0 t\,dt$$
$$+ \sum_{n=1}^{\infty} \left(a_n \int_{-T/2}^{T/2} \cos n\omega_0 t \cos k\omega_0 t\,dt \right.$$
$$\left. + b_n \int_{-T/2}^{T/2} \sin n\omega_0 t \cos k\omega_0 t\,dt \right).$$

However, here we recognize

$$\int_{-T/2}^{T/2} \cos n\omega_0 t \cos k\omega_0 t\,dt = 0,\ n \neq k,$$
$$\int_{-T/2}^{T/2} \sin n\omega_0 t \cos k\omega_0 t\,dt = 0,\ \text{all } n \text{ and } k, \tag{6}$$

which leads to

$$\int_{-T/2}^{T/2} f(t)\cos k\omega_0 t\,dt = a_k \int_{-T/2}^{T/2} \left(\cos k\omega_0 t \right)^2 dt = \frac{T}{2}a_k,$$

or

$$a_k = \frac{2}{T}\int_{-T/2}^{T/2} f(t)\cos k\omega_0 t\,dt,\ k = 1,2,3,.... \tag{7}$$

In a similar manner, if we multiply the series (4) by $\sin(k\omega_0 t)$ and integrate over one period, we will obtain

$$b_k = \frac{2}{T}\int_{-T/2}^{T/2} f(t)\sin k\omega_0 t\,dt,\ k = 1,2,3,.... \tag{8}$$

▶   **Fourier coefficients:** The constants defined by (5), (7), and (8), and summarized
    below in (9), are called *Fourier coefficients*.

It is customary to change the dummy index k in (7) and (8) back to n and define
$A_0 = (1/2)a_0$ so that (5) can be combined with (7).[2] In doing so, we now define the Fourier
coefficients and series by

$$a_n = \frac{2}{T}\int_{-T/2}^{T/2} f(t)\cos n\omega_0 t\,dt, \quad n = 0,1,2,\dots$$

$$b_n = \frac{2}{T}\int_{-T/2}^{T/2} f(t)\sin n\omega_0 t\,dt, \quad n = 1,2,3,\dots,$$

(9)

$$f(t) = \frac{1}{2}a_0 + \sum_{n=1}^{\infty}\left(a_n\cos n\omega_0 t + b_n\sin n\omega_0 t\right).$$

(10)

EXAMPLE 1 Find the Fourier series of the periodic ramp function (see Fig. 8.1)

$$f(t) = \begin{cases} 0, & -\pi < t < 0, \\ t, & 0 < t < \pi, \end{cases} \quad f(t+2\pi) = f(t).$$

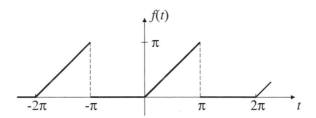

Figure 8.1

Solution: We first observe that the period of this function is $T = 2\pi$. Thus, the
series we seek has the form

$$f(t) = \frac{1}{2}a_0 + \sum_{n=1}^{\infty}\left(a_n\cos nt + b_n\sin nt\right),$$

where

$$a_0 = \frac{1}{\pi}\int_{-\pi}^{\pi} f(t)\,dt = \frac{1}{\pi}\int_{0}^{\pi} t\,dt = \frac{\pi}{2},$$

---

[2]Writing the constant A_0 in this fashion does not simplify the calculations in (9). That is,
the case $n = 0$ must be separately evaluated.

$$a_n = \frac{1}{\pi} \int_0^\pi t \cos nt \, dt = \begin{cases} -\dfrac{2}{\pi n^2}, & n = 1,3,5,\ldots \\ 0, & n = 2,4,6,\ldots, \end{cases}$$

and

$$b_n = \frac{1}{\pi} \int_0^\pi t \sin nt \, dt = \frac{(-1)^{n-1}}{n}, \quad n = 1,2,3,\ldots .$$

In the last result, we have used the relation $\cos n\pi = (-1)^n$. These values substituted into the above series is our intended result, which we write as

$$f(t) = \frac{\pi}{4} - \frac{2}{\pi} \sum_{\substack{n=1 \\ (\text{odd})}}^{\infty} \frac{\cos nt}{n^2} - \sum_{n=1}^{\infty} \frac{(-1)^n}{n} \sin nt.$$

8.2.2 Convergence of the series

The above treatment of Fourier series thus far has been purely formal, i.e., nonrigorous. Nonetheless, we can take the point of view that (10) defines the Fourier series of a periodic function with constants defined by the Fourier coefficients. For the series to *exist*, all we need is for the function $f(t)$ to be at least *piecewise continuous* on the closed interval $[-T/2, T/2]$. This means that discontinuities in $f(t)$ are permitted, but they must be *finite jumps* at isolated points. For *convergence* of the series, we need to also require the derivative to be piecewise continuous. This leads us to the following theorem which we state without proof (see [1] for a proof).

Theorem 8.1 (*Pointwise convergence*) If $f(t + T) = f(t)$ for all t, and $f(t)$ and $f'(t)$ are piecewise continuous functions, then the Fourier trigonometric series (10) with Fourier coefficients defined by (9) converges for all values of t. The sum of the series is $f(t)$ at points of continuity of $f(t)$ and equals the average value $(1/2)[f(t^+) + f(t^-)]$ at points of discontinuity of $f(t)$.

The conditions listed in Theorem 8.1 are only *sufficient conditions* for convergence, not necessary conditions. For example, the function $f(t) = |t|^{1/2}$, $-T/2 < t < T/2$, does not have a piecewise continuous derivative on any interval containing $t = 0$, and yet it has a convergent Fourier series. Also, it should come as no surprise that the series does not converge to the function value (if it exists!) at points of discontinuity. That is, we can always change the function value at a finite number of points without changing the Fourier series.

In order to illustrate the convergence of a Fourier series at a point of discontinuity, let us consider the periodic function known as a square wave (see Fig. 8.2), viz.,

$$f(t) = \begin{cases} 0, & -\pi < t < 0, \\ 1, & 0 < t < \pi, \end{cases} \qquad f(t + 2\pi) = f(t). \tag{11}$$

Some simple calculations reveal that the Fourier series for this function is

$$f(t) = \frac{1}{2} + \frac{2}{\pi} \sum_{\substack{n=1 \\ (odd)}}^{\infty} \frac{\sin nt}{n}. \tag{12}$$

Because of discontinuities in the function $f(t)$ at $t = 0$ and multiples of π, the series (12) converges to the average value one-half of the left-hand and right-hand limits. These values are illustrated by the crosses in Fig. 8.2. Elsewhere, the series converges to the value of the function $f(t)$. In particular, setting $x = \pi/2$ in (12) leads to

$$f(\pi/2) = 1 = \frac{1}{2} + \frac{2}{\pi}\left(1 - \frac{1}{3} + \frac{1}{5} - \frac{1}{7} + \cdots\right)$$

from which we deduce the interesting result

$$1 - \frac{1}{3} + \frac{1}{5} - \frac{1}{7} + \cdots = \frac{\pi}{4}. \tag{13}$$

Leibniz (around 1673) was the first to discover this relation between π and the reciprocal of the odd integers, but he obtained it from geometrical considerations alone. There are numerous other relations similar to (13) that can be deduced from known Fourier series relations.

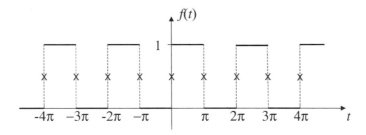

Figure 8.2 Square wave.

By taking only a finite number of terms in the series (12), we obtain what is called the *partial sum*, i.e.,

$$S_N(t) = \frac{1}{2} + \frac{2}{\pi} \sum_{\substack{n=1 \\ (odd)}}^{N} \frac{\sin nt}{n}. \tag{14}$$

If these partial sums are plotted, it can be observed that the sum tends to overshoot the function in the vicinity of the discontinuities and then approach the value one or zero (see Fig. 8.3). This feature, which can be directly attributed to a decrease in the bandwidth associated with a finite sum Fourier series (see Section 8.3.1), is typical of any partial sum in the vicinity of a finite discontinuity of a given function $f(t)$. It was first studied by J. Willard Gibbs (1839-1903) and is now widely known as the *Gibbs' phenomenon*.

In the use of Fourier series we may need to perform certain operations on the series, such as integration and differentiation. Integration is usually permitted and, when required in our analysis, we often make the assumption that *termwise differentiation* of the series is permitted. However, unless certain additional conditions are satisfied by the function $f(t)$, this may lead to some difficulties. For example, the square-wave function defined by (11) has the convergent Fourier series given by (12). Clearly, the derivative at every point of continuity is zero because the function is constant between the discontinuities. Note that termwise differentiation of the series (12) leads to

$$0 = 0 + \frac{2}{\pi} \sum_{\substack{n=1 \\ (\text{odd})}}^{\infty} \cos nt, \tag{15}$$

which is clearly an absurd result. For instance, at $t = 0$ this tells us

$$0 = 0 + \frac{2}{\pi}(1 + 1 + 1 + \cdots). \tag{16}$$

To be assured that we can perform certain operations on a series, we must ordinarily impose a stronger type of convergence known as *uniform convergence*. Basically, the original series and the differentiated series must both be uniformly convergent. Uniform convergence implies that the given function $f(t)$ is *continuous*, rather than just piecewise continuous as required for pointwise convergence.

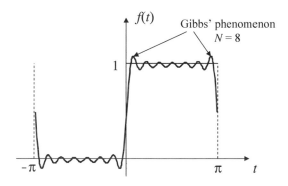

Figure 8.3 Gibbs' phenomenon illustrated for the square wave series (14) with $N = 8$.

8.2.3 Even and odd functions: cosine and sine series

Definition 8.1 If a given function $f(t)$ defined on $[-T/2, T/2]$ satisfies $f(-t) = f(t)$, we say the function is *even*. If $f(-t) = -f(t)$, we say the function is *odd*.

Products of even and odd functions have the following properties:

$$
\begin{aligned}
\text{Even} \times \text{Even} &= \text{Even} \\
\text{Even} \times \text{Odd} &= \text{Odd} \\
\text{Odd} \times \text{Odd} &= \text{Even}
\end{aligned}
$$

Because many functions that arise in practice are either even or odd, we find it useful to recognize the following—if $f(t)$ is an *even function*, then

$$\int_{-T/2}^{T/2} f(t)\,dt \;=\; 2\int_0^{T/2} f(t)\,dt; \tag{17}$$

and if $f(t)$ is an *odd function*, then

$$\int_{-T/2}^{T/2} f(t)\,dt \;=\; 0. \tag{18}$$

▸ **Cosine Series:** If $f(t)$ is an even function, the product $f(t)\cos n\omega_0 t$ is an even function and $f(t)\sin n\omega_0 t$ is an odd function. Hence, $b_n = 0$, $n = 1,2,3,\dots$, and the Fourier series reduces to a *cosine series*

$$\boxed{f(t) \;=\; \frac{1}{2}a_0 \;+\; \sum_{n=1}^{\infty} a_n \cos n\omega_0 t,} \tag{19}$$

$$\boxed{a_n \;=\; \frac{4}{T}\int_0^{T/2} f(t)\cos n\omega_0 t\,dt, \quad n = 0,1,2,\dots.} \tag{20}$$

▸ **Sine Series:** If $f(t)$ is an odd function, the product $f(t)\cos n\omega_0 t$ is an odd function and $f(t)\sin n\omega_0 t$ is an even function. Then, $a_n = 0$, $n = 0,1,2,\dots$, and the series reduces to a *sine series*

$$\boxed{f(t) \;=\; \sum_{n=1}^{\infty} b_n \sin n\omega_0 t,} \tag{21}$$

$$\boxed{b_n \;=\; \frac{4}{T}\int_0^{T/2} f(t)\sin n\omega_0 t\,dt, \quad n = 1,2,3,\dots.} \tag{22}$$

EXAMPLE 2 Find the Fourier series of the sawtooth function (see Fig. 8.4)

$$f(t) = t, \quad -1 < t < 1, \quad f(t+2) = f(t).$$

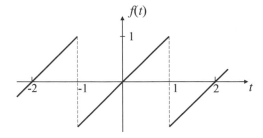

Figure 8.4

Solution: The period is $T = 2$ and, because $f(t)$ is an odd function, we seek only a sine series of the form

$$f(t) = \sum_{n=1}^{\infty} b_n \sin n\pi t,$$

where

$$b_n = 2 \int_0^1 f(t) \sin n\pi t \, dt = \frac{2}{n\pi} (-1)^{n-1}, \quad n = 1,2,3,\dots.$$

Here, we have again written $\cos n\pi = (-1)^n$, and thus, the Fourier series is

$$f(t) = \frac{2}{\pi} \sum_{n=1}^{\infty} \frac{(-1)^{n-1}}{n} \sin n\pi t.$$

Finally, we note that any function $f(t)$ can be written as the sum of an even function and an odd function, i.e.,

$$\begin{aligned}
f(t) &= \frac{1}{2}[f(t) + f(-t)] + \frac{1}{2}[f(t) - f(-t)] \\
&= f_e(t) + f_o(t).
\end{aligned} \tag{23}$$

8.2.4 Nonperiodic functions: extensions to other intervals

If a Fourier series converges, it must converge to a periodic function having the same period as the trigonometric functions. Suppose, however, we consider a *nonperiodic* function $f(t)$ defined only on the finite interval $[-T/2, T/2]$. Let $\hat{f}(t)$ denote the *periodic extension* of $f(t)$ over the entire axis. Then, provided it satisfies the conditions of Theorem 8.1, $\hat{f}(t)$ will have a convergent Fourier series that also represents the original function $f(t)$ over the finite interval $[-T/2, T/2]$. Hence, the only distinction between the two functions is the restricted interval of convergence of $f(t)$ to $-T/2 < t < T/2$. If the function $f(t)$ is defined on a nonsymmetric interval $[c, c + T]$, then we write the series as

$$f(t) = \frac{1}{2}a_0 + \sum_{n=1}^{\infty} \left(\cos \frac{2n\pi t}{T} + \sin \frac{2n\pi t}{T} \right), \quad c < t < c + T, \tag{24}$$

where

$$\begin{aligned}
a_n &= \frac{2}{T} \int_c^{c+T} f(t) \cos \frac{2n\pi t}{T} \, dt, \quad n = 0,1,2,\dots, \\
b_n &= \frac{2}{T} \int_c^{c+T} f(t) \sin \frac{2n\pi t}{T} \, dt, \quad n = 1,2,3,\dots.
\end{aligned} \tag{25}$$

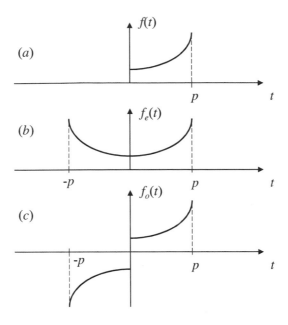

Figure 8.5 (*a*) The given function $f(t)$, (*b*) the even extension $f_e(t)$, and (*c*) the odd extension $f_0(t)$.

Given a function $f(t)$ defined only on the finite interval [0,p] as illustrated in Fig. 8.5(*a*), we can express it in either a cosine series or a sine series by imagining we "extend" it to the larger interval [$-p$,p] and define $T = 2p$. For example, if we let $f_e(t)$ denote the *even extension* of the function $f(t)$ as shown in Fig. 8.5(*b*), then $f_e(t)$ has a cosine series that converges to $f(t)$ on the interval [0,p]. Similarly, if we let $f_o(t)$ denote the *odd extension* of the function $f(t)$ as shown in Fig. 8.5(*c*), then $f_o(t)$ has a sine series that converges to $f(t)$ on the interval [0,p]. These type of series representations are sometimes called *half-range expansions*.

EXAMPLE 3 Find a Fourier series representation of

$$f(t) = t^2, \quad 0 < t < 1,$$

(*a*) as a sine series with period $T = 2$.
(*b*) as a cosine series with period $T = 2$.
(*c*) as a full trigonometric series with period $T = 1$.

Solution:
(*a*) The sine series has the form

$$f(t) = \sum_{n=1}^{\infty} b_n \sin n\pi t,$$

where

$$b_n = 2 \int_0^1 t^2 \sin n\pi t \, dt$$

$$= \frac{2}{n\pi}(-1)^{n-1} + \frac{4}{n^3\pi^3}[(-1)^n - 1], \quad n = 1,2,3,\dots.$$

(*b*) The corresponding cosine series is

$$f(t) = \frac{1}{2}a_0 + \sum_{n=1}^{\infty} a_n \cos n\pi t,$$

where

$$a_0 = 2 \int_0^1 t^2 \, dt = \frac{2}{3}$$

$$a_n = 2 \int_0^1 t^2 \cos n\pi t \, dt = \frac{4}{n^2\pi^2}(-1)^n, \quad n = 1,2,3,\dots.$$

(*c*) For the full trigonometric series with period $T = 1$, we use Eqs. (24) and (25) to write

$$f(t) = \frac{1}{2}a_0 + \sum_{n=1}^{\infty} (a_n \cos 2n\pi t + b_n \sin 2n\pi t),$$

where

$$a_0 = 2 \int_0^1 t^2 \, dt = \frac{2}{3},$$

$$a_n = 2 \int_0^1 t^2 \cos 2n\pi t \, dt = \frac{1}{n^2\pi^2}, \quad n = 1,2,3,\dots,$$

$$b_n = 2 \int_0^1 t^2 \sin 2n\pi t \, dt = -\frac{1}{n\pi}, \quad n = 1,2,3,\dots.$$

Hence, in this case we are led to the series

$$f(t) = \frac{1}{3} + \sum_{n=1}^{\infty} \left(\frac{\cos 2n\pi t}{n^2\pi^2} - \frac{\sin 2n\pi t}{n\pi} \right).$$

Remark: Note that part (*c*) in Example 3 is not simply the sum of parts (*a*) and (*b*). That is, the full trigonometric series (with $T = 1$) is not simply a sum of the cosine series and sine series (with $T = 2$) corresponding to even and odd extensions of the same function. In general, these three representations will be distinct.

8.3 Power Signals: Exponential Fourier Series

It is well known that a sinusoidal steady-state solution to an ac circuit problem is found most simply by inputting a complex time function $e^{i\omega t}$, where ω is angular frequency. In circuit problems and other related problems it is often useful to represent a given waveform or signal $f(t)$ in a complex exponential series rather than a trigonometric series. For example, the simple sinusoidal signal $f(t) = A\cos(\omega_0 t + \varphi)$ can be written as

$$f(t) = \text{Re}\left[A e^{i(\omega_0 t + \varphi)}\right], \tag{26}$$

where Re means the real part, ω_0 is the angular frequency, and φ is the phase. In this formulation, the signal is often called a *phasor*.

If the signal $f(t)$ is a periodic function having period $T = 2\pi/\omega_0$, then we have shown that we may represent it by a Fourier trigonometric series (called the *quadrature* form of the Fourier series)

$$f(t) = \frac{1}{2}a_0 + \sum_{n=1}^{\infty} \left(a_n \cos n\omega_0 t + b_n \sin n\omega_0 t\right), \tag{27}$$

$$a_n = \frac{2}{T}\int_{-T/2}^{T/2} f(t)\cos n\omega_0 t\, dt, \quad n = 0,1,2,\ldots,$$

$$b_n = \frac{2}{T}\int_{-T/2}^{T/2} f(t)\sin n\omega_0 t\, dt, \quad n = 1,2,3,\ldots. \tag{28}$$

However, we can replace $\cos n\omega_0 t$ and $\sin n\omega_0 t$ with complex exponential functions through use of the Euler identities, i.e.,

$$\cos n\omega_0 t = \frac{1}{2}(e^{in\omega_0 t} + e^{-in\omega_0 t}),$$

$$\sin n\omega_0 t = \frac{1}{2i}(e^{in\omega_0 t} - e^{-in\omega_0 t}). \tag{29}$$

By inserting Eqs. (29) into the series (27), collecting like terms, and noting that $1/i = -i$, we are led to

$$f(t) = \frac{1}{2}a_0 + \sum_{n=1}^{\infty}\left[\frac{1}{2}(a_n - ib_n)e^{in\omega_0 t} + \frac{1}{2}(a_n + ib_n)e^{-in\omega_0 t}\right]. \tag{30}$$

Now, if we define new (complex) constants

$$c_0 = \frac{1}{2}a_0, \quad c_n = \frac{1}{2}(a_n - ib_n), \quad c_{-n} = \frac{1}{2}(a_n + ib_n), \tag{31}$$

then (30) can be written as

$$f(t) = c_0 + \sum_{n=1}^{\infty}\left(c_n e^{in\omega_0 t} + c_{-n} e^{-in\omega_0 t}\right)$$

$$= c_0 + \sum_{n=1}^{\infty} c_n e^{in\omega_0 t} + \sum_{n=-1}^{-\infty} c_n e^{in\omega_0 t}, \tag{32}$$

or, more simply, as

$$f(t) = \sum_{n=-\infty}^{\infty} c_n e^{in\omega_0 t} = \sum_{n=-\infty}^{\infty} |c_n| e^{in\omega_0 t + i\varphi_n}, \tag{33}$$

where $c_n = |c_n|e^{i\varphi_n}$. The series (33) is called an *exponential* (or *complex*) *Fourier series* and the *Fourier coefficients* deduced from Eqs. (28) and (31) are defined by

$$c_n = \frac{1}{T}\int_{-T/2}^{T/2} f(t)e^{-in\omega_0 t}\,dt, \quad n = 0,\pm1,\pm2,\dots . \tag{34}$$

If the basic period of $f(t)$ is defined on the interval $[0,T]$ rather than on $[-T/2,T/2]$, then we can write

$$c_n = \frac{1}{T}\int_0^{T} f(t)e^{-in\omega_0 t}\,dt, \quad n = 0,\pm1,\pm2,\dots . \tag{35}$$

Remark: Note that the constant c_0 defined by (34) or (35) represents the *average value* of the signal $f(t)$.

The exponential form of the series (33) is often preferred over trigonometric series in analytical solutions because the Fourier coefficients are usually easier to calculate. Last, we note that $c_{-n} = c_n^*$ when $f(t)$ is real, where * denotes the complex conjugate.

EXAMPLE 4 A certain type of *full-wave rectifier* converts the input voltage $v(t)$ to its absolute value at the output, i.e., $|v(t)|$. Find the exponential Fourier series of the rectified sine wave (see Fig. 8.6)

$$f(t) = |v(t)| = A|\sin\pi t|, \quad A > 0.$$

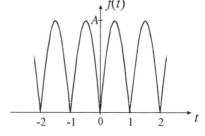

Figure 8.6

Solution: Here we see that $T = 1$ and $f(t) = \sin \pi t,\ 0 < t < 1.$ Thus, $\omega_0 = 2\pi/T = 2\pi$ and the series we seek is

$$f(t) = \sum_{n=-\infty}^{\infty} c_n e^{2ni\pi t},$$

where

$$c_n = A \int_0^1 \sin(\pi t) e^{-2ni\pi t}\, dt = \frac{A}{2i} \int_0^1 \left[e^{-(2n-1)i\pi t} - e^{-(2n+1)i\pi t} \right] dt,$$

which reduces to

$$c_n = -\frac{2A}{\pi(4n^2 - 1)},\quad n = 0, \pm 1, \pm 2, \dots.$$

The desired series then takes the form

$$f(t) = -\frac{2A}{\pi} \sum_{n=-\infty}^{\infty} \frac{e^{2ni\pi t}}{4n^2 - 1} = \frac{2A}{\pi} - \frac{4A}{\pi} \sum_{n=1}^{\infty} \frac{\cos 2n\pi t}{4n^2 - 1}.$$

The last result is a cosine series that arises because $f(t)$ is an even function.

8.3.1 Parseval's theorem and the power spectrum

Periodic signals are classified as *power signals*. In particular, if the period is T, then the *average power* across a 1-ohm resistance is found by integrating over one period, viz.,

$$P = \frac{1}{T} \int_{-T/2}^{T/2} |f(t)|^2\, dt. \tag{36}$$

The average power is related to the Fourier coefficients c_n through *Parseval's relation*

$$P = \frac{1}{T} \int_{-T/2}^{T/2} |f(t)|^2\, dt = \sum_{n=-\infty}^{\infty} |c_n|^2. \tag{37}$$

To formally derive Parseval's relation, let us consider multiplying Eq. (33) by its complex conjugate to find

$$|f(t)|^2 = \sum_{k=-\infty}^{\infty} \sum_{n=-\infty}^{\infty} c_n c_k^* e^{in\omega_0 t} e^{-ik\omega_0 t}. \tag{38}$$

Now, by integrating over one period, we have

$$\begin{aligned}
\int_{-T/2}^{T/2} |f(t)|^2\, dt &= \sum_{k=-\infty}^{\infty} \sum_{n=-\infty}^{\infty} c_n c_k^* \int_{-T/2}^{T/2} e^{i(n-k)\omega_0 t}\, dt \\
&= T \sum_{k=-\infty}^{\infty} \sum_{n=-\infty}^{\infty} c_n c_k^* \delta_{kn} = T \sum_{n=-\infty}^{\infty} |c_n|^2,
\end{aligned} \tag{39}$$

where $\delta_{kn} = 0$, $k \neq n$ and $\delta_{kn} = 1$, $k = n$ is the Kronecker delta (see Section 4.3.2).

Parseval's theorem (37) combined with the exponential Fourier series (33) shows that power in a periodic signal is distributed over discrete frequencies that are harmonically related to one another by $\omega_n = n\omega_0$ (called a *line spectrum*). We call ω_0 the *fundamental frequency,* and a graph of power P versus frequency ω is called a *two-sided power spectrum* of the signal $f(t)$. The zeroth coefficient c_0 represents the *dc component* in frequency of the signal $f(t)$, and $|c_n|$ is amplitude of the frequency component which has even symmetry. Also, φ_n provides the phase spectrum which has odd symmetry.

As a final comment, we note that the symbol $f(t)$ is a "time domain" representation of a given signal, whereas the Fourier series is a "frequency domain" representation of the same signal. The frequency domain representation is sometimes considered advantageous because it is essentially a "steady-state" description of the signal.

EXAMPLE 5 Calculate the average power of the periodic signal (period $T = 2$)

$$f(t) = 2\cos 5\pi t + \sin 6\pi t,$$

(*a*) using a time domain analysis.
(*b*) using a frequency domain analysis.

Solution:

(*a*) In the time domain, we calculate

$$P = \frac{1}{T}\int_{-T/2}^{T/2} |2\cos 5\pi t + \sin 6\pi t|^2\, dt.$$

Because the period $T = 2$, this expression reduces to

$$
\begin{aligned}
P &= \frac{1}{2}\int_{-1}^{1} \left(4\cos^2 5\pi t + 4\cos 5\pi t \sin 6\pi t + \sin^2 6\pi t\right) dt \\
&= \int_0^1 \left(4\cos^2 5\pi t + \sin^2 6\pi t\right) dt \\
&= \int_0^1 \left(2 + 2\cos 10\pi t + \frac{1}{2} - \frac{1}{2}\cos 12\pi t\right) dt,
\end{aligned}
$$

from which we deduce

$$P = 2 + \frac{1}{2} = 2.5\ \text{W}.$$

(*b*) For the frequency domain analysis, we start with Euler's formulas to write

$$
\begin{aligned}
f(t) &= 2\cos 5\pi t + \sin 6\pi t \\
&= e^{i5\pi t} + e^{-i5\pi t} + \frac{1}{2i}(e^{i6\pi t} + e^{-i6\pi t}).
\end{aligned}
$$

Hence, because this last expression is the Fourier series representation of the periodic signal $f(t)$, we see by inspection that

$$c_5 = c_{-5} = 1, \qquad c_6 = c_{-6} = \frac{1}{2i},$$

and all other $c_n = 0$. Hence, it follows that

$$
\begin{aligned}
P &= \sum_{n=-\infty}^{\infty} |c_n|^2 \\
&= |c_5|^2 + |c_{-5}|^2 + |c_6|^2 + |c_{-6}|^2 \\
&= 1 + 1 + \frac{1}{4} + \frac{1}{4} \\
&= 2.5 \text{ W}.
\end{aligned}
$$

8.4 Eigenvalue Problems and Orthogonal Functions

Eigenvalue problems are a special class of boundary value problems (BVPs) closely related to Fourier series through the orthogonal sets of functions that arise as solutions. In solving such problems, the differential equation (DE) is usually of the form

$$
\boxed{M[y] + \lambda y = 0, \quad a < x < b,}
\tag{40}
$$

where M is a differential operator (typically of second-order) defined by

$$
M \equiv A_2(x)D^2 + A_1(x)D + A_0(x), \quad D = \frac{d}{dx}.
\tag{41}
$$

The solution in such cases depends upon both x and the parameter λ. Thus, if y_1 and y_2 constitute linearly independent solutions of (40), the general solution is

$$
y = C_1 y_1(x,\lambda) + C_2 y_2(x,\lambda).
\tag{42}
$$

We now seek particular solutions of this general family (if any) that satisfy the *homogeneous boundary conditions* (BCs)

$$
\boxed{\text{BC:} \quad \begin{cases} B_1[y] \equiv y(a) + h_1 y'(a) = 0, \\ B_2[y] \equiv y(b) + h_2 y'(b) = 0, \end{cases}}
\tag{43}
$$

where B_1 and B_2 are boundary operators and h_1 and h_2 are constants.

Finding *nontrivial* solutions (i.e., nonzero solutions) of Eq. (40) subject to BCs (43) is what we call an *eigenvalue problem*. The values of λ that permit such solutions are called *eigenvalues* and the corresponding (nonzero) solutions are called *eigenfunctions*. The theory associated with this class of eigenvalue problems is similar to that presented in Chapter 3 with respect to matrices. In the present case, however, the number of eigenvalues (and corresponding eigenfunctions) is infinite rather than finite.

EXAMPLE 6 Find the eigenvalues and eigenfunctions of

$$y'' + \lambda y = 0, \quad 0 < x < p, \quad y(0) = 0, \ y(p) = 0.$$

Solution: By assuming a solution of the form $y = e^{mx}$, the resulting auxiliary equation $m^2 + \lambda = 0$ has roots

$$m = \pm\sqrt{-\lambda}.$$

Thus, the form of the general equation depends on the value of the parameter λ. We therefore take cases.

CASE I: Assume $\lambda = 0$, for which $m = 0, 0$ and

$$y = C_1 + C_2 x.$$

The prescribed BCs demand that $C_1 = C_2 = 0$, leading to the trivial solution $y = 0$. Hence, $\lambda = 0$ is not an eigenvalue.

CASE II: Assume $\lambda = -k^2 < 0$, for which $m = \pm k$ and

$$y = C_1 \cosh kx + C_2 \sinh kx.$$

Once again, the BCs demand that $C_1 = C_2 = 0$, leading to the trivial solution $y = 0$.

CASE III: Assume $\lambda = k^2 > 0$, which leads to $m = \pm ik$ and

$$y = C_1 \cos kx + C_2 \sin kx.$$

The first BC requires $C_1 = 0$, whereas the second condition yields

$$y(p) = C_2 \sin kp = 0.$$

By choosing $k = n\pi/p$, $n = 1,2,3, \dots$, we can leave C_2 arbitrary. Thus, we deduce that the eigenvalues are given by

$$\lambda_n = \frac{n^2 \pi^2}{p^2}, \quad n = 1,2,3,\dots,$$

and the corresponding eigenfunctions are any nonzero multiples of

$$\varphi_n(x) = \sin\frac{n\pi x}{p}, \quad n = 1,2,3,\dots .$$

EXAMPLE 7 Find the eigenvalues and eigenfunctions of

$$y'' + \lambda y = 0, \quad 0 < x < 1, \quad y(0) = 0, \ y(1) + y'(1) = 0.$$

Solution: The cases to consider are the same as those in the last example.

CASE I: Assume $\lambda = 0$, for which

$$y = C_1 + C_2 x.$$

The prescribed BCs demand that $C_1 = C_2 = 0$, leading to the trivial solution $y = 0$. Hence, $\lambda = 0$ is not an eigenvalue.

CASE II: Assume $\lambda = -k^2 < 0$, which leads to

$$y = C_1 \cosh kx + C_2 \sinh kx.$$

This time the BCs give us $C_1 = 0$ and the transcendental equation

$$y(1) + y'(1) = C_2(\sinh k + k \cosh k) = 0,$$

or

$$C_2(\tanh k + k) = 0.$$

By plotting the curves $u = \tanh k$ and $u = -k$ as shown in Fig. 8.7(a), we can see there are no intersections for $k > 0$. Hence, $C_2 = 0$ and we must conclude that there are no negative eigenvalues.

CASE III: Assume $\lambda = k^2 > 0$, which yields

$$y = C_1 \cos kx + C_2 \sin kx.$$

The first BC requires $C_1 = 0$, whereas the second condition yields

$$y(1) + y'(1) = C_2(\sin k + k \cos k) = 0.$$

If we require $C_2 \neq 0$, then we need solutions of the transcendental equation $\sin k + k \cos k = 0$, or

$$\tan k = -k.$$

By plotting the curves $u = \tan k$ and $u = -k$ as shown in Fig. 8.7(b), we can see that there are indeed infinitely-many roots of this equation which identify the eigenvalues. That is, the eigenvalues are given by

$$\lambda_n = k_n^2,$$

where the k_n's are solutions of the transcendental equation

$$\tan k_n + k_n = 0, \quad n = 1,2,3,\dots.$$

The corresponding eigenfunctions are

$$\varphi_n(x) = \sin k_n x, \quad n = 1,2,3,\ldots .$$

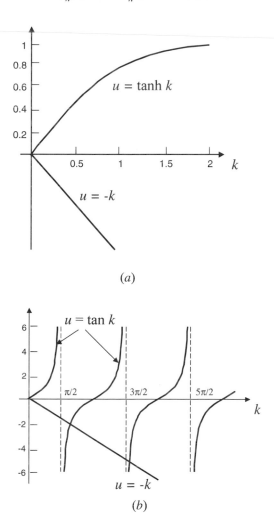

(a)

(b)

Figure 8.7 Graph of $(a)\, u = \tanh k$ and $u = -k$, and $(b)\, u = \tan k$ and $u = -k$.

Note that in both these last two examples we chose the arbitrary constant in front of the eigenfunctions to be unity as a matter of convenience. Any nonzero value for the constant will yield the same eigenfunction (i.e., eigenfunctions are unique only to within a multiplicative constant).

When the eigenequation is of a more general form, such as

$$ay'' + by' + \lambda y = 0, \tag{44}$$

we are led to the auxiliary equation

$$am^2 + bm + \lambda = 0. \tag{45}$$

The roots of (45) are

$$m = \frac{-b \pm \sqrt{b^2 - 4a\lambda}}{2a}, \tag{46}$$

and thus the three cases involved in the search for eigenvalues correspond to

$$\begin{array}{l} b^2 - 4a\lambda = 0, \\ b^2 - 4a\lambda > 0, \\ b^2 - 4a\lambda < 0. \end{array} \tag{47}$$

EXAMPLE 8 Find the eigenvalues and eigenfunctions of

$$y'' + y' + \lambda y = 0, \quad 0 < x < 3, \quad y(0) = 0, \, y(3) = 0.$$

Solution: The auxiliary equation leads to

$$m = -\frac{1}{2} \pm \frac{1}{2}\sqrt{1 - 4\lambda}.$$

We leave it to the reader to show that only the case in which $1 - 4\lambda = -4k^2 < 0$ yields any eigenvalues. In this case, $m = -1/2 \pm ik$, from which we obtain

$$y = e^{-x/2}(C_1 \cos kx + C_2 \sin kx).$$

The BCs require that $C_1 = 0$ and

$$y(3) = e^{-3/2} C_2 \sin 3k = 0.$$

For $C_2 \neq 0$, we choose $3k = n\pi$, $n = 1,2,3,...,$ to satisfy this last equation. Hence, from the assumed relation $\lambda = 1/4 + k^2$, the eigenvalues are found to be

$$\lambda_n = \frac{1}{4} + \frac{n^2\pi^2}{9}, \quad n = 1,2,3,...,$$

and the corresponding eigenfunctions are

$$\varphi_n(x) = e^{-x/2} \sin\frac{n\pi x}{3}, \quad n = 1,2,3,....$$

8.4.1 Regular Sturm-Liouville systems

Here we wish to examine some of the theory concerning eigenvalue problems. To start, let us consider the general eigenequation

$$M[y] + \lambda y = 0, \quad a < x < b,$$ (48)

where M is defined by (41) and here we assume $A_2(x) \neq 0$ on $[a,b]$.

▸ **Self-adjoint form:** We say a DE is in *self-adjoint form* if it can be expressed as

$$p(x)y'' + p'(x)y' + [q(x) + \lambda r(x)]y = 0,$$ (49)

or, equivalently, as

$$\frac{d}{dx}[p(x)y'] + [q(x) + \lambda r(x)]y = 0.$$ (50)

To put (48) in the form suggested by (49) or (50), we divide the DE in (48) by $A_2(x)$ and multiply by $p(x)$. This action leads to

$$p(x)y'' + p(x)\frac{A_1(x)}{A_2(x)}y' + \left[p(x)\frac{A_0(x)}{A_2(x)} + \lambda\frac{p(x)}{A_2(x)}\right]y = 0.$$ (51)

This last equation is in self-adjoint form (49) provided

$$p'(x) = p(x)\frac{A_1(x)}{A_2(x)}.$$ (52)

By separating variables in (52), we deduce that

$$p(x) = \exp\left[\int \frac{A_1(x)}{A_2(x)}dx\right].$$ (53)

Hence, virtually any equation can be put in self-adjoint form. Among other reasons, putting the DE in self-adjoint form identifies the functions $p(x)$ and $r(x)$, the latter called the *weighting function*.

EXAMPLE 9 Put the following DE in self-adjoint form:

$$x^2y'' + 3xy' + \lambda y = 0.$$

Solution: By first dividing the DE by x^2 and then using (53), we see that

$$p(x) = \exp\left(3\int \frac{1}{x} dx \right) = e^{3\ln x} = x^3.$$

Hence, the self-adjoint form is

$$x^3 y'' + 3x^2 y' + \lambda x y = 0.$$

From this last result, we note that the weighting function is $r(x) = x$.

In the following work it is useful to introduce the notion of a self-adjoint operator.

▸ **Self-adjoint operator:** A *self-adjoint operator* has the form

$$L = \frac{d}{dx}\left[p(x)\frac{d}{dx} \right] + q(x). \tag{54}$$

Using L, we can rewrite (50) together with a prescribed set of BCs more compactly as

$$\boxed{\begin{aligned} L[y] + \lambda r(x)y &= 0, \\ \text{BC:} \quad B_1[y] = 0, \quad B_2[y] &= 0, \end{aligned}} \tag{55}$$

Eigenvalue problems of the type (55) in which the BCs are *unmixed* are called *regular Sturm-Liouville systems,* named in honor of the two French mathematicians who investigated problems of this nature in the 1830s. Because we assume $A_2(x) \neq 0$ on $[a,b]$, it follows that $p(x) > 0$ on this interval. This class of eigenvalue problems is characterized by the following theorem.

Theorem 8.2 The eigenvalues of a regular Sturm-Liouville system have the properties:
- the eigenvalues are real
- only one eigenfunction belongs to each eigenvalue
- the eigenvalues form an infinite sequence ordered in increasing magnitude so that

$$\lambda_1 < \lambda_2 < \cdots \lambda_n < \cdots, \quad \lambda_n \to \infty \text{ as } n \to \infty.$$

It can be shown that the eigenfunctions are also real, but their most important property is that they are mutually orthogonal (see Theorem 8.3 below).

> **Definition 8.2** A set of real-valued functions $\{\varphi_n(x)\}$, $n = 1,2,3, \dots$, is said to be *orthogonal* on the interval $[a,b]$ with respect to a weighting function $r(x) > 0$ if
> $$\int_a^b r(x)\varphi_n(x)\varphi_k(x)\,dx = 0, \quad n \neq k.$$

▸ **Norm:** When $k = n$ in the integral above in Definition 8.2, we are led to

$$\|\varphi_n(x)\| = \sqrt{\int_a^b r(x)[\Phi_n(x)]^2\,dx}, \quad n = 1,2,3,\dots, \tag{56}$$

called the *norm* of the function $\varphi_n(x)$. In many cases, the weighting function $r(x)$ is simply unity.

The following theorem is a central result in the general theory concerning the eigenfunctions of Sturm-Liouville systems.

> **Theorem 8.3** (*Orthogonality*) If $\varphi_n(x)$ and $\varphi_k(x)$ are eigenfunctions of a regular Sturm-Liouville system (55) belonging to distinct eigenvalues λ_n and λ_k, respectively, then they are mutually orthogonal, i.e.,
> $$\int_a^b r(x)\varphi_n(x)\varphi_k(x)\,dx = 0, \quad n \neq k.$$

EXAMPLE 10 Find the norm of the eigenfunctions of

$$y'' + \lambda y = 0, \quad y(0) = 0, \; y(p) = 0.$$

Solution From Example 6, we know $\varphi_n(x) = \sin\dfrac{n\pi x}{p}$, $n = 1,2,3,\dots$. Thus,

$$\|\varphi_n(x)\|^2 = \int_0^p \sin^2\frac{n\pi x}{p}\,ds = \frac{p}{2},$$

and, consequently, by taking the square root, we obtain the norm

$$\|\varphi_n(x)\| = \sqrt{\frac{p}{2}}, \quad n = 1,2,3,\dots.$$

8.4.2 Generalized Fourier series

If $\{\varphi_n(x)\}$ denotes an orthogonal set of eigenfunctions on the interval $[a,b]$ with weighting function $r(x)$, and $f(x)$ is a suitable function on this interval, then we might consider series expansions of the form

$$f(x) = \sum_{n=1}^{\infty} c_n \varphi_n(x), \quad a < x < b, \tag{57}$$

where the c's are constants to be determined. By analogy with trigonometric Fourier series, let us multiply both sides of (57) by $r(x)\varphi_k(x)$ and integrate over the interval $[a,b]$. Doing so yields

$$\int_a^b r(x)f(x)\varphi_k(x)\,dx = \sum_{n=1}^{\infty} c_n \int_a^b r(x)\varphi_n(x)\varphi_k(x)\,dx, \tag{58}$$

where we have interchanged the order of summation and integration. Because of the orthogonality relation (Theorem 8.3), it follows that all terms of the series vanish except for that when $n = k$. Thus, we deduce that

$$\int_a^b r(x)f(x)\varphi_k(x)\,dx = c_k \|\varphi_k(x)\|^2. \tag{59}$$

If we solve (59) for c_k and change the dummy subscript back to n, we have

$$c_n = \|\varphi_n(x)\|^{-2} \int_a^b r(x)f(x)\varphi_n(x)\,dx, \quad n = 1,2,3\ldots. \tag{60}$$

Infinite series expansions like (57) are called *generalized Fourier series*, or *eigenfunction expansions*. The constants (60) are called *Fourier coefficients*. Without proof, we state the following important theorem concerning convergence.

Theorem 8.4 If $\{\varphi_n(x)\}$ is the set of eigenfunctions of a regular Sturm-Liouville system on $[a,b]$, and if $f(x)$ and $f'(x)$ are piecewise continuous functions on $[a,b]$, then the eigenfunction expansion (57) converges on the open interval (a,b) to $f(x)$ at points of continuity and to the average value $(1/2)[f(x^+) + f(x^-)]$ of the left-hand and right-hand limits at points of discontinuity.

Notice that Theorem 8.4 is remarkably similar to Theorem 8.1 concerning trigonometric series. In general, there are many similarities between trigonometric and generalized Fourier series. Let us illustrate with the simple eigenvalue problem

$$y'' + \lambda y = 0, \quad 0 < x < p, \quad y(0) = 0, \quad y(p) = 0, \tag{61}$$

in which the eigenvalues are $\lambda_n = n^2\pi^2/p^2$, $n = 1,2,3,...$, and the corresponding eigenfunctions are $\varphi_n(x) = \sin(n\pi x/p)$, $n = 1,2,3,...$ (recall Example 5). Therefore, based on Theorems 8.3 and 8.4, the eigenfunctions are orthogonal and can be used to find a generalized Fourier series of the form

$$f(x) = \sum_{n=1}^{\infty} c_n \sin\frac{n\pi x}{p}, \quad 0 < x < p, \tag{62}$$

where the Fourier coefficients are

$$c_n = \frac{2}{p}\int_0^p f(x)\sin\frac{n\pi x}{p}\,dx, \quad n = 1,2,3,.... \tag{63}$$

In arriving at (63) we have used the normalization factor found in Example 9. The series defined by (62) and (63) is recognized as simply a *sine series* of the type previously discussed in Section 8.2.3. Along similar lines, the eigenvalue problem

$$y'' + \lambda y = 0, \quad 0 < x < p, \quad y'(0) = 0, \; y'(p) = 0, \tag{64}$$

has eigenvalues $\lambda_0 = 0$, $\lambda_n = n^2\pi^2/p^2$, $n = 1,2,3...$, and eigenfunctions $\varphi_0(x) = 1$, $\varphi_n(x) = \cos(n\pi x/p)$, $n = 1,2,3,....$, which lead to the *cosine series*

$$f(x) = \frac{1}{2}a_0 + \sum_{n=1}^{\infty} a_n \cos\frac{n\pi x}{p}, \quad 0 < x < p, \tag{65}$$

where

$$a_n = \frac{2}{p}\int_0^p f(x)\cos\frac{n\pi x}{p}\,dx, \quad n = 0,1,2,.... \tag{66}$$

EXAMPLE 11 Find an expression for the Fourier coefficients associated with the generalized Fourier series arising from the eigenfunctions of

$$y'' + y' + \lambda y = 0, \quad 0 < x < 3, \quad y(0) = 0, \; y(3) = 0.$$

Solution: We first solve for the eigenvalues and eigenfunctions, which are readily found to be (recall Example 8)

$$\lambda_n = \frac{1}{4} + \frac{n^2\pi^2}{9}, \quad n = 1,2,3,...,$$

$$\varphi_n(x) = e^{-x/2}\sin\frac{n\pi x}{3}, \quad n = 1,2,3,....$$

Thus, the generalized Fourier series for these eigenfunctions is of the form

$$f(x) = \sum_{n=1}^{\infty} c_n e^{-x/2} \sin \frac{n\pi x}{3}, \quad 0 < x < 3.$$

At this point we need to put the DE in self-adjoint form to identify the weighting function $r(x)$. This we do by multiplying the DE by e^x to obtain

$$\frac{d}{dx}(e^x y') + \lambda e^x y = 0,$$

from which we deduce $r(x) = e^x$. The square of the norm is

$$\begin{aligned}
\|\varphi_n(x)\|^2 &= \int_0^3 e^x \left[e^{-x/2} \sin \frac{n\pi x}{3} \right]^2 dx \\
&= \int_0^3 \sin^2 \frac{n\pi x}{3} dx \\
&= \frac{3}{2}, \quad n = 1,2,3,\dots,
\end{aligned}$$

and the Fourier coefficients are therefore defined by

$$\begin{aligned}
c_n &= \frac{2}{3} \int_0^3 e^x f(x) \left(e^{-x/2} \sin \frac{n\pi x}{3} \right) dx \\
&= \frac{2}{3} \int_0^3 e^{x/2} f(x) \sin \frac{n\pi x}{3} dx, \quad n = 1,2,3,\dots.
\end{aligned}$$

EXAMPLE 12 Find a generalized Fourier series expansion of the function $f(x) = 1$, $0 < x < 1$, in terms of the eigenfunctions of

$$y'' + \lambda y = 0, \quad 0 < x < 1, \quad y(0) = 0, \ y(1) + y'(1) = 0.$$

Solution: We previously found in Example 7 that the eigenfunctions are given by $\varphi_n(x) = \sin k_n x$, $n = 1,2,3,\dots$, where the k's are the positive roots of

$$\sin k_n + k_n \cos k_n = 0, \quad n = 1,2,3,\dots.$$

The corresponding generalized Fourier series we seek is

$$f(x) = \sum_{n=1}^{\infty} c_n \sin k_n x, \quad 0 < x < 1.$$

In this case the DE is in self-adjoint form so we identify $r(x) = 1$. The square of the norm is

$$\|\varphi_n(x)\|^2 = \int_0^1 \sin^2 k_n dx = \frac{1}{2}\left(1 - \frac{\sin 2k_n}{2k_n} \right),$$

or, because $\sin 2k_n = 2\sin k_n \cos k_n$ and $\sin k_n = -k_n \cos k_n$, we can write

$$\|\varphi_n(x)\|^2 = \frac{1}{2}\left(1 + \cos^2 k_n\right).$$

Based on the above results, the Fourier coefficients are

$$c_n = \frac{2}{1 + \cos^2 k_n}\int_0^1 (1)\sin k_n x \, dx = \frac{2(1 - \cos k_n)}{k_n(1 + \cos^2 k_n)},$$

and, consequently,

$$f(x) = 2\sum_{n=1}^{\infty}\frac{1 - \cos k_n}{k_n(1 + \cos^2 k_n)}\sin k_n x, \quad 0 < x < 1.$$

8.4.3 Periodic Sturm-Liouville systems

In addition to regular Sturm-Liouville systems, we have the class of *periodic systems* defined by

$$
\boxed{
\begin{array}{l}
L[y] + \lambda y = 0, \quad a < x < b \\[4pt]
\text{BC:} \quad y(a) = y(b), \quad y'(a) = y'(b),
\end{array}
}
\tag{67}
$$

where $L = D[p(x)D] + q(x)$ and $p(a) = p(b)$. Theorem 8.3 is applicable for this class of problems and Theorem 8.2 also applies here with one possible exception. That is, it is "possible" for more than one eigenfunction to exist for each eigenvalue in a periodic system as illustrated in the following example.

EXAMPLE 13 Find the eigenvalues and eigenfunctions belonging to

$$y'' + \lambda y = 0, \quad -\pi < x < \pi$$
$$y(-\pi) = y(\pi), \quad y'(-\pi) = y'(\pi).$$

Solution: Here $p(x) = 1$ so $p(-\pi) = p(\pi)$. When $\lambda = 0$, the general solution is

$$y = C_1 + C_2 x.$$

The BCs require that $C_2 = 0$ while leaving C_1 arbitrary. Thus, we deduce that

$$\lambda_0 = 0, \quad \varphi_0(x) = 1$$

form an eigenvalue-eigenfunction pair. It can be shown that there are no negative eigenvalues. By assuming, therefore, that $\lambda = k^2 > 0$, the general solution is

$$y = C_1 \cos kx + C_2 \sin kx.$$

Upon simplification, the BCs lead to

$$C_1 \sin k\pi = 0, \quad C_2 \sin k\pi = 0,$$

from which we conclude that $k = n$, $n = 1,2,3,\dots$. The positive eigenvalues are

$$\lambda_n = n^2, \quad n = 1,2,3,\dots,$$

and because both C_1 and C_2 are arbitrary, it follows that there are *two* eigenfunctions corresponding to each positive eigenvalue. Although not unique, it is customary to choose the eigenfunctions

$$\varphi_n(x) = \begin{cases} \cos nx, & n = 1,2,3,\dots \\ \sin nx, & n = 1,2,3,\dots. \end{cases}$$

The eigenfunctions chosen in this fashion are all mutually orthogonal.

As a final comment here we note that the above set of eigenfunctions will lead to Fourier series of the form

$$f(x) = \frac{1}{2}a_0 + \sum_{n=1}^{\infty} \left(a_n \cos nx + b_n \sin nx\right),$$

where

$$a_n = \frac{1}{\pi}\int_{\pi}^{\pi} f(x)\cos nx\, dx, \quad n = 0,1,2,\dots,$$

$$b_n = \frac{1}{\pi}\int_{\pi}^{\pi} f(x)\sin nx\, dx, \quad n = 1,2,3,\dots.$$

Clearly, this is the general Fourier series representation for a periodic function with period $T = 2\pi$ that was first discussed in Section 8.2.

8.4.4 Singular Sturm-Liouville systems

Some of the more interesting eigenvalue problems found in practice are classified as *singular* as defined below. The basic DE still has the self-adjoint form

$$L[y] + \lambda y = 0, \quad a < x < b, \tag{68}$$

where $L = D[p(x)D] + q(x)$. Singularities associated with (68) change the general nature of the the system, especially in the form of BCs necessary to maintain the basic properties of the eigenvalues (Theorem 8.2) and orthogonality of the eigenfunctions (Theorem 8.3).

Definition 8.3 A Sturm-Liouville system is said to be *singular* if any of the following occur:
- ▸ $p(a) = 0$ and/or $p(b) = 0$.
- ▸ $p(x)$, $q(x)$, or $r(x)$ becomes unbounded at $x = a$ or $x = b$ (or both).
- ▸ either end point of the interval is infinite (or both).

The proper type of BCs to prescribe depends on the type of singularity. For example, if the singularity arises from $p(a) = 0$, then the proper type of BC at the endpoint $x = a$ is given by $|y(a)| < \infty$, or more formally, by

$$y(x),\ y'(x)\ \text{finite as}\ x \to a^+. \tag{69}$$

If there is no singularity at the endpoint $x = b$, any type of unmixed BC is acceptable.

In practice, *Legendre's equation* (see Sections 1.7 and 2.4.1) often appears in the form of an eigenequation given by

$$(1 - x^2)y'' - 2xy' + \lambda y = 0, \quad -1 < x < 1, \tag{70}$$

which we note is in self-adjoint form. Clearly, $p(x) = 1 - x^2$ vanishes at both endpoints $x = \pm 1$, which classifies this equation as *singular*. To ensure the orthogonality of the eigenfunctions, we therefore require BCs of the form

$$\text{BC:}\quad y(x),\ y'(x)\quad \text{finite as}\quad x \to -1^+,\ x \to 1^-. \tag{71}$$

We previously mentioned (Section 1.7) that bounded solutions of Legendre's DE exist only for the values

$$\lambda_n = n(n + 1), \quad n = 0,1,2,..., \tag{72}$$

which we now conclude are the eigenvalues. However, the proof that these are the only eigenvalues is too complicated to present here. Given the eigenvalues (72), the corresponding eigenfunctions are the *Legendre polynomials*, viz.,

$$\varphi_n(x) = P_n(x), \quad n = 0,1,2,.... \tag{73}$$

With $r(x) = 1$ from inspection of (70), it follows from Theorem 8.3 that the Legendre polynomials form an orthogonal set of eigenfunctions; i.e.,

$$\int_{-1}^{1} P_n(x)P_k(x)dx = 0, \quad n \neq k. \tag{74}$$

Also, when $k = n$, it can be shown that the norm is

$$\|P_n(x)\| = \sqrt{\int_{-1}^{1}[P_n(x)]^2 dx} = \sqrt{\frac{2}{2n + 1}}. \tag{75}$$

Based on the results (74) and (75), we seek generalized Fourier series of the form

$$f(x) = \sum_{n=0}^{\infty} c_n P_n(x), \quad -1 < x < 1, \tag{76}$$

where the Fourier coefficients are defined by

$$c_n = \left(n + \frac{1}{2}\right) \int_{-1}^{1} f(x) P_n(x) dx, \quad n = 1, 2, 3, \ldots. \tag{77}$$

Series of this type are also called *Legendre series*.

EXAMPLE 14 Find the Legendre series for

$$f(x) = \begin{cases} -1, & -1 < x < 0 \\ 1, & 0 < x < 1. \end{cases}$$

Solution: We first observe that $f(x)$ is an odd function, and because $P_n(x)$ is even or odd depending on index n, it follows that $c_n = 0$, $n = 0, 2, 4, \ldots.$ For odd index, we write

$$c_n = (2n + 1) \int_0^1 P_n(x) dx, \quad n = 1, 3, 5, \ldots.$$

Recalling

$$P_1(x) = x, \quad P_3(x) = \frac{1}{2}(5x^3 - 3x), \quad P_5(x) = \frac{1}{8}(63x^5 - 70x^3 + 15x),$$

we find that $c_1 = 3/2$, $c_3 = -7/8$, and $c_5 = 11/16$, and thus

$$f(x) = \frac{3}{2} P_1(x) - \frac{7}{8} P_3(x) + \frac{11}{16} P_5(x) + \cdots, \quad -1 < x < 1.$$

Finding an expression for the general Fourier coefficient c_n is usually a difficult task analytically. In practice, we might settle for just the first few coefficients, as done in Example 14. If the function $f(x)$ is a polynomial, then some simplifications take place as illustrated in the next example.

EXAMPLE 15 Find the Legendre series for $f(x) = 2x + x^2$.

Solution: Because the given polynomial is of degree two, we expect Legendre polynomials in the series only up to second degree. That is, because two polynomials are equal if and only if they are of the same degree, it follows that

$$2x + x^2 = c_0 P_0(x) + c_1 P_1(x) + c_2 P_2(x)$$

$$= c_0 + c_1 x + c_2 \frac{1}{2}(3x^2 - 1)$$

$$= (c_0 - \frac{1}{2} c_2) + c_1 x + \frac{3}{2} c_2 x^2.$$

By comparing like coefficients on each side of the last expression, we readily find that $c_0 = 1/3$, $c_1 = 2$, and $c_2 = 2/3$. Hence, we deduce that

$$2x + x^2 = \frac{1}{3}P_0(x) + 2P_1(x) + \frac{2}{3}P_2(x).$$

Another singular Sturm-Liouville problem that often arises involves *Bessel's equation* (see Sections 1.9 and 2.5), which in this setting is generally expressed as

$$x^2 y'' + xy' + (\lambda x^2 - v^2)y = 0, \quad 0 < x < b, \quad v \geq 0. \tag{78}$$

In self-adjoint form, Eq. (78) becomes

$$\frac{d}{dx}(xy') - \frac{v^2}{x}y + \lambda xy = 0, \tag{79}$$

where we identify $p(x) = x$ and $r(x) = x$. Clearly, $x = 0$ is a singular point, so typical BCs are generally of the form

$$\text{BC:} \quad y(x), \, y'(x) \quad \text{finite as} \quad x \to 0^+, \quad y(b) + h_2 y'(b) = 0, \tag{80}$$

where h_2 is a constant.

> **Remark:** Although the first BC in (80) requires that both y and y' be bounded, we find that it is generally sufficient to simply require $|y(0)| < \infty$.

For illustrative purposes, let us assume the BC (80) at $x = b$ is $y(b) = 0$. To solve for the eigenvalues, we take the following cases.

CASE I: By assuming $\lambda = 0$, Eq. (78) reduces to the Cauchy-Euler equation $x^2 y'' + xy' - v^2 y = 0$ with general solution (see Section 1.4.5)

$$y = \begin{cases} C_1 + C_2 \ln x, & v = 0 \\ C_1 x^v + C_2 x^{-v}, & v \neq 0. \end{cases} \tag{81}$$

The BC at $x = 0$ demands that $C_2 = 0$ in both cases above. Also, the BC at $x = b$ leads to $C_1 = 0$ in both cases. Thus, $\lambda = 0$ is *not* an eigenvalue.

CASE II: Assume $\lambda = -k^2 < 0$. In this case, the general solution is given in terms of *modified Bessel functions* (Section 2.5.2); i.e.,

$$y = C_1 I_\nu(kx) + C_2 K_\nu(kx). \tag{82}$$

Because $K_\nu(kx)$ is not bounded at $x = 0$, we must set $C_2 = 0$. The remaining BC requires that

$$y(b) = C_1 I_\nu(kb) = 0. \tag{83}$$

However, the modified Bessel function of the first kind does not have zeros away from the origin (recall Fig. 2.16); thus, we deduce there are no negative eigenvalues.

CASE III: Last, we set $\lambda = k^2 > 0$, and obtain the general solution

$$y = C_1 J_\nu(kx) + C_2 Y_\nu(kx). \tag{84}$$

The solution $Y_\nu(kx)$ is not bounded at the origin, so we set $C_2 = 0$. The remaining BC requires that

$$y(b) = C_1 J_\nu(kb) = 0. \tag{85}$$

Here now the situation is different owing to the fact that the standard Bessel function has infinitely many zeros on the positive axis, although we do not have explicit values for them all. We simply say that the eigenvalues of this problem are related to the zeros of the Bessel function through the relation

$$\lambda_n = k_n^2, \quad \text{where} \quad J_\nu(k_n b) = 0, \quad n = 1,2,3,..., \tag{86}$$

and the corresponding eigenfunctions are

$$\varphi_n(x) = J_\nu(k_n x), \quad n = 1,2,3,.... \tag{87}$$

EXAMPLE 16 Assuming $\lambda > 0$, determine the eigenvalues and eigenfunctions of the singular Sturm-Liouville problem

$$4x^2 y'' + (\lambda x^2 - 3)y = 0, \quad 0 < x < 4,$$
$$y(x),\ y'(x) \quad \text{finite as} \quad x \to 0^+, \quad y'(4) = 0.$$

Solution By setting $\lambda = k^2 > 0$ and writing the DE as

$$x^2 y'' + \left(\frac{k^2}{4} x^2 - \frac{3}{4} \right) y = 0,$$

we can identify $a = 1/2$, $b = k/2$, $c = 1$, and $\nu = 1$ by comparison with the general form discussed in Section 1.9.4. Hence, the general solution takes the form

$$y = \sqrt{x}\left[C_1 J_1\left(\frac{kx}{2}\right) + C_2 Y_1\left(\frac{kx}{2}\right)\right].$$

To investigate the behavior for x near the origin, we recall the asymptotic relations

$$J_1\left(\frac{kx}{2}\right) \sim \frac{kx}{4}, \quad Y_1\left(\frac{kx}{2}\right) \sim -\frac{4}{\pi kx}, \quad x \to 0^+,$$

and deduce that $C_2 = 0$ to avoid unbounded solutions. It follows that the only bounded solution is

$$y = C_1 \sqrt{x} J_1\left(\frac{kx}{4}\right).$$

The derivative of the solution is

$$y' = C_1\left[\frac{1}{2\sqrt{x}}J_1\left(\frac{kx}{2}\right) + \frac{k\sqrt{x}}{2}J_1'\left(\frac{kx}{2}\right)\right],$$

and by imposing the BC at $x = 4$, we see that

$$y'(4) = C_1\left[\frac{1}{4}J_1(2k) + kJ_1'(2k)\right] = 0.$$

We conclude that the eigenvalues are given by

$$\lambda_n = k_n^2, \quad J_1(2k_n) + 4k_n J_1'(2k_n) = 0, \quad n = 1,2,3,...,$$

with corresponding eigenfunctions

$$\varphi_n(x) = \sqrt{x} J_1(k_n x/2), \quad n = 1,2,3,....$$

The orthogonality property of the Bessel functions is given by

$$\int_0^b x J_\nu(k_n x) J_\nu(k_m x)\,dx = 0, \quad m \neq n, \tag{88}$$

but the norm depends on the form of the BC at the endpoint $x = b$. For the simple BC $y(b) = 0$, it has been shown that the norm is

$$\|\varphi_n(x)\| = \sqrt{\int_0^b x\left[J_\nu(k_n x)\right]^2 dx} = \sqrt{\frac{1}{2}b J_{\nu+1}(k_n b)}, \quad n = 1,2,3,.... \tag{89}$$

However, the norm is somewhat different for different BCs at $x = b$. The generalized Fourier series associated with the Bessel functions, also called a *Bessel series*, is defined by

$$f(x) = \sum_{n=1}^{\infty} c_n J_\nu(k_n x), \quad 0 < x < b, \tag{90}$$

where, based on (89), the Fourier coefficients are

$$c_n = \frac{2}{b^2 [J_{\nu+1}(k_n b)]^2} \int_0^b x f(x) J_\nu(k_n x) dx, \quad n = 1,2,3,\dots. \tag{91}$$

Clearly, the evaluation of such integrals (in general) is not an easy task!

EXAMPLE 17 Find the Bessel series for

$$f(x) = \begin{cases} x, & 0 < x < 1 \\ 0, & 1 < x < 2, \end{cases}$$

corresponding to $\{J_1(k_n x)\}$, where $J_1(2k_n) = 0$, $n = 1,2,3,\dots$.

Solution: The series we seek has the form

$$f(x) = \sum_{n=1}^{\infty} c_n J_1(k_n x), \quad 0 < x < 2,$$

where

$$c_n = \frac{1}{2[J_2(2k_n)]^2} \int_0^2 x f(x) J_1(k_n x) dx$$

$$= \frac{1}{2[J_2(2k_n)]^2} \int_0^1 x^3 J_1(k_n x) dx$$

By making the change of variable $t = k_n x$ and using the integral formula [recall property (BJ13) in Section 2.5]

$$\int t^\nu J_{\nu-1}(t) dt = t^\nu J_\nu(t) + C,$$

it follows that

$$c_n = \frac{J_2(k_n)}{2k_n [J_2(2k_n)]^2}, \quad n = 1,2,3,\dots.$$

8.5 Green's Function

In Chapter 1 we learned that *any* particular solution of a nonhomogeneous DE can be combined with the general homogeneous solution to produce the general solution of the nonhomogeneous DE. However, it can be useful in practice to use a method that leads to a unique choice of particular solution that also has special physical significance. Such a method is introduced here, based on the construction of a specific function called the Green's function.

8.5.1 One-sided Green's function

To begin, we consider the general initial value problem (IVP) described by

$$A_2(t)y'' + A_1(t)y' + A_0(t)y = F(t), \; t > t_0, \;\; y(t_0) = k_0, \; y'(t_0) = k_1. \tag{92}$$

The unknown function y is now considered a function of t. For purposes of developing the theory, we use the linearity of the problem to split it into the following two problems (recall Section 1.3.3):

PROBLEM (A):
$$\begin{aligned} A_2(t)y'' + A_1(t)y' + A_0(t)y &= 0, \\ y(t_0) = k_0, \; y'(t_0) &= k_1 \end{aligned} \tag{93}$$

PROBLEM (B):
$$y'' + a_1(t)y' + a_0(t)y = f(t), \;\; y(t_0) = 0, \; y'(t_0) = 0 \tag{94}$$

The significant feature in each problem is that either the DE or the initial conditions (ICs) are homogeneous. Also, in PROBLEM (B) we have put the DE in *normal form*, where $a_1(t) = A_1(t)/A_2(t)$, $a_0(t) = A_0(t)/A_2(t)$, and $f(t) = F(t)/A_2(t)$.

For PROBLEM (A) above, we denote the solution by y_H. This solution physically represents the response of the original system described by the IVP (92) that is entirely due to the ICs in the absence of an external disturbance $F(t)$. We denote the solution to PROBLEM (B) above by y_P. In this case the solution represents the response of the same system which is at rest until time $t = t_0$, at which time it is subject to the external input $f(t)$. We often refer to the external input as a forcing function.

The general solution of PROBLEM (A) can be written in the form

$$y_H = C_1 y_1(t) + C_2 y_2(t). \tag{95}$$

By imposing the prescribed ICs, we obtain

$$\begin{aligned} C_1 y_1(t_0) + C_2 y_2(t_0) &= k_0, \\ C_1 y_1'(t_0) + C_2 y_2'(t_0) &= k_1, \end{aligned}$$

the simultaneous solution of which is

$$C_1 = \frac{k_0 y_2{}'(t_0) - k_1 y_2(t_0)}{W(y_1,y_2)(t_0)}, \quad C_2 = \frac{k_1 y_1(t_0) - k_0 y_1{}'(t_0)}{W(y_1,y_2)(t_0)}. \tag{96}$$

The denominators in (96) are the nonvanishing Wronskian. Note that when $k_0 = k_1 = 0$, Eqs. (96) yield $C_1 = C_2 = 0$, and consequently we obtain the *trivial solution* $y_H = 0$.

From Section 1.4.4 we know a particular solution of the DE in PROBLEM (B) is given by (using the method of variation of parameters)

$$y_P = u(t)y_1(t) + v(t)y_2(t), \tag{97}$$

where

$$u(t) = -\int \frac{y_2(t)f(t)}{W(y_1,y_2)(t)}\,dt, \quad v(t) = \int \frac{y_1(t)f(t)}{W(y_1,y_2)(t)}\,dt. \tag{98}$$

Rather than define $u(t)$ and $v(t)$ by indefinite integrals, here we wish to choose them so that y_P satisfies the ICs in (94), viz.,

$$u(t) = -\int_{t_0}^{t} \frac{y_2(\tau)f(\tau)}{W(y_1,y_2)(\tau)}\,d\tau, \quad v(t) = \int_{t_0}^{t} \frac{y_1(\tau)f(\tau)}{W(y_1,y_2)(\tau)}\,d\tau. \tag{99}$$

The substitution of these results into (97) leads to the particular solution

$$y_P = \int_{t_0}^{t} g_1(t,\tau)f(\tau)\,d\tau, \tag{100}$$

where

$$g_1(t,\tau) = \frac{y_1(\tau)y_2(t) - y_1(t)y_2(\tau)}{W(y_1,y_2)(\tau)} = \frac{\begin{vmatrix} y_1(\tau) & y_2(\tau) \\ y_1(t) & y_2(t) \end{vmatrix}}{W(y_1,y_2)(\tau)}. \tag{101}$$

The complete solution of the IVP (92) is therefore

$$y = y_H + y_P = C_1 y_1(t) + C_2 y_2(t) + \int_{t_0}^{t} g_1(t,\tau)f(\tau)\,d\tau. \tag{102}$$

The function $g_1(t,\tau)$ defined by (101) is called the *one-sided Green's function*. Its construction depends only upon knowledge of the homogeneous solutions $y_1(t)$ and $y_2(t)$. Because these solutions always exist if the DE has continuous coefficients and $A_2(t) \neq 0$, the Green's function always exists under these conditions.

EXAMPLE 18 Use the method of Green's function to solve the IVP

$$y'' + y = \sin t, \quad t > 0, \quad y(0) = 1, \; y'(0) = -1.$$

Solution: We first identify the associated problems:

PROBLEM (A): $y'' + y = 0, \quad y(0) = 1, \; y'(0) = -1$

PROBLEM (B): $y'' + y = \sin t, \quad y(0) = 0, \; y'(0) = 0$

The general solution of PROBLEM (A) is $y_H = C_1 \cos t + C_2 \sin t$. By imposing the ICs, we deduce that $C_1 = 1$ and $C_2 = -1$, and thus,

$$y_H = \cos t - \sin t.$$

For PROBLEM (B), we first note that $W(\cos t, \sin t) = 1$, and consequently,

$$g_1(t,\tau) = \begin{vmatrix} \cos \tau & \sin \tau \\ \cos t & \sin t \end{vmatrix} = \cos \tau \sin t - \cos t \sin \tau.$$

Hence,

$$y_P = \int_0^t g_1(t,\tau) \sin \tau \, d\tau = \sin t \int_0^t \cos \tau \sin \tau \, dt - \cos t \int_0^t \sin^2 \tau \, dt$$

$$= \frac{1}{2}(\sin t - t \cos t).$$

Last, by combining solutions y_H and y_P, we get

$$y = y_H + y_P = \left(1 - \frac{1}{2}t\right)\cos t - \frac{1}{2}\sin t.$$

As a final comment, we note that the one-sided Green's function that we have defined here is the same (to within a multiplicative constant) as the *impulse response function* $h(t - \tau)$ associated with linear systems like electric circuits, oscillating systems, and so on (e.g., see Section 9.6). That is, it can be shown that the one-sided Green's function $g_1(t,\tau)$ is a solution of the nonhomogeneous IVP

$$y'' + a_1(t)y' + a_0(t)y = \delta(t - t_0), \quad y(t_0) = 0, \; y'(t_0) = 0, \tag{103}$$

in which the input function $f(t)$ has been replaced by the delta or *impulse function* (see Section 2.2.4).

8.5.2 Boundary value problems

When BCs are prescribed, the resulting boundary value problem (BVP) is given by

$$A_2(x)y'' + A_1(x)y' + A_0(x)y = F(x), \quad a < x < b, \quad B_1[y] = \alpha, \ B_2[y] = \beta. \tag{104}$$

Analogously to the IVP treated in Section 8.5.1, we split the problem into two problems:

PROBLEM (A):
$$\begin{array}{c} A_2(x)y'' + A_1(x)y' + A_0(x)y = 0, \\ B_1[y] = \alpha, \ B_2[y] = \beta \end{array} \tag{105}$$

PROBLEM (B):
$$L[y] = f(x), \quad B_1[y] = 0, \ B_2[y] = 0 \tag{106}$$

Note in PROBLEM (B) we have put the DE in self-adjoint form, where L is defined by (54). The physical interpretations of the solutions of each problem, denoted by y_H and y_P, are essentially the same as for the solutions defined in the previous section.

The solution of PROBLEM (A) is given by

$$y_H = C_1 y_1(x) + C_2 y_2(x), \tag{107}$$

where the arbitrary constants C_1 and C_2 can be determined by imposing the prescribed BCs in PROBLEM (A). Also, by analogy with IVPs discussed in the previous section, we assume the solution of PROBLEM (B) can be expressed in the form

$$y_P = -\int_a^b g(x,s)f(s)\,ds, \tag{108}$$

where $g(x,s)$ is the *Green's function* we wish to define. [The minus sign is chosen so that $g(x,s)$ has the proper physical interpretation, viz., the response of the system to a point source.] If we formally apply the operator L to both sides of (108), we see that

$$L[y_P] = -\int_a^b L[g]f(s)\,ds.$$

To be a solution of PROBLEM (B), the right-hand side must reduce to $f(x)$. This will happen provided

$$L[g] = -\delta(x - s), \ a < x < b. \tag{109}$$

Similarly, by applying the homogeneous BCs in PROBLEM (B) to (108), we see that the Green's function satisfies the prescribed BCs in PROBLEM (B); i.e.,

$$B_1[g] = 0, \qquad B_2[g] = 0. \tag{110}$$

Hence, the Green's function $g(x,s)$ is a solution of the BVP described by (109) and (110). As a consequence of this, the Green's function physically represents the solution of a system to a unit load (delta function) applied at position $x = s$. Also, it can be shown that (see [1])

$$g(s^+,s) = g(s^-,s),$$

$$\left.\frac{\partial g}{\partial x}\right|_{x=s^-}^{x=s^+} = -\frac{1}{p(s)}. \tag{111}$$

The first condition in (111) establishes the continuity of the Green's function at $x = s$ and the second condition shows that a jump discontinuity of magnitude $1/p(s)$ exists at $x = s$. The jump discontinuity is an essential feature of the Green's function.

Based on the above information concerning the Green's function, we can now construct a formula for it. Equation (109) suggests that if either $x < s$ or $x > s$, it follows that $L[g] = 0$. Thus, let z_1 and z_2 be two solutions of the homogeneous equation

$$A_2(x)y'' + A_1(x)y' + A_0(x)y = 0.$$

We then select these solutions in such a way that each satisfies a different BC in PROBLEM (B), viz.,

$$B_1[z_1] = 0, \qquad B_2[z_2] = 0. \tag{112}$$

Thus, to satisfy both (109) and (110), we assume the Green's function takes the form

$$g(x,s) = \begin{cases} u(s)z_1(x), & x < s \\ v(s)z_2(x), & x > s, \end{cases} \tag{113}$$

where $u(s)$ and $v(s)$ can be determined from the conditions in (111). Namely,

$$v(s)z_2(s) - u(s)z_1(s) = 0,$$

$$v(s)z_2'(s) - u(s)z_1'(s) = -\frac{1}{p(s)},$$

the simultaneous solution of which is

$$u(s) = -\frac{z_2(s)}{p(s)\,W(z_1,z_2)(s)}, \qquad v(s) = -\frac{z_1(s)}{p(s)\,W(z_1,z_2)(s)}. \tag{114}$$

However, it can be shown that (by differentiating both sides)

$$p(s)W(z_1,z_2)(s) = C \text{ (constant)},$$

so that the Green's function (113) reduces to

$$g(x,s) \;=\; \begin{cases} -z_1(x)z_2(s)/C, \ x < s \\ -z_1(s)z_2(x)/C, \ x > s. \end{cases} \tag{115}$$

Because C is constant, it follows that the Green's function (115) is *symmetrical* in x and s, i.e., $g(x,s) = g(s,x)$. This property can sometimes be useful in the construction of the Green's function by other methods (e.g., numerical techniques, etc.).

EXAMPLE 19 Construct the Green's function associated with the BVP

$$y'' + k^2 y \;=\; f(x), \quad 0 < x < 1, \quad y(0) = \alpha, \ y(1) = \beta \quad (k \neq 0).$$

Solution: Linearly independent solutions of the homogeneous DE $y'' + k^2 y = 0$ are $\cos kx$ and $\sin kx$. To construct solutions z_1 and z_2, we assume they are linear combinations of the fundamental solutions. For example, we assume

$$z_1(x) \;=\; c_1 \cos kx + c_2 \sin kx,$$

and by imposing the first homogeneous BC $z_1(0) = 0$, we set $c_1 = 0$ and $c_2 = 1$, among other choices. Hence,

$$z_1(x) \;=\; \sin kx.$$

Similarly, by writing

$$z_2(x) \;=\; c_3 \cos kx + c_4 \sin kx,$$

and imposing $z_2(1) = 0$, we are led to

$$c_3 \cos k + c_4 \sin k \;=\; 0.$$

By selecting $c_3 = \sin k$ and $c_4 = -\cos k$, we then obtain

$$z_2(x) \;=\; \sin k \cos kx - \cos k \sin kx \;=\; \sin k(1 - x).$$

It remains to determine the constant C. First observe that the operator $L = D^2 + k^2$ is self-adjoint with $p(x) = 1$. Hence, it follows that

$$C \;=\; p(x)W(z_1,z_2)(x) \;=\; \begin{vmatrix} \sin kx & \sin k(1-x) \\ k\cos kx & -k\cos k(1-x) \end{vmatrix} \;=\; -k \sin k.$$

Because the Green's function involves division by C, we must now restrict

$k \neq n\pi$ $(n = 1,2,3,...)$ to avoid division by zero. Under this condition, we deduce that

$$g(x,s) = \begin{cases} \dfrac{\sin ks \sin k(1-x)}{k \sin k}, & x < s \\[2mm] \dfrac{\sin kx \sin k(1-s)}{k \sin k}, & x > s. \end{cases}$$

If we allow $k = n\pi$ $(n = 1,2,3,...)$, then $C = 0$ in Example 19 and the Green's function does not exist.[3] In particular, we note that $z_1 = z_2 = \sin n\pi x$, which is not permitted in the construction of the Green's function. We can summarize this situation by the theorem below, known also as the *Fredholm alternative*. Formulated as an eigenvalue problem, it could be rephrased as saying—either the nonhomogeneous problem has a unique solution or else the associated homogeneous problem has an eigenfunction corresponding to a *zero eigenvalue*.

Theorem 8.5 If the coefficients of the self-adjoint operator L are continuous on the interval $[a,b]$, and if $f(x)$ is also continuous on this same interval, then the BVP

$$L[y] = f(x), \quad a < x < b, \quad B_1[y] = \alpha, \; B_2[y] = \beta,$$

has a unique solution or else the associated homogeneous problem

$$L[y] = 0, \quad a < x < b, \quad B_1[y] = 0, \; B_2[y] = 0,$$

has a nontrivial solution.

EXAMPLE 20 Use the method of Green's function to solve

$$2x^2 y'' + 3xy' - y = 4x\sqrt{x}, \quad 0 < x < 1, \quad y(0) = 0, \; y(1) = 0.$$

Solution: Dividing the problem into two problems, we obtain

PROBLEM (A): $2x^2 y'' + 3xy' - y = 0, \quad y(0) = 0, \; y(1) = 0$

---

[3] Although the Green's function defined by (115) does not always exist (e.g., when $C = 0$), it might be possible in such cases to construct a "modified" Green's function that does exist. However, the treatment of a modified Green's function is considered beyond the scope of this text.

PROBLEM (B): $\dfrac{d}{dx}(x\sqrt{x}\,y') - \dfrac{1}{2\sqrt{x}}y = 2x$, $y(0) = 0$, $y(1) = 0$.

Because both the DE and BCs are homogeneous in PROBLEM (A), the only solution is the trivial solution $y_H = 0$. For PROBLEM (B), we have already put the DE in self-adjoint form, obtained by first dividing the original equation by $2x^2$ and then multiplying by $x\sqrt{x}$.

To construct $g(x,s)$, we recognize that the equation is a Cauchy-Euler DE for which $\sqrt{x}$ and $1/x$ are linearly independent solutions of PROBLEM (A)   (see Section 1.4.5). Linear combinations of these solutions that satisfy the homogeneous BCs and are suitable for constructing $g(x,s)$ are readily found to be

$$z_1 = \sqrt{x}, \qquad z_2 = \sqrt{x} - \dfrac{1}{x}.$$

An easy calculation reveals that

$$C = p(x)\,W(z_1,z_2)(x) = x\sqrt{x}\left(\dfrac{3}{2x\sqrt{x}}\right) = \dfrac{3}{2},$$

and hence, we deduce that

$$g(x,s) = \begin{cases} -\dfrac{2}{3}\sqrt{x}(\sqrt{s} - 1/s), \ x < s \\[2mm] -\dfrac{2}{3}\sqrt{s}(\sqrt{x} - 1/x), \ x > s. \end{cases}$$

The solution we seek is therefore (here $y = y_P$)

$$y = -\int_0^1 g(x,s)(2s)\,ds = \dfrac{2}{3}(\sqrt{x} - 1/x)\int_0^x \sqrt{s}(2s)\,ds + \dfrac{2}{3}\sqrt{x}\int_x^1 (\sqrt{s} - 1/s)(2s)\,ds,$$

which reduces to

$$y = \dfrac{4}{5}\sqrt{x}(x - 1).$$

8.5.3 Bilinear formula

Generalized Fourier series can be used in the construction of a Green's function by an interesting and important approach to solving nonhomogeneous BVPs of the form

$$L[y] = f(x), \ a < x < b, \ B_1[y] = 0, \ B_2[y] = 0. \tag{116}$$

Closely associated with this problem is a regular Sturm-Liouville problem

$$L[y] + \lambda r(x)y = 0, \quad B_1[y] = 0, \ B_2[y] = 0, \tag{117}$$

where L, B_1, and B_2 are the same operators in both (116) and (117). The Sturm-Liouville problem (117) is not unique, but $r(x)$ is chosen on the basis of tractability. Let us assume the eigenvalues and eigenfunctions of (117) are λ_n and $\varphi_n(x)$, $n = 1, 2, 3, \dots$.

We start by assuming that the solution of (116) can be written as

$$y = \sum_{n=1}^{\infty} c_n \varphi_n(x), \quad a < x < b, \tag{118}$$

called an *eigenfunction expansion*. Clearly, this choice of solution satisfies the BCs because each of the eigenfunctions satisfies the BCs. We substitute (118) directly into the DE in (116), which leads to

$$L[y] = \sum_{n=1}^{\infty} c_n L[\varphi_n(x)] = f(x). \tag{119}$$

However, from (117) we recognize that $L[\varphi_n(x)] = -\lambda_n r(x) \varphi_n(x)$, and (119) becomes

$$-\sum_{n=1}^{\infty} c_n \lambda_n r(x) \varphi_n(x) = f(x),$$

which can be arranged as

$$\frac{f(x)}{r(x)} = \sum_{n=1}^{\infty} a_n \varphi_n(x), \quad a < x < b. \tag{120}$$

In the last step we introduced $a_n = -c_n \lambda_n$ for notational convenience. Equation (120) now represents a generalized Fourier series for the function $f(x)/r(x)$, and therefore

$$\begin{aligned}
a_n &= \|\varphi_n(x)\|^{-2} \int_a^b r(x) \frac{f(x)}{r(x)} \varphi_n(x)\, dx \\
&= \|\varphi_n(x)\|^{-2} \int_a^b f(x) \varphi_n(x)\, dx, \quad n = 1, 2, 3, \dots.
\end{aligned} \tag{121}$$

In summary, the formal solution of (116) is given by

$$\boxed{\; y = -\sum_{n=1}^{\infty} \left(\frac{a_n}{\lambda_n} \right) \varphi_n(x), \quad a < x < b, \;} \tag{122}$$

where the a's are defined by (121).

The solution (122) is *unique*, provided none of the eigenvalues vanish; i.e., $\lambda_n \neq 0$, $n = 1, 2, 3, \dots$. If one of the eigenvalues, say λ_j, is zero, the problem generally has *no solution*. However, if both $\lambda_j = 0$ and $a_j = 0$, it can be shown that the problem has *infinitely-many solutions*, but we will not pursue this special case.

EXAMPLE 21 Use the eigenfunction expansion method to solve

$$y'' + 2y' + y = e^x, \quad y(0) = 3, \quad y(1) = 0.$$

Solution: Because the BCs are nonhomogeneous, we begin by splitting the problem into two problems:

PROBLEM (A): $y'' + 2y' + y = 0$, $y(0) = 3$, $y(1) = 0$

PROBLEM (B): $y'' + 2y' + y = e^x$, $y(0) = 0$, $y(1) = 0$.

The general solution of PROBLEM (A) is $y_H = (C_1 + C_2 x)e^{-x}$, and by imposing the nonhomogeneous BCs, we find

$$y_H = 3(1 - x)e^{-x}.$$

For PROBLEM (B), we first multiply the DE by e^{2x} to put it in self-adjoint form; i.e.,

$$\frac{d}{dx}(e^{2x}) + e^{2x}y = e^{3x},$$

from which we find $r(x) = e^{2x}$ and $f(x) = e^{3x}$. The related eigenvalue problem is

$$\frac{d}{dx}(e^{2x}) + (\lambda + 1)e^{2x}y = 0, \quad y(0) = 0, \ y(1) = 0,$$

obtained by adding $\lambda e^{2x}y$ to the left-hand side. Now simplifying, we have

$$y'' + 2y' + (\lambda + 1)y = 0, \quad y(0) = 0, \ y(1) = 0.$$

The eigenvalues and corresponding eigenfunctions are, respectively, $\lambda_n = n^2\pi^2$ and $\varphi_n(x) = e^{-x}\sin n\pi x$, $n = 1,2,3,....$ Hence, the solution of PROBLEM (B) is

$$y_P = -\sum_{n=1}^{\infty}\left(\frac{a_n}{n^2\pi^2}\right)e^{-x}\sin n\pi x,$$

where $\|\varphi_n(x)\|^2 = 1/2$ and

$$a_n = 2\int_0^1 e^{2x}\sin n\pi x\,dx = \frac{2n\pi[1 - (-1)^n e^2]}{4 + n^2\pi^2}, \quad n = 1,2,3,....$$

By combining solutions y_H and y_P, we obtain

$$y = 3(1 - x)e^{-x} - \frac{2e^{-x}}{\pi}\sum_{n=1}^{\infty}\frac{[1 - (-1)^n e^2]}{n(4 + n^2\pi^2)}\sin n\pi x.$$

The above eigenfunction expansion method provides us with an alternative technique for constructing the Green's function discussed in Section 8.5.2. To see this, we simply rewrite the solution (122) in the form

$$y = -\int_a^b \left(\sum_{n=1}^{\infty} \frac{\varphi_n(x)\varphi_n(s)}{\|\varphi_n(x)\|^2 \lambda_n} \right) f(s)\,ds, \tag{123}$$

where we have replaced the a's by their integral representation (121), using s as a dummy variable, and interchanged the order of summation and integration. Now, by comparing (123) with (108), we deduce that the Green's function has the series representation

$$\boxed{\; g(x,s) = \sum_{n=1}^{\infty} \frac{\varphi_n(x)\varphi_n(s)}{\|\varphi_n(x)\|^2 \lambda_n}, \;} \tag{124}$$

called a *bilinear formula*. Thus, in addition to the method of Section 8.5.2, the Green's function can be constructed entirely from only knowledge of the eigenvalues and eigenfunctions of an associated Sturm-Liouville problem.

Suggested Reading

L. C. Andrews, *Elementary Partial Differential Equations with Boundary Value Problems* (Academic Press, Orlando, 1986).

R. V. Churchill and J. W. Brown, *Fourier Series and Boundary Value Problems*, 4th ed. (McGraw-Hill, New York, 1987).

L. W. Couch, II, *Digital and Analog Communication Systems*, 6th ed. (Prentice-Hall, Upper Saddle River, 2001).

A. Papoulis, *Signal Analysis* (McGraw-Hill, New York, 1977).

G. P. Tolstov, *Fourier Series* (Dover, New York, 1976).

R. E. Ziemer and W. H. Tranter, *Principles of Communications*, 5th ed. (Wiley, New York, 2002)

Chapter 8

Exercises

In Problems 1 and 2, find the Fourier trigonometric series of the periodic function $f(t)$, defined over one period.

1. $f(t)1 = |t|, \ -\pi < t < \pi$

Ans. $\dfrac{\pi}{2} - \dfrac{4}{\pi} \displaystyle\sum_{\substack{n=1 \\ (\text{odd})}}^{\infty} \dfrac{\cos nt}{n^2}$

2. $f(t) = t(1 - t), \ -1 < t < 1$

Ans. $-\dfrac{1}{3} - \dfrac{4}{\pi} \displaystyle\sum_{n=0}^{\infty} \dfrac{(-1)^n}{n}$
$\times (\cos n\pi t + \sin n\pi t)$

3. Show that the Fourier series of the periodic rectangular pulse (see Chapter 2)

$$f(t) = \sum_{n=-\infty}^{\infty} \Pi\left(\frac{t - nT}{\tau}\right),$$

is given by

$$f(t) = \frac{\tau}{T} \sum_{n=-\infty}^{\infty} \text{sinc}\left(\frac{n\tau}{T}\right) e^{2in\pi t/T}.$$

4. Given $f(t) = \pi - t, \ 0 < t < \pi$, find
 (a) a sine series (with $T = 2\pi$).
 (b) a cosine series (with $T = 2\pi$).
 (c) a full Fourier trigonometric series (with $T = \pi$).

Ans. (a) $2\displaystyle\sum_{n=1}^{\infty} \dfrac{\sin nt}{n}$

(b) $\dfrac{\pi}{2} + \dfrac{4}{\pi} \displaystyle\sum_{n=1}^{\infty} \dfrac{\cos(2n - 1)t}{(2n - 1)^2}$

(c) $\dfrac{\pi}{2} + \displaystyle\sum_{n=1}^{\infty} \dfrac{\sin 2nt}{n}$

5. Calculate the total average power in the signal

$$f(t) = 4\sin(3\pi t) + 6\sin(2\pi t + \pi/4),$$

using a
 (a) time domain analysis.
 (b) frequency domain analysis.

Ans. $P = 26$ Watts

6. Given that the instantaneous power in a particular voltage is

$$p(t) = A\cos^2 \omega_0 t,$$

calculate the total average power.

Ans. $A/2$

In Problems 7 and 8, find the exponential Fourier series of the given function.

7. $f(t) = \begin{cases} A, \ 0 < t < T/2 \\ 0, \ T/2 < t < T \end{cases}$

Ans. $\dfrac{A}{2} \displaystyle\sum_{n=-\infty}^{\infty} \dfrac{\sin(n\pi/2)}{n\pi/2} e^{2\pi int/T - in\pi/2}$

8. $f(t) = t(1 - t), \ -1 < t < 1$

In Problems 9 and 10, find all real eigenvalues and eigenfunctions.

9. $y'' + \lambda y = 0, \ y(0) = 0, \ y'(1) = 0$

Ans. $\lambda_n = \dfrac{n^2 \pi^2}{4}$,

$\varphi_n(x) = \sin \dfrac{n\pi x}{2}, \ n = 1, 3, 5, \ldots$

10. $y'' - 3y' + 2\lambda y = 0,$
 $y(0) = 0, \ y(1) = 0$

Ans. $\lambda_n = \dfrac{9}{8} + \dfrac{n^2 \pi^2}{2}$

$\varphi_n(x) = e^{3x/2} \sin n\pi x, \ n = 1, 2, 3, \ldots$

11. Given the eigenvalue problem

$$y'' + \lambda y = 0,$$
$$hy(0) + y'(0) = 0, \ y(1) = 0,$$

(*a*) show that if $\lambda = 0$ is an eigenvalue, then $h = 1$.

(*b*) If $h > 1$, show that there is exactly one negative eigenvalue.

(*c*) If $h < 0$, show that the eigenvalues are non-negative.

12. Find a generalized Fourier series for the function $f(x) = 1$ in terms of the eigenfunctions of

$$y'' + 4y' + (4 + 9\lambda)y = 0,$$
$$y(0) = 0, \ y(2) = 0.$$

Ans. $\dfrac{\pi}{2} e^{-2x} \displaystyle\sum_{n=1}^{\infty} \dfrac{n[1 - (-1)^n e^4]}{4 + n^2\pi^2/4} \sin\dfrac{n\pi x}{2}$

13. Determine the subset of the eigenvalues $\lambda_n = n(n + 1)$, $= 0, 1, 2, ...,$ and corresponding eigenfunctions of

$$(1 - x^2)y'' - 2xy' + \lambda y = 0,$$
$$y(0) = 0, \ |y(1)| < \infty.$$

14. Find the eigenvalues $(\lambda \geq 0)$ and eigenfunctions of

$$xy'' + y' + \lambda xy = 0,$$
$$|y(0)| < \infty, \ y'(1) = 0.$$

Ans. $\lambda_n = k_n^2$, $\varphi_n(x) = J_0(k_n x)$, where $J_0'(k_n) = 0$, $n = 1, 2, 3, ...$

In Problems 15 and 16, use the one-sided Green's function to solve the given IVP.

15. $y'' + y' - 2y = -4$,
$\quad y(0) = 2, \ y'(0) = 3$

Ans. $y = 2 + 2e^{-t/2} \sinh\dfrac{3t}{2}$

16. $y'' + 2y' + y = 3te^{-t}$,
$\quad y(0) = 4, \ y'(0) = 2$

Ans. $y = e^{-t}\left(\dfrac{1}{2}t^3 + 6t + 4\right)$

17. The small motions $y(t)$ of a spring-mass system are governed by the IVP

$$my'' + ky = F(t),$$
$$y(0) = y_0, \ y'(0) = v_0,$$

where m is the mass, k is the spring constant, and $F(t)$ is an external force.

(*a*) Show that the one-sided Green's function is $(\omega_0 = \sqrt{k/m})$

$$g_1(t,\tau) = \dfrac{1}{\omega_0}\sin[\omega_0(t - \tau)].$$

(*b*) Solve the IVP when the forcing function is $F(t) = P\cos\omega t$, $\omega \neq \omega_0$.

(*c*) Solve the IVP when the forcing function is $F(t) = P\cos\omega_0 t$.

In Problems 18 and 19, use the method of Green's function to solve the BVP.

18. $y'' - y = e^x$, $y(0) = 0$, $y'(1) = 1$

Ans. $y = \dfrac{1}{2}xe^x + \dfrac{(1 - e)\sinh x}{\cosh 1}$

19. $y'' + 2y' + y = x$,
$\quad y(0) = -3, \ y(1) = -1$

Ans. $y = (x - 1)e^{-x} + x - 2$

20. By direct construction, show that the Green's function for the singular BVP

$$y'' - k^2y = f(x), \ -\infty < x < \infty,$$
$$|y(\pm\infty)| < \infty$$

is given by

$$g(x,s) = \dfrac{1}{2k}e^{-k|x-s|}, \ k > 0.$$

In Problems 21-23, solve the BVP by the method of eigenfunction expansion.

21. $y'' = -x, \quad y(0) = 0, \ y(\pi) = 0$

$\quad$ *Ans.* $y = 2 \displaystyle\sum_{n=1}^{\infty} \frac{(-1)^{n-1}}{n^3} \sin nx$

22. $y'' + y = 1, \ y'(0) = 1, \ y'(1) = 1$

$\quad$ *Ans.* $y = 1 + \sin x + \left(\dfrac{\cos 1 - 1}{\sin 1} \right) \cos x$

23. $y'' - y = e^x, \ y(0) = 2, \ y'(1) = 0$

$\quad$ *Ans.* $y = 2\cosh(x - 1)$

$$+ 2 \sum_{n=1}^{\infty} \left(\frac{(-1)^{n-1} e + (2n - 1)\pi/2}{[1 + (2n-1)^2 \pi^2/4]^2} \right) \times \sin\left[\frac{(2n-1)\pi x}{2} \right]$$

24. Determine the bilinear formula for the
$\quad$ (*a*) Green's function in Example 19.
$\quad$ (*b*) What values of k lead to $\lambda_n = 0$?

$\quad$ *Ans.* (*a*) $g(x,s) = 2 \displaystyle\sum_{n=1}^{\infty} \frac{\sin n\pi x \sin n\pi s}{n^2 \pi^2 - k^2}$

In Problems 25 and 26, find a bilinear formula for the Green's function defined by the given operators and BCs.

25. $L = D^2 + 1; \ y(0) = 0, \ y'(1) = 0$

$\quad$ *Ans.* $8 \displaystyle\sum_{n=1}^{\infty} \frac{\cos(n - \frac{1}{2})\pi x \cos(n - \frac{1}{2})\pi s}{(2n - 1)^2 \pi^2 - 4}$

26. $L = D^2 - 1; \ y'(0) = \alpha, \ y'(\pi) = \beta$

$\quad$ *Ans.* $\dfrac{1}{\pi} + \dfrac{2}{\pi} \displaystyle\sum_{n=1}^{\infty} \frac{\cos nx \cos ns}{n^2 + 1}$

Chapter 9

Fourier and Related Transforms

9.1 Introduction / 454
9.2 Fourier Integral Representations / 454
    9.2.1   Cosine and sine integral representations / 457
9.3 Fourier Transforms in Mathematics / 458
    9.3.1   Fourier cosine and sine transforms / 460
9.4 Fourier Transforms in Engineering / 461
    9.4.1   Energy spectral density function / 462
    9.4.2   Table of Fourier transforms / 463
    9.4.3   Generalized Fourier transforms / 463
9.5 Properties of the Fourier Transform / 466
    9.5.1   Time and frequency shifting / 468
    9.5.2   Differentiation and integration / 469
    9.5.3   Convolution theorem / 469
9.6 Linear Shift-Invariant Systems / 471
9.7 Hilbert Transforms / 473
    9.7.1   Analytic signal representation / 475
    9.7.2   Kramers-Krönig relations / 476
    9.7.3   Table of transforms and properties / 476
9.8 Two-Dimensional Fourier Transforms / 477
    9.8.1   Linear systems in optics / 479
    9.8.2   Coherent imaging systems / 480
9.9 Fractional Fourier Transform / 483
    9.9.1   Application in optics / 486
9.10 Wavelets / 487
    9.10.1 Haar wavelets / 488
    9.10.2 Wavelet transform / 491
    Suggested Reading / 492
    Exercises / 493

Our objective in this chapter is to develop the basic properties associated with the Fourier transform and its inverse. This transform is particularly well-suited for determining the spectral character of energy signals such as those that arise in the analysis of linear shift-invariant systems. We also extend our analysis to the two-dimensional Fourier transform commonly used in optics applications, and to the related Hilbert transform, fractional Fourier transform, and wavelet transform.

9.1 Introduction

In Chapter 8 we found that periodic functions are classified as *power signals* that can be represented by a *Fourier series*. Nonperiodic functions that are also square-integrable are classified as *energy signals*, and they can be analyzed through application of the *Fourier transform*.

The use of Fourier transforms in mathematics, physics, and engineering applications dates back to the pioneering work of Joseph Fourier (see historical comments in Chapter 8). During the last few decades, however, there have been significant generalizations of the idea of integral transform, and many new uses of the transform method have evolved. Advances in computing power and software have also contributed to the increased feasibility of numerical evaluation of Fourier integrals and series.

Integral transforms of various types have become essential working tools of nearly every engineer and applied scientist. Of several varieties, the integral transforms of Fourier and Laplace are the most widely used in practice. In this chapter we consider the Fourier transform and briefly discuss some closely related transforms. The Fourier transform is basic to the spectral analysis of time-varying waveforms and linear systems. We introduce the Laplace transform in Chapter 10, which is basic to circuit analysis and control theory.

9.2 Fourier Integral Representations

In the last chapter we showed that a variety of periodic functions can be represented by a Fourier series. Here we wish to examine what happens when the period of a periodic function is allowed to become unbounded. The limiting result is clearly a function defined on the entire x-axis that is *not* periodic.

If $f(x)$ is a piecewise continuous periodic function, with period $T = 2p$ and a piecewise continuous derivative, it has the pointwise convergent Fourier series representation (see Section 8.2)

$$f(x) = \frac{1}{2}a_0 + \sum_{n=1}^{\infty} \left(a_n \cos \frac{n\pi x}{p} + b_n \sin \frac{n\pi x}{p} \right), \tag{1}$$

where

$$a_n = \frac{1}{p} \int_{-p}^{p} f(x) \cos \frac{n\pi x}{p} dx, \; n = 0,1,2,\ldots,$$

$$b_n = \frac{1}{p} \int_{-p}^{p} f(x) \sin \frac{n\pi x}{p} dx, \; n = 1,2,3,\ldots. \tag{2}$$

If we change the dummy variable of integration in Eqs. (2) to t and insert these integrals back into (1), the resulting expression is

$$f(x) = \frac{1}{2p}\int_{-p}^{p} f(t)\,dt + \frac{1}{p}\int_{-p}^{p} f(t)\sum_{n=1}^{\infty} \cos\left[\frac{n\pi(t-x)}{p}\right]dt, \tag{3}$$

where we have interchanged the order of summation and integration and used the trigonometric identity

$$\cos(A-B) = \cos A \cos B + \sin A \sin B. \tag{4}$$

We now examine what happens as we let p tend to infinity. To ensure the integral in (3) outside the summation vanishes in the limit, we require

$$\int_{-\infty}^{\infty} |f(t)|\,dt = \int_{-\infty}^{\infty} |f(x)|\,dx < \infty, \tag{5}$$

which says that $f(x)$ is *absolutely integrable*. For the remaining sum in (3), we let $\Delta s = \pi/p$ and then consider the limit

$$f(x) = \lim_{\Delta s \to 0} \frac{1}{\pi}\int_{-\pi/\Delta s}^{\pi/\Delta s} f(t)\sum_{n=1}^{\infty} \cos[n\Delta s(t-x)]\Delta s\,dt. \tag{6}$$

When Δs is a small positive number, the points $n\Delta s$ are evenly spaced along the s-axis. In this case, we expect the series in (6) to be approximated in the limit as $\Delta s \to 0$ by

$$\sum_{n=1}^{\infty} \cos[n\Delta s(t-x)]\Delta s \sim \int_{0}^{\infty} \cos[s(t-x)]\,ds. \tag{7}$$

Although the approximation (7) is nonrigorous, it does suggest that (6) approaches the double integral

$$\begin{aligned}
f(x) &= \frac{1}{\pi}\int_{-\infty}^{\infty}\int_{0}^{\infty} f(t)\cos[s(t-x)]\,ds\,dt \\
&= \frac{1}{\pi}\int_{0}^{\infty}\int_{-\infty}^{\infty} f(t)\cos[s(t-x)]\,dt\,ds,
\end{aligned} \tag{8}$$

where we have interchanged the order of integration in the last step.

The purely formal procedure we just went through has led us to an important and correct result known as the *Fourier integral theorem* (which we state without additional proof).

Theorem 9.1 (*Fourier integral theorem*) If $f(x)$ is piecewise continuous with a piecewise continuous derivative, and is absolutely integrable on $(-\infty,\infty)$, then

$$f(x) = \frac{1}{\pi}\int_{0}^{\infty}\int_{-\infty}^{\infty} f(t)\cos[s(t-x)]\,dt\,ds$$

at points of continuity of the function. At points of discontinuity, the integrals converge to the average value $\frac{1}{2}[f(x^{+}) + f(x^{-})]$.

The conditions listed in Theorem 9.1 are only sufficient conditions, not necessary conditions.[1] That is, there exist functions that have valid integral representations but which do not satisfy the conditions of this theorem. Moreover, the conditions stated in the theorem are not the most general set of sufficient conditions that have been established. Nonetheless, the conditions stated are broad enough to cover most of the functions commonly found in engineering applications.

To emphasize the analogy between Fourier series and the Fourier integral theorem, let us rewrite (8) as

$$f(x) = \frac{1}{\pi} \int_0^\infty \int_{-\infty}^\infty f(t)(\cos st \cos sx + \sin st \sin sx) \, dt \, ds, \tag{9}$$

or, equivalently, as

$$f(x) = \int_0^\infty [A(s)\cos sx + B(s)\sin sx] \, ds, \tag{10}$$

where

$$A(s) = \frac{1}{\pi} \int_{-\infty}^\infty f(t)\cos st \, dt, \tag{11}$$

$$B(s) = \frac{1}{\pi} \int_{-\infty}^\infty f(t)\sin st \, dt. \tag{12}$$

In the setting of (10)-(12), we call (10) the *Fourier integral representation* of the function $f(x)$ with *Fourier coefficients* (11) and (12) in analogy with Fourier series. The general theory of such integral representations closely parallels that of Fourier series.

EXAMPLE 1 Find the Fourier integral representation of the rectangular pulse centered at the origin (see Fig. 9.1)

$$f(x) = \begin{cases} 1, & |x| < 1, \\ 0, & \text{otherwise}. \end{cases}$$

Solution: From (11) and (12), we obtain

$$A(s) = \frac{1}{\pi} \int_{-\infty}^\infty f(x)\cos sx \, dx$$

$$= \frac{1}{\pi} \int_{-1}^1 \cos sx \, dx = \frac{2\sin s}{\pi s},$$

Figure 9.1

---

[1]The notations x^+ and x^- in Theorem 9.1 denote right-hand and left-hand limits.

$$B(s) = \frac{1}{\pi} \int_{-\infty}^{\infty} f(x) \sin sx \, dx = \frac{1}{\pi} \int_{-1}^{1} \sin sx \, dx = 0.$$

Thus, the integral representation (10) becomes

$$f(x) = \frac{2}{\pi} \int_{0}^{\infty} \left(\frac{\sin s}{s} \right) \cos sx \, dx.$$

Because $x = 0$ is a point of continuity of $f(x)$ in Example 1, it follows that

$$f(0) = 1 = \frac{2}{\pi} \int_{0}^{\infty} \frac{\sin s}{s} \, ds,$$

from which we deduce

$$\int_{0}^{\infty} \frac{\sin s}{s} \, ds = \frac{\pi}{2}. \tag{13}$$

At the points $x = \pm 1$, there is a jump discontinuity in $f(x)$, but from Theorem 9.1 the Fourier integral converges to 1/2, the average value of the left-hand and right-hand limits.

9.2.1 Cosine and sine integral representations

We take note of the fact that $f(x)$ in Example 1 is an *even* function. This led to a representation of $f(x)$ that only involved the cosine function in (10).

▸ **Cosine integral representation:** If $f(x)$ is an *even function*, the Fourier integral representation (10) reduces to the *cosine integral representation*

$$f(x) = \int_{0}^{\infty} A(s) \cos sx \, dx, \tag{14}$$

where

$$A(s) = \frac{2}{\pi} \int_{0}^{\infty} f(x) \cos sx \, dx. \tag{15}$$

▸ **Sine integral representation:** If $f(x)$ is an *odd function*, we obtain the *sine integral representation*

$$f(x) = \int_{0}^{\infty} B(s) \sin sx \, dx, \tag{16}$$

where

$$B(s) = \frac{2}{\pi} \int_{0}^{\infty} f(x) \sin sx \, dx. \tag{17}$$

9.3 Fourier Transforms in Mathematics

The notion of a Fourier transform can most easily be introduced here by starting with the Fourier integral relation (Theorem 9.1) written as

$$f(x) = \frac{1}{\pi} \int_0^\infty \int_{-\infty}^\infty f(t) \cos[s(x-t)] \, dt \, ds. \tag{18}$$

Then, by use of Euler's formula

$$\cos x = \frac{1}{2}\left(e^{ix} + e^{-ix}\right), \tag{19}$$

we can rewrite (18) as

$$\begin{aligned} f(x) &= \frac{1}{2\pi} \int_0^\infty \int_{-\infty}^\infty f(t)\left[e^{is(x-t)} + e^{-is(x-t)}\right] dt \, ds \\ &= \frac{1}{2\pi} \int_{-\infty}^\infty \int_{-\infty}^\infty f(t) e^{is(x-t)} \, dt \, ds, \end{aligned} \tag{20}$$

or

$$f(x) = \frac{1}{2\pi} \int_{-\infty}^\infty e^{isx} \int_{-\infty}^\infty e^{-ist} f(t) \, dt \, ds, \tag{21}$$

which is the exponential form of Fourier's integral theorem.

What we have established with (21) is the pair of transform formulas

$$F(s) = \int_{-\infty}^\infty e^{-ist} f(t) \, dt, \tag{22}$$

$$f(x) = \frac{1}{2\pi} \int_{-\infty}^\infty e^{isx} F(s) \, ds. \tag{23}$$

It is customary to use the same variable x in both transform relations and introduce the notation

$$\mathscr{F}\{f(x;s)\} = \int_{-\infty}^\infty e^{-isx} f(x) \, dx = F(s), \tag{24}$$

$$\mathscr{F}^{-1}\{F(s);x\} = \frac{1}{2\pi} \int_{-\infty}^\infty e^{isx} F(s) \, ds = f(x). \tag{25}$$

We call (24) the *Fourier transform* of the function $f(x)$ and (25) the *inverse Fourier transform* of $F(s)$.

The location of the constant $1/2\pi$ that appears in (25) can appear in either (24) or (25), and, for reasons of symmetry, it is also common in some cases to write $1/\sqrt{2\pi}$ in front of each integral. Moreover, there is some variation in the literature as to which integral actually represents the transform and which one is the inverse transform. In practice, all such differences are of little concern except when consulting tables or referring to their properties.

Because one is essentially the complex conjugate of the other, the Fourier transform and inverse transform integrals have many properties in common. Thus, if we establish some property for the transform, it is clear that a similar property holds for the inverse transform. In some sense, this makes the use of Fourier transforms easier than that of certain other integral transforms (such as the Laplace transform in Chapter 10).

EXAMPLE 2 Find the Fourier transform of the rectangular pulse (see also Example 1 and Section 2.2.2)

$$f(x) = \Pi(x/2) = \begin{cases} 1, & |x| < 1, \\ 0, & \text{otherwise.} \end{cases}$$

Solution: Directly from (24), we see that

$$\mathscr{F}\{f(x);s\} = \int_{-\infty}^{\infty} e^{-isx} f(x)\, dx = \int_{-1}^{1} e^{-isx}\, dx,$$

or

$$\mathscr{F}\{f(x);s\} = F(s) = \frac{2\sin s}{s}.$$

EXAMPLE 3 Find the Fourier transform of

$$f(x) = U(x)e^{-ax}, \quad a > 0,$$

where $U(x)$ is the unit step function (see Section 2.2.1).

Solution: From (24),

$$\mathscr{F}\{f(x);s\} = \int_{0}^{\infty} e^{-(a+is)x}\, dx = \frac{1}{a + is}, \quad a > 0.$$

There are a number of other functions whose Fourier transform can be readily deduced through elementary integration methods. However, more sophisticated integration techniques must be used in many cases, such as residue theory from complex variables (Chapter 7). As with other integral transforms, we can also rely on tables for determining the Fourier transform of a wide variety of functions that commonly occur in practice (e.g., see Table 9.1 in Section 9.4.2 or [13] for an extensive list of transform pairs).

9.3.1 Fourier cosine and sine transforms

From Section 9.2, we found that when the function $f(x)$ is even, the Fourier integral representation reduces to

$$f(x) = \int_0^\infty A(s)\cos sx\, ds = \frac{2}{\pi}\int_0^\infty \cos sx \int_0^\infty f(t)\cos st\, dt\, ds. \tag{26}$$

▸ **Fourier cosine transform:** Based on the integral relation (26), we define the *Fourier cosine transform* and *inverse cosine transform*, respectively, by

$$\mathscr{F}_C\{f(x);s\} = \sqrt{\frac{2}{\pi}}\int_0^\infty f(x)\cos sx\, dx = F_C(s), \quad s > 0, \tag{27}$$

$$\mathscr{F}_C^{-1}\{F_C(s);x\} = \sqrt{\frac{2}{\pi}}\int_0^\infty F_C(s)\cos sx\, ds = f(x), \quad x > 0. \tag{28}$$

The above results (27) and (28) are interesting in that they imply the equivalence of the "forward" and "backward" (inverse) cosine transforms. It is precisely for that reason in this case that we split the constant $2/\pi$ "symmetrically" between (27) and (28).

▸ **Fourier sine transform:** If $f(x)$ is an odd function, we are led to the *Fourier sine transform* and *inverse sine transform* defined by

$$\mathscr{F}_S\{f(x);s\} = \sqrt{\frac{2}{\pi}}\int_0^\infty f(x)\sin sx\, dx = F_S(s), \quad s > 0, \tag{29}$$

$$\mathscr{F}_S^{-1}\{F_S(s);x\} = \sqrt{\frac{2}{\pi}}\int_0^\infty F_S(s)\sin sx\, ds = f(x), \quad x > 0. \tag{30}$$

Hence, the sine transform and inverse sine transform are also of the same form.

If the given function $f(x)$ is neither even nor odd, but defined only for $x > 0$, then it may have both a cosine and a sine transform. Moreover, the even and odd "extensions" of $f(x)$ (see Section 8.2.4) will have exponential transforms. To relate these transforms, we point out that it can be shown that the Fourier transform of the *even extension* of $f(x)$ [denoted $f_e(x)$] is related to the cosine transform by

$$\mathscr{F}\{f_e(x);s\} = \sqrt{2\pi}\,\mathscr{F}_C\{f(x);s\}, \tag{31}$$

and that of the *odd extension* of $f(x)$ [denoted $f_o(x)$] is related to the sine transform by

$$\mathscr{F}\{f_o(x);s\} = -i\sqrt{2\pi}\,\mathscr{F}_S\{f(x);|s|\}\,\text{sgn}(s), \tag{32}$$

where the *signum function* is defined by (recall Section 2.2.1)

$$\text{sgn}(x) = \begin{cases} -1, & x < 0 \\ 1, & x > 0. \end{cases} \tag{33}$$

EXAMPLE 4 Given the integral relation

$$I = \int_0^\infty e^{-ax}\cos sx\,dx = \frac{a}{s^2+a^2}, \quad a > 0,$$

deduce the Fourier transform of $f(x) = e^{-a|x|}$, $a > 0$.

Solution: Because the given function is an even function, we use (31) to obtain

$$\mathscr{F}\{f(x);s\} = \sqrt{2\pi}\,\mathscr{F}_C\{f(x);s\} = 2\int_0^\infty e^{-ax}\cos sx\,dx$$
$$= \frac{2a}{s^2+a^2}, \quad a > 0.$$

9.4 Fourier Transforms in Engineering

The notation of Fourier transform introduced in Section 9.3 is commonly used in mathematics applications. In engineering applications, however, it is more common to use the notation below.

Periodic functions are classified as *power signals* and are represented by Fourier series (see Section 8.3.1). On the other hand, nonperiodic functions properly defined are called energy signals and can be represented by Fourier transforms. We define the total energy of a signal $f(t)$ by the expression

$$E = \int_{-\infty}^\infty |f(t)|^2\,dt < \infty. \tag{34}$$

In formally deriving the Fourier transform of the function $f(x)$ in Section 9.3 we made the assumption that $f(x)$ is absolutely integrable [recall Eq. (5)]. An *energy signal* is defined as one for which (34) is satisfied, i.e., the amount of energy in the signal is finite.

In the analysis of energy signals it is customary to define the *Fourier transform* of the time-varying signal $f(t)$ and the *inverse Fourier transform*, respectively, by

$$\boxed{F(\omega) = \int_{-\infty}^\infty e^{-i\omega t}f(t)\,dt = \mathscr{F}\{f(t)\}} \tag{35}$$

$$\boxed{f(t) = \frac{1}{2\pi}\int_{-\infty}^\infty e^{i\omega t}F(\omega)\,d\omega = \mathscr{F}^{-1}\{F(\omega)\}.} \tag{36}$$

Note that the transform (35) now involves the variable ω in the complex exponential function (kernel). In this setting the variable ω is interpreted as *angular frequency* and $\omega/2\pi$ is *linear frequency*, the latter specified in Hertz (Hz).

If $f(t)$ is a voltage signal, then the Fourier transform $F(\omega)$ has dimension of voltage per unit frequency, which describes the distribution of the signal voltage among the various frequencies. Therefore, the transform function $F(\omega)$ is called the *Fourier spectrum* of the signal function $f(t)$. In general, the spectrum $F(\omega)$ is a complex function which we denote by

$$F(\omega) = R(\omega) + iX(\omega) = |F(\omega)| e^{i\theta(\omega)}, \tag{37}$$

where $R(\omega)$ and $X(\omega)$ are the real and imaginary parts, respectively. The quantity $|F(\omega)|$ is called the *magnitude (amplitude) spectrum* and $\theta(\omega)$ is the *phase spectrum*.

EXAMPLE 5 If $f(t)$ is a real time signal, show that $R(\omega)$ is an even function of ω and that $X(\omega)$ is an odd function of ω.

Solution: By use of Euler's formula $e^{-i\theta} = \cos\theta - i\sin\theta$, we write

$$F(\omega) = \int_{-\infty}^{\infty} e^{-i\omega t} f(t)\, dt = \int_{-\infty}^{\infty} f(t)\cos\omega t\, dt - i\int_{-\infty}^{\infty} f(t)\sin\omega t\, dt$$

and thus, identify

$$R(\omega) = \int_{-\infty}^{\infty} f(t)\cos\omega t\, dt,$$
$$X(\omega) = -\int_{-\infty}^{\infty} f(t)\sin\omega t\, dt.$$

From inspection, we note that $R(-\omega) = R(\omega)$, and therefore conclude that $R(\omega)$ is an even function. Similarly, $X(-\omega) = -X(\omega)$, which is the condition for an odd function.

Based on the result of Example 5, it follows that when the signal $f(t)$ is a real function, its magnitude spectrum $|F(\omega)|$ is an even function, whereas its phase spectrum $\theta(\omega)$ is an odd function. Functions for which this is true are sometimes called *Hermitian*.

9.4.1 Energy spectral density function

The total energy (34) of the (real or complex) signal $f(t)$ can be written as

$$E = \int_{-\infty}^{\infty} |f(t)|^2\, dt = \int_{-\infty}^{\infty} f(t) f^*(t)\, dt, \tag{38}$$

where * denotes conjugation. Replacing $f^*(t)$ by the complex conjugate of (36) yields

$$E = \frac{1}{2\pi} \int_{-\infty}^{\infty} f(t) \int_{-\infty}^{\infty} e^{-i\omega t} F^*(\omega) d\omega \, dt$$

$$= \frac{1}{2\pi} \int_{-\infty}^{\infty} F^*(\omega) \int_{-\infty}^{\infty} e^{-i\omega t} f(t) dt \, d\omega$$

$$= \frac{1}{2\pi} \int_{-\infty}^{\infty} F^*(\omega) F(\omega) d\omega,$$

where we have interchanged the order of integration. Therefore, it follows that

$$E = \int_{-\infty}^{\infty} |f(t)|^2 dt = \frac{1}{2\pi} \int_{-\infty}^{\infty} |F(\omega)|^2 d\omega. \tag{39}$$

Equation (39) is known as *Parseval's* (or *Rayleigh's*) *energy theorem* and compares directly with Parseval's power principle for Fourier series [Eq. (37) in Chapter 8].

The quantity

$$S_E(\omega) = |F(\omega)|^2 \tag{40}$$

is the *energy spectral density* associated with the signal $f(t)$. It expresses how the total energy in the signal is distributed among the various frequencies. In terms of this quantity, we can write the total energy as

$$E = \frac{1}{2\pi} \int_{-\infty}^{\infty} S_E(\omega) d\omega. \tag{41}$$

9.4.2 Table of Fourier transforms

Recorded below in Table 9.1 is a selected list of transform pairs. Entries in this table are based on the transform relations (35) and (36), viz.,

$$F(\omega) = \mathscr{F}\{f(t)\} = \int_{-\infty}^{\infty} e^{-i\omega t} f(t) dt,$$

$$f(t) = \mathscr{F}^{-1}\{F(\omega)\} = \frac{1}{2\pi} \int_{-\infty}^{\infty} e^{i\omega t} F(\omega) d\omega. \tag{42}$$

When using Eqs. (24) and (25) to define the transform pairs, the variables t and ω in the entries in Table 9.1 must be replaced with x and s, respectively.

9.4.3 Generalized Fourier transforms

Many of the functions commonly appearing in the mathematical analysis of various systems do not satisfy the condition of having finite energy such as simple sinusoids, step functions, impulse (delta) functions, and so on. In such cases we need to extend our notion of Fourier transform to accommodate this class of functions.

Based on the "sifting" property (see Section 2.2.4), we note that the "evaluation" of the Fourier transform of the impulse function $\delta(t - t_0)$ leads to

$$\mathscr{F}\{\delta(t - t_0)\} = \int_{-\infty}^{\infty} e^{-i\omega t} \delta(t - t_0) dt = e^{-i\omega t_0}. \tag{43}$$

Hence, a shift of the impulse by time t_0 is a phase shift for each frequency ω by the amount ωt_0. Note also that the magnitude spectrum $|F(\omega)| = |e^{-i\omega t_0}| = 1$ is uniform over all frequencies and, by analogy with white light, this spectrum is called "white." The phase spectrum $\theta(\omega) = -t_0\omega$ is linear with negative slope. The inverse transform associated with (43) is

$$\mathscr{F}^{-1}\{e^{-i\omega t_0}\} = \frac{1}{2\pi}\int_{-\infty}^{\infty} e^{i\omega t} e^{-i\omega t_0} d\omega = \delta(t - t_0). \tag{44}$$

From a purely formal point of view, we can simply interchange the roles of t and ω in the above results to obtain the Fourier transform relation

$$\mathscr{F}\{e^{i\omega_0 t}\} = \int_{-\infty}^{\infty} e^{-i\omega t} e^{i\omega_0 t} dt = 2\pi\delta(\omega - \omega_0), \tag{45}$$

from which we deduce the Fourier transform pairs

Table 9.1 Selected Fourier Transform Pairs

	$f(t) = \mathscr{F}^{-1}\{F(\omega)\}$	$F(\omega) = \mathscr{F}\{f(t)\}$
1.	$U(t)e^{-at}$	$1/(a + i\omega)$
2.	$U(t)te^{-at}$	$1/(a + i\omega)^2$
3.	$e^{-a|t|}$	$2a/(a^2 + \omega^2)$
4.	$e^{-t^2/(2\sigma^2)}$	$\sigma\sqrt{2\pi}\, e^{-\sigma^2\omega^2/2}$
5.	$\mathrm{sgn}(t)$	$2/(i\omega)$
6.	$1/t$	$-i\pi\,\mathrm{sgn}(\omega)$
7.	$U(t)$	$\pi\delta(\omega) + 1/(i\omega)$
8.	$\delta(t - t_0)$	$e^{-i\omega t_0}$
9.	1	$2\pi\delta(\omega)$
10.	$e^{\pm i\omega_0 t}$	$2\pi\delta(\omega \mp \omega_0)$
11.	$\cos\omega_0 t$	$\pi\big[\delta(\omega - \omega_0) + \delta(\omega + \omega_0)\big]$
12.	$\sin\omega_0 t$	$-i\pi\big[\delta(\omega - \omega_0) - \delta(\omega + \omega_0)\big]$
13.	$\Pi(t/2)$	$\dfrac{2\sin\omega}{\pi\omega}$
14.	$\Pi(t/b)\cos(\pi t/b)$	$\dfrac{2b}{\pi}\dfrac{\cos(b\omega/2)}{1 - (b\omega/\pi)^2}$
15.	$\mathrm{comb}(t/T)$	$2\pi\,\mathrm{comb}(\omega/\omega_0), \quad \omega_0 = 2\pi/T$

$$\mathscr{F}\{e^{\pm i\omega_0 t}\} = 2\pi\,\delta(\omega \mp \omega_0),$$

$$\mathscr{F}^{-1}\{\delta(\omega \mp \omega_0)\} = \frac{1}{2\pi}e^{\pm i\omega_0 t}. \tag{46}$$

Equations (46) suggests that the spectra of the complex exponential functions $e^{\pm i\omega_0 t}$ can be represented by line spectra located at $\omega = \pm\omega_0$. We noted in Chapter 8 that periodic signals are classified as power signals, not energy signals, and that they have line spectra. Nonetheless, with the above formalism involving the impulse function, we can use the Fourier transform to deduce the spectrum of a simple sinusoidal function. For example, if $f(t) = \cos\omega_0 t$, then (see Figure 9.2)

$$\mathscr{F}\{\cos\omega_0 t\} = \frac{1}{2}\mathscr{F}\{e^{i\omega_0 t}\} + \frac{1}{2}\mathscr{F}\{e^{-i\omega_0 t}\} = \pi\,\delta(\omega - \omega_0) + \pi\,\delta(\omega + \omega). \tag{47}$$

Similarly,

$$\mathscr{F}\{\sin\omega_0 t\} = \frac{1}{2i}\mathscr{F}\{e^{i\omega_0 t}\} - \frac{1}{2i}\mathscr{F}\{e^{-i\omega_0 t}\} = -i\pi\,\delta(\omega - \omega_0) + i\pi\,\delta(\omega + \omega). \tag{48}$$

Thus, both $\cos\omega_0 t$ and $\sin\omega_0 t$ have the same magnitude spectrum but not the same phase spectrum.

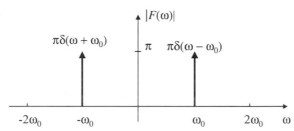

Figure 9.2 Magnitude line spectrum of $f(t) = \cos\omega_0 t$.

Next, we consider the Fourier transform of $f(t) = \mathrm{sgn}(t)e^{-a|t|}$, $a > 0$, where $\mathrm{sgn}(t)$ is the signum function (Section 2.2.1). Here,

$$\begin{aligned}
\mathscr{F}\{\mathrm{sgn}(t)e^{-a|t|}\} &= \int_{-\infty}^{\infty} e^{-i\omega t}\,\mathrm{sgn}(t)e^{-a|t|}\,dt \\
&= -\int_{-\infty}^{0} e^{-i\omega t}e^{at}\,dt + \int_{0}^{\infty} e^{-i\omega t}e^{-at}\,dt \\
&= -\int_{0}^{\infty} e^{-(a-i\omega)t}\,dt + \int_{0}^{\infty} e^{-(a+i\omega)t}\,dt,
\end{aligned}$$

where we have replaced t by $-t$ in the first integral. The evaluation of these integrals now yields

$$\mathscr{F}\{\mathrm{sgn}(t)e^{-a|t|}\} = -\frac{1}{a - i\omega} + \frac{1}{a + i\omega} = -\frac{2i\omega}{a^2 + \omega^2}. \tag{49}$$

In the limit $a \to 0$, we formally deduce that

$$\mathscr{F}\{\mathrm{sgn}(t)\} = \frac{2}{i\omega}. \tag{50}$$

By expressing the step function (Section 2.2.1) in terms of the signum function, we are led to

$$\mathscr{F}\{U(t)\} = \mathscr{F}\left\{\frac{1}{2} + \frac{1}{2}\mathrm{sgn}(t)\right\} = \pi\delta(\omega) + \frac{1}{i\omega}. \tag{51}$$

The presence of an impulse function at dc ($\omega = 0$) is a consequence of the fact that the unit step function $U(t)$ has a nonzero average value (viz., $1/2$).

Any attempt to calculate transform relations like (50) and (51) directly from definition of the signum and step functions would lead to incorrect results since neither function is an energy signal. Such transform relations, called "generalized transforms," are only formal results and care must be exercised in their evaluation.

Last, we take note of the fact that we can extend these ideas of generalized Fourier transform to include general periodic functions. That is, let $f(t)$ denote a periodic signal with period T and Fourier series representation

$$f(t) = \sum_{n=-\infty}^{\infty} c_n e^{in\omega_0 t}, \quad \omega_0 = 2\pi/T. \tag{52}$$

By calculating the Fourier transform of this series termwise, relying on the formal results of (46), we arrive at

$$\mathscr{F}\{f(t)\} = F(\omega) = 2\pi \sum_{n=-\infty}^{\infty} c_n \delta(\omega - n\omega_0). \tag{53}$$

Hence, the Fourier transform of a periodic signal can be interpreted as a sequence of impulse functions located at the harmonic frequencies of the signal (i.e., a *comb* function with amplitudes c_n). This shows that we can now analyze both periodic and nonperiodic signals with a unified treatment involving the Fourier transform.

9.5 Properties of the Fourier transform

The Fourier transform $F(\omega)$ and inverse transform $f(t)$ represent frequency domain and time domain descriptions of the same function. We will assume such relations are unique and further, that $\mathscr{F}^{-1}\mathscr{F}\{f(t)\} = f(t)$ and $\mathscr{F}\mathscr{F}^{-1}\{F(\omega)\} = F(\omega)$ for all t and ω.

The direct calculation of Fourier transforms is often tedious and can be quite complicated in some cases. Once a few transform relations are known, however, we may be able to deduce the transform of other functions in a relatively simple way by application of certain *operational properties* of the transform. For easy reference, these properties are tabulated below in Table 9.2. All properties in Table 9.2 are based on transform definitions (42).

The *linearity property* in Table 9.2 is simply a property of integrals and is therefore also applicable for the inverse transform. The *duality property* follows from a simple interchange of t and ω in the transform relation [e.g., recall our derivation of Eq. (45)]. To derive the *scaling property*, we first observe that for $\alpha > 0$,

$$\mathscr{F}\{f(\alpha t)\} = \int_{-\infty}^{\infty} e^{-i\omega t} f(\alpha t) dt = \frac{1}{\alpha} \int_{-\infty}^{\infty} e^{-i\omega x/\alpha} f(x) dx, \tag{54}$$

where $x = \alpha t$. Hence, we deduce that

$$\mathscr{F}\{f(\alpha t)\} = \frac{1}{\alpha} F(\omega/\alpha), \quad \alpha > 0.$$

Similarly, if $\alpha < 0$ then

$$\mathscr{F}\{f(\alpha t)\} = -\frac{1}{\alpha} F(\omega/\alpha), \quad \alpha < 0,$$

and, consequently, by combining these results we obtain the *scaling property*

$$\mathscr{F}\{f(\alpha t)\} = \frac{1}{|\alpha|} F(\omega/\alpha), \quad \alpha \neq 0. \tag{55}$$

If $\alpha > 1$, the function $f(\alpha t)$ is a compressed version of $f(t)$ and its spectrum $F(\omega)$ is expanded in frequency by $1/\alpha$. The magnitude of the spectrum also changes to maintain the same energy balance. If $0 < \alpha < 1$, then the function $f(\alpha t)$ is an expanded version of $f(t)$ and its spectrum is compressed. Negative values of α reverse time compared with $f(t)$.

Table 9.2 Operational Properties

1. Linearity:	$\mathscr{F}\{C_1 f(t) + C_2 g(t)\} = C_1 F(\omega) + C_2 G(\omega)$
2. Duality:	$\mathscr{F}\{F(t)\} = 2\pi f(-\omega)$
3. Scaling:	$\mathscr{F}\{f(\alpha t)\} = \dfrac{1}{|\alpha|} F(\omega/\alpha), \quad \alpha \neq 0$
4. Time shift:	$\mathscr{F}\{f(t - t_0)\} = e^{-i\omega t_0} F(\omega)$
5. Frequency shift:	$\mathscr{F}\{f(t) e^{i\omega_0 t}\} = F(\omega - \omega_0)$
6. Differentiation (time):	$\mathscr{F}\{f^{(n)}(t)\} = (i\omega)^n F(\omega), \quad n = 1, 2, 3, \dots$
7. Differentiation (freq.):	$\mathscr{F}\{t f(t)\} = i F'(\omega)$
8. Integration:	$\mathscr{F}\left\{\int_{-\infty}^{t} f(\tau) d\tau\right\} = \pi F(0) \delta(\omega) + \dfrac{1}{i\omega} F(\omega)$
9. Convolution:	$\mathscr{F}\{f(t) * g(t)\} = F(\omega) G(\omega)$ $\mathscr{F}\{f(t) g(t)\} = \dfrac{1}{2\pi} F(\omega) * G(\omega)$

9.5.1 Time and frequency shifting

If a given signal is delayed in time by an amount t_0, the corresponding effect on the spectrum (following a change of variable $x = t - t_0$) is given by

$$\mathscr{F}\{f(t - t_0)\} = \int_{-\infty}^{\infty} e^{-i\omega t} f(t - t_0)\,dt$$

$$= e^{-i\omega t_0} \int_{-\infty}^{\infty} e^{-i\omega x} f(x)\,dx,$$

or

$$\mathscr{F}\{f(t - t_0)\} = e^{-i\omega t_0} F(\omega). \tag{56}$$

Hence, a *time shift* delay of a signal leads to the same magnitude spectrum $|F(\omega)|$ but causes a phase shift $-\omega t_0$. Similarly, a *frequency shift* arises when the time function is multiplied by a complex exponential function $e^{i\omega_0 t}$; i.e.,

$$\mathscr{F}\{f(t)e^{i\omega_0 t}\} = \int_{-\infty}^{\infty} e^{-i(\omega - \omega_0)t} f(t)\,dt = F(\omega - \omega_0). \tag{57}$$

In particular, if $f(t)$ is a real signal, it follows that

$$\mathscr{F}\{f(t)\cos\omega_0 t\} = \mathscr{F}\left\{f(t)\frac{1}{2}\left(e^{i\omega_0 t} + e^{-i\omega_0 t}\right)\right\} = \frac{1}{2}\left[F(\omega - \omega_0) + F(\omega + \omega_0)\right]. \tag{58}$$

The process of multiplying a real time signal by a sine or cosine function is called *amplitude modulation*. Note that the modulation process in (58) has moved the spectrum $F(\omega)$ up in frequency by the amount ω_0 and down in frequency by the same amount, both with amplitude reduced by $1/2$ (see Fig. 9.3).

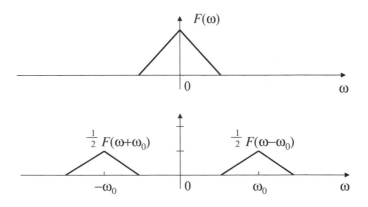

Figure 9.3 Amplitude modulation.

9.5.2 Differentiation and integration

The *differentiation property* in Table 9.2 is obtained by first noting

$$f'(t) = \frac{1}{2\pi} \int_{-\infty}^{\infty} e^{i\omega t}(i\omega)F(\omega)d\omega;$$

hence,

$$\mathscr{F}\{f'(t)\} = i\omega F(\omega), \tag{59}$$

whereas in general,

$$\mathscr{F}\{f^{(n)}(t)\} = (i\omega)^n F(\omega), \quad n = 1,2,3,\dots. \tag{60}$$

Also, it also follows that

$$\mathscr{F}\{tf(t)\} = iF'(\omega). \tag{61}$$

If we integrate both sides of the inverse transform relation (36) from $-\infty$ to t, we find

$$\int_{-\infty}^{t} f(\tau)d\tau = \frac{1}{2\pi} \int_{-\infty}^{\infty} F(\omega) \int_{-\infty}^{t} e^{i\omega\tau} d\tau d\omega. \tag{62}$$

However,

$$\int_{-\infty}^{t} e^{i\omega\tau} d\tau = \int_{-\infty}^{\infty} e^{i\omega\tau} d\tau - \int_{-\infty}^{\infty} e^{-i(-\omega)\tau} U(\tau - t)d\tau$$

$$= 2\pi\delta(\omega) - e^{-i\omega t}\left[\pi\delta(\omega) - \frac{1}{i\omega}\right], \tag{63}$$

where we have used (44), (50), and (56). Thus, from (62) we are now led to

$$\int_{-\infty}^{t} f(\tau)d\tau = \frac{1}{2\pi} \int_{-\infty}^{\infty} F(\omega)\left\{2\pi\delta(\omega) - e^{-i\omega t}\left[\pi\delta(\omega) - \frac{1}{i\omega}\right]\right\}d\omega$$

$$= F(0) - \mathscr{F}^{-1}\left\{\left[\pi\delta(\omega) - \frac{1}{i\omega}\right]F(\omega)\right\}, \tag{64}$$

and by taking the transform of each side, we deduce that

$$\mathscr{F}\left\{\int_{-\infty}^{t} f(\tau)d\tau\right\} = \pi F(0)\delta(\omega) + \frac{1}{i\omega}F(\omega). \tag{65}$$

9.5.3 Convolution theorem

Perhaps the most useful of all the properties associated with the Fourier transform is that of *convolution* given in Table 9.2. To prove the convolution property, we first introduce the *convolution integral*

$$f(t) * g(t) = \int_{-\infty}^{\infty} f(\xi)g(t - \xi)d\xi. \tag{66}$$

By making the change of dummy variable $u = t - \xi$, we get

$$f(t) * g(t) = \int_{-\infty}^{\infty} f(t - u)g(u)du = g(t) * f(t). \tag{67}$$

That is, the convolution integral satisfies the commutative relation. Let us now take the Fourier transform of (66), which leads to

$$\mathscr{F}\{f(t) * g(t)\} = \int_{-\infty}^{\infty} e^{-i\omega t} \int_{-\infty}^{\infty} f(\xi)g(t - \xi)d\xi\,dt$$
$$= \int_{-\infty}^{\infty} \int_{-\infty}^{\infty} e^{-i\omega t} f(\xi)g(t - \xi)d\xi\,dt. \tag{68}$$

We treat (68) as an iterated integral for which the order of integration can be interchanged. Doing so and making the change of variable $t = x + \xi$, we find

$$\mathscr{F}\{f(t) * g(t)\} = \int_{-\infty}^{\infty} \int_{-\infty}^{\infty} e^{-i\omega(x + \xi)} f(\xi)g(x)dx\,d\xi$$
$$= \int_{-\infty}^{\infty} e^{-i\omega\xi} f(\xi)d\xi \int_{-\infty}^{\infty} e^{-i\omega x} g(x)dx \tag{69}$$
$$= F(\omega)G(\omega),$$

and thus, through properties of the Fourier transform, conclude that

$$\boxed{\mathscr{F}^{-1}\{F(\omega)G(\omega)\} = \int_{-\infty}^{\infty} f(\xi)g(t - \xi)d\xi.} \tag{70}$$

Equation (70), or variations of it, are commonly referred to as the *convolution theorem*.

In Section 9.6 we illustrate a conventional use of the convolution theorem (70), whereas in Example 6 below we illustrate a somewhat novel use of the theorem.

EXAMPLE 6 Use the convolution theorem to evaluate the integral

$$I = \int_{-\infty}^{\infty} \frac{dx}{(x^2 + a^2)(x^2 + b^2)}, \quad a, b > 0.$$

Solution: We first note that the integral has the form

$$I = \int_{-\infty}^{\infty} F(\omega)G(\omega)d\omega,$$

where $F(\omega) = 1/(\omega^2 + a^2)$ and $G(\omega) = 1/(\omega^2 + b^2)$. From entry #3 in Table 9.1, we recognize these functions as Fourier transforms, respectively, of

$$f(t) = \frac{1}{2a} e^{-a|t|}, \quad g(t) = \frac{1}{2b} e^{-b|t|}.$$

Hence, using (70) leads to

$$\frac{1}{2\pi} \int_{-\infty}^{\infty} e^{i\omega t} F(\omega) G(\omega) d\omega = \int_{-\infty}^{\infty} f(\xi) g(t - \xi) d\xi,$$

and by setting $t = 0$, we get

$$\int_{-\infty}^{\infty} F(\omega) G(\omega) d\omega = 2\pi \int_{-\infty}^{\infty} f(\xi) g(-\xi) d\xi.$$

The left-hand side of this result is the integral I, and, because the functions $f(t)$ and $g(t)$ are even functions, the above integral can be written as

$$I = 4\pi \int_{0}^{\infty} f(\xi) g(\xi) d\xi$$
$$= \frac{\pi}{ab} \int_{0}^{\infty} e^{-(a+b)\xi} d\xi,$$

from which we deduce

$$I = \int_{-\infty}^{\infty} \frac{dx}{(x^2 + a^2)(x^2 + b^2)} = \frac{\pi}{ab(a + b)}, \quad a, b > 0.$$

9.6 Linear Shift-Invariant Systems

In this section we provide a brief review of general linear systems analysis and terminology such as that associated with electric circuits. In Fig. 9.4 we illustrate the basic diagram of a linear system with input $f(t)$ and output $g(t)$, related by $g(t) = L\{f(t)\}$. In such problems we consider the input and output functions in the time domain, and L is a *linear operator* which mathematically describes the particular system under study.

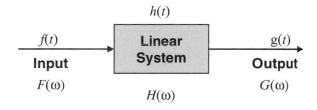

Figure 9.4 Input and output functions of a linear system.

We say a system is *linear* if it is true that

$$L\{C_1 f_1(t) + C_2 f_2(t)\} \;=\; C_1 g_1(t) + C_2 g_2(t), \tag{71}$$

where C_1 and C_2 are arbitrary constants. The system is further called *time-invariant*, or *shift-invariant*, if a time-delayed input $f(t - t_0)$ always produces a time-delayed output $g(t - t_0) = L\{f(t - t_0)\}$.

When the input to a linear system is a delta (impulse) function, the output $h(t)$ is called the *impulse response* of the system, which is directly related to the *Green's function* (see Section 8.5.1). In mathematical symbols we write $f(t) = \delta(t)$ and the output function is then represented by

$$g(t) \;=\; L\{\delta(t)\} \;=\; h(t). \tag{72}$$

To develop a useful relation in the time domain between the input function $f(t)$, the output function $g(t)$, and the impulse response function $h(t)$, we start with the delta function sifting property (see Section 2.2.4)

$$f(t) \;=\; \int_{-\infty}^{\infty} f(\tau)\delta(t - \tau)d\tau \tag{73}$$

and apply the linear operator L to both sides to find

$$L\{f(t)\} \;=\; g(t) \;=\; \int_{-\infty}^{\infty} f(\tau) L\{\delta(t - \tau)\}\, d\tau. \tag{74}$$

However, based on the result of Eq. (72), we can express (74) in the form of a convolution integral, viz.,

$$g(t) \;=\; \int_{-\infty}^{\infty} f(\tau)h(t - \tau)d\tau \;=\; f(t) * h(t). \tag{75}$$

Thus, we see that the mathematical analysis of *linear shift-invariant* (LSI) *systems* is ideally suited for Fourier transform methods. In particular, by taking the Fourier transform of each side of (75) through use of the convolution property in Table 9.2, we obtain the frequency domain relation between input and output functions given by

$$G(\omega) \;=\; F(\omega)H(\omega), \tag{76}$$

where $G(\omega) = \mathscr{F}\{g(t)\}$, $F(\omega) = \mathscr{F}\{f(t)\}$, and $H(\omega) = \mathscr{F}\{h(t)\}$. The function $H(\omega)$ is called the *system transfer function* of the linear system, which is the Fourier transform of the impulse response function $h(t)$. From (76), the transfer function can also be determined from the ratio of output transform to input transform, i.e.,

$$H(\omega) \;=\; \frac{G(\omega)}{F(\omega)}. \tag{77}$$

The transfer function defines the system frequency throughput or bandwidth—a wide bandwidth function allows high frequencies through the system. In general the transfer function is complex and the magnitude $|H(\omega)|$ is called the *gain factor*. The system amplifies the frequencies where the gain factor is high and attenuates those frequencies where the gain factor is low.

EXAMPLE 7 The input/output of the RC lowpass filter shown below in Fig. 9.5 are related by the differential equation [$g(t)$ = electric charge]

$$R\frac{dg}{dt} + \frac{1}{C}g(t) = f(t).$$

Find the transfer function and corresponding impulse response function.

Solution: By taking the Fourier transform of the governing equation, we obtain

$$\left(i\omega R + \frac{1}{C}\right)G(\omega) = F(\omega),$$

from which we deduce

$$H(\omega) = \frac{G(\omega)}{F(\omega)} = \frac{C}{1 + i\omega RC}.$$

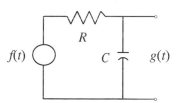

Figure 9.5 RC network.

Using entry #1 in Table 9.1, we obtain the corresponding impulse response function

$$h(t) = \frac{1}{R}e^{-t/RC}U(t).$$

9.7 Hilbert Transforms

In applications involving systems analysis of electric circuits it is sometimes necessary to determine the spectrum $F(\omega)$ defined by (37) when only its real part $R(\omega)$ or its imaginary part $X(\omega)$ is known. If the signal $f(t)$ is a *causal function* [i.e., $f(t) \equiv 0$ for $t < 0$], then this mathematical problem can be solved by what is called the Hilbert transform.

In the special case in which $f(t)$ is an even function, it follows from definition that (recall Example 5)

$$R(\omega) = 2\int_0^\infty f(t)\cos\omega t\,dt \tag{78}$$

and $X(\omega) = 0$. Through the inverse transform relation, it can then be shown that

$$f(t) = \frac{1}{\pi}\int_0^\infty R(\omega)\cos\omega t\,d\omega. \tag{79}$$

Similarly, when $f(t)$ is an odd function, we have $R(\omega) = 0$ and

$$X(\omega) = -2\int_0^\infty f(t)\sin\omega t\,dt, \tag{80}$$

and consequently,

$$f(t) = -\frac{1}{\pi}\int_0^\infty X(\omega)\sin\omega t\, d\omega. \tag{81}$$

Hence, if $f(t)$ is either an even or odd function, it can be recovered entirely from the real or imaginary part of its spectrum $F(\omega)$.

In the general case where $f(t)$ is neither even nor odd, it can always be expressed as the sum of an even and an odd function. That is [recall Eq. (23) in Chapter 8],

$$f(t) = f_e(t) + f_o(t),$$

where $f_e(t)$ is an even function and $f_o(t)$ is an odd function defined, respectively, by

$$f_e(t) = \frac{1}{2}\big[f(t) + f(-t)\big],$$
$$f_o(t) = \frac{1}{2}\big[f(t) - f(-t)\big]. \tag{82}$$

Thus, we recognize $R(\omega)$ as the Fourier transform of $f_e(t)$ and $X(\omega)$ as the Fourier transform of $f_o(t)$. In the case where $f(t)$ is causal, then $f(-t) = 0$ for $t > 0$, and from (82) we deduce that

$$f(t) = 2f_e(t) = 2f_o(t), \quad t > 0.$$

Under this condition we can write

$$f(t) = \frac{2}{\pi}\int_0^\infty R(\omega)\cos\omega t\, d\omega = -\frac{2}{\pi}\int_0^\infty X(\omega)\sin\omega t\, d\omega. \tag{83}$$

A causal function, therefore, can always be recovered from knowledge of either $R(\omega)$ or $X(\omega)$.

Equation (83) suggests there exists some direct relation between the real and imaginary parts of the spectrum of a causal function. To derive this relation, we start with

$$f_e(t) = f_o(t)\mathrm{sgn}(t),$$
$$f_o(t) = f_e(t)\mathrm{sgn}(t). \tag{84}$$

From properties of even and odd functions, we write

$$R(\omega) = \int_{-\infty}^\infty e^{-i\omega t} f_e(t)\, dt$$
$$= \int_{-\infty}^\infty e^{-i\omega t} f_o(t)\mathrm{sgn}(t)\, dt \tag{85}$$

Here we can use the convolution integral [see entry #9 in Table 9.2] to evaluate this last expression, recognizing that

$$\mathscr{F}\{f_o(t)\} = iX(\omega), \qquad \mathscr{F}\{\mathrm{sgn}(t)\} = \frac{2}{i\omega}. \tag{86}$$

Thus, Eq. (85) is equivalent to

$$R(\omega) = \frac{1}{2\pi} \int_{-\infty}^{\infty} i X(y) \left[\frac{2}{i(\omega - y)} \right] dy = \frac{1}{\pi} \int_{-\infty}^{\infty} \frac{X(y)}{\omega - y} \, dy. \tag{87}$$

Similarly, because

$$iX(\omega) = \int_{-\infty}^{\infty} e^{-i\omega t} f_o(t) \, dt$$

$$= \int_{-\infty}^{\infty} e^{-i\omega t} f_e(t) \operatorname{sgn}(t) \, dt,$$

we likewise deduce that

$$X(\omega) = -\frac{1}{\pi} \int_{-\infty}^{\infty} \frac{R(y)}{\omega - y} \, dy. \tag{88}$$

Equations (87) and (88) form a *Hilbert transform pair*. That is, the real and imaginary parts of the spectrum of a causal function are Hilbert transforms of one another. This means that if we know only the real part of the spectrum of a causal function, we can in theory always find the imaginary part through the use of (88), and conversely, if we know only the imaginary part of a causal function, the real part is determined from (87).

EXAMPLE 8 A filter that simply phase-shifts all frequency components of its input by $-1/2\pi$ radians has a transfer function defined by (see Section 9.6)

$$H(\omega) = -i \operatorname{sgn}(\omega).$$

Show that the output function $g(t)$ of this filter is the Hilbert transform of the input function $f(t)$.

Solution: The input and output frequency domain representations are related by

$$G(\omega) = H(\omega)F(\omega) = -i \operatorname{sgn}(\omega)F(\omega).$$

Consequently, by use of the transform relation (50), coupled with the duality principle (entry #2 in Table 9.2) and convolution property (entry #9 in Table 9.2), we are led to the Hilbert transform relation

$$g(t) = \frac{1}{\pi} \int_{-\infty}^{\infty} \frac{f(\tau)}{t - \tau} \, d\tau.$$

9.7.1 Analytic signal representation

A related problem involving Hilbert transforms is to find a complex representation of a real signal. For instance, if $f(t)$ is a real signal and $\hat{f}(t)$ is its Hilbert transform, the *analytic signal representation* of this function is given by

$$z(t) = f(t) + i\hat{f}(t). \tag{89}$$

The main advantage in this representation is that it leads to a one-sided frequency spectrum for the function $z(t)$.

To illustrate the one-sided spectrum characteristic of an analytic signal, let us take the Fourier transform of (89), i.e., $\mathscr{F}\{z(t)\} = \mathscr{F}\{f(t)\} + i\mathscr{F}\{\hat{f}(t)\}$, which leads to (see Example 8)

$$Z(\omega) = F(\omega) + i[-i\,\text{sgn}(\omega)F(\omega)] = F(\omega)[1 + \text{sgn}(\omega)].$$

From definition of the signum function, we see that this last result is equivalent to

$$Z(\omega) = 2F(\omega)U(\omega) = \begin{cases} 2F(\omega), & \omega > 0 \\ 0, & \omega < 0. \end{cases} \tag{90}$$

The factor of 2 in this last equation accounts for the fact that the spectrum $F(\omega)$ for the real signal $f(t)$ is two-sided. Specifically, if $F(\omega)$ is a bandpass spectrum, then $Z(\omega)$ is just twice the positive frequency portion of $F(\omega)$.

The analytic signal representation (89) is particularly useful in the analysis of narrowband signals such as random noise at the output of a bandpass filter. For example, a narrowband signal can be represented by $s(t) = \text{Re}[z(t)e^{i\omega_0 t}]$, which reduces to

$$s(t) = f(t)\cos\omega_0 t - \hat{f}(t)\sin\omega_0 t. \tag{91}$$

9.7.2 Kramers-Krönig relations

The Hilbert transforms (87) and (88) are also known as the Kramers-Krönig relations after the work of Kramers and Krönig on dispersion in optics. Optical dispersion is the result of a dependence of the index of refraction on wavelength or angular frequency. That is, the index of refraction n may have a real part determined by the phase velocity and a (negative) imaginary part determined by absorption, which are related through the *Kramers-Krönig relations* (see [3])

$$\text{Re}[n^2(\omega_0) - 1] = \frac{2}{\pi}\int_0^\infty \frac{\omega\,\text{Im}[n^2(\omega) - 1]}{\omega^2 - \omega_0^2}\,d\omega,$$

$$\text{Im}[n^2(\omega_0) - 1] = -\frac{2}{\pi}\int_0^\infty \frac{\omega_0\,\text{Re}[n^2(\omega) - 1]}{\omega^2 - \omega_0^2}\,d\omega. \tag{92}$$

A simple change of variable reduces Eqs. (92) to the form given by (87) and (88).

9.7.3 Table of transforms and properties

The Hilbert transform is basically convolution. For that reason it does not produce a change of *domain* as is the case with the Fourier transform. That is, if $f(t)$ is a function of time, so is its Hilbert transform $\hat{f}(t)$. One of the interesting properties associated with the Hilbert transform is given by

$$\int_{-\infty}^{\infty} f(t)\hat{f}(t)\,dt \;=\; 0. \tag{93}$$

In words, this says a signal and its Hilbert transform are *orthogonal*. This property can be readily established with the use of Parseval's relation for Fourier transforms.

Listed below in Table 9.3 are some basic properties and transforms associated with the Hilbert transform.

$$\hat{f}(t) \;=\; H\{f(t)\} \;=\; \frac{1}{\pi}\int_{-\infty}^{\infty}\frac{f(\tau)}{t-\tau}\,d\tau,$$

$$f(t) \;=\; H^{-1}\{\hat{f}(t)\} \;=\; -\frac{1}{\pi}\int_{-\infty}^{\infty}\frac{\hat{f}(\tau)}{t-\tau}\,d\tau, \tag{94}$$

Table 9.3 Table of Hilbert Transforms

	$f(t) = H^{-1}\{\hat{f}(t)\}$	$\hat{f}(t) = H\{f(t)\}$
1.	$f(at+b)$	$\hat{f}(at+b)$
2.	$f(t)+g(t)$	$\hat{f}(t)+\hat{g}(t)$
3.	$\dfrac{d^{n}f(t)}{dt^{n}},\; n=1,2,3,\dots$	$\dfrac{d^{n}\hat{f}(t)}{dt^{n}}$
4.	1	0
5.	$\dfrac{1}{t}$	$-\pi\delta(t)$
6.	$\delta(t)$	$\dfrac{1}{\pi t}$
7.	$e^{\pm i\omega_0 t}$	$\mp i e^{\pm i\omega_0 t}$
8.	$\sin(\omega_0 t + \theta)$	$-\cos(\omega_0 t + \theta)$
9.	$\dfrac{a}{t^2+a^2}$	$\dfrac{t}{t^2+a^2}$
10.	$\Pi(t/T)$	$\dfrac{1}{\pi}\ln\left|\dfrac{2t+T}{2t-T}\right|$

9.8 Two-Dimensional Fourier transforms

Because images in optical systems are inherently two-dimensional in nature, their (spatial) frequency analysis requires Fourier transforms in two dimensions. The spatial frequency can be interpreted as z-direction propagating plane waves whose directions are measured with respect to the x and y axes. Thus, if $f(x,y)$ is a piecewise continuous function of x and y such that

$$\int_{-\infty}^{\infty}\int_{-\infty}^{\infty} |f(x,y)|^2\,dx\,dy < \infty,$$

we can (by analogy with one-dimensional transforms) consider the two-dimensional Fourier transform pair

$$F(\xi,\eta) = \int_{-\infty}^{\infty}\int_{-\infty}^{\infty} e^{-i(x\xi+y\eta)} f(x,y)\,dx\,dy, \tag{95}$$

$$f(x,y) = \frac{1}{(2\pi)^2}\int_{-\infty}^{\infty}\int_{-\infty}^{\infty} e^{i(x\xi+y\eta)} F(\xi,\eta)\,d\xi\,d\eta. \tag{96}$$

We call (95) the *two-dimensional* (*double*) *Fourier transform* of the function $f(x,y)$ and (96) the *inverse Fourier transform*. We can interpret (95) as an iterated Fourier transform applied sequentially to the variable x and then to y. That is, we define

$$\hat{f}(\xi,y) = \int_{-\infty}^{\infty} e^{-ix\xi} f(x,y)\,dx,$$

$$F(\xi,\eta) = \int_{-\infty}^{\infty} e^{-iy\eta} \hat{f}(\xi,y)\,dy.$$

The inverse transform (96) follows in a similar manner. We also denote the double Fourier transform and its inverse, respectively, by the notation

$$\mathscr{F}_{(2)}\{f(x,y); x \to \xi, y \to \eta\} = F(\xi,\eta),$$

$$\mathscr{F}_{(2)}^{-1}\{F(\xi,\eta); \xi \to x, \eta \to y\} = f(x,y). \tag{97}$$

Operational properties of the one-dimensional Fourier transform carry over in a natural sort of way to the double Fourier transform. For example, the *shift properties* 4 and 5 in Table 9.2 now become

$$\mathscr{F}_{(2)}\{f(x-a,y-b); x \to \xi, y \to \eta\} = e^{-i(a\xi+b\eta)} F(\xi,\eta), \tag{98}$$

$$\mathscr{F}_{(2)}\{e^{i(ax+by)} f(x,y); x \to \xi, y \to \eta\} = F(\xi-a,\eta-b). \tag{99}$$

Also, the *convolution theorem* (70) for a two-dimensional Fourier transform takes the form

$$\mathscr{F}_{(2)}^{-1}\{F(\xi,\eta)G(\xi,\eta); \xi \to x, \eta \to y\} = \int_{-\infty}^{\infty}\int_{-\infty}^{\infty} f(x-u,y-v)g(u,v)\,du\,dv. \tag{100}$$

Other properties found in Table 9.2 can be similarly extended to the double Fourier transform featured here, but we leave the details of doing so to the reader (see [15]).

If we treat x and y as spatial variables, then by writing $\xi = 2\pi v_x$ and $\eta = 2\pi v_y$, we can interpret the quantities v_x and v_y as *spatial frequencies* (also called *wave numbers*). That is, our treatment of the two-dimensional Fourier transform in this setting can be interpreted as analogous to the one-dimensional treatment provided in the "frequency analysis" of time-varying waveforms. Thus, whereas the Fourier transform of a temporal function leads to frequency specified in *cycles per second* or *Hertz*, the two-dimensional Fourier transform of a spatial function leads to spatial frequencies specified in *cycles per meter* with respect to the x and y axes.

We say a function of two variables is separable if it can be written as a product of two functions, each one depending on only one variable. That is, the function $f(x,y)$ is *separable* if we can write

$$f(x,y) = f_1(x)f_2(y). \tag{101}$$

In this case, the double Fourier transform of $f(x,y)$ reduces to a product of one-dimensional Fourier transforms, i.e.,

$$\mathscr{F}_{(2)}\{f(x,y); x \to \xi, y \to \eta\} = \int_{-\infty}^{\infty} e^{-ix\xi} f_1(x)dx \int_{-\infty}^{\infty} e^{-iy\eta} f_2(y)dy = F_1(\xi)F_2(\eta), \tag{102}$$

where $F_1(\xi) = \mathscr{F}\{f_1(x); x \to \xi\}$ and $F_2(\eta) = \mathscr{F}\{f_2(y); y \to \eta\}$. Separable functions can be important to recognize because they simplify the analysis. In particular, in optics it is common to find circularly symmetric functions separable in polar coordinates, viz., $f(r,\theta) = f_1(r)f_2(\theta)$. Functions of this latter type can often be handled best by the *Fourier-Bessel transform*, more commonly called the *Hankel transform* (see Chapter 10).

9.8.1 Linear systems in optics

The analysis of *LSI systems* in two dimensions is a powerful tool commonly used in electro-optics (EO) systems such as an optical imaging system. We previously (Section 9.6) discussed the one-dimensional analog of this type of system with respect to electrical networks in which the inputs and outputs are real functions of time (e.g., voltages and currents). In the case of an imaging system, the inputs/outputs can be real functions (intensity) or complex valued functions representing field amplitude.

A LSI system in two dimensions is characterized by the *impulse response function* $h(x,y)$ and its Fourier transform $H(\xi,\eta)$ in the same fashion as for the one-dimensional system. Hence, if we denote the input function to such a system by $f(x,y)$, the output function $g(x,y)$ is related by the convolution integral [recall Eq. (75)]

$$g(x,y) = \int_{-\infty}^{\infty}\int_{-\infty}^{\infty} f(u,v)h(x-u, y-v)du\,dv. \tag{103}$$

By taking the two-dimensional Fourier transform of (103), we are led to

$$G(\xi,\eta) = H(\xi,\eta)F(\xi,\eta), \tag{104}$$

from which we deduce

$$H(\xi,\eta) = \frac{G(\xi,\eta)}{F(\xi,\eta)}. \tag{105}$$

As before, we refer to the function $H(\xi,\eta)$ as the *transfer function* of the system.

9.8.2 Coherent imaging systems

A *coherent imaging system* is one in which the object being imaged is illuminated by coherent, monochromatic (single-frequency) light. The LSI system analysis for this case is applied directly to field quantities.

Many simplifications take place in the analysis of imaging systems if we adopt the *ABCD* ray-matrix approach (see Section 15.7). That is, we can represent the entire imaging system by one *ABCD* matrix, including complex elements that may arise from finite aperture stops. In this case, the input and output functions in free space, $U(\mathbf{r},0)$ and $U(\mathbf{r},L)$, are related by the *generalized Huygens-Fresnel integral*

$$
\begin{aligned}
U(\mathbf{r},L) &= -\frac{ik}{2\pi B} \exp(ikL) \iint_{-\infty}^{\infty} d^2s\, U(\mathbf{s},0) \\
&\quad \times \exp\left[\frac{ik}{2B}(As^2 - 2\mathbf{s}\cdot\mathbf{r} + Dr^2)\right] \\
&= \iint_{-\infty}^{\infty} d^2s\, U(\mathbf{s},0)\, h(\mathbf{s},\mathbf{r}),
\end{aligned}
\tag{106}
$$

where L is propagation distance, k is optical wave number, and $h(\mathbf{s},\mathbf{r})$ is the impulse response function expressed in vector notation, i.e., $\mathbf{r} = \langle x,y \rangle$ and $\mathbf{s} = \langle u,v \rangle$. We are also adopting the standard notation $d^2s = du\, dv$. In arriving at this result we have tacitly assumed circular symmetry for all optical elements in the system. Let us further assume the optical wave at the input plane is a unit amplitude Gaussian-beam wave described by

$$
U(\mathbf{r},0) = \exp\left(-\frac{1}{2}\alpha k r^2\right) , \quad \alpha = \frac{2}{kW_0^2} + i\frac{1}{F_0} .
\tag{107}
$$

From (106), it follows that the field of the wave at the output plane is

$$
U(\mathbf{r},L) = \frac{1}{A + i\alpha B} \exp(ikL) \exp\left[-\frac{1}{2}\beta(L)k r^2\right],
\tag{108}
$$

where

$$
AD - BC = 1,
\tag{109}
$$

$$
\beta(L) = \frac{\alpha D - iC}{A + i\alpha B} = \frac{2}{kW^2} + i\frac{1}{F} .
\tag{110}
$$

The quantities W_0 and F_0 are, respectively, the beam radius and phase front radius of curvature of the wave at the input plane, and W and F are the same at the output plane.

Let us assume the input function is a point source on the optical axis described by the two-dimensional delta function

$$
U(\mathbf{s},0) = \delta(\mathbf{s}).
\tag{111}
$$

We can interpret this two-dimensional delta function in much the same way as for the one-dimensional counterpart. In particular, the sifting property is

$$\int_{-\infty}^{\infty} \int_{-\infty}^{\infty} f(\mathbf{s}) \delta(\mathbf{r} - \mathbf{s}) d^2 s = f(\mathbf{r}). \tag{112}$$

Also, in rectangular coordinates, for example, we can write $\delta(\mathbf{r} - \mathbf{s}) = \delta(x - u)\delta(y - v)$, which is simply a product of one-dimensional delta functions. The substitution of (111) into (106) then leads to the free-space propagation impulse response given by

$$U(\mathbf{r},L) = h(0,\mathbf{r}) = -\frac{ik}{2\pi B} \exp\left(ikL + \frac{ikDr^2}{2B} \right). \tag{113}$$

We can interpret this quantity as a Gaussian-beam wave under the paraxial approximation that reduces to a spherical wave in the absence of optical elements (i.e., $B = L$ and $D = 1$).

To derive the associated transfer function we use the two-dimensional Fourier transform defined by

$$H(\mathbf{v}) = \mathcal{F}_{(2)}\{h(0,\mathbf{r})\} = \int_{-\infty}^{\infty}\int e^{-2\pi i \mathbf{r} \cdot \mathbf{v}} h(0,\mathbf{r}) d^2 r, \tag{114}$$

where $\mathbf{v} = (v_x, v_y)$ is a two-dimensional spatial frequency vector. Substituting Eq. (113) into (114), and dividing by $H(0)$ to normalize the expression, we obtain the *coherent transfer function* (CTF)

$$\hat{H}(\mathbf{v}) = \frac{H(\mathbf{v})}{H(0)} = \exp\left(-\frac{2\pi^2 iBv^2}{kD} \right), \tag{115}$$

where $v = (v_x^2 + v_y^2)^{1/2}$ is the magnitude of the spatial frequency vector.

Note that by using the generalized Huygens-Fresnel integral, we obtain the Gaussian approximation (115) for the CTF, which is simpler than that obtained by the analysis of "physical hard apertures" (i.e., those with a finite diameter, not a Gaussian approximation). For example, the single-lens system illustrated in Fig. 9.6 leads to the expressions for the impulse response and CTF given by

$$h(0,\mathbf{r}) = \frac{k^2 D_G^2}{32\pi L_f^2} \exp[ik(L + L_f)] \exp\left(-\frac{k^2 D_G^2 r^2}{32 L_f^2} + \frac{ikr^2}{L_f} \right), \tag{116}$$

$$\hat{H}(\mathbf{v}) \cong \exp\left[-8\left(\frac{\lambda L_f v}{D_G} \right)^2 \right]. \tag{117}$$

On the other hand, when the analysis is based on a hard aperture cutoff, we obtain the corresponding well-known expressions (see [16] or [27])

$$h(0,\mathbf{r}) = \frac{k^2 D_G^2}{32\pi L_f^2} \frac{J_1(k D_G r / 2 L_f)}{k D_G r / 2 L_f}, \tag{118}$$

$$\hat{H}(v) = U(D_G / 2\lambda L_f - |v|) = \begin{cases} 1, & v < D_G / 2\lambda L_f \\ 0, & v > D_G / 2\lambda L_f. \end{cases} \tag{119}$$

In these expressions, L denotes propagation distance from the transmitter to the lens, L_f denotes distance behind the lens to the photodetector, and D_G is the diameter of the lens. Note that both the impulse response (116) and CTF (117) are Gaussian functions, whereas the hard aperture results (118) and (119) involve a Bessel function for the impulse response and a step function for the CTF. In Fig. 9.7 below we compare the Gaussian model (116) for the impulse response with the conventional result (118). The first zero of the Bessel function in (118) occurs at $r = 1.22\lambda L_f/D_G$, which corresponds to the spatial resolution of the imaging system, whereas the $1/e^2$ value of the normalized impulse response (116) occurs at $r = 1.27\lambda L_f/D_G$, nearly the same value.

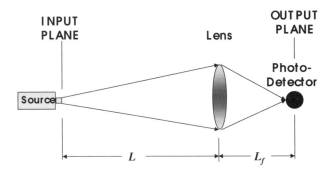

Figure 9.6 Propagation geometry for a single-lens imaging system.

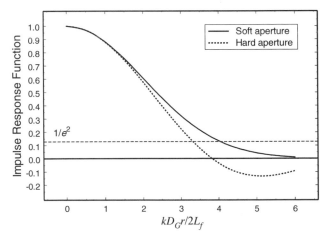

Figure 9.7 Normalized impulse response defined by Eqs. (116) (soft aperture) and (118) (hard aperture). Also shown is the $1/e^2$ cutoff for the soft aperture.

9.9 Fractional Fourier Transform

Mathematicians have concerned themselves for many years with the notion of taking an operation originally defined for integer order and generalizing it to fractional order, or even into the continuum. For example, the gamma function $\Gamma(x)$ is such a generalization of the factorial function $n!$. Many years before the introduction of the gamma function, however, Leibniz raised the question in 1695 concerning derivatives of noninteger order. In the ordinary sense, derivatives are restricted to those of integer order—the first and second derivative, and so on. Nonetheless, it is possible to extend the definition to noninteger order. One way to do this is based on the relation

$$D^n x^p = \frac{\Gamma(p+1)}{\Gamma(p-n+1)} x^{p-n}, \quad n = 1, 2, 3, \ldots,$$

where $D = d/dx$. Because the right-hand side of this expression is meaningful for any real number n for which $\Gamma(p - n + 1)$ is defined, we might argue that the left-hand side can also be meaningful for noninteger n. In particular, by replacing n with a parameter α, not restricted to integer values, we can define a "fractional-order" derivative by

$$D^\alpha x^p = \frac{\Gamma(p+1)}{\Gamma(p-\alpha+1)} x^{p-\alpha}, \quad \alpha \geq 0.$$

Fractional-order derivatives of other types of functions can also be defined as can fractional-order integrals, but we will not do so (see [11]).

The *fractional Fourier transform* (FRFT) is a generalization of the ordinary Fourier transform similar to the notion of a fractional derivative. The fractional Fourier transform has become the focus of many studies in recent years because of its applications in fields like quantum mechanics, optics, and signal processing. However, the FRFT has been known for roughly 70 years and has been reinvented at least twice over that period by different groups of scientists (see [20] and [21]).

To begin, we find it convenient here to reintroduce the Fourier transform in terms of the following variables

$$\begin{aligned}
F(v) &= \mathscr{F}\{f(x)\} = \int_{-\infty}^{\infty} e^{-2\pi i v x} f(x)\, dx \\
f(x) &= \mathscr{F}^{-1}\{F(v)\} = \int_{-\infty}^{\infty} e^{2\pi i v x} F(v)\, dv,
\end{aligned} \tag{120}$$

obtained from Eqs. (42) by letting $t = x$ and $\omega = 2\pi v$. This form has appeal in that the multiplicative factor $1/2\pi$ does not explicitly appear in front of either integral.

Some interesting observations arise if we take successive Fourier transforms of the transformed function. For example, consider

$$\begin{aligned}
\mathscr{F}_1\{f(x)\}(x_1) &= \int_{-\infty}^{\infty} e^{-2\pi i x x_1} f(x)\, dx = F(x_1), \\
\mathscr{F}_2\{f(x)\}(x_2) &= \int_{-\infty}^{\infty} e^{-2\pi i x_1 x_2} F(x_1)\, dx_1 = f(-x_2).
\end{aligned} \tag{121}$$

Continuing in this fashion, we have

$$
\mathcal{F}_3\{f(x)\}(x_3) = \int_{-\infty}^{\infty} e^{-2\pi i x_2 x_3} f(-x_2)\, dx_2 = F(-x_3),
$$

$$
\mathcal{F}_4\{f(x)\}(x_4) = \int_{-\infty}^{\infty} e^{-2\pi i x_3 x_4} f(-x_2)\, dx_3 = f(x_4),
$$

(122)

the last of which returns us to the original function, i.e., $\mathcal{F}_4\{f(x)\}(x) = f(x)$. This exercise brings up the interesting question—rather than take consecutive Fourier transforms by integer jumps between 1 and 4, can we take "fractional" jumps? That is, can we define the expression $\mathcal{F}_\alpha\{f(x)\}(x)$ for noninteger values of α?

In order to develop the fractional Fourier transform, let us first consider the differential equation

$$
f''(x) + 4\pi^2\left[\left(\frac{2n+1}{2\pi}\right) - x^2\right]f(x) = 0,
$$

(123)

whose solutions are the *Hermite-Gaussian functions* (e.g., see Problem 13 in Chapter 2)

$$
\Psi_n(x) = \frac{2^{1/4}}{\sqrt{2^n n!}} H_n(\sqrt{2\pi}\, x) e^{-\pi x^2}, \quad n = 0, 1, 2, \dots.
$$

(124)

If we apply the Fourier transform in (120) to the above differential equation, the transformed equation becomes

$$
F''(v) + 4\pi^2\left[\left(\frac{2n+1}{2\pi}\right) - v^2\right]F(v) = 0.
$$

(125)

In arriving at this result, we have used Properties 6 and 7 in Table 9.2. Because Eq. (125) is identical in form with Eq. (123), it follows that (although we will not show it) the eigenfunctions (124) satisfy the related "eigenvalue problem"

$$
\mathcal{F}\{\Psi_n(x)\} = \lambda_n \Psi_n(v),
$$

(126)

where the eigenvalues are $\lambda_n = i^{-n} = e^{-in\pi/2}$.

Because the eigenfunctions (124) form an *orthonormal* set, we can express certain functions in a *generalized Fourier series* of the form

$$
f(x) = \sum_{n=0}^{\infty} A_n \Psi_n(x),
$$

(127)

where the Fourier coefficients are defined by

$$
A_n = \int_{-\infty}^{\infty} f(x)\Psi_n(x)\, dx.
$$

(128)

The Fourier transform of (127) then yields

$$
\mathcal{F}\{f(x)\} = \sum_{n=0}^{\infty} A_n \mathcal{F}\{\Psi_n(x)\} = \sum_{n=0}^{\infty} A_n e^{-in\pi/2} \Psi_n(v).
$$

(129)

Based on Eq. (129), we now introduce the FRFT of order α by the expression

$$\mathcal{F}_\alpha\{f(x)\}(v) = \sum_{n=0}^\infty A_n e^{-i\alpha n\pi/2}\Psi_n(v) = \sum_{n=0}^\infty A_n e^{-in\varphi}\Psi_n(v), \tag{130}$$

where we have defined

$$\varphi = \frac{\alpha\pi}{2}. \tag{131}$$

The standard Fourier transform corresponds to $\alpha = 1$, but this is in the form of a Fourier series rather than an integral. To put (130) in the form of an integral, we can replace A_n in (130) with its integral definition (128), which leads to

$$\begin{aligned}
\mathcal{F}_\alpha\{f(x)\}(v) &= \int_{-\infty}^\infty f(x)\sum_{n=0}^\infty e^{-in\varphi}\Psi_n(x)\Psi_n(v)dx \\
&= \sqrt{2}\int_{-\infty}^\infty f(x)e^{-\pi(x^2+v^2)}\sum_{n=0}^\infty \frac{(e^{-i\varphi}/2)^n}{n!}H_n(\sqrt{2\pi}\,x)H_n(\sqrt{2\pi}\,v)dx.
\end{aligned} \tag{132}$$

By use of the identity (called Mehler's formula)

$$\sum_{n=0}^\infty H_n(X)H_n(Y)\frac{(s/2)^n}{n!} = \frac{1}{\sqrt{1-s^2}}\exp\left[\frac{2XYs - (X^2+Y^2)s^2}{1-s^2}\right], \tag{133}$$

Eq. (132) can be written in the form

$$\boxed{\mathcal{F}_\alpha\{f(x)\}(v) = \int_{-\infty}^\infty f(x)K_\alpha(x,v)dx,} \tag{134}$$

where (after some algebra)

$$K_\alpha(x,v) = \begin{cases} c(\varphi)\exp\{ia(\varphi)[(x^2+v^2)-2b(\varphi)vx]\}, & \text{if } \alpha \ne 0,1,2 \\ \delta(x-v), & \text{if } \alpha = 0 \\ \delta(x+v), & \text{if } \alpha = 2 \\ e^{-2\pi ivx}, & \text{if } \alpha = 1 \end{cases} \tag{135}$$

$$a(\varphi) = \frac{\pi}{2}\cot\varphi, \quad b(\varphi) = \frac{2}{\cos\varphi}, \quad c(\varphi) = \sqrt{1 - i\cot\varphi}. \tag{136}$$

Equation (134) is the integral representation of the FRFT. Some essential properties of the FRFT are the following: (*i*) it is *linear*, (*ii*) the first-order transform ($\alpha = 1$) corresponds to the standard Fourier transform, and (*iii*) it is additive in index, i.e., $\mathcal{F}_\alpha\mathcal{F}_\beta = \mathcal{F}_{\alpha+\beta}$.

Because the Hermite-Gaussian functions (124) are closely associated with the harmonic oscillator problem of quantum mechanics, early application of the FRFT occurred in quantum mechanics. Later, the FRFT was discovered by other scientists who recognized its utility in optics and signal processing. That this is so is a consequence of the fact that Hermite-Gaussian functions are also fundamental in optical wave propagation in graded index (GRIN) media. Of course, in optics it is the two-dimensional analog of (134) that must be used.

9.9.1 Application in optics

Let us consider the geometry illustrated in Fig. 9.8 involving a thin lens of focal length f. By using the *ABCD* matrix representation from Section 15.7, we can represent the path in Fig. 9.8 by the matrix

$$\mathbf{M}_1 = \begin{pmatrix} 1 & z_2 \\ 0 & 1 \end{pmatrix}\begin{pmatrix} 1 & 0 \\ -\dfrac{1}{f} & 1 \end{pmatrix}\begin{pmatrix} 1 & z_1 \\ 0 & 1 \end{pmatrix} = \begin{bmatrix} 1 - \dfrac{z_2}{f} & (z_1 - z_2) + z_2\left(2 - \dfrac{z_1}{f}\right) \\ -\dfrac{1}{f} & 1 - \dfrac{z_1}{f} \end{bmatrix}. \tag{137}$$

When $z_1 = z_2$, then $\mathbf{M}_1$ takes the form

$$\mathbf{M}_1 = \begin{pmatrix} \cos\varphi & f\sin^2\varphi \\ -\dfrac{1}{f} & \cos\varphi \end{pmatrix}, \tag{138}$$

where we define

$$\cos\varphi = 1 - \frac{z_1}{f}, \qquad \sin\varphi = \frac{1}{f}\sqrt{z_1(2f - z_1)}. \tag{139}$$

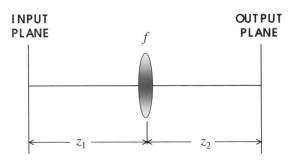

INPUT PLANE f **OUTPUT PLANE**

z_1 z_2

Figure 9.8 Geometry involving a lens of focal length f.

Under this configuration, the optical system in Fig. 9.8 realizes a FRFT. In the special case in which $z_1 = z_2 = f$, corresponding to $\varphi = \pi/2$, the geometry of the system is that associated with the standard Fourier transform in the output plane of an object in the input plane.

The basic property of lenses that is responsible for their focusing capability is that the optical path across them is a quadratic function of the distance $r = \sqrt{x^2 + y^2}$ from the z-

axis. A quadratic GRIN medium is a quadratic lenslike medium that has an index of refraction n defined by

$$n^2 = n_1^2 \left[1 - (n_2/n_1)r^2 \right], \tag{140}$$

where n_1 and n_2 are the GRIN medium parameters. The *ABCD* matrix associated with propagation in a quadratic GRIN medium over distance z is

$$\mathbf{M}_G = \begin{pmatrix} \cos\left(\sqrt{\dfrac{n_2}{n_1}}\, z \right) & \sqrt{\dfrac{n_1}{n_2}}\, \sin\left(\sqrt{\dfrac{n_2}{n_1}}\, z \right) \\[2em] -\sqrt{\dfrac{n_2}{n_1}}\, \sin\left(\sqrt{\dfrac{n_2}{n_1}}\, z \right) & \cos\left(\sqrt{\dfrac{n_2}{n_1}}\, z \right) \end{pmatrix}. \tag{141}$$

At distance $z = \alpha L$, where $L^2 = \pi^2(n_1/n_2)$, the matrix (141) can be written as

$$\mathbf{M}_G = \begin{pmatrix} \cos\varphi & \sqrt{\dfrac{n_1}{n_2}}\, \sin\varphi \\[2em] -\sqrt{\dfrac{n_2}{n_1}}\, \sin\varphi & \cos\varphi \end{pmatrix}, \tag{142}$$

where $\varphi = \alpha\pi/2$, and α is the order of the FRFT realized by the matrix (142). In particular, by introducing the parameter

$$\frac{1}{f} = \sqrt{\frac{n_2}{n_1}}\, \sin\varphi, \tag{143}$$

the matrix $\mathbf{M}_G$ is the same as $\mathbf{M}_1$ defined by (138).

Note in the configuration illustrated in Fig. 9.8 that if the lens is removed (i.e., if $1/f = 0$), the matrix $\mathbf{M}_1$ defined by (138) reduces to a standard two-dimensional rotation matrix similar to that introduced in the study of conic sections. Because the *ABCD* matrix relates the position and derivative coordinates r_1, s_1 at the input plane to r_2, s_2 at the output plane, where $s = dr/dz = r'$, we recognize that the angle φ represents a rotation in the rs-plane. The standard Fourier transform corresponds to a rotation angle $\varphi = \pi/2 + 2n\pi$, but the angle $\varphi = \alpha\pi/2$ for fractional α is associated with the FRFT. Other methods of introducing FRFT into optics exist, but we will not pursue them.

9.10 Wavelets

Fourier series and the Fourier transform have been around since the beginning of the nineteenth century. By contrast, the development of wavelets is much more recent—its origins go back several decades, but the popular use of wavelets in signal analysis and other areas extends only over the last two decades (e.g., see [6] and [25]) A wavelet looks like a wave that travels for one or more periods, but is nonzero only over a short interval rather than propagating forever like sines and cosines. Whereas Fourier analysis is appropriate for filtering or compressing signals that have time-dependent wavelike features, wavelets are more useful in analyzing signals that have "localized" features where standard Fourier analysis is not very efficient.

The first application of wavelets involved the data analysis of seismic surveys such as used in oil and mineral exploration. These surveys are made up of many two-dimensional pictures or slices, sewn together to obtain a three-dimensional image of the structure of rock below the surface. Each slice is obtained by placing "seismic microphones" (geophones) at equally spaced intervals along a line. Explosives set off at one end of the line create a seismic wave in the ground that is recorded by each microphone along the line. The Fourier transform is not a good tool in this application because it can only provide frequency information (oscillations)—it gives no direct information about *when* an oscillation occurred. Wavelets can keep track of time and frequency information. They can be used to "zoom in" on short bursts, or be used to "zoom out" to detect long, slow oscillations.

9.10.1 Haar wavelets

In wavelet analysis there are two functions that play a primary role—the *scaling function* φ called the "father wavelet" and the *wavelet* ψ called the "mother wavelet." The building blocks of wavelet analysis are *translations* and *dilations* (both in height and width) of the basic scaling function. The simplest scaling function is the *Haar scaling function* defined by (see Fig. 9.9)

$$\varphi(x) = U(x) - U(x-1) = \begin{cases} 1, & 0 \le x < 1 \\ 0, & \text{elsewhere}. \end{cases} \tag{144}$$

Note that the function $\varphi(x-n)$ for integer n has the same graph as $\varphi(x)$ but translated to the right by n units if $n > 0$ (and to the left if $n < 0$). Whereas the scaling function (144) is discontinuous at $x = 0$ and $x = 1$, the translates $\varphi(x-n)$ are discontinuous at integers $x = n$ and $x = n + 1$.

The rectangular blocks associated with the Haar scaling function are not sufficiently "thin" to analyze signals of high frequency. However, we note that the function $\varphi(2x)$ has width one half that of the Haar function (144), and that the graph of the function $\varphi(2x-n)$ is the same as that of $\varphi(2x)$ except shifted by $n/2$ units. To achieve narrower or wider building blocks, therefore, we can consider dilates of the scaling functions of the general form $\varphi(2^m x)$, $m = 0, \pm 1, \pm 2, \ldots$. The graph of $\varphi(2^m x)$ is a pulse of unit height and width $1/2^m$. It is easy to show that translations of such functions form an orthogonal set on the real axis, i.e., they do not overlap so that

$$\int_{-\infty}^{\infty} \varphi(2^m x - n)\varphi(2^m x - k)dx = \begin{cases} 0, & k \ne n \\ 1/2^m, & k = n. \end{cases} \tag{145}$$

Having an orthogonal basis is not the complete picture required for representing a given function $f(x)$. We also need a way of isolating the "pulses" that are generated by the father wavelets. This can be accomplished with the introduction of another function $\psi(x)$ called the mother wavelet, or simply, the wavelet. Analogous to the Haar scaling function, we define the *Haar wavelet* by (see Fig. 9.10)

$$\psi(x) = \varphi(2x) - \varphi(2x - 1) = \begin{cases} 1, & 0 \le x < \dfrac{1}{2} \\ -1, & \dfrac{1}{2} \le x < 1 \\ 0, & \text{otherwise}. \end{cases} \tag{146}$$

Note that the Haar wavelet $\psi(x)$ has both positive and negative parts, and further, both parts are of the same size in the sense that

$$\int_{-\infty}^{\infty} \psi(x)\,dx = 0. \tag{147}$$

Also, the dilates $\psi(2^m x)$, $m = 0, \pm 1, \pm 2, \ldots$, are either narrower or wider, depending on whether m is positive or negative, respectively.

Given a continuous function $f(x)$ that is absolutely integrable and $f(x) \to 0$ as $x \to \pm\infty$, we wish to construct a "good" discrete wavelet approximation to it. We start with a *frame* defined by

$$f_m(x) = \sum_{n=-\infty}^{\infty} \alpha_m[n]\varphi(2^m x - n), \tag{148}$$

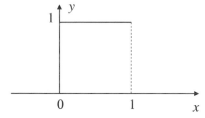

Figure 9.9 Graph of the Haar scaling function.

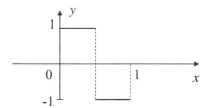

Figure 9.10 Graph of the Haar wavelet function.

where

$$\alpha_m[n] = 2^m \int_{-\infty}^{\infty} f(x)\varphi(2^m x - n)\,dx = 2^m \int_{n/2^m}^{(n+1)/2^m} f(x)\,dx. \tag{149}$$

In the last step in (149) we are recognizing that the function $\varphi(2^m x - n)$ vanishes everywhere except over the given interval. From this it can be shown that

$$|f(x) - f_m(x)| \to 0, \quad m \to \infty,$$

and consequently, we find that (148) can produce an arbitrarily good approximation to the given function $f(x)$ with increasing values of m.

An important idea in wavelet analysis is that the "wide" dilates $\varphi(x/2)$, $\psi(x/2)$ can be written as linear combinations of the "narrow" translates $\varphi(x)$, $\varphi(x-1)$, viz.,

$$\varphi(x/2) = \varphi(x) + \frac{1}{2}\psi(x/2),$$

$$\psi(x/2) = \varphi(x) - \frac{1}{2}\psi(x/2),$$

or, by inverting the system,

$$\varphi(x) = \frac{1}{2}\varphi(x/2) + \frac{1}{2}\psi(x/2),$$

$$\varphi(x-1) = \frac{1}{2}\varphi(x/2) - \frac{1}{2}\psi(x/2). \tag{150}$$

For illustrative purposes, we first take $m = 0$ and constants α_0 and α_1, noting that

$$\alpha_0\,\varphi(x) + \alpha_1\,\varphi(x-1) = \frac{\alpha_0+\alpha_1}{2}\varphi(x/2) + \frac{\alpha_0-\alpha_1}{2}\psi(x/2),$$

whereas for general m, it follows that

$$\alpha_0\,\varphi(2^m x - 2n) + \alpha_1\,\varphi(2^m x - 2n - 1)$$
$$= \frac{\alpha_0+\alpha_1}{2}\varphi(2^{m-1}x - n) + \frac{\alpha_0-\alpha_1}{2}\psi(2^{m-1}x - n). \tag{151}$$

Using this approach for the whole series (148), we obtain a *splitting* of the frame given by

$$f_m(x) = f_{m-1}(x) + d_{m-1}(x), \tag{152}$$

where

$$f_{m-1}(x) = \sum_{n=-\infty}^{\infty} \alpha_{m-1}[n]\varphi(2^{m-1}x - n),$$

$$\alpha_{m-1}[n] = \frac{1}{2}\alpha_m[2n] + \frac{1}{2}\alpha_m[2n+1] \tag{153}$$

$$= 2^{m-1}\int_{-\infty}^{\infty} f(x)\varphi(2^{m-1}x - n)\,dx,$$

and

$$d_{m-1}(x) = \sum_{n=-\infty}^{\infty} \beta_{m-1}[n]\psi(2^{m-1}x - n),$$

$$\beta_{m-1}[n] = \frac{1}{2}\alpha_m[2n] - \frac{1}{2}\alpha_m[2n+1] \tag{154}$$

$$= 2^{m-1}\int_{-\infty}^{\infty} f(x)\psi(2^{m-1}x - n)\,dx.$$

Thus, we see that the frame $f_m(x)$ can be expressed as a sum of the previous frame $f_{m-1}(x)$ and the previous *detail* $d_{m-1}(x)$. In particular, because we can write

$$f_{m+1} = f_m + d_m = f_{m-1} + d_{m-1} + d_m = f_{m-k} + d_{m-k} + \cdots + d_m, \tag{155}$$

we see that we can represent a given frame f_{m+1} in terms of any previous frame f_{m-k} plus the sum of all details from d_{m-k} to d_m.

As shown above, the basic building blocks $\varphi(x)$ and $\psi(x)$ of the Haar wavelets are quite simple. The disadvantage is that these wavelets are discontinuous and therefore do not generally approximate continuous signals very well. Nonetheless, the Haar wavelets still illustrate the general ideas underlying a continuous version of the building blocks $\varphi(x)$ and $\psi(x)$ that leads to a *multiresolution analysis* (which we do not present).

9.10.2 Wavelet transform

The wavelet transform and associated inversion formula decompose a signal into a weighted sum of its various frequency components, much like a Fourier transform. However, the weights in this case involve a given wavelet instead of the exponential e^{isx}.

The wavelet function $\psi(x)$ used in the definition of the transform satisfies the following two requirements:

1. $\psi(x)$ is continuous and $\psi(x) \leq M e^{-C|x|}$, for positive constants M and C.

2. $\int_{-\infty}^{\infty} \psi(x)dx = 0$.

An example of a wavelet function satisfying these two requirements is $\psi(x) = xe^{-x^2}$ (see Fig. 9.11). However, it is often assumed that the wavelet function vanishes outside some interval $-T \leq x \leq T$, which is a stronger requirement than the first condition above.

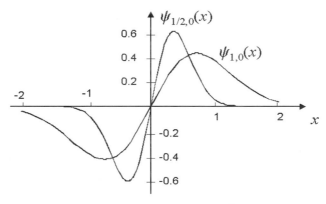

Figure 9.11 Graph of $\psi_{1,0}(x) = \psi(x) = xe^{-x^2}$ and $\psi_{1/2,0}(x)$.

Given a wavelet $\psi(x)$ satisfying the two requirements above, the *wavelet transform* of a suitable function $f(x)$ is defined by

$$W_f(a,b) = \frac{1}{\sqrt{|a|}} \int_{-\infty}^{\infty} f(x)\psi^*\left(\frac{x-b}{a}\right) dx = \sqrt{|a|} \int_{-\infty}^{\infty} f(at+b)\psi^*(t)dt. \tag{156}$$

The second integral in (156) is obtained from the first by a simple change of variable. The transform $W_f(a,b)$ in this case is a function of two variables. As a decreases, the wavelet function, which we also define by the notation

$$\psi_{a,b}(x) = \frac{1}{\sqrt{|a|}} \psi\left(\frac{x-b}{a}\right),$$
(157)

becomes "tall and narrow." As an illustration of this, we plot both $\psi_{1,0}(x)$ and $\psi_{1/2,0}(x)$ for the function $\psi(x) = xe^{-x^2}$ in Fig. 9.11. On the other hand, the nonzero portion of the wavelet is shifted from $x = 0$ to $x = b$ for nonzero b. Hence, we can interpret $W_f(a,b)$ as measuring the frequency component of the function $f(x)$ that oscillates with frequency proportional to $1/a$ near the point $x = b$.

The inversion formula for the wavelet transform is the double integral given by

$$f(x) = \frac{1}{C_s} \int_{-\infty}^{\infty}\int_{-\infty}^{\infty} |a|^{-1/2} W_f(a,b)\psi\left(\frac{x-b}{a}\right)\frac{db\,da}{a^2},$$
(158)

where

$$C_s = 2\pi \int_{-\infty}^{\infty} \frac{|\Psi(s)|}{|s|} ds.$$
(159)

The function $\Psi(s)$ is the Fourier transform of the wavelet defined by

$$\Psi(s) = \int_{-\infty}^{\infty} e^{isx}\psi(x)dx.$$
(160)

Suggested Reading

L. C. Andrews and B. K. Shivamoggi, *Integral Transforms for Engineers* (SPIE Press, Bellingham, 1999).

R. N. Bracewell, *The Fourier Transform and Its Applications* (McGraw-Hill, New York, 1978).

R. R. Goldberg, *Fourier Transforms* (Cambridge University Press, New York, 1961).

A. Papoulis, *The Fourier Integral and Its Applications* (McGraw-Hill, New York, 1963).

I. N. Sneddon, *Fourier Transforms* (McGraw-Hill, New York, 1951).

I. N. Sneddon, *The Use of Integral Transforms* (McGraw-Hill, New York, 1972).

R. E. Ziemer and W. H. Tranter, *Principles of Communications*, 5th ed. (Wiley, New York, 2002)

Chapter 9

Exercises

1. Use the result of (13) to show that

$$\int_0^\infty \frac{\sin s \cos s}{s} \, ds = \frac{\pi}{4}.$$

2. Use the Fourier integral (10) to deduce that

$$e^{-|x|} = \frac{2}{\pi} \int_0^\infty \frac{\cos sx}{1 + s^2} \, ds.$$

In Problems 3 and 4, use (24) and (25) to obtain the Fourier transform of the given function.

3. $f(x) = \begin{cases} 0, & x < 0 \\ e^{-ax}, & x > 0 \end{cases}$ $(a > 0)$

Ans. $\dfrac{1}{a + is}$

4. $f(x) = xe^{-a|x|}$, $(a > 0)$

Ans. $-\dfrac{4ias}{(s^2 + a^2)^2}$

In Problems 5 and 6, use relations (35) and (36) to obtain the Fourier transform of the given function.

5. $f(t) = e^{-t^2/(2\sigma^2)}$

Ans. $\sigma\sqrt{2\pi}\, e^{-\sigma^2\omega^2/2}$

6. $f(t) = t^2 e^{-t^2/2}$

Ans. $\sqrt{2\pi}(1 - \omega^2)e^{-\omega^2/2}$

7. Verify that the following are energy signals.

(a) $f(t) = \Pi(t/12)\cos(6\pi t)$.
See Section 2.2.2.
(b) $f(t) = e^{-|t|}\cos(12\pi t)$.

8. Find the amplitude spectrum of

$$f(t) = A\,\Pi\!\left(\frac{t - t_0}{T}\right).$$

Ans. $T\,|\mathrm{sinc}(\omega/2\pi T)|$

9. Calculate the Fourier transform of the triangle function (Section 2.2.2)

$$\Lambda(t/T) = \begin{cases} 1 - |t|/T, & |t| < T \\ 0, & |t| > T. \end{cases}$$

Ans. $T\mathrm{sinc}^2(\omega/2\pi T)$

In Problems 10-12, use the convolution theorem to show the given result.

10.
$$\int_{-\infty}^\infty \Pi\!\left(\frac{t - \tau}{T}\right)\Pi\!\left(\frac{\tau}{T}\right) d\tau = T\Lambda\!\left(\frac{t}{T}\right).$$

11. $\displaystyle\int_{-\infty}^\infty \sin ax \sin bx \, \frac{dx}{x^2} = \frac{\pi a}{2}$,

$(0 < a < b)$

12. $\displaystyle\int_{-\infty}^\infty e^{-i\omega t - |t|}\,\frac{\sin t}{t}\, dt = \tan^{-1}\frac{2}{\omega^2}$

In Problems 13 and 14, use residues (Chapter 7) to verify the given Fourier transform relation.

13. $\mathscr{F}\!\left\{\dfrac{1}{t^2 + k^2}\right\} = \dfrac{\pi}{k}e^{-k|\omega|}$, $(k > 0)$

14. $\mathscr{F}_s\!\left\{\dfrac{x^3}{x^4 + k^4}; x \to s\right\}$

$$= \sqrt{\frac{\pi}{2}}\cos\!\left(\frac{ks}{\sqrt{2}}\right)e^{-ks/\sqrt{2}},\quad (k > 0)$$

15. Use residue theory from Section 7.6.3 to establish the Fourier transform integral relation $(a > 0)$

$$\int_{-\infty}^\infty \frac{e^{-i\omega t}}{(a - i\omega)^n}\, d\omega = \frac{2\pi t^{n-1}}{(n-1)!}e^{-at}U(t),$$

$$n = 1, 2, 3, \ldots.$$

16. From the result of Problem 15, deduce that $(a > 0)$

$$\mathcal{F}\{t^n e^{-at} U(t)\} = \frac{n!}{(a + i\omega)^n},$$
$$n = 0, 1, 2, \ldots.$$

17. Given the impulse response function in Example 7, determine the output of an *RC* lowpass filter for the input function

$$f(t) = A\Pi\left(\frac{t - T/2}{T}\right), \quad A = \text{const.}$$

Ans.

$$g(t) = \begin{cases} 0, & t < 0 \\ A(1 - e^{-t/RC}), & 0 < t < T \\ A[e^{-(t-T)/RC} - e^{-t/RC}], & t > T. \end{cases}$$

18. Given that

$$\mathcal{F}\{(a^2 - t^2)^{-1/2} U(a - |t|)\} = \frac{1}{2} J_0(a\omega),$$

show that $(0 < a < b)$

$$\int_0^\infty J_0(ax)J_0(bx)\,dx = \frac{2}{\pi b} K\left(\frac{a}{b}\right),$$

where $K(m)$ is the *complete elliptic integral*

$$K(m) = \int_0^{\pi/2} \frac{d\theta}{\sqrt{1 - m^2 \sin^2\theta}}.$$

19. Verify that the given functions satisfy the Hilbert transform relations.

$$R(\omega) = -\frac{a}{\omega^2 + a^2}, \quad X(\omega) = -\frac{\omega}{\omega^2 + a^2}.$$

20. By use of the convolution theorem, show that the inverse Fourier transform of the Hilbert transform result (90) leads to

$$\mathcal{F}^{-1}\{Z(\omega)\} = f(t) + \frac{i}{\pi} \int_{-\infty}^\infty \frac{f(\tau)}{t - \tau}\,d\tau,$$

and therefore deduce that

$$\mathcal{F}^{-1}\{Z(\omega)\} = z(t) = f(t) + i\hat{f}(t).$$

21. By use of Parseval's relation, derive the Hilbert transform orthogonality relation

$$\int_{-\infty}^\infty f(t)\hat{f}(t)\,dt = 0.$$

In Problems 22-24, show that the double Fourier transform satisfies the given property.

22. $\mathcal{F}_{(2)}\{f(ax,by); x \to \xi, y \to \eta\}$
$$= \frac{1}{|ab|} F\left(\frac{\xi}{a}, \frac{\eta}{b}\right)$$

23. $\mathcal{F}_{(2)}\{f(x - a, y - b); x \to \xi, y \to \eta\}$
$$= e^{i(a\xi + b\eta)} F(\xi, \eta)$$

24. $\mathcal{F}_{(2)}\{f(x,y)\cos \omega x; x \to \xi, y \to \eta\}$
$$= \frac{1}{2}[F(\xi + \omega, \eta) + F(\xi - \omega, \eta)]$$

Chapter 10

Laplace, Hankel and Mellin Transforms

10.1 Introduction / 496
10.2 Laplace Transform / 496
    10.2.1 Table of transforms and operational properties / 500
    10.2.2 Inverse transforms—I / 503
    10.2.3 Inverse transforms—II / 506
10.3 Initial Value Problems / 508
    10.3.1 Simple electric circuits / 509
    10.3.2 Impulse response function / 510
    10.3.3 Stability of linear systems / 511
10.4 Hankel Transform / 513
    10.4.1 Operational properties and table of transforms / 516
10.5 Mellin Transform / 519
    10.5.1 Operational properties and table of transforms / 520
    10.5.2 Complex variable methods / 523
10.6 Applications Involving the Mellin Transform / 526
    10.6.1 Products of random variables / 526
    10.6.2 Electromagnetic wave propagation / 527
10.7 Discrete Fourier Transform / 529
    10.7.1 Discrete transform pair / 529
10.8 Z-transform / 533
    10.8.1 Operational properties / 535
    10.8.2 Difference equations / 536
10.9 Walsh Transform / 538
    10.9.1 Walsh functions / 539
    10.9.2 Walsh series and the discrete Walsh transform / 540
    Suggested Reading / 542
    Exercises / 542

In this chapter we introduce several transforms that are commonly used in engineering applications. Except for the Laplace transform, which can be used in a variety of applications, the other integral transforms are considered more specialized. We also briefly discuss the notion of a discrete Fourier transform, a discrete Laplace transform (called the Z-transform), and a discrete Walsh transform.

10.1 Introduction

Integral transforms are common working tools of every engineer and scientist. The Fourier transform studied in Chapter 9 is basic to frequency spectrum analysis of time-varying waveforms. Here, we study the Laplace transform used in control theory and in the analysis of initial-value problems like those associated with electric circuits. In addition, we introduce the Hankel transform (directly related to a two-dimensional Fourier transform) and the Mellin transform. The Hankel transform is essential to the analysis of diffraction theory and image formation in optics (see [15]), and the Mellin transform is useful in probability theory and in optical wave propagation (see [28]).

▸   **Integral transform:** An *integral transform* is a relation of the form

$$F(s) = \int_{-\infty}^{\infty} K(s,t)f(t)dt, \tag{1}$$

such that a given function $f(t)$ is transformed into another function $F(s)$ by means of an integral.

We call the function $F(s)$ the *transform* of $f(t)$, and $K(s,t)$ is the *kernel* of the transformation. If the kernel is defined by

$$K(s,t) = e^{-st}U(t) = \begin{cases} 0, & t < 0 \\ e^{-st}, & t \geq 0, \end{cases}$$

where $U(t)$ is the *unit step function* (Section 2.2), we obtain the *Laplace transform* (Section 10.2), and if the kernel takes on one of the forms $K(s,t) = e^{\pm ist}$, we obtain the *Fourier transform* (Chapter 9). Also, if $K(s,t) = tJ_0(st)U(t)$, where $J_0(x)$ is a *Bessel function* (Section 2.5), we are led to the *Hankel transform* (Section 10.4), and the *Mellin transform* (Section 10.5) is associated with the kernel $K(s,t) = t^{s-1}U(t)$.

10.2 Laplace Transform

Whereas the Fourier transform grew out of the pioneering work on heat conduction by Joseph Fourier (1768-1830), the Laplace transform can be traced back to the work of Oliver Heaviside (1850-1925) who used operational calculus to solve circuit problems with discontinuous input functions. The Laplace transform is ordinarily defined outright but it can be instructive to develop it directly from the Fourier integral theorem.

To begin, we assume $f(t)$ and its derivative $f'(t)$ are both piecewise continuous functions on $t \geq 0$. The function $f(t)$ is also assumed to be zero for $t < 0$, in which case it is called a *causal function*. Although we do not require $f(t)$ to be absolutely integrable, we do assume the related function

$$g(t) = e^{-ct}f(t)U(t) \tag{2}$$

has this property, where c is a positive constant. It therefore follows that

$$\int_0^\infty e^{-ct} |f(t)| \, dt < \infty. \tag{3}$$

Because the function (2) satisfies the conditions of the Fourier integral theorem [see Theorem 9.1 and Eq. (21) in Chapter 9], we can write

$$g(t) = \frac{1}{2\pi} \int_{-\infty}^{\infty} e^{ist} \int_{-\infty}^{\infty} g(x)e^{-isx} \, dx \, ds,$$

or equivalently,

$$f(t)U(t) = \frac{e^{ct}}{2\pi} \int_{-\infty}^{\infty} e^{ist} \int_0^\infty f(x)e^{-(c+is)x} \, dx \, ds. \tag{4}$$

If we now introduce the change of variable $p = c + is$, then (4) becomes

$$f(t)U(t) = \frac{1}{2\pi i} \int_{c-i\infty}^{c+i\infty} e^{pt} \int_0^\infty f(x)e^{-px} \, dx \, dp. \tag{5}$$

Hence, in a heuristic manner we have derived the pair of transformation formulas

$$\mathcal{L}\{f(t)\} = \int_0^\infty e^{-pt} f(t) \, dt = F(p), \tag{6}$$

$$\mathcal{L}^{-1}\{F(p)\} = \frac{1}{2\pi i} \int_{c-i\infty}^{c+i\infty} e^{pt} F(p) \, dp = f(t)U(t), \tag{7}$$

called the *Laplace transform* of $f(t)$ and the *inverse Laplace transform*, respectively. Note that Eq. (7) represents the inverse Laplace transform in integral form, but where integration takes place in the complex p-plane.

To ensure the existence of the Laplace transform (6), we need to impose a condition on the function $f(t)$ in addition to being piecewise continuous. That is, we need to specify that $f(t)$ also be of exponential order [e.g., see Eq. (3)].

▸   **Exponential order:** We say $f(t)$ is of *exponential order* if there exists a constant $c > 0$ such that $\lim_{t \to \infty} f(t)e^{-ct} = 0$.

Theorem 10.1 If $f(t)$ is piecewise continuous on $t \geq 0$ and of exponential order, it has a Laplace transform defined by Eq. (6).

Routine integration techniques will produce the Laplace transform of many elementary functions. In some cases, however, we may have to resort to various "tricks" and manipulations in the evaluation of a particular transform.

EXAMPLE 1 Find $\mathcal{L}\{e^{at}\}$.

Solution: By definition,

$$\mathcal{L}\{e^{at}\} = \int_0^\infty e^{-pt} e^{at}\, dt = \int_0^\infty e^{-(p-a)t}\, dt,$$

from which we deduce

$$\mathcal{L}\{e^{at}\} = \frac{e^{-(p-a)t}}{-(p-a)}\Big|_0^\infty = \frac{1}{p-a}, \quad p > a.$$

Remark: Although the transform variable p was assumed to be complex in our derivation of (6) and (7), in many applications it is convenient to treat p as a real parameter as we did in Example 1. We will generally follow such treatment throughout this section.

Note that by allowing $a \rightarrow 0^+$ in the result of Example 1, we get the limiting case

$$\mathcal{L}\{1\} = \frac{1}{p}, \quad p > 0. \tag{8}$$

It is common in practice to impose a restriction like that in (8) and in Example 1 on the transform variable p to ensure convergence of the defining integral. Restrictions of this kind, however, have little effect in most applications of the Laplace transform.

EXAMPLE 2 Find $\mathcal{L}\{t^n\}$, $n = 0, 1, 2, \dots$.

Solution: By definition,

$$\mathcal{L}\{t^n\} = \int_0^\infty e^{-pt} t^n\, dt.$$

In this case we note that by making the change of variable $x = pt$, the integral becomes

$$\mathcal{L}\{t^n\} = \frac{1}{p^{n+1}} \int_0^\infty e^{-x} x^n \, dx,$$

which, through properties of the gamma function (Section 2.3.1), reduces to

$$\mathcal{L}\{t^n\} = \frac{n!}{p^{n+1}}, \quad n = 0,1,2,\dots.$$

By using the gamma function, we can generalize the result of Example 2 to include

$$\mathcal{L}\{t^x\} = \int_0^\infty e^{-pt} t^x \, dt = \frac{\Gamma(x+1)}{p^{x+1}}, \quad x > -1. \tag{9}$$

In particular, if $x = -1/2$, Eq. (9) yields the result

$$\mathcal{L}\{t^{-1/2}\} = \frac{\Gamma(\frac{1}{2})}{\sqrt{p}} = \sqrt{\frac{\pi}{p}}, \quad p > 0. \tag{10}$$

Our last example here involves the *error function* and *complementary error function* defined, respectively, by (see Section 2.3.4)

$$\mathrm{erf}(t) = \frac{2}{\sqrt{\pi}} \int_0^t e^{-x^2} \, dx, \tag{11}$$

$$\mathrm{erfc}(t) = \frac{2}{\sqrt{\pi}} \int_t^\infty e^{-x^2} \, dx. \tag{12}$$

EXAMPLE 3 Find the Laplace transform $\mathcal{L}\{\mathrm{erf}(t)\}$.

Solution: From the defining integral, we have

$$\begin{aligned} \mathcal{L}\{\mathrm{erf}(t)\} &= \int_0^\infty e^{-pt} \mathrm{erf}(t) \, dt \\ &= \frac{2}{\sqrt{\pi}} \int_0^\infty \int_0^x e^{-pt-x^2} \, dx \, dt. \end{aligned}$$

Recharacterizing the region of integration $0 \le x \le t$, $0 \le t < \infty$ (see Fig. 10.1) by $x \le t < \infty$, $0 \le x < \infty$, we can interchange the order of integration above to get

$$\mathcal{L}\{\mathrm{erf}(t)\} \;=\; \frac{2}{\sqrt{\pi}}\int_0^\infty e^{-x^2}\int_x^\infty e^{-pt}\,dt\,dx$$

$$=\; \frac{2}{p\sqrt{\pi}}\int_0^\infty e^{-x^2 - px}\,dx$$

$$=\; \frac{2}{p\sqrt{\pi}}\,e^{p^2/4}\int_0^\infty e^{-(x + p/2)^2}\,dx,$$

where we have written

$$x^2 + px = \left(x + \frac{1}{2}p\right)^2 - \frac{1}{4}p^2.$$

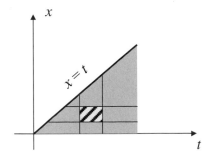

Figure 10.1 Region of integration.

Finally, by making the change of variable $u = x + p/2$, we see that

$$\mathcal{L}\{\mathrm{erf}(t)\} \;=\; \frac{1}{p\sqrt{\pi}}\,e^{p^2/4}\int_{p/2}^\infty e^{-u^2}\,du,$$

from which we deduce

$$\mathcal{L}\{\mathrm{erf}(t)\} \;=\; \frac{1}{p\sqrt{\pi}}\,e^{p^2/4}\,\mathrm{erfc}(p/2), \quad p > 0.$$

10.2.1   Table of transforms and operational properties

After a transform is calculated it is usually placed in a table like Table 10.1 for reference purposes (much like an integral table). A large number of Laplace transforms are known (e.g., see [13]), but we list only a few of those in Table 10.1.

Like the Fourier transform, the calculation of Laplace transforms directly from the integral definition is often tedious and difficult. However, once we have developed the transforms of several functions, many other transforms can be obtained from these through use of certain properties called *operational properties* (see Table 10.2).

The *linearity* property in Table 10.2 is a simple consequence of the linearity property of integration. The *shift* property follows directly from observing

$$\mathcal{L}\{e^{at}f(t)\} \;=\; \int_0^\infty e^{-pt}\big[e^{at}f(t)\big]dt \;=\; \int_0^\infty e^{-(p-a)}f(t)\,dt, \tag{13}$$

and recognizing that the last integral is the Laplace transform of $f(t)$ with transform variable $p - a$. To show the *derivative* property, we first observe that

$$\mathcal{L}\{f'(t)\} \;=\; \int_0^\infty e^{-pt}f'(t)\,dt \;=\; f(t)e^{-pt}\Big|_0^\infty + p\int_0^\infty e^{-pt}f(t)\,dt,$$

Table 10.1 Table of Selected Laplace Transforms

	$F(p) = \mathcal{L}\{f(t)\}$	$f(t) = \mathcal{L}^{-1}\{F(p)\}$
1.	$\dfrac{1}{p}$	1
2.	$\dfrac{1}{p^n}$ $(n = 1,2,3\ldots)$	$\dfrac{t^{n-1}}{(n-1)!}$
3.	$\dfrac{1}{p^x}$, $x > 0$	$\dfrac{t^{x-1}}{\Gamma(x)}$
4.	$\dfrac{1}{(p-a)^n}$ $(n = 1,2,3,\ldots)$	$\dfrac{t^{n-1}e^{at}}{(n-1)!}$
5.	$\dfrac{1}{(p-a)^x}$, $x > 0$	$\dfrac{t^{x-1}e^{at}}{\Gamma(x)}$
6.	$\dfrac{1}{p^2+k^2}$	$\dfrac{1}{k}\sin kt$
7.	$\dfrac{p}{p^2+k^2}$	$\cos kt$
8.	$\dfrac{1}{p(p^2+k^2)}$	$\dfrac{1}{k^2}(1-\cos kt)$
9.	$\dfrac{1}{p^2(p^2+k^2)}$	$\dfrac{1}{k^3}(kt-\sin kt)$
10.	$\dfrac{1}{\sqrt{p^2+k^2}}$	$J_0(kt)$
11.	e^{-ap}, $a > 0$	$\delta(t-a)$
12.	$\dfrac{1}{p}e^{-ap}$, $a > 0$	$U(t-a)$
13.	$\dfrac{1}{p}e^{-k/p}$, $k > 0$	$J_0(2\sqrt{kt})$
14.	$\dfrac{1}{p}e^{k/p}$, $k > 0$	$I_0(2\sqrt{kt})$

Table 10.2 Operational Properties of the Laplace Transform

1. Linearity:	$\mathscr{L}\{C_1 f(t) + C_2 g(t)\} = C_1 F(p) + C_2 G(p)$
2. Shift:	$\mathscr{L}\{e^{at} f(t)\} = F(p - a)$
3. Transform of Derivative:	$\mathscr{L}\{f'(t)\} = pF(p) - f(0)$ $\mathscr{L}\{f''(t)\} = p^2 F(p) - pf(0) - f'(0)$
4. Derivative of Transform:	$\mathscr{L}\{t^n f(t)\} = (-1)^n F^{(n)}(p),\ n = 1,2,3,\dots$

where we have performed an integration by parts, assuming that $f'(t)$ is piecewise continuous and of exponential order. By simplifying this result, we obtain the first derivative relationship (Property 3) in Table 10.2. Similarly, the second derivative property follows from application of the first, viz.,

$$\mathscr{L}\{f''(t)\} = p\mathscr{L}\{f'(p)\} - f'(0) = p[pF(p) - f(0)] - f'(0)$$
$$= p^2 F(p) - pf(0) - f'(0).$$

To formally verify Property 4 in Table 10.2, we start with the transform relation

$$F(p) = \int_0^\infty e^{-pt} f(t)\,dt$$

and differentiate both sides with respect to p to obtain

$$F'(p) = \int_0^\infty e^{-pt}(-t)f(t)\,dt = -\int_0^\infty e^{-pt}[tf(t)]\,dt. \tag{14}$$

Thus, we deduce that

$$\mathscr{L}\{tf(t)\} = -F'(p). \tag{15}$$

Repeated differentiation under the integral yields Property 4.

EXAMPLE 4 Find $\mathscr{L}\{\cos kt\}$ and $\mathscr{L}\{\sin kt\}$.

Solution: Based on Euler's formula $\cos kt + i\sin kt = e^{ikt}$, we find

$$\mathscr{L}\{\cos kt\} + i\mathscr{L}\{\sin kt\} = \mathscr{L}\{e^{ikt}\} = \frac{1}{p - ik},$$

where we have used the result of Example 1 and the linearity property. By separating this last result into real and imaginary parts, we obtain

$$\mathcal{L}\{\cos kt\} + i\mathcal{L}\{\sin kt\} = \frac{1}{p - ik} \cdot \frac{1 + ik}{1 + ik} = \frac{p}{p^2 + k^2} + i\frac{k}{p^2 + k^2},$$

and by matching up the real and imaginary parts, we deduce that

$$\mathcal{L}\{\cos kt\} = \frac{p}{p^2 + k^2}, \qquad \mathcal{L}\{\sin kt\} = \frac{k}{p^2 + k^2}.$$

EXAMPLE 5 Evaluate $\mathcal{L}\{5e^{-2t}\cos 3t\}$.

Solution: From Example 4 we have

$$\mathcal{L}\{5\cos 3t\} = \frac{5p}{p^2 + 9}.$$

Hence, by use of the shift property, it follows that

$$\mathcal{L}\{5e^{-2t}\cos 3t\} = \frac{5(p + 2)}{(p + 2)^2 + 9} = \frac{5p + 10}{p^2 + 4p + 13}.$$

EXAMPLE 6 Evaluate $\mathcal{L}\{2te^{-3t}\sin t\}$.

Solution: Based on the transform relation (see Example 4)

$$\mathcal{L}\{\sin t\} = \frac{1}{p^2 + 1}$$

and Eq. (15), we first obtain

$$\mathcal{L}\{t\sin t\} = -\frac{d}{dp}\left(\frac{1}{p^2 + 1}\right) = \frac{2p}{(p^2 + 1)^2}.$$

If we now apply the shift and linearity properties, we find that

$$\mathcal{L}\{2te^{-3t}\sin t\} = \frac{4(p + 3)}{[(p + 3)^2 + 1]^2} = \frac{4p + 12}{(p^2 + 6p + 10)^2}.$$

10.2.2 Inverse transforms—Part I

Generally, the use of Laplace transforms is effective only if we can also solve the inverse problem; i.e., given $F(p)$, what is $f(t)$? In symbols, we write

$$\mathcal{L}^{-1}\{F(p)\} = f(t), \tag{16}$$

called the *inverse Laplace transform*. To evaluate inverse transforms, we rely on the following methods:

- The use of *tables* (Table 10.1)
- The use of *operational properties* (Table 10.3)
- The method of *partial fractions* (from calculus)
- The method of *residues* (see Chapter 7)

Table 10.3 Operational Properties of the Inverse Laplace Transform

1. Linearity:	$\mathcal{L}^{-1}\{C_1 F(p) + C_2 G(p)\} = C_1 f(t) + C_2 g(t)$
2. Shift:	$\mathcal{L}^{-1}\{F(p - a)\} = e^{at} f(t)$
3. Convolution:	$\mathcal{L}^{-1}\{F(p)G(p)\} = \displaystyle\int_0^t f(\tau)g(t - \tau)d\tau$ $= \displaystyle\int_0^t f(t - \tau)g(\tau)d\tau$

The linearity property and shift property appearing in Table 10.3 are simply restatements of Properties 1 and 2 in Table 10.2. In general, each property established for the forward transform is also a property for the inverse transform. The convolution property in Table 10.3 is often stated as a separate property called the convolution theorem. It is an important property in the analysis of linear systems, providing a direct link between input and output functions (see Section 10.3.2).

EXAMPLE 7 Find $\mathcal{L}^{-1}\left\{\dfrac{p - 5}{p^2 + 6p + 13}\right\}$.

Solution: First, we complete the square in the denominator to get

$$\frac{p - 5}{p^2 + 6p + 13} = \frac{p - 5}{(p + 3)^2 + 4} = \frac{(p + 3) - 8}{(p + 3)^2 + 4}.$$

Then using Properties 1 and 2 in Table 10.3, we obtain

$$\mathcal{L}^{-1}\left\{\frac{p-5}{p^2+6p+13}\right\} = \mathcal{L}^{-1}\left\{\frac{(p+3)-8}{(p+3)^2+4}\right\}$$

$$= e^{-3t}\mathcal{L}^{-1}\left\{\frac{p-8}{p^2+4}\right\}$$

$$= e^{-3t}\left[\mathcal{L}^{-1}\left\{\frac{p}{p^2+4}\right\} - 4\mathcal{L}^{-1}\left\{\frac{2}{p^2+4}\right\}\right],$$

from which we deduce

$$\mathcal{L}^{-1}\left\{\frac{p-5}{p^2+6p+13}\right\} = e^{-3t}(\cos 2t - 4\sin 2t).$$

To prove the convolution property in Table 10.3, we begin by considering the product

$$F(p)G(p) = \int_0^\infty e^{-p\tau}f(\tau)d\tau \int_0^\infty e^{-px}g(x)dx$$
$$= \int_0^\infty \int_0^\infty e^{-p(\tau+x)}f(\tau)g(x)dx\,d\tau. \tag{17}$$

The change of variable $x = t - \tau$ then leads to

$$F(p)G(p) = \int_0^\infty \int_\tau^\infty e^{-pt}f(\tau)g(t-\tau)dt\,d\tau, \tag{18}$$

which we can interpret as an iterated integral over the region $\tau \le t < \infty$, $0 \le \tau < \infty$, similar to that shown in Fig. 10.1. Following the approach (in reverse order) in Example 3, we get

$$F(p)G(p) = \int_0^\infty \int_0^t e^{-pt}f(t)g(t-\tau)d\tau\,dt$$
$$= \int_0^\infty e^{-pt}\int_0^t f(\tau)g(t-\tau)d\tau\,dt. \tag{19}$$

The inside integral in (19) is called the *convolution integral*, which we also denote by

$$\boxed{f(t)*g(t) = \int_0^t f(\tau)g(t-\tau)d\tau.} \tag{20}$$

Note that the Laplace convolution is not the same as that defined for the Fourier transform [recall Eq. (66) in Chapter 9]. By making the change of dummy variable $u = t-\tau$, we see that

$$f(t)*g(t) = -\int_t^0 f(t-u)g(u)du = \int_0^t g(u)f(t-u)du = g(t)*f(t). \tag{21}$$

That is, the convolution integral satisfies the commutative relation. Now, by formally taking the inverse Laplace transform of (19), we obtain the important result known as the *convolution theorem*, viz.,

$$\mathcal{L}^{-1}\{F(p)G(p)\} = f(t) * g(t) = \int_0^t f(t - \tau)g(t)d\tau. \tag{22}$$

EXAMPLE 8 Find $\mathcal{L}^{-1}\left\{\dfrac{7}{p^2(p^2 + 9)}\right\}$,

(*a*) by the convolution theorem.
(*b*) by the use of partial fractions.

Solution:
(*a*) Let us select $F(p) = 7/p^2$ and $G(p) = 1/(p^2 + 9)$, whose inverse transforms are, respectively,

$$\mathcal{L}^{-1}\left\{\frac{7}{p^2}\right\} = f(t) = 7t, \qquad \mathcal{L}^{-1}\left\{\frac{1}{p^2 + 9}\right\} = g(t) = \frac{1}{3}\sin 3t.$$

Hence, by use of the convolution property (22), we are led to

$$\mathcal{L}^{-1}\left\{\frac{7}{p^2(p^2 + 9)}\right\} = f(t) * g(t) = \frac{7}{3}\int_0^t (t - \tau)\sin 3\tau\, d\tau$$

$$= \frac{7}{27}(3t - \sin 3t).$$

(*b*) By the method of partial fractions, we first observe

$$\frac{7}{p^2(p^2 + 9)} = \frac{A_1}{p} + \frac{A_2}{p^2} + \frac{B + Cp}{p^2 + 9}.$$

Calculating, we find $A_1 = 0$, $A_2 = 21/27$, $B = 21/27$, and $C = 0$, which yields

$$\mathcal{L}^{-1}\left\{\frac{7}{p^2(p^2 + 9)}\right\} = \frac{21}{27}\mathcal{L}^{-1}\left\{\frac{1}{p^2}\right\} + \frac{7}{27}\mathcal{L}^{-1}\left\{\frac{3}{p^2 + 9}\right\} = \frac{7}{27}(3t - \sin 3t).$$

10.2.3 Inverse transforms—Part II

The techniques introduced in Section 10.2.2 for evaluating inverse Laplace transforms are adequate for a wide variety of routine applications. In some cases, however, the desired inverse transform may not be available in tables or not amenable to the operational properties alone. For such problems we may need to use complex variable methods to directly evaluate the inverse transform defined by the integral [recall Eq. (7)]

$$\mathscr{L}^{-1}\{F(p)\} = f(t)U(t) = \frac{1}{2\pi i}\int_{c-i\infty}^{c+i\infty} e^{pt}F(p)\,dp. \tag{23}$$

The complex inversion integral in Eq. (23) can be evaluated by the use of residue theory from Chapter 7. That is, if $F(p)$ is an analytic function in the complex p-plane except for a finite number of *isolated singularities* a_k, $k = 1,2,...,N$, we consider the evaluation of the integral in (23) around the closed contour shown in Fig. 10.2, which encloses *all* isolated singularities of $F(p)$. As the radius R of the circular arc C_R becomes unbounded, we are led in the limit to the *complex inversion formula*

$$\boxed{\mathscr{L}^{-1}\{F(p)\} = f(t)U(t) = \sum_{k=1}^{N} \mathrm{Res}\{e^{pt}F(p);a_k\}.} \tag{24}$$

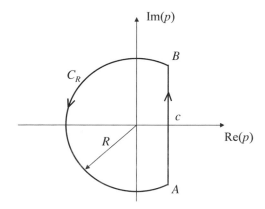

Figure 10.2 Contour of integration.

EXAMPLE 9 Evaluate $\mathscr{L}^{-1}\left\{\dfrac{1}{p(p^2 + k^2)}\right\}$.

Solution: Here we note that the function has simple poles at $p = 0, \pm ik$. Thus,

$$\mathscr{L}^{-1}\left\{\frac{1}{p(p^2 + k^2)}\right\} = \mathrm{Res}\left\{\frac{e^{pt}}{p(p^2 + k^2)};0\right\}$$

$$+ \mathrm{Res}\left\{\frac{e^{pt}}{p(p^2 + k^2)};ik\right\} + \mathrm{Res}\left\{\frac{e^{pt}}{p(p^2 + k^2)};-ik\right\}$$

$$= \frac{1}{k^2} - \frac{1}{2k^2}(e^{ikt} + e^{-ikt})$$

$$= \frac{1}{k^2}(1 - \cos kt).$$

A rather useful result emerges if we consider the case in which

$$F(p) = \frac{P(p)}{Q(p)},$$

where $P(p)$ and $Q(p)$ are both polynomials with no common factors and the degree of $Q(p)$ exceeds that of $P(p)$. Now, suppose that $Q(p) = K(p - a_1)(p - a_2)\cdots(p - a_N)$, where K is a constant and $a_j \neq a_k$ if $j \neq k$. Thus, the function $F(p)$ has simple poles at the zeros of $Q(p)$ and, using residues, it follows from (24) that

$$\mathscr{L}^{-1}\{F(p)\} = f(t)U(t) = \sum_{k=1}^{N} \frac{P(a_k)}{Q'(a_k)} e^{a_k t}. \qquad (25)$$

Equation (25) is known as the first *Heaviside expansion formula* and is usually derived from a partial fraction expansion of $F(p)$.

10.3 Initial Value Problems

The Laplace transform is a powerful tool for solving linear differential equations (DEs) with constant coefficients—in particular, *initial value problems* (IVPs). The usefulness of the transform for this class of problems rest primarily on the fact that the transform of the DE, together with prescribed initial conditions, reduces the problem to an algebraic equation in the transformed function. This algebraic problem is easily solved, and the inverse Laplace transform of its solution is the solution of the IVP that we seek.

   In the solution of DEs we ordinarily denote the unknown function by $y(t)$ and its transform by $Y(p)$.

EXAMPLE 10 Solve the IVP

$$y'' - 6y' + 9y = t^2 e^{3t}, \quad y(0) = 2, \ y'(0) = 6.$$

Solution: By defining $\mathscr{L}\{y(t)\} = Y(p)$, the term-by-term Laplace transform of the DE by the use of Property 2 in Table 10.2 yields

$$\mathscr{L}\{y'(t)\} = pY(p) - y(0) = pY(p) - 2,$$

$$\mathscr{L}\{y''(t)\} = p^2 Y(p) - py(0) - y'(0) = p^2 Y(p) - 2p - 6,$$

$$\mathscr{L}\{t^2 e^{3t}\} = \frac{2}{(p - 3)^3}.$$

Hence, the transformed IVP takes the form

$$[p^2 Y(p) - 2p - 6] - 6[pY(p) - 2] + 9Y(p) = \frac{2}{(p - 3)^3},$$

or, upon simplification,

$$(p^2 - 6p + 9)Y(p) = 2(p - 3) + \frac{2}{(p - 3)^3}.$$

Now, solving this last equation for $Y(p)$, we obtain

$$Y(p) = \frac{2}{p - 3} + \frac{2}{(p - 3)^5},$$

and upon taking the inverse transform, we get (from Table 10.1)

$$y(t) = 2\mathcal{L}^{-1}\left\{\frac{1}{p - 3}\right\} + \frac{2}{4!}\mathcal{L}^{-1}\left\{\frac{4!}{(p - 3)^5}\right\} = 2e^{3t} + \frac{1}{12}t^4 e^{3t}.$$

10.3.1 Simple electric circuits

The circuit shown below in Fig. 10.3 is composed of a *resistor* with a resistance of R ohms (Ω), *capacitor* with a capacitance of C farads (F), and an *inductor* with an inductance of L henrys (H), connected in series with a *voltage source* of $E(t)$ volts (V). IVPs involving simple electric circuits are readily handled by the Laplace transform. The DE governing the electric charge q on the capacitor is given by (see Section 15.2.4)

$$Lq'' + Rq' + C^{-1}q = E(t), \quad t > 0. \tag{26}$$

Let the *RLC* circuit shown in Fig. 10.3 have components $L = 0.1\,\mathrm{H}, R = 2\,\Omega$, $C = 1/260\,\mathrm{F}$, and $E(t) = 100\sin 60t$ V. If, initially, $i(0) = q(0) = 0$, we wish to calculate the charge q on the capacitor at all later times. Using these values, the IVP (26) can be formulated as

$$\frac{1}{10}q'' + 2q' + 260q = 100\sin 60t, \quad q(0) = 0, \ i(0) = q'(0) = 0. \tag{27}$$

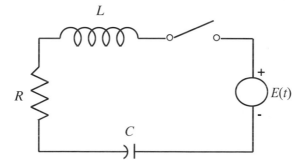

Figure 10.3 *RLC* circuit.

To begin, we multiply by 10 to put the problem in the more convenient form

$$q'' + 20q' + 2600q = 1000\sin 60t, \quad q(0) = 0, \quad i(0) = q'(0) = 0. \tag{28}$$

The term-by-term Laplace transform of this equation leads to

$$(p^2 + 20p + 2600)Q(p) = \frac{60,000}{p^2 + 3600},$$

which has solution

$$Q(p) = \frac{60,000}{(p^2 + 3600)[(p + 10)^2 + 2500]}. \tag{29}$$

We have written the second factor in the denominator in proper form to invoke the shift property in Table 10.3. In this case it is helpful to use a partial fraction expansion and write

$$Q(p) = \frac{Ap + B}{p^2 + 3600} + \frac{C(p + 10) + D}{(p + 10)^2 + 2500},$$

the inversion of which yields the solution form

$$q(t) = A\cos 60t + \frac{B}{60}\sin 60t + e^{-10t}\left(C\cos 50t + \frac{D}{50}\sin 50t\right). \tag{30}$$

By the method of partial fractions, we find that

$$A = -\frac{30}{61}, \quad B = -\frac{1500}{61}, \quad C = \frac{30}{61}, \quad \text{and} \quad D = \frac{1800}{61},$$

from which we deduce

$$q(t) = -\frac{5}{61}(6\cos 60t + 5\sin 60t) + \frac{6}{61}e^{-10t}(5\cos 50t + 6\sin 50t). \tag{31}$$

In some cases, it is also useful to know the *steady-state response*, which we get from (31) by allowing $t \to \infty$, i.e.,

$$q(t)\Big|_{t \to \infty} = -\frac{5}{61}(6\cos 60t + 5\sin 60t). \tag{32}$$

10.3.2 Impulse response function

In the study of *linear systems*, such as electric circuits, the systems engineer characterizes the system under study by the *transfer function*[1]

$$G(p) = \frac{Y(p)}{F(p)}, \tag{33}$$

---

[1] The transfer function $G(p)$ is the same as $H(\omega)$ introduced in Section 9.6 for $p = i\omega$.

where $\mathcal{L}\{y(t)\} = Y(p)$ and $\mathcal{L}\{f(t)\} = F(p)$.

To illustrate, let us assume the linear system under study is governed by the IVP

$$ay'' + by' + cy = f(t), \quad y(0) = 0, \ y'(0) = 0, \tag{34}$$

where a, b, c are constants. The Laplace transform of (34) leads to

$$(ap^2 + bp + c)Y(p) = F(p), \tag{35}$$

from which we deduce the transfer function (see also Section 9.6)

$$G(p) = \frac{Y(p)}{F(p)} = \frac{1}{ap^2 + bp + c}. \tag{36}$$

By defining the inverse Laplace transform

$$g(t) = \mathcal{L}^{-1}\{G(p)\} = \mathcal{L}^{-1}\left\{\frac{1}{ap^2 + bp + c}\right\}, \tag{37}$$

we can use the convolution theorem to write the solution of (34) in the form

$$y(t) = \int_0^t f(\tau)g(t - \tau)d\tau. \tag{38}$$

We see, therefore, that knowledge of the transfer function or its inverse Laplace transform allows us to solve the IVP (34) for any input function $f(t)$. In particular, if the input function is the unit impulse function, i.e., if $f(t) = \delta(t)$, then (38) reduces to

$$y(t) = \int_0^t \delta(\tau)g(t - \tau)d\tau = g(t). \tag{39}$$

In words, $g(t)$ is "the response of the linear system at rest (34) that is subject to a unit impulse input at time $t = 0$." For this reason, the function $g(t) = \mathcal{L}^{-1}\{G(p)\}$ is known as the *impulse response function*, or *Green's function* (recall Section 8.5.1), of the system described by (34).

In particular, the transfer function of the *RLC* circuit defined by Eq. (27) is

$$G(p) = \frac{10}{p^2 + 20p + 2600} = \frac{10}{[(p + 10)^2 + 2500]}. \tag{40}$$

The impulse response function, defined as the inverse Laplace transform of (40), is then

$$g(t) = 10e^{-10t}\sin 50t. \tag{41}$$

10.3.3 Stability of linear systems

One of the needs of a design engineer is to distinguish between functions in time that are "stable" and those that are "unstable" (recall the discussion in Section 3.5.3). Basically, the distinction is based on whether the given function remains finite or becomes unbounded.

To study the stability of certain functions, we need to establish the relationship between bounded and unbounded functions and their Laplace transforms. In control theory, for example, it is necessary to deal with electrical or mechanical systems subjected to an input or excitation that produces some kind of output or response, e.g., a linear system. For a linear system, we define a stable system in the following way.

▸ **Stable system:** A *stable system* is one that produces a bounded output for every bounded input.

If the input function is of the form $f(t) = e^{at}\cos(kt + \varphi)$, where a, k, φ are constants, then this function is bounded as $t \to \infty$ only if $a \le 0$. That is, if $a > 0$, the input function represents an unbounded function that is oscillatory if $k \ne 0$ and nonoscillatory if $k = 0$. Hence, for $a > 0$, the input function $F(p)$ has a pole (see Section 7.4.4) in the right-half p-plane. If $a = 0$ the pole is pure imaginary and for $a < 0$ the pole is in the left-half p-plane. In general it can be shown that the input function is bounded for all t if and only if $F(p)$ has no poles to the right of the imaginary axis, and any poles on the imaginary axis are simple.

Based on the above definition of stable system, we can study the stability of a linear system by analyzing the transfer function

$$G(p) = \frac{Y(p)}{F(p)}, \tag{42}$$

introduced in Section 10.3.2 as the ratio of the Laplace transform of the output function to the Laplace transform of the input function. By use of the complex inversion formula (24), the solution of the linear system with transfer function (42) is given by

$$\mathcal{L}^{-1}\{Y(p)\} = y(t)U(t) = \sum_{k=1}^{N} \mathrm{Res}\{e^{pt}F(p)G(p); a_k\}, \tag{43}$$

where a_k, $k = 1, 2, \dots, N$ denote the isolated singularities of the product $F(p)G(p)$. If the input function is bounded, then the stability of the system is entirely determined by the location of the poles of the transfer function (42). Specifically, we can say the system is stable if all the poles are to the left of the imaginary axis.

Remark: The transfer function of a stable system has all its poles lying to the left of the imaginary axis—this is so because, if $G(p)$ has a simple pole on the imaginary axis, and if also $F(p)$ has the same simple pole on the imaginary axis, the product $F(p)G(p)$ would have a second-order pole at this point. Nonetheless, some authors refer to the case where $G(p)$ has simple poles on the imaginary axis as only "marginally unstable" rather than unstable.

EXAMPLE 11 Discuss the stability of a system for real constants a and b, if the input $f(t)$ and output $y(t)$ are related by

$$y''' - ay'' + b^2 y' - ab^2 y = f(t).$$

Solution: With initial values set to zero (for the purpose of this analysis), the Laplace transform of the given equation becomes

$$(p^3 - ap^2 + b^2p - ab^2)Y(p) = F(p),$$

from which we readily deduce the transfer function

$$G(p) = \frac{1}{p^3 - ap^2 + b^2p - ab^2} = \frac{1}{(p-a)(p^2+b^2)}.$$

Thus, the transfer function has simple poles at $p = a$ and at $p = \pm ib$ ($b \neq 0$), in which case the system is always *unstable*. To better understand this conclusion, we look at specific cases:

- If $a > 0$, the system has a pole in the right-half plane and the system is unstable.

- If $a \leq 0$, and $b \neq 0$, the poles $\pm ib$ are simple and lie on the imaginary axis. The system is unstable, but is sometimes referred to as only "marginally unstable."

- If $b = 0$ and $a \neq 0$, the pole 0 on the imaginary axis is of second-order and the system is unstable.

10.4 Hankel Transform

Hankel transforms arise naturally in solving boundary-value problems formulated in cylindrical coordinates and they are used in diffraction theory to describe the image-forming process of a linear filtering operation for coherent and incoherent imaging. For circularly symmetric functions, the Hankel transform and its inversion formula are mathematically equivalent to a two-dimensional Fourier transform and its inversion formula (see Section 9.8).

To formally derive the Hankel transform, we start with the two-dimensional Fourier transform pair[2]

$$F(\xi,\eta) = \frac{1}{2\pi} \iint_{-\infty}^{\infty} e^{-i(x\xi+y\eta)} f(x,y)\,dx\,dy, \tag{44}$$

$$f(x,y) = \frac{1}{2\pi} \iint_{-\infty}^{\infty} e^{i(x\xi+y\eta)} F(\xi,\eta)\,d\xi\,d\eta. \tag{45}$$

---

[2] In some cases the two dimensional Fourier transform may be expressed by a product of one-dimensional transforms defined by [*i.e.*, if $f(x,y) = \hat{f}(x)\hat{g}(y)$, then $F(\xi,\eta) = \hat{F}(\xi)\hat{G}(\eta)$]

$$\hat{F}(\xi) = \frac{1}{\sqrt{2\pi}} \int_{-\infty}^{\infty} e^{-ix\xi} \hat{f}(x)\,dx, \quad \hat{G}(\eta) = \frac{1}{\sqrt{2\pi}} \int_{-\infty}^{\infty} e^{-iy\eta} \hat{g}(y)\,dy,$$

where we "split" the constant $1/2\pi$ between the transform and the inverse transform.

We call (44) the *two-dimensional Fourier transform* of the function $f(x,y)$ and (45) the *inverse two-dimensional Fourier transform*. In this case we have divided the multiplicative constant $1/(2\pi)^2$ equally between the transform and the inverse transform. In applications involving these transform pairs, such as in optics, the problem often exhibits circular symmetry for natural reasons. When this happens some simplification takes place because one radial variable will suffice in place of two independent rectangular coordinates. Let us assume the function being transformed has circular symmetry, viz., $f(x,y) \equiv f(\sqrt{x^2 + y^2}) \equiv f(r)$, and represent (44) and (45) in polar coordinates. That is, if $f(x,y) \equiv f(r)$, we expect the transform function to also exhibit circular symmetry, $F(\xi,\eta) \equiv F(\rho)$, where the polar coordinates are defined by

$$
\begin{aligned}
x &= r\cos\theta, & y &= r\sin\theta, \\
\xi &= \rho\cos\varphi, & \eta &= \rho\sin\varphi.
\end{aligned}
\tag{46}
$$

Under these transformations, we can write (44) as

$$
F(\rho,\varphi) = \frac{1}{2\pi} \int_0^\infty \int_0^{2\pi} e^{-i\rho r\cos(\theta-\varphi)} rf(r)\,d\theta\,dr,
\tag{47}
$$

where we recognize $x\xi + y\eta = \rho r\cos(\theta - \varphi)$. The innermost integral yields

$$
\frac{1}{2\pi} \int_0^{2\pi} e^{-\rho r\cos(\theta-\varphi)}\,d\theta = J_0(\rho r),
$$

where $J_0(x)$ is a Bessel function of order zero. Because the resulting integral in (47) is a function of ρ alone (independent of φ), we see that the transform function is

$$
\boxed{F(\rho) = \mathcal{H}_0\{f(r)\} = \int_0^\infty rf(r)J_0(\rho r)\,dr,}
\tag{48}
$$

called the *Hankel transform of order zero*. Performing similar transformations on the inverse transform integral (45) give us the *inversion formula*

$$
\boxed{f(r) = \mathcal{H}_0^{-1}\{F(\rho)\} = \int_0^\infty \rho F(\rho)J_0(r\rho)\,d\rho.}
\tag{49}
$$

Hankel transforms of higher order are also sometimes used in practice. For example, the general *Hankel transform of order v* is defined by

$$
\mathcal{H}_v\{f(r)\} = \int_0^\infty f(r)J_v(\rho r)\,dr = F(\rho), \quad v > -1/2
\tag{50}
$$

and its *inversion formula* by

$$
\mathcal{H}_v^{-1}\{F(\rho)\} = \int_0^\infty \rho F(\rho)J_v(r\rho)\,d\rho = f(r), \quad v > -1/2.
\tag{51}
$$

A heuristic argument for the special case $v = n/2 - 1$ can be presented to derive (50) and (51), which parallels the development of the Hankel transform of order zero, by considering the multiple Fourier transform of order n applied to a radially symmetric function.

The basic requirement for the existence of the Hankel transform (48) is that the function $\sqrt{r}f(r)$ be piecewise continuous and absolutely integrable on the real line. Let us illustrate the calculation of some elementary Hankel transforms.

EXAMPLE 12 Find $\mathcal{H}_0\{U(a - r)\}$, $a > 0$.

Solution: This magnitude squared of this transform represents the far-field diffraction pattern of a circular aperture. From definition of the step function $U(x)$, we have

$$\mathcal{H}_0\{U(a - r)\} = \int_0^a rJ_0(\rho r)dr = \int_0^{a\rho} xJ_0(x)dx,$$

where the last step follows a change of variable. The direct evaluation of this last integral using Property (BJ13) in Section 2.5.1 leads to

$$\mathcal{H}_0\{U(a - r)\} = \frac{a}{\rho}J_1(a\rho), \quad a > 0.$$

EXAMPLE 13 Find $\mathcal{H}_0\left\{\frac{1}{r}e^{-ar}\right\}$, $a > 0$.

Solution: From definition,

$$\mathcal{H}_0\left\{\frac{1}{r}e^{-ar}\right\} = \int_0^\infty e^{-ar}J_0(\rho r)\ dr = \sum_{n=0}^\infty \frac{(-1)^n (\rho/2)^{2n}}{(n!)^2} \int_0^\infty r^{2n}e^{-ar}dr,$$

where we have written the Bessel function in its series representation and interchanged the order of integration and summation. The integral under the sum is recognized as a Laplace transform of r^{2n}, and thus we deduce that

$$\mathcal{H}_0\left\{\frac{1}{r}e^{-ar}\right\} = \frac{1}{a}\sum_{n=0}^\infty \frac{(-1)^n (\rho/2a)^{2n}}{(n!)^2}(2n)!.$$

By writing the factorials in Pochhammer notation (see Section 2.6.1) as

$$\frac{(2n)!}{n!} = 2^{2n}(\tfrac{1}{2})_n,$$

we see that

$$\mathcal{H}_0\left\{\frac{1}{r}e^{-ar}\right\} = \frac{1}{a}\sum_{n=0}^\infty \frac{(-1)^n (\frac{1}{2})\rho^{2n}}{n!} = \frac{1}{a}\,{}_1F_0\left(\frac{1}{2};\text{—};-\rho^2/a^2\right),$$

where $_1F_0$ denotes a generalized hypergeometric function. However, by using Property (GH2) in Section 2.6.4, we can express the Hankel transform as

$$\mathcal{H}_0\left\{\frac{1}{r}e^{-ar}\right\} = \frac{1}{\sqrt{a^2 + \rho^2}}, \quad a > 0.$$

The Gaussian function $f(r) = e^{-a^2r^2}$ arises in many applications, particularly in optics. By following a procedure similar to that in Example 13, it can be shown that the Fourier transform of a Gaussian function is another Gaussian function, i.e.,

$$\mathcal{H}_0\left\{e^{-a^2r^2}\right\} = \frac{1}{2a^2}e^{-\rho^2/4a^2}, \quad a > 0. \tag{52}$$

10.4.1 Operational properties and table of transforms

As in the case of other transforms, new Hankel transforms can often be obtained from the use of certain *operational properties* in conjunction with known transforms. However, there are only a few operational properties (see Table 10.4) associated with the Hankel transform, e.g., there is *no* convolution theorem. A short table of Hankel transforms is given in Table 10.5.

Table 10.4 Operational Properties of the Hankel Transform

1. Linearity:	$\mathcal{H}_0\{C_1f(r) + C_2g(r)\} = C_1F(\rho) + C_2G(\rho)$
2. Scaling:	$\mathcal{H}_0\{f(ar)\} = (1/a^2)F(\rho/a), \ a > 0$
3. Derivative:	$\mathcal{H}_0\left\{\frac{1}{r}\frac{d}{dr}[rf'(r)]\right\} = -\rho^2F(\rho)$

The *linearity* property follows directly from the linearity of the defining integral. The *scaling* property can be deduced from the simple change of variable $x = ar$, which yields

$$\mathcal{H}_0\{f(ar)\} = \int_0^\infty rf(ar)J_0(\rho r)dr = \frac{1}{a^2}\int_0^\infty xf(x)J_0(\rho x/a)dx. \tag{53}$$

From (53), our result in Table 10.4 directly follows. To develop the derivative relation (Property 3 in Table 10.4), it is necessary to first show that

$$\mathcal{H}_1\{f'(r)\} = -\rho \mathcal{H}_0\{f(r)\} \tag{54}$$

and

$$\mathcal{H}_0\left\{\frac{1}{r}\frac{d}{dr}[rf(r)]\right\} = \rho \mathcal{H}_1\{f(r)\}. \tag{55}$$

Equation (54) requires integration by parts and the identity (see Section 2.5.1)

$$\frac{d}{dr}[rJ_1(\rho r)] = \rho r J_0(\rho r).$$

Equation (55) follows in a similar manner, but we leave the details in both cases to the reader.

Table 10.5 Table of Selected Hankel Transforms

	$F(\rho) = \mathcal{H}_0\{f(r)\}$	$f(r) = \mathcal{H}_0^{-1}\{F(\rho)\}$
1.	$\dfrac{a}{\rho}J_1(a\rho), \quad a>0$	$U(a-r)$
2.	$\dfrac{1}{\rho}$	$\dfrac{1}{r}$
3.	$\dfrac{1}{\sqrt{\rho^2+a^2}}, \quad a>0$	$\dfrac{1}{r}e^{-ar}$
4.	$\dfrac{a}{(\rho^2+a^2)^{3/2}}, \quad a>0$	e^{-ar}
5.	$\dfrac{1}{2a^2}e^{-\rho^2/4a^2}, \quad a>0$	$e^{-a^2r^2}$
6.	$\dfrac{U(a-\rho)}{\sqrt{a^2-\rho^2}}, \quad a>0$	$\dfrac{1}{r}\sin ar$
7.	$\sin^{-1}\dfrac{a}{\rho}, \quad a>0$	$\dfrac{1}{r^2}\sin ar$

EXAMPLE 14 (*Diffraction through a circular slit*) A plane wave is passed through a circular slit of inside radius a and outside radius $b > a$. Find the far-field diffraction pattern of this aperture function.

Solution: The far-field diffraction, also called the Fraunhofer diffraction pattern, can be found from the Hankel transform of order zero of the circular slit function defined by

$$f(r) = \begin{cases} 1, & a < r < b \\ 0, & \text{otherwise}. \end{cases}$$

We recognize that this function is simply the difference of two step functions, i.e.,

$$f(r) = U(b - r) - U(a - r).$$

Hence, based on Example 13 and the linearity property (Property 1 in Table 10.4), we deduce that

$$\mathcal{H}_0\{U(b - r) - U(a - r)\} = \frac{bJ_1(b\rho) - aJ_1(a\rho)}{\rho}.$$

In Fig. 10.4(*a*), (*b*) below we plot the field of the far-field diffraction pattern for the slit aperture, which corresponds to a circular aperture and a circular (centered) obscuration. We also show in Fig. 10.4(*c*) the standard far-field diffraction pattern for a circular slit without the obscuration.

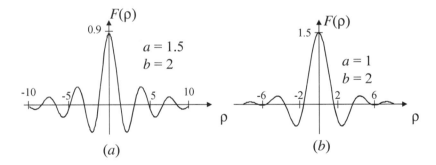

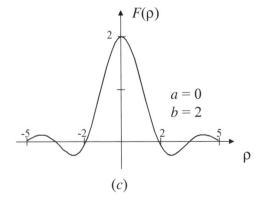

Figure 10.4 Far-field diffraction pattern for (*a*) a circular slit aperture (*a* = 1.5, *b* = 2), (*b*) a circular slit aperture (*a* = 1, *b* = 2), and (*c*) a circular aperture (*a* = 0, *b* = 2).

10.5 Mellin Transform

Like the Hankel transform, the Mellin transform is a specialized tool in that its utility is somewhat limited. However, in the proper setting it can be a very effective tool.

The Mellin transform and its inversion formula can be formally derived from the Fourier integral theorem (see Theorem 9.1) in a manner similar to that of the Laplace transform (see Section 10.2). If we assume that $g(u)$ satisfies the conditions of the Fourier integral theorem, then we can write [see Eq. (21) in Chapter 9]

$$g(u) = \frac{1}{2\pi} \int_{-\infty}^{\infty} e^{i\xi u} \int_{-\infty}^{\infty} g(t) e^{-i\xi t} \, dt \, d\xi.$$

By introducing the change of variables $x = e^t$, $y = e^u$, and $s = c - i\xi$, where c is a fixed constant, we obtain Fourier's integral theorem in the form

$$g(\ln y) y^{-c} = \frac{1}{2\pi i} \int_{c-i\infty}^{c+i\infty} y^{-s} \int_{0}^{\infty} g(\ln x) x^{-c} x^{s-1} \, dx \, ds. \tag{56}$$

If we now define

$$f(x) = g(\ln x) x^{-c},$$

then (56) leads to the pair of transform relations

$$\boxed{M\{f(x)\} = F(s) = \int_{0}^{\infty} x^{s-1} f(x) \, dx,} \tag{57}$$

$$\boxed{M^{-1}\{F(s)\} = f(x) U(x) = \frac{1}{2\pi i} \int_{c-i\infty}^{c+i\infty} x^{-s} F(s) \, ds.} \tag{58}$$

We call (57) the *Mellin transform* of the function $f(x)$ and (58) the *inverse Mellin transform* of $F(s)$.

If the integrals

$$\int_{0}^{\infty} x^{a-1} f(x) \, dx, \quad \int_{0}^{\infty} x^{b-1} f(x) \, dx,$$

both converge for real a and b ($a < b$), then the Mellin transform of $f(x)$ converges uniformly to $F(s)$ in any finite domain interior to the infinite vertical strip $a < \sigma < b$, where $\sigma = \text{Re}(s)$. In such cases, the function $F(s)$ is analytic in this vertical strip.

EXAMPLE 15 Find $M\{e^{-ax}\}$, $a > 0$.

Solution: From definition, we have

$$M\{e^{-ax}\} = \int_0^\infty x^{s-1} e^{-ax} dx = \frac{\Gamma(s)}{a^s}, \quad \text{Re}(s) > 0,$$

where we are using properties of the gamma function (see Section 2.3.1). Also, the form of the integral above suggests that the Mellin transform of this function is related to the Laplace transform through the relation

$$M\{e^{-ax}; x \to s\} = \mathcal{L}\{x^{s-1}; x \to a\},$$

where we are explicitly listing the transform variable in each case.

Note that for $a = 1$ in Example 15, we obtain the special case

$$M\{e^{-x}\} = \Gamma(s), \quad \text{Re}(s) > 0. \tag{59}$$

EXAMPLE 16 Find $M\left\{\dfrac{1}{1+x}\right\}$.

Solution: In this case the defining integral is

$$M\left\{\frac{1}{1+x}\right\} = \int_0^\infty \frac{x^{s-1}}{1+x} dx.$$

We recognize this integral as a beta function (Section 2.3.2), but also that in Example 24 in Chapter 7. Hence, it follows directly from previous results that

$$M\left\{\frac{1}{1+x}\right\} = \Gamma(s)\Gamma(1-s) = \frac{\pi}{\sin \pi s}, \quad 0 < \text{Re}(s) < 1.$$

10.5.1 Operational properties and table of transforms

The basic *operational properties* of the Mellin transform are similar to those of the Laplace and Hankel transforms. Some of the basic properties are listed in Table 10.6, and in Table 10.7 we provide a short list of Mellin transforms.

Table 10.6 Operational Properties of the Mellin Transform

1. Linearity:	$\mathrm{M}\{C_1 f(x) + C_2 g(x)\} = C_1 F(s) + C_2 G(s)$
2. Scaling:	$\mathrm{M}\{f(ax)\} = (1/a^s)F(s),\ a > 0$
3. Translation:	$\mathrm{M}\{x^a f(x)\} = F(s + a),\ a > 0$
4. Derivative:	$\mathrm{M}\{f^{(n)}(x)\} = (-1)^n \dfrac{\Gamma(s)}{\Gamma(s-n)} F(s-n),$ $n = 1, 2, 3, \ldots$
5. Convolution:	$\mathrm{M}\left\{ \displaystyle\int_0^\infty f(x/u)g(u)\dfrac{du}{u} \right\} = F(s)G(s)$

Once again, the *linearity* property is a simple consequence of the property of integrals. The *scaling* property follows from the change of variable $x = t/a$; i.e.,

$$\mathrm{M}\{f(ax)\} = \int_0^\infty x^{s-1} f(ax)\,dx = \frac{1}{a^s}\int_0^\infty t^{s-1} f(t)\,dt. \tag{60}$$

From the observation

$$\mathrm{M}\{x^a f(x)\} = \int_0^\infty x^{s+a-1} f(x)\,dx, \tag{61}$$

we deduce the *translation* property. Also, if $f'(x)$ is a continuous function on $x \geq 0$, then we see that

$$\begin{aligned}
\mathrm{M}\{f'(x)\} &= \int_0^\infty x^{s-1} f'(x)\,dx \\
&= x^{s-1} f(x)\Big|_0^\infty - (s-1)\int_0^\infty x^{s-2} f(x)\,dx,
\end{aligned}$$

from which we deduce

$$\mathrm{M}\{f'(x)\} = -(s-1)F(s-1),\quad \sigma_1 < \mathrm{Re}(s) < \sigma_2. \tag{62}$$

In arriving at (62), we require σ_1 and σ_2 to be real constants such that

$$\lim_{\substack{x \to 0 \\ x \to \infty}} x^{s-1} f(x) = 0,\quad \sigma_1 < \mathrm{Re}(s) < \sigma_2.$$

Continued application of these results leads to Property 4 in Table 10.6.

Last, to derive the *convolution* property, we start with the product

$$F(s)G(s) = \int_0^\infty \int_0^\infty (xy)^{s-1} f(x)g(y)\,dx\,dy. \tag{63}$$

By making the change of variables $x = t/u$, $y = u$, and switching the order of integration, we are led to

$$F(s)G(s) = \int_0^\infty t^{s-1} \int_0^\infty f(t/u)g(u)\,du\,dx, \tag{64}$$

from which Property 5 in Table 10.6 follows.

Table 10.7 Table of Selected Mellin Transforms

	$F(s) = M\{f(x)\}$	$f(x) = M^{-1}\{F(s)\}$
1.	$\dfrac{\pi}{\sin \pi s}$	$\dfrac{1}{1 + x}$
2.	$\dfrac{\Gamma(s)\Gamma(m - s)}{a^s \Gamma(m)}$, $\quad a > 0,\ m > 0$	$\dfrac{1}{(1 + ax)^m}$
3.	$\dfrac{\Gamma(s/a)\Gamma(m - s/a)}{a\Gamma(m)}$, $\quad a > 0,\ m > 0$	$\dfrac{1}{(1 + x^a)^m}$
4.	$\dfrac{\Gamma(s)}{a^s}$, $\quad a > 0$	e^{-ax}
5.	$\dfrac{\Gamma(s/2)}{2a^s}$, $\quad a > 0$	$e^{-a^2 x^2}$
6.	$\dfrac{\Gamma(s)}{a^s}\cos(\pi s/2)$, $\quad a > 0$	$\cos ax$
7.	$\dfrac{\Gamma(s)}{a^s}\sin(\pi s/2)$, $\quad a > 0$	$\sin ax$
8.	$-\dfrac{\sqrt{\pi}}{8}\dfrac{\Gamma(s/4)}{\Gamma(1/2 - s/4)}$	$\sin^2(x^2)$
9.	$\dfrac{2^{s-1}\Gamma(s/2 + v/2)}{a^s \Gamma(1 + v/2 - s/2)}$, $\quad a > 0,\ v > -\dfrac{1}{2}$	$J_v(ax)$
10.	$\dfrac{1}{2\sqrt{\pi}}\dfrac{\Gamma(v + s/2 + 1/2)\Gamma(1/2 - s/2)}{\Gamma(1 + v - s/2)\Gamma(1 - s/2)}$, $\quad v > -\dfrac{1}{2}$	$J_v^2(x)$

10.5.2 Complex variable methods

Direct evaluation of the Mellin transform and/or its inverse can often be accomplished more readily by the direct use of residue theory from complex variables (see Chapter 7). To illustrate, let us assume that $f(z)$ is a rational function having no poles on the positive real axis, but having N poles off the real axis at $z = a_1, a_2, ..., a_N$. We further assume there exists real constants σ_1 and σ_2 such that

$$\lim_{z \to 0} z^s f(z) = 0, \qquad \sigma_1 < \mathrm{Re}(s) < \sigma_2,$$

$$\lim_{|z| \to \infty} z^s f(z) = 0, \qquad \sigma_1 < \mathrm{Re}(s) < \sigma_2.$$

Then, by integrating the function $z^{s-1} f(z)$ around the closed contour shown in Fig. 10.5, which encloses all poles of $f(z)$, we find by application of the residue theorem that

$$\int_0^\infty x^{s-1} f(x)\,dx = \frac{\pi e^{-i\pi s}}{\sin \pi s} \sum_{k=1}^N \mathrm{Res}\{z^{s-1} f(z); a_k\}. \qquad (65)$$

Equation (65) can be readily modified to include the case where $f(z)$ has simple poles on the real axis, although we do not include that case here.

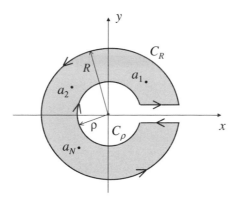

Figure 10.5 Contour of integration.

EXAMPLE 17 Find $M\left\{\dfrac{1}{1+x^2}\right\}$.

Solution: The Mellin transform we wish to calculate is similar to that in Example 16, but here we illustrate the method of residues. Note that $f(z) = 1/(1 + z^2)$ has simple poles located at $z = \pm i$, with residues given by

$$\text{Res}\left\{\frac{z^{s-1}}{1+z^2};i\right\} = \frac{i^{s-1}}{2i} = -\frac{1}{2}e^{i\pi s/2},$$

$$\text{Res}\left\{\frac{z^{s-1}}{1+z^2};-i\right\} = \frac{(-i)^{s-1}}{-2i} = -\frac{1}{2}e^{3i\pi s/2},$$

where we have used $i = e^{i\pi/2}$ and $-i = e^{3i\pi/2}$. Thus, from (85) we deduce that

$$\int_0^\infty \frac{x^{s-1}}{1+x^2}dx = \frac{\pi e^{-i\pi s}}{\sin \pi s}\frac{1}{2}(e^{i\pi s/2} + e^{3i\pi s/2}) = \frac{\pi}{\sin \pi s}\frac{1}{2}(e^{-i\pi s/2} + e^{i\pi s/2}),$$

which upon further reduction yields

$$\int_0^\infty \frac{x^{s-1}}{1+x^2}dx = \frac{\pi\cos(\pi s/2)}{\sin \pi s} = \frac{\pi}{2\sin(\pi s/2)}, \quad 0 < \text{Re}(s) < 2.$$

In calculating the inverse Mellin transform by complex variable methods, we start with the complex inversion formula

$$M^{-1}\{F(s)\} = f(x)U(x) = \frac{1}{2\pi i}\int_{c-i\infty}^{c+i\infty} x^{-s}F(s)ds. \tag{66}$$

The use of residue calculus in evaluating (66) is similar to that illustrated in Section 10.2.3 for the inverse Laplace transform. In the case of the inverse Mellin transform, however, we may need to consider more than one contour in the complex s-plane as shown below in Fig. 10.6. That is, if $F(s)$ has poles in the left-half plane, we use the contour shown in Fig. 10.6(a), which yields the inverse transform for the restricted interval $0 < x < x_0$, where x_0 is determined such that the integral along the circular arc in Fig. 10.6(a) vanishes as the radius tends to infinity. In some cases the function $F(s)$ may also have poles in the right-half plane, which we then enclose by a contour as shown in Fig. 10.6(b). This time the result will be valid for $x > x_0$. Let us illustrate with an example.

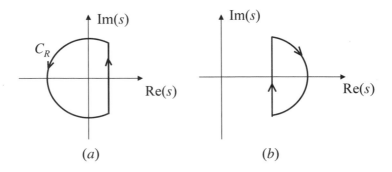

Figure 10.6 Contours of integration for inverse Mellin transform.

EXAMPLE 18 Use the method of residues to find

$$\mathrm{M}^{-1}\left\{\frac{\Gamma(s)\Gamma(m-s)}{a^s\Gamma(m)}\right\}, \quad a>0, \quad 0<\mathrm{Re}(s)<m.$$

Solution: The function $F(s) = \Gamma(s)\Gamma(m-s)/a^s\Gamma(m)$ has two sets of simple poles. Namely, the function $\Gamma(s)$ has simple poles at $s = -n$ ($n = 0, 1, 2, ...$) whereas $\Gamma(m-s)$ has simple poles at $s = m+n$ ($n = 0, 1, 2, ...$).

If we first choose the contour shown in Fig. 10.6(a), we find that

$$\mathrm{M}^{-1}\{F(s)\} = \sum_{n=0}^{\infty} \mathrm{Res}\{x^{-s}F(s); -n\} = \sum_{n=0}^{\infty}\frac{(-1)^n\Gamma(m+n)}{n!\Gamma(m)}(ax)^n$$

$$= \sum_{n=0}^{\infty}(m)_n\frac{(-ax)^n}{n!},$$

where $(m)_n$ is the Pochhammer symbol (Section 2.6.1). Thus,

$$\mathrm{M}^{-1}\{F(s)\} = {}_1F_0(m; -; -ax),$$

and from (GH2) in Section 2.6.4, we obtain

$$\mathrm{M}^{-1}\left\{\frac{\Gamma(s)\Gamma(m-s)}{a^s\Gamma(m)}\right\} = \frac{1}{(1+ax)^m}, \quad 0<x<1/a.$$

If we now choose the second contour shown in Fig 10.6(b), we are led to

$$\mathrm{M}^{-1}\{F(s)\} = -\sum_{n=0}^{\infty}\mathrm{Res}\{x^{-s}F(s); m+n\},$$

where the negative sign in front of the summation is due to integrating in the negative direction along the closed contour. In this case, it can be shown that the residue leads to

$$\mathrm{Res}\{x^{-s}F(s); m+n\} = -\frac{(-1)^n\Gamma(m+n)}{n!\Gamma(m)}(ax)^{-(m+n)} = -(m)_n\frac{(-1)^n}{n!}(ax)^{-(m+n)},$$

and hence, the inverse transform yields

$$\mathrm{M}^{-1}\left\{\frac{\Gamma(s)\Gamma(m-s)}{a^s\Gamma(m)}\right\} = \frac{1}{(ax)^m}{}_1F_0\left(m; -; -\frac{1}{ax}\right) = \frac{1}{(1+ax)^m}, \quad x>1/a.$$

Combining results, we deduce that

$$\mathrm{M}^{-1}\left\{\frac{\Gamma(s)\Gamma(m-s)}{a^s\Gamma(m)}\right\} = \frac{1}{(1+ax)^m}, \quad 0<x<\infty.$$

10.6 Applications Involving the Mellin Transform

Applications involving the Laplace transform and Hankel transform, in addition to those presented thus far, will be taken up in Chapter 12. Here we briefly examine some applications in which the Mellin transform is well suited.

10.6.1 Products of random variables

Unlike the distribution of sums of independent random variables, the distribution of products of independent random variables has received relatively little attention. It has been shown that the Mellin transform is an efficient tool for obtaining the distribution function of such products (see [31]). The general theory is too complicated for our purposes, but we can illustrate the utility of the Mellin transform by examining certain products involving only two random variables.

To begin, let us consider the product $\mathbf{z} = \mathbf{xy}$, where $\mathbf{x}$ and $\mathbf{y}$ are real, independent, positive, random variables. From basic probability theory (see Chapter 13), it is known that the probability density function (PDF) of $\mathbf{z}$ is related to the PDFs of $\mathbf{x}$ and $\mathbf{y}$ by

$$p_{\mathbf{z}}(z) = \frac{d}{dz} \int_0^\infty \int_0^{z/x} p_{\mathbf{x}}(x) p_{\mathbf{y}}(y) \, dy \, dx = \int_0^\infty p_{\mathbf{x}}(x) p_{\mathbf{y}}(z/x) \frac{dx}{x}. \tag{67}$$

We recognize the last integral in (67) as the convolution integral (see Property 5 in Table 10.6) of the Mellin transform. Hence, we can express (67) in the alternate form

$$p_{\mathbf{z}}(z) = M^{-1}\{F(s)G(s); s = z\}, \tag{68}$$

where $F(s)$ and $G(s)$ are the Mellin transforms, respectively, of $p_{\mathbf{x}}(x)$ and $p_{\mathbf{y}}(y)$. In solving for the PDF of $\mathbf{z}$, therefore, we can use either the integral in (67) directly or evaluate (68) as an inverse Mellin transform.

EXAMPLE 19 Find the PDF of the product $\mathbf{z} = \mathbf{xy}$, where $\mathbf{x}$ and $\mathbf{y}$ are independent Cauchy random variables, each having PDF of the form

$$p_{\mathbf{x}}(x) = \frac{2\,U(x)}{\pi(1 + x^2)}.$$

Solution: In this example we will use (67) directly, which leads to

$$p_{\mathbf{z}}(z) = \frac{4}{\pi^2} \int_0^\infty \frac{x\,dx}{(1 + x^2)(z^2 + x^2)}.$$

We can interpret this integral as the Mellin transform

$$p_{\mathbf{z}}(z) = \frac{4}{\pi^2} M\left\{ \frac{1}{(1 + x^2)(z^2 + x^2)}; s = 2 \right\}.$$

Using the variable ζ here for a complex variable, we find the function $f(\zeta) = 1/(1 + \zeta^2)(z^2 + \zeta^2)$ has simple poles at $\zeta = \pm i$ and at $\zeta = \pm iz$. Thus, using the method of residues, we can write

$$
M\left\{\frac{1}{(1 + x^2)(z^2 + x^2)}\right\} = \text{Res}\left\{\frac{1}{(1 + \zeta^2)(z^2 + \zeta^2)}; i\right\} + \text{Res}\left\{\frac{1}{(1 + \zeta^2)(z^2 + \zeta^2)}; -i\right\}
$$

$$
+ \text{Res}\left\{\frac{1}{(1 + \zeta^2)(z^2 + \zeta^2)}; iz\right\} + \text{Res}\left\{\frac{1}{(1 + \zeta^2)(z^2 + \zeta^2)}; -iz\right\}
$$

$$
= \frac{\pi(z^{s-2} - 1)}{2(z^2 - 1)\sin(\pi s/2)}.
$$

Therefore, the result we need is

$$
p_z(z) = \lim_{s \to 2} \frac{2(z^{s-2} - 1)}{\pi(z^2 - 1)\sin(\pi s/2)} = \frac{4\ln z\, U(z)}{\pi^2(z^2 - 1)},
$$

where the last step follows an application of L'Hôpital's rule.

10.6.2 Electromagnetic wave propagation

The study of wave propagation in a random medium leads to difficult integrals to solve using conventional methods. Even numerical techniques for many of these integrals are not easy to apply because the integrand is often expressed as the difference of two quantities that each produce a divergent integral, or the integrand involves a product of one function that goes to infinity and one that tends to zero at one of the integration limits. Moreover, numerical solutions do not provide the kind of insight into general behavior that analytic solutions provide. For example, an analytic solution identifies the natural parameters of the problem, which is difficult to do with numerical methods.

The Mellin transform provides a straightforward method for evaluating many of the integrals that commonly arise in wave propagation problems (see [28]). To illustrate the technique, we will rely primarily on entries found in Table 10.7 in Section 10.5.1. To begin, we examine the integral

$$
\sigma_I^2 = 8\pi^2 k^2 L \int_0^1 \int_0^\infty \kappa \Phi_n(\kappa)\left(1 - \cos\frac{L\kappa^2\xi}{k}\right) d\kappa\, d\xi, \tag{69}
$$

which describes the intensity fluctuations of a plane wave propagating through a random refractive medium that generates weak intensity fluctuations. The quantity k is the optical wave number, L is the propagation path length, ξ is a normalized distance variable, and $\Phi_n(\kappa)$ is the spatial spectrum of refractive-index fluctuations, which we will assume is given by the Kolmogorov power-law spectrum

$$\Phi_n(\kappa) = 0.033 \, C_n^2 \, \kappa^{-11/3}. \tag{70}$$

Here, the parameter C_n^2 is the refractive-index structure parameter that provides a measure of the strength of fluctuations of the random medium. By inserting (70) into (69) and using the trigonometric identity $2\sin^2(x/2) = 1 - \cos x$, we can express (69) as

$$
\begin{aligned}
\sigma_I^2 &= 16\pi^2 k^2 L(0.033)C_n^2 \int_0^1 \int_0^\infty \kappa^{-8/3} \sin^2\left(\frac{L\kappa^2\xi}{2k}\right) d\kappa \, d\xi \\
&= 5.211 \, C_n^2 k^2 L \left(\frac{L}{2k}\right)^{5/6} \int_0^1 \xi^{5/6} d\xi \int_0^\infty x^{-8/3} \sin^2(x^2) dx,
\end{aligned}
\tag{71}
$$

where the second step arises after a change of variables. The first integral is elementary and leads to

$$\int_0^1 \xi^{5/6} d\xi = \frac{6}{11}, \tag{72}$$

whereas the second integral is the Mellin transform (see Table 10.7)

$$\int_0^\infty x^{-8/3} \sin^2(x^2) dx = M\left\{\sin^2(x^2); s = -\frac{5}{3}\right\} = -\lim_{s \to -\frac{5}{3}} \frac{\sqrt{\pi}}{8} \frac{\Gamma(s/4)}{\Gamma(1/2 - s/4)}. \tag{73}$$

By combining the above results, we get the classic result

$$\sigma_I^2 = 1.23 \, C_n^2 k^{7/6} L^{11/6}. \tag{74}$$

We should point out that direct application of standard software packages on the integral in (69) may lead to an error message.

Tilt jitter in an imaging system is a significant effect caused by atmospheric turbulence. It has been shown that the tilt angle variance on a mirror for a two-axis Zernike tilt is given by (see also Section 15.8)

$$\sigma_{\text{tilt}}^2 = 1.303 k^2 \int_0^L C_n^2(z) \int_0^\infty \kappa^{-8/3} \left(\frac{16}{kD}\right)^2 \left[\frac{J_2(\kappa D/2)}{\kappa D/2}\right]^2 d\kappa \, dz, \tag{75}$$

where D is the diameter of the mirror and we have expressed the structure parameter as a function of propagation distance z. Because the Zernike tilt is essentially a phase quantity, diffraction effects have been ignored in (75).

By making the change of variable $x = \kappa D/2$, we find

$$\sigma_{\text{tilt}}^2 = \frac{105.1}{D^{1/3}} \int_0^L C_n^2(z) dz \int_0^\infty x^{-14/3} J_2^2(x) dx, \tag{76}$$

where the second integral is the Mellin transform (see Table 10.7)

$$\int_0^\infty x^{-14/3} J_2^2(x)\,dx \;=\; M\left\{J_2^2(x); s = -\frac{11}{3}\right\}$$

$$= \lim_{s \to -\frac{11}{3}} \frac{1}{2\sqrt{\pi}} \frac{\Gamma(s/2)\Gamma(1/2 - s/2)}{\Gamma^2(1 - s/2)}.$$ (77)

If we designate the first integral in (76) by

$$\mu_0 \;=\; \int_0^L C_n^2(z)\,dz,$$

the Zernike-tilt variance reduces to

$$\sigma_{\text{tilt}}^2 \;=\; \frac{6.08\,\mu_0}{D^{1/3}}.$$ (78)

The separate x and y tilt variances are each one-half of this value. Last, if the turbulence is constant along the path, i.e., if C_n^2 is constant, then (78) reduces to

$$\sigma_{\text{tilt}}^2 \;=\; \frac{6.08\,C_n^2 L}{D^{1/3}}.$$ (79)

10.7 Discrete Fourier Transform

In many engineering applications the function (signal) under consideration is a continuous function of time that needs to be processed by a digital computer. To do this the continuous time-domain signal $x(t)$ must be sampled at discrete intervals of time. The sampled signal $\tilde{x}(t)$ is then processed as an approximation to the true signal $x(t)$.

The relation between a continuous function $x(t)$ and its sample values $x(kT)$, $k = 0, \pm 1, \pm 2, \ldots$, where T is a fixed interval of time, is one of prime importance in digital processing techniques. If the Fourier transform (Chapter 9) of $x(t)$ is nonzero only over a finite range of frequencies, it turns out that the continuous function $x(t)$ can always be recovered from knowledge of its sample values $x(kT)$, provided that the sampling rate is "fast enough," i.e., at a rate that is at least twice the highest significant frequency of the signal. This remarkable result is known as the *Sampling Theorem* and plays a central role in digital processing techniques. Functions whose transform is zero everywhere except for a finite interval are known as *band-limited* waveforms in signal analysis. Such signals do not actually exist in the real world, but theoretical considerations of band-limited waveforms are fundamental to the digital field.

10.7.1 Discrete transform pair

In an ideal situation we can assume that sampling is performed instantaneously and thus represent the sampled waveform by (see Fig. 10.7)

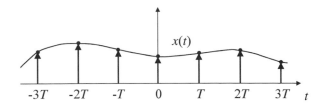

Figure 10.7 Sampled function.

$$\tilde{x}(t) = \sum_{k=-\infty}^{\infty} x(t)\delta(t-kT) = \sum_{k=-\infty}^{\infty} x(kT)\delta(t-kT), \qquad (80)$$

where $\delta(t-kT)$ is the impulse function. The sampled function is really a train of impulse functions in this sense, but it is otherwise treated as if it were a continuous function of t. We recognize (80) as a *comb function* (see Section 2.2.4) where the impulses are weighted by the sample values $x(kT)$. In reality, we cannot obtain an infinite number of samples as suggested in (80). That is, we must always settle for N samples over a total time duration NT, and in this case, Eq. (80) is approximated by

$$\tilde{x}(t) = \sum_{k=0}^{N-1} x(kT)\delta(t-kT). \qquad (81)$$

To obtain a Fourier transform pair for sampled functions, we let $X(f)$ denote the Fourier transform of the continuous waveform $x(t)$. Hence, in terms of linear frequency f we write the standard Fourier pair in the form (see Chapter 9)

$$X(f) = \int_{-\infty}^{\infty} x(t)e^{-2\pi i f t}\, dt,$$
$$x(t) = \int_{-\infty}^{\infty} X(f)e^{2\pi i f t}\, df. \qquad (82)$$

This particular formulation of the Fourier transform pair has a certain appeal because the multiplicative constant $1/2\pi$ has been absorbed.

Next, let us take the Fourier transform of the sampled function (81), which yields

$$\tilde{X}(f) = \sum_{k=0}^{N-1} \int_{-\infty}^{\infty} x(kT)\delta(t-kT)e^{-2\pi i f t}\, dt = \sum_{k=0}^{N-1} x(kT)e^{-2\pi i f k T}. \qquad (83)$$

At this point the transform function $\tilde{X}(f)$ is a continuous function of frequency f, and represents our approximation to the true frequency function $X(f)$. Owing to the periodic nature of the complex exponential function in (83), we see that $\tilde{X}(f)$ is a periodic function with period $1/T$, the reciprocal of the sampling frequency. Ordinarily we select discrete frequencies $f = j/NT$, $j = 0, 1, ..., N-1$, which covers one period. It is also customary to introduce the simplifying notation

$$x(k) \equiv x(kT), \qquad X(j) \equiv \tilde{X}(j/NT),$$

so that (83) becomes

$$X(j) = \sum_{k=0}^{N-1} x(k)e^{-2\pi ijk/N}, \quad j = 0, 1, ..., N - 1.$$
(84)

We call (84) a *discrete Fourier transform* (DFT).

We can easily derive the inverse DFT by first making the observation

$$\sum_{j=0}^{N-1} e^{-2\pi i(k-m)j/N} = \begin{cases} 0, & k \neq m \\ N, & k = m. \end{cases}$$

Then, by multiplying both sides of (84) by $e^{2\pi imj/N}$ and summing from $j = 0$ to $N - 1$, we find

$$\sum_{j=0}^{N-1} X(j)e^{2\pi imj/N} = \sum_{k=0}^{N-1} x(k) \sum_{j=0}^{N-1} e^{-2\pi i(k-m)j/N} = Nx(m).$$
(85)

If we now change the free index m to k, we finally deduce the *inverse DFT* relation

$$x(k) = \frac{1}{N} \sum_{j=0}^{N-1} X(j)e^{2\pi ijk/N}, \quad k = 0, 1, ..., N - 1.$$
(86)

EXAMPLE 20 Find the DFT of the four point sequence $\{x(k)\} = \{1, 1, 0, 0\}$, and then calculate the inverse DFT of the result.

Solution: The DFT in this case is simply

$$X(j) = \sum_{k=0}^{3} x(k)e^{-i\pi jk/2}, \quad j = 0, 1, 2, 3,$$

which gives us

$$X(0) = \sum_{k=0}^{3} x(k) = 2,$$

$$X(1) = \sum_{k=0}^{3} x(k)e^{-i\pi k/2} = 1 - i,$$

$$X(2) = \sum_{k=0}^{3} x(k)e^{-i\pi k} = 0,$$

$$X(1) = \sum_{k=0}^{3} x(k)e^{-3i\pi k/2} = 1 + i.$$

From these four values, we readily calculate the inverse DFT defined by

$$x(k) = \frac{1}{4}\sum_{k=0}^{3} X(j)e^{i\pi jk/2}, \quad k = 0,1,2,3,$$

which leads back to the original sample values

$$x(0) = 1, \quad x(1) = 1, \quad x(2) = 0, \quad x(3) = 0.$$

As pointed out above, one of the major distinctions in the DFT compared with the standard Fourier transform is that both $\{x(k)\}$ and $\{X(j)\}$ form *periodic sequences* with period N. However, the DFT (84) and inverse DFT (86) have operational properties analogous to those associated with the standard Fourier transform and its inverse (see Section 9.5). For example, the *shift* properties of the DFT take the form

$$\sum_{k=0}^{N-1} \left[x(k)e^{-2\pi ijk/N} \right] = X(j + m), \tag{87}$$

$$\sum_{k=0}^{N-1} \left[x(k + m)e^{-2\pi ijk/N} \right] = X(j)e^{2\pi ijm/N}. \tag{88}$$

Also, the *convolution theorem* can be expressed as

$$\frac{1}{N}\sum_{k=0}^{N-1} X(j)Y(j)e^{2\pi ijk/N} = x(k) * y(k) = \sum_{m=0}^{N-1} x(m)y(k - m), \tag{89}$$

and *Parsevel's relation* is given by

$$\frac{1}{N}\sum_{k=0}^{N-1} |X(j)|^2 = \sum_{m=0}^{N-1} |x(m)|^2. \tag{90}$$

It is easy to see that N^2 complex multiplications and additions are required to compute the complete DFT spectrum of a signal. To do such calculations in real time is generally prohibitive because in practice the number of necessary samples N may be large. However, there exist algorithms that allow the computation of the spectrum of a signal using approximately $N \log_2 N$ complex multiplications and additions. This gives significant computational savings for large N. Various algorithms for doing these computations are referred to as *fast Fourier transform* (FFT) *algorithms*, the first being developed in 1965 by J. W. Cooley and J. W. Tukey [9]. Fortunately, today, FFT algorithms are included in most computer mathematics packages, so it is not necessary to write special programs to use them.

10.8 Z-Transform

Communication systems using pulse modulation techniques, where several different messages may be interlaced on a time-sharing basis, rely on sample values of the signal taken at regular spaced intervals of time. In this case the sample values constitute the full available information about the signals. Another area of application based on sample values includes control systems in which feedback is applied on a basis of sample values of some quantity which is to be controlled. The Z-transform is used heavily in these areas of application, although it has also proven useful in other applications as well (see [18]).

The Z-transform is an operation that converts a discrete signal into a complex frequency domain representation. In this regard, it is the discrete analog of the Laplace transform, and thus will have many properties in common with the Laplace transform. Because the Laplace transform is ordinarily associated with causal functions, we will consider only that class of functions in the present discussion. Let $x(t)$ be a continuous function on $t \geq 0$ and of exponential order. The related sampled function $\tilde{x}(t)$ then has the representation

$$\tilde{x}(t) = \sum_{n=0}^{\infty} x(nT)\delta(t - nT), \qquad (91)$$

where T denotes the time interval between samples. Unlike the DFT, in developing properties of the Z-transform we normally assume that the sampled function consists of an infinite number of samples as indicated by (91). The Laplace transform of the sampled function (91) formally leads to

$$\mathcal{L}\{\tilde{x}(t)\} = \sum_{n=0}^{\infty} \int_0^{\infty} e^{-pt} x(nT)\delta(t - nT)dt = \sum_{n=0}^{\infty} x(nT)e^{-pnT}. \qquad (92)$$

By introducing the notation

$$z = e^{pT}, \qquad X(z) = \mathcal{L}\{\tilde{x}(t)\},$$

we can rewrite (92) as $[\sigma = \operatorname{Re}(p)]$

$$X(z) = \sum_{n=0}^{\infty} x(nT)z^{-n}, \qquad |z| > e^{\sigma T}. \qquad (93)$$

We say that $X(z)$ is the *Z-transform* of the sampled function $x(nT)$, but will also call it the Z-transform of the continuous function $x(t)$. It has been shown that the Z-transform $X(z)$ is an analytic function of z outside the circle $|z| = e^{\sigma T}$. Hence, the series on the right in (93) is the Laurent series expansion of $X(z)$ about the origin (see Section 7.4.3). Because the spacing between samples has no effect on developing properties and on the use of the Z-transform, it is customary to set $T = 1$ and write (93) as

$$\boxed{Z\{x(t)\} = X(z) = \sum_{n=0}^{\infty} x(n)z^{-n}.} \qquad (94)$$

EXAMPLE 21 Find the Z-transform of $x(t) = 1$.

Solution: Clearly, the sample values are $x(n) = 1$, $n = 0, 1, 2, ...$, and, hence,

$$Z\{1\} = \sum_{n=0}^{\infty} z^{-n} = \frac{1}{1 - z^{-1}} = \frac{z}{z - 1}, \quad |z| > 1,$$

where we have summed the series as a geometric series.

EXAMPLE 22 Find the Z-transform of $x(t) = a^t$, $a > 0$.

Solution: Here, $x(n) = a^n$, $n = 0, 1, 2, ...$, so that

$$Z\{a^t\} = \sum_{n=0}^{\infty} \left(\frac{a}{z} \right)^n = \frac{z}{z - a}, \quad |z| > a.$$

EXAMPLE 23 Find the Z-transform of $x(t) = t$.

Solution: With $x(n) = n$, $n = 0, 1, 2, ...$, we get

$$Z\{t\} = \sum_{n=0}^{\infty} nz^{-n} = z \sum_{n=0}^{\infty} nz^{-(n+1)},$$

which we can write as

$$Z\{t\} = -z \frac{d}{dz} \sum_{n=0}^{\infty} z^{-n} = \frac{z}{(z - 1)^2}, \quad |z| > 1.$$

Additional Z-transforms can be derived in a similar manner. Of course, like other transforms, it is necessary in practice to also find the *inverse Z-transform*,

$$\boxed{x(n) = Z^{-1}\{X(z)\}, \quad n = 0, 1, 2, ...,} \tag{95}$$

which can be obtained analogously to the Laplace transform, i.e., from a table, by the method of partial fractions, or by the use of residue theory. A short table of Z-transforms and inverse transforms is given below in Table 10.8.

Table 10.8 Table of Selected Z-Transforms

	$X(z) = Z\{x(t)\}$	$x(t)$	$x(n) = Z^{-1}\{X(z)\}$
1.	$\dfrac{z}{z-1}$	1	$U(n)$
2.	$\dfrac{z}{(z-1)^2}$	t	n
3.	$\dfrac{z}{z-a}, \quad a>0$	a^t	a^n
4.	$\dfrac{az}{(z-a)^2}, \quad a>0$	ta^t	na^n
5.	$\dfrac{1}{z-1}$	$U(t-1)$	$U(n-1)$
6.	$e^{1/z}$	$\dfrac{1}{\Gamma(t+1)}$	$\dfrac{1}{n!}$

10.8.1 Operational properties

The Z-transform has *operational properties* analogous to those of the Laplace transform, including the convolution theorem. Some of these properties are listed below in Table 10.9 for easy reference. Because the proofs of these properties are similar to those presented for the Laplace transform, we leave the details to the reader this time.

Table 10.9 Operational Properties of the Z-Transform

1. Linearity:	$Z\{C_1 x(t) + C_2 y(t)\} = C_1 X(z) + C_2 Y(z)$
2. Shift-1:	$Z\{x(t+1)\} = zX(z) - zx(0)$
3. Shift-2:	$Z\{x(t-a)U(t-a)\} = z^{-a}X(z)$
4. Derivative of Transform:	$Z\{tx(t)\} = -zX'(z)$
5. Convolution:	$Z^{-1}\{X(z)Y(z)\} = x(n)*y(n) = \displaystyle\sum_{k=0}^{n} x(k)y(n-k)$

EXAMPLE 24 Find the inverse Z-transform of $X(z) = \dfrac{z + 3}{z - 2}$.

Solution: By writing

$$X(z) = \frac{z}{z - 2} + \frac{3}{z - 2},$$

it follows from entry #3 in Table 10.8 and entry #3 in Table 10.9 that

$$x(n) = 2^n + 3 \cdot 2^{n-1} U(n - 1)$$

$$= \begin{cases} 1, & n = 0 \\ 5 \cdot 2^{n-1}, & n = 1, 2, 3, \dots. \end{cases}$$

EXAMPLE 25 Use the convolution theorem to calculate

$$Z^{-1}\left\{ \frac{z^2}{(z - 2)(z - 3)} \right\}.$$

Solution: By setting

$$X(z) = \frac{z}{z - 2}, \qquad Y(z) = \frac{z}{z - 3},$$

we obtain from Table 10.8 the inverse transform relations

$$Z^{-1}\{X(z)\} = x(n) = 2^n, \qquad Z^{-1}\{Y(z)\} = y(n) = 3^n.$$

Thus, the convolution theorem in Table 10.9 (see entry #5) yields

$$Z^{-1}\{X(z)Y(z)\} = \sum_{k=0}^{n} 2^k 3^{n-k} = 3^n \sum_{k=0}^{n} \left(\frac{2}{3} \right)^k.$$

We recognize this last result as a geometric series, from which we deduce

$$Z^{-1}\left\{ \frac{z^2}{(z - 2)(z - 3)} \right\} = 3^{n+1}\left[1 - \left(\frac{2}{3} \right)^{n+1} \right], \quad n = 0, 1, 2, \dots.$$

10.8.2 Difference equations

The Laplace transform is well-suited for solving differential equations with initial conditions, i.e., IVPs; whereas the Z-transform is ideally suited for solving *difference equations* with initial conditions. The general theory of difference equations with constant coefficients is quite similar to that of differential equations with constant coefficients. To better understand this, let us recall the definition of derivative

$$y'(t) = \lim_{T \to 0} \frac{y(t + T) - y(t)}{T}. \tag{96}$$

Rather than passing to the limit, let us consider the case when $T = 1$ and define

$$\Delta y(t) = y(t + 1) - y(t). \tag{97}$$

We refer to (97) as the *first-order difference* of the function $y(t)$. Higher-order differences can be deduced by repeated use of this relation. For example, the second-order difference is

$$\begin{aligned} \Delta^2 y(t) &= \Delta[y(t + 1) - y(t)] \\ &= [y(t + 2) - y(t + 1)] - [y(t + 1) - y(t)], \end{aligned} \tag{98}$$

or

$$\Delta^2 y(t) = y(t + 2) - 2y(t + 1) + y(t). \tag{99}$$

Continuing in this fashion, we can construct $\Delta^3 y(t)$, $\Delta^4 y(t)$, and so forth.

Although we can consider t as a continuous variable above, it is generally regarded as a discrete variable in most applications. Also, for notational convenience it is customary to adopt the notation $y(n) = y_n$, $y(n + 1) = y_{n+1}, \dots$. Adopting this notation, therefore, we will now define the Z-transform by

$$Z\{y_n\} = Y(z) = \sum_{n=0}^{\infty} y_n z^{-n}. \tag{100}$$

In solving difference equations by the method of Z-transform, we find the following properties to be quite useful:

$$\boxed{\begin{aligned} Z\{y_{n+1}\} &= z[Y(z) - y_0], \\ Z\{y_{n+2}\} &= z^2[Y(z) - y_0] - zy_1. \end{aligned}} \tag{101}$$

The reader may recognize the relations (101) as simple modifications of similar properties associated with the Laplace transform.

EXAMPLE 26 Use the Z-transform to solve the difference equation

$$y_{n+1} - 2y_n = 0, \quad y_0 = 3.$$

Solution: We apply the first property in (101) to deduce the transformed equation

$$z[Y(z) - 3] - 2Y(z) = 0,$$

which has the solution

$$Y(z) = \frac{3z}{z-2}.$$

The inversion of this function leads to the solution

$$y_n = 3 \cdot 2^n, \qquad n = 0, 1, 2, \ldots.$$

EXAMPLE 27 Solve the difference equation

$$y_{n+2} + 3y_{n+1} + 2y_n = 0, \qquad y_0 = 1, \ y_1 = 2.$$

Solution: This time the Z-transform yields

$$z^2[Y(z) - 1] - 2z + 3z[Y(z) - 1] + 2Y(z) = 0.$$

Now solving for $Y(z)$, we find that

$$Y(z) = \frac{z^2 + 5z}{(z+1)(z+2)} = \frac{4z}{z+1} - \frac{3z}{z+2},$$

where we have used a partial fraction decomposition. Inverting this result gives the desired solution

$$y_n = 4(-1)^n - 3(-2)^n$$

$$= (-1)^n[4 - 3 \cdot 2^n], \qquad n = 0, 1, 2, \ldots.$$

10.9 Walsh Transform

In 1923, Joseph L. Walsh published a paper in which he described a complete system of orthogonal functions on the unit interval which take on only the values $+1$ or -1. These functions, now known as Walsh functions, exhibit many properties that are comparable with the complex exponential functions associated with Fourier series and Fourier transforms. However, in spite of the similarities between the Fourier and Walsh functions, there is a fundamental difference between the two groups, which suggests that the transforms are not directly interchangeable in general.

Since the introduction of the fast Fourier transform (FFT) in 1965 (see Section 10.7), much attention has been focused on Walsh functions because of the increased processing speed associated with algorithms involving these functions. The general class of Walsh functions have important practical applications in various fields like signal processing, filtering, coding and multiplexing, pattern recognition and image processing, spectroscopy, and statistical analysis (e.g., see [38]).

10.9.1 Walsh functions

The *Walsh functions*, denoted by wal(n,t), form a complete system of orthogonal functions on the interval [0,1]. They are defined so that they assume only the values +1 or -1 on this interval, and have a denumerable number of discontinuities at the points where the sign changes. The first six Walsh functions are pictured in Fig. 10.8.

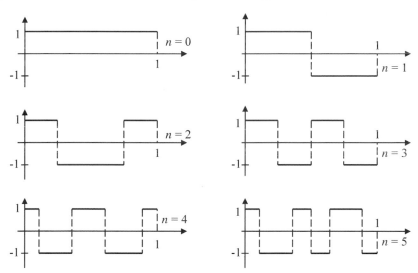

Figure 10.8 The first six Walsh functions.

The Walsh functions can be expressed analytically over the domain [0,1] in the following form:

$$\text{wal}(0,t) = 1,$$

$$\text{wal}(n,t) = 1 + 2\sum_{m=1}^{M} (-1)^m U(t - t_{nm}), \quad n = 1,2,3,..., \tag{102}$$

where $U(t)$ is the unit step function (see Section 2.2.1) and t_{nm} is a set of numbers that can be represented by the matrix

$$[t_{nm}] = \begin{pmatrix} 1/2 & 0 & 0 & 0 & 0 & 0 & \cdots \\ 1/4 & 3/4 & 1 & 0 & 0 & 0 & \cdots \\ 1/4 & 1/2 & 3/4 & 0 & 0 & 0 & \cdots \\ 1/8 & 3/8 & 5/8 & 7/8 & 1 & 0 & \cdots \\ \vdots & \vdots & \vdots & \vdots & \vdots & \vdots & \end{pmatrix}. \tag{103}$$

For comparison with the trigonometric functions $\cos \omega t$ and $\sin \omega t$ associated with a Fourier series, it is sometimes useful to introduce the corresponding functions

$$\text{cal}(n,t) = \text{wal}(2n,t),$$

$$\text{sal}(n,t) = \text{wal}(2n+1,t). \tag{104}$$

With respect to the midpoint of the Walsh interval $[0,1]$, it is clear that $\text{cal}(n,t)$ is an *even function* and $\text{sal}(n,t)$ is an *odd function* (see Fig. 10.8).

10.9.2 Walsh series and the discrete Walsh transform

The Walsh functions (102) form an *orthonormal* set on the unit interval, i.e.,

$$\int_0^1 \text{wal}(n,t)\text{wal}(k,t)dt = \begin{cases} 0, & k \neq n, \\ 1, & k = n. \end{cases} \tag{105}$$

Hence, if a given function $f(t)$ is piecewise continuous, it can be represented in a *Walsh series* of the form

$$f(t) = \sum_{n=0}^{\infty} W(n)\text{wal}(n,t), \tag{106}$$

where the "Fourier coefficients" are defined by (see Section 8.2)

$$W(n) = \int_0^1 f(t)\text{wal}(n,t)dt, \quad n = 0,1,2,.... \tag{107}$$

Generally speaking, functions that have a convergent Fourier series will also have a convergent Walsh series.

Equations (106) and (107) can be used as the basis for defining a discrete Walsh transform. That is, we let $t = k\tau$ in the integral (107) which can then be approximated by the finite sum (e.g., see Section 10.7.1)

$$W(n) = \frac{1}{N}\sum_{k=0}^{N-1} f(k\tau)\text{wal}(n,k\tau), \quad n = 0,1,...,N-1, \tag{108}$$

where $N = 2^m$ and $N\tau = 1$. Equation (108) defines the *discrete Walsh transform* (DWT), which maps a set of N points in the time domain into a sequence of points in a domain called the *sequency domain* (analogous to *frequency domain* associated with the conventional Fourier transform). The sequency domain measures the numbers of zeros per second (zps) of the time function. The *inverse discrete Walsh transform* is obtained from (106) by retaining only a finite number of terms in the series to obtain

$$f(k\tau) = \sum_{k=0}^{N-1} W(n)\text{wal}(n,k\tau), \quad k = 0,1,...,N-1. \tag{109}$$

Like the discrete Fourier transform (see Section 10.7), it is customary here to drop the parameter τ in the arguments of the Walsh transform and its inverse. Consequently, we now write (108) and (109) in the form

$$W(n) = \frac{1}{N}\sum_{k=0}^{N-1} f(k)\,\text{wal}(n,k), \quad n = 0,1,...,N-1, \tag{110}$$

$$f(k) = \sum_{k=0}^{N-1} W(n)\,\text{wal}(n,k), \quad k = 0,1,...,N-1. \tag{111}$$

In this notation, the discrete Walsh function wal(n,k) lends itself to a matrix representation. For example, when $N = 2$ we have

$$[\text{wal}(n,k)]_2 = \begin{pmatrix} 1 & 1 \\ 1 & -1 \end{pmatrix}, \tag{112}$$

and when $N = 4$,

$$[\text{wal}(n,k)]_4 = \begin{pmatrix} 1 & 1 & 1 & 1 \\ 1 & 1 & -1 & -1 \\ 1 & -1 & -1 & 1 \\ 1 & -1 & 1 & -1 \end{pmatrix}. \tag{113}$$

In practice it is found to be more convenient to relate the discrete Walsh functions like (112) and (113) to a group of easily generated matrices known as the *Hadamard matrices*. The first Hadamard matrix is defined by

$$\mathbf{H}_1 = \mathbf{H}(0) = 1, \tag{114}$$

and successive Hadamard matrices by

$$\mathbf{H}_N = \mathbf{H}(m) = \begin{pmatrix} \mathbf{H}(m-1) & \mathbf{H}(m-1) \\ \mathbf{H}(m-1) & -\mathbf{H}(m-1) \end{pmatrix}, \quad m = 1,2,3,.... \tag{115}$$

This gives a very systematic process for generating a Hadamard matrix of any order corresponding to $N = 2^m$. Then we note the number of sign interchanges in each row of the matrix $\mathbf{H}_N$, and rearrange the rows to form a new matrix $\mathbf{H}_N(W)$, called the Walsh-Hadamard matrix, which is exactly the matrix representation of the discrete Walsh functions.

As a final comment, we point out that *fast Walsh transform* (FWT) algorithms have been developed similar to the FFT algorithms associated with Fourier transforms. The total number of operations required by the FWT is the same as that for the FFT, but each operation of the FWT is simply an addition or subtraction rather than a complex multiplication followed by addition as required by the FFT. Thus, there is considerable savings in time and hardware realized by the FWT as compared with the FFT.

Suggested Reading

L. C. Andrews and B. K. Shivamoggi, *Integral Transforms for Engineers* (SPIE Press, Bellingham, 1999).

E. O. Brigham, *The Fast Fourier Transform* (Prentice-Hall, Upper Saddle River, 1974).

H. S. Carslaw and J. C. Jaeger, *Operational Methods in Applied Mathematics* (Oxford, University Press, Oxford, 1941).

R. V. Churchill, *Operational Mathematics* (McGraw-Hill, New York, 1972).

W. R. LePage, *Complex Variables and the Laplace Transform for Engineers* (Dover, New York, 1980).

I. N. Sneddon, *The Use of Integral Transforms* (McGraw-Hill, New York, 1972).

C. J. Tranter, *Integral Transforms in Mathematical Physics* (Wiley & Sons, New York, 1951).

H. J. Weaver, *Applications of Discrete and Continuous Fourier Analysis* (Wiley, New York, 1983).

Chapter 10

Exercises

In Problems 1-4, evaluate the Laplace transform from the defining integral.

1. $f(t) = t^{\frac{5}{2}}$ *Ans.* $\dfrac{15}{8p^3}\sqrt{\dfrac{\pi}{p}}$

2. $f(t) = U(t - a)$ *Ans.* $\dfrac{e^{-ap}}{p}$

3. $f(t) = e^{-t^2/4}$ *Ans.* $\sqrt{\pi}\,e^{p^2}\mathrm{erfc}(p)$

4. $f(t) = \mathrm{erf}\left(1/\sqrt{t}\right)$ *Ans.* $\dfrac{1}{p}(1 - e^{-2\sqrt{p}})$

5. Given $F(p) = \mathcal{L}\{f(t)\}$, show for constants a and b that

$$\mathcal{L}^{-1}\{F(ap + b)\} = \frac{1}{a}e^{-bt/a}f(t/a), \ a > 0.$$

6. Use Tables 10.1 and 10.2 to show that

$$\mathcal{L}\{L_n(t)\} = \frac{1}{p}\left(\frac{p-1}{p}\right)^n,$$

where the *Laguerre polynomials* are defined by

$$L_n(t) = \frac{e^t}{n!}\frac{d^n}{dt^n}(t^n e^{-t}), \ n = 0, 1, 2, \ldots.$$

In Problems 7 and 8, use Tables 10.1-3 and/or partial fractions to deduce the inverse Laplace transform of the given function.

7. $F(p) = \dfrac{p}{p^2 - 6p + 13}$

Ans. $e^{3t}(\cos 2t + \frac{3}{2}\sin 2t)$

8. $F(p) = \dfrac{p + 1}{p^3 + p^2 - 6p}$

Ans. $\dfrac{3}{10}e^{2t} - \dfrac{1}{6} - \dfrac{2}{15}e^{-3t}$

In Problems 9 and 10, use the convolution theorem to deduce the inverse Laplace transform.

9. $Y(p) = \dfrac{1}{p^4(p^2 + 1)}$

Ans. $\dfrac{1}{6}t^3 - t + \sin t$

10. $Y(p) = \dfrac{1}{(p + 1)^2(p^2 + 4)}$

Ans. $\dfrac{1}{25}e^{-t}(2 + 5t)$

$- \dfrac{1}{50}(4\cos 2t + 3\sin 2t)$

11. Use $\mathscr{L}\{J_0(t)\} = 1/\sqrt{p^2 + 1}$ and the convolution theorem to show that

$\int_0^\infty J_0(t - \tau)J_0(\tau)d\tau = \sin t$.

12. Show that

$\mathscr{L}^{-1}\left\{\dfrac{1}{\sqrt{p}(p - 1)}\right\} = e^t \operatorname{erf}(\sqrt{t})$.

In Problems 13 and 14, use the Laplace transform to solve the given problem.

13. Find the current in an RLC-circuit with $R = 100\ \Omega$, $L = 0.1$ H, $C = 10^{-3}$ F, and $E(t) = 155\sin 377t$. Assume zero charge and current at time $t = 0$.

Ans. $i(t) = 0.52e^{-990t} - 0.042e^{-10t}$
$- 0.484\cos 377t + 1.380\sin 377t$

14. Determine the transfer function and

impulse response for the system described by

$9y'' - 12y' + 4y = f(t)$.

Ans. $G(p) = \dfrac{1}{9p^2 - 12p + 4}$

$g(t) = \dfrac{1}{9}te^{2t/3}$

In Problems 15-18, use integration to verify the Hankel transform relation.

15. $\mathcal{H}_0\{e^{-ar^2}\} = \dfrac{1}{2a}e^{-p^2/4a}$, $a > 0$

16. $\mathcal{H}_0\{r^2 e^{-ar^2}\}$

$= \dfrac{1}{2a^2}\left(1 + p - \dfrac{p^2}{4a}\right)e^{-p^2/4a}, a > 0$

17. $\mathcal{H}_0\{(a^2 - r^2)^{\mu-1}U(a - r)\}$

$= 2^{\mu-1}\Gamma(\mu)\left(\dfrac{a}{p}\right)^\mu J_\mu(ap), a > 0$

18. $\mathcal{H}_0\left\{\dfrac{U(a - r)}{\sqrt{a^2 - r^2}}\right\}$

$= \sqrt{\dfrac{a\pi}{2p}}J_{1/2}(ap), a > 0$

19. With respect to a, integrate both sides of the integral $(0 < a < b)$

$\int_0^\infty \cos(ax)J_0(bx)dx = \dfrac{1}{\sqrt{b^2 - a^2}}$

to deduce that

$\mathcal{H}_0\left\{\dfrac{1}{r^2}\sin r\right\} = \sin^{-1}\dfrac{1}{p}, p > 1$.

20. Use the integral representation

$\dfrac{1}{\sqrt{r^2 + a^2}} = \dfrac{1}{\sqrt{\pi}}\int_0^\infty e^{-(r^2 + a^2)x}\dfrac{dx}{\sqrt{x}}$

to deduce that

$\mathcal{H}_0\left\{\dfrac{1}{\sqrt{r^2 + a^2}}\right\} = \dfrac{1}{p}e^{-ap}, a > 0$.

In Problems 21-26, verify the given Mellin transform relation.

21. $M\left\{\dfrac{1}{(1 + ax)^m}\right\} = \dfrac{\Gamma(s)\Gamma(m - s)}{a^s\Gamma(m)}$,

$$0 < \text{Re}(s) < m$$

22. $M\{x^{-\nu}J_\nu(ax)\} = \dfrac{a^{\nu-s}2^{s-\nu-1}\Gamma(s/2)}{\Gamma(\nu + 1 - s/2)}$,

$$a > 0, \ \nu > -1/2$$

23. $M\{\ln x f(x)\} = F'(s)$

24. $M\{x^n f^{(n)}(x)\} = (-1)^n \dfrac{\Gamma(s + n)}{\Gamma(s)} F(s),$

$$n = 1, 2, 3, \ldots$$

25. $M\left\{\displaystyle\int_x^\infty f(u)\,du\right\} = \dfrac{1}{s}F(s + 1)$

26. $M^{-1}\{\Gamma(s)F(1 - s)\} = \mathcal{L}\{f(t); t \to x\}$

In Problems 27 and 28, use residues to calculate the Mellin transform.

27. $f(x) = \dfrac{1}{1 + x^4}$

  Ans. $\dfrac{\pi}{4\sin(\pi s/4)}$

28. $f(x) = \dfrac{1}{(1 + x)^3}$

  Ans. $\dfrac{\pi(s - 1)(s - 2)}{2\sin(\pi s)}$

29. For the discrete Fourier transform (DFT),

  (a) derive the *convolution* property

$$\frac{1}{N}\sum_{j=0}^{N-1} X(j)Y(j)e^{2\pi ijk/N}$$
$$= \sum_{m=0}^{N-1} x(m)y(k - m).$$

  (b) From (a), deduce *Parseval's relation*

$$\frac{1}{N}\sum_{j=0}^{N-1} |X(j)|^2 = \sum_{k=0}^{N-1} |x(k)|^2.$$

In Problems 30-33, verify the given Z-transform relation.

30. $Z\left\{\dfrac{1}{\Gamma(t + 1)}\right\} = e^{1/z}$

31. $Z\{t^2\} = \dfrac{z(z + 1)}{(z - 1)^3}$

32. $Z\{ta^t\} = \dfrac{az}{(z - a)^2}$, $a > 0$

33. $Z^{-1}\left\{\dfrac{z^2}{(z - 1)(z - 3)}\right\}$

$$= \begin{cases} 1, & n = 0 \\ \dfrac{1}{2}(3^{n+1} - 1), & n = 1, 2, 3, \ldots \end{cases}$$

34. The sequence of numbers

$$0, 1, 1, 2, 3, 5, 8, 13, 21, \ldots,$$

where each number after the first is the sum of the two preceeding numbers, is called a *Fibonocci sequence.*

  (a) Use the Z-transform to find a formula for the general term y_n of the sequence.

  (b) Show that

$$\lim_{n \to \infty} \frac{y_{n+1}}{y_n} = \frac{1 + \sqrt{5}}{2}.$$

  Ans. $y_n = \dfrac{1}{\sqrt{5}}\left[\left(\dfrac{1 + \sqrt{5}}{2}\right)^n - \left(\dfrac{1 - \sqrt{5}}{2}\right)^n\right]$

Chapter 11

Calculus of Variations

11.1   Introduction / 546
11.2   Functionals and Extremals / 547
    11.2.1 Euler-Lagrange equation / 548
    11.2.2 Special cases of the Euler-Lagrange equation / 550
11.3   Some Classical Variational Problems / 552
    11.3.1 Shortest arc connecting two points / 552
    11.3.2 Surface of revolution with minimum area / 552
    11.3.3 Brachistochrone problem / 553
11.4   Variational Notation / 555
    11.4.1 Natural boundary conditions / 557
11.5   Other Types of Functionals / 559
    11.5.1 Functionals with several dependent variables / 559
    11.5.2 Functionals with higher-order derivatives / 561
    11.5.3 Functionals with several independent variables / 562
11.6   Isoperimetric Problems / 564
    11.6.1 Constraints and Lagrange multipliers / 565
    11.6.2 Sturm-Liouville problem / 567
11.7   Rayleigh-Ritz Approximation Method / 567
    11.7.1 Eigenvalue problems / 570
11.8   Hamilton's Principle / 572
    11.8.1 Generalized coordinates and Lagrange's equations / 573
    11.8.2 Linear theory of small oscillations / 575
11.9   Static Equilibrium of Deformable Bodies / 579
    11.9.1 Deflections of an elastic string / 580
    11.9.2 Deflections of an elastic beam / 580
11.10 Two-Dimensional Variational Problems / 581
    11.10.1 Forced vibrations of an elastic string / 581
    11.10.2 Equilibrium of a stretched membrane / 583
    Suggested Reading / 584
    Exercises / 584

Historical Comments: The earliest problems of elementary calculus that gained wide attention of mathematicians involved the determination of a maximum or minimum of some quantity. This interest stemmed from the fact that, prior to the invention of the calculus by Sir Isaac Newton (1642-1727) and Gottfried Wilhelm von Leibniz (1646-1716), there existed no systematical procedure for solving such extremal problems. Similar types of extremal problems form the central core of the theory of the calculus of variations.

In 1696, John Bernoulli (1667-1748) proposed his now famous *brachistochrone problem*—that is, finding the path in a vertical plane down which a particle will fall from one point to another in minimal time. The brachistochrone problem gained the attention of many mathematicians like Newton, Leibniz, Guillaume F. A. de L'Hôpital (1661-1704), and the older Bernoulli brother James (1654-1705), all of whom provided a correct solution.

Up until the time of the brachistochrone problem, the problems we associate today with the calculus of variations originated as certain isolated maximum or minimum problems not treatable by techniques of the elementary calculus. People such as Leonhard Euler (1707-1783) continued the development of the subject and introduced the variational notation for which the subject is now named. Today the calculus of variations plays an increasingly important role in the fields of analysis, physics, and engineering.

The purpose of this chapter is to develop an optimization technique known as the calculus of variations. This powerful method can be applied to a variety of mathematical and physical problems to derive the governing differential equation of the problem. Among others, such problems include those that arise from Hamilton's principle and in applying the principle of minimum potential energy to determine the equilibrium configuration of a deformable body. Higher-dimensional problems often lead to some of the classic partial differential equations of mathematical physics.

11.1 Introduction

The *calculus of variations* deals with the optimization problem of finding an extremal (maxima or minima) of a quantity in the form of an integral. The simplest example of such a problem is to show that the shortest path between two points in space is a straight line. A similar problem formulated from Fermat's principle leads to *Snell's law*.

From the beginning, the calculus of variations had its development closely interlaced with that of the differential calculus. Recall from calculus that a *necessary condition* for a continuous point function[1] $y = f(x)$ to have a maximum or minimum at a point x_0 in the interval $a < x_0 < b$ is that $f'(x_0) = 0$. *Sufficient conditions* that y be a maximum (or a minimum) at x_0 are that $f'(x_0) = 0$ *and* $f''(x_0) < 0$ [or $f''(x_0) > 0$].

The calculus of variations is fascinating in that it provides a unification of several diverse areas of mathematics and physics, the latter often using energy as a key concept. For example, variational principles are used in Hamiltonian and Lagrangian mechanics, modern quantum theory, and in optimal control theory (see [35]).

11.2 Functionals and Extremals

Most of the elementary variational problems involve an integral of the form

$$I(y) = \int_a^b F(x,y,y')dx, \tag{1}$$

where $y' = dy/dx$ and $F(x,y,y')$ is a continuous function in each of its three arguments x, y, y'. For a given function $F(x,y,y')$, the function $y = y(x)$ will determine the real number I. In this setting, we refer to $I(y)$ as a *functional*, i.e., a "function of a function." Sometimes the quantity $F(x,y,y')$ is also referred to as a functional because of its dependency on y. However, the dependency of y on the variable x is unknown, which means that the path between $x = a$ and $x = b$ is unknown.

The typical problem in the calculus of variations is to find a function $y = y(x)$ from a suitable class of "admissible functions" for which the quantity I defined by Eq. (1) is a *stationary value* (i.e., a maximum, minimum, or saddle point).

▸ **Admissible function:** We say $y = y(x)$ belongs to the *class of admissible functions* Γ if it is single-valued with continuous first- and second-order derivatives on the interval $a \le x \le b$.

In addition to specifying that $y = y(x)$ be an admissible function, we ordinarily impose the "forced" boundary conditions (BCs)

$$y(a) = y_1, \quad y(b) = y_2, \tag{2}$$

which fixes the location of the endpoints of y. Another type of endpoint condition in place of (2) can also occur, called a "natural" BC (e.g., see Section 11.4.1). In other problems, a combination of forced and natural BCs might arise.

Let us assume that $y = \Phi(x)$ is the true function which generates a stationary value of $I(y)$. In this setting, the optimizing curve $y = \Phi(x)$ is often called an *extremal*. We then construct a family of curves having the form (see Fig. 11.1)

---

[1]We use the term "point" function in this chapter to distinguish between a function from elementary calculus and a more general type of function studied here called a "functional."

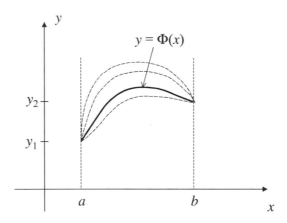

Figure 11.1 Family of admissible curves.

$$y = \Phi(x) + \varepsilon\eta(x), \tag{3}$$

where $\eta(x)$ belongs to the class of admissible functions Γ, and ε is an arbitrary parameter independent of x. We refer to the change $\varepsilon\eta(x)$ as a *variation* of the true function $\Phi(x)$—hence the name "calculus of variations." If the forced BCs (2) are imposed, then $\eta(x)$ must satisfy the homogeneous BCs

$$\eta(a) = 0, \quad \eta(b) = 0. \tag{4}$$

That is, all admissible curves in Fig. 11.1 must pass through the same endpoints.

11.2.1 Euler-Lagrange equation

Based on the above formulation, we wish to derive a necessary condition for $y = \Phi(x)$ to be an extremal of I. To begin, note that the substitution of (3) into (1) leads to

$$I(\Phi + \varepsilon\eta) = \int_a^b F(x, \Phi + \varepsilon\eta, \Phi' + \varepsilon\eta')dx. \tag{5}$$

If we momentarily think of η as being fixed, then I can be viewed as a continuous "point function" of ε; i.e., $I = I(\varepsilon)$. To find an extremal of I, therefore, we can treat (5) as a simple point function and set $I'(\varepsilon) = 0$. However, unlike most extremal problems for point functions, here we know beforehand that the extremal of (5) occurs when $\varepsilon = 0$.

If we assume that differentiation is permitted under the integral, we formally obtain

$$\begin{aligned}
I'(\varepsilon) &= \frac{d}{d\varepsilon} \int_a^b F(x, \Phi + \varepsilon\eta, \Phi' + \varepsilon\eta')dx \\
&= \int_a^b \left(\frac{\partial F}{\partial y} \frac{dy}{d\varepsilon} + \frac{\partial F}{\partial y'} \frac{dy'}{d\varepsilon} \right) dx.
\end{aligned} \tag{6}$$

However, because

$$\frac{dy}{d\varepsilon} = \eta(x), \quad \frac{dy'}{d\varepsilon} = \eta'(x),$$

we have in the limit $\varepsilon \to 0$ that

$$I'(0) = \int_a^b \left[\eta(x)\frac{\partial F}{\partial y} + \eta'(x)\frac{\partial F}{\partial y'} \right] dx = 0. \tag{7}$$

If we now integrate the second term in (7) by parts, we find

$$\int_a^b \eta'(x)\frac{\partial F}{\partial y'} dx = \eta(x)\frac{\partial F}{\partial y'}\bigg|_a^b - \int_a^b \eta(x)\frac{d}{dx}\left(\frac{\partial F}{\partial y'} \right) dx$$
$$= -\int_a^b \eta(x)\frac{d}{dx}\left(\frac{\partial F}{\partial y'} \right) dx, \tag{8}$$

the last step of which is a consequence of (4). Thus, Eq. (7) becomes

$$\int_a^b \eta(x)\left[\frac{\partial F}{\partial y} - \frac{d}{dx}\left(\frac{\partial F}{\partial y'} \right) \right] dx = 0, \tag{9}$$

and, because $\eta(x)$ is arbitrary, we deduce from (9) that

$$\boxed{\frac{\partial F}{\partial y} - \frac{d}{dx}\left(\frac{\partial F}{\partial y'} \right) = 0.} \tag{10}$$

Equation (10), known as the *Euler-Lagrange equation*, is a *necessary condition* for $y = \Phi(x)$ to be an extremal of Eq. (1). That is, any extremal of (1) must necessarily satisfy (10). Nonetheless, we do not know if this is a maximum, minimum, or horizontal inflection. *Sufficient conditions* for the variational problem do exist but are quite complicated to develop, so we ordinarily seek only stationary functions.

▸ **Stationary function:** Any admissible function $y = \Phi(x)$ satisfying the Euler-Lagrange equation (10) and BCs (2) is a *stationary function* of I. In that case we say that $I(\Phi)$ is a *stationary value.*

EXAMPLE 1 Find the stationary function and stationary value of

$$I(y) = \int_0^{\pi/2} \left[y^2 + (y')^2 - 2y\sin x \right] dx, \quad y(0) = 0, \ y(\pi/2) = \frac{1}{2}.$$

Solution: With $F(x,y,y') = y^2 + (y')^2 - 2y\sin x$, we obtain

$$\frac{\partial F}{\partial y} = 2y - 2\sin x, \quad \frac{\partial F}{\partial y'} = 2y', \quad \frac{d}{dx}\left(\frac{\partial F}{\partial y'}\right) = 2y''.$$

Hence, the Euler-Lagrange equation (10) becomes

$$y'' - y = -\sin x.$$

We recognize this differential equation (DE) as a second-order, linear, nonhomogeneous equation with general solution (see Section 1.4)

$$y = C_1 \cosh x + C_2 \sinh x + \frac{1}{2}\sin x.$$

Applying the prescribed BCs leads to

$$y(0) = C_1 = 0,$$
$$y(\pi/2) = C_1 \cosh\frac{\pi}{2} + C_2 \sinh\frac{\pi}{2} + \frac{1}{2}\sin\frac{\pi}{2} = \frac{1}{2},$$

from which we deduce $C_1 = C_2 = 0$. The stationary function is therefore

$$y = \Phi(x) = \frac{1}{2}\sin x,$$

which yields the stationary value

$$I(\Phi) = \frac{1}{4}\int_0^{\pi/2}(1 - 4\sin^2 x)dx = -\frac{\pi}{8}.$$

At this point, however, we do not know if this is a maximum or minimum of I.

11.2.2 Special cases of the Euler-Lagrange equation

By completing the indicated differentiations in the Euler-Lagrange equation (10), we obtain the expanded form

$$\frac{\partial^2 F}{\partial y'^2}y'' + \frac{\partial^2 F}{\partial y\partial y'}y' + \frac{\partial^2 F}{\partial x\partial y'} - \frac{\partial F}{\partial y} = 0, \tag{11}$$

which exhibits the fact that it is in general a second-order, nonlinear DE. In its general form, Eq. (11) can be very difficult or impossible to solve exactly. In some applications, however, one or more of the arguments x, y, y' may be missing from the functional F. When this happens the first integral of Eq. (11) can readily be found, which reduces the DE to a *first-order DE* for which there exists a number of standard solution techniques. Special cases of the Euler-Lagrange equation in which one or more of the arguments x, y, y' is missing are listed below.

CASE I: If $F = F(y')$ and $\dfrac{\partial^2 F}{\partial y'^2} \neq 0$, then Eq. (11) reduces to simply

$$y'' = 0.$$

(12)

Because F does not depend explicitly on variables x and y, all partial derivative with respect to these variables are zero and Eq. (11) becomes

$$\frac{\partial^2 F}{\partial y'^2} y'' = 0.$$

We have also assumed $\partial^2 F / \partial y'^2 \neq 0$, so this last result reduces to Eq. (12).

CASE II: If $F = F(x, y')$, then the Euler-Lagrange equation (10) becomes

$$\frac{d}{dx}\left(\frac{\partial F}{\partial y'} \right) = 0.$$

In particular, by integrating this result with respect to x, we obtain

$$\frac{\partial F}{\partial y'} = C \qquad \text{(constant)}.$$

(13)

CASE III: If $F = F(x, y)$, then Eq. (11) reduces to

$$\frac{\partial F}{\partial y} = 0.$$

(14)

Equation (14) follows by simply setting all derivatives of F with respect to y' equal to zero in Eq. (11).

CASE IV: If $F = F(y, y')$, then Eq. (11) becomes

$$F - y' \frac{\partial F}{\partial y'} = C \qquad \text{(constant)}.$$

(15)

To derive Eq. (15), we first note the equivalence of the expressions

$$\frac{d}{dx}\left(F - y'\frac{\partial F}{\partial y'} \right) = \frac{\partial F}{\partial x} + y'\left[\frac{\partial F}{\partial y} - \frac{d}{dx}\left(\frac{\partial F}{\partial y'} \right) \right] = \frac{\partial F}{\partial x}.$$

Because F does not depend explicitly on x, we have $\partial F / \partial x = 0$. Consequently, the first term above is zero and our intended result (15) follows immediately.

11.3 Some Classical Variational Problems

Below we consider some of the problems that were among the first to be treated by variational methods, but which still serve as excellent examples.

11.3.1 Shortest arc connecting two points

Perhaps the simplest of all variational problems is that of determining the smooth plane curve that has the "shortest length" between two fixed points. If we let (x_1, y_1) and (x_2, y_2) denote the two fixed points, then the functional we wish to minimize is the arc length relation from calculus (see also Section 4.6.1)

$$I(y) = \int_{x_1}^{x_2} \sqrt{1 + (y')^2}\, dx, \quad y(x_1) = y_1, \ y(x_2) = y_2, \tag{16}$$

where $ds = \sqrt{1 + (y')^2}\, dx$ is the element of arc length and $I(y)$ represents the total path length in this problem. We recognize the functional $F(x, y, y') = \sqrt{1 + (y')^2}$ as one of the type discussed in CASE I above, and hence, the governing DE is

$$y'' = 0. \tag{17}$$

The general solution of (17) is the *linear function* $y = C_1 + C_2 x$. Thus, by imposing the BCs, we find that the stationary function is the *line* between the fixed points (x_1, y_1) and (x_2, y_2), viz.,

$$y = \Phi(x) = m(x - x_1) + y_1, \tag{18}$$

where $m = (y_2 - y_1)/(x_2 - x_1)$ is the slope of the line. In this case, it is clear that the line (18) is a *minimizing extremal* of the functional (16).

11.3.2 Surface of revolution with minimum area

When a circular wire ring is dipped into a soap solution and withdrawn, a soap film bounded by the wire is formed. If then a second smaller circular wire is made to touch the first soap film and moved away, the two rings will be joined by a surface of film which is a surface of revolution in the particular case when the rings have their centers on the same axis perpendicular to their planes. It can be shown by the principles of mechanics that the surface of revolution so formed must be one of "minimum surface area." The problem is to determine the shape of such a surface by identifying the curve in the xy-plane that is revolved around the x-axis to form the surface.

   The functional to be minimized in this problem is the surface area integral from calculus given by

$$I(y) = 2\pi \int_a^b y\sqrt{1 + (y')^2}\, dx, \quad y(a) = y_1, \ y(b) = y_2. \tag{19}$$

Here we see that the functional F does not depend explicitly on x, so the Euler-Lagrange equation given by CASE IV above yields

$$\frac{y(y')^2}{\sqrt{1 + (y')^2}} - y\sqrt{1 + (y')^2} = C_1,$$

or, after rearranging terms,

$$-y = C_1\sqrt{1 + (y')^2}. \tag{20}$$

Equation (20) is a first-order DE that can be solved by separation of variables (see Section 1.3.1). That is, by writing (20) in the form

$$C_1\frac{dy}{\sqrt{y^2 - C_1^2}} = dx,$$

we are led through integration of both sides to the general solution

$$C_1 \cosh^{-1}(y/C_1) + C_2 = x, \tag{21}$$

where $\cosh^{-1}x$ is the inverse hyperbolic cosine and C_2 is a constant of integration. By solving (21) for y, we obtain the more useful form

$$y = C_1\cosh\left(\frac{x - C_2}{C_1}\right). \tag{22}$$

Recall from calculus that the hyperbolic cosine is defined by $\cosh x = 1/2(e^x + e^{-x})$. Equation (22) is a two-parameter family of *catenaries*. If the constants C_1 and C_2 can be determined so that y passes through the two fixed points (x_1,y_1) and (x_2,y_2), then this catenary is a stationary function. However, for two arbitrary points there may exist no catenary, and hence, no extremal of I.

> **Remark:** A *catenary* is also the curve that is formed by a hanging telephone line between two poles.

11.3.3 Brachistochrone problem

The *brachistochrone problem* is the most interesting variational problem from an historical point of view because it provided the first impetus to systematic research in the calculus of variations. In this problem the smooth curve is sought along which a point mass moves, under only the action of gravity (friction and resistance being neglected), from one point to a nearby lower point in the "shortest time." The term "brachistochrone" comes from the Greek words *brachistos* (meaning "shortest") and *chronos* (meaning "time").

If we designate the initial and final points by (x_1,y_1) and (x_2,y_2), then the total time of descent is given by

$$T = \int_0^T dt = \int_0^L \frac{ds}{v}, \tag{23}$$

where $v = ds/dt$ is the speed of descent and L is the total path length. By utilizing the arc length relation

$$ds = \sqrt{1 + (y')^2}\, dx,$$

we can write the total time of descent in the form

$$T = \int_{x_1}^{x_2} \frac{\sqrt{1 + (y')^2}}{v}\, dx. \tag{24}$$

By virtue of the *conservation of energy principle* (i.e., the sum of kinetic and potential energies is a constant), we can express the speed v as a function of y. Indeed, if the initial speed at (x_1, y_1) is v_0, then the conservation of energy principle at any other point along the path reads

$$\frac{1}{2} mv^2 + mgy = \frac{1}{2} mv_0^2 + mgy_1, \tag{25}$$

where m is the mass of the particle and g is the acceleration due to gravity. By solving (25) for v, we obtain

$$v = \sqrt{2g(y_1 - y) + v_0^2} = \sqrt{2g}\sqrt{k - y},$$

where $k = y_1 + v_0^2/2g$. The total time of descent (24) now becomes

$$T = \frac{1}{\sqrt{2g}} \int_{x_1}^{x_2} \frac{\sqrt{1 + (y')^2}}{\sqrt{k - y}}\, dx, \quad y(x_1) = y_1,\ y(x_2) = y_2. \tag{26}$$

The functional F is independent of x and thus, we can use Eq. (15) to obtain

$$\frac{\sqrt{1 + (y')^2}}{\sqrt{k - y}} - \frac{(y')^2}{\sqrt{k - y}\sqrt{1 + (y')^2}} = C,$$

or, after some algebraic manipulation,

$$(k - y)\left[1 + (y')^2\right] = \frac{1}{C^2} = C_1. \tag{27}$$

By separating variables, we get

$$dx = -\frac{\sqrt{k - y}}{C_1 - (k - y)}\, dy, \tag{28}$$

where the minus sign reflects the fact that the slope of the curve must always be negative. Next, by writing $k - y = C_1 \sin^2 \alpha$, we can rewrite (28) as

$$dx = C_1(1 - \cos 2\alpha)\, d\alpha,$$

the integral of which leads to

$$x(\alpha) = \frac{1}{2}C_1(2\alpha - \sin 2\alpha) + C_2,$$

$$y(\alpha) = k - C_1\sin^2\alpha = y_1 + \frac{v_0^2}{2g} - \frac{1}{2}C_1(1 - \cos 2\alpha), \tag{29}$$

where C_2 is a constant of integration. Last, by making the change of parameter $\theta = 2\alpha$, and letting $C_2 = a$ and $C_1 = 2b$, we obtain the parametric equations

$$x(\theta) = a + b(\theta - \sin\theta),$$

$$y(\theta) = y_1 + \frac{v_0^2}{2g} - b(1 - \cos\theta). \tag{30}$$

Equations (30) are recognized as the parametric equations of a *cycloid*. This curve can be generated by the motion of a fixed point on the circumference (rim) of a circle of radius b, which rolls on the negative side (i.e., underside) of the line $y = y_1$ (see Fig. 11.2). Although we won't prove so, it can be shown that values of a and b exist so that the cycloid will pass through any pair of points (x_1,y_1) and (x_2,y_2). Also, it can be shown that (30) indeed produces a minimum for the variational problem.

The formulation (24) for the shortest time is the same functional suggested in the principle by Pierre de Fermat (1601-1665): *light rays travel along the quickest paths*. In *Fermat's principle*, however, the velocity v may be a function of both x and y and written as $v(x,y) = c/n(x,y)$, where c is the speed of light in a vacuum and $n(x,y)$ is the index of refraction (e.g., see Section 15.5). Among other results, Fermat's principle leads to *Snell's law* (see [6] or [32]).

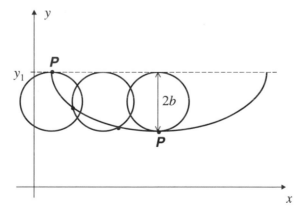

Figure 11.2 One arch of a cycloid.

11.4 Variational Notation

Recall that a necessary condition for a continuous point function of one variable $f(x)$ to have an extremal at some point x_0 in a specified open interval $a < x_0 < b$ is that $f'(x_0) = 0$. Similarly, if a point function of two variables $f(x,y)$ is defined throughout a domain D, and if the partial derivatives $\partial f/\partial x$ and $\partial f/\partial y$ exist and are continuous everywhere in D, then *necessary conditions* that $f(x,y)$ possess an extremal at some point (x_0,y_0) in D are that

$$\frac{\partial f(x_0,y_0)}{\partial x} = 0, \quad \frac{\partial f(x_0,y_0)}{\partial y} = 0.$$

These two requirements are equivalent to specifying that the differential df satisfy

$$df = \frac{\partial f}{\partial x} dx + \frac{\partial f}{\partial y} dy = 0, \quad \text{at } (x_0,y_0). \tag{31}$$

In order to establish more clearly an analogy between "differentials" associated with a point function and the notion of a "variation" associated with a functional, it is customary to introduce a new symbol. In the expression $y = \Phi(x) + \varepsilon\eta(x)$, the change $\varepsilon\eta(x)$ is called the variation of y, which we now denote by

$$\delta y = \varepsilon\eta(x). \tag{32}$$

The symbol δy plays a role in the discussion of a functional that is similar to that of the differential df in the discussion of a point function. However, from a geometrical viewpoint, the differential df of a point function of one variable is a first-order approximation to the change in that function "along a particular curve." But, the variation δy represents a first-order approximation to the change in y "from curve to curve." In the following we provide a more complete definition of the variation of a functional.

Definition 11.1 The *first variation* of the functional $I(y)$ is defined by

$$\delta I = \varepsilon \lim_{\varepsilon \to 0} \frac{d}{d\varepsilon} I(\Phi + \varepsilon\eta),$$

whenever this limit exists for all η belonging to the class Γ.

If we apply Def. 11.1 to the functional $F(x,y,y')$, then through use of the chain rule, we find

$$\delta F = \frac{\partial F}{\partial y}\delta y + \frac{\partial F}{\partial y'}\delta y', \tag{33}$$

which bears a great resemblance to the differential expression in (31).

Remark: Because F depends on three arguments x, y, y', we might expect Eq. (33) to read

$$\delta F = \frac{\partial F}{\partial x}\delta x + \frac{\partial F}{\partial y}\delta y + \frac{\partial F}{\partial y'}\delta y'.$$

But, x is *not varied* here (only y), so that $\delta x = 0$.

To illustrate the use of the variational notation in finding stationary values, let us apply the first variation to the functional

$$I(y) = \int_a^b F(x,y,y')dx. \tag{34}$$

This action leads to

$$\delta I = \int_a^b \delta F(x,y,y')dx = \int_a^b \left(\frac{\partial F}{\partial y}\delta y + \frac{\partial F}{\partial y'}\delta y' \right) dx = 0. \tag{35}$$

If we write $\delta y' = \dfrac{d}{dx}(\delta y)$ and integrate by parts, then

$$\delta I = \frac{\partial F}{\partial y'}\delta y \bigg|_a^b + \int_a^b \left[\frac{\partial F}{\partial y} - \frac{d}{dx}\left(\frac{\partial F}{\partial y'} \right) \right]\delta y\,dx = 0. \tag{36}$$

By assuming the variation $\delta y = 0$ at $x = a,b$ [i.e., $\eta(a) = \eta(b) = 0$], the term outside the integral in (36) vanishes. Hence, by analogy with Eq. (9) in Section 11.2, we see that a *stationary function* for the integral $I(y)$ is one for which the first variation of the integral is zero, and this will lead once again to the Euler-Lagrange equation.

11.4.1 Natural boundary conditions

Thus far we have concentrated on extremal problems in which the BCs are "forced" [recall Eq. (2)]. When this is the case we require $\eta(a) = \eta(b) = 0$, or $\delta y = 0$ at $x = a,b$. In some applications, however, we find that BCs are not prescribed in this fashion at one or both endpoints (perhaps because we do not know the proper physical endpoint conditions). In such cases the variational calculus can help us identify the proper "natural BCs" to prescribe. Moreover, it may happen in some problems that the endpoint values for x are not prescribed and we merely require the admissible functions to begin and end along certain curves. Here we wish to address only the special case in which the endpoints $x = a$ and $x = b$ are both fixed, but the value of y is not prescribed at either endpoint.

Recall from Eq. (36) we showed that the first variation of $I(y)$ yields

$$\delta I = \frac{\partial F}{\partial y'}\delta y \bigg|_a^b + \int_a^b \left[\frac{\partial F}{\partial y} - \frac{d}{dx}\left(\frac{\partial F}{\partial y'} \right) \right]\delta y\,dx = 0.$$

Now, if $\delta y \neq 0$ at either endpoint, then the term outside the integral will vanish only if we require

$$\frac{\partial F}{\partial y'}\bigg|_{x=a} = 0, \qquad \frac{\partial F}{\partial y'}\bigg|_{x=b} = 0. \tag{37}$$

Equations (37) are called *natural boundary conditions* of the problem. In some instances it may happen that a BC is prescribed at one endpoint, say at $x = a$, but not the other. In that case we have a forced BC at $x = a$ and a natural BC at $x = b$.

EXAMPLE 2 For positive constants T, ρ, and ω, determine the stationary functions of

$$I(y) = \int_0^1 [T(y')^2 - \rho\omega^2 y^2]dx,$$

subject to the prescribed BCs

(*a*) $y(0) = 0$, $y(1) = 1$.
(*b*) $y(0) = 0$.

Solution: The Euler-Lagrange equation is readily found to be

$$y'' + k^2 y = 0,$$

where $k = \rho\omega^2/T$. The general solution is

$$y = C_1 \cos kx + C_2 \sin kx.$$

(*a*) By imposing the first set of BCs on the general solution, it follows that $C_1 = 0$ and $C_2 \sin k = 1$, and consequently,

$$y = \frac{\sin kx}{\sin k}, \quad k \neq n\pi, \; n = 0, 1, 2, \dots.$$

(*b*) In the second case above, we only have one prescribed BC, which yields

$$y = C_2 \sin kx.$$

Here must find an appropriate natural BC for the endpoint $x = 1$. From (37), we see that $\partial F/\partial y' = 2Ty'$, and thus, the natural BC is

$$y'(1) = 0.$$

By imposing this condition on our solution, we find

$$y'(1) = kC_2 \cos k = 0.$$

We conclude, therefore, that there can exist no solution in this case unless we select the constant $k = \pi/2, 3\pi/2, 5\pi/2, \dots$. However, for these values of k the constant C_2 is left arbitrary and we obtain an infinite collection of stationary functions given by $y = C_2 \sin kx$, $k = \pi/2, 3\pi/2, 5\pi/2, \dots$.

11.5 Other Types of Functionals

In this section we wish to briefly consider three extensions of the general theory of extremals as presented above. These extensions include (*i*) functionals with several dependent variables, (*ii*) functionals with higher-order derivatives, and (*iii*) functionals with several independent variables.

11.5.1 Functionals with several dependent variables

Let $y_1, y_2, ..., y_m$ be a set of linearly independent point functions, all of which belong to the class of admissible functions Γ. We then consider functionals of the form

$$I(y_1, y_2, ..., y_m) = \int_a^b F(x, y_1, y_2, ..., y_m, y_1', y_2', ..., y_m')dx. \tag{38}$$

We can facilitate the discussion of such functionals by formally introducing the vector notation

$$\mathbf{y}(x) = [y_1(x), y_2(x), ..., y_m(x))] \tag{39}$$

and then rewriting (38) in the vector form

$$I(\mathbf{y}) = \int_a^b F(x, \mathbf{y}, \mathbf{y}')dx. \tag{40}$$

For simplicity, we assume the vector $\mathbf{y}$ satisfies the forced BCs

$$\mathbf{y}(a) = \boldsymbol{\mu}, \quad \mathbf{y}(b) = \mathbf{v}, \tag{41}$$

where $\boldsymbol{\mu}$ and $\mathbf{v}$ are fixed vectors.

Here we use the vector variation

$$\mathbf{y} = \boldsymbol{\Phi}(x) + \delta\mathbf{y}, \tag{42}$$

and set $\delta I = 0$ to find (after some manipulation)

$$\delta I = \sum_{j=1}^m \int_a^b \left[\frac{\partial F}{\partial y_j} - \frac{d}{dx}\left(\frac{\partial F}{\partial y_j'} \right) \right] \delta y_j \, dx = 0. \tag{43}$$

Because of the linear independence of the various δy_j, each term of the series in (43) must vanish. Thus, we are led to the *system of Euler-Lagrange equations*

$$\frac{\partial F}{\partial y_j} - \frac{d}{dx}\left(\frac{\partial F}{\partial y_j'} \right) = 0, \quad j = 1, 2, ..., m. \tag{44}$$

The classic area of application for problems of this nature is in the use of Hamilton's principle to problems in mechanics (see Section 11.8)

EXAMPLE 3 Find the stationary functions of

$$I(y,z) = \int_0^{\pi/2} \left[(y')^2 + (z')^2 + 2yz \right] dx,$$
$$y(0) = 3, \; y(\pi/2) = 1, \; z(0) = -3, \; z(\pi/2) = -1.$$

Solution: Because F does not depend explicitly on x, we can write each Euler-Lagrange equation in the form of CASE IV in Section 11.4.2. Doing so, we obtain the system of *nonlinear* equations

$$(z')^2 - (y')^2 + 2yz = C_1,$$
$$(y')^2 - (z')^2 + 2yz = C_2,$$

where C_1 and C_2 are arbitrary constants. However, by using (44) directly, we are led to the system of *linear* equations

$$y'' - z = 0,$$
$$z'' - y = 0.$$

Clearly, this second set of equations is the easier system to solve. From the first equation we have $z = y''$, which substituted into the second equation yields

$$y^{(4)} - y = 0,$$

where $y^{(4)} = d^4 y/dx^4$. The general solution of this DE is given by

$$y = C_1 e^x + C_2 e^{-x} + C_3 \cos x + C_4 \sin x,$$

and, consequently,

$$z = y'' = C_1 e^x + C_2 e^{-x} - C_3 \cos x - C_4 \sin x.$$

By imposing the prescribed BCs on the above solutions, we obtain

$$y(0) = C_1 + C_2 + C_3 = 3,$$
$$y(\pi/2) = C_1 e^{\pi/2} + C_2 e^{-\pi/2} + C_4 = 1,$$
$$z(0) = C_1 + C_2 - C_3 = -3,$$
$$z(\pi/2) = C_1 e^{\pi/2} + C_2 e^{-\pi/2} - C_4 = -1,$$

which requires $C_1 = C_2 = 0$, $C_3 = 3$ and $C_4 = 1$. Hence, the stationary functions are

$$y = 3 \cos x + \sin x,$$

$$z = -(3 \cos x + \sin x).$$

11.5.2 Functionals with higher-order derivatives

In the study of elastic deformation of beams, for example, it becomes necessary to determine extremals of functionals of the form (see Section 11.9.2)

$$I(y) = \int_a^b F(x,y,y',y'')dx, \tag{45}$$

where F satisfies general continuity and differentiability requirements in each of its arguments x, y, y', y''. We seek extremals of (45) subject to the forced BCs

$$y(a) = y_1, \quad y(b) = y_2. \tag{46}$$

In accordance with the variational notation, we have

$$\delta I = \int_a^b \left(\frac{\partial F}{\partial y} \delta y + \frac{\partial F}{\partial y'} \delta y' + \frac{\partial F}{\partial y''} \delta y'' \right) dx. \tag{47}$$

By writing $\delta y' = \dfrac{d}{dx}(\delta y)$, $\delta y'' = \dfrac{d}{dx}(\delta y')$, and then integrating by parts, we obtain

$$\int_a^b \frac{\partial F}{\partial y'} \delta y' \, dx = \frac{\partial F}{\partial y'} \delta y \Big|_b^a - \int_a^b \frac{d}{dx}\left(\frac{\partial F}{\partial y'} \right) \delta y \, dx,$$

$$\int_a^b \frac{\partial F}{\partial y''} \delta y'' \, dx = -\frac{d}{dx}\left(\frac{\partial F}{\partial y''} \right) \delta y \Big|_b^a + \frac{\partial F}{\partial y''} \delta y' \Big|_b^a + \int_a^b \frac{d^2}{dx^2}\left(\frac{\partial F}{\partial y''} \right) \delta y \, dx.$$

Combining the above results and setting $\delta I = 0$ yields

$$\begin{aligned}
\delta I = \int_a^b &\left[\frac{\partial F}{\partial y} - \frac{d}{dx}\left(\frac{\partial F}{\partial y'} \right) + \frac{d^2}{dx^2}\left(\frac{\partial F}{\partial y''} \right) \right] \delta y \, dx \\
&+ \left[\frac{\partial F}{\partial y'} - \frac{d}{dx}\left(\frac{\partial F}{\partial y''} \right) \right] \delta y \Big|_b^a + \frac{\partial F}{\partial y''} \delta y' \Big|_b^a = 0.
\end{aligned} \tag{48}$$

Here we require the integral to vanish as well as both terms outside the integral. Thus, for the integral to vanish with arbitrary δy, we impose the Euler-Lagrange equation

$$\boxed{\frac{\partial F}{\partial y} - \frac{d}{dx}\left(\frac{\partial F}{\partial y'} \right) + \frac{d^2}{dx^2}\left(\frac{\partial F}{\partial y''} \right) = 0.} \tag{49}$$

The first term in (48) outside the integral vanishes by virtue of the forced BCs (46) which require $\delta y = 0$ at $x = a, b$. The remaining term in (48) must also go to zero, but we cannot assume $\delta y' = 0$ at $x = a, b$. Thus, we are led to the natural BCs

$$\left.\frac{\partial F}{\partial y''}\right|_{x=a} = 0, \qquad \left.\frac{\partial F}{\partial y''}\right|_{x=b} = 0. \tag{50}$$

The necessity of prescribing two additional BCs also follows from the fact that the Euler-Lagrange equation (49) in general is a fourth-order DE, thereby requiring four BCs. However, had we specified the forced BCs $y'(a) = y_3$, $y'(b) = y_4$ in addition to those in (46), there would be no natural BCs like (50). That is, the last term outside the integral in (48) would vanish in this case because $\delta y' = 0$ at $x = a, b$. On the other hand, if no forced BCs are specified, then in addition to the natural BCs (50) we would require the two natural BCs

$$\left.\left[\frac{\partial F}{\partial y'} - \frac{d}{dx}\left(\frac{\partial F}{\partial y''}\right)\right]\right|_{x=a} = 0, \qquad \left.\left[\frac{\partial F}{\partial y'} - \frac{d}{dx}\left(\frac{\partial F}{\partial y''}\right)\right]\right|_{x=b} = 0. \tag{51}$$

EXAMPLE 4 Find the stationary functions of

$$I(y) = \int_0^1 \left[\frac{1}{2}(y'')^2 - y\right]dx, \quad y(0) = 0, \ y(1) = 0.$$

Solution: Based on (49), the Euler-Lagrange equation is readily found to be

$$y^{(4)} = 1.$$

For this problem the natural BCs (50) are given by

$$y''(0) = 0, \quad y''(1) = 0.$$

Four successive integrations of the governing DE yields the general solution

$$y = \frac{1}{24}x^4 + \frac{1}{6}C_1 x^3 + \frac{1}{2}C_2 x^2 + C_3 x + C_4,$$

and by imposing the forced and natural BCs, we find $C_1 = -1/2$, $C_2 = C_4 = 0$, and $C_3 = 1/24$, from which we deduce the stationary function

$$y = \frac{1}{24}(x^4 - 2x^3 + x).$$

11.5.3 Functionals with several independent variables

Let us now consider a functional that is a function of a point function of two independent variables. We denote the independent variables by x and y, and the point function by $u = u(x,y)$. It is assumed that u is continuous with continuous first and second partial derivatives in some bounded, simply connected (i.e., "no holes") domain D with smooth boundary curve C. The functional we wish to study is then defined by

$$I(u) = \iint_D F(x,y,u,u_x,u_y)\,dx\,dy, \tag{52}$$

where $u_x = \partial u/\partial x$ and $u_y = \partial u/\partial y$, and where F is continuous and has continuous first and second partial derivatives in each of its arguments.

If we let

$$u = \Phi(x,y) + \varepsilon\eta(x,y) = \Phi(x,y) + \delta u, \tag{53}$$

then the first variation of (52) gives us

$$\begin{aligned}
\delta I &= \iint_D \left(\frac{\partial F}{\partial u}\delta u + \frac{\partial F}{\partial u_x}\delta u_x + \frac{\partial F}{\partial u_y}\delta u_y \right) dx\,dy \\
&= \iint_D \left[\frac{\partial F}{\partial u} - \frac{\partial}{\partial x}\left(\frac{\partial F}{\partial u_x} \right) - \frac{\partial}{\partial y}\left(\frac{\partial F}{\partial u_y} \right) \right]\delta u\,dx\,dy \\
&\quad + \iint_D \left[\frac{\partial}{\partial x}\left(\frac{\partial F}{\partial u_x}\delta u \right) + \frac{\partial}{\partial y}\left(\frac{\partial F}{\partial u_y}\delta u \right) \right] dx\,dy,
\end{aligned} \tag{54}$$

where we have used the identity

$$\begin{aligned}
\frac{\partial}{\partial x}\left(\frac{\partial F}{\partial u_x}\delta u \right) + \frac{\partial}{\partial y}\left(\frac{\partial F}{\partial u_y}\delta u \right) &= \left[\frac{\partial}{\partial x}\left(\frac{\partial F}{\partial u_x} \right) + \frac{\partial}{\partial y}\left(\frac{\partial F}{\partial u_y} \right) \right]\delta u \\
&\quad + \frac{\partial F}{\partial u_x}\delta u_x + \frac{\partial F}{\partial u_y}\delta u_y.
\end{aligned} \tag{55}$$

By application of Green's theorem in the plane (see Theorem 4.7) to the last integral in (54), we get

$$\begin{aligned}
\delta I &= \iint_D \left[\frac{\partial F}{\partial u} - \frac{\partial}{\partial x}\left(\frac{\partial F}{\partial u_x} \right) - \frac{\partial}{\partial y}\left(\frac{\partial F}{\partial u_y} \right) \right]\delta u\,dx\,dy \\
&\quad + \oint_C \left(-\frac{\partial F}{\partial u_y}dx + \frac{\partial F}{\partial u_x}dy \right)\delta u\,ds.
\end{aligned} \tag{56}$$

For stationary values of $I(u)$, we set $\delta I = 0$, which leads to the two-dimensional Euler-Lagrange equation

$$\boxed{\frac{\partial F}{\partial u} - \frac{\partial}{\partial x}\left(\frac{\partial F}{\partial u_x} \right) - \frac{\partial}{\partial y}\left(\frac{\partial F}{\partial u_y} \right) = 0.} \tag{57}$$

If we impose the forced BC

$$u = f(x,y) \quad \text{on } C, \tag{58}$$

then $\delta u = 0$ on C and the line integral in (56) vanishes. If, however, no forced BC is prescribed, then the vanishing of the line integral in (56) requires the integrand to vanish, which leads to the natural BC

$$\frac{\partial F}{\partial u_y} - \frac{\partial F}{\partial u_x}\frac{dy}{dx} = 0 \quad \text{on } C. \tag{59}$$

EXAMPLE 5 Determine the Euler-Lagrange equation associated with finding stationary values of

$$I(u) = \iint_D \left[(u_x)^2 + (u_y)^2\right] dx\,dy, \quad u = f(x,y) \text{ on } C.$$

Solution: By calculating the relevant partial derivatives, we find

$$\frac{\partial F}{\partial u} = 0, \quad \frac{\partial F}{\partial u_x} = 2u_x, \quad \frac{\partial F}{\partial u_y} = 2u_y.$$

Thus, the Euler-Lagrange equation (57) becomes

$$u_{xx} + u_{yy} = 0,$$

which is *Laplace's equation* of potential theory. Solving this equation subject to the prescribed forced BC is known as a *Dirichlet problem* (see Sections 4.7.2, 7.8, and 12.5). The natural BC in this case is

$$u_y - u_x \frac{dy}{dx} = 0,$$

or $\nabla u \cdot \hat{\mathbf{N}} = 0$, where $\hat{\mathbf{N}} = <-dy/ds, dx/ds>$ is the outward unit normal (see Chapter 4) to the boundary curve C. In this latter instance the resulting potential problem is called a *Neumann problem*.

11.6 Isoperimetric Problems

Historically, the oldest problem associated with the calculus of variations was proposed by the ancients. The problem is that of determining the plane, closed, smooth curve of fixed length which encloses the largest area. This problem was intuitively solved by Queen Dido of Carthage and for that reason is sometimes called the "Problem of Dido."

We now easily recognize that a circle provides the fixed length curve that encloses the largest plane area. This problem is a member of a larger class of problems that are customarily referred to as *isoperimetric problems*. They involve the determination of a stationary function of a given functional, subject to the condition that the value of another functional remain constant (e.g., the arc length being a fixed value in the problem of Dido).

11.6.1 Constraints and Lagrange multipliers

We can formulate the general isoperimetric problem we wish to address in the following way. That is, we seek a stationary function (or stationary value) of

$$I(y) = \int_a^b F(x,y,y')dx, \quad y(a) = y_1, \; y(b) = y_2, \tag{60}$$

subject to the constraint

$$J(y) = \int_a^b G(x,y,y')dx = k \;\text{(constant)}. \tag{61}$$

To find the stationary function in this case, we consider the related problem of finding a stationary function of the quotient

$$Q(y) = \frac{I(y)}{J(y)}, \tag{62}$$

known as the *Rayleigh quotient*. If we once again designate $\Phi(x)$ as the true solution (stationary function), then we can define λ as the *stationary value*

$$\lambda = Q(\Phi) = \frac{I(\Phi)}{J(\Phi)}. \tag{63}$$

If we assume the rule of differentiating a quotient also applies to the variation notation (which it does), we can formally calculate the first variation of Q, which yields

$$\delta Q = \frac{J(\Phi)\delta I - I(\Phi)\delta J}{J^2(\Phi)} = \frac{\delta I - \lambda \delta J}{J(\Phi)}. \tag{64}$$

Thus, the condition $\delta Q = 0$ requires that $\delta I - \lambda \delta J = 0$, from which we deduce the Euler-Lagrange equation

$$\boxed{\frac{\partial F}{\partial y} - \frac{d}{dx}\left(\frac{\partial F}{\partial y'}\right) - \lambda\left[\frac{\partial G}{\partial y} - \frac{d}{dx}\left(\frac{\partial G}{\partial y'}\right)\right] = 0.} \tag{65}$$

Rather than use the Rayleigh quotient (62), the conventional way to find a stationary function of $I(y)$ subject to the constraint $J(y) - k = 0$ is to form the functional

$$H(y) = I(y) - \lambda[J(y) - k]$$

and then set $\delta H = \delta I - \lambda \delta J = 0$. Formulated in this fashion, the parameter λ is called a

Lagrange multiplier. Note that the condition $\delta H = 0$ leads to the same Euler-Lagrange equation (65) as found above. Thus, the stationary functions of $H(y) = I(y) - \lambda[J(y) - k]$ are the same as those of the Rayleigh quotient $Q(y) = I(y)/J(y)$. The Rayleigh quotient, however, has the advantage that it provides an interpretation for the parameter λ that is useful in approximation techniques for eigenvalue problems (see Section 11.7).

EXAMPLE 6 (*Problem of Dido*) Find the smooth curve in the *xy*-plane of fixed length k which passes through the points $(-a,0)$ and $(a,0)$, and for which the area bounded by the curve and the *x*-axis is a maximum.

Solution: As formulated, we wish to find a stationary function of

$$I(y) = \int_{-a}^{a} y\,dx, \quad y(-a) = 0, \ y(a) = 0,$$

subject to the constraint

$$J(y) = \int_{-a}^{a} \sqrt{1 + (y')^2}\,dx = \pi a.$$

We have chosen a symmetrical interval (without loss of generality) and fixed the value of the constraint (i.e., the arc length) for mathematical convenience.

By identifying $F(x,y,y') = y$ and $G(x,y,y') = \sqrt{1 + (y')^2}$, the Euler-Lagrange equation (65) becomes

$$1 + \lambda \frac{d}{dx}\left[\frac{y'}{\sqrt{1 + (y')^2}}\right] = 0,$$

which can be rearranged as

$$\frac{y''}{[1 + (y')^2]^{3/2}} = -\frac{1}{\lambda}.$$

The reason for writing the Euler-Lagrange equation in this last form is that we can now recognize the left-hand side of the equation as the expression for the *curvature* of the function y. Because the right-hand side is constant, we deduce that the curvature of the desired curve is a constant. This identifies the curve as a circular arc through the two fixed points $(-a,0)$ and $(a,0)$. Hence, the solution must be of the form

$$y - C = \pm\sqrt{a^2 - x^2},$$

where we can choose either branch (upper or lower) of the circle and C is a constant chosen so that the prescribed constraint is satisfied. In our case, this means that $C = 0$, i.e., the circular arc is a semicircle (either upper half or lower half).

To solve the Euler-Lagrange equation directly, we note that the functional $f(x,y,y') = F(x,y,y') - \lambda G(x,y,y')$ is missing the variable x; hence, we can use CASE IV in Section 11.2.2 to obtain the reduced form of the Euler-Lagrange equation given by $f - y' \,\partial f/\partial y' = C_1$, which is

$$\frac{y\sqrt{1 + (y')^2} - \lambda}{\sqrt{1 + (y')^2}} = C_1,$$

where C_1 is a constant. Separation of variables yields the solution

$$(x - C_2)^2 + (y - C_1)^2 = \lambda^2,$$

which we recognize as that of a circle (or circular arc). We leave the remaining details of applying the prescribed conditions to the reader.

11.6.2 Sturm-Liouville problem

We now wish to consider a class of optimization problems that involve finding extremals of the *quadratic functional*

$$I(y) = \int_a^b \left[p(x)(y')^2 - q(x)y^2\right]dx, \tag{66}$$

where $p(x)$, $p'(x)$, and $q(x)$ are continuous functions on $a \le x \le b$, and $p(x) > 0$ on $a < x < b$. We seek extremals of (66) subject to the constraint

$$J(y) = \int_a^b r(x)y^2\,dx = C \text{ (constant)}, \tag{67}$$

where $r(x)$ is continuous on $a \le x \le b$, and $r(x) > 0$ on $a < x < b$. In some cases it is convenient to select the constant on the right-hand side of (67) as unity.

The above extremal problem is an example of an isoperimetric problem for which the governing Euler-Lagrange equation (65) takes the form (after simplification)

$$\frac{d}{dx}\left[p(x)y'\right] + \left[q(x) + \lambda r(x)\right]y = 0, \quad a < x < b. \tag{68}$$

If we also impose the forced BCs

$$y(a) = 0, \quad y(b) = 0, \tag{69}$$

then the problem described by (68) and (69) is recognized as a regular *Sturm-Liouville system* of the type studied in Section 8.4. Also, we see that the *eigenvalues* are actually *stationary values* of the Rayleigh quotient $I(y)/J(y)$, i.e., $\lambda = I(\Phi)/J(\Phi)$. For further discussion of this class of equations, see Chapter 8.

11.7 Rayleigh-Ritz Approximation Method

The method used thus far for solving variational problems necessitates solving an ordinary DE (nonlinear in many cases) subject to certain forced or natural BCs. However, the Euler-Lagrange equation is often difficult to solve exactly, so it may become necessary to employ some approximation or numerical method.

The method we introduce here is an approximation method, first proposed by W. Ritz in 1908, which makes use of *Rayleigh's principle*:

"... the potential and kinetic energies in an elastic system are distributed such that the frequencies of the components are a minimum."

The method, now widely known as the *Rayleigh-Ritz method*, is applied directly to the variational integral, rather then to the associated Euler-Lagrange equation as in the case of other techniques (see Section 11.5.2).

To develop the Rayleigh-Ritz method, let us assume the given functional has the more general form given by

$$I(y) = \int_a^b F(x,y,y',y'')dx. \tag{70}$$

We attempt to approximate this extremal with a *trial solution* ψ defined by

$$\psi = g_0(x) + \sum_{j=1}^N c_j g_j(x), \tag{71}$$

where the c's are the unknowns and the functions $g_0(x), g_j(x), j = 1,2,...,N$, are linearly independent continuous functions chosen in the following way. The function $g_0(x)$ is chosen to satisfy the forced BCs exactly and the remaining N functions are chosen to satisfy homogeneous versions of the forced BCs. If the forced BCs are homogeneous, then we set $g_0(x) \equiv 0$. Also, if the functional $F(x,y,y',y'')$ exhibits any symmetries, it is useful to build such symmetries into the choice of $g_0(x), g_j(x), j = 1,2,...,N$.

One advantage of the Rayleigh-Ritz method is that we only need to satisfy the forced BCs with our choice of $g_0(x), g_j(x), j = 1,2,...,N$. If natural BCs exist, we do not have to choose the linearly independent functions $g_0(x), g_j(x), j = 1,2,...,N$, to satisfy them also, i.e., satisfying natural BCs is simply part of the extremal process.

Historically, methods like the Rayleigh-Ritz approximation (called weighted residual methods) evolved into what are now called *finite element* methods. The primary difference is that the approximating functions (polynomial splines) in a finite element method have compact support (i.e., a finite set of values over which they are nonzero).

EXAMPLE 7 Use the Rayleigh-Ritz method to find an approximate solution for the extremal of

$$I(y) = \frac{1}{2}\int_{-1}^1 \left[(2 - x^2)(y'')^2 + 40y^2 - 2(2 - x^2)y\right]dx.$$

Solution: Because no forced BCs are prescribed, we assume all BCs are natural. Although we don't need them to find our approximate solution, it can readily be shown that these natural BCs are $y''(\pm 1) = 0$, $y'''(\pm 1) = 0$.

To begin, we note that the functional $F(x,y,y',y'')$ is symmetric about the line $x = 0$, as are the natural BCs above. Thus, it is useful to select a trial solution that is symmetric about this line. In particular, let us assume

$$\psi = \sum_{j=1}^N c_j x^{2(j-1)}.$$

The direct substitution of the trial solution into the given functional yields

$$I(c_1, ..., c_N) = \frac{1}{2} \int_{-1}^{1} \left\{ (2-x)^2 \left[\sum_{j=1}^{N} c_j(2j-2)(2j-3)x^{2j-4} \right]^2 \right.$$

$$\left. + 40 \left(\sum_{j=1}^{N} c_j x^{2j-2} \right)^2 - 2(2-x^2) \sum_{j=1}^{N} c_j x^{2j-2} \right\} dx.$$

We can interpret this last expression as a point function of N variables $c_1, ..., c_N$. Hence, we can find stationary values of I by differentiating with respect to c_k and setting the result to zero. Doing so, we obtain

$$\frac{\partial I}{\partial c_k} = \sum_{j=1}^{N} c_j \int_{-1}^{1} \left[(2-x^2)(2j-2)(2j-3)(2k-2)(2k-3)x^{2j+2k-8} + 40x^{2j+2k-4} \right] dx$$

$$- \int_{-1}^{1} (2-x^2)x^{2k-2} dx = 0, \quad k = 1, 2, ..., N.$$

If we now introduce the simplifying notation

$$a_{jk} = (2j-2)(2j-3)(2k-2)(2k-3) \int_{-1}^{1} (2-x^2)x^{2j+2k-8} dx + 40 \int_{-1}^{1} x^{2j+2k-4} dx$$

$$= (2j-2)(2j-3)(2k-2)(2k-3) \left[\frac{4(j+k)-10}{4(j+k-3)^2-1} \right] + \frac{80}{2j+2k-3},$$

$$f_k = \int_{-1}^{1} (2-x^2)x^{2k-2} dx = \frac{4k+6}{4k^2-1},$$

then the above system of equations can be written in the form

$$\sum_{j=1}^{N} a_{jk} c_j = f_k, \quad k = 1, 2, ..., N.$$

This last system of equations can readily be solved by matrix methods (see Chapter 3). However, if $N = 1$, it follows that $c_1 \cong 0.042$, and our approximate solution is simply the constant

$$\psi = 0.042.$$

For $N = 3$, it can be shown that $c_1 \cong 0.045$, $c_2 \cong -0.012$, $c_3 \cong 0.002$, and consequently,

$$\psi = 0.045 - 0.012x^2 + 0.002x^4.$$

Note that ψ satisfies the natural BCs exactly when $N = 1$, whereas for $N = 3$, we obtain

$$\psi''(\pm 1) \cong 0.000, \quad \psi'''(\pm 1) \cong \pm 0.048.$$

In general, we don't expect our approximate solution to satisfy exactly the natural BCs, but it should approximate them reasonably well.

11.7.1 Eigenvalue problems

When the eigenvalue problem arises from a variational composition, the Rayleigh-Ritz method can be a very effective technique for approximating the eigenvalues and corresponding eigenfunctions. Let us again consider the regular Sturm-Liouville system formulated by the isoperimetric problem (see Section 11.6.1)

$$I(y) = \int_a^b \left[p(x)(y')^2 - q(x)y^2 \right] dx, \quad y(a) = 0, \; y(b) = 0, \tag{72}$$

$$J(y) = \int_a^b r(x)y^2 \, dx = C. \tag{73}$$

Recall that finding stationary functions of this isoperimetric problem is equivalent to finding stationary functions of the Rayleigh quotient $Q(y) = I(y)/J(y)$, and that the eigenvalues λ represent stationary values given by $\lambda = I(\Phi)/J(\Phi)$.

In particular, if λ_1 represents the smallest eigenvalue (i.e., $\lambda_1 < \lambda_2 < \lambda_3 \cdots$), we can use the Rayleigh quotient to obtain an estimate of it. Let us assume that $g_1(x)$ is a continuous function such that

$$g_1(a) = 0, \quad g_1(b) = 0. \tag{74}$$

A general property of eigenfunctions is that the jth eigenfunction $\varphi_j(x)$ has exactly $(j-1)$ zeros on the open interval $a < x < b$. In constructing an approximation to one of the eigenfunctions, therefore, we must take this into account. Hence, we choose $g_1(x)$ so it has no zeros on the open interval $a < x < b$. In that case it follows that

$$\Lambda_1 = Q[g_1(x)] = \frac{I[g_1(x)]}{J[g_1(x)]} \tag{75}$$

is an approximation to the first eigenvalue λ_1. Higher-order eigenvalues can be estimated in a similar manner by selecting suitable functions $g_j(x)$, $j = 1, 2, 3, \dots$.

To find approximations to the first N eigenvalues and eigenfunctions, we start with a trial solution of the form

$$\psi = \sum_{j=1}^N c_j g_j(x), \tag{76}$$

where the constants c_j, $j = 1, 2, \dots, N$ are the unknowns. We form this expression by selecting suitable linearly independent functions $g_j(x)$, $j = 1, 2, 3, \dots$ so that each one satisfies the homogeneous forced BCs like (74). Otherwise, the choice of $g_j(x)$ is somewhat arbitrary, but chosen to depict the general nature of the solution if that is known. The direct substitution of (76) into (72) and (73) yields

$$I(c_1,...,c_N) = \int_a^b \left\{ p(x)\left[\sum_{j=1}^N c_j g_j{}'(x) \right]^2 - q(x)\left[\sum_{j=1}^N c_j g_j(x) \right]^2 \right\} dx, \tag{77}$$

$$J(c_1,...,c_N) = \int_a^b r(x)\left[\sum_{j=1}^N c_j g_j(x) \right]^2 dx. \tag{78}$$

The stationary values of (77) and (78) are determined by setting $\delta I - \Lambda \delta J = 0$, which, because I and J are now point functions of the c's, can be expressed as

$$\frac{\partial I}{\partial c_k} - \Lambda \frac{\partial J}{\partial c_k} = 0, \quad k = 1,2,...,N. \tag{79}$$

The direct calculation of the appropriate partial derivatives in (79) gives us

$$\frac{\partial I}{\partial c_k} = 2\sum_{j=1}^N c_j \int_a^b \left[p(x)g_j{}'(x)g_k{}'(x) - q(x)g_j(x)g_k(x) \right] dx = 2\sum_{j=1}^N c_j a_{jk}, \tag{80}$$

$$\frac{\partial J}{\partial c_k} = 2\sum_{j=1}^N c_j \int_a^b \left[r(x)g_j(x)g_k(x) \right] dx = 2\sum_{j=1}^N c_j b_{jk}, \tag{81}$$

and, therefore, we can express (79) as

$$\sum_{j=1}^N (a_{jk} - \Lambda b_{jk})c_j = 0, \quad k = 1,2,...,N. \tag{82}$$

Nontrivial solutions for the c's of this system of homogeneous equations are possible only for values of Λ satisfying the *secular equation*

$$\det(a_{jk} - \Lambda b_{jk}) = 0, \tag{83}$$

which has exactly N solutions $\Lambda_1, \Lambda_2, ..., \Lambda_N$. Once again, the solution of (82) and/or (83) can best be treated by using matrix methods. However, the method is not uniformly accurate for all eigenvalues. Greater error is always observed in the larger eigenvalues for any N.

EXAMPLE 8 Use the Rayleigh-Ritz method to approximate the first two eigenvalues of

$$y'' + \lambda y = 0, \quad y(0) = 0, \ y(1) = 0.$$

Solution: Note that this eigenequation arises from the isoperimetric problem (72) and (73) in which $p(x) = r(x) = 1$ and $q(x) = 0$. Among other choices, let us select the trial solution

$$\psi = c_1 x(1 - x) + c_2 x^2(1 - x).$$

In this case, we find

$$a_{jk} = \int_0^1 g_j{}'(x)g_k{}'(x)dx, \quad b_{jk} = \int_0^1 g_j(x)g_k(x)dx, \quad j,k = 1,2,$$

which leads to

$$a_{11} = \frac{1}{3}, \; a_{12} = a_{21} = \frac{1}{6}, \; a_{22} = \frac{2}{15}; \quad b_{11} = \frac{1}{30}, \; b_{12} = b_{21} = \frac{1}{60}, \; b_{22} = \frac{2}{105}.$$

Hence, the secular equation becomes

$$\begin{vmatrix} \dfrac{1}{3} - \dfrac{\Lambda}{30} & \dfrac{1}{6} - \dfrac{\Lambda}{60} \\[2mm] \dfrac{1}{6} - \dfrac{\Lambda}{60} & \dfrac{2}{15} - \dfrac{\Lambda}{105} \end{vmatrix} = 0,$$

from which we calculate $\Lambda_1 = 10$ and $\Lambda_1 = 42$. The exact eigenvalues in this example are $\lambda_1 = \pi^2 \cong 9.87$ and $\lambda_2 = 4\pi^2 \cong 39.48$ (see Example 6 in Chapter 8). Last, the corresponding eigenfunctions are $\psi_1 = x(1 - x)$, $\psi_2 = x(1 - 3x + 2x^2)$.

11.8 Hamilton's Principle

In this section we wish to consider an area of application from mechanics which involves the equation(s) of motion for various systems of particles. The material presented here is based upon a knowledge of only the most rudimentary concepts of particle mechanics. Also, our discussion makes no attempt to treat relativistic phenomena, but is confined to problems in "classical mechanics."

One of the most significant principles of mathematical physics is named in honor of Sir William Rowan Hamilton (1805-1865). If we let T denote kinetic energy and V denote potential energy, then *Hamilton's principle* states:

"... among all motions consistent with the constraints that carry a conservative system from an initial configuration at time t_1 to a final configuration at time t_2, that which actually occurs provides a stationary value to the 'action'

$$I = \int_{t_1}^{t_2} L \, dt = \int_{t_1}^{t_2} (T - V) dt."$$

The energy difference

$$L = T - V \tag{84}$$

is called the *Lagrangian*. Stationary values of the action integral are found by setting

$$\delta I = \delta \int_{t_1}^{t_2} L dt = 0. \tag{85}$$

It can be shown that the integral I is actually a minimum when compared with that corresponding to any neighboring path having the same terminal configuration, provided the time interval is "sufficiently short."

11.8.1 Generalized coordinates and Lagrange's equations

A system of n particles is said to have $3n$ *degrees of freedom* if there are no constraints imposed on the motion of the particles. It is usually possible then to choose $3n$ linearly independent coordinates $x_1, y_1, z_1, ..., x_n, y_n, z_n$, which uniquely specify the position of all components of the system.

For a single particle, the *kinetic energy* expressed in terms of standard Cartesian coordinates assumes the form

$$T = \frac{1}{2}mv^2 = \frac{1}{2}m(\dot{x}^2 + \dot{y}^2 + \dot{z}^2), \tag{86}$$

where m is the mass, v is the velocity (speed), and we are using the notation

$$\dot{x} = \frac{dx}{dt}, \quad \dot{y} = \frac{dy}{dt}, \quad \dot{z} = \frac{dz}{dt}. \tag{87}$$

Here, the parameter t represents time. For a system of n particles, we write

$$T = \frac{1}{2}\sum_{j=1}^{n} m_j(\dot{x}_j^2 + \dot{y}_j^2 + \dot{z}_j^2), \tag{88}$$

where m_j denotes the mass of the jth particle.

If we confine our attention to only *conservative fields* (see Section 4.6.2), then the *potential energy* for a system of n particles is a function of the form

$$V = V(x_1, y_1, z_1, ..., x_n, y_n, z_n),$$

which depends only on the position coordinates but not on their time derivatives. The Lagrangian $L = T - V$ is therefore a function of the $3n$ position coordinates and their time derivatives, but does not depend explicitly on the time variable t—a fundamental characteristic of conservative fields.

When constraints are imposed upon the system of n particles, they have the effect of reducing the number of independent coordinates describing the simultaneous positions, or configurations, of the particles. A constraint exists when a particle is attached to a rod, or if a particle is confined to move only on a particular surface, etc. If the constraints are completely specified by the k ($k < 3n$) consistent and independent equations

$$g_j(x_1, y_1, z_1, ..., x_n, y_n, z_n) = 0, \quad j = 1, 2, ..., k, \tag{89}$$

the number of degrees of freedom is reduced to $m = 3n - k$. This means that we should be able to express the $3n$ original coordinates in terms of m linearly independent variables $q_1, q_2, ..., q_m$, called *generalized coordinates*. The choice of generalized coordinates is not unique.

In terms of generalized coordinates, the action integral of Hamilton's principle takes the form

$$I(q_1, ..., q_m) = \int_{t_1}^{t_2} L(q_1, ..., q_m, \dot{q}_1, ..., \dot{q}_m) dt. \tag{90}$$

We recognize this functional as the same form given by (38); hence, by setting $\delta I = 0$, we are led to the system of equations

$$\frac{\partial L}{\partial q_j} - \frac{d}{dt}\left(\frac{\partial L}{\partial \dot{q}_j}\right) = 0, \quad j = 1, 2, ..., m, \tag{91}$$

called *Lagrange's equations of motion*.

> **Remark:** Note that for a single particle confined to one degree of freedom, the kinetic energy is $T = 1/2 m\dot{q}^2$ and the Lagrange equation of motion becomes (because L does not depend explicitly on t)
>
> $$L - \dot{q}\frac{\partial L}{\partial \dot{q}} = \text{constant},$$
>
> or $L - 2T = \text{constant}$. From this relation we obtain the *conservation of energy principle* $E = T + V = \text{constant}$.

EXAMPLE 9 Determine the equation of motion in generalized coordinates for the simple plane pendulum of mass m with connecting rod of length b and negligible weight (see Fig. 11.3).

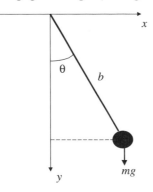

Figure 11.3 Pendulum.

Solution: By inspection we see that the pendulum has only one degree of freedom represented by $q_1 = \theta$. That is, if we specify the position of the mass m by x and y, we initially have two degrees of freedom. However, because x and y are related through the constraint

$$x^2 + y^2 = b^2,$$

the number of degrees of freedom is therefore reduced to one.

In terms of x and y, the position of the mass m is

$$\mathbf{r} = x\hat{\mathbf{i}} + y\hat{\mathbf{j}} = b \sin \theta \hat{\mathbf{i}} + b \cos \theta \hat{\mathbf{j}},$$

where $\hat{\mathbf{i}}$ and $\hat{\mathbf{j}}$ are unit vectors along the x- and y-axes, respectively (see Chapter 4). From this relation, we obtain

$$\dot{\mathbf{r}} = b(\cos \theta \hat{\mathbf{i}} - \sin \theta \hat{\mathbf{j}})\dot{\theta},$$

so that $v^2 = \dot{\mathbf{r}} \cdot \dot{\mathbf{r}} = b^2(\cos^2\theta + \sin^2\theta)\dot{\theta}^2 = (b\dot{\theta})^2$. Hence, the kinetic energy becomes

$$T = \frac{1}{2}m(b\dot{\theta})^2.$$

By using the x-axis as a reference level, the potential energy associated with the mass is therefore

$$V = -mgy = -mgb \cos \theta,$$

where g is the gravitational constant. The Lagrangian is given by

$$L = T - V = \frac{1}{2}m(b\dot{\theta})^2 + mgb \cos \theta,$$

from which we find

$$\frac{\partial L}{\partial \theta} = -mgb \sin \theta, \quad \frac{\partial L}{\partial \dot{\theta}} = mb^2 \dot{\theta}.$$

These expressions lead to the well-known equation of motion (after simplification)

$$\frac{d^2\theta}{dt^2} + \frac{g}{b} \sin \theta = 0,$$

which is independent of the mass m.

11.8.2 Linear theory of small oscillations

The Lagrangian formulation provides us with a systematic way of obtaining the equations of motion for systems with several degrees of freedom. Unfortunately, the resulting equations are often intractable due to their nonlinear nature and coupling. For a wide class of dynamical problems, however, the true behavior of the *nonlinear* system can be accurately approximated by solutions of a related *linear* system. These are systems with a stable equilibrium configuration such that motions with small displacements can usually persist near the equilibrium state. For example, in the pendulum problem of Example 9 we can often make the approximation $\sin \theta \cong \theta$ to reduce the equation of motion to a linear equation. In this case the linear DE provides very accurate results about the motion—even for quite large angles!

If the original system is specified by m generalized coordinates $q_1, q_2, ..., q_m$, these coordinates can usually be selected in such a way that all have zero value in the equilibrium configuration state. For simplicity, here we will discuss only those systems with two degrees of freedom specified by generalized coordinates q_1 and q_2.

For a conservative system with two degrees of freedom, the potential energy function will depend only on q_1 and q_2, but not on their time derivatives. Let us assume the equilibrium point is specified by $q_1 = q_2 = 0$, and expand the potential energy $V(q_1,q_2)$ in a power series about this point, i.e.,

$$V(q_1,q_2) = V(0,0) + \left(\frac{\partial V(0,0)}{\partial q_1}\right) q_1 + \left(\frac{\partial V(0,0)}{\partial q_2}\right) q_2$$
$$+ \frac{1}{2}\left[\left(\frac{\partial^2 V(0,0)}{\partial q_1^2}\right) q_1^2 + 2\left(\frac{\partial^2 V(0,0)}{\partial q_1 \partial q_2}\right) q_1 q_2 + \left(\frac{\partial^2 V(0,0)}{\partial q_2^2}\right) q_2^2\right] + \cdots. \tag{92}$$

We can choose $V(0,0) = 0$ as the arbitrary reference level and, because V must be stationary at equilibrium (i.e., $\delta V = 0$), the linear terms in (92) must vanish. Neglecting terms of order greater than two, the potential energy function can then be approximated by the *quadratic form*

$$V(q_1,q_2) = \frac{1}{2}(b_{11} q_1^2 + 2b_{12} q_1 q_2 + b_{22} q_2^2), \tag{93}$$

where the b's are constants defined by (92).

It has been shown that the potential energy function for systems that exhibit some sort of oscillatory behavior can be reasonably approximated by a quadratic function (93) in the generalized coordinates. We also write the kinetic energy of the system in the form

$$T(q_1,q_2,\dot{q}_1,\dot{q}_2) = \frac{1}{2}(a_{11} \dot{q}_1^2 + 2a_{12} \dot{q}_1 \dot{q}_2 + a_{22} \dot{q}_2^2), \tag{94}$$

where the a's may depend on q_1 and q_2 (and the masses). For small departures from equilibrium, however, these coefficients may be replaced by their values at equilibrium (viz., $q_1 = q_2 = 0$).

By using (93) and (94), Lagrange's equations (91) become

$$\frac{d}{dt}\left(\frac{\partial T}{\partial \dot{q}_j}\right) + \frac{\partial V}{\partial q_j} = 0, \quad j = 1,2, \tag{95}$$

or, specifically,

$$a_{11}\frac{d^2 q_1}{dt^2} + a_{12}\frac{d^2 q_2}{dt^2} + b_{11} q_1 + b_{12} q_2 = 0,$$

$$a_{12}\frac{d^2 q_1}{dt^2} + a_{22}\frac{d^2 q_2}{dt^2} + b_{12} q_1 + b_{22} q_2 = 0. \tag{96}$$

The equations (96) are linear with constant coefficients and can therefore be easily solved by matrix methods. That is, by introducing the matrices

$$\mathbf{Q} = \begin{pmatrix} q_1 \\ q_2 \end{pmatrix}, \quad \mathbf{A} = \begin{pmatrix} a_{11} & a_{12} \\ a_{12} & a_{22} \end{pmatrix}, \quad \mathbf{B} = \begin{pmatrix} b_{11} & b_{12} \\ b_{12} & b_{22} \end{pmatrix}, \tag{97}$$

we can express (96) as

$$\mathbf{A}\frac{d^2\mathbf{Q}}{dt^2} + \mathbf{BQ} = \mathbf{0}. \tag{98}$$

By analogy with solutions of the simple harmonic oscillator $m(d^2x/dt^2) + kx = 0$ (e.g, the spring-mass system in Section 15.2 , we look for solutions of (98) of the form

$$\mathbf{Q} = \mathbf{X}\cos(\omega t - \varphi), \tag{99}$$

where $\mathbf{X}$ denotes amplitude, ω is angular frequency, and φ is phase. The substitution of (99) into (98) leads to the eigenvalue problem (see Section 3.3.1)

$$(\mathbf{B} - \omega^2\mathbf{A})\mathbf{X} = \mathbf{0}, \tag{100}$$

in which the eigenvalues ω are solutions of

$$\det(\mathbf{B} - \omega^2\mathbf{A}) = \begin{vmatrix} b_{11} - \omega^2 a_{11} & b_{12} - \omega^2 a_{12} \\ b_{12} - \omega^2 a_{12} & b_{22} - \omega^2 a_{22} \end{vmatrix} = 0. \tag{101}$$

Because we have assumed symmetric matrices $\mathbf{A}$ and $\mathbf{B}$, it follows that the solutions ω^2 of (101) are all real and the positive square roots are the eigenvalues ω_1 and ω_2 which represent the allowed (angular) frequencies of the system. To illustrate the basic solution procedure, we consider the classic double pendulum below in Example 10.

EXAMPLE 10 Determine the natural frequencies of oscillation for the double pendulum shown below in Fig. 11.4. Assume the masses are equal and the length of each rod is the same.

Solution: Clearly, the angles θ_1 and θ_2 are the generalized coordinates representing two degrees of freedom. If we take the rectangular coordinates of each mass to be (x_1, y_1) and (x_2, y_2), respectively, then

$$x_1 = b\sin\theta_1, \quad y_1 = b\cos\theta_1,$$
$$x_2 = b\sin\theta_1 + b\sin\theta_2,$$
$$y_2 = b\cos\theta_1 + b\cos\theta_2.$$

The total potential energy of the masses is given by

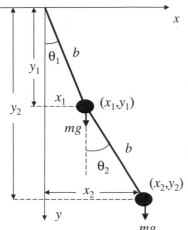

$$V(\theta_1,\theta_2) = -mgy_1 - mgy_2$$
$$= -mgb(2\cos\theta_1 + \cos\theta_2).$$

Using the small angle approximation

$$\cos x \cong 1 - \frac{1}{2}x^2 + \cdots,$$

we obtain (ignoring the additive constant)

$$V(\theta_1,\theta_2) \cong mgb\left(\theta_1^2 + \frac{1}{2}\theta_2^2\right).$$

Figure 11.4 Double pendulum.

Similarly, for the kinetic energy we find

$$T(\theta_1,\theta_2,\dot\theta_1,\dot\theta_2) = \frac{1}{2}m(\dot x_1^2 + \dot y_1^2) + \frac{1}{2}m(\dot x_2^2 + \dot y_2^2)$$
$$= mb^2\dot\theta_1^2 + mb^2\dot\theta_1\dot\theta_2\cos(\theta_1 - \theta_2) + \frac{1}{2}mb^2\dot\theta_2^2,$$

from which we deduce the quadratic approximation

$$T(\theta_1,\theta_2,\dot\theta_1,\dot\theta_2) \cong mb^2\dot\theta_1^2 + mb^2\dot\theta_1\dot\theta_2 + \frac{1}{2}mb^2\dot\theta_2^2.$$

Based on (95) and (96), the governing linear equations are

$$2\frac{d^2\theta_1}{dt^2} + \frac{d^2\theta_2}{dt^2} + 2\frac{g}{b}\theta_1 = 0,$$

$$\frac{d^2\theta_1}{dt^2} + \frac{d^2\theta_2}{dt^2} + \frac{g}{b}\theta_2 = 0.$$

By introducing the matrices

$$\mathbf{Q} = \begin{pmatrix} \theta_1 \\ \theta_2 \end{pmatrix}, \qquad \mathbf{A} = \begin{pmatrix} 2 & 1 \\ 1 & 1 \end{pmatrix}, \qquad \mathbf{B} = \begin{pmatrix} 2 & 0 \\ 0 & 1 \end{pmatrix},$$

we find

$$\det(\mathbf{B} - \lambda\mathbf{A}) = \begin{vmatrix} 2 - 2\lambda & -\lambda \\ -\lambda & 1 - \lambda \end{vmatrix} = 0,$$

where $\lambda = (b/g)\omega^2$. The eigenvalues λ are solutions of $\lambda^2 - 4\lambda + 2 = 0$, from which we calculate $\lambda_1 = 0.586$, $\lambda_2 = 3.414$. The corresponding (angular) frequencies of oscillation are found to be

$$\omega_1 = \sqrt{\lambda_1 g/b} = 0.765\sqrt{b/g}, \quad \omega_2 = \sqrt{\lambda_2 g/b} = 1.848\sqrt{b/g}.$$

Also, we note that the eigenvectors are

$$\mathbf{Q}_1 = \begin{pmatrix} 0.414 \\ 0.586 \end{pmatrix}, \quad \mathbf{Q}_2 = \begin{pmatrix} -2.414 \\ 3.414 \end{pmatrix},$$

and, hence, the general solution of the original linear system is

$$\begin{pmatrix} \theta_1 \\ \theta_2 \end{pmatrix} = \begin{pmatrix} 0.414 \\ 0.586 \end{pmatrix} C_1 \cos(\omega_1 t - \varphi_1) + \begin{pmatrix} -2.414 \\ 3.414 \end{pmatrix} C_2 \cos(\omega_2 t - \varphi_2),$$

where $C_1, C_2, \varphi_1, \varphi_2$ are constants to be determined by specifying a set of initial conditions.

11.9 Static Equilibrium of Deformable Bodies

An important special case of motion of a body takes place when the net external force acting on the body is zero. In this case we say the body is *at rest*, or in a *state of static equilibrium*. If, when a body is displaced slightly from its equilibrium position P it tends to return to position P, we say that P is a point of *stable equilibrium*. One of the most important principles of mechanics concerning the static equilibrium of elastic bodies is the *principle of minimum potential energy*:

"... of all the stable equilibrium configurations of an elastic body satisfying the prescribed displacement boundary conditions, the true configuration makes the total potential energy of the system, expressed in terms of displacements, a *minimum*."

If the equilibrium state is not stable, the work "minimum" is replaced by "stationary" in the above energy principle.

Remark: Also implied by the minimum energy principle is that the external forces must be *conserved*, i.e., derivable from a potential energy function.

Actually, in practice it is the *converse of the principle of minimum potential energy* that is most often used:

"... of all displacements of an elastic body satisfying prescribed boundary conditions, those which make the potential energy a minimum satisfy $\delta V = 0$, where V is the total potential energy of the body."

11.9.1 Deflections of an elastic string

It is perhaps instructive to consider the simple problem of finding the static equilibrium configuration of an elastic string supporting a distributed load. We assume the string is stretched tightly with tension τ (per unit length) between two supports located at $x = 0$ and $x = b$, and subjected to a distributed vertical force of intensity $q(x)$ per unit length. It is further assumed that the deflections are "small" and that gravity can be neglected.

The energy (per unit length) due to the distributed load $q(x)$ is given by the product $-q(x)y$ (force times displacement), where the minus sign is conventional. Thus, the expression

$$d\Omega = -q(x)y\,ds = -q(x)y\sqrt{1 + (y')^2}\,dx$$

represents the *external energy* per incremental arc length ds along the string. The *elastic energy* (also called the *strain energy*) due to internal forces is the product of the tension τ and the increase in length due to displacement; hence,

$$dU = \tau(ds - dx) = \tau\left[\sqrt{1 + (y')^2} - 1\right]dx \cong \frac{1}{2}\tau(y')^2\,dx,$$

where the last step follows by expanding the square root term in a binomial series and retaining only the first nonzero term in brackets (i.e., $|y'(x)| \ll 1$). The total potential energy of the string is the sum of external and elastic energies, which leads to

$$V = U + \Omega = \int_0^b\left[\frac{1}{2}\tau(y')^2 - q(x)y\right]dx. \tag{102}$$

Last, we assume the ends of the string are attached to fixed supports, which gives us the forced BCs

$$y(0) = 0, \quad y(b) = 0. \tag{103}$$

For stationary values of the potential energy V, we set $\delta V = 0$, which yields the Euler-Lagrange equation

$$y'' = -\frac{1}{\tau}q(x), \quad y(0) = 0, \ y(b) = 0. \tag{104}$$

The general solution of this linear DE is easily obtained by repeated integration of the right-hand side. Subjecting this general solution to the prescribed BCs then produces the desired stationary function $y = \Phi(x)$. For example, in the special case of a uniform loading in which $q(x) = q_0$ (constant), we obtain the well-known parabolic deflection shape given by

$$y = \frac{q_0}{2\tau}x(b - x). \tag{105}$$

11.9.2 Deflections of an elastic beam

A similar but more interesting case arises when a uniform beam of length b is subject to a distributed load of intensity $q(x)$. As for the elastic string problem, the load causes the beam to bend or be displaced from its equilibrium configuration along the x-axis. To determine the displacement curve (of the resulting centroidal axis), we assume the displacements are small and well within the elastic limit of the beam.

Like the elastic string, the total potential energy of the beam is also composed of the external energy plus the elastic (strain) energy. In the present problem, this total energy leads to (see [30])

$$V = \int_0^b \left[\frac{1}{2} EI(y'')^2 - q(x)y \right] dx, \tag{106}$$

where E is *Young's modulus* and I is the moment of inertia. The product EI is physically a measure of the "stiffness" of the beam. We recognize (106) as a specialization of the functional given by Eq. (45), and thus the governing Euler-Lagrange equation is

$$\frac{d^2}{dx^2}[EIy''] = q(x), \tag{107}$$

which is a fourth-order linear equation. For a uniform beam, the quantity EI is constant and the general solution can be found through four successive integrations of the right-hand side.

If we prescribe the forced BCs

$$y(0) = 0, \quad y(b) = 0, \tag{108}$$

it is still necessary to consider the natural BCs because four BCs are required. In this case, the natural BCs are

$$EIy''(0) = 0, \quad EIy''(b) = 0. \tag{109}$$

11.10 Two-Dimensional Variational Problems

In this section we apply Hamilton's principle to systems that involve a *continuous* distribution of mass—to be distinguished from a system composed of a discrete set of mass particles. The means for effecting this application is to replace sums over discrete particles by integrals over the continuous mass distributions. We will also extend the principle of minimum potential energy to continuum systems where more than one spatial coordinate is involved.

Our primary purpose in this section is to derive the governing partial differential equation (PDE) for several two-dimensional variational problems. Solution techniques are presented in Chapter 12.

11.10.1 Forced vibrations of an elastic string

Let us once again consider the stretched string problem of Section 11.9.1, but this time we assume the load acting on the string is also time-dependent. The string is then permitted to vibrate in the plane containing the x-axis in such a fashion that each particle of the string moves in a direction transverse to the x-axis. The amplitude of vibration u is assumed to be "small" (compared with the length of the string) in the sense that $|\partial u/\partial x| \ll 1$. Thus, the resulting PDE governing the vibrational modes will be *linear*. We further assume that no resistive forces (damping) are present, i.e., the system is *conservative*.

Let us denote the external load acting on a string of length b by $q(x,t)$. The transverse displacement $u(x,t)$ describes the shape of the string at all time t. By examining a small element of the string of length dx, we see that the kinetic energy of this element is given by

$$dT = \frac{1}{2}\rho(x)dx(u_t)^2,$$

where $\rho(x)$ is the mass density (per unit length) and u_t is the velocity of the small element of the string. Accordingly, the total kinetic energy is

$$T = \frac{1}{2}\int_0^b \rho(x)(u_t)^2\,dx. \tag{110}$$

Based on results given in Section 11.9.1, the total potential energy is

$$V = \int_0^b \left[\frac{1}{2}\tau(u_x)^2 - q(x,t)u\right]dx, \tag{111}$$

where τ is the (constant) tension in the string. Thus, the Lagrangian takes the form

$$L = T - V = \int_0^b \left[\frac{1}{2}\rho(x)(u_t)^2 - \frac{1}{2}\tau(u_x)^2 + q(x,t)u\right]dx. \tag{112}$$

From Hamilton's principle, the integral whose stationary values we seek is

$$I = \int_{t_1}^{t_2} L\,dt = \int_{t_1}^{t_2}\int_0^b \left[\frac{1}{2}\rho(x)(u_t)^2 - \frac{1}{2}\tau(u_x)^2 + q(x,t)u\right]dx\,dt, \tag{113}$$

which is of the form discussed in Section 11.5.3. Hence, the governing Euler-Lagrange equation leads to

$$\rho(x)u_{tt} = \tau u_{xx} + q(x,t),\quad 0 < x < b,\ t > 0, \tag{114}$$

which is the *one-dimensional wave equation* with forcing function. For the special case in which the forcing function is identically zero and the mass density is constant, i.e., $\rho(x) = \rho_0$, this equation reduces to the standard form of the wave equation given by

$$u_{xx} = c^{-2}u_{tt},\quad 0 < x < b,\ t > 0, \tag{115}$$

where $c = \sqrt{\tau/\rho_0}$ is a constant with the dimension of velocity.

To be a "properly posed problem," the solution of (114) or (115) must satisfy certain prescribed BCs and initial conditions (ICs). If we assume the ends of the string are fixed, then we prescribe the forced BCs

$$u(0,t) = 0,\quad u(b,t) = 0,\quad t > 0. \tag{116}$$

The associated ICs take the general form

$$u(x,0) = f(x), \quad u_t(x,0) = g(x), \quad 0 < x < b, \tag{117}$$

which describes the initial configuration of the string and the velocity at each point along its length. We solve this problem in Chapter 12.

11.10.2 Equilibrium of a stretched membrane

We now wish to determine the equation of equilibrium of a stretched membrane (in the xy-plane) with a large isotropic tension τ applied around the boundary curve C. We assume the tension is essentially constant and that a distributed load $q(x,y)$, independent of time, is applied over the surface of the membrane. The equilibrium position of the membrane is obtained by minimizing the total potential energy V.

The tension τ is specified in units of energy per unit area, commonly called "surface tension." The elastic energy stored during a small displacement is $\tau \Delta A$, where ΔA represents the change in surface area. If dS represents the surface area after deformation of a small element dA, then

$$\Delta A = dS - dA = \left[\sqrt{1 + (u_x)^2 + (u_y)^2} - 1 \right] dA.$$

Under the assumption of small displacements, we can expand the above square root in a binomial series to obtain

$$\Delta A \cong \frac{1}{2} \left[(u_x)^2 + (u_y)^2 \right] dA,$$

and thus, the total strain energy of the membrane becomes

$$U = \frac{\tau}{2} \iint_D \left[(u_x)^2 + (u_y)^2 \right] dA, \tag{118}$$

where D is the domain inside the boundary curve C. The external potential energy due to the distributed load is similarly given by

$$\Omega = -\iint_D q(x,y) u \, dA, \tag{119}$$

and, consequently, the sum of strain and external potential energies yields

$$V = U + \Omega = \iint_D \left\{ \frac{\tau}{2} \left[(u_x)^2 + (u_y)^2 \right] - q(x,y) u \right\} dA. \tag{120}$$

Based on this expression, the principle of minimum potential energy leads to the governing PDE

$$u_{xx} + u_{yy} = -\frac{1}{\tau} q(x,y), \tag{121}$$

known as *Poisson's equation*. In the absence of the distributed load, this PDE reduces to *Laplace's equation*

$$u_{xx} + u_{yy} = 0. \tag{122}$$

Because neither (121) nor (122) depend on time t, ICs are not appropriate to prescribe. The standard BC that is prescribed is the *Dirichlet condition*

$$u = f \text{ on } C. \tag{123}$$

Hence, the solution of (122) subject to the BC (123) is called a *Dirichlet problem*. It is considered the oldest boundary value problem and thus, is also called the *boundary value problem of the first kind* (see also Chapter 12).

Suggested Reading

G. A. Bliss, *Calculus of Variations* (The Mathematical Association of America: Open Court Publishing, Chicago, 1925).

B. A. Finlayson, *Method of Weighted Residuals and Variational Principles* (Academic Press, New York, 1972).

I. M. Gelfand and S. V. Fomin, *Calculus of Variations*, trans. by R. Silverman (Prentice-Hall, Englewood Cliffs, 1963).

C. Lanczos, *The Variational Principles of Mechanics* (U. Toronto Press, Toronto, 1970).

H. Sagan, *Introduction to the Calculus of Variations* (McGraw-Hill, New York, 1969).

Chapter 11

Exercises

In Problems 1-4, find the stationary function and stationary value of the given variational problem.

1. $\int_0^1 \dfrac{\sqrt{1 + (y')^2}}{y} dx$, $y(0) = 1$, $y(1) = 2$

Ans. $(x - 2)^2 + y^2 = 5$, $\dfrac{1}{2} \ln \dfrac{3 + \sqrt{5}}{3 - \sqrt{5}}$

2. $\int_0^{\pi/8} [(y')^2 + 2yy' - 16y^2] dx$,
$y(0) = 0$, $y(\pi/8) = 2$

Ans. $y = 2\sin 4x$, 4

3. $\int_0^1 \dfrac{1 + y^2}{(y')^2} dx$, $y(0) = 0$, $y(1) = \sinh 1$

Ans. $y = \sinh x$, 1

4. $\int_1^2 [2y^2 + x^2(y')^2] dx$

Ans. $y = 4/x^2$, 0

5. Find the Euler-Lagrange equation and natural BC for

$$I = \int_0^1 [(y')^2 + 4\cos y] dx.$$

Ans. $y'' + 2\sin y = 0$,
$\quad\quad y'(0) = 0, \; y'(1) = 0$

6. A particle starts from rest at $(0,0)$ and traverses a brachistochrone to the line $x + y = 1$.

(*a*) Determine the curve and point of intersection with the given line.

(*b*) Find the time of travel.

Ans. (*a*) $x = \dfrac{2}{\pi}(\theta - \sin\theta)$,

$\quad\quad y = \dfrac{2}{\pi}(1 - \cos\theta)$;

$\quad\quad \left(1 - \dfrac{2}{\pi}, \dfrac{2}{\pi}\right)$

(*b*) $\sqrt{\pi/2g}$

In Problems 7 and 8, find the stationary function and value of the given variational problem.

7. $\displaystyle\int_0^1 [(y')^2 + (z')^2 + y'z']dx$,
$\quad y(0) = 1, y(1) = 0, z(0) = -1, z(1) = 2$

Ans. $y = 1 - x, \; z = 3x - 1$

8. $\displaystyle\int_0^1 [(y')^2 + (z')^2 + y^2 + z^2]dx$,
$\quad y(0) = 1, y(1) = \sinh 1,$
$\quad z(0) = 0, z(1) = \sinh 1$

Ans. $y = z = \sinh x$

9. Find a stationary function of
$\displaystyle\int_0^1 [(y')^2 + 2y]dx, \; y(0) = 0, y(1) = \dfrac{1}{2}$

subject to the constraint

$$\int_0^1 2y\,dx = \dfrac{1}{6}.$$

Ans. $y = x^2 - \dfrac{1}{2}x$

10. Find the Euler-Lagrange equation for the isoperimetric problem

$$\int_{-1}^1 (1 - x^2)(y')^2\,dx,$$

$$y(-1) = 1, \; y(1) = 1,$$

$$\int_{-1}^1 y^2\,dx = 1.$$

Ans. $(1 - x^2)y'' - 2xy' + \lambda y = 0$

11. Find the Euler-Lagrange equation for the isoperimetric problem

$$\int_0^1 x(y')^2\,dx, \; |y(0)| < \infty, y(1) = 0$$

$$\int_0^1 xy^2\,dx = 1.$$

Ans. $xy'' + y' + \lambda y = 0$

In Problems 12 and 13, find the eigenvalues and eigenfunctions of the given isoperimetric problem.

12. $\displaystyle\int_0^\pi (y')^2\,dx, \; y(\pi) = 0$,
$\displaystyle\int_0^\pi y^2\,dx = $ const.

13. $\displaystyle\int_0^1 e^{2x}(y')^2\,dx, \; y(0) = 0$,
$\displaystyle\int_0^1 y^2\,dx = $ const.

14. Given the reduced variational problem

$$\int_0^1 (y'' + xy + x)\delta y\,dx = 0,$$

$$y(0) = 0, \; y(1) = 0,$$

(*a*) find a one-term Ritz approximation $\psi = c_1 x(1 - x)$.

(*a*) find a two-term Ritz approximation $\psi = c_1 x(1 - x) + c_2 x^2(1 - x)$.

Ans. (*a*) $c_1 = 0.263$
$\quad\quad$ (*b*) $c_1 = 0.177, \; c_2 = 0.173$

15. Use the trial solution

$$\psi = 2x - 1 + c_1 \sin \pi x$$

and the Ritz method to approximate the minimizing curve of

$$I = \frac{1}{2} \int_0^1 [(y')^2 - y^2 + 2(x - 1)] dx,$$

$$y(0) = -1, \ y(1) = 1.$$

Ans. $c_1 = \dfrac{2}{\pi(\pi^2 - 1)} \cong 0.0718$

16. The equilibrium configuration of a uniform beam resting on an elastic foundation leads to the potential energy variational integral

$$V = \frac{1}{2} \int_0^1 [(y'')^2 - y^2 - x^3 y] dx,$$

$$y(0) = 0, \ y''(0) = 0.$$

(*a*) Solve by the Ritz method and trial solution $y = c_1 x + c_2 x^3$.
(*b*) Find the exact solution.

Ans. (*a*) $c_1 = -\dfrac{315}{1048}, \ c_2 = \dfrac{1}{1048}$
(*b*) $y = 1.362 \sinh x - 1.663 \sin x$
$\qquad - 0.5x^3$

17. Use the Ritz method and trial solution $\psi = c_1 x(1 - x) + c_2 x^2(1 - x)$ to find approximations to the first *two* eigenvalues and eigenfunctions of the isoperimetric problem

$$\int_0^1 (y')^2 dx, \ y(0) = 0 \ y(1) = 0,$$

$$\int_0^1 xy^2 dx = 1.$$

Ans. $\Lambda_1 = 19.19, \ \Lambda_2 = 102.1$
$\psi^{(1)} = 5.21 x(1 - x) + 4.37 x^2(1 - x)$
$\psi^{(2)} = 25.94 x(1 - x) - 44.92 x^2(1 - x)$

In Problems 18 and 19, determine the Lagrange equation(s) for the given system.

18. (Compound pendulum):

$$T = \frac{I}{2} \dot\theta^2, \ V = mgb(1 - \cos \theta),$$

where I is moment of inertia.

Ans. $I \dfrac{d^2\theta}{dt^2} + mgb \sin \theta = 0$

19. (Particle acted on by springs with spring constants k_1 and k_2):

$$T = \frac{m}{2} (\dot x^2 + \dot y^2),$$

$$V = \frac{1}{2} (k_1^2 x^2 + k_2^2 y^2)$$

Ans. $m \dfrac{d^2 x}{dt^2} + k_1 x = 0,$

$\qquad m \dfrac{d^2 y}{dt^2} + k_2 y = 0$

In Problems 20 and 21, find a complete solution to the system of Lagrange equations associated with the given variational problem.

20. $I = \int_\pi^{2\pi} (2xy + \dot x^2 + \dot y^2) dt$

Ans. $\begin{pmatrix} x \\ y \end{pmatrix} = C_1 \begin{pmatrix} 1 \\ 1 \end{pmatrix} e^{-t} + C_2 \begin{pmatrix} 1 \\ 1 \end{pmatrix} e^{t}$

$\qquad + C_3 \begin{pmatrix} 1 \\ -1 \end{pmatrix} \cos t + C_4 \begin{pmatrix} 1 \\ -1 \end{pmatrix} \sin t$

21. $I = \int_2^8 (6xy + 5x^2 + 5y^2$
$\qquad\qquad + 2\dot x^2 + 2\dot y^2) dt$

Ans. $\begin{pmatrix} x \\ y \end{pmatrix} = C_1 \begin{pmatrix} 1 \\ -1 \end{pmatrix} e^{-t} + C_2 \begin{pmatrix} 1 \\ -1 \end{pmatrix} e^{t}$

$\qquad + C_3 \begin{pmatrix} 1 \\ 1 \end{pmatrix} e^{-2t} + C_4 \begin{pmatrix} 1 \\ 1 \end{pmatrix} e^{t}$

22. A rod of length b vibrates longitudinally. Assume $u(x,t)$ is the displacement of any point on the rod originally at distance x from one end, and the kinetic and potential energies are given by

$$T = \frac{1}{2}\rho \int_0^b (u_t)^2 \, dx,$$

$$V = EA \int_0^b (u_x)^2 \, dx,$$

where ρ is the mass density, E is the modulus of elasticity, and A is the cross sectional area. Show that the governing Euler-Lagrange equation is

$$\rho u_{tt} = EA u_{xx}.$$

23. Let an elastic membrane cover a region in the xy-plane and let it vibrate in a direction perpendicular to the plane. Given that the kinetic and potential energies are

$$T = \frac{1}{\rho} \iint_R (u_t)^2 \, dxdy$$

$$V = \frac{1}{2}\tau \iint_R [(u_x)^2 + (u_y)^2] \, dxdy,$$

show that the governing partial differential equation is

$$u_{tt} = a^2(u_{xx} + u_{yy}).$$

24. Consider the double pendulum problem in Example 10 for the case where the masses satisfy $m_2 = 2m_1 = 2m$ and the rod lengths are $b_2 = 3b_1 = 3b$.

(a) Find the angular frequencies of vibration.

(b) What are the corresponding eigenvectors?

$Ans.\,(a)\;\omega_1 = 0.518\sqrt{\dfrac{g}{b}},\;\; \omega_2 = 1.932\sqrt{\dfrac{g}{b}}$

$(b)\;\mathbf{Q}_1 = \begin{pmatrix} 0.536 \\ 0.732 \end{pmatrix},\;\; \mathbf{Q}_2 = \begin{pmatrix} -7.464 \\ 2.732 \end{pmatrix}$

25. Consider the simple pendulum shown in the figure below whose support m_1 is free to translate in the x-direction along a frictionless surface. Using x and θ as generalized coordinates,

(a) show that the potential energy and kinetic energy for "small movements" are given by

$$T_1 = \frac{1}{2}m_1 \dot{x}^2,$$

$$T_2 = \frac{1}{2}m_2(\dot{x}^2 + 2b\dot{x}\dot{\theta} + b^2\dot{\theta}^2),$$

$$V = -m_2 gb.$$

(b) find the governing Lagrange equations.

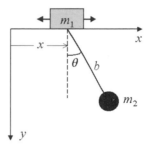

Ans. (b) $(m_1 + m_2)\dot{x} + m_2 b\dot{\theta} = \text{const.}$

$\frac{1}{2}(m_1 + m_2)\dot{x} - \frac{1}{2}m_2 b^2\dot{\theta}^2 + m_2 b\dot{\theta} = \text{const.}$

26. The free vibrations of a linear triatomic molecule has potential energy and kinetic energy described by

$$V = \frac{k}{2}[(x_2 - x_1)^2 + (x_3 - x_4)^2],$$

$$T = \frac{m}{2}(\dot{x}_1^2 + \dot{x}_3^2) + \frac{M}{2}\dot{x}_2^2,$$

where m is the mass of the first and third molecule, M is the mass of the second molecule, and k is an equivalent "spring constant." Find the angular frequencies of vibration.

$$Ans. \ \omega_1 = 0, \ \omega_2 = \sqrt{\frac{k}{m}},$$

$$\omega_3 = \sqrt{\frac{k}{m}\left(1 + \frac{2m}{M}\right)}$$

27. The anaharmonic oscillator is a classical problem in quantum mechanics. A particular form of the problem leads to the equation

$$\frac{h^2}{2m}y'' - \frac{1}{2}m\omega^2 x^2 y - \varepsilon x^4 y + \lambda y = 0,$$

where all parameters are constants. The parameter h is Planck's constant, m is the mass, ω is angular frequency, λ denotes energy level, and ε is a small perturbation constant. We seek bounded solutions (as $x \to \pm\infty$) of this equation. The eigenvalues λ_n represent the "allowed" energy levels of the system, the ground state being the first eigenvalue. When $\varepsilon = 0$, the ground state is $\lambda_0 = 1/2h\omega$. For $\varepsilon \neq 0$, use the Ritz method with trial solution $\psi = \exp(-m\omega x^2/2h)$ to find an approximation to the ground state energy.

$$Ans. \ \Lambda_0 = \frac{1}{2}h\omega + \varepsilon\frac{3h^2}{4m^2\omega^2}$$

28. The path of a ray of light in the neighborhood of the sun is the minimizing curve of

$$I = \int_{\theta_1}^{\theta_2} \sqrt{gr^2 + g^2(dr/d\theta)^2}\, d\theta,$$

where $g = 1/(1 - k/r)$ (k constant). Show that this path satisfies the DE

$$. \ \frac{d^2R}{d\theta^2} + R = \frac{3}{2}kR^2, \ R = 1/r.$$

Chapter 12

Partial Differential Equations

12.1 Introduction / 591
12.2 Classification of Second-Order PDEs / 591
12.3 The Heat Equation / 592
    12.3.1 Homogeneous boundary conditions / 593
    12.3.2 Nonhomogeneous boundary conditions / 596
    12.3.3 Derivation of the heat equation / 599
12.4 The Wave Equation / 600
    12.4.1 d'Alembert's solution / 602
12.5 The Equation of Laplace / 604
    12.5.1 Rectangular domain / 605
    12.5.2 Circular domain / 608
    12.5.3 Maximum-minimum principle / 610
12.6 Generalized Fourier Series / 611
    12.6.1 Convective heat transfer at one endpoint / 611
    12.6.2 Nonhomogeneous heat equation / 612
    12.6.3 Nonhomogeneous wave equation / 615
12.7 Applications Involving Bessel Functions / 617
    12.7.1 Vibrating membrane / 617
    12.7.2 Scattering of plane waves by a circular cylinder / 618
12.8 Transform Methods / 621
    12.8.1 Heat conduction on an infinite domain: Fourier transform / 622
    12.8.2 Heat cnduction on a semi-infinite domain: Laplace transform / 624
    12.8.3 Nonhomogeneous wave equation / 626
    12.8.4 Poisson integral formula for the half-plane / 628
    12.8.5 Axisymmetric Dirichlet problem for a half-space: Hankel transform / 630
Suggested Reading / 631
Exercises / 632

Historical Comments: In 1727, John Bernoulli (1667-1748) treated the vibrating string problem by imagining the string to be a thin thread having a number of equally spaced weights placed along it. However, because his governing equation was not time-dependent, it was not truly a partial differential equation. The French mathematician Jean Le Rond d'Alembert (1717-1783) derived the one-dimensional wave equation as we know it today by letting the number of weights in Bernoulli's model become infinite while at the same time allowing the space between them to go to zero. His famous solution, widely known as d'Alembert's solution, appeared around 1746, six years before the separation of variables technique was introduced by Daniel Bernoulli (1700-1782). Leonhard Euler (1707-1783), D. Bernoulli, and Joseph Louis Lagrange (1736-1813) all solved the wave equation in the mid-1700's by the method of separating the variables, also called the Bernoulli product method.

Laplace's equation arose in the study of gravitational attraction by Pierre-Simon Laplace (1749-1827) and Adrien-Marie Legendre (1752-1833), both of whom were professors of mathematics at the Ecole Militaire in France. In fact, it was Legendre's famous 1782 study of gravitational attraction of spheroids that introduced what are now called Legendre polynomials (see Chapter 2).

The first major step toward developing a general method of solution of partial differential equations began in the early 1800's when Joseph Fourier (1768-1830) made his famous study of the heat equation. He is best known today for his celebrated 1822 book *Théorie analytique de la chaleur*. The chief contribution of Fourier was the idea that almost any function can be represented in a sine series.

Mathematicians in the nineteenth century worked vigorously on problems associated with Laplace's equation, trying to extend the work of Legendre and Laplace. Despite this great effort by well-known mathematicians, very little was known about the general properties of the solutions of Laplace's equation until 1828 when George Green (1793-1841), a self-taught British mathematician whose main interest was in electricity and magnetism, and the Russian mathematician Michel Ostrogradsky (1801-1861) independently studied properties of a class of solutions known as harmonic functions. In the second half of the nineteenth century, mathematicians began to make progress on problems concerning the existence of solutions to partial differential equations.

Historical Comments (Cont'd):

      Throughout the years, a host of mathematicians have played a role in the general development of the theory of ordinary and partial differential equations, but their names are too numerous to mention here. Motivated in part by the many areas of application in mathematics, engineering, and science, and an ever-increasing need to solve more and more complicated equations (both linear and nonlinear), partial differential equations will remain an active area of research for many more years.

Our objective in this chapter is to discuss the basic solution techniques associated with the heat equation, wave equation, and Laplace's equation. Each of these three partial differential equations is a simple representative of a certain class of equations, and hence, provides much insight into the general behavior of other equations in the class. Integral transform techniques are utilized for problems in which the spatial domain becomes unbounded.

12.1 Introduction

In general, *partial differential equations* (PDEs) are prominent in those physical and geometrical problems involving functions that depend on more than one independent variable. These variables may involve time and/or one or more spatial coordinates. Our treatment of such equations is intentionally brief, however, focusing on only the basic equations of *heat conduction*, *wave phenomena*, and *potential theory*. These equations are simple representatives of three important classes of PDEs that include most of the equations of mathematical physics.

12.2 Classification of Second-Order PDEs

PDEs are classified as to order and linearity in much the same way as ordinary DEs (ODEs) in Chapter 1. For example, the *order* of a PDE is the order of the partial derivative of highest order appearing in the equation; the equation is *linear* if the unknown function and all its derivatives are of the first degree algebraically. The general linear PDE of second order in two variables has the form

$$Au_{xx} + Bu_{xy} + Cu_{yy} + Du_x + Eu_y + Gu = F, \tag{1}$$

where $A, B, C, ..., F$ may be functions of x and y, but not u. Here, we are using the notation u_x for $\partial u/\partial x$, u_{xy} for $\partial^2 u/\partial y \partial x$, and so on. If $F(x,y) \equiv 0$, we say that (1) is *homogeneous*; otherwise, it is *nonhomogeneous*.

      By *solution* of a PDE, we mean a function u that has all partial derivatives occurring in the PDE that, when substituted into the equation, reduces it to an identity. A *general solution* of a PDE is a collection of all the solutions. However, the general solution of a PDE involves *arbitrary functions* rather than *arbitrary constants* as is the case for ODEs. For example, the general solution of $2u_x - u_y = 0$ is $u(x,y) = G(x + 2y)$, where G is any differentiable function of $x + 2y$. Because of this, the methods of solution of a PDE are generally very different than those used in solving ODEs.

The general second-order PDE (1) is classified according to the following scheme:

$$
\begin{aligned}
B^2 - 4AC &= 0 \quad \text{(parabolic)} \\
B^2 - 4AC &> 0 \quad \text{(hyperbolic)} \\
B^2 - 4AC &< 0 \quad \text{(elliptic)}
\end{aligned}
$$

Equations belonging to the same classification exhibit many features in common, and thus, by solving a particular member of any class we are exposed to certain fundamental characteristics shared by all members of that class. This classification scheme is also important when numerical techniques are used for finding solutions.

When the coefficients in (1) are functions of x and y, the classification of an equation can vary from region to region. For instance, the PDE

$$u_{xx} - xu_{yy} + u = 0,$$

where $B^2 - 4AC = 4x$, is of hyperbolic type for $x > 0$, elliptic for $x < 0$, and parabolic along the y-axis.

The PDEs of greatest interest in mathematical physics are simple examples from each category of the above classification scheme. These equations are

$$
\begin{aligned}
\text{Heat equation:} \quad & u_{xx} = a^{-2} u_t \quad \text{(parabolic)} \\
\text{Wave equation:} \quad & u_{xx} = c^{-2} u_{tt} \quad \text{(hyperbolic)} \\
\text{Laplace equation:} \quad & u_{xx} + u_{yy} = 0 \quad \text{(elliptic)} \\
\text{Helmholtz equation:} \quad & u_{xx} + u_{yy} + k^2 u = 0 \quad \text{(elliptic)}
\end{aligned}
$$

The *heat equation* governs diffusion processes, whereas the *wave equation* is fundamental in oscillation problems like vibrating strings and membranes. *Laplace's equation*, also called the *potential equation*, is basic to steady-state heat conduction involving homogeneous solids; gravitational, magnetic, and electrostatic potential problems; steady flow of currents in solid conductors; and the velocity potential of inviscid, irrotational fluids; among other applications. The *Helmholtz equation*, sometimes called the reduced wave equation, is associated with the spatial behavior of both the heat and wave equations.

12.3 The Heat Equation

In the study of the flow of heat in thermally conducting regions, the governing PDE is the *heat equation*. This same equation arises, however, in other diffusion processes and is therefore also widely known as the *diffusion equation*.

In its three-dimensional form, the heat equation is written as

$$\nabla^2 u = a^{-2} u_t, \tag{2}$$

where u is the temperature, a^2 is a physical constant known as the *diffusivity*, and ∇^2 is the *Laplacian operator*. In rectangular coordinates we have

$$\nabla^2 u = u_{xx} + u_{yy} + u_{zz}. \tag{3}$$

Heat transfer is the science that predicts energy transfer in material bodies owing to temperature differences. It takes place in one of three fundamental modes: *conduction*, *convection*, or *radiation*. Of these, we are primarily concerned with conduction. The fundamental problem in the mathematical theory of heat conduction is to solve (2) for the temperature u in a homogeneous solid when the distribution of temperature throughout the solid is known at time $t = 0$, called an *initial condition* (IC). In addition, the solution u must satisfy a *boundary condition* (BC) prescribed at each exposed point of the solid which is ordinarily one of three distinct kinds:

- *Dirichlet condition*—specifies the temperature along each finite surface

- *Neumann condition*—specifies the heat flow across the boundary surface

- *Robin's condition*—specifies the rate at which heat is lost from the solid due to convection (*Newton's law of cooling*)

In this section we concern ourselves with the one-dimensional analogue of (2) in which the temperature $u = u(x,t)$ depends only on one spatial variable x and on time t (see Section 12.3.3 for a derivation of the governing PDE). For example, this form of the heat equation governs the temperature distribution in a long rod or wire of length p whose lateral surface is insulated (see Fig. 12.1). The *one-dimensional* form of (2) for this case and the temperature of the rod at time $t = 0$ are prescribed, respectively, by

$$u_{xx} = a^{-2} u_t, \quad 0 < x < p, \, t > 0, \tag{4}$$

$$\text{IC:} \quad u(x,0) = f(x), \quad 0 < x < p. \tag{5}$$

12.3.1 Homogeneous boundary conditions

To begin, let us assume the initial temperature is described by (5) and the ends of the rod are "immediately" placed in contact with ice packs at 0°C and maintained at this temperature at all later times. The BCs are therefore described by the Dirichlet conditions

$$\text{BC:} \quad u(x,0) = 0, \, u(x,p) = 0, \quad t > 0. \tag{6}$$

$x = 0$ $x = p$

x

Figure 12.1 Rod of finite length.

It is customary to solve PDEs of the type (4) by the technique of *separation of variables*; that is, we seek solutions of the product form

$$u(x,t) = X(x)W(t). \tag{7}$$

The substitution of (7) into the heat equation (4), followed by some algebraic manipulation, leads to the separated form $[X'' = d^2X/dx^2,\ W' = dW/dt]$

$$\frac{X''(x)}{X(x)} = \frac{W'(t)}{a^2 W(t)}. \tag{8}$$

In (8), we note that the left-hand side is a function of x alone and the right-hand side is a function of t alone. Since these are independent variables, we conclude that each ratio must be a constant, which we can equate to a common constant, $-\lambda$; that is,

$$\frac{X''(x)}{X(x)} = -\lambda, \quad \frac{W'(t)}{a^2 W(t)} = -\lambda. \tag{9}$$

The negative sign in the separation constant is not required, but conventional. These last results lead to separate ODEs for the unknown factors, which are

$$X'' + \lambda X = 0, \quad 0 < x < p,$$
$$W' + \lambda a^2 W + 0, \quad t > 0. \tag{10}$$

Under the assumption (7), the prescribed BCs take the form

$$u(0,t) = X(0)W(t) = 0, \quad u(p,t) = X(p)W(t) = 0, \tag{11}$$

or

$$X(0) = 0, \quad X(p) = 0. \tag{12}$$

The first ODE in (10), subject to the BCs (12), forms an *eigenvalue problem* of the type studied in Chapter 8. In particular, based on Example 6 in Section 8.4, we know that the eigenvalues and eigenfunctions are given by

$$\lambda_n = \frac{n^2\pi^2}{p^2}, \quad X(x) \equiv \varphi_n(x) = \sin\frac{n\pi x}{p}, \quad n = 1,2,3,\dots. \tag{13}$$

With the separation constant λ restricted to the values in (13), it follows that the solution of the second DE in (10) is given by

$$W_n(t) = c_n e^{-a^2 n^2 \pi^2 t/p^2}, \quad n = 1,2,3,\dots, \tag{14}$$

where the c's are arbitrary constants.

From (13) and (14), we generate the family of solutions

$$u_n(x,t) = \varphi_n(x)W_n(t), \quad n = 1,2,3,\ldots, \tag{15}$$

and by summing over all solutions (15), we obtain the series solution

$$u(x,t) = \sum_{n=1}^{\infty} \varphi_n(x)W_n(t) = \sum_{n=1}^{\infty} c_n \sin\left(\frac{n\pi x}{p}\right) e^{-a^2 n^2 \pi^2 t/p^2}. \tag{16}$$

Equation (16) represents a solution of the heat equation (4) and prescribed BCs (6) for any choice of constants c_n, $n = 1,2,3,\ldots$. Summing solutions in this fashion is a generalization of the *superposition principle* from ODEs (see Section 1.4.1) to an infinite number of solutions. Note that $u(x,t) \to 0$ as $t \to \infty$ for any choice of constants, in agreement with physical conditions (i.e., the BCs demand that the steady-state solution go to zero). Last, to satisfy the prescribed IC (5), we set $t = 0$ in (16) to find

$$u(x,0) = f(x) = \sum_{n=1}^{\infty} c_n \sin\frac{n\pi x}{p}, \quad 0 < x < p, \tag{17}$$

which we recognize as a Fourier *sine series* (see Section 8.2). Therefore, the constants in (17) are defined by

$$c_n = \frac{2}{p} \int_0^p f(x)\sin\frac{n\pi x}{p}\,dx, \quad n = 1,2,3,\ldots. \tag{18}$$

Equation (16) is what we call a *formal solution* of the heat equation with prescribed IC and BCs. However, it has been shown that the series (16) will converge to a *strict solution* of the heat equation, provided the initial data $f(x)$ is merely *piecewise continuous*—a consequence of the "smoothing process" of heat conduction (or diffusion).

EXAMPLE 1 Solve the heat conduction problem

$$u_{xx} = a^{-2}u_t, \quad 0 < x < \pi, \, t > 0,$$

BC: $\quad u_x(0,t) = 0, \, u_x(\pi,t) = 0, \quad t > 0,$

IC: $\quad u(x,0) = x, \quad 0 < x < \pi.$

Solution: This problem is a variation of that described above, although the BCs are still homogeneous. We can interpret this problem physically as that corresponding to the situation where the rod is initially heated in a linear fashion prescribed by the temperature $f(x) = x$, while the ends at $x = 0$ and $x = \pi$ are insulated (i.e., there is no heat flow into the surrounding medium at the ends).

The separation of variables this time leads to the eigenvalue problem and time-dependent equation given, respectively, by

$$X'' + \lambda X = 0, \quad X'(0) = 0, \, X'(\pi) = 0,$$

$$W' + \lambda a^2 W = 0.$$

Next, following the method illustrated in Section 8.4, we readily find that the above eigenvalue problem has eigenvalues and eigenfunctions given by

$$\lambda_0 = 0, \quad \varphi_0(x) = 1,$$
$$\lambda_n = n^2, \quad \varphi_n(x) = \cos nx, \quad n = 1, 2, 3, \dots .$$

With this restriction on the separation constant, the time-dependent term yields

$$T_0(t) = c_0, \quad T_n(t) = c_n e^{-a^2 n^2 t}, \quad n = 1, 2, 3, \dots ,$$

and hence, by use of the superposition principle we get

$$u(x,t) = c_0 + \sum_{n=1}^{\infty} c_n \cos(nx) e^{-a^2 n^2 t}.$$

By imposing the prescribed IC, this gives us

$$u(x,0) = x = c_0 + \sum_{n=1}^{\infty} c_n \cos nx,$$

which we recognize as a cosine series. Hence,

$$c_0 = \frac{\pi}{2}, \quad c_n = \begin{cases} -\dfrac{4}{\pi n^2}, & n = 1,3,5,\dots \\ 0, & n = 2,4,6,\dots \end{cases}$$

and the solution we seek is

$$u(x,t) = \frac{\pi}{2} - \frac{4}{\pi} \sum_{\substack{n=1 \\ (\text{odd})}}^{\infty} \frac{\cos nx}{n^2} e^{-a^2 n^2 t}.$$

12.3.2 Nonhomogeneous boundary conditions

Suppose we now consider the problem described by

$$
\begin{array}{ll}
 & u_{xx} = a^{-2} u_t, \quad 0 < x < p, \, t > 0, \\
\text{BC:} & u(0,t) = T_1, \, u(p,t) = T_2, \quad t > 0, \\
\text{IC:} & u(x,0) = f(x), \quad 0 < x < p,
\end{array} \tag{19}
$$

where T_1 and T_2 are constants. When these constants are both zero, the problem reduces to that in the previous section so we assume that at least one of them is nonzero. Physically, this problem corresponds to holding the endpoints at constant temperatures other than 0°C.

In this problem the separation of variables will not work in a direct manner. For this reason, we use a modified approach here. Based on physical observations, we expect the temperature $u(x,t)$ to approach a *steady-state temperature function $S(x)$* in the limit as $t \to \infty$.

Hence, the steady-state solution must satisfy the simple DE $u_{xx} = S'' = 0$, which the heat equation reduces to in the steady-state situation. Therefore, we assume the solution of (19) can be expressed in the form

$$u(x,t) = S(x) + v(x,t),\tag{20}$$

where $v(x,t)$, called the *transient solution*, must vanish in the limit $t \to \infty$. The substitution of (20) into (19) leads to

$$S'' + v_{xx} = a^{-2}v_t,$$
$$\text{BC:} \quad S(0) + v(0,t) = T_1,\ S(p) + v(p,t) = T_2,\tag{21}$$
$$\text{IC:} \quad S(x) + v(x,0) = f(x).$$

By taking the limit $t \to \infty$ of the system (21), the transient terms all vanish in the DE and BC, and we are left with the *steady-state problem*

$$S'' = 0, \quad S(0) = T_1,\ S(p) = T_2.\tag{22}$$

The solution of (22) is readily found to be the linear function

$$S(x) = T_1 + (T_2 - T_1)\frac{x}{p}.\tag{23}$$

Having found the steady-state solution (23), the remaining transient problem deduced from (21) is

$$v_{xx} = a^{-2}v_t,$$
$$\text{BC:} \quad v(0,t) = 0,\ v(p,t) = 0,\tag{24}$$
$$\text{IC:} \quad v(x,0) = f(x) - T_1 - (T_2 - T_1)\frac{x}{p}.$$

Here we see that the transient solution $v(x,t)$ satisfies a heat conduction problem of the type for which the BCs are homogeneous. Hence, the separation of variables method applied to (24) leads to the formal solution

$$v(x,t) = \sum_{n=1}^{\infty} c_n \sin\left(\frac{n\pi x}{p}\right) e^{-a^2 n^2 \pi^2 t/p^2}.\tag{25}$$

If we impose the IC in (24), it follows that

$$c_n = \frac{2}{p}\int_0^p \left[f(x) - T_1 - (T_2 - T_1)\frac{x}{p}\right]\sin\frac{n\pi x}{p}\,dx$$
$$= \frac{2}{p}\int_0^p f(x)\sin\frac{n\pi x}{p}\,dx + \frac{2}{n\pi}\left[(-1)^n T_2 - T_1\right], \quad n = 1,2,3,\dots.\tag{26}$$

Combining the steady-state and transient solutions, we get

$$\boxed{\begin{aligned} u(x,t) &= S(x) + v(x,t) \\ &= T_1 + (T_2 - T_1)\frac{x}{p} + \sum_{n=1}^{\infty} c_n \sin\left(\frac{n\pi x}{p}\right) e^{-a^2 n^2 \pi^2 t/p^2}, \end{aligned}}$$

(27)

where the c's are defined by (26).

EXAMPLE 2 Solve the heat conduction problem

$$u_{xx} = a^{-2}u_t, \quad 0 < x < 10, \ t > 0$$
BC: $\quad u(0,t) = 10, \ u(10,t) = 30, \quad t > 0$
IC: $\quad u(x,0) = 0, \quad 0 < x < 10.$

Solution: Because the BCs are nonhomogeneous, we assume the solution can be written as $u(x,t) = S(x) + v(x,t)$, where the steady-state solution satisfies

$$S'' = 0, \quad S(0) = 10, \ S(10) = 30.$$

Clearly, the steady-state solution is once again a linear function of x given by $S(x) = 10 + 2x$. In this case the transient solution satisfies

$$v_{xx} = a^{-2}v_t,$$
BC: $\quad v(0,t) = 0, \ v(10,t) = 0,$
IC: $\quad v(x,0) = -S(x) = -10 - 2x,$

which leads to

$$v(x,t) = \sum_{n=1}^{\infty} c_n \sin\left(\frac{n\pi x}{10}\right) e^{-a^2 n^2 \pi^2 t/100}.$$

Imposing the IC prescribed in the above transient problem, we find that

$$c_n = \frac{20}{n\pi}[(-1)^n 3 - 1], \quad n = 1,2,3,...,$$

from which we deduce the solution

$$\begin{aligned} u(x,t) &= S(x) + v(x,t) \\ &= 10 + 2x + \frac{20}{\pi}\sum_{n=1}^{\infty} \frac{[(-1)^n 3 - 1]}{n}\sin\left(\frac{n\pi x}{10}\right) e^{-a^2 n^2 \pi^2 t/100}. \end{aligned}$$

12.3.3 Derivation of the heat equation

To derive the governing equation of heat flow in a uniform rod from first principles, we apply the law of conservation of thermal energy to a slice of the rod between x and $x + \Delta x$. The *law of conservation of thermal energy* states that

"... the rate of heat entering a region plus that which is generated inside the region equals the rate of heat leaving the region plus that which is stored."

Suppose we let $Q(x,t)$ be the rate of heat flow at the point x and time t, also known as the *heat flux*. If A denotes the cross-sectional area of a uniform rod, then $AQ(x,t)$ represents the rate at which heat enters the slice at x, and similarly, $AQ(x + \Delta x,t)$ represents the rate at which heat leaves the slice at $x + \Delta x$ (see Fig. 12.2). If a heat source is present within the region, the rate at which heat is generated in the slice is $A \Delta x F(x,t)$, where $F(x,t)$ is the rate of heat generation per unit volume. Lastly, the rate of heat energy storage in the slice is proportional to the time rate change of temperature $u(x,t)$. To find the constant of proportionality necessitates the introduction of the *specific heat* constant c, defined as the heat energy that must be supplied to a unit mass of the rod to raise it one unit of temperature. Thus, if ρ is the (constant) *mass density* of the rod (i.e., mass per unit volume), the rate of change of heat energy is approximately $A \Delta x \rho c u_t$.

If we now invoke the law of conservation of thermal energy, it follows that

$$AQ(x,t) + A \Delta x F(x,t) = AQ(x + \Delta x,t) + A \Delta x \rho c u_t, \tag{28}$$

which can be rearranged as (note that the area A drops out)

$$-\left[\frac{Q(x + \Delta x,t) - Q(x,t)}{\Delta x} \right] = \rho c u_t - F(x,t). \tag{29}$$

By allowing $\Delta x \to 0$, we obtain from (29)

$$-Q_x = \rho c u_t - F(x,t). \tag{30}$$

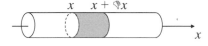

Figure 12.2 Small slice of a finite rod with uniform cross-sectional area A.

We can relate the heat flux to the temperature by use of *Fourier's law*, which is

$$Q = -k u_x, \quad k > 0, \tag{31}$$

where k is a positive constant known as the *heat conductivity*. Combining (30) and (31), we are led to the governing *nonhomogeneous* heat equation

$$u_{xx} = a^{-2} u_t - \frac{1}{k} F(x,t), \tag{32}$$

where $a^2 = k/\rho c$ is the *thermal diffusivity*. In the absence of the heat source, (32) reduces to the *homogeneous* form of the heat equation discussed above.

12.4 The Wave Equation

Imagine a violin string under a large tension τ that is "plucked" and allowed to vibrate freely. If the string has mass density (per unit length) ρ and initially lies along the x-axis, then the transverse movement $u(x,t)$ after it is plucked is a solution of the one-dimensional *wave equation* (see Section 11.10.1)

$$u_{xx} = c^{-2}u_{tt}, \tag{33}$$

where $c = \sqrt{\tau/\rho}$ is a physical constant having the dimension of velocity.

Like the heat equation, the general form of the wave equation in three dimensions is given by

$$\nabla^2 u = c^{-2}u_{tt}, \tag{34}$$

where the Laplacian in rectangular coordinates is defined by (3). In our analysis, however, we will consider only the one-dimensional case (33) and discuss its solutions in terms of the displacements from equilibrium of a vibrating string along the portion of the axis $0 < x < p$. The motion of the string will depend upon its initial deflection and velocity (speed) at the time of release. Hence, the problem we wish to solve is formulated by

$$
\begin{array}{ll}
& u_{xx} = c^{-2}u_{tt}, \quad 0 < x < p, \, t > 0 \\
\text{BC:} & u(0,t) = 0, \, u(p,t) = 0, \quad t > 0 \\
\text{IC:} & u(x,0) = f(x), \, u_t(x,0) = g(x), \quad 0 < x < p.
\end{array}
\tag{35}
$$

Because the boundary conditions in (35) are homogeneous, we can resort immediately to the separation of variables method. Hence, as before, we assume solutions exist of the form $u(x,t) = X(x)W(t)$, which leads to the system of ODEs

$$X'' + \lambda X = 0, \quad X(0) = 0, \, X(p) = 0,$$

$$W'' + \lambda c^2 W = 0. \tag{36}$$

The eigenvalue problem in (36) is one we have solved many times; thus,

$$\lambda_n = \frac{n^2\pi^2}{p^2}, \quad \varphi_n(x) = \sin\frac{n\pi x}{p}, \quad n = 1,2,3,\dots. \tag{37}$$

With the separation constant λ defined in this fashion, the solution to the second ODE in (36) is

$$W_n(t) = a_n\cos\frac{nc\pi t}{p} + b_n\sin\frac{nc\pi t}{p}, \quad n = 1,2,3,\dots, \tag{38}$$

where the a's and the b's are arbitrary constants.

By combining solutions (36) and (37), we obtain the set of solutions

$$u_n(x,t) = \left(a_n\cos\frac{nc\pi t}{p} + b_n\sin\frac{nc\pi t}{p}\right)\sin\frac{n\pi x}{p}, \quad n = 1,2,3,\dots. \tag{39}$$

The individual solutions in (39) are called *standing waves* since, for a fixed value of t, each can be viewed as having the shape $\sin(n\pi x/p)$ with amplitude $W_n(t)$ (see Fig. 12.3). The points where $\sin(n\pi x/p) = 0$ are called *nodes* and physically correspond to zero displacement of the string. The number of nodes is equal to $n - 1$. Also, if we now think of x as fixed, each solution in (39) represents the motion of a point on the string with abscissa x. Moreover, by writing $W_n(t)$ in the form

$$W_n(t) = \sqrt{a_n^2 + b_n^2} \cos\left(\frac{nc\pi t}{p} - \alpha_n\right), \tag{40}$$

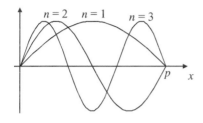

Figure 12.3 Standing waves.

where $\alpha_n = \tan^{-1}(b_n/a_n)$, we see that the solutions (39) represent *simple harmonic motion* (SHM) of (angular) *frequency* $\omega_n = nc\pi/p$ and *amplitude* $\sqrt{a_n^2 + b_n^2} \sin(n\pi x/p)$. The smallest frequency ω_1 denotes the *fundamental frequency* of the system and all other frequencies are integer multiples (or harmonics) of the fundamental frequency $\omega_1 = c\pi/p$.

In order to satisfy the prescribed ICs, we must impose the superposition principle to get

$$u(x,t) = \sum_{n=1}^{\infty} \left(a_n \cos\frac{nc\pi t}{p} + b_n \sin\frac{nc\pi t}{p} \right) \sin\frac{n\pi x}{p}. \tag{41}$$

The first IC requires

$$u(x,0) = f(x) = \sum_{n=1}^{\infty} a_n \sin\frac{n\pi x}{p}, \quad 0 < x < p, \tag{42}$$

from which we deduce

$$a_n = \frac{2}{p}\int_0^p f(x)\sin\frac{n\pi x}{p}dx, \quad n = 1,2,3,\dots. \tag{43}$$

The second IC yields

$$u_t(x,0) = g(x) = \sum_{n=1}^{\infty} \left(\frac{n\pi c}{p}\right) b_n \sin\frac{n\pi x}{p}, \quad 0 < x < p, \tag{44}$$

and, consequently,

$$b_n = \frac{2}{n\pi c}\int_0^p g(x)\sin\frac{n\pi x}{p}dx, \quad n = 1,2,3,\dots. \tag{45}$$

EXAMPLE 3 Solve the vibrating string problem

$$u_{xx} = c^{-2}u_{tt}, \quad 0 < x < 1, \ t > 0$$
$$\text{BC:} \quad u(0,t) = 0, \ u(1,t) = 0$$

$$\text{IC:} \quad u(x,0) = \begin{cases} Ax, & 0 < x < 1/2, \\ A(1-x), & 1/2 < x < 1, \end{cases} \quad u_t(x,0) = 0.$$

Solution: We assume A is a small positive constant. Following the separation of variables method illustrated above, we obtain the solution

$$u(x,t) = \sum_{n=1}^{\infty} (a_n \cos nc\pi t + b_n \sin nc\pi t) \sin n\pi x.$$

Because the initial velocity is zero, it follows that $b_n = 0$, $n = 1,2,3,\ldots$. Thus, setting $t = 0$ in the solution, we are led to the first IC

$$u(x,0) = \sum_{n=1}^{\infty} a_n \sin n\pi x,$$

where

$$a_n = 2A \int_0^{1/2} x \sin n\pi x \, dx + 2A \int_{1/2}^{1} (1-x) \sin n\pi x \, dx$$

$$= \frac{4A}{n^2 \pi^2} \sin(n\pi/2), \quad n = 1,2,3,\ldots .$$

The solution then becomes

$$u(x,t) = \frac{4A}{\pi^2} \sum_{\substack{n=1 \\ (\text{odd})}}^{\infty} \frac{\sin(n\pi/2)}{n^2} \cos(n\pi ct) \sin(n\pi x).$$

12.4.1 d'Alembert's solution

Suppose now we consider a string that is so long we consider it to be of "infinite extent." In this case the actual mathematical form of the BCs is no longer important but only the initial deflection and velocity. Thus, we now consider the problem described by

$$u_{xx} = c^{-2}u_{tt}, \quad -\infty < x < \infty, \ t > 0$$

$$\text{IC:} \quad u(x,0) = f(x), \ u_t(x,0) = g(x), \quad -\infty < x < \infty. \tag{46}$$

The solution technique we are about to illustrate is attributed to the famous French mathematician, d'Alembert. At first, d'Alembert thought he had developed a technique that could be applied to a large class of equations, but he soon realized that this was not the case. In fact, the wave equation is one of the rare examples for which d'Alembert's technique proves fruitful. However, in addition to being of historical value, the solution form developed by d'Alembert provides us with a more comprehensive understanding of the solutions of the wave equation in general.

We begin by introducing the change of independent variables

$$\begin{aligned} r &= x + ct, \\ s &= x - ct, \end{aligned} \tag{47}$$

which, through use of the chain rule, reduces the wave equation to the *canonical form*

$$u_{rs} = 0. \tag{48}$$

By integrating (48) twice, once with respect to s and then with respect to r, we find that

$$u(r,s) = F(r) + G(s), \tag{49}$$

where F and G are arbitrary (but differentiable) functions. In terms of the original variables, we have the general solution

$$u(x,y) = F(x+ct) + G(x-ct). \tag{50}$$

To determine the functions in (50), we impose the ICs in (46), which lead to

$$\begin{aligned} u(x,0) &= F(x) + G(x) = f(x), \\ u_t(x,0) &= cF'(x) - cG'(x) = g(x). \end{aligned} \tag{51}$$

These equations are equivalent to

$$\begin{aligned} F(x) + G(x) &= f(x), \\ F(x) - G(x) &= \frac{1}{c}\int_0^x g(z)\,dz + C_1, \end{aligned} \tag{52}$$

where C_1 is an arbitrary constant. The simultaneous solution of these equations is

$$\begin{aligned} F(x) &= \frac{1}{2}f(x) + \frac{1}{2c}\int_0^x g(z)\,dz + \frac{1}{2}C_1, \\ G(x) &= \frac{1}{2}f(x) - \frac{1}{2c}\int_0^x g(z)\,dz - \frac{1}{2}C_1. \end{aligned} \tag{53}$$

Because we permit the arguments in F and G to extend over all numbers, the results (53) can be generalized to

$$\begin{aligned} F(x+ct) &= \frac{1}{2}f(x+ct) + \frac{1}{2c}\int_0^{x+ct} g(z)\,dz + \frac{1}{2}C_1, \\ G(x-ct) &= \frac{1}{2}f(x-ct) - \frac{1}{2c}\int_0^{x-ct} g(z)\,dz - \frac{1}{2}C_1. \end{aligned} \tag{54}$$

Finally, by combining the results in (54), we obtain *d'Alembert's solution*:

$$u(x,t) = \frac{1}{2}[f(x+ct) + f(x-ct)] + \frac{1}{2c}\int_{x-ct}^{x+ct} g(z)\,dz. \qquad (55)$$

The interesting observation here is that once we know the initial displacement $f(x)$ and initial velocity $g(x)$, the solution (55) is immediately determined. This makes d'Alembert's solution easy to apply as compared with the infinite series solution obtained from separation of variables. For instance, if the initial deflection shape were described by the "Gaussian" function

$$f(x) = e^{-x^2}, \quad -\infty < x < \infty,$$

and $g(x) \equiv 0$, then (55) leads to

$$u(x,t) = \frac{1}{2}\{\exp[-(x+ct)^2] + \exp[-(x-ct)^2]\}.$$

The functions $\exp[-(x+ct)^2]$ and $\exp[-(x-ct)^2]$ in the above solution have interesting physical interpretations as wave phenomena. The first function, plotted as a function of x alone, is exactly the same in shape as the original function $\exp(-x^2)$, but with every point on it displaced a distance ct to the left of the corresponding point in $\exp(-x^2)$. Thus, the function $\exp[-(x+ct)^2]$ represents a wave of displacement traveling to the left with velocity c. In the same manner, we can interpret the function $\exp[-(x-ct)^2]$ as a wave of displacement traveling to the right with velocity c. The above solution $u(x,t)$ is then simply the superposition of these two *traveling waves*.

12.5 The Equation of Laplace

Perhaps the most important PDE in mathematical physics is *Laplace's equation*, also known as the *potential equation*. Its importance stems from the many diverse areas of application that lead to this equation, ranging from a variety of potential problems to steady-state heat conduction. In its general form, Laplace's equation appears as

$$\nabla^2 u = 0, \qquad (56)$$

whereas in two dimensions it has the rectangular coordinate representation

$$u_{xx} + u_{yy} = 0. \qquad (57)$$

Because the fields of application of Laplace's equation do not involve time, ICs are not prescribed for the solutions of (57). Rather, we find it is proper to simply prescribe a single BC (at each boundary point of the region of interest). The most common BCs that might be prescribed for solutions of (57) fall mainly into two categories, giving us two primary types of boundary value problems. If R denotes a region in the xy plane and C its boundary curve, then one type of problem is

$$\begin{aligned} u_{xx} + u_{yy} &= 0 \quad \text{in } R, \\ \text{BC:} \quad u &= f \quad \text{on } C, \end{aligned} \qquad (58)$$

which is called a *Dirichlet problem* or *boundary value problem of the first kind*. In this problem we specify the value of u at each boundary point by the function f. The second most common problem is characterized by

$$u_{xx} + u_{yy} = 0 \text{ in } R,$$
$$\text{BC:} \quad \frac{\partial u}{\partial n} = f \text{ on } C, \tag{59}$$

called a *Neumann problem* or *boundary value problem of the second kind*. The derivative $\partial u/\partial n$ denotes the *normal derivative* of u and is positive in the direction of the outward normal to the boundary curve C. By definition, $\partial u/\partial n = \nabla u \cdot \hat{\mathbf{n}}$, where $\hat{\mathbf{n}}$ is the *outward* unit normal vector to C (see Chapter 4).

12.5.1 Rectangular domain

To begin, we wish to consider solutions of Laplace's equation (58) in a rectangular domain. A simple example of this is formulated by the Dirichlet problem (see Fig. 12.4):

$$u_{xx} + u_{yy} = 0, \quad 0 < x < a, \, 0 < y < b,$$
$$\text{BC:} \quad \begin{cases} u(0,y) = 0, & u(a,y) = 0, \\ u(x,0) = 0, & u(x,b) = f(x). \end{cases} \tag{60}$$

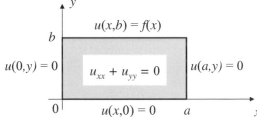

Figure 12.4 Rectangular domain.

A problem like (60) would physically arise if three edges of a thin rectangular isotropic plate with insulated flat surfaces were maintained at 0°C and the fourth edge was maintained at temperature $f(x)$ until steady-state conditions prevailed throughout the plate. In general, two-dimensional heat flow is governed by the two-dimensional heat equation

$$u_{xx} + u_{yy} = a^{-2}u_t, \tag{61}$$

where a^2 is the thermal diffusivity. When steady-state conditions prevail, we have $u_t = 0$ throughout the region of interest and (61) reduces to Laplace's equation. Thus, we can say the solution of (60) is the steady-state temperature distribution interior to the plate in this case. The same mathematical problem, however, describes the steady-state temperature distribution in a long rectangular bar bounded by the planes $x = 0$, $x = a$, $y = 0$, $y = b$, whose temperature variation in the z direction may be neglected. Other physical situations also lead to the same mathematical problem.

Our solution technique for (60) is basically the same as that for the heat and wave equations. That is, we seek solutions of (60) in the product form $u(x,y) = X(x)Y(y)$ which, when substituted into the PDE leads to the system of ODEs

$$X'' + \lambda X = 0,$$
$$Y'' - \lambda Y = 0,$$
(62)

where λ is the separation constant. Because the first three BCs in (60) are homogeneous, they reduce immediately to

$$X(0) = 0, \quad X(a) = 0, \quad Y(0) = 0.$$
(63)

The fourth condition, which is nonhomogeneous, must be handled separately.

The solution of the first ODE in (62), subject to the first two BCs in (63), leads to the eigenvalues and eigenfunctions described by

$$\lambda_n = \frac{n^2\pi^2}{a^2}, \quad \varphi_n(x) = \sin\frac{n\pi x}{a}, \quad n = 1,2,3,\dots.$$
(64)

Corresponding to these values of the separation constant, the solutions of the remaining ODE in (62) subject to the last BC in (63) is

$$Y_n(y) = \sinh\frac{n\pi y}{a}, \quad n = 1,2,3,\dots.$$
(65)

Thus, for any choice of the c's, the function

$$u(x,y) = \sum_{n=1}^{\infty} c_n \sin\frac{n\pi x}{a}\sinh\frac{n\pi y}{a}$$
(66)

satisfies Laplace's equation and the three homogeneous BCs prescribed in (60). The remaining task is to determine the c's in (66). We do this by setting $y = b$ in (66), which yields

$$u(x,b) = f(x) = \sum_{n=1}^{\infty} c_n \sinh\frac{n\pi b}{a}\sin\frac{n\pi x}{a}, \quad 0 < x < a.$$
(67)

From the theory of Fourier series, it follows that

$$c_n = \frac{2}{a\sinh(n\pi b/a)}\int_0^a f(x)\sin\frac{n\pi x}{a}\,dx, \quad n = 1,2,3,\dots,$$
(68)

and the problem is formally solved.

A more realistic situation occurs when the temperature is prescribed by nonzero values along all four edges of the plate, rather than along just one. To solve this more general problem we simply superimpose two solutions, each of which corresponds to a problem in which temperatures of 0°C are prescribed along parallel edges of the plate (see Example 4 below).

The Neumann problem for the rectangle is solved similar to that above, as are problems featuring a mix of Dirichlet and Neumann conditions. The following example features a mix of BCs.

EXAMPLE 4 Solve

$$u_{xx} + u_{yy} = 0, \quad 0 < x < \pi, \ 0 < y < 1,$$

BC: $$\begin{cases} u_x(0,y) = 0, \quad u_x(\pi,y) = 0, \\ u(x,0) = T_0 \cos x, \quad u(x,1) = T_0 \cos^2 x. \end{cases}$$

Solution: Physically, this problem corresponds to a steady-state temperature distribution problem for a rectangular plate with temperatures prescribed along its edges $y = 0$ and $y = 1$, but whose other two edges are insulated.

Proceeding with the separation of variables leads to

$$X'' + \lambda X = 0, \quad X'(0) = 0, \ X'(\pi) = 0,$$
$$Y'' - \lambda Y = 0.$$

The problem for X is a standard eigenvalue problem for which

$$\lambda_0 = 0, \quad \varphi_0(x) = 1,$$
$$\lambda_n = n^2, \quad \varphi_n(x) = \cos nx, \quad n = 1,2,3,\ldots .$$

Thus, the solution for Y becomes

$$Y_n(y) = \begin{cases} a_0 + b_0 y, & n = 0 \\ a_n \cosh ny + b_n \sinh ny, & n = 1,2,3,\ldots, \end{cases}$$

from which we deduce

$$u(x,y) = a_0 + b_0 y + \sum_{n=1}^{\infty} (a_n \cosh ny + b_n \sinh ny) \cos nx.$$

The BC at $y = 0$ requires that

$$u(x,0) = T_0 \cos x = a_0 + \sum_{n=1}^{\infty} a_n \cos nx,$$

which suggests, by matching like terms, that $a_0 = 0$, $a_1 = T_0$, and $a_n = 0, n = 2,3,4,\ldots$. Therefore, our solution reduces to

$$u(x,y) = b_0 y + T_0 \cosh y \cos x + \sum_{n=1}^{\infty} b_n \sinh ny \cos nx.$$

By imposing the last BC at $y = 1$, we find

$$T_0 \cos^2 x = b_0 + T_0 \cosh 1 \cos x + \sum_{n=1}^{\infty} b_n \sinh n \cos nx,$$

or, upon arranging terms and writing $\cos^2 x = (1/2)(1 + \cos 2x)$, we get

$$\frac{1}{2}T_0(1 + \cos 2x) - T_0\cosh 1 \cos x = b_0 + \sum_{n=1}^{\infty} b_n \sinh n \cos nx.$$

Matching like terms once again, we have

$$b_0 = \frac{T_0}{2}, \quad b_1 = -\frac{T_0\cosh 1}{\sinh 1}, \quad b_2 = \frac{T_0}{2\sinh 2},$$

with all other constants zero. Thus, we finally obtain (after some algebraic manipulation)

$$u(x,y) = T_0\left[\frac{y}{2} + \frac{\sinh(1-y)}{\sinh 1}\cos x + \frac{\sinh 2y}{2\sinh 2}\cos 2x\right].$$

Notice in Example 4 that the final solution is not in the form of an infinite series. This makes it particularly easy to calculate numerical values of the solution throughout the rectangle. The reason for this situation here, of course, is that the temperature functions prescribed along the edges $y = 0$ and $y = 1$ are actually composed of linear combinations of eigenfunctions belonging to the set $\varphi_n(x)$, $n = 0,1,2,\dots$. Thus, only a finite number of eigenfunctions are required in the series solution.

12.5.2 Circular domain

In solving a potential problem by the separation of variables method in a circular domain, it is necessary to express the problem in polar coordinates. By setting

$$x = r\cos\theta, \quad y = r\sin\theta, \tag{69}$$

it can be shown through use of the chain rule that

$$u_{xx} + u_{yy} = u_{rr} + \frac{1}{r}u_r + \frac{1}{r^2}u_{\theta\theta}. \tag{70}$$

Let us consider the steady-state heat conduction problem for a flat plate in the shape of a circular disk with boundary curve $x^2 + y^2 = \rho^2$. We assume that the plate is isotropic, that the flat surfaces are insulated, and that the temperature is known everywhere on the circular boundary. The temperature inside the disk is then a solution of the Dirichlet problem (see Fig. 12.5)

$$\boxed{\begin{aligned} &u_{rr} + \frac{1}{r}u_r + \frac{1}{r^2}u_{\theta\theta} = 0, \quad 0 < r < \rho, \ -\pi < \theta < \pi, \\ &\text{BC:} \quad u(\rho,\theta) = f(\theta), \quad -\pi < \theta < \pi. \end{aligned}} \tag{71}$$

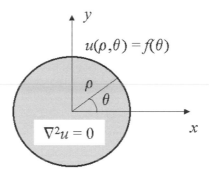

Figure 12.5 Circular domain.

Before we attempt to solve (71), some observations about the solution are in order. First, we need to recognize that $r = 0$ is a "mathematical boundary" of the problem, although it is clearly not a physical boundary. Moreover, to obtain physically meaningful solutions we need to impose at $r = 0$ the *implicit* BC

$$\text{BC:} \quad |u(0,\theta)| < \infty, \tag{72}$$

which simply states that the solution is bounded at $r = 0$. Second, in order to allow θ to assume any value rather than be restricted to the 2π interval cited in (71), we specify that $f(\theta)$, and consequently $u(r,\theta)$, be *periodic* functions in the variable θ with period 2π. Hence, it is necessary to also specify the periodic BCs

$$\text{BC:} \quad u(r,-\pi) = u(r,\pi), \quad u_\theta(r,-\pi) = u_\theta(r,\pi), \tag{73}$$

which in essence are continuity requirements along the "mathematical" slit $\theta = \pi$.

By assuming a product solution $u(r,\theta) = R(r)\Theta(\theta)$, we find that the potential equation in (71) separates according to

$$\frac{r^2 R'' + rR'}{R} = -\frac{\Theta''}{\Theta} = \lambda. \tag{74}$$

Therefore, we are led to the ODEs

$$\Theta'' + \lambda\Theta = 0, \quad \Theta(-\pi) = \Theta(\pi), \ \Theta'(-\pi) = \Theta'(\pi), \tag{75}$$

$$r^2 R'' + rR' - \lambda R = 0, \quad |R(0)| < \infty, \tag{76}$$

where we have incorporated BCs (72) and (73).

The periodic eigenvalue problem described by (75) has solution (recall Example 13 in Section 8.4.3)

$$\begin{aligned}
\lambda_0 &= 0, \quad \varphi_0(x) = 1, \\
\lambda_n &= n^2, \quad \varphi_n(\theta) = a_n\cos n\theta + b_n\sin n\theta, \quad n = 1,2,3,...,
\end{aligned} \tag{77}$$

where the a's and b's are arbitrary constants. Equation (76) is a Cauchy-Euler equation (see Section 1.4.5) with solution

$$R_n(r) = \begin{cases} C_1 + C_2 \ln r, & n = 0, \\ C_3 r^n + C_4 r^{-n}, & n = 1,2,3,\dots. \end{cases} \tag{78}$$

However, to keep the solution bounded we must select $C_2 = C_4 = 0$. Collecting solutions, we obtain the general family of solutions

$$u_n(r,\theta) = \begin{cases} 1, & n = 0, \\ \left(\dfrac{r}{\rho}\right)^n (a_n \cos n\theta + b_n \sin n\theta), & n = 1,2,3,\dots, \end{cases} \tag{79}$$

each of which satisfies Laplace's equation and the implicit BCs prescribed by (72) and (73). In arriving at (79), we set $C_1 = 1$ and $C_3 = 1/\rho^n$ for mathematical convenience. By invoking the superposition principle, we obtain the solution form

$$\boxed{u(r,\theta) = \frac{1}{2}a_0 + \sum_{n=1}^{\infty} \left(\frac{r}{\rho}\right)^n (a_n \cos n\theta + b_n \sin n\theta),} \tag{80}$$

where we have expressed the constant term as $a_0/2$, again for mathematical convenience. At $r = \rho$ the BC in (71) leads to

$$f(\theta) = \frac{1}{2}a_0 + \sum_{n=1}^{\infty} (a_n \cos n\theta + b_n \sin n\theta), \tag{81}$$

from which it follows

$$a_n = \frac{1}{\pi}\int_{-\pi}^{\pi} f(\theta)\cos n\theta\, d\theta, \quad n = 0,1,2,\dots, \tag{82}$$

$$b_n = \frac{1}{\pi}\int_{-\pi}^{\pi} f(\theta)\sin n\theta\, d\theta, \quad n = 1,2,3,\dots. \tag{83}$$

12.5.3 Maximum-minimum principle

Continuous solutions of Laplace's equation, which also have continuous first- and second-order derivatives, are called *harmonic functions* (see also Sections 4.7.2, 6.4.3, and 7.8). Because of the numerous properties of harmonic functions, this is considered an important branch of applied mathematics in its own right, commonly called *potential theory*. However, we won't present a general development here of the basic properties of harmonic functions (see Section 4.7.2), but consider only one particular property.

If we set $r = 0$ in the solution (80) above, we get

$$u(0,\theta) = \frac{1}{2}a_0 = \frac{1}{2\pi}\int_{-\pi}^{\pi} f(\theta)\,d\theta. \tag{84}$$

Thus, we see that the temperature at the center of the disk is simply the average of the temperature on the boundary of the disk. A similar result holds for any fixed radius from the center of the plate; viz., it can be shown that

$$u(0,\theta) = \frac{1}{2\pi} \int_{-\pi}^{\pi} u(r,\theta)d\theta, \ 0 < r \le \rho. \tag{85}$$

This remarkable property is useful in numerical methods and also for establishing the following significant theorem in the study of harmonic functions concerning the maximum and minimum values.

Theorem 12.1 (*Maximum-minimum principle*) If u is a harmonic function in a region R, then both the maximum and minimum values of u are attained on the boundary C of R.

The maximum-minimum principle is actually a consequence of the mean value property of harmonic functions described in (85). That is, if we imagine that the solution u is the steady-state temperature distribution in a region R, such as a metal plate, then the temperature at any interior point cannot exceed all other nearby points, for if it did, heat would flow from the hot points to the cooler points. But then the temperature would be changing in time, and this is a contradiction that steady-state conditions prevail. Also, an interior maximum or minimum could never be the average of its neighbors as required by the mean value property (85).

12.6 Generalized Fourier Series

In this section we briefly examine some applications involving eigenfunction expansions that lead to *generalized Fourier series* (recall Section 8.4.2).

12.6.1 Convective heat transfer at one endpoint

In solving the heat equation we have limited our BCs thus far to either prescribing the unknown function u or its normal derivative u_x on the boundary. However, when heat is lost from an end of the rod into the surrounding medium through convection, the BC involves a linear combination of u and u_x.

Suppose, for example, we consider a long thin rod of length p that is initially heated to the temperature distribution $f(x)$. At time $t = 0$ the end at $x = 0$ is exposed to ice packs at 0°C while heat exchange into the surrounding medium through convection takes place at the end $x = p$. Mathematically, this problem is characterized by

$$\begin{aligned} &u_{xx} = a^{-2}u_t, \ \ 0 < x < p, \ t > 0, \\ \text{BC:} \ \ &u(0,t) = 0, \ hu(p,t) + u_x(p,t) = 0, \ \ t > 0, \\ \text{IC:} \ \ &u(x,0) = f(x), \ \ 0 < x < p, \end{aligned} \tag{86}$$

where the constant h is the heat transfer coefficient.

By using the separation of variables technique, we are led to the eigenvalue problem and time-dependent equation

$$X'' + \lambda X = 0, \quad X(0) = 0, \quad hX(p) + X'(p) = 0, \tag{87}$$

$$W' + \lambda a^2 W = 0. \tag{88}$$

The above eigenvalue problem is similar to that in Example 7 in Section 8.4. Thus, we will consider only the case $\lambda = k^2 > 0$, which yields the solution

$$X(x) = C_1 \cos kx + C_2 \sin kx. \tag{89}$$

The BC at $x = 0$ requires $C_1 = 0$, whereas the second condition leads to $(C_2 \neq 0)$

$$h \sin kp + k \cos kp = 0. \tag{90}$$

If we denote the solutions of this transcendental equation by k_n, $n = 1,2,3,...$ (which incorporates the heat transfer coefficient h), then the eigenvalues and eigenfunctions are

$$\lambda_n = k_n^2, \quad \varphi_n(x) = \sin k_n x, \quad n = 1,2,3,.... \tag{91}$$

The solution of (88) is therefore

$$W_n(t) = c_n \exp(-a^2 k_n^2 t), \quad n = 1,2,3,..., \tag{92}$$

and hence,

$$u(x,t) = \sum_{n=1}^{\infty} c_n \exp(-a^2 k_n^2 t) \sin k_n x. \tag{93}$$

From the initial condition

$$u(x,0) = f(x) = \sum_{n=1}^{\infty} c_n \sin k_n x, \tag{94}$$

we now obtain the arbitrary constants

$$c_n = \|\varphi_n(x)\|^{-2} \int_0^P f(x)\sin k_n x \, dx = \frac{\int_0^P f(x)\sin k_n x \, dx}{\int_0^P \sin^2 k_n x \, dx}, \quad n = 1,2,3,.... \tag{95}$$

12.6.2 Nonhomogeneous heat equation

When a heat source is present within the domain of interest, the governing PDE is *nonhomogeneous*. Thus, the problem we wish to discuss can be formulated by

$$
\begin{array}{ll}
& u_{xx} = a^{-2}u_t - q(x,t), \quad 0 < x < p, \ t > 0, \\
\text{BC:} & u(0,t) = 0, \ u(p,t) = 0, \quad t > 0, \\
\text{IC:} & u(x,0) = f(x), \quad 0 < x < p,
\end{array}
\tag{96}
$$

where $q(x,t)$ is proportional to the heat source.

Here we begin by looking for a solution of the form

$$
u(x,t) = \sum_{n=1}^{\infty} E_n(t)\varphi_n(x),
\tag{97}
$$

where the eigenfunctions $\varphi_n(x)$ belong to the eigenvalue problem associated with the homogeneous heat equation [i.e, when $q(x,t) \equiv 0$]. Consequently, they are solutions of

$$
X'' + \lambda X = 0, \quad X(0) = 0, \ X(p) = 0,
\tag{98}
$$

which yields $\lambda_n = n^2\pi^2/p^2$ and $\varphi_n(x) = \sin(n\pi x/p)$. By substituting the assumed solution form (97) into the nonhomogeneous equation in (96) and rearranging terms, we obtain

$$
\begin{aligned}
a^2 q(x,t) &= u_t - a^2 u_{xx} \\
&= \sum_{n=1}^{\infty}\left[E_n'(t) + \frac{a^2 n^2 \pi^2}{p^2} E_n(t) \right] \sin\frac{n\pi x}{p}.
\end{aligned}
\tag{99}
$$

For a fixed value of t, we can interpret (99) as a Fourier sine series with Fourier coefficients defined by

$$
E_n'(t) + \frac{a^2 n^2 \pi^2}{p^2} E_n(t) = \frac{2a^2}{p}\int_0^p q(x,t)\sin\frac{n\pi x}{p}\,dx, \quad n = 1,2,3,\dots .
\tag{100}
$$

Thus, (100) represents a system of first-order, linear, nonhomogeneous ODEs which has the general solution (see Section 1.3.2)

$$
E_n(t) = \left[c_n + a^2 \int_0^t \exp\left(\frac{a^2 n^2 \pi^2 \tau}{p^2} \right) Q_n(\tau)\,d\tau \right] \exp\left(-\frac{a^2 n^2 \pi^2 t}{p^2} \right),
\tag{101}
$$

$$
n = 1,2,3,\dots,
$$

where, for notational convenience, we have introduced

$$
Q_n(t) = \frac{2}{p}\int_0^p q(x,t)\sin\frac{n\pi x}{p}\,dx, \quad n = 1,2,3,\dots .
\tag{102}
$$

The substitution of (101) into the solution form (97) provides us with the formal solution

$$u(x,t) = \sum_{n=1}^{\infty} \left[c_n + a^2 \int_0^t \exp\left(\frac{a^2 n^2 \pi^2 \tau}{p^2} \right) Q_n(\tau) d\tau \right]$$
$$\times \exp\left(-\frac{a^2 n^2 \pi^2 t}{p^2} \right) \sin\frac{n\pi x}{p}.$$

(103)

Finally, by imposing the IC in (96) we have

$$u(x,0) = f(x) = \sum_{n=1}^{\infty} c_n \sin\frac{n\pi x}{p},$$

(104)

where

$$c_n = \frac{2}{p} \int_0^p f(x) \sin\frac{n\pi x}{p} dx, \quad n = 1,2,3,\dots.$$

(105)

EXAMPLE 5 Solve the nonhomogeneous problem

$$u_{xx} = u_t + (1-x)\cos t, \quad 0 < x < 1, \ t > 0,$$
BC: $u(0,t) = 0, \ u(1,t) = 0, \quad t > 0,$
IC: $u(x,0) = 0, \quad 0 < x < 1.$

Solution: The eigenvalue problem associated with this example has solution

$$\lambda_n = n^2\pi^2, \quad \varphi_n(x) = \sin n\pi x, \quad n = 1,2,3,\dots.$$

Hence, we seek solutions of the form

$$u(x,t) = \sum_{n=1}^{\infty} E_n(t) \sin n\pi x,$$

which, when substituted into the nonhomogeneous PDE, yields

$$-(1-x)\cos t = \sum_{n=1}^{\infty} [E_n{}'(t) + n^2\pi^2 E_n(t)] \sin n\pi x.$$

Therefore, we see that

$$E_n{}'(t) + n^2\pi^2 E_n(t) = -2\int_0^1 (1-x)\cos t \sin n\pi x \, dx$$
$$= -\frac{2}{n\pi}\cos t, \quad n = 1,2,3,\dots,$$

with solution

$$E_n(t) = \left(c_n - \frac{2}{n\pi} \int_0^t e^{n^2\pi^2\tau} \cos\tau \, d\tau \right) e^{-n^2\pi^2 t}$$
$$= c_n e^{-n^2\pi^2 t} + \frac{2}{n\pi(1 + n^4\pi^4)} \left[n^2\pi^2\left(e^{-n^2\pi^2 t} - \cos t \right) - \sin t \right].$$

Last, by putting this expression back into the assumed solution form and imposing the IC, we find $c_n = 0$ and thus,

$$u(x,t) = \frac{2}{\pi} \sum_{n=1}^{\infty} \frac{\sin n\pi x}{n(1+n^4\pi^4)} \left[n^2\pi^2 \left(e^{-n^2\pi^2 t} - \cos t \right) - \sin t \right].$$

12.6.3 Nonhomogeneous wave equation

The motion of an elastic string under the influence of an external stimulus is called *forced*. It is characterized by a nonhomogeneous wave equation. For illustration purposes, let us consider the problem of a guy wire stretched tightly between two fixed supports at $x = 0$ and $x = \pi$. We assume the wire is initially at rest until a "gust" of wind comes along. If the wind can be modeled as a simple sinusoidal function of time with constant amplitude applied normally to the wire, the subsequent motions of the wire are described by solutions of

$$
\begin{aligned}
&u_{xx} = c^{-2}u_{tt} - P\sin\omega t, \quad 0 < x < \pi, \ t > 0, \\
\text{BC:} \quad &u(0,t) = 0, \ u(\pi,t) = 0, \quad t > 0, \\
\text{IC:} \quad &u(x,0) = 0, \ u_t(x,0) = 0, \quad 0 < x < \pi,
\end{aligned}
\tag{106}
$$

where P is a constant, $P\sin\omega t$ is proportional to the wind force, and ω is angular frequency.

We can apply a method similar to that in Section 12.6.2 for the nonhomogeneous heat equation. Namely, we seek a solution of (106) of the form

$$u(x,t) = \sum_{n=1}^{\infty} E_n(t)\varphi_n(x), \tag{107}$$

where the eigenfunctions $\varphi_n(x) = \sin nx$ belong to the eigenvalue problem

$$X'' + \lambda X = 0, \quad X(0) = 0, \ X(\pi) = 0. \tag{108}$$

Once again, we recognize this eigenvalue problem as that belonging to the associated homogeneous wave equation. By substituting the assumed solution form (107) into the nonhomogeneous wave equation in (106), we obtain

$$
\begin{aligned}
Pc^2\sin\omega t &= u_{tt} - c^2 u_{xx} \\
&= \sum_{n=1}^{\infty} \left[E_n''(t) + n^2 c^2 E_n(t) \right] \sin nx.
\end{aligned}
\tag{109}
$$

For a fixed value of t, we can interpret (109) as a Fourier sine series of the function on the left with Fourier coefficients defined by

$$E_n''(t) + n^2 c^2 E_n(t) = \frac{2Pc^2}{\pi} \sin \omega t \int_0^\pi \sin nx\, dx$$

$$= \begin{cases} \dfrac{4Pc^2}{n\pi} \sin \omega t, & n = 1,3,5,\ldots \\[2mm] 0, & n = 2,4,6,\ldots. \end{cases} \tag{110}$$

Equation (110) is a nonhomogeneous ODE that we can solve by use of the one-sided Green's function (see Section 8.5.1). That is,

$$E_n(t) = \begin{cases} a_n \cos nct + b_n \sin nct + \dfrac{4Pc}{\pi n^2} \displaystyle\int_0^t \sin[nc(t-\tau)] \sin \omega \tau\, d\tau, & n = 1,3,5,\ldots \\[4mm] a_n \cos nct + b_n \sin nct, & n = 2,4,6,\ldots, \end{cases} \tag{111}$$

and, consequently,

$$u(x,t) = \sum_{n=1}^\infty (a_n \cos nct + b_n \sin nct) \sin nx$$

$$+ \frac{4Pc}{\pi} \sum_{\substack{n=1 \\ (\text{odd})}}^\infty \frac{\sin nx}{n^2} \int_0^t \sin[nc(t-\tau)] \sin \omega \tau\, d\tau. \tag{112}$$

Last, because the prescribed ICs are homogeneous, we find $a_n = b_n = 0$, $n = 1,2,3,\ldots$, and, upon evaluating the integral, we are led to

$$\boxed{u(x,t) = \frac{4Pc}{\pi} \sum_{\substack{n=1 \\ (\text{odd})}}^\infty \frac{\sin nx}{n^2(\omega^2 - n^2 c^2)} (\omega \sin nct - nc \sin \omega t),} \tag{113}$$

where we assume $\omega \neq nc$ $(n = 1,3,5,\ldots)$.

If the input frequency is one of the values $\omega = nc$ $(n = 1,3,5,\ldots)$, the amplitude of the resulting motion (113) of the guy wire becomes unbounded, and *resonance* is said to occur. Of course, we recognize that in any physical situation the amplitude cannot truly become unbounded. A certain amount of natural damping is always present (although we have ignored such effects) and this tends to limit the magnitude of the amplitude variations. Nontheless, large amplitudes do sometimes arise and guy wires can sometimes "snap."

Resonance as described here has also played a deleterious role in the collapse of certain bridges. For example, on November 7, 1940, only four months after its grand opening, the Tacoma Narrows bridge at Puget Sound in the state of Washington collapsed from large undulations caused by the wind blowing across the superstructure. Because of such possibilities, it is an important aspect of design that the natural frequency of the structure be different from the frequency of any probable forcing function.

12.7 Applications Involving Bessel Functions

In solving certain heat conduction problems or wave motion problems in circular or cylindrical domains, it often happens that *Bessel functions* (Section 2.5.1) arise in the solution process. Below we illustrate two typical physical problems.

12.7.1 Vibrating membrane

For illustrative purposes, let us consider the problem of determining the small transverse displacements u of a thin circular membrane (such as a drumhead) of unit radius whose edge is rigidly fixed. If the displacements depend only on the radial distance r from the center of the membrane and on time t, the problem we wish to solve is given by

$$
\begin{aligned}
& u_{rr} + \frac{1}{r}u_r = c^{-2}u_{tt}, \quad 0 < r < 1,\ t > 0, \\
& \text{BC:} \quad u(1,0) = 0, \quad t > 0, \\
& \text{IC:} \quad u(r,0) = f(r),\ u_t(r,0) = 0, \quad 0 < r < 1.
\end{aligned}
\tag{114}
$$

In (114), we have assumed the initial displacement of the drumhead is given by $f(r)$ with zero velocity. To ensure a bounded solution, we must also impose the implicit BC

$$
\text{BC:} \quad |u(0,t)| < \infty.
\tag{115}
$$

We begin by assuming (114) has a product solution of the general form $u(r,t) = R(r)W(t)$. By substituting this assumed solution into the PDE and separating the variables, we are led to the eigenvalue problem

$$
rR'' + R' + \lambda rR = 0, \quad |R(0)| < \infty,\ R(1) = 0,
\tag{116}
$$

whereas the time-dependent equation is

$$
W'' + \lambda c^2 W = 0.
\tag{117}
$$

We recognize the DE in (116) as Bessel's equation (see Section 2.5.1), for which

$$
R(r) = C_1 J_0(kr) + C_2 Y_0(kr),
\tag{118}
$$

where C_1 and C_2 are arbitrary constants and $\lambda = k^2 > 0$. To satisfy the boundedness condition in (116), we set $C_2 = 0$ and the remaining BC yields

$$
R(1) = C_1 J_0(k) = 0.
\tag{119}
$$

Hence, we let k_n denote the zeros of the zero-order Bessel function and claim that the eigenvalues and eigenfunctions are

$$\lambda_n = k_n^2, \; \varphi_n(r) = J_0(k_n r); \; J_0(k_n) = 0, \; n = 1, 2, 3, \ldots . \tag{120}$$

For these eigenvalues, the solution of (117) is

$$W_n(t) = a_n \cos k_n ct + b_n \sin k_n ct, \quad n = 1, 2, 3, \ldots, \tag{121}$$

and consequently,

$$u(r,t) = \sum_{n=1}^{\infty} (a_n \cos k_n ct + b_n \sin k_n ct) J_0(k_n r). \tag{122}$$

The individual solutions in (122) can be viewed as *standing waves* with fixed shape $J_0(k_n r)$ and amplitude given by (121). Thus, the situation is similar to that of the vibrating string problem in Section 12.4. However, because the zeros of the Bessel function are not regularly spaced (in contrast with the zeros of the sine function appearing in the vibrating string problem), the sound emitted from a drum is quite different from that of a violin or guitar. In musical tones the zeros are evenly spaced and the frequencies are integral multiples of the fundamental frequency.

By imposing the ICs, we obtain

$$u(r,0) = f(r) = \sum_{n=1}^{\infty} a_n J_0(k_n r), \quad 0 < r < 1, \tag{123}$$

$$u_t(r,0) = 0 = \sum_{n=1}^{\infty} k_n c b_n J_0(k_n r), \quad 0 < r < 1. \tag{124}$$

These series are Bessel series for which we deduce $b_n = 0$, $n = 1, 2, 3, \ldots$, and

$$a_n = \frac{2}{[J_1(k_n)]^2} \int_0^1 r f(r) J_0(k_n r) dr, \quad n = 1, 2, 3, \ldots . \tag{125}$$

12.7.2 Scattering of plane waves by a circular cylinder

We now introduce Hankel functions (Section 2.5.3) to study the effects of a stationary object on an incident plane wave. In particular, we consider the problem of a circular cylinder of radius $r = a$ standing vertically in a shallow sea or stream. We assume an infinite plane wave of complex displacement η_i is advancing along the positive x-axis, i.e.,

$$\eta_i(r,\theta) = A e^{ikx} = A e^{ikr\cos\theta}, \tag{126}$$

where A is the amplitude, k is the wave number of the wave, and we have introduced polar coordinates. The total wave behind the cylinder can be expressed as the sum of incident and scattered waves so that $\eta = \eta_i + \eta_s$. Although the incident wave η_i satisfies the *Helmholtz equation* (reduced wave equation) everywhere, the scattered wave η_s satisfies the Helmholtz equation only outside the cylinder. Hence, we write

$$\nabla^2 \eta_s + k^2 \eta_s = 0, \quad r > a, \tag{127}$$

where $\nabla^2 \eta_s = \partial^2 \eta_s / \partial r^2 + (1/r)\partial \eta_s / \partial r + (1/r^2)\partial^2 \eta_s / \partial \theta^2$. Far from the cylinder $(r \to \infty)$ the scattered wave must be outgoing (radiation condition). Also, on the cylinder the normal velocity $\partial \eta / \partial r$ vanishes so that

$$\frac{\partial \eta}{\partial r}(a,\theta) = \frac{\partial \eta_i}{\partial r}(a,\theta) + \frac{\partial \eta_s}{\partial r}(a,\theta) = 0. \tag{128}$$

By separation of variables $\eta_s(r,\theta) = R(r)\Theta(\theta)$, the Helmholtz equation (127) reduces to the ODEs

$$\Theta'' + \lambda\Theta = 0, \quad \Theta(-\pi) = \Theta(\pi), \quad \Theta'(-\pi) = \Theta'(\pi),$$

$$r^2 R'' + rR' + (k^2 r^2 - \lambda)R = 0, \quad r > a, \tag{129}$$

$$\text{BC:} \lim_{r \to \infty} R(r) \sim e^{ikr},$$

where we have incorporated periodic BCs in the first equation. Based on Example 13 in Section 8.4.3, the eigenvalues of the first equation are $\lambda_n = n^2$, $n = 0,1,2,...$, and the corresponding eigenfunctions are $\cos n\theta$ and $\sin n\theta$. However, from the symmetry of the problem in this case we retain only the even eigenfunctions $\cos n\theta$.

We recognize the second equation in (129) as Bessel's DE of order n. However, in this case it is best to select the general solution in the form

$$R_n(r) = \alpha_n H_n^{(1)}(kr) + \beta_n H_n^{(2)}(kr), \tag{130}$$

which involves *Hankel functions* of the first and second kind defined by

$$H_n^{(1)}(kr) = J_n(kr) + i Y_n(kr),$$
$$H_n^{(2)}(kr) = J_n(kr) - i Y_n(kr). \tag{131}$$

For $r \to \infty$, the asymptotic behavior of these function is given by (see [34])

$$H_n^{(1)}(kr) \sim \sqrt{\frac{2}{\pi kr}} e^{ikr - \pi/4 - n\pi/2},$$

$$\tag{132}$$

$$H_n^{(2)}(kr) \sim \sqrt{\frac{2}{\pi kr}} e^{-ikr - \pi/4 - n\pi/2},$$

which represent outgoing and incoming waves, respectively. In order for the wave to be outgoing at large distances from the cylinder, the scattered wave can only consist of Hankel functions of the first kind. Consequently, by imposing the superposition principle, we are led to the scattered wave

$$\eta_s(r,\theta) = \frac{1}{2}\alpha_0 H_0^{(1)}(kr) + \sum_{n=1}^{\infty} \alpha_n H_n^{(1)}(kr)\cos n\theta. \tag{133}$$

To combine (133) with the incident wave (126), it is helpful at this point to expand the incident wave in a Fourier series of the polar angle θ. That is, we write

$$\eta_i(r,\theta) = A e^{ikr\cos\theta} = \frac{1}{2}C_0 + \sum_{n=1}^{\infty} C_0 \cos n\theta, \tag{134}$$

where

$$C_n = \frac{2A}{\pi}\int_0^\pi e^{ikr\cos\theta}\cos n\theta\, d\theta$$
$$= \begin{cases} 2AJ_0(kr), & n = 0 \\ 2Ai^{\,n}J_n(kr), & n = 1,2,3,\dots. \end{cases} \tag{135}$$

Thus,

$$\eta_i(r,\theta) = AJ_0(kr) + 2A\sum_{n=1}^{\infty} i^{\,n}J_n(kr)\cos n\theta, \tag{136}$$

and by combining the results of (133) and (136), we obtain

$$\begin{aligned}\eta(r,\theta) &= \eta_i(r,\theta) + \eta_s(r,\theta) \\ &= AJ_0(kr) + \frac{1}{2}\alpha_0 H_0^{(1)}(kr) + \sum_{n=1}^{\infty}\left[2Ai^{\,n}J_n(kr) + \alpha_n H_n^{(1)}(kr)\right]\cos n\theta.\end{aligned} \tag{137}$$

To satisfy the BC (128), we must have

$$\begin{aligned}\frac{\partial\eta}{\partial r}(a,\theta) &= AkJ_0'(ka) + \frac{1}{2}\alpha_0 kH_0^{(1)\prime}(ka) + \sum_{n=1}^{\infty}\left[2Ai^{\,n}kJ_n'(ka) + \alpha_n kH_n^{(1)\prime}(kr)\right]\cos n \\ &= 0.\end{aligned} \tag{138}$$

It follows that

$$\alpha_0 = -\frac{2AJ_0'(ka)}{H_0^{(1)\prime}(ka)}, \quad \alpha_n = -\frac{2Ai^{\,n}J_n'(ka)}{H_n^{(1)\prime}(ka)}, \tag{139}$$

from which we deduce

$$\begin{aligned}\eta(r,\theta) = A&\left[J_0(kr) - \frac{J_0'(ka)}{H_0^{(1)\prime}(ka)}H_0^{(1)}(kr)\right] \\ &+ 2A\sum_{n=1}^{\infty} i^{\,n}\left[J_n(kr) - \frac{J_n'(ka)}{H_n^{(1)\prime}(ka)}H_n^{(1)}(kr)\right]\cos n\theta.\end{aligned} \tag{140}$$

As a physical deduction of our result, we note that the spatial factor of the dynamic pressure on the cylinder is

$$p(a,\theta) = \rho g \eta(r,\theta) = A\rho g \left[J_0(ka) - \frac{J_0{}'(ka)}{H_0^{(1)'}(ka)} H_0^{(1)}(ka) \right]$$
$$+ 2A\rho g \sum_{n=1}^{\infty} i^n \left[J_n(ka) - \frac{J_n{}'(ka)}{H_n^{(1)'}(ka)} H_n^{(1)}(ka) \right] \cos n\theta , \tag{141}$$

where ρ is the mass density of the fluid and g is the gravitational acceleration constant. We can simplify this expression by using the Wronskian identity

$$W(J_n, H_n)(z) = J_n(z)H_n^{(1)'}(z) - H_n^{(1)}(z)J_n{}'(z) = \frac{2i}{\pi z}, \tag{142}$$

which arises from the Wronskian relation $W(J_n, Y_n)(z) = 2/\pi z$ and recognizing that

$$W(J_n, H_n)(z) = i\left[J_n(z)Y_n{}'(z) - Y_n(z)J_n{}'(z) \right] = \frac{2i}{\pi z}.$$

Therefore, using (142) we find that (141) reduces to

$$p(a,\theta) = \frac{2Ai\rho g}{\pi k a H_0^{(1)'}(ka)} + \frac{4Ai\rho g}{\pi k a} \sum_{n=1}^{\infty} \frac{i^n \cos n\theta}{H_n^{(1)'}(ka)}. \tag{143}$$

Last, the amplitude of the total horizontal force on the cylinder is in the direction of the x-axis and is given by

$$F = ah \int_0^{2\pi} p(a,\theta) \cos \theta \, d\theta = \frac{4A\rho g h}{k H_1^{(1)'}(ka)}, \tag{144}$$

where h is the depth of the fluid (sea).

12.8 Transform Methods

Most of the problems discussed thus far involve finite domains. For certain situations, however, it may be convenient to think of a long rod or string as if it were of infinite extent. For example, if we are interested in the temperature distribution in the middle portion of a long rod, prior to the time when such temperatures are greatly influenced by the BCs of the rod, we often model the problem as if the domain extended over $-\infty < x < \infty$. In other instances, we might model the problem so that the rod extends over $0 < x < \infty$ when we are interested in the temperature distribution only near the finite boundary at the origin.

Problems on infinite domains are usually best handled by the use of integral transforms. In our treatment here, we will illustrate the use of Fourier, Laplace, and Hankel transforms.

12.8.1 Heat conduction on an infinite domain: Fourier transform

We begin by considering the problem of heat conduction in an infinite rod when the initial temperature distribution is known. The problem is mathematically characterized by

$$
\begin{array}{ll}
& u_{xx} = a^{-2}u_t, \quad -\infty < x < \infty, \, t > 0, \\
\text{BC:} & u(x,t) \to 0, \, u_x(x,t) \to 0 \quad \text{as } |x| \to \infty, \, t > 0, \\
\text{IC:} & u(x,0) = f(x), \quad -\infty < x < \infty.
\end{array}
\tag{145}
$$

Because of the extent of the spatial variable x, we use the Fourier exponential transform. In particular, we will use the transform definition from Section 9.3. Hence, by introducing

$$
\mathscr{F}\{u(x,t); x \to s\} = \int_{-\infty}^{\infty} e^{-isx} u(x,t)\,dx = U(s,t),
\tag{146}
$$

it follows that

$$
\begin{aligned}
\mathscr{F}\{u_{xx}(x,t); x \to s\} &= -s^2 U(s,t), \\
\mathscr{F}\{u_t(x,t); x \to s\} &= U_t(s,t).
\end{aligned}
\tag{147}
$$

We should point out that the strict BCs appearing in (145) are necessary to ensure the existence of various Fourier transforms. Once we have produced a solution of the problem, we may be able to relax these conditions to only requiring a bounded solution.

Using the above relations, the transformed problem takes the form

$$
\begin{array}{ll}
& U_t + a^2 s^2 U = 0, \quad t > 0, \\
\text{IC:} & U(s,0) = F(s),
\end{array}
\tag{148}
$$

where $\mathscr{F}\{f(x); s\} = F(s)$. The solution of the initial value problem (148) is

$$
U(s,t) = F(s)e^{-a^2 s^2 t}, \quad -\infty < s < \infty.
\tag{149}
$$

The solution of the original problem is now recovered by taking the inverse Fourier transform, which by use of the convolution theorem yields

$$
u(x,t) = \int_{-\infty}^{\infty} f(\xi)g(x - \xi, t)\,d\xi,
\tag{150}
$$

where

$$
g(x,t) = \mathscr{F}^{-1}\{e^{-a^2 s^2 t}; s \to x\} = \frac{1}{2a\sqrt{\pi t}}e^{-x^2/4a^2 t}.
\tag{151}
$$

This last result can be obtained from Table 9.1. The formal solution of this problem is therefore

$$
u(x,t) = \frac{1}{2a\sqrt{\pi t}}\int_{-\infty}^{\infty} f(\xi)e^{-(x-\xi)^2/4a^2 t}\,d\xi.
\tag{152}
$$

By making the change of variable $z = (x - \xi)/2a\sqrt{t}$, we can express (152) in the alternate form

$$u(x,t) = \frac{1}{\sqrt{\pi}} \int_{-\infty}^{\infty} f(x - 2az\sqrt{t}) e^{-z^2} dz. \tag{153}$$

This last form of the solution is particularly useful if the initial temperature distribution takes on constant values along the axis. For example, if $f(x) = T_0$, then

$$u(x,t) = \frac{T_0}{\sqrt{\pi}} \int_{-\infty}^{\infty} e^{-z^2} dz = T_0. \tag{154}$$

Physically, this tells us that the temperature distribution remains at the constant value T_0, which is intuitively correct since there is no reason for the temperature to change.

The result given by (154) does not satisfy the BCs in (145) and, moreover, this function does not have a Fourier transform. In spite of this, the solution (154) is the correct result. Therefore, in handling problems by the integral transform method we usually take the approach of finding a "tentative solution" through formal reasoning, and then test the solution to see if it satisfies all "necessary" physical constraints of the original problem.

Suppose that the problem described by (145) is modified to the semi-infinite domain

$$
\begin{aligned}
&u_{xx} = a^{-2} u_t, \quad 0 < x < \infty, \ t > 0, \\
\text{BC:} \quad &u(0,t) = 0, \quad u(x,t) \to 0, \ u_x(x,t) \to 0 \quad \text{as } x \to \infty, \ t > 0, \\
\text{IC:} \quad &u(x,0) = f(x), \quad 0 < x < \infty.
\end{aligned} \tag{155}
$$

The semi-infinite spatial interval together with the BC at $x = 0$ suggests the use of the Fourier sine transform in this case. Hence, if we define

$$\mathscr{F}_S\{u(x,t); x \to s\} = \sqrt{\frac{2}{\pi}} \int_0^{\infty} u(x,t) \sin sx \, dx = U(s,t), \tag{156}$$

$$\mathscr{F}_S\{u_t(x,t); x \to s\} = \sqrt{\frac{2}{\pi}} \frac{\partial}{\partial t} \int_0^{\infty} u(x,t) \sin sx \, dx = U_t(s,t), \tag{157}$$

and, through two integrations by parts,

$$
\begin{aligned}
\mathscr{F}_S\{u_{xx}(x,t); x \to s\} &= -s^2 U(s,t) + \sqrt{\frac{2}{\pi}} s u(0,t) \\
&= -s^2 U(s,t),
\end{aligned} \tag{158}
$$

the transformed problem takes the form

$$
\begin{aligned}
&U_t + a^2 s^2 U = 0, \quad t > 0, \\
\text{IC:} \quad &U(s,0) = F(s),
\end{aligned} \tag{159}
$$

where $\mathscr{F}_S\{f(x); s\} = F(s)$. The solution of the initial value problem (159) is

$$U(s,t) = F(s)e^{-a^2s^2t}, \quad 0 < s < \infty, \tag{160}$$

and the inverse sine transform applied to (160) yields a solution in the integral form

$$u(x,t) = \sqrt{\frac{2}{\pi}} \int_0^\infty e^{-a^2s^2t} F(s)\sin sx\,dx. \tag{161}$$

The solution form given by (161) is not as convenient to use as that developed for the exponential transform using the convolution theorem. A convolution theorem for the sine transform does exist but it takes a more complicated form and may not be a good alternative in some cases. Last, we mention that we did not subscript the transformed functions with *S* in this problem. Because the sine transform is the only one used in this example, such additional notation is redundant, and possible confusing.

12.8.2 Heat conduction on a semi-infinite domain: Laplace transform

Let us now consider a very long rod, one end of which is exposed to a time-varying heat reservoir. We assume the initial temperature distribution is 0°C along the rod, and thus, the subsequent temperatures are solutions of

$$\boxed{\begin{aligned} u_{xx} &= a^{-2}u_t, \quad 0 < x < \infty,\ t > 0, \\ \text{BC:} \quad u(0,t) &= f(t), \quad u(x,t) \to 0, \quad \text{as } x \to \infty,\ t > 0, \\ \text{IC:} \quad u(x,0) &= 0, \quad 0 < x < \infty. \end{aligned}} \tag{162}$$

This problem is similar to that described by (155), but the end condition at $x = 0$ suggests that the Laplace transform (see Chapter 10) is the proper tool. Here we define

$$\mathcal{L}\{u(x,t);t \to p\} = \int_0^\infty e^{-pt}u(x,t)dt = U(x,p), \tag{163}$$

from which it follows

$$\begin{aligned} \mathcal{L}\{u_{xx}(x,t);t \to p\} &= U_{xx}(x,p), \\ \mathcal{L}\{u_t(x,t);t \to p\} &= pU(x,p). \end{aligned} \tag{164}$$

Consequently, the transformed problem becomes

$$U_{xx} - \left(\frac{p}{a^2}\right)U = 0, \quad x > 0, \tag{165}$$

$$\text{BC:} \quad U(0,p) = F(p), \quad U(x,p) \to 0 \text{ as } x \to \infty,$$

where $F(p) = \mathcal{L}\{f(t)\}$. The bounded solution of (165) is readily found to be

$$U(x,p) = F(p)e^{-x\sqrt{p}/a}, \tag{166}$$

which through use of the convolution theorem yields

$$
\begin{aligned}
u(x,t) &= \mathcal{L}^{-1}\{F(p)e^{-x\sqrt{p}/a}; p \to t\} \\
&= \int_0^t f(\tau)g(x,t-\tau)d\tau,
\end{aligned}
$$

(167)

where

$$
g(x,t) = \mathcal{L}^{-1}\{e^{-x\sqrt{p}/a}; p \to t\} = \frac{x}{2a\sqrt{\pi}\, t^{3/2}}e^{-x^2/4a^2 t}.
$$

(168)

Last, the substitution of (168) into (167) leads to the formal solution

$$
u(x,t) = \frac{x}{2a\sqrt{\pi}}\int_0^t \frac{f(\tau)}{(t-\tau)^{3/2}}\exp\left[-\frac{x^2}{4a^2(t-\tau)}\right]d\tau.
$$

(169)

EXAMPLE 6 Solve (162) for the special case in which $f(t) = T_0$ (constant).

Solution: The formal solution is that given by (169). By making the change of variable $z = x/2a\sqrt{t-\tau}$, we find that (169) becomes

$$
u(x,t) = \frac{2}{\sqrt{\pi}}\int_{x/2a\sqrt{t}}^{\infty} f(t-x^2/4a^2 z^2)e^{-z^2}dz = \frac{2T_0}{\sqrt{\pi}}\int_{x/2a\sqrt{t}}^{\infty} e^{-z^2}dz.
$$

Recalling the definition of the complementary error function (Section 2.3.4)

$$
\mathrm{erfc}(x) = \frac{2}{\sqrt{\pi}}\int_x^{\infty} e^{-z^2}dz,
$$

we can express the above solution as

$$
u(x,t) = T_0\,\mathrm{erfc}\left(\frac{x}{2a\sqrt{t}}\right).
$$

The physical implication of the solution suggests that for any fixed value of x, the temperature in the rod at that point will eventually approach T_0 if we wait long enough $(t \to \infty)$. Also, at any instant of time t the temperature $u(x,t)$ will tend to zero as $x \to \infty$. Finally, we recognize that the temperature remains constant along any member of the family of parabolas in the xt-plane described by

$$
\frac{x}{2a\sqrt{t}} = \text{constant}.
$$

12.8.3 Nonhomogeneous wave equation

We now wish to use the Fourier and Laplace transform methods together to solve the nonhomogeneous wave equation described by

$$u_{xx} = c^{-2}u_{tt} - q(x,t), \quad -\infty < x < \infty, \ t > 0$$

$$\text{IC:} \quad u(x,0) = f(x), \ u_t(x,0) = g(x), \quad -\infty < x < \infty. \tag{170}$$

To begin, it is convenient to split the problem into two problems. We let $u_H(x,t)$ denote the solution of the homogeneous equation and specified ICs given by

$$u_{xx} = c^{-2}u_{tt}, \quad -\infty < x < \infty, \ t > 0$$

$$\text{IC:} \quad u(x,0) = f(x), \ u_t(x,0) = g(x), \quad -\infty < x < \infty, \tag{171}$$

and $u_P(x,t)$ is the solution of the nonhomogeneous equation and homogeneous ICs[1]

$$u_{xx} = c^{-2}u_{tt} - q(x,t), \quad -\infty < x < \infty, \ t > 0$$

$$\text{IC:} \quad u(x,0) = 0, \ u_t(x,0) = 0, \quad -\infty < x < \infty. \tag{172}$$

By use of the linearity property, the solution we seek is $u(x,t) = u_H(x,t) + u_P(x,t)$.

The infinite extent on x suggests the use of the exponential Fourier transform. Hence, by using the relations (146) and (147), the transformation of the problem (171) leads to

$$U_{tt} + c^2 s^2 U = 0, \quad t > 0,$$

$$\text{IC:} \quad U(s,0) = F(s), \ U_t(s,0) = G(s). \tag{173}$$

By standard solution techniques, we find

$$U_H(s,t) = F(s)\cos cst + \frac{G(s)}{cs}\sin cst, \tag{174}$$

the inverse Fourier transform of which yields

$$u_H(x,t) = \frac{1}{2\pi}\int_{-\infty}^{\infty} e^{isx}\left[F(s)\cos cst + \frac{G(s)}{cs}\sin cst\right]ds. \tag{175}$$

By the use of Euler's formulas

$$\cos x = \frac{1}{2}(e^{ix} + e^{-ix}), \quad \sin x = \frac{1}{2i}(e^{ix} - e^{-ix}), \tag{176}$$

we can write

---

[1]The use of subscripts H and P for the solutions of (171) and (172) should not be confused with partial derivatives involving the variables x and t.

$$u_H(x,t) = \frac{1}{4\pi}\int_{-\infty}^{\infty}\left[e^{is(x+ct)} + e^{is(x-ct)}\right]F(s)\,ds + \frac{1}{4\pi}\int_{-\infty}^{\infty}\left[e^{is(x+ct)} - e^{is(x-ct)}\right]\frac{G(s)}{ics}\,ds. \quad (177)$$

From the inverse transform relation

$$f(x) = \frac{1}{2\pi}\int_{-\infty}^{\infty}e^{isx}F(s)\,ds, \quad (178)$$

we recognize the first integral in (177) as

$$\frac{1}{4\pi}\int_{-\infty}^{\infty}\left[e^{is(x+ct)} + e^{is(x-ct)}\right]F(s)\,ds = \frac{1}{2}\left[f(x+ct) + f(x-ct)\right]. \quad (179)$$

The second integral in (177) can be similarly identified if we start with

$$g(z) = \frac{1}{2\pi}\int_{-\infty}^{\infty}e^{isz}G(s)\,ds \quad (180)$$

and integrate both sides to find

$$\int_{x-ct}^{x+ct}g(z)\,dz = \frac{1}{2\pi}\int_{-\infty}^{\infty}\left[e^{is(x+ct)} - e^{is(x-ct)}\right]\frac{G(s)}{is}\,ds. \quad (181)$$

By combining the results of (179) and (181), we obtain the solution

$$u_H(x,t) = \frac{1}{2}\left[f(x+ct) + f(x-ct)\right] + \frac{1}{2c}\int_{x-ct}^{x+ct}g(z)\,dz, \quad (182)$$

which we recognize as *d'Alembert's solution* once again (see Section 12.4.1).

The Fourier transform applied to the second problem (172) leads to

$$U_{tt} + c^2s^2U = Q(s,t), \quad t > 0,$$

$$\text{IC:} \quad U(s,0) = 0, \ U_t(s,0) = 0. \quad (183)$$

To solve (183), we can resort to the Laplace transform and define

$$\mathcal{L}\{U(s,t); t \to p\} = \hat{U}(s,p), \quad \mathcal{L}\{U_{tt}(s,t); t \to p\} = p^2\hat{U}(s,p), \quad (184)$$

where we have incorporated the homogeneous initial conditions. The transformed problem takes the simple form

$$(p^2 + c^2s^2)\hat{U}(s,p) = \hat{Q}(s,p), \quad (185)$$

where $\hat{Q}(s,p) = \mathcal{L}\{Q(s,t); t \to p\}$. The solution of (185) is

$$\hat{U}(s,p) = \frac{\hat{Q}(s,p)}{p^2 + c^2 s^2}, \tag{186}$$

and by invoking the convolution theorem of the Laplace transform, we obtain

$$U_P(s,t) = \frac{1}{cs} \int_0^t \sin[cs(t-\tau)]Q(s,\tau)d\tau. \tag{187}$$

To invert (187) by the inverse Fourier transform, we formally write

$$\begin{aligned}
u_P(x,t) &= \frac{1}{2\pi} \int_0^t \int_{-\infty}^{\infty} e^{isx} \sin[cs(t-\tau)] \frac{Q(s,\tau)}{cs} ds \, d\tau \\
&= \frac{1}{4\pi} \int_0^t \int_{-\infty}^{\infty} \left\{ e^{is[x+c(t-\tau)]} - e^{is[x-c(t-\tau)]} \right\} \frac{Q(s,\tau)}{ics} ds \, d\tau,
\end{aligned} \tag{188}$$

where we have utilized Euler's formula (176) once again. Similar to the result of Eq. (181), we now deduce that

$$\int_{x-c(t-\tau)}^{x+c(t-\tau)} q(z,\tau)dz = \frac{1}{2\pi} \int_{-\infty}^{\infty} \left\{ e^{is[x+c(t-\tau)]} - e^{is[x-c(t-\tau)]} \right\} \frac{Q(s,\tau)}{is} ds, \tag{189}$$

and hence, (188) can be written in the form

$$u_P(x,t) = \frac{1}{2c} \int_0^t \int_{x-c(t-\tau)}^{x+c(t-\tau)} q(z,\tau)dz \, d\tau. \tag{190}$$

The sum of (182) and (190) leads to our solution (now independent of transforms)

$$\boxed{\begin{aligned}
u(x,t) &= u_H(x,t) + u_P(x,t) \\
&= \frac{1}{2}\left[f(x+ct) + f(x-ct) \right] + \frac{1}{2c} \int_{x-ct}^{x+ct} g(z)dz \\
&\quad + \frac{1}{2c} \int_0^t \int_{x-c(t-\tau)}^{x+c(t-\tau)} q(z,\tau)dz \, d\tau.
\end{aligned}} \tag{191}$$

12.8.4 Poisson integral formula for the half-plane

Suppose we wish to find the steady-state temperature distribution in a very large rectangular plate when the temperature is prescribed by the function f along one edge of the plate and tends to zero along each of the other edges. Such a problem is mathematically described by

$$\boxed{\begin{aligned}
& u_{xx} + u_{yy} = 0, \quad -\infty < x < \infty, \; y > 0, \\
\text{BC:} \quad & \begin{cases} u(x,0) = f(x), & -\infty < x < \infty, \\ u(x,y) \to 0 & \text{as} \quad x^2 + y^2 \to \infty. \end{cases}
\end{aligned}} \tag{192}$$

In this case we have a choice of integral transform to use. We can either use the Fourier exponential transform on the variable x or the Fourier sine transform on the variable y. Here we use the former. The transformed problem becomes

$$U_{yy} - s^2 U = 0, \quad y > 0,$$
$$\text{BC:} \quad U(s,y) = F(s), \tag{193}$$

from which we deduce the general solution

$$U(s,y) = A(s)e^{-sy} + B(s)e^{sy}, \tag{194}$$

where $A(s)$ and $B(s)$ are arbitrary functions. From the BC in (193), we see that

$$U(s,y) = \begin{cases} F(s)e^{sy}, & s < 0, \\ F(s)e^{-sy}, & s > 0, \end{cases} \tag{195}$$

or, equivalently,

$$U(s,y) = F(s)e^{-|s|y}, \quad -\infty < s < \infty. \tag{196}$$

Based on the inverse Fourier transform relations

$$u(x,y) = \mathscr{F}^{-1}\{U(s,y); s \to x\} = \mathscr{F}^{-1}\{F(s)e^{-|s|y}; s \to x\}, \tag{197}$$

$$\mathscr{F}^{-1}\{e^{-|s|y}; s \to x\} = \frac{1}{\pi}\frac{y}{x^2 + y^2}, \tag{198}$$

and the convolution theorem, it follows that

$$u(x,y) = \frac{y}{\pi}\int_{-\infty}^{\infty} \frac{f(\xi)\,d\xi}{(\xi - x)^2 + y^2}, \quad y > 0. \tag{199}$$

Equation (199) is called the *Poisson integral formula for the half-plane*. In addition to representing solutions for all Dirichlet problems formulated in the upper half-plane, it is a useful result for developing certain properties of harmonic functions in this domain (see also Section 7.8.1).

EXAMPLE 7 Use the Poisson integral formula (199) to solve the Dirichlet problem (192) when

$$f(x) = \begin{cases} T_0, & x < 0 \\ 0, & x > 0. \end{cases}$$

Solution: The substitution of this function directly into (199) yields

$$u(x,y) = \frac{yT_0}{\pi} \int_{-\infty}^{0} \frac{d\xi}{(\xi-x)^2 + y^2}$$

$$= \frac{T_0}{\pi}\left(\frac{\pi}{2} - \tan^{-1}\frac{x}{y}\right).$$

By using the trigonometric identity

$$\frac{\pi}{2} - \tan^{-1}\alpha = \tan^{-1}\frac{1}{\alpha},$$

the above solution can be put in the more useful form

$$u(x,y) = \frac{T_0}{\pi}\tan^{-1}\frac{y}{x}.$$

Curves in the upper half-plane for which the steady-state temperature is constant are called *isotherms*. For our particular problem, these curves are defined by a family of lines through the origin defined by (C constant)

$$y = Cx.$$

12.8.5 Axisymmetric Dirichlet problem for a half-space: Hankel transform

One of the principal uses of the Hankel transform is in the solution of boundary-value problems involving cylindrical coordinates. For example, let us consider the axisymmetric Dirichlet problem for the half-space $z > 0$, which is described by

$$
\boxed{
\begin{array}{l}
u_{rr} + \dfrac{1}{r}u_r + u_{zz} = 0, \quad 0 < r < \infty, \ z > 0 \\[2mm]
\text{BC:} \quad \begin{cases} u(r,0) = f(r), \quad 0 < r < \infty, \\ u(r,z) \to 0 \quad \text{as} \quad r^2 + z^2 \to \infty, \ z > 0 \end{cases}
\end{array}
}
\tag{200}
$$

If we apply the Hankel transform of order zero (see Chapter 10) to the radial variable r, we find that

$$\mathcal{H}_0\{u(r,z); r \to \rho\} = U(\rho), \quad \mathcal{H}_0\{f(r)\} = F(\rho),$$

$$\mathcal{H}_0\left\{u_{rr} + \frac{1}{r}u_r\right\} = \mathcal{H}_0\left\{\frac{1}{r}\frac{d}{dr}(ru_r)\right\} = -\rho^2 U(\rho),
\tag{201}$$

and the problem described by (200) becomes

$$U_{zz} - \rho^2 U = 0, \quad z > 0$$

BC: $\quad \begin{cases} U(\rho,0) = F(\rho), \\ U(\rho,z) \to 0 \quad \text{as} \quad z \to \infty. \end{cases}$ \hfill (202)

Clearly, the solution of (202) is $U(\rho,z) = F(\rho)e^{-\rho z}$, and inverting it, we obtain

$$u(r,z) = \mathcal{H}_0^{-1}\{F(\rho)e^{-\rho z}; \rho \to r\} = \int_0^\infty \rho F(\rho) e^{-\rho z} J_0(\rho r) d\rho. \tag{203}$$

EXAMPLE 8 Solve the problem described by (200) for the special case where

$$u(r,0) = f(r) = \frac{1}{\sqrt{r^2 + a^2}}, \quad a > 0.$$

Solution: We first observe that

$$\mathcal{H}_0\left\{\frac{1}{\sqrt{r^2 + a^2}}\right\} = F(\rho) = \frac{1}{\rho}e^{-a\rho},$$

which we obtain from Table 10.5 by interchanging the roles of r and ρ. By substituting this result into the solution formula (203), we obtain the Laplace transform integral

$$u(r,z) = \int_0^\infty e^{-(z+a)\rho} J_0(\rho r) d\rho$$
$$= \mathcal{L}\{J_0(\rho r); \rho \to (z+a)\},$$

from which we deduce (see Table 10.1)

$$u(r,z) = \frac{1}{\sqrt{r^2 + (z+a)^2}}.$$

Suggested Reading

L. C. Andrews, *Elementary Partial Differential Equations with Boundary Value Problems* (Academic Press, Orlando, 1986).

L. C. Andrews and B. K. Shivamoggi, *Integral Transforms for Engineers* (SPIE Press, Bellingham, 1999).

R. V. Churchill and J. W. Brown, *Fourier Series and Boundary Value Problems*, 4th ed. (McGraw-Hill, New York, 1987).

M. D. Greenberg, *Foundations of Applied Mathematics* (Prentice-Hall, Upper Saddle River, 1978).

R. Haberman, *Elementary Applied Partial Differential Equations with Fourier Series and Boundary Value Problems*, 3 rd ed. (Prentice-Hall, Upper Saddle River, 1998).

I. N. Sneddon, *The Use of Integral Transforms* (McGraw-Hill, New York, 1972).

E. Zauderer, *Partial Differential Equations of Applied Mathematics*, 2nd ed. (Wiley, New York, 1989).

Chapter 12

Exercises

In Problems 1 and 2, solve the given heat equation.

1. $u_{xx} = a^{-2}u_t$, $0 < x < p$, $t > 0$
BC: $u_x(0,t) = 0$, $u_x(p,t) = 0$,
IC: $u(x,0) = T_0 \sin^2(\pi x/p)$

Ans. $\dfrac{T_0}{2}\left[1 - \cos\left(\dfrac{2\pi x}{p}\right)e^{-4a^2n^2\pi^2t/p^2}\right]$

2. $u_{xx} = a^{-2}u_t$, $0 < x < 1$, $t > 0$
BC: $u(0,t) = T_0$, $u(1,t) = T_0$,
IC: $u(x,0) = T_0 + x(1 - x)$

Ans. $T_0 + \dfrac{8}{\pi^3}\displaystyle\sum_{\substack{n=1 \\ (odd)}}^{\infty} \dfrac{\sin n\pi x}{n^3}e^{-a^2n^2\pi^2t}$

3. The ends $x = 0$ and $x = 100$ of a rod 100 cm in length, with insulated lateral surface, are held at temperature $0°$ and $100°C$, respectively, until steady-state conditions prevail. Then, at time $t = 0$, the temperatures of the two ends are interchanged. What is the resulting temperature distribution?

Ans. $100 - x$
$-\dfrac{200}{\pi}\displaystyle\sum_{n=1}^{\infty} \dfrac{1}{n}\sin\dfrac{n\pi x}{50}e^{-a^2n^2\pi^2t/2500}$

In Problems 4 and 5, solve the given wave equation.

4. $u_{xx} = c^{-2}u_{tt}$, $0 < x < 1$, $t > 0$
BC: $u(0,t) = 0$, $u(1,t) = 0$,
IC: $u(x,0) = 0$, $u_t(x,0) = v_0$

Ans. $\dfrac{4v_0}{c\pi^2}\displaystyle\sum_{\substack{n=1 \\ (odd)}}^{\infty} \dfrac{1}{n^2}\sin nc\pi t \sin n\pi x$

5. $u_{xx} + 2u_x + u = u_{tt}$, $0 < x < \pi$, $t > 0$
BC: $u(0,t) = 0$, $u(\pi,t) = 0$,
IC: $u(x,0) = e^{-x}$, $u_t(x,0) = 0$

Ans. $\dfrac{4}{\pi}e^{-x}\displaystyle\sum_{\substack{n=1 \\ (odd)}}^{\infty} \dfrac{1}{n}\cos nt \sin nx$

6. A string π meters long is stretched between fixed supports until the wave speed $c = 40$ m/s. If the string is given an initial velocity $4\sin x$ from its equilibrium position, find the maximum displacement and state its location.

Ans. 10 cm at $x = \pi/2$
and $t = \pi/80$ s.

7. A uniform string occupying the domain $0 \le x < \infty$ begins its motion with initial displacement $f(x)$ and initial velocity $g(x)$. Show that the later motion is that of the right-half of a doubly infinite string provided

$$f(x) + f(-x) = -\frac{1}{c} \int_{-x}^{x} g(z)dz.$$

In Problems 8-10, solve the given potential problem.

8. $u_{xx} + u_{yy} = 0,\ 0<x<\pi,\ 0<y<1,$
BC: $u(0,y) = 0,\ u(\pi,y) = 0,$
$u(x,0) = 0,\ u(x,1) = \sin x$

Ans. $\dfrac{\sinh y}{\sinh 1} \sin x$

9. $u_{xx} + u_{yy} = 0,\ 0<x<1,\ 0<y<1,$
BC: $u_x(0,y) = 0,\ u_x(1,y) = 0,$
$u(x,0) = x^2,\ u_y(x,1) = 0$

Ans. $\dfrac{1}{3} + \dfrac{4}{\pi^2} \sum_{n=1}^{\infty} \dfrac{(-1)^n}{n^2 \cosh n\pi}$
$\times \cosh[n\pi(1-y)] \cos n\pi x$

10. $u_{rr} + \dfrac{1}{r}u_r + \dfrac{1}{r^2}u_{\theta\theta} = 0,\ 0<r<1,$

BC: $u(1,\theta) = \begin{cases} 0, & -\pi<\theta<0 \\ T_0, & 0<\theta<\pi \end{cases}$

Ans. $\dfrac{1}{2}T_0$
$+ \dfrac{2T_0}{\pi} \sum_{n=1}^{\infty} \dfrac{r^{2n-1}}{2n-1} \sin[(2n-1)\theta]$

11. Determine the steady-state temperature distribution in a semi-circular disk bounded by $y = \sqrt{1-x^2}$, $y = 0$, with 0°C prescribed along the diameter $y = 0$, while along the circumference the temperature is
(a) $u(1,\theta) = \sin\theta$.
(b) $u(1,\theta) = T_0$ (const.).

Ans. (a) $r\sin\theta$
(b) $\dfrac{4T_0}{\pi} \sum_{\substack{n=1 \\ (\text{odd})}}^{\infty} \dfrac{r^n}{n} \sin n\theta$

12. Solve the heat conduction problem

$u_{xx} = a^{-2}u_t,\ 0<x<1,\ t>0$
BC: $u(0,t) = T_1,\ u(1,t) + u_x(1,t) = T_2,$
IC: $u(x,0) = T_1$

Ans. $T_1 + \dfrac{1}{2}(T_2 - T_1)x$
$+ 2(T_2 - T_1) \sum_{n=1}^{\infty} \dfrac{\cos k_n \sin k_n x}{k_n(1 \cos^2 k_n)} e^{-a^2 k_n^2 t},$
where: $\sin k_n + k_n \cos k_n = 0$

In Problems 13-15, solve the given nonhomogeneous equation.

13. $u_{xx} = a^{-2}u_t - A\cos\omega t,$
BC: $u_x(0,t) = 0,\ u_x(\pi,t) = 0,$
IC: $u(x,0) = 0$

Ans. $\dfrac{a^2 A}{\omega} \sin\omega t$

14. $u_{xx} = u_t - e^{-t},$
BC: $u_x(0,t) = 0,\ u(1,t) = 0,$
IC: $u(x,0) = 0$

Ans. $\dfrac{16}{\pi} \sum_{n=1}^{\infty} \dfrac{(-1)^{n-1} \cos(n-\frac{1}{2})\pi x}{(2n-1)[(2n-1)^2\pi^2 - 4]}$
$\times\left[e^{-t} - e^{-(n-\frac{1}{2})^2\pi^2 t}\right]$

15. $u_{xx} = c^{-2}u_{tt} - P\cos\omega t,$
BC: $u(0,t) = 0,\ u(\pi,t) = 0,$
IC: $u(x,0) = 0,\ u_t(x,0) = 0$

Ans. $\dfrac{4Pc^2}{\pi} \sum_{\substack{n=1 \\ (\text{odd})}}^{\infty} \dfrac{\sin nx}{n(\omega^2 - n^2c^2)}$
$\times(\cos nct - \cos\omega t)$

16. The wind is blowing over a suspension cable on a bridge that is 15 m long. The distributed force caused by the wind is roughly $0.02\sin(21\pi t)$. Assume the tension in the cable is

40,000 N and the cable has a mass of 10 kg/m. Determine the subsequent motion. Will resonance occur?

Ans. $-\dfrac{2.448}{40,000}\displaystyle\sum_{\substack{n=1 \\ (\text{odd})}}^{\infty} \sin\dfrac{n\pi x}{15}$

$\times \left[\dfrac{21\sin(4.216n\pi t) - 4.216n\sin(21\pi t)}{n^2(17.77n^2 - 441)}\right]$

Some resonance is likely because the harmonic $\omega_5 = 21.08\pi$ is close to that of the input function $\omega = 21\pi$.

17. Given

$$u_{rr} + \frac{1}{r}u_r = c^{-2}u_{tt},\ \ 0 < r < 1,\ t > 0$$

BC: $u(1,t) = 0,\ u_x(\pi,t) = 0,$

IC: $u(r,0) = 0,\ u_t(r,0) = 1,$

show that

$$u(r,t) = \frac{2}{c}\sum_{n=1}^{\infty}\frac{\sin(k_n ct)}{k_n^2 J_1(k_n)}J_0(k_n r),$$

where $J_0(k_n) = 0,\ n = 1,2,3,\dots$.

18. Over a long, solid cylinder of unit radius and uniform temperature distribution T_1 is fitted a long, hollow cylinder $1 \le r \le 2$ of the same material at uniform temperature T_2. Find the temperature distribution throughout the two cylinders.

Ans. $T_2 + \dfrac{1}{2}(T_1 - T_2)$

$\times \displaystyle\sum_{n=1}^{\infty}\dfrac{J_1(k_n)}{k_n[J_1(2k_n)]^2}J_0(k_n r)e^{-a^2 k_n^2 t},$

where $J_0(2k_n) = 0,\ n = 1,2,3,\dots$.

19. Find the steady-state temperature distribution in a solid homogeneous cylinder of unit radius for which

$$u_{rr} + \frac{1}{r}u_r + u_{zz} = 0,\ 0 < r < 1,\ 0 < z < \pi,$$

BC: $u(1,z) = T_0,\ u(r,0) = 0,\ u_z(r,\pi) = 0.$

Ans.

$\dfrac{4T_0}{\pi}\displaystyle\sum_{n=1}^{\infty}\dfrac{I_0[(n-1/2)r]}{(2n-1)I_0(n-1/2)}\sin(n-1/2)z$

In Problems 20-22, use the Fourier transform to solve the given BVP.

20. Solve the problem described by (145) when $u(x,0) = e^{-x^2/4a^2}$.

Ans. $\dfrac{1}{\sqrt{1+t}}e^{-x^2/4a^2(1+t)}$

21. Solve the problem described by (145) when

$$u(x,0) = \begin{cases} T_0, & |x| < 1 \\ 0, & |x| > 1. \end{cases}$$

Ans.

$\dfrac{T_0}{2}\left[\text{erf}\left(\dfrac{x+1}{2a\sqrt{t}}\right) - \text{erf}\left(\dfrac{x-1}{2a\sqrt{t}}\right)\right]$

22. $u_{xx} + u_{yy} = 0, 0 < x < \infty, y > 0$

BC: $u(0,y) = 0,\ u(x,0) = T_0,$

Hint: Use the sine transform.

Ans. $\dfrac{2T_0}{\pi}\arctan\dfrac{x}{y}$

In Problems 22 and 23, use the Laplace transform to solve the given BVP.

23. $u_{xx} = a^{-2}u_t, 0 < x < \infty, t > 0$

BC: $u_x(0,t) = -T_0,$

IC: $u(x,0) = 0$

Ans.

$T_0\left[2a\sqrt{\dfrac{t}{\pi}}e^{-x^2/4a^2 t} - x\,\text{erfc}\left(\dfrac{x}{2a\sqrt{t}}\right)\right]$

24. Consider the motions of a string fastened at the origin but whose far end is looped around a "frictionless" peg that exerts no vertical force on the loop. Determine the displacements given that the string is moving under the action of gravity and the problem is characterized by (K = const.)

$$c^2 u_{xx} = u_{tt} - K, 0 < x < \infty, t > 0$$

BC: $u(0,t) = 0,$

IC: $u(x,0) = 0, u_t(x,0) = 0.$

Ans. $\dfrac{K}{2}\left[t^2 - (t - x/c)^2 U(t - x/c)\right]$

25. Given the BVP

$$u_{xx} = u_t, 0 < x < 1, t > 0$$

BC: $u(0,t) = 0, u(1,t) = T_0,$

IC: $u(x,0) = T_0,$

(*a*) use the Laplace transform to obtain the solution of the transformed problem

$$U(x,p) = \frac{T_0}{p}\left\{1 - e^{-\sqrt{p}}\left[\frac{1 - e^{-2(1-x)\sqrt{p}}}{1 - e^{-2\sqrt{p}}}\right]\right\}.$$

(*b*) By expanding $1/(1 - e^{-2\sqrt{p}})$ in a series of ascending powers of $e^{-2\sqrt{p}}$, show that the solution in (*a*) becomes

$$U(x,p) = \frac{T_0}{p}\Big[1 - e^{-x\sqrt{p}}$$
$$+ e^{-(2-x)\sqrt{p}} - e^{-(2+x)\sqrt{p}} + \cdots\Big].$$

(*c*) Given that

$$\mathcal{L}\left\{\mathrm{erf}\left(\frac{1}{\sqrt{t}}\right)\right\} = \frac{1}{p}(1 - e^{-2\sqrt{p}}),$$

invert the solution in (*b*) to deduce that

$$u(x,t) = T_0\left[\mathrm{erf}\left(\frac{x}{2\sqrt{t}}\right) + \mathrm{erfc}\left(\frac{2-x}{2\sqrt{t}}\right)\right.$$
$$\left. - \mathrm{erfc}\left(\frac{2-x}{2\sqrt{t}}\right) + \cdots\right].$$

Chapter 13

Probability and Random Variables

13.1 Introduction / 638
13.2 Random Variables and Probability Distributions / 640
    13.2.1 Cumulative distribution function / 640
    13.2.2 Probability density function / 642
    13.2.3 Discrete random variables / 644
13.3 Examples of Density Functions / 646
    13.3.1 Gaussian (or normal) distribution / 646
    13.3.2 Uniform distribution / 648
    13.3.3 Rayleigh distribution / 648
    13.3.4 Gamma distribution / 649
13.4 Expected Values / 649
    13.4.1 Higher-order moments / 650
    13.4.2 Characteristic functions / 653
13.5 Conditional Probability / 655
    13.5.1 Conditional CDF and PDF / 656
    13.5.2 Expected values / 657
13.6 Functions of One Random Variable / 658
    13.6.1 Continuous random variables / 659
    13.6.2 Expected values / 662
    13.6.3 Characteristic function method / 663
13.7 Two Random Variables / 665
    13.7.1 Joint distribution and density functions / 665
    13.7.2 Marginal density functions / 667
    13.7.3 Conditional distributions and densities / 668
    13.7.4 Independent random variables / 672
    13.7.5 Expected values / 675
    13.7.6 Moments and correlation / 673
    13.7.7 Bivariate Gaussian distribution / 675
13.8 Functions of Two or More Random Variables / 677
    13.8.1 Sums of two random variables / 678
    13.8.2 Rician distribution / 681
    13.8.3 Products of random variables / 683
    13.8.4 Quotients of random variables / 685
    13.8.5 Two functions of two random variables / 686
    13.8.6 Sums of several random variables / 688
13.9 Limit Distributions / 690
    13.9.1 Gaussian density function / 690
    13.9.2 Gamma density function / 691
    Suggested Reading / 692
    Exercises / 693

> **Historical Comments:** The origins of probability theory can be traced back to the correspondence between Blaise Pascal (1623-1662) and Pierre Fermat (1601-1665) concerning gambling games. Their theory, considered the first foundation of probability theory, remained largely a tool reserved for use in games of chance until Pierre S. Laplace (1749-1827) and Karl Friedrich Gauss (1777-1855) applied it to other problems.
>
> Further interest in probability was generated when it was recognized that the probability of an event, e.g., in the kinetic theory of gases and in many social and biological phenomena, often depends on preceding outcomes. In Russia, for example, the study of such linked chains of events (now known as Markov chains or Markov processes) was initiated in 1906-1907 by Andrei A. Markov (1856-1922), a student of Chebyshev. Important advances in Markov processes were made by Andrei N. Kolmogorov (1903-1987) in 1931. Kolmogorov is also credited with establishing modern probability theory in 1933 by his use of the theory of measure and integration advanced in the early twentieth century by Henri Lebesgue (1875-1941) and Félix E. E. Borel (1871-1956).

> The objective of this chapter is to provide an overview of probability theory that is useful in physical applications. In particular, the theory presented here makes it possible to mathematically describe or model random signals that arise, for example, in the analysis of communication systems. Such random signals, called random processes, are discussed in further detail in Chapter 14.

13.1 Introduction

The mathematical theory developed in basic courses in engineering and the physical sciences is usually based on *deterministic* phenomena. As an example, the input to a linear filter is often presumed to be a deterministic quantity, such as a sine wave, step function, impulse function, and so on, leading to a deterministic output. However, in practice the input to a filter may contain a fluctuating or "random" quantity (noise) that yields some uncertainty about the output. In general, an unpredictable noise signal always appears at the input to any communication receiver and thus interferes with the reception of incoming radio or radar "signals." Situations like this that involve uncertainty or randomness in some form cannot be analyzed by deterministic methods but must be treated by *probabilistic methods*. Probability theory has become an indispensable tool in engineering and scientific analysis involving electron emission, radar detection, quality control, statistical mechanics, turbulence, and noise, among other areas.

Some of the terms used in discussing random happenings include the following:

- *random experiment*—one in which the outcome is uncertain before the experiment is performed (e.g., flipping a coin, etc.). A single instance of an experiment is called a *trial*.
- *event*—a collection of possible outcomes or a random experiment (also called an *outcome* or *sample point*).
- *sample space* —the whole set of possible outcomes (called the *universal set*).

When the outcomes of a random experiment are countable, we say they are *discrete* (e.g., tossing a coin, rolling a die, etc.). Outcomes that are not countable are said to form a *continuum*. Fundamentally, there are two approaches to the study of probability—the *relative-frequency approach* and the *axiomatic approach*.

The relative-frequency approach to probability is closely linked to the frequency of occurrence of a defined "event." In the classical sense we let N denote the number of outcomes of a given experiment, all of which are assumed to be "equally likely," and let N_A denote the number of outcomes "favorable" to event A. The *relative frequency* of event A is then defined by the ratio

$$r(A) = \frac{N_A}{N}. \tag{1}$$

If N is sufficiently large, we can roughly associate the relative frequency (1) with the probability $\Pr(A)$ of the event A. If we let S denote the universal set (entire space of possible outcomes), it follows that $\Pr(S) = 1$. That is, a "sure thing" has probability one.

The relative-frequency approach is useful when first learning about probability because it attempts to attach some physical significance to the probabilities involved in simple engineering and physical science applications. However, the relative-frequency approach to probability is severely limited and cannot handle the complexity of many physical situations. For these complex situations it is much easier to use the axiomatic approach that we describe below.

Again we let S denote our universal sample space. With each event $A \in S$ we associate a number $\Pr(A)$, called the *probability* of A, such that the following *axioms of probability* are satisfied:

Axiom I. For every $A \in S$,

$$0 \leq \Pr(A) \leq 1.$$

Axiom II. The entire universal sample space S has the probability

$$\Pr(S) = 1.$$

Axiom III. For mutually exclusive events A and B,

$$\Pr(A \text{ and } B) = \Pr(A) + \Pr(B).$$

The *axiomatic approach* to probability, which emphasizes the deductive character of the theory, is a rigorous approach where the probabilities of an event satisfy the above axioms (postulates). Other probabilities are then consequences of these axioms. Although rigorous mathematical treatments based on Kolmogorov's classic work in 1933 often require knowledge of measure theory, this is usually not necessary in understanding the basic ideas and applying them to engineering and physics problems. In our approach to probability we will assume the reader has some fundamental intuitive understanding of probability concepts. Hence, we will start with the notion of a *random variable* and build on the above axioms of probability.

13.2 Random Variables and Probability Distributions

A *probability distribution* shows the probabilities that we observe in an experiment. The quantity that we observe in a given trial of an experiment is a number called a *random variable* because the value it will assume in the next trial is "random" and, consequently, depends on chance. Owing to its random nature, we may assign probabilities to the possible values (events) of the variable. In the following, random variables will be designated by boldface letters such as **x** and **y**.

▸ **Discrete random variable:** If the random variable can only take on certain discrete values, we say it is a *discrete random variable*.

▸ **Continuous random variable:** If a random variable can assume any value within a specified range (possibly infinite), it is called a *continuous random variable*. In our discussions we concentrate primarily on continuous random variables.

13.2.1 Cumulative distribution function

For a given random variable **x**, there are three primary types of events to consider involving probabilities. These events are, respectively,

$$\{\mathbf{x} \le a\}, \quad \{a < \mathbf{x} \le b\}, \quad \{\mathbf{x} > b\}.$$

For the general event $\{\mathbf{x} \le x\}$, where x is any real number, we introduce the function

$$\boxed{F_{\mathbf{x}}(x) = \Pr(\mathbf{x} \le x), \quad -\infty < x < \infty,} \tag{2}$$

called the *cumulative distribution function* (CDF) or, more briefly, the *distribution function*. In words, the CDF is the probability of the event:

> "The random variable **x** is less than or equal to the value x in a trial of some chance experiment."

Observe that the events $\{\mathbf{x} \le a\}$ and $\{a < \mathbf{x} \le b\}$ are mutually exclusive events, and also that

$$\{\mathbf{x} \le a\} + \{a < \mathbf{x} \le b\} = \{\mathbf{x} \le b\}.$$

Hence, in terms of the CDF (2), it follows that

$$F_{\mathbf{x}}(a) + \Pr(a < \mathbf{x} \le b) = F_{\mathbf{x}}(b),$$

or, equivalently,

$$\Pr(a < \mathbf{x} \le b) = F_{\mathbf{x}}(b) - F_{\mathbf{x}}(a). \tag{3}$$

Because the distribution function is a probability, it must satisfy the basic axioms of probability. As a consequence, the CDF $F_{\mathbf{x}}(x)$ has the following properties:

$$
\begin{array}{l}
1. \quad 0 \le F_{\mathbf{x}}(x) \le 1, \quad -\infty < x < \infty \\[2mm]
2. \quad F_{\mathbf{x}}(a) \le F_{\mathbf{x}}(b), \quad \text{for } a < b \\[2mm]
3. \quad F_{\mathbf{x}}(-\infty) = 0, \quad F_{\mathbf{x}}(\infty) = 1
\end{array}
\tag{4}
$$

Finally, because

$$\{\mathbf{x} > x\} = \{\mathbf{x} < \infty\} - \{\mathbf{x} \le x\},$$

we deduce further that

$$
\begin{aligned}
\Pr(\mathbf{x} > x) &= F_{\mathbf{x}}(\infty) - F_{\mathbf{x}}(x) \\
&= 1 - F_{\mathbf{x}}(x).
\end{aligned}
\tag{5}
$$

In terms of the CDF, we can now provide a more precise definition of a continuous random variable. Namely, when the CDF $F_{\mathbf{x}}(x)$ is continuous everywhere, we say that $\mathbf{x}$ is a *continuous random variable*. A typical continuous CDF is illustrated in Fig. 13.1.

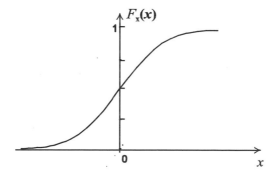

Figure 13.1 A typical cumulative distribution function (CDF) of a continuous random variable.

13.2.2 Probability density function

If **x** is a continuous random variable, we define its *probability density function* (PDF) by

$$f_{\mathbf{x}}(x) = \frac{dF_{\mathbf{x}}(x)}{dx}. \qquad (6)$$

This being the case, the CDF can be recovered from the PDF by evaluating the integral

$$F_{\mathbf{x}}(x) = \int_{-\infty}^{x} f_{\mathbf{x}}(u)\,du, \qquad (7)$$

where u is a dummy variable of integration. The shaded area shown in Fig. 13.2 under the PDF represents the CDF (7). Based on properties of integrals, it follows that

$$\Pr(a < \mathbf{x} \le b) = F_{\mathbf{x}}(b) - F_{\mathbf{x}}(a)$$

$$= \int_{a}^{b} f_{\mathbf{x}}(x)\,dx. \qquad (8)$$

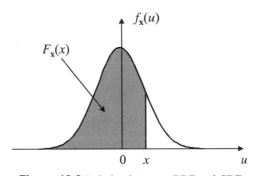

Figure 13.2 Relation between PDF and CDF.

In words, Eq. (8) represents the area under the PDF curve between $x = a$ and $x = b$. And because the probability (7) is nondecreasing, the PDF must be nonnegative; i.e.,

$$f_{\mathbf{x}}(x) \ge 0, \quad -\infty < x < \infty.$$

Finally, by virtue of Axiom II, we see that

$$\int_{-\infty}^{\infty} f_{\mathbf{x}}(x)\,dx = 1. \qquad (9)$$

That is, the total area under the PDF curve always equals unity.

EXAMPLE 1 (*Uniform distribution*) Determine the associated CDF (see Fig. 13.3) for the uniform distribution defined by

$$f_\mathbf{x}(x) = \begin{cases} 1, & 0 < x < 1 \\ 0, & \text{otherwise.} \end{cases}$$

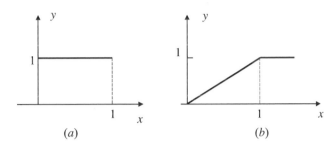

Figure 13.3 Uniform distribution: (*a*) PDF and (*b*) CDF.

Solution: Based on Eq. (7), we see that

$$F_\mathbf{x}(x) = \int_{-\infty}^{x} f_\mathbf{x}(u)\,du = \begin{cases} 0, & x \le 0 \\ x, & 0 < x < 1, \\ 1, & x > 1. \end{cases}$$

EXAMPLE 2 (*Negative exponential distribution*) For $b > 0$, determine the value of the constant C so that the following is a proper PDF (called the negative exponential distribution). Also, find the associated CDF.

$$f_\mathbf{x}(x) = \begin{cases} C e^{-x/b}, & x \ge 0 \\ 0, & x < 0 \end{cases}$$

Solution: Here we simply evaluate the integral given by (9) to find

$$C \int_{0}^{\infty} e^{-x/b}\,dx = Cb = 1.$$

Hence, we see that $C = 1/b$ so that

$$f_\mathbf{x}(x) = \begin{cases} \dfrac{1}{b} e^{-x/b}, & x \ge 0 \\ 0, & x < 0. \end{cases}$$

To calculate the associated CDF, we now simply evaluate

$$F_\mathbf{x}(x) = \frac{1}{b} \int_{0}^{x} e^{-u/b}\,du = 1 - e^{-x/b}, \; x \ge 0,$$

whereas $F_x(x) = 0$ for $x < 0$.

13.2.3 Discrete random variables

Although our primary interest concerns continuous random variables, we wish to illustrate in this section that discrete random variables can often be treated by the methods of continuous random variables by introducing the notions of step function and delta function (see Section 2.2). To more closely follow standard convention, we use the notation $U(x)$ for the step function and define

$$U(x-a) \equiv \text{step}(x-a) = \begin{cases} 1, & x \geq a \\ 0, & x < a. \end{cases} \tag{10}$$

The definition of CDF given in Section 13.2.1 is applicable whether or not the random variable $\mathbf{x}$ is continuous. If the distribution function $F_x(x)$ is discontinuous, it then represents either a discrete or a mixed random variable. That is, when the CDF has one or more finite jump discontinuities, but takes on constant values between jumps (i.e., it is made up entirely of "steps"), we say that $\mathbf{x}$ is a discrete random variable. On the other hand, if the CDF has one of more finite jumps but rises continuously over some finite interval on the x-axis, we say that $\mathbf{x}$ is a *mixed random variable*. In this case it has characteristics of both a discrete and a continuous random variable. Both cases are illustrated in Fig. 13.4.

Let $\mathbf{x}$ denote a discrete random variable that takes on only values x_k with probabilities $\Pr(\mathbf{x} = x_k)$, $k = 1, 2, 3, \dots$. Based on properties of the step function (10), it follows that the CDF can always be written in the form

$$F_x(x) = \sum_{k=1}^{\infty} \Pr(\mathbf{x} = x_k) U(x - x_k). \tag{11}$$

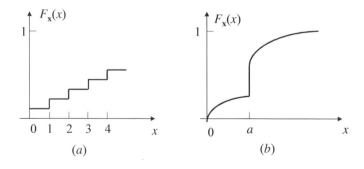

Figure 13.4 (*a*) Discrete random variable and (*b*) mixed random variable.

Thus, through Eq. (6) we formally find that the associated PDF takes the form

$$f_x(x) = \sum_{k=1}^{\infty} \Pr(\mathbf{x} = x_k) \delta(x - x_k), \tag{12}$$

where we are using the relation $dU(x - a)/dx = \delta(x - a)$.

An important special case of (12) occurs when to each outcome of an experiment, the same number $\mathbf{x} = a$ is assigned. That is, the "random variable" is really a *constant* (also called deterministic). The CDF (11) in this case reduces to

$$F_{\mathbf{x}}(x) = U(x - a), \tag{13}$$

where $\Pr(\mathbf{x} < a) = \Pr(\mathbf{x} > a) = 0$, and $\Pr(\mathbf{x} = a) = 1$. The PDF of a constant, or deterministic quantity, is therefore

$$f_{\mathbf{x}}(x) = \delta(x - a). \tag{14}$$

One of the basic discrete distribution functions for a random variable $\mathbf{n}$ is the *Poisson distribution* given by

$$\Pr(\mathbf{n} = k) = \frac{m^k}{k!} e^{-m}, \quad m > 0, \tag{15}$$

where m is the *mean* (see Section 13.4) and $k = 0, 1, 2, \ldots$. Among other applications, the Poisson distribution arises in determining the number of photo-electrons emitted in a fixed time interval, which is proportional to the total radiant energy of light incident on a surface. The total probability associated with (15) is given by the sum

$$\sum_{k=0}^{\infty} \frac{m^k}{k!} e^{-m} = e^{-m} e^{m} = 1, \tag{16}$$

where we recognize the series as that of the exponential function. Equation (16) is simply a discrete analog of Eq. (9).

EXAMPLE 3 (*Poisson distribution*) Write the CDF and PDF for the Poisson distribution (15) as a continuous random variable.

Solution: Because $\mathbf{n}$ is restricted to nonnegative integers, it follows that

$$F_{\mathbf{n}}(x) = \Pr(\mathbf{n} \le x) = 0, \quad x < 0.$$

For $x \ge 0$, we use Eq. (11) to write

$$F_{\mathbf{n}}(x) = \sum_{k=0}^{\infty} \Pr(\mathbf{n} = x) U(x - k) = \sum_{k=0}^{\infty} \frac{m^k}{k!} e^{-m} U(x - k), \quad x \ge 0.$$

Hence, following Eq. (12), the Poisson density function can be written as

$$f_{\mathbf{n}}(x) = \sum_{k=0}^{\infty} \frac{m^k}{k!} e^{-m} \delta(x - k).$$

> **Remark:** One of the observations we need to make here concerns the distinction between probabilities associated with discrete and continuous random variables. Namely, for a discrete random variable we discuss the probability that the random variable equals a specific value, e.g., $\mathbf{x} = a$, whereas the probability associated with a continuous random variable must involve an "interval" $a < \mathbf{x} \leq b$ rather than just a "point."

13.3 Examples of Density Functions

It is customary to describe distributions not by their CDF, but by their PDF or *density function*. There are numerous examples of PDFs that arise in engineering applications. In Table 13.1 we provide a list of some of these density functions and their basic properties. A few of these PDFs are discussed below in greater detail.

13.3.1 Gaussian (or normal) distribution

Perhaps the most important of all density functions is the *normal distribution*, also called the *Gaussian distribution*. Its importance stems from the fact that it is a limit distribution for large sums of random variables (see Section 13.9). Its functional form is given by

$$f_{\mathbf{x}}(x) = \frac{1}{\sigma\sqrt{2\pi}} \exp\left[-\frac{(x-m)^2}{2\sigma^2} \right], \tag{17}$$

where m and σ^2 are parameters (identified as the *mean* and *variance*, respectively, in Section 13.4). In communication devices where noise is a random variable, it is often assumed that the noise is a Gaussian variate with $m = 0$.

To calculate the CDF, we first note that

$$F_{\mathbf{x}}(x) = \frac{1}{\sigma\sqrt{2\pi}} \int_{-\infty}^{x} \exp\left[-\frac{(u-m)^2}{2\sigma^2} \right] du = \frac{1}{\sqrt{2\pi}} \int_{-\infty}^{(x-m)/\sigma} e^{-s^2/2}\, ds, \tag{18}$$

the last step following the change of variable $s = (u - m)/\sigma$. We now write

$$F_{\mathbf{x}}(x) = \frac{1}{\sqrt{2\pi}} \int_{-\infty}^{0} e^{-s^2/2}\, ds + \frac{1}{\sqrt{2\pi}} \int_{0}^{(x-m)/\sigma} e^{-s^2/2}\, ds,$$

and, by recalling the *error functions* (see Section 2.3.4)

$$\mathrm{erf}(x) = \frac{2}{\sqrt{\pi}} \int_{0}^{x} e^{-t^2}\, dt, \qquad \mathrm{erfc}(x) = \frac{2}{\sqrt{\pi}} \int_{x}^{\infty} e^{-t^2}\, dt, \tag{19}$$

we are led to[1]

$$F_x(x) = 1 - \frac{1}{2}\operatorname{erfc}\left(\frac{x-m}{\sigma\sqrt{2}}\right) = \frac{1}{2}\left[1 + \operatorname{erf}\left(\frac{x-m}{\sigma\sqrt{2}}\right)\right]. \tag{20}$$

Table 13.1 Table of Selected Distributions and Their Properties

Name and Type of Distribution	PDF	CDF	Mean	Variance
Binomial (Discrete)	$f_n(x) = \sum_{k=0}^{\infty}\binom{n}{k}p^k q^{(n-k)}\delta(x-k)$	$F_n(x) = \sum_{k=0}^{\infty}\binom{n}{k}p^k q^{(n-k)}U(x-k)$	np	npq
Poisson (Discrete)	$f_n(x) = \sum_{k=0}^{\infty}\frac{m^k}{k!}e^{-m}\delta(x-k)$	$F_n(x) = \sum_{k=0}^{\infty}\frac{m^k}{k!}e^{-m}U(x-k)$	m	m
Uniform (Continuous)	$f_x(x) = \begin{cases}\dfrac{1}{b-a}, & a<x<b \\ 0, & \text{otherwise}\end{cases}$	$F_x(x) = \begin{cases}0, & x\le a \\ \dfrac{x-a}{b-a}, & a<x\le b \\ 1, & x>b.\end{cases}$	$\dfrac{a+b}{2}$	$\dfrac{(b-a)^2}{12}$
Rayleigh (Continuous)	$f_x(x) = \dfrac{x}{\sigma^2}e^{-x^2/2\sigma^2}U(x)$	$F_x(x) = (1 - e^{-x^2/2\sigma^2})U(x)$	$\sqrt{\dfrac{\pi}{2}}\sigma$	$\dfrac{(4-\pi)\sigma^2}{2}$
Rician (Continuous)	$f_x(x) = \dfrac{x}{\sigma^2}e^{-(x^2+m^2)/2\sigma^2}$ $\times I_0\!\left(mx/\sigma^2\right)U(x)$	$1 - Q(m, x/\sigma)^{\dagger}$	‡	‡
Neg. Exp. (Continuous)	$f_x(x) = \dfrac{1}{b}e^{-x/b}U(x), b>0$	$F_x(x) = (1 - e^{-x/b})U(x)$	b	b^2
Gamma (Continuous)	$f_x(x) = \dfrac{x^{c-1}}{\Gamma(c)}e^{-x}U(x), c>0$	$F_x(x) = \dfrac{1}{\Gamma(c)}\gamma(c,x)U(x)$	c	c
Gaussian (Continuous)	$f_x(x) = \dfrac{1}{\sigma\sqrt{2\pi}}\exp\left[-\dfrac{(x-m)^2}{2\sigma^2}\right]$	$F_x(x) = \dfrac{1}{2}\left[1 + \operatorname{erf}\left(\dfrac{x-m}{\sigma\sqrt{2}}\right)\right]$	m	σ^2

See Eq. (144) in Section 13.8.2.   ‡See Example 9 in Chapter 2

---

[1]Another definition of the error function commonly used in statistics is

$$\operatorname{Erfc}(x) = \frac{1}{\sqrt{2\pi}}\int_x^{\infty}e^{-t^2/2}\,dt.$$

In terms of this definition it follows that $F_x(x) = 1 - \operatorname{Erfc}[(x-m)/\sigma]$.

> **Remark:** For easy calculations involving the error function, the following approximation (among others) is often used:
>
> $$\mathrm{erfc}(x) \cong \frac{e^{-x^2}}{\sqrt{\pi(1+x^2)}}, \quad x > 1.$$

13.3.2 Uniform distribution

A special case of the uniform distribution was introduced in Example 1. A more general form of the *uniform distribution* is given by

$$f_{\mathbf{x}}(x) = \begin{cases} \dfrac{1}{b-a}, & a < x \le b \\ 0, & \text{otherwise}. \end{cases} \tag{21}$$

Through integration, we readily determine the CDF

$$F_{\mathbf{x}}(x) = \begin{cases} 0, & x \le a \\ \dfrac{x-a}{b-a}, & a < x \le b \\ 1, & x > b. \end{cases} \tag{22}$$

An important case of a uniform distribution occurs in dealing with random noise for which the random phase θ has the PDF

$$f_{\theta}(\theta) = \frac{1}{2\pi}, \quad -\pi < \theta \le \pi, \tag{23}$$

and equals zero elsewhere.

13.3.3 Rayleigh distribution

The distribution for pointing errors in a radar system and that for the amplitude of random noise in which the phase is uniformly distributed is the *Rayleigh distribution*

$$f_{\mathbf{x}}(x) = \frac{x}{\sigma^2} e^{-x^2/2\sigma^2} U(x), \tag{24}$$

where σ^2 is a parameter and $U(x)$ is the unit step function. The associated CDF is

$$F_{\mathbf{x}}(x) = \frac{1}{\sigma^2} \int_0^x u e^{-u^2/2\sigma^2} \, du = \int_0^{x^2/2\sigma^2} e^{-s} \, ds,$$

which reduces to

$$F_{\mathbf{x}}(x) = (1 - e^{-x^2/2\sigma^2}) U(x). \tag{25}$$

13.3.4 Gamma distribution

The *gamma distribution,* which is often associated with a sum of negative exponential random variables (see Section 13.9.2), is defined by

$$f_{\mathbf{x}}(x) = \frac{x^{c-1}}{\Gamma(c)} e^{-x} U(x), \quad c > 0,$$

(26)

where $\Gamma(c)$ is the *gamma function* (Section 2.3.1). In this case, the CDF is

$$F_{\mathbf{x}}(x) = \frac{1}{\Gamma(c)} \int_0^x u^{c-1} e^{-u} du, \quad x > 0,$$

(27)

which, in terms of the *incomplete gamma function* (Section 2.3.1), becomes

$$F_{\mathbf{x}}(x) = \frac{1}{\Gamma(c)} \gamma(c,x) U(x).$$

(28)

Note that the special case $c = 1$ reduces the gamma distribution (26) to the *negative exponential distribution* (see Table 13.1)

$$f_{\mathbf{x}}(x) = e^{-x} U(x).$$

(29)

EXAMPLE 4 For the special case in which $c = 1/2$, express the CDF of the gamma distribution in terms of the error function.

Solution: With $c = 1/2$, the CDF (27) becomes

$$F_{\mathbf{x}}(x) = \frac{1}{\sqrt{\pi}} \int_0^x u^{-1/2} e^{-u} du = \frac{2}{\sqrt{\pi}} \int_0^{\sqrt{x}} e^{-s^2} ds,$$

from which we deduce

$$F_{\mathbf{x}}(x) = \operatorname{erf}(\sqrt{x}) U(x).$$

13.4 Expected Values

If $\mathbf{x}$ is a random variable, the *expected value* of any function of $\mathbf{x}$, say $g(\mathbf{x})$, is defined by

$$E[g(\mathbf{x})] = \int_{-\infty}^{\infty} g(x) f_{\mathbf{x}}(x) dx.$$

(30)

The most important expected value occurs when $g(\mathbf{x}) = \mathbf{x}$, which leads to

$$E[\mathbf{x}] = m = \int_{-\infty}^{\infty} x f_{\mathbf{x}}(x)\,dx. \tag{31}$$

We call m the *mean (value)*, or *expected value*, of the random variable $\mathbf{x}$. The mean is also denoted by $\bar{x}$ and by m_1, among other notations. The mean is that value around which most other values of the random variable tend to cluster. Note that (31) is the mathematical expression used in calculus for computing the first moment or "center of mass" along a line where $f_{\mathbf{x}}(x)$ is the mass density. Equation (31) is used for both discrete and continuous random variables, although in the former case we write

$$\begin{aligned} E[\mathbf{x}] &= \sum_{k=1}^{\infty} \Pr(\mathbf{x} = x_k) \int_{-\infty}^{\infty} x \delta(x - x_k)\,dx \\ &= \sum_{k=1}^{\infty} x_k \Pr(\mathbf{x} = x_k). \end{aligned} \tag{32}$$

If a PDF satisfies $f_{\mathbf{x}}(-x) = f_{\mathbf{x}}(x)$, we say the PDF is an *even function* (or symmetric about the origin), and in this case the mean value is zero. More generally, if a PDF is symmetric about a point $x = \mu$, then μ is the mean.

The "expectation" symbol $E[.]$ introduced in (30) is a *linear operator*; i.e.,

$$E[C_1 g_1(\mathbf{x}) + C_2 g_2(\mathbf{x})] = C_1 E[g_1(\mathbf{x})] + C_2 E[g_2(\mathbf{x})], \tag{33}$$

where C_1 and C_2 are any constants, which is a consequence of the linearity of integrals.

> **Remark:** The bracket notation $\langle g(\mathbf{x}) \rangle$ is often used in place of the expectation $E[g(\mathbf{x})]$ (e.g., we use the bracket notation $\langle \rangle$ in Chapter 14).

13.4.1 Higher-order moments

For the case when $g(\mathbf{x}) = \mathbf{x}^n$, $n = 2, 3, 4, \ldots$, we obtain the expectation

$$E[\mathbf{x}^n] = m_n = \int_{-\infty}^{\infty} x^n f_{\mathbf{x}}(x)\,dx, \quad n = 2, 3, 4, \ldots, \tag{34}$$

called the *higher-order moments* of the distribution. Mathematically, the higher-order moments describe the characteristic behavior in the "tails" of the distribution. The *second moment* m_2 is the most important of these higher-order moments— it corresponds to the moment of inertia of a mass distribution with respect to an axis through the origin $x = 0$. The second moment also is interpreted physically as being equal to the time average of the square of a random voltage (or current). In this latter case, it is proportional to the average power.

Also significant are the *central moments* defined by

$$\mu_n = E[(\mathbf{x} - m)^n] = \int_{-\infty}^{\infty} (x - m)^n f_{\mathbf{x}}(x)\,dx, \quad n = 2, 3, 4, \ldots. \tag{35}$$

These moments describe the distribution around the mean value $m = m_1$. Of these, the most important is the second central moment

$$\mu_2 = \text{Var}(\mathbf{x}) = \int_{-\infty}^{\infty} (x - m)^2 f_{\mathbf{x}}(x)\,dx, \tag{36}$$

called the *variance* of the random variable $\mathbf{x}$. It is commonly designated by the symbol $\sigma_{\mathbf{x}}^2$ and corresponds to a moment of inertia with respect to an axis through the center of mass. The quantity $\sigma_{\mathbf{x}}$ is called the *standard deviation* and roughly measures the dispersion or width of the PDF about the mean value.

The central moments (35) can be related to the standard moments (34) by first writing $(x - m)^n$ in a binomial series to get

$$(x - m)^n = \sum_{k=0}^{n} \binom{n}{k} (-1)^k m^k x^{n-k}, \qquad \binom{n}{k} = \frac{n!}{k!(n-k)!},$$

which, when substituted into (35), yields

$$\mu_n = \sum_{k=0}^{n} \binom{n}{k} (-1)^k m^k \int_{-\infty}^{\infty} x^{n-k} f_{\mathbf{x}}(x)\,dx = \sum_{k=0}^{n} \binom{n}{k} (-1)^k m^k E[\mathbf{x}^{n-k}]. \tag{37}$$

When $n = 2$, for example, we find that the variance becomes

$$\begin{aligned}
\sigma_{\mathbf{x}}^2 &= E[\mathbf{x}^2 - 2m\mathbf{x} + m^2] \\
&= E[\mathbf{x}^2] - 2mE[\mathbf{x}] + m^2,
\end{aligned} \tag{38}$$

which in general we write as

$$\boxed{\sigma_{\mathbf{x}}^2 = m_2 - m_1^2.} \tag{39}$$

> **Remark:** When only one random variable appears in the discussion, we often write σ^2 for $\sigma_{\mathbf{x}}^2$.

EXAMPLE 5 Calculate the central moments for the Gaussian distribution

$$f_{\mathbf{x}}(x) = \frac{1}{\sigma\sqrt{2\pi}} \exp\left[-\frac{(x - m)^2}{2\sigma^2} \right].$$

Solution: Using (35), we find that

$$\begin{aligned}
\mu_n &= \frac{1}{\sigma\sqrt{2\pi}} \int_{-\infty}^{\infty} (x - m)^n \exp\left[-\frac{(x - m)^2}{2\sigma^2} \right] dx \\
&= \frac{1}{\sigma\sqrt{2\pi}} \int_{-\infty}^{\infty} u^n e^{-u^2/2\sigma^2}\,du.
\end{aligned}$$

Clearly, all moments with n odd are zero. Hence, we let $n = 2k$ and use properties of even functions to obtain

$$\mu_{2k} = \frac{2}{\sigma\sqrt{\pi}}\int_0^\infty u^{2k} e^{-u^2/2\sigma^2}\,du \quad (\text{Let } t = u^2/2\sigma^2)$$

$$= \frac{2^k \sigma^{2k}}{\sqrt{\pi}}\int_0^\infty e^{-t} t^{k-1/2}\,dt$$

$$= \frac{2^k \sigma^{2k}}{\sqrt{\pi}}\Gamma(k+1/2), \quad k = 1,2,3,\dots,$$

where we have made use of the gamma function. Writing the gamma function in terms of factorials [see property (G4) in Section 2.3.1], we have

$$\mu_{2k} = \frac{(2k)!\,\sigma^{2k}}{2^k k!}, \quad k = 1,2,3,\dots.$$

In particular, when $k = 1$, we obtain $\mu_2 = \sigma^2$, justifying the notation used in (17).

EXAMPLE 6 Calculate the moments $E[\mathbf{x}^n]$, $n = 1,2,3,\dots$, of the Rayleigh distribution

$$f_{\mathbf{x}}(x) = \frac{x}{b^2} e^{-x^2/2b^2} U(x), \ b > 0.$$

Solution: Here we are led to

$$E[\mathbf{x}^n] = \frac{1}{b^2}\int_0^\infty x^{n+1} e^{-x^2/2b^2}\,dx \quad (\text{Let } t = x^2/2b^2)$$

$$= 2^{n/2} b^n \int_0^\infty e^{-t} t^{n/2}\,dt$$

$$= 2^{n/2} b^n \Gamma(1+n/2), \quad n = 1,2,3,\dots.$$

For $n = 1$ and $n = 2$, we find

$$m_1 = \sqrt{\frac{\pi}{2}}\,b, \quad m_2 = 2b^2,$$

and, consequently, the variance is

$$\sigma_{\mathbf{x}}^2 = \left(2 - \frac{\pi}{2}\right)b^2.$$

13.4.2 Characteristic functions

One of the most useful functions arising in the calculation of expected values of a random variable **x** is the *characteristic function* defined by

$$\Phi_{\mathbf{x}}(s) = E[e^{is\mathbf{x}}] = \int_{-\infty}^{\infty} e^{isx} f_{\mathbf{x}}(x) dx. \tag{40}$$

We recognize (40) as 2π times the *inverse Fourier transform* of the PDF (see Section 9.3). Hence, it follows from the theory of Fourier transforms that the PDF is simply $1/2\pi$ times the *Fourier transform* of the characteristic function; i.e.,

$$f_{\mathbf{x}}(x) = \frac{1}{2\pi} \int_{-\infty}^{\infty} e^{-isx} \Phi_{\mathbf{x}}(s) ds. \tag{41}$$

Between the two, the characteristic function is generally easier to determine; hence, in practice we often find $\Phi_{\mathbf{x}}(s)$ and then calculate the PDF through the use of (41).

EXAMPLE 7 Find the characteristic function associated with the Gaussian PDF

$$f_{\mathbf{x}}(x) = \frac{1}{\sigma\sqrt{2\pi}} \exp\left[-\frac{(x - m)^2}{2\sigma^2}\right].$$

Solution: From definition, we have

$$\Phi_{\mathbf{x}}(s) = \frac{1}{\sigma\sqrt{2\pi}} \int_{-\infty}^{\infty} e^{isx} \exp\left[-\frac{(x - m)^2}{2\sigma^2}\right] dx,$$

which, after the change of variable $t = x - m$, reduces to

$$\Phi_{\mathbf{x}}(s) = \frac{e^{ims}}{\sigma\sqrt{2\pi}} \int_{-\infty}^{\infty} e^{ist - t^2/2\sigma^2} dt.$$

From Table 9.1, we readily find that the above expression gives us

$$\Phi_{\mathbf{x}}(s) = e^{ims - \sigma^2 s^2/2}.$$

Perhaps the most practical property of the characteristic function of a random variable **x** is its relation to the moments of **x**. By setting $s = 0$ in the defining integral (40), we first observe that

$$\Phi_{\mathbf{x}}(0) = \int_{-\infty}^{\infty} f_{\mathbf{x}}(x) dx = 1. \tag{42}$$

This property is always satisfied by the characteristic function. To see the connection with the statistical moments, we now write

$$e^{isx} = \sum_{n=0}^{\infty} \frac{(isx)^n}{n!}.$$ (43)

By substituting this series into the integral in (40), and interchanging the order of summation and integration, we obtain

$$\begin{aligned} \Phi_{\mathbf{x}}(s) &= \sum_{n=0}^{\infty} \frac{(is)^n}{n!} \int_{-\infty}^{\infty} x^n f_{\mathbf{x}}(x)\,dx \\ &= \sum_{n=0}^{\infty} \frac{(is)^n}{n!} E[\mathbf{x}^n]. \end{aligned}$$ (44)

Hence, we see that the moments $E[\mathbf{x}^n]$, $n = 1,2,3,...$, are simply the coefficients of $(is)^n/n!$ in the Maclaurin series expansion of the characteristic function $\Phi_{\mathbf{x}}(s)$. Moreover, from properties of the Maclaurin series, we also know that

$$\Phi_{\mathbf{x}}(s) = \sum_{n=0}^{\infty} \Phi_{\mathbf{x}}^{(n)}(0) \frac{s^n}{n!},$$ (45)

from which we deduce

$$E[\mathbf{x}^n] = (-i)^n \Phi_{\mathbf{x}}^{(n)}(0), \quad n = 1,2,3,....$$ (46)

EXAMPLE 8 Use the characteristic function method to calculate the moments of a zero-mean Gaussian distribution.

Solution: A zero-mean Gaussian distribution is described by

$$f_{\mathbf{x}}(x) = \frac{1}{\sigma\sqrt{2\pi}} e^{-x^2/2\sigma^2},$$

and its associated characteristic function is (recall Example 7)

$$\Phi_{\mathbf{x}}(s) = e^{-\sigma^2 s^2/2}.$$

By expanding this characteristic function in a Maclaurin series, we have

$$\Phi_{\mathbf{x}}(s) = \sum_{k=0}^{\infty} \frac{(is)^{2k}}{k!} (\sigma^2/2)^k = \sum_{\substack{n=0 \\ (\text{even})}}^{\infty} \frac{(is)^n\, n!}{(n/2)!\, n!} (\sigma^2/2)^{n/2}.$$

Because only even-order terms appear in the above series, we conclude that all odd-order moments are zero. Also, from Eqs. (44)-(46) we further see that the even-order moments of the Gaussian distribution are

$$E[\mathbf{x}^{2n}] = \frac{(2n)!}{2^n \, n!} \sigma^{2n}, \quad n = 1,2,3,....$$

In particular, when $n = 2$, we get $E[\mathbf{x}^4] = 3\sigma^4$.

Remark: Notice that the moments found in Example 8 are exactly the same as the central moments found in Example 5.

Not all characteristic functions have a Maclaurin series expansion of the form (45). For example, the characteristic function associated with the *Cauchy distribution*

$$f_{\mathbf{x}}(x) = \frac{\pi/a}{x^2 + a^2}, \quad a > 0, \tag{47}$$

is

$$\Phi_{\mathbf{x}}(s) = e^{-a|s|}. \tag{48}$$

Clearly, the characteristic function (48) does not have a Maclaurin series about $s = 0$ because it has no derivatives at $s = 0$ (at least in the usual sense). That is, all moments of the Cauchy distribution are divergent.

13.5 Conditional Probability

In practice, we often need to find the probability of an event B under the condition that event A has occurred. This type of probability is called the *conditional probability of B given A*, and denoted by $\Pr(B|A)$. The probability that both A and B occur is defined by

$$\Pr(A \cap B) = \Pr(B|A)\Pr(A), \tag{49}$$

where the symbol $\cap$ denotes *intersection* (i.e., common to both A and B). By solving (49) for the conditional probability, we obtain

$$\Pr(B|A) = \frac{\Pr(A \cap B)}{\Pr(A)}, \quad \Pr(A) \neq 0. \tag{50}$$

If A and B are two events in a sample space S and $\Pr(A) \neq 0$, $\Pr(B) \neq 0$, then

$$\Pr(A \cap B) = \Pr(B|A)\Pr(A) = \Pr(A|B)\Pr(B). \tag{51}$$

Equation (51) is known as the *multiplication rule* of conditional probabilities. Also, if events A and B are such that

$$\Pr(A \cap B) = \Pr(A)\Pr(B), \tag{52}$$

they are called *independent events*. In this case it also follows that

$$\Pr(A|B) = \Pr(A), \quad \Pr(B|A) = \Pr(B). \tag{53}$$

13.5.1 Conditional CDF and PDF

Based on Eq. (50), we define the *conditional distribution function*

$$F_{\mathbf{x}}(x|A) = \Pr(\mathbf{x} \le x|A) = \frac{\Pr[(\mathbf{x} \le x) \cap A]}{\Pr(A)}. \tag{54}$$

The conditional CDF (54) satisfies all properties previously stated for any CDF.

Assuming that $\mathbf{x}$ is a continuous random variable, we define the associated *conditional probability density function* by

$$f_{\mathbf{x}}(x|A) = \frac{dF_{\mathbf{x}}(x|A)}{dx}. \tag{55}$$

The conditional PDF (55) also has all the properties of an ordinary PDF.

If A is an event that can be expressed in terms of the random variable $\mathbf{x}$, then $F_{\mathbf{x}}(x|A)$ can be found from $F_{\mathbf{x}}(x)$. For example, let A be the event

$$A = \{\mathbf{x} \le a\},$$

where a is a constant; then from (54) we deduce that

$$F_{\mathbf{x}}(x|A) = \Pr(\mathbf{x} \le x|A) = \frac{\Pr[(\mathbf{x} \le x) \cap (\mathbf{x} \le a)]}{\Pr(A)}. \tag{56}$$

The intersection of the two events in (56) leads to

$$(\mathbf{x} \le x) \cap (\mathbf{x} \le a) = \begin{cases} \mathbf{x} \le x, & x \le a, \\ \mathbf{x} \le a, & x > a, \end{cases} \tag{57}$$

from which we deduce

$$F_{\mathbf{x}}(x|\mathbf{x} \le a) = \frac{\Pr(\mathbf{x} \le x)}{\Pr(\mathbf{x} \le a)} = \frac{F_{\mathbf{x}}(x)}{F_{\mathbf{x}}(a)}, \quad x \le a, \tag{58}$$

$$F_{\mathbf{x}}(x|\mathbf{x} \le a) = \frac{\Pr(\mathbf{x} \le a)}{\Pr(\mathbf{x} \le a)} = 1, \quad x > a. \tag{59}$$

The associated conditional PDF is

$$f_{\mathbf{x}}(x|\mathbf{x} \le a) = \frac{dF_{\mathbf{x}}(x)/dx}{F_{\mathbf{x}}(a)} U(a-x) = \frac{f_{\mathbf{x}}(x)}{F_{\mathbf{x}}(a)} U(a-x),$$ (60)

where $U(a-x)$ is the unit step function.

Along similar lines, if A is the event defined by

$$A = \{a < \mathbf{x} \le b\},$$

then the conditional CDF and PDF are, respectively,

$$F_{\mathbf{x}}(x|a < \mathbf{x} < b) = \begin{cases} 0, & x \le a, \\ \dfrac{F_{\mathbf{x}}(x) - F_{\mathbf{x}}(a)}{F_{\mathbf{x}}(b) - F_{\mathbf{x}}(a)}, & a < x \le b, \\ 1, & x > b, \end{cases}$$ (61)

$$f_{\mathbf{x}}(x|a < \mathbf{x} \le b) = \frac{f_{\mathbf{x}}(x)}{F_{\mathbf{x}}(b) - F_{\mathbf{x}}(a)} [U(x-a) - U(x-b)].$$ (62)

Because the event A in this case tells us the outcome of the "experiment" lies between a and b, it follows that the conditional PDF (62) is zero outside this interval.

13.5.2 Expected values

When using a conditional PDF to calculate expected values, the resulting expectation is called a *conditional expected value*. In general, we write

$$E[g(\mathbf{x})|A] = \int_{-\infty}^{\infty} g(x) f_{\mathbf{x}}(x|A) dx.$$ (63)

For the case when $g(\mathbf{x}) = \mathbf{x}^n$, $n = 1,2,3,...$, we obtain the conditional moments

$$E[\mathbf{x}^n|A] = \int_{-\infty}^{\infty} x^n f_{\mathbf{x}}(x|A) dx, \quad n = 1,2,3,....$$ (64)

Suppose, for example, that A is the event

$$A = \{a < \mathbf{x} \le b\}.$$

Here, we use (62) to deduce that

$$\begin{aligned} E[\mathbf{x}|a < \mathbf{x} \le b] &= \int_{-\infty}^{\infty} x f_{\mathbf{x}}(x|a < \mathbf{x} \le b) dx \\ &= \frac{\int_{a}^{b} x f_{\mathbf{x}}(x) dx}{F_{\mathbf{x}}(b) - F_{\mathbf{x}}(a)}. \end{aligned}$$ (65)

EXAMPLE 9 Given that $\mathbf{x}$ is a Gaussian random variable with mean m and variance σ^2, calculate the conditional moment $E[\mathbf{x}|\mathbf{x} \leq m]$.

Solution: The Gaussian PDF for the random variable $\mathbf{x}$ is

$$f_{\mathbf{x}}(x) = \frac{1}{\sigma\sqrt{2\pi}} \exp\left[-\frac{(x-m)^2}{2\sigma^2}\right].$$

Based on Eq. (60), the conditional PDF takes the form

$$f_{\mathbf{x}}(x|\mathbf{x} \leq m) = \frac{f_{\mathbf{x}}(x)}{F_{\mathbf{x}}(m)} U(m-x),$$

where

$$F_{\mathbf{x}}(m) = \frac{1}{\sigma\sqrt{2\pi}} \int_{-\infty}^{m} \exp\left[-\frac{(x-m)^2}{2\sigma^2}\right] dx = \frac{1}{\sigma\sqrt{2\pi}} \int_{-\infty}^{0} e^{-u^2/2\sigma^2} du,$$

or

$$F_{\mathbf{x}}(m) = \frac{1}{2}.$$

Therefore, the conditional PDF becomes

$$f_{\mathbf{x}}(x|\mathbf{x} \leq m) = 2f_{\mathbf{x}}(x)U(m-x),$$

and, consequently,

$$E[\mathbf{x}|\mathbf{x} \leq m] = \int_{\infty}^{\infty} x f_{\mathbf{x}}(x|\mathbf{x} \leq m)dx = \frac{2}{\sigma\sqrt{2\pi}} \int_{-\infty}^{m} x \exp\left[-\frac{(x-m)^2}{2\sigma^2}\right] dx,$$

or

$$E[\mathbf{x}|\mathbf{x} \leq m] = \sigma\sqrt{\frac{2}{\pi}}.$$

13.6 Functions of One Random Variable

When a signal and accompanying random noise pass through a nonlinear device, such as a square-law detector, the random output of the device will have different statistical properties. In many cases the output of the nonlinear device can be represented by some function

$$\mathbf{y} = g(\mathbf{x}). \tag{66}$$

Generally, it is assumed we know the PDF associated with the random variable $\mathbf{x}$ from which we wish to determine the PDF for $\mathbf{y}$.

13.6.1 Continuous random variables

Suppose $\mathbf{x}$ is a continuous random variable with known PDF given by $f_{\mathbf{x}}(x)$. Let $y = g(x)$ be a given continuous *monotone function* for which there is a unique inverse $x = g^{-1}(y)$. We first make the observation

$$g(x + \Delta x) = y + \Delta y,$$

or, equivalently,

$$x + \Delta x = g^{-1}(y + \Delta y).$$

Hence, it follows that

$$\Pr(y < \mathbf{y} \le y + \Delta y) = \Pr\!\left[g^{-1}(y) < \mathbf{x} \le g^{-1}(y + \Delta y)\right]$$
$$= \Pr(x < \mathbf{x} \le x + \Delta x), \tag{67}$$

from which we deduce

$$F_{\mathbf{y}}(y + \Delta y) - F_{\mathbf{y}}(y) = F_{\mathbf{x}}(x + \Delta x) - F_{\mathbf{x}}(x). \tag{68}$$

By dividing both sides of (68) by $\Delta x \, \Delta y$ and taking the limit as $\Delta x \to 0$, $\Delta y \to 0$, we obtain (upon simplification)

$$f_{\mathbf{y}}(y)dy = f_{\mathbf{x}}(x)dx, \tag{69}$$

which we rearrange as

$$f_{\mathbf{y}}(y) = f_{\mathbf{x}}(x)\frac{dx}{dy}. \tag{70}$$

The sign of the derivative dx/dy in (70) will depend upon whether the function $y = g(x)$ is *increasing* or *decreasing* (see Fig. 13.5). Since a PDF must always be positive, we replace this derivative with its absolute value, which leads to

$$\boxed{f_{\mathbf{y}}(y) = f_{\mathbf{x}}(x)\left|\frac{dx}{dy}\right| = \frac{f_{\mathbf{x}}(x)}{|g'(x)|}.} \tag{71}$$

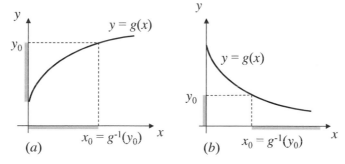

Figure 13.5 The function $y = g(x)$ as (a) a monotonically increasing function and (b) a monotonically decreasing function.

From definition, the CDF associated with the random variable **y** is given by

$$F_y(y) = \int_{-\infty}^{y} f_y(u)\,du.$$ (72)

If we want to relate the CDF $F_y(y)$ directly to $F_x(x)$, we first take the case where $g(x)$ is a *monotonically increasing* function. From Fig. 13.5(a), it follows that the events $\{\mathbf{y} \le y\}$ and $\{\mathbf{x} \le g^{-1}(y)\}$ are equal and, hence, in this case we can write

$$F_y(y) = \int_{-\infty}^{g^{-1}(y)} f_x(u)\,du = F_x[g^{-1}(y)].$$ (73)

Based on Fig. 13.5(b) where the function is *monotonically decreasing*, we see that the event $\{\mathbf{y} \le y\}$ is the same as $\{\mathbf{x} \ge g^{-1}(y)\}$, and consequently,

$$F_y(y) = 1 - \int_{-\infty}^{g^{-1}(y)} f_x(u)\,du = 1 - F_x[g^{-1}(y)].$$ (74)

EXAMPLE 10 (*Linear transformation*) Given the PDF $f_x(x)$, find the PDF $f_y(y)$, where **y** is defined by the linear transformation

$$\mathbf{y} = a\mathbf{x} + b, \quad a > 0.$$

Solution: With $a > 0$, the function $g(x) = ax + b$ is monotonically increasing with derivative $g'(x) = a$. Also, solving for x yields

$$x = g^{-1}(y) = \frac{y - b}{a},$$

from which we deduce

$$f_y(y) = \frac{1}{a} f_x\!\left(\frac{y - b}{a}\right), \quad a > 0.$$

It follows now that the associated CDF is given by

$$F_y(y) = F_x\!\left(\frac{y - b}{a}\right), \quad a > 0.$$

In cases where the function $y = g(x)$ does not have a unique inverse, we need to modify the relation (71). Here, we simply add the probabilities of all the "infinitesimal" events $\{x < \mathbf{x} \le x + dx\}$ that go into the same event described by $\{y < \mathbf{y} \le y + dy\}$. If $x_1, x_2, \dots, x_n$ are all the values for which

$$y = g(x_1) = g(x_2) = \dots = g(x_n),$$

then it can be shown that (see [24])

$$f_y(y) = \frac{f_x(x_1)}{|g'(x_1)|} + \frac{f_x(x_2)}{|g'(x_2)|} + \dots + \frac{f_x(x_n)}{|g'(x_n)|}. \tag{75}$$

Of course, the values $x_1, x_2, \dots, x_n$ all depend upon y (e.g., see Example 11).

EXAMPLE 11 (*Square-law device*) The output of a square-law device is defined by the quadratic transformation

$$\mathbf{y} = a\mathbf{x}^2, \quad a > 0,$$

where $\mathbf{x}$ is the random input variable. Find an expression for the PDF $f_y(y)$ given that we know $f_x(x)$.

Solution: We first observe that if $y < 0$, then $y = ax^2$ has no real solutions; hence, it follows that $f_y(y) = 0$ for $y < 0$.

For $y > 0$, there are two solutions given by

$$x_1 = \sqrt{\frac{y}{a}}, \quad x_2 = -\sqrt{\frac{y}{a}},$$

where

$$g'(x_1) = 2ax_1 = 2\sqrt{ay},$$
$$g'(x_2) = 2ax_2 = -2\sqrt{ay}.$$

Based on (75), we deduce that

$$f_y(y) = \frac{1}{2\sqrt{ay}}\left[f_x\left(\sqrt{\frac{y}{a}}\right) + f_x\left(-\sqrt{\frac{y}{a}}\right)\right]U(y),$$

where $U(y)$ is the unit step function. We leave it to the reader to show that the associated CDF is (which can be verified through differentiation)

$$F_y(y) = \left[F_x\left(\sqrt{\frac{y}{a}}\right) - F_x\left(-\sqrt{\frac{y}{a}}\right)\right]U(y).$$

If $\mathbf{x}$ is a continuous random variable and $g(x)$ is *constant* over some interval, then $\mathbf{y} = g(\mathbf{x})$ is a random variable of mixed type. Basically, this means that the CDF $F_y(y)$ is discontinuous and the PDF $f_y(y)$ contains one or more delta functions. Consider the following example.

EXAMPLE 12 (*Half-wave rectifier*) The output of a half-wave rectifier is defined by $\mathbf{y} = g(\mathbf{x}) = \mathbf{x}\,U(\mathbf{x})$, or

$$y = \begin{cases} x, & x \geq 0 \\ 0, & x < 0. \end{cases}$$

Given that $f_{\mathbf{x}}(x) = \dfrac{1}{2}e^{-|x|}$, determine the CDF and PDF for $\mathbf{y}$.

Solution: Note that $g(x) = x\,U(x)$ is constant over the interval $x < 0$. This means that all probability associated with negative values of $\mathbf{x}$ will "pile up" at the point $y = 0$, leading to a delta function in the PDF for $\mathbf{y}$ at the origin.

Directly from definition, we first calculate

$$F_{\mathbf{x}}(x) = \frac{1}{2}\int_{-\infty}^{x} e^{-|u|}\,du = \begin{cases} 1 - \dfrac{1}{2}e^{-x}, & x \geq 0 \\[2mm] \dfrac{1}{2}e^{x}, & x < 0. \end{cases}$$

Because the function $g(x) = x\,U(x)$ is monotonically increasing for $x \geq 0$, we find

$$F_{\mathbf{y}}(y) = F_{\mathbf{x}}(y) = 1 - \frac{1}{2}e^{-y}, \quad y \geq 0.$$

If $y < 0$ we have the impossible event and $F_{\mathbf{y}}(y) = 0$. Hence,

$$F_{\mathbf{y}}(y) = \left(1 - \frac{1}{2}e^{-y}\right) U(y),$$

and through differentiation

$$f_{\mathbf{y}}(y) = \frac{dF_{\mathbf{y}}(y)}{dy} = \frac{1}{2}e^{-y}U(y) + \left(1 - \frac{1}{2}e^{-y}\right)\delta(y).$$

However, from the property $f(x)\delta(x) = f(0)\delta(x)$, our PDF simplifies to

$$f_{\mathbf{y}}(y) = \frac{1}{2}e^{-y}U(y) + \frac{1}{2}\delta(y).$$

13.6.2 Expected values

If the function $\mathbf{y} = g(\mathbf{x})$ is any integrable function of a random variable $\mathbf{x}$, then we can calculate its expected value from either

$$E[\mathbf{y}] = \int_{-\infty}^{\infty} y f_{\mathbf{y}}(y)\,dy \tag{76}$$

or

$$E[g(\mathbf{x})] = \int_{-\infty}^{\infty} g(x) f_{\mathbf{x}}(x)\,dx. \tag{77}$$

Observe that in the latter case, it is not necessary to first calculate the PDF for **y** but simply use that for **x**.

EXAMPLE 13 The output **y** of a half-wave rectifier is $\mathbf{y} = \mathbf{x}U(\mathbf{x})$ (see Example 12). Calculate the mean and variance of **y**, given that the input variable **x** has a Gaussian PDF with zero mean and variance σ^2.

Solution: Given that

$$f_{\mathbf{x}}(x) = \frac{1}{\sigma\sqrt{2\pi}} e^{-x^2/2\sigma^2},$$

we have

$$E[\mathbf{y}] = \frac{1}{\sigma\sqrt{2\pi}} \int_{-\infty}^{\infty} xU(x)e^{-x^2/2\sigma^2} dx$$

$$= \frac{1}{\sigma\sqrt{2\pi}} \int_{0}^{\infty} xe^{-x^2/2\sigma^2} dx,$$

which yields

$$E[\mathbf{y}] = \frac{\sigma}{\sqrt{2\pi}}.$$

Similarly, the second moment gives us

$$E[\mathbf{y}^2] = \frac{1}{\sigma\sqrt{2\pi}} \int_{0}^{\infty} x^2 e^{-x^2/2\sigma^2} dx = \frac{\sigma^2}{\sqrt{\pi}} \int_{0}^{\infty} t^{1/2} e^{-t} dt = \frac{1}{2}\sigma^2,$$

which follows from properties of the gamma function (Section 2.3.1). The variance of the output **y** is therefore given by

$$\mathrm{Var}(\mathbf{y}) = E[\mathbf{y}^2] - E^2[\mathbf{y}]$$

$$= \frac{1}{2}\left(1 - \frac{1}{\pi}\right)\sigma^2.$$

13.6.3 Characteristic function method

Given that $\mathbf{y} = g(\mathbf{x})$, a useful method for calculating the PDF of **y** from that of **x** involves the use of the characteristic function and the expected values discussed in Section 13.6.2. Namely, the characteristic function for **y** is defined by

$$\Phi_{\mathbf{y}}(s) = E[e^{isg(\mathbf{x})}] = \int_{-\infty}^{\infty} e^{isg(x)}f_{\mathbf{x}}(x)dx. \tag{78}$$

The PDF for **y** is then related through the Fourier transform relation

$$f_y(y) = \frac{1}{2\pi} \int_{-\infty}^{\infty} e^{-isy} \Phi_y(s)\,ds. \tag{79}$$

Let us illustrate with some examples.

EXAMPLE 14 Find the PDF for **y** from Example 12 using the characteristic function method.

Solution: From Example 12, the PDF for **x** is given by $f_x(x) = \dfrac{1}{2}e^{-|x|}$. If we substitute this function into Eq. (78), we get

$$
\begin{aligned}
\Phi_y(s) &= \frac{1}{2}\int_{-\infty}^{\infty} e^{ixU(x)s}\, e^{-|x|}\,dx \\
&= \frac{1}{2}\int_{0}^{\infty} e^{ixs}\, e^{-x}\,dx \;+\; \frac{1}{2}\int_{-\infty}^{0} e^{x}\,dx \\
&= \frac{1}{2}\int_{0}^{\infty} e^{-(1-is)x}\,dx \;+\; \frac{1}{2}\int_{0}^{\infty} e^{-x}\,dx,
\end{aligned}
$$

which yields (see Table 9.1)

$$\Phi_y(s) = \frac{1}{2(1-is)} + \frac{1}{2}.$$

By inverting the characteristic function through the use of (79), we are led to

$$f_y(y) = \frac{1}{4\pi}\int_{-\infty}^{\infty} e^{-isy}\left(\frac{1}{1-is} + 1\right)ds = \frac{1}{2}e^{-y}U(y) + \frac{1}{2}\delta(y).$$

EXAMPLE 15 If the input variable **x** to a square-law device is a zero-mean Gaussian PDF with variance σ^2, find the PDF for the output variable **y**.

Solution: The square-law device is characterized by $\mathbf{y} = a\mathbf{x}^2$, $a > 0$ (recall Example 11). Hence, the characteristic function for **y** is

$$
\begin{aligned}
\Phi_y(s) &= \frac{1}{\sigma\sqrt{2\pi}} \int_{-\infty}^{\infty} e^{isax^2}\, e^{-x^2/2\sigma^2}\,dx \\
&= \frac{1}{\sqrt{1 - 2ia\sigma^2 s}}.
\end{aligned}
$$

Inverting this function by the Fourier transform we obtain

$$f_{\mathbf{y}}(y) = \frac{1}{2\pi} \int_{-\infty}^{\infty} \frac{e^{-isy}}{\sqrt{1 - 2ia\sigma^2 s}} ds = \frac{e^{-y/2a\sigma^2}}{2a\sigma^2} \frac{1}{2\pi i} \int_{1-i\infty}^{1+i\infty} \frac{e^{py/2a\sigma^2}}{\sqrt{p}} dp,$$

where $p = 1 - 2ia\sigma^2 s$ in the last step. From entry #3 in Table 10.1 of Laplace transforms (i.e., $\mathcal{L}^{-1}\{1/\sqrt{p}; p \to y/2a\sigma^2\} = \sqrt{2a\sigma^2/\pi y}$), this last integral yields

$$f_{\mathbf{y}}(y) = \frac{1}{\sigma\sqrt{2\pi ay}} e^{-y/2a\sigma^2} U(y),$$

which is known as the *chi-square distribution* with one degree of freedom (e.g., see [24]).

13.7 Two Random Variables

Up to this point we have considered PDFs and CDFs associated with a single random variable. In some cases, however, the single random variable may be a function of two or more other random variables such as the product $\mathbf{z} = \mathbf{xy}$ (e.g., see Section 13.8.3). We often make calculations concerning the random variable $\mathbf{z}$ in such cases by developing a joint PDF and/or CDF model of $\mathbf{x}$ and $\mathbf{y}$. In doing so, it is helpful to visualize an outcome of an experiment as a point in the $\xi\eta$-plane with coordinates $(\mathbf{x},\mathbf{y})$.

13.7.1 Joint distribution and density functions

If $\mathbf{x}$ and $\mathbf{y}$ are two random variables, we associate the events $\{\mathbf{x} \le x\}$ and $\{\mathbf{y} \le y\}$ with the respective probabilities

$$\Pr(\mathbf{x} \le x) = F_{\mathbf{x}}(x), \qquad \Pr(\mathbf{y} \le y) = F_{\mathbf{y}}(y), \tag{80}$$

where $F_{\mathbf{x}}(x)$ and $F_{\mathbf{y}}(y)$ are the cumulative distribution functions of $\mathbf{x}$ and $\mathbf{y}$. For any pair of random variables $\mathbf{x}$ and $\mathbf{y}$, the event

$$\{\mathbf{x} \le x\} \cap \{\mathbf{y} \le y\} = \{\mathbf{x} \le x, \mathbf{y} \le y\} \tag{81}$$

is represented by a quadrant in the $\xi\eta$-plane having its vertex at the point (x,y) as illustrated in Fig. 13.6. The probability of this event is given by

$$\boxed{F_{\mathbf{xy}}(x,y) = \Pr(\mathbf{x} \le x, \mathbf{y} \le y),} \tag{82}$$

called the *joint (cumulative) distribution function* of $\mathbf{x}$ and $\mathbf{y}$.

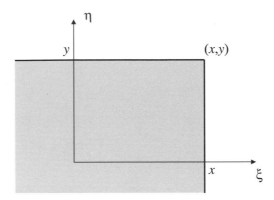

Figure 13.6 Defining region for the joint CDF (82).

The joint CDF (82) has properties analogous to those previously discussed for a single random variable, viz.,

$$
\begin{aligned}
&\text{1.} \quad 0 \le F_{\mathbf{xy}}(x,y) \le 1, \quad -\infty < x < \infty, \ -\infty < y < \infty \\
&\text{2.} \quad F_{\mathbf{xy}}(-\infty,y) = F_{\mathbf{xy}}(x,-\infty) = F_{\mathbf{xy}}(-\infty,-\infty) = 0 \\
&\text{3.} \quad F_{\mathbf{xy}}(\infty,\infty) = 1 \\
&\text{4.} \quad F_{\mathbf{xy}}(\infty,y) = F_{\mathbf{y}}(y), \quad F_{\mathbf{xy}}(x,\infty) = F_{\mathbf{x}}(x)
\end{aligned}
\tag{83}
$$

Also, $F_{\mathbf{xy}}(x,y)$ is a nondecreasing function as either (or both) x and y increase. Property 4 above is a consequence of the fact, for example, that the probability of the event $\{\mathbf{x} \le x\}$ is the same as that for the event $\{\mathbf{x} \le x, \ \mathbf{y} < \infty\}$. In this context, we say that $F_{\mathbf{x}}(x)$ is a *marginal distribution function* of the random variable $\mathbf{x}$. Similarly, $F_{\mathbf{y}}(y)$ is a *marginal distribution function* of the random variable $\mathbf{y}$. Although both marginal distributions $F_{\mathbf{x}}(x)$ and $F_{\mathbf{y}}(y)$ can be determined from the joint distribution function $F_{\mathbf{xy}}(x,y)$, the joint distribution $F_{\mathbf{xy}}(x,y)$ cannot in general be constructed from only knowledge of $F_{\mathbf{x}}(x)$ and $F_{\mathbf{y}}(y)$.

The *joint density function* of the random variables $\mathbf{x}$ and $\mathbf{y}$ is defined by

$$
f_{\mathbf{xy}}(x,y) = \frac{\partial^2 F_{\mathbf{xy}}(x,y)}{\partial x\, \partial y},
\tag{84}
$$

provided the joint CDF is continuous and differentiable. The joint PDF is also commonly called a *bivariate probability density function*. Given the joint PDF, the joint CDF can be recovered by integrating the latter over the rectangular domain described by the integral (see Fig. 13.6)

$$F_{xy}(x,y) = \int_{-\infty}^{x}\int_{-\infty}^{y} f_{xy}(\xi,\eta)\,d\xi\,d\eta. \tag{85}$$

Because $F_{xy}(\infty,\infty) = 1$, it follows that

$$F_{xy}(\infty,\infty) = \int_{-\infty}^{\infty}\int_{-\infty}^{\infty} f_{xy}(x,y)\,dx\,dy = 1. \tag{86}$$

If we define the event $A = \{(x,y) \in D\}$, where D is some domain of the $\xi\eta$-plane, then the probability assigned to this event is

$$\Pr(A) = \iint_{D} f_{xy}(x,y)\,dx\,dy. \tag{87}$$

[When confusion is not likely, we often write integrals in terms of x and y, as in (86) and (87), instead of using dummy variables ξ and η.]

13.7.2 Marginal density functions

From Property 4 in (83) for the joint CDF $F_{xy}(x,y)$, we have

$$F_{x}(x) = F_{xy}(x,\infty) = \int_{-\infty}^{x}\int_{-\infty}^{\infty} f_{xy}(\xi,\eta)\,d\xi\,d\eta,$$

the derivative of which with respect to x yields the result

$$f_{x}(x) = \int_{-\infty}^{\infty} f_{xy}(x,\eta)\,d\eta = \int_{-\infty}^{\infty} f_{xy}(x,y)\,dy. \tag{88}$$

The PDF $f_x(x)$ is called the *marginal density function* of the random variable x. Similarly, the marginal density function of the random variable y is defined by

$$f_{y}(y) = \int_{-\infty}^{\infty} f_{xy}(x,y)\,dx. \tag{89}$$

In general, the marginal PDF of a random variable is obtained by integrating the joint PDF over the range of the other random variable.

EXAMPLE 16 Given the joint PDF

$$f_{xy}(x,y) = \begin{cases} \dfrac{2}{\pi}(1 - x^2 - y^2), & 0 < x^2 + y^2 \le 1 \\ 0, & x^2 + y^2 > 1, \end{cases}$$

(a) determine the marginal PDFs for x and y.
(b) determine $\Pr(x^2 + y^2 < b^2)$, $0 < b < 1$.

Solution:

(a) From (88), we obtain $(0 < x^2 + y^2 \le 1)$

$$f_{\mathbf{x}}(x) = \int_{-\infty}^{\infty} f_{\mathbf{xy}}(x,y)dy$$

$$= \frac{2}{\pi} \int_{-\sqrt{1-x^2}}^{\sqrt{1-x^2}} (1 - x^2 - y^2)dy$$

$$= \frac{4}{\pi} \left[(1 - x^2)^{3/2} - \frac{1}{3}(1 - x^2)^{3/2} \right],$$

which leads to

$$f_{\mathbf{x}}(x) = \begin{cases} \frac{8}{3\pi}(1 - x^2)^{3/2}, & |x| \le 1 \\ 0, & |x| > 1. \end{cases}$$

By symmetry, the marginal PDF for **y** is the same with x replaced by y.

(*b*) By using (87) and converting the integrand to polar coordinates, it follows that

$$\Pr(\mathbf{x}^2 + \mathbf{y}^2 < b^2) = \frac{2}{\pi} \int_0^b \int_0^{2\pi} (1 - r^2) r d\theta \, dr,$$

the evaluation of which yields

$$\Pr(\mathbf{x}^2 + \mathbf{y}^2 < b^2) = b^2(2 - b^2), \quad 0 < b < 1.$$

13.7.3 Conditional distributions and densities

To discuss conditional distributions involving a pair of random variables **x** and **y**, we consider only the case where the conditioning event depends on **y**. In particular, given the event $A = \{\mathbf{y} \le y\}$, then

$$\Pr(\mathbf{y} \le y) = F_{\mathbf{y}}(y),$$

$$\Pr(\mathbf{x} \le x, \mathbf{y} \le y) = F_{\mathbf{xy}}(x,y) = F_{\mathbf{x}}(x|\mathbf{y} \le y)F_{\mathbf{y}}(y),$$

from which we deduce the conditional CDF

$$F_{\mathbf{x}}(x|\mathbf{y} \le y) = \frac{F_{\mathbf{xy}}(x,y)}{F_{\mathbf{y}}(y)}. \tag{90}$$

The corresponding conditional PDF is

$$f_{\mathbf{x}}(x|\mathbf{y} \le y) = \frac{1}{F_{\mathbf{y}}(y)} \int_{-\infty}^{y} f_{\mathbf{xy}}(x,\eta)d\eta. \tag{91}$$

More generally, given the event $A = \{a < \mathbf{y} \le b\}$, it readily follows that

$$F_{\mathbf{x}}(x|a < \mathbf{y} \le b) = \frac{F_{\mathbf{xy}}(x,b) - F_{\mathbf{xy}}(x,a)}{F_{\mathbf{y}}(b) - F_{\mathbf{y}}(a)}, \tag{92}$$

$$f_{\mathbf{x}}(x|a<\mathbf{y}\le b) \;=\; \frac{1}{F_{\mathbf{y}}(b)-F_{\mathbf{y}}(a)}\int_a^b f_{\mathbf{xy}}(x,y)dy. \tag{93}$$

One of the most common types of conditional probability is one in which we consider the event $A = \{\mathbf{y} = y\}$. In this case, by setting $a = y$ and $b = y + \Delta y$ in (92) and taking the limit as $\Delta y \to 0$, we are led to

$$F_{\mathbf{x}}(x|\mathbf{y}=y) \;=\; \frac{1}{f_{\mathbf{y}}(y)}\int_{-\infty}^x f_{\mathbf{xy}}(\xi,y)d\xi. \tag{94}$$

Hence, by differentiating both sides with respect to x, we find the conditional PDF

$$f_{\mathbf{x}}(x|\mathbf{y}=y) \equiv f_{\mathbf{x}}(x|y) \;=\; \frac{f_{\mathbf{xy}}(x,y)}{f_{\mathbf{y}}(y)}, \tag{95}$$

where it is customary to write $f_{\mathbf{x}}(x|y)$ only when there is no confusion. By interchanging the roles of x and y, we also obtain

$$f_{\mathbf{y}}(y|\mathbf{x}=x) \equiv f_{\mathbf{y}}(y|x) \;=\; \frac{f_{\mathbf{xy}}(x,y)}{f_{\mathbf{x}}(x)}. \tag{96}$$

Note that because the marginal PDF for $\mathbf{x}$ is defined by

$$f_{\mathbf{x}}(x) \;=\; \int_{-\infty}^{\infty} f_{\mathbf{xy}}(x,y)dy$$

and Eq. (95) can be expressed as

$$f_{\mathbf{xy}}(x,y) \;=\; f_{\mathbf{x}}(x|y)f_{\mathbf{y}}(y), \tag{97}$$

we can also write

$$f_{\mathbf{x}}(x) \;=\; \int_{-\infty}^{\infty} f_{\mathbf{x}}(x|y)f_{\mathbf{y}}(y)dy, \tag{98}$$

called the *total probability*. Thus, to remove the condition $\mathbf{y} = y$, we simply take the average of the conditional PDF over the marginal PDF of $\mathbf{y}$. Finally, by interchanging the roles of x and y in (97) and equating the result with (97), we are led to *Bayes' theorem*

$$f_{\mathbf{y}}(y|x) \;=\; \frac{f_{\mathbf{x}}(x|y)f_{\mathbf{y}}(y)}{f_{\mathbf{x}}(x)}. \tag{99}$$

EXAMPLE 17 At a particular instant of time we observe the random variable $\mathbf{y}$, which consists of a signal $\mathbf{x}$ plus random noise $\mathbf{n}$; i.e.,

$$\mathbf{y} \;=\; \mathbf{x} + \mathbf{n}.$$

If **x** and **n** are both Gaussian variates with densities defined by

$$f_x(x) = \frac{1}{b\sqrt{2\pi}}\exp(-x^2/2b^2), \quad f_n(n) = \frac{1}{\sigma\sqrt{2\pi}}\exp(-n^2/2\sigma^2),$$

where b^2 is signal power and σ^2 is noise power, determine the conditional PDF of the signal **x** given that we observed a particular value of **y**, say $\mathbf{y} = y$.

Solution: From Bayes' theorem (99), we have

$$f_x(x|y) = \frac{f_y(y|x)f_x(x)}{f_y(y)},$$

where to use this result we must determine the conditional density $f_y(y|x)$ and the marginal density $f_y(y)$. However, when **x** is given, the randomness of **y** is due entirely to the noise **n**. Hence, by the transformation $\mathbf{n} = \mathbf{y} - \mathbf{x}$, we get

$$f_y(y|x) = f_n(y-x) = \frac{1}{\sigma\sqrt{2\pi}}\exp\left[-\frac{(y-x)^2}{2\sigma^2}\right].$$

Now, based on (98) with the roles of **x** and **y** reversed, we find

$$f_y(y) = \int_{-\infty}^{\infty} f_y(y|x)f_x(x)\,dx$$

$$= \frac{1}{2\pi b\sigma}\int_{-\infty}^{\infty}\exp\left[-\frac{(y-x)^2}{2\sigma^2}\right]\exp\left(-\frac{x^2}{2b^2}\right)dx,$$

the integration of which (after completing the square in x and making a change of variable) yields

$$f_y(y) = \frac{1}{\sqrt{2\pi(\sigma^2 + b^2)}}\exp\left[-\frac{y^2}{2(\sigma^2 + b^2)}\right].$$

Observe that this is just another Gaussian distribution with zero mean and variance $\sigma^2 + b^2$. By invoking Bayes' theorem, we deduce that

$$f_x(x|y) = \frac{\dfrac{1}{2\pi\sigma b}\exp\left[-\dfrac{(y-x)^2}{2\sigma^2}\right]\exp\left(-\dfrac{x^2}{2b^2}\right)}{\dfrac{1}{\sqrt{2\pi(\sigma^2 + b^2)}}\exp\left[-\dfrac{y^2}{2(\sigma^2 + b^2)}\right]},$$

which simplifies to

$$f_x(x|y) = \frac{1}{s\sqrt{2\pi}}\exp\left\{-\frac{[x - b^2 y/(\sigma^2 + b^2)]^2}{2s^2}\right\},$$

where $s^2 = \sigma^2 b^2/(\sigma^2 + b^2)$. Note that this is a Gaussian PDF with mean value equal to $b^2 y/(\sigma^2 + b^2)$ and variance s^2.

The conditional PDF $f_x(x|y)$ is sometimes called the *posterior density function* of the signal **x**. The *principle of maximum likelihood*, which is equivalent to Bayes' rule, gives us the best estimate of the signal **x**, given the observation **y** = *y*. In particular, the *maximum-likelihood estimate* $\hat{\mathbf{x}}$ is that value for which the conditional PDF $f_x(x|y)$ is maximum. In the case of Example 17, for instance, the best estimate for the signal value **x** = *x* is the mean value

$$\hat{\mathbf{x}} = \frac{b^2 y}{\sigma^2 + b^2}.$$

As the noise vanishes (i.e., $\sigma^2 \to 0$), our estimate above becomes $\hat{\mathbf{x}} \to y$.

When the conditioning event A is imposed on a joint distribution, we can generalize the above results to develop the necessary relationships. For example, if $A = \{a < \mathbf{x} \le b\}$, we deduce from (92) and (93) that

$$F_{xy}(x,y \,|\, a < \mathbf{x} \le b) = \begin{cases} 0, & x \le a \\[2mm] \dfrac{F_{xy}(x,y) - F_{xy}(a,y)}{F_x(b) - F_x(a)}, & a < x \le b \\[3mm] \dfrac{F_{xy}(b,y) - F_{xy}(a,y)}{F_x(b) - F_x(a)}, & x > b, \end{cases} \tag{100}$$

$$f_{xy}(x,y \,|\, a < \mathbf{x} \le b) = \frac{f_{xy}(x,y)}{F_y(b) - F_y(a)}, \quad a < x < b, \tag{101}$$

and $f_{xy}(x,y \,|\, a < \mathbf{x} \le b) = 0$ otherwise. More generally, if the conditioning event is $A = \{(\mathbf{x},\mathbf{y}) \in D\}$, then the conditional joint PDF becomes

$$f_{xy}[x,y \,|\, (\mathbf{x},\mathbf{y}) \in D] = \frac{f_{xy}(x,y)}{\iint_D f_{xy}(x,y)\,dx\,dy}, \quad (\mathbf{x},\mathbf{y}) \in D, \tag{102}$$

and $f_{xy}[x,y \,|\, (\mathbf{x},\mathbf{y}) \in D] = 0$ for $(\mathbf{x},\mathbf{y}) \notin D$.

EXAMPLE 18 Determine the conditional PDF $f_{xy}(x,y \,|\, \mathbf{x}^2 + \mathbf{y}^2 < b^2)$, given that

$$f_{xy}(x,y) = \frac{1}{2\pi\sigma^2} \exp\left(-\frac{x^2 + y^2}{2\sigma^2} \right).$$

Solution: Based on (102), we must first calculate the probability

$$\Pr(\mathbf{x}^2 + \mathbf{y}^2 < b^2) = \frac{1}{2\pi\sigma^2} \iint_D \exp\left(-\frac{x^2 + y^2}{2\sigma^2}\right) dx\,dy.$$

Changing to polar coordinates, we see that

$$\Pr(\mathbf{x}^2 + \mathbf{y}^2 < b^2) = \frac{1}{2\pi\sigma^2} \int_0^{2\pi} \int_0^b e^{-r^2/2\sigma^2} r\,dr\,d\theta = 1 - e^{-b^2/2\sigma^2},$$

and thus deduce that

$$f_{xy}(x,y \mid \mathbf{x}^2 + \mathbf{y}^2 < b^2) = \frac{\exp\left[-(x^2 + y^2)/2\sigma^2\right]}{2\pi\sigma^2\left[1 - \exp(-b^2/2\sigma^2)\right]}, \quad x^2 + y^2 < b^2,$$

and $f_{xy}(x,y \mid \mathbf{x}^2 + \mathbf{y}^2 < b^2) = 0$ otherwise.

13.7.4 Independent random variables

We say two random variables $\mathbf{x}$ and $\mathbf{y}$ are *statistically independent* if

$$\Pr(\mathbf{x} \le x, \mathbf{y} \le y) = \Pr(\mathbf{x} \le x)\Pr(\mathbf{y} \le y). \tag{103}$$

From (103), it follows that

$$F_{xy}(x,y) = F_x(x)F_y(y), \tag{104}$$

$$f_{xy}(x,y) = f_x(x)f_y(y). \tag{105}$$

Also, because $f_{xy}(x,y) = f_x(x \mid y)f_y(y)$, for example, we see that

$$f_x(x \mid y) = f_x(x). \tag{106}$$

which is simply another statement that $\mathbf{x}$ and $\mathbf{y}$ are statistically independent.

13.7.5 Expected values

If $\mathbf{x}$ and $\mathbf{y}$ are random variables, the *expected value* of the function $g(\mathbf{x},\mathbf{y})$ is

$$E[g(\mathbf{x},\mathbf{y})] = \int_{-\infty}^{\infty} \int_{-\infty}^{\infty} g(x,y)f_{xy}(x,y)dx\,dy. \tag{107}$$

Because the expectation operator defined by (107) is a linear operator (see Section 13.4), we can use this property to simplify certain calculations. To illustrate, let us consider the particular cases in which $g(\mathbf{x},\mathbf{y}) = \mathbf{x} + \mathbf{y}$ and $g(\mathbf{x},\mathbf{y}) = (\mathbf{x} + \mathbf{y})^2$. In the first case, we find

$$E[\mathbf{x} + \mathbf{y}] = \int_{-\infty}^{\infty}\int_{-\infty}^{\infty} (x + y)f_{\mathbf{xy}}(x,y)\,dx\,dy$$

$$= \int_{-\infty}^{\infty} x \int_{-\infty}^{\infty} f_{\mathbf{xy}}(x,y)\,dy\,dx + \int_{-\infty}^{\infty} y \int_{-\infty}^{\infty} f_{\mathbf{xy}}(x,y)\,dx\,dy$$

$$= \int_{-\infty}^{\infty} x f_{\mathbf{x}}(x)\,dx + \int_{-\infty}^{\infty} y f_{\mathbf{y}}(y)\,dy,$$

where we have interchanged the order of integration in the first term. Hence, we conclude that

$$E[\mathbf{x} + \mathbf{y}] = E[\mathbf{x}] + E[\mathbf{y}]. \tag{109}$$

Similarly,

$$E[(\mathbf{x} + \mathbf{y})^2] = \int_{-\infty}^{\infty} (x^2 + 2xy + y^2)f_{\mathbf{xy}}(x,y)\,dx\,dy$$

$$= E[\mathbf{x}^2] + 2E[\mathbf{xy}] + E[\mathbf{y}^2]. \tag{110}$$

13.7.6 Moments and correlation

Moments like those appearing on the right hand sides of (109) and (110) are special cases of joint moments. In general, we define the *joint moments* of two random variables $\mathbf{x}$ and $\mathbf{y}$ by

$$m_{jk} \equiv E[\mathbf{x}^j\mathbf{y}^k] = \int_{-\infty}^{\infty}\int_{-\infty}^{\infty} x^j y^k f_{\mathbf{xy}}(x,y)\,dx\,dy, \quad j,k = 1,2,3,\dots. \tag{111}$$

Of special importance is the moment m_{11} defined by

$$E[\mathbf{xy}] = \int_{-\infty}^{\infty}\int_{-\infty}^{\infty} xy f_{\mathbf{xy}}(x,y)\,dx\,dy, \tag{112}$$

called the *correlation* of the random variables $\mathbf{x}$ and $\mathbf{y}$. If $\mathbf{x}$ and $\mathbf{y}$ are statistically independent, then $f_{\mathbf{xy}}(x,y) = f_{\mathbf{x}}(x)f_{\mathbf{y}}(y)$ and (112) leads to

$$E[\mathbf{xy}] = E[\mathbf{x}]E[\mathbf{y}]. \tag{113}$$

On the other hand, if $E[\mathbf{xy}] = 0$, we say that $\mathbf{x}$ and $\mathbf{y}$ are *orthogonal*.

The *covariance* of the random variables $\mathbf{x}$ and $\mathbf{y}$ is defined by the central moment

$$\mathrm{Cov}(\mathbf{x},\mathbf{y}) = E[(\mathbf{x} - \bar{x})(\mathbf{y} - \bar{y})]$$

$$= \int_{-\infty}^{\infty}\int_{-\infty}^{\infty} (x - \bar{x})(y - \bar{y})f_{\mathbf{xy}}(x,y)\,dx\,dy$$

$$= \int_{-\infty}^{\infty}\int_{-\infty}^{\infty} (xy - \bar{x}y - \bar{y}x + \bar{x}\,\bar{y})f_{\mathbf{xy}}(x,y)\,dx\,dy, \tag{114}$$

from which we deduce

$$\boxed{\mathrm{Cov}(\mathbf{x},\mathbf{y}) = E[\mathbf{xy}] - E[\mathbf{x}]E[\mathbf{y}].} \tag{115}$$

If $\mathrm{Cov}(\mathbf{x},\mathbf{y}) = 0$, we say that $\mathbf{x}$ and $\mathbf{y}$ are *uncorrelated*. If $\mathbf{x}$ and $\mathbf{y}$ are statistically independent, it follows that they are also uncorrelated; however, the converse may not be true. That is, if $\mathbf{x}$ and $\mathbf{y}$ are uncorrelated they need not be statistically independent.

EXAMPLE 19 If $\text{Cov}(\mathbf{x},\mathbf{y}) = 0$, show that

$$\text{Var}(\mathbf{x} + \mathbf{y}) = \text{Var}(\mathbf{x}) + \text{Var}(\mathbf{y}) = \sigma_x^2 + \sigma_y^2.$$

Solution: From definition, we have

$$\text{Var}(\mathbf{x} + \mathbf{y}) = E[(\mathbf{x} + \mathbf{y})^2] - E^2[\mathbf{x} + \mathbf{y}]$$

$$= E[\mathbf{x}^2] + 2E[\mathbf{xy}] + E[\mathbf{y}^2]$$
$$- E^2[\mathbf{x}] - 2E[\mathbf{x}]E[\mathbf{y}] - E^2[\mathbf{y}]$$

$$= \text{Var}(\mathbf{x}) + \text{Var}(\mathbf{y}) + 2\text{Cov}(\mathbf{x},\mathbf{y}).$$

However, because $\text{Cov}(\mathbf{x},\mathbf{y}) = 0$, our intended result follows.

The ratio

$$\rho = \frac{\text{Cov}(\mathbf{x},\mathbf{y})}{\sigma_x \sigma_y} \tag{116}$$

is called the *correlation coefficient* of the random variables $\mathbf{x}$ and $\mathbf{y}$. It provides a measure of the degree to which $\mathbf{x}$ and $\mathbf{y}$ are statistically related. Based on the inequality

$$|\text{Cov}(\mathbf{x},\mathbf{y})| \leq \sigma_x \sigma_y, \tag{117}$$

which we state without proof, it follows from (116) that

$$-1 \leq \rho \leq 1. \tag{118}$$

EXAMPLE 20 Determine the correlation coefficient between random variables $\mathbf{x}$ and $\mathbf{y}$ whose joint density function is

$$f_{\mathbf{xy}}(x,y) = \begin{cases} x + y, & 0 \leq x \leq 1, \ 0 \leq y \leq 1 \\ 0, & \text{elsewhere}. \end{cases}$$

Solution: From the symmetry of the PDF, it follows that

$$E[\mathbf{x}] = E[\mathbf{y}] = \int_0^1 \int_0^1 x(x + y)\,dx\,dy = \frac{7}{12}.$$

Also,

$$E[\mathbf{xy}] = \int_0^1 \int_0^1 xy(x + y)\,dx\,dy = \frac{1}{3},$$

and consequently,

$$\text{Cov}(\mathbf{x},\mathbf{y}) = E[\mathbf{xy}] - E[\mathbf{x}]E[\mathbf{y}] = \frac{1}{3} - \left(\frac{7}{12}\right)^2 = -\frac{1}{144}.$$

Similarly,

$$\sigma_x^2 = \sigma_y^2 = \int_0^1 \int_0^1 \left(x - \frac{7}{12}\right)^2 (x+y)\,dx\,dy = \frac{11}{144},$$

from which we deduce that

$$\rho = \frac{\text{Cov}(\mathbf{x},\mathbf{y})}{\sigma_x \sigma_y} = -\frac{1}{11}.$$

Because $\rho < 0$ in Example 20, we say that $\mathbf{x}$ and $\mathbf{y}$ are negatively correlated. That is, when one variable is large the other tends to be small.

13.7.7 Bivariate Gaussian distribution

Perhaps the most important joint PDF is the *bivariate Gaussian density function*

$$f_{\mathbf{xy}}(x,y) = \frac{1}{2\pi\sigma_x\sigma_y\sqrt{1-\rho^2}} \times \exp\left[-\frac{\dfrac{(x-\bar{x})^2}{\sigma_x^2} - \dfrac{2\rho(x-\bar{x})(y-\bar{y})}{\sigma_x\sigma_y} + \dfrac{(y-\bar{y})^2}{\sigma_y^2}}{2(1-\rho^2)} \right], \tag{119}$$

where ρ is the correlation coefficient (see below). To simplify notation, it is sometimes useful to scale the variables by introducing the zero-mean, unit variance, variates

$$x' = (x - \bar{x})/\sigma_x, \quad y' = (y - \bar{y})/\sigma_y,$$

and then redefine x' and y' by x and y. Doing so yields the simpler functional form

$$f_{\mathbf{xy}}(x,y) = \frac{1}{2\pi\sqrt{1-\rho^2}} \exp\left[-\frac{x^2 - 2\rho xy + y^2}{2(1-\rho^2)} \right], \tag{120}$$

which we will use in our discussion below.

The marginal density function of $\mathbf{x}$ can be found by writing

$$x^2 - 2\rho xy + y^2 = (y - \rho x)^2 + (1 - \rho^2)x^2,$$

and then evaluating the integral

$$f_{\mathbf{x}}(x) = \int_{-\infty}^{\infty} f_{\mathbf{xy}}(x,y)\,dy$$

$$= \frac{1}{2\pi\sqrt{1-\rho^2}} \int_{-\infty}^{\infty} \exp\left[-\frac{(y-\rho x)^2 + (1-\rho^2)x^2}{2(1-\rho^2)}\right]dy, \tag{121}$$

to deduce that

$$f_{\mathbf{x}}(x) = \frac{e^{-x^2/2}}{2\pi\sqrt{1-\rho^2}} \int_{-\infty}^{\infty} \exp\left[-\frac{(y-\rho x)^2}{2(1-\rho^2)}\right]dy = \frac{1}{\sqrt{2\pi}}e^{-x^2/2}. \tag{122}$$

By symmetry, we also conclude that

$$f_{\mathbf{y}}(y) = \frac{1}{\sqrt{2\pi}}e^{-y^2/2}. \tag{123}$$

EXAMPLE 21 Use the joint Gaussian PDF given by (120) to show that

$$E[\mathbf{xy}] = \rho.$$

Solution: Directly from definition we find

$$E[\mathbf{xy}] = \int_{-\infty}^{\infty} xy f_{\mathbf{xy}}(x,y)\,dy$$

$$= \frac{1}{2\pi\sqrt{1-\rho^2}} \int_{-\infty}^{\infty}\int_{-\infty}^{\infty} xy\exp\left[-\frac{(y-\rho x)^2 + (1-\rho^2)x^2}{2(1-\rho^2)}\right]dx\,dy$$

$$= \frac{1}{2\pi\sqrt{1-\rho^2}} \int_{-\infty}^{\infty} xe^{-x^2/2}\int_{-\infty}^{\infty} y\exp\left[-\frac{(y-\rho x)^2}{2(1-\rho^2)}\right]dy\,dx,$$

where we have reversed the order of integration in the last expression. By making the change of variable $t = y - \rho x$, the inside integral leads to

$$E[\mathbf{xy}] = \frac{\rho}{\sqrt{2\pi}} \int_{-\infty}^{\infty} x^2 e^{-x^2/2}\,dx,$$

the evaluation of which gives us

$$E[\mathbf{xy}] = \rho.$$

Remark: In the case of the general form of the bivariate Gaussian distribution (119), it can be shown that $\text{Cov}(\mathbf{x},\mathbf{y}) = \rho\sigma_x\sigma_y$.

13.8 Functions of Two or More Random Variables

We now wish to discuss the case when the random variable **z** is related to random variables **x** and **y** by an equation of the form

$$\mathbf{z} = g(\mathbf{x},\mathbf{y}). \tag{124}$$

Our primary concern is with the development of the CDF and PDF of the random variable **z**, given that we know the joint PDF for **x** and **y**. There are several methods available for making such calculations, which we discuss below.

METHOD I: Here we first calculate the CDF of $\mathbf{z} = g(\mathbf{x},\mathbf{y})$ by determining the probability

$$F_{\mathbf{z}}(z) = \Pr(\mathbf{z} \le z) = \Pr[g(\mathbf{x},\mathbf{y}) \le z]. \tag{125}$$

The event $g(\mathbf{x},\mathbf{y}) \le z$ is represented by a certain region in the $\xi\eta$-plane, denoted by D. Hence, Eq. (125) leads to the integral

$$F_{\mathbf{z}}(z) = \iint_{D} f_{\mathbf{xy}}(x,y)\,dx\,dy, \tag{126}$$

where $f_{\mathbf{xy}}(x,y)$ is the joint distribution of **x** and **y**. It now follows that the PDF of the random variable **z** is found from calculating

$$\boxed{f_{\mathbf{z}}(z) = \frac{d}{dz}\iint_{D} f_{\mathbf{xy}}(x,y)\,dx\,dy.} \tag{127}$$

METHOD II: In this case we use conditional statistics. Thus, if we fix **y**, say $\mathbf{y} = y$, the random variable **z** depends only upon the random variable **x** and we can use the method discussed in Section 13.6. Namely, if $g(x,y)$ is a monotone function of x with inverse $x = g^{-1}(x,y)$, we first find

$$f_{\mathbf{z}}(z|y) = \frac{f_{\mathbf{x}}(x|y)}{\left|\dfrac{\partial g(x,y)}{\partial x}\right|}\Bigg|_{x = g^{-1}(x,y)}, \tag{128}$$

and then make the calculation

$$\boxed{f_{\mathbf{z}}(z) = \int_{-\infty}^{\infty} f_{\mathbf{z}}(z|y)f_{\mathbf{y}}(y)\,dy,} \tag{129}$$

where $f_{\mathbf{y}}(y)$ is the marginal PDF of **y**.

METHOD III: The third method, based on the characteristic function, is often the easiest procedure to apply. The characteristic function of the random variable **z** is defined by

$$\Phi_{\mathbf{z}}(s) = \int_{-\infty}^{\infty}\int_{-\infty}^{\infty} e^{isg(x,y)} f_{\mathbf{xy}}(x,y)\,dx\,dy, \tag{130}$$

which we invert to obtain

$$f_{\mathbf{z}}(z) = \frac{1}{2\pi}\int_{-\infty}^{\infty} e^{-isz}\Phi_{\mathbf{z}}(s)\,ds. \tag{131}$$

13.8.1 Sums of two random variables

Perhaps the most important case involving a sum is the simple relation

$$\mathbf{z} = \mathbf{x} + \mathbf{y}. \tag{132}$$

We will illustrate how to find the PDF of **z** by both METHODS I and III above. METHOD II in this instance is really the same as METHOD I.

To use METHOD I, we note from Fig. 13.7 that the domain of integration D is defined by $x + y \le z$. Hence, from Eq. (126) we have

$$F_{\mathbf{z}}(z) = \int_{-\infty}^{\infty}\int_{-\infty}^{z-y} f_{\mathbf{xy}}(x,y)\,dx\,dy, \tag{133}$$

the derivative of which with respect to z yields

$$f_{\mathbf{z}}(z) = \int_{-\infty}^{\infty} f_{\mathbf{xy}}(z-y,y)\,dy. \tag{134}$$

If **x** and **y** are independent, then (134) becomes

$$f_{\mathbf{z}}(z) = \int_{-\infty}^{\infty} f_{\mathbf{x}}(z-y)f_{\mathbf{y}}(y)\,dy = \int_{-\infty}^{\infty} f_{\mathbf{x}}(x)f_{\mathbf{y}}(z-x)\,dx, \tag{135}$$

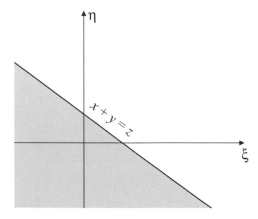

Figure 13.7 Domain defined by $x + y \le z$.

which is simply a *convolution* of the respective marginal densities. However, if **x** and **y** take on only positive values, then this last result becomes

$$f_z(z) = \int_0^z f_x(z-y) f_y(y) dy = \int_0^z f_x(x) f_y(z-x) dx, \quad z > 0. \tag{136}$$

Without knowing the specific joint PDF $f_{xy}(x,y)$, we cannot in general for METHOD III construct a formula for the density $f_z(z)$ like that in METHOD I. However, in the special case in which **x** and **y** are independent, it follows that

$$\Phi_z(s) = E[e^{is(x+y)}] = \Phi_x(s) \Phi_y(s). \tag{137}$$

Hence, the inverse Fourier transform of this product once again leads to the convolution integral given by (135).

EXAMPLE 22 Given that **x** and **y** are independent with marginal PDFs

$$f_x(x) = ae^{-ax} U(x), \quad f_y(y) = be^{-by} U(y), \quad a,b > 0,$$

determine the PDF for $z = x + y$.

Solution: Using the result of (135), we deduce from METHOD I that

$$f_z(z) = ab \int_0^z e^{-a(z-y)} e^{-by} dy = abe^{-az} \int_0^z e^{(a-b)y} dy.$$

Evaluating this integral, we see that

$$f_z(z) = \begin{cases} \dfrac{ab}{b-a} (e^{-az} - e^{-bz}), & b \neq a \\ a^2 z e^{-az}, & b = a, \end{cases} \quad z > 0,$$

and $f_z(z) = 0$, $z < 0$.
To use METHOD III, we first observe that

$$\Phi_z(s) = ab \int_0^\infty \int_0^\infty e^{is(x+y)} e^{-(ax+by)} dx dy$$
$$= ab \int_0^\infty e^{-(a-is)x} dx \int_0^\infty e^{-(b-is)y} dy,$$

or

$$\Phi_z(z) = \frac{ab}{b-a} \left(\frac{1}{a-is} - \frac{1}{b-is} \right).$$

For $b \neq a$, we can use Table 9.1 (or the technique of Example 15) to solve

$$f_z(z) = \frac{ab}{2\pi(b-a)}\left[\int_{-\infty}^{\infty}\frac{e^{-isz}}{a-is}\,ds - \int_{-\infty}^{\infty}\frac{e^{-isz}}{b-is}\,ds\right]$$

$$= \frac{ab}{b-a}(e^{-az} - e^{-bz})U(z).$$

Similarly, for $b = a$, we obtain

$$f_z(z) = \frac{a^2}{2\pi}\int_{-\infty}^{\infty}\frac{e^{-isz}}{(a-is)^2}\,ds = a^2 z e^{-az}U(z).$$

Another important sum in applications is that defined by

$$\mathbf{z} = \mathbf{x}^2 + \mathbf{y}^2. \tag{138}$$

In this case the random variable $\mathbf{z}$ is always nonnegative. We illustrate this situation for the case in which $\mathbf{x}$ and $\mathbf{y}$ are independent Gaussian random variables in Example 23 below.

EXAMPLE 23 Find the PDF for the sum $\mathbf{z} = \mathbf{x}^2 + \mathbf{y}^2$ given that $\mathbf{x}$ and $\mathbf{y}$ are independent Gaussian random variables defined by the joint PDF

$$f_{\mathbf{xy}}(x,y) = f_{\mathbf{x}}(x)f_{\mathbf{y}}(y) = \frac{1}{2\pi}e^{-(x^2+y^2)/2}.$$

Solution: For $z > 0$, the region D of the $\xi\eta$-plane for which $x^2 + y^2 \le z$ is a circle with radius $\sqrt{z}$. Hence, by METHOD I we have

$$F_{\mathbf{xy}}(z) = \frac{1}{2\pi}\iint_{x^2+y^2 \le z}e^{-(x^2+y^2)/2}\,dx\,dy$$

$$= \frac{1}{2\pi}\int_0^{\sqrt{z}}\int_0^{2\pi}r e^{-r^2/2}\,d\theta\,dr,$$

the last step of which follows from a change to polar coordinates. Completing the integration, we are led to

$$F_z(z) = (1 - e^{-z/2})U(z),$$

the derivative of which yields the negative exponential distribution

$$f_z(z) = \frac{1}{2}e^{-z/2}U(z) + (1 - e^{-z/2})\delta(z) = \frac{1}{2}e^{-z/2}U(z),$$

where we have used the property (Section 2.2.4) $g(z)\delta(z) = g(0)\delta(z)$.

To use METHOD III, we first calculate

$$\begin{aligned}
\Phi_z(s) &= \frac{1}{2\pi} \int_{-\infty}^{\infty}\int_{-\infty}^{\infty} e^{is(x^2+y^2)} e^{-(x^2+y^2)/2}\, dx\, dy \\
&= \frac{1}{2\pi} \int_0^{\infty}\int_0^{2\pi} r e^{-(1/2-is)r^2}\, d\theta\, dr \\
&= \frac{1}{1-2is},
\end{aligned}$$

where we have switched to polar coordinates again. The inversion of this last result (by the use of Table 9.1) gives us

$$f_z(z) = \frac{1}{2\pi} \int_{-\infty}^{\infty} \frac{e^{-isz}}{1-2is}\, ds = \frac{1}{2} e^{-z/2} U(z).$$

13.8.2 Rician distribution

The Rician distribution, also called the Rice-Nakagami distribution, is commonly associated with detection problems in both communications and radar. In particular, the envelope of a signal embedded in additive narrowband noise is governed by this distribution. In accordance with the notation used here, the Rician distribution is the PDF associated with the random variable

$$\mathbf{z} = \sqrt{\mathbf{x}^2 + \mathbf{y}^2}, \tag{139}$$

where $\mathbf{x}$ and $\mathbf{y}$ are independent Gaussian random variables with joint PDF

$$f_{\mathbf{xy}}(x,y) = \frac{1}{2\pi\sigma^2} \exp\left[-\frac{(x-m)^2 + y^2}{2\sigma^2}\right]. \tag{140}$$

In this case the mean value of $\mathbf{y}$ is taken as zero and the mean value of $\mathbf{x}$ is m. We have assumed the variance of both $\mathbf{x}$ and $\mathbf{y}$ is σ^2.

Following the approach used in Example 23, we first determine the CDF by making the change of variables to polar coordinates $x = r\cos\theta$, $y = r\sin\theta$. This action yields the integral for the CDF given by

$$\begin{aligned}
F_z(z) &= \frac{1}{2\pi\sigma^2} \iint_{\sqrt{x^2+y^2}\le z} \exp\left[-\frac{(x-m)^2+y^2}{2\sigma^2}\right] dx\, dy \\
&= \frac{1}{2\pi\sigma^2} \int_0^z\int_0^{2\pi} r\exp\left[-\frac{(r\cos\theta - m)^2 + r^2\sin^2\theta}{2\sigma^2}\right] d\theta\, dr \\
&= \frac{1}{2\pi\sigma^2} e^{-m^2/2\sigma^2} \int_0^z\int_0^{2\pi} r\exp\left(-\frac{r^2}{2\sigma^2}\right) \exp\left(\frac{mr\cos\theta}{\sigma^2}\right) d\theta\, dr,
\end{aligned}$$

the derivative of which, with respect to z, yields

$$f_z(z) = \frac{U(z)}{2\pi\sigma^2} e^{-(z^2+m^2)/2\sigma^2} \int_0^{2\pi} \exp\left(\frac{mz\cos\theta}{\sigma^2}\right) d\theta.$$

We recognize this last integral as that defining the modified Bessel function [see property (BI4) in Section 2.5.2], and therefore

$$f_z(z) = \frac{z}{\sigma^2} e^{-(z^2+m^2)/2\sigma^2} I_0\left(\frac{mz}{\sigma^2}\right) U(z), \tag{141}$$

where $I_0(x)$ is a modified Bessel function of the first kind.

Equation (141) represents the *Rician* or *Rice-Nakagami distribution* [also recall Example 7 in Section 2.5.2]. Because it is commonly associated with the output of an envelope detector in a coherent receiver system (see Section 15.4.2), we can interpret the input to the detector as a signal with amplitude m and additive narrowband noise. In this case, the variance σ^2 represents the noise power. When the signal is absent, (i.e., $m = 0$), (141) reduces to the *Rayleigh distribution*

$$f_z(z) = \frac{z}{\sigma^2} e^{-z^2/2\sigma^2} U(z). \tag{142}$$

The CDF associated with the Rician distribution (141) is defined by the integral

$$F_z(z) = \frac{1}{\sigma^2} \int_0^z x e^{-(x^2+m^2)/2\sigma^2} I_0\left(\frac{mx}{\sigma^2}\right) dx, \tag{143}$$

which cannot be evaluated in closed form. In early radar applications it was common to express the CDF (143) in terms of the *Marcum Q-function* defined by

$$Q(a,b) = \int_b^\infty x e^{-(x^2+a^2)/2} I_0(ax) dx. \tag{144}$$

To derive (141) by the use of METHOD III, it is preferable to first determine the PDF for the random variable $\mathbf{w}$ defined by

$$\mathbf{w} = \mathbf{x}^2 + \mathbf{y}^2, \tag{145}$$

and then let $\mathbf{z} = \sqrt{\mathbf{w}}$. In this case we have that

$$f_\mathbf{w}(w) = \frac{1}{2\pi} \int_{-\infty}^\infty e^{-isw} \Phi_{x^2}(s)\Phi_{y^2}(s) ds = \frac{1}{2\pi} \int_{-\infty}^\infty e^{-isw} \frac{\exp\left(\dfrac{ims}{1-2i\sigma^2 s}\right)}{1-2i\sigma^2 s} ds$$

$$= \frac{e^{-(w+m^2)/2\sigma^2}}{4\pi i\sigma^2} \int_{1-i\infty}^{1+i\infty} e^{pw/2\sigma^2} \frac{e^{m^2/2\sigma^2 p}}{p} dp,$$

where we have made the change of variable $p = 1 - 2i\sigma^2 s$ and performed some algebra.

Because we now recognize this as an inverse Laplace transform integral, the use of entry #14 in Table 10.1 gives us the result

$$f_{\mathbf{w}}(w) = \frac{1}{2\sigma^2} e^{-(w+m^2)/2\sigma^2} I_0\left(\frac{m\sqrt{w}}{\sigma^2}\right) U(w). \tag{146}$$

Finally, this reduces to the Rician PDF (141) through the transformation

$$f_{\mathbf{z}}(z) = 2z f_{\mathbf{w}}(z^2). \tag{147}$$

13.8.3 Products of random variables

Simple products of the form

$$\mathbf{z} = \mathbf{xy} \tag{148}$$

arise in a variety of situations. In particular, this product is associated with the correlation coefficient between random variables $\mathbf{x}$ and $\mathbf{y}$. Here we discuss two examples in which the variables $\mathbf{x}$ and $\mathbf{y}$ are assumed independent and belong to either the family of gamma distributions or the family of Gaussian distributions.

To begin, let us consider the case in which $\mathbf{x}$ and $\mathbf{y}$ are positive-definite random variables, both governed by gamma distributions

$$f_{\mathbf{x}}(x) = \frac{\alpha(\alpha x)^{\alpha-1}}{\Gamma(\alpha)} e^{-\alpha x} U(x), \ \alpha > 0,$$

$$f_{\mathbf{y}}(y) = \frac{\beta(\beta y)^{\beta-1}}{\Gamma(\beta)} e^{-\beta y} U(y), \ \beta > 0, \tag{149}$$

where $\Gamma(x)$ is the gamma function. In each case the mean has been set to unity and the parameter α or β then represents the reciprocal of the corresponding variance. By first holding $\mathbf{x}$ fixed, we can use METHOD II with

$$f_{\mathbf{z}}(z|x) = \frac{\beta(\beta z/x)^{\beta-1}}{x\Gamma(\beta)} e^{-\beta z/x} U(z), \tag{150}$$

to obtain

$$f_{\mathbf{z}}(z) = \int_0^\infty \left[\frac{\beta(\beta z/x)^{\beta-1}}{x\Gamma(\beta)} e^{-\beta z/x} \right]\left[\frac{\alpha(\alpha x)^{\alpha-1}}{\Gamma(\alpha)} e^{-\alpha x} \right] dx$$

$$= \frac{(\alpha\beta)^{\alpha\beta}}{\Gamma(\alpha)\Gamma(\beta)} \int_0^\infty x^{\alpha+\beta-1} e^{-\alpha x - \beta z/x} dx. \tag{151}$$

This last integral leads to a modified Bessel function of the second kind, i.e.,

$$f_{\mathbf{z}}(z) = \frac{2(\alpha\beta)^{(\alpha+\beta)/2}}{\Gamma(\alpha)\Gamma(\beta)} z^{(\alpha+\beta)/2-1} K_{\alpha-\beta}\left(2\sqrt{\alpha\beta z}\right) U(z). \tag{152}$$

The distribution (152) is sometimes termed the *generalized K distribution* because, for $\beta = 1$, it reduces to the *K distribution*

$$f_z(z) = \frac{2\alpha^{(\alpha+1)/2}}{\Gamma(\alpha)} z^{(\alpha-1)/2} K_{\alpha-1}\left(2\sqrt{\alpha z}\right) U(z). \tag{153}$$

Now let us consider the case in which **x** and **y** are independent Gaussian variables with joint density function

$$f_{\mathbf{xy}}(x,y) = \frac{1}{2\pi\sigma^2} e^{-(x^2+y^2)/2\sigma^2}. \tag{154}$$

Hence, we are assuming that **x** and **y** have zero means and equal variances. The characteristic function (METHOD III) for $\mathbf{z} = \mathbf{xy}$ is defined by the integral

$$\begin{aligned}
\Phi_{\mathbf{z}}(s) &= \frac{1}{2\pi\sigma^2} \int_{-\infty}^{\infty}\int_{-\infty}^{\infty} e^{isxy} e^{-(x^2+y^2)/2\sigma^2} \, dx\, dy \\
&= \frac{1}{2\pi\sigma^2} \int_{-\infty}^{\infty} e^{-x^2/2\sigma^2} \int_{-\infty}^{\infty} e^{isxy - y^2/2\sigma^2} \, dy\, dx \\
&= \frac{1}{\sigma\sqrt{2\pi}} \int_{-\infty}^{\infty} e^{-(1/\sigma^2 + \sigma^2 s^2)x^2/2} \, dx.
\end{aligned} \tag{155}$$

By using properties of the error function, the last integral in (155) reduces to

$$\Phi_{\mathbf{z}}(s) = \frac{1}{\sqrt{1 + \sigma^4 s^2}}. \tag{156}$$

Finally, the inversion of the characteristic function (156) yields

$$f_z(z) = \frac{1}{2\pi} \int_{-\infty}^{\infty} \frac{e^{-isz}}{\sqrt{1 + \sigma^4 s^2}} \, ds = \frac{1}{\pi} \int_{0}^{\infty} \frac{\cos sz}{\sqrt{1 + \sigma^4 s^2}} \, ds, \tag{157}$$

the evaluation of which gives us (see [13])

$$f_z(z) = \frac{1}{\pi\sigma^2} K_0\left(\frac{|z|}{\sigma^2}\right). \tag{158}$$

EXAMPLE 24 Calculate the variance of the random variable **z** in which

$$f_z(z) = \frac{1}{\pi\sigma^2} K_0\left(\frac{|z|}{\sigma^2}\right).$$

Solution: Because of symmetry of the PDF, we readily deduce that $E[\mathbf{z}] = 0$. Thus, the variance in this case leads to

$$\text{Var}(\mathbf{z}) = E[\mathbf{z}^2] = \frac{1}{\pi\sigma^2} \int_{-\infty}^{\infty} z^2 K_0\left(\frac{|z|}{\sigma^2}\right) \, dz = \sigma^4.$$

However, because $\mathbf{z} = \mathbf{xy}$ [see Eqs. (154)-(158) above], we can more easily calculate the variance, not by the above integral, but through the relation

$$\text{Var}(\mathbf{z}) = E[\mathbf{x}^2]E[\mathbf{y}^2] = \sigma^4 .$$

13.8.4 Quotients of random variables

In the case of a quotient like

$$\mathbf{z} = \frac{\mathbf{y}}{\mathbf{x}}, \tag{159}$$

we find that either METHODS II or III work best. For METHOD II, we fix the random variable $\mathbf{x}$ and note that $\mathbf{y} = \mathbf{xz}$. Hence, with $z = g(x,y) = y/x$, we see that

$$\frac{\partial g(x,y)}{\partial y} = \frac{1}{x} \tag{160}$$

and

$$f_\mathbf{y}(y) = |x| f_\mathbf{y}(xz|x). \tag{161}$$

It now follows that

$$f_\mathbf{z}(z) = \int_{-\infty}^{\infty} |x| f_\mathbf{y}(xz|x) f_\mathbf{x}(x) dx = \int_{-\infty}^{\infty} |x| f_\mathbf{xy}(x,xz) dx . \tag{162}$$

EXAMPLE 25 Given that $\mathbf{x}$ and $\mathbf{y}$ are independent with marginal PDFs

$$f_\mathbf{x}(x) = ae^{-ax}U(x), \quad f_\mathbf{y}(y) = be^{-by}U(y), \quad a,b > 0,$$

determine the PDF for $\mathbf{z} = \mathbf{y}/\mathbf{x}$.

Solution: Because $\mathbf{x}$ and $\mathbf{y}$ are independent, it follows that

$$f_\mathbf{y}(xz|x) f_\mathbf{x}(x) = abe^{-(ax+bxz)}U(x)U(xz),$$

and using the result of (162), we deduce from METHOD II that

$$f_\mathbf{z}(z) = ab\int_0^{\infty} xe^{-(a+bz)x}dx = \frac{ab}{(a+bz)^2}U(z).$$

EXAMPLE 26 Find the PDF for **z** = **y**/**x**, given that **x** and **y** are independent Gaussian random variables defined by the joint PDF

$$f_{xy}(x,y) = f_x(x)f_y(y) = \frac{1}{2\pi\sigma^2}e^{-(x^2+y^2)/2\sigma^2}.$$

Solution: In this case we will use METHOD III. By definition

$$\Phi_z(s) = \int_{-\infty}^{\infty}\int_{-\infty}^{\infty}e^{isy/x}f_{xy}(x,y)dxdy$$

$$= \frac{1}{2\pi\sigma^2}\int_{-\infty}^{\infty}e^{-x^2/2\sigma^2}\int_{-\infty}^{\infty}e^{isy/x-y^2/2\sigma^2}dydx,$$

where we have changed the order of integration in the last step. The inside integral is recognized as an inverse Fourier transform integral, which yields (see Table 9.1)

$$\int_{-\infty}^{\infty}e^{isy/x-y^2/2\sigma^2}dy = \sigma\sqrt{2\pi}e^{-\sigma^2s^2/2x^2},$$

and thus the characteristic function reduces to

$$\Phi_z(s) = \frac{1}{\sigma\sqrt{2\pi}}\int_{-\infty}^{\infty}e^{-x^2/2\sigma^2-\sigma^2s^2/2x^2}dx.$$

By use of the integral formula

$$\int_{-\infty}^{\infty}e^{-a^2x^2-b^2/x^2}dx = \frac{\sqrt{\pi}}{a}e^{-2a|b|}, \quad a>0,$$

it follows that the above integral defining the characteristic function gives

$$\Phi_z(s) = e^{-|s|}.$$

Inverting this characteristic function leads to the Cauchy distribution

$$f_z(z) = \frac{1}{2\pi}\int_{-\infty}^{\infty}e^{-isz}e^{-|s|}ds = \frac{1}{\pi(z^2+1)}.$$

13.8.5 Two functions of two random variables

In some situations we want the joint distribution of random variables **z** and **w** which are related to random variables **x** and **y** through equations of transformation given by

$$\mathbf{z} = g(\mathbf{x},\mathbf{y}), \quad \mathbf{w} = h(\mathbf{x},\mathbf{y}). \tag{163}$$

An important example is given by the *linear transformation*

$$\begin{aligned}\mathbf{z} &= a\mathbf{x}+b\mathbf{y},\\ \mathbf{w} &= c\mathbf{x}+d\mathbf{y},\end{aligned} \quad ad-bc\neq 0. \tag{164}$$

It may also be necessary in some cases to obtain the inverse transformation. However, transformations of the form (163) or (164) have a unique *inverse transformation* only if the Jacobian is not zero.

The *Jacobian of transformation* for the general transformation (163) is defined by the 2×2 determinant of partial derivatives

$$J(x,y) = \begin{vmatrix} \dfrac{\partial g}{\partial x} & \dfrac{\partial g}{\partial y} \\[2mm] \dfrac{\partial h}{\partial x} & \dfrac{\partial h}{\partial y} \end{vmatrix}, \tag{165}$$

which, for the linear transformation (164), reduces to $J(x,y) = ad - bc$ (see also Chapters 3 and 5). Hence, the condition $ad - bd \neq 0$ given in (164) is a consequence of requiring that the Jacobian of transformation not vanish. Of course, for the more general transformation (163), the Jacobian will be a function of x and y that may vanish at some points but not at others.

EXAMPLE 27 Calculate the Jacobian of transformation for

$$\mathbf{x} = \mathbf{r}\cos\theta, \quad \mathbf{y} = \mathbf{r}\sin\theta.$$

Solution: The given transformation is simply the transformation from polar coordinates to rectangular coordinates. Because the transformation is not linear in this case, we must use the determinant in (165) to calculate the Jacobian. Doing so yields

$$J(r,\theta) = \begin{vmatrix} \dfrac{\partial x}{\partial r} & \dfrac{\partial x}{\partial \theta} \\[2mm] \dfrac{\partial y}{\partial r} & \dfrac{\partial y}{\partial \theta} \end{vmatrix} = \begin{vmatrix} \cos\theta & -r\sin\theta \\ \sin\theta & r\cos\theta \end{vmatrix},$$

which reduces to the well-known result

$$J(r,\theta) = r(\cos^2\theta + \sin^2\theta) = r.$$

To determine the joint density function of $\mathbf{z}$ and $\mathbf{w}$ from that of $\mathbf{x}$ and $\mathbf{y}$, we use a generalization of the relation given in Eq. (71). Specifically, the desired relation is

$$f_{zw}(z,w) = \frac{f_{xy}(x,y)}{|J(x,y)|}. \tag{166}$$

Let us illustrate with an example.

EXAMPLE 28 Find the joint PDF $f_{zw}(z,w)$ for the linear transformation

$$\mathbf{z} = \mathbf{x}\cos\theta + \mathbf{y}\sin\theta,$$

$$\mathbf{w} = -\mathbf{x}\sin\theta + \mathbf{y}\cos\theta,$$

given that **x** and **y** are correlated jointly Gaussian random variables with joint PDF

$$f_{xy}(x,y) = \frac{1}{2\pi\sqrt{1-\rho^2}}\exp\left[-\frac{x^2 - 2\rho xy + y^2}{2(1-\rho^2)}\right].$$

Solution: In this case the Jacobian of transformation is

$$J(x,y) = \begin{vmatrix} \cos\theta & \sin\theta \\ -\sin\theta & \cos\theta \end{vmatrix} = 1,$$

and the inverse transformation is

$$\mathbf{x} = \mathbf{z}\cos\theta - \mathbf{w}\sin\theta,$$
$$\mathbf{y} = \mathbf{z}\sin\theta + \mathbf{w}\cos\theta.$$

Based on Eq. (166), we now deduce that

$$f_{zw}(z,w) = \frac{1}{2\pi\sqrt{1-\rho^2}}$$
$$\times\exp\left[-\frac{z^2 + w^2 - 2\rho(z^2 - w^2)\sin\theta\cos\theta - 2\rho zw(\cos^2\theta - \sin^2\theta)}{2(1-\rho^2)}\right],$$

which is another bivariate Gaussian density function. Note that **z** and **w** become statistically independent (hence, also uncorrelated) for the particular choice $\theta = \pi/4$. That is, the above bivariate distribution in this case reduces to

$$f_{zw}(z,w) = \frac{1}{2\pi\sqrt{1-\rho^2}}\exp\left[-\frac{z^2}{2(1+\rho)}\right]\exp\left[-\frac{w^2}{2(1-\rho)}\right],$$

where both **z** and **w** have zero means and variances given by

$$\text{Var}(\mathbf{z}) = 1 + \rho, \quad \text{Var}(\mathbf{w}) = 1 - \rho.$$

13.8.6 Sums of several random variables

When a random process (see Chapter 14) is sampled, such as the random output of a detection device, the sum of the samples leads to an expression of the form

$$\mathbf{z} = \mathbf{x}_1 + \mathbf{x}_2 + \cdots + \mathbf{x}_n. \tag{167}$$

Of particular interest in this situation is the case in which each of the random variables $\mathbf{x}_k$ has the same density function and they are all mutually independent. This can be accomplished by sampling the same random process at sufficiently large intervals of time.

Given the premise that each of the random variables $\mathbf{x}_k$ in (167) has the same density function and they are all mutually independent, the easiest way to determine the density function for the sum $\mathbf{z}$ is through the use of characteristic functions. Namely, if each random variable $\mathbf{x}_k$ has the characteristic function $\Phi_{\mathbf{x}_k}(s)$, then the characteristic function of the sum is simply the product of each characteristic function. This leads to the expression

$$\Phi_{\mathbf{z}}(s) = \prod_{k=1}^{n} \Phi_{\mathbf{x}_k}(s) = \left[\Phi_{\mathbf{x}}(s)\right]^n, \tag{168}$$

from which we can then calculate

$$f_{\mathbf{z}}(z) = \frac{1}{2\pi} \int_{-\infty}^{\infty} e^{-isz} \left[\Phi_{\mathbf{x}}(s)\right]^n ds. \tag{169}$$

EXAMPLE 29 If each random variable $\mathbf{x}_k$, $k = 1, 2, \ldots, n$, has the same PDF

$$f_{\mathbf{x}}(x) = \frac{1}{\sqrt{2\pi}\,\sigma} e^{-(x-m)^2/2\sigma^2},$$

determine the PDF for the sum of statistically independent random variables $\mathbf{z} = \mathbf{x}_1 + \mathbf{x}_2 + \cdots + \mathbf{x}_n$.

Solution: Each random variable $\mathbf{x}_k$ is Gaussian distributed and thus has the characteristic function (recall Example 7)

$$\Phi_{\mathbf{x}}(s) = e^{ims - \sigma^2 s^2/2},$$

the nth power of which yields

$$\Phi_{\mathbf{z}}(s) = e^{inms - n\sigma^2 s^2/2}.$$

The characteristic function for $\mathbf{z}$ is clearly that associated with another Gaussian random variable, this time with mean nm and variance $n\sigma^2$; hence,

$$f_{\mathbf{z}}(z) = \frac{1}{\sqrt{2n\pi}\,\sigma} e^{-(x-nm)^2/2n\sigma^2}, \quad n = 1, 2, 3, \ldots.$$

13.9 Limit Distributions

Sums of independent Poisson random variables have a Poisson distribution, and sums of independent Gaussian random variables have a Gaussian distribution. Distributions with this property, called *infinite divisibility*, are often of special significance in applications.

The ubiquity of the Gaussian distribution in science and engineering applications is not only attributed to its occurrence with sums of independent Gaussian random variables (see Example 29), but also with large sums of independent variables with any distribution of the individual random variables in the sum.

13.9.1 Gaussian density function

To understand how the Gaussian distribution is reached in a limiting manner, let us consider the particular sum of *zero-mean* random variables

$$\mathbf{z} = \frac{1}{\sqrt{n}}(\mathbf{x}_1 + \mathbf{x}_2 + \cdots + \mathbf{x}_n), \tag{170}$$

where we scale each random variable by $1/\sqrt{n}$ so that the resulting variance of $\mathbf{z}$ remains finite in the limit as $n \to \infty$. Other than the fact that each random variable $\mathbf{x}_k$ in the sum (170) has the same variance σ^2 and zero mean value, we make no further assumptions regarding their distribution. Therefore, let us consider the limit of the characteristic function given by

$$
\begin{aligned}
\lim_{n \to \infty} \Phi_{\mathbf{z}}(s) &= \lim_{n \to \infty} E^n[e^{is\mathbf{x}/\sqrt{n}}] \\
&= \lim_{n \to \infty} \left(1 + \frac{is E[\mathbf{x}]}{\sqrt{n}} - \frac{s^2 E[\mathbf{x}^2]}{2n} + \cdots \right)^n,
\end{aligned}
\tag{171}
$$

where we have retained only the most significant terms of the Maclaurin expansion for the characteristic function. By assumption, $E[\mathbf{x}] = 0$ and $E[\mathbf{x}^2] = \sigma^2$, so we can write

$$\lim_{n \to \infty} \Phi_{\mathbf{z}}(s) = \lim_{n \to \infty} \left(1 - \frac{s^2 \sigma^2}{2n} + \cdots \right)^n. \tag{172}$$

We now simplify (172) by use of the binomial formula

$$(1 + x)^n = \sum_{k=0}^{n} \binom{n}{k} x^k, \tag{173}$$

which leads to

$$
\begin{aligned}
\lim_{n \to \infty} \Phi_{\mathbf{z}}(s) &= \lim_{n \to \infty} \sum_{k=0}^{n} \binom{n}{k} \left(-\frac{s^2 \sigma^2}{2n} \right)^k \\
&= \lim_{n \to \infty} \sum_{k=0}^{n} \frac{n!}{k!(n-k)!n^k} \left(-\frac{s^2 \sigma^2}{2} \right)^k.
\end{aligned}
\tag{174}
$$

However, by the use of *Stirling's formula,*[2] it can be shown that

$$\lim_{n \to \infty} \frac{n!}{(n-k)! n^k} = 1,$$ (175)

and thus (174) reduces to

$$\lim_{n \to \infty} \Phi_z(s) = \sum_{k=0}^{\infty} \frac{(-s^2 \sigma^2/2)^k}{k!} = e^{-s^2 \sigma^2/2}.$$ (176)

We recognize the characteristic function arrived at in (176) as that of the Gaussian distribution with zero mean (recall Example 7). Hence, the limit density function is the Gaussian distribution

$$f_x(x) = \frac{1}{\sqrt{2\pi}\,\sigma} e^{-x^2/2\sigma^2}.$$ (177)

This important result that the limit distribution of a large number of random variables is a Gaussian density function is widely known as the *Central Limit Theorem* (see also [24]).

13.9.2 Gamma density function

When the sum (170) is composed of only causal or positive-definite random variables, the limit distribution under suitable conditions leads to the gamma PDF, not the Gaussian distribution. To see this, let us first consider the sum of complex random variables of the form

$$\begin{aligned}
z &= \frac{1}{\sqrt{n}}[(x_1 + iy_1) + (x_2 + iy_2) + \cdots + (x_n + iy_n)] \\
&= \frac{1}{\sqrt{n}}(x_1 + x_2 + \cdots + x_n) + \frac{i}{\sqrt{n}}(y_1 + y_2 + \cdots + y_n).
\end{aligned}$$ (178)

If we consider z as shown in the last step as separate sums of real variables x_k and y_k which all have zero means and equal variances, then application of the central limit theorem above leads to the conclusion that the real part X and imaginary part Y of z each approach a Gaussian distribution with zero mean. The magnitude squared of $z = X + iY$ is then a positive random variable which we can denote by

$$u = |z|^2 = X^2 + Y^2.$$ (179)

The PDF for u is the negative exponential distribution (e.g., see Example 23)

$$f_u(u) = \frac{1}{b} e^{-u/b} U(u),$$ (180)

where b is the mean. We now consider the finite sum of positive random variables

$$w = u_1 + u_2 + \cdots + u_m,$$ (181)

---

[2]Stirling's formula is often written as $n! \sim \sqrt{2\pi n}\, n^n e^{-n}$, $n \to \infty$.

where each variate $\mathbf{u}_k$ is governed by a negative exponential distribution (180), all with the same mean. Under this condition, we can use the characteristic function method to write the characteristic function of $\mathbf{w}$ in the form

$$\Phi_{\mathbf{w}}(s) = \prod_{k=1}^{m} \Phi_{\mathbf{u}_k}(s) = \frac{1}{(1 - ibs)^m} . \tag{182}$$

Inverting (182) through use of the Fourier transform yields the *gamma distribution*

$$f_{\mathbf{w}}(w) = \frac{1}{b^m \Gamma(m)} w^{m-1} e^{-w/b} U(w) . \tag{183}$$

In the above fashion we see that the gamma distribution is associated with a finite sum of negative exponential random variables, which themselves evolve from the magnitude squared[3] of the limit distribution of a sum of complex random variables. The gamma distribution is also used as an approximation for certain other finite sums of positive random variables. In some cases, the discrete parameter m is formally replaced with a continuous positive parameter c (see Table 13.1).

Suggested Reading

P. Beckmann, *Probability in Communication Engineering* (Harcourt, Brace, & World, New York, 1967).

H. Cramér, *Mathematical Methods of Statistics* (Princeton University Press, Princeton, 1946).

B. R. Frieden, *Probability, Statistical Optics, and Data Testing*, 2nd ed. (Springer-Verlag, New York, 1991).

C. W. Helstron, *Probability and Stochastic Processes for Engineers*, 2nd ed. (Macmillan, New York, 1991).

M. G. Kendall and A. Stuart, *The Advanced Theory of Statistics*, Vol. 3 (Hafner, New York, 1968).

A. Papoulis, *Probability, Random Variables, and Random Processes*, 2nd ed. (McGraw-Hill, New York, 1984).

---

[3]If we considered only the magnitude $\mathbf{w} = |\mathbf{z}|$, not the magnitude squared, the resulting PDF would be the Rayleigh distribution in place of the negative exponential distribution. A finite sum of Rayleigh variates can then be approximated by the m distribution of Nakagami, obtained formally from (183) by the change of variables $w = r^2$.

Chapter 13

Exercises

1. A certain class of diodes are tested to see how long they last before burning out. Assume the CDF for a diode as a function of time $\mathbf{t}$ is

$$F_{\mathbf{t}}(t) = \begin{cases} 0, & t < 0 \\ 1 - e^{-\mu t}, & t \geq 0, \end{cases}$$

where μ is a constant. What is the probability that

(*a*) a given diode burns out between times a and b $(0 \leq a \leq b)$?

(*b*) a given diode lifetime is longer than time c?

(*c*) What is the PDF for the diode?

Ans. (*a*) $\Pr(a < \mathbf{t} < b) = e^{\mu a} - e^{-\mu b}$
(*b*) $\Pr(\mathbf{t} > c) = e^{-\mu c}$
(*c*) $f_{\mathbf{t}}(t) = \begin{cases} 0, & t < 0 \\ \mu e^{-\mu t}, & t \geq 0 \end{cases}$

2. The PDF of a random variable $\mathbf{x}$ is

$$f_{\mathbf{x}}(x) = K e^{-2x} U(x).$$

(*a*) Find the value of K.
(*b*) Find the probability $\mathbf{x} > 1$.
(*c*) Find the probability $\mathbf{x} \leq 0.5$.

Ans. (*a*) 2.0 (*b*) 0.1353
(*c*) 0.6321

3. Verify the expressions for the mean and variance in Table 13.1 for the

(*a*) binomial distribution.
(*b*) Poisson distribution.
(*c*) uniform distribution.

4. For the PDF in Problem 2, calculate

(*a*) the mean value.
(*b*) the variance.

Ans. (*a*) 1/2 (*b*) 1/4

5. For the gamma distribution in Table 13.1, find the

(*a*) characteristic function.
(*b*) moments $E[\mathbf{x}^n]$, $n = 1, 2, 3, \ldots$.
(*c*) variance $\sigma_{\mathbf{x}}^2$.

Ans. (*a*) $\Phi_{\mathbf{x}}(s) = \dfrac{1}{(1 - is)^c}$
(*b*) $<E[\mathbf{x}^n] = \dfrac{\Gamma(n + c)}{\Gamma(c)}$

6. Given the characteristic function

$$\Phi_{\mathbf{x}}(s) = e^{-a|s|},$$

show that the related PDF is

$$f_{\mathbf{x}}(x) = \dfrac{a/\pi}{x^2 + a^2}.$$

7. If $\mathbf{x}$ is a Gaussian random variable [see Eq. (17)], show that

(*a*) $\Pr(|\mathbf{x} - m| \leq k\sigma) = \mathrm{erf}\left(\dfrac{k}{\sqrt{2}}\right)$.

(*b*) $f_{\mathbf{x}}(x \mid |\mathbf{x} - m| \leq k\sigma) = \dfrac{e^{-(x-m)^2/2\sigma^2}}{\sqrt{2\pi}\,\sigma\,\mathrm{erf}\left(\dfrac{k}{\sqrt{2}}\right)}$.

8. A random variable $\mathbf{x}$ has the PDF

$$f_{\mathbf{x}}(x) = e^{-2|x|}.$$

(*a*) Find the PDF of the random variable $\mathbf{y}$, where $\mathbf{y} = 3\mathbf{x} - 4$.

(*b*) Find the probability that $\mathbf{y} < 0$.
(*c*) Find the probability that $\mathbf{y} > \mathbf{x}$.

Ans. (*a*) $f_{\mathbf{y}}(y) = \dfrac{1}{3} e^{-2|y+4|/3}$
(*b*) 0.9653 (*c*) 9.158×10^{-3}

9. Let $\mathbf{x}$ have the PDF in Problem 8.
(*a*) If $\mathbf{y} = \mathbf{x}^2$, find the PDF for $\mathbf{y}$.
(*b*) What is $\Pr(\mathbf{y} < 2)$?

Ans. (*a*) $f_{\mathbf{y}}(y) = \dfrac{1}{\sqrt{y}} e^{-2\sqrt{y}} U(y)$
(*b*) 0.9409

10. A random variable θ is uniformly distributed over $(0, 2\pi)$. If $\mathbf{x} = \cos\theta$,

(*a*) find the PDF of $\mathbf{x}$.

(b) find the mean of $\mathbf{x}$.

(c) find the variance of $\mathbf{x}$.

(d) What is the probability $\mathbf{x} > \frac{1}{2}$?

Ans. (a) $f_{\mathbf{x}}(x) = \dfrac{1}{\pi\sqrt{1 - x^2}}$

(b) 0   (c) 1/2   (d) 1/3

11. If $\mathbf{x}$ is a Gaussian random variable with zero mean, show that

$$\mathrm{Var}(\mathbf{x}|\mathbf{x} > 0) = \left(1 - \frac{2}{\pi}\right)\sigma^2 .$$

12. The actual time to failure of a communications satellite is governed by $f_t(t) = \frac{1}{3}e^{-t/3}U(t)$. What is the conditional mean lifetime of the satellite, given that it has survived for 2 years?

Ans. 5 years

13. If $\mathbf{x}$ is a Rayleigh variable (see Table 13.1) and $\mathbf{y} = a + b\mathbf{x}^2$, find $\mathrm{Var}(\mathbf{y})$.

Ans. $\mathrm{Var}(\mathbf{y}) = 4b^2\sigma^4$

14. Given the joint PDF

$$f_{\mathbf{xy}}(x,y) = Cx^2e^{-x-xy}U(x)U(y),$$

find C so this is a proper PDF.

Ans. $C = 1$

15. Given the joint PDF in Problem 14,

(a) find $f_{\mathbf{y}}(y)$.

(b) find $f_{\mathbf{x}}(x|y)$.

(c) find $E[\mathbf{x}^4|y]$.

(d) find $f_{\mathbf{xy}}(x,y|\mathbf{y} > 3/\mathbf{x})$.

Ans. (a) $f_{\mathbf{y}}(y) = 2U(y)/(1 + y)^3$

(b) $f_{\mathbf{x}}(x|y)$

$\qquad = \frac{1}{2}x^2(1 + y)^3 e^{-x-xy}U(x)U(y)$

(c) $E[\mathbf{x}^4|y] = 360/(1 + y)^4$

(d) $f_{\mathbf{xy}}(x,y|\mathbf{y} > 3/\mathbf{x}) = x^2e^{3-x-xy}$

16. The random variables $\mathbf{x}$ and $\mathbf{y}$ have means 0 and 1, respectively. If also $\sigma_{\mathbf{x}}^2 = 4$, $\sigma_{\mathbf{y}}^2 = 9$, and $\mathrm{Cov}(\mathbf{x,y}) = 2$, find

(a) the normalized correlation

coefficient ρ.

(b) the mean and variance of the random variable $\mathbf{z} = 5\mathbf{x} - 2\mathbf{y}$.

Ans. (a) 1/3   (b) −2, 96

17. If $\mathbf{x}$ and $\mathbf{y}$ are independent, Gaussian random variables with means m_1, m_2 and variances σ_1^2, σ_2^2, respectively, show that

$$\mathrm{Pr}(\mathbf{xy} < 0) = \frac{1}{2} - 2\,\mathrm{erf}\!\left(\frac{m_1}{\sigma_1}\right)\mathrm{erf}\!\left(\frac{m_2}{\sigma_2}\right).$$

18. Given the joint PDF

$$f_{\mathbf{xy}}(x,y) = \begin{cases} \dfrac{8}{\pi}(1 - x^2 - y^2), \\ \qquad 0 < x^2 + y^2 < 1 \ (x,y > 0) \\ 0, \ \text{elsewhere} \end{cases}$$

calculate $\mathrm{Pr}(\mathbf{x} > 1/2)$.

Ans. 0.2532

19. Given the joint PDF

$$f_{\mathbf{xy}}(x,y) = \begin{cases} C(3xy + x + 3y + b), \\ \qquad 0 < x \le 1, \ 0 < y < 1 \\ 0, \ \text{elsewhere} \end{cases}$$

(a) find values of C and b so that $\mathbf{x}$ and $\mathbf{y}$ are statistically independent.

(b) find $E[\mathbf{xy}]$.

Ans. (a) $C = 4/15$, $b = 1$ (b) 1/3

20. Given $\mathbf{y} = \mathbf{x} + \mathbf{n}$ and the joint PDF

$$f_{\mathbf{xy}}(x,y) = \frac{\sqrt{15}}{4\pi}e^{-(x^2 + y^2 + xy/2)},$$

(a) find $f_{\mathbf{x}}(x|y)$.

(b) find the best estimate for $\mathbf{x}$ given that $\mathbf{y} = 3$.

Ans. (a) $f_{\mathbf{x}}(x|y) = \dfrac{1}{\sqrt{\pi}}e^{-(x+y/4)^2}$

(b) $\hat{\mathbf{x}} = -\dfrac{3}{4}$

21. If $\mathbf{z} = \mathbf{xy}$, where $\mathbf{x}$ and $\mathbf{y}$ are independent random variables with

$$f_{\mathbf{x}}(x) = \begin{cases} \dfrac{1}{\pi\sqrt{1 - x^2}}, & |x| < 1 \\ 0, & |x| > 1 \end{cases}$$

$$f_{\mathbf{y}}(y) = \frac{y}{\sigma^2} e^{-y^2/2\sigma^2} U(y),$$

show that **z** is Gaussian with zero mean and variance σ^2.

22. If $\mathbf{z} = \mathbf{x}/\mathbf{y}$, and **x** and **y** are statistically independent random variables for which

$$f_{\mathbf{x}}(x) = a e^{-ax} U(x),$$
$$f_{\mathbf{y}}(y) = b e^{-by} U(y),$$

(*a*) find $f_{\mathbf{z}}(z)$.
(*b*) Does $E[\mathbf{z}]$ exist.

Ans. (*a*) $f_{\mathbf{z}}(z) = \dfrac{ab\,U(z)}{(az + b)^2}$

(*b*) No

23. If $\mathbf{z} = \mathbf{x}/\mathbf{y}$, and **x** and **y** are statistically independent random variables for which

$$f_{\mathbf{x}}(x) = \frac{x}{\alpha^2} e^{-x^2/2\alpha^2} U(x),$$

$$f_{\mathbf{y}}(y) = \frac{y}{\beta^2} e^{-y^2/2\beta^2} U(y),$$

(*a*) find $f_{\mathbf{z}}(z)$.
(*b*) Show that $(k > 0)$

$$\Pr(\mathbf{x} \le k\mathbf{y}) = \frac{k^2}{k^2 + \alpha^2/\beta^2}.$$

Ans. $f_{\mathbf{z}}(z) = \dfrac{2\alpha^2 z}{\beta^2(z^2 + \alpha^2/\beta^2)^2} U(z)$

24. Given $\mathbf{z} = \mathbf{x} + \mathbf{y}$, $\mathbf{w} = \mathbf{x} - \mathbf{y}$, and the joint PDF

$$f_{\mathbf{xy}}(x,y) = \begin{cases} 1/\pi, & 0 < x^2 + y^2 \le 1 \\ 0, & x^2 + y^2 > 1, \end{cases}$$

(*a*) find $f_{\mathbf{zw}}(z,w)$.
(*b*) find $f_{\mathbf{z}}(z)$.
(*c*) find Var(**z**).

Ans. (*a*) $f_{\mathbf{zw}}(z,w) = \dfrac{1}{2\pi}$, $0 < z^2 + w^2 \le 2$

(*b*) $f_{\mathbf{z}}(z) = \dfrac{1}{\pi}\sqrt{2 - z^2}$, $-\sqrt{2} < z < \sqrt{2}$

(*c*) $\text{Var}(\mathbf{z}) = \dfrac{1}{2}$

25. If **x** and **y** are statistically independent with densities as given in Problem 22 with $a = b = 1$, and

$$\mathbf{z} = \mathbf{x} + \mathbf{y}, \quad \mathbf{w} = \frac{\mathbf{x}}{\mathbf{x} + \mathbf{y}},$$

find the PDFs for **z** and **w**.

Ans. $f_{\mathbf{z}}(z) = z e^{-z} U(z)$ and **w** is uniform on $(0,1)$

Chapter 14

Random Processes

14.1   Introduction / 698
14.2   Probabilistic Description of Random Process / 698
    14.2.1   First- and second-order statistics / 699
    14.2.2   Stationary random processes / 700
14.3   Autocorrelation and Autocovariance Functions / 700
    14.3.1   Time averages and ergodicity / 703
    14.3.2   Basic properties / 405
    14.3.3   Structure functions / 707
14.4   Cross-Correlation and Cross-Covariance / 708
    14.4.1   Basic properties / 709
    14.4.2   Cross-correlation techniques in radar / 710
14.5   Power Spectral Density Functions / 711
    14.5.1   Riemann-Stieltjes integral / 712
14.6   Transformations of Random Processes / 716
    14.6.1   Memoryless nonlinear transformations / 716
    14.6.2   Linear systems / 718
14.7   Stationary Gaussian Processes / 722
    14.7.1   Multivariate Gaussian distributions / 723
    14.7.2   Detection devices / 725
    14.7.3   Zero-crossing problem / 726
Suggested Reading / 729
Exercises / 729

The mathematical background reviewed in Chapter 13 provides the basis for developing the statistical description of random functions (signals) known as random processes. Basically, we mathematically describe such signals by either a correlation function or a power spectral density, which are related through the Fourier transform. Among many other areas, the theory of random processes is essential to the field of statistical communication theory.

14.1 Introduction

In Chapter 13 we introduced the concept of random variable and its related probability distribution. A natural generalization of the random variable concept is that of random process. A *random process*, also called a *stochastic process*, is a collection of time functions and an associated probability description. The entire collection of such functions is called an *ensemble*. Ordinarily, we represent any particular member of the ensemble by simply $\mathbf{x}(t)$, called a *sample function* or *realization*. For a fixed value of time, say t_1, the quantity $\mathbf{x}_1 = \mathbf{x}(t_1)$ can then be interpreted as a *random variable*.

A *continuous* random process is one in which the random variables $\mathbf{x}_1$, $\mathbf{x}_2$, ..., can assume any value within a specified range of possible values. A *discrete* random process is one in which the random variables can assume only certain isolated values (possibly infinite in number).

One of the most common random process occurring in engineering applications is *random noise*, e.g., a "randomly" fluctuating voltage or current at the input to a receiver that interferes with the reception of a radio or radar signal, or the current through a photo-electric detector, and so on. Although we limit our discussion to random processes of time t, it is possible to readily extend these ideas to the notion of a *random field*, which in general is a function of both time t and space $\mathbf{R} = (x,y,z)$. Atmospheric wind velocity, temperature, and index of refraction fluctuations are typical examples of random fields.

14.2 Probabilistic Description of Random Process

If we imagine "sampling" the random process $\mathbf{x}(t)$ at a finite number of times $t_1, t_2, ..., t_n$, then we obtain the collection of random variables $\mathbf{x}_k = \mathbf{x}(t_k)$, $k = 1, 2, ..., n$. The probability measure associated with these random variables is described by the *joint probability density function* (PDF) of order n:

$$f_{\mathbf{x}}(x_1, t_1; x_2, t_2; ...; x_n, t_n).$$

In principle, we can develop the theory of a continuous random process by describing the joint probability density function of all orders. However, this is generally an impossible task, so we usually settle for only first-order and/or second-order distributions. We denote the first-order PDF by $f_{\mathbf{x}}(x,t)$ and the second-order PDF by $f_{\mathbf{x}}(x_1, t_1; x_2, t_2)$.

14.2.1 First- and second-order statistics

The function defined by

$$F_{\mathbf{x}}(x,t) = \Pr[\mathbf{x}(t) \le x] \tag{1}$$

is called the first-order *distribution function* of the random process $\mathbf{x}(t)$. The corresponding first-order PDF is

$$f_{\mathbf{x}}(x,t) = \frac{\partial F_{\mathbf{x}}(x,t)}{\partial x}. \tag{2}$$

Similarly, the second-order distribution function and corresponding PDF are defined, respectively, by

$$F_{\mathbf{x}}(x_1,t_1;x_2,t_2) = \Pr[\mathbf{x}(t_1) \le x_1, \mathbf{x}(t_2) \le x_2], \tag{3}$$

$$f_{\mathbf{x}}(x_1,t_1;x_2,t_2) = \frac{\partial^2 F_{\mathbf{x}}(x_1,t_1;x_2,t_2)}{\partial x_1 \partial x_2}. \tag{4}$$

We note that $F_{\mathbf{x}}(x_1,t_1;\infty,t_2) = F_{\mathbf{x}}(x_1,t_1)$ and

$$f_{\mathbf{x}}(x_1,t_1) = \int_{-\infty}^{\infty} f_{\mathbf{x}}(x_1,t_1;x_2,t_2)dx_2. \tag{5}$$

Conditional PDFs and distributions associated with random processes can be defined in much the same manner as done in Chapter 13 for random variables. For example, the conditional PDF of $\mathbf{x}_2 = \mathbf{x}(t_2)$, given the process took on value $\mathbf{x}_1$ at time t_1, is defined by

$$f_{\mathbf{x}_2}(x_2,t_2 | x_1,t_1) = \frac{f_{\mathbf{x}}(x_1,t_1;x_2,t_2)}{f_{\mathbf{x}_1}(x_1,t_1)}. \tag{6}$$

EXAMPLE 1 Calculate the joint PDF $f_{\mathbf{x}}(x_1,t_1;x_2,t_2)$ given that $\mathbf{x}_1$ and $\mathbf{x}_2$ are obtained from the random process

$$\mathbf{x}(t) = \mathbf{A}\cos(\omega t + \boldsymbol{\varphi}),$$

in which the amplitude $\mathbf{A}$ and phase $\boldsymbol{\varphi}$ are random variables.

Solution: At times t_1 and t_2 we obtain

$$\mathbf{x}_1 = \mathbf{A}\cos(\omega t_1 + \boldsymbol{\varphi}),$$

$$\mathbf{x}_2 = \mathbf{A}\cos(\omega t_2 + \boldsymbol{\varphi}).$$

We can interpret these as transformation equations from random variables $\mathbf{A}$ and $\boldsymbol{\varphi}$ to $\mathbf{x}_1$ and $\mathbf{x}_2$. Hence, their joint PDFs are related by [see (166) in Section 13.8.5]

$$f_{\mathbf{x}}(x_1,t_1;x_2,t_2) = \frac{f_{\mathbf{A}\varphi}(A,\varphi)}{|J(A,\varphi)|},$$

where J is the Jacobian of transformation defined by

$$J(A,\varphi) = \begin{vmatrix} \cos(\omega t_1 + \varphi) & -A\sin(\omega t_1 + \varphi) \\ \cos(\omega t_2 + \varphi) & -A\sin(\omega t_2 + \varphi) \end{vmatrix} = A\sin[\omega(t_1 - t_2)].$$

14.2.2 Stationary random processes

Suppose the first-order PDF does not depend on time, i.e., $f_{\mathbf{x}}(x,t) = f_{\mathbf{x}}(x)$, and further, that the second-order PDF has the form

$$f_{\mathbf{x}}(x_1,t_1;x_2,t_2) = f_{\mathbf{x}}(x_1,x_2;t_2 - t_1) \tag{7}$$

for all t_1 and t_2. That is, the second-order or joint PDF depends only on the time difference $\tau = t_2 - t_1$ but not on the specific times t_1 and t_2. If all marginal and joint PDFs depend only on the time difference $\tau = t_2 - t_1$, but not on the specific time origin, we have what is called a *stationary random process*. Such a process can also be described as one in which its moments are invariant under translations in time.

Random noise produced by an electronic device is usually considered a stationary process during the (finite) interval of observation, as are many other random processes that occur in engineering applications. In general, if the parameters producing a random process do not change significantly during the finite observation time, we can often treat that process as stationary. Of course, if any of the PDFs associated with a random process do change with the choice of time origin, we say that process is *nonstationary*.

The random process in Example 1 above is not in general a stationary process, because the joint PDF will usually depend on the specific times t_1 and t_2, rather than just on their difference. However, if the joint PDF for the amplitude and phase has the form

$$f_{\mathbf{A}\varphi}(A,\varphi) = \frac{1}{2\pi}f_{\mathbf{A}}(A), \quad A > 0, \ 0 \le \varphi < 2\pi,$$

where the phase φ is uniformly distributed over 2π, then it readily follows that the random process $\mathbf{x}(t)$ in Example 1 is stationary (at least through the second-order PDF); i.e., the solution for $\mathbf{A}$ in terms of $\mathbf{x}_1$ and $\mathbf{x}_2$ depends only on $\tau = t_2 - t_1$.

14.3 Autocorrelation and Autocovariance Functions

In this chapter we will use the notation $\langle\ \rangle$ to denote an ensemble average instead of the expectation symbol $E[.]$ used in Chapter 13. We define the *mean*, also called the *expected value* or *ensemble average*, of the random process $\mathbf{x}(t)$ by

$$\boxed{<\mathbf{x}(t)> = m(t) = \int_{-\infty}^{\infty} x f_\mathbf{x}(x,t)\,dx,}$$ (8)

where we are emphasizing the fact that the mean value in general may depend on time. Similarly, the *variance* $\mathrm{Var}[\mathbf{x}(t)] \equiv \sigma_\mathbf{x}^2(t) = <\mathbf{x}^2(t)> - m^2(t)$ is also a function of time in the general case. However, if the random process is stationary, then its mean value and variance are independent of time; in particular, we write $<\mathbf{x}(t)> = m$.

▸ **Autocorrelation function:** Let $\mathbf{x}_1$ and $\mathbf{x}_2$ denote random variables taken from a real stationary random process $\mathbf{x}(t)$ at times t_1 and $t_2 = t_1 + \tau$, respectively. We define the *autocorrelation function* $R_\mathbf{x}(t_1,t_2) \equiv R_\mathbf{x}(\tau)$ (also called simply the *correlation function*) by the expression

$$\boxed{R_\mathbf{x}(\tau) = <\mathbf{x}(t_1)\mathbf{x}(t_2)> = \iint_{-\infty}^{\infty} x_1 x_2 f_\mathbf{x}(x_1,x_2;\tau)\,dx_1\,dx_2.}$$ (9)

If $\mathbf{x}(t)$ is a *complex* stationary random process, then $R_\mathbf{x}(\tau) = <\mathbf{x}(t_1)\mathbf{x}^*(t_2)>$, where the asterisk * denotes the complex conjugate of the quantity.

▸ **Autocovariance function:** The *autocovariance function* (or *covariance function*) of a real stationary process is defined by the ensemble average

$$C_\mathbf{x}(t_1,t_2) \equiv C_\mathbf{x}(\tau) = <[\mathbf{x}(t_1)-m][\mathbf{x}(t_2)-m]>,$$ (10)

from which we deduce

$$\boxed{C_\mathbf{x}(\tau) = R_\mathbf{x}(\tau) - m^2.}$$ (11)

Hence, when the mean is zero, the correlation and covariance functions are identical. Also, when $t_1 = t_2$ ($\tau = 0$), the covariance function (11) reduces to the variance $\sigma_\mathbf{x}^2 = C_\mathbf{x}(0)$ of the random variable $\mathbf{x}$. The autocovariance $C_\mathbf{x}(\tau)$ is generally considered more informative, but the autocorrelation $R_\mathbf{x}(\tau)$ is more directly measurable.

Analogous to the correlation coefficient defined by (116) in Chapter 13, it is sometimes useful to introduce the *normalized covariance function*

$$\rho_\mathbf{x}(\tau) = \frac{C_\mathbf{x}(\tau)}{C_\mathbf{x}(0)}.$$ (12)

Because the maximum of the covariance function occurs at $\tau = 0$ [see Eq. (23) below], it follows that

$$-1 \leq \rho_\mathbf{x}(\tau) \leq 1.$$ (13)

To be considered a *strict* stationary process, we require all marginal and joint density functions to be independent of the choice of time origin. However, this requirement is more stringent than necessary in most practical situations.

▸ **Wide-sense stationary:** If all we know is that the mean value $<x(t)>$ and variance σ_x^2 are constant, and that the covariance function $C_x(\tau)$ depends only on the time interval $\tau = t_2 - t_1$, we say the random process $\mathbf{x}(t)$ is stationary in the *wide sense*.

Strict stationary processes are automatically wide-sense stationary, but the converse is not necessarily true. For most wide-sense stationary processes, it is usually (but not always) the case that

$$C_x(\tau) \to 0, \quad |\tau| \to \infty. \tag{14}$$

EXAMPLE 2 Calculate the mean and covariance of the random process defined by $\mathbf{x}(t) = \mathbf{A}\cos(\omega t + \varphi)$, in which the amplitude $\mathbf{A}$ and phase φ are random variables with joint PDF (see also Example 1)

$$f_{\mathbf{A}\varphi}(A,\varphi) = \frac{1}{2\pi} f_{\mathbf{A}}(A), \quad A > 0, \ 0 \le \varphi < 2\pi.$$

Solution: By definition, the mean value is

$$\begin{aligned}
<\mathbf{x}(t)> &= \int_0^\infty \int_0^{2\pi} A\cos(\omega t + \varphi) f_{\mathbf{A}\varphi}(A,\varphi)d\varphi\,dA \\
&= \frac{1}{2\pi}\int_0^\infty A f_{\mathbf{A}}(A)dA \int_0^{2\pi}\cos(\omega t + \varphi)d\varphi = 0,
\end{aligned}$$

which follows directly from integration over φ. Similarly, the correlation function leads to

$$\begin{aligned}
R_x(t_1,t_2) &= \frac{1}{2\pi}\int_0^\infty A^2 f_{\mathbf{A}}(A)dA\int_0^{2\pi}\cos(\omega t_1 + \varphi)\cos(\omega t_2 + \varphi)d\varphi \\
&= \frac{1}{4\pi}<\mathbf{A}^2>\int_0^{2\pi}\left\{\cos[\omega(t_1 - t_2)] + \cos[\omega(t_1 + t_2 + 2\varphi)]\right\}d\varphi \\
&= \frac{1}{2}<\mathbf{A}^2>\cos[\omega(t_1 - t_2)].
\end{aligned}$$

Hence, the process $\mathbf{x}(t)$ is (wide-sense) stationary and, because the mean is zero, we can write the covariance as

$$C_x(\tau) = R_x(\tau) = \frac{1}{2}<\mathbf{A}^2>\cos\omega\tau, \quad \tau = t_2 - t_1.$$

EXAMPLE 3 Consider the random process

$$\mathbf{x}(t) = \mathbf{a}\cos\omega t + \mathbf{b}\sin\omega t,$$

where ω is a constant and $\mathbf{a}$ and $\mathbf{b}$ are independent Gaussian random variables satisfying

$$<\mathbf{a}> = <\mathbf{b}> = 0, \ <\mathbf{a}^2> = <\mathbf{b}^2> = \sigma^2.$$

Determine
(a) the autocorrelation function for $\mathbf{x}(t)$, and
(b) the second-order PDF for $\mathbf{x}_1$ and $\mathbf{x}_2$.

Solution:

(a) Because $\mathbf{a}$ and $\mathbf{b}$ are statistically independent random variables with zero means, it follows that $\langle \mathbf{ab} \rangle = \langle \mathbf{a} \rangle \langle \mathbf{b} \rangle = 0$, and thus

$$
\begin{aligned}
R_x(t_1,t_2) &= \langle (\mathbf{a}\cos \omega t_1 + \mathbf{b}\sin \omega t_1)(\mathbf{a}\cos \omega t_2 + \mathbf{b}\sin \omega t_2) \rangle \\
&= \langle \mathbf{a}^2 \rangle \cos \omega t_1 \cos \omega t_2 + \langle \mathbf{b}^2 \rangle \sin \omega t_1 \sin \omega t_2 \\
&= \sigma^2 \cos[\omega(t_2 - t_1)],
\end{aligned}
$$

or

$$
R_x(\tau) = \sigma^2 \cos \omega \tau, \quad \tau = t_2 - t_1 .
$$

(b) The mean value of $\mathbf{x}(t)$ is $\langle \mathbf{x}(t) \rangle = \langle \mathbf{a} \rangle \cos \omega t + \langle \mathbf{b} \rangle \sin \omega t = 0$. Hence, $\sigma_x^2 = R_x(0) = \sigma^2$, and the first-order PDF of $\mathbf{x}(t)$ is given by

$$
f_x(x,t) = \frac{1}{\sqrt{2\pi}\,\sigma} e^{-x^2/2\sigma^2} .
$$

The second-order PDF depends on the correlation coefficient between $\mathbf{x}_1$ and $\mathbf{x}_2$, which, because the mean is zero, can be calculated from

$$
\rho_x(\tau) = \frac{R_x(\tau)}{R_x(0)} = \cos \omega \tau,
$$

and consequently (see Section 13.7.7),

$$
f_x(x_1,t_1;x_2,t_2) = \frac{1}{2\pi\sigma^2 |\sin \omega \tau|} \exp\left(-\frac{x_1^2 - 2x_1 x_2 \cos \omega \tau + x_2^2}{2\sigma^2 \sin^2 \omega \tau} \right).
$$

Last, based on the above results, we deduce that the random process $\mathbf{x}(t)$ is stationary in the wide sense (at least).

14.3.1 Time averages and ergodicity

Up to this point we have considered what are called *ensemble averages*, such as the mean value and covariance of a given random process $\mathbf{x}(t)$. Such averages depend upon knowledge of the various marginal and joint PDFs associated with the random process. In practice, however, we must usually deal with a single realization of a random process, such as a noisy radar signal, over some time period T that we will assume extends over $(-T/2,T/2)$. If $x(t)$ is a particular realization of a given random process, we can define its *finite-time-average* by the integral

$$\overline{\mathbf{x}_T(t)} = \frac{1}{T}\int_{-T/2}^{T/2}\mathbf{x}(t)dt, \tag{15}$$

where the subscript T denotes that we observe the sample function only over time interval T. In the limit $T \to \infty$, we obtain the *long-time-average*

$$\overline{\mathbf{x}(t)} = \lim_{T \to \infty}\frac{1}{T}\int_{-T/2}^{T/2}\mathbf{x}(t)dt. \tag{16}$$

In the same manner, if $\mathbf{x}(t)$ is a realization of a stationary random process, we define the *long-time-average* correlation function by

$$\Re_{\mathbf{x}}(\tau) = \overline{\mathbf{x}(t)\mathbf{x}(t+\tau)} = \lim_{T \to \infty}\frac{1}{T}\int_{-T/2}^{T/2}\mathbf{x}(t)\mathbf{x}(t+\tau)dt. \tag{17}$$

Random processes for which ensemble averages like (8) and (9) can be replaced with time averages like (16) and (17) are said to be *ergodic*. That is, for an ergodic process, we have

$$\overline{\mathbf{x}(t)} = \langle\mathbf{x}(t)\rangle, \tag{18}$$

$$\Re_{\mathbf{x}}(\tau) = R_{\mathbf{x}}(\tau). \tag{19}$$

Although we will not prove it, it has been shown that (18) is valid, for example, if and only if (see [24])

$$\lim_{T \to \infty}\frac{1}{T}\int_{-T}^{T}\left(1 - \frac{|\tau|}{T}\right)C_{\mathbf{x}}(\tau)d\tau = 0. \tag{20}$$

To better understand the concept of ergodicity, let us imagine we have available a large number of identical noise generators that make up the ensemble. One method of determining the average voltage output of such generators would be to sample the output of a single generator (one realization) for a long period of time and then calculate its average. Or, we could sample the voltage output of all generators at a particular instant of time and then calculate the average value. If the process is ergodic, these two methods of computing averages always give the same value. If a random process is ergodic, it must also be stationary; however, if a random process is stationary, it may or may not be ergodic.

The theory of random processes is usually formulated in terms of ensemble averages, but actual measurements are ordinarily based on time averages. Therefore, the assumption that a stationary process is also ergodic turns out to be essential in most applications (whether or not it can be proved!). In general, it is a difficult task to decide whether a given stationary process is also ergodic because there are no simple conditions that guarantee it. The validity of Eq. (14) shows only that the mean is ergodic, not necessarily other parameters. However, for a stationary *Gaussian random process* with zero mean and continuous correlation function, one simple condition that implies ergodicity is

$$\int_{-\infty}^{\infty}|R_{\mathbf{x}}(\tau)|\,d\tau < \infty.$$

There is no comparable condition for a non-Gaussian process.

EXAMPLE 4 Calculate the time average mean and covariance of the random process defined by $\mathbf{x}(t) = \mathbf{A}\cos(\omega t + \varphi)$, in which the amplitude $\mathbf{A}$ and phase φ are *random variables*.

Solution: By using the definition given by (16), we have that

$$\overline{\mathbf{x}(t)} = \mathbf{A}\lim_{T\to\infty}\frac{1}{T}\int_{-T/2}^{T/2}\cos(\omega t + \varphi)\,dt$$

$$= \mathbf{A}\lim_{T\to\infty}\frac{1}{T}\frac{\sin(\omega t + \varphi)}{\omega}\bigg|_{-T/2}^{T/2} = 0.$$

Thus, the covariance and time average correlation function are both the same. From (17), therefore, we find

$$\Re_x(\tau) = \mathbf{A}^2\lim_{T\to\infty}\frac{1}{T}\int_{-T/2}^{T/2}\cos(\omega t + \varphi)\cos[\omega(t + \tau) + \varphi]\,dt$$

$$= \frac{1}{2}\mathbf{A}^2\lim_{T\to\infty}\frac{1}{T}\int_{-T/2}^{T/2}\{\cos\omega\tau + \cos[\omega(2t + \tau) + 2\varphi]\}\,dt,$$

which, upon taking the limit, reduces to

$$\Re_x(\tau) = \frac{1}{2}\mathbf{A}^2\cos\omega\tau.$$

However, because the amplitude $\mathbf{A}$ is a random variable, we see that this process is not ergodic (recall Example 2). In Example 2 the random variation in $\mathbf{x}(t)$ (due to random variables $\mathbf{A}$ and φ) is over the *ensemble*, not with respect to time.

14.3.2 Basic properties

Here we wish to present a few basic properties shared by the correlation and covariance functions of a real stationary random process $\mathbf{x}(t)$. Because the proofs for covariance functions are essentially the same as for correlation functions, we only present proofs of these properties for the former. Moreover, if the random process $\mathbf{x}(t)$ is also ergodic, these properties hold for time averages as well as ensemble averages. The fundamental properties are the following:

$$R_x(0) = \langle\mathbf{x}^2(t)\rangle \geq 0, \quad C_x(0) = \sigma_x^2 \geq 0, \tag{21}$$

$$R_x(-\tau) = R_x(\tau), \quad C_x(-\tau) = C_x(\tau), \tag{22}$$

$$|R_x(\tau)| \leq R_x(0), \quad |C_x(\tau)| \leq C_x(0). \tag{23}$$

Equations (21) follow immediately from definition, e.g.,

$$R_x(0) = \langle \mathbf{x}(t)\mathbf{x}(t+\tau) \rangle |_{\tau=0} = \langle \mathbf{x}^2(t) \rangle. \tag{24}$$

Equations (22) follow from the observation that $R_x(t_1,t_2) = R_x(t_2,t_1)$. Basically, these relations imply that the correlation and covariance functions are *even functions* of τ. Finally, the validity of Eqs. (23) can be established from making the observation that

$$\langle [\mathbf{x}(t) - \mathbf{x}(t+\tau)]^2 \rangle = \langle \mathbf{x}^2(t) \rangle + \langle \mathbf{x}^2(t+\tau) \rangle - 2\langle \mathbf{x}(t)\mathbf{x}(t+\tau) \rangle$$

$$= 2[R_x(0) - R_x(\tau)]. \tag{25}$$

Because the left-hand side of this last expression is nonnegative, it follows that the right-hand side is also nonnegative. Consequently, $R_x(\tau)$ has its maximum value at the origin.

Suppose $\mathbf{x}(t)$ is a stationary process of the form

$$\mathbf{x}(t) = m + \mathbf{n}(t), \tag{26}$$

in which m is deterministic and $\mathbf{n}(t)$ is a random "noise process" for which $\langle \mathbf{n}(t) \rangle = 0$. From definition,

$$\begin{aligned} R_x(\tau) &= \langle [m + \mathbf{n}(t)][m + \mathbf{n}(t+\tau)] \rangle \\ &= m^2 + \langle \mathbf{n}(t)\mathbf{n}(t+\tau)] \rangle \\ &= m^2 + R_n(\tau). \end{aligned} \tag{27}$$

Thus, if $\mathbf{x}(t)$ has a nonzero mean value (i.e., a dc component), then the autocorrelation function $R_x(\tau)$ has a constant component. We deduce, therefore, that if $\langle \mathbf{x}(t) \rangle = m$ and $\mathbf{x}(t)$ has no periodic component, then in the limit $|\tau| \to \infty$ we find that $\mathbf{n}(t)$ and $\mathbf{n}(t+\tau)$ become independent, and consequently,

$$R_x(\infty) = m^2 + \lim_{|\tau| \to \infty} \langle \mathbf{n}(t) \rangle \langle \mathbf{n}(t+\tau) \rangle = m^2. \tag{28}$$

Remark: Note that whereas the autocorrelation function satisfies the relation (27), the autocovariance of the same random process yields $C_x(\tau) = C_n(\tau)$. This simply illustrates the fact that the autocovariance function describes characteristics of only the "noise."

The normalized covariance function $\rho_x(\tau)$ defined by (12) is often called the "correlation coefficient," similar to that defined by Eq. (116) in Section 13.7.6. However, here it is actually a function of the time difference τ, not a constant. For a zero mean process, the normalized covariance function defines the correlation between $\mathbf{x}(t)$ and $\mathbf{x}(t+\tau)$. That is, the more rapidly the process $\mathbf{x}(t)$ changes in time, the more rapidly the function $\rho_x(\tau)$ will decrease from its maximum value of unity (at $\tau = 0$) with increasing τ. This decrease in $\rho_x(\tau)$ is often characterized by a *correlation time* τ_0, but its definition is not universal. One common definition of τ_0 used in engineering applications identifies the correlation time τ_0 as the value for which $\rho_x(\tau_0) = 1/e$.

EXAMPLE 5 A stationary random process $x(t)$ has an autocorrelation function given by

$$R_x(\tau) = 25e^{-4|\tau|} + 16e^{-|\tau|}\cos 20\tau + 9.$$

Calculate
(a) $\langle x^2(t)\rangle$.
(b) $\langle x(t)\rangle$.
(c) $\mathrm{Var}[x(t)]$.

Solution: (a) From definition,

$$\langle x^2(t)\rangle = R_x(0) = 50.$$

(b) Because $R_x(\tau)$ has a constant component, that constant identifies the mean. That is, $R_x(\infty) = 9$, from which we deduce

$$\langle x(t)\rangle = \pm 3.$$

(c) Based on parts (a) and (b) above, we have that

$$\mathrm{Var}[x(t)] = \langle x^2(t)\rangle - \langle x(t)\rangle^2 = 41.$$

14.3.3 Structure functions

As previously mentioned, actual random processes in practice are often approximated with sufficient accuracy by stationary random functions. Of course, there are many instances in which the assumption of stationarity is not warranted. For example, atmospheric parameters such as wind velocity fluctuations, temperature fluctuations, and so on, are not strictly stationary because their mean values are constant only over relatively short time periods. This difficulty can often be alleviated if the random process has *stationary increments*. In other words, rather than work directly with the random process $x(t)$ itself, we concentrate on the function $x(t + t_1) - x(t_1)$, which often behaves very much like a stationary process even though $x(t)$ may not be stationary. Such functions have what we consider a slowly varying mean and can be described most conveniently in terms of structure functions rather than covariance functions.

It is customary in the study of turbulence, among other areas, to write

$$x(t) = m(t) + x_1(t), \tag{29}$$

where $m(t) = \langle \mathbf{x}(t) \rangle$ is the mean of $\mathbf{x}(t)$ and $\mathbf{x}_1(t)$ is the fluctuating part with mean value $\langle \mathbf{x}_1(t) \rangle = 0$. The *structure function* associated with the random process $\mathbf{x}(t)$ is

$$
\begin{aligned}
D_{\mathbf{x}}(t_1, t_2) &= \langle [\mathbf{x}(t_1) - \mathbf{x}(t_2)]^2 \rangle \\
&= [m(t_1) - m(t_2)]^2 + \langle [\mathbf{x}_1(t_1) - \mathbf{x}_1(t_2)]^2 \rangle.
\end{aligned}
\tag{30}
$$

Here we see the utility of the structure function approach. If the mean value is not constant but is "slowly varying," then the difference in means in Eq. (30) is still nearly zero and the structure function can be approximated by

$$
\boxed{D_{\mathbf{x}}(t_1, t_2) \cong \langle [\mathbf{x}_1(t_1) - \mathbf{x}_1(t_2)]^2 \rangle.}
\tag{31}
$$

To put the above ideas into a more precise mathematical framework, let $\mathbf{x}(t)$ be a random process such that the ensemble average of $\mathbf{x}(t + \tau) - \mathbf{x}(t)$ is independent of t. If it is also true that the ensemble average of $[\mathbf{x}(t + \tau) - \mathbf{x}(t)]^2$ is independent of t, we then call $\mathbf{x}(t)$ a random process with *stationary increments*. We ordinarily characterize a random process with stationary increments by the structure function rather than by the covariance function.

A stationary process may be considered a special case of a process with stationary increments. For instance, if $\mathbf{x}(t)$ is a stationary process, then its structure function and covariance function are directly related as are their respective spectra. In particular, it follows from definition that

$$
\begin{aligned}
D_{\mathbf{x}}(\tau) &= \langle [\mathbf{x}(t + \tau) - \mathbf{x}(t)]^2 \rangle \\
&= \langle \mathbf{x}^2(t + \tau) \rangle + \langle \mathbf{x}^2(t) \rangle - 2 \langle \mathbf{x}(t + \tau)\mathbf{x}(t) \rangle \\
&= 2[C_{\mathbf{x}}(0) - C_{\mathbf{x}}(\tau)].
\end{aligned}
\tag{32}
$$

14.4 Cross-Correlation and Cross-Covariance

In practice we often have to consider more than one random process in the same experiment. For a communication engineer, this might be simply the sum of a random signal and random noise. Another common example arises when the two random processes are the input $\mathbf{x}(t)$ and output $\mathbf{y}(t)$ to a linear filter. In all such cases it may be useful to consider the *cross-correlation function*

$$
\begin{aligned}
R_{\mathbf{xy}}(t_1, t_2) &= \langle \mathbf{x}(t_1)\mathbf{y}(t_2) \rangle \\
&= \iint_{-\infty}^{\infty} x_1 y_2 f_{\mathbf{xy}}(x_1, t_1; y_2, t_2) \, dx_1 \, dy_2,
\end{aligned}
\tag{33}
$$

where $f_{\mathbf{xy}}(x_1, t_1; y_2, t_2)$ is the joint PDF of $\mathbf{x}(t_1)$ and $\mathbf{y}(t_2)$. Similarly, the *cross-covariance function* is defined by

$$
C_{\mathbf{xy}}(t_1, t_2) = R_{\mathbf{xy}}(t_1, t_2) - \langle \mathbf{x}(t_1) \rangle \langle \mathbf{y}(t_2) \rangle.
\tag{34}
$$

The cross-correlation and cross-covariance functions describe how one random process is related to the other. If the processes are *statistically independent*, then $C_{xy}(t_1, t_2) \equiv 0$. In this case, the random processes $x(t)$ and $y(t)$ are also called "incoherent," but incoherent processes are not necessarily statistically independent. The random processes are said to be *uncorrelated* if $R_{xy}(t_1, t_2) = \langle x(t_1) \rangle \langle y(t_2) \rangle$ for all t_1 and t_2. Only in the case of a Gaussian processes does this last condition imply that $x(t)$ and $y(t)$ are also statistically independent. We say two processes $x(t)$ and $y(t)$ are *jointly stationary* if their joint PDF does not depend on time and their cross-correlation function (33) depends only on $\tau = t_2 - t_1$.

14.4.1 Basic properties

The basic properties of cross-correlation and cross-covariance functions are quite different in general from those of autocorrelation and autocovariance functions. Here we assume that the random processes $x(t)$ and $y(t)$ are jointly stationary in the wide sense (not simply individually stationary). Under this condition we have the following properties:

$$R_{xy}(-\tau) = R_{yx}(\tau), \quad C_{xy}(-\tau) = C_{yx}(\tau), \tag{35}$$

$$|R_x(\tau)| \leq \sqrt{R_x(0)R_y(0)}, \quad |C_x(\tau)| \leq \sqrt{C_x(0)C_y(0)}, \tag{36}$$

$$2|R_{xy}(\tau)| \leq R_x(0) + R_y(0). \tag{37}$$

Equation (35) tells us that cross-correlation and cross-covariance functions are generally *not* "even functions" of τ. In fact, the functions $R_{xy}(\tau)$ and $R_{yx}(\tau)$ are "mirror images" of one another, so we need to distinguish between them in most cases. Also, the maximum value of a cross-correlation function need not occur at $\tau = 0$, and, unlike autocorrelation/autocovariance functions, there is no physical interpretation of either $R_{xy}(0)$ or $C_{xy}(0)$. Last, if $x(t)$ and $y(t)$ are Gaussian random processes, then it can be shown that the condition $C_{xy}(\tau) = 0$ is both a necessary and a sufficient condition for $x(t)$ and $y(t)$ to be *statistically independent*.

EXAMPLE 6 Two stationary and jointly stationary random processes are described by

$$x(t) = 5\cos(\omega t + \theta),$$

$$y(t) = 20\sin(\omega t + \theta),$$

where θ is a random variable uniformly distributed over $(0, 2\pi)$. Find the cross-correlation function $R_{xy}(\tau)$.

Solution: By definition,

$$R_{xy}(\tau) = \langle \mathbf{x}(t)\mathbf{y}(t + \tau) \rangle$$
$$= 50 \langle \sin(2\omega t + \omega\tau + 2\theta) + \sin \omega\tau \rangle,$$

where we have used the identity

$$\sin A \cos B = \frac{1}{2}\left[\sin(A + B) + \sin(A - B)\right].$$

The ensemble average of the first sinusoid above with the random phase term 2θ is zero, reducing the cross-correlation function to

$$R_{xy}(\tau) = 50 \sin \omega\tau.$$

If $\mathbf{z}(t) = \mathbf{x}(t) + \mathbf{y}(t)$ is a stationary random process, then the autocorrelation function of $\mathbf{z}(t)$ has the form

$$
\begin{aligned}
R_z(\tau) &= \langle [\mathbf{x}(t) + \mathbf{y}(t)][\mathbf{x}(t + \tau) + \mathbf{y}(t + \tau)] \rangle \\
&= \langle \mathbf{x}(t)\mathbf{x}(t + \tau) \rangle + \langle \mathbf{y}(t)\mathbf{y}(t + \tau) \rangle + \langle \mathbf{x}(t)\mathbf{y}(t + \tau) \rangle + \langle \mathbf{y}(t)\mathbf{x}(t + \tau) \rangle \\
&= R_x(\tau) + R_y(\tau) + R_{xy}(\tau) + R_{yx}(\tau).
\end{aligned}
\tag{38}
$$

In general, the autocorrelation function of a sum of stationary random processes is a sum of all autocorrelation and cross-correlation functions. If $\mathbf{x}(t)$ and $\mathbf{y}(t)$ are statistically independent and at least one of them has a zero mean, then it follows from (38) that

$$R_z(\tau) = R_x(\tau) + R_y(\tau). \tag{39}$$

14.4.2 Cross-correlation techniques in radar

Among other uses, cross-correlation techniques provide a practical way of extracting a small known signal from a combination of signal and noise such as that which occurs in certain radar systems. Let us imagine that a radar signal $\mathbf{x}(t)$ is transmitted in the direction of a given target. The echo or return signal from the target is generally "much weaker" than the original signal and will be delayed in time by the propagation time to the target and back, and this is directly related to the distance to the target. In addition, the echo signal will also be accompanied by random noise $\mathbf{n}(t)$ at the input to the receiver system. We denote the return signal by

$$\mathbf{y}(t) = a\mathbf{x}(t - T) + \mathbf{n}(t), \tag{40}$$

where a is a positive constant much smaller than unity and T is the round-trip delay of the signal. The cross-correlation function of the transmitted signal and echo signal at the receiver input is

$$R_{xy}(\tau) = \langle \mathbf{x}(t)\mathbf{y}(t + \tau)\rangle = aR_x(\tau - T) + R_{xn}(\tau), \tag{41}$$

where $R_{xn}(\tau)$ is the cross-correlation between the signal and noise. However, the signal and noise are generally statistically independent, and the noise usually has zero mean. Hence, in this case the cross-correlation between the signal and noise is zero and (41) reduces to

$$R_{xy}(\tau) = aR_x(\tau - T). \tag{42}$$

Equation (42) shows that the cross-correlation function of the transmitted and echo signals is proportional to the autocorrelation function of the transmitted signal alone shifted by the time delay T. Because the autocorrelation function assumes its maximum value at the origin, the delay time T (and hence, the target distance) can be found by adjusting τ so that the cross-correlation function (42) is a maximum.

14.5 Power Spectral Density Functions

A wide variety of engineering applications involving random processes are concerned with the determination of a particular correlation function or its Fourier transform, the latter of which is called a *power spectral density function*. The correlation function quantifies the similarity of the fluctuations with a time difference of τ. The power density quantifies these fluctuations as a sinusoidal component at the frequency defined by $1/\tau$. Essentially, both of these functions provide the same information about the random process being studied, but historically they evolved from different groups of scientists— correlation functions were primarily a product of mathematicians and statisticians, while spectral density functions were developed mostly as a tool of engineering.

Before discussing the definition of power spectral density for a random function it may be useful to first review the basic notion of a Fourier transform and its inverse for deterministic functions (see Section 9.4). Let us suppose that $f(t)$ represents a *deterministic* (nonrandom) function of time, such as a signal or voltage, and assume that $f(t)$ is absolutely integrable; i.e.,

$$\int_{-\infty}^{\infty} |f(t)|\, dt < \infty. \tag{43}$$

The frequency content of the time function $f(t)$ can then be obtained from its *Fourier transform*

$$F(\omega) = \int_{-\infty}^{\infty} e^{-i\omega t} f(t)\, dt, \tag{44}$$

where ω represents *angular frequency*. In this setting we say that $F(\omega)$ is the *spectrum* associated with the time signal $f(t)$. Based on its definition, it is clear that $F(\omega)$ is a complex function except in the special case when $f(t)$ is real and even. The value $F(0)$ is called the *dc* (direct current) *component* of the spectrum and $F(\omega)$, $\omega \neq 0$, is called the *ac* (alternating current) *component*. The original time function $f(t)$ can be fully recovered from the spectrum by the *inverse Fourier transform*

$$f(t) = \frac{1}{2\pi} \int_{-\infty}^{\infty} e^{i\omega t} F(\omega) d\omega. \tag{45}$$

Equation (45) is also considered an integral representation of the time function $f(t)$ in terms of its frequency function $F(\omega)$. We should point out that these definitions of Fourier transform are not universal. That is, the signs of the complex exponential functions in Eqs. (44) and (45) are sometimes interchanged and the multiplicative constant $1/2\pi$ may appear in front of either integral or its square root in front of each expression.

14.5.1 Riemann-Stieltjes integral

Let us assume that $\mathbf{x}(t)$ is a *complex stationary random process* with mean value zero. Clearly, random processes do not satisfy the basic condition (43) for the existence of a Fourier transform. However, a stationary random process can be represented in the form of a stochastic (random) *Riemann-Stieltjes integral* (also called a *Fourier-Stieltjes integral* or *spectral representation*)

$$\mathbf{x}(t) = \frac{1}{2\pi} \int_{-\infty}^{\infty} e^{i\omega t} d\mathbf{v}(\omega), \tag{46}$$

where $d\mathbf{v}(\omega)$ is a random complex amplitude. Of course, the random functions $\mathbf{x}(t)$ and $d\mathbf{v}(\omega)$ are not the same for each realization of the random process.[1] Because we have assumed the mean value of $\mathbf{x}(t)$ is zero, its covariance and correlation functions are the same. Thus, by using the Riemann-Stieltjes integral (46), the covariance (or correlation) function is given by

$$\begin{aligned} C_{\mathbf{x}}(\tau) &= \langle \mathbf{x}(t_1)\mathbf{x}^*(t_2) \rangle \\ &= \left(\frac{1}{2\pi}\right)^2 \iint_{-\infty}^{\infty} \exp\left[i(\omega_1 t_1 - \omega_2 t_2)\right] \langle d\mathbf{v}(\omega_1) d\mathbf{v}^*(\omega_2) \rangle, \end{aligned} \tag{47}$$

where ω_1 and ω_2 are dummy variables of frequency and the asterisk $*$ denotes the complex conjugate quantity. In order that the covariance function (47) be a function of the time difference $\tau = t_1 - t_2$, characteristic of stationary functions, we must require that the random amplitude satisfy

$$\langle d\mathbf{v}(\omega_1) d\mathbf{v}^*(\omega_2) \rangle = 2\pi \delta(\omega_2 - \omega_1) S_{\mathbf{x}}(\omega_1) d\omega_2 d\omega_1, \tag{48}$$

where $\delta(\omega_2 - \omega_1)$ is the *Dirac delta function* (see Chapter 2) and $S_{\mathbf{x}}(\omega_1) \geq 0$. By inserting Eq. (48) into Eq. (47), and utilizing the *sifting property* of the Dirac delta function

$$\int_{-\infty}^{\infty} \delta(\omega - a) g(\omega) d\omega = g(a), \tag{49}$$

---

[1]Note that Eq. (46) expresses the relation between a given realization of a random process and its spectrum in the same way that Eq. (45) does for a deterministic function provided we relate $d\mathbf{v}(\omega)$ to $F(\omega)d\omega$.

we obtain the Fourier transform relation (now replacing ω_1 by ω)

$$C_x(\tau) = \frac{1}{2\pi} \int_{-\infty}^{\infty} e^{i\omega\tau} S_x(\omega) d\omega. \tag{50}$$

We call $S_x(\omega)$ the *power spectrum* or *power spectral density* of the random process $x(t)$. It has also been shown that the power spectrum $S_x(\omega)$ is the inverse transform of the covariance function $C_x(\tau)$, i.e., that these functions are Fourier transform pairs. In this case, we can also write

$$S_x(\omega) = \int_{-\infty}^{\infty} e^{-i\omega\tau} C_x(\tau) d\tau. \tag{51}$$

The general transform relations (50) and (51) are widely known as the *Wiener-Khintchine theorem*.

If the random process $x(t)$ is *real*, then the covariance function is a real, even function and it can be shown that the power spectrum is also a real, even function. Thus, these two transform pairs in this special case can also be expressed as *cosine transforms*

$$C_x(\tau) = \frac{1}{\pi} \int_0^{\infty} S_x(\omega)\cos\omega\tau \, d\omega, \tag{52}$$

$$S_x(\omega) = 2 \int_0^{\infty} C_x(\tau)\cos\omega\tau \, d\tau. \tag{53}$$

Note that the constants in the integrals in (52) and (53) are sometimes interchanged.

It is common practice to define the power spectral density as the Fourier transform of the *autocorrelation function* $R_x(\tau)$ rather than the *autocovariance function* $C_x(\tau)$. However, these functions are the same for zero-mean processes, and, for non–zero-mean processes, the relationship between the two functions is given by [recall Eq. (11)]

$$R_x(\tau) = C_x(\tau) + m^2.$$

Hence, the Fourier transform of the autocorrelation function in this case will still be essentially the same as that of the autocovariance function, except the former will always contain an impulse function at dc ($\omega = 0$) originating from the mean. For this reason, it makes little difference which definition of power spectral density is used in practice.

The power spectrum $S_x(\omega)$ is a frequency domain or spectral decomposition of the temporal correlations of the complex random process $x(t)$. Because we are assuming that $x(t)$ is an ergodic process, the power spectrum can be estimated by time averages in practice. In Table 14.1 we illustrate some typical autocovariance functions and their corresponding spectral density.

Table 14.1 Autocovariance Function and Spectral Density

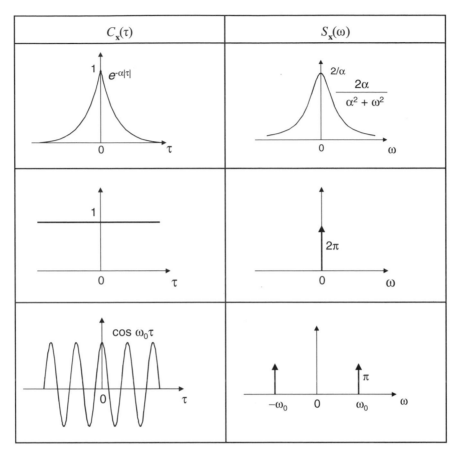

EXAMPLE 7 The covariance function associated with a random telegraph wave is

$$C_x(\tau) = e^{-\alpha|\tau|},$$

where $\alpha > 0$. Calculate the associated power spectral density.

Solution: From the Fourier transform relation (51), we obtain

$$S_x(\omega) = \int_{-\infty}^{\infty} e^{-i\omega\tau} e^{-\alpha|\tau|} d\tau$$
$$= 2\int_0^{\infty} e^{-\alpha\tau} \cos \omega\tau \, d\tau,$$

which reduces to (see Table 14.1)

$$S_x(\omega) = \frac{2\alpha}{\alpha^2 + \omega^2}.$$

EXAMPLE 8 Given the stationary random process

$$\mathbf{x}(t) = A \cos(\omega_0 t + \varphi),$$

where φ is a random variable uniformly distributed over $(0, 2\pi)$, and A and ω_0 are constants, determine the spectral density function.

Solution: In this case we must first calculate the autocovariance function. Based on Example 2, we recall the mean value of $\mathbf{x}(t)$ is zero and

$$C_x(\tau) = \frac{A^2}{2} \cos \omega_0 \tau.$$

Thus, by writing $\cos \omega_0 \tau = \frac{1}{2}(e^{i\omega_0 \tau} + e^{-i\omega_0 \tau})$, we find that (see Table 14.1)

$$S_x(\omega) = \frac{A^2}{4} \int_{-\infty}^{\infty} \left[e^{-i(\omega - \omega_0)\tau} + e^{-i(\omega + \omega_0)\tau} \right] d\tau$$

$$= \frac{\pi A^2}{2} \left[\delta(\omega - \omega_0) + \delta(\omega + \omega_0) \right].$$

Regardless of whether the random process $\mathbf{x}(t)$ is real or complex, the power spectral density $S_x(\omega)$ is a *real* function. Moreover, if $\mathbf{x}(t)$ is a *real* stationary process the power spectral density is *real* and *even*, i.e., $S_x(\omega) = S_x(-\omega)$. However, we should remember that only positive frequencies have physical meaning—illustrating the power spectrum with positive *and* negative frequencies is simply a mathematical convenience.

Because the autocovariance $C_x(\tau)$ and power spectral density $S_x(\omega)$ are Fourier transform pairs, it follows that a shorter correlation time τ_0 corresponds to a wider spectral density. From physical considerations this is obvious, because as a random process $\mathbf{x}(t)$ varies more rapidly with time, it requires a higher-frequency content, which results in a wider spectrum.

From an engineering point of view, we have the following interpretations of statistical averages of ergodic processes (e.g., a voltage or current signal).

1. The *mean value* $<\mathbf{x}(t)>$ is the *dc component* of the signal.
2. The *mean-squared value* $<\mathbf{x}(t)>^2 = R_x(\infty)$ is the *power in the dc component* of the signal.
3. The *second moment* $<\mathbf{x}^2(t)> = R_x(0)$ is the total *average power* of the signal.
4. The *variance* $\sigma_x^2 = C_x(0) = \frac{1}{2\pi} \int_{-\infty}^{\infty} S_x(\omega) d\omega$ is the total *average power in the ac component* of the signal.
5. The *standard deviation* σ_x is the *root-mean-square* (rms) *value of the ac component* of the signal.

14.6 Transformations of Random Processes

Signals embedded in random noise are ordinarily passed through some filtering devices and analyzed by nonlinear operations like mixing and rectification. In these cases, we can imagine a random process $\mathbf{x}(t)$ as the input to some receiver device and $\mathbf{y}(t)$ as the output random process. The relationship between $\mathbf{x}(t)$ and $\mathbf{y}(t)$ can be described by a *transformation* which we represent in general by

$$\mathbf{y}(t) = T[\mathbf{x}(t)]. \tag{54}$$

The symbol T is an operator that describes the relation between input and output processes. We will briefly consider two cases: (*i*) transformations by a *memoryless nonlinear device* and (*ii*) transformations by a *linear system* (i.e., a filter).

14.6.1 Memoryless nonlinear transformations

Let us consider a system in which the output $\mathbf{y}(t_1)$ at time t_1 depends only on the input $\mathbf{x}(t_1)$ and not upon any other past or future values of $\mathbf{x}(t)$. If we designate the system by the relation

$$\mathbf{y}(t) = g[\mathbf{x}(t)], \tag{55}$$

where $y = g(x)$ is a function assigning a unique value of y to each value of x, we then say the system effects a *memoryless* transformation. Because the function $g(x)$ does not depend explicitly on time t, we also say the system is *time-invariant*.

EXAMPLE 9 (*Square-law device*) A stationary, zero-mean, narrowband, Gaussian random process is passed through a square-law device (also called a quadratic rectifier). Let us represent the input to the square-law device by the complex quantity

$$\mathbf{w}(t) = \mathbf{x}(t) + i\,\mathbf{y}(t),$$

where $\mathbf{x}(t)$ and $\mathbf{y}(t)$ are independent Gaussian random processes with zero means and $\langle \mathbf{x}^2(t)\rangle = \langle \mathbf{y}^2(t)\rangle = 1$. The output of the square-law device is given by

$$\mathbf{z}(t) = |\mathbf{w}(t)|^2 = \mathbf{x}^2(t) + \mathbf{y}^2(t).$$

(*a*) Determine the first-order PDF of the output $\mathbf{z}(t)$.
(*b*) Find the autocorrelation function $R_z(\tau)$ in terms of $R_x(\tau)$ and $R_y(\tau)$.

Solution:
(*a*) The joint PDF of the input processes $\mathbf{x}(t)$ and $\mathbf{y}(t)$ is

$$f_{\mathbf{xy}}(x,t;y,t) = f_{\mathbf{x}}(x,t)f_{\mathbf{y}}(y,t) = \frac{1}{2\pi}e^{-(x^2+y^2)/2}.$$

The characteristic function for $\mathbf{z}(t)$ is (see Section 13.4.2)

$$\begin{aligned}
\Phi_{\mathbf{z}}(s) &= \int_{-\infty}^{\infty}\int_{-\infty}^{\infty}e^{is(x^2+y^2)}f_{\mathbf{xy}}(x,t;y,t)dx\,dy \\
&= \frac{1}{2\pi}\int_{-\infty}^{\infty}\exp\bigl[-(1/2-is)x^2\bigr]dx\int_{-\infty}^{\infty}\exp\bigl[-(1/2-is)y^2\bigr]dy \\
&= \frac{1}{1-2is}.
\end{aligned}$$

Consequently, the PDF for $\mathbf{z}(t)$ is

$$f_{\mathbf{z}}(z,t) = \frac{1}{2\pi}\int_{-\infty}^{\infty}\frac{e^{-isz}}{1-2is}ds = \frac{1}{2}e^{-z/2}U(z),$$

where $U(z)$ is the unit step function.

(b) To calculate the autocorrelation function we start with

$$\begin{aligned}
R_{\mathbf{z}}(\tau) &= \langle\mathbf{z}(t)\mathbf{z}(t+\tau)\rangle \\
&= \langle[\mathbf{x}^2(t)+\mathbf{y}^2(t)][\mathbf{x}^2(t+\tau)+\mathbf{y}^2(t+\tau)]\rangle \\
&= \langle\mathbf{x}^2(t)\mathbf{x}^2(t+\tau)\rangle + \langle\mathbf{y}^2(t)\mathbf{y}^2(t+\tau)\rangle + \langle\mathbf{x}^2(t)\mathbf{y}^2(t+\tau)\rangle + \langle\mathbf{y}^2(t)\mathbf{x}^2(t+\tau)\rangle \\
&= \langle\mathbf{x}^2(t)\mathbf{x}^2(t+\tau)\rangle + \langle\mathbf{y}^2(t)\mathbf{y}^2(t+\tau)\rangle \\
&\qquad + \langle\mathbf{x}^2(t)\rangle\langle\mathbf{y}^2(t+\tau)\rangle + \langle\mathbf{y}^2(t)\rangle\langle\mathbf{x}^2(t+\tau)\rangle.
\end{aligned}$$

However, for Gaussian statistics it can be shown that (e.g., see Problem 22)

$$\begin{aligned}
\langle\mathbf{x}^2(t)\mathbf{x}^2(t+\tau)\rangle &= \langle\mathbf{x}^2(t)\rangle\langle\mathbf{x}^2(t+\tau)\rangle + 2\langle\mathbf{x}(t)\mathbf{x}(t+\tau)\rangle^2 \\
&= R_{\mathbf{x}}^2(0) + 2R_{\mathbf{x}}^2(\tau),
\end{aligned}$$

$$\langle\mathbf{y}^2(t)\mathbf{y}^2(t+\tau)\rangle = R_{\mathbf{y}}^2(0) + 2R_{\mathbf{y}}^2(\tau).$$

Consequently, we deduce that

$$R_{\mathbf{z}}(\tau) = R_{\mathbf{x}}^2(0) + R_{\mathbf{y}}^2(0) + 2R_{\mathbf{x}}^2(\tau) + 2R_{\mathbf{y}}^2(\tau) + 2R_{\mathbf{x}}(0)R_{\mathbf{y}}(0),$$

or, because $R_{\mathbf{x}}(\tau) = R_{\mathbf{y}}(\tau)$, this reduces to

$$R_{\mathbf{z}}(\tau) = 4R_{\mathbf{x}}^2(0) + 4R_{\mathbf{x}}^2(\tau) = 4R_{\mathbf{x}}^2(0)\bigl[1 + \rho^2(\tau)\bigr],$$

where $\rho(\tau) = R_{\mathbf{x}}(\tau)/R_{\mathbf{x}}(0)$ [recall that $\mathbf{x}(t)$ and $\mathbf{y}(t)$ have zero means].

14.6.2 Linear systems

Relationships involving correlation/covariance functions and power spectral density functions between the input and output of various *linear systems* (see Fig. 14.1) is important to a wide class of engineering applications. In some cases we consider only an "ideal physical system," defined as one that (*i*) is physically realizable, (*ii*) has constant parameters, (*iii*) is stable, and (*iv*) is linear. We define these terms below.

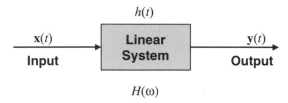

Figure 14.1 Linear system with impulse response $h(t)$.

In Section 9.6 we discussed the basics of linear systems with deterministic input and output functions. For such systems, the response properties are described by the *unit impulse response function* $h(t)$, or its Fourier transform $H(\omega)$ called the *system transfer function*. Analogously, if $\mathbf{x}(t)$ is a sample function of a stationary random process, the system output or response $\mathbf{y}(t)$ is related in the same fashion by the integral

$$\mathbf{y}(t) = \int_{-\infty}^{\infty} h(\xi)\mathbf{x}(t-\xi)d\xi = \int_{-\infty}^{\infty} h(t-\eta)\mathbf{x}(\eta)d\eta. \tag{56}$$

However, because of the random nature of $\mathbf{x}(t)$ and $\mathbf{y}(t)$, we must use a statistical approach in describing input/output relations of such systems.

A *physically realizable* system is one that is *causal*, i.e., it cannot respond to an input until that input has been applied. This condition requires that

$$h(t) = 0, \quad t < 0, \tag{57}$$

in which case Eq. (56) becomes

$$\mathbf{y}(t) = \int_{0}^{\infty} h(\xi)\mathbf{x}(t-\xi)d\xi = \int_{-\infty}^{t} h(t-\eta)\mathbf{x}(\eta)d\eta. \tag{58}$$

In some of the analysis below we find it is not always necessary to assume the system is causal. If the system has *constant parameters*, then the impulse response $h(t)$ is independent of the time at which a unit impulse is applied. Also, stationary inputs will always lead to stationary outputs. We say the system is *stable* if bounded inputs produce only bounded outputs. To ensure this last condition, we require that

$$\int_{-\infty}^{\infty} |h(t)| \, dt < \infty. \tag{59}$$

Finally, a system is called *linear* if, when the input $\mathbf{x}_1$ produces the output $\mathbf{y}_1$ and $\mathbf{x}_2$ produces $\mathbf{y}_2$, then the input $C_1\mathbf{x}_1 + C_2\mathbf{x}_2$ produces the output $C_1\mathbf{y}_1 + C_2\mathbf{y}_2$ for any constants C_1 and C_2. An important property of a linear system is that, if the input is a Gaussian random process, the output is also a Gaussian random process.

The *expected value* of the output of a linear system can be found by forming the expected value of the integral in (56). That is, if we assume that the expected value and integration operations can be interchanged, then

$$\langle \mathbf{y}(t) \rangle = \int_{-\infty}^{\infty} h(\xi) \langle \mathbf{x}(t - \xi) \rangle d\xi. \tag{60}$$

Under the assumption that $\mathbf{x}(t)$ is stationary, it follows that

$$\langle \mathbf{x}(t - \xi) \rangle = \langle \mathbf{x}(t) \rangle = m,$$

where m is constant; hence, we deduce that

$$\langle \mathbf{y}(t) \rangle = m \int_{-\infty}^{\infty} h(\xi) d\xi = mH(0), \tag{61}$$

where the system transfer function $H(\omega)$ is the Fourier transform of the input response $h(t)$. Basically, Eq. (61) tells us that the dc component of the output is equal to the dc component of the input times the *dc gain* of the system $H(0)$. Therefore, if the input process has zero mean, the output process also has zero mean.

For the *second moment* of the output $\mathbf{y}(t)$, we start with

$$\begin{aligned}
\langle \mathbf{y}^2(t) \rangle &= \left\langle \int_{-\infty}^{\infty} h(\xi)\, \mathbf{x}(t - \xi) d\xi \int_{-\infty}^{\infty} h(\eta) \mathbf{x}(t - \eta) d\eta \right\rangle \\
&= \int_{-\infty}^{\infty} \int_{-\infty}^{\infty} h(\xi) h(\eta) \langle \mathbf{x}(t - \xi) \mathbf{x}(t - \eta) \rangle d\xi \, d\eta,
\end{aligned} \tag{62}$$

which yields

$$\langle \mathbf{y}^2(t) \rangle = \int_{-\infty}^{\infty} \int_{-\infty}^{\infty} h(\xi) h(\eta) R_{\mathbf{x}}(\xi - \eta) d\xi \, d\eta. \tag{63}$$

Hence, the second moment of the output process (i.e., the total average power of the output signal) depends on the autocorrelation function of the input process.

EXAMPLE 10 Consider a linear system with inpulse response function

$$h(t) = 5e^{-3t} U(t),$$

where $U(t)$ is the unit step function. If $\mathbf{x}(t) = 4\cos(2t + \varphi)$, where φ is uniformly distributed over $(0, 2\pi)$, find
(a) the mean of $\mathbf{y}(t)$, and
(b) the variance of $\mathbf{y}(t)$.

Solution:
(a) From Example 2 we know $\langle \mathbf{x}(t) \rangle = 0$, and therefore $\langle \mathbf{y}(t) \rangle = 0$.

(b) Here we first calculate the autocorrelation function of $\mathbf{x}(t)$. From Example

8, it follows that the autocorrelation and autocovariance functions are both equal to

$$R_x(\tau) = 8\cos 2\tau.$$

The direct substitution of this result into (63) leads to

$$<\mathbf{y}^2(t)> = 200 \int_0^\infty \int_0^\infty e^{-3(\xi+\eta)} \cos[2(\xi - \eta)]\, d\xi\, d\eta,$$

which, upon evaluation, yields the total average power given by

$$\text{Var}[\mathbf{y}(t)] = <\mathbf{y}^2(t)> = \frac{200}{13}.$$

In practice it can be important to establish relations between the correlation and spectral density functions for the output of a linear system in terms of similar quantities for the input processs. To begin, let us consider the correlation function of the output process $\mathbf{y}(t)$. Thus, we start with the expression

$$<\mathbf{y}(t)\mathbf{y}(t + \tau)> = \int_{-\infty}^\infty \int_{-\infty}^\infty h(\xi)h(\eta)<\mathbf{x}(t - \xi)\mathbf{x}(t + \tau - \eta)>\, d\xi\, d\eta, \qquad (64)$$

which is the same as

$$R_y(\tau) = \int_{-\infty}^\infty \int_{-\infty}^\infty h(\xi)h(\eta)R_x(\tau + \xi - \eta)\, d\xi\, d\eta. \qquad (65)$$

In a likewise manner, it follows that the cross-correlation leads to

$$<\mathbf{x}(t)\mathbf{y}(t + \tau)> = <\mathbf{x}(t)\int_{-\infty}^\infty h(\xi)\, \mathbf{x}(t + \tau - \xi)>\, d\xi, \qquad (66)$$

from which we deduce the input-output relation

$$R_{xy}(\tau) = \int_{-\infty}^\infty h(\xi)R_x(\tau - \xi)\, d\xi. \qquad (67)$$

Notice that (67) is a convolution-type integral. The remaining cross-correlation function is found from the association $R_{xy}(\tau) = R_{yx}(-\tau)$.

EXAMPLE 11 (*White noise*) If $\mathbf{x}(t)$ is "white noise," its spectral density is $S_x(\omega) = S_0$, where S_0 is a constant. Find the autocorrelation function $R_y(\tau)$ and cross-correlation function $R_{xy}(\tau)$ for a white noise input.

Solution: Because the Fourier transform of a constant is the delta function, i.e., $R_x(\tau) = S_0\delta(\tau)$, it follows that a white noise process is uncorrelated at distinct times ($\tau \neq 0$). By putting this autocorrelation function in (65) and using the sifting property of the delta function (see Chapter 2), we are led to the expression

$$R_y(\tau) \;=\; S_0 \int_{-\infty}^{\infty} h(\eta - \tau)h(\eta)d\eta \;=\; S_0 \int_{-\infty}^{\infty} h(\eta)h(\eta + \tau)d\eta.$$

Thus, the autocorrelation function of the output is proportional to the time correlation function of the impulse response function.

For the cross-correlation function defined by (67), we see that

$$R_{xy}(\tau) \;=\; S_0 \int_{-\infty}^{\infty} h(\xi)\delta(\tau - \xi)d\xi \;=\; S_0\,h(\tau).$$

That is, for a white noise input the cross-correlation function $R_{xy}(\tau)$ is directly proportional to the impulse response of the linear system.

The result $R_{xy}(\tau) = S_0\,h(\tau)$ obtained in Example 11 provides us with a useful scheme for measuring the impulse response of any linear system. Rather than rely on the output of a unit impulse applied to the linear system to determine $h(t)$, which has certain inherent difficulties associated with it, an alternative is to measure the cross-correlation function between a white noise input and the corresponding output.

The spectral density of a random process is a measure of how the average power of the process is distributed in the frequency domain. Because the spectral density is simply the Fourier transform of the appropriate covariance (or correlation) function, we have, for example,

$$\begin{aligned}
S_y(\omega) \;=\; \mathscr{F}\{R_y(\tau)\} &= \int_{-\infty}^{\infty}\int_{-\infty}^{\infty}\int_{-\infty}^{\infty} h(\xi)h(\eta)R_x(\tau + \xi - \eta)e^{-i\omega\tau}\,d\tau\,d\xi\,d\eta \\
&= \int_{-\infty}^{\infty}\int_{-\infty}^{\infty} h(\xi)h(\eta)S_x(\omega)e^{i\omega(\xi - \eta)}\,d\xi\,d\eta \qquad (68) \\
&= S_x(\omega)\int_{-\infty}^{\infty} h(\xi)e^{i\omega\xi}\,d\xi \int_{-\infty}^{\infty} h(\eta)e^{-i\omega\eta}\,d\eta,
\end{aligned}$$

which reduces to

$$S_y(\omega) \;=\; S_x(\omega)H(\omega)H(-\omega) \;=\; S_x(\omega)|H(\omega)|^2. \qquad (69)$$

Along the same lines,

$$S_{xy}(\omega) \;=\; S_x(\omega)H(\omega), \qquad (70)$$

$$S_{yx}(\omega) \;=\; S_x(\omega)H(-\omega). \qquad (71)$$

EXAMPLE 12 Consider a linear system with impulse response function

$$h(t) \;=\; 5e^{-3t}U(t).$$

If the input to the system is $x(t) = 4\cos(2t + \varphi)$, where φ is uniformly distributed over $(0, 2\pi)$, find:

(*a*) the power spectral density of the output $\mathbf{y}(t)$.

(*b*) the variance of the output $\mathbf{y}(t)$.

Solution:

(*a*) We begin by calculating the transfer function

$$H(\omega) = \int_{-\infty}^{\infty} e^{-i\omega t} h(t) dt = 5\int_{0}^{\infty} e^{-(3+i\omega)t} dt,$$

which yields

$$H(\omega) = \frac{5}{3 + i\omega}, \qquad |H(\omega)|^2 = \frac{25}{9 + \omega^2}.$$

Also, based on Example 10, we have $R_x(\tau) = 8\cos 2\tau$, and consequently,

$$S_x(\omega) = 8\int_{-\infty}^{\infty} e^{-i\omega\tau}\cos 2\tau\, d\tau = 8\pi[\delta(\omega - 2) + \delta(\omega + 2)].$$

Hence, by the use of Eq. (69) we deduce that

$$S_y(\omega) = 8\pi[\delta(\omega - 2) + \delta(\omega + 2)]\frac{25}{9 + \omega^2} = \frac{200\pi}{13}[\delta(\omega - 2) + \delta(\omega + 2)],$$

where we are using property (*c*) of Eqs. (8) in Section 2.2.4.

(*b*) The variance of the output (which represents the average power in this case because the mean is zero) can be calculated directly from the spectral density found in part (*a*), viz.,

$$\text{Var}[\mathbf{y}(t)] = C_y(0) = \frac{1}{2\pi}\int_{-\infty}^{\infty} S_y(\omega) d\omega = \frac{200}{13}.$$

Example 12 illustrates that the spectral or frequency domain analysis may be easier to perform than the corresponding time domain analysis. However, to be proficient at the frequency domain analysis one must have a good understanding of the fundamentals of Fourier analysis (see Chapters 8 and 9).

14.7 Stationary Gaussian Processes

Because normal, or Gaussian, random processes are so common in practice, we wish to develop a few properties peculiar to this class of processes. We say a real random process $\mathbf{x}(t)$ is a *Gaussian process* if the random variables

$$\mathbf{x}(t_1), \mathbf{x}(t_2), \dots, \mathbf{x}(t_n)$$

are jointly Gaussian for any n and times $t_1, t_2, ..., t_n$. One important consequence of such random processes is that sums (or differences) of Gaussian processes are also Gaussian.

14.7.1 Multivariate Gaussian distributions

If **x** and **y** are jointly Gaussian random variables with zero means, their joint PDF is given by the *bivariate distribution* [see Eq. (119) in Chapter 13]

$$f_{xy}(x,y) = \frac{1}{2\pi\sigma_1\sigma_2\sqrt{1-\rho^2}} \exp\left[-\frac{1}{2(1-\rho^2)}\left(\frac{x^2}{\sigma_1^2} - \frac{2\rho xy}{\sigma_1\sigma_2} + \frac{y^2}{\sigma_2^2}\right)\right], \tag{72}$$

where

$$\sigma_1^2 = <x^2>, \qquad \sigma_2^2 = <y^2>, \qquad \rho = \frac{<xy>}{\sigma_1\sigma_2}. \tag{73}$$

An important property of this density function is that

$$<x^2 y^2> = <x^2><y^2> + 2<xy>^2 = \sigma_1^2\sigma_2^2(1 + 2\rho^2), \tag{74}$$

which we leave for the reader to verify (see Problem 22 in the EXERCISES and [24]). Also, because a bivariate Gaussian distribution is completely specified by its first- and second-order moments, a consequence of this is that a wide-sense stationary Gaussian process is also stationary in the *strict* sense.

The case of n random variables $x_1, x_2, ..., x_n$ that are jointly Gaussian leads to the *multivariate Gaussian distribution*

$$f_x(x_1, x_2, ..., x_n) = \frac{1}{(2\pi)^{n/2} |\det(\mathbf{C})|^{1/2}} \exp\left[-\frac{1}{2}(\mathbf{x} - \mathbf{m})^T \mathbf{C}^{-1}(\mathbf{x} - \mathbf{m})\right], \tag{75}$$

where **x** and **m** are column vectors[2]

$$\mathbf{x} = \begin{pmatrix} x_1 \\ x_2 \\ \vdots \\ x_n \end{pmatrix}, \qquad \mathbf{m} = \begin{pmatrix} m_1 \\ m_2 \\ \vdots \\ m_n \end{pmatrix}, \tag{76}$$

and where $(\mathbf{x} - \mathbf{m})^T$ denotes the transpose of the column vector $(\mathbf{x} - \mathbf{m})$. The column vector **m** is the *mean vector* and **C** is a *covariance* square matrix defined by

---

[2]To avoid confusion in notation here, we will not designate random variables that appear as matrix elements [like x_i in (76)] by bold letters.

$$
\mathbf{C} = \begin{pmatrix} c_{11} & c_{12} & \cdots & c_{1n} \\ c_{21} & c_{22} & \cdots & c_{2n} \\ \vdots & \vdots & & \vdots \\ c_{n1} & c_{n2} & \cdots & c_{nn} \end{pmatrix}, \tag{77}
$$

where the matrix elements are defined by the expected value (see also Chapter 13)

$$
c_{ij} = \text{Cov}(\mathbf{x}_i \mathbf{x}_j) = \langle (x_i - m_i)(x_j - m_j) \rangle; \quad i,j = 1,2,\ldots,n. \tag{78}
$$

By definition, the covariance matrix (77) is symmetrical, i.e., $c_{ij} = c_{ji}$, and the diagonal elements are the variances

$$
c_{ii} = \text{Var}(\mathbf{x}_i) = \sigma_i^2, \quad i = 1,2,\ldots,n. \tag{79}
$$

Last, when the random variables $\mathbf{x}_1, \mathbf{x}_2, \ldots, \mathbf{x}_n$ are uncorrelated, the covariance matrix (77) reduces to the diagonal matrix

$$
\mathbf{C} = \begin{pmatrix} \sigma_1^2 & 0 & \cdots & 0 \\ 0 & \sigma_2^2 & \cdots & 0 \\ \vdots & \vdots & & \vdots \\ 0 & 0 & \cdots & \sigma_n^2 \end{pmatrix}. \tag{80}
$$

For the special case in which $n = 2$, we find that the covariance matrix becomes

$$
\mathbf{C} = \begin{pmatrix} \sigma_1^2 & \rho\sigma_1\sigma_2 \\ \rho\sigma_1\sigma_2 & \sigma_2^2 \end{pmatrix}, \tag{81}
$$

where σ_1^2 and σ_2^2 are the variances of $\mathbf{x}_1$ and $\mathbf{x}_2$, and ρ is the correlation coefficient between $\mathbf{x}_1$ and $\mathbf{x}_2$. When the correlation coefficient $\rho = 0$, we say the random variables $\mathbf{x}_1$ and $\mathbf{x}_2$ are uncorrelated. In the general case of (81), however, it follows that

$$
\mathbf{C}^{-1} = \frac{1}{\det(\mathbf{C})} \begin{pmatrix} \sigma_2^2 & -\rho\sigma_1\sigma_2 \\ -\rho\sigma_1\sigma_2 & \sigma_1^2 \end{pmatrix}, \tag{82}
$$

$$
\det(\mathbf{C}) = (1 - \rho^2)\sigma_1^2\sigma_2^2. \tag{83}
$$

This case, of course, leads to the bivariate distribution (72).

14.7.2 Detection devices

Communication systems may be broadly classified in terms of *linear operations*, such as *amplification* and *filtering*, and *nonlinear operations*, such as *modulation* and *detection*. Common detection devices include the following (see also Section 15.4.2):

1. *Square-law detector:* $\quad\quad\quad\quad\quad$ $y(t) = x^2(t)$

2. *Linear detector (full-wave):* $\quad$ $z(t) = |x(t)|$

3. *Linear detector (half-wave):* $\quad$ $w(t) = x(t)\,U[x(t)]$

The quantity $U(x)$ is the unit step function (see Section 2.2).

For the square-law detector, we derived an expression in Example 11 of Chapter 13 for the output PDF in terms of the PDF for the input process $x(t)$. For a specified PDF at the input, we did the same in Example 12 in Chapter 13 for the half-wave detector (rectifier). Here, we assume the input $x(t)$ in all three cases above is a zero-mean stationary Gaussian process with PDF

$$f_x(x) = \frac{1}{\sqrt{2\pi R_x(0)}}\exp\left[-\frac{x^2}{2R_x(0)}\right], \tag{84}$$

where $R_x(0) = \sigma_x^2$. Given that this is so, we wish to determine the statistics of the detector output.

To begin, we note that the PDF for the outputs of the detection devices are

$$f_y(y) = \frac{1}{\sqrt{2\pi R_x(0)y}}\exp\left[-\frac{y}{2R_x(0)}\right]U(y), \tag{85}$$

$$f_z(z) = 2f_x(z)\,U(z), \tag{86}$$

$$f_w(w) = \frac{1}{2}\delta(w) + f_x(w)\,U(w), \tag{87}$$

which can be readily deduced using the method of Section 13.6. The expected value of the output in each case is

$$\langle y(t)\rangle = \langle x^2(t)\rangle = R_x(0), \tag{88}$$

$$\langle z(t)\rangle = \langle |x(t)|\rangle = \int_{-\infty}^{\infty}|x|f_x(x)dx = \sqrt{\frac{2R_x(0)}{\pi}}, \tag{89}$$

$$\langle w(t)\rangle = \langle x(t)U[x(t)]\rangle = \int_0^{\infty}xf_x(x)dx = \sqrt{\frac{R_x(0)}{2\pi}}. \tag{90}$$

In all three cases, the mean of the detector output is determined by the autocorrelation function of the input process.

For higher-order statistics, we only consider the square-law detector. Here, the autocorrelation function of the output is [see Eq. (74)]

$$R_y(\tau) = \langle \mathbf{x}^2(t)\mathbf{x}^2(t+\tau) \rangle = \langle \mathbf{x}^2(t) \rangle \langle \mathbf{x}^2(t+\tau) \rangle + 2\langle \mathbf{x}(t)\mathbf{x}(t+\tau) \rangle^2, \tag{91}$$

from which we deduce

$$R_y(\tau) = R_x^2(0) + 2R_x^2(\tau). \tag{92}$$

In particular, the power in the dc component of the output is

$$\langle \mathbf{y}^2(t) \rangle = R_y(0) = 3R_x^2(0), \tag{93}$$

and the total average power in the ac component is

$$\sigma_y^2 = \langle \mathbf{y}^2(t) \rangle - \langle \mathbf{y}(t) \rangle^2 = 2R_x^2(0). \tag{94}$$

14.7.3 Zero-crossing problem

In many applications we are interested in the frequency of positive and negative crossings through zero (or some nonzero threshold value) of a random signal like the output current from a detector. It has been shown that the frequency of either positive or negative crossings is based on the joint PDF of the signal $\mathbf{x}(t)$ and its time derivative $\mathbf{x}'(t)$. The time derivative is defined by the standard limit process

$$\mathbf{x}'(t) = \lim_{\Delta t \to 0} \frac{\mathbf{x}(t+\Delta t) - \mathbf{x}(t)}{\Delta t}. \tag{95}$$

If we assume that $\mathbf{x}(t)$ is a Gaussian random process, it follows that the difference in (95) is also Gaussian and we conclude that the derivative $\mathbf{x}'(t)$ is Gaussian. Hence, the derivatives of *any* order of a Gaussian process are also Gaussian.

Next, let us find the autocorrelation function $R_{x'}(\tau)$ of the derivative $\mathbf{x}'(t)$ of a stationary process (not necessarily Gaussian). To begin, we define $\mathbf{Y}(t) = \mathbf{x}(t+\Delta t) - \mathbf{x}(t)$ and then calculate the correlation function for $\mathbf{Y}(t)$. Thus,

$$\begin{aligned} R_Y(\tau) &= \langle \mathbf{Y}(t)\mathbf{Y}(t+\tau) \rangle \\ &= \langle [\mathbf{x}(t+\Delta t) - \mathbf{x}(t)][\mathbf{x}(t+\tau+\Delta t) - \mathbf{x}(t+\tau)] \rangle \\ &= \langle \mathbf{x}(t+\Delta t)\mathbf{x}(t+\tau+\Delta t) \rangle - \mathbf{x}(t)\mathbf{x}(t+\tau+\Delta t) \\ &\quad - \mathbf{x}(t+\Delta t)\mathbf{x}(t+\tau) + \mathbf{x}(t)\mathbf{x}(t+\tau) \\ &= 2R_x(\tau) - R_x(\tau+\Delta t) - R_x(\tau-\Delta t), \end{aligned} \tag{96}$$

from which we deduce

$$R_{Y/\Delta t}(\tau) = \frac{2R_x(\tau) - R_x(\tau + \Delta t) - R_x(\tau - \Delta t)}{(\Delta t)^2}$$

$$= -\frac{1}{(\Delta t)^2}\left[R_x''(\tau)(\Delta t)^2 + 2R_x^{(4)}(\tau)\frac{(\Delta t)^4}{4!} + \cdots\right]. \tag{97}$$

In attaining the last line in (97) we have expanded both $R_x(\tau + \Delta t)$ and $R_x(\tau - \Delta t)$ in Taylor series about $\Delta t = 0$. Consequently, in the limit $\Delta t \to 0$ we see that $Y/\Delta t \to x'$, and deduce that the autocorrelation function of the derivative $x'(t)$ is given by

$$\boxed{R_{x'}(\tau) = -R_x''(\tau).} \tag{98}$$

Based on the result of (98), we conclude that the average power in the derivative $x'(t)$ is given by the second moment

$$<[x'(t)]^2> = -R_x''(0). \tag{99}$$

Also, it follows from (95) that the mean of any differentiable stationary process is

$$<x'(t)> = 0. \tag{100}$$

Without providing the details, it has been shown that the cross-correlation function of $x(t)$ and its derivative $x'(t)$ is (see [4] or [24])

$$R_{xx'}(\tau) = R_x'(\tau). \tag{101}$$

In particular, at $\tau = 0$ we see that

$$R_{xx'}(0) = R_x'(0) = <x(t)x'(t)> = 0, \tag{102}$$

which follows from the fact that $R_x(\tau)$ is an even function, and thus its derivative at $\tau = 0$ must vanish. For a stationary process, therefore, we see that $x(t)$ and its derivative $x'(t)$ are uncorrelated, and, therefore, for a Gaussian process they are also statistically independent.

> **Remark:** Recall from Section 14.4 that we stated two random variables (processes) are not necessarily independent when they are uncorrelated, unless they are also Gaussian.

To obtain the power spectrum of the derivative $x'(t)$, we use (98) to write

$$S_{x'}(\omega) = -2\int_0^\infty R_x''(\tau)\cos \omega\tau \, d\tau$$

$$= 2\omega^2 \int_0^\infty R_x(\tau)\cos \omega\tau \, d\tau, \tag{103}$$

where we have performed integration by parts twice. Thus, the power spectrum of the derivative $\mathbf{x}'(t)$ is related to the power spectrum of the original process $\mathbf{x}(t)$ by

$$S_{\mathbf{x}'}(\omega) = \omega^2 S_{\mathbf{x}}(\omega), \tag{104}$$

and, by combining this last result with (99), we further deduce that

$$\langle [\mathbf{x}'(t)]^2 \rangle = -R_{\mathbf{x}}''(0) = \frac{1}{2\pi} \int_{-\infty}^{\infty} \omega^2 S_{\mathbf{x}}(\omega) d\omega. \tag{105}$$

Thus far, we have only developed some basic properties involving correlation functions associated with the derivative $\mathbf{x}'(t)$ of a random process $\mathbf{x}(t)$, such as the random current at the output of a detector. However, the derivative (or slope) of $\mathbf{x}(t)$ is related to the number of positive and negative crossings of such a current through a given (threshold) level. The frequency of positive and negative crossings is therefore identical with the concept of frequency of fades and surges. A *fade* happens when the desired current $\mathbf{x}(t)$ drops below a given level and a *surge* happens when it goes above this level.

In the 1940s, Rice [26] showed that the *mean frequency of surges* (positive crossings of level x_0) is given by

$$\langle n^+(x_0) \rangle = \int_0^{\infty} x' p_{\mathbf{x}\mathbf{x}'}(x_0, x') dx', \tag{106}$$

where $p_{\mathbf{x}\mathbf{x}'}(x, x')$ is the joint PDF between $\mathbf{x}(t)$ and $\mathbf{x}'(t)$. Similarly, the *mean frequency of fades* (negative crossings of level x_0) is

$$\langle n^-(x_0) \rangle = -\int_{-\infty}^{0} x' p_{\mathbf{x}\mathbf{x}'}(x_0, x') dx'. \tag{107}$$

Consequently, the mean frequency of either positive or negative crossings of a threshold value x_0 by a current is given by the expected number of crossings per second defined by

$$\langle n(x_0) \rangle = \frac{1}{2} \int_{-\infty}^{\infty} |x'| p_{\mathbf{x}\mathbf{x}'}(x_0, x') dx'. \tag{108}$$

We have shown above that when $\mathbf{x}(t)$ is Gaussian, so is its derivative. Moreover, these are statistically independent random processes so that their joint PDF is simply

$$p_{\mathbf{x}\mathbf{x}'}(x, x') = \frac{1}{\sqrt{2\pi}\,\sigma} \exp\left[-\frac{(x-m)^2}{2\sigma^2} \right] \frac{1}{\sqrt{2\pi}\,b} \exp\left(-\frac{x'^2}{2b^2} \right), \tag{109}$$

where $m = \langle \mathbf{x}(t) \rangle$,

$$\sigma^2 = \frac{1}{2\pi} \int_{-\infty}^{\infty} S_{\mathbf{x}}(\omega) d\omega, \tag{110}$$

$$b^2 = -R_{\mathbf{x}}''(0) = \frac{1}{2\pi} \int_{-\infty}^{\infty} \omega^2 S_{\mathbf{x}}(\omega) d\omega. \tag{111}$$

Last, by inserting the PDF (109) into (108) and performing the integration, we are led to

$$\langle n(x_0) \rangle = v_0 \exp\left[-\frac{(x_0 - m)^2}{2\sigma^2}\right], \tag{112}$$

where $v_0 = b/2\pi\sigma = \sqrt{-R_x''(0)}/2\pi\sigma$. The quantity v_0 represents the expected number of fades or surges through the mean value m.

For further reading about the zero crossing problem, also see [22] and [24].

Suggested Reading

W. B. Davenport, Jr., *Probability and Random Processes: An Introduction for Applied Scientists and Engineers* (McGraw-Hill, New York, 1970).

E. Dougherty, *Random Processes for Image and Signal Processing* (SPIE Press, Bellingham, 1998).

C. W. Helstron, *Probability and Stochastic Processes for Engineers*, 2nd ed. (Macmillan, New York, 1991).

A. Papoulis, *Probability, Random Variables, and Random Processes*, 2nd ed. (McGraw-Hill, New York, 1984).

J. G. Proakis and M. Salehi, *Communication Systems Engineering*, 2nd ed. (Prentice-Hall, Upper Saddle River, 2002).

Chapter 14

Exercises

1. The random process $x(t) = e^{-At} U(t)$, depends on the random variable A with PDF $f_A(A)$. Calculate the PDFs

    (a) $f_x(x,t)$.

    (b) $f_x(x_1,t_1;x_2,t_2)$.

Ans. (a) $f_x(x,t) = \dfrac{1}{|tx|} f_A[-\ln(x)/t]$

    (b) $f_x(x_1,t_1;x_2,t_2)$

    $= \dfrac{1}{|t_1 x_1|} f_A[-\ln(x_1)/t_1]\delta(x_2 - x_1^{t_2/t_1})$

2. If $f_A(A) = 1/M$, $0 < A < M$, in Problem 1, show that

    (a) $\langle x(t) \rangle = \dfrac{1 - e^{-Mt}}{Mt}$.

    (b) $R_x(t_1,t_2) = \dfrac{1 - e^{-M(t_1 + t_2)}}{M(t_1 + t_2)}$.

3. Calculate the correlation function and covariance function of the random process

$$\mathbf{x}(t) = \mathbf{A} \cos \omega_0 t,$$

where $\mathbf{A}$ is a random variable for which $\langle \mathbf{A} \rangle = 1$ and $\langle \mathbf{A}^2 \rangle = 4$.

Ans. $R_\mathbf{x}(t_1, t_2) = 4 \cos \omega_0 t_1 \cos \omega_0 t_2$
$C_\mathbf{x}(t_1, t_2) = 3 \cos \omega_0 t_1 \cos \omega_0 t_2$

4. Given the random process

$$\mathbf{x}(t) = \mathbf{a} \cos \omega_0 t + \mathbf{b} \cos \omega_0 t,$$

where $\mathbf{a}$ and $\mathbf{b}$ are statistically independent random processes with

$$\langle \mathbf{a} \rangle = \langle \mathbf{b} \rangle = 0,$$
$$\langle \mathbf{a}^2 \rangle = \langle \mathbf{b}^2 \rangle = \sigma^2,$$

find the time-average correlation process. Is the process ergodic?

Ans. $\Re_\mathbf{x}(\tau) = \dfrac{1}{2}(\mathbf{a}^2 + \mathbf{b}^2) \cos \omega_0 \tau$
No

5. Find the structure function associated with a stationary random process $\mathbf{x}(t)$, given the covariance function

(a) $C_\mathbf{x}(\tau) = b_0 \exp(-|\tau/\tau_0|)$.
(b) $C_\mathbf{x}(\tau) = \exp[-a^2(\tau/\tau_0)^2]$.

6. If $\mathbf{x}(t)$ is a complex random process with mean $\langle \mathbf{x}(t) \rangle = m$ and correlation function defined by

$$R_\mathbf{x}(\tau) = \langle \mathbf{x}(t)\mathbf{x}^*(t + \tau) \rangle,$$

show that

(a) $R_\mathbf{x}(-\tau) = R_\mathbf{x}^*(\tau)$.
(b) $C_\mathbf{x}(\tau) = R_\mathbf{x}(\tau) - |m|^2$.

7. A random process is given by

$$\mathbf{x}(t) = \begin{cases} \mathbf{a}t + \mathbf{b}, & t \geq 0 \\ 0, & t < 0 \end{cases}$$

where $\mathbf{a}$ and $\mathbf{b}$ are statistically independent random variables. If $\mathbf{a}$ is a Gaussian random variable with mean zero and $\sigma^2 = 9$, and $\mathbf{b}$ is uniform on the interval $(0,6)$, find the

(a) mean value of $\mathbf{x}(t)$.
(b) the variance of $\mathbf{x}(t)$.

Ans. (a) 3   (b) $9t^2 + 3$

8. If a particular realization of the random process in Problem 7 yields

$$\mathbf{x}(2) = 10, \quad \mathbf{x}(4) = 20,$$

what is $\mathbf{x}(8)$?

Ans. 40

9. Determine the mean and variance of the random process $\mathbf{x}(t)$, given that

(a) $R_\mathbf{x}(\tau) = 100 e^{-\tau^2} \cos 2\pi\tau + 36$.
(b) $R_\mathbf{x}(\tau) = 10\left(\dfrac{\tau^2 + 8}{\tau^2 + 4}\right)$.

Ans. (a) ± 6, 100
(b) $\pm\sqrt{10}$, 10

10. The correlation function of a random process is given by

$$R_\mathbf{x}(\tau) = \begin{cases} 1 - \dfrac{|\tau|}{T}, & |\tau| < T \\ 0, & |\tau| > T. \end{cases}$$

Show that its power spectrum is given by

$$S_\mathbf{x}(\omega) = \frac{4 \sin^2(\omega T/2)}{T\omega^2}.$$

11. The impulse response of a linear system is

$$h(t) = \frac{1}{2T} U(T - |t|).$$

Find

(a) the system function.
(b) the power spectrum of the output as a function of the input power spectrum.

(*c*) the cross-correlation function of the input and output.

(*d*) the autocorrelation function of the output as a function of the input autocorrelation function.

Ans.

(*a*) $H(\omega) = \dfrac{\sin \omega T}{\omega T}$

(*b*) $S_y(\omega) = S_x(\omega)\left(\dfrac{\sin \omega T}{\omega T}\right)^2$

(*c*) $R_{xy}(\tau) = \dfrac{1}{2T}\displaystyle\int_{-T}^{T} R_x(\tau - \xi)\,d\xi$

(*d*) $R_y(\tau) = \dfrac{1}{2T}\displaystyle\int_{-2T}^{2T} R_x(\tau - \xi)\left(1 - \dfrac{|\xi|}{2T}\right)d\xi$

12. Two statistically independent stationary random processes $\mathbf{x}(t)$ and $\mathbf{y}(t)$ have autocorrelation functions

$$R_x(\tau) = 25e^{-10|\tau|}\cos 100\pi\tau,$$
$$R_y(\tau) = \frac{16\sin 50\pi\tau}{50\pi\tau}.$$

If
$$\mathbf{u}(t) = \mathbf{x}(t) + \mathbf{y}(t),$$
$$\mathbf{w}(t) = \mathbf{x}(t) - \mathbf{y}(t),$$

(*a*) find $R_u(\tau)$.
(*b*) find $R_w(\tau)$.

Ans. (*a*) $R_u(\tau) = 25e^{-10|\tau|}\cos 100\pi\tau$
$$+ \frac{16\sin 50\pi\tau}{50\pi\tau}$$
(*b*) $R_w(\tau) = R_u(\tau)$

13. For the random processes given in Problem 12,

(*a*) find $R_{uw}(\tau)$.
(*b*) find $R_{wu}(\tau)$.
(*c*) If $\mathbf{z}(t) = \mathbf{x}(t)\mathbf{y}(t)$, find $R_z(\tau)$.

Ans. (*a*) $R_{uw}(\tau) = R_x(\tau) - R_y(\tau)$
(*b*) $R_{wu}(\tau) = R_{uw}(-\tau) = R_{uw}(\tau)$
(*c*) $R_z(\tau) = R_x(\tau)R_y(\tau)$

14. Two jointly stationary random processes are described by

$$\mathbf{x}(t) = A\cos(\omega_0 t + \mathbf{\theta}),$$
$$\mathbf{y}(t) = B\sin(\omega_0 t + \mathbf{\theta}),$$

where $\mathbf{\theta}$ is a random variable uniformly distributed on $(0, 2\pi)$. Find the cross-correlation function $R_{xy}(\tau)$.

Ans. $R_{xy}(\tau) = \dfrac{1}{2}(AB)\sin \omega_0 \tau$

15. A stationary random process $\mathbf{x}(t)$ has power spectral density

$$S_x(\omega) = \frac{9}{\omega^2 + 64}.$$

Find a spectral density for bandlimited white noise $\mathbf{y}(t)$ that has the same value at zero frequency (dc) and the same average power as $\mathbf{x}(t)$.

Ans. $S_y(\omega) = \begin{cases} \dfrac{9}{64}, & -4\pi < \omega < 4\pi \\ 0, & \text{otherwise} \end{cases}$

16. Find the autocorrelation functions for the (zero mean) random processes $\mathbf{x}(t)$ and $\mathbf{y}(t)$ in Problem 15.

Ans. $R_x(\tau) = \dfrac{9}{16}e^{-8|\tau|}$
$$R_y(\tau) = \frac{9}{16}\left(\frac{\sin 4\pi\tau}{4\pi\tau}\right)$$

17. Given that $\mathbf{x}(t)$ is a Gaussian random process with PDF (84), derive the PDFs given by Eqs. (85)-(87) for the square-law detector, full-wave linear detector, and half-wave linear detector, respectively.

18. Based on PDF models (85)-(87),
(*a*) verify the mean values given in Eqs. (88)-(90).
(*b*) Find the autocorrelation function of the output for the full-wave linear detector.

19. If the input to a square-law device is $x(t) = n(t) + m$, where $n(t)$ is a zero-mean, stationary, Gaussian process, show that the output $y(t) = x^2(t)$ has the covariance function

$$C_y(\tau) = 2[C_x(\tau)]^2 + 4m^2 C_x(\tau).$$

20. What is the spectral density of the output of the square-law device in Problem 19
    (a) if $m = 0$?
    (b) if $m \neq 0$?

21. Given that x and y are zero-mean Gaussian random processes for which

$$\langle x^2(t) \rangle = \langle y^2(t) \rangle = \sigma^2,$$
$$\langle x(t)y(t) \rangle = \rho\sigma^2,$$

and that z and w are related by

$$z(t) = \frac{1}{\sqrt{2}}[x(t) + y(t)],$$

$$w(t) = \frac{1}{\sqrt{2}}[-x(t) + y(t)],$$

show that

$$\langle z^2(t) \rangle = \sigma^2(1 + \rho),$$
$$\langle w^2(t) \rangle = \sigma^2(1 - \rho).$$

22. For the inverse transformation of the random processes in Problem 21 given by

$$x(t) = \frac{1}{\sqrt{2}}[z(t) - w(t)],$$

$$y(t) = \frac{1}{\sqrt{2}}[z(t) + w(t)],$$

use the results of Problem 21 to establish that
    (a) $z(t)$ and $w(t)$ are statistically independent random processes, i.e.,

$$\langle z(t)w(t) \rangle = \langle z(t) \rangle \langle w(t) \rangle = 0.$$

    (b) Based on part (a), show that

$$\langle z^2(t)w^2(t) \rangle = \langle z^2(t) \rangle \langle w^2(t) \rangle$$
$$= \sigma^4(1 - \rho^2).$$

(c) Use results from parts (a) and (b) to deduce that

$$\langle x^2(t)y^2(t) \rangle = \frac{3}{2}\sigma^4(1 + \rho^2)$$

$$- \frac{1}{2}\sigma^4(1 - \rho^2)$$

$$= \sigma^4(1 + 2\rho^2).$$

Chapter 15

Applications

15.1 Introduction / 734
15.2 Mechanical Vibrations and Electric Circuits / 734
    15.2.1 Forced oscillations—I / 734
    15.2.2 Damped motions / 737
    15.2.3 Forced oscillations—II / 738
    15.2.4 Simple electrical circuits / 740
15.3 Buckling of a Long Column / 742
15.4 Communication Systems / 745
    15.4.1 Frequency modulated signals / 745
    15.4.2 Nonlinear devices / 747
    15.4.3 Coherent detection optical receiver / 751
    15.4.4 Threshold detection / 754
15.5 Applications in Geometrical Optics / 756
    15.5.1 Eikonal equation / 758
    15.5.2 Frenel-Serret formulas revisited / 759
    15.5.3 The heated window / 759
15.6 Wave Propagation in Free Space / 762
    15.6.1 Hankel transform method / 762
    15.6.2 Huygens-Fresnel integral: lowest-order Gaussian mode / 764
    15.6.3 Hermite-Gaussian modes / 765
15.7 *ABCD* Matrices for Paraxial Systems / 767
    15.7.1 Generalized Huygens-Fresnel integral / 769
    15.7.2 Gaussian lens / 770
    15.7.3 Fourier-transform plane / 772
15.8 Zernike Polynomials / 773
    15.8.1 Application in optics / 774
    15.8.2 Atmospheric effects on imaging systems / 775
    15.8.3 Aperture filter functions / 777
Exercises / 780

15.1 Introduction

In this final chapter we wish to present a collection of applications that make use of the various mathematical techniques introduced in the preceding chapters. For this purpose we have selected problems in mechanical vibrations, communication systems, and optics, among others.

15.2 Mechanical Vibrations and Electric Circuits

To begin, we consider some *initial value problems* (IVPs) involving DEs of the second order in connection with mechanical vibrations and electric circuits. Problems in both of these application areas are mathematically similar, the general problem being to solve the linear DE

$$A_2(t)y'' + A_1(t)y' + A_0(t)y = F(t), \quad t > 0, \tag{1}$$

where $y' = dy/dt$, $y'' = d^2y/dt^2$, subject to the prescribed *initial conditions* (ICs)

$$y(0) = k_0, \quad y'(0) = k_1. \tag{2}$$

15.2.1 Forced oscillations—I

When a given weight (having mass m) is attached to an elastic spring suspended from a fixed support (like a ceiling beam), the spring will stretch to an equilibrium position by an amount s that varies with the weight mg, where $g = 9.8$ m/s^2 (32 ft/s^2) is the gravitational constant. To remain in equilibrium, Hooke's law states that the spring will exert an upward restoring force f proportional to the amount of stretch; that is,

$$f = mg = ks,$$

where the spring constant k depends on the "stiffness" of the spring. If the spring-mass system is also subjected to a downward (positive direction) external force $F(t)$, the body or mass will move in the vertical direction. In addition to the external force, there may exist a retarding force caused by resistance of the medium in which the motion takes place, or possibly by friction. For example, the mass could be suspended in a viscous medium (like oil), connected to a dashpot damping device (like a shock absorber), and so on. In practice, many such retarding forces are approximately proportional to the velocity of the moving body and act in a direction opposing the motion.

    As a consequence of *Newton's second law of motion* ($F = ma$), the sum of forces acting on the spring-mass system leads to the governing DE

$$my'' + cy' + ky = F(t), \quad t > 0, \tag{3}$$

where c is a positive retarding force constant and y represents the displacement from equilibrium of the mass alone. Such a spring-mass system is illustrated in Fig. 15.1. The motion of the mass is said to be *undamped* when $c = 0$ and *damped* when $c \neq 0$. The motion is further classified as *free* when $F(t) \equiv 0$ and *forced* otherwise.

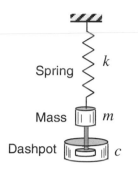

Figure 15.1 Mass on a spring.

For the special case in which c is sufficiently small compared with mk, it may be acceptable to neglect the damping term in (3). If this is done and no external force acts on the mass, then the free oscillations of the mass are described by solutions of the IVP

$$my'' + ky = 0, \quad y(0) = y_0, \; y'(0) = v_0, \quad (4)$$

where y_0 is the initial displacement of the mass from equilibrium and v_0 is the initial velocity at that point. By setting $\omega_0 = \sqrt{k/m}$, it is easy to show that the motion of the mass is described by the sinusoidal solution (e.g., see Section 1.4.2)

$$y = y_0 \cos \omega_0 t + \frac{v_0}{\omega_0} \sin \omega_0 t. \qquad (5)$$

It is often convenient to rewrite the solution (5) in the form

$$y = A \cos(\omega_0 t - \varphi), \qquad (6)$$

where A is the *amplitude* of the motion defined by

$$A = \sqrt{y_0^2 + \left(\frac{v_0}{\omega_0}\right)^2}, \qquad (7)$$

and φ is the phase angle that satisfies the relations

$$\cos \varphi = \frac{y_0}{A}, \quad \sin \varphi = \frac{v_0}{A\omega_0}. \qquad (8)$$

Motion described by (6) is called *simple harmonic motion*. It is clearly periodic motion with the *period* of the motion defined by

$$T = \frac{2\pi}{\omega_0} = 2\pi\sqrt{\frac{m}{k}}. \qquad (9)$$

The amplitude A gives the maximum displacement of the mass from equilibrium. The reciprocal of the period T is the *natural frequency* of the system measured in hertz (Hz); that is,

$$f_0 = \frac{1}{T} = \frac{\omega_0}{2\pi}. \qquad (10)$$

EXAMPLE 1 Suppose a 4-kg mass is attached to a spring with $k = 100$ N/m. If the mass is released 10 cm below equilibrium with velocity (speed) of 1 m/s directed upward, find the equation of motion of the mass. Also find the amplitude, phase, period, and frequency.

Solution: The governing IVP is described by

$$4y'' + 100y = 0, \quad y(0) = 0.1, \ y'(0) = -1,$$

or, upon simplifying,

$$y'' + 25y = 0, \quad y(0) = 0.1, \ y'(0) = -1.$$

The equation of motion deduced from this IVP is

$$y = 0.1\cos 5t - 0.2\sin 5t$$

$$= 0.224\cos(5t + 1.107),$$

for which the amplitude is $A = 22.4$ cm and the phase is $\varphi = -1.107$ rad. Also, the period of motion is $T = 2\pi\sqrt{m/k} = 2\pi/5$ s and the natural frequency of the spring-mass system is $f = 1/T = 5/2\pi \cong 0.8$ Hz.

In the presence of a simple sinusoidal external force $F(t) = P\cos\omega t$, but negligible damping, the IVP for a system initially at rest is

$$my'' + ky = P\cos\omega t, \quad y(0) = 0, \ y'(0) = 0. \tag{11}$$

Under the assumption that the frequency of the forcing function is different from the natural frequency of the spring-mass system, i.e., $\omega \neq \omega_0$, the solution of (11) is readily found to be (see Section 1.4.3)

$$y = \frac{P}{m(\omega_0^2 - \omega^2)}(\cos\omega t - \cos\omega_0 t), \quad \omega \neq \omega_0. \tag{12}$$

In this case the motion consists of the superposition of two modes of vibration—the *natural mode* at angular frequency ω_0 and the *forced mode* at angular frequency ω.

An interesting phenomenon occurs when the forcing frequency ω in (12) is close to the natural frequency ω_0, i.e., when $|\omega_0 - \omega|$ is "small." By the use of trigonometric identities $\cos(a \pm b) = \cos a \cos b \mp \sin a \sin b$ we can rewrite the solution (12) in the more revealing form

$$y = \frac{2P}{m(\omega_0^2 - \omega^2)}\sin\left[(\omega_0 - \omega)\frac{t}{2}\right]\sin\left[(\omega_0 + \omega)\frac{t}{2}\right]. \tag{13}$$

The period of the first sine wave in (13) is large compared with the period of the second sine wave, producing motion visualized as a rapid oscillation with angular frequency $(\omega_0 + \omega)/2$, but with a slowly varying sinusoidal amplitude known as the *envelope* (see Fig. 15.2). Motion of this type exhibits what is called a *beat*. The phenomenon of beats can most easily be demonstrated with acoustic waves—for example, when the strings of two guitars of nearly the same frequency are plucked at the same time.

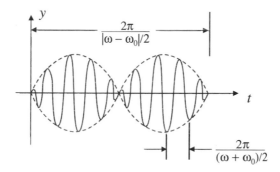

Figure 15.2 Phenomenon of beats.

Last, in the limit as $\omega \to \omega_0$, i.e., when the system is excited at its natural frequency, we can obtain the response of the system directly from (12) through the use of L'Hôpital's rule. Doing so yields

$$y = \frac{P}{2m\omega_0} t \sin \omega_0 t. \tag{14}$$

Because the amplitude now increases with time, such motion leads to the phenomenon of *resonance*.

15.2.2 Damped motions

When damping effects cannot be reasonably ignored, the *free motions* of the spring-mass system are described by solutions of

$$my'' + cy' + ky = 0, \; t > 0, \; c > 0. \tag{15}$$

Assuming solutions of the form $y = e^{\lambda t}$, we obtain the auxiliary equation

$$m\lambda^2 + c\lambda + k = 0, \tag{16}$$

with roots

$$\lambda_1, \lambda_2 = \frac{-c \pm \sqrt{c^2 - 4mk}}{2m}. \tag{17}$$

The solution clearly takes on three distinct forms, depending on the magnitude of the damping term c.

CASE I. *Overdamping*: $c^2 > 4mk$

In this case the damping is large and leads to roots λ_1, λ_2 that are *real*, *negative*, and *distinct*. The general solution is

$$y = e^{-ct/2m}(C_1 e^{\alpha t} + C_2 e^{-\alpha t}), \qquad \alpha = \frac{\sqrt{c^2 - 4mk}}{2m}. \tag{18}$$

The solution (18) is a smooth *nonoscillatory* type of motion that goes to zero as $t \to \infty$.

CASE II. *Critical damping*: $c^2 = 4mk$

The roots of the auxiliary equation are equal, i.e., $\lambda_1 = \lambda_2 = -c/2m$, so the general solution of (15) takes the form

$$y = e^{-ct/2m}(C_1 + C_2 t). \tag{19}$$

The possible motions in this case are similar to the overdamped case.

CASE III. *Underdamping*: $c^2 < 4mk$

Here, the roots of the auxiliary equation (17) are complex conjugates, from which we deduce

$$y = e^{-ct/2m}(C_1 \cos \mu t + C_2 \sin \mu t), \qquad \mu = \frac{\sqrt{4mk - c^2}}{2m}. \tag{20}$$

In this case the mass will *oscillate* back and forth across the equilibrium position with a steadily decreasing amplitude [i.e., $y(t) \to 0$ as $t \to \infty$].

15.2.3 Forced oscillations—II

Fourier series can be a useful tool in the analysis of steady-state solutions of certain types of differential equation (DE) with general periodic forcing function. In particular, let us consider the same spring-mass system discussed in the previous sections, i.e.,

$$my'' + cy' + ky = F(t). \tag{21}$$

We could use the one-sided Green's function (Section 8.5.1) to solve this problem, but here we will take $F(t)$ to be a periodic function (other than a simple sinusoid) for which the method of Fourier series is better suited (Section 8.2).

Let us assume the external force is the periodic square wave defined for $t > 0$ by

$$F(t) = \begin{cases} 100, & 0 < t < 1 \\ -100, & 1 < t < 2 \end{cases}, \quad F(t + 2) = F(t). \tag{22}$$

We further assume $m = 10$ kg, $c = 0.5$ kg/s, and $k = 1000$ N/m, for which the natural (angular) frequency is roughly $\sqrt{k/m} = 10$ rad/s. Over the extended interval $-\infty < t < \infty$, the periodic function (22) is an odd function, and thus, we consider a sine series

$$F(t) = \sum_{n=1}^{\infty} b_n \sin n\pi t, \qquad (23)$$

where

$$b_n = 2\int_0^1 100\sin n\pi t\, dt = \begin{cases} \dfrac{400}{n\pi}, & n = 1,3,5,\dots \\ 0, & n = 2,4,6,\dots. \end{cases} \qquad (24)$$

The resulting DE can now be expressed as

$$10y'' + 0.5y' + 1000y = \frac{400}{\pi} \sum_{\substack{n=1 \\ (\text{odd})}}^{\infty} \frac{\sin n\pi t}{n}. \qquad (25)$$

We assume the steady-state solution (i.e., after transient effects diminish) of (25) can be represented by the Fourier series

$$y(t) = \frac{1}{2}A_0 + \sum_{n=1}^{\infty} A_n \cos n\pi t + B_n \sin n\pi t. \qquad (26)$$

The direct substitution of this assumed solution into the DE yields (after simplification)

$$10\sum_{n=1}^{\infty}(-n^2\pi^2)(A_n \cos n\pi t + B_n \sin n\pi t)$$
$$+ 0.5\sum_{n=1}^{\infty} n\pi(-A_n \sin n\pi t + B_n \cos n\pi t) \qquad (27)$$
$$+ 500A_0 + 1000\sum_{n=1}^{\infty}(A_n \cos n\pi t + B_n \sin n\pi t) = \frac{400}{\pi}\sum_{\substack{n=1 \\ (\text{odd})}}^{\infty}\frac{\sin n\pi t}{n}.$$

At this point we simply match like coefficients on each side of the DE. For example, for $n = 0$ we have $A_0 = 0$, whereas for general n we obtain the simultaneous equations

$$(1000 - 10n^2\pi^2)A_n + 0.5n\pi B_n = 0,$$
$$-0.5n\pi A_n + (1000 - 10n^2\pi^2)B_n = \begin{cases} \dfrac{400}{n\pi}, & n = 1,3,5,\dots \\ 0, & n = 2,4,6,\dots. \end{cases} \qquad (28)$$

By solving these equations simultaneously, we note that only those coefficients with odd index n will be nonzero because the right-hand side is nonzero only for odd index n. For odd index n, therefore, we obtain the result

$$A_n = -\frac{200}{\Delta_n}, \quad n = 1,3,5,\dots,$$

$$\qquad (29)$$

$$B_n = \frac{400}{n\pi\Delta_n}(1000 - 10n^2\pi^2), \quad n = 1,3,5,\dots,$$

where

$$\Delta_n = (1000 - 10n^2\pi^2)^2 + 0.25n^2\pi^2. \tag{30}$$

From these expressions we are led to the result

$$\begin{aligned}
y(t) = &-2.5\times10^{-4}\cos\pi t + 0.141\sin\pi t \\
&- 1.56\times10^{-2}\cos 3\pi t + 0.379\sin 3\pi t \\
&- 9.29\times10^{-5}\cos 5\pi t - 0.0174\sin 5\pi t + \cdots.
\end{aligned} \tag{31}$$

The solution (31) is plotted in Fig. 15.3. Note that the amplitude of the sine term $\sin 3\pi t$ dominates the solution. Also, the angular frequency of this dominant term is 3π rad/s, whereas the fundamental frequency of the periodic input was π rad/s. This happened because the natural frequency of the undamped system is 10 rad/s, very close to the frequency 3π rad/s of the dominant term. We can deduce from this that the overtone 3π rad/s, not the fundamental, resonated with the system. Overtones may dominate the steady-state response for any underdamped system that is forced with a periodic input having a frequency smaller than the natural frequency of the system.

As a final comment here we take note that similar results could be deduced for other types of physical systems. For instance, an electric circuit problem arises as a simple analog of the above mechanical system if we simply replace the mass m by inductance L, the resistive constant c by the resistance R, the spring constant k by the inverse capacitance $1/C$, and the input forcing function by the electromotive force $E(t)$. In such cases we can interpret the unknown function y as representing the charge q on the capacitor (see the following section).

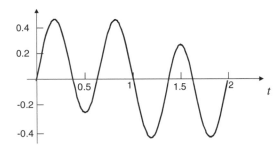

Figure 15.3 Steady-state solution given by Eq. (31) to a square-wave periodic input.

15.2.4 Simple electrical circuits

Let us consider the electric circuit shown in Fig. 15.4, which is composed of a *resistor*, *inductor*, and *capacitor* connected in series with a *voltage source* $E(t)$ measured in volts (V). When the switch is closed at time $t = 0$, a *current* of $i(t)$ amperes (A) will flow in the loop and a *charge* of $q(t)$ coulombs (C) will accumulate on the capacitor at time t. The voltage drop across a resistor is $E_R(t) = Ri(t)$, where R is resistance measured in ohms (Ω), whereas that across an inductor is $E_L(t) = Ldi/dt$, where L is inductance measured in henrys (H). The voltage drop across a capacitor is $E_C(t) = q(t)/C$, where C is capacitance measured in farads (F). The current and charge are related by

$$i = \frac{dq}{dt}. \tag{32}$$

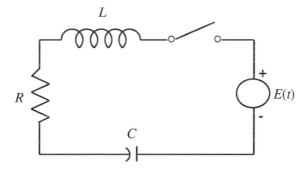

Figure 15.4 *RLC* circuit.

The network featured here is called an *RLC* circuit. Its behavior at each point in the network is determined by solving appropriate DEs that result from applying *Kirchoff's current* and *voltage laws* (see below).

Kirchoff's Laws
1. The sum of the currents into (or away from) any point is zero.
2. The sum of the instantaneous voltage drops in a specified direction is zero around any closed path.

By applying the above laws, the governing DE for the *RLC* circuit becomes

$$L\frac{di}{dt} + Ri + \frac{q}{C} = E(t). \tag{33}$$

To express (33) in terms of only the charge, we can use (32) to eliminate the current, which leads to

$$Lq'' + Rq' + C^{-1}q = E(t), \quad t > 0. \tag{34}$$

We recognize (34) as the same DE in form as that which governs the damped motions of a forced spring-mass system. Based on this comparison, we can make the following analogies between mechanical and electrical systems:

Charge q corresponds to **position** y.
Current i corresponds to **velocity** $v = y'$.
Inductance L corresponds to **mass** m.
Resistance R corresponds to **damping constant** c.
Inverse capacitance C^{-1} corresponds to **spring constant** k.
Electromotive force $E(t)$ corresponds to **forcing function** $F(t)$.

EXAMPLE 2 The circuit shown in Fig. 15.4 has an impressed voltage of $E(t) = E_0 \cos \omega t$. Determine the steady-state current in the circuit.

Solution: The governing IVP is

$$Lq'' + Rq' + \frac{1}{C}q = E_0 \cos \omega t.$$

By using variation of parameters (Section 1.4.3) and assuming a solution of the form $q_{s.s.} = A \cos \omega t + B \sin \omega t$, we find that A and B must satisfy

$$\left(\frac{1}{C} - L\omega^2 \right) A + \omega R B = E_0,$$

$$-\omega R A + \left(\frac{1}{C} - L\omega^2 \right) B = 0.$$

Solving for A and B, we are led to

$$
\begin{aligned}
q_{s.s.}(t) &= \frac{(1/C - L\omega^2)E_0}{(1/C - L\omega^2)^2 + \omega^2 R^2} \cos \omega t + \frac{\omega R E_0}{(1/C - L\omega^2)^2 + \omega^2 R^2} \sin \omega t \\
&= \frac{E_0}{\sqrt{(1/C - L\omega^2)^2 + \omega^2 R^2}} \sin(\omega t + \alpha),
\end{aligned}
$$

where $\sin \alpha = (1/C - L\omega^2)/\Delta$, $\cos \alpha = \omega R/\Delta$, and $\Delta = \sqrt{(1/C - L\omega^2)^2 + \omega^2 R^2}$. The corresponding steady-state current is therefore (after some algebraic rearranging)

$$i_{s.s.}(t) = q_{s.s.}'(t) = \frac{E_0}{\sqrt{(\omega L - 1/\omega C)^2 + R^2}} \cos(\omega t + \alpha),$$

where $\omega L - 1/\omega C$ is the *reactance* and $\sqrt{(\omega L - 1/\omega C)^2 + R^2}$ is the *impedence*.

15.3 Buckling of a Long Column

Vertical columns have been used extensively in architectural structures throughout the centuries. One of the oldest engineering problems concerns the buckling of such a column under a compressive load. Euler developed the first truly mathematical model that can rather accurately predict the *critical compressive load* that a column can withstand before deformation or *buckling* takes place.

Consider a long slender column or rod of length L that is simply supported at each end and is subject to an axial compressive load P applied at the top, as illustrated in Fig. 15.5.

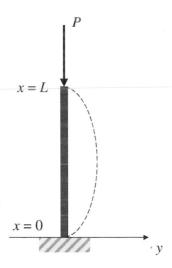

P

$x = L$

$x = 0$

y

Figure 15.5 Buckling column.

By "long," we mean that the length of the column is much greater than the largest dimension in its cross section. From the elementary theory of small deflections of beams and columns, the departure y from the vertical x axis for such a column is governed by the second-order differential equation (see [30])

$$EIy'' = M, \qquad (35)$$

where M is the *bending moment*, E is the *modulus of elasticity*, and I is the *moment of inertia* of the cross section of the column. When the column is deflected a small amount y from the vertical axis due to the load P (shown by the dashed curve in Fig. 15.5), the bending moment is

$$M = -Py.$$

Because the column is assumed to be simply supported at the endpoints, there can be no displacement at these points. Thus, Eq. (35) together with boundary conditions (BCs) takes the form

$$EIy'' + Py = 0, \ y(0) = 0, \ y(L) = 0. \qquad (36)$$

Clearly, the problem (36) has the trivial solution $y = 0$, corresponding to the column not bending away from the x axis. However, if P is sufficiently large, the column may suddenly bow out of its equilibrium state into a state of "buckling." The smallest load P_1 that leads to buckling is called the *Euler critical load,* and the corresponding deflection mode is called the *fundamental buckling mode.*

The classic example of a buckling column involves the case when E, I, and P are all constant, and then it is easy to show that the Euler critical load and corresponding deflection mode are given, respectively, by (recall Example 6 in Section 8.4)

$$P_1 = \frac{\pi^2 EI}{L^2}, \qquad y = C \sin \frac{\pi x}{L}, \qquad (37)$$

where C is an arbitrary constant. The constant C remains undetermined in this analysis because we chose a *linear* model (although the buckling problem is fundamentally nonlinear). Generally, only the critical load is of interest so the value of the constant C may be of little concern.

Let us consider the case where the column is tapered so the moment of inertia is not constant, but is given by $I(x) = \alpha x$, where α is a positive constant. In this case, the buckling problem (36) becomes

$$E\alpha x y'' + Py = 0, \ y(0) = 0, \ y(L) = 0. \qquad (38)$$

In solving the equation, we first rewrite it in the form

$$x^2 y'' + k^2 xy = 0,$$
(39)

where $k^2 = P/EI\alpha > 0$. Then, by comparing (39) with the general form given by Eq. (146) in Chapter 1, we see that $a = 1/2$, $b = 2k$, $c = 1/2$, and $v = 1$, from which we deduce the solution in terms of Bessel functions

$$y = \sqrt{x}\left[C_1 J_1(2k\sqrt{x}) + C_2 Y_1(2k\sqrt{x})\right],$$
(40)

where C_1 and C_2 are arbitrary constants.

To apply the first BC in (38), we use the small argument forms for the Bessel functions (see Section 2.5.1), which yields

$$
\begin{aligned}
y(0) &= \lim_{x \to 0} \sqrt{x}\left[C_1 J_1(2k\sqrt{x}) + C_2 Y_1(2k\sqrt{x})\right] \\
&= \lim_{x \to 0} \sqrt{x}\left[C_1 k\sqrt{x} + C_2 \frac{1}{\pi k \sqrt{x}}\right] \\
&= -C_2 \frac{1}{\pi k}.
\end{aligned}
$$

Clearly, this last expression equals zero only if we set $C_2 = 0$. The remaining BC then requires that

$$y(L) = C_1 \sqrt{L} J_1(2k\sqrt{L}) = 0.$$
(41)

There are infinitely many solutions of this last equation [determined by the zeros of $J_1(x)$], the first of which leads to the Euler critical load

$$P_1 = E\alpha k_1^2,$$
(42)

where k_1 is the smallest value for which $J_1(2k\sqrt{L}) = 0$. The corresponding fundamental deflection mode is therefore

$$y = C_1 \sqrt{x} J_1(2k_1\sqrt{x}).$$
(43)

EXAMPLE 3 Given a simply supported column of length $L = 400$ cm, $\alpha = 10^{-4}$ cm^3, and elastic modulus $E = 2 \times 10^9$ g/cm^2, calculate the Euler critical load and determine the point along the vertical axis where the maximum displacement takes place.

Solution: With $L = 400$, we first determine k_1 based on Eq. (41); i.e.,

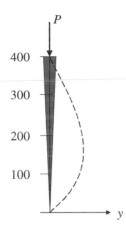

Figure 15.6 Buckling of a tapered column.

$$J_1(40k_1) = 0.$$

The first zero of the Bessel function $J_1(.)$ occurs at 3.832; hence, we set $40k_1 = 3.832$, which yields $k_1 = 0.0958$. The Euler critical load is then given by

$$P_1 = E\alpha k_1^2 = 1836 \text{ kg}.$$

The maximum displacement of the beam takes place at the point where the slope $y' = 0$, which from (43) and the identity [see (BJ7) in Section 2.5.1]

$$\frac{d}{du}[uJ_1(u)] = uJ_0(u),$$

leads to

$$y' = \frac{d}{dx}\left[C_1\sqrt{x}J_1(2k_1\sqrt{x})\right] = C_1 k_1 J_0(2k_1\sqrt{x}) = 0.$$

The first zero of $J_0(.)$ occurs at 2.405, and thus, by setting $2k_1\sqrt{x} = 2.405$, we find that $x \cong 158$ cm, which is closer to the bottom (more tapered end) of the column as illustrated in Fig. 15.6.

15.4 Communication Systems

Communication systems are broadly classified in terms of *linear operations* like *amplification* and *filtering*, and *nonlinear operations* like *modulation* and *detection*.

15.4.1 Frequency modulated signals

Historically, the first type of continuous-wave (CW) modulation to evolve was *amplitude modulation* (AM) associated with standard radio broadcasting. AM is that process whereby the amplitude of a carrier sinusoidal signal (with fixed frequency and phase) is altered. The modulated waveform in this case has the form

$$f(t) = A(t)\cos(\omega_0 t + \theta_0), \tag{44}$$

where ω_0 is the fixed angular frequency of the carrier, θ_0 is the phase, and $A(t)$ is the message signal; the latter also called the modulating signal. Other common modulation schemes involve either *phase* or *frequency modulation*, collectively referred to as *angle modulation*. In either of these cases, the modulated waveform is described by

$$f(t) = A\cos[\theta(t)] = A\cos[\omega_0 t + \gamma(t)], \tag{45}$$

where $\theta(t)$ is the "phase angle" and $\gamma(t)$ contains the message signal.

The time-domain description of a frequency modulated (FM) waveform is provided by writing (see [37])

$$\theta(t) = \omega_0 t + k_f \int_0^t g(\tau) d\tau + \theta_0, \tag{46}$$

where k_f and θ_0 are constants. The instantaneous angular frequency is obtained by differentiating the phase angle, which yields

$$\omega_i = \theta'(t) = \omega_0 + k_f g(t). \tag{47}$$

The simplest analysis of FM waveforms occurs when the message signal is a simple sinusoid, i.e., *tone modulation*. In this case we write

$$g(t) = a \cos \omega_m t, \tag{48}$$

where a is a constant and ω_m is the fixed angular frequency associated with the tone. The instantaneous angular frequency (47) in this case becomes

$$\omega_i = \omega_0 + \Delta\omega \cos \omega_m t, \tag{49}$$

where $\Delta\omega = ak_f$ denotes the *peak frequency deviation* from the carrier frequency ω_0. The resulting FM waveform (45) with $\theta_0 = 0$ can then be expressed as

$$f(t) = A \cos(\omega_0 t + \beta \sin \omega_m t), \tag{50}$$

where

$$\beta = \frac{\Delta\omega}{\omega_m}. \tag{51}$$

The parameter β is called the FM *modulation index* and represents the maximum phase deviation (in radians). It is defined only in the case of tone modulation.

To obtain a spectral analysis (i.e., a Fourier series) of the FM signal, we begin by writing it in the complex exponential representation, i.e.,

$$\begin{aligned} f(t) &= A \cos(\omega_0 t + \beta \sin \omega_m t) \\ &= \mathrm{Re}\{A \exp[i(\omega_0 t + \beta \sin \omega_m t)]\} \\ &= \mathrm{Re}\{A \exp(i\omega_0 t) \exp(i\beta \sin \omega_m t)\}. \end{aligned} \tag{52}$$

The second exponential in (52) is a periodic function with period $T = 2\pi/\omega_m$. Hence, we write the Fourier expansion (see Section 8.3)

$$\exp(i\beta \sin \omega_m t) = \sum_{n=-\infty}^{\infty} c_n e^{in\omega_m t}, \tag{53}$$

where

$$c_n = \frac{1}{T} \int_{-T/2}^{T/2} \exp(i\beta \sin \omega_m t) e^{-in\omega_m t} dt = \frac{1}{T} \int_{-T/2}^{T/2} \exp\left[i(\beta \sin \omega_m t - n\omega_m t)\right] dt. \quad (54)$$

By making the change of variable $x = \omega_m t$, this last expression becomes

$$c_n = \frac{1}{2\pi} \int_{-\pi}^{\pi} e^{i(\beta \sin x - nx)} dx = J_n(\beta), \ n = 0, \pm 1, \pm 2, ..., \quad (55)$$

where we have used property (BJ5) in Section 2.5.1 for the Bessel function. Therefore, the representation of the FM signal (52) takes the form

$$f(t) = \text{Re}\left\{ A \sum_{n=-\infty}^{\infty} J_n(\beta) \exp\left[i(\omega_0 t + n\omega_m t)\right] \right\}$$

$$= A \sum_{n=-\infty}^{\infty} J_n(\beta) \cos\left[(\omega_0 t + n\omega_m t)\right]. \quad (56)$$

Last, from (56) it is clear that the general FM tone modulation waveform has an infinite number of sidebands of magnitude $AJ_n(\beta)$ at angular frequencies $\omega_0 \pm n\omega_m$. All lines are equally spaced by the modulating frequency ω_m. However, the magnitude of such spectral components quickly diminishes away from the carrier frequency ω_0 because of the relative behavior of the Bessel function. For all practical purposes, the bandwidth of the FM signal is finite, i.e., the power is essentially contained within the first few sidebands. The effective bandwidth (98% of total power) is often approximated by *Carson's rule*

$$BW \cong 2f_m(1 + \beta), \ f_m = \omega_m/2\pi. \quad (57)$$

15.4.2 Nonlinear devices

Random noise, which appears at the input to any communications receiver, interferes with the reception of incoming radio and radar signals. When this noise is channeled through a passband filter whose bandwidth is narrow compared with the center frequency ω_0 of the filter, the output is called *narrowband noise*. If the filter bandwidth is sufficiently narrow, such noise at the output of the filter is essentially a random sinusoid (single frequency) that can be mathematically represented by (see [22])

$$\mathbf{n}(t) = \mathbf{x}(t)\cos \omega_0 t - \mathbf{y}(t)\sin \omega_0 t, \quad (58)$$

where $\mathbf{x}(t)$ and $\mathbf{y}(t)$ are independent, stationary, Gaussian random processes with zero means and equal variances σ^2.

Let us assume the incoming signal to the receiver system is a simple sinusoid $s(t) = A \cos \omega_0 t$, where the amplitude A is constant. In this case, the total output of the linear filter is the sum of signal and noise, which takes the form

$$\mathbf{u}(t) = A \cos \omega_0 t + \mathbf{n}(t) = \mathbf{r}(t)\cos[\omega_0 t + \boldsymbol{\theta}(t)], \quad (59)$$

where $\mathbf{r}(t)$ and $\theta(t)$ are the random *envelope* and random *phase*, respectively, of the output waveform given by

$$\mathbf{r}(t) = \sqrt{[A + \mathbf{x}(t)]^2 + \mathbf{y}^2(t)},$$

$$\theta(t) = \tan^{-1} \frac{\mathbf{y}(t)}{A + \mathbf{x}(t)}.$$

Statistical properties of the envelope $\mathbf{r}(t)$ are described by the *Rician distribution* (e.g., see Example 7 in Section 2.5.2)

$$f_\mathbf{r}(r) = \frac{r}{\sigma^2} e^{-(r^2 + A^2)/2\sigma^2} I_0\left(\frac{Ar}{\sigma^2}\right) U(r), \tag{60}$$

where $U(r)$ is the unit step function (Section 2.2.1). In the absence of the signal ($A = 0$), Eq. (60) reduces to the *Rayleigh distribution*

$$f_\mathbf{r}(r) = \frac{r}{\sigma^2} e^{-r^2/2\sigma^2} U(r). \tag{61}$$

If the output (59) of a passband filter is now fed into a memoryless nonlinear device whose output $\mathbf{w}(t) = g[\mathbf{u}(t)]$ is a function of the input $\mathbf{u}(t)$ at the same instant of time, say, at $t = t_1$, the output of the nonlinear device in this case is $g(r \cos \varphi)$, where $\varphi = \omega_0 t_1 + \theta$. We recognize this as an *even, periodic* function of φ which has the *Fourier cosine series* representation (see Section 8.2)

$$g(r \cos \varphi) = \frac{1}{2} g_0(r) + \sum_{n=1}^{\infty} g_n(r) \cos n\varphi, \tag{62}$$

where the Fourier coefficients are defined by

$$g_n(r) = \frac{2}{\pi} \int_0^\pi g(r \cos \varphi) \cos n\varphi \, d\varphi, \quad n = 0, 1, 2, \dots. \tag{63}$$

Each term of the series (62) can be thought of as the output of a zonal filter centered at the nth harmonic of ω_0. For example, the first term of the series $(1/2)g_0(r)$, which represents the low-pass component of the output called the *audio* or *video output*, is the only term ordinarily used in *detection* devices like a square-law detector. The next term of the series $g_1(r) \cos(\omega_0 t_1 + \theta)$, which has the same frequency as the input waveform, is the only significant term in the case of a *nonlinear amplifier* or *limiter*. Note that the term $g_1(r) \cos(\omega_0 t_1 + \theta)$ reproduces the input phase exactly but distorts the input amplitude.

The so-called *vth-law device* ($v \geq 0$), which falls into the categories of *half-wave, full-wave even*, and *full-wave odd*, is a class of nonlinear devices commonly used in communication systems. The output of such a device can be expressed as a function $\mathbf{w}(t)$ in terms of envelope and phase by (see Fig. 15.7)

$$\mathbf{w}(t) = \mathbf{u}^v(t) = \mathbf{r}^v(t)\left\{\frac{1}{2}c_0(v) + \sum_{n=1}^{\infty} c_n(v)\cos\left[n\omega_0 t + n\theta(t)\right]\right\}, \tag{64}$$

where the specific Fourier coefficients $c_n(v)$ for each type of device are given below.

Figure 15.7 Block diagram of vth-law device.

The specific vth-law devices mentioned above have the following transfer characteristics and Fourier coefficients (see [10]):

1. *Half-wave device*: $\quad \mathbf{w}(t) = \mathbf{u}^v(t)\,U[\mathbf{u}(t)]$

$$c_n(v) = \frac{\Gamma(1+v)}{2^v\,\Gamma\!\left(1+\dfrac{v+n}{2}\right)\Gamma\!\left(1+\dfrac{v-n}{2}\right)}, \quad n = 0,1,2,\ldots$$

2. *Full-wave even device*: $\quad \mathbf{w}(t) = |\mathbf{u}(t)|^v$

$$c_n(v) = \frac{\Gamma(1+v)}{2^{v-1}\,\Gamma\!\left(1+\dfrac{v+n}{2}\right)\Gamma\!\left(1+\dfrac{v-n}{2}\right)}, \quad n = 0,2,4,\ldots$$

$$c_n(v) = 0, \quad n = 1,3,5,\ldots$$

3. *Full-wave odd device*: $\quad \mathbf{w}(t) = \begin{cases} \mathbf{u}^v(t), & \mathbf{u}(t) \geq 0 \\ -|\mathbf{u}(t)|^v, & \mathbf{u}(t) < 0 \end{cases}$

$$c_n(v) = 0, \quad n = 0,2,4,\ldots$$

$$c_n(v) = \frac{\Gamma(1+v)}{2^{v-1}\,\Gamma\!\left(1+\dfrac{v+n}{2}\right)\Gamma\!\left(1+\dfrac{v-n}{2}\right)}, \quad n = 1,3,5,\ldots$$

As an illustration, let us consider the case of a *bandpass limiter*, which is a type of full-wave odd device $(v < 1)$ used in frequency modulation (FM) receivers to remove unwanted amplitude variations. We choose the zonal filter in Fig. 15.7 to correspond with the frequency of the input waveform (i.e., $n = 1$). The output $\mathbf{z}(t)$ of the zonal filter then takes the form

$$\mathbf{z}(t) = c_1(v)\mathbf{r}^v(t)\cos\left[\omega_0 t + \theta(t)\right]. \tag{65}$$

A measure of system performance is provided by the time and ensemble averaged *signal-to-noise ratio* (SNR) at the output of the zonal filter $\mathbf{z}(t)$, which we define by

$$\overline{\mathrm{SNR}_0} = \frac{\overline{S_0}}{\overline{N_0}} = \frac{(m_1 - m_1|_{A=0})^2}{m_2 - m_1^2},$$ (66)

where $\overline{S_0}$ and $\overline{N_0}$ are time-averaged signal and noise powers, respectively, and m_1 and m_2 are the first and second ensemble moments of the output $\mathbf{z}(t)$ defined by

$$m_k = <\mathbf{z}^k(t)> = c_1^k(\nu) \int_0^\infty \int_0^{2\pi} r^{k\nu} \cos^k(\omega_0 t + \theta) f_{\mathbf{r}\theta}(r,\theta) d\theta\, dr, \quad k = 1, 2.$$ (67)

Here, $f_{\mathbf{r}\theta}(r,\theta)$ is the joint PDF of the envelope and phase given by (see Section 13.8.2)

$$f_{\mathbf{r}\theta}(r,\theta) = \frac{r}{2\pi\sigma^2} \exp\left[-\frac{1}{2\sigma^2}(r^2 + A^2 - 2Ar\cos\theta)\right].$$ (68)

The first moment corresponding to $k = 1$ in (67) now leads to

$$m_1 = \frac{c_1(\nu)}{2\pi\sigma^2} \int_0^\infty r^{\nu+1} e^{-(r^2+A^2)/2\sigma^2} \int_0^{2\pi} \cos(\omega_0 t + \theta) \exp\left(\frac{Ar}{\sigma^2}\cos\theta\right) d\theta\, dr.$$ (69)

At this point we will further restrict the analysis to the special case $\nu = 0$, which is called an *ideal limiter*. Essentially, an ideal bandpass limiter reduces the random output $\mathbf{w}(t)$ of the νth-law device to that of a square wave with random phase. To evaluate the integrals in (69) for this case, we use $c_1(0) = 4/\pi$ and the integral relation

$$\frac{1}{2\pi} \int_0^{2\pi} \cos(n\omega_0 t + n\theta) \exp(a\cos\theta) d\theta = I_n(a)\cos n\omega_0 t,$$ (70)

to deduce

$$m_1 = \frac{4}{\pi\sigma^2} e^{-A^2/2\sigma^2} \int_0^\infty r e^{-r^2/2\sigma^2} I_1\left(\frac{Ar}{\sigma^2}\right) dr,$$ (71)

where $I_n(x)$ is a modified Bessel function. Next, by employing the integral formula

$$\int_0^\infty x^\mu e^{-a^2 x^2} I_n(bx) dx = \frac{b^n}{2^{n+1} a^{n+\mu+1}} \Gamma\left(\frac{n+\mu+1}{2}\right)$$
$$\times {}_1F_1\left(\frac{n+\mu+1}{2}; 1+n; \frac{b^2}{4a^2}\right),$$ (72)

along with the identities (CH3) and (GH12) in Section 2.6, we arrive at

$$m_1 = 2\sqrt{\frac{S}{\pi}} \, _1F_1\left(\frac{1}{2}; 2; -S\right)\cos\omega_0 t = 2\sqrt{\frac{S}{\pi}} e^{-S/2}[I_0(S/2) + I_1(S/2)]\cos\omega_0 t, \quad (73)$$

where $S = A^2/2\sigma^2$ is the input SNR at the output of the bandpass filter, and $_1F_1(a;c;x)$ is a confluent hypergeometric function. The time-averaged signal power is therefore

$$\begin{aligned}\overline{S_0} &= \overline{(m_1 - m_1|_{S=0})^2} = \frac{4S}{\pi} e^{-S}[I_0(S/2) + I_1(S/2)]^2 \overline{\cos^2\omega_0 t} \\ &= \frac{2S}{\pi} e^{-S}[I_0(S/2) + I_1(S/2)]^2.\end{aligned} \quad (74)$$

In a similar manner, the second moment deduced from (67) with $k = 2$ and $v = 0$ leads to

$$\begin{aligned}m_2 &= \frac{c_1^2(0)}{2\pi\sigma^2}\int_0^\infty r e^{-(r^2+A^2)/2\sigma^2}\int_0^{2\pi}\cos^2(\omega_0 t + \theta)\exp\left(\frac{Ar}{\sigma^2}\cos\theta\right)d\theta\,dr \\ &= \frac{8}{\pi^2}\left[1 + \frac{1}{S}(S - 1 + e^{-S})\cos 2\omega_0 t\right],\end{aligned} \quad (75)$$

where we have used (GH14) in Section 2.6. The time-averaged noise power is

$$\overline{N_0} = \overline{m_2 - m_1^2} = \frac{8}{\pi^2}\left[1 + \frac{1}{S}(S - 1 + e^{-S})\overline{\cos 2\omega_0 t}\right] = \frac{8}{\pi^2}, \quad (76)$$

and, consequently, the output SNR (66) becomes

$$\overline{SNR_0} = \frac{Se^{-S}[I_0(S/2) + I_1(S/2)]^2}{\frac{4}{\pi} - Se^{-S}[I_0(S/2) + I_1(S/2)]^2} \sim \begin{cases}\frac{\pi}{4}S, & S \ll 1 \\ 2S, & S \gg 1.\end{cases} \quad (77)$$

15.4.3 Coherent detection optical receiver

Optical receivers are broadly divided into two basic types—*direct* (or *power*) *detector* and *coherent detector*. Coherent detection is also commonly called *heterodyne detection*. It involves the addition of a single-frequency strong local oscillator (LO) signal with the incoming signal before photo-detection (see Fig. 15.8). The objective of this mixing process is to convert a weak incoming signal to an intermediate frequency (IF) in the microwave regime for improved detection and processing (see [14] and [19]).

We assume the received signal and the local oscillator are both infinite plane waves described by

$$E_S(t) = A_S e^{j(\omega_S t + \varphi_S)}, \qquad E_{LO}(t) = A_{LO}e^{j(\omega_{LO}t + \varphi_{LO})}, \quad (78)$$

where A_S and A_{LO} are the amplitudes, ω_S and ω_{LO} are angular frequencies, φ_S and φ_{LO} are

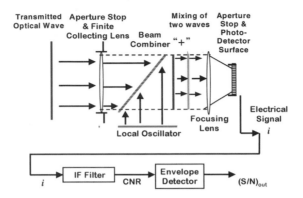

Figure 15.8 Coherent detection system.

the phases, and we use $j^2 = -1$ in this section. The intensity of the combined fields incident on the photo-detector is

$$I = |E_S(t) + E_{LO}(t)|^2$$
$$= A_S^2 + A_{LO}^2 + 2A_S A_{LO} \cos(\omega_{IF} t), \tag{79}$$

where we set $\omega_{IF} = \omega_S - \omega_{LO}$ and assumed zero phase difference between the signal and LO (i.e., $\varphi_{LO} \cong \varphi_S$). The signal current coming from the photo-detector into the filter has a form similar to (79) and, after filtering at the IF frequency, leads to

$$\mathbf{i}(t) = i_{IF} \cos(\omega_{IF} t) + \mathbf{n}(t), \tag{80}$$

where i_{IF} is proportional to the product $A_S A_{LO}$, and $\mathbf{n}(t)$ is taken to be a random *narrowband Gaussian noise process* with zero mean. In coherent detection the LO shot noise usually dominates all other noise sources. Because the signal information A_S that we desire is at the IF rather than at baseband, it is necessary to "detect" the signal a second time. This is often accomplished with the use of an envelope detector, which can be in the form of a linear or quadratic rectifier followed by a low-pass filter.

System performance in this case can be based on the *carrier-to-noise ratio* (CNR) at the output of the IF filter, which is CNR $= i_{IF}^2 / \sigma_\mathbf{n}^2$, or based on the SNR at the output of the low-pass filter. To determine this last quantity, we rewrite the current (80) as (see Section 15.4.2)

$$\mathbf{i}(t) = \mathbf{r}(t) \cos[\omega_{IF} t + \mathbf{\theta}(t)], \tag{81}$$

where $\mathbf{r}(t)$ is the random envelope of the total IF current and $\mathbf{\theta}(t)$ is the random phase. Because we have assumed the additive noise in (80) is Gaussian distributed, it follows that the statistics of the *envelope* $\mathbf{r}(t)$ in the absence of signal are governed by a *Rayleigh distribution*

$$p_{\mathbf{n}}(r) = \frac{r}{\sigma_n^2} \exp\left(-\frac{r^2}{2\sigma_n^2} \right) U(r), \tag{82}$$

where $U(r)$ is the unit step function and the noise power is defined by the ensemble average $\sigma_n^2 = \langle \mathbf{n}^2 \rangle$. Similarly, the envelope of the sum of signal (s) plus noise is governed by the *Rician distribution* (see Sections 13.8 and 15.4.2)

$$p_{s+\mathbf{n}}(r) = \frac{r}{\sigma_n^2} \exp\left(-\frac{i_{IF}^2 + r^2}{2\sigma_n^2} \right) I_0\left(\frac{r i_{IF}}{\sigma_n^2} \right) U(r), \tag{83}$$

where $I_0(x)$ is a modified Bessel function.

The concept of SNR at the output of the low-pass filter is somewhat ambiguous because of the mixing of signal and noise terms. In particular, the mean noise power will have components of the type $n \times n$ (those due to the noise beating with itself) and of the type $s \times n$ (those due to the noise beating with the signal). Fortunately, most definitions of SNR lead to similar results. For example, a common definition of SNR for an unmodulated signal at the output of a quadratic envelope detector is

$$\mathrm{SNR}_0 = \frac{[\langle r^2 \rangle - \langle r^2 | s = 0 \rangle]^2}{\langle r^4 \rangle - \langle r^2 \rangle^2}, \tag{84}$$

where s is the signal, the numerator represents the square of the mean output power minus the mean value when the signal is absent, and the denominator is simply the variance of the output power. By using the result of Example 4 below for the nth moment of the envelope $\mathbf{r}(t)$, we are led to the well-known expression (see [10] or [22])

$$\mathrm{SNR}_0 = \frac{\mathrm{CNR}^2}{1 + 2\mathrm{CNR}} \sim \begin{cases} \mathrm{CNR}^2, & \mathrm{CNR} \ll 1 \\ \mathrm{CNR}/2, & \mathrm{CNR} \gg 1. \end{cases} \tag{85}$$

In the CNR $\ll 1$ case in (85), the $n \times n$ terms tend to dominate, leading to the quadratic output SNR_0 in terms of input CNR. This case demonstrates the so-called *suppression characteristic* of envelope detection. For CNR $\gg 1$, however, it is the $s \times n$ terms that dominate, leading to the linear relationship for output SNR_0.

EXAMPLE 4 For the Rician density function defined by

$$p_{\mathbf{r}}(r) = \frac{r}{\sigma^2} e^{-(r^2 + A^2)/2\sigma^2} I_0\left(\frac{Ar}{\sigma^2} \right) U(r),$$

where A and σ^2 are constants, calculate the nth statistical moment $\langle r^n \rangle$, $n = 1, 2, 3, \dots$.

Solution: From definition,

$$\langle \mathbf{r}^n \rangle = \int_0^\infty r^n p_\mathbf{r}(r)\, dr = \frac{1}{\sigma^2} e^{-A^2/2\sigma^2} \int_0^\infty r^{n+1} e^{-r^2/2\sigma^2} I_0\!\left(\frac{Ar}{\sigma^2}\right) dr.$$

By representing the Bessel function by its power series representation [Eq. (54) in Chapter 2], we are led to

$$\langle \mathbf{r}^n \rangle = \frac{1}{\sigma^2} e^{-A^2/2\sigma^2} \sum_{k=0}^\infty \frac{(A/2\sigma^2)^{2k}}{(k!)^2} \int_0^\infty r^{2k+n+1} e^{-r^2/2\sigma^2}\, dr$$

$$= (2\sigma^2)^{n/2} e^{-A^2/2\sigma^2} \sum_{k=0}^\infty \frac{\Gamma(k + 1 + n/2)(A^2/2\sigma^2)^k}{k!\,k!}$$

$$= (2\sigma^2)^{n/2} e^{-A^2/2\sigma^2} \Gamma\!\left(1 + \frac{n}{2}\right) \sum_{k=0}^\infty \frac{(1 + n/2)_k (A^2/2\sigma^2)^k}{(1)_k\, k!},$$

where we have used properties of the gamma function and Pochhammer symbol (Chapter 2) in arriving at our result. Summing the remaining series, we obtain

$$\langle \mathbf{r}^n \rangle = (2\sigma^2)^{n/2} \Gamma\!\left(1 + \frac{n}{2}\right) {}_1F_1\!\left(1 + \frac{n}{2}; 1; \frac{A^2}{2\sigma^2}\right) e^{-A^2/2\sigma^2}.$$

However, by using Kummer's transformation [(CH3) in Section 2.6.3] in this last expression, we can write the moments in the form

$$\langle \mathbf{r}^n \rangle = (2\sigma^2)^{n/2} \Gamma\!\left(1 + \frac{n}{2}\right) {}_1F_1\!\left(-\frac{n}{2}; 1; -\frac{A^2}{2\sigma^2}\right), \quad n = 1,2,3,\dots.$$

15.4.4 Threshold detection

The need to determine the presence of a "signal" embedded in "noise" is fundamental in communication systems. There are several techniques for detecting the signal, which ordinarily rely on a *threshold* device of some kind. Only when the output of the detector exceeds the set threshold value do we say a *signal is present*. *False alarms* arise when noise alone exceeds the threshold value and is consequently interpreted as signal. On the other hand, if the signal plus noise does not exceed the threshold, it is called *missed detection*. Threshold detection concepts are illustrated in Fig. 15.9.

Let us assume a detector is followed by a filter of bandwidth B, where the bandwidth is chosen to match the waveform or frequency spread of the incoming signal-power envelope. The output current from the filter has the form

$$\mathbf{i}(t) = s(t) + \mathbf{n}(t), \tag{86}$$

where $s(t)$ is the signal current and $\mathbf{n}(t)$ is the random noise current. The total noise power in the detector current is defined by

$$\sigma_{\mathbf{n}}^2 = \langle \mathbf{i}^2(t) \rangle - \langle \mathbf{i}(t) \rangle^2 = \langle \mathbf{n}^2(t) \rangle. \tag{87}$$

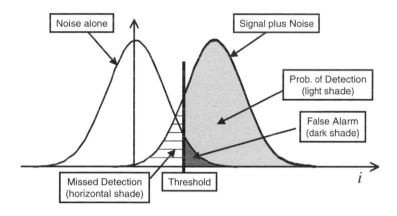

Figure 15.9 Probability of detection and false alarm.

False alarm rates and threshold detection (or missed detection) involve the SNR. We define the output SNR by the ratio of the power of the detector signal current s^2 to the power of the noise current (87), which yields

$$\text{SNR} = \frac{s^2}{\sigma_{\mathbf{n}}^2}. \tag{88}$$

The reliability of a communication system can be determined in terms of the probability density function of the output signal from the detector. In terms of the output PDF, we define the *probability of detection* in terms of the cumulative distribution above a prescribed threshold value. The *probability of miss* or *fade* below threshold is therefore defined by $P_m = 1 - P_d$, where P_d is the probability of detection (see Fig. 15.9). The fade probability provides us with an estimate of how likely the output current $\mathbf{i}(t)$ from the detector is to drop below a prescribed threshold i_T. Related quantities of interest are the *frequency of surges* and *frequency of fades* of the output current, which are identical with the frequency of positive and negative crossings of the threshold level. The *expected number of crossings per second* is defined by Eq. (108) in Section 14.7.3 (see also [4] and [26]).

A typical source of noise in the receiver system is that called *detector noise* or *shot noise*. The PDF model for such noise is commonly assumed to be a zero-mean Gaussian process defined by

$$f_{\mathbf{n}}(i) = \frac{1}{\sqrt{2\pi}\,\sigma_{\mathbf{n}}} \exp\left(-\frac{i^2}{2\sigma_{\mathbf{n}}^2} \right). \tag{89}$$

The total output current from the filter containing both signal and noise [see (86)] therefore has mean value s. Hence, the total current $\mathbf{i}(t)$ is governed by the non-zero mean Gaussian PDF

$$f_{s+\mathbf{n}}(i) = \frac{1}{\sqrt{2\pi}\,\sigma_\mathbf{n}} \exp\left[-\frac{(i-s)^2}{2\sigma_\mathbf{n}^2}\right],$$

(90)

and the probabilities of detection and false alarm deduced from these PDF models are given explicitly by

$$P_d = \int_{i_T}^{\infty} f_{s+\mathbf{n}}(i)\,di = \frac{1}{2}\,\mathrm{erfc}\left(\frac{i_T - s}{\sqrt{2}\sigma_\mathbf{n}}\right),$$

(91)

$$P_{fa} = \int_{i_T}^{\infty} f_\mathbf{n}(i)\,di = \frac{1}{2}\,\mathrm{erfc}\left(\frac{i_T}{\sqrt{2}\sigma_\mathbf{n}}\right),$$

(92)

where $\mathrm{erfc}(x)$ is the *complementary error function* (see Section 2.3.4).

Last, the performance measure in digital communications is provided not directly by the SNR, but rather by the *probability of error*, also called the *bit error-rate* (BER). The most basic form of pulsed modulation in digital communications is on-off keying (OOK). Each bit symbol is transmitted by pulsing the source either on or off during each bit time. Because of detector noise, a transmitted "0" may be mistaken for a "1" and vice-versa. Assuming each symbol is equally likely to be sent, the BER is given by

$$\Pr(E) = \frac{1}{2}\Pr(1|0) + \frac{1}{2}\Pr(0|1) = \frac{1}{2}\,\mathrm{erfc}\left(\frac{\mathrm{SNR}}{2\sqrt{2}}\right),$$

(93)

where we set $i_T = s/2$ and SNR is defined by (88). Here, $\Pr(1|0)$ is the probability of false alarm (92) and $\Pr(0|1)$ is missed detection defined by $1 - P_d$.

15.5 Applications in Geometrical Optics

That branch of optics which is characterized by the neglect of diffraction effects (i.e., corresponding to the limiting case in which wavelength $\lambda \to 0$) is called *geometrical optics*. Under the approximation of geometrical optics, the optical laws can be formulated through geometric considerations alone. In particular, the energy may then be regarded as being transported along the ray paths (see [32]).

As pointed out in Section 11.3.3, the path for a ray of light in a medium is a curve for which the integral

$$I = \int_{\mathbf{R}_1}^{\mathbf{R}_2} n(\mathbf{R})\,ds$$

(94)

is an extremum. Here, $n(\mathbf{R})$ is the index of refraction, $\mathbf{R} = (x,y,z)$ is a position vector in an inhomogeneous but isotropic medium, and s denotes arc length. The points $\mathbf{R}_1$ and $\mathbf{R}_2$ are two points connected by the ray path.

Although formulated in terms of arc length s, we have no *a priori* knowledge of arc length so we introduce a parameterization for all curves connecting the points $\mathbf{R}_1$ and $\mathbf{R}_2$ in terms of an arbitrary parameter t. Using Eq. (55) in Chapter 4, we can then write (94) in the form

$$I = \int_{t_1}^{t_2} n(\mathbf{R}) \| \mathbf{R}'(t) \| \, dt,$$

where t_1 and t_2 are defined such that $\mathbf{R}_1 = \mathbf{R}(t_1)$ and $\mathbf{R}_2 = \mathbf{R}(t_2)$. In vector form, it can be shown that Lagrange's equations of motion [see Eq. (91) in Chapter 11] take the form

$$\| \mathbf{R}'(t) \| \nabla n(\mathbf{R}) - \frac{d}{dt} \left[n(\mathbf{R}) \frac{\mathbf{R}'(t)}{\| \mathbf{R}'(t) \|} \right] = 0. \tag{95}$$

However, through the use of Eqs. (55) and (56) in Chapter 4, we can express this result in the more convenient form

$$\frac{d}{dt} \left(n \frac{d\mathbf{R}}{ds} \right) = \frac{ds}{dt} \nabla n,$$

where s again represents arc length. Through use of the chain rule, this result further reduces to

$$\frac{d}{ds} \left(n \frac{d\mathbf{R}}{ds} \right) = \nabla n. \tag{96}$$

Equation (96) is commonly called the *ray equation*. In some cases, the identity $(d/ds)n(\mathbf{R}) = \nabla n \cdot \mathbf{R}'(s)$ is used to write (96) in the expanded form

$$n\mathbf{R}'' + (\nabla n \cdot \mathbf{R}')\mathbf{R}' = \nabla n. \tag{97}$$

In a *homogeneous medium* we have $n = $ constant, so that (96) reduces to

$$\frac{d\mathbf{R}}{ds} = 0, \tag{98}$$

which has the solution

$$\mathbf{R}(s) = \mathbf{a}s + \mathbf{b}. \tag{99}$$

Here, $\mathbf{a}$ and $\mathbf{b}$ are constant vectors and (99) simply represents the equation of a *line*. That is, in a homogeneous medium the light rays travel in straight lines.

We should point out that our derivation of the ray equation (96) is based on the assumption that the refractive index n is a *continuously differentiable function* of position. Of course, this precludes consideration of the important practical problem of determining the ray path through a lens, which is usually considered a medium with a *discontinuous* refractive index. Because this latter case requires additional analytic tools, it will not be considered here.

15.5.1 Eikonal equation

Recall from Theorem 4.3 in Section 4.6.2 that saying $\mathbf{F}$ is a *conservative field* implies that the quantity $\mathbf{F} \cdot d\mathbf{R}$ is an *exact differential*, i.e., there is a scalar potential function φ such that $d\varphi = \mathbf{F} \cdot d\mathbf{R}$. We now consider the analogous situation in optics. Namely, if the vector $\mathbf{R}(s)$ represents an aggregate of rays that satisfy the ray equation (96), then under what circumstances does there exist a scalar potential function φ such that

$$d\varphi = n\mathbf{R}' \cdot d\mathbf{W}, \tag{100}$$

where $\mathbf{W}$ represents a *wavefront*? The answer is that

$$\nabla \times (n\mathbf{R}') = \mathbf{0}, \tag{101}$$

although we will not prove it. Given that (101) is true, then, in particular, because the family of rays is orthogonal to the wavefront, it follows that $n\mathbf{R}' \cdot d\mathbf{W} = 0$, and the result (100) leads to

$$\varphi = \text{const.} \tag{102}$$

That is, the wavefront can be described by the family of level surfaces (102).

In addition, the condition (101) implies that the vector $n\mathbf{R}'$ is directly related to the scalar potential by

$$\nabla \varphi = n\mathbf{R}' = n\hat{\mathbf{T}}. \tag{103}$$

By squaring this last result, we arrive at

$$(\nabla \varphi)^2 = n^2, \tag{104}$$

which relates the orthogonal surface (wavefront) directly to the index of refraction. Equation (104), which is independent of the family of rays, is commonly referred to as the *eikonal equation*. By forming the scalar product of each side of (103) with the unit vector $\mathbf{R}' = \hat{\mathbf{T}}$, we obtain $\nabla \varphi \cdot \hat{\mathbf{T}} = n$, which is simply a directional derivative; i.e.,

$$\frac{d\varphi}{ds} = n. \tag{105}$$

This last relation provides a direct equation for calculating the *eikonal* or wavefront φ (also called the phase when multiplied by the electromagnetic wave number $k = \lambda/2\pi$) through integration of the index of refraction n. Equations (104) and (105) are considered central to the study of geometrical optics (e.g., see [36]).

15.5.2 Frenet-Serret formulas revisited

Note that the Euler vector equation (97) is in general a pair of second-order ordinary differential equations. Under appropriate conditions, there exists a general solution that depends on four arbitrary constants. In optics terminology, this means that any medium with a continuous refractive index has associated with it a four-parameter family of ray paths constituting the general solution—this requires the specification of two coordinates on a plane and two direction cosines. A ray connecting any two points in the medium must be a member of that family.

In Section 4.4.2 we developed the *Frenet-Serret formulas*, which define a triad of unit vectors at each point of a space curve, viz.,

$$\begin{aligned}
\hat{\mathbf{T}}'(s) &= \kappa\hat{\mathbf{N}}(s), \\
\hat{\mathbf{N}}'(s) &= -\kappa\hat{\mathbf{T}}(s) + \tau\hat{\mathbf{B}}(s), \\
\hat{\mathbf{B}}'(s) &= -\tau\hat{\mathbf{N}}(s),
\end{aligned} \tag{106}$$

where τ is the torsion and $\kappa = 1/\rho$ is the curvature. The quantity ρ is the radius of curvature. Thus, we see that the derivative of each unit vector with respect to arc length can be written as a linear combination of the three unit vectors. From the first equation in (106), which we can also express as $\mathbf{R}''(s) = \kappa\hat{\mathbf{N}}(s)$, we obtain

$$\mathbf{R}''(s) = \frac{1}{\rho}\hat{\mathbf{N}}, \tag{107}$$

the derivative of which leads to

$$\mathbf{R}'''(s) = -\frac{1}{\rho^2}\hat{\mathbf{T}} - \frac{\rho'}{\rho}\hat{\mathbf{N}} + \frac{\tau}{\rho}\hat{\mathbf{B}}. \tag{108}$$

Based on the Frenet-Serret formulas, we can rewrite Eq. (97) for a ray path in the form

$$\nabla n = (\nabla n \cdot \hat{\mathbf{T}})\hat{\mathbf{T}} + \frac{n}{\rho}\hat{\mathbf{N}}. \tag{109}$$

This expression shows that the gradient of the refractive index at any point along a ray is a linear combination of the unit tangent and normal vectors, similar to the acceleration vector associated with an object moving through space (recall Section 4.4.3). Consequently, the binormal vector must be perpendicular to the gradient of the refractive index; i.e., $\nabla n \cdot \hat{\mathbf{B}} = 0$.

15.5.3 The heated window

Let us consider a window of thickness d between two plane parallel surfaces, as shown in Fig. 15.10, which separates two media of vastly differing temperatures. In this case a thermal gradient exists across the window. The material of the window is an inhomogeneous optical medium for which the index of refraction at a point depends on the temperature of the medium at that point, which in turn depends on the position of that point to the heating and cooling surfaces.

Let us assume that the temperature is a *linear function* of position and, although not quite correct, we will assume the refractive index is a *linear function* of temperature. Hence, we write

$$n(\mathbf{R}) = 1 + \mathbf{A} \cdot \mathbf{R}, \qquad (110)$$

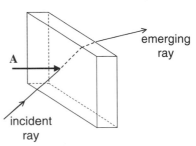

where $\mathbf{R}$ is the position vector and $\mathbf{A}$ is a vector in the direction of the thermal gradient, i.e., normal to the two parallel surfaces. We see, therefore, that the vector $\mathbf{A}$ is simply the gradient of the refractive index

$$\mathbf{A} = \nabla n. \qquad (111)$$

Figure 15.10 The heated window.

If we now substitute the above results (110) and (111) into the ray equation (96), we obtain

$$\frac{d}{ds}\left[(1 + \nabla n \cdot \mathbf{R})\mathbf{R}'\right] = \nabla n, \qquad (112)$$

the integral of which yields

$$(1 + \nabla n \cdot \mathbf{R})\mathbf{R}' = (\nabla n)s + \mathbf{C}, \qquad (113)$$

where $\mathbf{C}$ is a constant of integration. By "squaring" each side of (113) and noting that $\mathbf{R}' \cdot \mathbf{R}' = \hat{\mathbf{T}} \cdot \hat{\mathbf{T}} = 1$, we deduce that

$$n = \|(\nabla n)s + \mathbf{C}\|. \qquad (114)$$

Similarly, by taking the dot product of each side of (113) with $\mathbf{R}'$, we get

$$n = 1 + \nabla n \cdot \mathbf{R} = [(\nabla n)s + \mathbf{C}] \cdot \mathbf{R}'. \qquad (115)$$

We can determine the arbitrary vector $\mathbf{C}$ by imposing some initial conditions. Namely, let $\mathbf{R}(0) = \mathbf{R}_0$ denote the initial point with initial direction described by the unit tangent vector $\hat{\mathbf{T}}_0$ and let $n(0) = n_0$ denote the initial index of refraction. From (114) we obtain

$$n_0 = \|\mathbf{C}\| = \mathbf{C} \cdot \hat{\mathbf{T}}_0, \qquad (116)$$

and conclude that $\mathbf{C} = n_0 \hat{\mathbf{T}}_0$. Consequently, Eq. (115) can be rearranged as

$$\mathbf{R}' = \hat{\mathbf{T}} = \frac{(\nabla n)s + n_0 \hat{\mathbf{T}}_0}{\|(\nabla n)s + n_0 \hat{\mathbf{T}}_0\|}. \qquad (117)$$

The principal unit normal and unit binormal vectors can now be shown to equal

$$\hat{\mathbf{N}} = \frac{[(\nabla n)s + n_0\hat{\mathbf{T}}_0] \times [(\nabla n) \times \hat{\mathbf{T}}_0]}{\| (\nabla n)s + n_0\hat{\mathbf{T}}_0 \| \, \| (\nabla n) \times \hat{\mathbf{T}}_0 \|}, \tag{118}$$

$$\hat{\mathbf{B}} = -\frac{(\nabla n) \times \hat{\mathbf{T}}_0}{\| (\nabla n) \times \hat{\mathbf{T}}_0 \|}. \tag{119}$$

Note that the unit binormal (119) is constant, and thus the torsion is zero. This means that the ray path is a plane curve in agreement with (109). If we take into account the initial values of the three vectors (117)-(119), it follows that we can write these last relations more succinctly as

$$\hat{\mathbf{T}} = (n_0/n)\hat{\mathbf{T}}_0 + (s/n)\nabla n,$$
$$\hat{\mathbf{N}} = (n_0/n)\hat{\mathbf{N}}_0 - (s/n)\nabla n \times \hat{\mathbf{B}}_0, \tag{120}$$
$$\hat{\mathbf{B}} = \hat{\mathbf{B}}_0.$$

We can now solve the differential equation (117) for the position vector of the ray path by integrating each side to obtain

$$\mathbf{R}(s) = \frac{\| (\nabla n)s + n_0\hat{\mathbf{T}}_0 \| \nabla n}{\| \nabla n \|^2} + \frac{n_0}{\| \nabla n \|^3} \nabla n \times (\hat{\mathbf{T}}_0 \times \nabla n) \ln \left\{ \nabla n \cdot [(\nabla n)s + n_0\hat{\mathbf{T}}_0] \right.$$
$$\left. + \| \nabla n \| \, \| (\nabla n)s + n_0\hat{\mathbf{T}}_0 \| \right\} + \mathbf{K}, \tag{121}$$

where $\mathbf{K}$ is an arbitrary constant vector of integration. We can simplify this result by using the above relations, which yields

$$\mathbf{R}(s) = \frac{n\nabla n}{\| \nabla n \|^2} + \frac{n_0^2(\nabla n \times \hat{\mathbf{B}}_0)}{\rho_0 \| \nabla n \|^3} \ln \left\{ n_0(\nabla n \cdot \mathbf{T}_0) + n \| \nabla n \| + s \| \nabla n \|^2 \right\} + \mathbf{K}. \tag{122}$$

Finally, by setting $s = 0$ in this last equation, we deduce that

$$\mathbf{K} = \mathbf{R}_0 - \frac{n_0\nabla n}{\| \nabla n \|^2} + \frac{n_0^2(\nabla n \times \hat{\mathbf{B}}_0)}{\rho_0 \| \nabla n \|^3} \ln \left\{ n_0(\nabla n \cdot \mathbf{T}_0) + n_0 \| \nabla n \| \right\}, \tag{123}$$

and thus, the solution becomes

$$\mathbf{R}(s) = \mathbf{R}_0 + \frac{(n - n_0)\nabla n}{\| \nabla n \|^2} + \frac{n_0^2(\nabla n \times \hat{\mathbf{B}}_0)}{\rho_0 \| \nabla n \|^3} \ln \left\{ \frac{n_0(\nabla n \cdot \mathbf{T}_0) + n \| \nabla n \| + s \| \nabla n \|^2}{n_0(\nabla n \cdot \mathbf{T}_0) + n_0 \| \nabla n \|} \right\}. \tag{124}$$

Some additional simplification can take place in terms of the window thickness d and the incident angle θ_i of the light ray, but we will not pursue it (see [32]).

15.6 Wave Propagation in Free Space

Early theories of free-space optical wave propagation were based on geometric or ray optics (e.g., see Section 15.5). To include diffraction effects (deviations of light from ray paths), it is necessary to solve the wave equation (Section 12.4) for the optical field $U(x,y,z)$ which, for a monochromatic wave, reduces to the *Helmholtz equation*

$$\nabla^2 U + k^2 U = 0. \tag{125}$$

Under the *paraxial approximation*, which assumes the transverse spreading of the wave in the *xy*-plane is much less than the propagation path length along the positive *z*-axis, Eq. (125) reduces further to

$$\frac{\partial^2 V}{\partial x^2} + \frac{\partial^2 V}{\partial y^2} + 2ik\frac{\partial V}{\partial z} = 0, \tag{126}$$

where we have written $U(x,y,z) = V(x,y,z)e^{ikz}$. Here, $k = 2\pi/\lambda$ is the optical wave number, λ is wavelength, and $i^2 = -1$. Various methods for solving (126) are available, some of which we discuss below.

15.6.1 Hankel transform method

A useful model for the optical wave generated by a laser is known as a lowest-order *Gaussian-beam wave*, also called a transverse electromagnetic (or TEM$_{00}$) wave. Let us assume that $U(\mathbf{r},0)$ is a lowest-order Gaussian-beam wave defined by (see [2])

$$U(\mathbf{r},0) = a_0 \exp\left(-\frac{1}{2}\alpha k r^2\right), \tag{127}$$

where a_0 is the amplitude, $\mathbf{r} = <x,y>$, and α is a complex parameter describing the beam spot radius W_0 and phase front radius of curvature F_0 according to

$$\alpha = \frac{2}{k W_0^2} + i\frac{1}{F_0}. \tag{128}$$

Because of circular symmetry of the Gaussian-beam wave at the transmitter, we assume the wave retains this symmetry as it propagates along the positive *z*-axis. Therefore, let us write the paraxial wave equation (126) in cylindrical coordinates as

$$\frac{1}{r}\frac{\partial}{\partial r}\left(r\frac{\partial V}{\partial r}\right) + 2ik\frac{\partial V}{\partial z} = 0, \quad V(r,0) = a_0 \exp\left(-\frac{1}{2}\alpha k r^2\right), \tag{129}$$

where we have also incorporated the boundary condition (BC) at $z = 0$. By introducing the *Hankel transform* (Section 10.4) relations

$$\mathcal{H}_0\{V(r,z); r \to \rho\} = \hat{V}(\rho,z),$$

$$\mathcal{H}_0\left\{\frac{1}{r}\frac{\partial}{\partial r}\left(r\frac{\partial V}{\partial r}\right); r \to \rho\right\} = -\rho^2\hat{V}(\rho,z), \tag{130}$$

the transformed problem (126) becomes

$$\frac{d\hat{V}}{dz} = \frac{\rho^2}{2ik}\hat{V}(\rho,z), \quad \hat{V}(\rho,0) = \frac{a_0}{k\alpha}\exp\left(-\frac{\rho^2}{2k\alpha}\right). \tag{131}$$

We recognize (131) as a first-order linear DE that can be readily solved by separating variables (Section 1.3.1) to get

$$\hat{V}(\rho,z) = \frac{a_0}{k\alpha}\exp\left(-\frac{\rho^2}{2k\alpha}\right)\exp\left(-\frac{iz\rho^2}{2k}\right)$$

$$= \frac{a_0}{k\alpha}\exp\left[-\frac{1}{2k\alpha}(1 + i\alpha z)\rho^2\right], \tag{132}$$

which satisfies the imposed BC in (131). Now, by taking the inverse Hankel transform of (132), we are led to the solution

$$V(r,z) = \frac{a_0}{1 + i\alpha z}\exp\left[-\frac{k\alpha r^2}{2(1 + i\alpha z)}\right]. \tag{133}$$

Last, multiplying (133) by the unimportant phase term e^{ikz} gives us the field $U(r,z)$ at any position z along the propagation path.

The solution (133) illustrates that the optical wave at any distance z is always a Gaussian-beam wave, which is the main reason this mathematical model is so useful. By writing the argument in (133) in terms of its real and imaginary parts (see Chapter 6), we can express the field in a more revealing form given by

$$U(r,z) = e^{ikz}V(r,z) = \frac{a_0 W_0^2}{W^2}\exp\left(ikz - i\varphi - \frac{r^2}{W^2} - i\frac{kr^2}{2F}\right), \tag{134}$$

where W is the spot radius, F is the phase front radius of curvature, and φ is a longitudinal phase. These parameters are defined by

$$W = W_0\sqrt{(1 - z/F_0)^2 + (2z/kW_0^2)^2},$$

$$F = \frac{zW^2}{W^2 - W_0^2(1 - z/F_0)}, \tag{135}$$

$$\varphi = \tan^{-1}\left(\frac{2z/kW_0^2}{1 - z/F_0}\right).$$

15.6.2 Huygens-Fresnel integral: lowest-order Gaussian mode

The *Huygens-Fresnel principle* assumes that each point on a wavefront acts like a source of a secondary spherical wave. This principle leads to the Huygens-Fresnel integral (given below) which provides another solution form for the paraxial approximation to the field of a propagating optical wave in free space.

Written in terms of the free-space Green's function, which is a spherical wave under the paraxial approximation given by $G(\mathbf{s},\mathbf{r};z) = (1/4\pi)\exp[ikz + (ik/2z)|\mathbf{s} - \mathbf{r}|^2]$, the *Huygens-Fresnel integral* takes the form (see [16])

$$U(\mathbf{r},z) = -\frac{ik}{2\pi z}\int_{-\infty}^{\infty}\int_{-\infty}^{\infty}\exp\left[ikz + \frac{ik}{2z}|\mathbf{s} - \mathbf{r}|^2\right]U(\mathbf{s},0)d^2s, \tag{136}$$

where $U(\mathbf{s},0)$ is the optical wave field at the source plane and transverse position $\mathbf{s}$, and $U(\mathbf{r},z)$ is the optical field at propagation distance z and transverse position $\mathbf{r} = <x,y>$. We leave it to the reader to verify that (136) indeed satisfies the paraxial wave equation (126).

The substitution of $U(\mathbf{s},0)$ given by (127) into the integral (136) yields

$$U(\mathbf{r},z) = -\frac{ik}{2\pi z}a_0\exp\left(ikz + \frac{ik}{2z}r^2\right)\int_{-\infty}^{\infty}\int_{-\infty}^{\infty}\exp\left(-\frac{ik}{z}\mathbf{r}\cdot\mathbf{s}\right)$$
$$\times\exp\left[\frac{ik}{2z}(1 + i\alpha z)s^2\right]d^2s. \tag{137}$$

By writing $d^2s = ds_1 ds_2$, and then converting to polar coordinates through the transformation

$$s_1 = s\cos\theta, \quad s_2 = s\sin\theta,$$

where $s = \sqrt{s_1^2 + s_2^2}$ and $\tan\theta = s_2/s_1$, the double integral in (137) becomes

$$U(\mathbf{r},z) = -\frac{ik}{2\pi z}a_0\exp\left(ikz + \frac{ik}{2z}r^2\right)\int_0^{\infty}\int_0^{2\pi}\exp\left(-\frac{ik}{z}rs\cos\theta\right)$$
$$\times\exp\left[\frac{ik}{2z}(1 + i\alpha z)s^2\right]s\,d\theta\,ds. \tag{138}$$

By using property (BJ4) in Section 2.5.1, the inside integration yields

$$\int_0^{2\pi}\exp\left(-\frac{ik}{z}rs\cos\theta\right)d\theta = 2\pi J_0\left(\frac{krs}{z}\right).$$

Next, we can use a technique like that in Example 5 in Chapter 2 to perform the remaining integration on s. Doing so gives us our result

$$
\begin{aligned}
U(\mathbf{r},z) &= -\frac{ik}{z} a_0 \exp\left(ikz + \frac{ik}{2z} r^2\right) \int_0^\infty s J_0(krs/z) \exp\left[\frac{ik}{2z}(1 + i\alpha z) s^2\right] ds \\
&= \frac{a_0}{1 + i\alpha z} \exp\left[ikz - \frac{k\alpha r^2}{2(1 + i\alpha z)}\right],
\end{aligned}
\tag{139}
$$

which is the same as that obtained by the Hankel transform method.

15.6.3 Hermite-Gaussian modes

The majority of studies of wave propagation has involved the lowest-order Gaussian beam or TEM_{00} mode (see, for example, Section 15.6.2). In practice, however, there are certain scenarios when it is desirable to minimize the excitation of nonlinearities within the crystal of a laser or when the received optical wave needs to have a multiple spot pattern. In those cases, the higher-order solutions of the paraxial wave equation (126) can be used to generate *higher-order Gaussian-beam modes* with Hermite polynomials (CO_2 laser) in rectangular coordinates or Laguerre polynomials (HeNe laser) in cylindrical coordinates (see [29]).

The higher-order *Hermite-Gaussian modes* TEM_{mn} of a collimated beam at the exit aperture of a laser are described by

$$
U_{mn}(x,y,0) = H_m\left(\frac{\sqrt{2}x}{W_x(0)}\right) H_n\left(\frac{\sqrt{2}y}{W_y(0)}\right) \exp\left(-\frac{x^2}{W_x^2(0)} - \frac{y^2}{W_y^2(0)}\right),
\tag{140}
$$

where $m,n = 0,1,2,\dots$, the TEM_{00} spot size along the x and y axes at the transmitter is given by $W_x(0)$ and $W_y(0)$, respectively, and $H_n(x)$ is the nth Hermite polynomial (see Section 2.4.2). However, the higher-order modes always form a pattern of spots (see Fig. 15.11) rather than a single spot as exhibited by the TEM_{00} mode.

We can use the Huygens-Fresnel integral (136) to find the intensity (or irradiance) of the Hermite-Gaussian beam wave along the propagation path. The field at position z along the positive z-axis is defined by the double integral

$$
U_{mn}^0(\mathbf{r},L) = -\frac{ik}{2\pi L} \exp(ikL) \int\int_{-\infty}^\infty d^2s \, U_{mn}^0(\mathbf{s},0) \exp\left(\frac{ik|\mathbf{s} - \mathbf{r}|^2}{2L}\right),
\tag{141}
$$

where $U_{mn}^0(\mathbf{s},0)$ is the optical wave (140) at the transmitter. By defining $\mathbf{s} = \langle\xi,\eta\rangle$ and $\mathbf{r} = \langle x,y\rangle$, each integral in (141) has the form

$$
I_1 = -\frac{ik}{2\pi z} e^{ikz} \int_{-\infty}^\infty H_m\left(\frac{\sqrt{2}\xi}{W_x(0)}\right) \exp\left[\frac{ik}{2z}(\xi - x)^2\right] \exp\left(-\frac{\xi^2}{W_x^2(0)}\right) d\xi,
$$

$$
I_2 = -\frac{ik}{2\pi z} e^{ikz} \int_{-\infty}^\infty H_n\left(\frac{\sqrt{2}\eta}{W_y(0)}\right) \exp\left[\frac{ik}{2z}(\eta - x)^2\right] \exp\left(-\frac{\eta^2}{W_y^2(0)}\right) d\eta.
$$

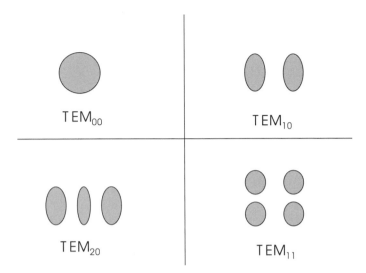

Figure 15.11 Illustrations depicting various patterns (light spots) associated with some of the lower-order Hermite-Gaussian beams.

From integral tables, we find that

$$\int_{-\infty}^{\infty} H_k(ax)e^{-(x-y)^2}\,dx = \sqrt{\pi}(1 - a^2)^{k/2} H_k\left(\frac{ay}{\sqrt{1-a^2}}\right), \tag{142}$$

and thus, by rearranging I_1 and I_2 above to fit this integral, it can be shown that the intensity of the optical wave deduced from $I(x,y,z) = |I_1 I_2|^2$ is

$$I(x,y,z) = \frac{W_x(0)\,W_y(0)}{W_x(z)\,W_y(z)}\,H_m^2\left(\frac{\sqrt{2}x}{W_x(z)}\right) H_n^2\left(\frac{\sqrt{2}y}{W_y(z)}\right)\exp\left(-\frac{2x^2}{W_x^2(z)} - \frac{2y^2}{W_y^2(z)}\right), \tag{143}$$

where $W_x(z)$, $W_y(z)$ are defined by

$$
\begin{aligned}
W_x^2(z) &= W_x^2(0) + \frac{4z^2}{k^2 W_x^2(0)}, \\
W_y^2(z) &= W_y^2(0) + \frac{4z^2}{k^2 W_y^2(0)}.
\end{aligned}
\tag{144}
$$

Because higher-order modes do not form a simple spot of light, the effective spot size characterizing the illuminated area of the higher-order modes is not well defined. However, it can be defined by the enclosed area $\sigma_x^2(z)_m \times \sigma_y^2(z)_n$, where

$$\sigma_x^2(z)_m = \frac{2\iint_{-\infty}^{\infty} x^2 I(x,y,z)\,dx\,dy}{\iint_{-\infty}^{\infty} I(x,y,z)\,dx\,dy} = \sqrt{2m+1}\,W_x(z), \tag{145}$$

and, similarly, $\sigma_y^2(z)_n = \sqrt{2n+1}\,W_y(z)$ (see also [29]).

15.7 *ABCD* Matrices for Paraxial Systems

An effective way to discuss optical wave propagation through various optical structures like lenses and apertures is by use of 2×2 matrices known as *ABCD ray matrices* (e.g., see [2] and [29]). The use of such matrices allows us to describe the propagation of an optical wave through a train of optical elements by utilizing the cascade scheme of multiplying successive matrix representations of each optical element, including those that describe the free-space path between optical elements. In this fashion the entire propagation path, consisting of various optical elements at arbitrary positions along the propagation path, can be represented by a single *ABCD* ray matrix. In doing so, we again invoke the paraxial approximation, which is valid when the separation distance between optical elements is large compared with the transverse extent of the beam.

The notion of a ray matrix is most easily understood by examining a one-dimensional analog. Consider a ray of light propagating between two points denoted by r_1 and r_2 in parallel transverse planes at $z = z_1$ and $z = z_2$ separated by distance L (see Fig. 15.12). The angle that the light ray makes from point r_1 to point r_2 is ϕ, for which

$$\tan\phi = \frac{r_2 - r_1}{L} = \frac{dr_1}{dz}. \tag{146}$$

Rewriting Eq. (146), we have

$$r_2 = r_1 + L\frac{dr_1}{dz} = r_1 + Lr_1', \tag{147}$$

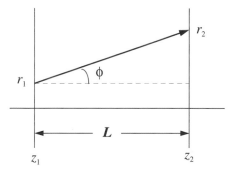

Figure 15.12 Line-of-sight section of length L.

and recognizing that the slope at r_1 is the same as that at r_2 yields

$$r_2' = r_1'. \tag{148}$$

By combining the last two equations, we obtain the matrix equation

$$\begin{pmatrix} r_2 \\ r_2' \end{pmatrix} = \begin{pmatrix} 1 & L \\ 0 & 1 \end{pmatrix} \begin{pmatrix} r_1 \\ r_1' \end{pmatrix}. \tag{149}$$

The 2×2 matrix in (149) is the *ABCD* ray matrix depicting free-space propagation over a path of length *L*. If we formally replace r_1 and r_2 with vectors $\mathbf{r}_1$ and $\mathbf{r}_2$, the same relation holds for rotationally symmetric systems. Other *ABCD* matrices for a thin lens and finite aperture stop with rotational symmetry are listed in Table 15.1. Note that the finite aperture stop, representing a "soft aperture," is similar to that of the thin lens with an imaginary focal length. This finite aperture is also called a *Gaussian aperture*.

An important property of all ray matrices listed in Table 15.1 is that

$$AD - BC = 1, \tag{150}$$

which is valid as long as input and output planes are in the same medium. Let us now consider an optical ray propagating through a sequence of optical elements (plus straight line sections) all properly aligned and arranged in cascade fashion as illustrated in Fig. 15.13. By representing each optical element and straight line section along the propagation path by its *ABCD* matrix, the overall *ABCD* matrix for *N* such matrices is

$$\begin{pmatrix} A & B \\ C & D \end{pmatrix} = \begin{pmatrix} A_N & B_N \\ C_N & D_N \end{pmatrix} \begin{pmatrix} A_{N-1} & B_{N-1} \\ C_{N-1} & D_{N-1} \end{pmatrix} \cdots \begin{pmatrix} A_1 & B_1 \\ C_1 & D_1 \end{pmatrix}. \tag{151}$$

Note that the matrices must be arranged in reverse order from the order of occurrence.

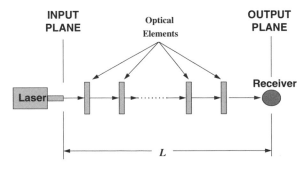

Figure 15.13 A ray-matrix optical system in cascade.

Table 15.1 Ray Matrices for Various Optical Elements

Structure	*Matrix*
Line-of sight section (length L)	$\begin{pmatrix} 1 & L \\ 0 & 1 \end{pmatrix}$
Thin lens (focal length F_G)	$\begin{pmatrix} 1 & 0 \\ -1/F_G & 1 \end{pmatrix}$
Finite aperture stop (aperture radius W_G)	$\begin{pmatrix} 1 & 0 \\ 2i/k W_G^2 & 1 \end{pmatrix}$
Gaussian lens (thin lens and aperture stop)	$\begin{pmatrix} 1 & 0 \\ -1/F_G + 2i/k W_G^2 & 1 \end{pmatrix}$

15.7.1 Generalized Huygens-Fresnel integral

Line-of-sight propagation of optical waves in the absence of optical elements between the input (transmitter) and output (receiver) planes can be described by the *Huygens-Fresnel integral* (see Section 15.6.2)

$$U(\mathbf{r},L) = -\frac{ik}{2\pi L}\int_{-\infty}^{\infty}\int_{-\infty}^{\infty} U(\mathbf{s},0)\exp\left(ikL + \frac{ik}{2L}|\mathbf{s}-\mathbf{r}|^2\right) d^2s, \qquad (152)$$

where L denotes total propagation distance, $\mathbf{s}$ and $\mathbf{r}$ are vectors in the transverse direction to the propagation path at the input and output planes, $k = 2\pi/\lambda$ is the optical wave number proportional to the reciprocal of wavelength λ, and $i^2 = -1$. Here, $U(\mathbf{s},0)$ and $U(\mathbf{r},L)$ denote the optical wave at the input and output planes, respectively. If we now consider an optical system composed of various optical elements exhibiting rotational symmetry between input and output planes as illustrated in Fig. 15.13, then, using the *ABCD* matrix representation (151) for the propagation channel between input and output, the Huygens-Fresnel integral for such an optical system assumes the more general form[1]

$$U(\mathbf{r},L) = -\frac{ik}{2\pi B}\exp(ikL)\int\int_{-\infty}^{\infty} d^2s\, U(\mathbf{s},0)$$
$$\times \exp\left[\frac{ik}{2B}(As^2 - 2\mathbf{s}\cdot\mathbf{r} + Dr^2)\right], \qquad (153)$$

---

[1]It is the Green's function that changes in the generalized Huygens-Fresnel integral. In particular, the generalized Green's function is $G(\mathbf{s},\mathbf{r};L) = (1/4\pi B)\exp\left[ikL + (ik/2B)(As^2 - 2\mathbf{s}\cdot\mathbf{r} + Dr^2)\right]$, which reduces to the standard Green's function when $A = D = 1$ and $B = L$.

called the *generalized Huygens-Fresnel integral*. When $A = D = 1$ and $B = L$, Eq. (153) reduces to the standard form of the Huygens-Fresnel integral (152).

Let us assume the optical wave at the input plane ($z = 0$) is a lowest-order Gaussian-beam wave described by

$$U(\mathbf{s},0) = a_0 \exp\left(-\frac{1}{2}\alpha k s^2\right) = a_0 \exp\left[\frac{ik}{2z}(i\alpha z)s^2\right], \qquad (154)$$

where a_0 is the amplitude on the optical axis and α is a complex parameter describing the beam spot radius W_0 and phase front radius of curvature F_0 according to

$$\alpha = \frac{2}{k W_0^2} + i\frac{1}{F_0}. \qquad (155)$$

By substituting (154) into (153) and performing an evaluation similar to that illustrated in Section 15.6.2, the field of the wave at the output plane is the Gaussian-beam wave

$$U(\mathbf{r},L) = \frac{a_0}{A + i\alpha B} \exp(ikL)\exp\left[-\frac{1}{2}\beta(L)k r^2\right], \qquad (156)$$

where

$$\beta(L) = \frac{\alpha D - iC}{A + i\alpha B} = \frac{2}{k W^2} + i\frac{1}{F}. \qquad (157)$$

The quantities $W = \{2/\text{Re}[k\beta(L)]\}^{1/2}$ and $F = 1/\text{Im}[\beta(L)]$ are, respectively, the beam radius and phase front radius of curvature of the wave at the output plane, and Re and Im denote the real and imaginary parts of the argument (see Chapter 6).

15.7.2 Gaussian lens

For a specific example, let us consider the optical system shown in Fig. 15.14 consisting of a thin lens and a Gaussian limiting aperture (known as a *Gaussian lens*) located at distance L_1 from a laser source and distance L_2 from the output plane. We assume the Gaussian lens has effective focal length F_G (m) and "soft aperture" radius W_G (m), the latter often related to a "hard aperture" diameter[2] D_G according to $D_G^2 = 8W_G^2$.

By characterizing the Gaussian lens by the complex parameter (see the last entry in Table 15.1)

$$\alpha_G = \frac{2}{k W_G^2} + i\frac{1}{F_G}, \qquad (158)$$

the overall *ABCD* ray matrix for the optical system shown in Fig. 15.14 is described by

---

[2]In characterizing a Gaussian beam or Gaussian lens, it is customary to refer to its "radius," whereas for a "hard" aperture stop one ordinarily refers to its "diameter."

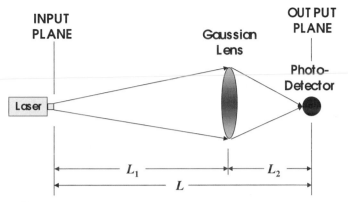

Figure 15.14 Propagation geometry for a Gaussian beam originating at distance L_1 to the left of a thin Gaussian lens of real focal length F_G and effective transmission radius W_G.

$$
\begin{pmatrix} A & B \\ C & D \end{pmatrix} = \begin{pmatrix} 1 & L_2 \\ 0 & 1 \end{pmatrix} \begin{pmatrix} 1 & 0 \\ i\alpha_G & 1 \end{pmatrix} \begin{pmatrix} 1 & L_1 \\ 0 & 1 \end{pmatrix}
$$

$$
= \begin{pmatrix} 1 + i\alpha_G L_2 & L_1 + L_2(1 + i\alpha_G L_1) \\ i\alpha_G & 1 + i\alpha_G L_1 \end{pmatrix}.
$$

(159)

For the special case of a spherical wave (i.e., a point source) at the input plane we use the delta function (Section 2.2.4) to write $U(\mathbf{s},0) = \delta(\mathbf{s})$, which leads to

$$
U(\mathbf{r},L) = \frac{ika_0}{2\pi B} \exp\left(ikL + \frac{ikDr^2}{2B} \right).
$$

(160)

Thus, the matrix elements B and D in this case reduce to

$$
B = L_2\left(1 + \frac{L_1}{L_2} - \frac{L_1}{F_G} + i\Omega_G \right),
$$

$$
D = 1 - \frac{L_1}{F_G} + i\Omega_G,
$$

(161)

where $\Omega_G = 2L_1/kW_G^2$ characterizes the finite size of the lens. Equation (160) now describes a Gaussian beam at the output plane for which $\beta(L) = -iD/B$.

EXAMPLE 5 For the optical system illustrated in Fig. 15.14, calculate the spot radius and phase front radius of curvature in the focal plane of the lens for a transmitted collimated beam given the following beam and optical system characteristics: $W_0 = 1$ cm, $F_0 = \infty$, $\lambda = 1.06$ μm, $W_G = 2$ cm, $F_G = 20$ cm, and $L_1 = 1$ km. Repeat the calculations for a point source.

Solution: The focal plane is characterized by the relationship $L_2 = F_G$. The matrix elements for this system with a collimated beam can be obtained from Eq. (159). Then, by using the relation given by (157), we find that $\beta(L) \cong 13{,}346 + 10{,}989i$, from which we deduce $W = \sqrt{\dfrac{2}{13{,}346k}}$, $F = 1/10{,}989$, or

$$W \cong 5 \,\text{μm}, \quad F \cong 91 \,\text{μm}.$$

In the case of a point source at the transmitter, we use Eqs. (160) and (161) to obtain $\beta(L) \cong 12{,}321 + 14{,}602i$. Here we find nearly the same results, viz.,

$$W \cong 5.2 \,\text{μm}, \quad F \cong 68 \,\text{μm}.$$

15.7.3 Fourier-transform plane

One of the most remarkable properties of a converging lens is its inherent ability to perform two-dimensional Fourier transforms. That is, if a small object is placed in the front (or left-hand) focal plane of a "sufficiently large" thin converging lens, its Fourier transform will appear in the back (or right-hand) focal plane (see Fig. 15.15).

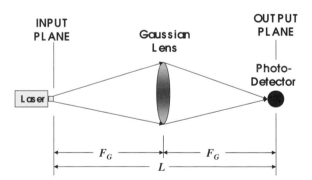

Figure 15.15 Propagation geometry for a Fourier-transform plane analysis. We assume a thin Gaussian lens of real focal length F_G and effective transmission radius W_G.

For the configuration shown in Fig. 15.14 we set $L_1 = L_2 = F_G$ in the *ABCD* matrix of Eq. (159), which leads to

$$\begin{pmatrix} A & B \\ C & D \end{pmatrix} = \begin{pmatrix} i\Omega_G & F_G(1 + i\Omega_G) \\ -\dfrac{1}{F_G}(1 + i\Omega_G) & i\Omega_G \end{pmatrix}, \tag{162}$$

where once again we set $\Omega_G = 2F_G/kW_G^2$. If the lens is sufficiently larger than the incident beam, then we can set $\Omega_G = 0$ in (162) to obtain

$$\begin{pmatrix} A & B \\ C & D \end{pmatrix} = \begin{pmatrix} 0 & F_G \\ -\dfrac{1}{F_G} & 0 \end{pmatrix}. \tag{163}$$

If the beam at the transmitter is a collimated beam with effective radius W_0, then the beam in the output plane described by (156) takes the form

$$U(\mathbf{r},L) = -\frac{ika_0 W_0^2}{2F_G}\exp(2ikF_G)\exp\left(-\frac{k^2 W_0^2 r^2}{4F_G^2}\right), \tag{164}$$

which, by defining $v = kr/F_G$, is proportional to the Fourier transform of the input beam (see also Chapters 9 and 10).

> **Remark:** The difference between (164) and a conventional two-dimensional Fourier transform is a phase term that is usually considered insignificant (see [16]).

15.8 Zernike Polynomials

The *Zernike polynomials* (see [7] and [23]) represent a set of functions of two variables that are orthogonal over a circle with unit radius. It is customary to define these polynomials as a product of two functions, one depending only on a radial coordinate r and the other depending only on the angular coordinate θ, i.e.,

$$Z_n^m(r,\theta) = R_n^m(r)e^{im\theta}, \tag{165}$$

where both m and n are integers, $n \geq 0$, $-n \leq m \leq n$, and $n \pm |m|$ is even. The radial polynomial $R_n^m(r)$ is a special case of Jacobi or hypergeometric polynomial that is normalized so that $R_n^m(1) = 1$. It is defined in general by

$$R_n^m(r) = \sum_{k=0}^{(n-|m|)/2} \frac{(-1)^k(n-k)!}{\left(\dfrac{n+m}{2}-k\right)!\left(\dfrac{n-m}{2}-k\right)!}\frac{r^{n-2k}}{k!}. \tag{166}$$

Clearly, the polynomial (166) has degree n and contains no power of r less than $|m|$, where m is the angular dependence. Also, the polynomial is even if m is even and odd if m is odd. The *orthogonality* property of these polynomials is

$$\int_0^1 \int_0^{2\pi} Z_n^m(r,\theta) Z_k^{m*}(r,\theta) r\, d\theta\, dr \;=\; \frac{\pi}{n+1}\,\delta_{nk} \;=\; \begin{cases} \pi/(n+1), & k=n \\ 0, & k\neq n \end{cases}, \tag{167}$$

where * denotes complex conjugate, and δ_{nk} is the Kronecker delta (see Section 4.3.2) defined by $\delta_{nk} = 0$ $(k \neq n)$ and $\delta_{nk} = 1$ $(k = n)$.

15.8.1 Application in optics

One of the principal uses of the Zernike polynomials is to represent fixed aberrations in optical systems in the form of a *generalized Fourier series* in Zernike polynomials. Lower-order Zernike polynomials are then referred to by such names as piston, tilt, focus, astigmatism, coma, and so forth. By the use of these polynomials, researchers have been able to study how aberrations affect various imaging systems. They are also useful in adaptive optics systems designed for atmospheric turbulence decomposition (see [27] and [33]).

Virtually any realistic wave front $\Phi(r,\theta)$ can be represented in a two-dimensional series of Zernike polynomials by one of the following expressions:

$$\Phi(r,\theta) \;=\; \sum_{n=0}^{\infty} \sum_{m=-n}^{n} C_{mn} R_n^m(r) e^{im\theta}, \tag{168}$$

or, alternatively,

$$\begin{aligned}
\Phi(r,\theta) \;=\;& A_{00} + \frac{1}{\sqrt{2}} \sum_{n=2}^{\infty} A_{0n} R_n^0(r) \\
& + \sum_{n=0}^{\infty} \sum_{m=1}^{n} \left(A_{mn}\cos m\theta + B_{mn}\sin m\theta \right) R_n^m(r).
\end{aligned} \tag{169}$$

The factor $1/\sqrt{2}$ is often introduced in the second term in (169) because it simplifies the final formulas for the Fourier coefficients. The Fourier coefficients in (169) are related to those in (168) by

$$\begin{aligned}
A_{00} &= C_{00}, \qquad A_{0n} = \sqrt{2}\,C_{0n}, \\
A_{mn} &= \frac{1}{2}(C_{mn} - iC_{-mn}), \qquad B_{mn} = \frac{1}{2i}(C_{mn} + iC_{-mn}),
\end{aligned} \tag{170}$$

which can be obtained by using the method illustrated in Section 8.3.

To obtain the Fourier coefficients in (168), we first multiply both sides of the expression by $rR_k^{|m|}(r)e^{-im\theta}$ and then integrate over the unit disk. This action leads to

$$\begin{aligned}
\int_0^1 \int_0^{2\pi} r\Phi(r,\theta) R_k^{|m|}(r) e^{-im\theta}\, d\theta\, dr \;=\;& \sum_{n=0}^{\infty} \sum_{m=-n}^{n} C_{mn} \int_0^1 \int_0^{2\pi} r R_n^{|m|}(r) R_k^{|m|}(r)\, d\theta\, dr \\
=\;& \frac{\pi}{k+1} \sum_{m=-k}^{k} C_{mk}.
\end{aligned} \tag{171}$$

For $k = 0$, we obtain the coefficient for piston given by

$$C_{00} = \frac{1}{\pi} \int_0^1 \int_0^{2\pi} r\Phi(r,\theta)\,d\theta\,dr = \bar{\Phi}(r,\theta). \qquad (172)$$

That is, this coefficient is simply the average value of the wave front $\bar{\Phi}(r,\theta)$. Similarly, the coefficient for focus is deduced by setting $k = 2$ and $m = 0$, which gives us

$$C_{02} = \frac{3}{\pi} \int_0^1 \int_0^{2\pi} r(2r^2 - 1)\Phi(r,\theta)\,d\theta\,dr. \qquad (173)$$

A useful property of the Zernike series (168) [or (169)] is that the root-mean-square (rms) wave front error due to primary (lower-order) aberrations can readily be calculated. That is, if all the Fourier coefficients in (168) are known, the geometric sum of the non-piston terms yields the wave front variance (or "mean-square" deformation)

$$(\Delta\Phi)^2 = \frac{1}{\pi} \int_0^1 \int_0^{2\pi} r[\Phi(r,\theta) - \bar{\Phi}(r,\theta)]^2\,d\theta\,dr = \sum_{n=1}^{\infty} \sum_{m=-n}^{n} \frac{|C_{mn}|^2}{n+1}, \qquad (174)$$

the square-root of which gives the *rms wave front error*. Note that this expression is a kind of Parseval's relation (see Section 8.3.1).

15.8.2 Atmospheric effects on imaging systems

The utility of the Zernike polynomials for analyzing the wave front of an optical wave that has passed through an imaging system with fixed aberrations has been briefly discussed in the previous section. This approach has provided a deep understanding of the lower-order aberrations on such a system. Moreover, in diagnosing the effects of atmospheric turbulence on a propagating optical wave, it can also be useful to express the turbulence-induced random phase perturbations on $\Phi(r,\theta)$ in a series of Zernike polynomials.

A modification of the Zernike polynomials (165) that is commonly used in studying atmospheric effects on imaging systems leads to Zernike functions defined by

$$\begin{aligned} Z_i(r) &\equiv Z_i[0,n] = \sqrt{n+1}\,R_n^0(r), & m = 0, \\ Z_{i,\text{even}}(r,\theta) &\equiv Z_{i,\text{even}}[m,n] = \sqrt{n+1}\,R_n^m(r)\sqrt{2}\cos m\theta, & m \neq 0, \\ Z_{i,\text{odd}}(r,\theta) &\equiv Z_{i,\text{odd}}[m,n] = \sqrt{n+1}\,R_n^m(r)\sqrt{2}\sin m\theta, & m \neq 0. \end{aligned} \qquad (175)$$

The ordering scheme for these polynomials uses the rule that, for a given n, the modes with smaller m are counted first. For $m > 0$, there are two Zernike functions for each (m,n) pair as given in (175). Following this ordering scheme, we list the first few polynomials $Z_i[m,n]$ in Table 15.2. Here, the quantity $Z_1[0,0]$ represents *piston* (constant retardation or advancement of the phase over the entire beam), $Z_2[1,1]$ and $Z_3[1,1]$ represent *tilt* in the x and y directions, respectively, $Z_4[0,2]$ is *focus* (or *defocus*), and $Z_5[2,2]$ and $Z_6[2,2]$ are two components of *astigmatism*.

Table 15.2 Table of Zernike Polynomials Through $(m,n) = (3,3)$

i	m	n	Zernike Polynomial $Z_i[m,n]$
1	0	0	1
2	1	1	$2r\cos\theta$
3	1	1	$2r\sin\theta$
4	0	2	$\sqrt{3}(2r^2 - 1)$
5	2	2	$\sqrt{6}r^2\sin 2\theta$
6	2	2	$\sqrt{6}r^2\cos 2\theta$
7	1	3	$\sqrt{8}(3r^3 - 2r)\sin\theta$
8	1	3	$\sqrt{8}(3r^3 - 2r)\cos\theta$
9	3	3	$\sqrt{8}r^3\sin 3\theta$
10	3	3	$\sqrt{8}r^3\cos 3\theta$

Up to this point we have considered only the case in which the telescope aperture of the imaging system has unit radius. To account for an aperture of diameter D rather than of unit radius, we can simply make the change of variable

$$r = 2\rho/D$$

in the above Zernike polynomials and corresponding integrals. Also, we can avoid complicated expressions like (171) for the Fourier coefficients by representing the wave front in a single series called a *modal expansion*, viz.,

$$\Phi(2\rho/D,\theta) = \sum_{i=1}^{\infty} a_i Z_i(2\rho/D,\theta). \tag{176}$$

In this case the Fourier coefficient a_i, which depends on only one index corresponding to the Zernike mode Z_i, is given by

$$a_i = \frac{4}{\pi D^2} \int_0^{D/2} \int_0^{2\pi} \rho\,\Phi(2\rho/D,\theta) Z_i(2\rho/D,\theta)\,d\theta\,d\rho, \quad i = 1,2,3,.... \tag{177}$$

As an optical wave propagates through the atmosphere, small index of refraction fluctuations cause random phase perturbations on the wave front, the mean value of which is zero. In this case the Fourier coefficients a_i, $i = 1,2,3,...$ defined by (177) are taken to be *Gaussian random variables* with zero mean, i.e.,

$$<a_i> = \frac{4}{\pi D^2} \int_0^{D/2} \int_0^{2\pi} \rho <\Phi(2\rho/D,\theta)> Z_i(2\rho/D,\theta) d\theta \, d\rho = 0. \tag{178}$$

The angle brackets $< >$ denote an ensemble average (see Chapter 14). Thus, we consider higher-order statistics of the Fourier coefficients like the *covariance* defined by

$$<a_i a_j> = \left(\frac{4}{\pi D^2}\right)^2 \int_0^{D/2} \int_0^{2\pi} \int_0^{D/2} \int_0^{2\pi} \rho_1 \rho_2 \, C_\Phi(\rho_1,\theta_1;\rho_2,\theta_2)$$
$$\times Z_i(2\rho_1/D,\theta_1) Z_j(2\rho_2/D,\theta_2) d\theta_1 \, d\rho_1 \, d\theta_2 \, d\rho_2, \tag{179}$$

where $C_\Phi(\rho_1,\theta_1;\rho_2,\theta_2) = <\Phi(2\rho_1/D,\theta_1)\Phi(2\rho_2/D,\theta_2)>$ is the covariance of the phase (wave front). Except for diagonal terms in which $i = j$, most other terms in (179) are zero. In particular, it has been shown that $<a_i a_j> = 0$ for $i - j = \text{odd}$. However, the fact that $<a_i a_j> \neq 0$ for all $j \neq i$, means that correlation exists between some of the various Fourier coefficients. Consequently, the Zernike polynomials are not considered an "optimal basis." It has been shown that an optimal basis can be found in the form of a Karhunen-Loève series. Moreover, the orthogonal functions forming the basis of such a Karhunen-Loève series are linear combinations of the Zernike polynomials, but we will not pursue that approach here (see, for example, [27]).

Building on the above results, the mean-square phase value is defined by

$$\varepsilon^2 = \frac{4}{\pi D^2} \int_0^1 \int_0^{2\pi} r <\Phi^2(2\rho/D,\theta)> d\theta \, dr = \sum_{i=1}^\infty <(a_i)^2>, \tag{180}$$

which is simply the sum of mean-square values of the random Fourier coefficients. For a piston-removed phase, the tilt modes Z_2 and Z_3 are the major contributors to the mean-square phase value (over 86%). If it is possible to use adaptive optics techniques to remove the first N Zernike modes in the wave front, the residual phase can be represented by

$$\Phi_N(2\rho/D,\theta) = \Phi(2\rho/D,\theta) - \sum_{i=1}^N a_i Z_i(2\rho/D,\theta), \tag{181}$$

and, consequently, the *mean-square phase error* for the residual phase becomes

$$\varepsilon_N^2 = \varepsilon^2 - \sum_{i=1}^N <(a_i)^2> = \sum_{i=N+1}^\infty <(a_i)^2>. \tag{182}$$

A good imaging system in the presence of atmospheric turbulence can be defined as one in which the mean square error $\varepsilon_N^2 \ll 1$.

15.8.3 Aperture filter functions

In the analysis of optical wave propagation through atmospheric turbulence, the statistical quantities of interest are often calculated for a single point in the transverse plane of a receiving aperture. One way to include the finite size of the receiver aperture in the analysis is by the use of an "aperture filter function." Basically, this means that we formulate the effect of a finite receiver aperture in terms of wave number (m^{-1}) instead of physical spatial distance (m). The relation between the wave number space and the actual physical space is the two-dimensional Fourier transform.

For adaptive optics systems, the use of aperture filter functions can be effective in the theoretical analysis of lower-order aberrations like piston and tilt. In this short section, however, we will not attempt to perform that kind of analysis but limit our treatment to calculating the filter functions associated with various Zernike modes.

The two-dimensional Fourier transform of the Zernike polynomials (175), scaled by the area of the aperture, is given by

$$
\begin{aligned}
G_i(\kappa,\varphi) &= \frac{1}{\pi} \int_{-\infty}^{\infty}\int_{-\infty}^{\infty} e^{\,i\mathbf{r}\cdot\boldsymbol{\kappa}} Z_i(r,\theta) U(1 - |r|)\,dx\,dy \\
&= \frac{1}{\pi} \int_0^1 \int_0^{2\pi} e^{\,i r\kappa\cos(\theta-\varphi)} Z_i(r,\theta)\, r\,d\theta\,dr,
\end{aligned}
\tag{183}
$$

where $\mathbf{r} = \langle x,y \rangle$ and $\boldsymbol{\kappa} = \langle \kappa_x, \kappa_y \rangle$ are two-dimensional vectors in the spatial and wave number domains, respectively, $U(x)$ is the step function depicting the finite size of the aperture (unit radius), and in the last step we introduced the polar coordinates

$$
\begin{aligned}
x &= r\cos\theta, & y &= r\sin\theta, \\
\kappa_x &= \kappa\cos\varphi, & \kappa_y &= \kappa\sin\varphi.
\end{aligned}
\tag{184}
$$

If we substitute the Zernike polynomials (175) directly into (183), we are led to three types of filter function described by

$$
\left\{
\begin{array}{l}
G_i(\kappa) \\
G_{i,\text{even}}(\kappa,\varphi) \\
G_{i,\text{odd}}(\kappa,\varphi)
\end{array}
\right\}
= \frac{\sqrt{n+1}}{\pi} \int_0^1 \int_0^{2\pi} e^{\,i\kappa r\cos(\theta-\varphi)} r R_n^m(r)
\left\{
\begin{array}{l}
1 \\
\sqrt{2}\cos m\theta \\
\sqrt{2}\sin m\theta
\end{array}
\right\} d\theta\,dr.
\tag{185}
$$

Integrals of the type (185) are difficult to evaluate in general, but can be readily evaluated for some of the lower-order Zernike modes. For example, in the case of piston in which $m = 0$ and $R_0^0(r) = 1$, i.e., $Z_1[0,0] = 1$, the above integral reduces to

$$
G_1(\kappa) = \frac{1}{\pi} \int_0^1 r \int_0^{2\pi} e^{\,i\kappa r\cos(\theta-\varphi)} d\theta\,dr = 2\int_0^1 r J_0(\kappa r)\,dr,
\tag{186}
$$

or,

$$
G_1(\kappa) = \frac{2 J_1(\kappa)}{\kappa},
\tag{187}
$$

where we are utilizing the identities (BJ4) and (BJ13) in Section 2.5.1. For the case of tilt corresponding to $Z_2[1,1] = 2r\cos\theta$, we find that (185) becomes

$$G_{2,\text{even}}(\kappa,\varphi) = \frac{2}{\pi}\int_0^1 r^2 \int_0^{2\pi} e^{i\kappa r\cos(\theta-\varphi)}\cos\theta\,d\theta\,dr. \qquad (188)$$

To evaluate the inside integral, we use the identity (BJ2) in Section 2.5.1 written as

$$e^{i\kappa r\cos(\theta-\varphi)} = J_0(\kappa r) + 2\sum_{k=0}^{\infty} i^k J_k(\kappa r)\cos k(\theta-\varphi) \qquad ,$$

and then obtain

$$\int_0^{2\pi} e^{i\kappa r\cos(\theta-\varphi)}\cos\theta\,d\theta = J_0(\kappa r)\int_0^{2\pi}\cos\theta\,d\theta$$
$$+ 2\cos k\varphi\sum_{k=1}^{\infty} i^k J_k(\kappa r)\int_0^{2\pi}\cos k\theta\cos\theta\,d\theta \qquad (189)$$
$$+ 2\sin k\varphi\sum_{k=1}^{\infty} i^k J_k(\kappa r)\int_0^{2\pi}\sin k\theta\cos\theta\,d\theta.$$

Except for the term $k = 1$ in the first summation, all other integrals above go to zero because of the orthogonality property of the trigonometric functions (see Section 8.2). Hence, we are left with

$$\int_0^{2\pi} e^{i\kappa r\cos(\theta-\varphi)}\cos\theta\,d\theta = 2iJ_1(\kappa r)\cos\varphi\int_0^{2\pi}\cos^2\theta\,d\theta$$
$$= 2\pi iJ_1(\kappa r)\cos\varphi, \qquad (190)$$

and, consequently, the integral (188) reduces to

$$G_{2,\text{even}}(\kappa,\varphi) = 4i\cos\varphi\int_0^1 r^2 J_1(\kappa r)\,dr$$
$$= 4i\,\frac{J_2(\kappa)}{\kappa}\cos\varphi. \qquad (191)$$

The above integral was evaluated with the use of identity (BJ13) in Section 2.5.1 once again. Following along similar lines, it has been shown that the aperture filter functions in the general case are given by

$$\left.\begin{array}{l} G_i(\kappa) \\ G_{i,\text{even}}(\kappa,\varphi) \\ G_{i,\text{odd}}(\kappa,\varphi) \end{array}\right\} = \sqrt{n+1}\,\frac{2J_{n+1}(\kappa)}{\kappa}\left\{\begin{array}{l} (-1)^{n/2} \\ (-1)^{(n-m)/2}\sqrt{2}i^m\cos m\varphi \\ (-1)^{(n-m)/2}\sqrt{2}i^m\sin m\varphi. \end{array}\right. \qquad (192)$$

The values of m and n are determined by the choice of Zernike polynomial in Table 15.2.

Because of the presence of the term i^m, the filter functions (192) are sometimes called the *complex filter functions*. In practice, it is the absolute value squared of these expressions that we want, called simply the *filter functions*. Thus, the filter functions are defined by

$$\left.\begin{array}{c} F_i(\kappa) \\ F_{i,even}(\kappa,\varphi) \\ F_{i,odd}(\kappa,\varphi) \end{array}\right\} = (n+1)\left[\frac{2J_{n+1}(\kappa)}{\kappa}\right]^2 \begin{cases} 1\ (m=0) \\ 2\cos^2 m\varphi \\ 2\sin^2 m\varphi. \end{cases} \qquad (193)$$

Remark: The above results are based on an aperture of unit radius. To account for an aperture of diameter D, we can make the change of variable $r = 2\rho/D$ in the definition of the Zernike polynomials or, alternatively, we can simply replace κ in Eqs. (192) and (193) with $\kappa D/2$.

Chapter 15

Exercises

1. A 2-lb weight stretches a spring 6 in. An impressed force $16\sin 8t$ is acting upon the spring, and the weight is pulled down 3 in. below the equilibrium position and released from rest. Determine the subsequent motion if damping is neglected.

Ans. $y = \left(\frac{1}{4} - t\right)\cos 8t + \frac{1}{8}\sin 8t$

2. A 20-N weight is suspended by a frictionless spring with $k = 98$ N/m. An external force $2\cos 7t$ N acts on the weight and damping is neglected.

(*a*) Find the frequency of the beat.

(*b*) Find the amplitude of the motion given that it starts from rest in equilibrium.

(*c*) How many seconds are there between beats?

Ans. (*a*) 0.0056 Hz (*b*) 2
(*c*) ~ 89.76 s

3. If $m = 1$ gm, $c = 0.02$ gm/s,

and $k = 25$ gm/s^2, find the steady-state solution of the resulting spring-mass system (21) when $F(t + 2\pi) = F(t)$, and

$$F(t) = \begin{cases} t + \pi/2, & -\pi < t < 0 \\ -t + \pi/2, & 0 < t < \pi. \end{cases}$$

Ans. $y = \sum_{n=1}^{\infty}(A_n\cos nt + B_n\sin nt),$

$$A_n = \frac{4(25 - n^2)}{n^2\pi\Delta},\quad B_n = \frac{0.08}{n\pi\Delta},$$
$$\Delta = (25 - n^2)^2 + (0.02n)^2$$

4. Find the current in an RLC circuit for $q(0) = i(0) = 0$, and

(*a*) $R = 6$ Ω, $L = 1$ H, $C = 0.04$ F, $E(t) = 24\cos 5t$ V.

(*b*) $R = 80$ Ω, $L = 20$ H, $C = 0.01$ F, $E(t) = 100$ V.

Ans. (*a*) $i = e^{-3t}(3\sin 4t - 4\cos 4t)$
$+ 4\cos 5t$
(*b*) $i = -5e^{-t}\sin t$

5. The amplitudes of two successive maximum currents in a series circuit with $L = 10^{-4}$ H and $C = 10^{-6}$ F are measured to be 0.02 and 0.1 A. What is the resistance R?

Ans. $R = 2.19$ Ω

6. When the moment of inertia is given

by $I(x) = \alpha\sqrt{x}$, show that the solution of the buckling problem is

$$y = C_1 \sqrt{x} J_{2/3}(2kx\sqrt{x}).$$

7. The small deflections of a uniform column of length b bending under its own weight are governed by

$$\theta'' + K^2 x\theta = 0, \quad \theta'(0) = 0, \quad \theta(b) = 0,$$

where θ is the angle of deflection from the vertical and $K > 0$.

(a) Show that the solution satisfying the first BC (at $x = 0$) is

$$\theta(x) = C\sqrt{x} J_{-1/3}\left(\frac{2}{3}Kx\sqrt{x}\right), \quad C = \text{const.}$$

(b) Find the shortest column length for which buckling can occur.

Ans. (b) $b \cong 1.99K^{-2/3}$

8. A certain FM signal is described by

$$f(t) = A\cos(20\pi \times 10^6 t + 2\sin 1000\pi t).$$

(a) Use Carson's rule to calculate the bandwidth in hertz.

(b) If the total average power is 10 watts, determine the amount of power in the first two sidebands.

Ans. (a) BW $= 3000$ Hz
(b) $P = 10[J_1^2(2) + J_{-1}^2(2)] = 6.7$ W

9. Solve for the output SNR of the full-wave odd device in Section 15.4.2 for the case $0 < v < 1$. Show that your answer reduces to (77) as $v \to 0$.

10. By directly solving for the coefficients $c_n(v)$ in the Fourier series (64), verify the results given in the text

(a) for a half-wave device.
(b) for a full-wave even device.
(c) for a full-wave odd device.

11. By defining the SNR (84) for a linear detector, we have

$$\text{SNR} = \frac{[\langle r \rangle - \langle r | s = 0 \rangle]^2}{\langle r^2 \rangle - \langle r \rangle^2}.$$

(a) Show that

$$\text{SNR} = \frac{S_0}{N_0},$$

$$S_0 = \frac{\pi}{4}[{}_1F_1(-1/2; 1; -S) - 1]^2,$$

$$N_0 = {}_1F_1(-1; 1; -S)$$
$$- \frac{\pi}{4}\,{}_1F_1^2(-1/2; 1; -S).$$

(b) Use asymptotic formulas for ${}_1F_1$ functions in part (a) to derive the asymptotic results

$$\text{SNR} \sim \begin{cases} 0.92S^2, & S \ll 1 \\ 2S, & S \gg 1. \end{cases}$$

12. Under the geometrical optics approximation, the *variance of log-intensity* of a plane wave propagating through atmospheric turbulence over a path of length L is defined by the integral

$$\sigma_\chi^2 = \frac{\pi^2 L^3}{3}\int_0^\infty \Phi_n(\kappa)\kappa^5\,d\kappa.$$

Given the spatial power spectrum

$$\Phi_n(\kappa) = 0.033 C_n^2 \kappa^{-11/3} e^{-\kappa^2/\kappa_m^2},$$

where C_n^2 is the refractive-index structure parameter and κ_m is a parameter (related to the inner scale of turbulence), show that

$$\sigma_\chi^2 \cong 0.054\Gamma(7/6)C_n^2 L^3 \kappa_m^{7/3}.$$

13. Under the same conditions as stated in Problem 12, the *covariance of log-intensity* of a plane wave is defined by the integral

$$B_\chi(\rho) = \frac{\pi^2 L^3}{3}\int_0^\infty J_0(\kappa$$

References

[1] Andrews, L. C.: *Elementary Partial Differential Equations with Boundary Value Problems* (Academic Press, Orlando, 1986).

[2] Andrews, L. C. and R. L. Phillips: *Laser Beam Propagation through Random Media* (SPIE Optical Engineering Press, Bellingham, 1998).

[3] Arfken, G.: *Mathematical Methods for Physicists*, 3rd ed. (Academic Press, San Diego, 1985).

[4] Beckmann, P.: *Probability in Communication Engineering* (Harcourt Brace & World, New York, 1967).

[5] Boggess, A. and F. J. Narcowich, *A First Course in Wavelets with Fourier Analysis* (Prentice-Hall, Upper Saddle River, 2001).

[6] Born, M. and E. Wolf: *Principles of Optics*, 6th ed. (Cambridge University Press, Cambridge, 1980).

[7] Bell, W. W.: *Special Functions for Scientists and Engineers* (Van Nostrand, London, 1968).

[8] Boyce, W. and R. DiPrima: *Elementary Differential Equations*, 5th ed. (Wiley, New York, 1992).

[9] Cooley, J. W. and J. W. Tukey, "An algorithm for the machine calculation of complex Fourier series," *Math. Comp.* **19**, 297-301 (1965).

[10] Davenport, Jr., W. B. and W. L. Root: *An Introduction to the Theory of Random Signals and Noise* (IEEE Press, New York, 1987).

[11] Debnath, L. "Generalized calculus and its applications," *Int. J. Math. Educ. Sci. Technol.* **9**, 399-416 (1978).

[12] Debnath, L. and P. Mikusinski: *Introduction to Hilbert Spaces with Applications* (Academic Press, San Diego, 1990).

[13] Erdélyi, A., W. Magnus, F. Oberhettinger, and F. G. Tricomi: *Tables of Integral Transforms* (in two volumes) (McGraw-Hill, New York, 1954).

[14] Gagliardi, R. M. and S. Karp.: *Optical Communications*, 2nd ed. (John Wiley & Sons, New York, 1995).

[15] Gaskill, J. D.: *Linear Systems, Fourier Transforms, and Optics* (Wiley & Sons, New York, 1978).

[16] Goodman, J. W.: *Introduction to Fourier Optics* (McGraw-Hill, New York, 1968).

[17] Ince, E. I.: *Ordinary Differential Equations*, (Dover, New York, 1956).

[18] Jury, E. I.: *Theory and Application of the Z-Transform* (Wiley & Sons, New York, 1964).

[19] Kingston, R. H.: *Optical Sources, Detectors, and Systems: Fundamentals and Applications* (Academic Press, San Diego, 1995).

[20] Lohmann, A. W., D. Mendlovic, and Z. Zalevsky.: "Fractional transformations in optics," in *Progress in Optics XXXVIII*, E. Wolf, ed., Elsevier Science B. V. (1998).

[21] Mendlovic, D. and H. M. Ozaktas: "Fractional Fourier transforms and their optical implementation: I," *J. Opt. Soc. Am. A* **10**, 1875-1881 (1993); "Fractional Fourier transforms and their optical implementation: II," *J. Opt. Soc. Am. A* **10**, 2522-2531 (1993).

[22] Middleton, D.: *An Introduction to Statistical Communication Theory* (McGraw-Hill, New York, 1960).

[23] Noll, R. J.: "Zernike polynomials and atmospheric turbulence," *J. Opt. Soc. Am.* **66**, 207-211 (1976).

[24] Papoulis, A.: *Probability, Random Variables, and Stochastic Processes* (McGraw-Hill, 1965).

[25] Resnikoff, H. L. and R. O. Wells, Jr.: *Wavelet Analysis: The Scalable Structure of Information* (Springer-Verlag, New York, 1998).

[26] Rice, S. O.: "The mathematical analysis of random noise," *Bell Sys. Tech. J.* **23**, 282-332 (1944); "Statistical properties of a sine wave plus random noise," *Bell Sys. Tech. J.* **27**, 109-158 (1948).

[27] Roggemann, M. C. and B. Welsh: *Imaging Through Turbulence* (CRC Press, Boca Raton, 1996).

[28] Sasiela, R. J.: *Electromagnetic Wave Propagation in Turbulence* (Springer, New York, 1994).

[29] Siegman, A. E.: *Lasers* (University Science, Mill Valley, 1986).

[30] Sokolnikoff, I. S.: *Mathematical Theory of Elasticity*, 2nd ed. (McGraw-Hill, New York, 1956).

[31] Springer, M. D. and W. E. Thompson, "The distribution of products of beta, gamma, and Gaussian random variables," *SIAM J. Appl. Math.* **18**, 721-737 (1970).

[32] Stavroudis, O. N.: *The Optics of Rays, Wavefronts, and Caustics* (Academic Press, New York, 1972).

[33] Tyson, R. K.: *Principles of Adaptive Optics* (Academic Press, San Diego, 1991).

[34] Watson, G. N.: *A Treatise on the Theory of Bessel Functions*, 2nd ed. (Cambridge University Press, London, 1952).

[35] Weinstock, R.: *Calculus of Variations with Applications to Physics and Engineering* (Dover, New York, 1974).

[36] Wheelon, A. D.: *Electromagnetic Scintillation: I Geometrical Optics* (Cambridge University Press, Cambridge, 2001).

[37] Ziemer, R. E. and W. H. Tranter: *Principles of Communications: Systems Modulation, and Noise*, 5th ed. (Wiley & Sons, New York, 2002).

[38] *Applications of Walsh Functions, 1970 Proceedings, Symposium and Workshop*, sponsored by Naval Res. Lab. and U. Maryland, Doc. AD 707431, Nat. Tech. Infor. Ser., Virginia.

Index

A

ABCD matrices, 767
Absolute derivative, 258
Absolutely integrable, 455
Acceleration, 166
    centripetal, 167
    coriolis, 167, 170
    gravitational, 14
Addition of
    complex numbers, 274
    matrices, 111
    random variables, 678, 688
    vectors, 147
Admissible function, 547
Airy
    equation, 56, 93
    functions, 94
Ampère's law, 208
Amplitude spectrum, 462
Analytic
    function, 291
    signal representation, 475
Angular frequency, 416, 462
Annular region, 327
Annulus, 347
Arc length, 163, 182
Argument, 277
    principal value, 277
Associated Laguerre polynomial, 80
Asymptotic stability, 130
Autocorrelation function. *See*
    correlation function
Autocovariance function. *See*
    covariance function
Auxiliary condition, 6
    See also Boundary conditions *and*
    Initial conditions
Auxiliary equation, 20, 123
Average,
    ensemble, 700
    power, 405, 715
    time, 703
    (*See also*, mean value)
Axioms of probability, 639

B

Base vectors,
    Cartesian, 148, 226
    curvilinear, 248
    reciprocal, 227
    rectilinear, 226
Basis, 134
Bayes's theorem, 669
Beam. *See* Elastic beam
Bernoulli,
    Daniel, 62, 590
    James, 2, 546, 590
    John, 2
Bernoulli equation, 387
Bessel equation, 50, 84
    related to, 56
Bessel, Friedrich W., 62
Bessel functions,
    Airy, 94
    first-kind, 52, 84, 617
    Hankel, 93
    modified, 88
    spherical, 92
    second-kind, 55, 86
Bessel series, 438
Beta function, 70
Bilinear formula, 446
Bilinear transformation, 311
Binomial
    distribution, 647
    series, 101
Binormal vector, 164
Bit error rate, 756
Bivariate distribution, 665
Blasius, theorem of, 395
Borel, Félix E. E., 638
Boundary conditions, 420
    forced, 547
    implicit, 609
    natural, 557, 561
    periodic, 431, 609
Boundary value problem, 442
    eigenvalue problem, 420
Brachistochrone problem, 553

Branch
  cut, 302
  point, 302
Buckling problem, 742
  Euler load, 743

C

Canonical form, 603
Carson's rule, 747
Carrier-to-noise ratio, 752
  (*See also*, signal-to-noise ratio)
Cartesian
  coordinates, 146
  tensor notation, 156, 216
Catenary, 553
Cauchy, Augustin-Louis, 272
Cauchy
  distribution, 655
  inequality, 337
  integral formula, 332, 336
  integral theorem, 325
  principal value, 365
Cauchy-Euler equation, 31
Cauchy-Riemann equations, 198, 289
  polar form, 291
Causal function, 473
Central limit theorem, 691
Central moments, 650
Centripetal acceleration, 167
Characteristic function, 653
  normal distribution, 653
Characteristic
  value. *See* Eigenvalue
Chebyshev
  equation, 82
  polynomials, 81
Chi-square distribution, 665
Christoffel symbols, 251
Circle of convergence, 342
Circulation, 182, 394
Cis, 278
Classification
  of ODEs, 3
  of PDEs, 591
Clifford, W. L., 144
Coefficients of a DE, 17
Column,

buckling, 742
  vector, 121
Comb function, 65
Compatibility condition, 199
Complementary error function, 72
Complex
  conjugate, 275
  function, 281
  line integral, 321
  number, 273
  plane, 275
  potential, 388
  series, 339
  velocity, 387
Component
  contravariant, 222
  covariant, 222
  vector, 146
Conditional
  density functions, 656, 668
  expected value, 657
  probability, 655, 668
Conductivity constant, 599
Confluent hypergeometric
  equation, 97
  functions, 97
Conformal mapping, 313
Conjugate complex number, 275
Connected set, 183, 282
  simply, 183
Conservation of
  energy principle, 554
  mass, 175
  thermal energy, 599
Conservative field, 183
Constraints, 565
Contiguous functions, 96
Continuous
  distribution, 641
  random variable, 640
Contour, 321
  closed, 322
  indented, 371
  simple, 321
Contour integral, 321
Contraction, 224
Contravariant tensor, 222

Convective heat transfer, 611
Convergence, 339
    circle (radius) of, 342
    pointwise, 340, 409
    uniform, 340
Convolution theorem,
    discrete Fourier, 532
    Fourier transform, 470
    Laplace transform, 505
    Mellin, 522
Coordinates,
    Cartesian, 146
    curvilinear, 245
    cylindrical, 246
    polar, 608
    rectilinear, 218
    spherical, 268
Coriolis acceleration, 167
Correlation, 673
    coefficient, 674
    function, 701
Cosine,
    integral representation, 457
    inverse, 304
    series, 412
    transform, 460
Covariance function, 701
    normalized, 701
    table of, 714
Covariance, 673
    matrix, 723
Covariant
    derivative, 255
    tensor, 222
Critical
    damping, 738
    point, 129, 314
Cross
    -correlation function, 708
    -correlator, 710
    -covariance function, 708
    -spectral density, 721
Cross product, 150, 235
Cumulative distribution function, 640
    conditional, 655
Curl, 176
Curvature, 164

Curve
    arc length of, 163, 182
    rectifiable, 324
    simple, 321
Curvilinear coordinates, 245
Cycloid, 555
Cylinder, flow around, 396
Cylindrical coordinates, 246

D

d'Alembert, Jean Le Rond, 590
d'Alembert solution, 602
Damped motion, 735
dc component, 419, 715
Deformation of path, 326
Degrees of freedom, 573
Deleted neighborhood, 282
Delta,
    function. *See* Impulse function
    Kronecker, 157, 217
DeMoivre formula, 279
Density function, 642
    table of, 647
Dependent, linearly, 19, 135
Derivative of a
    complex function, 287
    vector function, 161
Detection,
    probability of, 756
    threshold, 754
Determinant, 112
Dido, problem of, 566
Difference equations, 536
Differential equation, 3
    Airy, 93
    Bessel, 50, 84
    Cauchy-Euler, 31
    Cauchy-Riemann, 289
    confluent hypergeometric, 97
    constant coefficients, 20
    elliptic, 592
    Helmholtz, 618, 762
    Hermite, 79
    homogeneous, 9, 17
    hyperbolic, 592
    hypergeometric, 96
    Laguerre, 80

Laplace, 177, 604
Legendre, 40, 76
linear, 4, 8
nonhomogeneous, 9, 24, 28
nonlinear, 4
ordinary, 4
parabolic, 592
partial, 591
separable, 7, 594
Sturm-Liouville, 425, 431, 432
Differential operator, 177
Diffraction pattern,
    circular slit, 517
    square aperture, 75
Diffusion equation. *See* Heat equation
Diffusivity constant, 593
Digamma (psi) function, 71
Dirac delta function. *See* Impulse
    function
Dirac, Paul A. M., 65
Direction angles, 147
Directional derivative, 171
Dirichlet
    condition, 584, 593
    problem, 198
Discrete
    Fourier transform, 529
    random variable, 640, 644
    Walsh transform, 540
    Z-transform, 533
Distribution,
    conditional, 656, 668
    continuous, 641
    discrete, 644
    joint, 665, 698
    marginal, 666
    table of, 647
Divergence, 173
    theorem, 202
Domain, 183, 282
Dot product, 148, 235
Doublet, 392
Drag, 396

E

Eigenfunction, 420
    expansion, 428, 447

orthogonality of, 427
Eigenvalue, 116, 241, 420
    problem, 116, 124, 420, 570
Eigenvector, 117, 124, 241
Eikonal equation, 758
Einstein, A., 144
Einstein summation rule, 156
Elastic
    beam, 580
    membrane, 583
    string, 580, 600
Electromechanical analogies, 741
Elliptic equation. *See* Laplace's
    equation
Elliptic integral, complete, 494
Energy, 405
    signal, 406, 461
    spectral density, 463
Ensemble, 698
    average, 700
Entire function, 295
Envelope, 737, 748
Equidimensional equation. *See*
    Cauchy-Euler equation
Equilibrium point. *See* Critical point
Equipotential
    curves, 306, 383
    surfaces, 170
Ergodic process, 704
Error function, 12, 72
    complementary, 72
Essential singularity, 353
Euclidean space, 136
Euler
    buckling load, 743
    constant, 71
    formulas, 278
Euler-Lagrange equation, 548, 561, 563
    system of, 559
Euler, Leonhard, 62, 404, 546, 590
Even
    function, 411
    periodic extension, 414
Event, 639
Existence-uniqueness theorem, 6, 18
Expansion theorem, 137, 428
Expected value, 649

conditional, 657
Exponential
    function, complex, 295
    order, 497
Extension, periodic, 413
Extremal, 547

F

Fade, 728
False alarm, 754
    probability of, 756
Faraday's law, 207
Fermat, Pierre de, 555, 638
Fermat's principle, 555
Fibonacci sequence, 544
Field,
    conservative, 183
    irrotational, 387
    scalar, 170
    vector, 173
Fluid
    flow, 175, 386
    ideal, 387
    incompressible, 387
Flux, 189, 599
Forced oscillations, 734
Forcing function, 8
    See also Input function
Fourier
    -Bessel series, 438
    coefficients, 408
    cosine integral, 457
    cosine series, 412
    integral representation, 456
    integral theorem, 455
    -Legendre series, 434
    law, 599
    series, 405
    series, complex, 416
    series, generalized, 428, 611, 774
    sine integral, 457
    sine series, 412
Fourier, Joseph, 404, 590
Fourier transform, 458, 461, 622
    cosine, 460
    discrete, 529
    fractional, 483

inverse, 458, 461
    properties of, 467
    sine, 460
    table of, 464
    two-dimensional, 477
Fractional-order
    derivatives, 483
    Fourier transform, 483
Free oscillations, 735
Frenet-Serret formulas, 164, 759
Frequency,
    angular, 416, 462
    fundamental, 419
    modulation, 745
    spectrum, 419
Fresnel integrals, 73
Frobenius, Georg, 44
Frobenius method, 44
Function,
    analytic, 291
    complex, 281
    multivalued, 300
    vector, 173
Functional, 547
    quadratic, 567
Fundamental
    frequency, 419, 601
    matrix, 122
    solutions, 18
    theorem of calculus, 330

G

Galilei, Galileo, 144
Gamma distribution, 69, 647, 649
    moments of, 69
Gamma function, 50, 67
    incomplete, 68
Gauss, Carl F., 62, 95, 144
Gauss divergence theorem, 202
Gauss's law, 208
Gaussian
    distribution. *See* Normal
        distribution
    function, 65
General solution,
    first-order, 9
    second-order, 19, 24

partial DE, 591
    system of linear DEs, 124, 132
Generalized
    coordinates, 573
    force components, 263
    Fourier series, 428, 611, 774
    hypergeometric functions, 98
Generating function, 77, 84
Geometric series, 340
Geometrical optics, 756
    eikonal equation, 758
    ray equation, 757
Gibbs, J. Williard, 144
Gibbs' phenomenon, 410
Gradient, 171
Green
    formulas, 199
    theorem, 194, 325
Green, George, 144, 404, 590
Green's function, 12, 438, 444
    bilinear formula for, 446
    one-sided, 439

H

Half-plane,
    Dirichlet problem for, 379
    Neumann problem for, 379
Half-range expansions, 413
Half-wave rectifier, 662
Hamilton, William Rowan, 144, 274
Hamilton principle, 572
Hankel functions, 93, 619
Hankel transform, 513, 630, 762
    inverse, 514
    properties of, 516
    table of, 517
Harmonic
    conjugate, 294
    function, 197, 293
    motion, 601
    oscillator, 577
Heat
    conductivity constant, 599
    equation, 592
    flux, 599
Heaviside expansion formula, 508
Heaviside, Oliver, 144

Heaviside unit function. *See* step
    function
Helmholtz equation, 618, 762
Hermite, Charles, 62
Hermite
    equation, 79
    polynomials, 78
Hermite-Gaussian function, 484, 765
Hermitian, 462
Hilbert space, 138
Hilbert transform, 473
    table of, 477
Homogeneous
    DE, 9, 17, 18
    solution, 9
    systems of DEs, 122, 123
Huygens, Christian, 2
Huygens-Fresnel integral, 764
    generalized, 480, 769
Hyperbolic
    equation. *See* Wave equation
    functions, 299
Hypergeometric
    equation, 96
    functions, 95

I

Ideal fluid, 387
Identity
    matrix, 113
    transformation, 220
Image
    point, 283
    plane, 283
Imaginary
    axis, 275
    part of, 274
Implicit boundary condition, 609
Improper integral, 365
Impulse function, 65
    sifting property of, 66
Impulse response function, 12, 472, 511
Incomplete gamma function, 68
Incompressible flow, 175
Inconsistent equations, 115
Indented contour, 371
Independence of path, 183, 328

Independent
  events, 656
  random variables, 672
  solutions, 19
Indicial equation, 45
Inequality
  Cauchy, 337
  Schwarz, 137
  triangle, 137
Infinite divisibility, 690
Infinity in complex plane, 310
Initial
  condition, 6, 11
  value problem, 6, 439, 508, 734
Inner product, 136, 139
  space, 135
Input function, 8, 17
Integral,
  contour, 321
  Fourier, 456
  improper, 365
  line, 179
  surface, 185
  transform, 496
  volume, 200
Interior point, 282
Intrinsic derivative, 259
Invariant, 222
Inverse
  hyperbolic functions, 304
  of a matrix, 114
  trigonometric functions, 304
Irregular singular point, 44
Irrotational fluid, 387
Isolated singularity, 353
Isoperimetric problem, 564
Isotherms, 383

J

Jacobian, 191, 201, 219, 245, 314, 687
Joint
  distribution, 665, 698
  moments, 673
Jointly stationary, 709

K

K distribution, 684

Kepler, Johannes, 168
Kepler's laws, 169
Kernel, 496
Kirchoff's laws, 741
Kolmogorov, Andrei N., 638
Kramers-Krönig relations, 476
Kronecker delta, 157, 217
Kummer, Ernst E., 62
Kummer functions. *See* Confluent
  hypergeometric functions
Kutta-Joutkowski lift formula, 399

L

Lagrange
  equation, 80
  multiplier, 566
Lagrange, Joseph Louis, 62, 590
Lagrangian, 572
Laguerre
  associated polynomials, 80
  equation, 80
  polynomials, 80
Laplace equation, 177, 604
  Dirichlet problem, 605
  Neumann problem, 605
Laplace, Pierre Simon de, 590
Laplace transform, 496, 624
  complex inversion formula, 507
  inverse, 497, 503
  properties of, 502, 504
  table of, 501
Laplacian, 177, 593
Laurent
  series, 346
  theorem, 349
Lebesgue, Henri, 638
Legendre, Adrien M., 62, 590
Legendre
  equation, 40, 76
  functions of the second kind, 42
  polynomials, 41, 76
  series, 434
Leibniz, Gottfried Wilhelm von, 2, 546
Length
  of a curve, 163
  of a vector, 145
Level curve, 306, 383

Levi-Civita, T., 144
L'Hôpital, G. F. A., 546
L'Hôpital's rule, 55
Lift, 396
Limit,
    complex function, 284
    distribution, 690
Linear
    dependence, 19, 135
    differential equation, 8, 17
    independence, 19, 135
    mapping, 307
    shift-invariant system, 472, 479
    system, 114, 471, 718,
    vector space, 133
Line integral,
    complex, 321
    vector, 179
Liouville, Joseph, 404
Liouville theorem, 338
Logarithm function, complex, 301
    principal branch, 301

M

Maclaurin series, 344
Magnitude of a
    complex number, 276
    vector, 145
Mapping, 283
    bilinear, 311
    conformal, 313
    fixed point, 312
    linear, 307
    one-to-one, 296
    reciprocal, 308
Marcum Q-function, 682
Marginal
    density, 666, 667
    distribution, 666
Markov, Andrei A., 638
Matrix, 110
    eigenvalue problem, 116
    inverse, 114
    nonsingular, 114
    properties, 111
    singular, 114
    skew-symmetric, 119

    symmetric, 113, 119
    transpose, 113
Maximum likelihood,
    estimate, 671
    principle of, 671
Maximum modulus theorem, 338, 611
Maxwell, James C., 144
Maxwell's equations, 206
Mean value, 650, 700
Mellin transform, 519
    inverse, 519
    properties of, 521
    table of, 522
Metric
    space, 135
    tensor, 228, 248
Mixed
    random variables, 644
    triple product, 151, 235
ML inequality, 323
Möbius transformation. *See* Bilinear
    transformation
Modified Bessel functions, 88
Modulus of complex number, 276
Moments,
    central, 650
    of a distribution, 650
    generating function for, 654
    joint, 673
Morera's theorem, 337
Multiplication rule, probability, 655
Multivalued function, 300
Multivariate distribution, 723

N

Narrowband noise, 747, 752
Necessary condition, 546, 549, 556
Negative exponential distribution, 643
Neighborhood, 282
Neumann
    condition, 593
    problem, 198
Newton, Sir Isaac, 2, 546
Newton second law of motion, 2
Noise,
    narrowband, 747, 752
    white, 720

Nonhomogeneous
    DE, 9, 24, 28
    solution. *See* Particular solution
    systems of DEs, 131
Nonlinear DE, 4
Nonsingular matrix, 114
Nontrivial solution, 420
Norm, 136, 139, 145, 427
Normal distribution, 73, 646
    bivariate, 665
    moments of, 651, 654
    multivariate, 723
Normal form, 8, 17
Normalized
    covariance function, 701
    vector, 145
Normal random variable, 646, 647
    characteristic function for, 653
    density function for, 73, 646
    moments of, 651, 654
Normal vector, 164, 186

O

Odd
    function, 411
    periodic extension, 414
One-sided Green's function, 439
One-to-one mapping, 296
Open set, 183, 282
Order of DE, 4
Ordinary
    DE, 4
    point, 36
Oriented surface, 204
Orthogonal
    coordinates, 253
    curves, 306
    eigenfunctions, 427
    eigenvectors, 137
    functions, 42, 76, 427
    random variables, 673
    vectors, 148
Orthonormal, 137
Oscillations,
    damped, 735
    forced, 735
    free, 735

    small, 575
    undamped, 735
Ostrogradsky, Michel, 144, 590
Output function, 8, 471
Overdamping, 738

P

Parabolic equation. *See* Heat equation
Parallelogram law, 147
Parametric representation, 162
Paraxial approximation, 762
Parseval's relation for
    energy signals, 463
    power signals, 418
Partial
    DE, 591
    fractions, 506
    sum, 339, 410
Particular solution, 5, 9, 132
Path. *See* Contour.
Pascal, Blaise, 638
Pendulum problem, 574
    compound, 586
    double, 577
Period, 296, 406
Periodic
    boundary condition, 431
    extension, 414
    function, 296, 406
    Sturm-Liouville system, 431
Permutation symbols,
    Cartesian coordinates, 158, 218
    rectilinear coordinates, 232
Phase, 748
    plane, 129
    spectrum, 462
Physical components, 239
Picard's theorem, 356
Piecewise continuous function, 409
Pochhammer symbol, 94
Point at infinity, 310
Pointwise convergence, 340, 409
Poisson
    distribution, 645
    equation, 209, 583
Poisson integral formula for
    circular domain, 381

half plane, 379, 628
Polar
    coordinates, 608
    form of complex numbers, 276
Pole
    of order *m*, 353
    simple, 354
Polygamma function, 71
Polynomials,
    Chebyshev, 81
    Hermite, 78
    Laguerre, 80
    Legendre, 41, 76
Positive direction, 322
Posterior density function, 671
Potential,
    complex, 388
    equation. *See* Laplace's equation
    energy principle, 579
    function, 183
Power
    series, 34, 341
    series method, 34
    signal, 406, 461
    spectral density, 711
Principal
    axes, 120, 241
    branch, 302
    value, 277
Principle of
    maximum likelihood, 671
    minimum potential energy, 579
Probability
    as relative frequency, 639
    axioms of, 639
    conditional, 655, 668
    density function, 642
    total, 669
Probability density functions, 642
    binomial, 647
    Cauchy, 655
    chi-square, 665
    conditional, 656
    gamma, 647, 649
    K, 684
    negative exponential, 643
    normal, 646, 647

Poisson, 645, 647
Rayleigh, 647, 648
Rician, 647
table of, 647
uniform, 92, 643, 647, 648
Product of
    complex numbers, 274
    random variables, 526, 683
    vectors, 148-151
Projection of a vector, 149
Proper transformation, 219
Psi function. *See* Digamma function
Pure imaginary number, 274

Q

Q-function, 682
Quotient
    law, 224
    of random variables, 685

R

Radius of convergence, 342
Random process, 698
    stationary, 700
Random variable,
    continuous, 640
    discrete, 640, 644
    of mixed type, 644
    products of, 526, 683
    quotients of, 685
    statistically independent, 672
    sums of, 678, 688
Rayleigh distribution, 92, 647, 648
    moments of, 652
Rayleigh
    energy theorem, 463
    principle, 568
    quotient, 565
Rayleigh-Ritz method, 567
Real
    axis, 275
    part of, 274
Realization, 698
Reciprocal mapping, 388
Rectangle function, 64
Rectifiable curve, 324
Rectilinear coordinates, 218

Recurrence formula, 38, 50
Region, 282
Regular
    singular point, 44
    Sturm-Liouville system, 425
Relative frequency, 639
Removable singularity, 353
Residue, 357
Residue theorem, 361
Resonance, 616
Ricci, G., 144
Ricci's theorem, 257
Rician distribution, 92, 681, 647
    moments of, 753
Rice-Nakagami distribution. *See* Rician
    distribution
Riemann, Georg. F. B., 272
Riemann
    -Christoffel tensor, 259
    -Stieltjes integral, 712
    zeta function, 71
Right-hand rule, 150
RLC circuit, 509, 741
Robin condition, 593
Root-mean-square (rms) value, 715

S

Sample space, 639
Sampled function, 529
Scalar, 145, 216
    field, 170
    multiplication, 148
    triple product, 151
Schwarz inequality, 137
Self-adjoint equation, 425
Separation of variables,
    ODEs, 7
    PDEs, 594
Sequency domain, 540
Series,
    geometric, 340
    of Bessel functions, 438
    convergence of, 339, 342, 409
    of eigenfunctions, 428, 611
    Fourier, 405
    Laurent, 346
    Maclaurin, 344

partial sum of, 339, 410
    power, 341
    Taylor, 344
Sifting property, 66
Signal-to-noise ratio, 92, 750
Signum function, 63
Simple
    curve, 321
    harmonic function, 601, 735
    pole, 353
    zero, 352
Simply connected, 183
Sinc function, 65
Sine,
    integral representation, 457
    inverse, 304
    series, 412
    transform, 460
Singular
    matrix, 114
    Sturm-Liouville system, 432
Singular point, 36, 292, 352
    branch, 302
    essential, 353
    irregular, 44
    isolated, 353
    pole, 353
    regular, 44
    removable, 353
Sink, 391
Skew-symmetric tensor, 225
Smooth
    curve, 180, 321
    surface, 185
Solution, 4
    general, 5, 9, 591
    homogeneous, 9
    particular, 5, 9
    steady-state, 596
    trivial, 9
Source, 391
Spectral density, 711
    cross, 721
Spectrum,
    line, 419
    magnitude, 462
    phase, 462

Spherical Bessel functions, 92
Spherical coordinates, 201, 268
Square
    -integrable, 406
    matrix, 110
    wave, 410
Square-law device, 661, 716
Stable system, 129, 511
    asymptotically, 130
Stagnation point, 388
Standard deviation, 651
Standing waves, 601
Stationary,
    function, 549
    increments, 707
    in the wide sense, 702
    random process, 700
    value, 549
Steady-state solution, 596
Step function, 63
Stereographic projection, 310
Stirling's formula, 691
Stochastic variable. *See* Random
    variable
Stokes, George G., 144
Stokes's theorem, 204
Stream function, 388
Streamline, 387
Structure function, 707
Sturm, Jacques C. F., 404
Sturm-Liouville system, 425, 431, 432,
    567
Sum of random variables, 678, 688
Summation convention, 156
Superposition principle, 18, 595
Surface,
    area, 189
    integral, 185
    normal to, 186
Surge, 728
Symmetric
    matrix, 113, 119
    tensor, 225
System
    function, 472
    linear, 471
    of DEs, 121

T
Table of
    covariance functions, 714
    distributions, 647
    Fourier transforms, 464
    Laplace transforms, 501
    Hankel transforms, 517
    Mellin transforms, 522
    power spectral densities, 714
    Z-transforms, 535
Tangent vector, 162
Taylor
    series, 344
    theorem, 344
Tensor,
    components, 156
    contravariant, 222
    covariant, 222
    definition of, 221
    metric, 228, 248
    mixed, 222
    notation, 156, 216
    physical components, 239
    transformation laws, 221
Thermal
    conductivity, 599
    diffusivity, 593
Time
    average, 703
    invariant, 716
Torsion of a curve, 165
Total probability, 669
Trajectories, 129
Transfer function, 472, 479, 481, 510
Transient solution, 597
Transpose of a matrix, 113
Traveling waves, 604
Triangle
    function, 64
    inequality, 137
Trigonometric
    form of complex numbers, 277
    functions, complex, 297
    series. *See* Fourier series
Triple
    integral, 200
    product, 151, 235

Trivial solution, 9
Two-dimensional
    distribution, 665
    Fourier transform, 477

U

Uncorrelated random variables, 673
Undamped motion, 735
Underdamping, 738
Undetermined coefficients, 24
Uniform
    convergence, 340
    distribution, 92, 643, 647, 648
Unit
    binormal vector, 164
    circle, 363
    impulse response. *See* Impulse
        response function
    normal vector, 164
    step function. *See* step function
    tangent vector, 163
    vector, 145
Unitary space, 136
Unstable system, 130, 512
    marginally, 512

V

Variance of a distribution, 651
Variation of parameters, 9, 28, 132
Vector, 145, 216
    components, 146
    cross product, 150, 235
    differential operator, 177
    dot product, 148, 235
    field, 173
    magnitude, 145
    space, 133
    triple product, 151, 235
    unit, 145
    zero, 145
Velocity,
    complex, 387
    potential, 388
    vector, 166
Vibrating
    membrane, 583
    string, 580, 600

Volume integral, 200
Vortex flow, 391

W

Walsh
    functions, 539
    transform, 538
Wave equation, 582, 600
    d'Alembert solution of, 602
Wavelet, 487
    transform, 491
Weierstrass, Karl, 62, 272
Weierstrass *M*-test, 341
Weighted residual, 568
Weighting function, 425
White noise, 720
Wide-sense stationary, 702
Wiener-Khinchin theorem, 713
Work, 182, 184
Wronskian, 19, 122

Y

Young's modulus, 581

Z

Zernike polynomials, 773
    aperture filter functions, 777
    modal expansion, 776
    table of, 776
Zero
    of analytic function, 352
    crossings, 726
    vector, 145
Zeta function. *See* Riemann zeta
    function
Z-transform, 533
    properties of, 535
    table of, 535

Larry C. Andrews is a professor of mathematics at the University of Central Florida and an associate member of the School of Optics/CREOL. He is also an associate member of the Florida Space Institute (FSI). Previously, he held a faculty position at Tri-State University and was a staff mathematician with the Magnavox Company, antisubmarine warfare (ASW) operation. He received a doctoral degree in theoretical mechanics in 1970 from Michigan State University. Dr. Andrews has been an active researcher in optical wave propagation through random media for more than 20 years and is the author or co-author of 10 textbooks on differential equations, boundary value problems, special functions, integral transforms, and wave propagation through random media. Along with wave propagation through random media, his research interests include special functions, random variables, atmospheric turbulence, and signal processing.

Ronald L. Phillips is Director of the Florida Space Institute (FSI) and a professor in the Department of Electrical and Computer Engineering at the University of Central Florida. Dr. Phillips is also a member of the Department of Mathematics and an associate member of the Center for Research and Education in Optics and Lasers (CREOL). He has held positions on the faculties at Arizona State University and the University of California, San Diego. He received a doctoral degree in Electrical Engineering in 1970 from Arizona State University. Dr. Phillips has been an active researcher in wave propagation through random media for more than 22 years. He was awarded a Senior NATO Postdoctoral Fellow in 1977 and the American Society for Engineering Education 1983 Medal for outstanding contributions in research. Dr. Phillips is co-author of two textbooks on wave propagation through random media and, in addition to optical wave propagation, his research interests include optical communications and imaging through atmospheric turbulence.

AS
NATURE
MADE HIM

AS
NATURE
MADE HIM

≡

The Boy Who Was Raised as a Girl

JOHN COLAPINTO

HarperCollins*Publishers*Ltd

AS NATURE MADE HIM: THE BOY WHO WAS RAISED AS A GIRL

Copyright © 2000 by John Colapinto.

http://www.harpercanada.com

HarperCollins books may be purchased for educational,
business, or sales promotional use. For information please write:
Special Markets Department, HarperCollins Canada, 55 Avenue
Road, Suite 2900, Toronto, Ontario, Canada M5R 3L2.

First HarperCollins hardcover ed. ISBN 0-00-200047-4
First HarperCollins trade paper ed. ISBN 0-00-638614-8

Canadian Cataloguing in Publication Data
Colapinto, John, 1958–
    As nature made him : the boy who was raised as a girl.
        ISBN 0-00-200047-4
        1. Transsexuals — Canada — Biography. 2. Sex change
    — Case studies. 3. Gender identity. 4. Nature and nurture.
    I. Title.
    RC560.G45C64 2000   305.9'066   C99-932789-5

99 00 01 02 03 04 RRD 8 7 6 5 4 3 2 1

Printed and bound in the United States

To Donna

I have entered on an enterprise which is without precedent, and will have no imitator. I propose to show my fellows a man as nature made him, and this man shall be myself.

—ROUSSEAU, *Confessions*

How could I not be glad to know my birth?

—SOPHOCLES, *Oedipus Rex*

The difficulty is to detach the framework of fact—of absolute undeniable fact—from the embellishments of theorists and reporters. Then, having established ourselves upon this sound basis, it is our duty to see what inferences may be drawn and what are the special points upon which the whole mystery turns.

—*Memoirs of Sherlock Holmes*

Author's Note

≡

THIS IS A WORK OF NONFICTION. All passages of dialogue are taken verbatim from tape transcripts of psychological interviews, from contemporaneous psychiatric session notes, or from the direct recollection of witnesses to, or participants in, these events. No dialogue or scenes have been invented for the purposes of "narrative flow," "atmosphere," or any other quasi-novelistic purpose. The account of Dr. Money's appearance on the Canadian Broadcasting Corporation television show in 1967 is taken from a videotape of that program—a tape that was, miraculously, not destroyed in the thirty years since its broadcast. Direct dialogue from the Psychohormonal Research Unit sessions (published here for the first time) is taken from tape transcripts which Dr. Money supplied to the patient in June 1998 upon the request of the patient's local physician.

Preface

=

ON THE MORNING of 27 June 1997 I paid my first visit to David Reimer's home, a small, nondescript dwelling in a working-class neighborhood of Winnipeg, Manitoba. There was nothing about the house to suggest that its owner might arouse the interest of a journalist from New York City—not to mention the fascination of scientists and doctors the world over. On the well-tended lawn, a child's bicycle lay on its side. At the curb was parked an eight-year-old secondhand Toyota. Inside the house, a handmade wooden cabinet in a corner of the living room held the standard emblems of family life: wedding photos and school portraits, china figurines, and souvenirs from family trips. There was a knock-off antique coffee table, a worn easy chair, and a sofa—which was where my host, a wiry young man dressed in a jean jacket and scuffed work boots, seated himself.

At thirty-one years of age David Reimer could have passed for more than a decade younger. Partly it was the sparseness of his facial hair—just a few blond wisps that sprouted from his jawline; partly it was a certain delicacy to his prominent cheekbones and tapering chin. Otherwise he looked and sounded like what he was: a blue-collar factory worker, a man of high school education, whose fondest pleasures were to do a little weekend fishing with his dad in the local river or to have a backyard barbecue with his wife and kids. He was the kind of rough-edged but affable young man whose conversation ran to such topics as his tinkering with his car engine, his work woes, or the challenges of raising three kids on less than forty thousand a year.

I had come to Winnipeg to learn all I could about David Reimer, but my chief interest was in his childhood—a subject that, when I raised it, brought an immediate and dramatic change in him. Gone was the smile on his face and the bantering tone in his voice. Now his brows gathered together above his small straight nose, his eyes began to blink with startled rapidity, and he thrust his chin forward like someone who'd just been challenged to a fight. His voice—a deep, burred baritone—took on a new pitch and rhythm, an insistent, hammering rhythm, which for all its obvious aggrievement and anger also carried the pleading edge of someone desperate to communicate emotions that he feared others could never understand. How well even *he* understood these emotions was not immediately clear. I noticed that when David described events that had occurred prior to his fifteenth birthday, he tended to drop the pronoun *I* from his speech, replacing it with the distancing *you*—almost as if he were speaking about someone else altogether. Which, in a sense, he was.

"It was like brainwashing," he was saying as he lit the first in an unbroken chain of cigarettes. "I'd give just about anything to go to a hypnotist to black out my whole past. Because it's torture. What they did to you in the body is sometimes not near as bad as what they did to you in the *mind*—with the psychological warfare in your head."

He was referring to the events that had begun to unfold on an April morning three decades earlier when, at eight months of age, he lost his entire penis to a botched circumcision. As a result of that irreparable injury, his parents had taken him to see a famed expert in sex research at the renowned Johns Hopkins Hospital in Baltimore where they were convinced to submit their son to a surgical sex change. The process involved clinical castration and other genital surgery when he was a baby, followed by a twelve-year program of social, mental, and hormonal conditioning to make the transformation take hold in his psyche. The case was reported in the medical literature as an unqualified success, and he became one of the most famous (though unnamed) patients in the annals of modern medicine.

It was a fame that derived not only from the fact that his medical and surgical metamorphosis from boy to girl was the first infant sex reassignment ever reported on a developmentally normal child, but also from a stunning statistical long shot that lent special significance to the case. He had been born an identical twin. His brother and sole sibling provided to the experiment a built-in matched control—a genetic clone who, with penis and testicles intact, was raised as a male. That the twins were reported to have grown into happy, well-adjusted children of opposite sex seemed unassailable proof of the primacy of environment over biology in the differentiation of the sexes. Textbooks in medicine and the social sciences were rewritten to include the case, and a precedent for infant sex reassignment as standard treatment in cases of newborns with injured or irregular genitals was established. The case also became a touchstone for the feminist movement in the 1970s, when it was widely cited as proof that the gender gap was purely a result of cultural conditioning, not biology. For Dr.

John Money, the medical psychologist who was the architect of the experiment, the so-called "twins case" became the most publicly celebrated triumph of a forty-year career that in 1997 earned him the accolade "one of the greatest sex researchers of the century."

But as the mere existence of the young man who sat in front of me on that morning in June 1997 would suggest, the experiment was a failure—a fact not publicly revealed until that spring, in the medical journal *Archives of Pediatrics and Adolescent Medicine*. There, authors Dr. Milton Diamond, a biologist at the University of Hawaii, and Dr. Keith Sigmundson, a psychiatrist from Victoria, British Columbia, had documented how David had struggled against his imposed girlhood from the start and how, at age fourteen, he had reverted to the sex written in his genes and chromosomes. The paper had set off shock waves in medical circles around the world, generating furious debate about the ongoing practice of infant sex reassignment (a procedure more common than a layperson might think). The paper also raised troubling questions about the way the case had been reported in the first place, why it had taken almost twenty years for a follow-up to reveal the actual outcome, and why that follow-up was conducted not by Dr. Money or Johns Hopkins, but by outside researchers. The answers to these questions, fascinating for what they suggest about the mysteries of sexual identity, also brought to light a thirty-year rivalry between eminent sex researchers, a rivalry whose very bitterness not only dictated how this most unsettling of medical tragedies was exposed, but also may have been the impetus behind the experiment in the first place.

What was shaping up for medicine to be a highly public scandal involving some of the biggest names in the world of sex research was for David Reimer a purely private catastrophe.

Apart from two television interviews that he granted in the summer of 1997 (his face obscured, his voice disguised), he had never told his story in full to a journalist. He had agreed to speak to me, for an article I was preparing on the case for *Rolling Stone* magazine, on the condition that I withhold crucial details of his identity. Accordingly, in the article I did not reveal where he was born, raised, and continued to live, and I invented pseudonyms for his parents, Ron and Janet, and for his identical twin brother, Brian. The physicians who treated him in Winnipeg, I identified by initials. David himself I called, variously, "John" and "Joan," the pseudonyms given to him by Diamond and Sigmundson in their journal article describing the macabre double life he had been obliged to lead. So careful was I to provide not even the most oblique clue to David's geographic whereabouts that I omitted even to mention the historic blizzard that paralyzed Winnipeg on the morning of his circumcision accident—a freak late April snowfall eerily evocative of those reversed natural wonders that always presage horror in Shakespearean and Greek tragedies.

My *Rolling Stone* article appeared in December 1997. At nearly twenty thousand words, it was as thorough a job as could be managed under the space constraints and deadline pressures of magazine journalism. But even as the piece went to press, it was clear that David's life and the scientific machinations that played such a decisive part in shaping it were of sufficient complexity, scientific import, and human drama to require, for their fullest telling, a book. David, it transpired, had been thinking along the same lines and wanted me to be the book's author. At which point I was obliged to reveal to him an important condition for my taking on the project: that he abandon the mask of John/Joan.

Quite apart from the fact that I could not imagine writing a book in which the central character, his family, friends, physi-

cians, and others exist as pseudonyms moving against the indeterminate background of a "city somewhere in the North
American Midwest," I also knew that a knowlege of his specific
geographic location and the people who inhabited it were considered vital to a proper understanding of the case. In a tale at
the very heart of which lies the debate surrounding nature versus nurture, genetics versus environment, biology versus rearing, it was imperative that I be permitted to describe in detail
the sociocultural milieu in which David was raised. Finally, as a
writer, I knew how many of the story's peculiarly poetic reso-
nances would be lost should David insist upon anonymity. To
stick to pseudonyms would mean forfeiting the story of how
David, when beginning his laborious switch back to boyhood
at age fourteen, had rechristened himself with a male name different from his original birth name of Bruce—one that not only
had the kind of down-to-earth masculine directness he favored,
but also evoked for him his accomplishment in triumphing over
the array of forces that had conspired, for the first fourteen
years of his life, to convince him that he was someone other
than the person he felt himself inwardly to be. It was owing to
this unlikely victory that he had decided to name himself after
the child in the Bible story who slew the seemingly invincible
giant Goliath. Here, and in myriad other instances, to retain
pseudonyms was to sacrifice a fact that reflected not only on the
saga as a whole, but on David's own understanding of it.

Beginning with the interviews he had first granted to Diamond and Sigmundson in early 1993 for their journal article
and continuing with the interviews he gave to me for *Rolling
Stone*, David had been moving by degrees out of the shadows
of shame and secrecy in which he had been living. By the time I
spoke to him about abandoning the mask of John/Joan, he had
already come a long way in that journey. After discussing it
with his wife, parents, and brother, and sleeping on it for one

night, David told me that he was ready to step forward as his true self.

For the purposes of my reconstructing his past, David closed no doors, shut down no avenue of inquiry. He granted me over one hundred hours of interviews spread over twelve months, and he signed confidentiality waivers giving me access to an array of private legal papers, therapy notes, Child Guidance Clinic reports, IQ tests, medical records, and psychological workups that had accumulated over the course of his remarkable childhood. He assisted me in finding the schoolteachers and classmates who had known him in childhood—a difficult task of sleuthing since he had kept no school yearbooks, remembered few of his peers' last names, and had spent the previous decade and a half trying to forget and avoid anyone who had known him in his previous incarnation as a girl. Most crucially, David helped me obtain interviews with all of his family members, including his father, who because of the painfulness of these events had not spoken of them to anyone in more than twenty years. It is only through the Reimer family's rare candor that the full story of John/Joan can finally be told. Although that story is primarily about David Reimer and his experience of living on both sides of the gender divide, it is also about the young couple who, barely out of their teens, made the momentous decision to submit one of their twin baby boys to this unprecedented, and ultimately doomed, experiment in psychosexual engineering.

"My parents feel very guilty, as if the whole thing was their fault," David explained to me during my first visit to Winnipeg. "But it wasn't like that. They did what they did out of *kindness* and love and desperation. When you're desperate, you don't necessarily do all the right things."

PART ONE

A Game of Science Fiction

I

≡

THE IRONY WAS that Ron and Janet Reimer's life together had begun with such special promise. That it would survive its trials is attributable perhaps in part to their shared heritage in an ethnic and religious background virtually defined by the hardiness of its people in the face of suffering.

Both Ron Reimer and Janet Schultz were descended from families who were Mennonite, the Anabaptist sect founded in sixteenth-century Holland. Like the Amish, Ron's and Janet's Mennonite ancestors were pacifists who followed a simple, nonworldly life based directly on Christ's teachings in the Sermon on the Mount. During the Inquisition, Mennonites were tortured and slaughtered in the thousands, the survivors escaping to begin a three-hundred-year search for a country that would allow them to live as a culture and religion apart. The majority went to Russia and farmed, but in the late 1800s, large numbers began to migrate to the New World, some settling in Nebraska and Kansas. The densest concentrations, however, settled in Canada, where the federal government, eager to populate its empty western plains, offered to the Mennonites complete religious freedom, their own schools, and exemption from military service. The first Mennonites arrived in southern Manitoba in 1874. Within five years, over ten thousand had followed, transplanting entire Russian villages to the Canadian prairie. It was in this wave of immigrants that both Ron's and Janet's great-grandparents, who were Dutch Mennonites directly descended from the earliest followers of the sect, came to Manitoba.

Their arrival coincided with that moment when the Canadian Pacific Railway reached Winnipeg, and transformed the once tiny and isolated fur-trapping settlement and Hudson's Bay trading post. Within three decades the settlement had become a major grain capital of the North American middle west. "All roads lead to Winnipeg," the *Chicago Record Herald* reported in 1911. "It is destined to become one of the greatest distributing commercial centers of the continent as well as a manufacturing community of great importance."

Though the city failed to live up to those grand predictions, Winnipeg did grow rapidly in size, sophistication, and importance over the first half of the twentieth century, establishing the country's first national ballet company and symphony orchestra. Today its population is over 600,000, and the city's downtown core, built around the meandering curves of the Red River, boasts an impressive stand of modern high-rises to complement its fine Victorian buildings.

The Mennonites on the surrounding prairies had long felt the lure of Winnipeg's affluence, and after World War II the more assimilated families began to move into the city to take jobs in manufacturing, trucking, and construction. Among them were Ron Reimer's parents, Peter and Helen, who in 1949 sold their farm in nearby Deloraine and moved to the Winnipeg neighborhood of St. Boniface, where Peter took a job in a slaughterhouse and Helen raised their four young children, of whom Ron was the eldest.

Even as a small child, he was dutiful and hardworking, a boy whose combination of personal privacy and dogged industry often amazed his own mother. "He was always so shy and quiet," Helen Reimer recalls, "but he was also such a *busy* little boy. I had to think up ways to keep him out of trouble. I would show him how to cook. He always wanted to be doing something with food and cooking." It was a passion that would stay with Ron. As an adult he would eventually support his wife and two children by running his own business as the operator of a coffee truck, supplying sandwiches and other prepared foods to construction sites around Winnipeg.

By 1957, when Ron was in his early teens, the music of Elvis Presley, Chuck Berry, and Little Richard had reached Winnipeg. Cars, girls, beer, and rock 'n' roll music soon had strong claims on his attention. For Mennonites of Ron's parents' generation, the swift cultural changes of the late 1950s were threatening. Though not themselves especially devout, they had only a decade earlier moved from an almost exclusively Mennonite farm community where some of the day-to-day values and assumptions were still closer to those of nineteenth-century rural Russia than late-twentieth-century urban North America. In what would prove to be a kind of reverse migration, the Reimers were among many Mennonite families who, in an effort to resist the seismic cultural shifts taking place in the city, returned their families to their roots on the prairie. In 1959, Ron's father bought a farm some sixty miles from the city, near the town of Kleefeld, in Mennonite country, and moved his family there.

Ron, fifteen years old at the time, hated the move. Kleefeld itself was little more than a ramshackle scattering of stores along a few hundred yards of gravel highway (grain store, post office, grocery), with nowhere for Ron to channel his formidable work ethic. He would pick two hundred pounds of saskatoons and sell them for twenty-five cents a pound—grueling labor for little pay; nothing like the money he was able to make in the city. And his father insisted on taking even those paltry sums from Ron for upkeep of the old clapboard farmhouse on its patch of scrubby land.

It was in this state of boredom, penury, and growing friction with his strict and authoritarian father that Ron, at seventeen, accepted the invitation of his friend Rudy Hildebrandt to visit Rudy's girlfriend in the nearby town of Steinbach. Rudy's girlfriend had a nice-looking roommate, a girl named Janet, whom Ron might like.

Like Ron, Janet Schultz was raised in Winnipeg, the eldest child of Mennonite parents who had joined the postwar migration from the prairie to the city. Growing up in the Winnipeg neighborhood of St. Vital, Janet was a lively and inquisitive girl whose passion for reading—first Nancy Drew and Hardy Boys books, then thrillers, and eventually books on psychology—opened up for her a perspective on life beyond the traditional Mennonite values of her parents—and in particular her mother, with whom she constantly clashed. "I wanted an education, but my mother wanted me to get out to work and bring home money," Janet says. Eventually she was convinced to quit school after ninth grade and take a job at a sewing factory. Janet gave her paychecks to her mother, which did little to foster goodwill between them. A further gulf opened between mother and daughter when Janet, in her early teens, stopped attending the Mennonite church. "I found it was so restrictive," she says. "I didn't think it was biblical. They said it was a sin to smile. I didn't think that way." In fact, by age fifteen, Janet was given to joking about her parents' religion. "Why don't Mennonites ever make love standing up?" she liked to ask her friends. "Because someone might think they were *dancing*!" Janet herself loved to go dancing and roller-skating, and as an exceptionally pretty hazel-eyed brunet with a shapely figure, she never lacked for dates.

Convinced that their eldest child and only daughter was slipping dangerously from their control, Janet's parents, like Ron Reimer's, joined the migration of city Mennonites back to the farm. In 1960, when Janet was fourteen, the Schultzes relocated to New Bothwell, a tiny settlement amid the silos and grain fields forty-five minutes from Winnipeg. Janet missed the city's movie theaters, restaurants, roller rinks, and dance halls—and soon began accepting dates from any boy who had a car and thus could offer her escape from the farm. Janet's mother tried to curb her daughter's social life but to no avail. Shortly after Janet's fifteenth birthday, her mother told her to move out. Janet went gladly. She moved to the nearby city of Steinbach, where she found work at a sewing factory and shared a small apartment in a rooming house with her cousin Tina. Not long after that, Tina's boyfriend brought a young man over to meet Janet. He was a tall blond boy of seventeen with large blue eyes and a shy way of glancing at her. His name was Ron Reimer. "I was flirting with Ron," she says, laughing, "and I was thinking he wasn't flirting back, so I figured he didn't like me."

Ron *did* like her, but was too shy to reveal his feelings in front of the other couple. He invited Janet to have a look at his car on the street, then asked her out to see a movie on the weekend. He raised money for their date by taking the transmission out of a junkyard Ford and selling it to a friend for ten dollars. That weekend, Ron and Janet went to see *Gidget Goes Hawaiian*. "I don't think I watched five minutes of that movie," Janet laughs. "I was too busy eyeballing him. Oh, he was so sexy!"

Over the course of the summer they saw a lot of each other,

joining Tina and Rudy on double dates—usually just a drive out to one of the isolated country roads where they would park, drink a six-pack, make out, and talk. As Ron and Janet compared their backgrounds, they were amazed to discover how much they had in common. Their similarities drew them together, but paradoxically enough so did their differences. Janet could compensate for Ron's sometimes passive reluctance to take decisive action; Ron, on the other hand, with his slow, considered approach to life, could rein Janet in from her more reckless enthusiasms and impulses. Together they made up a single entity stronger than either one of them.

When Janet decided to move back into Winnipeg, there was never any question but that Ron would follow her. Though they did not rent an apartment together—this was the early 1960s, and such boldness would have been unthinkable for a pair not yet out of their teens—Ron did spend much of his time with Janet in her rooming house. It was there that they slept together for the first time. Both had been virgins. And not long after that, Janet missed her period. She had just turned eighteen. Ron was nineteen, soon to turn twenty. It was young to marry, but they had talked about marriage before. This was simply a sign that they should bless their union sooner rather than later. The two were married on 19 December 1964 in the city of Steinbach. In acknowledgment of the emancipation they now felt from their disapproving parents, they deliberately declined to be married in one of the city's twenty Mennonite churches.

The newlyweds moved into a tiny cold-water flat in downtown Winnipeg. They couldn't afford better. Janet was getting minimum wage working as a waitress at the Red Top diner; Ron was toiling for low pay at a factory that made windows. That they would have to bring in more money was obvious—especially when, during one of Janet's checkups with her obstetrician, she learned that she was pregnant with twins. Ron was nervous, but Janet refused to be anything but optimistic. "I was so excited," she says, "because all my life I'd been dreaming, Oh wouldn't it be wonderful to have twins?"

That June, when Janet was five months pregnant, Ron landed a union job at one of the city's biggest slaughterhouses, and his pay more than doubled, enabling them to move into a two-bedroom apartment on the corner of Dubuc and Des Meurons Streets. Then the couple had a scare. When she was in the latter stages of her pregnancy, Janet developed a serious case of toxemia—a pregnancy-related form of high blood pressure that, untreated, can be harmful to the fetus. Her doctor recommended that she have her labor and delivery induced.

On 22 August 1965, some four weeks before her projected due date, Janet was admitted to St. Boniface Hospital. During his wife's labor, Ron sat in the visitor's lounge nervously awaiting the outcome. After several hours, a nurse came and announced that everything had gone fine and that he was the father of identical twins. In his relief and excitement at hearing that Janet and the babies were alive and well, Ron failed to take in anything else. So as he hurried through the doorway toward the nursery to see his children, he was brought up short by a smiling nurse who called out to him, "Boy or girl?"

"I don't know!" Ron called back. "I just know there's *two* of 'em!"

They named the twins Bruce and Brian. They were so similar in appearance that people could not tell them apart, but Janet and Ron, like the parents of most identical twins, could soon distinguish the children easily. Bruce, the elder of the two by twelve minutes, had been born slightly underweight and as a result had had to stay in the hospital a few days to be fattened up. But by the time he joined his twin brother at home, it was clear that he was the more active child, tending to writhe and wriggle and to wake in the night with greater frequency than his brother Brian, a peaceful, less rowdy baby. Both bore a striking resemblance to Janet, with their upturned noses and small round mouths.

By the time the boys were six months of age, Janet felt like an old hand at pacifying, feeding, and changing them. Ron had received another raise, and the family moved to a still bigger and nicer place to live—an actual house on Metcalfe Street, not far from their former apartment. Life seemed to be shaping up beautifully for the young family.

Which is what made it so unsettling when, shortly after the twins were seven months old, Janet noticed that they seemed to be in distress when they urinated. At first she thought it was just the wet diapers that made them cry; then she noticed that even after a diaper change they would scream and complain. She examined their penises and noticed that their foreskins seemed to be sealing up at the tip and making it difficult for the boys to pass water. She took the babies to see her pediatrician, who explained that they were suffering from a condition called phimosis. It was not rare, he said, and was easily remedied by circumcision. After talking about it with Ron, Janet agreed to have the children circumcised at St. Boniface Hospital.

The operations were scheduled for the morning of 27 April, but because Ron was working the late shift at the slaughterhouse, he and Janet decided that he should drive the kids in to be admitted the night before. Apart from the normal concern any parent would feel on the eve of such an operation, Ron and Janet felt no particular trepidation about the circumcisions. Nor should they have. St. Boniface was an excellent, fully modern general teaching hospital. Housed in a seven-story building, it had seven hundred beds, a cardiac care unit, and a children's hospital where, in the mid-1960s, some 2,600 babies were delivered annually and roughly a thousand circumcisions performed each year, all without mishap.

"We weren't worried," Janet says. "We didn't know we had anything to worry about."

Ordinarily, pediatricians experienced in circumcisions performed the procedure at St. Boniface Hospital, but on the morning of 27 April 1966 the usual attending physician, for reasons lost to history, was not available when the Reimer twins were scheduled for their operations. The duty fell to Dr. Jean-Marie Huot, a forty-six-year-old general practitioner.

When a nurse was dispatched to collect the first of the children, it was pure happenstance that she lifted baby Bruce from the bassinet first.

With the baby fixed and draped on the operating table, Dr. Max Cham, the anesthesiologist, administered gas to put Bruce to sleep. (Though newborns were routinely circumcised without anesthesia, a child of eight months, like baby Bruce, could not be operated on while conscious.) Sources differ slightly on what happened next. Court papers later filed against the surgeon, hospital, and three attending nurses refer to an "artery clamp" that was used to secure the piece of foreskin that was meant to be cut away. An artery clamp, however, would be a most unusual choice for such a procedure. According to Dr. Cham, with whom I spoke in the winter of 1997, Dr. Huot used the standard Gomco clamp. Designed specifically for circumcisions, the clamp is used to prevent excessive bleeding: the foreskin is stretched over a bell-shaped metal sheath; a round clamping device then closes over the stretched foreskin and compresses it against the bell, squeezing the foreskin and thus making it blood-free for excision by scalpel.

Regardless of which clamp was used, it is not in doubt that Dr. Huot elected to use not a scalpel to cut away Bruce's foreskin, but a Bovie cautery machine. This device employs a generator to deliver an electric current to a sharp, needlelike cutting instrument, which burns the edges of an incision as it is made, sealing the blood vessels to prevent bleeding—a quite superfluous consideration if Huot had indeed used a Gomco clamp, and a dangerous one, since it would bring perilously close to the penis a current that could be conducted by the metal bell encasing the organ. If, at the same time, the current to the needle were to be turned up almost to the maximum, the results could be cataclysmic.

According to the later testimony of operating room personnel, the electrocautery machine was turned on, and the hemostat dial, which controlled the amount of heat in the needle, was set at the minimum. Dr. Huot lowered the needle and touched it to Bruce's foreskin. Subsequent testing of the machine revealed that it was in proper working order. Whether through temporary mechanical malfunction, user error, or some combination of the two, the needle failed to sever the flesh on the first pass. The hemostat control was turned up. Once again the instrument was applied to the foreskin; again it failed to cut. The cautery machine's current was increased. The needle was once again brought into contact with the foreskin.

"I heard a sound," recalls Dr. Cham, "just like steak being seared."

A wisp of smoke curled up from the baby's groin. An aroma as of cooking meat filled the air.

A urologist was quickly summoned. On duty that morning was Dr. Earl K. Vann. He cleared the instruments and inspected the organ. It appeared oddly blanched in color. He felt the penis with his gloved hand and noticed that it had an unusual firmness. Vann took a probe and attempted to pass it through the urinary meatus—the hole at the end of the penis. The probe would not pass through. Vann told the operating room personnel that he would have to perform an emergency suprapubic cystotomy to place a catheter and thus enable the baby to pass urine. He made an incision below Bruce's belly button, then threaded a length of tubing into the incision, through the muscle wall, and into the bladder. This was sewn into place. A bag to catch the child's urine was affixed to the free end of the catheter. The baby was then wheeled out to the burn ward.

It was decided not to attempt to circumcise his twin brother.

On nights when Ron worked the late shift, the Reimers' normal routine was for Janet to prepare dinner, which they would eat together when Ron got home from work shortly after midnight. They would talk about their day, maybe watch a little TV, and often not make it to bed until two or three in the morning. They'd usually sleep until noon or one. They were sleeping on the morning of 27 April when the phone rang.

Janet answered. It was the hospital calling.

"They said to come in and see the doctor," Janet recalls. "They said there was a slight accident, and they needed to see us right away." Ron took the phone and asked the person on the other end of the line what was going on. "They just said they wanted to see us," Ron says. "They didn't say there was anything wrong."

But Ron and Janet could tell by the person's tone of voice that something unusual was happening. They dressed and headed out to their car. Opening their front door, they discovered that the city, which for some weeks had been in the full delicious flood of early spring, had been hit by a freak blizzard. The pathway to the curb was completely obliterated by snow; the car was buried up to its bumpers. Flakes continued to sift down thickly from a bleached sky.

Ron dug the car out, and they began the slow journey through streets clogged and snarled with snowbound traffic. Five blocks north on St. Mary's Road, then a right turn onto Tache Avenue and the eight blocks up to the hospital. Over the car radio they heard that the airport had been closed down. Seven inches were expected to fall over the course of the day. Already the weathermen were proclaiming it one of the worst blizzards in the city's history. Longtime residents would recall the storm clearly more than thirty years later.

Having finally made it the one mile from their house to St. Boniface Hospital, Ron and Janet rushed inside, only to wait in the doctor's office for what seemed a very long time. Dr. Huot entered. In a businesslike voice he told the Reimers that there had been an accident while circumcising baby Bruce.

"What do you mean, an accident?" Janet said.

Dr. Huot said that Bruce's penis had been burned.

"I sort of froze," Janet recalls. "I didn't cry. It was just like I turned to stone." When she finally gathered her wits enough to speak, Janet found herself asking if they had also burned her *other* child.

"No," Dr. Huot replied. "We didn't touch Brian."

Ron and Janet asked to see their injured baby right away. The doctor said that Bruce was recovering from a surgical procedure to install a catheter. The Reimers were told not to worry, that they could see the child tomorrow. They collected their uninjured son, Brian, and drove home through the steadily falling snow.

The next day Ron and Janet returned to the hospital. Dr. Vann took them to see the baby. Janet's first glimpse of her son is a memory that even three decades later causes her face to drain of blood. Standing over Bruce's bassinet in the burn unit, she looked at his penis—or what was left of it.

"It was blackened, and it was sort of like a little string. And it was right up to the base, up to his body." To Ron the penis looked "like a piece of charcoal. I knew it wasn't going to come back to life after that."

Nevertheless, Janet asked the urologist, "Will it still grow, and he'll just have a *little* penis?"

The doctor shook his head. "I don't think so. That's not the way it works."

Over the next few days, baby Bruce's penis dried and broke away in pieces. It was not very long before all vestiges of the organ were gone completely.

Bruce remained in the hospital while Ron and Janet watched a parade of the city's top local specialists examine him. The doctors gave little hope. Phallic reconstruction, a crude and makeshift expedient even today, was in its infancy in the 1960s—a fact made plain by the plastic surgeon, Dr. Desmond Kernahan, when he described the limitations of a penis that would be constructed from flesh farmed from Bruce's thigh or abdomen. "Such a penis would not, of course, resemble a normal organ in color, texture, or erectile capability," Kernahan wrote in his consultation report. "It would serve as a conduit for urine, but that is all." Even that was optimistic, according to Dr. M. Schwartz, a urologist who also examined the child: "Insofar as the future outlook is concerned," he wrote, "restoration of the penis as a functional organ is out of the question." Dr. G. L. Adamson, head of the Department of Neurology and Psychiatry at the Winnipeg Clinic, evaluated Bruce's projected psychological and emotional future. "One can predict," Adamson wrote, "that he will be unable to live a normal sexual life from the time of adolescence: that he will be unable to consummate marriage or have normal heterosexual relations, in that he will have to recognize that he is incomplete, physically defective, and that he must live apart."

Pediatrician Dr. Harry Medovoy was also called in to consult on the case. Though Medovoy had spent his entire career practicing in Manitoba, he had an international reputation. He was a member of the editorial board of the American journal *Pediatrics* and founder of a children's hospital at the Winnipeg Health Sciences Center, which bears his name today. Though he was a relentless booster of Canadian medicine, it was Medovoy's opinion that the child should be seen at one of the major American medical centers. He recommended the Mayo Clinic, a mere half-day's train ride away in Rochester, Minnesota. Thus, upon Bruce's release from the hospital on 7 June—six weeks after he was first admitted to St. Boniface—Ron and Janet took him on the train to Rochester.

At the Mayo Clinic the baby was examined by a team of doctors. They recommended that Bruce have an artificial phallus constructed at some time shortly before he began school. Like the Winnipeg doctors, the Mayo Clinic physicians explained that phalloplasties were by no means foolproof: they required multiple surgeries through childhood, and the cosmetic and functional results were not promising.

Ron and Janet could hardly believe that this was all the Mayo Clinic doctors could offer them. They wondered why they had bothered to go to the expense and trouble of coming all the way to this famous medical center merely to hear what they had heard back in Canada.

Feeling that they had now exhausted all their options, Ron

and Janet returned to Winnipeg and tried to reconcile themselves to raising a son who, no matter how successful the phalloplasty, "must live apart."

The *Winnipeg Free-Press* and its rival, the *Tribune,* soon got wind of the story. The newspapers each ran an article about a child whose penis was burned off at St. Boniface Hospital. The press did not print the Reimers' name, however, so Ron and Janet were able to keep secret from their neighbors the dreadful accident that had befallen their child. When Janet accepted invitations from other young mothers in the neighborhood to come over for coffee, she sat silently while the others happily traded information about their babies. Only when she got home did she burst into tears and wail, "I hate you, God!" Her taciturn husband, typically, permitted himself no such outpouring of emotion. Ron had once tried to confide in a couple of friends at work about the accident, but the guys joked about it. "I stopped talking to those people," Ron says. "I stopped talking to everybody, pretty much." It only added to the young couple's misery that Brian's phimosis had long since cleared up by itself, his healthy penis a constant reminder that the disastrous circumcision on Bruce had been utterly unnecessary in the first place.

The twins' first birthday, on 22 August 1966, passed in gloom for Ron and Janet. By January they felt like prisoners in their house. They could not even go out together to see a movie (if they had felt so inclined), since they were afraid to hire a baby-sitter who might gossip about the tragedy. By February, Ron began to wake in the middle of the night from dreams that he was strangling Dr. Huot.

Then on a Sunday evening in mid-February—some ten

17

months after Bruce's accident—Ron and Janet saw something that jolted them from their despondency. Their small black-and-white TV happened to be tuned to the Canadian Broadcasting Corporation's popular current affairs program *This Hour Has Seven Days*, where a man identified as Dr. John Money was a guest. A suavely charismatic individual in his late forties, bespectacled and with the long, elegantly cut features of a matinee idol, Dr. Money was talking about the wonders of gender transformation taking place at Johns Hopkins Hospital in Baltimore.

Today, with the subject of transexual surgeries a staple of daytime talk shows, it is difficult to imagine just how alien the concept seemed on that February evening in 1967. Fifteen years earlier, in 1952, a spate of publicity had attended the announcement by American ex-GI George Jorgensen that he had undergone surgical transformation to become Christine. That operation, performed in Denmark, had been roundly criticized by American hospitals, which refused to perform the surgeries. The subject had faded from public view—until now, when Johns Hopkins announced that it had not only performed two male-to-female sex changes, but had established the world's first clinic devoted solely to the practice of converting adults from one sex to the other. The driving force behind the renowned hospital's adoption and promotion of the controversial procedure was the man who now appeared on the Reimers' TV screen: Dr. John Money.

The name rang a distant bell for Ron and Janet. Shortly after Bruce's accident, one of the Winnipeg plastic surgeons had said that he had mentioned Bruce's case to a leading sex researcher at a medical meeting in the United States; the man had suggested that Ron and Janet raise Bruce as a girl. The doctors at the Mayo Clinic had also said something to Ron and Janet about a man in Baltimore who could help them raise Bruce as a girl. While the Mayo Clinic doctors had not themselves recommended the procedure, they had said that the Reimers might like to get a second opinion. At the time, Ron and Janet had not even considered the idea of a sex change. Or so they had thought. As they watched Dr. Money on television, they realized that the idea had never completely left them; it had lodged in the backs of their minds, as Ron puts it, "like a seed that had been planted." Now, as they watched and listened to Dr. Money speak, it was as if that seed had grown and burst into full flower.

It was his confidence that was most striking. Even under the pressure of the staring television cameras and live studio audience, Dr. Money's words, tinged with a highly cultured, British-sounding accent, issued forth with uncanny fluency. He did not stumble over a single syllable, even when the show's interviewer—a bulldoglike young man named Alvin Davis—asked pointedly why psychiatrists were "so opposed" to the practice that Dr. Money was promoting.

"Well," Money said, "I suppose it's a self-evident fact that there are many people who feel that this is not the psychiatric way to treat these patients, since the usual definition of psychiatry is in terms of psychotherapy and the talking treatment. However, there are a small group of people who, like myself, believe that it's thoroughly justified in an attempt to constantly increase our ability to help human beings and to see exactly what the outcome is when, let's say, twenty or thirty people can be followed for five to ten years after having received this kind of treatment."

"But isn't it a fact," Davis said, "that a homosexual will come to you and say, 'I want to be castrated.' And then *you* will make the judgment—or you and a board, a panel at Johns Hopkins will make the judgment—about whether to castrate that person?"

"Yes," Dr. Money said, mildly. "If you want to state it that way, it's true."

"Not only to castrate that person," Davis continued, his voice taking on the rising tone of a prosecutor, "but to inject hormones into the person and virtually change the person—*not* into a female, but into a male with female parts. Aren't you arrogating to yourselves certain decisions that not only psychiatrists don't want to have, but perhaps God doesn't want to have?"

"Well," Dr. Money said, the flicker of a smile underlining the martini-dry sarcasm in his tone, "would *you* like to argue on God's side?"

"No," Davis said. "I would like to know whether *you* believe God doesn't belong in this."

"Well," Money replied, returning to his tone of unflappable calm, a tone ever so slightly shaded by patient condescension, "I'm not sure that's really a particularly relevant question— although I'm aware that many would. May I," he continued, "give you the answer of the group of ministers in Baltimore who were interrogated by the press at the time of the announcement in the papers there? The thirteen of them agreed that in terms of the magnitude of the problem—especially in terms of its magnitude in the lives of the people concerned—that this was ethically justifiable as an attempt to help them. There was one person who withheld an answer until a later date, and that was a representative of the Roman church."

"Why isn't the work being done here in Canada?" Davis demanded to know. And he repeated his earlier query: "And why are so many psychiatrists here so opposed to it?"

"Oh," Money said, almost languidly, "I would think for the same reason that there tends to be a traditionalism in most places. I don't need to tell you that in many branches of medicine, science—or even housekeeping or farming—there is a tendency to hang on to the past, to cling to the past."

"And *you're* the pioneer?" Davis asked.

"Well," Dr. Money said, "perhaps in a small way."

At this point the camera cut from Dr. Money and his questioner to a blond woman who walked out onto the set. Dressed in a narrow skirt, high heels, and a matching close-fitting jacket, she took a seat in the chair across from the two men. A close-up shot revealed that her round, pretty face was expertly made up, in the style of the mid-1960s, with heavy eyeliner, mascara, and foundation, her mouth thickly painted with lipstick.

"This is Mrs. Diane Baransky," the show's announcer said. "Until four years ago, her name was Richard."

Ron and Janet gaped at the TV screen. It was their first glimpse, ever, of a transexual. It was one thing to hear Dr. Money talk about sex change in the abstract; it was another to see it with your own eyes. Ron and Janet could hardly believe it. If they hadn't been told that Mrs. Baransky was born a man, they would never have guessed it. Even *knowing* it, it was hard to believe. She looked like an attractive, even *sexy* woman. The way she moved, walked, sat—even her voice, despite an ever so slight huskiness, had the timbre of a woman's as she said hello to her host and fellow guest.

After a few preliminary questions from Davis, Dr. Money spoke up, deftly seizing the reins of the interview.

"Diane," he said, "I think people would be extremely interested if you could give us a short sketch of the difference that it makes to have had this procedure—to compare the old life with the new."

"Well, there is a tremendous difference," Mrs. Baransky said. "It's a way of finding yourself. You actually fit into society, you're more accepted in a more normal society." She explained that the discrepancy between her anatomic self as a male and her inward sense of herself as a female had been a trial to her growing up. "As a teenager—or being young—when you're different from anybody else, it's very hard." Becoming a woman, Mrs. Baransky explained, had solved all her problems of being teased and "singled out." Until her sex change, she had felt completely alone. Now she was accepted as a woman and had recently married her husband, a fellow hairdresser. "I was different," she said. "I was never complete. I was neither a man nor a woman."

"And now you feel complete as a woman?" Davis asked.

Her response was unequivocal: "Oh, yes, definitely. Yes. Completely—body and mind."

The audience was then invited to ask questions. It was near the end of the segment that a young man asked the question that had been forming in Janet's mind. He asked about "the other group of sex patients" whom Dr. Money treated—newborn babies with what Dr. Money had earlier called "unfinished genitals," babies whose private parts were neither male nor female at birth. In replying to this question, Dr. Money explained that he and his colleagues at Johns Hopkins could, through surgery and hormone treatments, make such children into whichever sex seemed best, and that the child could be raised happily in that sex. "The psychological sex in these circumstances," as Dr. Money put it, "does not always agree with the genetic sex nor with whether the sex glands are male or female."

Despite the big words and the rapidity with which Dr. Money spoke them, Janet and Ron caught their meaning. Dr. Money was saying that the sex a baby was born with didn't matter; you could convert a baby from one sex to the other.

Janet turned to Ron. "I think I should write to this Dr. Money," she said.

Ron agreed. When the segment ended a few minutes later, Janet wrote a letter to Dr. Money describing what had happened to Bruce. Dr. Money's reply was prompt. He expressed great optimism about what could be done for the Reimers' baby at Johns Hopkins and urged them to bring the child to Baltimore without delay.

After so many months of grim predictions, bleak prognoses, and hopelessness, Dr. Money's words, Janet says, felt like a balm. "Someone," she says, "was finally *listening*."

2

≡

Dʀ. Mᴏɴᴇʏ ᴡᴀs indeed listening. In a sense, Janet's cry for help was one that he might have been waiting for his entire professional life.

At the time the Reimer family's plight became known to him, John Money was already one of the most respected, if controversial, sex researchers in the world. Born in 1921 in New Zealand, he had come to America at the age of twenty-five, received his Ph.D. in psychology from Harvard, then joined Johns Hopkins, where his rise as a researcher and clinician specializing in sexuality was meteoric. Fifteen years after joining Johns Hopkins, he was already widely credited as the man who coined the term *gender identity* to describe a person's inner sense of himself or herself as male or female. He was also known as the world's undisputed authority on the psychological ramifications of ambiguous genitalia and was making headlines around the world for his establishment of the pioneering Johns Hopkins clinic for transexual surgeries.

As his unflappable appearance on *This Hour Has Seven Days* would suggest, Money was also a formidable promoter of his ideas. "He's a terribly good speaker, very organized, and very persuasive in his recital of the facts regarding a case," says Dr. John Hampson, a child psychiatrist who, with his wife, Joan, coauthored a number of Money's groundbreaking papers on sexual development in the mid-1950s. "I think a lot of people were envious. He's kind of a charismatic person, and some people dislike him."

Money's often overweening confidence actually came to him at some cost. His childhood and youth in rural New Zealand had been beset by anxieties, personal tragedies, and early failure. The son of an Australian father and an English mother who belonged to the Brethren church, he was a thin, delicate child raised in an atmosphere of strict religious observance—or what he would later derisively call "tightly sealed, evangelical religious dogma." His sense of intellectual superiority developed early. On his first day of school at age five, he was set upon by bullies and took refuge with a female cousin in the girls' play-shed, where boys would not be caught dead. "Having not measured up as a fighter," Money would later write, "I was set on the pathway of outwitting other kids by being an intellectual achiever. That was easier for me than for most of them."

Money's childhood difficulties were compounded by his vexed relationship with his father. Six decades later he would write with barely controlled venom of this father, portraying him as a brutal man who heartlessly shot and killed the birds that infested his fruit garden, and administered to his four-year-old son an "abusive interrogation and whipping" over a broken window. This incident, Money wrote, helped establish his lifelong rejection of "the brutality of manhood."

Money was eight years old when his father died of a chronic

kidney ailment. "My father died without my being able to forget or forgive his unfair cruelty," Money wrote. Not told of his father's death until three days after seeing him carried off to the hospital, Money's shock was compounded by the experience of being informed by an uncle that now *he* would have to be the man of the household. "That's rather heavy duty for an eight-year-old," Money wrote. "It had a great impact on me." As an adult, Money would forever avoid the role of "man of the household." After one brief marriage ended in divorce in the early 1950s, he never remarried, and has never had children.

After his father's death, Money was raised in an exclusively feminine atmosphere by his mother and spinster aunts, whose anti-male diatribes also had a lasting effect on him. "I suffered from the guilt of being male," he wrote. "I wore the mark of man's vile sexuality"—that is, the penis and testicles. In light of Money's future fame in both adult and infant sex change, his next comment has an unsettling tenor: "I wondered if the world might really be a better place for women if not only farm animals but human males also were gelded at birth."

A solitary adolescent with a passion for astronomy and archaeology, Money also harbored youthful ambitions as a musician, a goal doomed to disappointment once Money realized that he would never be more than a skilled amateur. As an undergraduate at Victoria University, in the New Zealand capital city of Wellington, Money discovered a new passion into which he rechanneled his thwarted creativity: the science of psychology. Like so many students drawn to the study of the mind and emotions, Money's interest in the discipline was in large part as a means for solving certain troubling questions about himself. His first serious work in psychology, his master's thesis, concerned "creativity in musicians," in which, Money writes, "I began to investigate my relative lack of success in comparison with that of other music students."

His decision soon after that to narrow his studies to the psychology of sex had a similarly personal basis. Having departed sharply from his parents' faith, Money grew increasingly to react against what he saw as the repressive religious strictures of his upbringing. The academic study of sexuality, which removed even the most outlandish sexual practices from moral considerations into the "pure" realm of scientific inquiry, was for Money an emancipation. From his twenties on, he would be a fierce proselytizer for sexual curiosity and exploration. By the mid-1970s, with the sexual revolution in full rampage, Money would step out publicly as a champion of open marriage, nudism, and other more rarefied manifestations of the culture's sexual unbuttoning. "There is plenty of evidence that bisexual group sex can be as personally satisfying as a paired partnership, provided each partner is 'tuned in' on the same wavelength," he wrote in his book *Sexual Signatures*. Elsewhere, he has described his own private life as casual and eclectic—"a give-and-take of sexual visitations and friendly companionships with compatible partners, some women, some men."

Reveling in his role as "agent provocateur of the sexual revolution" (as the *New York Times* dubbed him in 1975), Money rarely missed an opportunity to spread his gospel of sexual emancipation: extolling the heightened pleasures of sex under a black light to a student after a speaking engagement at the University of Nebraska; appearing in court as an expert witness to defend the 1973 pornographic film *Deep Throat*, which he praised as a "cleansing" movie that would help keep marriages together; penning op-ed pieces for the *New York Times* in which he called for a "new ethic of recreational sex." A patient treated by Money in the 1970s for a rare endocrine disorder recalls the psychologist once casually asking him if he'd ever experienced a "golden shower." A sexually inexperienced youth at the time, the patient did not know what Money was talking about. "Getting pissed on," Money airily announced with the twinkling, slightly insinuating smile with which he liked to deliver such deliberately provocative comments.

Convinced that embargoes on certain words promoted prudery, Money inserted the words *fuck*, *cock*, and *cunt* into his regular conversation with colleagues and patients. Dr. Fred Berlin, a professor of psychiatry at the Johns Hopkins School of Medicine and a colleague who considers Money one of his most important mentors, defends Money's penchant for sexual outspokenness. "Because he thinks it's important to desensitize people in discussing sexual issues," Berlin says, "he will sometimes use four-letter words that others might find offensive. Perhaps he could be a little more willing to compromise on that, but John is an opinionated person who isn't looking necessarily to do things differently than the way he's concluded is best."

While Money's conclusions about the best approach to sexual matters merely raised eyebrows in the mid-1970s, they provoked outrage at the dawn of the more conservative 1980s, when Money ventured into areas of which even some of the most adventurous sexual explorers were leery. In 1986, Money published *Lovemaps,* an exhaustive study of such practices as sadomasochism, coprophilia, amputation fetishes, autostrangulation, and various other behaviors that he called, not perversions, but "paraphilias," in an effort to destigmatize and decriminalize them. The topic of pedophilia became a particular interest, and one that Money took obvious delight in publicly espousing.

"A childhood sexual experience," he explained to *Time* magazine in April 1980, "such as being the partner of a relative or of an older person, need not necessarily affect the child adversely." He granted an interview to *Paidika*, a Dutch journal of pedophilia, which carries ads for the North American Man-Boy Love Association and other pro-pedophile groups.

"If I were to see the case of a boy aged ten or twelve who's intensely attracted toward a man in his twenties or thirties, and the relationship is totally mutual, and the bonding is genuinely totally mutual, then I would not call it pathological in any way," he told the journal, and added, "It's very important once a relationship has been established on such positive and affectionate grounds that it should not be broken up precipitously." In 1987, Money wrote an admiring foreword to an unusual volume published in Denmark entitled *Boys and Their Contacts with Men*. By Dutch professor Theo Sandfort, the book presented what purported to be verbatim testimonials of boys as young as eleven years old rhapsodically describing the delights of sex with men as old as sixty. "For those born and educated after the year 2000," Money wrote, "we will be their history, and they will be mystified by our self-important, moralistic ignorance of the principles of sexual and erotic development in childhood." Money concluded his foreword with the proclamation "It is a very important book, and a very positive one."

Money's response to criticism for the public airing of such views was always to launch counterattacks of his own, ridiculing his critics for their adherence to an outmoded sexual Puritanism. In an autobiographical essay included in his 1985 book of collected writings, *Venuses Penuses,* Money dubbed himself a "missionary" of sex, proudly proclaiming, "It has not been as easy for society to change as it had been for me to find my own emancipation from the 20th-century legacy of fundamentalism and Victorianism in rural New Zealand."

Money's experimental, taboo-breaking attitude to sex found its echo in the way he pursued his professional research career. Eschewing the more trammeled byways of sex

30

research, Money deliberately sought out exotic corners of the field. He found just such a relatively undiscovered realm of human sexuality in 1948, while in the first year of study for his Ph.D. in psychology at Harvard. In a tutorial called Fieldwork and Seminar in Clinical Psychology, Money was presented with the case of a fifteen-year-old genetic male born not with a penis, but with a tiny, nublike phallus resembling a clitoris. At puberty, the boy had developed breasts. It was Money's first exposure to hermaphroditism—also known as intersexuality—a term of classification for a variety of birth anomalies of the internal and external sex organs. Often described in lay terms as a condition of being half-man, half-woman, the syndrome derives its name from a combination of the names of the Greek gods of love, Hermes and Aphrodite, and occurs as often as one in two thousand births (by some estimates). The symptoms vary from the extreme manifestation of a genetic female born with a penis-sized clitoris and fused labia resembling a scrotum, to a male whose genital resemblance to a girl at birth is so total that his true biological sex is not suspected until puberty when "she" fails to menstruate—to anything in between.

Money was fascinated by hermaphroditism and wrote his doctoral dissertation on the subject. Until then the syndrome had been studied almost solely from a biological perspective. Money approached it from a psychological angle, investigating the mental and emotional repercussions of growing up as anatomically neither boy nor girl. His thesis, entitled "Hermaphroditism: An Inquiry into the Nature of a Human Paradox," was completed in 1952 and led to his invitation to join Johns Hopkins, where the world's first and largest clinic for studying and treating intersexual conditions had been established. The clinic's director, pioneering pediatric endocrinologist Lawson Wilkins, teamed Money with two married psychiatrists, Drs. Joan and John Hampson, to study the mental and emotional makeup of the intersexual patients treated in the clinic. The three researchers made up the newly created Psychohormonal Research Unit.

Over the next six years, Money and the Hampsons studied some 131 intersexuals ranging in age from toddlers to adults. Money (who was lead investigator and author of the team's published reports) claimed to observe a striking fact about intersexes who had been diagnosed with identical genital ambiguities and chromosomal makeups but raised in the opposite sex from one another: more than 95 percent of them reportedly fared equally well psychologically whether they had been raised as boys or girls. Money called these groupings of patients "matched pairs" and said they were proof that the primary factor determining an intersexual child's gender identity was not biology, but rather the way the child was raised. He concluded that these children were born wholly undifferentiated in terms of their psychological sex and that they formed a conception of themselves as masculine or feminine solely through rearing.

This theory was the foundation on which Money based his recommendation to Johns Hopkins surgeons and endocrinologists that they could surgically and hormonally steer intersexual newborns into whichever sex, boy or girl, they wished. Such surgeries would range from cutting down enlarged clitorises on mildly intersexual girls to full sex reversal on intersexual boys born with undeveloped penises. These conversions to girlhood were foreordained by the state of surgical technology: it was easier for surgeons to construct a synthetic vagina than to create an artificial penis. Money's only provisos were that such "sex assignments" and reassignments be done within the first two and a half years of life (after which time, Money theorized, a child's psychosexual orientation ceased to be as malleable) and that once the sex had been decided upon, doctors and parents never waver in their decision lest they risk introducing fatal ambiguities into the child's mind.

By providing a seemingly solid psychological foundation for such treatments, Money had offered physicians a relatively simple surgical solution to one of the most vexing and emotionally fraught conundrums in medicine: how to deal with the birth of an intersexual child. "One can hardly begin to imagine what it's like for a parent when the first question—'Is it a boy or a girl?'—results in a response from the physician that they're just not sure," says Dr. Fred Berlin. "John Money was one of those folks who, years ago, before this was even talked about, was out there doing his best trying to help families, trying to sort through what's obviously a difficult circumstance."

Money, however, was not interested chiefly in intersexes. As he stated as early as his Harvard thesis, he recognized the scientific worth of intersexes primarily as what he called "experiments of nature"—as a cohort of research subjects who could shed light on the question of sexual development in *normal* humans—who could, in fact, resolve one of the longest-running debates in science; namely, whether it is primarily nature or nurture that shapes our sexual sense of self. It was in his first published papers at Johns Hopkins that Money generalized the theory of psychosexual neutrality at birth from hermaphrodites to include *all* children, even those born without genital irregularity.

"From the sum total of hermaphroditic evidence," he wrote in 1955, "the conclusion that emerges is that sexual behavior and orientation as male or female does not have an innate, instinctive basis. In place of a theory of instinctive masculinity or femininity which is innate, the evidence of hermaphroditism lends support to a conception that, psychologically, sexuality is undifferentiated at birth and that it becomes differentiated as masculine or feminine in the course of the various experiences of growing up." In short, Money was advancing a view that human beings form a sense of themselves as boy or girl according to whether they are dressed in blue or pink, given a masculine or feminine name, clothed in pants or dresses, given guns or Barbies to play with. Many years later, Money would describe how he arrived at some of his more radical theories about human sexual behavior. "I frequently find myself toying with concepts and working out potential hypotheses," he mused. "It is like playing a game of science fiction."

While Money's theory of human newborns as total psychosexual blank slates may strike a contemporary reader as science fiction, such was not the case in the mid-1950s, when it was met with almost universal acceptance by clinicians and scientists—an acceptance not difficult to understand in the context of the time. Explanations for sex differences had been moving toward a nurturist view for decades. Prior to that, the pendulum had been pointing in the naturist direction—thanks to the discovery at the end of the nineteenth century of the so-called male and female hormones, testosterone and estrogen. The discovery of these chemical-based internal secretions had led biologists to proclaim the riddle of sex differences solved: testosterone was the masculinizing agent; estrogen, the feminizing. They confidently predicted that male homosexuals would be discovered to possess an excess of the "female" hormone in their bloodstream and a deficiency of the "male" hormone. Minute analysis of the urine and blood of adult homosexual men, however, revealed no such hormonal imbalances. Under the microscope, a straight and a gay man's internal secretions are identical. Other experiments meant to show the hormonal basis of sexual identity also failed, and as the failures mounted, enthusiasm for a biological explanation of sexual differences gradually waned. Simultaneously, the first half of the twentieth century and the advent of Freud and modern psychology saw a rapid increase in social learning models for human behavior. Against this background, the Johns Hopkins team's conclusions that sexual identity and orientation were solely shaped by parents and society fit perfectly into an intellectual zeitgeist in thrall to behaviorist theories. Nor did it detract from the papers' reception that they carried the imprimatur of Johns Hopkins Hospital, one of the premier medical research institutions in the world.

The Johns Hopkins team's 1955 intersex papers were proclaimed instant classics and won that year's Hofheimer Prize from the American Psychiatric Association. The Hampsons soon left Johns Hopkins for Washington State University and by 1961 had drifted out of gender identity research. As a result, Money alone became heir to the award-winning papers' reputation. And as sole director of the Psychohormonal Research Unit (after Lawson Wilkins's death in 1962), he was also the lone beneficiary of the unit's success. In 1963 Money was awarded a grant of $205,920 from the National Institutes of Health—a considerable sum in early-1960s dollars, but merely the first of several NIH grants that would sustain Money and his unit for the next thirty-five years. In 1965 he served as Mead Johnson visiting professor of pediatrics at the University of Buffalo Children's Hospital, and was awarded the Children's Hospital of Philadelphia Medal "for contributions to the study of the psychological development of children." A year later he would begin to garner fame outside the academic realm when he finally succeeded in persuading Johns Hopkins to establish the clinic for the treatment and study of adult transexuals.

Money had been galvanized by transexualism since 1952, when the revelations about Christine Jorgensen first hit the press. In Jorgensen's case, Money saw tantalizing proof of his theory that environment, not biology, determines psychological sex, for here was a person born with apparently normal male biological makeup and genitals whose inner sense of self had differentiated as female—in direct contradiction to his chromosomal, gonadal, hormonal, reproductive, and anatomic sex. What greater evidence could there be that gender identity is determined not by biology but by environment? Determined to study such individuals in the greatest number possible, Money set out to get Johns Hopkins into transexual research and treatment, which was still a repellent idea for the majority in the American medical establishment.

In his campaign to establish Johns Hopkins as the first hospital in America to embrace transexual surgeries, Money knew that he would first have to bring on board a respected medical man. (Money himself was a psychologist and did not possess a medical degree of any kind.) He turned first to Dr. Howard Jones, the Johns Hopkins gynecologist who had perfected the surgical techniques for sex assignment on Money's infant intersexual subjects. "I can recall," Jones says, "that for a number of months, maybe even *years*, John kept raising the question of whether we shouldn't get into the transexual situation." While Jones was interested in experimental medicine (he would eventually leave Johns Hopkins for the University of Virginia where he would found the nation's first in vitro fertilization clinic), he was resistant to the idea of performing elective castrations and genital reconstruction on adults.

But Money was persistent. He turned for help to Dr. Harry Benjamin. The acknowledged grandfather of transexual study in America, Benjamin had for the previous ten years been quietly referring transexual patients to doctors in Casablanca and Morocco for sex change surgery. Money enlisted three of Benjamin's postoperative transexuals to come to Johns Hopkins and meet with Jones and pediatric endocrinologist Milton Edgerton. Eventually Jones and Edgerton were convinced. "John finally marshaled enough evidence," as Jones puts it, "to indicate that this was something that maybe should be done." Fittingly enough, Money was given the job of naming the new clinic for adult transexual surgeries. He dubbed it the Gender Identity Clinic.

The first complete transexual surgery at Johns Hopkins was performed by Dr. Jones on 1 June 1965, when a New Yorker named Phillip Wilson became Phyllis Avon Wilson. But it still remained for Johns Hopkins to sell the idea to the American public. While some members of the sex change committee argued for keeping the existence of the clinic quiet, Money pushed for a preemptive strike and argued in favor of creating a press release that would circumvent leaked rumors about what the team had done. Money's argument prevailed, and he helped concoct a press release with the hospital's public relations department. The statement was issued on 21 November 1966. Money later revealed that a strategic decision had been made to issue the press release to the *New York Times* alone. The prestige of the *Times*, the Johns Hopkins team hoped, would set the tone for all other media coverage. "The plans," Money later wrote, "worked out exactly as hoped."

The *Times* treated the revelations with none of the scandalized outrage that had greeted the Jorgensen case in 1952. The front-page story used verbatim quotations from Gender Identity Clinic chairman John Hoopes, culled directly from the Johns Hopkins press release, and presented the procedure as a humane and effective solution to an intractable psychosexual problem. Similarly approving stories followed in all three news weeklies, *Time, Newsweek,* and *U.S. News & World Report.* In April 1967 *Esquire* magazine published an exhaustive feature on the Johns Hopkins clinic, in which Money was admiringly quoted. Indeed, of all the coverage in late 1966 and early 1967 of Johns Hopkins' pioneering foray into transexual surgery, by far the hardest edged was CBC's *This Hour Has*

Seven Days, in which Alvin Davis sharply challenged Money on the ethics and efficacy of switching people's sex. Except for the single stinging rejoinder ("Would *you* like to argue on God's side?"), Money had refused to rise to the bait, and thus, for his fellow Gender Identity Clinic committee members, set the standard for how to handle direct attacks. Money's calm, judicious performance was a masterpiece of public relations, and all the more impressive to those who knew the ferocity with which, in ordinary life, he responded to even the mildest opposition to his opinions.

As Money himself would admit in an essay written in 1990, "In the practice of my psychohormonal research, I do not suffer fools gladly." This was an understatement. The psychologist's violent reactions to intellectual challenge were legendary. "John was unusually brilliant," says Dr. Donald Laub, a pioneer in adult transexual surgical techniques who has known Money for thirty years. "He may be the smartest person I've ever met. He was so smart that it was a *problem*—because he knew everybody *else* was dumb." By all accounts, Money had no compunction about letting others know his low opinion of their intellectual firepower. "Even when John asked for feedback, what he was looking for was agreement," says Dr. Howard Devore, a psychologist who earned his Ph.D. under Money in the Psychohormonal Research Unit in the mid-1980s. Should that agreement fail to be forthcoming, Money was never afraid to let his displeasure be known. As early as the mid-1950s, Money had a reputation for tantrums among his coworkers, underlings, and students that preceded him throughout the academic world.

"Every center that I trained at after [Johns Hopkins]," says Devore, "when people saw on my résumé that I had worked with John Money, they would ask me to comment, off the record, what it was like working with him and was he 'as bad as people say?' I was just amazed at how consistent his world-wide reputation actually was. And frankly, John didn't do that much to hide it. I once saw him stand up at an academic meeting and shout a presenter down because he didn't agree with what she was saying."

By February 1967—when Ron and Janet Reimer first saw John Money on television—his reputation was for all intents and purposes unassailable. Dr. Benjamin Rosenberg, himself a leading psychologist who specialized in sexual identity, says that Money was "the leader—the front-runner on everything having to do with mixed sex and hermaphrodites and the implications for homosexuality and on and on and on."

Money's reach and influence throughout the academic and scientific world would help to define the scientific landscape for decades to come—indeed, to the present day: many of his students and protégés, trained in his theories of psychosexual differentiation, have gone on to occupy the top positions at some of the most respected universities, research institutions, and scientific journals in the country. His former students include Dr. Anke Ehrhardt, now a senior professor at Columbia University; Dr. Richard Green, director of the Gender Identity Clinic in London, England; Dr. June Reinisch, who for years was head of the famed Kinsey Institute; and Dr. Mark Schwartz, director of the influential Masters and Johnson Clinic.

On the clinical side, Money's influence was perhaps even more remarkable. His theories on the psychosexual flexibility at birth of humans form the cornerstone of an entire medical speciality—pediatric endocrinology. Professor Suzanne Kessler, in her 1998 book, *Lessons from the Intersexed*, suggests that Money's views and their implications for the treatment of ambiguously sexed babies form among physicians "a consensus that is rarely encountered in science."

There was, however, at least one researcher in the mid-1960s who was willing to question John Money. He was a young graduate student fresh from the University of Kansas.

The son of struggling Ukrainian Jewish immigrant parents, Milton Diamond, whom friends called Mickey, was raised in the Bronx, where he had sidestepped membership in the local street gangs for the life of a scholar. As an undergraduate majoring in biophysics at the City College of New York, Diamond had become fascinated by the role hormones played in human behavior. Seeking a place to do graduate work, he chose Kansas, where anatomist William C. Young (famous for his hallmark studies of the 1930s on the role of hormones in the estrus cycle) ran a laboratory. In a stroke of serendipity, Diamond's arrival in Kansas in the fall of 1958 coincided with the time when a trio of researchers on Young's staff—Charles Phoenix, Robert Goy, and Arnold Gerall—stood on the brink of a discovery about the sex-differentiating role of hormones that would change the science and study of sexual development forever.

Disillusionment with earlier hormone studies had led many sex researchers, including Young's team, to shift their focus from the role played by hormones in the mature organism to the role played by hormones in the womb. Working from guinea pig studies done two decades earlier by Soviet sex researcher Vera Dantchakoff, the Kansas team sought to learn the role played by the hormones that bathe a developing fetus's brain and nervous system. Earlier researchers had shown that, in humans, in the early stages of gestation, the male and female fetus's internal and external sex organs are identical to one another. Between six and eight weeks, however, changes start to take place. If the fetus's cells bear the male (XY) chromosome, the fetal gonads differentiate as testicles, which begin to pump out testosterone. This prenatal androgen is the agent that masculinizes the developing fetus's external genitals—turning the undifferentiated genital tubercle into a penis, causing the open genital sinus to fuse along the midline and form the scrotum, into which the testicles descend—and at the same time masculinizes the internal reproductive system by spurring the growth of the seminal ducts (another testicular secretion suppresses growth of the rudimentary female internal structures). If, on the other hand, the fetus bears the female (XX) chromosome, the gonads develop as ovaries, no testosterone is produced, in the absence of which the external genitals and internal anatomy differentiate as female, the genital tubercle develops as a clitoris, the genital sinus remains open and becomes the entrance to the vagina, and the internal structures develop as fallopian tubes and uterus.

The question for the Kansas team was whether these prenatal hormonal effects on the anatomy were mirrored in the brain. To find out, they set about creating a cohort of hermaphrodite guinea pigs by injecting large doses of testosterone into the wombs of pregnant mothers. When exposed to testosterone at a critical stage in fetal development, the female guinea pigs were born, as expected, with clitorises enlarged to the size of penises. The researchers then set out to learn if the masculinization of a treated female's anatomy was matched by a corresponding masculinization of her sexual behavior.

In observing the treated females as they grew from childhood to maturity, the team noticed something extraordinary. Not only did the treated females demonstrate an increased physical activity distinct from that of their untreated sisters, they also did not, in the presence of normal males, present their hindquarters for sexual penetration in the normal female in-heat posture known as lordosis. Instead, the testosterone-treated females (even those that showed no clitoral enlargement) attempted to mount their untreated sisters.

I spoke with team member Robert Goy, shortly before his death in 1999, about the breakthrough moment of his research career. His voice was charged with an excitement that suggested he had just made the discovery the night before. "We couldn't schedule tests fast enough," he told me. "We were testing every night—night after night after night—and getting data, and analyzing it, and reanalyzing it."

Milton Diamond was in the thick of the research, performing adjunct experiments on the pregnant mothers to learn what, if any, influence the testosterone had on their functioning. Having come to Kansas hoping to learn something new and interesting about the action of hormones on behavior, Diamond found himself present at one of the most significant biological breakthroughs in sex research of the twentieth century.

There was concern among members of the team about how their professor, William Young, would react to the results. They knew him to be an adherent of the theories of psychosexual neutrality advanced just four years earlier by John Money's team at Johns Hopkins. "Young was a great follower of John Money and the Hampsons," Goy told me. "He had been thinking all this time that the organizing principle for sexual behavior was experience. So his world was shaken by these results. But he was wonderfully adaptable, and the truth was more important to him than anything else. It's very unusual in a scientist. Most scientists fall in love with their own ideas and theories, and you can't shake them out of it. Will Young wasn't like that."

In fact it was Young who settled the debate that flared among the research team members when it came time to write up the results. Unsure precisely how to label the behavior of the treated female guinea pigs—the team toyed with calling it "masculine mimicry" or "pseudodifferentiation"—they were overruled by Young, who told them they had discovered not the role played by prenatal testosterone in creating a *simulation* of masculine behavior, but masculine behavior itself. Accordingly, Young advised the team to state unequivocally that they had discovered, in the fetal guinea pig, the *organizing principle* for adult masculine sexual behavior.

"Young was an anatomist," Goy explained, "and if you understand the way anatomists use the term *organization,* it makes that choice of word inevitable. Anatomists believe that the organs of the body are organized by a set of tissues that are differentiated in a special way and combined so that they carry out a definite function or malfunction of that organ. And that's the way he used the word *organization.* He meant that all of the tissues underlying sexual behavior—whether peripheral structures, brain tissues, blood, or muscles—are organized into a whole; and that that organization is imposed by exposure to hormones before birth; and that that organization is either masculine or feminine. And he believed that we had discovered the principle that organizes the tissues in a masculine form."

Still, when the team came to write up their results, which would appear in a 1959 issue of the journal *Endocrinology,* Young urged caution in how directly they should extrapolate their experimental animal work to sexual differentiation in humans—largely out of Young's respect for Money's work with the Hampsons. The team agreed to soften their statements on the applicability of their research to humans. "We said there may be some way that the guinea pig picture will 'complement' or 'supplement' the human picture by accounting for 'discrepancies,' " Goy said.

Not everyone in the lab was satisfied with that decision. The youngest member, Mickey Diamond, felt that Young and the others were being too cautious in failing to link their animal findings directly to the human situation. "I believe in evolution," Diamond says, chuckling. "I didn't see any reason that human beings would be different from other mammals in that regard." He felt so strongly that when he was applying for a research grant in his final year at Kansas and was required to submit an original paper, he decided to write an essay taking on Money and the Hampsons' theory of psychosexual neutrality at birth.

In that paper, entitled "A Critical Evaluation of the Ontogeny of Human Sexual Behavior," Diamond rejected outright the Johns Hopkins team's theory. Citing the guinea pig findings, Diamond described as "specious" a theory that said man is "completely divested of his evolutionary heritage," and stated that prebirth factors "set limits" on how far culture, learning, and environment can direct gender identity in humans. Marshaling evidence from biology, psychology, psychiatry, anthropology, and endocrinology to argue that gender identity is hardwired into the brain virtually from conception, the paper was an audacious challenge to Money's authority (especially coming from an unknown graduate student at the University of Kansas).

Addressing the theory about the psychosexual flexibility of intersexes, Diamond pointed out that such individuals had experienced "a genetic or hormonal imbalance" in the womb, and he argued that even if human hermaphrodites *could* be steered into one sex or the other as newborns (as Money claimed), this was not necessarily evidence of their gender neutrality at birth. It might simply suggest that the organization of their nervous systems and brains had undergone in utero a similar ambiguous organization as their genitals. In short, they had an inborn neurological capability to go both ways—a capability, Diamond hastened to point out, that genetically normal children certainly would not share. As for transexuals, who showed no observable anatomic ambiguity of sex, Diamond postulated that they, too, might possess an as yet undiscovered biological condition that hardwired their brains to a program opposite to the evidence of their bodies—a possibility that Diamond was able to back up with evidence from no less an authority than Dr. Harry Benjamin himself, who had recently reported that in forty-seven out of eighty-seven of his patients, he "could find no evidence that childhood conditioning" was involved in their conviction that they were living in the wrong sex.

Had he known of it at the time, Diamond might also have drawn upon an obscure paper in the foreign literature for his critique—a paper that had questioned the Johns Hopkins team's protocols for intersex treatment some six years earlier. In a 1959 edition of *The Canadian Psychiatric Association Journal*, three Toronto physicians, Dr. Daniel Cappon, Dr. Calvin Ezrin, and Dr. Patrick Lynes, had pointed out serious flaws in the Hopkins team's statistical and research methods. "[T]hese workers," the Canadians wrote, "failed to relate the physical and psychological wholes of the person and only compared component parts without submitting these comparisons to mathematical validation." In conducting their own research on a cohort of seventeen intersexual patients, the Canadian doctors took precautions that the Johns Hopkins team had not. To prevent subjective tainting of their results, the Canadians split their research team in two: one to study the patients from an endocrinologic perspective, the other to study the patients from a psychological perspective. For comparative purposes, the Canadian team also carried out research on a control group of nonhermaphrodites, as well as on a series of homosexuals and transvestites.

The team's results showed that it was dangerous indeed to suppose that no link existed between an intersexual child's biological makeup and its gender identity; that in fact the status of the chromosomes, gonads, or hormones might predispose a hermaphrodite child to identify more with one sex than the other in adulthood. Stating that the Johns Hopkins team had based its recommendations to surgeons on "shaky theory," the Canadians had expressed particular unease about the recommendation that males born with tiny or nonexistent penises should, without exception, be castrated and converted into girls. Such sex-changed children, the Canadians had warned, "were liable to be brought up tragically incongruously with the main somatic sex."

The Canadian team's findings would have made a strong addition to Diamond's exhaustive theoretical critique, but he did not learn of the paper's existence until after his own was published (at which point he began to cite it in his own papers). "The Canadian paper got lost somewhere," Diamond says. "It just died. I think it was maybe Hopkins compared to Podunk." But in 1965, Diamond's paper was published in a high-profile, well-respected American journal, the *Quarterly Review of Biology,* where it could not be missed—least of all by John Money, considering that the *Quarterly Review* was at that time published out of Johns Hopkins.

I was sitting with Diamond in his cluttered, windowless office on the campus of the University of Hawaii Medical School as he reminisced about these origins of his thirty-year-long scientific debate with John Money. It was June of 1997, just two months after Diamond and Sigmundson's "John/Joan" paper had delivered a blow to his old rival. A mild-mannered sixty-four-year-old with frizzy graying hair and beard, Diamond was clearly exhausted from fielding the unending stream of phone calls, faxes, and letters from both reporters and fellow scientists requesting more information about, or an interview with, John/Joan. Dressed in a pale blue overlaundered T-shirt riddled with holes, a pair of jeans, and battered running shoes, Diamond told me that professors at the University of Hawaii are "paid in sunshine." His putty-colored pallor suggested that he had not been drawing his full wages. Diamond had, in fact, spent the majority of his thirty years in Honolulu doing experiments or hunched over his computer in the tiny office he calls his "cave," pumping out more than one hundred journal articles and eight books on sexuality. On the wall beside him was tacked a snapshot of his four daughters; on the messy desk in front of him were heaped papers, books, open journals, and boxed sets of both Robert Johnson and Bach tapes.

Diamond insists that he bore John Money no personal animus at the time of writing his 1965 article and that his intent was not to embarrass him. He says that his paper had merely been an effort to advance the field of knowledge in the time-honored scientific tradition of assertion and challenge. Diamond points out that after the article's publication, he actually made an overture to Money, suggesting that they collaborate on an article. Though he recognized that they stood on opposite sides of the nature-nurture debate, Diamond believed this was precisely why their collaboration would be of particular value. He shakes his head and smiles at the naïveté that compelled him, a mere graduate student, to suggest a collaboration with one of the leading scientists in the field—a scientist whom, furthermore, he had just publicly challenged in a leading journal. "I really believed that it was an intellectually good thing to do," Diamond says. Money evidently felt otherwise. "His attitude was, Why should I do anything with *you*?" Diamond says. "Who knows *you*?" Diamond admits that he was not completely surprised by the reaction. "I had challenged his theory, which he took as an argument against *him*. Which it wasn't."

Yet even a scientist less thin-skinned than John Money might have been stung by the calm, relentless logic of Diamond's critique—which, near the end, raised the most rudimentary Science 101 objection to the unquestioning acceptance of Money's theory of psychosexual neutrality in normal children. "To support [such a] theory," Diamond wrote, "we have been presented with no instance of a normal individual appearing as an unequivocal male and being reared successfully as a female." And Diamond had added: "If such an individual is available he has not been referred to by proponents of a 'neutrality-at-birth' theory. It may be assumed that such an individual will be hard to find."

Hard—but not, as events transpired, impossible. For it was just one year and eight months after Diamond threw down this gauntlet in the *Quarterly Review of Biology* that Dr. John Money received a letter from a young mother in Winnipeg, Canada, describing the terrible circumcision accident that had befallen one of her identical twin baby boys.

3

≡

Ron and Janet Reimer made their first trip to Johns Hopkins in early 1967, shortly after seeing Dr. Money on TV. The young couple—aged twenty and twenty-one respectively—were awestruck by the vast domed medical center dominating the top of a rise on Baltimore's Monument Street. Dr. Money's Psychohormonal Research Unit was located in the Henry Phipps Psychiatric Clinic, a gloomy Victorian building tucked away off a back courtyard. The unit's offices, located on the fourth floor, were reached by way of a rickety turn-of-the-century elevator. Money's own inner sanctum (where most of his meetings with the Reimers would take place over the next eleven years) reflected the psychologist's eccentric tastes in interior decoration. Furnished with a couch, Oriental rugs, and a profusion of potted plants, the room also featured brightly colored afghans thrown over the backs of armchairs, a collection of carved aboriginal sculptures of erect phalluses, vaginas, and breasts on a mantel, and a collection of primitive blowguns, darts, and masks hanging on the walls. The Reimers had certainly never seen anything like this before, but Dr. Money, with his smoothly confident, professional manner—not to mention the diplomas on his wall—made the Reimers feel that they were, finally, in the best possible hands. "I looked up to him like a god," says Janet. "I accepted whatever he said." And what Dr. Money had to say was exactly what the Reimers ached to hear.

In his many published versions of this first interview, Money has recounted how he spelled out to the young couple the advantages of sex reassignment for their baby—"using non-technical words, diagrams, and photographs of children who had been reassigned." He explained to Ron and Janet that their baby could be given a perfectly functional vagina—"adequate for sexual intercourse and for sexual pleasure, including orgasm." He also explained to them that although their child would not, if changed into a girl, be able to bear children, she would develop psychologically as a woman and would find her erotic attraction to men. As a married woman she would be perfectly capable of adopting children of her own.

What is not clear from Money's written accounts of this meeting is whether Janet and Ron, whose education at the time did not go beyond ninth and seventh grades, respectively, understood that such a procedure was in fact purely experimental—that while Money and his colleagues at Johns Hopkins had performed sex reassignments on hermaphrodite children, no such infant sex change had ever been attempted on a child born, like their Bruce, with normal genitals and nervous system. Today Ron and Janet say that this was a distinction they did not fully grasp until many years later. The crucial point they gleaned from Dr. Money was his conviction that the procedure had every chance for success. "I see no reason," Janet recalls him saying, "that it shouldn't work."

Money's eagerness to begin seemed evident in his recounting of the interview almost ten years later. "If the parents stood by their decision to reassign the child as a girl," he wrote in *Sexual Signatures*, "surgeons could remove the testicles and construct feminine external genitals immediately. When she was eleven or twelve years old, she could be given the female hormones."

If Dr. Money seemed to be in a hurry, he was. He explained to Ron and Janet that they would have to make up their minds quickly. For according to one of the finer points of his theory, the *gender identity gate*—Money's term for that point after which a child has locked into an identity as male or female—comes at two and a half to three years of age. Bruce was now nineteen months. "The child was still young enough so that whichever assignment was made, erotic interest would almost certainly direct itself toward the opposite sex later on," Money wrote, "but the time for reaching a final decision was already short."

Ron and Janet, however, were not prepared to have Bruce immediately admitted to the hospital. They needed time to decide on something as momentous as having their child undergo a surgical sex change. They told Dr. Money that they would have to go home and think about it. Janet says that he made no secret of his impatience with the delay. Upon their return to Winnipeg, the couple received letters from him urging them to reach a final decision. "He wrote in a letter that we were 'procrastinating,'" Janet recalls, "but we wanted to move slow because we had never heard of anything like this."

Back home, Ron and Janet canvassed opinions. Their pediatrician recommended against such drastic treatment and stuck by his earlier advice that Ron and Janet wait until the child was of preschool age before beginning the long process of phalloplasty. Janet's mother, Betty, was inclined to trust the expert from Baltimore but had no real opinion of her own. Ron decided not even to bring it up to his parents since he felt sure they would be against it.

Finally Ron and Janet realized that only they could decide the fate of their child. They alone were the ones living with the reminder, at each diaper change, of his terrible injury. Janet saw the benefits of changing their son into a daughter. "I didn't know much back then," she says, "and I thought women were the gentler sex. Mistakenly. I have since learned that women are the hard-core knockabout tough guys. *Men* are the gentler sex, by far, from my experience. But I thought, with his injury, it would be easier for Bruce to be raised as a girl—to be raised gently. He wouldn't have to prove anything like a man had to."

Ron, too, could see the benefits of changing Bruce's sex. "You know how little boys are," Ron says. "*Who can pee the furthest?* Whip out the wiener and whiz against the fence. Bruce wouldn't be able to do that, and the other kids would wonder why." And then, of course, there was the entire question of Bruce's sex life. Ron could not even imagine the humiliations and frustrations that would entail. As a girl and woman, though, Bruce wouldn't face all that, Ron reasoned. If what Dr. Money told them was true, she could live a normal life, she could get married, she could be happy.

Within days of their return from Baltimore, Ron and Janet stopped cutting the baby's hair, allowing the soft, light brown locks to curl down past the ears. Janet used her sewing machine to turn his pajamas into girlish granny gowns. Their son had become, for Ron and Janet, their daughter. Dr. Money had counseled them, when deciding what to call their new daughter, to select a name beginning with the same letter as her former name and to avoid calling her after any female family members with whom her identity could become confused. Janet, following Dr. Money's instructions, called her new baby daughter Brenda Lee.

There was, of course, still one more step to take. That summer, Ron and Janet left Brenda's twin brother, Brian, with an aunt and uncle, then flew back to Baltimore with their daughter. Now twenty-two months old, she was still within the window that Money had established as safe for infant sex change. On Monday, 3 July 1967, Brenda underwent surgical castration in a gynecologic operating room at Johns Hopkins Hospital. The surgeon was Money's Gender Identity Clinic cofounder, Dr. Howard Jones. Today Jones says he can recall few specifics about the case. He says that all decisions regarding reassignment of sex were the responsibility of Money and pediatric endocrinologist Dr. Robert Blizzard.

"My chief interest was the physical situation and the surgical potential," Jones says. "Was the patient healthy and able to withstand the operation?—all that kind of stuff. The case was pretty well worked up before I ever got involved." For Jones, the surgery on Brenda Reimer was like the routine castrations he had been performing on hermaphrodite babies over the previous twelve years—and apparently Johns Hopkins Hospital viewed the operation the same way. Officials of the hospital have declined all comment on the case, but a Johns Hopkins public relations person, JoAnne Rodgers, told me in the winter of 1998, "In all surgeries that were considered, in the sixties, to be experimental, there were protocols in place to have those approved by appropriate committees and boards." Dr. Jones cannot recall that the hospital convened any special committee or board in the case of Bruce Reimer's historic conversion to girlhood.

The main procedure was a bilateral orchidectomy—removal of both testicles. As Jones's operating room notes reveal, the baby, under general anesthesia, was placed on his back on the operating table, each foot secured in a stirrup so that the groin was exposed for the doctors. Three clamps were placed on the scrotum, and two incision lines were drawn on either side of the midline. With a pair of scissors, Dr. Jones cut away the demarcated scrotal flesh in a strip 1.5 centimeters long to lay bare the testicles and seminal vesicles within. With a scalpel, Jones cut away both the right and left testicles, then used a length of catgut thread to tie off the cord and vessels that in adulthood would have carried sperm to the severed urethra.

In closing the scrotum, Dr. Jones then fashioned a rudimentary exterior vagina using the remaining scrotal skin, which he pulled up from its lower edge to meet the top edge of the incision and sewed in a manner that left the scrotum not as a single empty sack, but as two symmetrical flaps. "A rolled piece of gauze covered with telfa was then placed in the midline to effect a midline furrow leaving constructed labia majora on either side," Jones's operating room note concludes.

Ron and Janet say that by the time they decided to have their baby undergo clinical castration, they had eradicated any doubts they might have had about the efficacy of the treatment. This was a crucial turnabout since according to Dr. Money it was a "vital consideration" that the parents of a sex-reassigned child harbor no doubts that could weaken the child's identification as a girl and woman.

Whether Dr. Money himself was able to eradicate his own doubts about the child's future development is debatable. In a letter he wrote on 28 August 1967, more than a month after Brenda's sex change surgery, his tone admitted of considerable caution regarding the child's prognosis. This was perhaps to be expected, since the letter was addressed to the Winnipeg lawyer whom Ron and Janet had hired to sue St. Boniface Hospital and the doctor who had botched the circumcision.

"The reassignment of a baby's sex is usually undertaken only in cases of a birth defect of the genitalia," Money wrote. "Then one usually expects that the child's psychosexual differentiation will be congruous with the sex of rearing. In any given case, however, it is not possible to make an absolute prediction."

And indeed, by the summer of 1967, when Bruce Reimer underwent his castration, Dr. Money had special reason to be particularly reluctant to make an "absolute prediction" about the patient's future psychosexual development. Two years earlier he had undertaken to discover if the findings of the Kansas team about the masculinizing effects on behavior of prenatal testosterone in guinea pigs could be observed in humans. Under Money's direction, one of his graduate students, Anke Ehrhardt, had studied a group of ten girls, ranging in age from three to fourteen, who had been subjected to excesses of testosterone in the womb when their pregnant mothers had taken a synthetic steroid called progestin to prevent threatened miscarriage. Like the guinea pigs in the Kansas team's study, nine of the ten girls had been born with masculinized genitals—an oversized clitoris and in some cases partially fused labia. As interviews with the children and their parents revealed, all nine of those girls demonstrated what Money and Ehrhardt called (in an article published six months before Bruce Reimer's castration) "tomboyishness." This, the authors explained, included marked preferences for "masculine-derived" clothes and "outdoor pursuits," a "strong interest in boys' toys" (these included guns and toy soldiers), a "high incidence of interest and participation in muscular exercise and recreation," and a "minimal concern for feminine frills, doll play, baby care, and household chores."

≡

Central to Dr. Money's program for the sex assignment of hermaphrodites was his edict that the children, when very young, know nothing of their ambiguous sexual status at birth. Money put the same stricture into effect with baby Brenda Reimer. "He told us not to talk about it," Ron says. "Not to tell Brenda the whole truth and that she shouldn't know she wasn't a girl."

It was shortly after the Reimers' return from Baltimore, and not long before the twins' second birthday, when Janet first put Brenda in a dress. It was a special dress that Janet had sewn herself, using the white satin from her own wedding gown. "It was pretty and lacy," Janet recalls. "She was ripping at it, trying to tear it off. I remember thinking, Oh my God, she knows she's a boy and she doesn't want girls' clothing. She doesn't want to be a girl. But then I thought, Well, maybe I can *teach* her to want to be a girl. Maybe I can train her so that she wants to be a girl."

Ron and Janet tried their best to do just that. They furnished her with dolls to play with; they tried to teach her to be neat and tidy; and they tried, whenever possible, to reinforce her identity as a girl. So when, for instance, the twins had just turned four, and Brian was watching Ron shave and asked if he could shave, too, Ron gave him an empty razor and some shaving cream to play with. When Brenda also clamored for a razor, Ron refused. "I told her girls don't shave," Ron says. "I told her girls don't have to." Janet offered to put makeup on Brenda, but Brenda didn't want to wear makeup.

"I remember saying, 'Oh, can I shave, too?' " David says of this incident, which forms his earliest childhood memory of life as Brenda. "My dad said, 'No, no. You go with your mother.' I started crying, 'Why can't *I* shave?' "

Brian says that the episode was typical of the way their parents tried to steer him and his sister Brenda into opposite sexes—and how such efforts were inevitably doomed to failure. "I recognized Brenda as my sister," Brian says. "But she never, ever acted the part."

Today, with the twins having rejoined each other on the same side of the gender divide, the stark physical differences between them eerily testify to all that David has been through. When David first introduced me to Brian in the summer of 1997, I instinctively assumed that the man who took my hand in a firm grip was an *older* brother, so different did this balding, dark-bearded, bearlike man look from his youthfully thin, smooth-faced brother. It was only when I looked a little closer at Brian's face and recognized the startling familiarity of the eyes, nose, and distinctively shaped mouth that I realized I was meeting David's identical twin, and that he was in fact the *younger* of the two (albeit by a scant twelve minutes).

As children, their physical differences were, if less pronounced, equally deceptive. Photographs of them as preschoolers show a pair of exceptionally attractive children: a puppy-eyed little boy with a crew cut, and a slim, brown-eyed girl with wavy chestnut hair framing a face of delicate prettiness. However, by all accounts of family, teachers, guidance clinic workers, and relatives, this illusion of two children of opposite sexes disappeared the second Brenda moved, spoke, walked, or gestured.

"When I say there was nothing feminine about Brenda," Brian laughs, "I mean there was *nothing* feminine. She walked like a guy. Sat with her legs apart. She talked about guy things, didn't give a crap about cleaning house, getting married, wearing makeup. We both wanted to play with guys, build forts and have snowball fights and play army. She'd get a skipping rope for a gift, and the only thing we'd use *that* for was to tie people up, whip people with it. She played with *my* toys: Tinkertoys, dump trucks. This toy sewing machine she got just sat." That is, David recalls, until the day when Brenda, who loved to take things apart to see how they worked, sneaked a screwdriver from her dad's tool kit and dismantled the toy.

Enrolled in Girl Scouts, Brenda was miserable. "I remember making daisy chains and thinking, If this is the most exciting thing in Girl Scouts, forget it," David says. "I kept thinking of the fun stuff my brother was doing in Cubs." Given dolls at Christmas and birthdays, Brenda simply refused to play with them. "What can you *do* with a doll?" David says today, his voice charged with remembered frustration. "You *look* at it. You *dress* it. You *undress* it. Comb its hair. It's boring! With a car, you can drive it somewhere, *go* places. I wanted cars." Brenda also wanted toy guns. Once, around age eight, she went to the store to buy an umbrella. Waiting in line to pay, she saw a nearby display of toy machine guns. After a moment's hesitation, she put down the umbrella and bought one of the guns. At age ten, Brenda would prove to be a crack shot with the pellet rifle Ron and Janet bought for Brian—a rifle in which, ironically enough, Brian himself evinced little interest.

Brenda had always tried to co-opt Brian's toys and clothes—a habit that would invariably initiate fights. "There were knock-down-drag-out wrestling matches all the time," Janet says. "Brian was a weakling compared to Brenda. She was wiry. More often than not, Brenda won. Poor Brian felt so bad getting beat up by a girl."

Ron and Janet were troubled by Brenda's masculine behavior, but having been admonished by Dr. Money not to entertain any doubts about their daughter, they felt that to do so would only increase the problem. Instead they tried to focus on those moments when Brenda's behavior *could* be construed as stereotypically feminine. "She could be sort of feminine sometimes, when she wanted to please me," Janet says. "She'd be less rough, keep herself clean and tidy, and help a little bit in the kitchen."

In her letters to Dr. Money describing Brenda's progress, Janet made sure to emphasize those moments so that the psychologist would know that Janet and Ron were doing everything they could to implement his plans. She also informed Money of their daughter's masculine leanings, but the psychologist assured her that this was mere "tomboyism." This was an explanation that Janet found comforting, and she would cling to it for many years to come. "I have seen all kinds of women in my life," she says, "and some of them, you'd swear they were men. So I thought, Well, maybe it won't be a problem, because there are lots of women who aren't very effeminate. Maybe it could work. I *wanted* it to work."

Ron's and Janet's parents were struck by Brenda's behavior. "When a girl would come to play with her," says Ron's mother, Helen, "she would not play like a girl, and then she would say to her mother that she wanted the girl to go home."

"I noticed it when she had that fight with the boy across the street," says Janet's mother, Betty. "This boy tried to beat her up. And Brenda beat back."

Janet's uncle Johnny and aunt Evelyn were also unable to ignore certain realities about their niece. They might have surmised that Brenda was simply imitating Brian, but they knew better. They knew Brian particularly well because they were the ones who had baby-sat him for the three weeks while Ron, Janet, and Brenda were in Baltimore for Brenda's operation. Without his sister around, Brian had been a quiet, gentle, sensitive boy—quite unlike the little terror who was tearing up Ron and Janet's home with Brenda. Johnny and Evelyn formed the private opinion that, if anything, *Brenda* was the leader of the pair, and it was Brian who followed her lead into boyish mayhem and mischief. "She was the instigator," Johnny says. Neither Johnny nor Evelyn ever voiced this to Ron and Janet, of course. "We were trying to go along with this," Evelyn explains. "We were not going to start looking for trouble."

Brenda, meanwhile, was having her own doubts. "You don't wake up when you're four and a half years old, look at the clock, and say, 'Yup, I feel like a boy,'" David explains. "You're too young." At the same time, he says he knew something was amiss, even before he fully understood the concept of boy and girl. "I thought I was very similar to my brother. It's not so much me being a guy, it's more that we were *brothers*. It didn't matter that I was in a dress."

Brian didn't question his sister's boyish ways until they went off to school. "I was in grade one or two," he says, "and I saw all these other girls doing their thing, combing their hair, holding their dolls. Brenda was not like that. Not at all." At that time Brenda voiced the ambition to be a garbage man. "She'd say, 'Easy job, good pay,'" Brian explains. "I thought it was kinda bizarre—my *sister* a garbage man?" Brian would finally grow so perplexed with his sister's unconventional behavior that he went to his mother about it. "Well, that's Brenda being a tomboy," Janet told him, which he accepted.

It was not an explanation that Brenda's schoolmates were prepared to accept, however. Upon entering kindergarten at Woodlawn, a small school near their house, Brenda became the object of instant ridicule from both boys and girls. "As you'd walk by, they'd start giggling," David recalls. "Not one, but almost the whole class. It would be like that every day. The whole school would make fun of you about one thing or another."

"It started the first day of kindergarten," Janet says. "Even the teacher didn't accept her. The teacher knew there was something different."

She did indeed. Contacted twenty-six years later, the twins' kindergarten teacher, Audrey McGregor, said she had never seen a girl like Brenda before or since. At first glance the child looked like the thousands of other girls who have passed through McGregor's classroom, but there was a rough-and-tumble rowdiness, an assertive, pressing dominance, and a complete lack of any demonstrable feminine interests that were unique to Brenda in McGregor's experience. And there was something else. McGregor mentions an incident that occurred shortly after the school year began. "A female classmate of Brenda's came up to me," McGregor recalls, "and she asked, 'How come Brenda stands *up* when she goes to the bathroom?' "

Ever since setting out to toilet train the twins, Janet had been grappling with Brenda, trying to convince her daughter not to stand and face the toilet bowl when she peed. No amount of coaxing seemed to work. Janet had mentioned the problem to Dr. Money, who had assured her that it was common for girls to insist on standing up to urinate and that the problem would correct itself in time. It had not. For Janet, Brenda's stubborn insistence on standing created a housekeeping nightmare, since Brenda's urine stream, which shot out almost perpendicular to her body from her severed urethra, splashed all over the back of the toilet seat. As for any suggestion that Brenda's stubborn insistence on standing up to urinate indicated that the treatment was not working—this was not, Janet says, something she could afford to believe.

Kindergarten teacher McGregor, unschooled in Money's theories of child development, formed a quite different opinion about Brenda. "She was more a *boy*," McGregor says, "in the *nature* of things." Furthermore, McGregor was convinced that Brenda herself, on some unconscious level, knew this. "I don't think she *felt* she was a little girl," the teacher says.

McGregor's surmise was correct. Plunged into the sexually polarized world of school, Brenda now knew there was something seriously different about her. "You know generally what a girl is like," David says, "and you know generally what a guy is like. And everyone is telling you that you're a girl. But you say to yourself, I don't *feel* like a girl. I liked to do guy stuff. It didn't match. So you figure, Well, there's something *wrong* here. If I'm supposed to be like this girl over here, but I'm acting like this guy, I guess I gotta be an *it*."

Brenda's personal difficulties were obvious in her functioning in the classroom. On her year-end Kindergarten Inventory of Skills, she was rated unsatisfactory in category after category: Social Living, Work Habits, Listening Skills, Speaking Skills, Reading Skills. The school threatened to hold Brenda back to repeat kindergarten. Janet complained to Dr. Money during a follow-up visit to Johns Hopkins. Dr. Money responded by giving the child an IQ test. Over two days, his research assistant, Nanci Bobrow, administered the Wechsler Intelligence Scale (the standard IQ test). Brenda scored in the low 90s, which placed her in the middle 50 percent of the population, indicating an average intelligence. Three weeks later Dr. Money sent the results to Brenda's school. In an accompanying letter he painted a portrait of Brenda as a girl whose problems were temporary and well on the way to clearing up once she got over what he called her "playful negativism," which was the result of "the bad emotional situation created by her early hospitalizations."

"In such a case as the present one," he concluded, "I very strongly favor promotion, because the degree of under achievement observed is a function of an emotional interference-factor which will definitely not improve by retaining the child at the kindergarten level."

The school authorities in Winnipeg, upon receipt of

Dr. Money's letter, reversed their recommendation, and in September 1971, Brenda was advanced to first grade at a new school called Minnetonka.

Brenda's problems only got worse. On 29 October, less than two months after Brenda started first grade, her teacher, Sharyn Froome, filed a report with the district's Child Guidance Clinic. "I have had an extremely hard time interesting Brenda in any games or activities," wrote Froome, who saw Brenda's negative behavior as anything but "playful." Describing her simply as "very negativistic," and noting the child's total isolation from her peers, Froome wrote that Brenda "has been doing just the opposite of anything the other children do."

Child Guidance Clinic worker Joan Nebbs was among those who observed Brenda's functioning at this time. "Her mother would send her to school very clean and cutely dressed, in little fancy tops and things like that," Nebbs recalls. "She was quite fine-featured, with curly hair, and was a very pretty child with big brown eyes. It was her manner more than anything else that got in the way. She was always grubby. She'd always just been fighting with the kids and playing in the dirt. Brenda was *really* a rough little kid. She didn't want to sit down with a book. She'd rather play knock-'em-down-shoot-'em-up cop games." Nebbs says that Brenda sometimes tried to play with girls, but with little success. "She'd be trying to organize the girls to do things her way—trying to be the boss. She'd want them to play cowboys and Indians, chasing everyone around, general mayhem—and they didn't want any part of that."

Ron and Janet, who had hoped to keep Brenda's medical history confidential, had no choice but to recognize that that would be impossible. After repeated queries from both the school and the Child Guidance Clinic for information about Brenda that might throw light on her academic and social difficulties, Ron and Janet signed a confidentiality waiver authorizing their local pediatrician, Dr. Mariano Tan, to contact the Child Guidance Clinic.

"I hope you will keep this letter strictly confidential," Dr. Tan wrote to the clinic's director. "Both of these children have been under my care since Oct., 1966. They are identical twins—both *male*—however, because of an unfortunate accident during circumcision on Bruce (now Brenda), the penis was amputated." Tan went on to explain about Brenda's sex change at Johns Hopkins.

The revelations in Tan's letter seemed to explain, for both the clinic and the school, much about Brenda Reimer. "I just agreed it was a girl until I heard different," says Brian's first grade teacher, June Hunnie. "Once we knew the *background,* we thought to ourselves, Well, no wonder. What can you do to have a child sit down and quietly concentrate on classwork if there's all this horrible stuff going on in the background? It's impossible."

Indeed it was—at least for Brenda. At the end of that school year, Minnetonka informed Ron and Janet that while Brian would be promoted to the next grade in the fall of 1972, Brenda (despite Dr. Money's sanguine predictions) would have to remain behind.

4

≡

On 28 December 1972, four months after Brenda Reimer began her second attempt at first grade, John Money unveiled his "twins case."

The occasion was the annual meeting of the American Association for the Advancement of Science in Washington, D.C. There Money delivered to a capacity crowd of over one thousand scientists, feminists, students, and reporters the first speech in a two-day series of talks devoted to "Sex Role Learning in Childhood and Adolescence." The symposium, held in the Ambassador Ballroom of the Shoreham Hotel, featured an impressive roster of leading researchers in the field of sexual development. Only Money's appearance at the meeting would make headlines, however—thanks to the remarkable case he cited that morning, a still fuller account of which (he informed his audience) could be found in his book *Man & Woman, Boy & Girl* (coauthored with Anke Ehrhardt)—a book that happened to have been published, in an early example of cross-promotional marketing, the very day of Money's appearance at the symposium.

Man & Woman, Boy & Girl had been in the making for the previous four years. Culling data from the hundreds of hermaphrodites who had passed through his Psychohormonal Research Unit since the early 1950s, and drawing (as Money announced in the book's preface) on scientific specialties as diverse as "genetics, embryology, neuroendocrinology, endocrinology, neurosurgery, social, medical and clinical psychology, and social anthropology," the book was a daunting, ambitious-looking effort of scholarship—all the more so for its often impenetrable Latinate terminology and convoluted syntactic structures. Its thematic thrust, however, was surprisingly straightforward and was reducible to one organizing idea stated again and again in its three hundred pages. It was the same idea Money had first advanced in his mid-1950s papers on intersexes: namely, that the primary factors driving human psychosexual differentiation are learning and environment, not biology.

Appearing five years after Money and Ehrhardt's data showing that female humans exposed to excesses of testosterone in utero displayed "tomboyism" in later life, *Man & Woman, Boy & Girl* had little choice but to acknowledge what Money called "a determining influence" of prenatal hormones on adult sexual behavior. Money explained that these influences were not decisive, however. Describing them as merely adding "a certain special flavor" to the girls' behavior, Money stated that in the formation of gender identity, prebirth biological influences are secondary to the power of postbirth environmental factors, which override them. To prove this nurturist bias, Money repeatedly evoked his principle of "matched pairs" of hermaphrodites—intersexual patients who shared a similar syndrome yet had been raised successfully, he claimed, as opposite sexes.

But the careful reader might have been struck by what looked like an uncharacteristic admission that hermaphrodites could not tell the whole story of human sexual development. For midway through the book, Money confessed to the frustrating constraints that prevented sex researchers from conducting the kinds of experiments that would provide truly conclusive answers to the riddle of psychosexual differentiation in humans. "The ultimate test of the thesis that gender identity differentiation is not preordained in toto by either the sex chromosomes, the prenatal hormonal pattern, or the postnatal hormonal levels would be undertaken, if one had the same ethical freedom of working in experiments with normal babies as with animals," he wrote. "Since planned experiments are ethically unthinkable, one can only take advantage of unplanned opportunities, such as when a normal boy baby loses his penis in a circumcision accident."

Then Money revealed that just such an "unplanned opportunity" to experiment on a developmentally normal infant had come his way—and that he had seized it. Describing how the injured baby's parents had allowed their son to be surgically reassigned as a girl, Money also pointed out what he called an "extreme unusualness" to the case: the child in question was one of a pair of identical male twins. The momentous import of this would not have been lost on either Money's readers or his AAAS audience. Money was saying that he had used for his experiment a pair of children whose biology was as close to identical as any two human beings could be: a pair of children whose lives had begun with the same primordial zygote cell, whose DNA bore the same genetic blueprint, and whose brains and nervous systems had developed in the womb within the same bath of prenatal hormones. In short, the ultimate matched pair.

That Money recognized the very special place Brenda Reimer's case occupied in his work—and indeed within the entire history of sex research—was clear from the emphasis he gave it in *Man & Woman, Boy & Girl*. First mentioned in the book's introduction, it was then cited at various key points throughout the text: in Chapter 8 on "Gender Identity Differentiation," in Chapter 9 on "Developmental Differentiation," in Chapter 10 on "Pubertal Hormones." It was in Chapter 7, on "Gender Dimorphism in Assignment and Rearing," that Money explored the case at greatest length, his account having been assembled from firsthand observation of Brenda during the family's annual visits to his Psychohormonal Research Unit and from letters and phone calls with Janet during the year.

Money made mention in passing of Brenda's "tomboyish traits" but dismissed these as insignificant next to the myriad ways she conformed to the stereotypes of female behavior—examples of which were selected from Janet's hopeful cataloging over the years of Brenda's fitful attempts to act more like a girl. Money did make reference to Brenda's extraordinary bathroom habits, but as he had done with Janet, he assured his readers that "many girls" attempt standing to urinate like boys, and he hinted that by age five Brenda no longer stood to pee—and that any sporadic reversion to her old habits was merely her effort at "copying her brother." No mention was made of the academic, social, and emotional difficulties that had obliged Money to intervene on Janet's behalf with the Winnipeg school authorities a year and a half before the book's publication.

By any measure, the account portrayed the experiment as an unqualified success. In comparison with her twin brother, Brenda provided what Money variously described as an "extraordinary" and a "remarkable" contrast. Brian's interest in "cars and gas pumps and tools" was compared with Brenda's avid interest in "dolls, a doll house and a doll carriage"; Brenda's cleanliness was characterized as wholly different from Brian's total disregard for such matters; Brenda's interest in kitchen work was placed alongside Brian's disdain for it. Money did describe Brenda as always the "dominant twin," though he gave the impression that this was changing over time. By age three, he reported, her dominance over Brian had become "that of a mother hen." All in all, the twins embodied an almost miraculous division of taste, temperament, and behavior along gender lines and seemed the "ultimate test" that boys and girls are made, not born.

The significance of the case was not lost on the then-burgeoning women's movement, which had been arguing against a biological basis for sex differences for decades. Money's own papers from the 1950s on the psychosexual neutrality of newborns had already been used as one of the main foundations of modern feminism. Kate Millet, in her bestselling 1970 feminist bible, *Sexual Politics*, had quoted the 1950s papers as scientific proof that the differences between men and women reflect not biological imperatives, but societal expectations and prejudices. The twins case offered still more dramatic, and apparently irrefutable, evidence to support that view.

"This dramatic case," *Time* magazine duly reported on January 8, 1973, the week after Money debuted the case at the AAAS meeting in Washington, "provides strong support for a major contention of women's liberationists: that conventional patterns of masculine and feminine behavior can be altered. It also casts doubt on the theory that major sexual differences, psychological as well as anatomical, are immutably set by the genes at conception."

The *New York Times Book Review* hailed *Man & Woman,*

Boy & Girl as "the most important volume in the social sciences to appear since the Kinsey reports." It summed up the book's argument on the power of nurture to override nature thus: "[I]f you tell a boy he is a girl, and raise him as one, he will want to do feminine things."

The twins case was quickly enshrined in myriad textbooks ranging from the social sciences to pediatric urology and endocrinology. "The clear message here is that even if biologically based sex differences in behavioral predispositions exist, social factors such as the sex which the child is assigned and in which the child is reared can substantially override and obscure them," wrote Alice G. Sargent about the case in her 1977 women's studies text, *Beyond Sex Roles*. Sociologists were equally enthralled by the case and cited it as the premier example of society's power to mold the most fundamental building block of human identity. Typical was the textbook *Sociology*, first published in 1977, in which Ian Robertson wrote that Money's work "indicates that children can easily be raised as a member of the opposite sex" and that what few inborn sex differences might exist in humans "are not clear-cut and can be overridden by cultural learning." The 1979 volume *Textbook of Sexual Medicine*, by Robert Kolodny and renowned sex researchers Masters and Johnson, cited the case as compelling evidence of the power of nurture over nature: "The childhood development of this (genetically male) girl has been remarkably feminine and is very different from the behavior exhibited by her identical twin brother. The normality of her development can be viewed as a substantial indication of the plasticity of human gender identity and the relative importance of social learning and conditioning in this process."

Money meanwhile did his part to ensure that the case got maximum exposure in both the academic and lay press.

Through the 1970s he made the case the centerpiece of his public addresses, rarely giving a speech in which he did not mention
it. He soon introduced refinements into his crowd-pleasing presentation. At a March 1973 address at the prestigious
Nebraska Symposium on Motivation, Money included a slide
show in which he displayed a close-up photograph of Bruce's
groin following the loss of his penis and a shot of the twins
standing near a doorway. Brian is dressed in a short-sleeved
shirt and dark trousers, Brenda in a sleeveless dress, white ankle
socks, and white shoes. Money also showed a shot of Brenda
alone, taken by Money himself. The child is seated awkwardly
on the patterned upholstery of his office sofa. She wears a
floral dress and running shoes, her bare left knee lifted self-
protectively against the lens, her left hand deliberately obscuring her face. "In the last illustration," Money told his audience,
"you have a pretty persuasive example of feminine body talk."

At his Nebraska lecture, Money also dropped a telling comment in summing up the case, when he told his listeners that
Brenda's successful sex change refuted charges that "Money
studies only odd and atypical cases, not normal ones." To those
in the know, this was a not very veiled allusion to Money's principal theoretical rival, Milton Diamond.

In fact, Diamond did not object to Money's use of "odd
and atypical cases" to study gender identity formation. He
merely questioned the theoretical conclusions that Money
drew from them. Since publishing his challenge to Money in
1965, Diamond had taken a teaching post at the University
of Louisville in Kentucky, where he set to work studying
intersexes himself. In his own interviews with intersexual
patients, whom he met at the Louisville Children's Hospital,
Diamond found that an imposed sex assignment in early
infancy was by no means the magical panacea Money's writings suggested.

71

Instead, Diamond met several patients who contradicted the claim that rearing in a particular sex will always make a child accept that designation. There was the female baby exposed to excessive testosterone in utero, who was reared from birth as a girl but at age six stated to her mother that she was "a boy." There was the genetic male born with a tiny penis and raised as a girl, who at age seventeen voluntarily came to Louisville Children's Hospital requesting a change of sex to male—and was willing to endure more than twenty-five surgeries to construct an artificial penis, so vehemently was "she" determined to live in the sex of her genes and chromosomes. Even in those instances when an intersexual child *did* seem to accept a sex in contradiction to his or her biology, Diamond was not convinced that they had undergone a transformation in their core sexual identity. Such cases "should be considered a credit to human role flexibility and adaptability rather than an indelible feature of upbringing," he warned in the book *Perspectives in Reproduction and Sexual Behavior*, published in 1968.

In the years following publication of that book, Diamond was heartened to see that his views were beginning to be noticed by a scattering of scientists, researchers, and clinicians. In England, a pair of physicians, Dewhurst and Gordon, who had been treating intersexual patients for a decade, published their book, *The Intersexual Disorders*, in which they specifically questioned Money's assertion that rearing in a particular sex invariably led to a child's identifying with that sex. They not only cited a nationwide survey of British physicians whose clinical experience with intersexes contradicted Money's claim, but also referred to Diamond's work with intersexes in Louisville. A year later, in 1970, a fellow American joined Diamond for the first time in challenging Money's theory of human psychosexual differentiation.

Dr. Bernard Zuger was a Manhattan-based child psychiatrist whose work treating young male homosexuals and their families had caused him to question the prevailing view that sexual orientation results from rearing and environment. By exploring the family dynamics of his gay patients, Zuger discovered that in many cases the stereotypical pattern of an overbearing mother and a detached, hostile father did pertain; but by actually observing children in their family settings, Zuger came to believe that such a dynamic was not a *cause* of the child's homosexuality, but an *effect*. Long-term interviews with some fifty-five children (some of whom Zuger would follow for thirty years) showed that in virtually every case the boys demonstrated very early feminine play preferences, interests, and behavior. The father's efforts to bond over masculine interests were rebuffed by the child, and the father—rejected—would emotionally withdraw; the mother would move in to fill the vacuum, thus creating the observed pattern of a distant father and overbearing mother. Zuger suspected a biological basis for homosexuality that contradicted the universally accepted nurturist view—a view, as Zuger later wrote, that was founded to a remarkable degree on Money and the Hampsons' prize-winning 1950s papers on hermaphrodites. It was in an effort to learn how the Johns Hopkins team had arrived at those findings that Zuger submitted their work to close review.

Like the Canadian team more than a decade earlier, Zuger found serious problems with the Johns Hopkins team's methodology, interpretation of the clinical data, and statistical analysis. Noting that the papers were "lacking in such data as the ages when individual cases were observed, their subsequent course, and the part substitution therapy played in maintaining their gender role," Zuger also referred to new biological evidence, which had arisen in the intervening fifteen years, that cast further doubt on the Hopkins team's conclusions. Unlike the Canadian team, however, Zuger actually reanalyzed the

Johns Hopkins data using what he considered proper statistical methods and in light of the new biological findings. In doing so, he meticulously dismantled case after case cited by Money and the Hampsons and showed how children who, according to the team, had been raised in contradiction to their prevailing biological sex had in fact accepted a gender assignment in keeping with one or another of the factors that constitute a person's biological makeup as male or female: the chromosomes, the gonads, or the hormones. Summing up, Zuger wrote that of the sixty-five instances given as evidence for the dominance of rearing over biology, only four cases could be said to have escaped challenge—and even those were questionable. "The four cases," Zuger wrote, "might be explained on the basis of the 'flexibility' which Diamond attributes to human sexuality, or perhaps even by specific biologic factors which more detailed studies might have brought to light."

Slated for publication in a 1970 issue of the journal *Psychosomatic Medicine,* a prepublication copy of Zuger's paper was shown by the journal's editors to Money, who fired off a blistering response.

"It is difficult for the seeing to give art instruction to the blind," Money began, before proceeding to accuse Zuger of "intentionally biased sampling" and lambasting his work as "argumentative," and "very conjectural." Declining to address any of the specific scientific, methodological, and statistical unorthodoxies Zuger had highlighted, Money instead issued a threat to the journal editors: "I am sure you have ascertained, by now, the strength of my feeling about Dr. Zuger's manuscript. I do not want to take the easy way out and recommend simply that you do not publish it, because I know it would be equally easy, these days, to journal-shop and get the manuscript into print in another journal. What I really want is to ask Dr. Zuger to subject his manuscript to a very radical, total revision." A revision, in Money's exhaustive spelling out, that would bring Zuger's conclusions into agreement with Money's.

It was a measure of Money's academic power that the editors took his advice. They asked Zuger to revise his paper along the lines suggested by Money. Zuger declined, pointing out that Money had made no criticism "carrying any substance whatever" and adding, "Dr. Money's notion of a total revision, way beyond the scope of the paper, amounts to, of course, stalling it forever." Instead, through an arrangement agreed upon by both researchers, Zuger's paper and Money's letter of rebuttal were published in their entirety in the September-October 1970 edition of the journal.

Whatever larger debate might have been stimulated by the cumulative weight of the critiques by the Canadians in 1959, Diamond in 1965 and 1968, the British team in 1969, and Zuger in 1970 was effectively quashed by the fanfare that attended the publication, in late 1972, of Money's magnum opus, *Man & Woman, Boy & Girl*, and in particular its remarkable chapter on the twins case.

Dr. Mel Grumbach, a pediatric endocrinologist at the University of California, San Francisco, and a world authority on the subject, says that Money's twins case was decisive in the universal acceptance not only of the theory that human beings are psychosexually malleable at birth, but also of sex reassignment surgery as treatment of infants with ambiguous or injured genitalia. Once confined principally to Johns Hopkins, the procedure soon spread and today is performed in virtually every major country with the possible exception of China. While no annual tally of infant sex reassignments has ever been made, one physician conservatively estimates that three to five cases of babies with incongruous genitalia requiring sex change crop up annually in every major American city—giving the United States alone a total of at least one hundred such operations a year. Globally that figure could be as high as a thousand a year.

"Doctors were very influenced by the twin experience," Grumbach explains. "John stood up at a conference and said, 'I've got these two twins, and one of them is now a girl, and the other is a boy.' They were saying they took this normal boy and changed him over to a girl. That's powerful. That's *really* powerful. I mean, what is your response to that? This case was used to reinforce the fact that you can really do *anything*. You can take a normal XY male and convert it into a female in the neonatal period, and it won't make any difference." Grumbach adds, "John Money is a major figure, and what he says gets handed down and accepted as gospel by some."

But not all. Mickey Diamond had continued his research into how the sexual nervous system is organized before birth, and his studies had only strengthened his conviction that neither intersexual nor normal children were born psychosexually neutral—a conviction that would make him view with alarm the burgeoning practice of infant sex reassignment. And he was more convinced than ever that converting a normal infant from one sex to the other would be impossible. "But I didn't have any evidence to disprove the twins case at the time," Diamond says. "I didn't have anything except a theoretical argument to challenge it." He vowed to follow the case closely—a decision, he says, that was made from purely scientific motives. If, however, Diamond also by now felt a degree of personal involvement in his theoretical dispute with Money, that was perhaps understandable. For in the chapter directly following his account of the twins case in *Man & Woman, Boy & Girl*, Money had lashed out at Diamond and the others who had challenged his classic papers. Restating his own position,

Money had acidly observed: "It would not have been necessary to belabor this point, except that some writers still don't understand it," and he went on to say that the work of Diamond and the others was "instrumental in wrecking the lives of unknown numbers of hermaphroditic youngsters."

At the time of *Man & Woman, Boy & Girl*'s publication, Money and Diamond had limited their debate solely to published papers and books. That was shortly to change.

In September 1973, some nine months after the book's publication, John Money chaired the Third Annual International Symposium on Gender Identity, held at the Hotel Libertas in Dubrovnik, Yugoslavia. The symposium brought together a number of the leading authorities in the field of sexual development. These included Money's coauthor Dr. Anke Ehrhardt, who had taken a position as clinical associate professor in psychiatry at the State University of New York at Buffalo; Dr. Donald Laub, the Stanford Medical School professor and plastic surgeon who specialized in sex change surgery; and Dr. Ira Pauly, a psychiatrist who today is still a leader in the field of transexualism. Milton Diamond, not invited as either presenter or panelist, had nevertheless come to Yugoslavia to attend the conference. After the first day of speeches, during which Money had given the keynote address, the scientists gathered at an evening cocktail reception. The convivial gathering took place in a large room with vast windows that framed a view of the sunset over the Aegean Sea.

"I was sitting with some people over at one end of the room," Diamond recalls, "and Money was sitting over in another part of the room with Anke Ehrhardt. And all of a sudden he gets up and shouts at the top of his voice, 'Mickey Diamond, I hate your fucking guts!' "

An altercation ensued.

"They were arguing over the twins case," says Vern Bullough, then a professor at the State University of New York at Buffalo, and a friend of both men. "Mickey pointed out to John that all the data was not there, that it was too early to draw definitive conclusions about the kid. John suddenly slugged Mickey. Hit him. Mickey did not fight back. He just repeated, 'The data is not there.' John yelled at him, 'We have to stick together as sex researchers and not challenge one another!' " (Diamond says that he cannot recall any physical contact during this encounter.)

The combatants were separated, but the incident, Bullough says, threw a considerable pall over the party. Still, it did not inhibit Money's ongoing promotion of the twins case in lectures, published papers, and the press. The following June, Baltimore's *News American* newspaper ran a long profile on Money, in which the twins case was highlighted as his most impressive accomplishment in sex and gender research. "There isn't any question which one is the boy and which one the girl," Money told the newspaper. "It's just plain obvious."

"Such findings," the story continued, "could have an effect on future attitudes about sex roles that could prove comparable to that of Darwin's theory of Evolution."

5

≡

In 1967, at the time of Brenda's castration, Dr. Money had stipulated to the Reimers that he see the child once a year for follow-up consultations. The trips, which were sometimes separated by as many as eighteen months, were meant to "guard against the psychological hazards" associated with growing up as a sex-reassigned child, as Money said in a letter to the Reimers' lawyer. According to the Reimers, however, and to contemporaneous clinical notes, the family's trips to the Psychohormonal Research Unit only exacerbated the confusion and fear that Brenda was already suffering. As Money's private case files show, Brenda reacted with terror on her first follow-up trip to Johns Hopkins at age four. "[T]here was something almost maniacal about her refusals [to be tested]," Money wrote in his notes, "and the way she hit, kicked and otherwise attacked people."

"You get the idea *something* happened to you," David says, explaining the dread that engulfed him during those mysterious annual visits to the Psychohormonal Research Unit, "but you don't know what—and you don't want to know." Brian, who was also required to submit to sessions with Dr. Money on each visit, found the trips equally bewildering and unsettling. "For the life of me I couldn't understand why, out of all the kids in my class, I'm the only one going with my sister to Baltimore to talk to this Dr. Money? It made us feel like we were aliens." The twins soon developed a conviction that everyone, from their parents to Dr. Money and his colleagues, was keeping something from them. "There was something not adding up," Brian says. "We knew that at a very early age. But we didn't make the connection. We didn't know."

All they did know was that Dr. Money and his associates seemed to take an inordinate amount of interest in everything about them. Some of the questions they were asked were relatively innocuous—"What's your favorite food?" "Who do you like more, Mom or Dad?" Others were less so. Dr. Money repeatedly asked the children about the differences between boys' and girls' genitalia and about what they knew about how babies were made. For Brenda, there were also private sessions with Dr. Money in which she was asked minutely detailed, numbingly repetitious questions about the toys she liked to play with, whether she fought with boys, whether she liked to play with girls. David says that Dr. Money and his coworkers dismissed Brenda's concerns about her boyish behavior and feelings. "They'd tell me, 'You shouldn't be *ashamed* of being a girl,'" David says. "They'd say, 'Girls can do the same things as boys.' One woman—an associate of Dr. Money's—told me, 'That's a typical tomboy thing; I did the same thing. You're just a tomboy.' But I was saying to myself, No, it's not *quite* like that. I don't think that's *quite* it."

Money's Psychohormonal Research Unit files corroborate

David's claim that Money and his colleagues seemed unwilling or unable to see and hear Brenda's efforts to tell them of her sexual confusion. At her earliest visits to the unit, Brenda could not consciously articulate her feelings of not being a girl, but as Money's notes show, those feelings were clear in her interviews and in the psychological tests Money and his students administered to her.

On a 19 June 1972 visit to the Psychohormonal Research Unit, when Brenda was six, she was given the Draw a Person Test, a standard test in which children demonstrate the primacy of their own gender identity by representing their own sex when instructed to draw a person. But Brenda did not draw a girl. Instead she produced the standard childish representation of a boy, which her tester, Money's student R. Clopper, called a "stick figure." Asked who it was, Brenda said, "Me." Asked to draw a figure of the sex *opposite* to herself, Brenda refused. Only after what the notes describe as "considerable coaxing" did she draw another stick figure, which she called "Brenda with a ponytail." Then she changed her response to "Brian," then changed again and said it was Brenda herself. Asked what the "opposite sex" figure to herself was wearing, Brenda said, "A dress."

David says he quickly learned to try to tell Money and his coworkers what they wanted to hear. And indeed, in Money's notes Brenda can sporadically be seen making sober avowals to her love of "sewing, cleaning, dusting and doing dishes." As Money's notes also show, however, Brenda often slipped up in her pose of serene and dutiful femininity. In one instance during the June 1972 visit, she can be seen actually feeling out Money for the correct way to answer him, and readjusting her response—on the fly—to fit her questioner's expectations.

The exchange began when Money asked if Brenda fought

back or ran away when boys started to fight her. Brenda at first blurted out, "Fight back," but then immediately reversed herself. "No," she said. "I just run away." Money, clearly noting this transparent attempt to tell him what he wanted to hear, asked the question again. Now Brenda could not be budged. She insisted that she did not fight boys—"Because I'm a girl."

"You're a girl?" the psychologist asked.

"And not a boy," Brenda felt compelled to assure him. Then, apparently unsure whether she had given the correct answer, she asked, "Girls don't fight, do they?"

Minutes later, when Money asked the question from yet another angle (Did she use her hands to fight people?), Brenda promptly contradicted her earlier avowals with the exclamation that she hit hard—"with my *fist*."

By the following year, when Brenda was seven, Money's notes show that she was less prone to such childish mistakes of inconsistency. When Money conducted his standard "schedule of inquiry" with her, she dutifully snapped out her answers with the swiftness of a call-and-response routine.

"Do you like to play house sometimes?" Money asked.

"Yes."

"Who plays mother?"

"Me."

"And who plays father?"

"My brother."

"And who is the baby?"

"My doll."

"How do you play with the dolls?"

"Feed them and, uh, give them milk. That's all."

At this same visit, however, Brenda's subconscious conviction that she was a boy emerged. For it was after the above exchange that Money asked Brenda to describe a "good dream." She began by describing a child on a farm with a horse. Before announcing the sex of the child, Brenda (as Money dictated in his notes) "paus[ed], and search[ed] for the next word" before revealing that the child was a boy.

"It was nice," Brenda continued, "and he wanted to eat, and he wanted to drink. He wanted to go to bed, and he wanted to sleep. That's all."

The presence of Brian in joint interviews with his sister did little to soften the impression Brenda gave of a scrappy, headstrong, dominant little pugilist. Indeed, Money's transcripts of his joint interviews with the twins only serve to reinforce an impression that family members, teachers, Child Guidance Clinic personnel, and others in Winnipeg described to me: that Brenda was the more traditionally masculine of the two children. When Money questioned the six-year-old twins about how to play with a doll, it was Brian who first spoke up, talking excitedly of how you hold, feed, and nurse them. Only when Brenda was pressed to respond did she try to parrot her brother's answer.

In this same interview, Money asked a question he would put to the twins repeatedly over the years: who was "the boss." Brian at first claimed that he was, but Money was clearly dubious. (Two years earlier, he had already noted that Brian "does a lot of copying of her.") He now repeated his question to Brian, asking if he truly was the boss. Brian's bravado instantly collapsed.

"I don't know," he admitted.

Brenda pounced. "*Are* you the boss?" she challenged him. "Do you *want* to be the boss? I don't think so. OK, *I'll* be the boss."

In this same joint interview, Money questioned the twins

about their respective fighting habits. Brian said he fought—but only with girls, and in particular a little girl with orange hair who picked on him.

"Do you fight with other boys?" Money asked.

"No," Brian said. "I fight with girls." Brenda then explained that she defended Brian against his female antagonists, telling them, "You better not hit my brother."

Brenda was still more explicit about her role as Brian's protector a year later. In an interview alone with Money, she once again described how she rescued Brian from bullies. At the same time she let slip that she sometimes bullied Brian herself.

"Do you and Brian fight sometimes?" Money asked.

"Yep," Brenda said.

"Do you fight with your hands, or fists, or feet—or how?"

"Fists and feet and hands."

"Can you beat Brian up, or does he beat you?"

"I could beat him up."

"Who wins?"

"I do."

It was at this same visit that Money compared how the twins threw a ball. Brian had privately told the psychologist that Brenda threw "like a girl." To test this exciting thesis, Money gave Brenda a ball of modeling clay and asked her to throw it. "[S]he pitched the ball in a fairly straightforward way from her left hand (both children are left handed)," Money dictated in his private notes. "It was a standard overhand throw."

David says that at seven he was still too young to be able to formulate in words his inner sense of being identical to his brother in every way but in the anatomy of his genitals. Brenda, though, can be seen clearly struggling to articulate this concept in an exchange with Money over the difference between boys' and girls' private parts. It was a topic that Money had quizzed the twins on since their earliest visits to the unit and one that clearly caused both children acute embarrassment. In the present instance, Brenda dodged and weaved for several minutes, too mortified and frightened to say the words *penis* and *vagina*. Instead she employed a number of stalling tactics that she had perfected over the years in her verbal dueling matches with Money. To his question about how to tell boys and girls apart, she first offered that a boy has short hair, a girl long. Money asked the question again. Brenda said that boys wear pants, girls dresses. This went on for several more minutes until Money, clearly growing impatient, said, "Well, I'll help you. You have a look down here, between the legs. How is a boy and how is a girl down there? What's the difference?"

"You mean it's flat?" Brenda said.

"A boy has a penis—for peeing through," Money said. "It is just like a little sausage, huh? What does a girl have?"

"I don't know."

"Well, she has it flat," Money said. He continued: "A boy doesn't have that. They are both different." Money repeated: "They are both different. Now we know, don't we?"

At this point Money's private notes continue: "Spontaneously she adds: 'But we're twins. We're twins.'"

Money, obviously taken aback by the vehemence of this rare outburst from the ordinarily tight-lipped girl, asked, "What does it mean when you say you're twins?"

Brenda helplessly cataloged several of the physical things that made her identical to her brother: their left-handedness, their voices, their eyes. Too ashamed to speak directly of her genitals, she left it up to Money to settle the mystery of how two such completely similar children could be "both different" in their anatomic sex. But Money failed, or declined, to catch Brenda's meaning, and instead returned to his standard schedule of inquiry—the list of prepared questions about toys, friends, school, and fighting that he worked through at each visit.

As the twins got older, Money's questioning grew more explicit. "Dr. Money would ask, 'Do you ever dream of having sex with women?'" Brian says. "'Do you ever get an erection?' And the same with Brenda. 'Do you think about this? About that?'"

While attempting to probe the twins' sexual psyches, Money also tried his hand at programming Brenda's and Brian's respective sense of themselves as girl and boy. One of his theories of how children form their different *gender schemas*—Money's term—was that they must understand at a very early age the differences between male and female sex organs. Pornography, he believed, was ideal for this purpose. "[E]xplicit sexual pictures," he wrote in his book *Sexual Signatures,* "can and should be used as part of a child's sex education." Such pictures, he said, "reinforce his or her own gender identity/role."

"He would show us pictures of kids—boys and girls—with no clothes on," Brian says. David recalls that Dr. Money also showed them pictures of adults engaged in sexual intercourse. "He'd say to us, 'I want to show you pictures of things that moms and dads do.'"

Money had two sides to his personality, according to the twins: "One when Mom and Dad weren't around," Brian says, "and another when they were." When their parents were present, Money was avuncular, mild-mannered. Alone with the children he could be irritable or worse, especially when they defied him. They were particularly resistant, the twins say, to Money's requests that they remove their clothes and inspect each other's genitals. David recalls an occasion when he attempted to defy the psychologist. "He told me to take my clothes off," David says, "and I just did not do it. I just stood there. And he screamed, '*Now!*' Louder than that. I thought he was going to give me a whupping. So I took my clothes off and stood there, shaking." In a separate conversation with me, Brian recalls that same incident. "'Take your clothes off—*now*!'" Brian shouts.

Though the children could not know this, the genital inspections that Dr. Money demanded they perform were central to his theory of how children develop a sense of themselves as boy or girl—and thus, in Money's mind, crucial to the successful outcome of Brenda's sex reassignment. For as Money stressed in his writings of the period, "The firmest possible foundations for gender schemas are the differences between male and female genitals and reproductive behavior, a foundation our culture strives mightily to withhold from children. All young primates explore their own and each others' genitals, masturbate, and play at thrusting movements and copulation—and that includes human children everywhere, as well as subhuman primates. The only thing wrong about these activities is not to enjoy them."

But the children did not enjoy these enforced activities—particularly those involving "play at thrusting movements and copulation," which Brian recalls that Dr. Money first introduced when the twins were six years old. Money, he says, would make Brenda assume a position on all fours on his office sofa and make Brian come up behind her on his knees and place his crotch against her buttocks. Variations on the therapy included Brenda lying on her back with her legs spread and Brian lying on top of her. On at least one occasion, Brian says, Dr. Money took a Polaroid photograph of them while they were engaged in this part of the therapy.

Of all the therapy the children received, this particular

form of counseling left the deepest impression on both twins. Today David is still unwilling to speak about it. "There are some things I don't *want* to remember," he says. In 1989 he did describe the sessions to Jane Fontane, the woman who would become his wife. The two had just watched a TV documentary on CIA torture involving electroshock to people's genitals. "He cried hysterically," Jane told me. "He was crying about John Money. I'd never seen him like that. I tried to comfort him. David said Dr. Money made him go on all fours and made Brian go up behind his butt. They were being photographed. He mentioned that very act."

Brian speaks of the coital mimicry only with the greatest emotional turmoil. "It's very hard to— I don't understand why to this day we were forced to do that," Brian says.

Brian's perplexity would have instantly been eradicated had he ever made a study of John Money's theory of childhood sexual rehearsal play, articulated repeatedly in books, papers, speeches, and press interviews published over a quarter century, and its supposed critical importance in the establishment of healthy gender identity.

Money's fascination with the topic of coital mimicry in children had its origins in a trip he made in late 1969 to the northern coast of Australia with three professors from the University of New South Wales. There Money visited for two weeks in a village of coastal aborigines called the Yolngu—a tribe Money would later describe as wholly heterosexual and entirely free of any psychosexual gender confusions or dysfunction whatsoever. While visiting one of the tribal elementary schools, Money heard a secondhand report from an eight-year-old child "that two six-year-old relatives at the camp-fire the previous night had given a demonstration of nigi-nigi"—a term Money understood to mean, through his preteen interpreter, "sexual intercourse." This incident, coupled with Money's belief in the tribe's lack of any gender confusion, was the foundation for his theory that childhood "sexual rehearsal play" was vital to the formation of a healthy adult gender identity—a theory he first articulated in a 1970 paper on the Yolngu published in the *British Journal of Medical Psychology.*

"The straightforward attitude of the Yolngu towards nudity and sex play in young children allows these children to grow up with a straightforward attitude towards sex differences, towards the proper meaning and eventual significance of the sex organs, and towards their own reproductive destiny and sense of identity as male or female," he wrote. Conversely, Money hypothesized, Western society's restrictions on such sex play in young children was highly detrimental and was the root cause of such things as homosexuality, pedophilia, and lust murders.

One of Money's colleagues on the trip, Professor J. E. Cawte, who has studied the Yolngu for almost thirty years, says that he has never witnessed sexual rehearsal play among the tribal children and knows of no researcher who has. Professor Cawte is similarly mystified at the claim that adults of the Yolngu manifest no sexual difficulties. As a psychiatrist who has ministered to the needs of the tribe for decades, Cawte says he has treated many of the Yolngu adults for a wide variety of what he calls "sexual neuroses" and dysfunctions of every variety.

Nevertheless, the Yolngu's purported habit of childhood sexual rehearsal play and their alleged freedom from any psychosexual confusion became a constant reference in almost every public utterance of Money's for the next three decades. He included a section on sexual rehearsal play in *Man & Woman, Boy & Girl* and published an article on the theory in *The Sciences* magazine in 1975. By the time he came to write

Sexual Signatures in the mid-1970s (a time concurrent with his treatment of the Reimer twins), the issue of childhood sexual rehearsal play had assumed the dimensions of a crusade—and one that could move Money to shrill flights of rhetoric. "[W]hat happens in our culture?" he wrote. "Children's sex explorations are treated like a contagious disease. . . . [D]on't let them see the incontrovertible differences in their genitals, and don't, at all costs, let them rehearse copulation—the one universal human activity that still imperatively demands that the two sexes behave differently and harmoniously!"

In an interview with the pornographic magazine *Genesis* in April 1977, Money vented his frustration against the prohibition against childhood sexual rehearsal play and a psychologist's right to observe it. "The number of studies of the effects of depriving human infants and juveniles of sexual rehearsal play is exactly and precisely zero," he said, "because anyone who tried to conduct such a study would risk imprisonment for contributing to the delinquency of minors, or for being obscene. Just imagine the headlines and the fate of a research-grant application requesting funds to watch children playing fucking games!" He sounded the same theme in a 1984 speech, lamenting that it was a "crime" for a sexologist "to make a pictorial record of children's normal, healthy sexual rehearsal play" and returned to this theme in *Psychology Today* when he showed a book with pictures of young children engaged in sexual intercourse to interviewer Constance Holden and said, "You have just become a criminal by looking at those pictures of children."

In a 1988 appearance on *The Oprah Winfrey Show*, Money unexpectedly veered from the show's main topic (intersexuality) and put in a plug for his pet theory. "I worked among the aboriginal people on the north coast of Australia in the early seventies," Money told Oprah's audience. "I was very interested in the fact that they don't impose a sexual taboo on themselves, and they don't punish children for doing normal sexual rehearsal play . . . and I was very surprised to find out that there were no bisexual or no gay people in there." Oprah, who had clearly not been briefed on this particular aspect of Money's research, tried to deflect the remark. "I'm almost afraid to ask what all that means, Dr. Money," she interjected. Money, however, was not to be put off, and continued with an explicit description of the sexual rehearsal play he now claimed to have directly witnessed among the Yolngu. A year later, Money could still be heard trying to promote his theory on an episode of the Canadian TV show *The Originals,* where he scoffingly referred to the prudery of a society that prohibits such childhood exploration. "It has become very obvious to me," he said, "that sexual rehearsal play is part of nature's absolute intention, in order to allow children to grow up to be sexually normal."

But never having heard of Money's theory of sexual rehearsal play, Brian and Brenda Reimer could only perform the ritualized poses obediently, in complete perplexity about their meaning and wholly unaware of the critical role their counselor understood the episodes to have in the successful outcome of his most famous experiment in infant sex change.

Not surprisingly perhaps, Brenda, at age seven, began strongly to resist going to Baltimore. Money suggested to Ron and Janet that they sweeten the pill of the annual visits by blending the trips to Johns Hopkins with a family vacation. "Soon," Janet says, "we were promising Disneyland and side trips to New York just to get her to go."

It was also at this time that Dr. Money began increasingly to focus on the issue of vaginal surgery in his sessions with

Brenda. When she underwent her castration at the age of twenty-two months, Brenda was only at the first stage of the feminizing process. Dr. Jones had elected to wait until Brenda's body was closer to fully grown before performing the two remaining surgeries: the first to lower her urethra into the female position, the second to excavate a full vaginal canal. For Dr. Money, there was an increasingly urgent need for Brenda to prepare for these operations. Because genital appearance was critical to his theory of how one "learns" a sexual identity, he believed that Brenda's psychological sex change could not be complete until her physical sex change was finished.

There was only one problem. Brenda was determined not to have the surgery—ever. As Money's private clinical notes reveal, he first raised the issue of vaginal surgery with Brenda on her visit of 24 April 1973. He segued into the subject with deceptive casualness.

"That reminds me of something else I wanted to tell you about," Money said after interrogating her at length on the usual range of topics: fighting, how to tell boys and girls apart. "You know already the way you are made down there, between your legs, you are not exactly the same as other girls, eh?"

"Yes," Brenda said. She was understating the case considerably. Her vagina, with its small stumplike protrusion under the skin and its apparent scarring, caused her such confusion and anxiety that she could not even bring herself to look, or touch, between her own legs.

"Well, I have a message for you about that," Money said. "Here in this hospital we can fix it up for you and make it look like it's supposed to look."

"Huh?" Brenda said.

Money went on to explain that the doctors could operate on her so that she could urinate properly. (It was Money's

theory that Brenda's continuing unorthodoxies in the bathroom resulted solely from the condition of her uncompleted
vaginal surgery.) "How old will you be when you're ready
for that [operation]?" Money asked.

Brenda resorted to the reply she so often gave to Money's
queries. "I don't know."

Money suggested that Brenda would be ready at her next
visit, when she was eight—one year away. Brenda said nothing.
Money talked on at length about the "doctor in the white coat"
who would "fix it up down there." Finally Brenda found her
voice.

"I wouldn't do that," she said.

This was a position from which Brenda would refuse to shift.

Today David explains that his refusal to undergo vaginal
surgery was not only a result of his deep fear of hospitals, doctors, and needles. It had to do with certain realizations he came
to around this time—realizations that convinced him he was
not a girl and never would be, no matter what his parents, his
doctor, his teachers, or anyone else said. For as David explains,
when seven-year-old Brenda daydreamed of an ideal future, she
saw herself as a twenty-one-year-old male with a mustache, a
sports car, and surrounded by admiring friends. "He was somebody I wanted to *be*," David says today, reflecting on those
childhood fantasies. Based on those fantasies, Brenda was convinced that to submit to vaginal surgery would lock her into a
gender that was not her own.

Dr. Money, with the fate of his famous case hanging in the
balance, spared no effort to break down the child's resistance. The transcript of their encounter on 24 April 1973
continues with Money taking a new tack. Hoping to teach
Brenda about the vaginal opening and canal, which she did
not yet possess, Money asked, "How much do you know
about where babies come from?"

Brenda said, "From their mother's tummy."

"Now," Money said, circling closer to the issue at hand, "do you know how the baby gets out?"

Brenda, clearly tumbling to Money's tactic, stalled, mumbling incoherent syllables.

"When it's ready to get born," Money repeated, "how does it get out?"

Again Brenda stalled.

"I'll ask my question one more time," Money said. "When the baby is ready to get born, how does it get out from inside the mother? Where does it get out?"

Brenda, aware that she had driven Money to the limits of his patience, feigned not to have understood. "Oh!" she now exclaimed. "The mother gets her out."

Money was not to be put off so easily. "How does the mother get it out?" he repeated.

"Um, I don't know," Brenda finally said. "I didn't learn that at school."

"Would you like me to show you some pictures?" Money said.

Brenda made no recorded response.

"This is a book called *Two Births*," Money continued, opening a large coffee-table book for Brenda to look at.

Published one year earlier, *Two Births* is a vintage artifact of the early 1970s. Photographed by Ed Buryn, it is a record of two hippie women having home births. The large black-and-white photographs are expertly and beautifully made but are at the same time unsparingly graphic in their depiction of the moments before, during, and after birth. Intense close-ups show both women naked, grimacing, their bare breasts swollen, their vaginas distended as the babies' heads begin to push through the stretched orifices.

"See, there is the lady with the baby inside," Money said

as he leafed through the pages for Brenda. "Getting ready, almost ready to come out. . . . See, here's the baby just getting ready to come out and here it's really coming out. See, there's his head beginning to poke through. . . . There, it got all the way out."

"Now," Money continued, "I wanted to show you that picture of a baby being born because I wanted to tell you that, down there, the way you are, you can't find the baby hole yet." And suddenly Money was once again talking about the "doctor in the hospital here" who could give her a "baby hole."

Neither the pictures of the grimacing women with the spread legs and stretched vaginas nor Money's explanations of the pictures convinced Brenda to submit to the vaginal surgery. Nor did what followed—a description from Dr. Money of sexual intercourse.

"A lot of kids don't know that story," Money said when he had finished describing how the penis goes into the vagina, "because they don't have a doctor to tell them. The lucky kids who know about it are best if they don't talk about it too much."

"Yes," Brenda said.

"You are pretty wise, aren't you?"

"No," Brenda said.

"I think so."

"No," Brenda said. "I'm not."

"Aren't you?" Money persisted.

Brenda did not reply.

"How are you?" Money asked.

Brenda said nothing.

"I think you're a wise girl," Money said.

"No," Brenda repeated, "I'm not."

"You're one of my favorite girls."

According to David, Money's supposed affection for Brenda turned to increasing frustration, impatience, and anger as she continued to resist his blandishments. Brenda meanwhile reacted badly to the increasing pressure to submit to the operation. In the spring of 1974, facing another summer visit to Dr. Money's Psychohormonal Research Unit and yet another battle of wits and wills with him, Brenda found that the pressure was simply too much.

"I had a nervous breakdown," David says. "Because I knew, also, that right after I saw this guy on the summer holidays it would be *school*. It was a double whammy. I remember the summer I turned nine just huddling in a corner and shaking and crying."

Seeing their daughter's distress, Ron and Janet postponed that summer's visit. Finally, however, it was Ron, convinced that only Dr. Money could help their daughter, who insisted that Brenda return to Johns Hopkins in the fall. And so on 19 November 1974 the family again visited the Psychohormonal Research Unit. The two-day visit was a trial for all concerned—but especially for Brenda. In a one-on-one taped interview, Money tried in vain to get her to speak. She would only mumble monosyllables. When Money tried to raise the topic of vaginal surgery, Brenda scurried from the room, found her father in the hallway, and refused to leave his side.

Today David recognizes that if he had told his parents what went on between Brenda and the psychologist behind closed doors—the pressure tactics, cajoling, pornography, and unorthodox inspections and posings—Janet and Ron would never have made her return to Johns Hopkins. But the thought never occurred to her—for a simple and chilling reason.

"I thought my parents *knew*," David says. "I figured, they're responsible for me. They brought me here. They *must* know what's going on."

6

≡

Ron and Janet did not know what went on in the twins' sessions with Dr. Money. "The twins would be whisked off somewhere, I didn't know where," Janet says. "Dr. Money spent some of the time in a little office talking mostly with me, some to Ron." They had no reason to think that the psychologist was any different with Brenda and Brian than he was with them, and with Ron and Janet he was unfailingly polite and kind. Only once did they have any reason to suspect that there might be another side to Dr. Money. "One time we came into his office when he wasn't expecting us," Ron says, "and he was giving all holy shit to his secretary. Just *chewing* her out for something small—she forgot to mail a letter or something. When he saw us, he let it drop."

This unsettling glimpse was never repeated, so Ron and Janet wrote it off as a rare moment when the psychologist lost control. Otherwise they continued to think of Money as their closest confidant and friend. And he considered them important allies in his ongoing struggles with Brenda. In fact, the end of their fraught November 1974 visit, Dr. Money took Ron and Janet aside and gave them what he called a "homework assignment," telling them to find opportunities to talk with Brenda explicitly about her genitalia and vaginal surgery and impressing upon them how important it was that she agree, at the very next visit, to a vaginal inspection.

In a private note to himself after this meeting, Money was still more emphatic: "Next year it will be imperative for a physical examination to be done," he wrote. "There is an optimal length of time for dealing with a difficult issue by avoiding it, and that optimum will be passed next year, if it is not already passed this year." Something of Money's growing frustration with Brenda's stalwart resistance also crept into this note. "When Brenda is tense and hyperkinetic, she does not give an exactly endearing impression nor a particularly feminine one."

Back in Winnipeg, Ron and Janet got to work on their homework assignment. Told to impress upon Brenda the differences between male and female sex organs, Ron and Janet had been instructed by Money to allow her to see them naked. In *Sexual Signatures*, Money emphasized the importance of such parental genital displays for correct heterosexual child development, and even went so far as to recommend that parents engage in sexual intercourse in front of their children. "With a little calm guidance," he wrote, "the experience can be integrated into the child's sex education and serve to reinforce his or her own gender identity/role."

Janet and Ron drew the line at having sex in front of the twins, but Janet did try to follow the other parts of the homework assignment. She appeared naked, as often as possible, in front of Brenda. This only embarrassed the child, who seemed startled to see her mother walking around the house unclothed. "All of a sudden," David recalls, "right after we go on one of

John Money's trips, she's walking around stark naked." Desperate for the treatment to work, and afraid to contravene Money's orders in the least, Janet persevered. "He encouraged us to go to a nude beach," she says. "We knew of a river where there was nobody for miles around. Ron and I went in the buff, but the twins wouldn't." Janet also tried, in conversation, to "break the ice" with Brenda about the vaginal surgery, but with similar dismal results. "The minute I went anywhere near that topic," Janet says, "she'd leave the room."

The atmosphere in the Reimer home grew steadily more tense as Brenda realized that her parents were now working in collusion with Dr. Money to force her into the surgery. She began to rebel against her parents openly. Even the supposedly happy occasion of Christmas became an ordeal. Brenda raged against having to get into the party dress her parents insisted she wear when they went to see Ron's family in Kleefeld. Brenda had always hated going to see her extended family because this always meant that her parents would put special pressure on her to dress and act like a little lady. Making matters worse, her grandparents, aunts, and uncles would constantly scrutinize her. "They'd be studying me like a bug, to see how much I'd changed throughout the year," David says. "And as soon as I'd catch them staring at me, they'd look the other way. I told my dad, 'I don't know why, but I always feel like an oddball around my own family.' He said, very quietly, 'I know.'"

Ron's family, like the rest of the relatives, knew about Brenda's sex change, so Ron understood why they studied his daughter so closely. He also recognized in his heart of hearts what they were seeing. "*I* sort of knew it wasn't working after Brenda was seven or something," Ron says. "But what were we going to do?"

Neither Ron nor Janet could entertain the notion that they

had made the wrong decision. The only option was to put distance between themselves and anyone who seemed bent on making them face such a realization. From now on, Ron decided, they would see as little of his parents as possible.

But it proved difficult to segregate themselves from all reminders of Brenda's problems. Just that fall the Child Guidance Clinic had once again contacted them to say that her behavioral problems in school had worsened and that she was "hyper and defiant" and looked "unhappy." Furthermore, the clinic reported, Brian was also showing signs of increasingly serious emotional problems related to Brenda's predicament.

"At that point my main emotion toward my sister was jealousy," Brian explains. "She got all the attention. I was just the normal one. Mom and Dad were so worried about Brenda that they neglected me. I felt I was unimportant. I started to act up a bit, try to get some attention." He succeeded that March, when he was caught trying to shoplift from a local store and the proprietor threatened to press charges. For Janet and Ron, this proved to be the last straw.

At the time of Brenda's sex reassignment almost eight years earlier, their local pediatrician had advised them to move away from the area so that they could make a fresh start in a place where there was no lingering memory of their former son. They had refused his advice at the time. Now they saw the wisdom of it. It was imperative that they get away from the ghosts and doubters who haunted Winnipeg; it was imperative that they put as much distance as possible between themselves and Ron's parents, the Child Guidance Clinic—everyone.

That spring of 1975, Ron and Janet sold their house, their furniture, their appliances, and their '66 Pontiac. They bought a half-ton Chevy truck with a camper on the back. They packed up what few belongings they still owned and headed west for British Columbia. Ron had a friend out there who'd told him there was plenty of work. Yet so little had Ron planned this move—so completely had he failed to look ahead—that he would later castigate himself for having sold all their possessions and thus put himself in the position of having to buy everything when they got to BC.

"I remember thinking when we got there, Oh God, what did I *do*?" Ron says. "How could I have just picked up and moved? What an *idiot* thing to do!" Only much later, Janet says, did she and Ron fully face why they had so precipitately uprooted their lives and headed off to BC. "We were trying to escape."

That Dr. Money already understood this motive was clear from a note he made to himself at this time. "The plan to move to British Columbia may include a bit too much geographical magic," he wrote, "especially with regard to solving problems with the grandparental families. However, it could also turn out to be a perfectly satisfactory move."

Their destination was British Columbia's mountainous, wooded, sparsely populated interior. They settled in a tiny place called Ashton Creek. The nearest town was Enderby with a population of just 2,500. Ron bought a house trailer, which they parked in an encampment. The twins were enrolled in grade four at tiny Ashton Creek School.

"It was more of a country school," David says. "But it didn't matter what kind of school it was. If you're not comfortable, you're not going to be comfortable no matter what school you go to. You can go to a thousand schools, and it's always the same. Because the standard rule of thumb is: There's the girls over here, and there's the boys over there. Separated. Which direction [do I go]? There's no belonging. So you're an outcast. It doesn't change. School to school to school. It doesn't change."

In April the family returned to Baltimore for another visit with Dr. Money. By now, Brenda, almost ten years old, had developed a new attitude toward Dr. Money. Frowning, sullen, and almost completely mute, she refused to answer his questions in anything but grunting monosyllables. She also imagined that she had succeeded in keeping secret certain shameful impulses she had started to have, but she was wrong. According to Ron, it was during this visit that Dr. Money informed him of an issue that had arisen during his private sessions with Brenda.

"Money told us that he had asked Brenda what partner she would rather have, a boy or a girl," Ron recalls. "Brenda had said, 'A girl.'" Ron says that Dr. Money wanted to know how they felt about raising a lesbian. At a loss for how to respond, but relieved that Dr. Money did not seem to think it significant, Ron said what he honestly believed about homosexuality: "It's not the most important thing in life."

Money evidently agreed, for this clinical finding was not included in his next report on the twins, which appeared later that year in *Archives of Sexual Behavior*. Entitled "Ablatio Penis" (the Latin term for the medical condition of complete amputation of the organ), the paper recapped earlier data about the sex reassignment's success and added one new piece of evidence of the girl's happy femininity. Money recounted an exchange he had had with Brenda about the family's recent trip to the Washington Zoo: "I resorted to the standard question of which animal she'd want to be if she could change into one," Money reported. "She elected to be a monkey. . . . 'Would you want to be a boy monkey or a girl monkey?' I asked. 'A girl one,' she replied, and gave as the reason for this choice, 'I'm already a girl!' "

A question remains about this seemingly unequivocal statement of female gender identity (even if not dismissed as one of Brenda's typical efforts at telling the psychologist what he wanted to hear). In that interview session of 24 April 1973, Money had threaded his reel-to-reel tape recorder incorrectly. Thus Brenda's statements on the tape were virtually inaudible, and Money had to make a special effort to hear anything at all of the interview. "I'm pressing the earpiece closer into my ear," he dictated in his notes while listening to the playback of this exchange, "and hearing a little more now. . . . I ask her why she wants to be a monkey, but I cannot hear the reply on tape. I remember that it was something that I did not immediately understand until she demonstrated with her hands that she meant climbing and swinging. I then asked her if she would want to be a boy monkey or a girl monkey. Her reply is audible on tape, 'a girl one.' I inquire as to why. . . . Again, the reply is audible on the tape, 'I'm already a girl.' The pronunciation of *girl* is as if it were spelled g-r-i-r-l."

Upon seeing these interview transcripts for the first time, in 1998, David insists that he did not say "girl" but in fact used one of the standard evasion tactics he had by then developed. Instead of answering Money's question about the *sex* of the monkey, Brenda had instead answered what *kind* of monkey she would like to be. "I said 'gorilla,' " David says. Considering the particular clipped Canadian prairie accent in which all the Reimers speak, it is easy to see how the word *gorilla,* heard through a faulty recording, would come out "grirl." That it should then be interpreted by Money as "girl" is perhaps illustrative not of Brenda's gender identity but rather of the role the subjective hopes of scientific researchers can play in the gathering and interpretation of their data.

Epistemological vagaries notwithstanding, Money's "Ablatio Penis" paper ended on a note of high optimism. "No one [outside the family] knows [that she was born a boy]," Money wrote. "Nor would they ever conjecture. Her behavior is so normally that of an active little girl, and so clearly different by contrast from the boyish ways of her twin brother, that it offers nothing to stimulate anyone's conjectures."

Later that year, Money published yet another account of Brenda's successful metamorphosis. This time the intended audience was not only Money's scientific and medical colleagues, but the general public. The new account appeared in *Sexual Signatures*. Coauthored with journalist Patricia Tucker, the book was couched in the language of commercial pop-psych bestsellers, and it represented Money's bid for a wider audience. It also provided his most detailed and readable account of Brenda's sex change to date. Stripped of the often impenetrable medical jargon that characterized his earlier accounts of the sex reassignment, *Sexual Signatures* offered an unrelievedly upbeat, almost triumphant version of Brenda's story.

"Although the girl had been the dominant twin in infancy," Money wrote, "by the time the children were four years old there was no mistaking which twin was the girl and which the boy. At five, the little girl already preferred dresses to pants, enjoyed wearing her hair ribbons, bracelets and frilly blouses, and loved being her daddy's little sweetheart. Throughout childhood, her stubbornness and the abundant physical energy she shares with her twin brother and expends freely have made her a tomboyish girl, but nonetheless a girl." Describing Brenda's sex reassignment as "[d]ramatic proof that the gender identity option is open at birth for normal infants," Money went on to claim that the child's subsequent history was proof of how well the family had adjusted to the original decision in favor of castration.

Even as *Sexual Signatures* appeared in bookstores, the Reimer family's adjustment to their decision was growing more imperiled by the day in British Columbia.

Ron, who had found work in a sawmill after a grueling period of unemployment, was no better than he had ever been at talking about his daughter and the major medical decisions that were ahead of her. "I'm a workaholic," Ron admits. "If I'm worried about something, I just work harder." Rather than face what was happening with Brenda—her nervous breakdown before the previous year's trip to Johns Hopkins, her intense boyishness, her refusal to discuss the surgery, her continued dismal performance in school, her "lesbianism"—Ron would simply stagger home from an overtime shift at the sawmill, silently shovel in his dinner, then stare into the TV while he drained a six-pack. Often failing to join Janet in bed, he would simply slope off into unconsciousness in front of the TV.

Janet was faring little better. After six months in British Columbia she was feeling dangerously alienated. "I had no family to talk to," she says. "I had some friends, but they didn't know the real me—or the real Brenda." The only person who *did* know the real Brenda was Ron, and he refused to talk about her. David recalls vividly the chaos that engulfed their small trailer: "Mom crying and screaming," he says, "Dad drinking."

That summer Janet's condition deteriorated. She sank into a serious depression and found herself obsessively thinking about all that had happened to them. "Sometimes it didn't feel real," she says. "What I especially found difficult was all those years I had a strong sense of Christianity, of a living God—a God who laid out a path for you. I remember thinking, What kind of a purpose could this *possibly* have in life? All this pain after pain after pain? What possible purpose could there be in this *horrible* life?" She began to suffer wild mood swings, from volatile anger to weeping despondency. At times, she felt that her very sanity was in jeopardy. Janet says that she experienced periods of "psychosis"—episodes when she could not tell reality from fantasy.

"She was unpredictable," Brian says. "It was like walking on eggs. You didn't know what you were going to come home to." Janet consulted some doctors in the area. "I couldn't get help," she says. "One lady doctor said to me, 'Oh, all you need is another baby to keep you busy.' I said, 'The last thing I need is another child!' "

Ron and Janet were almost completely estranged. That summer Janet drifted into an affair with one of the local men to spite Ron. He found out and was devastated. Janet, guilt-ridden, swallowed a bottle of sleeping pills. Ron found her in time to drive her to the hospital in Enderby. Upon Janet's release, the couple talked about divorce but decided to soldier on somehow together. In the early fall, more bad luck arrived: their house trailer caught fire and burned to the ground, destroying all the family pictures and most of their possessions.

In November 1976, one year and five months after their flight from Winnipeg, the family packed up the few belongings that had escaped the fire and began the long drive back to Winnipeg. Janet and Ron were obliged to admit that their attempt to escape had only exacerbated the problems they were trying to flee. Janet's depression and suicide attempt, the near breakup of the marriage, and the acceleration in Ron's alcohol intake had all taken a visible toll on the twins. Brian was now beginning to have troubling outbursts of frustrated violence and anger against other kids. Brenda turned her feelings inward and became increasingly anxiety-ridden and depressed. She was also demonstrating overt hostility and distrust toward both parents, especially her mother—although, as David points out, he tried to hide these feelings because he was now trying with everything he had to hold together his parents' disintegrating marriage. "I thought it was all *my* fault," David explains. "So I would try to make them happy. I would try to be more ladylike."

Those efforts proved more difficult than ever for Brenda—especially now that she had passed her eleventh birthday, and certain physiological changes began to occur: her shoulders had started to widen and grow more muscular; her neck and biceps, too, began to thicken; and sometimes now her voice would crack into a strange squeaking sound.

All in all, the Reimers' sojourn in British Columbia formed a bitterly ironic contrast to the way their lives were portrayed in the *New York Times Book Review* of May 1975, when reviewer Linda Wolfe, working from the evidence presented by John Money in *Sexual Signatures,* wrote of "the identical twin boy whose penis was cauterized at birth and who, now that his parents have opted for surgical reconstruction to make him appear female, has been sailing contentedly through childhood as a genuine girl."

Janet with her twin baby boys, Brian and Bruce. The cataclysmic accident that would lead to Bruce's sex change was still some months in the future. *(Photo courtesy of the Reimer family)*

Brenda at age two, shortly after her surgical sex reassignment.
(Photo courtesy of the Reimer family)

Brenda and Brian as toddlers.
(Photo courtesy of the Reimer family)

Brian and Brenda around the time that they started kindergarten. *(Photo courtesy of the Reimer family)*

Brian and Brenda Reimer—the ultimate "matched pair."
(Photo courtesy of the Reimer family)

Brenda at age ten. "Everyone is telling you that you're a girl," David would later say, "but you say to yourself, 'I don't feel like a girl.'" *(Photo courtesy of the Reimer family)*

Christmas, mid-1970s, with Brenda dressed in her best clothes for visiting her paternal grandparents. *(Photo courtesy of the Reimer family)*

Dr. John Money
(Photo by Mike Mitchell)

Dr. Milton Diamond
(Photo by Sara Diamond)

Brian, Ron, and Janet on their way to what would be the family's final trip to Baltimore, in May 1978. Brenda, the photographer, was careful to stay on the opposite side of the lens at this stage of her life. *(Photo courtesy of the Reimer family)*

Brian and Brenda at age fourteen, shortly before they were told the truth of Brenda's birth. *(Photo courtesy of the Reimer family)*

Brenda at age twelve, shortly after she started estrogen therapy to promote the feminization of her figure.
(Photo courtesy of the Reimer family)

David and Brian in May 1980 at their uncle's wedding, where David made his public debut as a boy. David still carries the evidence of the binge-eating he had done, as Brenda, to disguise his breasts. They would be surgically removed five months later, and testosterone injections would soon make David catch up in height to his twin.
(Photo courtesy of the Reimer family)

David and Brian, May 1980, at their uncle's wedding.
(Photo courtesy of the Reimer family)

David at age eighteen.
(Photo courtesy of the Reimer family)

David and Jane on their wedding day, 22 September 1990. It was Jane's reaction to the truth about David's past that convinced him of her "true heart." *(Photo courtesy of the Reimer family)*

PART TWO

To Know My Birth

7

≡

THE REIMERS ARRIVED back in Winnipeg in mid-November 1976 and began trying to rebuild their lives. Janet found work as a cashier at a five-and-dime, and Ron took a job with a food company driving a lunch truck—a job he would soon leave to start his own business. Living temporarily out of the Capri Motel in the city's East End, the Reimers enrolled Brenda and Brian in Agassiz Drive Elementary, a small school located nearby on the edge of the comfortable middle-class neighborhood of College Heights. Up to this point Brenda's sole psychological therapy had been the counseling sessions during her annual visits to Johns Hopkins. This changed when she entered Agassiz Drive, where her anxiety, social isolation, and fear immediately drew the attention of the school's principal, Mr. Bergmann, who once again notified the city's Child Guidance Clinic. Joan Nebbs, the reading clinician who had handled Brenda's case a year and a half earlier, interviewed Brenda again in the fall of 1976.

"Brenda's interests are strongly masculine," Nebbs wrote in her new report on Brenda at age eleven. "She has marvelous plans for building tree houses, go-carts with CB radios, model gas airplanes. . . . [S]he appears to be more competitive and aggressive than her brother and is much more untidy both at home and at school." A session with the clinic's psychologist revealed that Brenda had "strong fears that something has been done to her genital organs" and that she had "some suicidal thoughts."

Brenda's case was referred to Dr. Keith Sigmundson, an amiable thirty-four-year-old who was then head of the clinic's Psychiatry Department. Born and raised in the small fishing town of Gimli, an hour's drive north of Winnipeg, Sigmundson had taken his medical and psychiatric degrees at the University of Manitoba in Winnipeg, then joined the Child Guidance Clinic, where his career ascent had been rapid. "Because I was just ahead of the baby boomers," Sigmundson says with typical self-deprecation, "I got a position that I was too young for and probably didn't deserve in the first place."

Even the most seasoned psychiatrist might have found Brenda Reimer's case a unique challenge. Sigmundson read Dr. Money's published accounts of the unequivocal female gender identity that Brenda had reportedly established, but from his very first meeting with the girl, Sigmundson was struck by her appearance. "She was sitting there in a skirt with her legs apart, one hand planted firmly on one knee," he recalls. "There was nothing feminine about her."

Sigmundson decided to establish a record of the girl's behavior in comparison with that of her twin brother and arranged for a clinical video to be made. Shot through a two-way mirror in a room at the Child Guidance Clinic, the videotape showed psychiatrist Dr. Doreen Moggey interviewing the twins. Or rather, *trying* to interview the twins. Brenda, whose yearly trips to the Psychohormonal Research Unit had made her acutely distrustful of any unfamiliar people or situations, immediately grew wary of the undertaking.

"It was a big room," David recalls, "nothing in it—just three chairs, one for me and one for my brother and one for this lady who was there talking. She had a notepad and she was writing. She was trying to get me to go over and sit down. But I was suspicious. I was checking all the nooks and crannies, checking the place out. I went up to the glass, and I saw the camera." Yelling at her brother that they were being filmed, Brenda immediately stalked out of the room and refused to come back.

Despite the abortive nature of the video, it did provide an accurate record of Brenda's mood, movements, and manner-isms, so that when Sigmundson convened a group of the city's senior psychiatrists, endocrinologists, and pediatricians to con-sult on the case, he showed them the video. "Everyone who saw Brenda that day identified that she looked like a boy," says Moggey, who attended the meeting. But in the conversation and debate that ensued among the assembled physicians, a con-sensus soon emerged that they had little choice but to continue the treatment Money had begun. It had simply gone too far to turn back. Nor was it lost upon Sigmundson that Brenda's case was famous in the medical literature. "I felt I had a responsibil-ity," Sigmundson says. "This was *the* case. The idea was that we were going to try to make this work."

To promote Brenda's female identification, Sigmundson decided that she should be treated by a woman psychiatrist. He enlisted Dr. Moggey. A keen-eyed woman with a brisk, take-charge manner, Moggey was, like Sigmundson, troubled by the case from the outset. In an early meeting with Brenda, on 30 December 1976, Moggey noted the fitful girlishness that

Brenda (especially when under observation) could incorporate into her mannerisms and utterances—"a mixture of masculine and feminine gestures and characteristics," as Moggey put it in her notes. The psychiatrist soon grew skeptical about the degree to which Brenda's sporadic feminine adaptations indicated her sense of herself as a girl. In her notes, Moggey wrote, "One gets impression that [Brenda] sometimes says what she thinks you want to hear—'I am a girl.'"

As the sessions progressed, Moggey's doubts quickly deepened. She documented the way Brenda repeatedly voiced the conviction that she was "just a boy with long hair in girl's clothes," and that people looked at her and said she "looks like a boy, talks like a boy." At the same time Moggey noted that Brenda was vehemently opposed even to talking about undergoing femininizing surgery on her genitals and flat-out refused to return to Johns Hopkins where, Brenda complained, people looked at her and "a man show[ed] her pictures of nude bodies."

Moggey had read John Money's accounts of the case and was mystified. "When you read the papers and when you saw the kid, they didn't go together," she says. "That wasn't the child he was describing." Nor, Moggey says, did Ron and Janet seem to be the parents Money was describing. In *Sexual Signatures*, Money had portrayed Brenda's parents as blissfully content with the difficult decision they had been forced to make in authorizing the sex change of their baby. Yet in her own interviews with Ron and Janet, Moggey heard of the couple's recent near split, about Ron's drinking, and about Janet's depression and suicide attempt. Far from a husband and wife happily raising their daughter, Ron and Janet seemed to be a couple barely holding themselves together as they anxiously tried to comply with Dr. Money's directives on how to raise Brenda.

Just how slavishly Ron and Janet were following Dr.

Money's program was borne home to Moggey upon her first visit with Brenda at the Agassiz Drive school. It was a frigid December day, and all the girls were dressed in pants; Brenda was the only girl in a skirt. When Moggey asked Janet why she did not put Brenda in pants, Janet helplessly replied, "Because Dr. Money said to put her in dresses." Moggey had to remind Janet that Winnipeg experienced some of the harshest winters on the continent (unlike Baltimore) and that Brenda should be in pants like the other girls. Only then did Janet agree to dress Brenda more appropriately. Brenda began to wear jeans like the other girls in her class.

On 3 January 1977, a month and a half after she took on the case, Moggey wrote to John Money. She informed him of Brenda's behavioral difficulties and requested some background information "to help Brenda and her family make a more appropriate adjustment to her problems." She asked what surgery had already been done on Brenda (Money's published accounts had always left this vague), what operations were planned, and what efforts Money had made to help both Brenda and her parents "adjust to the sex change."

Money replied on 17 January and in a breezy tone professed himself very pleased that Dr. Moggey had become involved in Brenda's case. He explained that the second stage of Brenda's vaginal surgery had not yet been performed due to the child's "fanatical fear of hospitals"—a fear, Money wrote, "that I have encountered on only one other occasion in 25 years of work at Johns Hopkins." He added that mention of hormone treatments or surgery induced in Brenda a "panic so intense that it's impossible to broach any conversation on such matters without the child fleeing from the room, screaming." Nevertheless, Money continued, there was now "an urgency" that Brenda's fears be overcome, because the need for hormone therapy and surgery was rapidly increasing with her approaching adolescence. "It will be one of the best things you can do for her if you can help her break down this extraordinary veto," he wrote. Dismissing Moggey's suggestion that Ron and Janet had not adjusted to Brenda's sex change, Money added, "With regard to the help that you can give the parents, I think it is not so much with regard to helping them adjust to the sex reassignment, as it is to helping them adjust to one another"—specifically, Money added, by helping Janet control her depressive mood swings.

Moggey (having already found a psychiatrist for Janet, whose moods would soon stabilize with antidepressants) was nonplused by Money's reply. "I thought, There are more problems here than are controlled by the mother," she says. Yet inclined to defer to the famous psychologist's apparently greater knowledge and experience of the case, Moggey forged on in her sessions with Brenda, reassuring the child that she was indeed a girl and impressing upon her the necessity that she return to Johns Hopkins to undergo surgery on her genitals.

"But the resistance!" Moggey says. Sullen, angry, unresponsive, Brenda often simply refused to speak, scowling and directing her gaze to the floor. Mere mention of the word *penis* or *vagina* would induce in the child an explosive panic. Brenda remained immovable on the subject of vaginal surgery or returning to the Psychohormonal Research Unit. "Won't look at pictures of female bodies with Dr. Money," Moggey noted on 20 January. "Won't accept going to see Dr. Money."

Struck by the obvious depth of the child's aversion and concerned that Money had not yet grasped the severity of the problems that Brenda and her family had been laboring under, Moggey wrote to him again on 2 February. This time she wrote at greater length and was explicit about Brenda's problems: she explained that the child was two years behind her peers academically and had not progressed well in school "from day one"; that Brian was "embarrassed" by Brenda's "tomboy like behavior"; that Brenda was having trouble making friends and talked openly about being different from other girls; that she was "not interested in developing female shape," felt anything to do with her body was "dirty," was still refusing surgery, and had expressed embarrassment about her trips to Baltimore where "a man [was] showing her pictures of nude bodies"; and that furthermore "she is saying no to returning to see you and threatens to run away if this is necessary." Moggey ended the letter by saying that despite Brenda's resistance, she was continuing to work toward making the girl return to Baltimore, "as I do feel Brenda needs to start her hormone therapy and start developing female characteristics."

Money's reply, dated 9 February, was brief. Failing to allude even glancingly to the many issues raised by Moggey, he simply reiterated his great pleasure that the psychiatrist had become involved, his happiness in working "in closest collaboration" with her, and his "great relief" that Moggey was helping to prepare Brenda for her return to Johns Hopkins.

Money's relief was to be short-lived. After two more months of trying—in vain—to convince Brenda to return to Johns Hopkins, Moggey decided that no amount of therapy could remove the girl's fierce resistance. That spring Moggey spoke to Ron and Janet about the alternative of Brenda's undergoing the vaginal surgery not in Baltimore, but in Winnipeg—a plan that would not only save considerable time, energy, and money for the Reimers, but also remove for Brenda the deep anxiety associated with her visits to Johns Hopkins.

Still Ron and Janet expressed trepidation about deviating in any way from Money's program. They would agree to it only if Dr. Money himself approved the plan. On 18 April, Moggey wrote to him again and apprised him of the proposal

117

to have Brenda's surgery performed in Winnipeg—"unless," she added, "there is some particular reason for returning to Johns Hopkins."

As it happened, Money considered there to be many reasons for the child to return to Johns Hopkins. He outlined these reasons in a two-page, single-spaced letter that seemed to carry, under its surface smoothness, a strong undertone of desperation at the prospect of losing control of, and contact with, his most famous research subject.

"It goes without saying the Reimers have freedom of choice with respect to location of professional services," Money began. "I shall go along with their decision, whatever it is. Nonetheless, I believe their wisest decision will be not to lose their Johns Hopkins contact, but to work out a joint program of cooperation between us and Brenda's local specialists." Citing the "unique" benefits of the "close collaboration between medical psychology, endocrinology and surgery" at Johns Hopkins, which, Money claimed, "eliminates the possibility of conflicting opinions among experts not accustomed to working together," he also extolled the vagina-building skills of Howard Jones, the surgeon who had castrated Brenda a decade earlier and who was scheduled to construct her artificial genitals. Yet apparently aware that even his own formidable powers of persuasion were not quite up to the task of explaining why it was more practicable for the child to travel two thousand miles for a complicated operation requiring lengthy follow-up care (rather than have the surgery in a hospital just minutes from home), Money concluded his letter with a plea that regardless of where the surgery was done, he not lose contact with Brenda. "I would like to continue seeing her," he wrote, "on approximately an annual basis in the future as in the past."

Only when the Reimers saw that Dr. Money would not stand in the way of Brenda's having the surgery in Winnipeg did they agree to the change in plan. But on the issue of Brenda's returning to Baltimore for her annual follow-ups, Ron and Janet found themselves once again persuaded by Money's eloquence. They decided that if Brenda's fears could be overcome by the local psychiatrists, they would continue to bring her to the Psychohormonal Research Unit for her yearly counseling sessions with Dr. Money. "We felt like we had nowhere else to turn," says Janet. "No one knew us, or Brenda, as well as Dr. Money did."

8

=

In the spring of Brenda's sixth-grade year at Agassiz Drive, her case was transferred to a new psychiatrist when Dr. Moggey, for family reasons, had to leave Winnipeg and go to live in Brandon, a small town some fifty miles north of the city. Moggey referred Brenda to Dr. Janice Ingimundson, a thirty-two-year-old alumna of the University of Manitoba Medical School. Ingimundson was a coolly rational Freudian whose considerable wit and warmth were carefully concealed under a scrupulously correct analytic detachment. Not a member of the city's Child Guidance Clinic, Ingimundson had her own private practice, which she operated out of an office in downtown Winnipeg. Her first session with Brenda was on 6 May 1977. She recalls being taken aback at her first glimpse of the patient.

"All the documentation claimed that this child had accepted her gender identity as a female," Ingimundson says. "Yet my one visual memory of this youngster is of kind of a"—she curls

her hands into fists and bends her elbows in a boxer's pose—
"*tough girl.* A rather boyish-looking girl. Rugged."

Little that Brenda said in her sessions contradicted
Ingimundson's first impression. Though at times Brenda said
what she thought the psychiatrist wanted to hear ("I want to be
pretty; I'm a girl, not a boy"), in the same breath she would
inevitably reveal contradictory feelings about herself. She
defended her preference for boy's clothing, telling Ingimundson, "I like dressing like this. It doesn't feel right to be in a
dress, like I shouldn't be in one." Asked about her feelings for
boys, Brenda said she "wants to beat up on [them]" and then
added a comment that offered almost an embarrassment of
riches for a devotee of Freudian symbolism. If a boy "laid a finger on her," Brenda told the psychiatrist, "she would take her
father's ax and cut it off." When Ingimundson tried to touch on
the topic of vaginal surgery, Brenda grew especially truculent.
"I have decided not to have that," she informed the psychiatrist, "and I don't want to talk about it."

Troubled by such statements, Ingimundson nevertheless
took Dr. Money's word that Brenda had formed a female gender identity. The psychiatrist felt she had little choice. "I
thought, The decision has been made," Ingimundson says. "If
you open up this can of worms now and say, 'Maybe this was
the wrong decision'—well, who is going to *do* that?" Accordingly, Ingimundson (like Moggey before her) resolved to work
hard to assure Brenda that she was a girl and urge her to submit as soon as possible to the vaginal surgery. Yet from their
first session on, Ingimundson was uncomfortable with every
aspect of the case—and in particular her sense that on some
subconscious level the child knew that she was a boy, yet knew
she must not speak of it.

"You'd talk to her," Ingimundson explains, "and conventionally masculine interests would come forward. Which is not

surprising; you see that in girls. But her embarrassment—not embarrassment, but her *difficulty*—in talking about those kinds of interests was acute. She didn't want to expose them. She used to say she had a 'secret.' She talked about wanting to be a 'detective.' She wanted to solve the mystery. That's the therapeutic puzzle: she wants to know, but she doesn't want to know. People in therapy want to know, but they *don't* want to know. They want to know only if it's good news."

And according to Ingimundson, it was clear that Brenda had come to believe that the truth was anything but good news. Ingimundson continues: "Brenda was basically saying, 'There's nothing wrong with me, why do these people want to cut into me?' And in retrospect, she was *right*. 'I'm a boy, I'm a male, so in that sense, there's nothing wrong with me; but anatomically, if I'm a girl, then there *is* something wrong with me.' So this is the bind that she's in. 'If I admit that I'm a boy, then I have to admit that there's something wrong with me anatomically. And if I admit that there's something wrong with me anatomically, *what happened*?' "

Thus convinced that Brenda's resistance stemmed from her inkling that she was not being told the whole truth about herself, Ingimundson, after their second session, talked to Brenda's parents. She urged Ron and Janet to begin preparing Brenda against the day when she would be informed about the circumstances of her birth, the accident she had suffered, and her subsequent sex change. "She has to be told," Ingimundson said, "and she has to be helped in accepting it." Only then, the psychiatrist believed, would Brenda recognize that there was no choice but to go forward with the vaginal surgery. In her notes, Ingimundson registered Ron's and Janet's terror at the prospect of telling Brenda the truth.

"I kind of wish that *you* could tell her," Janet said. But Ingimundson assured her that it was more appropriate for

Brenda to hear the truth from her parents. They should go slowly at first, laying the groundwork. Meanwhile Ingimundson would prepare Brenda emotionally for the final revelation.

A few days later, Ron had a private talk with Brenda. Sitting on the edge of her bed, he managed to choke out the statement that when Brenda was a baby, a doctor had "made a mistake down there" and that the surgery she was going to have was so that "other doctors" could fix up the "first doctor's mistake." Ron found it impossible to make any mention of Brenda's true birth status, her sex change, or in any way to explain what this "mistake" had been. Nor did Brenda evince any interest in these matters; in fact, her reaction seemed to suggest that she wanted to hear nothing more about them. As Ron later reported to Ingimundson, Brenda's sole response to his mysterious utterances about the "doctor who made a mistake" was to ask, "Did you beat him up?"

Today David explains that Ron's halting talk of an "accident" did nothing to tip Brenda off to the fact that she was a male who had been surgically changed into a girl. "You're not going to think of that in a million years," David says. "So I didn't know *what* he was talking about—and I didn't want to know."

Meanwhile Brenda was having her usual difficulties at school. "She has only been in the school for 4 days, and principal related peer problems," scribbled the school's social worker in Brenda's file. "Teasing—'looks like a boy, etc.'"

As the school year progressed, however, Brenda was gradually accepted into a small clique of misfit tomboys led by a girl named Heather Legarry, a short-haired brunet with an open smile. "I had been the subject of shunning several times throughout those early school years," says Heather today. "I knew what it was like, and I never would do that to another person—ever." Brenda, she says, seemed a natural candidate for inclusion in the group of tomboys who played soccer and dodge ball, climbed the jungle gym, and rode bikes. Though distrustful, at first, of Heather's overtures, Brenda finally dropped her guard and began to hang around with Heather and her friends. "Heather was the first friend I ever had," David says. "I didn't know what a friend *was*."

Heather, for her part, valued Brenda as a girl devoid of the duplicity and backstabbing that had poisoned so many of her relations with girls in the past and that even threatened the harmony of her current clique of tomboys. "Brenda didn't speak much," Heather says, "but when she did she was never vindictive or false. She was very honest. If she told me something was black or white, it was."

At the same time, Heather was not blind to what she calls Brenda's "oddity." Partly this was a feature of what Heather describes as Brenda's acute anxiety. "Brenda was always nervous about doing things that were the least bit unusual—like cutting through the university grounds on our bikes, which I did all the time. She was very nervous, very unsure of herself." Heather says that this nervousness even affected Brenda's speech. "Just making a sentence sometimes seemed hard for her."

And there was something still more odd about Brenda. "As far as I knew, Brenda was a girl—physically," Heather explains. "But from everything that she did and said, she indicated that she didn't want to be a girl. The other girls in our group were competitive against the boys; we wanted to prove we could do whatever they could do. We wanted to *show* them. We might get in arguments with the guys, but we wouldn't have gone as far as to fight with them physically. I wouldn't want a bruise on my face, for example. But Brenda fought with the boys. Brenda would take the bruises." Heather pauses and thinks about this for a moment. "I myself was a tomboy," she resumes, "but I never wanted to *be* a boy. Brenda did."

Heather's impression was only strengthened when, one day on the school playground, she noticed a small bald patch near the crown of Brenda's head. The hairless area was the result of an accident when Brenda was a baby—she'd pulled the cord of the electric frying pan while her father was cooking and been hit by hot grease. But that's not the story Brenda told to Heather. "She said she had deliberately taken a hot frying pan to her head to burn off the hair," Heather recalls, "because she 'wanted to be bald like a man.' "

It was through her friendship with Heather that Brenda became increasingly aware of a new and perilous undercurrent in the life of the classroom. She first noticed it when overhearing bits of conversation among the other girls—talk of "crushes" and "going steady" and "kissing." Then she saw boys and girls passing notes when the teacher's back was turned. Once she saw one. It was a love note, and it was signed "xoxoxox."

Brenda recognized in these developments a fertile field for fresh embarrassments and humiliations. She had resolved to give wide berth to the burgeoning dating scene—a resolution that she did not think would be difficult to keep, since none of the boys showed any romantic interest in her. Still, it soon became clear to her that she could not avoid the awakening sexuality of her peers entirely. That fall Heather took Brenda along to a birthday party for one of their classmates. The party began innocuously enough. With their host's parents acting as chaperones, the children dutifully played games of pin the tail on the donkey and Twister while a children's album played on the stereo. But when the parents left the room and went downstairs, everything changed.

"One of the kids took the record off—*zzzzt!*—and put on a makeout tune," David recalls. "Another guy put the lights out. Suddenly everyone's slow dancing and making out. I'm looking at Heather, and Heather was looking at me. We were the only two left over." The pair beat a quick retreat downstairs. "But you could hear through the vents," David says. "I could sense what was going on."

I asked David how he had felt as Brenda, watching his classmates pair off romantically. He thought for a moment. "I guess envious," he said finally. "These people looked like they knew where they belonged. There was no place for me to feel comfortable with anybody or anything."

Brenda's escalating alienation was clear in her sessions with Dr. Ingimundson, who continued doggedly to try to get Brenda to open up and discuss her genitals and to agree, finally, to surgery. But Brenda could not be budged.

"Not responsive to my efforts to engage," the psychiatrist wrote, when she and Brenda were three months into their sessions. "Silent . . . staring off into space—head turned away. . . . Telling me she feels trapped in office—wants to get out—or feels trapped inside self."

9

≡

I**N THE SUMMER OF** 1977, Brenda suddenly had to fend off an attack on a new front. On her last several trips to Baltimore, Dr. Money had spoken to her about the medication she would soon need in order to become a "normal girl." He was talking about estrogens, the female hormones that would simulate the effects of female puberty on Brenda's broad-shouldered, narrow-hipped boy's physique. Like the vaginal surgery, the prospect of growing breasts struck Brenda as a nightmare. So she was suspicious when one day soon after the end of her sixth-grade year at Agassiz Drive, her father produced a package of pills and told her to start taking them.

"What's this medicine for?" she asked.

Ron, struggling for the best way to put it, finally came up with, "It's to make you wear a bra."

"I said, 'I don't wanna wear a bra!' " David recalls. "I threw a fit."

The depths of Brenda's resistance to the hormones was clear in her dealings with the doctor whose specific job it was to prescribe and regulate her estrogen therapy—a pediatric endocrinologist named Jeremy Winter. A thirty-four-year-old professor at the Children's Hospital of Winnipeg, Winter had trained in Philadelphia under the respected endocrinologist Alfred Bongiovanni, who himself had trained at Johns Hopkins under Lawson Wilkins.

Given this academic pedigree, it was perhaps not surprising that Winter was the Winnipeg doctor least inclined to question the methods or conclusions of Money's twins case. Before meeting Brenda, he anticipated no problems with the treatment.

"I got the chart and looked at all the background information that was available," says Winter. "I read *Man & Woman, Boy & Girl*, and I believed it. I said, 'That makes sense, and everything fits, and I'm going to see this kid, and this is what we're going to do' "—namely, put the child on a course of estrogens and commence vaginal surgery immediately. But things did not work out that way.

"It was easily the most frustrating case we had in the clinic," Winter says. "We prided ourselves on excellent rapport with patients, being able to sit down with kids and talk and listen in a warm atmosphere. And here was this absolutely silent, angry child who didn't want to be there. I'd ask, 'Will you allow a blood test?' 'No.' 'Will you allow me to examine you?' 'No.' So I would have these monologues about the importance of taking the estrogen and having the vaginal surgery and how successful and wonderful this was going to be."

According to Winter, Brenda was especially adamant about never returning to Johns Hopkins. "I'd never seen a patient in my life who behaved that way about going to another doctor—who showed that depth of emotion," Winter recalls. Meanwhile, Winter had no choice but to try to get her to take the estrogen pills—an increasingly urgent need, since her twelfth birthday was approaching in late August.

Brenda continued trying to resist, but after continued entreaties from Winter, her parents, and Dr. Ingimundson (not to mention the threat that Dr. Money had once introduced into Brenda's head, that she would grow disproportionate limbs if she failed to take the drugs), she finally—on the eve of her twelfth birthday—began to take the pills. Or rather, she pretended to. When her parents were not looking, she would throw the small tablets into the toilet. "I remember the pink dye running out of them," David says. "I had to flush fast before my parents saw." Ron and Janet soon caught on, however, and took to standing over Brenda while she swallowed the daily medication—0.02 milligrams of ethinyl estradiol, later increased to 0.75 milligrams.

Soon enough, a pair of breasts sprouted on Brenda's chest along with a padding of fat around her waist and hips. The changes caused her deep mortification. In a bid to disguise the increasing feminization of her figure, she began prodigious bouts of eating. With several ice cream cones every day, her waistline swelled to forty inches. The added fat helped to camouflage her breasts and hips, but no amount of binge eating could hide certain other physiological changes that began to accelerate within her that fall. "Spontaneously expressed anxiety about her voice," Ingimundson wrote in her September session notes. "Starting to crack."

The dramatic deepening of Brenda's voice was a phenomenon endocrinologist Winter was at a loss to explain. Given her absence of testicles (the primary male hormone-secreting endocrine gland) and her estrogen therapy, her voice, by all known medical criteria, should not have undergone a virilizing change at puberty. Today Winter suggests that Brenda's vocal cords and larynx were perhaps thickening because of increased androgen secretion from her adrenal glands. Whatever the cause, one thing was not in doubt. Brenda's voice now began to change in a manner identical to her brother Brian's. She asked her mother why.

Thinking fast, Janet mentioned the deep-voiced actress Marlo Thomas from the TV situation comedy *That Girl*. "She has a raspy voice," Janet told her daughter. "It's normal for some girls to have voices like that."

Armed with this explanation, Brenda started seventh grade that fall at her new school: Glenwood Junior High, a large public school some five minutes' walk from her house. She was instantly exiled to the farthest periphery of Glenwood's social life, where she took her place among a haphazard collection of the school's misfits. One girl was an intersex. Another wore a complicated metal leg brace and built-up shoe to accommodate a right leg some three inches shorter than the left. Another, Esther Haselhauer, suffered from Poland's syndrome, a congenital birth anomaly that had stunted her growth, partially withered one hand, and completely retarded the growth of her right breast. Esther remembers that she sensed an immediate kinship with Brenda.

"She was hard to connect with," Esther says. "But there was something that I just responded to. It was . . . I don't know, a sadness. She reminded me of *me*." At the same time, Esther was aware that there was a big difference between the two of them, and indeed, between Brenda and any other girl she'd ever met. "Brenda wasn't a *girl* girl," Esther says. Furthermore, being with Brenda evoked a feeling in her that was curiously like that of being with a member of the *opposite* sex. "It was a feeling of security," Esther says. "When I was with Brenda, I felt so safe. Kids would beat me up because I was so small. But when I was with Brenda, if anyone tried picking on me, she let them have it."

David says that he was grateful for Esther's friendship, but that their differences made it impossible for them to get close. "She was always talking about guys," David says. Asked if Brenda ever expressed interest in romance with boys, Esther laughs. "Oh no!" she says. "That would be unthinkable; as unthinkable as me, at four feet tall, going out for the basketball team."

Six months after Brenda began seventh grade, Dr. Ingimundson received a letter from John Money requesting a progress report. Money's letter arrived at a particularly inopportune moment—just two days after a disastrous family session Ingimundson had held with the Reimers. Her notes on the 20 February 1978 session make it clear that Brenda's resistance to the surgery had abated not at all in the fourteen months since she had begun psychiatric treatment in Winnipeg. Refusing to utter a word, Brenda had pulled up the hood of her winter coat and folded her arms across her chest. Pressed to say something about the surgery, she finally burst into tears, while the adults looked on helplessly. "Parents & I at a loss for words," Ingimundson wrote in her session notes. The session ended soon after that.

With this fraught scene still fresh in her mind, Ingimundson wrote back to Money. She explained that despite their success in getting Brenda to take hormones, the local team had made no further progress in the case. "[N]o plans for surgery have been formalized," she wrote, "nor, for that matter, have they been discussed in a tentative fashion." She added that Brenda remained "resistant to medical attention" and was "still refusing examination of her genitals."

How Money greeted this letter is difficult to tell. His reply was written by his secretary—and it was brief. "Dr. Money has asked me to write you thanking you for your letter of March 8, 1978 giving a progress report on Brenda Reimer. We are very pleased to have it," the note read in its entirety.

A more expansive articulation of Money's opinion on the case soon appeared in a chapter of the book *Biological Determinants of Sexual Behavior*, an anthology of writings on gender identity published later that year in Britain. Once again the outlook was sunny. A full-page photograph of Brenda and Brian (taken at their last visit to the Psychohormonal Research Unit) showed them standing side by side against a white background. Brenda wears a short patterned dress with cap sleeves, her long, well-brushed hair falling to her shoulders; Brian is dressed in a boy's short-sleeved shirt and dark jeans, his close-cut hair exposing his ears. A pair of large black dots cover their faces, obscuring all but their identical jaws, chins, and eyebrows. Nothing in the accompanying text suggested that, beneath the black dots, either child was anything but smiling and happy. "Now prepubertal in age," Money wrote, "the girl has . . . a feminine gender identity and role, distinctly different from that of her brother."

An attentive reader might have noticed in this update certain evidence to suggest a less sanguine prognosis for the sex-changed twin. Elsewhere in the chapter, Money wrote of further research he had done into the role played by excessive testosterone exposure in genetic females in the womb. Money now revealed that there was reason to think that such exposure affected not only masculinization of play preference, toy preference, and career goals (as he and Ehrhardt had reported eleven years earlier), but other behavior as well. "The preliminary evidence indicates the possibility," Money wrote, "that there is a greater incidence of bisexuality and homosexuality [among such girls] than would be expected by chance."

That Money's famous sex-reassigned twin had spent her

entire prenatal life awash in a full complement of testosterone produced by the fetal testicles (a complement of testosterone some ten times the amount experienced by normal female fetuses) might have led some readers to conclude that the twin, at puberty, would in all likelihood manifest an erotic attraction to females; but on the all-important question of Brenda's sexual orientation, Money (perhaps forgetting what he had told Ron and Janet about her lesbianism) professed himself unable to venture even an educated guess as to what her partner preference might one day be. He wrote in his concluding remarks on Brenda's gender identity, "The final and conclusive evidence awaits the appearance of romantic interest and erotic imagery."

In the eleven years that had elapsed since Bruce Reimer's conversion to Brenda, none of her local doctors had ever met with John Money in person to discuss her case. But in the spring of 1978, Jeremy Winter was invited to deliver a lecture at the Johns Hopkins Medical School's Reproductive Biology Seminar. While in Baltimore, Winter arranged to speak with Money. Their meeting took place at the Psychohormonal Research Unit on 4 April. Winter detailed for Money the extreme difficulty the local treatment team was having in implementing his plans for Brenda: she continued to refuse to submit to a genital exam; refused even to discuss the issue of vaginal surgery; refused to return to Baltimore; and often refused to take her hormone pills. According to Winter, Dr. Money was wholly unconcerned by the issues he raised.

"He was supremely confident," Winter says. "Everything was perfect; there were no problems—and any concerns that I was raising were my naïveté and youth coming to the fore; and I would learn, in time, that everything was fine."

Money's own notes on the encounter confirm Winter's

impression. Money refers confidently to the time when Brenda will be able to "negotiate the decision [for surgery] herself"; alludes to his belief that Brenda's "intense phobia of white coats and doctors" reflects only a deep-seated sense memory of her circumcision accident at eight months of age; and opines, "I rather strongly suspect that Brenda already knows that she once had a penis and probably that she had been considered [sic] a boy." Still, this suspicion did not diminish Money's belief that Brenda would soon agree to vaginal surgery—possibly even at Johns Hopkins. As his notes show, Money also told Winter about Brenda's "intense rejection of any conversation regarding matters sexual, and of looking at books pertaining to any aspect of sex education." According to Winter, Money showed him some of the sex education materials. "He showed me photographs that he would use, dirty pictures, to see whether Brenda was homosexual or bisexual or heterosexual," Winter says.

Though unsettled by Money's seeming unconcern about the problems he had raised and troubled by the materials Money had shown him, Winter nevertheless resolved to feel encouraged by the visit. It was a relief that the world-renowned expert on gender identity did not consider Brenda's resistance to be an insurmountable obstacle to the eventual success of the sex reassignment. It was similarly a relief that Money, the world's leading authority on sex change, had endorsed the local team's approach to Brenda's case. "I was a very junior person going to the expert," Winter says, "and I was happy to get some reassurance."

But the dramatic depth of Brenda's resistance to Johns Hopkins and to the surgery was soon to be brought home to all concerned—especially John Money. On 2 May 1978, one month after Winter's trip to Baltimore, the Reimers returned with Brenda and Brian for a counseling session with Dr. Money. Brenda had fought hard against the visit, agreeing to go only when Ron and Janet promised an expensive side trip to New York City as a bribe. Yet even with Manhattan as a pill sweetener, the visit would prove so traumatic for Brenda that it marked the last time she would ever consent to go to Baltimore.

That something remarkable had occurred during Brenda's visit was obvious from a letter Money wrote to Winter several weeks after the encounter. Stating that "Brenda talked more extensively on this occasion than on her last visit," Money went on to say, "She was especially at ease with two youthful students doing an elective with me. She was quite explicit, however, about avoiding references to sex and sex-related topics, and to prospective surgery. . . . [S]he could not tolerate further continuance of such talk, and went into the next room to join her brother. I followed, and in bringing the session to a close, put my hand on her shoulder in what most youngsters would accept as a reassurance. She fled in panic. One of the students followed and helped her recover her composure. They walked, saying little, for about a mile." In concluding this oddly elliptical-sounding account of the events, Money referred to the student as "a woman." What he did not mention was that the woman had begun life as a man. She was a male-to-female transexual whom Money had enlisted to speak to Brenda about the positive aspects of surgical construction of a vagina.

The Reimers' trip to the unit had begun typically enough, which is to say, with Brenda displaying intense anxiety, anger, and depression—emotions that were reflected in the Sentence Completion Test she was made to fill out. "Compared to most families mine's . . ." "a loser," Brenda wrote. "I think most girls . . ." "aren't very nice." "I believe most women . . ." "aren't very nice either." "My feelings about married life are . . ." "rotten." "If I had sex relations . . ." "I wouldn't like it. Same if a boy would kiss me." "To me the future looks . . ." "bad."

But it was when Dr. Money introduced her to the transexual that Brenda's typically despairing mood turned to pure, deep-running panic.

"Dr. Money said, 'I've got someone for you to talk to who's been through what you're going to be going through,'" David recalls. Brenda was ushered into the presence of a person whom she immediately identified as a man wearing makeup, dressed in women's clothing, with a woman's hairstyle. When the person spoke, it was in a breathy, artificially high-pitched voice. "He's telling me about the surgery," David says, "how fantastic it was for him, and how his life turned out beautifully."

Brenda sat immobile, silent, apparently listening. But the words reached her through a clamoring, rising terror in her mind. "I was thinking, '*I'm* going to end up like that?'" David says.

When the transexual finished speaking, she led Brenda back into Dr. Money's office, where he sat waiting for her at his desk. Brenda sat in the armchair beside Money's desk. The transexual sat on the adjacent sofa. Money's transcripts of the meeting record what happened next.

"You do not have to have the operation for your sex organs if you don't want it," Money said. "And you can also change your mind and have it anytime you want to, whether you're in your twenties, or your thirties, or whatever. But from now on you're old enough to sign your own operative permit, and nobody can make you have an operation. As a matter of fact, nobody can make you take pills if you don't want. And you know that very well, because all you have to do is tell lies about them, hmmm?"

Dr. Money talked on in this vein for almost ten minutes, shifting back and forth from trying to sound friendly and supportive to sounding threatening and angry. He said that no one should make her feel as if she were having things forced upon her—even as he relentlessly tried to convince her to have the surgery. He spoke about her "gender identity," saying that she could not be a person unless she had one, and then he was talking about the operation again, about "sex organs for a female."

Brenda tried to interrupt, but Dr. Money said he wanted to tell her "a very nice story" about a patient who had been born with "a birth defect of the sex organs." Money began to talk about "clitorises" and "penises." Brenda again tried to interrupt him. "Let me finish," Money snapped. Recovering himself, he talked on about how this patient had always refused, like Brenda, to discuss his sexuality when he was growing up. Money said that he had learned from this patient not to force children to talk about things that disturbed them. Yet at the same time, Money continued to press her to speak. "I want you to know that I'm going to be the one person in the world that you can tell anything to, because I'm not going to yell at you," Money said. "And I'm not going to tell you you're crazy. I'm just going to listen and be helpful and find the answer to it. And you can tell me anything."

When Money finally fell silent, Brenda had only one question.

"Are you finished?" she said.

"We're finished."

Brenda got up and hurried toward Money's office door. Money and the transexual moved toward her. The transexual was saying something about taking Brenda up to the fifth floor where they could be alone. Dr. Money reached out for her. She felt the psychologist's fingers grasp her shoulder. Convinced that they were going to drag her off to the operating room, Brenda wrenched free of Money's grasp. Today

David cannot recall how he got out of Money's office. "I remember running," David says, "that's all."

"I heard the door slam open," says Brian, who was sitting in the waiting room, "and—*whoosh!*—there goes Brenda. Bolted. I hear John Money yelling. I see a bunch of people with lab coats running after her."

Janet and Ron, who were being interviewed in a nearby office, heard the commotion and came out into the corridor. "Dr. Money took off," Janet says. "We stayed with his assistants, waiting, while he went chasing."

Brenda ran blindly until she reached a set of stairs, which she dashed up, emerging onto a rooftop. The transexual had followed. Brenda crouched by a low brick wall that ran around the perimeter of the roof, trying to hide. David cannot recall what happened next. A report filed by the transexual (whose name has been whited out in the Psychohormonal Research Unit record) reveals that Brenda, with her pursuer close behind, fled down four flights of stairs and ran out of the hospital's back exit into a parking lot. The transexual searched the hospital grounds—then spotted Brenda running into the main entrance. She gave chase but once again lost sight of Brenda.

At the front desk, the transexual phoned Money at his office, gave a progress report on the search, then staked out the exit. Brenda appeared two minutes later heading for the door. The transexual intercepted her and offered to walk with her to calm her down. Brenda agreed only if they did not talk or come close to each other. "We walked," the transexual's notes continue, "Brenda about 4 feet behind me." It was in this strange configuration that Brenda and her would-be counselor proceeded silently some eight blocks from the hospital, then back again.

On their return to the hospital, they were met at the main

entrance by Viola Lewis—one of the few unit workers Brenda even remotely trusted. Lewis escorted the child to the nearby Sheraton Hotel, where Ron and Janet had been convinced to go and wait for her return.

Reunited finally with her parents and brother in their hotel room, Brenda told Janet that if ever again forced to see Dr. Money, she would kill herself.

10

═

Upon their return to Winnipeg from Baltimore, the Reimers found themselves enmeshed in a new crisis involving their daughter—although this time the drama did not directly include Brenda herself but rather the members of her local treatment team. Several weeks earlier, Dr. Ingimundson had terminated treatment with Brenda to take a leave of absence from her practice and have a baby. Ingimundson had referred Brenda to another psychiatrist, Dr. Sheila Cantor. An aggressive and outspoken woman, Cantor had taken a view of Brenda's case quite out of synch with that held by the rest of the local treatment team. After taking a look at Brenda's medical records and Child Guidance Clinic reports and having one joint session with Brenda and her parents, Cantor abruptly announced to the Reimers that Brenda's sex reassignment was a dismal failure and that the child must be allowed to switch sex immediately to boyhood.

Sigmundson says that such bluntness was typical of Cantor (who has since died of cancer). "She was a good psychiatrist, but so strongly opinionated about anything she touched that she would alienate people," says Sigmundson. Cantor certainly alienated the Reimers, who still labored under Dr. Money's instructions to suppress all doubts about the treatment.

It was the ordinarily taciturn Ron who spoke up. "My husband got very angry," Janet recalls. "He said, 'First of all, we have to be sure that she wants to be a boy; don't just *assume* this.' He hadn't yet accepted that Brenda was not to be." Nor had Janet. Nor had Dr. Winter, who sided with the Reimers in their dispute with the psychiatrist. While Winter admits that, with hindsight, Cantor was correct in her assessment of Brenda's condition, he thinks the psychiatrist erred in her approach to the problem. "Even if you've got the right answer in medicine," he says, "part of this whole business is that you've got to wait for people to catch up and come along with you. And if you don't do that, the best plans don't work."

With Winter's support, Ron and Janet appealed to Sigmundson, demanding that he remove Cantor from the case. Sigmundson did so, which left him in a serious bind. Having now run through three of the city's senior female psychiatrists, and still determined to assign Brenda's case to a woman in order to increase her feminine identification, Sigmundson was running out of qualified women.

Even as the doctors struggled to find a way forward with the case, Brenda had settled on her own strategy for coping with her predicament. When Brenda started eighth grade at Glenwood Junior High that fall, Esther Haselhauer noted the stunning change that had come over her friend. Ordinarily Brenda was never seen in anything but jeans and a T-shirt, wearing no makeup. But something had clearly happened over the summer.

"I remember she came into the classroom," Esther says, "and she was wearing this matching checkered beige pantsuit with stripes, her hair was brushed, and she was wearing lipstick, rouge, and mascara, and she was carrying a purse. It was obvious that she was trying very hard to fit in as a girl."

Indeed she was. Following her last traumatic trip to the Psychohormonal Research Unit, Brenda had become convinced that the only way to avoid the surgery was to play along to the best of her ability; she would try to act the part of a girl; she would try to convince everyone that she was happy. That way, she reasoned, they might not force her to have the operation. And who knew? Perhaps they were all right: perhaps if she made a true effort at living as a girl, she would begin to feel like one. As David puts it, "I decided to play ball. I tried my guts out. I was miserable. I was unhappy. I was uncomfortable. I felt awkward as hell. But the pressure was on me. And I tried my hardest."

For Ron and Janet and the members of the local treatment team, Brenda's behavior that fall was initially a cause for considerable joy. "Parents have found her to be much more enthusiastic about school this year," social worker Downey noted in Brenda's Child Guidance Clinic file in early September, "and she has apparently been out shopping with some other girls." This expedition to a local department store had actually been at the suggestion of their teacher, Mrs. Bailey, who had taken some of the more sympathetic girls aside and asked them to help Brenda a little with her dress and grooming, which left something to be desired. Ordinarily her mother would have helped Brenda with her makeup and clothes, but Janet had recently taken a job as a parking lot attendant and was gone early in the mornings, leaving Brenda to fend for herself. Otherwise it is unlikely that Brenda would have been allowed to go to school in the ill-applied makeup and unfortunate beige pantsuit, which she had bought for herself while on a shopping expedition with Ron.

"The pantsuit wasn't exactly in style," Esther says, "and the rouge was in circles on her cheeks. She came off looking more like a clown."

David concurs. "I remember these girls took one look at how I was dressed, and they said, 'We gotta take you shopping!' "

On a trip to the Hudson's Bay department store, the girls combed the racks for Brenda and picked out a feminine-styled blue turtleneck sweater and a pair of designer jeans. After school, when her mother was back from work, Brenda asked Janet for some lessons on how to alter her angular, gunslinger's stride, which since kindergarten had been a source of such hilarity among her peers. Janet showed Brenda how to balance a book on her head to practice straightening her spine and smoothing her stride. "It was unnatural," David recalls. "I'd get all tensed up after a while. But you were expected to walk like that, so I tried to do it."

Glenwood Junior High held dances on Friday nights in the school gym. All the kids went; Brenda did, too. "I'd get dressed up in my unisex disco clothes—jeans and high-heeled boots—and tell my parents, 'I'm going out to forget about my worries and my troubles.' " But Brenda discovered that her worries and troubles did not dissipate at the school dances, where the gym walls reverberated to the taped strains of Rod Stewart's "Do Ya Think I'm Sexy?"

"I'd mostly dance with girls, in groups of four or five," David says. But Brenda also danced with the occasional boy. "I was *expected* to dance with the boys, so I'd dance with a boy," David says. "I'd ask some guy, 'You wanna dance?' They kinda look at you"—he wrinkles his nose—"and say 'No, no, it's

OK.'" On a few occasions, however, boys *did* agree to accompany Brenda to the dance floor. "Sometimes it was a fast dance," David says. "Sometimes a slow dance." Circling the dance floor in the arms of a boy, it was painfully apparent to Brenda that she was not having the right sensations. Instead of any romantic flickering, she felt only embarrassment and excruciating awkwardness. One Friday night, David recalls, a boy in ninth grade who was one of the school's main heartthrobs defended Brenda when she was being teased by a group of girls. Jokingly, the boy kissed her. "It was a peck on the cheek," David says. "I went home and thought about it. I thought, 'It doesn't seem right. I don't like this. This is how it's supposed to happen, but it doesn't feel *right*.'"

Such feelings of doubt, which Brenda had long expressed in her therapy sessions, had always been dismissed by her therapists. "These psychiatrists and psychologists kept saying that these were 'normal confused feelings,'" David says. "You have a sense that it's beyond that, but people are telling you what you're thinking. And only *you* know what you're thinking. It makes you feel even more crazy."

Still, Brenda persevered. That fall a group of girls invited her to a pajama party. "Someone sneaked in a mickey of booze," David says. "I didn't drink any. I faked it. I put it to my mouth, but I put my tongue on the hole so I didn't get any. Everyone was talking about *boys*—'Who do you have a crush on?'and all this. So I said, 'Oh yeah, OK, I got a crush on this person or that person.' I mean, what do gay people do when they're in hiding? They pretend they're straight. You toe the line, like everybody else. Act like everybody else, and you'll be treated like everybody else."

It was clear to Brenda that she did not feel like everybody else in the room. Especially when the other girls began to get ready for bed.

"A girl got undressed in front of me," he says. "I was so embarrassed I looked the other way. She said, 'That's OK, you don't have to be embarrassed, you're one of us.' She showed me her bra and asked, 'What do you think of this?' I said, 'I don't know. It looks beautiful; I like the lace.' And I'm sitting there turning red. I felt like Mrs. Doubtfire in that movie." Asked about any other involuntary physiological reactions that he might have experienced in the company of so many half-dressed girls, David smiles—and answers with an analogy. "If you lose your arm," he says, "and you're dying of thirst, that stump is still going to move toward that glass of water to try to get it. It's instinct. It's in you."

However, Brenda was now living a life in which every instinct had to be denied, repressed, hidden: at dances, at parties, in the classroom, and on the street. "I was like a robot," he says, describing the playacting that his day-to-day, moment-to-moment survival now entailed. "You're so careful to look *normal*, but you don't want to go overboard. You're saying to yourself, This looks like an appropriate time to smile. So you smile. This looks like an appropriate time to cross your legs. So you cross your legs. You're always thinking one step ahead, like in a chess game."

It was a chess game Brenda was losing. Despite those few girls who had obeyed their teacher's orders to be nice to Brenda, the majority of her peers continued to jeer and laugh at her. "It was sad, because it was like the harder she tried, the more she failed," says Esther Haselhauer. "The ridicule became worse."

Of Brenda's eighth-grade tormentors, Wendy Holderston, the popular and pretty daughter of a local singing star, remembers Brenda as an "odd duck," a "tomboy" who had "a deep voice and really deep-set eyes"—characteristics that resulted in Brenda's being dubbed "Cavewoman," a name that soon caught on among their classmates. At first Brenda tried to absorb her peers' ridicule in "ladylike" fashion, but one day she'd had enough. In the school hallway she rounded on Wendy, David recalls, grabbed her by the front of her shirt, smashed her against the lockers, and threw her onto the ground. Boys who teased her got similar treatment. "That's what always impressed me about Brenda," Esther says. "She'd actually fight with the boys who teased her. She'd haul off and punch them. I always wished I could do that."

David is more rueful when he recalls his habit as Brenda of beating up the boys who teased her. "It only made people think I was a bigger weirdo and ostracize me more."

By late November, the earlier note of optimism about Brenda's social and academic life had vanished from her Child Guidance Clinic file. "She has no friends in class," Downey wrote on November 27. More frighteningly, Brenda was also showing disturbing signs of deterioration in her intellectual functioning—a precipitous descent into a helpless, childlike regression. "She cannot spell the days of the week or the months of the year," Downey noted in Brenda's file. "As 5th grade spelling was too difficult for her, she had been placed back in a 4th grade program. She has not learned certain set routines in the classroom. . . . She has been waiting for other students to unlock her combination lock as she cannot remember her number."

David recalls the humiliation associated with his helplessness. "When you're going through what I was going through, schoolwork is kind of low on your list," he says. "The last thing on your mind is a *test*. You're thinking about survival." The Child Guidance Clinic's Joan Nebbs uses the same term when she describes Brenda's predicament: "It was survival," she says. "To get through the day. To get through the hour."

As eighth grade progressed, Brenda's continued ability to

get through the hour became increasingly imperiled. The more her sense of the disjunction between her physical and mental selves increased and her feelings of entrapment and confusion escalated, the more her thoughts turned to the ultimate escape: suicide. "I kept visualizing a rope thrown over a beam," says David.

It was at this critical stage of Brenda's adolescence that Keith Sigmundson finally succeeded in placing her in the care of a new psychiatrist, a particularly gifted and empathetic one named Dr. Mary McKenty.

Upon entering Dr. McKenty's office at the Child Guidance Clinic for her first session on 2 January 1979, Brenda was surprised by the sight of the smiling, elderly, gray-haired woman who welcomed her. She was dressed in a tweed skirt and matching jacket and stood no more than five feet one inch tall. "She looked like she'd be baking cookies for her grandchildren," David says. "She didn't look like a typical psychiatrist."

And indeed, Mary McKenty was not a typical representative of the profession. McKenty had always eschewed the strict Freudian rules governing psychiatrist-patient relations—rules that frowned on excessive personal contact, or even warmth, between doctor and patient. She had always preferred to use a nurturing approach that was, in part, a reaction to the circumstances of her own childhood. Born in 1916, McKenty had been reared in an affluent home (her Scottish immigrant father worked for Richardson's, the province's most profitable grain company) but one lacking in introspection and overt displays of warmth. "Her Scottish upbringing was not attuned to psychiatry," says Evelyn Loadman, who first met McKenty in 1934 when the two were among the first women pre-med students in Manitoba history. "But Mary happened to be the kind of person who was very sensitive to others and couldn't really figure life out until she started to think about what we are made of."

It was in the early 1940s, in her first job at the Children's Hospital of Winnipeg, that McKenty's unique gifts as a child psychiatrist emerged. McKenty made no apparent attempts to delve into the protected realms of her patient's unconscious. Instead she would come down to the child's level, playing games, encouraging her patient to draw, paint, and write. Thus lulled into a state of distracted absorption, children could reveal themselves by word and deed. Loadman recalls that McKenty's nonpressuring approach lent an almost magical quality to her ability to rid her patients of neurotic behaviors—bed-wetting, hair-pulling—that had resisted the efforts of other therapists. "She was the only one who ever *cured* anybody as a psychiatrist at the Children's Hospital," Loadman says with a touch of wonder in her voice.

McKenty was hired by the Winnipeg Child Guidance Clinic at its inception in the late 1960s and quickly established a reputation for success with even the most difficult children. Indeed, when Keith Sigmundson first inherited Brenda Reimer's fraught and complex case in late 1976, he initially tried to assign it to McKenty, but she was recovering from a double mastectomy and thinking about retirement, and she turned down the case. Sigmundson returned to her only in the fall of 1978, when he was unable to find any other female psychiatrist willing to take Brenda on. Recognizing that Sigmundson was desperate, McKenty finally gave in and took the case.

Ron and Janet formed an immediate liking for the calm, grandmotherly psychiatrist. Brenda was, by necessity, more circumspect.

On the one hand, Brenda was drawn to McKenty. She liked how the psychiatrist spoke in a kindly, conversational voice and refrained from scribbling on a pad. Brenda also liked that the bulk of her sessions with McKenty were taken up playing board games—Clue, Cootie, Basquetball, Sorry!—or drawing, painting, or playing tongue twisters. At times, Dr. McKenty would drop in an occasional question about how Brenda was feeling and sometimes *would* jot something on a pad, but somehow it was different when Dr. McKenty did it. "She'd be writing something down," David recalls, "and I'd say, 'Oh, you're writing something down again—spying on me, huh?' And she'd say, 'Oh no, dear, this is just so I can make sure that we covered all the bases, so that I can help you.'"

Yet on the other hand, Brenda could not afford to trust McKenty completely, since she had to remain alert to the nightmare possibility that the psychiatrist's friendliness might simply be a more diabolical tactic to trick her into the surgery. As McKenty's therapy notes reveal, Brenda used an array of strategies to see if Dr. McKenty's friendliness was genuine. She drew a cruel caricature of the psychiatrist and showed it to her; she seized one of the clinic's toy machine guns and took the doctor "hostage"; she wrote out a "Death Warrant for Mary McKenty." The psychiatrist offered no resistance. She carefully preserved Brenda's nasty caricature of her, even obligingly signing her name underneath it; taken hostage, she allowed herself to be marched at gunpoint around the office; and on Brenda's "Death Warrant," she dutifully filled in the areas where Brenda had left room for the condemned woman's vital statistics.

"I was testing her," David says. "She passed the test." Brenda dropped her provoking tactics and began actually looking forward to her psychiatric sessions with McKenty. "We didn't see each other as patient and doctor," David says. "It was a friendship."

Brenda was now in particular need of a friend, since an old nemesis was about to reappear in her life. That January,

Brenda confided to McKenty that she had snooped in her mother's mail and seen a letter from Dr. Money announcing that he was coming to Winnipeg. Invited to give a talk at the local medical school, he had made plans to visit Brenda and her family—a visit, McKenty wrote, that Brenda was dreading.

Over the next two months, McKenty recorded Brenda's escalating anxiety as Money's impending visit drew nearer. On 31 January, Brenda recounted to McKenty a nightmare that she'd had about Money, her twin brother, and herself. "Dr. Money was a magician with a cape," Brenda told McKenty, "and he said he could make us disappear—pouf!—like that. I woke up and thought we had disappeared." To allay Brenda's fears, McKenty said that she did not have to see the psychologist when he visited. Together they created the "Don't Want to See Dr. Money Club." McKenty signed her name on the membership list. Brenda added her own signature and the words "Join my Club!" But a month later, with Money's visit less than two weeks away, Brenda had another dream. "I had on a fancy blue dress and my good shoes, too," she told McKenty. "The skirt was split because it's too narrow. Everything was cleaned up—the floor was washed. Something big seemed to be going to happen. There was a closet nearby. We have a closet like that in our house. I was scared because it seemed I could maybe get put in that closet."

The week of 22 March 1979 proved to be one of those gray Winnipeg periods when the weather has stalled in the indefinable season between late winter and early spring. The snowbanks, having lingered too long, were blackened with car exhaust, and the sky was the color of cement. Ron, driving his Dodge Dart to pick up Dr. Money at the city's Health Sciences Center (where the psychologist had just given his speech), looked with chagrin at the dirty snowbanks. He'd been hoping for better weather so that Money might take home a more favorable impression of their city.

Back at the Reimers' house, Janet was having similar feelings of anxiety about the impression Money would carry away of his visit to Winnipeg and, specifically, the Reimers' home. They had moved into the house two years earlier and put two thousand dollars into fixing the place up. Ordinarily Janet felt quite proud of what she'd been able to achieve on her budget. Now she was seeing the place through the eyes of Dr. Money, whom she considered the most aesthetically refined person she had ever met. "The house had an old gold rug down when we moved in," Janet says. "It had been cleaned, but it just looked dirty. The walls hadn't been painted in a long time, and we had a cheap couch, an old beige sofa with an orange thread running through it."

By the time she heard the car pull up in the driveway, Janet had worked herself into a state of considerable apprehension. When Dr. Money stepped through the front door into the Reimers' modest living room, however, he seemed to take particular pains to project a manner of friendly and accepting approval of everything that met his eye. If memories lingered in Money's mind of the last time he had seen the Reimers—during the disastrous Baltimore trip the previous spring—he showed no signs of it. Certainly Ron and Janet made every effort to put the episode out of their own minds.

"He was like a friend—or an uncle—who had been away a long time and had come for a visit," Janet says. Dr. Money admired Ron's homemade wall cabinet and complimented Janet on her pen-and-ink drawings, which hung on the walls. Meanwhile the twins had disappeared into the basement and refused to come up and meet Dr. Money.

The adults sat at the dining room table, and Ron offered Dr.

Money a beer. Sipping one of Ron's Canadian lagers, Money relaxed and spoke about his childhood in rural New Zealand, where he said he had once seen a fireball and where the earthquakes were so frequent that his mother had strung thread across her kitchen shelves to prevent bottles from falling during the tremors. Money talked about the dark beer that was popular in his native country and asked Ron if they had dark beer in Winnipeg.

Today, neither Ron nor Janet can remember precisely how it was that Dr. Money, who had planned merely to drop over for an hour or two, ended up spending the night. Janet recalls that Money glanced at the clock and announced that he had missed his flight; Ron thinks that perhaps a snowstorm canceled Money's flight back to Baltimore. In any case, once it was apparent that Dr. Money was stranded, Ron and Janet, out of politeness, invited him to stay with them—although they had only an air mattress in the front room for him to sleep on. To their surprise, the eminent psychologist accepted their offer. Ron phoned out for a bucket of chicken to accommodate their unexpected house guest. The children continued to hide in the basement. That is, until their parents forced them to come up.

During the stiff encounter in the living room, Money asked how the twins were doing in school. Brian did the talking. He said something noncommittal about their academic accomplishments, then asked Dr. Money how he liked their city and how long he was staying. "Then," Brian says, "we wanted to go." Before the two could escape to the basement, Money pulled out his wallet. Saying that he would have spent the money on a hotel room anyway, he bestowed on the children fifteen dollars each. The twins then hurried back downstairs. They did not emerge until the next morning, when Dr. Money had left for the airport. It was the last time the Reimers would see him in person.

It was not the last the city of Winnipeg would hear about John Money, however. After his departure, the *Winnipeg Free-Press* carried accounts, on two successive days, of Money's standing-room-only appearance at the university's Human Sexuality Conference. STUDENTS, MDs DEBATING WORTH OF SEX EXERCISE, ran the banner headline on the first day. MORAL VALUES OF LECTURERS QUESTIONED. Money had upset some students, the paper reported, by showing graphic slides of a range of unusual sexual behaviors. The slide show was in fact part of a standard lecture Money had devised to "desensitize" medical students to various sexual perversions, and had already generated fierce controversy in the local Baltimore newspapers when Money introduced the films to the Johns Hopkins Medical School curriculum in 1971. The show featured explicit photographs of people engaged in bestiality, urine-drinking, feces-eating, and various amputation fetishes. During the second day of his Winnipeg lecture (the paper reported), Money had also screened a stag film of five women and three men having group sex, then followed the screening with a speech in which he informed the assembled professors and first-year medical students that marriage was simply an economic compact in which the "heart follows the wallet"; that incest should not be prosecuted as a criminal offense; and that in cases where stepfathers sleep with their stepdaughters, the mother is often "happy" because she "is glad to have [her husband] off her back."

Dr. Robert Martin, a clinical psychologist and member of the University of Manitoba's Psychiatry Department, attended Money's talks. "He was a personification of the style of the time," Martin says. "He liked to shock, play the devil's advocate, and was very cocky and very self-assured. His attitude was very much one of bringing 'enlightenment' to the 'boonies.' He radiated that particular lack of anxiety that, personally, sets alarm bells off, and he gave the impression that he'd plunge into anything. He was not the sort of person that you would forget."

Steve Whysall, who also attended the lectures, agrees. A seasoned journalist who had worked in London's Fleet Street, Whysall was the *Free-Press* medical reporter who wrote the paper's accounts of Money's controversy-stirring visit. "I'd been around and thought I'd seen and heard quite a few things," he says. "But I was surprised that this [sexual material] came up in that form." Whysall was especially surprised by the deliberate casualness with which Money spoke about such outlandish sexual fetishes as feces-eating. He interviewed Money briefly after the lecture. "I asked him, 'Are you telling these doctors-to-be that they shouldn't be alarmed if they meet someone who comes to them with that kind of request or condition?' " Money, Whysall says, dismissed such queries. "He was pretending that *he* was shocked that I was so narrow-minded, so Puritanical."

Since the Reimers were not in the habit of reading the *Free-Press* (they preferred the tabloid *Tribune,* which did not cover the event), they failed to learn about the controversy their overnight guest had ignited. Brian alone happened to notice that Dr. Money's visit generated some media interest; while watching TV, he caught a snippet of Dr. Money on CKND Channel 9, but the glimpse was fleeting, and Brian failed to hear the gist of the report.

In the days directly following Dr. Money's visit to Winnipeg, Brenda began to make increasing strides in her therapy with Dr. McKenty—as if finally freed from the last vestiges of suspicion that Dr. McKenty was working in collusion with the psychologist from Baltimore. Arriving at her 4 April session from an Easter celebration with her family, and thus outfitted in full feminine attire—black cowl-necked top, garnet pendant, and mascara—Brenda pointedly rejected McKenty's compliments on her appearance, denying that she was even wearing makeup. At her next session, Brenda announced, "I hate dresses. I only wear them to funerals and weddings." On 4 May, McKenty wrote, "Showed me her purse and contents, which were a hairbrush, mascara, lip gloss and rouge given to her by her mother, but she remarked pleasantly, 'I hate that stuff.' "

It was the session of 8 June that marked Brenda's most dramatic psychotherapeutic breakthrough to date. That this was to be an unusual encounter is clear from the opening sentence of McKenty's notes. "Did not want to play any games," McKenty wrote. "Soon began to ask some questions about her medical condition." This marked the first time in Brenda's ten-year standoff with the medical profession that she ever voluntarily raised the issue of her genitals and the fact that they did not resemble those of other girls. Brenda told McKenty how her father had explained that a doctor "did something that was a mistake." McKenty asked Brenda what *she* thought had happened.

"I used to think," Brenda said, "that my mother had beaten me between the legs."

Keith Sigmundson was immediately informed, by McKenty, of Brenda's comment. The two psychiatrists discussed it, and they agreed that Brenda's statement about her mother seemed to fit with almost eerie neatness a central tenet of psychoanalysis: Freud's theory of the Oedipus complex—the developmental stage that supposedly marks every human being's psychosexual differentiation into boy or girl.

Named for the unlucky hero of Greek tragedy who unwittingly slept with his mother and murdered his father, the Oedi-

pus complex was founded on Freud's conviction that all children, both male and female, develop in earliest infancy an
erotic attachment to their mother—an attachment that eventually pits them against the father in competition for the mother's
erotic favors. In boys, Freud postulated, the Oedipus complex
gives rise to "castration anxiety": the terror that their father will
neutralize the son's sexual threat by castrating him. In girls,
Freud stated, the Oedipus complex breeds "penis envy"—a conviction that the castration *has already been performed*, that she
once had a penis and has had it removed by one of her parents.

In normal female development, Freud argued, the girl's
urgent desire to reclaim her missing penis compels her to
reroute her infantile erotic desires for her mother and direct
them toward her father so that she might, in sexual intercourse
with him, take back the penis stolen from her, and it is by this
means that she forms a "normal" heterosexual orientation. In
Freud's view, psychotherapy was primarily concerned with cur-
ing the mental illnesses and neuroses that result in patients
who, for a plethora of reasons, fail properly to resolve their
Oedipal dramas in childhood. According to psychoanalytic theory, a crucial step in the resolution is to face the initial castration anxiety and voice it—as Brenda had apparently done when
she described her childhood fear that her mother had damaged
her sex organs.

On this basis, Sigmundson explains, he and McKenty
hoped that Brenda's comment might be the articulation of a
universal Oedipal fear shared by all females. "So we thought,"
Sigmundson says, "we were getting somewhere."

At the same time, Sigmundson admits that he was also
forced to consider another interpretation of Brenda's comment—one that not only took into account the awkward fact of
Brenda's intensely masculine behavior, but also acknowledged
that Brenda *was* born a boy with a normal penis and testicles,

which had (at least partly on her mother's authorization) been lost. Bearing these factors in mind, Brenda's comment could be seen to signal not an Oedipal breakthrough, but something less abstract—namely, the gruesome but emotionally logical explanation that a young child had used to explain the scarred state of her genitals and the bouts of depression suffered by her guilt-ridden mother. Viewed thus, Brenda's comment to McKenty could be interpreted not as her imminent acceptance of herself as a girl, but its opposite: her recognition that her earlier fears of maternal castration were incorrect and that now she wanted to know what had *really* happened to her; a sign, perhaps, that she was approaching the point at which she was ready to embrace the boy she had always instinctively known herself to be.

Whichever interpretation proved correct, Sigmundson and McKenty were now convinced that Brenda's therapy was reaching a critical stage.

The events of that June also brought Janet to a critical stage. Told of Brenda's statement, Janet was aghast. Already feeling nearly insupportable grief and guilt over her daughter, she found this latest piece of news almost unbearable. "I was just stunned," she says. "I couldn't believe that Brenda thought I could do such a horrible thing. I wondered, What must she think of me that I would do a thing like that to my own child?"

Today Janet cannot recall if this incident was instrumental in undermining her confidence in the entire experiment. One thing is certain, however. That June she and Ron were fast approaching the time when they would ordinarily start planning their annual visit to Johns Hopkins. Dr. Money had contacted them recently and pressed them to make an appointment for July. Yet when July arrived, Janet and Ron did not follow through on the plan to go to Baltimore. When

Brenda anxiously inquired of Janet whether there were any plans to return to Johns Hopkins that summer, Janet responded with a question of her own.

"Would it do any good?" she asked.

"No," Brenda said.

"Then we're not going." Janet did not bother to contact Dr. Money to cancel the appointment. The Reimers simply did not show up. Nor would they ever again.

11

≡

IN THE FALL, when Brenda's psychotherapy resumed with Dr. McKenty (who, like many psychiatrists, always took a summer vacation), she told the psychiatrist that she had passed a boring summer at home watching TV and doing a paper route. Now, however, she was uncharacteristically excited about starting classes at a new school. The previous spring Dr. McKenty and the other members of the local treatment team had discussed with Ron and Janet the option of removing Brenda from the academic path and putting her into a vocational school where she could learn a trade. Ron and Janet, having come to recognize that Brenda would never be a scholar, approved this plan, and Brenda herself was enthusiastic about it. She told McKenty that she would like to become an auto mechanic—a job whose only drawback was that "no guy would hire a girl to fix his car."

In September 1979, Brenda, now fourteen, began ninth grade at R. B. Russell Vocational School, which was located across town in the scrappy industrial area of Winnipeg's West End. The brochures for the school had featured shots of a pleasant city campus; the reality proved somewhat different. A school that ministered to children with behavioral and family problems (some of the teenage girl students were reportedly already moonlighting as prostitutes), the campus, David says, was a forbidding concrete complex covered with graffiti.

The school had a rigorous hazing initiation week for freshmen. It was as part of this hazing process that Brenda was selected by her upper-year schoolmates as "Freshie Queen." On the day when she was to have the twenty-five-dollar prize bestowed on her, Brenda, as instructed, wore her best dress, a full-length gown with puffed sleeves and a ruffled lace neck. She then learned that she could not collect her cash prize until she danced with the "Freshie King." Her royal counterpart proved to be a small, stooped boy with a brush cut and a pained expression. The two got up and danced in front of the school. "It damn near killed me," David says.

The incident marked another turning point for Brenda. That September, after starting at R. B. Russell School, she took her sexual destiny into her own hands and simply stopped living as a female. Gone were the cowl-necked sweaters, garnet pendants, and purses she had adopted the previous year at Glenwood. She now wore a boy's denim jacket, torn at both elbows, dirty corduroy pants frayed at the cuffs, a pair of what McKenty described as "men's leather gloves," and heavy construction boots on her feet. She stopped washing her hair, which grew matted. Her voice had settled into the rumbling register in which David speaks today. Physically, her condition was such that "strangers turn to look at her," as McKenty wrote in her notes on Brenda. To the close observer, however, it was Brenda's mental state that would have drawn particular scrutiny and pity. For as photographs from this period reveal,

Brenda, for all her attempts to smile, had the eyes of a cornered animal.

"That was the worst time of her life," says Sigmundson. "R. B. Russell really brought things to a head in a way that may have taken much longer had she been in a more cultured society where the kids might have been more prepared to play the game. At the other schools they called her 'Cavewoman.' At R. B. Russell they looked at her and said, 'You're a fucking *gorilla.*' "

Despite the brutal intensification in her peers' taunts, Brenda refused to change. "I won't walk funny like girls do," she told McKenty—and she jumped up and did a caricature of a girl walking: "mincing along," McKenty noted, "with bent arm and 5th finger prominently displayed." Brenda enrolled in Appliance Repair—the first and only girl ever to take the course in the twelve years it was offered at R. B. Russell. The teacher, Hillel Taylor, was at first concerned about how a girl would fare with the boys in his class, but his fears were soon allayed. "Brenda could relate to the boys on a very equal basis," says Taylor, who has never been informed of Brenda's medical history. "I could picture someone like her making it in the military or something. I remember being questioned by my principal and other people involved—guidance counselors and so on— 'How is she fitting in?' 'How is she handling the boys?' " Taylor let them know that Brenda was adapting as if she were "just one of the guys."

Ron and Janet were not happy with Brenda's behavior, but that was fine with Brenda. "I was at that age where you rebel," David says. "I got so sick to death of doing what everyone wanted me to do. I got to that point in my life, I knew I was an oddball, I was willing to live my life as an oddball. If I wanted to wear my hair in a mess, that's how I wore it. I wore my clothes the way *I* wanted to."

And Brenda had more private ways of rebelling. Since earliest childhood she had been instructed both by her parents and by Dr. Money to urinate in the sitting position—despite her strong, overriding urge to face the toilet bowl standing up. Ever since she had been spotted by a kindergarten classmate peeing this way, Brenda had tried to refrain from standing up. No more. "If no one was around, I'd stand up," David says. "It was easier for me to do that. I figured, what difference did it make?" It made a difference to her peers at R. B. Russell. Caught one day urinating like a boy, Brenda was barred from the girls' bathroom. She tried sneaking into the boys' but was kicked out and threatened with a beating if she returned. With nowhere else to go, Brenda was reduced to sneaking out to a back alley near the school to urinate.

It was on one such trip that Brenda became conscious of a car idling in the gap between the houses that lined the alleyway. She noticed that the car had rental plates. The man at the wheel seemed to be looking at her. She zipped up her pants and moved away, but the car followed. Then she saw that the man behind the wheel was pointing a camera at her.

"I ran back to the school," David says. "I didn't know what he was up to. I *wondered* if maybe he was a reporter. You know that you're different. You go to the United States to see all these important people, so it's feasible a reporter would want to see you, but you don't know why—or why he's so anxious to get a photo of you."

The British Broadcasting Corporation's interest in John Money's famous twins case dated to some eight months prior to the day Brenda spotted the man trying to photograph her. Edward Goldwyn, an award-winning documentary filmmaker with the BBC series *Horizon,* had begun researching a film about gender identity in late 1978. A tenacious reporter with a background in science, Goldwyn had burrowed into all aspects of the subject, traveling around the globe to interview experts in the Dominican Republic, East Germany, Los Angeles, New York, and London. He inevitably heard much in his travels about Money's landmark case, which still stood as the single most compelling piece of evidence to prove the primacy of rearing over biology in the formation of gender identity. Yet when he discussed the experiment with experts, Goldwyn was surprised to hear rumblings that the case was not quite as it appeared in Money's writings.

"I was getting vibes from people in Baltimore being quite embarrassed by Money and the prominence of this case in the literature," Goldwyn says. "I could tell that these people were getting increasingly worried." They urged him not to put too much stock in the experiment until he had talked to the doctors in charge of the twin's care. Tipped off by a source whom he declines to name, Goldwyn learned that the child was being treated by Jeremy Winter.

Goldwyn contacted Winter in late 1978 and told him about the documentary he was making on gender identity.

"I was incredibly suspicious of some guy wanting to produce a show for TV entertainment," Winter says. "I was very frosty at the beginning." But Goldwyn quickly established his bona fides with Winter and showed him the extent of his reading and research. "He totally brought me around," says Winter. Having thus allayed Winter's fears, Goldwyn questioned him about the twins case. Winter cannot recall the precise words he used, but he says that he did disabuse the reporter of the notion that the case was a success. "At the very least, I'm sure I would have said, 'Look, I wouldn't take that case too seriously, because the reality of the child's psychological adjustment is really quite different.' "

Goldwyn wanted to know how different and was struck by Winter's reply: "He told me that the twin would have been suicidal if it hadn't been for Mary McKenty," Goldwyn says.

In January 1979, Goldwyn even visited the Reimers at home—a visit none of the family members can recall twenty years later, and with reason. Eager to see the family but concerned not to disrupt them by revealing that their identity had been learned by a journalist, Goldwyn settled on a ruse to gain access to their house. He declines to say precisely what his cover story was, but he says, "I came in as if I were asking if they could move their car because it was in my way. I was being, I suppose, a bit immoral, but I thought it was important for me to go and look and see for myself." Ron and Janet, he says, were "worried, lonely looking people." Brenda was surly, distinctly sexually ambiguous, and "somebody who I thought was really quite angry." In short, the family little conformed to Money's sunny portrait in *Sexual Signatures*. "Having found that the case wasn't a good data point—that Money's study actually didn't prove anything," Goldwyn says, "I felt the only thing to do was to leave it all out of my film. The only reason to put it in was to rubbish it."

This decision did not preclude Goldwyn's discussing what he had learned with a BBC colleague known for producing programs of a more controversial bent. Freelance TV journalist Peter Williams had recently been placed under contract by the BBC as executive producer on a new series called *Open Secret*, which was to deal specifically with medical scandals. Williams was fascinated by what Goldwyn told him about Money's famous case and asked freelance documentary filmmaker Martin Smith to direct a projected half-hour program on the case.

In late September 1979, Williams; his wife, Jo Taylor; Smith; and a small BBC-TV crew arrived in Winnipeg. Within days of their arrival, Dr. McKenty notified Sigmundson of Brenda's description of the strange incident in the alley near the school, where a man had tried to photograph her. Sigmundson immediately recognized that reporters had gotten wind of Brenda's location. As head of the clinic's psychiatry unit, Sigmundson had the most experience with the press, so it was agreed that he would handle the reporters.

"By that time," Sigmundson says, "there were *clear* doubts in my mind that this [sex reassignment] *ever* should have happened. At that point, I think I really wanted the world to know." Sigmundson agreed to speak to the reporters only under conditions that guaranteed the Reimers' anonymity. He demanded that the reporters agree in writing not to broadcast the photographs they had taken of Brenda; make no further effort to capture her on film; obscure the Reimers' location by omitting the names of all local treatment personnel; and finally, that the program not be sold in Canada or the United States. With these conditions agreed to by Williams and Smith, Sigmundson allowed himself to be interviewed at his home on 30 September.

Although appearing as an unnamed psychiatrist, Sigmundson nevertheless looked distinctly nervous as he faced the BBC cameras. Glancing frequently at a set of notes in his lap, he described the "significant psychological problems" from which Brenda had been suffering when she first came to his attention at the Child Guidance Clinic. He related the litany of Brenda's masculine appearance, her difficulties at school, and her failure to sustain friendships with peers. It was when Williams asked about the prognosis for the sex reassignment that Sigmundson paused. Several seconds passed before he spoke.

"When I took that long, long pause," Sigmundson says today, "I was wondering if I was really going to tell the truth or just fudge it. After all, it was still Hopkins. Money was the guru." When he finally answered, Sigmundson picked his words carefully, like a man tiptoeing through a minefield.

"I don't think all the evidence is in," he began. "And it may not be until she is a young adult that we're going to know everything about this particular case. At the present time, however, she does display certain features which would make me be very suspicious that she will ever make an adjustment as a woman."

Brenda's former psychiatrist, Doreen Moggey, also agreed to be interviewed. "I felt it needed to be done," Moggey says. "*Somebody* needed to say that this was not the rosy success story that was presented in the literature." On camera Moggey described the extreme difficulties she had faced with Brenda's case and recounted how she had notified Dr. Money of these difficulties by letter.

Ron and Janet Reimer learned of the BBC's presence in Winnipeg from Mary McKenty, who had declined to speak to the filmmakers.

"She called to say that there were reporters who wanted to see us," Janet recalls. McKenty told her that they should not feel they had to be interviewed, but Ron saw no reason to refuse—as long as they were not filmed or recorded. At that point Ron and Janet still remained convinced that Dr. Money's treatment would work, and they thought their testimony would be a help to other parents who might find themselves in a similar predicament. "We were too close to the situation," Janet says. "I had brainwashed myself. I couldn't *afford* to believe anything else."

In the Reimers' living room, Williams and his wife, Jo Taylor, asked how Brenda's treatment was working out. "I said I was still hopeful," Janet recalls. The reporters began to ask about Sigmundson's and Moggey's observations regarding Brenda's school performance and social life. The mood of the encounter changed. Janet started to cry, and Ron sank into a characteristic mute melancholy. The reporters asked to meet Brenda. Janet called her daughter in from outside and introduced the British visitors as editors of a poetry magazine that wished to publish one of Janet's poems.

Brenda, dressed in tattered jeans and a torn jacket, her unwashed hair falling in tangles around her face, stalked into the living room and said an awkward hello in her deep voice, then quickly disappeared. Her appearance seemed to make a strong impression on the reporters.

"When Brenda left the room," Janet says, "the woman got up and said, 'We're going to get to the bottom of this!' She seemed quite angry."

The BBC crew were headed for Baltimore. They had notified Money some weeks earlier that they were doing a documentary on the twins case. "Money initially showed considerable willingness and interest in being involved," says Smith, but that was before he learned of the reporters' investigative efforts in Winnipeg.

Williams and Smith arrived at Money's house in the early evening of 3 October 1979. At the time of his divorce more than twenty-five years earlier, Money had moved from the suburbs to an address just minutes by foot from Johns Hopkins, in a gritty, inner-city Baltimore neighborhood, where he continues to live to this day. "It was not the sort of place where you would expect a well-heeled academic or scientist to be living," says Smith. Money occupied the upper floors of a run-down corner store. Williams and Smith, admitted through a front door that boasted three locks, were no less surprised at the interior of Money's residence, which was decorated with the masks, totems, and sexual artifacts that also bedecked his office. Money himself was a convivial host—at least initially.

"There were a couple of his mature students around," Smith says. "We were having a drink quite casually in front of the fireplace and talking about the preparations for the interview, which was to take place the next day."

The reporters then eased toward revealing to Money the full scope of their documentary. "We think the case is very interesting," Smith remembers saying, "and we do want to do a documentary on it, but—"

"We should warn you," Williams recalls cutting in, "that we have heard other things about it."

Smith informed Money that they had spoken to the child's psychiatrists and that all was not what it appeared to be from Money's published writings. "At that point I think it's fair to say that he got extremely angry and annoyed," Smith says. "I think he felt that he'd been sandbagged into a corner. Which wasn't the case. In fact we quite deliberately told him that we had made contact with the psychiatrists *before* we did the filming."

Money, however, appeared to be in no state to appreciate such fine distinctions of journalistic etiquette. "His anger might have been that he felt that the child was being investigated or put at risk," Smith continues. "Or it might have been personal anger that someone should challenge his work. I don't know. But our relationships changed dramatically, and we were shortly out the door."

The telephone call to the Reimers' house in Winnipeg came later that evening. Janet and Ron had already gone to bed. Janet answered. It was Dr. Money calling from Baltimore, and he was in a panic. The content of the call has been preserved in notes taken the next day by Mary McKenty, to whom Janet recounted Money's conversation. Speaking of "persons unknown" but "suspected to be a Mr. Smith of the BBC" and another man— "a friend of Mr. Goldwyn"—Money told a wild tale of files possibly stolen from him and of reporters who had somehow learned of Brenda's whereabouts.

"He was all freaked out," Janet says. "He said, 'Don't

speak to any reporters.'" At which point Janet had no choice but to tell him the truth—that both she and Ron had already spoken to a man and woman from the BBC.

The extent of Money's displeasure was clear from a letter he wrote the next day to Sir Charles Curran, then director general of the BBC. After laying out his history with Williams and Smith, Money delivered a threat. "I would appreciate it," he informed Curran, "if you perused the contents of the program most carefully in light of the BBC's moral and legal obligation not to violate the privacy of a family which is at present particularly vulnerable to the possible effects of an invasion of privacy. I need hardly tell you that my concern is for the protection of this family. However, I must advise you that if their privacy is not appropriately protected I will counsel them to take legal steps to obtain compensation for any harm the BBC has caused them, and I trust that this will not be necessary."

But the BBC stood behind Williams and Smith, and the reporters moved on to the final stage of their reporting: to find a scientist who could comment on the significance of their findings. One name in particular kept coming up—that of the scientist who had inspired Money's wrath fourteen years earlier when he first questioned Money's conclusions and with whom Money had later clashed at the gender identity symposium in Dubrovnik.

"When we got onto Dr. Diamond," Smith says, "it was then quite interestingly obvious that we were getting into what is best described as scientific warfare—and that warfare can get quite bloody." Indeed, given Money and Diamond's long history as doctrinal adversaries, the BBC reporters were at first wary about using Diamond as an expert commentator on the case, fearing that any opinions he expressed might not reflect an objective scientific viewpoint. "You have to be careful to find out: Was this something personal, or was it not?" Williams says. "I was satisfied that Diamond was actually raising something which deeply troubled him ethically. Whether or not he liked Money is quite another matter."

Diamond says that he had no special dislike for Money. Their altercation of six years earlier he had forgiven as a by no means rare eccentricity in a scientist and perhaps a result of the bibulous nature of the Dubrovnik cocktail party. Even after that encounter, Diamond had tried to communicate with Money. "I asked John several times in the late 1970s about the twin," Diamond says. "He didn't want to talk about it. He said that the kid was going through some troubles unrelated to the sex reassignment and that it would be inappropriate to bother her at this time. So I let it go."

But Diamond had never deviated from his conviction that sex reassignment of a developmentally normal infant was impossible, and he had not hesitated to publish this opinion—even as recently as a few months before the BBC contacted him. In the 1979 volume *Frontiers of Sex Research*, Diamond had cited the case, saying that on the evidence Money had so far published, it seemed to be "good fuel" for the power of rearing over biology, but in what today looks like a statement of extraordinary prescience, Diamond warned, "with puberty, the penectomized twin has a good likelihood of rebelling at the assignment of rearing which is in conflict with biological heritage."

Diamond agreed to be interviewed, and his segment was filmed on a rocky precipice overlooking the ocean. Asked by Williams what impact it would have on the field if the twin were shown to be having "severe and sustained" problems, Diamond said, "I think it depends on who you ask. There are those who believe in the [case] almost as a religious entity." He went on to say that if all the combined medical, surgical, and social efforts could not succeed in making the child accept a female gender identity, "then maybe we really have to think that there is something important in the individual's biological makeup; that we don't come to this world neutral; that we come to this world with some degree of maleness and femaleness which will transcend whatever the society wants to put into it."

The documentary, entitled *The First Question* (in reference to the universal query at birth, "Is it a boy or a girl?"), aired in Britain on 19 March 1980. An impressively clear overview of the complex issues involved, the program also sought a balance in its depiction of Money's work. Included in the program was an interview with the mother of one of Money's intersexual research subjects, an XY male born with a tiny penis and undescended gonads, who on Money's recommendation had been surgically reassigned as a girl. At eight years of age, the child, Paula, was described as successfully living in her female assignment. Yet the program was unsparing in its depiction of the far more theoretically important case of the developmentally normal twin—whose case seemed to be on the brink of collapse.

Williams and Smith expected their program to stir controversy and comment. It did not. "The reaction was curiously muted," says Smith. "I was a bit surprised that print journalists didn't take it up." Diamond was similarly mystified at the failure of the documentary to provoke comment or follow-up from American programs like *60 Minutes*.

Determined to disseminate the BBC's findings to North American physicians, Diamond wrote up the results of the documentary in a short scientific paper and submitted it to the American science journal *Archives of Sexual Behavior*. Titled "Sexual Identity, Monozygotic Twins Reared in Discordant Sex Roles and a BBC Follow-Up," the paper appeared in a 1982 issue of the journal. In it Diamond quoted Sigmundson's and Moggey's verbatim comments about Brenda's problems as well as Sigmundson's doubts that she would ever make the adjustment to being a woman. Speaking to the wider implications of the case and its apparently imminent failure, Diamond added, "As for the twin, it is scientifically regrettable that so much of a theoretical and philosophical superstructure has been built on the supposed results of a single, uncontrolled and unconfirmed case. It is further regrettable that we here in the United States had to depend for a clinical follow-up [on] a British investigative journalist team for a case originally and so prominently reported in the American literature."

Upon the article's publication, Diamond was frustrated to see it meet with a reaction similar to that of the documentary. While some feminist scholars quietly dropped the twins case from new editions of their women's studies textbooks, the academic, scientific, and medical communities were oddly silent about the findings. "They ignored it," Diamond says. "It's not what they wanted to hear."

In the days and weeks following her glimpse of the mysterious man with the camera in the alley near her school, Brenda grew increasingly outspoken in her rejection of girlhood during her sessions with McKenty. She complained of how Brian had more friends because he wasn't constantly being teased, and she railed about the fact that her brother could fight people without looking like a weirdo. "All girls do is make *babies,*" she said.

At school, her "out" boyishness provoked escalating taunts and threats from her peers. One day shortly before Christmas, she was threatened by a classmate who brandished a knife. "I told my mother, 'I'm not going to that school anymore. I'll run away,'" David says. Janet supported Brenda's decision, as did McKenty, who arranged for a private tutor paid for by the government.

Away from the taunts and threats of her R. B. Russell class-mates, Brenda continued to assert her boyishness in word, dress, and deed while at home. Janet, hoping that Brenda's behavior was simply a "stage," continued to look for signs of femininity in her daughter. "Any little sign—and I was in seventh heaven," she says. "I misinterpreted a lot of things she did."

David remembers an incident from this period when Brenda found a pair of her mother's black kid gloves in a closet. "They felt nice and soft inside," David says. "I put them on. They reminded me of those cool Italian race car gloves that you see in the movies. I was thinking, These would give a good grip on the steering wheel. All of a sudden I realized my mother was behind me. I looked around and she was smiling at me, and she said, 'Go ahead. If you want to wear them, *go ahead*.' She thought I was trying to be feminine."

As the winter progressed, Janet found it increasingly diffi-cult to sustain such fantasies. One night she had a dream that years later she recognized as a sign of all she was struggling to repress about her daughter. In the dream Janet was visiting a woman whose boyfriend had just moved out. The woman, distraught, opened a trunk and reverently lifted out of it a huge stuffed penis. "She held it out to me like it was a brick of gold," Janet recalls. "And she said, 'That's what I've got to remember him by.'"

That winter, Janet began to feel returning the mood of deso-late hopelessness that had engulfed her during the family's ill-fated sojourn in British Columbia. In late January her psy-chiatrist, Dr. Nona Doupe, recognized that Janet was descend-ing into a serious depression and was once again a threat to herself. Dr. Doupe had Janet admitted to Victoria Hospital, where she remained for a month. Soon after her release, Janet spiraled back into despair—and life at the Reimer home fell into chaos. For Brenda, her mother's continued vigilance for signs of girlishness became intolerable; for Janet, the sight of Brenda in her boy's clothes and matted hair seemed an unspoken and unbearable rebuke for the decision she and Ron had made almost thirteen years earlier. In early March, McKenty noted Brenda's complaints "that nothing she does pleases her mother who criticizes and yells at her all the time." A few days later things reached a climax.

"She told me to clean up the fridge," David says. "I used as much elbow grease as possible, but it wasn't to her liking. I shouted, 'I'm doing the best I can!' She threw a box of cereal at my face. I threw it back at her. She was ready to hit me. I grabbed her hand and shoved her. My mother said, 'I'm going to tell your father!' "

Ron, pleading exhaustion, withdrew from the fray, turned on the television, and poured himself a drink. Janet recognized a return to the fatal pattern that had trapped the family in British Columbia. She called Dr. McKenty, whose notes on the conversation register the turmoil into which the family had plunged: Ron was drinking unsettling amounts of whiskey; Janet and Brenda were constantly at each other's throats; and now Brenda and Brian, too, were in open warfare, fighting all the time.

By now it was impossible for McKenty and the other members of the local treatment team to ignore the obvious: after almost four years of trying to implement Dr. Money's plan, Brenda and her family were only worse off. Dr. Winter was the only physician who still held out any hope. Convinced that the appearance of Brenda's uncompleted vagina was the chief stumbling block to her psychological acceptance of herself as a girl, he had long been the most vocal advocate for the surgery. But now, even he began to waver. "Early on, I had . . . pushed for early surgery," he wrote in a letter to Dr. McKenty. "I am not as convinced now that this is a good idea and therefore at the present time have no specific plans or opinions as to the proper time for the operation."

Ultimately Brenda forced the endocrinologist to come down off the fence. During an appointment at his office in mid-March, she refused to remove her hospital gown for a breast exam. The doctor asked again. She refused. The stand-off lasted twenty minutes. "It comes to a point in your life where you say, 'I've had enough,'" David says. "There's a limit for everybody. This was my limit."

Dr. Winter had reached *his* limit, too. "Do you want to be a girl or not?" he demanded.

It was the question Dr. Money had been asking her since the dawn of her consciousness, a question the local treatment team had badgered her with for years. It was a question she'd heard once too often.

She raised her head and bellowed into Winter's face, "*No!*"

To Brenda's surprise, Winter did not get angry. Instead he simply left his office for a moment, then returned. "OK," he told her. "You can get dressed and go home."

Only later would Brenda learn that Winter had, in step-ping out into the hallway, spoken with Dr. McKenty. He told her that in his opinion it was time the teenager was told the truth about who she was and what had happened to her.

It was Ron's custom to pick Brenda up in the car after her weekly sessions with Dr. McKenty. The afternoon of 14 March 1980 was no exception. The only difference was that when Brenda climbed into the car, Ron said that instead of driving straight home, they would get an ice cream cone.

Immediately Brenda was suspicious. "Usually when there was some kind of disaster in the family, good old dad takes you out in the family car for a cone or something," David says. "So

I was thinking, Is mother dying? Are you guys getting a divorce? Is everything OK with Brian?"

"No, no," Ron said to Brenda's nervous questioning. "Everything's fine."

It was not until Brenda had bought her ice cream and Ron had pulled the car into the family's driveway that he found the words he needed.

"He just started explaining, step by step, everything that had happened to me," David says. "He told me that I was born a boy, and about the accident when they were trying to circumcise me, and how they saw all kinds of specialists, and they took the best advice they had at the time, which was to try to change me over. My dad got very upset." It was the first time Brenda ever saw her father cry. She remained dry-eyed, however, staring straight ahead through the windshield, the ice cream cone melting in her hand.

"She just sat there listening, real quiet," Ron says, almost two decades after this extraordinary encounter between father and child. "I guess she was so fascinated with this *unbelievable* tale that I was telling her."

Today David says that the revelations awoke many emotions within him—anger, disbelief, amazement. But one emotion overrode all the others. "I was *relieved,*" he says. "Suddenly it all made sense why I felt the way I did. I wasn't some sort of weirdo. I wasn't *crazy.*"

Brenda did have a question for her father. It concerned that brief charmed span of eight months directly after her birth, the only period of her life when she ever had been, or ever would be, fully intact.

"What," she asked, "was my name?"

12

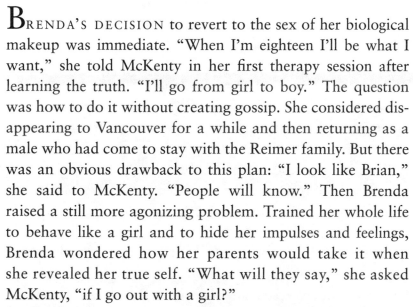

BRENDA'S DECISION to revert to the sex of her biological makeup was immediate. "When I'm eighteen I'll be what I want," she told McKenty in her first therapy session after learning the truth. "I'll go from girl to boy." The question was how to do it without creating gossip. She considered disappearing to Vancouver for a while and then returning as a male who had come to stay with the Reimer family. But there was an obvious drawback to this plan: "I look like Brian," she said to McKenty. "People will know." Then Brenda raised a still more agonizing problem. Trained her whole life to behave like a girl and to hide her impulses and feelings, Brenda wondered how her parents would take it when she revealed her true self. "What will they say," she asked McKenty, "if I go out with a girl?"

A month and a half later, the Reimers attended a large family gathering to celebrate Janet's youngest brother's engagement. Still living socially as a girl, Brenda had no choice but to go to the party in female attire: a dress, red shoes, panty hose, makeup, and a stylish, short, imitation white mink coat which Janet had bought specially for the occasion—and, perhaps, as a last inducement to Brenda to remain in the sex they had chosen for her. But the humiliation of parading herself publicly as a girl, now that she knew the truth, was too much for Brenda. Having vowed to change sex in three years, she now moved up the deadline. "In *two* years," she told McKenty a day after the party, "I want to look like a boy. I'd like a mustache."

At her next session, Brenda again moved up the deadline for becoming a boy. She wanted to do it *now*, and she told McKenty that she had been thinking about a boy's name for herself. She did not want to revert to her birth name, Bruce, which she considered a name for "geeks and nerds." She'd come up with two options. She liked Joe because it had no pretensions; it was a name for Everyman. She also thought of calling herself David, after the biblical king and giant-slayer. "It reminded me of the guy with the odds stacked against him," David says, "the guy who was facing up to a giant eight feet tall. It reminded me of courage."

Brenda left the final decision up to her parents, who chose the name David. Ron says it was easy to make the transition from calling their child Brenda and he cannot recall ever accidentally calling his son Brenda after that. Others, too, found Brenda's transformation to David easier to accept than they had anticipated. David's tutor, Dorothy Troop, says that she had initially been nervous when notified of the change, but when David arrived for his first tutoring session, Troop found that his maleness was far from an obstacle between them. Brenda had always been a sullen, depressed, angry child; as David, everything was different. "He was happier," Troop says, "far more settled and *alive* to what was going on around him." Troop gave David a chain with his new name on it. In return, David gave his tutor a gift: the imitation mink jacket he had worn to the family party. "He seemed to want to get rid of anything that reminded him of when he was Brenda," Troop recalls.

That August, one week after his fifteenth birthday, David made his big public debut as a boy among his extended family. The occasion was the wedding ceremony and reception of his uncle Dale. Using tape to flatten the breasts that still protruded from his chest, David donned a starched white dress shirt, a dark tie, and a charcoal gray suit identical to his brother Brian's. It was not easy, David says, to step out as a boy for the first time in front of aunts, uncles, grandparents, and friends. He knew that the whole family had been informed long ago about his sex reassignment as a baby, but this knowledge did not make it any easier for him, trained for so long to play the little lady in front of relatives. Still, determined to get up in front of the crowd, he danced with the bride and several of her bridesmaids. "Happy," Dr. McKenty wrote in her session notes with David two days later, "wedding a success."

David began to receive injections of testosterone. He soon boasted a growth of peach fuzz on his cheeks and chin, and he grew over an inch in height. On 22 October 1980 he underwent a double mastectomy, an intensely painful procedure that left him in agony for weeks afterward. He decided to wait until the following summer—until he finished tenth grade—before having any further surgery.

In the intervening months, he fell to brooding on the accident that had set his life on its bewildering course. "At that stage in his life," Dr. Winter says, "all he wanted was a gun to kill the doctor who had done that to him." As the dismal Winnipeg winter progressed, David's fantasies of revenge began to take on the contours of reality. With two hundred dollars saved from his paper route money, David bought an unlicensed 1950 Russian Luger on the streets of downtown Winnipeg. One February day he went to the Winnipeg clinic where Dr. Jean-Marie Huot had an office.

"I had the gun in my pocket," David says. "I opened the door to his office. He looked at me and says, 'Yes, what can I do for you?' I said, 'Do you remember me?' He said, 'No. Should I remember you?' I said, 'Take a good look.' Then he knew who I was. He nodded his head. I was intending to pull out the gun and blow his brains out, but he started crying. I felt sorry for him. He had his head down. I said, 'Do you know the hell you put me through?' He didn't say anything, just sat there, crying. I walked out. I could hear him behind me saying, 'Wait! Wait!' But I left. I sat by the river, crying."

David smashed the gun with a rock and threw it into the Red River. A few days later he admitted to McKenty that he had gone to Huot's office and "blasted him about the accident." He did not say he had been carrying a gun in his pocket.

I contacted Dr. Huot in the summer of 1997. He refused to speak about this encounter. "That was seventeen years ago," he said, "a very long time ago." Nor did he care to discuss the incident that had brought the murderously depressed fifteen-year-old boy to his office in the first place. Asked about the circumcision accident, Huot said in his heavy French-Canadian accent, "I'm not in a situation to start talking about that now, for sure, for sure, for *sure*."

On 2 July 1981, a month before his sixteenth birthday, David underwent surgery to create a rudimentary penis. Constructed of muscles and skin from the inside of his thighs, the penis was attached to the small stump of remaining penile corpora under the skin. False testicles, made of light-colored plastic, were inserted into his reconstructed scrotum. The sensation of a penis hanging between his legs was odd and unfamiliar. And he soon learned the drawbacks to phalloplasties. Over that first year, he was hospitalized eighteen times for blockages and infections in his artificial urethra. He would continue to be hospitalized regularly over the next three years.

Meanwhile, David tried to come to terms with his new life, and to prepare for reentering the world. In some respects, he says, this proved less difficult than he had feared. For apart from her fleeting friendships with Heather Legarry and Esther Haselhauer, Brenda had suffered severe social rejection; this, along with her almost annual changes of school, had guaranteed that no one ever got close enough to her to remark on her sudden vanishing—and David's sudden materialization. Still, after his reversion to his biological sex, David (fearing that he might run into someone who would recognize him as the former Brenda) took the precaution of lying low in his parents' basement. He watched TV, listened to records, and mulled all that had happened to him, trying to absorb and process it. This period would ultimately extend to nearly two years, until gradually, around the time of his eighteenth birthday, he began to emerge from the house, hanging out at local fast-food joints, roller rinks, and bars with Brian and his friends. Brian's buddies immediately accepted David as one of the guys, but there were inevitably kids who vaguely recalled that Brian had once had a sister named Brenda.

Together the twins dreamed up a story to explain Brenda's disappearance. They claimed she had gone to live with her boyfriend in British Columbia and had died in a plane crash. David was Brian's long-lost cousin. As for David's frequent hospitalizations, the twins said that they were to treat injuries sustained in a motorcycle accident.

"We all knew they weren't telling us the entire truth," says

Lyle Denike, one of Brian and David's friends from that era. "But we didn't want to push things too far. We knew we were dealing with something very personal."

Heather Legarry, Brenda's friend from sixth grade, also had doubts. In July 1983 she was working for the summer at her brother's Go-Cart track after completing her freshman year of college. "I was selling tickets," Heather says. "Suddenly there was a familiar face at the counter. It was Brian Reimer—or so I thought. I said hi, but instead of smiling, he flushed and stammered, then stepped away and pointed at this other guy. Up steps the *real* Brian. I asked, 'Who was that? He looks just like you.' Brian said, 'That's my cousin David.' I wondered if it was Brenda, but I just brushed it off, telling myself, If he says it's his cousin, it's his cousin."

"I couldn't say anything to her," David says of this encounter with the one person from his childhood whom he had considered a true friend. "It would take too long to explain everything. It was easier just to avoid people."

Later that summer, when David turned eighteen, he reached another milestone, for it was then that he came into possession of the money that had been held in trust for him since he was two and a half years old—money awarded to him when St. Boniface Hospital settled out of court with Ron and Janet for a sum of sixty-six thousand dollars in 1967. This was far less than the millions that some had predicted they would receive in compensation for their son's penile ablation. But then urgently in need of funds, and warned by their lawyer that a judge might overturn a large jury award, the young couple had accepted the hospital's offer. In 1960s dollars, it had seemed a considerable sum to Ron, who at the time had an annual income of only six thousand dollars. Placed in trust for David, the settlement money was to be used by Ron and Janet only for treatment associated with David's injury and had financed the family's annual trips to Johns Hopkins. By 1983 the money had grown to over a hundred and seventy thousand dollars—a sum that instantly made David one of the best-heeled young men in his peer group. In the hopes of "lassoing some ladies" (as he would later tell Diamond), he bought a souped-up van. Equipped with a wet bar, TV, and wall-to-wall carpeting, it was quickly dubbed "The Shaggin' Wagon."

David did not do any shagging in the van, however. Indeed it was in his relations to girls that he felt the worst complications of his transition—complications that were only exacerbated by the fact that by age eighteen he was not merely a passably attractive young man, but an arrestingly handsome one. His sudden popularity with what was now the opposite sex introduced a terrible dilemma, because he knew his penis neither resembled nor performed like the real thing (it was incapable of becoming erect). "How do you even *start* dating?" David says, recalling this period of his life. "You *can't*. You're in such an embarrassing situation."

Eventually he did date a girl two years his junior, a pretty but flighty sixteen-year-old. For David there was an ever-present anxiety. "I would think, What the hell is going to happen when she wants to go further than a kiss? How am I going to handle *that*?" He developed a strategy for stopping their sexual encounters before they became too intimate: he would drink a lot and then say, *I'm tired, I'm going to pass out now.* But one evening he miscalculated and truly did drop unconscious after drinking too much. When he woke in the morning, his girlfriend was beside him in the bed, and he could tell from her expression that she had looked between his legs. He had no choice now but to tell her. He explained that he had suffered an "accident." Within days, he says, everyone knew. Just as in his childhood, he was suddenly the object of muttered comments, giggling, and ridicule. For

David, this proved unbearable. The next day, he swallowed a bottle of his mother's antidepressants and lay down on his parents' sofa to die.

Ron and Janet discovered him unconscious. "Me and Janet looked at each other," Ron recalls, "and we were wondering if we *should* wake him up."

Janet remembers saying to Ron, "I wonder if we should just leave him, because that kid has done nothing but suffer all his life. He really wants to die." Within seconds, however, she had made up her mind, and they lifted him and rushed him to the hospital, where his stomach was pumped. On his release one week later, he tried it again, ingesting another bottle of his mother's antidepressants, then running a bath with the intention of drowning himself. "I was thinking, *When you're dead you don't feel anything, no pain in your heart, no pain in your body, no humiliation—nothing,* but I couldn't make it to the bathtub. Every step was like I had a hundred-pound weight on each foot." With the overdose beginning to suck him under, he lay down on the sofa and dropped unconscious. This time Brian saved him.

David withdrew from the world. He spent sojourns of up to six months at a time alone in a cabin in the woods near Lake Winnipeg. By now he refused to see even Dr. McKenty, but she had convinced him to bring a tape recorder with him to the cabin and to speak his thoughts into it. One night in January 1985, he did so.

"This is David Reimer," he began, his voice slurred with alcohol. "I'm nineteen. Soon I'll be turning twenty. I'm halfway through grade twelve. What I plan to do in my life is"—after a pause, he continued in a new tone—"OK, by the time I'm twenty-five, I should be all fixed up. I don't plan to marry until I'm in my thirties, because I'm just not the—not the type to get married." He rambled for a few minutes before returning to the subject that was preying obsessively on his mind. "I want to marry a chick that's sort of shy," he said. "Not *too* shy. And I would prefer her to have kids of her own. Because I want to have kids. And I can't have kids." This statement seemed to trigger a new set of associations. "Oh yeah," he said, "I got some money, about a hundred grand, because of an accident I had a long time ago. When I was small." He paused again, as if trying to decide whether he had the energy, or inclination, to speak about this part of his life. He did not. "Well, that's just about it," he said. "I hope everybody out there has a great life." He turned off the recorder and never made another tape.

It was not until almost a year after David had retreated to his cabin, that two friends of his, Harold Normand and Ron Mandel, talked him into leaving the woods and indeed getting far from frigid mid-winter Manitoba. Given subsequent events, there was an irony in the destination the three young men chose: Hawaii. On 11 January 1986 they flew to Honolulu, where they stayed for a week in the Outrigger Hotel not ten minutes drive from Milton Diamond's house. The trip had a salubrious effect on David, but it was an incident in the airplane on the way *to* Hawaii that suggested he was finally coming out of his depression and beginning to come to terms with the secrets of his past.

In the plane over the Pacific Ocean, he turned to Harold. "He said to me, 'I always wanted to tell you about that sister of Brian's,' " Harold recalls. "I said, 'You don't have to. I already know.' "

Harold had heard the truth three years earlier when he first met David. Immediately suspicious about the tale of the twin sister who had died in a plane crash, Harold mentioned the mystery to his parents. They instantly recalled the short newspaper item from 1967 about a twin boy who had lost his penis while being circumcised at St. Boniface Hospital. They had later learned through the grapevine that the family was named Reimer and had even heard whispers that the boy had been raised as a girl. "My parents put two and two together," Harold recalls. A uniquely private person himself, Harold had never gossiped about David's secret among their friends and had never revealed to David that he knew the truth.

In the months after their trip to Hawaii, David confided much to his friend that he had never told anyone except Mary McKenty. "He said to me that he never felt like a girl, so that when he found out he was a boy, his mind was made up to switch back," says Harold. "Either that, or he was going to be a lesbian. Because that was his biggest problem when he was growing up. He had feelings about girls."

After his return from Hawaii, David heard from his doctor about a new type of artificial penis, one that would, the doctor said, be a vast improvement over his current one. His new penis would resemble the real thing, and through the use of advanced microsurgery could be supplied with sensation. Shortly before his twenty-second birthday, David underwent a second phalloplasty. In a twelve-stage operation, which took three surgeons thirteen hours to perform, David underwent a procedure known as microvascular right radial artery forearm flap reconstruction of the penis—an operation in which the flesh, nerves, and an artery from his right wrist to elbow were cut away and formed into a tube to build the new urethra and main body of his penis, and a segment of cartilage was grafted from one of his left ribs to give structural support to the organ. Despite the long recovery time, David was delighted with the results, which were immeasurably better than his former phalloplasty. "I was driving down the street afterward," David says, "and I just started crying."

190

Despite the marked improvement in both appearance and sensation of his new penis, it would be two more years before David used it for sex. The delay had less to do with his feelings of confidence about his penis, he says, than with the legacy of what had been done to him in the operating room at Johns Hopkins Hospital when he was twenty-two months old—his castration. "I kept thinking, What am I going to say to the woman I meet who I want to marry?" David remembers. "What am I going to say to her when she says she wants children and I can't give her children?" Even if he did meet a woman who said she did not want to have children, she might change her mind later in life and then resent him. "I thought it would be *unfair* for me to do that to somebody I love," David says.

Still, David could not put thoughts of marriage and children out of his mind. His brother had married at the age of nineteen, and by the summer of 1988, Brian was twice a father—and possessed everything David wanted for himself. "I got so terribly lonely," David says. "I did something I'd never done before. I wound up praying to God. I said, 'You know, I've had such a terrible life. I'm not going to complain to You, because You must have some idea of why You're putting me through this. But I could be a good husband if I was given the chance; I think I could be a good father, if I was given a chance.' "

Two months later, Brian and his wife introduced David to a young woman of their acquaintance. Twenty-five years old, Jane Fontane was a pretty woman with blue eyes and shoulder-length strawberry blond hair. At five feet one and one hundred and eighty pounds, she was sensitive about her weight, but she carried her generous size easily, and to those who knew her, it seemed merely a natural adjunct to her nurturing personality. When I first met Jane in the summer of 1997, her combination of unflappability, affectionate friendliness, and infectious laughter reminded me of no one so much as the central character in Joyce Cary's comic novel *Herself Surprised*—the unsinkable Sara Monday, the picaresque mother of five children, a woman whom Cary describes as a kind of force of nature, a woman whose earthy goodness and fundamental optimism see her through every scrape life can throw at her—including her own youthful poor judgments.

Like Sara, Jane possessed a guilelessness and innocence that helped to explain how, by the time she met David, she was herself the single mother of three children—by three different fathers. Unworldly to a fault, Jane was a lifelong nonsmoker and nondrinker, a homebody who did not go to bars and didn't approve of "cursing." Her chief flaw was a certain neediness, a result perhaps of her difficult childhood in Winnipeg, where she was raised by her mother and stepfather.

Jane was sixteen when she joined the civilian cadets—an army program offered at her school as an extracurricular activity. There she met Robert, a cadet a few years older than her. "He was the first guy I ever fell in love with," she says. Robert suggested that they leave Winnipeg and move across the country to his hometown of Bancroft, Ontario. To her parents' chagrin, Jane agreed to the plan. The couple stayed with Robert's parents for the summer, then moved on to Quebec, where Jane soon learned that she was pregnant. Robert talked about marriage, but then he started taking off. "He'd go out for cigarettes and he wouldn't come back for six hours," Jane recalls. One day she saw him on the street holding hands with another girl. Shortly after that, she left and returned home to Winnipeg on the train.

Her parents were furious to learn that she was pregnant, but she was jobless and broke and had no choice but to stay with them. Her daughter was born in 1982. Jane was twenty years old. She was an excellent mother, surrounding her daughter with the love that she felt she had never received from her own parents. Jane eventually moved out to a small apartment in the city's West End, where a friend of a friend introduced her to Dean, a handsome, dark-haired young man who worked as a security guard. They started dating, but he was too young to settle down, even when Jane discovered that she was pregnant again. Their daughter was born in 1984. Dean helped out financially when he could, but his visits gradually grew less frequent and finally stopped altogether. With two infants at home, Jane could not work, but eventually she got a job through a government program and started making money. Life was looking up. Then she met a young man who lived across the way. His name was Raymond. "Our apartment block was right across from each other," Jane says. "He said, 'If you ever want to use my washer and dryer . . .'" Jane took him up on this offer, and more besides.

"I'm not proud of it," Jane says. "But I was really looking for love in all the wrong places. I wanted a relationship. I wanted someone to love me." When Raymond learned that Jane was carrying her third child, he told her about his "common-law wife" who happened to be returning soon from British Columbia. "That's how I lost Raymond," Jane says. Her son was born in the early spring of 1988. Jane was at the lowest point in her life.

Three weeks after her son's birth, Jane got a call from her mother. Anne had some news. Lately she had been keeping house for a young woman recuperating from surgery. Anne had mentioned to this woman Jane's difficult situation, saying that she would probably never find someone to marry her now that she was saddled with three children. The woman had mentioned that she knew a young man who might like to meet Jane: he was her brother-in-law, the identical twin of her husband, Brian.

Jane had little hope for this long-shot matchmaking effort,

but she gave the woman a call. Brian's wife told Jane all about David's accident and how he had received a substantial sum of money as a settlement. "She said he's got this *van* and a *convertible*. I said, 'Does it really matter how much money he has or what he has between his legs? If he's not good to me or the kids, he can go his own way.'"

The two women arranged a day when Jane would go to Brian's house and meet David. The two hit it off right away. David, who was probably the more nervous of the two, says, "She had such a true heart."

The foursome made plans for a double date and that weekend went to a restaurant. At the end of the night, David held Jane's hand, and they made a date to meet each other alone. Soon they were dating regularly, and as they fell increasingly for each other, David began to worry about when, and how, to tell Jane about his injury. He finally got up the nerve one day while they were driving in his van up to his cabin in the woods. He had not got more than a few words out when Jane stopped him. She told him she already knew, and she didn't care about it. "She said that she had known all that time and she didn't want to tell me because she figured it would bother me," David recalls. "That's when I knew it was the real thing; I knew that she cared for *me*."

Asked her feelings about knowing her husband was raised to age fifteen as a girl, Jane treats it as a fact less to be marveled at than one to inspire outrage. "When I saw those pictures of him as Brenda, I just shook my head and thought, Poor child. He didn't look like a girl to me. He looked like Dave. I thought, Going to school must have been the hardest thing."

In the fall of 1989, they moved into an apartment together. David's phalloplasty allowed him to have sex with Jane. "You know how it is when you get into a relationship," Jane laughs. "You do it a lot in the first year."

David sold his "Shaggin' Wagon"—emblem of the reckless, oats-sewing youth that he had never actually had. With the money, he bought a diamond ring.

"I remember," Jane says, "he came into the bedroom and he said, in a very serious voice, 'I want to talk to you.' We were sitting on the bed. He took out this box and opened it up. There was a ring inside. My eyes were like saucers. He said, 'Will you marry me?' "

On 22 September 1990, two years and four months after they first met, David Peter Reimer and Jane Anne Fontane were married at a ceremony in Regents Park United Church in the city of Winnipeg. Jane's two daughters were bridesmaids. David wore a white tuxedo; Jane wore a white dress. Standing before the congregation of some one hundred and thirty guests, made up of friends and family, on an unseasonably warm fall morning, David and Jane spoke the vows that they had written for one another.

"Jane," David said, "I take you to be my wife; to laugh with you in joy, to grieve with you in sorrow, to grow with you in love, to be faithful to you alone, as long as we both shall live."

And Jane said to him, "David, I choose you to be my life's partner. I promise to respect you, to encourage you, to forgive you and instill hope in you. I give you my love for this day, and for all the days to come."

PART THREE

As Nature Made Him

13

=

KEITH SIGMUNDSON REMEMBERS his discomfort on seeing the advertisement. It appeared, he says, sometime in the
1980s in an American Psychiatric Society newsletter, and it
said: "Will whoever is treating the twins please report."
Below this entreaty was a name and address: Dr. Milton Diamond, University of Hawaii-Manoa, John A. Burns School of
Medicine, Honolulu.

"I saw it," Sigmundson says, "but I couldn't bring myself
to answer."

In the ten years that had elapsed since Brenda's switch to
David in 1980, Sigmundson had toyed with the idea of publishing the true outcome of the case. He hadn't done it, and for
a very simple reason. "I was shit-scared of John Money," he
admits. "I didn't know what it would do to my career." It had
been one thing for Sigmundson to cooperate with the BBC
documentary and appear as an unidentified psychiatrist speaking about "difficulties" in the twin's psychological adjustment.

It would be quite another to challenge a man of Money's power directly by publishing a signed article describing how his most widely publicized and influential case had failed from the outset. Sigmundson put the idea out of his head. Diamond's ad was an awkward reminder. At first, Sigmundson almost answered it, but he had resisted the urge.

Money himself also mastered any urge he might have felt to publicize the case's outcome. After his encounter with the BBC reporters in October 1979, he dropped all direct references to the case from his published papers, books, and public lectures. To many in the field of sexual development, his sudden silence on the subject was perplexing.

Virginia Prince, a pioneering transvestite activist who founded the first magazine for cross-dressers, *Transvestia*, in 1960, was one of those curious about the fate of the case, since it had played a significant role in her own acceptance of herself and her sexuality. Born a male, Charles Prince was in his early teens when he first started dressing in women's clothes for erotic gratification—a clandestine activity that continued even after his marriage and the birth of his son. In his forties, Prince began to live full-time in the role of a woman, divorcing his second wife and changing his name to Virginia. Though a fully "out" transvestite by the time she learned of the twins case, Prince says that the story of the sex-changed baby nevertheless had a profound impact on her.

She first learned of it at a meeting of the Society for the Scientific Study of Sex (or Quad-S), an association of sex researchers and activists for which Money served as president for two years in the early 1970s. It was at a November 1972 Quad-S meeting in Palm Springs, California, that then-president Money gave a sneak preview of the twins case—one month before its wider unveiling at the American Association for the Advancement of Science in Washington, D.C.

"John presented pictures of the twins," Prince recalls. "One photograph was of the two kids playing. The girl had a bow in her hair and was wearing a little dress; she's sitting in the front of a wheelbarrow, and her brother is driving her around. The other was a portrait—a snapshot, but it was posed. The little boy's got kind of a scowl on his face, and he's not attempting to make a good showing. But the little girl is sitting up straight and smiling and looking at the camera, just as if she's saying, 'I'm happy as a clam.' That's the picture that got stamped on everybody's mind."

For Prince, Money's twins case was encouraging proof that sex and gender were not an immutably preordained biological phenomenon. At every subsequent meeting with Money, Prince asked for an update on the twins' progress. Ordinarily Money was happy to oblige. "He was very upbeat and happy with the results and proud [of] what was done," says Prince. At a lecture in Los Angeles a decade later, that attitude had changed, Prince recalls. It had been some time since she had heard Money comment on the case, and Diamond's paper on the BBC's investigation was yet to be published. "I asked him, 'Whatever became of the twins?'" Prince says. "He wasn't very forthcoming. He seemed to be a little bit put out that I should ask the question. He was very short about it."

At Johns Hopkins, Money grew similarly tight-lipped about the case. Deflecting questions about the twins when the topic arose on the wards or in his classroom, Money told especially inquisitive students and colleagues that he had "lost track" of the experiment after a "media invasion" by reporters. "He said that this family had been victimized by the BBC," says Money's former student Howard Devore, and that "the family and the case had been irreparably damaged as a result." Money offered a similar explanation for his silence to his protégé, Dr. June Reinisch, a psychologist who had become head of the Kinsey Institute after studying with Money in the 1960s. "[He said] he'd been cut off from the family because they had somehow blamed him for the BBC," Reinisch says.

This interpretation of events does not accord with Janet Reimer's recollections. She says that even after the BBC's visit to Winnipeg, she stayed in touch with Money. "I wrote him a letter about David switching back to being a boy and about what was happening in his life," Janet says. "David had got his [settlement money from the hospital] by then. He was already dating girls—maybe not dating, but hanging out with girls." Janet says that Money wrote back. "He said he would like to hear from David and Ron. And I wrote him a letter stating the truth: 'Ron and David do not wish to communicate with you.' I said, 'I'm telling you that as a friend; I don't want to give you a feeling of rejection, but they just don't want to deal with you.'" Janet says that Money retained a studious neutrality in his letters about the news that Brenda was now David. "He never let on that he was disappointed," Janet says. They continued to exchange letters sporadically through the 1980s. Money wrote to Janet about a trip he had made to Zimbabwe; he informed her of his bout with prostate cancer; and he mentioned the removal of his Psychohormonal Research Unit from the main campus of Johns Hopkins in 1986.

Apparently forgetting this exchange of letters, Money continued to insist to his scientific, academic, and medical colleagues that the case was "lost to follow-up"—a surprising claim if for no other reason than that the Reimers continued to live in the very house that Money had visited in 1979 and with the same telephone number that he had used to contact them.

While Money refrained from mentioning the twins case directly in his public statements after 1980, he continued to lecture on the efficacy of infant sex reassignment for boys with no penis, and Johns Hopkins Hospital continued to perform the procedure, even when an alternative treatment was developed in the mid-1970s by Dr. Mel Grumbach at the University of California, San Francisco, for cases of boys born with small penises. He discovered that he could increase phallus size in babies born with micropenis by giving injections of testosterone to the organ shortly after birth. In patients whose bodies were responsive to the hormone, the penis could be made to grow to a length that permitted standing urination and conventional copulation.

Grumbach spoke about the procedure at the Seventh Annual Birth Defects Institute Symposium on Genetic Mechanisms of Sexual Development, held in November 1976 in Albany, New York. He was surprised to discover that he could not earn an endorsement for the procedure from Johns Hopkins, still the single most influential hospital for intersex treatment. The meeting's chairman, Johns Hopkins pediatric endocrinologist Robert Blizzard (who had worked as a consultant to Money on Bruce Reimer's conversion to Brenda back in 1966), articulated the hospital's decision not to adopt the California team's treatment plan. "I think that we will be able to answer the question concerning the preference of rearing of those with [micropenis] in a few years—although not immediately," Blizzard said in his closing remarks to the symposium. "I believe that Dr. Grumbach's group on the West Coast is going to do what they believe is correct; namely, raise these children as boys; and our group on the East Coast are going to do what they think is correct; namely, raise these children as girls."

John Money was Johns Hopkins's most tireless promoter of that decision in the years that followed, stating in interviews, speeches, books, and papers that sex reassignment to girlhood was the sole option for baby boys with micropenis—or boys

203

who, like David Reimer, had lost their penis to injury. At a meeting of the National Institute of Child Health and Human Development in September 1987, Money mentioned such infant sex changes as being among his most important clinical contributions to medical science. The occasion for Money's comments was a ceremony at which he was being honored as one of four scientists in the country who had, for twenty-five consecutive years, been funded with taxpayers' money by the National Institutes of Health. "In syndromes of male hermaphroditism and micropenis, and in cases of *ablatio penis* from circumcision trauma, when there is insufficient phallic tissue for surgical reconstruction of an adequate urinary and copulatory penis, a baby may be assigned and clinically habituated as a girl," Money told the NIH audience in his acceptance speech. "In adulthood, comparison of such cases with those living as males without a penis shows a higher prevalence of satisfactory outcome in those living as females."

Money's comments were curious for at least two reasons: (1) no systematic follow-up studies had ever been published by Money, or Johns Hopkins, that demonstrated the prevalence of this satisfactory outcome, nor have there been in the years since Money's remarks; and (2) at the time that he described the ability of doctors to successfully change the sex of *developmentally normal boys* to girls in cases of penis loss, the only such experiment he had followed from babyhood to adulthood was that of Brenda Reimer—an experiment that had failed fully seven years earlier, when Brenda had become David.

14

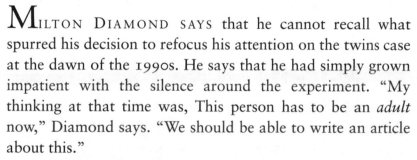

MILTON DIAMOND SAYS that he cannot recall what spurred his decision to refocus his attention on the twins case at the dawn of the 1990s. He says that he had simply grown impatient with the silence around the experiment. "My thinking at that time was, This person has to be an *adult* now," Diamond says. "We should be able to write an article about this."

Further incentive for returning to the subject was soon provided by Money himself, who in 1991 published *Biographies of Gender and Hermaphroditism in Paired Comparisons*, a career-summing monograph on his forty years of work at the Psychohormonal Research Unit. Presenting his largest collection to date of "matched pairs" the book was Money's latest defense of his theory that social learning overrides biological imperatives in the shaping of human sexual identity. Missing from the text was mention of the definitive test for his thesis—his ultimate matched pair: the sex-changed twin and her brother. In the book's introduction, Money explained the mysterious absence of the case from this otherwise comprehensive volume—an absence that he insinuated owed something to the machinations of his longtime challenger, Milton Diamond.

"On the international academic scene," Money wrote, "doctrinal rivalry regarding the origins of gender identity led to an alliance with an unscrupulous media"—here he inserted a parenthetical reference to Diamond's 1982 paper on the troubled case—"that prematurely terminated a unique longitudinal study of identical twins. A BBC crew of television sleuths, incited by the prospect of airing a doctrinal dispute, traced the whereabouts of the twins and their family and unethically invaded their privacy for programming purposes." Money provided no information about Brenda's 1980 decision to become David, and this brief reference, with its hint that Diamond was somehow connected to the case's premature termination, marked Money's last published comment on the case.

Perhaps understandably, Diamond was disinclined to allow this innuendo-steeped passage to stand as the final historical word on the twins experiment. That the academic community at large accepted Money's version of events was clear from yet another book published that year: *John Money: A Tribute*, a collection of essays written on the occasion of Money's seventieth birthday. Replete with paeans to Money's scholarship from longtime acolytes, including Anke Ehrhardt and June Reinisch, the volume also included a fulsome tribute from Dr. John Bancroft, a psychiatrist and clinical consultant at the Royal Edinburgh Hospital in Scotland, who is now director of the Kinsey Institute. A behaviorist who was a believer in the primacy of rearing over biology in sexual orientation, Bancroft had taken this nurturist view to its logical conclusion in his clinical work. As a sex therapist in Great Britain, he had experimented with trying (in vain) to convert adult homosexuals to heterosexuality through aversion therapy. In his tribute to Money, Bancroft referred with tart disapproval to the "recurring attack from Diamond" and went on to cast doubts on the veracity of the information Diamond had reported from the BBC about the psychological difficulties Brenda had suffered.

"Money has reported her development at various stages, consistent with his theoretical expectations," Bancroft wrote. "However, since the prepubertal stage, the scientific community has received no further authoritative reports, but rather rumors (*not* from Money) of troubled developments." He moved on to defend Money's decade-long silence on the case, casting it as evidence of Money's scrupulous care for the emotional health of his research subjects. "In a case such as this," wrote Bancroft, "when the attention of the scientific community (and in this case the media also) is focused on a particular individual, it is easy to see the need to withdraw and be silent to protect that individual; it must be extremely difficult to be the living test of a controversial theory!"

With his own academic integrity now being questioned, Milton Diamond did not have the luxury of withdrawing and being silent. Since the late 1970s and through the '80s, he had made periodic inquiries (and placed at least one ad) seeking information from endocrinologists and psychiatrists about the case. But now he resolved to redouble his efforts to learn the fate of the twin.

Through the BBC, Diamond found the name of a psychiatrist who had worked on the case—Dr. Doreen Moggey. That spring, he called her.

It had been fourteen years since Moggey terminated therapy with Brenda. She told Diamond that she did not know the final outcome of the experiment. She did, however, offer to give Diamond a phone number for the man who had overseen Brenda's psychiatric treatment: Keith Sigmundson.

"I remember the first words Sigmundson said to me when I called," Diamond recalls with a chuckle. "It was to the effect of 'I was wondering how long it would take for you to get here.'"

By that time Sigmundson was living in Victoria, British Columbia, where he had become head of the province's Division of Child Psychiatry. "Mickey said, 'Keith, we *gotta* do this,'" Sigmundson remembers. At first Sigmundson tried to beg off, but Diamond, he says, "kept on badgering me a little bit."

As someone who had seen firsthand the results of a reportedly successful sex reassignment, Sigmundson was inclined to agree with Diamond's thesis that the procedure of turning baby boys into girls was wrongheaded. Still, Sigmundson had been warned by colleagues that Diamond was a "fanatic" with an ax to grind. Further conversations with Diamond and a reading of his journal articles convinced Sigmundson otherwise. "I came to see that Mickey is a serious researcher and a caring guy who really believed that Money's theory had caused—and was continuing to cause—great harm to children." Sigmundson agreed to contact David Reimer and ask if he would be willing to cooperate with a follow-up article on his life.

"I wasn't sure what it was all about," David says about the call he received that spring from Sigmundson. At that time David had been married for less than a year and wanted nothing more than to put his tortured past behind him. Sigmundson was persistent, however, and David finally agreed to meet Diamond and see what happened.

Diamond flew to Winnipeg to meet David. Over lunch at a local diner, David learned for the first time about his own fame in the medical literature and how the reported success of his case stood as the precedent upon which thousands of sex reassignments had since been performed—and continued to be performed. "'There are people who are going through what you're going through every day,' " David recalls Diamond telling him, " 'and we're trying to stop that.' "

David was staggered. "I figured I was the only one," he says. "And here Diamond tells me they're doing all these surgeries based on *me*. That's why I decided to cooperate with Mickey." And there was another reason: David sensed in Diamond one of those people whose response to his sufferings was not purely detached and clinical. "When I told him a few things about my life," David says, "I saw that Mickey had tears on his cheeks."

Over the course of the following year, David and his wife and mother recounted to Diamond and Sigmundson the story of David's harrowing journey from boy to girl and back again. Using these interviews plus the detailed clinical records that had accumulated at the Child Guidance Clinic, Diamond set out, as the paper's lead author, to write up the results. He had promised the Reimers anonymity, agreeing to obscure their location, to omit the names of the local physicians, and to refer to David by pseudonym—or rather pseudonyms, since Diamond was faced with the narrative problem of retelling David's double life as both he and she. He settled on the solution of calling David variously Joan (for when he was Brenda) and John (following his switch back to his genetic sex). Only in a conversation with me two years later did Diamond notice that he had bestowed on John/Joan the Christian names of Money's two most important collaborators: Drs. John and Joan Hampson—an act that Diamond assured me was purely unconscious.

Written over the winter of 1994, the paper cast David's life as living proof of precisely the opposite of what Money had said it proved. Citing the Kansas team's classic work from the late 1950s, Diamond wrote that David's case was evidence that gender identity and sexual orientation are largely inborn, a result of prenatal hormone exposure and other genetic influences on the brain and nervous system, which set limits to the degree of cross-gender flexibility that any person can comfortably display. Diamond argued that while nurture may play a role in helping to shape a person's expressed degree of masculinity or femininity, nature is by far the stronger of the two forces in the formation of a person's private inner sense of self as man or woman, boy or girl.

Powerful as the paper was in presenting anecdotal evidence of the neurobiological basis of sexuality, it also served as a clear warning to physicians about the dangers of surgical sex reassignment for all newborns—not just those like David who are born with normal genitals and nervous system. Diamond argued that the procedure was equally misguided for intersexual newborns, since physicians have no way of predicting in which direction the infant's gender identity has differentiated. To change such children surgically into one sex or the other, he argued, was to consign at least half of them to lives as tortured as David's.

Accordingly, Diamond and Sigmundson offered a new set of guidelines for management of babies with ambiguous genitalia. Recognizing that a child must be raised as either a boy or a girl, they recommended that doctors continue to assign a firm sex to the baby—but only in terms of hair length, clothing, and name. Any irreversible surgical intervention, they said, must be delayed until the children were old enough to know, and be able to articulate, which gender they felt closest to. Or as Diamond put it to me, "To rear the child in a consistent gender—but keep away the knife."

Diamond was aware that writing the paper would inevitably raise the specter of a personal vendetta against Money. To minimize this danger, he removed from David's quoted utterances all reference to the famous psychologist. "In fact," Diamond says, "Money's name is only mentioned once.

I didn't want it to be an argument ad hominem. I wanted it to be a theoretical discussion."

Nevertheless it took Diamond and Sigmundson two years to find someone willing to publish their paper.

"We were turned down by all these journals that said it was too controversial," says Sigmundson. "*The New England Journal, The Journal of the American Medical Association.*" The article was finally accepted by the American Medical Association's *Archives of Pediatrics and Adolescent Medicine* in September 1996, with publication set for March 1997. In the intervening months, Diamond and Sigmundson felt considerable apprehension as they waited for their bombshell to go off. "We were basically telling all these physicians that they'd been doing the wrong thing for the past thirty years," Sigmundson says. "We knew we were going to be pissing a lot of people off."

Some critics, as expected, attempted to dismiss the paper on the grounds that Diamond was simply using David's history to embarrass a scientific rival, but at least one physician who saw a prepublication copy of the paper was inclined to agree strongly with its conclusions. Dr. William Reiner had two years earlier launched the first comprehensive long-term follow-up study of patients who had been sex reassigned. Trained as a pediatric urologist, Reiner had actually spent the first eighteen years of his medical career in California performing "normalizing" genital surgeries on intersexual children. It was early in his career that Reiner had his first glimmer of doubt about the Johns Hopkins treatment model. "I got babies and two-year-olds and four-year-olds and eight-year-olds and sixteen-year-olds," he says. "So I really saw a longitudinal view of all these urological conditions—all these birth defects—and I was therefore able to visualize in a relatively short period of time the kinds of effects that these conditions have on the lives of these kids and their families." Then, in 1986, Reiner met a patient who changed his life.

She was a fourteen-year-old girl—a Hmong immigrant—who had announced that she was dropping out of high school because she was "not a girl." To all outward appearances an anatomically normal female, she had nevertheless always rejected girls' play and had insisted on wearing gender-neutral clothes. At puberty she had arrived at the unshakable conviction that she wanted to change sex and live as a male. Referred to Reiner to discuss the possibility of reconstructive surgery, she was threatening suicide unless her wishes were met.

"I had a complete medical workup on the child done," says Reiner. Tests revealed that "she" was biologically a he—a 46XY male who suffered from a rare chromosomal condition that prevents masculine differentiation of the genitals. Reiner performed sex change surgery, after which the former girl effortlessly assumed the sex written in his DNA. The case convinced Reiner of what he had suspected for years: that the biological underpinnings for psychosexual identity are not so easily overridden by social and environmental rearing as he (and every other pediatric urologist, endocrinologist, psychiatrist, and psychologist) had been taught. This further forced Reiner to the uncomfortable conclusion that he had been doing the wrong thing in his surgical career in helping to steer intersexual children into one sex or the other at birth. In a 1996 edition of the *Journal of the American Academy of Child and Adolescent Psychiatry*, Reiner published a paper on the Hmong case, along with a warning to his fellow physicians about the long-accepted theory that rearing prevails over biology in shaping human sexuality.

Reiner also did something else. After eighteen years as a surgeon, he put down his scalpel. He began to retrain as a child psychiatrist specializing in psychosexual development and intersexual conditions. In 1995 he was hired by Johns Hopkins as an assistant professor in psychiatry. There he launched his study on the long-term psychosexual implications of sex reassignment. Reiner set out to follow sixteen patients, focusing particularly on six genetic males who were born without penises and as a result were castrated and raised as girls. Two years into his study, he noted that all six sex-changed boys were closer to males than to females in attitudes and behavior. Two had spontaneously reverted to being boys without being told of their male (XY) chromosome status.

"These are children who did not have penises," Reiner told me, "who had been reared as girls and yet *knew* they were boys. They don't say, 'I wish I was a boy' or 'I'd really rather be a boy' or 'I think I'm a boy.' They say, 'I *am* a boy.'" Reiner stressed the parallels between the children he was studying and David Reimer, who also "knew," despite his rearing as Brenda, that he was not a girl. Reiner wrote a supportive editorial in *Archives of Pediatrics and Adolescent Medicine* to accompany Diamond and Sigmundson's John/Joan paper.

Today Reiner says that both David's case and the trend in his own study support the findings that have emerged on the primacy of neurobiological influences on gender identity and sexual orientation. He cites the now-classic study done at Oxford University in 1971, which showed anatomic differences between the male and female brain in rats. Six years later, at UCLA, researchers narrowed these differences to a cluster of cells in the hypothalamus. A study done in the mid-1980s in Amsterdam located the corresponding area in the human hypothalamus, noting that it is twice as large in homosexual as in heterosexual men. Further studies have supported this finding. In 1993 and again in 1995, researcher Dean Hamer announced that in two separate studies of gay male brothers, he had found a certain distinctive pattern on their X chromosomes. The finding suggested that sexual orientation may have a genetic component.

Although Hamer's studies have failed to be replicated by other scientists, few sex researchers today dispute the mounting evidence of an inborn propensity for acting as, and inwardly identifying with, a particular sex. "It's quite clear that the vast majority of boys born with functioning testicles have masculine brains," Reiner says. He endorses Diamond and Sigmundson's recommendation to delay surgery in cases of penile loss or intersexuality and to impose only a provisional assignment that can be changed should the child voice a strong desire to live as the other sex. Reiner suggests that this treatment model is diametrically opposed to the one pioneered at Johns Hopkins by Money and his colleagues, in which a sexual identity is imposed on a child through unshakable fiat of physicians, and any doubts or confusions the child may express about the assignment are denied by caregivers. Reiner says that on the basis of David's case and the others he has studied, the decades-old Johns Hopkins treatment model needs to be reevaluated. "We have to learn to listen to the children themselves," he says. "They're the ones who are going to tell us what is the right thing to do."

Before Diamond and Sigmundson's journal article appeared in the *Archives of Pediatrics and Adolescent Medicine* in March 1997, the American Medical Association's public relations department alerted the media that something explosive was coming. On the day of the article's publication, the *New York Times* ran a front-page story headlined SEXUAL IDENTITY NOT PLIABLE AFTER ALL, REPORT SAYS, in which writer

Natalie Angier described David's life as having "the force of allegory." Twenty-four years after publishing news of the case's success, *Time* magazine ran a full-page story declaring, "The experts had it all wrong." Similar news accounts appeared around the world—and soon Diamond and Sigmundson were deluged with calls from reporters in several countries seeking an interview with the young man now known simply as John/Joan.

David agreed to appear on two television newsmagazine programs. He was shown in darkened silhouette on ABC-TV's *Primetime Live* and with his face obscured in a Canadian Broadcasting Corporation documentary. It was during the latter taping, which took place in New York City in June 1997, that I was introduced to David by Diamond and Sigmundson. The researchers had passed along to David the names of the many reporters who had requested an interview with him, but David (a rock'n' roll fan) had chosen the reporter from *Rolling Stone*.

At that first meeting with me, David was nervous and guarded. He explained that his childhood had made it difficult for him to trust strangers, but later, over a beer at the Hard Rock Cafe, he grew more relaxed. He spoke about how his parents and brother had been crucial supports in a childhood that he described as "a pit of darkness." I soon learned that a formidable sense of humor had also played a role in his survival. Describing the physical differences between himself and his heavier, slightly balding twin, he shouted over the pounding music, "I'm the young *cool* Elvis. He's the fat *old* Elvis."

But the strongest impression I was left with was of David's unequivocal masculinity. His gestures, walk, attitudes, tastes, vocabulary—none of them betrayed the least hint that he had been raised as a girl. And indeed, when I asked whether he thought his extraordinary childhood had given him a special insight into women, he dismissed the question. David had apparently never *been* a girl—not in his mind, where it counts. He insisted that his conversion from Brenda at age fourteen marked nothing more than a superficial switch in name—as if the double mastectomy, two phalloplasties, and lifelong course of testosterone injections he needs to compensate for his castration were mere details. "I've changed over," David said, "but mainly by name. The rest was all cosmetic. I just had repaired what was damaged. That's all."

Through the summer and fall of 1997, David's story continued to receive media coverage. With this coverage, another set of voices in the debate over the heretofore unexamined practice of infant sex reassignment began to be heard. These were the voices of those intersexes born after the publication of Money's 1955 protocols—people in their thirties and forties who as babies had undergone normalizing genital surgeries and sex assignments and who were now ready to speak on the record about their lives.

They had already begun to emerge as a public voice four years earlier, largely through the efforts of one person: a San Francisco–based activist named Cheryl Chase, who had been lobbying for changes in intersex treatment since the early 1990s. "I wasn't getting very far," admits Chase, a short-haired woman with a dry, rational manner that belies the passion driving her. "That changed overnight when the John/Joan case blew up."

At her birth in suburban New Jersey in 1956, Chase presented a classic case of ambiguous genitalia. Instead of a penis and testicles, there was a somewhat vaginalike opening behind her urethra, and a phallic structure of a size and shape that could be described as either an enlarged clitoris (if she were assigned as a girl) or a micropenis (if a boy). After three days of deliberation, the doctors told Chase's parents that their child should be reared as a boy. She was christened Charlie. But a year and a half later, her parents, still troubled by Charlie's unusual appearance, consulted another team of experts. They reassigned her as a girl and told her parents that she would grow up to be a happy, healthy, normal woman. Her parents changed her name from Charlie to Cheryl, and the doctors removed her clitoris.

Like David Reimer, Chase was then raised without knowledge of her true birth status. Thus, like David, she experienced a childhood punctuated with mysterious, unexplained surgeries and regular genital and rectal exams. Also like David, she grew up confused about her sex. "I was more interested in guns and radios," Chase says, "and if I tried to socialize with any kids, it was generally boys, and I would try to physically best my brother. I didn't fit with boys or girls, I was stigmatized and ostracized by my peers, and picked out for teasing all the time." At age ten, Cheryl's parents brought her to a psychiatrist, who attempted to prepare Cheryl for her role as wife and mother. As a preadolescent, she recognized that she was erotically attracted to females.

By age nineteen, Chase had done some of her own medical sleuthing and understood that she had been subjected to a clitorectomy as a child, and she began to search for her medical history. She was thwarted by her doctors, who refused to reveal the circumstances of her birth. It took three years for her to find a physician willing to disclose her medical records. It was then that Chase read that doctors had labeled her a "true hermaphrodite"—a term that refers to people whose gonads possess both ovarian and testicular tissue. This was also when she first learned that she had spent the first eighteen months of her life as a boy named Charlie, and that her parents, doctors, aunts, uncles, grandparents, and family friends had conspired to keep this secret from her. She also learned that the operation she had undergone at age eight (to relieve "stomachaches") had actually been to cut away the testicular part of her gonads.

Horrified and angered at the deceptions perpetrated upon her and aggrieved at the loss of her clitoris, which had rendered her incapable of orgasm, Chase began to seek out others like herself. Through letters to the editors of medical journals and magazines, news articles, listings with crisis hotlines, and ultimately on a website, she established a network of inter-sexes in cities across the country. In 1993, she dubbed the group the Intersex Society of North America, a peer support, activist, and advocacy group. By mid-1999, Chase had been contacted by nearly four hundred intersexes from around the world—many of whom told stories almost identical to her own.

To meet Chase and members of ISNA—as I did in the spring of 1997, when they held a peaceful demonstration outside Columbia Presbyterian Hospital in New York, where Chase's clitoral amputation was performed—is to enter a world where it is impossible to think of sex with the binary boy-girl, man-woman distinction we're accustomed to. There was Heidi Wal-cutt, genetically male with an XY chromosome constitution but born with a rudimentary uterus, fallopian tubes, internal sperm ducts, and a micropenis, who describes herself as a "true American patchwork quilt of gender." There was Martha Coventry, born with an enlarged clitoris but a fully functioning female reproductive system, who is the mother of two girls. There was Kiira Triea, assigned as a boy at age two, who did not learn of her intersexuality until puberty, when she began to menstruate through her phallus. At that stage she was referred as a patient to Dr. Money at the Psychohormonal Research Unit, where she was treated from age fourteen to seventeen, in the mid-1970s, concurrent with Brenda Reimer.

Kiira and David have never met or spoken, but Kiira's

story bears striking parallels to his. She describes how Dr. Money, evidently attempting to ascertain whether she possessed a male or a female gender identity, questioned her about her sex life—in the frank language for which he is well known. "Have you ever fucked somebody?" she remembers Money asking. "Wouldn't you like to fuck somebody?" She also describes how Money showed her a pornographic movie on a projector he kept in his office. "He wanted to know who I identified with in this movie," she says.

Contrary to Money's claim that an intersexual baby reared as a boy will develop an unequivocal male gender identity, Triea's sexuality and sense of self proved to be far more complicated than that. At fourteen she agreed to undergo feminizing surgery at Johns Hopkins to simulate female genitals, but when she became sexually active for the first time at age thirty-two, her erotic orientation was toward women.

The other intersexes in Chase's group show a similarly complex sexuality. Max Beck was first assigned and reared as a girl named Judy. Despite strong masculine thoughts, inclinations, behaviors, and attitudes, Judy tried to stick with her assignment in order to placate worried parents and relatives, even going so far as to marry in her early twenties. But at age twenty-seven Judy left the marriage and divorced. At the age of thirty-two she stopped taking estrogen, changed her name to Max, and began taking testosterone by patch. Yet even today, Max resists the simple designation of male. "I have always felt—and continue to feel myself to be—intersexed," he recently e-mailed me. " 'Masculine' is simply a more comfortable compromise, testosterone a tastier hormonal cocktail than estrogen." Not all the intersexes who joined Chase were sex-reassigned as babies. Dr. Howard Devore, the psychologist who studied under John Money in the 1980s, was born in 1958 with acute hypospadius (a penis open from base to tip) and with undescended, underdeveloped testicles, but was raised as a boy. Beginning at age three months he endured some sixteen "normalizing" surgeries through childhood, aimed at giving him a cosmetically convincing penis. The experience, Devore says, was emotionally devastating— and wholly unnecessary. His genitals still do not resemble those of a normal male and the sole result of his constant hospitalizations is a psychological scarring far worse than he would have experienced had he been raised with counseling to accept his atypical genitals. Devore refrained from making this argument to Money. "I learned very early that if you choose to do battle with John," he says, "you have to deal with a very, very angry man who's going to make you feel horrible for challenging him." (Devore says that only in the wake of the "John/Joan" revelations has he felt emboldened to make his intersexuality public—and to openly challenge his former professor.)

Armed with her own story and those of her fellow intersexes, Chase began trying to alert the medical establishment to the dangers of the protocols for intersex management initiated by Johns Hopkins. ISNA's stated aim was to abolish all cosmetic genital surgery on infants—not simply the castration and sex reversal of micropenis boys. While Chase did not oppose life-saving corrective surgery on genitals, she denounced as "barbaric" all medically unnecessary cosmetic treatments on newborns that could have an irreversible effect on their erotic or reproductive functioning. And ultimately, she said, she wanted to "end the idea that it's monstrous to be different."

Chase found it more difficult than she had anticipated to gain an audience with influential people in the field—including John Money. "I've written him several times, politely, asking if he would clarify his position for us," Chase told me. "Each time he would return my letters with a note scribbled on the corner saying he doesn't have enough time to talk to me."

Chase also wrote to the American Academy of Pediatrics— an association with a membership of over fifty-five thousand doctors in the United States, Canada, and Latin America. The AAP has long endorsed Money's protocols for intersex treatment. "I write to inform you that many who have been treated according to the model you outline have found that the treatment itself has rendered our lives an ordeal," Chase wrote to the AAP in 1995. "We who are intersexual have been discussing our experiences through the Intersex Society of North America . . . and we find that the current model of treatment does nothing to discourage the shame and secrecy surrounding intersexuality. . . . We would love to open a dialog with you, and we encourage you to mention, when you teach about treatment of intersex, the existence of a vocal, organized population of intersexual former patients who oppose the current model."

The AAP did not respond to this letter. Chase wrote to them again in 1996 and again received no response. That October, Chase and other ISNA members held a demonstration at the AAP's national conference in Boston. Academy officials refused to meet with the protesters, but they did distribute a press release among the journalists and protestors at the demonstration. "The American Academy of Pediatrics, a voice for children for over 60 years, is aware of the concerns and sensitive to the needs of intersexuals," the statement read. It went on to say that the AAP would not change its stance on intersex treatment and cited Money's work from the 1950s to defend its position.

Chase also appealed to former Surgeon General Joycelyn Elders, who prior to her appointment to Clinton's administration had practiced for over twenty years as a pediatric endocrinologist in Arkansas, where she regularly applied Money's protocols for intersex management to ambiguously sexed new-borns. Elders never acknowledged Chase's letters.

In 1996, Chase did succeed in persuading the *New York Times* to write a feature article about the burgeoning intersex activist movement, but in the story members of the medical establishment refused to discuss ISNA's complaints. Dr. John Gearhart, head of pediatric urology at Johns Hopkins, dismissed the group as "zealots." In a conversation with me in the summer of 1997, amid the media storm generated by Diamond and Sigmundson's article on the failed twins case, Gearhart was more politic when addressing the issues raised by ISNA and David's case. While he insisted that sex reassignment remains a viable option for boys born with micropenis or who lose their penises to injury, he noted that advances in penile reconstruction made him more hesitant to recommend the procedure today. "If John/Joan happened today," he told me, "I would sit down with those parents and say, 'The child has testicles; it's a normal male child.' I would suggest that you *could* change the child's gender, but I would not recommend that, because reconstructive genital surgery has come light-years since John/Joan's accident."

Gearhart also said that advances in medicine render ISNA's concerns obsolete. "When these people in ISNA were operated on, twenty-five and thirty years ago, there weren't really children's reconstructive surgeons around," he said. "So most of [these babies] had their clitoris or their penis amputated. That was wrong, OK? *That* was wrong. But the surgeons didn't know any better. Nowadays, people in modern reconstructive surgery are not cutting off little babies' clitorises or penises, or anything along those lines." Gearhart said that modern microsurgery retains sensation.

To hear the back-and-forth exchanges of doctors like Gearhart and activists like Cheryl Chase is to be convinced that the issues involved will not be settled anytime soon. For instance, Chase flatly rejects Gearhart's claim that surgeons maintain clitoral sensation after reducing the organ's size. Gearhart meanwhile continues to reject ISNA's call for change in the current treatment protocols, insisting that scores of intersexes live happily in the sex assigned to them in infancy and that Chase and the members of ISNA represent only the "disgruntled" few—a charge to which Chase and other ISNA members take particular exception. They insist that silence among intersexual adults does not reflect happiness with the decisions made for them as babies, but is instead a symptom of the shame and secrecy that are the legacy of the current treatment methods.

"It goes back to being completely isolated as children," Heidi Walcutt told me. "*Knowing* that there's this difference, but being silenced and being shamed about it. Some people never get to the point where they start looking for answers—let alone step out as an activist against what was done to them." Chase adds that there is also a strong disincentive for intersexes to speak out, since doing so often means undergoing a traumatic confrontation with parents who authorized the surgeries in the first place. Chase points out that more than a few ISNA members find themselves estranged from their families.

It is obviously difficult for an independent investigator to verify either Gearhart's or ISNA's conflicting claims about the relative happiness of adult intersexes who decline to speak about their lives; they are by definition invisible. Asked to provide a satisfied patient, every pediatric specialist I contacted voiced the Catch–22 that they "lose track" of their patients after young adulthood. Gearhart added, "And the ones I do know just want to live their lives in privacy."

I was able to locate and speak to one intersex who is not a member of ISNA or any other activist group. She is notable in her own right in that in the late 1970s she was repeatedly cited

223

as a particularly successful example of an intersex who was sexually reassigned in infancy. Her case was featured not only on an Emmy-winning ABC-TV science series documentary, but also in the BBC's investigative report on the twins case. She is Paula, the former John Money patient whom Peter Williams and Martin Smith included for balance in their twins case exposé. Living anonymously now in the Northeast, but located through an Internet search engine that lists census records, Paula agreed to speak with me on the condition that I not use her last name or otherwise reveal her identity.

In a series of phone conversations and a five-hour in-person interview with Paula and her mother, I learned the circumstances of Paula's birth. They were in many respects strikingly similar to the stories I had already heard from Chase and her colleagues in ISNA. Born in September 1971, the second of three children, Paula presented ambiguous genitalia with a scrotum but no testicles inside and a small penis of mostly empty skin. The local doctors recommended assignment as a boy, saying that the penis would grow and that the testicles would descend over time. The baby was duly christened Michael Edward. But Michael's mother remained upset by her baby's appearance and continued to consult doctors over the first year and a half of her child's life. When Michael was eighteen months old, a neighbor in whom Michael's mother had confided the dilemma brought over the current issue of *Time* magazine, which carried a story about one of a pair of twin baby boys who had lost his penis to circumcision and was later turned into a girl on the advice of Johns Hopkins psychologist John Money. According to the *Time* article, the sex change had been a complete success. Michael's mother immediately wrote to Dr. Money, who replied promptly and advised that she bring Michael to Baltimore for immediate sex reassignment as a girl.

"Within two days," Paula's mother says, "I was on my way down there with my husband and my child."

On 23 February 1973, Michael was operated on at Johns Hopkins by Dr. Howard Jones, who established that the baby (like Cheryl Chase) had gonads containing both ovarian and testicular cells. Jones removed the undescended gonads (to prevent spontaneous masculinization at puberty) and reconstructed the external genitalia so that they would appear more feminine. Full excavation of a vaginal canal was to wait until the baby was in her teens. In the meantime it was arranged that Paula would return periodically to Johns Hopkins for counseling with Dr. Money. And indeed, Paula's mother brought her daughter back to see Dr. Money several times a year throughout her childhood. Since Dr. Money had often said that Paula was one of his best patients, it came as little surprise to Paula's mother when, shortly after her daughter's seventh birthday, Dr. Money asked if she would be willing to put Paula on television to discuss her successful sex reassignment. "I said, 'If it would help one other person,' " Paula's mother recalls, " 'then that's all I want from it.' "

The program was part of the ABC-TV science series *The Body Human*. The episode, entitled "The Sexes," featured a scene of Paula, a freckle-faced, short-haired girl, during one of her trips to the Psychohormonal Research Unit. With the camera keeping a studious distance from the fertility sculptures arrayed around Money's plant-festooned office, the famous psychologist was shown sitting at his desk, in shirtsleeves and tie. He asked Paula questions as she faced him from a large, afghan-covered armchair. In a flowered dress with lace collar, her fingernails painted bright red, Paula smiled warily as she haltingly answered Money's questions about marriage and career. Meanwhile the narrator explained in voice-over, "At the

Johns Hopkins Hospital, under the enlightened care of specialist Dr. John Money, careful attention is paid to nurturing Paula's image of herself as a girl, preparing her for all the complete experiences of womanhood."

Shortly after this program aired in May 1979, Money again asked Paula's mother to put her daughter in front of the cameras. This time the reporters were with the BBC. Again Paula's mother agreed, and Dr. Money made the preliminary arrangements with Williams and Smith to interview Paula's mother. The filming took place in early October 1979. According to Paula's mother this interview proved less gratifying than her encounter with the ABC-TV producers. Within hours of doing the interview, she received an agitated call from Dr. Money, who told her he had learned that the BBC reporters had an "ulterior motive" in making their documentary. Money wanted Paula's mother to pull out of the interview. When he learned that the reporters had already done the interview, he was irate. "He was absolutely furious with those reporters," she recalls. "*Furious.*" The imbroglio did nothing to mar the relationship between Dr. Money and Paula's mother. She continued to bring her daughter to see Money for regular follow-up visits until Paula was eighteen years old, at which point Paula underwent the final stage of vaginal surgery and stopped going to Johns Hopkins.

Today, at twenty-seven, Paula is a slim woman with blue eyes and tawny, straight, side-parted hair that falls to her waist. Dressed in jeans, a blue shirt, and platform open-toed sandals, she passes easily as a woman, albeit a boyishly figured one. Her small breasts and hip shape are maintained only through a regular lifelong regimen of estrogen ingestion. She takes care to pitch her raspy voice in the upper part of its register, but it does at times dip into lower notes than would ordinarily be expected in even a deep-voiced woman. Paula takes assiduous care of her grooming, lavishing great attention on her long mane of hair;

she pays once-weekly visits to the pedicurist and manicurist for the maintenance of her nails; and in the course of our conversation, she frequently refreshed her makeup with the skill of a trained cosmetician.

Despite these obvious outward efforts to enhance her public femininity, Paula says that privately she has no choice but to think about her medical condition every day. Like the other intersexes I spoke with, Paula's surgically created vagina is a daily reminder that she was not born a typical woman. "I don't look like everyone else does," she says. "Not at all. So of course you're always going to have a constant reminder." Asked if her vagina carries sensation, Paula drags on her cigarette. "There's always lack of sensation where there's scarring," she says. Given these realities, it comes as little surprise to hear Paula say that despite the best efforts of medical science, she has a constant sense of living "with a secret." Asked if there are any close friends to whom she has felt comfortable divulging her secret, Paula's face hardens and she chuckles with brittle cynicism. "You have very few friends in this world—trust me." She takes another drag on her cigarette. "Yeah," she continues in the same tone. "That *anyone* can trust. There are very few people in this world." She says that she had a boyfriend for six years in her late teens and early twenties in whom she confided her secret, and he was understanding. Since then, however, she has preferred to keep her condition, and the circumstances of her birth, to herself.

A virgin at age twenty-seven, Paula says that she has never felt any sexual attraction to women. When I asked about this, she cut me off before I had even finished the question. "Not at all," she said. "Never. Never, at all. Not at all." I asked if, while growing up, she had ever thought, "Maybe I'm a boy," but again she spoke before I could get the question out. "Never," she said. "Never at all."

Paula is open about her desire not to upset her mother by voicing doubts about the decisions made on her behalf as a baby. She insists that she is happy with the choice made to reassign her as a girl—or rather, she expresses the view that no other decision could have been made at the time. "I can't see things being any other way," she tells me. "You know?" She pauses, then resumes. "As far as anybody was concerned at the time, this was, like, the only way."

Like the other former research subjects of John Money to whom I spoke, Paula has vivid memories of her counseling sessions with him. She was shown pictures of men and women engaged in sexual intercourse and queried about the most private aspects of her inner self. "He asked questions that a six-, seven-, nine-, *ten*-year-old would not *ever* be asked," Paula recalls: about masturbation, her private sex fantasies, how to deflect lesbian advances from other girls. "He would *press* you for answers," Paula says. "He would sit there and *press* you and *press* you and *press* you. It was way too much for a child. I always said to my mother, 'I don't know why the hell I have to go to him.'"

Appearing on network television as the world's first openly intersexual seven-year-old was also, Paula says, "traumatic." When she discusses this aspect of her dealings with Dr. Money, Paula's vocabulary and tone of voice lose their quality of studious feminine poise. Her voice drops several tones lower as she spits out a stream of angry expletives. "All that TV bullshit was garbage," she snarled. "It was *bullshit*, it was traumatic. I mean you have to understand, at that time of my life I was in *grade* school. People in my class were asking me about it the next day. I was too young to make my own decisions, but if I had had the choice I never would have done it. But that's my mama," she adds with a forgiving but exasperated smile. "She loves John

Money. She would do anything he said. She thought she was doing good. Helping other mothers. But I think it was bullshit."

Paula's mother has never questioned the decisions she made on Paula's behalf, nor has she questioned John Money's handling of the case. To this day she refers to Money as her "savior" and speaks in only the most glowing terms about him and about her former son's conversion to girlhood. "Everything worked out fantastic with Paula," she told me, speaking in ecstatically upbeat tones in a phone conversation before I met her daughter. "She is full of life and full of fun. She has never confided in me of any worries. She's a character! A real party girl. Loves life, parties, going out, oodles and oodles of friends. The phone never stops." She described Paula as the very quintessence of femininity. "*Loves* being a girl. Loves to shop. Buys the most expensive clothes. And jewelry. Everything is top designer." She said that Paula had never seemed even slightly tomboyish as a child (an observation in contrast to the recollections of one of the ABC-TV producers, whose impression of the seven-year-old Paula was "That little boy stayed a little boy, no matter *what* they did to him").

Paula's mother does voice one small concern about Paula's life today: her daughter's single status. When Paula's parents brought her to Johns Hopkins, Dr. Money had specifically explained that sex reassignment was being done to ensure that Paula could one day marry, have a normal heterosexual love life, and have children by adoption. "Dr. Money looked that far ahead," Paula's mother marvels. Marriage is the one area where Dr. Money's prognostications have yet failed to materialize, but Paula's mother has not given up hope. "For me," she says, "closure will come when Paula gets married." Paula, watching as her mother speaks, takes a hard drag on her cigarette, then looks away.

Paula is dubious that she will ever marry. For one thing there is the delicate circumstance of her unusual genitals—which Paula feels is a severe stumbling block to physical and emotional intimacy with a partner. There is also the fact that Paula views marriage as an outdated institution. So for now, she continues to live at home with her parents. Her father, who suffers from severe clinical depression, confines himself mostly to his room. Most of Paula's dealings are with her mother. It is a close but emotionally complex relationship in which Paula is totally devoted to her mother, despite Paula's spates of brittle snappishness. Having heard many times during the course of her growing up about the severe trauma her birth caused her mother, Paula lives a life dedicated to minimizing any further emotional upset or unease that her existence might cause. Paula has thus told her mother that she will live at home forever to look after her and will "never leave her."

In the meantime Paula devotes herself to her work. Her choice of career, while not unchallenging, does not reflect her extraordinary intelligence. Tested at Johns Hopkins at age ten, Paula's IQ was 132, placing her in the top 2.2 percent of the population. At the time, Money's associate Gregory K. Lehne wrote in Paula's file that her "future academic planning can include college and professional training, with every expectation of success." And in fact Paula had once planned to become a lawyer. She set those ambitions aside in her junior year of college, after her mother mentioned her wish that Paula pursue nursing. Though Paula had always expressed an understandable aversion to all aspects of the medical profession, she nevertheless quit college and enrolled in nursing school. Currently Paula is a registered nurse and is working toward her master's in nursing. Her mother is "ecstatic" about these developments and boasts of how Paula "has not looked back since."

Paula seems to be a young woman determined not to look back. She says that she has no criticism of those intersex activists who are lobbying the medical profession for change, but she takes the position, for herself, that it is better just to get on with life and not stir up the past. When her mother left the house, Paula admitted quietly, "Maybe they *should* wait and give kids the choice about surgery." That, however, is an opinion she is unlikely to pass on to the medical profession. She has turned down a recent request from Johns Hopkins to participate in a follow-up study on sex reassigned patients. She simply does not want to relive her childhood—which she nevertheless insists was a perfectly happy one. But that some degree of unresolved emotion around her childhood might linger is perhaps suggested by the fact that Paula chose obstetrics and gynecology as her nursing specialty. Today she helps to deliver babies in the same small hospital where she was born twenty-seven years ago, as Michael Edward.

The medical establishment's refusal to listen to those intersexes who *have* elected to speak about their experiences is no surprise to Cheryl Chase. "Our position implies that they have—unwittingly at best, and through willful denial at worst—spent their careers inflicting a profound harm from which their patients will never fully recover," Chase once wrote. She says that she does not expect the medical establishment to change its practices unless forced. Chase plans to force them. "I think a context will open up for surgeons who keep doing this to be vulnerable to lawsuits," she told me. "But it's going to take a while to create that context. Right now we can't sue because it's standard practice, and parents give permission. The first thing we want to have happen is that when they make their recommendation to parents, they tell them it's experimental and there's no evidence that it works and that there's plenty of people who've had it done to them who are mad as hell."

There are other needs as well. Anne Fausto-Sterling, an embryologist at Brown University, says that the medical establishment will have to provide education and emotional support to help parents with the difficult task of raising an infant whose genitals are atypical. "At the moment there is no ongoing counseling done by people skilled in psychosexual development," Fausto-Sterling says. "If there was really a wholesale change in this, the medical profession would have to do something like what they've done with genetic counseling—which is to develop a specialty of people who would work with these families long term and help them resolve both emotional and practical questions. The practical questions are very real: What do I do when it comes to undressing in gym? How do I intervene with the school system? There's a different infrastructure that needs to be built and put into place. I think it's the responsibility of the medical profession to do it."

Perhaps the biggest change that will have to take place is in the medical profession's current view of what it means to be reared with ambiguous genitals, since the Money and Hopkins guidelines are predicated on the belief that such a childhood would be psychologically and psychosexually devastating. Studies that would prove the truth of this intuitive observation are hard to come by; case histories of children reared with ambiguous genitals are rare because so few intersexual newborns have avoided surgical intervention. In 1989 a study did appear in the *Journal of Urology* on the lives of twenty males with micropenis who were reared in their biologic sex. Drs. Justine Reilly and C. R. J. Woodhouse of St. Peter's Hospital and The Hospital for Sick Children in London described how these patients, who ranged in age from ten to forty-three years, had all formed healthy male gender identities and "participated in normal male activities in childhood and adolescence." They also reported that nine (75 percent) of the older patients were sexually active and that "vaginal penetration usually is possible but adjustment of position or technique may be necessary." The researchers drew two main conclusions: "A small penis does not preclude normal male role and a micropenis or microphallus alone should not dictate a female gender assignment in infancy."

Reilly and Woodhouse's study, however, looked at the lives of only twenty patients, all of whom had the same syndrome. A much more exhaustive study exists on the lives of untreated intersexes who display a much wider range of conditions than micropenis alone. Written before the advent of the 1955 protocols, it is a unique and fascinating monograph that reviews over two hundred and fifty cases of intersexes who received no surgical intervention as babies. Furthermore, the study directly addresses the question of how children fare when they grow up with genitals of the sex opposite to that in which they are reared. "Do [these people], with such manifest sexual problems to contend with, break down under the strain, as psychiatric theory may lead one to believe," asked the study's author, "or do they make an adequate adjustment to the demands of life?"

Far from manifesting psychological traumas and mental illnesses, the study showed, the majority of patients rose above their genital handicap and not only made an "adequate adjustment" to life, but lived in a way virtually indistinguishable from people without genital difference—a result that clearly amazed the study's author.

"One would not have been surprised had the paradox of hermaphroditism been a fertile source of psychosis and neurosis," the investigator noted. "The evidence, however, shows that the incidence of the so-called functional psychoses in the most ambisexual of the hermaphrodites—those who could not help but be aware that they were sexually equivocal—was extraordinarily low. The incidence of neurotic psychopathology of the classic types, sufficiently severe and incapacitating to be unmistakable, was also conspicuously low." The study pointed out that genital ambiguity led to a "disheartenment" of mood in some patients and a social "reticence" in others but went on to say of these individuals, "there was no evidence that their disheartenment or reticence ordinarily accumulated to the proportions of psychopathology, seriously impairing their ability to cope with the essential business of life"—such as completing their education, going to the office each day, and earning a living each week.

Of particular interest are the study's in-depth interviews with ten intersexes who received no surgery or hormone treatments until they were old enough to make their own decision. Their lives only strengthened the investigator's impression that the condition of the genitalia plays a strikingly insignificant part in the way a person develops a stable and healthy gender identity, not to mention a secure and confident self-image. One patient with an enlarged clitoris at birth did not have the organ surgically reduced until the age of twelve, yet her childhood with masculinized genitalia left no wound on her psyche and did not impair her sense of herself as a girl. "[O]ne appreciates her remarkable stamina and the self-reliant way in which she had consolidated it," the author noted. A second girl with a similar medical history demonstrated a marked "social deftness and complete poise" and, despite her mother's depressions, "had emerged more stable than her adult sister or brother." About another girl whose masculinized genitals were not surgically altered until she consented to it at age twelve, "one would not be justified in saying that she is different from scores of other adolescents." A boy with an untreated micropenis had married at age twenty-four and, the study reported, "is meeting life most successfully without any suspicion of psychopathology. . . . His life is an eloquent and incisive testimony to the stamina of human personality." A "true hermaphrodite" with a micropenis, split scrotum, and breasts at puberty lived as a male with no surgery to correct these anomalies. "The youth is another living testimony . . . to the stamina of human personality in the face of sexual ambiguity of no mean proportions." A seventeen-year-old boy whose micropenis went untreated through childhood and adolescence "is making a stalwart and almost heroic adjustment to life." Likewise, a twenty-year-old born with a small, hypospadic penis that required him to sit to urinate and that went uncorrected until age nineteen; this patient "was almost a model of what the average citizen believes a healthy, well-adjusted American youth should be," the author noted: "confident, self-reliant, and optimistic."

Unfortunately, no experts in the debate on intersex treatment—including Milton Diamond, Bill Reiner, Anne Fausto-Sterling, or Cheryl Chase—has ever made reference to this valuable report. That such a rare and unique study has been overlooked is perhaps not surprising. Never commercially published or distributed, it can be obtained only through written application to the Widener Library at Harvard University, where it was submitted as a senior dissertation to the college's Ph.D. program in 1951. The author was a thirty-year-old doctoral candidate named John Money.

15

≡

JOHN MONEY HAS NEVER explained the shift that occurred in his thinking between the time he finished his Harvard thesis and the time he wrote his first papers on intersexes four years later, and he has never publicly commented on any aspect of his work since the revelations in Diamond and Sigmundson's paper.

Now seventy-eight years old and in semiretirement, he has nevertheless remained a prolific and opinionated writer on the subject of sex and sexuality. His latest book, *Unspeakable Monsters,* was published in the spring of 1999. Through the last two decades, his books and articles have continued to appear with regularity, and in the late 1980s he enjoyed an intense courting by the media over the publication of his book *Lovemaps*—Money's term for an individual's particular constellation of erotic tastes and impulses. Profiles and interviews with Money appeared in *Playboy, Cosmopolitan, Psychology Today, Omni,* and the *Atlantic Monthly.* In *Rolling Stone's*

1990 "Hot Issue," Money was celebrated as the "Hot Love Doctor," and he appeared on various TV programs.

Meanwhile Money was negotiating a subtle shift from his earlier extreme position on the primacy of rearing over biology in the making of boys and girls. In a May 1988 magazine profile, he seemed at some pains to characterize himself as a longtime champion of the role of *biology* in psychological sex differentiation, saying that when he was publishing papers on the behavioral influence of prenatal sex hormones in the 1950s, "many people in various branches of the social sciences were just enraged at the idea that hormones in the bloodstream before you were born could have a sex differentiating influence on you." In the same article, however, Money reiterated his claim that infant boys can, with surgery and hormone treatments, be turned into heterosexual women.

If the last two decades have seen the consolidation of Money's international reputation as one of the single most influential sexologists of the twentieth century, his career at Johns Hopkins has not been without its setbacks. The seeds for Money's problems were sown as early as 1975, when Dr. Joel Elkes, chairman of the Psychiatry Department and Money's longtime protector within the institution, was replaced by Dr. Paul McHugh.

By almost any measure, McHugh, a practicing Catholic and a sworn enemy of all fashions and fads in psychiatry, was John Money's diametrical opposite—save for the forcefulness of his opinions and his determination to put them into action. Today McHugh is famed as psychiatry's most outspoken scourge. In referring to McHugh's "ceaseless campaign to restore sanity to his own profession," a 1997 *Baltimore Sun* profile dubbed him "Dr. Iconoclast" and listed his "annihilating opinions on everything from doctor-assisted suicide (utterly wrong) to multiple personality disorder (it doesn't exist)." The profile also highlighted his excoriating disdain for "dubious practices—and practitioners—in the medical profession," including Dr. Jack Kevorkian, whom McHugh was quoted as calling "insane," and Dr. Bruno Bettelheim, the famous expert on children, whom McHugh called "a habitual liar, thankless friend, vicious bully, and brazen plagiarist."

McHugh has always reserved special scorn for the practice of sex-change surgery on adult transexuals. Classifying transexualism as merely one symptom in a larger complex of personality disorders, McHugh had long believed that psychiatrists should treat such patients with the talking cure, not radical, irreversible surgeries. In a 1992 article in the *American Scholar*, McHugh lambasted transexual surgery as "the most radical therapy ever encouraged by twentieth century psychiatrists" and likened its popularity to the once widespread practice of frontal lobotomy. "Johns Hopkins was one of the places in the United States where [transexual surgery] was given its start," McHugh pointed out in this article. "It was part of my intention, when I arrived in Baltimore in 1975, to help end it."

Two years after McHugh arrived at Johns Hopkins, Dr. Jon Meyer, a Hopkins psychiatrist and former director of the Gender Identity Clinic, produced a long-term follow-up of fifty postoperative and preoperative adult transexuals treated at Johns Hopkins since the clinic was founded in 1966. Meyer reported that none showed any measurable improvement in their lives and concluded that "sex reassignment surgery confers no objective advantage in terms of social rehabilitation." Presented at the American Psychiatric Association's Annual Convention in May 1977, the paper was published two years later in the *Archives of General Psychiatry*. The transgendered community reacted with outrage to the paper's alleged nonscientific methods and aims. To no avail. Its publication was heralded by an October 1979 press conference at Johns Hopkins, where it was announced to the assembled reporters that the Gender Identity Clinic was now closed. John Money was not notified about the press conference and was not consulted about the clinic's closing—an ignominious position for the man who had—virtually single-handed—spearheaded the movement to open it.

McHugh's tenure at Johns Hopkins also coincided with a sudden dramatic erosion in Money's once secure status as the institution's resident sexual revolutionary. In 1983, Money was informed that his controversial evening course in human sexology was being summarily dropped. Three years later, when Money turned sixty-five, he was notified that he would not be allowed to keep his Johns Hopkins office space—a privilege conferred upon some other retirement-age professors—but must remove himself from the campus. He was relocated to a shabby medical arts building four blocks from the hospital and university, across from an empty lot where the local homeless and addicts congregate. There, Money installed himself in one of the building's low-ceilinged basement offices. With a staff now reduced to a single graduate student, Money affixed to the cheap plywood door of his new space the sign he had removed from his former office door. It reads: JOHNS HOPKINS PSYCHO-HORMONAL RESEARCH UNIT.

Even after his physical removal from the institution, Money's problems with Johns Hopkins were not over. In the early 1990s one of his former research subjects raised a complaint against him and against Johns Hopkins. This patient, who wishes to remain anonymous, has asked me to refer to him as "Charlie Gordon"—a pseudonym that he did not choose randomly. It is the name of the protagonist in the 1960s Daniel Keyes novel *Flowers for Algernon*, which was later turned into the Cliff Robertson movie *Charly*. The fictional story of a retarded man who, as an experimental research subject, was turned into a genius, *Charly* bears striking parallels to the life of the man I have agreed to call Charlie Gordon.

Born in 1947, Gordon showed early signs of hypothyroidism, a congenital endocrine disorder whose symptoms include severely stunted growth and retarded intellectual development—syndromes then classified as the condition "cretinism." At age two, Gordon was referred to Lawson Wilkins's pediatric endocrine clinic at Johns Hopkins where he underwent experimental treatments of hormone replacement by ingesting cow thyroid glands in pill form. The treatment increased not only his physical stature, but also his intellectual powers. At age five he became a psychological research subject in the newly created Psychohormonal Research Unit, where John Money would, for the next twenty-five years, conduct adjunct studies on Gordon's adaptation to his changing bodily and intellectual stature. In an article published in the *Journal of Pediatrics* in September 1978, Money singled out Gordon as having demonstrated the largest increase in intelligence of all the research subjects. According to Money, he had gone from an IQ of 84 at age five to an IQ of 127 in adulthood: a 43-point gain that had taken him from low average to superior range—what Money called "a remarkable upgrading."

Over the course of their association, Gordon became one of Money's favorite research subjects and agreed to Money's request that he appear at medical school grand rounds, in which he was studied by scores of Johns Hopkins student doctors. At the same time, Gordon was making regular annual visits to the clinic for in-depth interviews with Money. Gordon found the encounters unsettling. "He was always saying 'fuck,' all the time," Gordon recalls. " 'Fuck this,' and 'fuck that.' As a kid I was raised in somewhat a religious background. When I'd do church things, he'd say, 'Oh, what do you do that shit for?' "

Money also questioned Gordon closely about sex. Believing that Dr. Money's interest in his erotic life was intended to help him cope with the difficulties associated with his condition, Gordon opened up without reserve, detailing the content of his sexual fantasies, describing his masturbation techniques, and recounting his experimental forays into ménage à trois and his childhood experiences of "playing doctor" with a neighborhood girl. Later, in his twenties, Gordon confessed to the insecurities that had gone along with his small stature, admitting that he had once sought relationships with girls many years his junior—some as young as fourteen. Only several years after he stopped treatment with Money did Gordon learn that Money's interest in his sex life was not simply therapeutic in nature.

This realization was brought home to him with particular force on a day in December 1989 when he was browsing in a bookstore and happened upon a copy of Money's latest volume, *Vandalized Lovemaps*. The book detailed Money's theory of how people develop sexual fetishes, perversions, and disorders, and it featured a number of case histories. The first one, entitled "Pedophilia in a Male with a History of Hypothyroidism," caught Gordon's eye. He began to scan the opening sentences and realized with amazement and horror that the case history being detailed was his own. He saw his sexual life laid out with extensive verbatim quotations culled from his taped interviews with Money and saw himself diagnosed as a pedophile on the evidence of his interest in teenage girls. More shocking still, says Gordon, was information published about his parents, which included a statement by Gordon's father, who had allegedly told Money that Gordon's mother had had an incestuous affair with her brother. Though Gordon and his

242

family were not mentioned by name, he felt that the details of the case would be unmistakable to anyone who knew him or his family. Stricken, Gordon phoned Money, but he could not reach him. "He wouldn't return my calls," Gordon says. "His associate said, 'He's busy.' "

In the spring of 1990, Gordon brought a formal complaint against Money and Johns Hopkins through the federal Department of Health and Human Services Office for Protection from Research Risks—a division of the National Institutes of Health. Gordon learned that scientists operating under federal research grants must adhere to stringent rules, which include gaining informed, signed consent from patients and research subjects about whom a researcher wishes to publish. Money had never secured consent from Gordon for the publication of the deeply private material in *Vandalized Lovemaps*. The Department of Health and Human Services launched an investigation and concluded that "given the nature of [the] information [disclosed by Money], the complainant could be indentifiable by persons acquainted with [him]." That fall, DHHS cited the Johns Hopkins University School of Medicine for "serious noncompliance" with federal regulations for the protection of human research subjects. Calling for "strong corrective action," DHHS required that the Johns Hopkins Psychiatry Department "republish departmental guidelines for safeguarding the identity of patients" and allow patients who did provide informed consent to review manuscripts before publication, and it demanded that Gordon receive an apology from Money in person and in the presence of his department chairperson, in this case Money's nemesis, Dr. Paul McHugh. Gordon says that this apology was never given, but he felt vindicated by the other sanctions. In a statement that Gordon prepared for an October 1997 meeting of President Clinton's National Bioethics Advisory Commission on human research subjects, he outlined his unhappy history with Dr. John Money and drew a parallel between his experiences as a research subject and those of the famous "John/Joan" whose story had broken in the press just eight months earlier.

Despite this string of professional reversals, embarrassments, and punishments, Money remained defiant, combative, and uncowed. Indeed the setbacks seemed only to fuel his contentious spirit. Increasingly his published work appeared to be as much an opportunity for Money to settle scores and air grievances as it was to elucidate the subject of human sexuality. His preface to the 1987 book *Gay, Straight and In Between,* a volume ostensibly about the origins of sexual orientation, included an unusual digression into his then-recent ouster from Johns Hopkins. "In the spring of 1986," he wrote, "I was delivered an edict: the space allotted to the Psychohormonal Research Unit . . . would be reallocated. The new space would be away from the hospital campus in a commercial building. No further explanation would be given. There would be no appeal. . . . My response was to write this book."

In a 1991 autobiographical essay included in the anthology *The History of Clinical Psychology in Autobiography,* Money continually veered from the subject of his contributions to sexology to revisit his battles with the Johns Hopkins administration. Angrily evoking the termination of his human sexology course, Money wrote, "What the students at Johns Hopkins have lost has become the gain of students around the world. For them I now have more time to write." In the same essay, Money excoriated the (unnamed) Paul McHugh as "the most contentiously destructive person I have ever known" and gloated that "[h]is clandestine efforts to get rid of me failed." Money portrayed his current diminished status in the grim and dangerous basement setting of the Psychohormonal Research Unit with not untypical grandiloquence. "Working as an off-campus exile," he wrote, "in a green subterranean jungle that flourishes under artificial light, I have a sense of kinship with dissidents like Galileo, who by order of the Vatican lived as an exile under house arrest."

Inevitably, perhaps, in the same essay Money addressed those in the field of sexual development who had challenged his scientific theories over the years. Though he did not mention Milton Diamond by name, there was little doubt that the University of Hawaii professor was high on the list of those whom Money now castigated for "shamelessly" attacking him. Yet after lambasting these critics, Money segued into a tone of lofty Olympian remove, finally dismissing his academic disputants as beneath his notice. "My personal impression," he wrote, "is that they are lacking in the special talent for original thinking, for formulating new concepts and hypotheses, and for making new discoveries." Of his continued academic survival, Money wrote, "I have survived by putting into practice my own maxim and have not been lured into declaring a war that I had no possible chance of winning. Instead of mounting a direct counterattack, I would adopt a policy of disengagement and redirect my energies into an alternative channel of achievement."

Money put into effect just such a strategy of disengagement six years later, in the spring of 1997, when his career and reputation suffered their greatest blow to date, from the worldwide media response to Diamond and Sigmundson's paper on the twins case. To the many news organizations that requested comment from him about the now-infamous case, the psychologist refused to speak, citing confidentiality laws. I was among the raft of reporters who sought an interview with Money (for the article I was preparing for *Rolling Stone*). In a letter, I urged him to speak with me, and assured him that I would treat the story with scrupulous objectivity. He declined, but over the ensuing weeks and months, we exchanged a number of e-mails in which he eventually offered to work with me as a kind of silent collaborator on what he called "a piece of investigative journalism." He offered to supply me with the requisite reprints from his published work and to vet my unpublished article to "check the accuracy of some data."

This invitation was withdrawn in late August. Having returned from a second trip to Winnipeg, I notified Money for the first time that I had located and interviewed the patient and his family and had furthermore secured David's promise of a signed confidentiality waiver freeing Money to speak to me about the case. Money's tone changed abruptly. From would-be silent collaborator, he now grew ice-cold. "Thank you for your e-mail of August 24th, to which my reply is that my position has not changed and will not change," he wrote. "I am not under any circumstances available for an interview regarding the Reimer case, and have no further comments to make. So please desist."

I did desist for the next two months while I wrote my *Rolling Stone* story. In early November, with the article going to press, I phoned Money's office to check some facts with his assistant, William Wang. I was surprised when Money got on the line. Although he refused to discuss David Reimer's case directly, he claimed that the media's reporting of it reflected nothing more than a conservative political bias. He was particularly incensed by the *New York Times* front-page story. "It's part of the antifeminist movement," he said. "They say masculinity and femininity are built into the genes so women should get back to the mattress and the kitchen." As to his failure to report the outcome of the case, Money was unapologetic, repeating his claim that he had lost contact with the Reimers when they did not return to Johns Hopkins and that the opportunity to conduct a follow-up had been denied to him. Money sounded affronted when I suggested to him what various of his defenders had hinted to me: that the misreporting of the case was all the Reimers' fault and that David's mother, in particular, in her zeal to believe in the experiment's success—and to please Money—had given him a "rosy picture."

"I was *not* being given a rosy picture," he said irritably, as if stung by the suggestion that he would have failed to factor in any such maternal bias in his assessment of the case's progress. "The only thing that was of importance to me was that I didn't get *any* picture at all after the family simply stopped coming to Johns Hopkins."

He stood by his original reporting of the case and dismissed my suggestion that he "misperceived" what was going on with the child in their one-on-one sessions. Furthermore, he implied that David's reversion to his biological sex might not have been entirely his own decision. "I have no idea," Money said, "how much he was coached in what he wanted, since I haven't seen the person." He also hinted that Diamond and Sigmundson's paper had a hidden agenda. "There is no reason I should have been excluded from the follow-up, was there?" he asked. "Someone had a knife in my back. But it's not uncommon in science. The minute you stick your head up above the grass, there's a gunman ready to shoot you." Told of these comments, Diamond says that he had repeatedly invited Money to share or publish information on the twins over the previous fifteen years, always to no avail.

When I asked Money about Diamond's appeal to delay surgery on intersexual babies until they are old enough to speak for themselves, Money grew angry. Apparently forgetting the conclusions he had reached in his own Harvard thesis review of over two hundred and fifty untreated intersexes, he emphatically rejected the idea that a person could survive a childhood with ambiguous genitalia. "I've seen the people who were the

247

victim of that," he said. "I've heard these poor people describe how they had to sit in a locked room and not go out for fear that someone would see them." Money insisted that surgical intervention at the earliest opportunity after birth was the only guarantee of the child's future happiness. "You cannot be an *it*," he declared, adding that Diamond's recommendations would lead intersexes back to the days when they locked themselves away in shame and worked as "circus freaks."

Money refused to discuss any aspect of his personal life. "You're trying to entrap me," he said, darkly. "Just like my patients try to entrap me."

At this point Money seemed determined to get off the phone. Before he did so, I reminded him that his now classic text, *Man & Woman, Boy & Girl*, was still in print and that it reports the twins case as a success. I asked if it would not be worthwhile for him to make changes in the text for a future edition. Money said flatly, "I'll be dead by then."

Despite its ring of finality, this proved not to be Money's last word on the case. After the publication of my *Rolling Stone* story in December 1997, he again broke his press silence when he granted an interview to a sympathetic writer and friend, Michael King, in the New Zealand magazine *The Listener*. There Money dismissed both Diamond and Sigmundson's John/Joan paper and my *Rolling Stone* article as part of a dark conspiracy against him. The *Listener* article furthermore hinted that David and his family were deliberately lying about Brenda's life for financial gain, since David had decided to collaborate on this book, and moviemakers had expressed interest in the saga. King's article described Money as "surprisingly resilient and sanguine" despite the controversy and ended with news of the undimmed status that

Money still enjoys among U.S. funding agencies. "He has recently been recommended for a grant from the National Institutes of Health for a major new project, a classification and consolidation of contemporary knowledge of paraphilias or 'perversions,' " King reported. I checked with the NIH in the summer of 1999 and learned that Money is still supported by the same NIH research grant that he was awarded in the mid-1950s. His most recent renewal was in the amount of $135,956.

Nor does Money lack for defenders within the academic community—and in particular among professors of psychology, many of whose tenured positions and clinical appointments have been built upon the promulgation and dissemination of Money's theories of psychosexual development. One of his more engaging and intelligent defenders is Dr. Kenneth Zucker, a psychologist at the Clarke Institute of Psychiatry in Toronto and a longtime adherent to Money's nurturist bias in gender identity formation. (In his clinical work, Zucker has for years attempted to modify homosexuality and transexualism in boy and girl children.) Several months after the publication of Diamond and Sigmundson's article, Zucker wrote a paper entitled "Experiment of Nurture," which was framed as a direct response to the John/Joan revelations. Showing his environmentalist leanings, Zucker suggested that the twins case had failed not because David possessed a male biology, but because of certain "psychosocial factors"—in particular "parental ambivalence regarding the initial decision to reassign the infant as a girl."

Efforts on the part of Money's defenders to blame the failure of the case on Ron's and Janet's supposed lack of commitment were by no means exclusive to Zucker; many of Money's acolytes have made the same charge to me in interviews. These charges might carry more weight if not for the fact that all the evidence shows that Ron and Janet were almost slavishly devoted to the experiment—not to mention that Money himself, in his reports on the case, repeatedly described Ron and Janet as particularly skilled and committed parents in the rearing of their daughter. To be sure, once news of the case's failure emerged, rumors apparently originating with Money leaked into the scientific community that Ron and Janet were rural fundamentalists whose restricted religious and cultural values had made it impossible for them to accept their child's sex change in the first place, and that therefore they unconsciously undermined it. In reality, Ron and Janet grew up and spent the majority of their lives in the modern metropolis of Winnipeg (except for the three teen years they spent on farms), and both had (like Money himself) thoroughly rejected the fundamentalist religion of their parents (so much so that they refused even to be married in a Mennonite church). All new claims to the contrary, neither Ron nor Janet labored under outmoded stereotypes of men's and women's roles which would have forbidden them from accepting a merely "tomboyish" daughter, nor was their surrounding community of 1970s Winnipeg—an eclectic, cosmopolitan mix of cultures, religions, backgrounds, races, and socioeconomic levels—predisposed to rejecting a girl who did not conform to rigid stereotypes of femininity.

Zucker's paper, however, did not concern itself solely with unnamed "psychosocial factors" that supposedly negatively influenced the case. He also presented a long-term follow-up on a second case of a developmentally normal baby boy who had been raised as a girl. In a shocking parallel to David's case, this child (also, coincidentally, a Canadian) had lost his penis to a bungled circumcision by electrocautery and had subsequently been castrated and reassigned as a girl at seven months of age in 1971. Now twenty-six years old, the patient was described by Zucker as still living in the female sex. "She denied any uncertainty about being a female from as far back as she could remember," Zucker wrote, "and did not report any dysphoric feelings about being a woman." At the same time, Zucker admitted that the case could not be deemed an unalloyed example of the efficacy of sex reassignment, for he was obliged to acknowledge that the patient, in childhood, had always enjoyed "stereotypically masculine toys and games"; that as an adult she works in a " 'blue collar' job practiced almost exclusively by men"; and that she is currently living with a woman, in her third significant sexual relationship with a member of the female sex.

Nevertheless, Zucker concluded, "In this case ... the experiment of nurture was successful regarding female gender identity differentiation," and he cited the case as convincing proof that her rearing as a girl "overrode any putative influences of a normal prenatal masculine sexual biology."

Struck by the seeming incongruity of these conclusions, I spoke with Zucker about the case at his office in Toronto in the summer of 1998. Our conversation only served to raise further doubts about the paper's conclusions, for Zucker was unable to answer any of my specific questions about whether the patient might not have been telling the researchers what they wanted to hear when she stated that she had never harbored any doubts about her gender. By now I understood that this is a phenomenon endemic to all areas of sex research that rely on patient testimony, but particularly so in the fraught and sensitive world of sex reassignment, where as one ISNA member told me, "You feel so embarrassed and ashamed to be talking to someone that you'll basically tell them *anything* so you can get the hell out of there." Zucker agreed that such scenarios are not unfamiliar, but he couldn't say whether such a dynamic was at work in the case in question. And for a simple reason. He had never met the patient and had based his reporting solely on information supplied to him by the people listed as coauthors of the paper. These included a gynecologist with no training in the assessment of gender identity, and a psychiatrist who had conducted only two interviews with the woman—the first when she was sixteen, the second when she was twenty-six.

There was, as well, further reason to feel uneasy about the paper's conclusions, and this had to do with its murky provenance. It was only in the closing moments of my interview with Zucker, after I had turned off my tape recorder, that he let fall that the paper had another silent collaborator—an investigator who, when notified of the researchers' efforts, had hastened to supply records he had gathered on the patient in her early childhood. The investigator was John Money, who had authorized and overseen the patient's sex reassignment in infancy and who had, true to practice, conducted a number of annual follow-ups with the child until she (for reasons unspecified in Zucker's paper) stopped returning to Johns Hopkins.

16

=

It has been twenty years since Brenda Reimer made her transformation to David. That metamorphosis marked a turning point in the family's fortunes. Ron, who had been struggling to get his business on its feet, began finally to build a faithful clientele of construction companies and factories. By the early 1980s he was earning forty thousand dollars a year, the best money he had made in his life. Janet continued to see her psychiatrist, and by the mid-1980s, with lithium treatment, her depressions had abated. "I found out what kind of person I really was," she says. "And I went to my children and apologized to them. I said, 'I know at times I was unreasonable and that you were wary of me sometimes because you didn't know what was going to come up next, and you didn't altogether trust me or feel you could take me into your confidence.' I told them that I felt great remorse."

The improvements in Ron's finances and Janet's emotional health brought a harmony the couple hadn't known since the earliest days of their marriage. "I would lay down my life for Ron," Janet told me in the summer of 1998. "Actually, I remember something Dr. Money once said to me. He said, 'I don't know why people always say *making love*; it's *making sex*.' Back then I didn't have an answer for him. Now I do. What I have with Ron is love. We make *love*."

Neither Ron nor Janet pretends that they can ever put the past behind them completely. Janet remains the more talkative on the subject of the guilt and grief that are the main emotions associated with their decision, thirty years ago, to turn their son into a daughter. Ron typically finds it more difficult to speak directly about these matters, but he communicates them nevertheless in his more spare and diffident speech.

"I wonder," I asked him in our first interview, "if you've ever got to a point where you forgot this had happened?"

Ron shook his head. "No," he said. "We never forget." Then he said it again. "Never forget." And once more: "Never forget."

I remembered a notation I had seen in Dr. Ingimundson's psychiatric notes from the spring of 1977 concerning a private meeting with Janet and Ron. Under the heading "Counter transference" (the psychoanalytic term for the emotions experienced by the therapist toward the patient), Ingimundson had written, "Have a need to protect them." I now felt something of the same need.

"I know David doesn't blame you at all," I told Ron. "He attributes all the best things in his life to you and Janet."

Ron smiled weakly and blinked away the moisture in his eyes. "I'm glad *he* feels that way," Ron said. "I don't know if *I* feel like that."

Perhaps the greatest insight that Ron gave me into his emotions concerning the failed experiment came when our formal interview was over, and we repaired from the backyard to the house. Ron poured us a pair of Crown Royal rye whiskeys, then invited me to watch a tape of his favorite movie. It had been a long day, and I told him I would probably head back to my Travelodge and turn in early. Ron was strangely, and uncharacteristically, insistent. "This is a really great movie," he said. "I got Brian to tape it for me off HBO. I've seen it maybe twenty times." The movie, he said, was called *Crossroads*. I soon realized that it was pointless to resist; by now Janet (who also loved the movie) had joined Ron in his entreaties. So I followed them to the basement, where we settled down in front of the television set.

I vaguely registered the movie's plot as it played on the Reimers' TV screen. Ralph Macchio plays a cocky young blues guitarist who befriends an eighty-year-old blues player, one Blind Willie Brown. Together the pair travel from New York City to Blind Willie's Mississippi home, where he has some "unfinished business" to take care of. As Ry Cooder's keening blues guitar soundtrack wailed over the opening credits, Janet turned to me and said, "We love this music. I think you have to have been to hell and back to love the blues." In its detail, and in the thorny, affectionate relationship between the older and the younger man, the movie was better than I expected; but I still found myself fighting off sleep as the saga reached its final act, which occurs at a stark, dusty crossroads in the depths of rural Mississippi—at which point I began to grow alert.

Drawing on the famous legend of blues guitarist Robert Johnson (who was said to have won his skills as a guitar player from a deal he signed with the devil at "the crossroads"), the movie now revealed that Blind Willie Brown had made a similar deal almost sixty years earlier, when he was seventeen years old. But Blind Willie had not become famous and celebrated. Instead he had ended up destitute in a Harlem old folks' home. Now he had come for a reckoning. Standing in the shadow of a leafless tree at the crossroads, he watched as the dapper, smooth-talking, grinning man with whom he had struck his deal all those years ago materialized from nowhere. The two men face each other. Ron, who was sitting in an armchair to my right, set down his rye and 7-Up and sat forward a little, bringing his face closer to the screen.

Confronting the man who had hoodwinked him into his bum deal, Willie Brown demands that the Man in Black tear up the contract between them—"and give me some peace."

The Man in Black laughs derisively. "Why on earth would I want to do that?" he asks.

Willie is outraged at the man's insouciance. "*You* sloughed up on *your* end of things," Willie shouts. "I didn't end up where I wanted. I didn't end up with *nothing*—didn't get nothing!"

But the grinning Man in Black offers no apologies. "Ain't nothing *ever* as good as we want it to be!"

Both Ron and Janet hung on every word of this dialogue— as if they expected that on this viewing, the scene might finally play out differently. When the scene ended, Ron sat back in his chair, then glanced quickly at me and away. Several times during our long interview that day, I had tried to get Ron to speak about how he now felt about John Money and the momentous decision he had convinced Ron and Janet to make. He had made a few halting, stumbling efforts to answer my question but had clearly failed to say all that was in him. Now I felt I had my answer. Along with Ron's grief and guilt there was an obvious admixture of outraged betrayal, which lay too deep for him to express in words.

Nor were those emotions solely confined to the way Ron and Janet felt about the son they had agreed to convert into a daughter. For David was by no means the only casualty of that doomed experiment. The matched control, too, had suffered, and suffered badly, with results that were still being felt. Brian's episode of shoplifting in the spring before the family's flight to British Columbia proved not to be an isolated incident, but a precursor of more serious transgressions to come.

"I was thirteen when I got involved with a bad crowd," Brian explains. "It started with drinking and smoking, and it eventually wound up into stealing cars and dope and fighting. For me, personally, I never got into armed robbery and I never really hurt anyone that bad. . . ." Then Brian thinks for a moment and amends that. "One person I hurt pretty bad." He is referring to a boy whose arm he broke so severely in a fight that he was called to court. Listening to the litany of Brian's brutal and criminal acts as a teen and young adult, I was mystified. Even in adulthood he clearly demonstrated, in comparison with David's more conventionally male attitudes, a greater aesthetic awareness and a sensitive side out of keeping with the endless tales of mayhem and brutality that filled his teens and twenties. "That's the side I couldn't show to people," he says. "The sensitive guy finished last. The tough guy gets the respect, and he gets left alone. That's bad to say, but that's the reality of it. He gets all the girlfriends; he gets invited to all the parties."

Being included by his friends was vitally important for Brian because of the abandonment he felt from his parents. "I had problems growing up, but they had to deal with my sister's problems, which were so much bigger," Brian says. "But try growing up all your life feeling that your problems are nothing."

Brian learned about his sister's true birth status from Janet on 14 March 1980—the same day that Ron told Brenda.

"My mother was working at the parking lot," Brian says. "She called me and said, 'Brian, I have to talk to you about something.' So I visited her at work, in this booth where she sat. We were having coffee. She said, 'It's about Brenda.' Then she says, 'Brenda was really your brother.' And I got upset."

Brian's reaction was typically explosive. He jumped from his chair and smashed his fist into the booth's reinforced glass window. "I broke both panes," Brian recalls. "I was pissed off. Then I cooled down, and mother told me everything that had happened—about the circumcision and everything. I said, 'Now I can understand. I can put the pieces of the puzzle together. It makes sense now.' But I felt, 'Shit, the first fourteen years of my life was a *lie.*'"

And there would be other emotional hurdles for Brian to get over—which became clear later that same day, when he saw his twin for the first time since hearing the news. "Dave was wearing a suit," Brian recalls. "He says, 'What d'you think?' I said, 'Hey. You look good. I'm happy for you.'" But Brian admits that David's transformation brought mixed emotions. In the past, Brian had always had his status as the family's only son to make him feel special. Now even that was gone. "I supported him one hundred percent," Brian says. "I felt a sense of relief because now he finally fitted into society. At the same time, I'm not big brother anymore."

Just how deeply affected Brian was by this swift and emotionally bewildering turn of events was clear a year later, shortly before the twins' sixteenth birthday and two weeks before David underwent his first phalloplasty, when on 17 June 1981, Brian was taken to the emergency room of St. Boniface Hospital to have his stomach pumped. He had drunk from a bottle of drain cleaner. At the time, Brian told his family that the suicide gesture was over a girl who had broken up with him. Today Brian admits that was not the truth. "Mom was worried sick about David," he says. "Every waking moment was David. It was 'Brian's OK, he can take care of himself.' Any problem I had seemed trivial compared to David's. So it was almost like I had to do *something* to get a little bit of attention."

At sixteen, to his parents' consternation, Brian quit school and took a job pumping gas. He moved out of the house and started living with a girlfriend. At nineteen, Brian married her, and had two children. The marriage proved tumultuous and ended in acrimony and divorce a few years later.

Brian's life reached its nadir after the divorce. Unemployed and trying to raise his children as a single father, he began to drink to excess and suffered bouts of severe depression. His children were temporarily removed from his custody and lived for six months with Ron and Janet. During that time Brian cleaned himself up and got control of his life. In the early 1990s he landed a well-paid union job working a lathe in a metal-spinning factory. He remarried, had a daughter, and moved into a house he bought in Winnipeg's West End. Prozac has helped with his mood swings.

Aside from his wife and children, Brian says that the person he knows best in the world is his brother. Despite the lingering rivalries that sometimes drive the twins, even in adulthood, to severe periods of feuding and dissension, they remain extraordinarily attuned to one another's inner lives. Yet Brian admits that, like his parents, he harbors guilt about David. It is a guilt that dates back roughly to sixth grade at Agassiz Drive school, when Brian pulled away from his social pariah sister. "I had a choice," Brian says. "I could be with my friends or with my sister. They made it quite clear—subtle but clear—that I had a choice." He chose to have friends—a choice for which he has never fully forgiven himself. "I wanted to have a life," he says. "I turned my back on Brenda." Only many years later, after David had been living in his true sex for almost five years, would the brothers again grow close.

At our first meeting, in late June 1997, Brian proudly cataloged the striking similarities between them, likening David and himself to those cases of identical twins separated at birth and who, when reunited in adulthood, discover that their lives bear uncanny parallels. "Both me and Dave married in September," he told me. "Both have one dog and one cat. Both are factory workers. Both make around the same amount of money. We both like to watch *Biography*, 20/20, *Fifth Estate*, 60 *Minutes*. Both love Elvis. In a way, it's always been me and my brother against the world."

More than two decades have passed since David Reimer had his final contact with Dr. John Money, when the famous sexologist slipped him fifteen dollars in his parents' living room. In the intervening years, David often fantasized about what he might say or do to the psychologist if they were ever to meet face-to-face. He admits that as a younger man his fantasies ran to violence. No more. Determined to get on with his life, he refuses to dwell on a past that he cannot change. In their paper, Diamond and Sigmundson describe David as a "forward-looking person." In conversation, Diamond calls him a true hero. And indeed, David's life today defies the dire prognosis of the psychiatrist who thirty-three years ago declared that he would never marry and "must live apart." At the same time, it has been impossible for him to put the past away entirely. Over the course of our interviews together, David spoke with a blunt and unvarnished honesty about his extraordinary childhood and youth. He spoke without self-pity. His sufferings were extreme, his survival almost miraculous: both lend to his unschooled speech an aura of oracular wisdom.

"I don't blame my parents," he told me. "A lot of people are going to be surprised by that. They'd have to put themselves in my situation and live out my life, knowing that my parents have sacrificed so much. My dad's a very special man. He's got a lot in his heart, and he doesn't know how to express himself. But you know, you can see in the soul of his eyes that he's hurting and that he cares and he loves you.

"My mother is a lot better, she's getting help. She admits that she did wrong things. Some people wouldn't even do that. You know, as a very little kid I had a crush on my mother. I used to pick dandelions for her. She was the most beautiful woman in the whole world to me.

"When I think of my brother as a kid, I see this little seven-year-old with a bean shave, puppy-dog eyes, asking for help. 'Help me! Help me!' He'd get into trouble, get into a fight, and I'd do my best to bail him out. He'd let the scrawniest guy beat the hell out of him. My brother hiding behind me! I'd look ridiculous fighting because of the way I was dressed—I didn't look the part to bail him out. My dad gave me shit when I fought because he thought it put Brian in a position where *he* would have to try to protect *me*. 'It's unfair to put your brother in a position like that.' I tried to explain: 'This is not my *brother's* fight, this is *my* fight.' It didn't do any good. I'm not going to take anything away from Brian: he had it rough because of me. But it was directed at *me*, not him. When they picked on him, they were making fun of me. 'There's your butch sister.'

"My childhood. It comes to me. I don't go and think about it. I'm trying to sleep, and these stupid thoughts come into my head, and I shake my head and I say, 'I'm going to think about something else,' but it will jump right back into my head. Memories of how I used to look. Memories of being belittled by my classmates. Memories of just trying to survive.

"If I had grown up as a boy without a penis? Oh, I would still have had my problems, but they wouldn't have been compounded the way they are now. If I was raised a boy, I would have been more accepted by other people. I would have been way better off if they had just left me alone, because when I switched back over, then I had *two* problems on my hands, not just one, because of them trying to brainwash me into accepting myself as a girl. So you got the *psychological* thing going in your head. When I'm intimate with my wife it sometimes *haunts* me. From time to time it gets to flashbacks of you as a kid, and it makes you— I admit, sometimes I have to get up out of the bed and go to the bathroom and throw up.

"You know, if I had lost my arms and my legs and wound up in a wheelchair where you're moving everything with a little rod in your mouth—would that make me less of a person? It just seems that they implied that you're nothing if your penis is gone. The second you lose *that,* you're nothing, and they've got to do surgery and hormones to turn you into something. Like you're a zero. It's like your whole personality, everything about you is all directed—all pinpointed— toward what's between the legs. And to me, that's ignorant. I don't have the kind of education that these scientists and doctors and psychologists have, but to me it's very ignorant. If a woman lost her breasts, do you turn her into a guy? To make her feel 'whole and complete'?

"I feel sorry for women. I've been there. 'You're a little lady—go into the kitchen.' Or 'We don't want you to chop wood—you might hurt yourself.' I remember when I was a kid and women were fighting like hell to get equal rights. I said, 'Good for them.' I kind of sensed what position women had in society. Way down there. And that's how I was portrayed. And I didn't want to go way down there. I felt, I can do what anybody else can! But 'Oh, you're a girl—you might get hurt playing ball.'

"At Agassiz Drive school there was this guy, Tubby Wayne.

He was a male chauvinist pig. 'Women are dirt; they can't do anything men can do.' He kept saying, 'You don't know anything. You're a *girl*; girls don't know anything.' So I finally said to him, 'You think you're so tough? Then hit me. C'mon hit me.' He says, 'No, I'm not going to hit you; you're a girl.' 'No. Hit me. I'm not going to put up with this.' He wouldn't, so I punched him—and he laughed at me. It was a good thing he didn't hit me, I guess. But I was thinking, Don't hide behind the 'I don't hit girls' excuse.

"The guys at work don't know about what happened to me. I mean, I work in a *slaughterhouse*. All men. Can you imagine?—'There's the freak who wore dresses as a kid.' They give you that male chauvinist crap all the time. Like they're always saying that they're the *boss* at home. They look at me and ask me, 'Who's the boss?' I say, 'Look, man, in my home it's a partnership. It doesn't mean I wimp out; sometimes I get my way, and sometimes I don't get my way. But either way, it's a partnership.' I mean, who wants a woman with no brains, who follows you blindly? That's more like a slave than a wife. You don't want a slave, you want somebody with her own opinions, somebody who puts you on the right track, someone to show you the right direction. It's very hard to talk to somebody who's stone-cold stupid, who follows you blindly.

"But you know, if I had had a normal life, and none of this had ever happened to me, I'd probably be one of these chauvinistic kind of guys, where the guy goes to work, breaks his back, comes home, and sucks down a beer and watches sports. And if I saw someone like *me* out on TV, I'd sit there saying, 'Oh God that's sick.' That's how I would be. So knowing that that person is *me,* you can realize how sick *I* feel looking back on all this. You wish to God you could switch places with anybody.

"After I tried to kill myself, they put me in one of those psycho wards. Right away they want to put you in a group meeting. You can't even face this by *yourself,* and they're going to stick you in a room full of people so you can discuss this with strangers? There was this doctor there who kept telling me that it was wrong for me to try to kill myself. Well, it's easy for him to sit there and say, 'Shame, shame, shame.' This guy's been putting himself through college, he's got a degree, probably living in a fancy house, he's got 2.2 kids, got a normal life. Don't compare that to me. Not the same.

"I'm happiest when I'm alone. Doesn't mean I'm not friendly. It's just I'm more comfortable when I'm by myself. It's not lonely. It's relaxing. It's soothing. It reminds me of my grandfather's farm. If I go for a walk there, I'm in total peace. I'm never alone there. You always feel like you're surrounded in a place like that. Surrounded by what, I don't know. But you're not alone.

"I still wonder from time to time what it would have been like to have a biological child: to see how much of me that child would have. It's not really an ego trip. It's just . . . you feel that way. But I love my kids, and they're my children—not just my *step*children. I'm going to want to tell my kids about what happened. I couldn't keep something like this away from them. I'd feel very uncomfortable about that. I already told my eldest, when she was fifteen. She had that look on her face, like, 'You wore *dresses*?' But all in all, she took it well. She said, 'I don't love you any less, Dad.' I know my middle child would understand, but I won't tell her for years. I have to wait until she's a lot older. My son? I'll tell him when he's about fifteen.

"I live my life through my son. Everything my son does, that's *me.* When my son has a little crush on a girl, and he's leaning on a fence, and the girl comes up and says, 'Hi,' and he's got that blush thing happening; when he's in Cub Scouts— that's *me.* I live through him. When he succeeds, that's like *me* succeeding. Some psychiatrist would disagree with me living my life through him, but I never had any kind of a childhood.

"It's going to be harder to tell my son than to tell my girls. You never know with boys and their fathers. He might think differently about me. You know: 'My father wore dresses, my father had a girl's name, my father lived as a girl.' I mean that's a lot to swallow. You can tell when it's hard for somebody to accept something or is embarrassed. And for him to look at me sideways . . .

"I'm sick to death of feeling ashamed of myself. That feeling will never go away. I did nothing wrong, but it's like you're conditioned to feel ashamed of yourself. The very thought that I was wearing a dress, having a girl's name—the long hair and everything—I'm going to have to carry that for the rest of my life. You can't erase memories like that. I just survive as best I can. Keep telling myself, 'It's not my fault. Not my fault what happened.'

"Mom and Dad wanted this to work so I'd be happy. That's every parent's dream for their child. But I couldn't be happy for my parents. I had to be happy for me. You can't be something that you're not. You have to be *you*."

Embarking on my interviews with David, I had been warned by his former psychiatrist Doreen Moggey that I was engaging in a dangerous psychological process. She feared that David could not negotiate the leap from his current life to his former one without risk of serious mental upset, and urged me to tread gingerly when guiding him back to the memories and events he had tried, for so long, to forget. I heeded her advice, but in my conversations with David—which often stretched to six hours at a sitting, going deep into the small hours of the morning—I found that he bridled at the cautious circumspection of my questioning. He seemed to want nothing more than to charge fearlessly into the past, to speak without restraint about all that had happened to him, to put his true life story on the permanent record, and thus to reclaim it as his own. I also discovered that he wished to reclaim his life not only in the abstract form of interviews with me.

In mid-January 1998 I located and spoke to his old friend Heather Legarry. She still lived in Winnipeg but was married now with a different last name, and she worked as a kindergarten teacher. As was the case in all the interviews I did with those who had known Brenda as a child (and to whom David had authorized me to tell his medical history), I did not at first tell her the truth of Brenda's birth, so as not to color her memories and impressions of the girl; but when I did finally reveal to her that Brenda had been born a boy and now lived as a man named David, she said that she would like to see him. Recalling how David had once dodged a confrontation with Heather at the Go-Cart track fifteen years earlier, I wondered how he would greet this news. I was surprised when David said that he would like to see her, too. That Sunday they met privately and had lunch, while I stayed home with Jane and helped her nurse her mostly-mock jealousy—a jealousy that became a little more real when David failed to return home for almost four hours. When he did arrive, he (with typical thoughtfulness) bore a red rose for his wife, but he also bore an expression, on his face, that I had never seen before; it was a look of serene and somehow euphoric peacefulness. Asked by Jane what they had talked about for *so long,* David even got off a joke about his horrendous past. "Oh," he said, "just *girl talk!*" In fact, he later said, he and Heather had discussed all aspects of his childhood, comparing notes on their shared memories, reexamining their old friendship in the light of what neither had known at the time. It was a meeting that had clearly removed from David a burden that had weighed on him for almost two decades.

None of which is to say that revisiting David's past with him for the purposes of researching this book was not without its painful, and precarious, moments—particularly when I stumbled upon one last family secret of the Reimers' that had never been exposed: one of the white lies that Janet had constantly been obliged to concoct for Brenda in order to quell her daughter's suspicions about her true identity; a white lie that had, through parental oversight, never been corrected with David, and which thus lay buried for almost twenty years, like an unexploded land mine.

On my first ever trip to Winnipeg, eight months earlier, one of the first things David had ever told me about his mother was that she was a published poet. He described to me, with touching pride, how he had actually met the editors of the journal that had printed his mother's work. I soon learned, of course, that the story of Janet as poet was a ruse. Wanting to learn from David his impressions of meeting the BBC reporters on that day in October 1979, I did not feel comfortable keeping up the fiction that they were poetry editors. I knew that I would have to explode the lie. I put it off for as long as I could, until close to the end of my monthlong stay in Winnipeg in early 1998. I tried to break the news gently, but David still reacted very badly to this jarring revelation.

"Why did I have to find out *this* way?" he raged. "I'm a thirty-two-year-old man!"

For the next two hours he railed against his mother's "gutlessness" in not telling him the truth; he shouted that he had been lied to all his life, and said, "You don't expect that from the people you trust and love!" His anger and upset were frighteningly intense, but thankfully David's fury soon burnt itself down, and the next day he visited his mother and spoke to

267

her about the deception. She apologized for having overlooked this final lie, and they made up over it. In the end, the exposure of that last, lingering falsehood had a positive effect, Janet said, in removing an obstacle that had persisted from David's childhood of secrets and lies. I was glad, and relieved, of this outcome, but at the same time, glimpsing David's apocalyptic rage made me wish, not for the first time, that he could bring himself to see a therapist, if only to vent the combustible anger that periodically built up in him over his past.

But David's dismal experience with the mental health profession as a child has guaranteed that he will never consult a psychiatrist in adulthood—unless it were his old friend, Mary McKenty, and that is impossible, given that she, at age eighty-three, has long since retired. Still, David and she have remained friends, and he accompanied me on a visit I made to her home in June 1998, on my final trip to Winnipeg. Though now in the grip of Alzheimer's disease, and often unable even to recognize her own children, McKenty knew David immediately. She was a tiny woman with long, gray hair and the masklike expression typical of Alzheimer's sufferers, but her face blossomed into a delighted smile when she saw David step over the threshold of her door.

While David chatted with Mary's live-in nurses, I retired with her to a small sitting room off the front hall. Fortunately, we had come on a day when Mary's memory was particularly good, and her answers to my questions, though abbreviated, were spoken in a strong clear voice that left no doubt that she knew precisely what we were talking about.

I asked Mary how she had handled Brenda's difficult case. She shrugged. "I tried to be sensitive and supportive," she said. I asked if Brenda had ever seemed at all like a girl. "No," she said, "not like a girl at all." I recalled to myself Money's claim that Brenda must have been "coached" in rejecting girlhood;

no evidence for this existed anywhere in the local treatment team's notes, and David had been unequivocal in stating that McKenty had in no way "coached" him. But I felt it was my obligation to ask. Had Mary always kept Brenda's true birth status from her? "Yes," she said. I asked if she had tried to steer her gently away from being a girl. "No," McKenty said. "The things were for *her* to realize and for *her* to do."

Impressed with what I had been able to glean about Mary's special human touch with her patients—and especially with Brenda—I asked her about her general approach to psychiatry. In reply, she said "You have a parental attitude to your patients." Then she glanced in the direction of the hallway, where David's voice could be heard. "Just like a parent," she added, "you often admire them." I was moved by this word *admire*—so different from what one ordinarily expects to hear a psychiatrist say about a patient, or to hear a highly educated doctor like McKenty say about a slaughterhouse sanitation worker like David Reimer. But it was clear that her admiration for David was total.

I asked if she had ever read any of John Money's work. Her face, which had been eerily immobile, crumpled into a scowl. "Yes," she said with clear distaste. I asked what she thought of it. "I thought it was *unusual*," she said dryly.

David came into the room. She immediately brightened. He crouched down on the carpet in front of where she sat on the sofa.

"It's been a long time, Mary," he said gently.

"How long has it been?"

"Ten years at least."

"And it's been almost twenty since Mary started treating you," I said.

David looked sheepish. Just that morning, he and I had been looking at Mary's therapy notes with Brenda, and

David had been aghast to be reminded of the cruel tricks he had played on her in their earliest sessions. He apologized to her now. "I drew a nasty picture of you," he said. "I took you hostage with a toy gun."

"I didn't mind," she said.

"I made a Death Warrant in your name!"

Mary laughed and pretended to reach for her cane. "You'd better watch it!" she said. Then Mary grew serious and looked at David. His smile faded, too. For a moment it might have been twenty years ago, and the two might have been patient and doctor again. And indeed, Mary's next utterance was a classic psychiatric inquiry, a question that went to the very heart of the psychotherapeutic enterprise, and to the heart, for that matter, of the universal enterprise of becoming a self-realized human being—the enterprise upon which we are all engaged.

"Is there anything you want to change?" she asked.

David looked down at his hands. He breathed a tired sigh. Then he looked at her. He smiled. "Everything I've wanted to do," he said, "I've done."

Among the things that David Reimer has done that give him greatest pride is his decision to speak out publicly about his ordeal, and the positive changes that have resulted from that act.

For despite the brave efforts of Cheryl Chase, despite the three decades that Milton Diamond spent trying to warn doctors about the dangers in current management of intersex conditions, despite the long-term follow-up of sex-reassigned youngsters in Bill Reiner's study, the medical establishment remained reluctant to address the issue. But in October 1998, amid the growing controversy ignited by Diamond and Sigmundson's "John/Joan" article, the American Academy of Pediatrics invited Diamond to address its prestigious annual meeting of urologists. Diamond spoke about the failed twins case and spelled out his and Sigmundson's revised protocols for the treatment of children with irregular or injured genitals. His speech was met with sustained applause from the physicians—the first tangible sign that the medical establishment as a whole might be prepared to alter what, for the past four decades, has been the accepted standard of care. It was a moment of triumph that Diamond declines to accept for himself, and instead directly attributes to the willingness of David Reimer to speak out about his extraordinary life as one of medicine's most famous, if unwitting, guinea pigs.

In speaking out, David has shaken to its foundations the clinical practice founded on John Money's work; he has also raised profound questions about a theory that has held sway for most of the twentieth century: Freud's theory that a child's healthy psychological development as boy or a girl rests on the presence or absence of the penis—the ultimate reason that David was converted to girlhood in the first place. It is a notion that today is also being called into question by neurobiological research, which is leading scientists toward the conclusion that, as Dr. Reiner says, "the most important sex organ is not the genitals; it's the brain."

David Reimer puts it another way when he speaks of his pride in his role as husband, father, and sole breadwinner in the family he never believed he would be lucky enough to have. "From what I've been taught by my father," he says, "what makes you a man is you treat your wife well, you put a roof over your family's head, you're a good father. Things like that add up much more to being a man than just *bang-bang-bang*—sex. I guess John Money would consider my children's biological fathers to be real men. But they didn't stick around to take care of the children. I did. That, to me, is a man."

Epilogue

=

IT IS ONE OF THE FIRST MAXIMS of science that no theory can be based on a single experiment. This truism applies as much to Milton Diamond's revelations about the failure of the twins case as it does (or *should* have) to Money's original reporting of its success. While it is also true that Diamond cites David's case *not* as an isolated one (he has repeatedly presented evidence to the same effect, both clinical and theoretical, in cases involving intersexes), it is only through continued study and follow-up on cases of developmentally normal boys turned into girls that medical science can confidently proclaim whether it is nature or nurture that predominates in the making of men and women, boys and girls.

Such cases, which rely exclusively on genital accidents resulting in loss of the penis, are necessarily rare. Yet with the high-profile debate generated by David's case, there are already signs that what cases *do* exist are being followed up and reported on with a greater rigor than previously. In February

1998 the *Urology Times* published a report by Dr. Bernardo Ochoa, former chief of pediatric and urologic surgery at the University of Antioquia in Medellín, Colombia, on the case of a baby boy who, like David, lost his penis in an injury and was subsequently reassigned to girlhood by castration, vaginal surgery, and hormone treatment—with results strikingly similar to David's. "She and her family received extensive psychosocial assistance," Ochoa reported. "However, when she became an adolescent, 14 years later, she demanded to be reassigned as a boy because she didn't feel she was a girl."

In pursuit of further data, doctors and researchers will almost certainly pay particularly close attention to events that began at Atlanta's Northside Hospital in 1985, when on a single day two developmentally normal newborn boys suffered severe penile burns from bungled electrocautery circumcisions. One boy lost his penis entirely, the other a significant portion. The parents of the child in the former case agreed to sex reassignment of their child as a female; the other parents opted to have their child receive plastic surgery to create an artificial penis.

Here was another of those one-in-a-million chances, another unplanned experiment of fate, in which two children became ideal matched controls in a living laboratory experiment, this one offering researchers a unique comparison study of how best to deal with the calamity of penile loss: sex reassignment or phalloplasty? With the children currently on the eve of their fourteenth birthdays, privacy issues have prevented much information on their condition from being released, and I was able to glean only minimal reports when I spoke with the several lawyers who represented the parents in their lawsuits against Northside Hospital.

Thomas Sampson is the lawyer who won a $22.8 million settlement for the parents of Antonio, the boy who is being raised as a male. According to Sampson, Antonio is doing better today than was ever expected at the time of his injury. He experienced a period of difficulty in his early school years, when he underwent several operations for phalloplasty and experienced some cruelty, both intentional and otherwise, from his peers. But now in tenth grade, his social situation is considerably better. He displays no difficulties with his identity as a male, though whether he will ever feel confident enough to have sexual intercourse with his artificial penis and to father his own children, only time will tell.

The fate of the other child is less well known and more carefully guarded. Castrated and converted to girlhood at less than two weeks of age, she is known in court documents simply as "Baby Doe." For the last year and a half, I have sought word of her mental and emotional state from her lawyers and physicians, but they have refused all comment. What few facts I was able to verify independently did not augur particularly well. Her parents divorced while she was still young. Doe lives with her mother—although in a relationship whose future is unsure. According to a lawyer involved in her case, she was recently placed under the care of a court-appointed guardian.

I spoke to David about Baby Doe during my last trip to see him in Winnipeg in the summer of 1998. Deeply upset, he insisted on dictating a letter to the child's parents, in which he offered himself for advice or support to them or their daughter. His letter went unanswered. "I wasn't surprised," David says. "I've been there, and so have my parents. They just need time."

Since then David has found himself dreaming of Doe, who has appeared to him in his sleep as a mute younger sister desperately trying to communicate something through enigmatic scribblings on a child's chalkboard. That David should feel a special kinship with Doe is not surprising, for not only has she undergone the same accident and treatment that have shaped

David's life, she also shares other uncanny similarities with him. Remarkably enough, *both* children injured at Atlanta's Northside Hospital in 1985 happen to have been born on 22 August—twenty years to the day after David and Brian Reimer came into the world. The psychologist who consulted on Baby Doe's case—five years after David announced his decision to live as a male—was Dr. John Money.

Acknowledgments

=

Besides the Reimer family and all others named in this book who granted me interviews, I'd also like to acknowledge the crucial help of Mel Myers, Josh Weinstein, John Danakas, and Keith Black in Winnipeg; Dave Amber at the American Academy for the Advancement of Science in Washington, D.C.; Sara Pinto at the BBC in London; Miriam Zuger, whom I happened first to contact just days after the passing of her husband, Bernard, and who supplied me with reprints of his fascinating papers; Holly Devor, who has written two books on intersexuality; Edward Eichel who gave me the difficult-to-find *Paidiki* interview with John Money; Drs. Mariano and Marvin Tan who fielded endless e-mail queries about the twins' early lives; and photographer Ed Buryn who kindly sent me a copy of his book *Twin Births*. Thanks, also, to the staffs of the New York Academy of Medicine Library, the library of the New York Psychoanalytic Institute, and the

New York Society Library, where much of this book was written.

This book would not exist if not for my friend and longtime editor at *Rolling Stone*, Bob Love. He helped me shape and fine-tune the structure of the original article (which formed the narrative blueprint for *As Nature Made Him*), and he was always a patient sounding board for the ideas herein. Many thanks, too, to *Rolling Stone* editor and publisher Jann Wenner, who was passionate about this project from the outset, and whose reaction to the article's 17,500-word first draft was to bark, "It's great. Now gimme more!" Thanks also to the other people at *Rolling Stone* whose work on the article bears a lasting imprint on this book: Erika Fortgang, Tom Conroy, and Marian Berelowitz.

Closer to home, I'd like to thank certain members of my family for the grounding they gave me in medical and scientific terminology and ideas which made my journeying in these areas a good deal less disorienting than it would otherwise have been: my mother Carol, a registered nurse; my brother Ted, a neurosurgeon; and my late father, Vincent, who was for many years the chief of Urology at St. Michael's Hospital in Toronto and who, nightly, regaled me and my three siblings with tales from the medical world—including, in the early 1970s, the amazing news that when newborn boys lose their penis to circumcision they are turned into girls. Not himself a pediatric urologist, my father did not perform infant sex reassignments, but he (like the majority of medical men) accepted the psychological rationale behind them. I hope he would be proud of my efforts, with this book, to disseminate to the medical community (and world at large) contradictory evidence not available for the past three decades.

Many thanks to my agent, Lisa Bankoff, who made sure my proposal came before the eyes of the superb Robert Jones at HarperCollins. Not only did Robert give this book the most meticulous and artful edit, purging it of *longueurs*, repetition, and other maladies, he also somehow found the time to write explanations for each suggested change in the margin. I took them all. Thanks, too, to Fiona Hallowell at HarperCollins. Any errors of fact or interpretation, meanwhile, are my own.

I want to thank my wife, Donna Mehalko, who is the first person to read everything I write and upon whose instincts I rely utterly. Thanks, too, to my son John Vincent. Born eleven months into the process of my writing this book, he gave me, with his beloved presence, a special insight into the unimaginable horror faced by Ron and Janet Reimer all those years ago, and whose newborn crying, at times a distraction, was also a happy goad to keep at the computer and press on to the end.